Cutting and Layering Propagation

Sticking softwood cuttings of carnation in cell packs.

Cross-section anatomy of adventitious root formation. Root primordia are emerging through a ring of phloem fibers.

Outdoor softwood cuttings under mist.

Mound layered apple stems in a stool bed ready to be cut as rooted understocks.

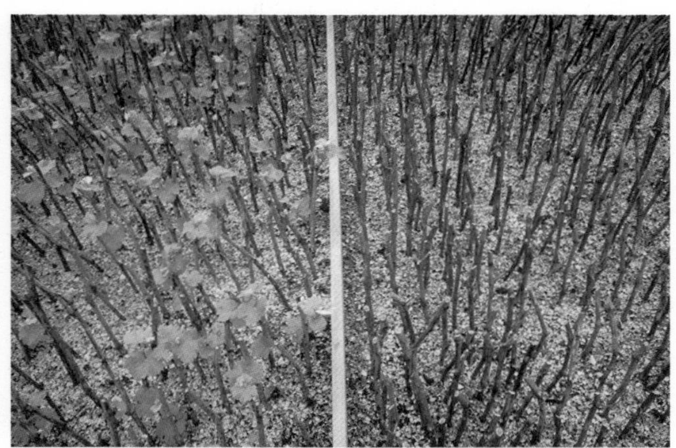

Hardwood cuttings of grape. Cuttings on the left have already rooted and are leafing out.

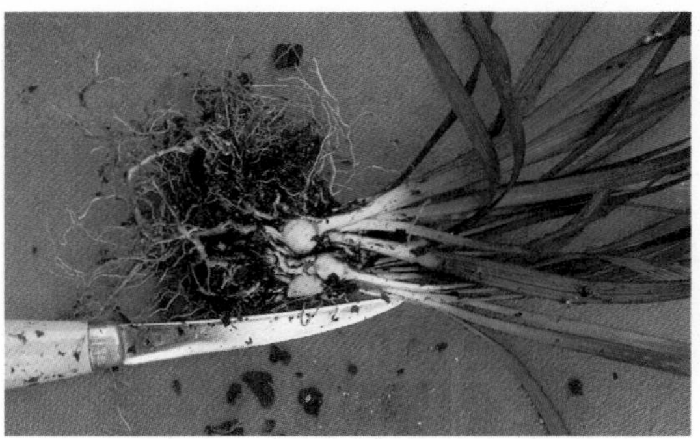

Liriope is a herbaceous perennial that is propagated by crown division.

PLANT PROPAGATION

Principles and Practices

SIXTH EDITION

Hudson T. Hartmann, Ph.D.
University of California, Davis

Dale E. Kester, Ph.D.
University of California, Davis

Fred T. Davies, Jr., Ph.D.
Texas A & M University
College Station

Robert L. Geneve, Ph.D.
University of Kentucky, Lexington

PRENTICE HALL
Upper Saddle River, New Jersey 07458

Library of Congress Cataloging-in-Publication Data

Plant propagation : principles and practices / Hudson T. Hartmann .
[et all.].—6th ed.
p. cm.
Rev. ed. of: Plant propagation / Hudson T. Hartmann, Dale E.
Kester, Fred T. Davies, Jr. 5th ed. c1990.

Includes bibliographical references and index.
ISBN 0-13-206103-1
1. Plant propagation. I. Hartmann, Hudson Thomas.
II. Hartmann, Hudson Thomas, Plant propagation.
SB119.P55 1997
631.5'3—dc20 96-28045
CIP

Acquisitions editor: *Charles Stewart*
Editorial production supervision: *Barbara Marttine Cappuccio*
Director of manufacturing and production: *Bruce Johnson*
Managing editor: *Mary Carnis*
Manufacturing buyer: *Ilene Sanford*
Design director: *Marianne Frasco*
Cover design: *Miguel Ortiz*
Marketing manager: *Debbie Yarnell*
Editorial assistant: *Kate Linsner*
Endpaper/cover photos: *Robert L. Geneve*

Simon & Schuster / A Viacom Company
Upper Saddle River, New Jersey 07458

Printed in the United States of America

10 9 8 7 6 5 4 3 2 1

ISBN 0-13-206103-1

Prentice-Hall International (UK) Limited, *London*
Prentice-Hall of Australia Pty. Limited, *Sydney*
Prentice-Hall Canada Inc., *Toronto*
Prentice-Hall Hispanoamericana, S.A., *Mexico*
Prentice-Hall of India Private Limited, *New Delhi*
Prentice-Hall of Japan, Inc., *Tokyo*
Simon & Schuster Asia Pte. Ltd., *Singapore*
Editora Prentice-Hall do Brasil, Ltda., *Rio de Janeiro*

DEDICATION

The sixth edition of *Plant Propagation* is dedicated to Dr. Hudson T. Hartmann. Dr. Hartmann died March 2, 1994 just as plans for the sixth edition were getting underway. Dr. Hartmann is remembered as a dedicated, hard-working, conscientious scientist, teacher, and human being. He conceived of the writing of this text about 1955 and asked the second author, Dr. Dale E. Kester, to join him. Following the publication of the first edition in 1959, four more editions followed in 1968, 1975, 1983, and 1990. Dr. Hartmann taught Plant Propagation at the University of California at Davis from 1945 to his retirement in 1980. His research in propagation involved early studies on hormones, mist propagation, and other aspects of cutting propagation particularly as they applied to fruit trees. He was also a specialist in olive research and development, attaining a worldwide reputation for this crop.

One of his primary accomplishments was his activity with the International Plant Propagation Society. He became a member in 1953 and then was instrumental in initiating the Western Region of the Society in 1960. He served as Western Region Editor for the Society from 1960 to 1993, serving also as International Editor from 1970 until 1991. During his career he published many scientific papers and popular articles. As well as the present text, he was senior author of *Plant Science: Growth, Development and Utilization of Cultivated Plants,* first edition (1981), second edition (1988) published by Prentice Hall.

Dr. Hartmann was a member of American Society for Horticultural Science, becoming a Fellow in 1974. As an undergraduate he was a member of Gamma Sigma Delta and Alpha Zeta. Dr. Hartmann has received many awards including the Charles G. Woodbury Award (1960), Joseph H. Gourley Award (1962), and Stark Brothers Award (1964) from ASHS. The American Association of Nurseryman awarded him its Norman J. Coleman Award (1970), The California Association of Nurseryman presented him with its Research award (1977). Pi Alpha Xi made him an honorary member (1981). The Western Region IPPS awarded Dr. Hartmann its Merit award (1979), Honorary Membership (1983), and established the Hudson T. Hartmann Western Region Research Grant in his honor. The International IPPS Board of Directors awarded him the International Award of Honor in 1990.

Dr. Hartmann was a close personal friend, a collaborator who made working together a pleasure, and a respected peer whose guidance and insight are missed.

Dr. Dale E. Kester

Contents

Preface *ix*

PART I
GENERAL ASPECTS
OF PROPAGATION

1

Introduction *1*

The Role of Plant Propagation in Human History *1*
Evolution of Plant Propagation *1*
The Development of Nurseries *4*
The Modern Plant Propagation Industry *5*
Plant Propagation Organizations *6*

2

Biology of Propagation *9*

Genetic Control in Propagation: Sexual Versus Asexual *10*
Plant Development, Gene Expression, and Epigenetic Variation *17*
The Concept of Hormonal Control of Plant Growth and Development *20*
Life Cycles in Plants *26*
Plant Nomenclature and the Concept of the Cultivar *31*

3

Environmental Factors of Propagation *40*

Fundamental Microclimatic and Edaphic Factors in the Propagation Environment *41*
Managing the Propagation Environment *45*
Containers for Propagating and Growing Young Liner Plants *62*
Management of Edaphic Factors in Propagation and Liner Production *68*
Management of Microclimatic Factors in Propagation and Liner Production *78*
Biotic Factors—Pathogen and Pest Management in Plant Propagation *83*
Post-Propagation Care of Liners *94*

4

Principles and Practices of Seed Selection *105*

Uses of Seeds in Propagation *106*
Seed Selection in Herbaceous Plant Species *106*

Seed Selection in Woody Plant Species *106*
Categories of Seed-Propagated Woody Plants *117*
Seed Selection and Production Systems *118*
Legal Controls to Genetic Purity *120*
Summary *121*

PART II
SEED PROPAGATION

5

The Development of Seeds *125*

Introduction *125*
What is a Seed? *126*
Formation of the Fruit, Seed, and Embryo *128*
Polyembryony and Apomixis *134*
Plant Hormones and Seed Development *141*
Ripening and Dissemination *143*

6

Techniques of Seed Production and Handling *147*

Sources of Seed *147*
Harvesting and Processing Seeds *149*
Seed Testing *153*
Seed Storage *167*

7

Principles of Propagation by Seed *177*

The Germination Process *177*
Dormancy: Regulation of Germination *194*

8

Techniques of Propagation by Seed *216*

Seed Propagation Systems *216*
Field Nurseries for Transplant Production *224*
Production of Transplants Under Protected Conditions *227*

PART III
VEGETATIVE PROPAGATION

9

Selection and Management of Clones in Vegetative Propagation *239*

History *239*
Reasons for Using Clonal Cultivars *240*
Genetic Basis of Clones *242*
Nongenetic Variation Within Clones *250*
Pathogens and Plant Propagation *258*
Management of Sources for Vegetative Propagation *259*
Propagation Sources and their Management *266*
Summary *270*

10

The Biology of Propagation by Cuttings *276*

Descriptive Observations of Adventitious Root and Bud Formation *277*
Correlative Effects: How Hormonal Control Affects Adventitious Root and Bud Formation *287*
The Biochemical Basis for Adventitious Root Formation *292*
Molecular/Biotechnological Advances in Asexual Propagation *295*
Factors Affecting Regeneration of Plants From Cuttings *298*
Treatment of Cuttings *308*
Environmental Manipulation of Cuttings *312*

11

Techniques of Propagation by Cuttings *329*

The Importance and Advantages of Propagation by Cuttings *329*
Types of Cuttings *330*
Stock Plants: Sources of Cutting Material *344*

Rooting Media *349*
Wounding *353*
Treating Cuttings With Auxins *355*
Preventive Disease Control *362*
Environmental Conditions for Rooting Leafy Cuttings *363*
Preparing the Propagation Bed, Bench, Rooting Flats, and Containers and Inserting the Cuttings *373*
Preventing Operation Problems with Mist Propagation *373*
Management Practices *374*
Cutting Nutrition *377*
Care of Cuttings During Rooting *378*
Cold Storage of Rooted and Unrooted Cuttings *380*
Handling Field-Propagated Plants *383*
Container-Grown Plants and Alternative Field Production Systems *385*

12

The Biology of Grafting *392*

The History of Grafting *393*
Terminology *393*
Reasons for Grafting and Budding *395*
Natural Grafting *400*
Formation of the Graft Union *401*
Graft Union Formation in T- and Chip Budding *406*
Factors Influencing Graft Union Success *409*
Polarity in Grafting *413*
Genetic Limits of Grafting *414*
Graft Incompatibility *416*
Scion-Rootstock (Shoot-Root) Relationships *425*

13

Techniques of Grafting *437*

Types of Grafts *438*
Detached Scion Graftage *439*
Approach Graftage *457*
Repair Graftage *458*
Production Processes of Graftage *461*
The Craftmanship of Grafting *469*
Aftercare of Grafted Plants *470*
Grafting Systems *472*

14

Techniques of Budding *481*

Rootstocks for Budding *482*
Time of Budding—Fall, Spring, or June *482*
Types of Budding *487*
Top-Budding (Topworking) *498*
Double-Working by Budding *498*
Microbudding *500*

15

Layering and its Natural Modifications *502*

Physiology of Regeneration by Layering *503*
Management of Plants During Layering *503*
Procedures in Layering *503*
Plant Modifications Resulting in Natural Layering *513*

16

Propagation by Specialized Stems and Roots *520*

Bulbs *520*
Corms *534*
Tubers *536*
Tuberous Roots and Stems *538*
Rhizomes *540*
Pseudobulbs *543*

PART IV
METHODS OF MICROPROPAGATION

17

Principles of Tissue Culture for Micropropagation *549*

Introduction *549*
Reproduction of Seedling Plants in Tissue Culture *552*
Micropropagation of Plantlets from Tissue Culture *556*

Types of Systems Used to Regenerate Plantlets by Micropropagation (Table 17-1) *563*
Callus, Cell, and Protoplast Culture Systems *567*
Somatic Embryogenesis and Synthetic Seed Production *572*
Control of the Tissue Culture Environment *576*
Special Problems Encountered by In Vitro Culture *577*
Variation in Micropropagated Plants *578*

18

Techniques of In Vitro Culture for Micropropagation and Biotechnology ***590***

Introduction *590*
Characteristics of Micropropagation *590*
Disadvantages of Micropropagation *592*
General Laboratory Facilities and Procedures *592*
Micropropagation Procedures *601*
Stage I—Establishment and Stabilization *601*
Stage II—Shoot Multiplication *604*
Stage III—Root Formation *605*
Stage IV—Acclimatization to Greenhouse Conditions *608*
Specific Protocols for Aseptic Culture *608*
Somatic Embryogenesis and the Production of Synthetic Seeds *609*
Methods for Micropropagating Representative Horticulture Crops *611*

PART V
PROPAGATION OF SELECTED PLANTS

19

Propagation Methods and Rootstocks for Important Fruit and Nut Species ***625***

20

Propagation of Ornamental Trees, Shrubs, and Woody Vines ***667***

21

Propagation of Selected Annuals and Herbaceous Perennials Used as Ornamentals ***725***

Indexes ***758***

Subject Index *758*
Plant Index, Scientific Names *765*
Plant Index, Common *768*

Preface

The sixth edition of *Plant Propagation* has been the result of many changes. First of all, as described in the dedication, Dr. Hartmann passed away as the plans for the sixth edition were starting. Prior to this, a third author, Dr. Fred T. Davies, had joined the author team and brought not only considerable expertise in propagation but also broadened the range of author experience in horticultural interest, background, age viewpoint, and geography, as well as providing continuity with a newer generation of researchers and teachers.

Following Dr. Hartmann's death, a reevaluation was made by the remaining authors in the areas of subject responsibility. The previous arrangement was that subject matter areas were divided between the authors who produced the initial drafts of their chapters. Chapters were then exchanged and changes made as agreed upon. In this manner, Dr. Hartmann had original responsibility for the introductory chapter, structures and management, cuttings, grafting and budding, and fruit tree and woody plant propagation. Dr. Kester had responsibility for seed selection and propagation, clonal selection, layering, separation and division, tissue culture, and herbaceous plant propagation. Dr. Davies later assumed responsibility for cuttings, layering, separation and division, and woody plant and herbaceous plant ornamentals.

Our first decision concerning the sixth edition was to add a fourth author who would also bring specific expertise in subject matter, geographical distribution, and experience. The new subject matter distribution was as follows—Dr. Kester: introduction (later divided into chapters on history and biology), seed selection, clonal selection, layering, tissue culture, and fruit tree propagation. Dr. Davies: structures and management, cuttings, grafting and budding, and woody ornamental propagation. The new author would take responsibility for seed propagation, separation and division, and herbaceous ornamentals. Dr. Robert Geneve, of the University of Kentucky, with expertise in seed propagation, tissue culture, and biotechnology was invited to join the team as the fourth author. The process of reevaluation involved a critical review of the entire text by five individuals with specific expertise in teaching and research in plant propagation.

Previous users of this text will note both close similarities to earlier editions as well as specific changes. First of all, we have retained the three-part objectives of the previous editions. These remain: (1) to present the scientific evidence that provides the theoretical framework upon which propagation is based, (2) to describe in detail procedures and techniques, and (3) to provide descriptions of up-to-date propagation methods for important horticultural plants. The book is intended for university-level students with some background in biology. We realize that in so doing the information presented may be somewhat more advanced than other students might find comfortable, particularly in technical programs where students may be more interested in techniques and procedures. The current format retains the flexibility to present techniques without the theoretical framework. The presence of the biological aspects, however, provides a challenge to the instructors who may need to be somewhat selective in the presentation of the material. To assist in this regard, a Teachers Manual has been prepared for this edition in which we have tried to sort out the key ideas and concepts.

There has been a substantial reorganization of the subject matter in the sixth edition. This has been the result of information from readers of the previous edition. Chapter 1 now is a discussion of the historical aspects of plant propagation and its role in human society. Chapter 2 is intended to provide the basic biological background that we believe is necessary for a comprehensive understanding of propagation in all of its aspects. The five biological concepts include the genetic basis of reproduction, the epigenetic control of gene expression, the hormonal control of growth and development, the concept of developmental cycles (seedling, clonal, and apomictic), and the concepts of botanical taxonomy and cultivars. We have chosen to include presentations of molecular biology and genetic engineering (in a simplified version) because we believe that this information will become increasingly important in future propagation as well as part of the core biological information modern horticultural students should have. Although some students will have covered this material in other classes, we believe that few texts integrate these topics in a manner that addresses the overall role of the propagator.

Chapter 3 deals with the environmental aspects of propagation and takes more of an ecological approach than earlier editions. It also incorporates integrated pest management (IPM) and best management practices (BPM).

The seed chapters have been rearranged. They start with a chapter on seed selection with the view that seed propagation also includes maintaining the integrity of the cultivar. Seed development follows, including newer information on its molecular basis. The seed handling chapter now includes seed testing and various aspects of pre-germination procedures. The next two chapters then provide the parallel presentation of the theory and practice of seed germination. Seed dormancy classification has been revisited with some modification of previous information.

Vegetative propagation chapters follow the previous format. Chapter 9 presents the concept of variation within clones with emphasis on trueness-to-cultivar, trueness-to-type, and freedom from pathogens. In the subsequent theory chapters, we have tried to integrate new and older concepts without resorting to encyclopedic listing of research studies.

Some reorganization has taken place in the tissue culture chapters but the primary format has been retained. The chapters on propagation of selected plants has been expanded, updated, and reorganized so that plants are listed by genus with cross-listing of common names.

As always, we have enjoyed the cooperation of many individuals who provided reviews, comments, and illustrative materials. We wish in particular to thank the following reviewers: S. Arulsekar, H.W. Barnes, L.W. Barnes, W. Barr, C. Baskin, M. Bennett, F.A. Blazich, F. Bliss, T.D. Davis, L. Ferguson, F.W. Garrett, T.H. Gradziel, S. Hottovy, B.H. Howard, E.G. Illa, M. Kane, R. Knight, T.D. Landis, H.J. Lang, M. Marcotrigiano, G. McGranahan, W. Micke, K.W. Mudge, J.R. Murray, W. Libby, D. Parfitt, W. Pill, J.E. Preece, W.M. Proebsting, F.S. Santamour, B. Rush, S. Southwick, J. Staub, J.B. Storey, D.K. Struve, E. Sutter, D. TeKrony, A. Walker, K. Warren, M.N. Westwood, J. Uyemoto, and R. Zimmerman.

Special thanks have to be made to our families and wives (Daphne Kester, Maritza Davies, and Pat Geneve) for their support and encouragement—and enduring the lack of companionship that resulted from heavy time commitments in writing the sixth edition.

We also thank Janet Williams and Carolyn Cobb, who have produced some of the artwork that is found in this edition, as well as our Production Editor, Barbara Cappuccio.

Dale E. Kester
Fred T. Davies, Jr.
Robert L. Geneve

About the Authors

Dale E. Kester is Professor of Pomology emeritus at the University of California, Davis. During his 40 years at University of California he taught courses in Plant Propagation and Pomology. He has had a lifelong collaboration with Dr. Hudson T. Hartmann which resulted in the publication of the first edition of *Plant Propagation-Principles and Practices* in 1959 followed by other editions in 1968, 1975, 1983 and 1990. He has authored over 100 research and popular publications in plant propagation and pomology. In 1996 he became Vice-President, program chair and President-elect of the International Plant Propagators Society-Western Region.

Fred T. Davies, Jr., Professor of Horticultural Sciences and Plant Physiology, Texas A&M University, has taught courses in plant propagation, nursery production, and management since 1979. He has co-authored some 87 research and technical publications. He was a Senior Fulbright Fellow to Mexico in 1993. He received the Distinguished Achievement Award for Nursery Crops from the American Society of Horticultural Sciences (1989), L.M. Ware Distinguished Research Award-ASHS Southern Region (1995), and S.B. Meadows Award of Merit-International Plant Propagators Society-Southern Region (1994). He was President and is currently Editor of the IPPS-Southern Region.

Robert L. Geneve is an Associate Professor of Horticulture at the University of Kentucky. He teaches courses in Plant Propagation, Seed Biology, and Plant Growth and Development. He has co-authored over 60 research and technical publications. He is currently the Associate Editor in the area of plant propagation for the American Society of Horticultural Sciences.

PART I—GENERAL ASPECTS OF PROPAGATION

"and the earth brought forth grass, and herb yielding seed after his kind, and the tree yielding fruit, whose seed was in itself, after his kind: and God saw that it was good."

Genesis 1:12, the Bible.

"man has become so utterly dependent on the plants he grows for food that, in a sense, the plants have "domesticated him." A fully domesticated plant cannot survive without the aid of man, but only a minute fraction of the human population could survive without cultivated plants."

from: J. P. Harlan, *Plants and Man,* 2nd edition.
Madison, WI: Amer Soc Agron. 1992.

1

Introduction

THE ROLE OF PLANT PROPAGATION IN HUMAN HISTORY

The propagation of plants is a fundamental occupation of humankind. Its discovery began what we now refer to as civilization and initiated human dominion over the earth. Agriculture began some 10,000 years ago when ancient peoples, who lived by hunting and gathering, began to cultivate plants and domesticate animals. These activities centered around stable communities and people began to select and propagate the kinds of plants that provided a greater and more convenient supply of food and perhaps other products for themselves and their animals (*18, 29*). Once this process began, humans could remain at the same site for long periods of time, thus creating centers of activity that eventually would become cities and countries.

Agriculture—the deliberate cultivation of crops and animals for use by humans—involves perhaps four kinds of activities: (a) selecting and (or) developing specific kinds of plants (*plant breeding*), (b) multiplying those plants and preserving their unique qualities (*plant propagation*), (c) growing them under controlled conditions for maximum yield (*crop production*), and (d) transforming and preserving the products of those plants for food or other uses, e.g., making bread, pressing oil, preparing wine, dehydration, etc. (*food technology*).

The relationship between man and domesticated plants (and animals) is symbiotic (*18, 22*). Human society cannot exist without the availability of the special kinds of food, fiber, and amenities provided by cultivated plants (and animals), most of which must be maintained under man's surveillance or they would not continue to exist. **Plant selection** provides the plants which are most useful to promote man's well-being. **Plant propagation** multiplies these plants and preserves their essential genetic characteristics.

EVOLUTION OF PLANT PROPAGATION

The pivotal role of plant propagation in the evolution of human society can be seen in terms of particular stages of agricultural development.

Hunting and Gathering

Most of the millions of years of human existence as hunters and gatherers were related to the presence of specific food resources including seeds, fruits, roots, and tubers, as well as animals that fed on the plants. The distribution and the characteristics of plant species was determined by the environment, that is, both the physical world (climate, soil, topography) and the biological interactions of plant, animal, and human populations (*18, 27, 29*). The interactions involved an **ecological climax** in which all of these components were more or less in equilibrium in different communities. Such a condition evidently existed for several millions of years, enabling humans to spread from their presumed place of origin in western Africa into Asia, Europe, and eventually into North and South America. Food supplies were abundant in the native vegetation although quite variable in different parts of the world. Apparently, primitive people were quite effective in searching out those kinds which were useful as well as in developing processes to utilize and preserve them.

What motivated humans to begin to propagate and grow specific kinds of plants near their abodes has been the subject of much scientific debate (*18, 29*). It is clear that the development of agriculture changed forever the relationship between humans and their surrounding environment. This event occurred in separate places of the world more or less simultaneously within a relatively short period of a few thousand years about 10,000 years ago: (a) the fertile cresent of southwest Asia and northeast Africa, extending from the valley of the Euphrates and Tigris rivers along the coasts of Syria, Turkey, Israel to the Nile Valley of Egypt, (b) China, including a northern segment and a tropical southern area, and (c) Central and South America, including areas in Mexico, and the coastal lowlands and highlands of Peru (*18, 20*).

The key activity bringing about this change must have been the deliberate propagation and cultivation of specific kinds of plants that were particularly useful to humans. As a result, a larger and more stable population could be supported which evolved into cities and countries. Human organization changed from subsistence existence, where everyone participated in the production of food and other items, to a division of labor among the population between agricultural and nonagricultural segments, and even to specialization within the agricultural segment. In this context, the plant propagator, who possessed specific knowledge and skills, has had to assume a key role.

Domestication

The number of domesticated plant species with which early civilization developed was relatively few, determined both by their usefulness in the primitive economy and the ease by which they could be propagated. The lists differed in the separate areas of the world where human societies evolved (*18, 27, 28, 29*). The most important food crops were seed plants, (cereals, such as wheat, barley, rice) which provided carbohydrates, and legumes (beans, peas) which provided protein. These seed-propagated plants could be subjected to genetic selection in consecutive propagation cycles for such agricultural characteristics as high yield, "non-shattering," large seed size, and reduced seed dormancy that were maintained more or less "fixed" because of their genetic tolerance to inbreeding (see Chapters 3 and 4). Highly desirable single plants of certain species, such as grape, fig, olive, pomegranate, potato, yams, banana, pineapple (*31*) could be selected directly from wild populations and "fixed" through vegetative propagation (see Chapters 2 and 9). Domestication of fruit plants, such as apple, pear, peach, apricot, citrus, and others occurred with the discovery of grafting methods (see Chapters 12, 13, and 14). By the time of recorded history (or that which can be reconstructed), most of the basic methods of propagation had been discovered. During domestication, crop plants had evolved beyond anything that existed in nature.

The establishment of specific crops and cropping systems resulted in some side effects that have continued to create problems (*18*). As the fields used to grow plants near human sites were disturbed and became depleted, certain aggressive plant species also were spontaneously established in these sites. These so-called "weedy" species have become a part of the agricultural system and more or less evolved along with cultivated plants.

Organization of Human Societies

Primitive. The initial phases of domestication probably involved plant breeding (selection), plant propagation, and plant production. With an increase in food supply, a larger population could be supported and division of labor began to occur. Classes of individuals may have included laborers, manufacturers, artisans, government bureaucrats associated with irrigation systems, religious groups, and soldiers as well as farmers and herdsmen. Historical records of early civilizations in Egypt and the Middle East (as well as archaeological investigations) have shown that the agricultural sector was well organized to produce food (cereals, vegetables, fruits, dates), fiber (flax, cotton) and other items for the nonagricultural components of society (*22*). Early Chinese writings indicate the knowledge of grafting, layering, and other techniques although rice and millet were the principal food sources. In the Americas, seed propagated crops (maize, beans, cucurbits, squash) as well as vegetatively propagated crops (potato, cassava, sweet potato, pineapple) were developed and grown.

Greek and Roman. Early writings (see supplementary references) described the agricultural world in detail with accounts of propagation techniques much as we know them today. Control of land and agricultural surplus were the keys to power and wealth (*29*). Small and large farms existed. Olive oil and wine were exported and grains were imported. Vegetables were grown near the home as were many fruits (fig, apple, pear, cherry, plum). Not only were food plants essential, but Romans developed ornamental gardening to a high level (*18*).

Medieval period of the Middle Ages. Society was organized around large estates, manor houses, and castles with landlords providing protection. Large areas of forest were kept as game preserves. Equally important, monasteries acted as independent agricultural and industrial organizations and preserved a great deal of the written and unwritten knowledge. In both kinds of institutions, a separation developed among those involved in the production of cereals, fibers, forages which were grown extensively in large fields (agronomy), vegetables, fruits, herbs, and flowers grown in "kitchen gardens" and orchards near the home (horticulture), and woody plants grown for lumber, fuel, and game preserves (forestry) (*22*).

The end of the medieval period and the beginnings of modern Europe brought a shift from a subsistence existence to a market economy and the emergence of land ownership (*29*). In Western Europe, both large landowners and owners of smaller individual plots emerged. In Eastern Europe, the shift was toward large wealthy estates with the populace being largely serfs.

Through these periods, the specific skills and knowledge of the plant propagator were possessed by specific individuals. These skills, considered as "trade secrets," were passed from father to son or to specific individuals. Often this knowledge was accompanied by superstition and sometimes attained religious significance.

Exploration, Science, and Learning

Major changes occurred as Europe moved from the subsistence phase of medieval agriculture towards the market-oriented economy of the sixteenth century (*29*).

Plant exchanges. The exchanges of plant material from the area of origin to developing countries of the world has been one of the major aspects of human development. Not only did the range of plants available for food, medicine, industrial uses, and gardening expand, but plant propagation methods to reproduce them was required. The Arab invasion in the ninth century introduced citrus and rice to southern Europe along with new concepts of cultivation and the use of irrigation. The voyages of Columbus opened the world to exploration and the interchange of plant materials from continent to continent—wheat, barley, many vegetables, fruits to the U.S.—potatoes, tomato, beans, corn, tobacco and other plants to Europe. A series of plant exploration trips were initiated with the voyages of Captain Cook in 1768 which included the plant explorers Sir Joseph Banks and Francis Masson who brought large numbers of exotic plants to England for the Royal Botanic Garden established at Kew outside of London (*20,26*). Centers of learning were established in many countries where scientific investigations began on all aspects of the biological and physical world. Linneaus established the Binomial System of Nomenclature and botanists began to catalogue the plants of the world. Plant collecting trips continued throughout the world, from Europe (David Douglas, 1823) and from the United States (David Fairchild, Frank Meyer, Joseph Rock, E.H. Wilson) (*11, 16, 20, 26*). Significant ornamental species which are mainstays of modern gardens were collected at that time, from the Orient (rhododendron, primula, lily, rose, chrysanthemum), Middle East (tulips, many bulb crops), and North America (evergreen and deciduous trees and shrubs). "Orangeries" and glasshouses (greenhouses) were invented to grow the exotic species in colder climates.

A London physician and amateur horticulturist Nathanial Ward invented the Wardian case early

FIGURE 1–1 A Wardian case invented by N.B. Ward in the early nineteenth century to use in transporting plants through long ocean voyages (*32*).

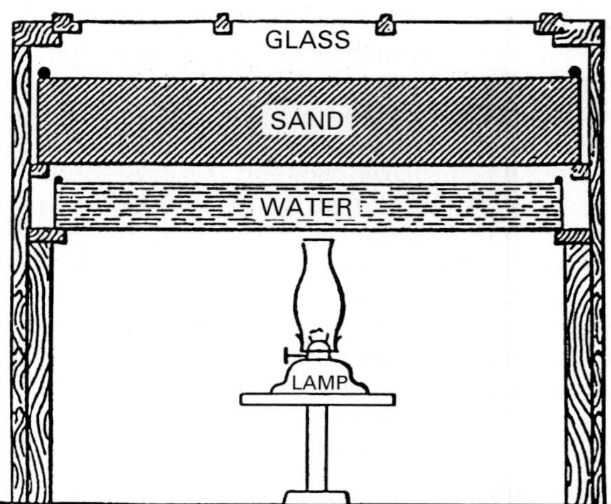

FIGURE 1–2 A simple device to provide bottom heat in propagation. Reproduced from Bailey, The Nursery-Manual (*5*).

in the 1800s to help preserve plant material on these long expeditions (*30*) (Figure 1–1).

Scientific and horticultural literature. Important books were written detailing the scientific and cultural developments of the day. Charles Darwin and his *Origin of Species* (*11*) as well as its important contemporary *Domestication of Plants* (*12*) introduced the concept of evolution and set the stage for the genetic discoveries following the rediscovery of Mendel's papers in 1900. The subsequent explosion in knowledge and application has provided the framework on which present day plant propagation is based as has the increase in knowledge of plant growth, anatomy, physiology, and other basics of biological science (*26*).

Books and articles on gardening and propagation began to appear (*14*). The first book on nurseries, *Seminarium,* was written by Charles Estienne in 1530. Later Charles Baltet, a practical nurseryman, published a famous book on "Grafting and Budding" in 1821, describing 180 methods of grafting (*9*). A book by Andrew J. Fuller—*Propagation of Plants*—was published in 1885 (*17*).

The Morrill Act. The passage of this Act by the U.S. Congress in 1862 was a landmark event that established land-grant colleges and fostered the scientific investigation of agriculture and mechanical arts. Departments of agronomy, horticulture, pomology, and related fields were established which became centers of scientific investigation, teaching, and extension. Liberty Hyde Bailey, a product of this system, published his first edition of *The Nursery Book* (*2*) later revised as the *Nursery Manual* in 1920 (*5*), which catalogued the known knowledge about plant propagation and the production of plants in the nursery (Figure 1–2). His *Cyclopedia of Horticulture* (*3*) in 1900–1902, *Standard Cyclopedia of Horticulture* (*4*) in 1914–1917, *Hortus* (*6*) in 1930, *Hortus Second* (*7*) in 1941, and *Manual of Cultivated Plants* (*8*) in 1940 and 1949 described the known plants in cultivation. An update—*Hortus Third* (*9*)—is a classic in the field.

M.G. Kains of Pennsylvania State College and later Columbia University, New York, published *Plant Propagation* (*23*), later revised by Kains and McQueston (*24*), which remained a standard text for many years. Several other books were written during this period including Adriance and Brison (*1*), Duruz (*15*), Hottes (*21*), and Malstede and Haber (*24*). The first edition of this book (*19*) was published in 1959 and has continued through five editions.

THE DEVELOPMENT OF NURSERIES

The concept of the nursery where plants are propagated to be transplanted to their permanent site either as part of the agricultural unit or to be sold to others has probably been a part of agriculture since its beginning. Nevertheless, the development of commercial nurseries is probably something that has developed largely within the recent era (*14*). Most agronomic crops (wheat, corn, etc.) and many vegetables were grown by seed with propagation being an integral part of the production cycle. A portion of the seed was retained each year to supply the seed for the next cycle. In regions of cold winters, starting vegetables as well as flowers in protected structures (cold frames, hotbeds) and later transplanting them to the open was an important part of production. The main reason was to extend the length of the growing season.

A number of important nurseries existed in France during the sixteenth and seventeenth centuries and eventually throughout Europe (*15*). Ghent, Belgium had a gardener's guild as early as 1366. The first glass house (greenhouse) was built in 1598. The Vilmorin family established a seed house and nursery business in 1815 which was maintained through seven generations.

Early plant breeding was often combined with a nursery as exemplified by Victor Lemoine (1850) who specialized in tuberous begonias, lilies, gladiolus, and other garden flowers. Nickolas Hardenpont and Jean Baptiste van Mons specialized in fruits, particularly pears. The Veitch family started a major nursery in England in 1832. Thomas Andrew Knight, a famous hybridizer of fruits, established the Royal Horticultural Society in 1804.

Early colonists brought seeds, scion, and plants to the United States from Europe and Spanish priests brought material to the West Coast. John Bartram is credited with providing a major impetus with his Botanical Garden in Philadelphia in 1728. The first nursery, however, was credited to William Prince and Son in 1730 on Long Island. These were followed by the expansion of nurseries throughout the eastern U.S. during the nineteenth century. To a large extent, the early nurseries specialized in selecting and grafting fruit trees although ornamentals and forest trees also began to be produced.

The establishment of the nursery industry in the Pacific Northwest was a unique accomplishment (*15*). In the summer of 1847, Henderson Lewelling of Salem, Iowa, established a traveling nursery of grafted nursery stock growing in a mixture of soil and charcoal in boxes on heavy wagons pulled by oxen which crossed the plains covering 2,000 miles to Portland, Oregon. Three hundred and fifty trees survived which were used to establish a nursery at Milwaukee, Oregon.

THE MODERN PLANT PROPAGATION INDUSTRY

The present day plant propagation industry is large and complex and involves not only the group that multiplies plants for sale and distribution but also a large group of industries that provides services, sells the product, is involved in regulation, provides consultation, carries on research, or is involved in teaching. The key person within this complex, however, is the *plant propagator* who possesses the knowledge and skills either to perform or to supervise the essential propagation task for specific plants (Table 1–1).

TABLE 1–1

Organizations and groups involved in the propagation industry

- Amateur propagators, hobbyists
- Nonprofit organizations
  - Arboreta
  - Botanical gardens
  - Research and teaching institutions
- Germ-plasm repositories
- Commercial wholesale nurseries
  - Ornamental landscape plants
    - bare root
    - container
  - Bedding plant producers
    - vegetables
    - flowers
  - Foliage plant producers
  - Fruit and nut tree nurseries
    - rootstock producers
    - scion varieties
  - Forest plants
    - reforestation
    - Christmas tree producers
- Tissue culture laboratories
  - Commercial nurseries
  - Research institution
    - private
    - institution
- Seed producers
  - Commercial companies
  - Certified seed growers

AREAS OF KNOWLEDGE OF PLANT PROPAGATION

1. First, successful propagation requires a knowledge of technical skills for manipulating plant growth which take a certain amount of skill, practice, and experience to master, such as how to bud and graft, how to make cuttings, and how to carry out tissue culture procedures. These may be considered the *art of propagation.*
2. Secondly, propagation requires a knowledge and insight into plant growth, development, and morphology as well as basic knowledge of physical, chemical, and ecological aspects of the propagation environment. This information may be considered the *science of propagation.* Much of this knowledge is acquired empirically by working with the plants themselves, but it is best supplemented by the formal study of chemistry, physics, botany, genetics, and plant physiology.

   Such knowledge aids propagators in understanding why they do the things they do.
3. Thirdly, a successful propagator must have a *knowledge of plants* and the propagation techniques that best reproduce them. To a large extent, the method selected must be related to the response of the kind of plant being propagated and to the conditions under which the propagation is being attempted.

Presentation of information in these three areas encompass the objectives of this book. This differentiation of subject matter is the basis for separating theory from techniques into different chapters and to provide an expanded section on the propagation of specific plant species and cultivars at the end of the book.

PLANT PROPAGATION ORGANIZATIONS

International Plant Propagators Society (IPPS)[1]. The society was organized in 1951 to recognize the special skills of the plant propagator and to foster the exchange of information among propagators. Since then, the organization has expanded to include an Eastern, Western, and Southern Region of the United States, Great Britain and Ireland, Australia, New Zealand, and Denmark with new groups arising in South America and Japan. Each Region holds an annual meeting whose papers are published in a *Combined Proceedings.* Membership requirements include activity in some phase of plant propagation and willingness to exchange information. The publications are only available to members and libraries.

[1]Washington Park Arboretum × D-P, University of Washington, Seattle, WA 98195 U.S.A.

American Association of Nurserymen (AAN)[2]. Organized in 1875, this association is a national trade organization of the U.S. nursery and landscape industry. It serves about 3,200 member firms involved in the nursery business—wholesale growers, garden center retailers, landscape firms, mail-order nurseries, and allied suppliers to the horticultural community.

AAN helps its members through information, educational meetings, an annual meeting, and publication of a twice monthly magazine *American Nurseryman.* Articles include information on propagation and plant materials.

American Society for Horticultural Science[3]. This organization has a membership of public and private scientists, educators, extension personnel, and industry members with an interest in horticulture. The organization holds annual national and regional meetings and publishes scientific reports in the *Journal of American Society for Horticultural Science, HortScience* and *HortTechnology.* It includes working groups in all propagation areas.

American Seed Trade Association (ASTA)[4]. This organization of seed companies has been serving the industry since 1883. ASTA holds a general meeting each year as well as sponsoring conferences on specific crops. It publishes a semi-monthly newsletter, an annual yearbook, and proceedings of individual conferences. It participates in regulatory activities that affect the seed industry.

Association of Official Seed Analysts[5]. Membership is by member seed laboratories, both private and governmental, mostly in the continental United States. The association holds an annual meeting and publishes the proceedings in the *Journal of Seed Tech-*

[2]Suite 500, 1250 I Street, N.W., Washington, DC 20005

[3]Business Office, ASHS, 600 Cameron Street, Alexandria, VA 22314-2562

[4]1030 15th Street, N.W., Washington, DC 20005 U.S.A.

[5]P.O. Box 4906, State Fairgrounds, Springfield, IL 62708-4906 U.S.A.

nology. A quarterly newsletter is published along with other publications, including *Rules for Testing Seeds*.

Association of Official Seed Certifying Agencies (AOSCA)[6]. Membership includes U.S. and Canadian agencies responsible for seed certification in their respective areas. This organization was organized originally in 1919 as the International Crop Improvement Association. These agencies maintain a close working relationship with the seed industry, seed regulatory agencies, governmental agencies involved in international seed market development and movement, and agricultural research and extension services.

Canadian Seed Trade Association[7]. This organization is similar to ASTA.

International Association for Plant Tissue Culture[8]. This organization is open to anyone interested in in-vitro plant cell, tissue, and organ culture (see Chapters 17, 18). The association publishes a newsletter three times a year and sponsors an international conference every four years. Members represent both academic and commercial organizations.

International Dwarf Fruit Tree Association[9]. This organization is for members interested in fruit tree rootstocks and propagation but also includes cultural aspects. An annual meeting is held whose papers are published in the proceedings *Compact Fruit Tree*.

International Seed Testing Association (ISTA)[10]. This is an intergovernmental association with worldwide membership accredited by the governments of 59 countries and involves 137 official seed-testing associations. The primary purpose is to develop, adopt, and publish standard procedures for sampling and testing seeds and to promote uniform application of these procedures for evaluation of seeds moving in international trade.

Secondary purposes are to promote research in all areas of seed science and technology, to encourage cultivar certification, and to participate in conferences and training courses promoting these activities.

A meeting is held every three years. A journal, *Seed Science and Technology,* is published monthly with a quarterly newsletter, *ISTA News Bulletin*. A number of technical handbooks are also published.

International Society for Horticultural Science[11]. This organization is an international society for horticultural scientists, educators, extension, and industry personnel. It sponsors an International Horticultural Congress every four years as well as numerous workshops and symposia. Proceedings are published in *Acta Horticulture*. A newsletter, *Chronica Horticulturae,* is published four times per year.

Professional Plant Growers Association, Inc. (Bedding Plants, Inc.—B.P.I.)[12]. This organization is directed towards bedding plant growers. Annual meetings are held and papers published in an annual proceedings. A *Buyer's Guide* is published every two years. In addition, a newsletter is published several times a year.

Society for in vitro Biology[13]. Organization is composed of biologists, both plant and animal, who do research on plant cellular and developmental biology. Includes use of plant tissue culture techniques. Publishes journal of scientific papers quarterly. Holds annual meeting.

REFERENCES

1. Adriance, G.W. and F.R. Brison. *Propagation of Horticultural Plants*. New York: McGraw Hill.
2. Bailey, L.H., 1891, revised, 1896, *The nursery book*. Harrisburg, PA: Mount Pleasant Press, J. Horace McFarland Co.
3. Bailey, L.H., 1900–1902, *Cyclopedia of American horticulture*. New York: MacMillan, 4th ed., 1906.
4. Bailey, L.H. 1914–1917. *Standard cyclopedia horticulture*. 3 Volumes. New York: MacMillan Company.

[6]2948 Hillsborough Street, Raleigh, NC 27607 U.S.A.

[7]2948 Baseline Road, #207, Ottawa, Ontario K2H 8T5, Canada

[8]Soil and Crop Sciences Dept., Texas A & M University, College Station, TX 77843 U.S.A.

[9]Dept. of Horticulture, Michigan State University, East Lansing, MI 48824 U.S.A.

[10]P.O. Box 412, CH-8046, Zurich, Switzerland

[11]ISHS Secretariat, K. Mercierlaan 92, 3001 Leuven, Belgium

[12]P.O. Box 27517, Lansing, MI 48909 U.S.A.

[13]8815 Centre Park Drive, Suite 210, Columbia, MD 21045 U.S.A.

5. Bailey, L.H. 1920 (revised). *The Nursery-Manual.* New York: MacMillan Company.

6. Bailey, L.H. 1930. *Hortus.* New York: MacMillan Company.

7. Bailey, L.H. and E.Z. Bailey. 1941. *Hortus Second.* New York: MacMillan Company.

8. Bailey, L.H., E.Z. Bailey, and staff of Bailey Hortorium. 1940, 1949. *Manual of cultivated plants.* New York: MacMillan.

9. Bailey, L.H., E.Z. Bailey, and staff of Bailey Hortorium. 1976. *Hortus Third.* New York: MacMillan.

10. Baltet, C. 1910. *The art of grafting and budding,* 6th ed. London: Crosby Lockwood (quoted by Hottes, 1922).

11. Cunningham, I.S. 1984. *Frank N. Meyer: plant hunter in Asia.* Ames, Iowa: Iowa State University Press.

12. Darwin, C., 1959. *The Origin of Species.* Philadelphia, PA: Univ. of Pennsylvania Press.

13. Darwin, C. 1868. *The variation of animals and plants under domestication.* London, Eng: J. Murray.

14. Davidson, H., Peterson and R. Mecklenburg. 1994. *Nursery management.* 3rd edition. Englewood Cliffs, N.J.: Prentice-Hall.

15. Duruz, W.P. 1st edition 1949, 2nd ed. 1953. *The principles of nursery management.* New York: A.T. de la Mare Company, Inc.

16. Fairchild, D. 1938. *The world was my garden.* New York: Schribner's.

17. Fuller, A.S. 1887. *Propagation of plants.* (quoted by Hottes, 1922).

18. Harlan, J.R. 1992. *Crops and Man.* 2nd edition. Madison, WI: Amer. Soc. of Agron., Inc. Crop Science of America, Inc.

19. Hartmann, H.T. and D. E. Kester. 1959. *Plant propagation—principles and practices.* Englewood Cliffs. N.J.: Prentice-Hall.

20. Hartmann, H.T., A.M. Kofranek, V.E. Rubatsky, and W. J. Flocker. 1988. *Plant Science: Growth, development and utilization of cultivated plants.* Second edition. Englewood Cliffs, N.J.: Prentice-Hall.

21. Hottes, A.C. 1917, 1922 (revised). *Practical Plant Propagation.* New York: A.T. de la Mare Co., Inc.

22. Janick, J., R.W. Shery, F.W. Woods, and V. W. Ruttan. 1969. *Plant Science.* San Francisco: W.H. Freeman and Company.

23. Kains, M.G. 1916, 1920. *Plant Propagation. Greenhouse and nursery practice.* New York: Orange Judd Company, Inc.

24. Kains, M.G. and L.M. McQuesten. 1938, 1942. *Propagation of plants.* New York: Orange Judd Publishing Co., Inc.

25. Mahlstede, J.P. and E.S. Haber. 1957. *Plant Propagation.* New York: Wiley.

26. Reed, H.S. 1942. *A short history of the plant sciences.* New York: The Ronald Press Company.

27. Sauer, C.O. 1969. *Agricultural origins and dispersal.* 2nd edition. Cambridge, Mass.: Massachusetts Institute of Technology Press.

28. Simmonds, N.W., ed. 1976. *Evolution of crop plants.* London: Longman Group Limited.

29. Solbrig, O.T. and D.J. Solbrig. 1994. *So shall you reap. Farming and crops in human affairs.* Washington, D.C.: Island Press.

30. Ward, N.B. 1842. *On the growing of plants in closely glazed cases.* 2nd. edition. London: J. van Voorst.

31. Zohary, D. and P. Spiegel-Roy. 1975. *Beginnings of fruit growing in the old world.* Science 187(4174): 319–27.

SUPPLEMENTARY READING

DeCandolle, A.L.P. 1902. *Origin of Cultivated Plants.* D. Appleton & Co.: New York.

Miller, P. 1768. *The gardener's dictionary.* London.

Raven, P.H., R.F. Evert and S.E. Eichhorn. 1992. *Biology of Plants.* 5th edition. Chapter 31. New York: Worth Publishers.

Spongberg, S.A. 1990. *A reunion of trees. The discovery of exotic plants and their introduction into North American and European landscapes.* Cambridge, Mass.: Harvard Univ. Press.

Viola, H.J. and C. Margolis. 1991. *Seeds of change.* Washington, DC: Smithsonian Institution Press.

2

Biology of Propagation

Introduction. Plant propagation involves the application of specific biological principles and concepts in the multiplication of plants for useful purposes. Implicit in this concept is maintaining the unique characteristics of the specific plants being propagated.

Any living organism can be described by genotype and phenotype. **Genotype** is the sum total of all of the genetic characteristics of the organisms controlled by genes. **Phenotype** is the overall appearance and performance of the organism, including size, vigor, color, seasonal patterns, shape, adaptations, and other traits that make the organism unique. The relationship between these two terms is a fundamental feature of biology in that the *phenotype results from the interaction of the genotype with the environment within which the organism is growing.*

Plant propagation deals with the multiplication and production of plants using propagules representing a specific genotype. *A propagule is any plant part used to produce a new plant or a population of plants.* Specific propagules include seeds, cuttings, layers, buds, scions, explants, and various kinds of specialized structures such as bulbs, corms and tubers. To maintain the genotype of the plant population being propagated, it is important to control the selection of the seed parents or the vegetative source of the propagule.

In this chapter, five fundamental concepts of biology are described which provide the fundamental framework upon which these dual plant propagation objectives depend. These include:

a) sexual versus asexual reproduction,
b) gene expression in the control of plant growth and development,
c) hormonal control of growth and development,
d) plant life cycles and phase change, and
e) plant nomenclature and the concept of the cultivar.

GENETIC CONTROL IN PROPAGATION: SEXUAL VERSUS ASEXUAL

The life cycle of flowering plants and ferns takes place in two biological generations (*24, 59*) (Figure 2–1). One—referred to as **sporophytic**—involves the vegetative growth of the plant body in size and structure (Figure 2–2). The sporophytic generation is asexual (without sex) in which the genotype is maintained among the cells of the plant through the type of cell division known as **mitosis.** The other generation—referred to as **gametophytic**—is much reduced in size and structure in the flowering plants (see Figure 2–4) or, as in ferns (see Figure 2–5), has a distinctly unique structure. The gametophytic generation involves special reproductive cells **(gametes)** and provides the opportunity to interchange genetic information each generation.

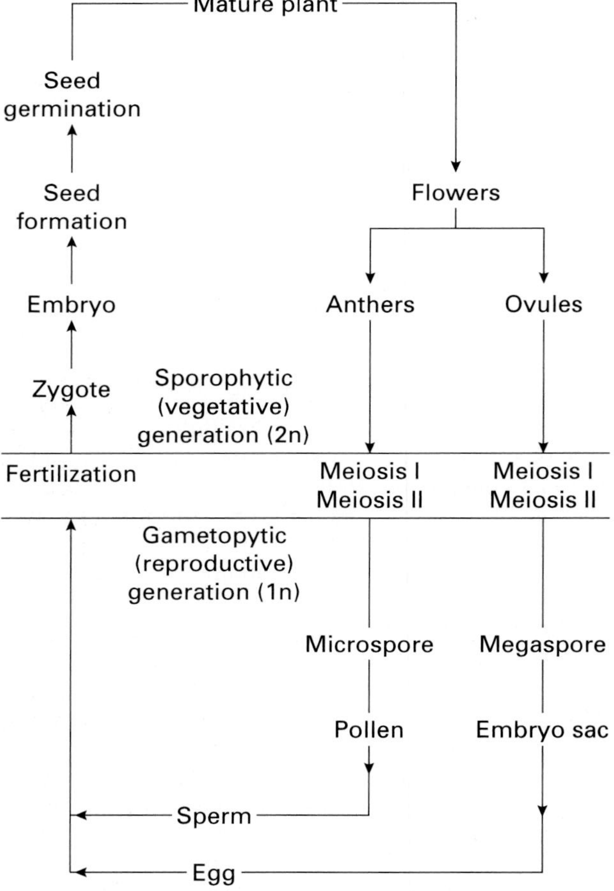

FIGURE 2–1 Alternating sporophyte and gametophyte generations in flowering plants. These two generations correspond to the vegetative (see Figure 2–2) and reproductive phases (see Figure 2–4) in plant development.

These two alternating cycles provide the opportunity for individual organisms with a specific genotype not only to multiply and increase their distribution but also to introduce variation through the reshuffling of genetic factors. In flowering plants and ferns, the sporophytic generation corresponds to the vegetative phase of the life cycle—i.e., growth of shoots and roots—and the gametophytic generation to the reproductive (flowering) phase.

In nature, the sporophytic (vegetative) generation allows the plant to occupy a specific environmental site. The gametophytic (reproductive) generation allows the population of plants to evolve new genotypes even more adapted to that or other sites.

During cultivation, genetic variation is controlled. First, plant breeders utilize genetic variation to create new kinds of plant materials. Secondly, propagators maintain and multiply the specific genotypes that are created.

Mitosis, Asexual Reproduction, and Vegetative Propagation

The sporophytic generation of flowering plants and ferns (nonflowering) is characterized by plant structures as stems, leaves, and roots which are composed of living and dead plant cells. A living plant cell contains a **nucleus,** and a **cytoplasm,** enclosed by a **cell wall**. The nucleus is surrounded by a membrane and the cytoplasm by the **plasma membrane (plasmalemma)** enclosing the cytoplasm, both of which have important functions in controlling movement of solutes.

The nucleus contains the **chromosomes,** which includes DNA that stores genetic information in **genes.** In the nondividing cell, the DNA is dispersed throughout the nucleus, but during the dividing phase becomes shortened and thickened into distinct **homologous pairs** of chromosomes with a specific morphology, size, and number for a particular genotype. The significance of the chromosome pairing is that each contains corresponding **complementary alleles** of single genes associated with specific locations on the chromosomes. The number of pairs of homologous chromosomes gives the **haploid** (n) number with the total number of chromosomes the **diploid** (2n) number. A species contains a specific number of chromosomes which may be dif-

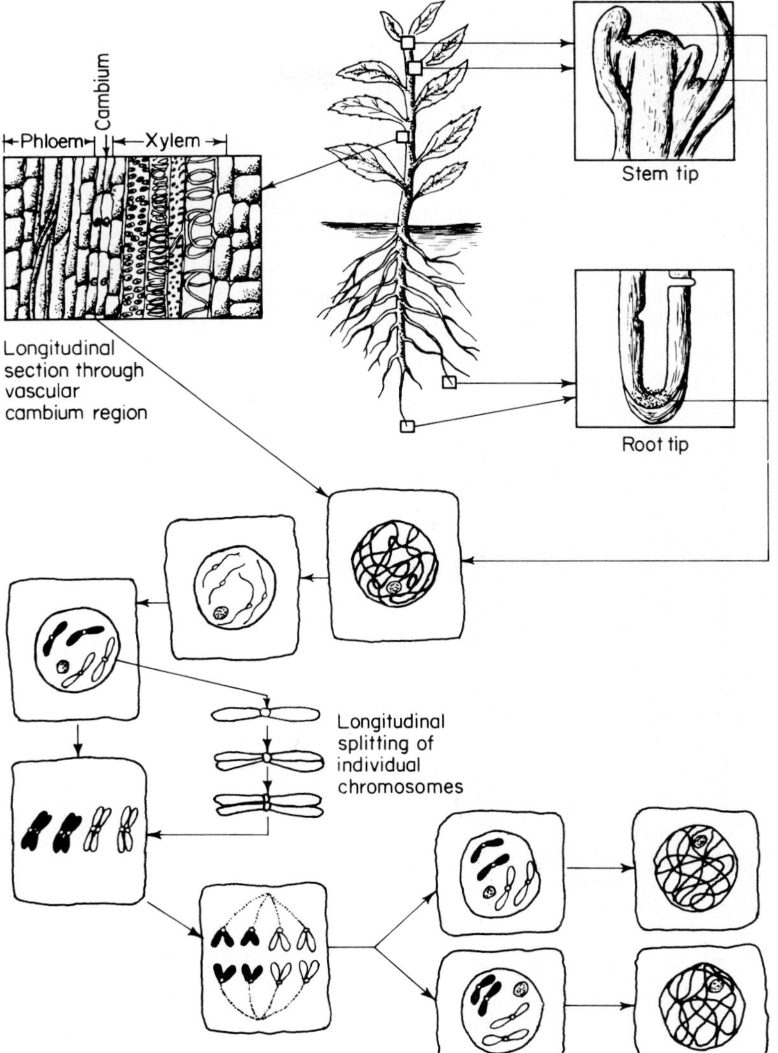

FIGURE 2–2 *Upper:* Meristematic (growing) regions in a dicotyledonous plant are located in the shoot tips (terminal and lateral), tips of primary and secondary roots, and the cambium. *Lower:* Mitosis in a meristematic cell involves the duplication of individual chromosomes to produce two daughter cells whose genotype is the same as the original cell.

ferent multiples of the basic n number. For example, the haploid (n) number for *Prunus* species (stone fruits) is 8. Vegetative cells of members of this genus, as peaches, apricots, and almonds, have 16 (2n = diploid); some cherries have 24 (3n = triploid); some plums have 32 (4n = tetraploid) and others 64 (8n = octoploid) chromosomes.

The cytoplasm contains **plastids, mitochrondria, vacuoles, ribosomes, dictyosomes,** and other structures (*59*). Some of these structures, plastids, mitochrondria, and ribosomes contain DNA and/or RNA which have "genetic" control.

Mitosis is the name given to cell division of vegetative cells in which each chromosome is duplicated (Figure 2–2). Growth in size and volume of the plant results from a combination of cell division **(mitosis),** cell enlargement, and cell differentiation. Mitosis takes place in primary **(shoot tip, root tip)** and secondary **(cambium, cork cambium)** growing regions **(meristems)** of the plant. Mitosis therefore is essential both for plant growth and vegetative propagation. Growth from the shoot apices allows the shoots and leaves to extend into the atmosphere and growth from the roots into the soil. Secondary meristems include the **cambium,** which increases the circumference of the stem and subsequently the volume, the **cork cambium,** which provides the outer bark protective area, and **intercalary zones** which extend shoot length at the base of internodes in monocots, such as grasses.

Mitosis is important in wound healing. The production of a mass of cells **(callus)** is the first step in the healing of the damaged area and is usually associated with the rooting of cuttings (Chapter 10). Wound induced callus formation is also involved in the formation of graft unions (Chapter 12) (Figure 2–3).

Mitosis initiates new (*de novo*) growing points on established vegetative structures such as a stem, root, or leaf. The term applied to these is **adventitious**. Some species develop natural adventitious roots (*7*), such as prop roots on monocots, aerial roots of some trees, tip layers of some blackberry (*Rubus*) species, and roots on bulbs, corms, and rhizomes. In nature, the ability to produce adventitious roots allows some species to adapt to specific environments and extend their range into new areas (*44*).

In plant propagation (*18, 60*), the ability to initiate adventitious roots on shoots and shoots on roots is the basis for the rooting of cuttings (Chapters 10 and 11), layering of plants (Chapter 15) and propagation by separation and division (Chapter 16). Adventitious organ formation is brought to a high level of sophistication and control in micropropagation and tissue culture (Chapters 17 and 18).

The significance of mitosis is that it is asexual. That is, no change in genotype normally takes place and the genotype of each propagule is maintained in the daughter cell during vegetative propagation. Individual plants of the progeny population are thus genetic copies both of the original source plant and each other. *The population of plants generated from a single individual which maintains the same genotype is known in biology as a* **clone** (*65, 66, 67, 69, 74, 75, 83*) *and the process as* **cloning** (*45*). In theory, all plants of a clone are identical in appearance; in practice, certain kinds of phenotypic and genetic variation may develop. For example, variation may occur at different planting sites due to environmental effects. Genetic changes **(mutations)** may occur spontaneously or result from exposure to specific factors in the environment. Whatever the cause, a certain amount of variation is associated with clones and may require special procedures to control during vegetative propagation (see Chapter 9).

Meiosis, Sexual Reproduction and Seedling Propagation

The gametophytic generation in the life cycle involves the formation of flowers, the production of male (pollen) and female (egg) gametes, pollination, and, after fertilization, the formation of a single-celled **zygote** which develops into the seed (Figure 2–4). In ferns, nonflowering plants, the structures are different but the process is similar (Figure 2–5).

Male and female **gametes** develop through a special kind of cell division known as **meiosis.**

FIGURE 2–3 Vegetative regeneration occurring in asexual propagation. *Left:* Adventitious shoots growing from a root cutting. *Center:* Adventitious roots developing from the base of a stem cutting. *Right:* Callus tissue produced from scion and stock in the formation of a graft union.

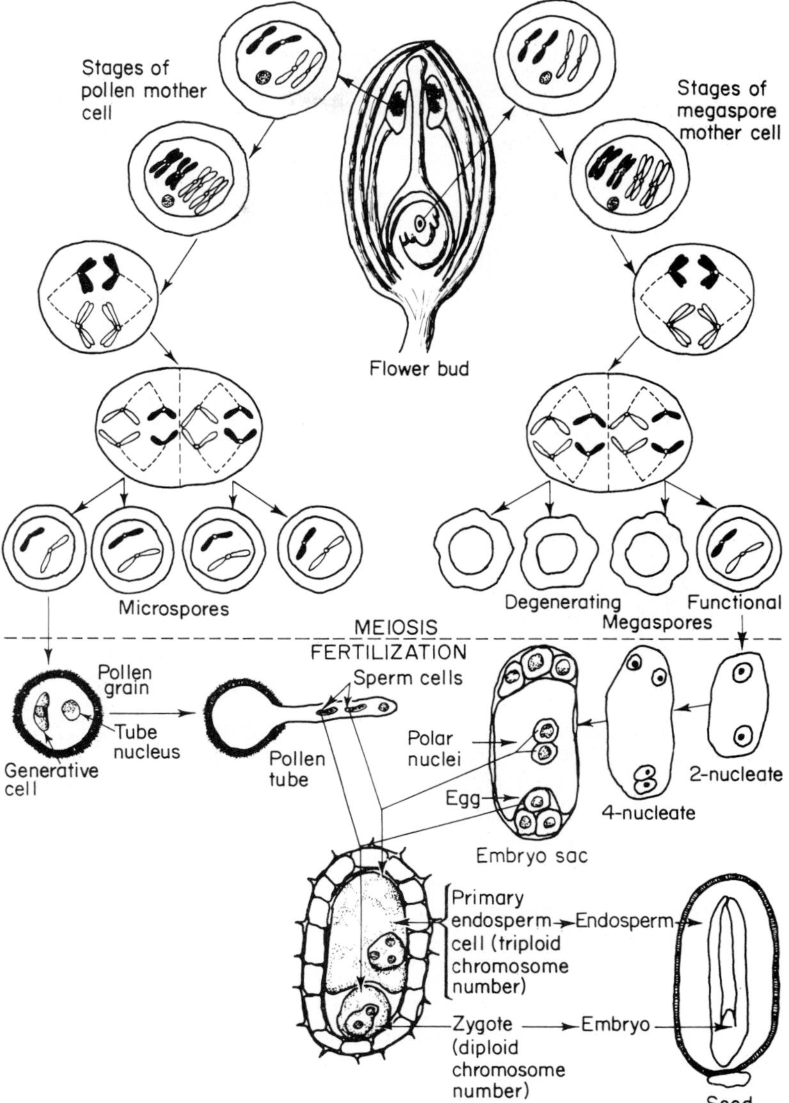

FIGURE 2–4 Sexual reproduction in a flowering plant. Meiosis occurs in the flower bud in the anther (male) and in the pistil (female) during the bud stage. During meiosis I the homologous chromosome pairs split to produce two attached chromatids, undergo crossing-over to exchange genetic material, and line up on opposite sides of the cell. During meiosis II, the two chromatids separate to produce four gametes, each with one half of the original number of chromosomes. Fertilization unites a male gamete (sperm) with a female gamete (egg) to produce a **zygote** which restores the diploid chromosome number. A second male gamete unites with two polar nuclei to produce the triploid **endosperm.**

Meiosis involves two steps. In meiosis I, homologous chromosomes thicken and pair together across the center of the cell. Each chromosome divides longitudinally into two **chromatids** which remain attached at one location. Exchange of genetic material can take place **(crossing-over)** between parts of individual chromatids at this point; each pair of **chromatids** then separates (meiosis II) to produce four separate gamete cells, each with half (i.e., n number) the number of chromosomes of the original diploid cell. This is called **reduction division** and results in haploid cells which develop into the **gametophytes,** i.e., the **pollen** and **embryo sac.** During flowering, **pollination** occurs to release the pollen, and the pollen tubes from each pollen grain grow into the flower where male gametes combine with the female gamete in the embryo sac **(fertilization).** This act restores the original diploid chromosome number but results in rearrangement of alleles. In higher plants, this single celled **zygote** is the beginning of the next sporophyte generation. In the process, chromosomes have segregated, alleles are rearranged through crossing-over, and the original diploid chromosome number is restored, resulting in a different genotype and a new organism.

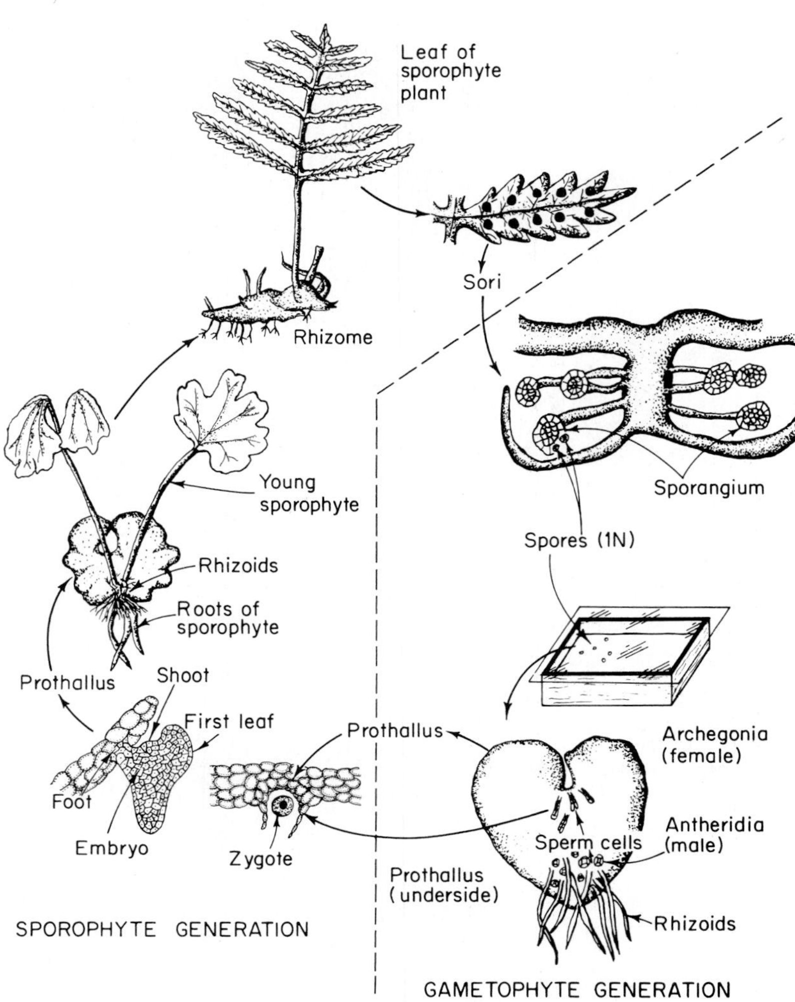

FIGURE 2–5 Growth and reproduction in a fern, a nonflowering plant showing the sporophyte and gametophyte generations. The sporophyte generation is the conspicuous vegetative portion that develops from a zygote, is diploid, and has roots, stems, and leaves. The gametophyte generation develops from haploid spores that develop through reduction division in special structures on the leaf. A prothallus is produced upon which are borne the male and female structures, which produce gametes that unite to produce a zygote that repeats the generation cycle.

The genotype of an individual seedling offspring determines its morphological appearance, physiological characteristics, and interaction with the environment. The degree of variability within this seedling population and the resemblance between parents and offspring depends upon the similarity of the paired alleles of individual genes. This relationship can be expressed by (a) the level of **heterozygosity** of the alleles on the two paired chromosomes, (b) the relative **dominance** and **recessiveness** of the individual alleles and, (c) the pattern of inheritance (see Figures 2–6 and 2–7). The principle utilized in maintaining genetic uniformity involves the relative **homozygosity** and **heterozygosity** of the parent plant(s). If corresponding alleles on homologous chromosomes are the same, the genotype is **homozygous** and the progeny will resemble the parent in respect to this gene; if they differ, the genotype is **heterozygous** and segregation for the two alleles takes place in the next generation. Since a large number of separate genes are on each chromosome, the relative heterozygosity of the parent genotype will determine the variability of the seedling progeny population.

In nature, seedling variability provides the opportunity for selection in the evolution of better adapted individuals (*29, 59, 70, 71*). In practice, genotypes become more or less fixed when grown over a long period of time in the same environmental site and evolve into **species** (see page 33). In cultivation, genetic variation provides the opportunity for plant breeders to develop populations of genotypes which have special value to humans: increased yield, special qualities, or unique appearance. These special groups of man-made and man-preserved populations of genotypes are referred to as **cultivated varieties** or **cultivars** (see page 34).

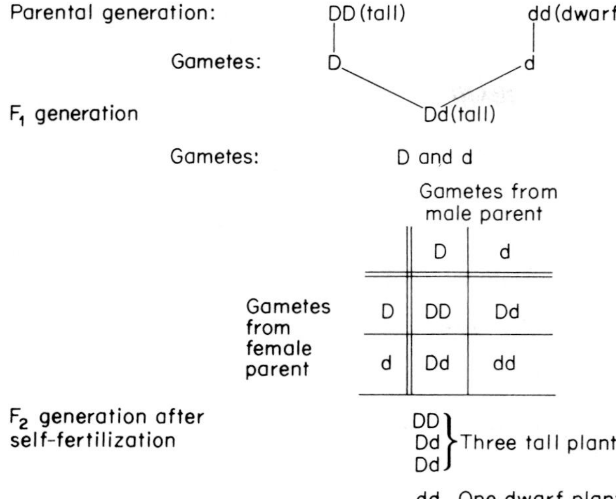

FIGURE 2–6 Inheritance involving a single pair of alleles in the gene controlling height of garden pea. *Tallness (D)* is dominant over *dwarf (d)*. A tall pea plant is either homozygous *(DD)* or heterozygous *(Dd)*. Segregation occurs in the F_2 generation to produce three genotypes *(DD, Dd, dd)* and two phenotypes (tall and dwarf).

Plant propagation and plant breeding utilize the same principles, many of the same practices but different objectives. Plant breeding generates variation by transferring genes through sexual reproduction and then stabilizes **(fixes)** the genotype (see Chapter 4). A large body of knowledge and methodology has developed over time to allow geneticists and plant breeders to manipulate genetic variation and improve the likelihood of creating new and improved cultivars (*1, 38, 56, 73*). Modern horticulture, as represented by commercial and private nurseries, has at its disposal an astonishing range of cultivars from which to choose (*6, 19, 49;* also see Chapters *19, 20, 21*).

Plant Breeding, Genetic Engineering and Biotechnology

Plant improvement has expanded from traditional plant breeding to include **genetic engineering** (*81*) and **biotechnology** (*34, 42*). This expansion has resulted from fundamental biological discoveries from a number of disciplines (genetics, bacteriology, biochemistry, physiology) (*34, 42*). The basic discoveries involve molecular biology of the gene; transfer systems of genetic information (see boxes, p. 17 and p. 20); the ability to manipulate plants, plant parts, and cells in culture (tissue culture) (see

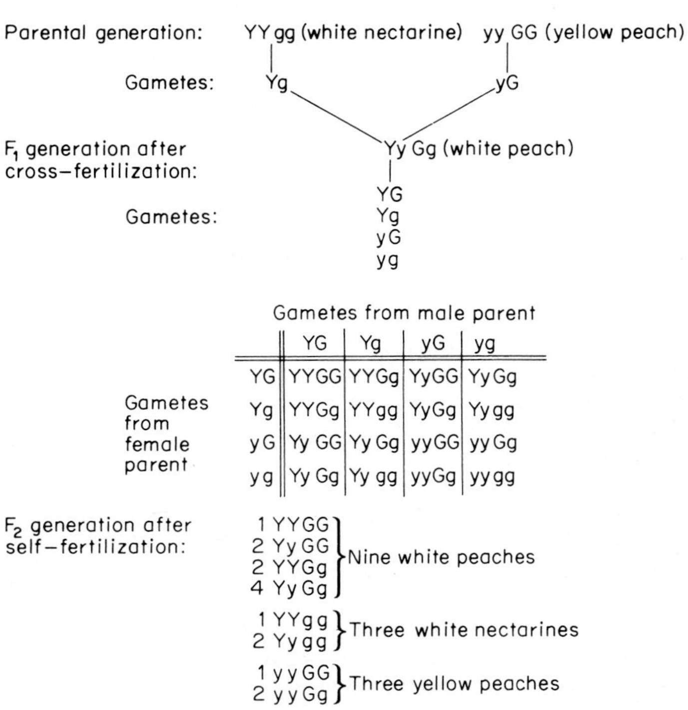

FIGURE 2–7 Simultaneous inheritance of two genes in a cross involving peach and nectarine (*Prunus persica*). *Fuzzy skin (G)* of a peach is dominant over the *smooth skin (g)* of the nectarine. *White flesh color (Y)* is dominant over *yellow flesh color (y)*. In the example shown, the phenotype of the F_1 generation is different from either parent. Segregation in the F_2 generation produces nine genotypes and four phenotypes.

Chapters 17, 18); the technology to clone genes and parts of chromosomes; and the capability to combine different kinds of DNA to **transform** individuals to produce **transgenic** plants.

The production of transgenic plants by **recombinant DNA technology** makes it possible to bypass normal sexual incompatibility that prevents genetic material from being transferred from one biological group (species, genera, and even plants and animals) to another, as illustrated in Figure 2–8. Traditional breeding combines entire sets of genes through hybridization among sexually compatible parents. Introduction of a single trait may require repeated backcrosses to the recipient parent, an effective method but one requiring time and generally large plant populations. Genetic engineering can create new genotypes outside of the traditional (whole-plant) methods of cross-fertilization (*16*) (Figure 2–8). Single genes affecting specific traits can be transferred into an already established cultivar utilizing genes from various biological sources (*16*).

A direct application of molecular biological research to plant propagation is the use of **molecular biochemical markers** to identify cultivars in propagation systems (Chapter 9). Culture techniques used in biotechnology propagate specific cultivars in aseptic culture under the term **micropropagation.** (Chapters 17 and 18).

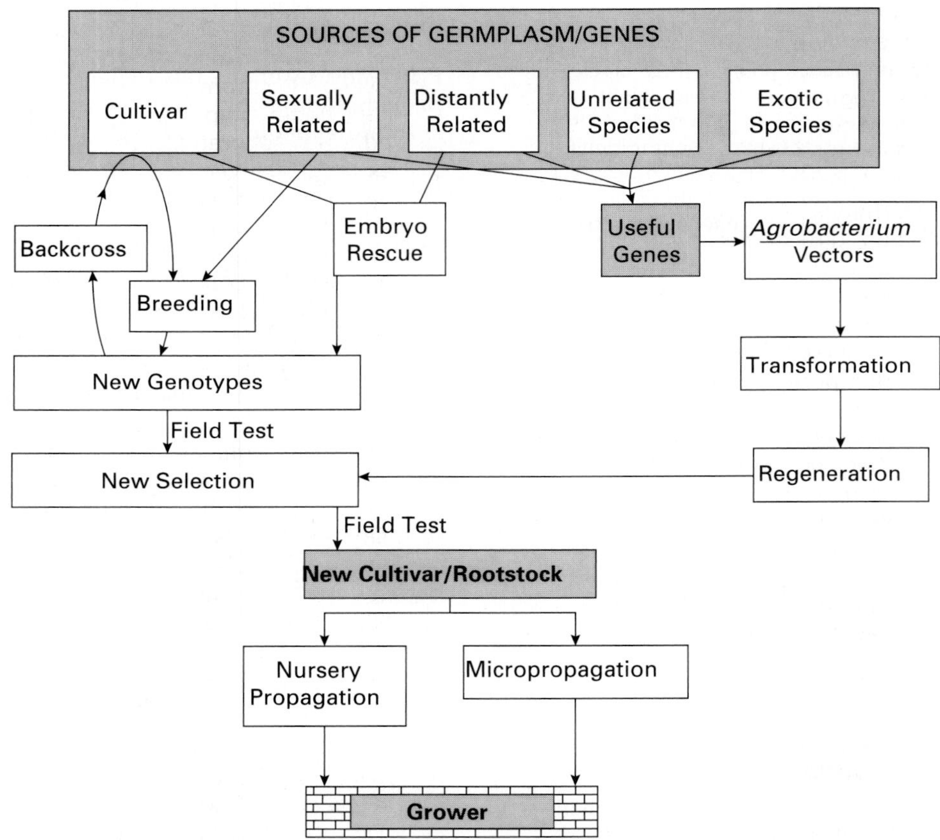

FIGURE 2–8 Relationship among conventional plant breeding, biotechnology, genetic engineering techniques, and propagation. Gene sources for conventional breeding are limited mostly to compatible species although the range can be extended by some techniques as embryo rescue. Recombinant gene technology can transfer useful genes from many biological sources to create new genotypes. New cultivars must be evaluated and introduced under the same conditions as conventional breeding. Reproduced by permission from Dandekar, et. al (16) 1992. Transgenic woody plants. In: Shaindow and R. WU (ed). *Transgenic plants.* 2. Academic Press: San Diego, CA. pp 129–151.

How is genetic information controlled? Three types of large macromolecules (Figure 2–9) essential to life are involved in inheritance and gene expression: **DNA** (deoxyribosenucleic acid), **RNA** (ribosenucleic acid) and **proteins** (*24, 59*). DNA is composed of long chains of **nucleotides,** which are combinations of four chemical **bases** (thymine, adenine, guanine, cytosine), a special sugar molecule (deoxyribose), and phosphoric acid (PO_4^-). Nucleotides in turn are joined into long chemical strands by PO_4^- radicles that connect the 5′ position of one sugar molecule to the 3′ position on the next. RNA has a similar structure but is a single strand, has a different sugar (ribose), and includes uracil instead of thymine. Proteins are large complex molecules made up of chains of different amino acids which are organized into groups of poly-peptides. Proteins may function as storage proteins, regulatory enzymes, and structural components of cells.

The specific combination of bases on chromosomes provides the genetic code that controls inheritance, determines the specific genotype of the organism, and directs the pattern of gene expression. A coding unit consists of three of the four bases and is known as a **codon.** A series of codons translates into a particular amino acid of which about 20 are known. These in turn combine to synthesize unique proteins which differ in their amino acids composition.

Chromosomes consist of proteins **(histones)** combined with two complementary chains of DNA. The two strands are arranged in a spiral, known as a **double helix,** connected to each other by a loosely attracted **hydrogen bond** between a base on one strand of DNA and a base on the other (adenine with thymine and guanine with cytosine). DNA has the unique capacity to replicate itself during mitosis when catalyzed by the enzyme **DNA polymerase.** During meiosis, DNA strands segregate and subsequently replicate to form four new combinations of DNA at each division.

From a molecular standpoint, a gene can be described as a linear piece of DNA which includes (a) an **initiation codon** (start site or promoter), (b) an **exon** (coding region), (c) an **intron** (noncoding region) and, (d) a **termination codon** (stop signal) (*61*). Adjoining the gene on the chromosome are additional areas which can function as **regulator sequences** and determine when a gene is turned on or off (Figure 2–10). Bacterial genes are somewhat similar but lack the introns.

Recombinant DNA technology. A piece of cloned DNA from one organism is inserted into the DNA on a chromosome of another organism to alter its genotype. Gene transfer is accomplished in the laboratory by several systems. One utilizes biological vectors (e.g., **plasmids** from certain bacteria, such as *Agrobacterium tumefaciens* and *A. rhizogenes* normally associated with the known plant diseases "crown gall" and "hairy root"). Bacterial DNA is joined to plant DNA to form **recombinant DNA.** A marker gene is incorporated into the DNA clone which can subsequently identify the recombinant gene in a cell. Examples of marker genes include *luciferase* from fireflies, which provides bioluminescence under specific conditions, *beta-glucuronidase* (GUS), which produces a blue color, and *kan* which confers resistance when grown in a medium with kanamycin (*42*). Other methods of DNA transfer include high voltage electric pulses **(electroporation)** and high-velocity microprojectiles coated with DNA or RNA **(particle bombardment)** (*42*).

PLANT DEVELOPMENT, GENE EXPRESSION, AND EPIGENETIC VARIATION

One of the principles in biology is that each living cell has the potential to reproduce an entire organism, since it possesses all of the necessary genetic information. The concept is known as **totipotency**(*31, 53, 45*). Experimental confirmation (*72*) of this concept was provided by producing adventitious embryos and subsequently entire plants from cell suspensions of wild carrot (Figure 2–11) and tobacco (*9, 80*).

In the intact plant, totipotency applies most directly to the zygote (Figure 2–4) and to the meristematic cells in growing points of the shoot and the root (Figure 2–2). Most cells differentiate into specialized kinds that determine structure or carry on specific functions. Many lose their capability to regenerate; others die and provide the structure of the plant.

Plant development and growth patterns are associated with seasonal and environmental cycles. Phenotypic variations that develop within a single individual result from the translation of genetic

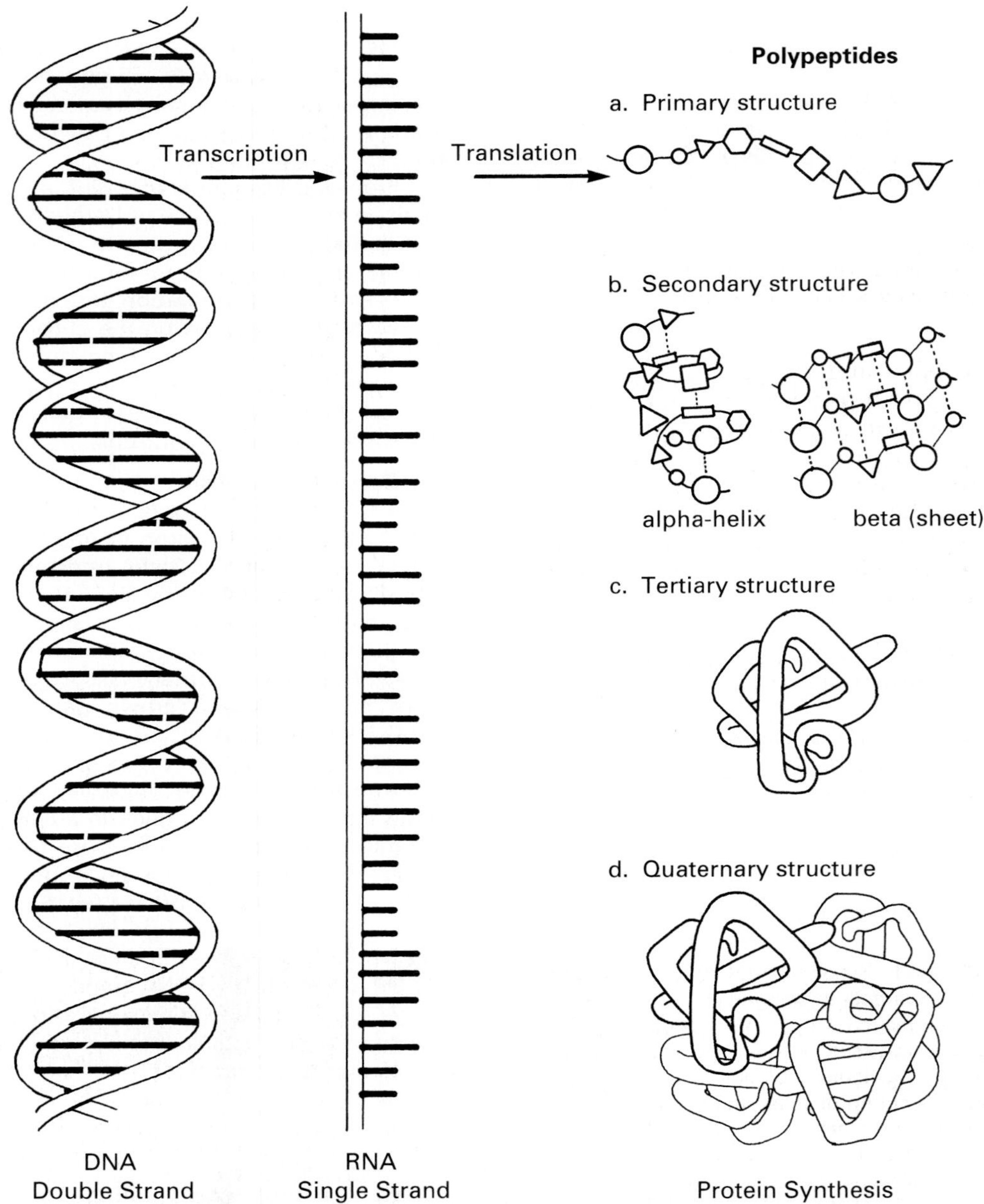

FIGURE 2–9 Three major macromolecules in the plant control growth and development. *Left:* DNA is composed of two long complementary chains of nucleotides that provide the genetic code which controls the composition of specific amino acids. *Center:* RNA molecules consist of a single long strand of nucleotides complementary to DNA. mRNA transcribes selected parts of the genetic code directly from DNA, and moves from the nucleus to the cytoplasm where genetic information is translated to direct the synthesis of specific proteins. *Right:* Proteins are large complex molecules. Four structural levels of a protein molecule are shown: (a) peptide composed of a sequence of amino acids, (b) polypeptide composed of peptides, which may have either a helical or flat structure, (c) combined polypeptides, which has a globular structure, and (d) protein shown with a quaternary structure. Proteins function for storage, as regulators (enzymes), or structural components of cells.

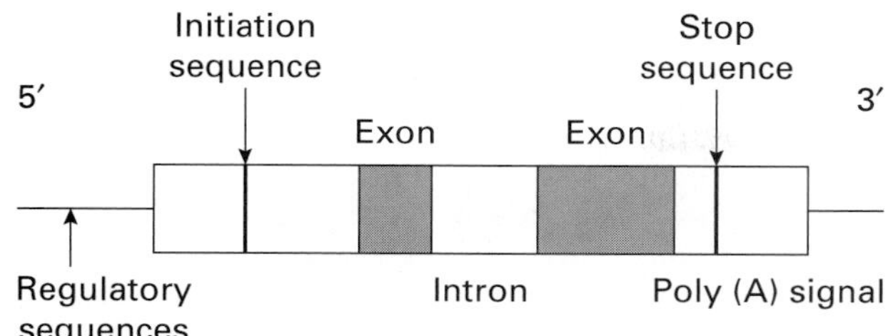

FIGURE 2–10 Schematic drawing of the structure of a gene as described in the text.

information encoded in genes to the formation of plant structure, growth patterns, and function. The phenomenon is known as **gene expression** and its control is the primary activity of plant propagation. Processes by which information in the DNA (i.e., genetic code) of the chromosome is channeled to direct growth, development, and plant structure are referred to as **epigenetic** (*24*) because change does not occur in the basic genotype of the plant during this process.

Variation related to epigenetic forces can be classed into three broad categories:

a) *Formation of specific structures, i.e., stems, roots, leaves, and flowers.* The process involves growth, enlargement, and differentiation of specialized cells within organs to produce the morphological and physiological variations that constitute the whole plant.

b) *Control of developmental cycles.* The primary cycles include seasonal cycles **(phenology)** and life cycles **(ontogeny)** and the unique morphological and physiological expressions associated with them. This category includes the phenomenon known as **phase change,** which will be described in more detail in the next section of this chapter. These cyclic patterns result from the fact that genes are turned on (i.e., expressed) at specific times in the life of the plant and/or in response to specific environmental or internal signals of the plant.

c) *Induction of adventitious shoots, roots, or embryos.* These events take place if specific cells retain the potential for regeneration during development or specific cells are induced to "dedifferentiate" and develop the capacity to regenerate. This potential to regenerate is the basis for vegetative propagation, particularly propagation by cuttings (Chapter 10) and micropropagation (Chapter 17).

Competency. The term **competency** is used to describe the potential of a given cell or tissue to develop in a particular way (*50, 51, 54*) (Figure 2–13), for instance, to initiate adventitious roots, buds, or embryos. This potential is a kind of internal "memory" within the cell mechanism. Development into a specific kind of cell, tissue, or organ may require a signal either from within the plant or

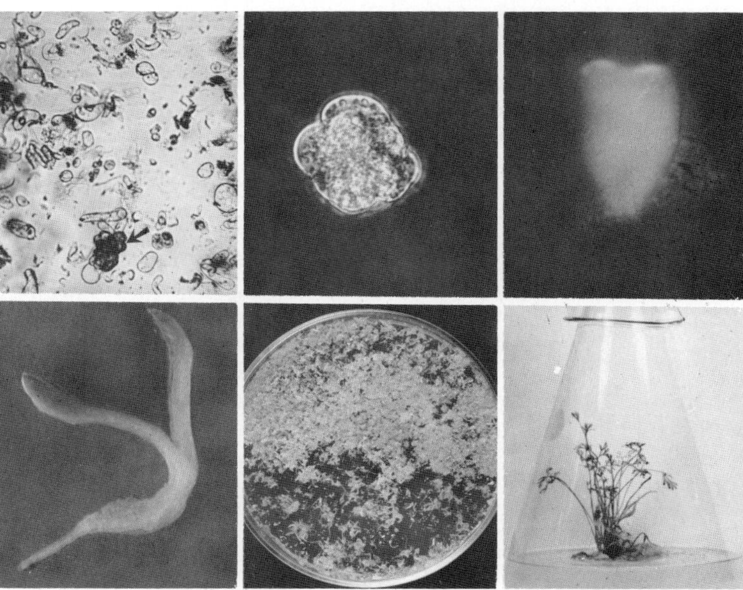

FIGURE 2–11 Totipotency of cells in carrot. *Top left:* Highly magnified view of suspended cells and cell clumps derived from tissue cultures of carrot growing in a liquid medium. Arrow points to a clump of cells—the beginning stage for a new plant. *Top center:* Single-cell clump (higher magnification) illustrating the globular stage of embryo development (see p. 132) *Top right:* A more advanced heart-shaped stage of embryo development. *Below left:* Mature embryoid that developed continuously from the globular stage, through the heart and torpedo (not shown) stages. *Below center:* Thousands of carrot plantlets that developed as embryoids. *Below right:* Single carrot plant that grew from a single embryoid transplanted from previous stage. (Courtesy F.C. Steward and M.O. Mapes.)

How does genetic information become expressed? DNA acts as a template to synthesize RNA molecules upon which the genetic code for specific amino acid sequences is transcribed. These new molecules are known as **messenger RNA** (mRNA) *(24, 59)*. Transcription starts with the recognition of genetic information encoded in DNA in the chromosomes (Figure 2–12). Recognition involves (a) regulation of the time or sequence in which a gene is turned on or off and, (b) transcription of specific DNA codons to messenger RNA's (mRNA). Transcription is mediated by a specific enzyme **(RNA polymerase)** which in plants is complicated (as compared to bacteria) by the fact that the intron (noncoding) sequences must be removed.

Single-stranded mRNA molecules—which may vary from 100 to 10,000 nucleotides in size—move from the nucleus into the cytoplasm (upper part of Figure 2–12). Two other kinds of RNA molecules in turn function to translate the "message" into information to produce proteins. Each transfer **RNA (tRNA),** which is small, about 80 nucleotides in size, is coded to a single kind of amino acid. The tRNA molecule has two important sites. One attaches itself to a specific amino acid at the 3′ end. The other, known as an **anticodon,** recognizes specific codons on mRNA at the 5′ end. RNA **(ribosomal RNA or rRNA)** in association with protein produce the **ribosome** within which **translation** (bottom of Figure 2–12) occurs and proteins are synthesized (middle of Figure 2–12). Specific codons on the mRNA initiate the process (*initiation*) and recognize a specific tRNA anticodon for a specific amino acid. As consecutive codons are revealed by the mRNA, a polypeptide begins to develop which continues to grow (*elongation*) until a special codon signal is reached (*termination*). As the process takes place, amino acids are released to the growing protein strand. tRNA and an RNA in turn are released to the surrounding cytoplasm.

from the environment. For example, cells in a growing plant may be induced to initiate flower buds in response to a particular length of day (*79*).

Determinism. The development of competency by a tissue requires a certain amount of time and/or exposure to an external or internal signal. **Determinism** is the term used to describe the degree of commitment that a group of cells have towards a specific direction at a specific time (*50, 51, 54*). Growth and development of stems, leaves, flowers, etc. involve a gradual change in cell structure and function which determines the kind of organ (i.e., stem, leaf, flower, etc.) that is produced.

Antisense technology. Only one of the two strands serves as the template for mRNA formation and is called the **sense strand.** It is possible for genetic engineers to reverse the copy of the sense mRNA, which is now called **antisense.** The specific segment becomes nonfunctional and cannot be translated, effectively turning off the gene associated with it. The technique has been used to decrease ethylene production in tomato fruits and control ripening.

At some point of development the direction becomes irreversible and the cells are said to be **determined.**

THE CONCEPT OF HORMONAL CONTROL OF PLANT GROWTH AND DEVELOPMENT

Plant hormones provide growth regulation within the plant as directed by the genetic code within the DNA of the chromosomes and the transcription and translation functions of mRNA, tRNA, and rRNA during protein synthesis. Hormones are naturally occurring chemicals of relatively low molecular weight present within plants (i.e., endogenous) in very low concentrations. These chemicals function to regulate plant growth and development (*24, 27, 37, 58, 59*). Although their existence and general functions have been known for a number of years, details of how hormones control growth within the plant are only now being revealed. The five major plant hormones are usually identified as **auxins, gibberellins, cytokinins, ethylene,** and **abscisic acid.** Each of these compounds have specific chemical structures (Figure 2–14).

In addition to these substances, certain chemicals, some natural and some synthetic, have hormonal effects when exogenously applied to plants. These substances, along with the natural hormones,

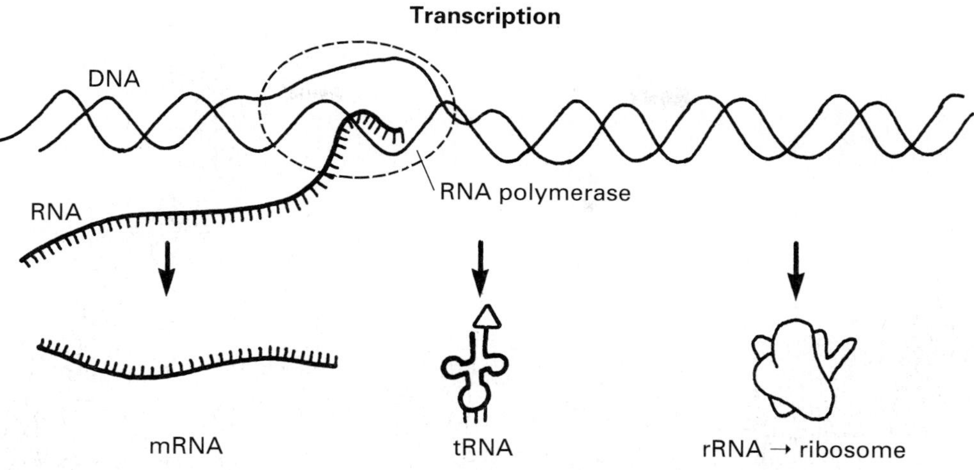

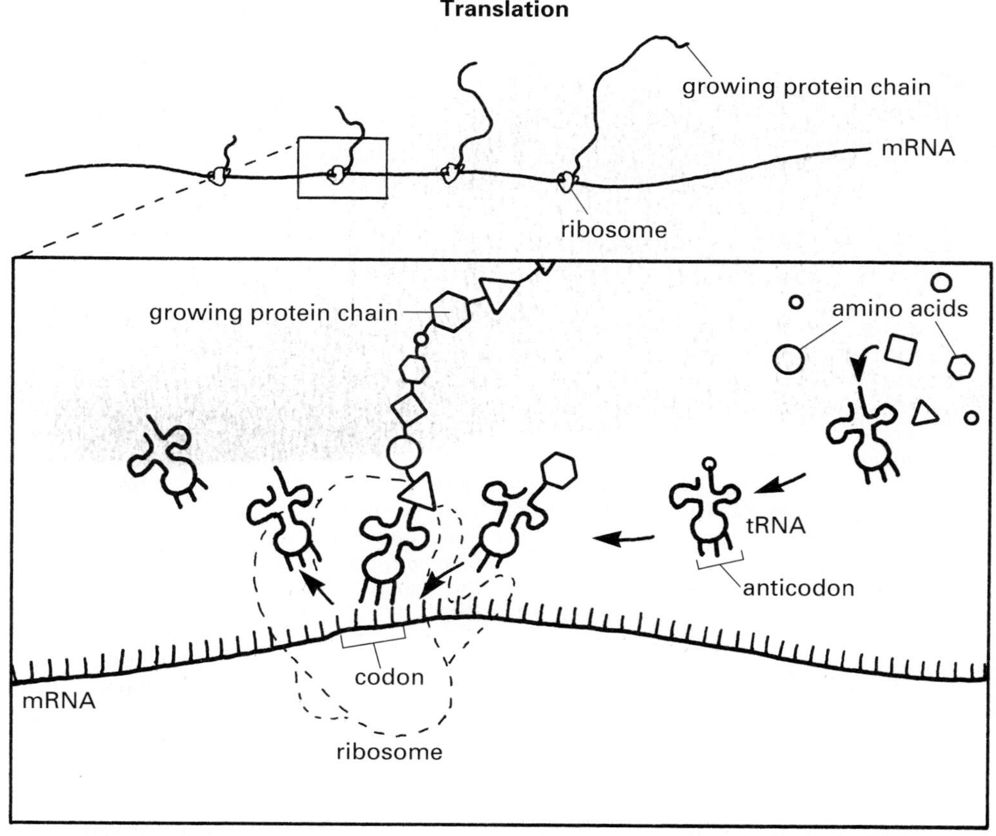

FIGURE 2–12 Schematic drawing illustrating **transcription** and **translation** of genetic codes for the synthesis of proteins. See text for details.

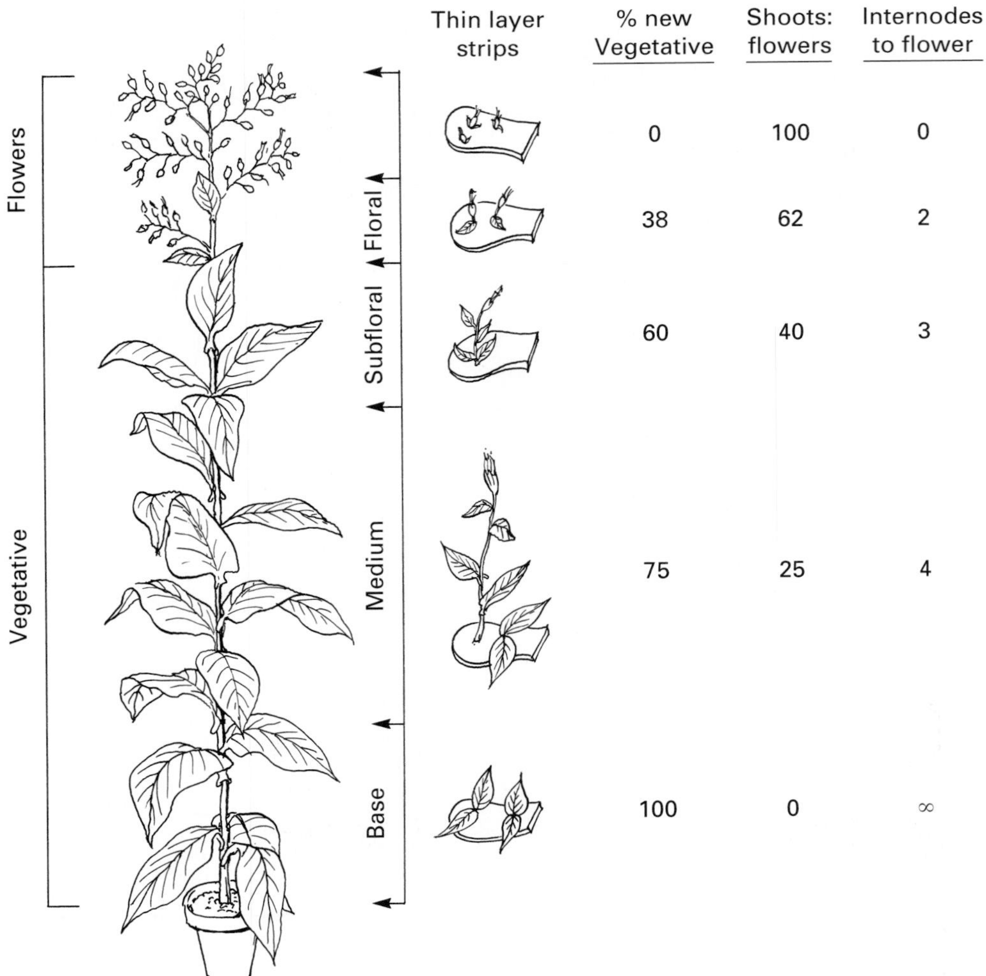

FIGURE 2–13 Principles of competency and determinism are illustrated by this experiment with tobacco. The plant is divided into four zones, basal (B), median (M), subflowering (SFZ) and flowering (FZ). "Thin layer explants" of sections of epidermis were excised and grown in sterile culture. Vegetative buds only were produced from B, vegetative shoots were produced from M and SFZ but grew into flowers; the number of nodes required decreasing with height. Flowers only were produced from FZ. These results showed that a gradient in competency for flowering existed with age and location. Once tissue potential is committed to flowering, it can be said to be determined. Redrawn with permission from Tran Thanh Van, M. 1973. Direct flower neoformation from superficial tissue of small explants of *Nicotiana tabacum L.* Planta (Berl.) 115:87–92.

are combined into the term **plant growth regulator (PGR)** (*27, 37, 58*). Much research and development effort has gone into a continuing search for PGR's that can be used in plant regulation.

Auxins. One of the first hormone responses in plants was the observation that light affected the direction of growth of plant coleoptiles in germinating seeds (*17*). Fritz Went (*84*) and a number of other researchers showed that these effects could be induced by plant extracts, which were subsequently shown to contain the plant hormone **indoleacetic acid** (IAA). IAA is synthesized from the amino acid L-tryptophan in leaf primordia, young leaves and developing seeds. IAA moves from cell to cell in a polar gradient (i.e., tip to base). Auxin is involved in

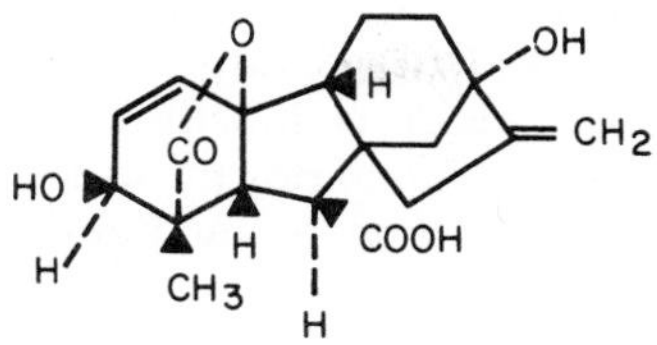

GA_3 (Gibberellic acid)

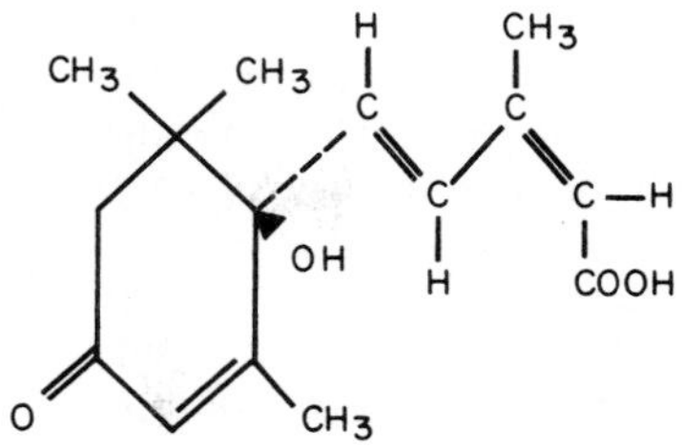

(S)–Abscisic acid

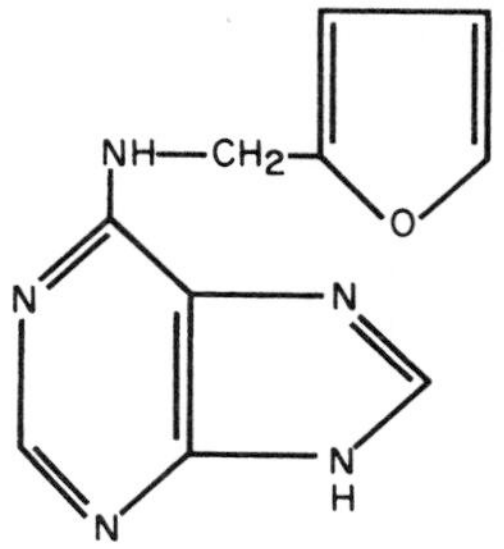

6–Furfurylamino purine (kinetin)

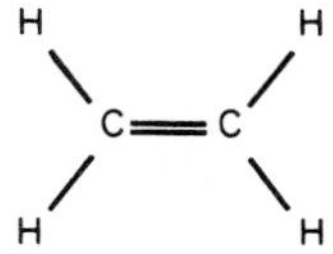

Ethylene

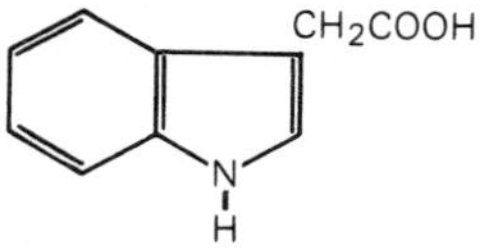

3–indoleacetic acid
beta–Indoleacetic acid
(IAA)

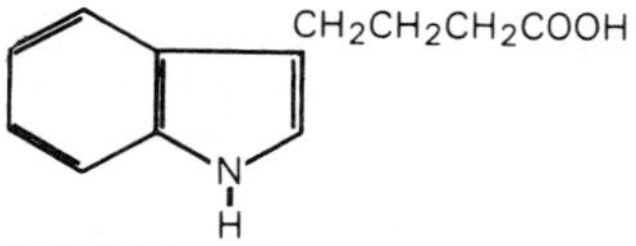

3–indolebutyric acid
gamma–(Indole–3)–butyric acid
(IBA)

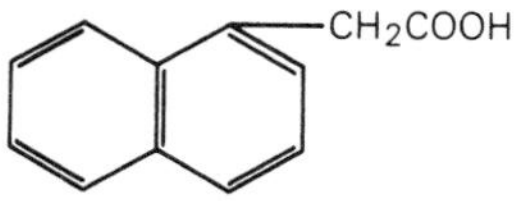

1–Naphthaleneacetic acid
alpha–naphthaleneacetic acid
(NAA)

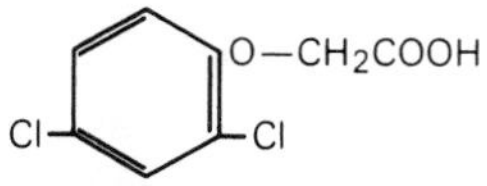

2,4–Dichlorophenoxyacetic acid
(2,4–D)

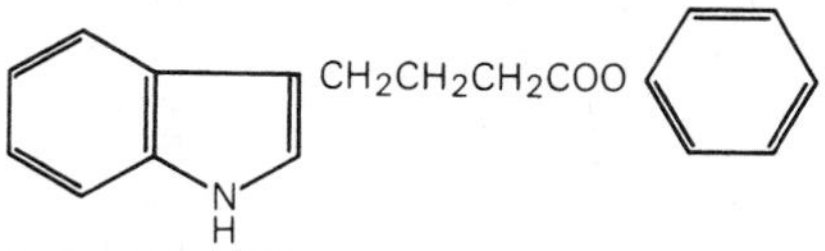

Aryl ester of IBA
(P–IBA)

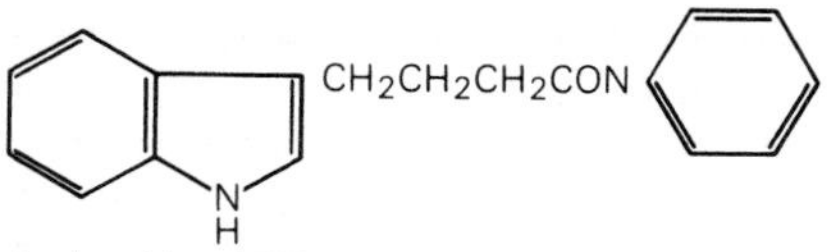

Aryl amide of IBA
(NP–IBA)

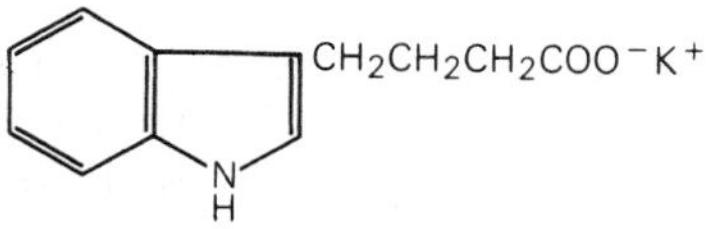

K-salt of IBA
(K–IBA)

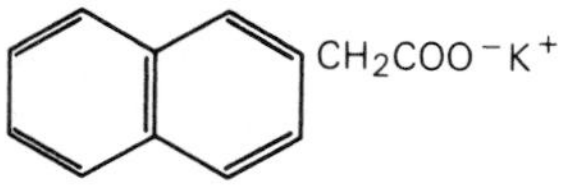

K-salt of NAA
(K–α–NAA)

FIGURE 2–14 Chemical structures of plant hormones and plant growth regulators with hormone activity.

many plant activities including coleoptile bending toward light, inhibition of lateral buds by terminal buds (apical dominance), formation of abscission layer on leaves and fruit, and activation of cambial cell growth. The most widely used function for auxin in plant propagation is the induction of adventitious roots on cuttings (Chapter 10) and the control of morphogenesis in micropropagation (Chapters 17, 18).

Plant physiologists have sought to define the **mode of action** of the chemicals on plants. Growth promotion by auxins is thought to take place by two mechanisms: (a) by promoting the transport of H^+ ions across cell walls, and increasing their extensibility, and (b) by inducing the transcription of specific mRNA's necessary for sustained growth. In morphogenesis, applied auxins appear capable of erasing cellular programs of differentiation, reverting cells to a dedifferentiated state and reinstating cell division (see Chapter 10).

Auxins may be regulated within the cell by rates of synthesis or by deactivation. One method of deactivation is by **conjugation** with small molecules such as alcohols, amino acids, or sugars. The conjugated auxin is inactivated but is protected from oxidative breakdown by IAA-oxidase. Conjugation may be a way to facilitate storage. On the other hand, complexing with proteins (referred to as **binding**) may be necessary for physiological activity to take place.

Synthetic auxins have been discovered which have the same functions as IAA in the plant but do not disintegrate as readily when applied to living tissue. The most useful synthetic auxins, discovered about 1935, are **indolebutyric acid** (IBA) and **naphthalene acetic acid** (NAA). IBA has since been found to occur naturally. Other auxins used in propagation are listed in Table 2–1.

Certain plant growth regulators have been shown to reduce auxin activity and are referred to as

TABLE 2–1

Characteristics of important plant growth regulators and hormones. Those marked with asterisk (*) occur naturally. Adapted from *Plant Cell Culture* 1993 catalogue. Sigma Chemical Co., St. Louis Mo.

Name	Chemical name	Mol. Wt.	Use[1]	Solvent	Sterilization[2]	Storage	
						Powder	Liquid
	A. Auxins						
IAA*	indole-3-acetic acid	175.2	CM;D,S	EtOH; 1N NaOH	CA\F	–0°C	–0°C
IBA*	indole-3-butyric acid	203.2	CM;D,S	EtOH; 1N NaOH	CA\F	0–5°C	–0°C
KIBA	indole-3-butyric acid-potassium salt	241.3	CM;D,S	Water	CA\F	0–5°C	–0°C
NAA	alpha-napthaleneacetic acid	186.2	CM;D,S	1N NaOH	CA	RT	0–5°C
2,4D	2,4dichloro-phenoxy-acetic acid	221.0	CM;Ap	EtOH; 1N NaOH	CA	RT	0–5°C
2,45T	2,4,5-tri-chloro-phenoxy-acetic acid	255.5	Ap	EtOH	CA	RT	0–5°C
	B. Cytokinins						
BA	6-benzyl-amino-purine	225.3	CM	1N NaOH	CA\F	RT	0–5°C
4CPPU	N-(2-chloro-4-pyridyl) n′-phenyl-urea	247.7	CM	DMSO	F	0–5°C	0–5°C
DPU	1,3diphenyl-urea	212.3	CM	DMSO	F	RT	0–5°C
2iP*	6(,-di-methyl-allyl-amino) purine	203.2	CM	1N NaOH	CA\F	–0°C	–0°C
kinetin		215.2	CM,D,S	1N NaOH	CA\F	–0°C	–0°C
TDZ	Thidiazuron	220.2	CM	DMSO; EthOH	CA	RT	0–5°C
Zeatin*		219.2	CM	1N NaOH	CA\F	–0°C	–0°C
	C. Gibberellins						
GA_3*	gibberellic acid	346.4	CM;D,S;A Ap	EtOH	CA\F	RT	0–5°C
KGA_3	gibberellic acid potassium salt	384.5	CM;D,S;A Ap	water	CA\F	0–5°C	–0°C
	D. Inhibitors						
ABA*	Abscisic acid	264.3	CM;D,S; Ap	1N NaOH	CA\F	–0°C	–0°C

[1]CM = culture medium; D,S = dip or soak; Ap = apply to plant

[2]CA = coautoclavable with other media; F = filter sterilize; CA/F = autoclavable with other components but some loss in activity may occur

anti-auxins. These are complex chemicals whose names have been abbreviated to TIBA, PTAA, naptalam (or NPA), 2,4,6,-T, and PCIB (*27*).

Cytokinins. This group of hormones is one of the most important in plant growth because they appear essential for cell division. These substances were discovered during research at the laboratory of Professor Skoog at the University of Wisconsin to develop methods for growing plant tissues aseptically in culture. Coconut milk, which is liquid endosperm, as well as yeast extract were found to promote cell division in callus tissue when supplied in an agar medium under sterile conditions. Adenine, a component of both RNA and DNA (see page 17), produced the same response. An extract from autoclaved fish sperm DNA yielded a compound which stimulated cell division in the presence of auxin. The compound given the name **kinetin** was subsequently shown to be a representative of a group of hormones which have been called **cytokinins** (*68*). Similar natural substances include **zeatin, isopentenyladenine (2iP)** and synthetic substances as **benzyladenine (BA** or **BAP).** Other substances with cytokinin activity include **thiourea, diphenyl urea, thidizuron (TDZ)** and **N-2-chloro-4-puridyl-N`-phenylurea (CPPU).**

The interaction of auxin and cytokinin is one of the primary relationships in plant propagation. A high auxin/cytokinin ratio favors rooting, a high cytokinin/auxin ratio favors shoot formation, and a high level of both favors callus development (see chapters on rooting and micropropagation). The interaction among cytokinins, abscisic acid, and gibberellin controls seed dormancy (see chapters on germination). Cytokinins promote cell division of galls and nodules produced from *Rhizobium* or *Rhizobactor,* nematodes, or infection by *Agrobacterium tumefaciens.* Cytokinin also acts in intact plants to delay or reduce senescence, slowing down the breakdown of chlorophyll and cellular protein.

The mode of action of cytokinins is not clear but is known to stimulate protein synthesis (*27*). Cytokinin has been found in tRNA but the specific mechanisms by which cellular cytokinins are synthesized are not specifically worked out.

Gibberellins. Gibberellins were discovered before World War II by Japanese scientists trying to explain the abnormally tall growth and reduced yield of rice infected with a fungus known as *Gibberella fukikuori* (sexual) or *Fusarium moniliformne* (asexual form). An active ingredient was extracted from the fungus and its chemical structure determined (Figure 2–14). More than 90 forms of gibberellin have since been found in plants but only a few appear to be physiologically active. Gibberellic acid (GA_3) (Table 2–1) is the most important commercial product but a combination of GA_4 and GA_7 is also available (*27*).

Gibberellins occur in high concentrations in developing seeds and have important functions in germination and control of dormancy. They also occur in high concentrations in stem apices, particularly in leaf primordia, roots, fruits, and tubers. They are transported within the plant in the xylem and phloem. Their function in the plant is to promote shoot elongation through the increase of both cell division and elongation. The application of GA can be dramatic as in the overcoming of specific genetic dwarfness in plants.

GA's regulate synthesis of seed enzymes in cereals, induce seed germination, and stimulate flowerings in long-day plants and biennials.

Some plant growth regulators used as growth retardants function by inhibiting the synthesis of gibberellin in plants and are called "antigibberellins" (*27*). Some of these include *chlormequat* (CCC), *AMO 1618* (Carvadan) and *ancymidol.* Another group of retardants with the same action include compounds with a triazole ring in their structure. An important compound of this group is *paclobutrazol.*

Abscisic acid (ABA) (*21, 27*). This naturally occurring material is one of a class of **growth inhibitors** which can play an important role in many plant activities. ABA in plants can be identified and quantified by gas chromatography, flame spectrometry, or ELISA (see page 263). The ABA molecule has two isomeric forms, *cis-* and *trans-*. The first is the more common and plants can convert the second to the first. The chemical structure also has a (+) and (–) form which cannot be interconverted. The (+) form is active and occurs in nature. Commercial products are mixtures of both.

ABA is synthesized from mevalonic acid either directly or from the breakdown of carotenoid pigments. Biosynthesis occurs in chloroplasts. ABA is present in all organs of higher plants and in phloem and xylem sap, but its function is apparently dependent upon concentration. At low concentrations ABA can stimulate elongation but it mostly functions as a growth inhibitor. Horticulturally, ABA is involved in the control of dormancy of buds and seeds, abscission (although ethylene has greater effects), and plant response to stress, particularly moisture. ABA regulates stomatal closure, controls water and ion uptake by roots, and affects leaf senescence and abscission.

In propagation, ABA is involved in germination and dormancy of seeds (see Chapter 7) and plays a role in the embryogenesis and production of seeds (see Chapter 5).

Ethylene (*27*). This compound is a gas with a very simple chemical structure (Figure 2–14) but which can have profound effects on plant growth, including epinasty in high concentrations, senescence and abscission in leaves and fruits, promotion of flowering, stimulation of lateral buds, stimulation of latex production, and induction of flowering. In propagation, ethylene can induce initiation of adventitious roots, stimulate germination in some seeds, and overcome dormancy. Auxin sometimes stimulates ethylene production but the effect is limited to vegetative organs. Wounding and stress however, consistently result in the increase of ethylene. Naturally occurring ethylene is involved in the ripening and maturation of fruits and is widely used to induce ripening under commercial storage conditions. A liquid chemical (2=chloroethylphosphonic acid, known commonly as ETHEPHON) is absorbed by plant tissue where it breaks down to produce ethylene. It is used on some crops to promote ripening, to act as a thinning agent, or to promote flowering. Ethylene gas is a natural byproduct of combustible fuels and escaping fumes can cause damage in commercial storages. Likewise, since ethylene is a byproduct of ripening fruit, it can cause damage to other plant materials when they are stored with fruit.

Ethylene is synthesized from the amino acid methionine via the pathway that includes 1-aminocylclopropane-1-carboxylic acid (also known as ACC). The process is enhanced by exogenous carbohydrates, light, cytokinins, auxins, and carbon dioxide. ACC can produce ethylene effects directly. Several chemical inhibitors of ethylene exist including the most well-known aminoethoxyvinylglycine (AVG) which inhibits the enzyme synthesis of ethylene.

Other possible natural hormones. (*27*). Certain other naturally occurring chemicals in plants are considered by some plant physiologists to be plant hormones. These include polyamines, some complex oligosaccharides, and inositol triphosphate. Polyamines include *putrescine, spermidine* and *spermine.* These are synthesized from the amino acids arginine and ornithine and are widespread in plant tissue. In some experiments, polyamines have shown auxin activity and inhibited ethylene synthesis. GA and cytokinins treatments have increased polyamine synthesis in some plants.

Myo-inositol is usually classed as a vitamin and is an important ingredient of some tissue cultures. It is part of the growth-promoting fraction of coconut milk as well as yeast extract.

LIFE CYCLES IN PLANTS

Plants typically exhibit consecutive periods of *vegetative growth,* e.g., formation of stems, leaves, and roots, and *reproductive development,* e.g., the formation of flowers, fruits, and seeds. These are generally organized into two kinds of cycles. One is determined by seasonal patterns based upon changes in climate dictated by where the plants live. Climatic cycles are based on seasonal patterns of temperature and/or rainfall. These are expressed by growth from shoot and root tips, and cambia, and represent species' adaptations to the environment (*37, 58*).

In addition, plants undergo life cycle changes beginning with the propagation of a specific propagule (seed, cutting, scion, explant, etc.) and ending with the death of the individual plant. The growth and development patterns taking place are referred to as **ontogenetic** and the variations that are expressed in this process are characteristic of the individual plant.

For example, **annual** plants complete the entire sequence from germination to dissemination of seeds and death in one growing season. **Biennial** plants have a two-year cycle. During the first season, the plants are vegetative and grow as a low clump or rosette of leaves. During the second season, the plants are reproductive, produce flowers and seeds, then die. The transition from the vegetative to the reproductive stage is often in response to an environmental trigger, such as a cold period or a certain length of day. **Perennial** plants live for more than two years and repeat the vegetative-reproductive change annually (or biennially) in new shoots produced by buds. These cycles also tend to be related to consecutive periods of warm-cold temperatures or to wet-dry environmental cycles. **Herbaceous perennials** produce shoots which grow during one season and die back at the end of the growing season. The plant survives during adverse periods by specialized underground stem structures with roots that remain perennial (bulbs, rhizomes, crowns; see Chapters 15 and 16). **Woody perennials** develop permanent woody stems that continue to increase annually from apical and lateral buds with characteristic growth and dormancy cycles.

Seedling Life Cycle (*43*)

In this text, an individual plant that develops from a seed is referred to as a **seedling** whether it is an annual, biennial, herbaceous, or woody perennial. Genetically this term applies throughout its life cycle. The life cycle of a plant (Figure 2–15) begins with the formation of the zygote which grows into an embryo, receives nutrients from the mother plant, and undergoes characteristic morphological and physiological stages of development. At first, growth involves cell division of the entire embryo as it increases in size. Later, growth potential develops with a polar orientation as the embryo develops its characteristic structure. These changes are described in Chapter 5.

Germination initiates a dramatic change from the embryonic pattern to the developmental patterns of the young seedling. Growth potential is now polarized between the shoot and root; cell division is concentrated in the root tips, shoot tips, and axillary buds. Subsequently, the extension of the root and shoot is accompanied by an increase in volume. New nodes are continuously laid down as leaves and axillary growing points are produced. Initially, lateral buds produce only vegetative shoots. Eventually, the plant reaches the reproductive stage at which the growing points become flowers as a response to specific signals within the plant or from the environment, such as chilling **(vernalization)** or **photoperiod.** The initial stage in the induction of flowering is an internal physiological change in the growing points that precedes morphological change. The second stage is initiation of flower parts followed by differentiation of the flower structure. Flowering and seed production follows to start a new seedling cycle. In many plants, growth and reproduction are geared to cold-warm seasonal cycles or wet-dry rainfall patterns.

Phase change. Superimposed upon the obvious variation in growth (vegetative) and development (reproductive) stages and growth habit patterns is another system that controls the expression of characteristics of the plant at different ontogenetic stages. Morphological and physiological traits expressed in different parts of the same plant may follow a distinct pattern depending upon the developmental "age" at which the cells are laid down (*11,*

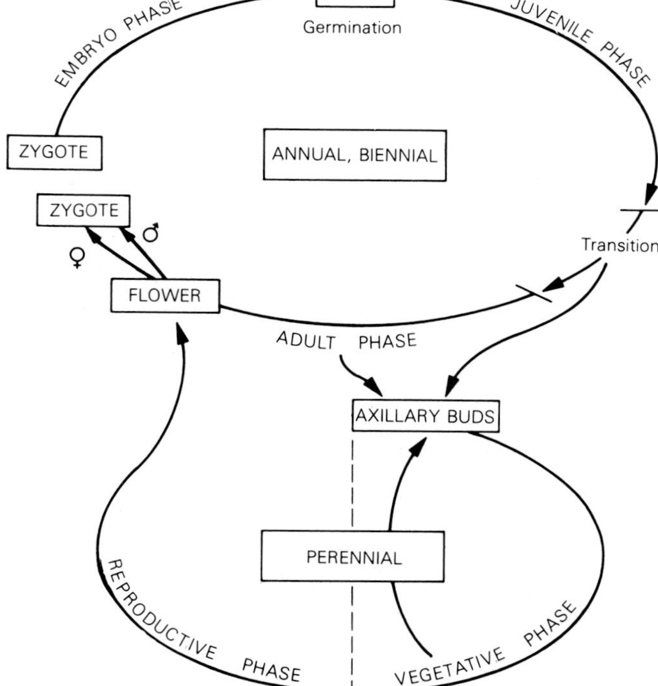

FIGURE 2–15 The **seedling cycle** in plants. Model illustrates change involving embryo development, juvenile, transition, and adult phases. In the annual or biennial, the apical meristem progresses more or less continuously through one (annual) or two (biennial) growing seasons *(upper circle).* In herbaceous and woody perennials *(lower circle)* the adult vegetative meristem is renewed continuously by seasonal cycles of growth and reproduction.

32, 33, 82). This is shown by the ivy plant of Figure 2–16. The name for this phenomenon is **phase change** and the shift from vegetative to reproductive is **maturation.**

Phase change has been recognized by botanists, foresters, and horticulturists (*28*) for many years and labeled with such terms as **juvenility** (*20, 25, 32, 62, 76, 82, 85*), **ontogenetic aging** (*23*) and **cyclophysis** (*55, 62, 63*). This phenomenon is not an effect of chronological aging due to crop overload, growth slowdown, senescence or other debilitating influences but is related to the shift from vegetative growth to reproductive development as it occurs in the seedling plant.

The phases of seedling plant development has traditionally been identified as follows:

1. *Juvenile.* In this initial stage of vegetative growth following germination, the plant size expands but the apical meristems are unable to induce flowers.
2. *Transitional.* Plant is outwardly vegetative but internally is in transition to the reproductive phase. This phase is sometimes called "adolescent" by foresters.
3. *Adult* (*mature*). This is the reproductive stage at which the plant produces flowers.

These phases are identified most readily in long lived perennial plants, such as, trees, shrubs and herbaceous perennials where differences in juvenile and mature traits are most conspicuous. Nonetheless, phase change has been demonstrated to occur in annual plants such as maize (57) and it must be recognized as a universal characteristic of plant development. In maize, phases not only may differ in morphology but are controlled by specific genes (Table 2–2).

The transition from juvenile to vegetative adult is gradual but the change from vegetative adult to reproductive adult is abrupt.

Phase change has been associated with the same aspects of plant development that were described on page 19.

1. *Flowering.* The pattern and age when flowering begins is the most characteristic aspect of phase change. The age of flowering may vary from a few months in some annuals to many years in some perennials. Certain tree species

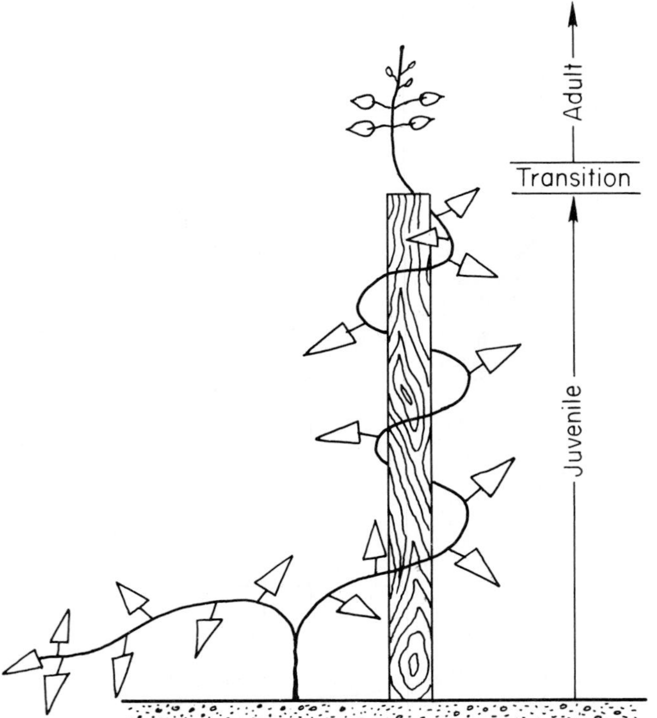

FIGURE 2–16 Phase changes in ivy *(Hedera helix)* in which the **juvenile** (nonflowering) phase is a vine which, as it grows into a vertical form, undergoes a **transition** into the **adult** (mature) phase in which flowers and fruits are produced.

TABLE 2–2

Morphological expression of maize as associated with phase change (59)

	Cuticle	*Epidermal Cells*	*Aerial Roots*	*Root Hairs*	*Epicuticular Wax*
1. Juvenile:	thin	circular	yes	yes	yes
2. Adult:					
A. Vegetative	thick	rectangular	no	no	no
B. Reproductive	Production of flower structures				

require 25 to 50 years to begin to flower (*12*) (Table 2–3). Usually, flowering begins in the upper and peripheral parts of the plant where shoots and branches have reached a specific ontogenetic age measurable by the specific number of nodes produced.

2. *Morphology and physiology.* Those parts of the plant in the juvenile phase often differ morphologically and physiologically (e.g., leaf form, thorniness, vigor). Ivy (*Hedera*) is a vine in the juvenile phase with distinct leaves whereas the adult phase is a shrub with rounded leaves. This aspect is discussed more fully in Chapter 9.
3. *Regeneration potential.* Each phase tends to have characteristic regenerative potential. Embryos in culture, for instance, tend to be **embryogenic,** i.e., proliferate to initiate new embryos. Cuttings from juvenile shoots tend to initiate adventitious roots more readily than cuttings from adult shoots.

TABLE 2–3

Age of flower development in some woody plants in the seedling phase (12)

Species	*Length of juvenile period*
Rose (Rosa spp.)	20–30 days
Grape (Vitis)	1 year
Stone fruits (Prunus spp.)	2–8 years
Apple (Malus spp.)	4–8 years
Citrus (Citrus spp.)	5–8 years
Scotch pine (Pinus sylvestris)	5–10 years
Ivy (Hedera helix)	5–10 years
Birch (Betula pubescens)	5–10 years
Pear (Pyrus spp.)	6–10 years
Sequioa (Sequioa sempervirens)	5–10 years
Pine (Pinus monticola)	7–20 years
Larch (Larix decidua)	10–15 years
Ash (Fraxinus excelsior)	15–20 years
Maple (Acer pseudoplatanus)	15–20 years
Arborvitae (Thuja plicata)	15–25 years
Douglas-Fir (Pseudotsuga menziesii)	20 years
Bristle cone pine (Pinus aristata)	20 years
Redwood (Sequoiadendron giganteum)	20 years
Norway spruce (Picea abies)	20–25 years
Hemlock (Tsuga heterophylla)	20–30 years
Sitka spruce (Picea sitchensis)	20–35 years
Oak (Quercus robur)	25–30 years
Fir (Abies amabilis)	30 years
Beech (Fagus sylvatica)	30–40 years

Juvenile traits are apparently adaptations to evolutionary pressure through control of the early stages of the seedling cycle, such as protecting young plants from predatory animals because of thorns or bitter tasting foliage, or enabling seedling to compete with other individuals in the same space because of vigorous vegetative growth and delayed initiation of flowering. Considering the wide range of environments in which plants grow, it is no wonder that juvenile and adult phases take many morphological and physiological forms.

Although biological control of phase change within an individual plant is described as epigenetic (p. 19) there is a genetic component that determines the overall pattern of phase change within populations of plants and in different species. Studies with model systems, including corn and *Arabidopsis* (*57*), have identified specific genes that determine the time that phase change events occur and the expression of specific morphological markers. Time of flowering and other traits are inherited in avocado and pear (*48, 81*) and long seedling cycles can limit progress in breeding programs (*35*).

Epigenetic variation means that the cells of an individual plant have the same genotype; yet they are capable of expressing strikingly different morphology and physiology patterns. An indication of epigenetic change is also indicated in that following the sexual reproduction (i.e. meiosis and recombination) the adult phase is reversed during the gametophyte generation to produce a zygote, embryo, and new seedling plant which recapitulates the embryonic and juvenile phases (*25, 43*). This relationship indicates that even though differences in phase change among plants may be genetically controlled, the phase control *within a single plant* is epigenetic (nongenetic). Differential expression of genetic information involves in part controlling the timing of expression of specific genes.

Clonal Life Cycle

The **clonal cycle** concept applies to *individual* perennial plants that are propagated as clones (Figure 2–17). Differences develop in the competency (see p. 19) of individual growing points to exhibit epigenetic variation during the seedling cycle. The difference in competency becomes strikingly apparent when individual growing points are removed and vegetatively propagated. A vegetative propagule (bud, scion, cutting, explant) removed from different locations in the plant, e.g., as a terminal or lateral bud, retains the specific level of juvenility (or maturation) when it is separated from the plant and propagated vegetatively. As a result, the morphology and physiology of individual progeny plants propagated from different parts of the seedling source may vary significantly in the three types of categories described previously.

This means that vegetatively propagated cultivars may show significantly different biological and horticultural characteristics than those of seedling plants from which they were obtained. Propagules removed from different parts of the same plant may show different expressions of the clonal cycle due to differences in their relative juvenility or maturation at the time of propagation.

This phenomenon of phenotypic variation among vegetative propagules of the same plant has been recognized by horticulturists and foresters for a long time and is widely utilized in propagation and plant growing as shown by the following examples.

a) Ivy (*Hedera*) is a classic example in which the juvenile vine form is readily propagated and maintained by cuttings whereas the adult flowering phase produces a treelike form (Figure 2–18).

b) Apple, pear, and citrus seedlings tend to be thorny, vigorous, and much delayed in flowering whereas plants grafted from buds of the adult (flowering) phase are more preco-

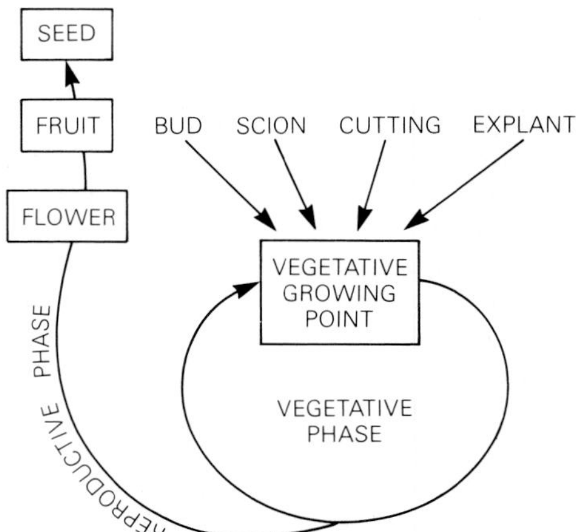

FIGURE 2–17 A **clonal life cycle** in a plant results when it is vegetatively propagated. The type of growth, time of flowering, and other characteristics may vary among different propagules depending upon the location on the seedling plant from which the propagule was taken. With continued propagation, the clone is stabilized into a mature phase with consecutive vegetative and reproduction stages.

FIGURE 2–18 Vegetative propagation from the mature phase of ivy *(Hedera helix)* produces a plant with a bushy structure and palmate leaves *(left)*, whereas vegetative propagation from the juvenile phase produces a vine with narrower and lobed leaves.

FIGURE 2–19 Nursery trees propagated from the juvenile *(left)* part of a 'Kara' mandarin citrus nucellar seedling shows thorns on the nursery tree. Tree will be vigorous, come into bearing late, and produce poor quality fruit. *Right:* Nursery tree from the mature part of the tree has no thorns and will be less vigorous, bear relatively young, and produce good quality fruit. (Courtesy H.B. Frost. From H.J. Webber and L.D. Batchelor, *The citrus industry,* vol. 1. University of California Press, Berkeley.)

> cious in age of flowering and tend to lose the undesirable thorny, vigorous growth habit of the juvenile phase (Figure 2–19). Consecutive propagation stabilizes these characteristics into the mature clones that are the cultivars used in commerce.

The application of this phenomenon to propagation is discussed throughout this text particularly in Chapters 9 and 10.

Apomictic Life Cycle

Apomixis is a process in which normal sexual process of zygote formation is replaced by systems that do not involve sexual recombination. In these systems, an embryo originates directly from a haploid or a diploid cell present within the reproductive structures (*5, 30, 56, 59, 73*). The primary types of apomixis include **adventitious embryony, recurrent apomixis** (diplospory), and **nonrecurrent apomixis.**

Adventitious Embryony (Nucellar Embryony, Nucellar Budding).

Embryos originate from cells of the nucellus or within the embryo sac or from the integument (Chapter 5). These embryos have the same genotype as the plant on which they were produced. When this phenomenon occurs, seeds may have more than one embryo **(polyembryony)** one being sexual, the remainder apomictic. This type of reproduction is typical for citrus and some mango species whose horticultural significance is described in Chapter 19.

Recurrent apomixis (diplospory). The egg mother cell is produced without undergoing reduction division. Consequently, the egg develops directly into the embryo and has the diploid chromosome number and same genotype as the parent mother cell. This type is known to occur in some species of *Poa* (bluegrass), *Taraxacum* (dandelion), *Allium* (onion), *Rubus* (blackberry), and others.

Nonrecurrent apomixis (aplospory). This type occurs if the embryo arises directly from the egg nucleus. It occurs rarely and is of genetic interest because the embryo is haploid. This type of growth cycle would not be considered apomictic because the haploid seedling is not a clone of the parent plant.

The significance of the apomictic cycle is that seeds produced by a nonsexual process duplicate the genotype of the parent plant. Apomixis can be considered a form of cloning which "fixes" the genetic potential of apomictic populations. The use in plant breeding is described in Chapter 4. Horticultural importance varies with individual species. Plants of the apomictic cycle go through the same phase changes as the seedling with characteristic juvenile and adult phenotypes. For some cultivars, as in citrus, the juvenile seedling product many not have horticulturally acceptable characteristics (see p. 254).

PLANT NOMENCLATURE AND THE CONCEPT OF THE CULTIVAR

Plant propagation involves the preservation of specific genotypes that are important to humans. It is essential therefore to have some means of labeling them (*26, 64*). Systems of nomenclature have gradually evolved through the efforts of botanists and horticulturists to provide uniform worldwide plant identification. The systems are embodied in the *International Code of Botanical Nomenclature* (*47*) and the *International Code of Nomenclature for Cultivated Plants* (*78*).

Botanical Classification

A botanical name includes the *genus, species,* both in Latin, and a designation of the individual who first provided the name. For example, the peach is *Prunus persica* L. Peaches are in the Rose family (Roseaceae) along with many other important horticultural plants, such as roses, strawberries, and apples. The genus *Prunus* includes the stone fruits (plum, cherry, almond) characterized by a fruit with both a fleshy mesocarp and a hard endocarp. The species is the designation of specific "kind" which reproduces itself as a unit. L. refers to Carolus Linneaus, the eighteenth-century botanist who originated the system of nomenclature and first bestowed the name.

The **species** is the fundamental unit in taxonomy used to designate a population of plants that can be recognized and which reproduces itself as a unit (*59, 69, 70, 71*). In nature, individuals within one species normally interbreed freely but do not interbreed with another species because of separation either by distance or by some physiological, morphological, or genetic barrier that prevents the interchange of genes between them. A true species can usually be propagated and maintained by seed.

Propagation of a species from different parts of its natural range may result in variation in appearance and adaptability among separate collections (*36*). Taxonomists recognize the *subspecies and botanical varieties* categories as naturally occurring subdivisions within the species resulting from geographical separation which has created a recognizable difference in morphology. The term *form* indicates a particular phenotypic difference, as a blue or white color. To get a total picture of species variability, one should sample all parts of the range rather than relying on a narrow range of selection.

Natural variation among native plants of a species can also be described with the terms **cline** and **ecotype** (*46, 59*). A cline refers to continuous differences in genetically controlled physiological and morphological characteristics that occur within a species in different parts of its range. Differences are related to continuous variation in the environment. Populations have evolved due to adaptations to the local environment. Where plant differences are distinct and discontinuous, the term *ecotype* is used. Seed collection of natural populations are described in Chapter 4.

Cultivated plants are designated by a scientific name but may be a complex hybrid rather than a distinct "natural" species. For example, peach cultivars are variations within a recognized species but

PLANT TAXONOMY

In addition to practical aspects of plant identification, taxonomy also involves an understanding of evolutionary relationships among living organisms (both plant and animal). Concepts of these relationships have themselves evolved over time utilizing increasingly sophisticated techniques of morphology, physiology, and genetics (*59*). Botanical (and animal) classification is based upon increasing specialization and complexity in structure and organization of the plant resulting from evolutionary processes. The most important separation is between the **prokaryotes** and **eukaryotes.** Prokaryotes include bacteria which are represented by two *groups* (*Archaebacteria and Eubacteria*). *Bacteria* are mostly single-celled organisms, do not have organelles, and have a single circular molecule of DNA not structurally associated with proteins. Most have small circular DNA fragments (plasmids) which provide a means of genetic transfer in nature from bacteria to host plants. Bacteria are the causal agent for some of the most serious propagation diseases. On the other hand, plasmids from specific bacteria are agents in genetic engineering (see p. 17).

All other plants are classed as Eukaryotes. Eukaryotes are single and multicelled organisms whose cells include a nucleus containing the chromosomes composed of DNA associated with histone proteins and a cytoplasm with organelles (**mitochondria, plastids,** etc). They exhibit sexual reproduction, i.e., alternation of generation. The classification of Eukaryotes is included in Table 2–4.

TABLE 2–4

Classification of prokaryotes and eukaryotes

- Prokaryotes—one-celled bacteria
  - Archaebacteria
  - Eubacteria
- Eukaryotes—multicelled plants
  - Protista
    - algae
    - slime molds
    - chytrids
  - Fungi
    - Zygomycetes
    - Ascomycetes
    - Basidiomycetes
  - Plantae
    - Bryophytes (no seeds)
      - liverworts
      - hornworts
      - mosses
    - Vascular plants (have stem structures)
      - Without seeds
        - psilotophytes
        - lycophytes
        - horsetails
        - ferns
      - Seed producers
        - Angiosperms—enclosed seeds
          - Monocots (single cotyledon)
          - Dicots (two cotyledons)
        - Gymnosperms—naked seeds
          - Ginkgo
          - Conifers
        - Gnetophytes (Gnetum, Welwitsia)

the European prune (*Prunus domestica* L.) is a complex hybrid which apparently developed in cultivation. Interspecific and intergeneric hybrid names include an X as shown in the following examples (*26*).

Genus and species: *Cercis canadensis* (Eastern redbud)

Botanical variety: *Cercis canadensis* var. *alba* (white flowered eastern redbud)

Cultivar: *Cercis canadensis* cv. Forest Pansy

Interspecific hybrid within a genus:

x Viburnum x *burkwoodii* (a group of cultivars originating as hybrids of *V. carlesi* x *V. utile*).

Intergeneric hybrid within a family:

x Cupressocyparis leylandii (a group of cultivars originating as fertile hybrids of *Chamaecyparis nootkatensis* and *Cypressus macrocarpa*).

Nomenclature of Cultivated Varieties (Cultivars)

The development of agriculture and horticulture as a human endeavor evolved over a long period of time largely through the ability to maintain special kinds of plants during propagation which were particularly useful to man. Although the Code of Botanical Nomenclature applies to cultivated plant materials, horticulturists eventually came to the conclusion that the categories described in the Code were not adequate to cover the special kinds of plants in cultivation.

An International Code of Nomenclature for Cultivated Plants was first formulated in 1930 (*10, 78*) but has been extensively revised.

a) The code recognizes the term **cultivar** (contraction of the two words "cultivated variety") as a special taxonomic category for cultivated plants. A cultivar must have arisen in cultivation, must be clearly distinguishable by any characters (morphological, physiological, cytological, chemical, or other) and, when reproduced sexually or asexually, must retain its distinguishing characters. Cultivar is synonymous with such older terms as *variety* (English), *variete* (French), *variedad* (Spanish), *sorte* (German), *sort* (Scandinavian), *ras* (Russian), or *razza* (Italian), many of which continue to be used in practice.

b) Cultivars (and varieties) are identified by a non-Latin name (after 1956), usually bestowed by the originator. The Code establishes guidelines by which cultivar names can be applied.

c) The Code establishes registration procedures for specific cultivated plant groups by which cultivar names can be recorded in an official system recognized worldwide by propagators of that particular plant group.

The scientific name of a cultivar includes the: (a) genus, (b) species, and (c) cultivar name, the first two in the usual Latin form and the latter in words of a common language. The cultivar name can be set off by the abbreviation **cv.**, or by single quotation marks, but not both. It can also just be attached to the common name.

Syringa vulgaris cv. Mont Blanc
Syringa vulgaris 'Mont Blanc'
Lilac 'Mont Blanc'

A cultivar should be given a proper name that is recognized by horticulturists everywhere and will not confuse its identity. Multiple names for the same cultivar, or one name for several similar cultivars, arising either by accident or from deliberate changes in name, can only lead to confusion and misrepresentation. The primary principle in naming cultivars is the same as with botanical names, i.e., the earliest name applied should have priority. Once a name is correctly applied, it should be changed only for exceptional reasons. Additional rules assist in choosing a plant name (*3, 78*).

Seed propagated cultivars are those whose uniformity and reproducibility can be achieved by genetic methods as described in Chapter 4.

a) **Lines, multilines, and inbred lines.** Uniformity is achieved largely by self-pollination either because of flower structure or by artificial controls.

Triticum aestivum 'Marquis' is a wheat cultivar in which the uniformity is maintained in propagation because it is homozygous and self-pollinated.

b) **Mixtures** and **composite** or **synthetic lines.** These are groups of cross-fertilized individuals which can be distinguished by one or more characters but are not necessarily genetically uniform.

'Ranger' alfalfa is a cultivar derived from intercrossing five seed-propagated lines each maintained in isolation. 'Sternenzauber' is a mixture of different color forms of *Phlox drummondi* but with the same starlike corolla shape.

c) **Hybrids.** These include groups of individuals which are reconstituted each generation and which are distinguishable from other cultivars by one or more characters.

'US 13' corn is produced by consecutive crossing of four inbred lines. 'Granex' onion is derived from crossing two onion inbred lines.

d) **Provenances** (see Chapter 4). Although not used this way in forestry, the Code allows for plants of specific provenance to be considered as a cultivar providing distinguishing characteristics that are retained when propagated.

e) **Obligate apomicts.** These are seedling plants which are reproduced through apomixis (see p. 31).

Asexually reproduced cultivars maintain their characteristics by vegetative propagation, as by cuttings, layering, division, grafting, or micropropagation.

a) **Clone.** This category includes a population of plants derived from a single individual. It may be a mixture of clones which can be distinguished from others but which are not necessarily genetically uniform.

Examples of clones are 'Redhaven' peach, 'King Alfred' daffodil, and 'Burbank Russet' potato.

b) **Bud mutation.** Individuals propagated from a distinguishable mutant from a cultivar and distinct from the parent plant.

'Starkrimson' is a red-skinned mutation of 'Delicious' apple.

c) Plants showing special or unique growth phases (see p. 30) within clones retaining those characters with vegetative propagation can be identified as a cultivar.

Examples include *Abies amabilis* 'Spreading Star,' *a prostrate* form maintained by propagation of lateral branches, *'Chamaecyparis lawsoniana 'Ellwoodii,'* a juvenile form, and *Picea abies* 'Pygmaea,' a witches' broom type of spruce.

Legal Protection of Cultivars

The right to propagate specific cultivars developed through controlled selection and/or breeding programs can be protected by a number of legal devices. These allow the originator of specific cultivars to control their distribution by preventing their propagation by others except through the payment of royalties, fees, or other arrangements. These protections have rewarded plant breeders and observant horticulturists, both private and institutional, with monetary compensations for their efforts as plant breeders (*15, 39, 52*).

Legal protection has been available in the U.S. since 1930 with the passage of the Townsend-Purnell Act which added vegetatively propagated plants to the general patenting law for inventions (*41*). This protection was extended to some seed propagated cultivars with the Plant Variety Protection Act passed in 1970 and revised in 1994 (*4*).

The range of plant materials covered was greatly expanded by the *Diamond* vs. *Chakrabarty* court decision that extended protection to "everything under the sun that is made by man." This now includes not only plant cultivars of all types but microorganisms, plant parts, biotechnical techniques, genes, and other plant innovations (*41*).

Many countries of the world have legal systems that grant protection to patents and breeders rights and a large network of such programs have developed (*8, 13, 14, 21, 52*). Guidelines for their use have been produced by the International Convention for the Protection of New Varieties of Plants in 1961, 1972, 1978, and 1991 (*79*).

Patents. A plant patent is a grant from the U.S. Patent Office which extends patent protection to plants. Exclusive rights are given to the inventor (or heirs or assigns) of a "distinct and new" kind of plant (cultivar) for a 17-year period. Only vegetatively propagated cultivars (i.e., by cuttings, grafting, layering, micropropagation), but not tuber propagated plants (potato and Jerusalem artichoke), can be patented. A plant found growing wild in nature is not considered patentable. Plant patents under this law are recognized in the United States and its territories and possessions. Agreements must be worked out with individuals in other countries to expand this protection throughout the world.

The applicant for a plant patent must be the inventor or discoverer, who subsequently has reproduced the new cultivar by propagation. If a person who was not the inventor applies for the patent, it would not be granted or, if discovered afterward, would become null and void. A description of the plant must be provided. There is no necessity to prove that the new cultivar has superior merit, only that it is "distinct and new."

To obtain direct information, contact the U.S. Patent and Trademark Office, Washington, DC 20231.

Plant variety protection (*4*). The U.S. Plant Variety Protection Act (PVPA) extends plant patent protection to certain seed propagated cultivars which can be maintained as "lines," including F_1 hybrids. Tuber propagated crops are also protected. The new plant must be novel (i.e., not identical to any previous plant), distinctive, uniform, and stable.

A plant breeding certificate allows breeders propagation protection for many agricultural and horticultural cultivars propagated by seed, including such crops as cotton, alfalfa, soybeans, and marigolds. A certificate protects these rights for 20 years. These rights may be sold or licensed.

To obtain specific information about U.S. patent matters, contact the Plant Variety Protection Office, USDA, National Agricultural Library Building, Room 500, 10301 Baltimore Blvd., Beltsville, MD 20705.

Trademarks (*22, 28, 41*). A registered trademark offers protection for a name that indicates the specific origin of a plant (or product). For instance, a nursery may obtain a trademark that indicates that a specific plant is grown by them. The mark is distinct from the cultivar name and both identities should be provided. The trademark is any word, symbol, device, logo, or similar distinguishing mark. A trademark is granted for ten years but can be renewed indefinitely as long as it remains in use.

One of the oldest trademarks "Star^R Roses" first used in 1907 designates roses produced by Conard-Pyle Co. (*39*).

Utility patents(*28, 41*). This protection is under the general patent law which uses the criteria novelty and utility. An application requires the same full description as a plant patent. It may include more than one claim that involves specific uses of the plant. The patent disclosure must allow the public to practice the invention upon expiration of the patent. Utility patents are used by commercial genetic engineering and biotechnology firms to control the use of specific genes or particular technologies.

Other methods. *Contracts* may be used to control the propagation of specific plants as well as the selling of their fruit and other products. Enforcement comes under contract law. *Trade secrets* are protected by law and can provide some protection for disclosure of certain technology. This may include information that is not disclosed to the public or temporary protection prior to disclosure for patent application (*40*). *Copyrights* have the purpose of preventing reproduction or copies of printed material. Although this device could apply to plant materials, they are usually used to control reproduction of pictures or printed material about the plant which is used in brochures or catalogs.

Summary. Plant propagation can be called "applied biology" because it requires an understanding of specific biological principles. Advances in the understanding of basic biology of plants has an impact on the practical art of plant propagation. In this chapter, we describe five basic concepts for which we believe a student in plant propagation needs to have at least a general understanding. (a) The first concept is the genetic basis of life and the control of variation both in plant breeding and in plant propagation. The traditional concept of genetic transfer through pollination and fertilization has been extended by the discovery of gene transfer and biotechnology. How much this potential will change the role of the plant propagator is difficult to foresee. (b) The second concept is control of plant development and gene expression through an epigenetic system within the plant. This aspect is truly within the realm of the plant propagator whose task is to manipulate the process through the selection of propagules, environmental control, and management. (c) The third concept is the hormonal control of plant growth and development either through endogenous substances within the plant (e.g., auxins, gibberellins, abscisic acid, cytokinins, or ethylene) or PGR's applied exogenously. (d) The interaction of these three basic types of control (genotype, gene expression, and hormone control) creates a range of life cycles in individual plants. The endogenous mechanism of phase change creates modifications of these patterns depending upon whether they are sexually, vegetatively, or apomictically produced. (e) The fifth concept is that of botanical nomenclature, particularly the species and cultivar. The end result of propagation is the production and distribution of specific cultivars whose "trueness-to-type" is preserved during propagation. In the modern world, identity of these cultivars can be protected by various legal conventions which must be respected by the propagator.

REFERENCES

1. Abbott, A.J. and R.K. Atkin, eds. 1987. *Improving vegetatively propagated crops.* New York: Academic Press.
2. Addicott, F. 1982. *Abscission.* Berkeley: Univ. of Calif. Press.
3. American Association of Nurserymen. 1988. *How to use, select and register cultivar names.* Washington, D.C.: Amer. Assn. Nurs.
4. Anonymous. 1994. Congressional passage of new PVP law a triumph for seed industry. *Diversity* 10(3): 34–35.
5. Asker, S. and L. Jerling. 1992. *Apomixis in plants,* Boca Raton, FL: CRC Press, Inc.
6. Bailey, L.H. and E.Z. Bailey. 1976. *Hortus Third.* New York: MacMillan.
7. Barlow, 1994. The origin, diversity and biology of shoot-bourne roots. In: T.M. Davis and B.E. Haussig, eds. *Biology of adventitious rooting.* New York and London: Plenum Press. pp. 1–24.
8. Barnaby, C.J. 1992. Plant variety rights and plant production—help or hindrance. *Comb. Proc. Intl. Plant Prop. Soc.* 42:269–272.
9. Braun, A.C. 1959. A demonstration of the recovery of crown-gall tumor cell with the use of complex tumors and single-cell origin. *Proc. Natl. Acad. Sci.* U.S.A. 45:932–938.
10. Brickell, C.D., ed. 1980. International code of nomenclature for cultivated plants—1980. *Regnum Vegetabile* 104:7–32. (Obtainable from Crop Science

Society of America, 677 South Segoe Road, Madison, Wis. 53771.)

11. Brink, R.A., Phase change in higher plants and somatic cell heredity. *Quart. Rev. Bio.* 37(1):1–22.
12. Clark, J.R. 1983. Age-related changes in trees. *Jour. Arbor.* 9:201–5.
13. Costin, J.J. 1990. Plant breeders rights. *Com. Proc. Intl. Plant Prop. Soc.* 40:321–323.
14. Craig, P. 1991. Canada catches up. *Amer. Nurs.* 22173(4):111.
15. Craig, R. 1994. Intellectual property protection of Pelargoniums. *HortTech.* 3:284–290.
16. Dandekar, A. M., G.H. McGranahan and D.J. James. 1993. Transgenic woody plants. In *Transgenic plants,* volume 2. Present status and social and economic impacts. pp. 129–151. New York: Academic Press.
17. Darwin, C. and F. Darwin. 1881. *The power of movement in plants.* New York, NY: Appleton Century Crofts.
18. Davies, F.T., T.M. Davis, D.E. Kester. 1944. Commercial importance of adventitious rooting to horticulture. In T.M. Davis and B.E. Haussig, eds. *Biology of adventitious rooting.* New York and London: Plenum Press. pp. 53–60.
19. Dirr, M. and C.W. Heuser, Jr. 1987. *The reference manual of woody plants, from seed to tissue culture.* Athens GA. Varsity Press.
20. Doorenbos, J. 1965. Juvenile and adult phases in woody plants. In *Handbuch der Pflanzenphysiologie,* Vol. 15 (Part 1). Berlin: Springer-Verlag, pp. 1222–35.
21. Durzan, D.J. and M.D. Durzan. 1993. Opportunities for biotechnology transfer to developing countries. *HortTech.* 3:268–277.
22. Elliott, Wm. J. 1991. Property rights and plant germplasm. *HortSci.* 26:364–365.
23. Fortanier, E.J. and H. Jonkers. 1976. Juvenility and maturity as influenced by their ontogenetical and physiological ageing. *Acta Hort.* 56:37–44.
24. Foskett, D.E. 1994. *Plant growth and development.* San Diego, CA: Academic Press, Inc.
25. Frost, H.B. 1938. Nucellar embryony and juvenile characters in clonal varieties of Citrus. *J. Hered.* 29:423–432.
26. Geneve, R. 1990. A rose by any other name would spell chaos. *American Nurseryman* 172(3):43–47.
27. George, E.F., 1993. *Plant propagation by tissue culture, Part 1. The Technology.* Edington, Wilts, England: Exergetics, Ltd.
28. Gioia, V.G. 1992. Plant variety protection. *Comb. Proc. Intl. Plant Prop. Soc.* 42:206–208.
29. Goebel, K. von. 1928–1938. *Organographie der Pflanzen.* 3 Anglage, Jena, Fischer Verlag.
30. Guftafsson, A. 1946–1947. *Apomixis in higher plants,* Parts I–III. Lunds Univ. Arsskrift, N.F. Avid, 2 Bd 42, Nr.3:42(2); 43(12).
31. Haberlandt, G. 1902. Kulturversuche mit isolierten Pflanzellen. *Sitzungsber. Akad. der Wiss. Wien, Math.—Naturwiss.* Ki. 111–69–92.
32. Hackett, W.P. 1985. Juvenility, maturation and rejuvenation in woody plants. In *Horticultural Reviews,* Vol. 7, J. Janick, ed. Westport, Conn.: AVI Publ. Co. pp. 109–55.
33. Hackett, W.P. and J.R. Murray. (in press). Approaches to understanding maturation or phase change. In: *Biotechnology of Horticultural crops.* R. Geneve, J. Preece Wallingford: CAB International.
34. Hammerslag, F.A. and R.E. Litz, eds. 1992. *Biotechnology of perennial fruit crops.* Wallingford, C.A.B. International.
35. Hansche, P.E. and Wm. Beres. 1980. Genetic remodeling of fruit and nut trees to facilitate cultivar improvement. *HortSci.* 15:710–715.
36. Harlan, J.R. 1992. *Crops and Man.* 2nd edition. Madison, WI: ASA, CSSA.
37. Hartmann, H.T., A.M. Kofranek, V.E. Rubatzky, and W. J. Flocker. 1988. *Plant science., Growth, development, and utilization of cultivated plants* (2nd ed.). Englewood Cliffs, N.J.: Prentice-Hall.
38. Hayward, M.D., N.O. Bosemark, and I Romagoza, eds. 1993. *Plant Breeding. Principles and Prospects.* London. England: Chapman Hall.
39. Hutton, R.J. 1994. Plant patents—a boon to horticulture and the American public. *HortTech.* 2:280–283.
40. Ihnen, J. et. al., 1989. *Protecting plant germplasm: Alternatives to patent and plant variety protection. Intellectual property rights associated with plants.* Amer. Soc. of Agron. Madison, WI. ASA Spec. Publ. 52. pp. 123–143.
41. Jondle, R.J. 1993. Legal protection for plant intellectual property. *HortTech.* 3:301–307.
42. Kays, S.J. 1991. Biotechnology—a horticulturist's perspective: introduction to the colloquium. *HortSci.* 26:1020–1021.
43. Kester, D.E. 1983. The clone in horticulture. *HortScience* 18(6):831–37.
44. Kovar, J.L. and R.O. Kuchenbach. 1994. Commercial importance of adventitious roots to agronomy. In T.M. Davis and B.E. Haussig, eds. *Biology of adventitious rooting.* New York and London: Plenum Press. pp. 25–36.
45. Krikorian, A.D. 1982. Cloning higher plants from aseptically cultured tissues and cells. *Biol. Rev.* 57:151–218.
46. Langlet, O. 1962. Ecological variability and taxonomy of forest trees. In *Tree growth,* T. T. Kozlowski, ed. New York: Ronald Press, pp. 357–69.
47. Lanjouw, J., ed. 1966. International code of botanical nomenclature. *Regnum Vegetabile* 46:402.
48. Lavi, U., Lahav, E., Degani C. and S. Gazit. 1992. The genetics of the juvenile phase in avocado and its applications for breeding. *J. Amer. Soc. Hort. Sci.* 117(6):981–984.

49. MacDonald, B. 1987. *Practical woody plant propagation for nursery growers.* Portland, OR: Timber Press.

50. Meins, F. 1986. Determination and morphogenetic competence in plant tissue culture, In Yeoman, M.M., ed. *Plant cell culture technology.* Boston, MS: Blackwell Sci. Pub.

51. Mohnen, D. 1994. Novel experimental systems for determining cellular competence and determination. In *Biology of adventitious root formation,* T.M. Davis and B.E. Hassig. New York and London: Plenum Press. pp. 87–98.

52. Moore, J.N. 1993. Plant patenting: a public fruit breeder's assessment. *HortTech.* 3:262–266.

53. Morgan, T.H. 1901. *Regeneration.* London: MacMillan (quoted by Krikorian, 1982).

54. Murray, J.R., M. C Sanchez, A.G. Smith and W.P. Hackett. 1994. Differential competence for adventitious root formation in histologically similar cell types. In T.D. Davis and B.E. Hassig, ed. *Biology of Adventitious Root Formation.* New York: Plenum Press. pp. 99–110.

55. Oleson, P.O. 1982. On cyclophysis and topophysis. *Silvia Genetica* 27:173–78.

56. Poehlman, J.M. and D.A. Sleper. 1995. *Breeding Field Crops.* 4th ed. Ames, Iowa: Iowa State University Press.

57. Poehig, R.S. 1990. Phase change and the regulation of shoot morphogenesis in plants. *Science* 250:923–929.

58. Preece, J.E. and P.E. Read. 1993. *The biology of horticulture.* New York: Wiley.

59. Raven, P.H., R.F. Evert and S.E. Eichhorn. 1992. *Biology of plants.* New York: Worth.

60. Ritchie, G.A. 1994. Commercial application of adventitious rooting to forestry. In *Biology of adventitious root formation,* T.M. Davis and B.E. Hassig, ed. New York and London: Plenum Press, pp. 37–52.

61. Schaff, D.A. 1992. Biotechnology—gene transfer: terminology, techniques and problems involved. *HortScience* 26:1021–1024.

62. Schaffelitzky de Muckadell, M. 1959. Investigations on aging of apical meristems in woody plants and its importance in silviculture. *Forstl. Forsogov. Damm.* 25:310–455.

63. Seelinger, R. 1924. Topophysis und Zyklophysis pflanzlicher organe und ihre bedeutung for die Pflanzenkultur. *Angew. Bot.* 6:191–200.

64. Sheldon, A. 1988. What's in a name? *Amer. Hort.* 67(4):34–36.

65. Shull, G.H. 1912a. 'Genotypes', 'biotypes,' 'pure lines' and 'clones.' *Science* 35:27–29.

66. Shull, G.H. 1912b. 'Phenotype' and 'clone.' *Science* 35:182–183.

67. Shull, G.H. 1915. Genetic definitions in the New Standard Dictionary. *Amer. Nat.* 49:52–59.

68. Skoog, F. and C.O. Miller. 1957. Chemical regulation of growth and organ formation in plant tissues cultured in vitro. *Symp. Soc. for Exp. Biol.* 11:118–31.

69. Stearn, W.T. 1949. The use of the term 'clone.' *Jour. Roy. Hort Soc.* 74:41–47.

70. Stebbins, G.L. 1950. *Variation and evolution in plants.* New York: Columbia Univ. Press.

71. ———. 1971. *Processes of organic evolution* (2nd ed.). Englewood Cliffs, N.J.: Prentice-Hall.

72. Steward, F.C. M.O. Mapes, A.E. Kent and R.D. Holstein. 1964. Growth and development of cultured plant cells. *Science* 143:20–27.

73. Stoskopf, N.C., D.T. Tomes and B.D. Christie. 1993. *Plant Breeding Theory and Practice.* Boulder, CO: Westview Press.

74. Stout, A.B. 1929. The clone in plant life. *J. New York Bot. Gard.* 30:25–37.

75. Stout, A.B. 1940. The nomenclature of cultivated plants. *Amer. J. Bot.* 27:339–347.

76. Stoutmyer, V.T. 1937. Regeneration in various types of apple wood. *Iowa Agr. Expt. Sta. Res. Bul.* 220:308–52.

77. Tran Thanh Van, M. 1973. Direct flower neoformation from superficial tissue of small explants of *Nicotiana tabacum* L. *Planta* (Berl.) 115:87–92.

78. Trehane, P.C., C.D. Bricknell, B.R. Baum, W.L.A. Hetterscheid, A.C. Leslie, J. McNeill, A.A. Spongberg and F. Vrugtman (eds). 1995. *The International Code of Nomenclature for Cultivated Plants—1995.* 6th edition. Wimborne, U.K.: Quarterjack Publishing.

79. UPOV, 1991. *International convention for the protection of new varieties of plants.* UPOV Publ. No. 221(E).

80. Vacil, I.K. and A.C. Hildebrandt. 1965. Differentiation of tobacco plants from single, isolated cells microcultures. *Science* 150:889–892.

81. Visser, T., J.J. Verhaegh and D.P. de Vries. 1976. A comparison of apple and pear seedlings with reference to the juvenile period. I. Seedling growth and yield. *Acta Hort.* 56:205–214.

82. Wareing, P.F. and V.M. Frydman. 1976. General aspects of phase change with special reference to *Hedera helix* L. *Acta Hort.* 56:57–68.

83. Webber, H.J. 1903. New horticultural and agricultural terms. *Science* 18:501–503.

84. Went, F.W. 1937, *Phytohormones.* New York, NY: MacMillan.

85. Zimmerman, R. 1972. Juvenility and flowering in woody plants: a review. *HortScience:* 7:447–455.

SUPPLEMENTARY READING

BAILEY, L.H., E.Z. BAILEY, and staff of L.H. Bailey Hortorium. 1976. *Hortus third.* New York: Macmillan.

COLLOQUIUM. 1991. The horticultural dilemma—Trademarks, patents, royalties and cultivars. *HortScience* 26:359–365.

HARTMANN, H.T., A.M. KOFRANEK, V.E. RUBATSKY and W.J. FLOCKER. 1988. *Plant science: Growth, development, and utilization of cultivated plants* (2nd ed.). Englewood Cliffs, N.J.: Prentice-Hall.

LINDSEY, K. and M.G.K. JONES. 1990. *Plant biotechnology in agriculture.* Englewood Cliffs: Prentice-Hall.

MACDONALD, B. 1987. *Practical woody plant propagation for nursery growers.* Portland, Oreg.: Timber Press.

RAVEN, P.H., R.F. EVERT and S.E. EICHHORN. 1992. *Biology of plants.* New York: Worth.

3

Environmental Factors of Propagation

This chapter reviews environmental factors, manipulation of the propagation microclimate, and the management of pests and beneficial organisms. Also discussed are propagation structures, containers and media, plant nutrition, preventative measures—including integrated pest management (IPM) and current best management practices (BMP), and post propagation handling of **liner**[1] plants.

To enhance the propagation of plants, commercial producers manipulate the environment of *propagules* (cuttings or seeds) by managing: (1) **microclimatic conditions** (relative humidity, temperature, light, and gases), (2) **edaphic factors** (propagation medium or soil, mineral nutrients, and water), and (3) **biotic factors**—interaction of propagules with other organisms (symbiotic mycorrhizal fungi, pathogens, insect pests, etc.) (Figure 3–1).

Unique ecological conditions exist during propagation. Commercial propagators may have to compromise to obtain an "average environment" in which a whole range of species are propagated either by cuttings, seed, or tissue culture explants (*90*). Propagators very often use a monoculture of one species or cultivar that optimizes the propagation environment, but can also increase susceptibility of plant propagules to disease and insect problems. "Family bulking" is a method of vegetatively propagating elite Douglas-Fir selections by cuttings. Different clones are mixed (bulked) together to avoid monoculture (monoclonal) conditions (Figure 3–2). The environmental conditions that are optimum for plant propagation are frequently conducive for pests (pathogenic fungi, viruses, bacteria, insect, and mite development). Astute propagators not only manage the environment during propagation, but also manipulate the environment of **stock plants** prior to selecting propagules, i.e. **shading** and **stooling** to maximize rooting potential of a propagule; and post propagation—**hardening-off** (weaning

[1]A liner traditionally refers to lining out nursery stock in a field row. The term has evolved to mean a small plant produced from a rooted cutting, seedling, plug, or tissue culture plantlet. *Direct sticking* or *direct rooting* into smaller pots is commonly done in U.S. propagation nurseries. Seedlings and rooted cuttings can also be transplanted into small *liner pots* and allowed to become established during liner production, before being transplanted to larger containers (upcanned) or outplanted into the field (Figure 3–27).

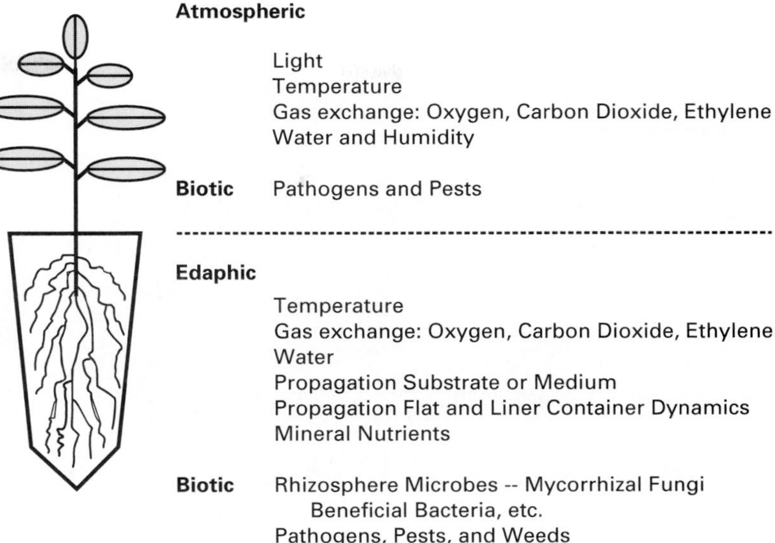

FIGURE 3–1 The propagation environment: manipulation of microclimatic, edaphic, and biotic factors. Modified from Landis (*90*).

rooted cuttings from the mist system and changing fertility regimes) to assure growth and survival of tender rooted liner plants after propagation.

FUNDAMENTAL MICROCLIMATIC AND EDAPHIC FACTORS IN THE PROPAGATION ENVIRONMENT

In propagating and growing young nursery plants, facilities and procedures are best arranged to optimize the response of plants to some fundamental microclimatic and edaphic factors influencing their growth and development: **light, water, temperature, gases,** and **mineral nutrients**. In addition, young nursery plants require protection from pathogens and other pests, as well as control of salinity levels in the growing media. The propagation structures, equipment, and procedures described in this chapter, if handled properly, maximize the plants' growth and development by controlling their environment.

Light

The management of light can be critical for rooting cuttings, germinating seeds, growing seedlings, or shoot multiplication of **explants** during tissue culture propagation. Light can be manipulated by controlling *irradiance*[2], *light duration* (photoperiod), and *light quality* (wave length). Relative comparisons of light units for propagation are found in the enclosed chart (see page 42).

Daylength (photoperiod). Higher plants are classified as long-day, short-day, or day-neutral, based on the effect of photoperiod on initiation of reproductive growth. **Long-day** plants, which flower chiefly in the summer, will flower when the **critical photoperiod** of light is equaled or exceeded; **short-day** plants, such as chrysanthemums, flower when

[2] *Irradiance* is the relative amount of light as measured by radiant energy per unit area. Irradiance, intensity and photon flux all measure the amount of light very differently. They are not interchangeable terms. Only *photosynthetic photon flux* (PPF) should really be used for plant propagation, since the process of photosynthesis relies on the number of photons intercepted, not light given off by a point source (intensity) or energy content (irradiance). *Photosynthetic active radiation* (PAR) is measured in the 400 to 700 nanometer (nm) waveband as PPF in micromoles of photons per unit area per time ($\mu mol\ m^{-2}s^{-1}$) with a *quantum sensor,* or as watts per square meter (W/m^2) with a *pyranometric sensor.* Some propagators still measure light intensity with a *photometric sensor,* which determines foot-candles or lux (1 foot-candle = 10.8 lux). A photometric sensor is relatively insensitive to wavelengths that are important for plant growth, i.e., it may record high light intensity from an artificial electric light source, but it does not take into account if the light source is rich in green and yellow, or poor in red and blue light—which would lead to poor plant growth. Quantum and radiometric (pyranometer) sensors can be purchased from instrument companies (i.e., Licor, Inc. of Lincoln, NE). For determining *light quality or wavelength,* the spectral distribution is measured with a portable spectroradiometer, which is a very expensive piece of equipment.

FIGURE 3–2 Vegetative propagation of Douglas-Fir by cuttings. This method, known as "family bulking," enables production of up to 25 trees from each individual control pollinated seed. In this manner, limited quantities of elite genetic material can be expanded while maintaining significant genetic diversity among families. *Top:* Over one million cuttings from twenty-four families in a propagation house. *Middle:* Hardened-off rooted cuttings in Multipot-104 trays ready for lining out in a bareroot nursery. *Bottom:* Over 1.5 million rooted Douglas-Fir cuttings following fall lining out into a bareroot nursery. (Photographs by G. A. Ritchie).

Relative comparison of light units for solar radiation and artificial lighting *(93,146)**

Light Source	*Energy [Photosynthetic Photon Flux]* (μmol $m^{-2}s^{-1}$)	*Radiation [Irradiance]* (watts m^{-2})	*Illumination [Light Intensity]* (lux)	(ft-candles)
Solar Radiation				
Full Sunlight	2000	450	108,000	10,037
Heavy Overcast	60	15	3,200	297
Artificial Light Source				
Metal halide (400 w) lamp @ 2 m height	19	4	1,330	124

*Photosynthetically active radiation (PAR): 400 to 700 nm. Conversions between energy, radiation, and illumination units are complicated and will be different for each light source. The spectral distribution curve of the radiant output must be known in order to make conversions.

the critical photoperiod is not exceeded. Reproductive growth in **day-neutral** plants, such as roses, is not triggered by photoperiod. The discovery of photoperiodism by Garner and Allard demonstrated that the dark period, not the light period, is most critical to initiation of reproductive growth, even though light cycles are traditionally used to denote a plant's photoperiod. In propagation, fresh seed collected in the fall from selected woody plant species, such as *Larix,* need long-day conditions to germinate. Dahlia cuttings need short-day conditions to trigger tuberous root formation.

Photoperiod is normally extended under short-day conditions of late fall and early winter by lighting with incandescent lights, or high intensity discharge lights (HID) (Figures 3–3 and 3–13). Conversely, photoperiod can be shortened under the long-day conditions of late spring and summer by covering stock plants and cuttings with black cloth or plastic that eliminates all light.

FIGURE 3–3 *Upper and lower left:* computer-control led environmental manipulation of propagation facilities; including: *upper right,* a mechanized traveling mist boom under high-pressure sodium vapor lights; *lower right:* automated shade material programmed to close along the top of the house when preset radiant energy levels are reached. This system works well with contact polyethylene propagation, *lower center:* in commercial English nurseries. *Lower left:* automated metering system for monitoring CO_2 injection.

Light Quality. Light quality is perceived by the human eye as color and corresponds to a specific range of wavelengths. Red light is known to enhance seed germination of selected lettuce cultivars, while far-red light inhibits germination. Far-red light can promote bulb formation on long-day plants, such as onion (*Allium cepa*). Blue light enhances in-vitro bud regeneration of tomato (*100*). Using greenhouse covering materials with different spectral light-transmitting characteristics, researchers at Clemson University (*105*) have been able to control the height and development of greenhouse grown plants, rather than relying on chemical application of growth regulators for height control. This has application for plant propagation, liner production, and plant tissue culture systems.

Water-Humidity Control

Water management and humidity control are critical for any propagation procedure. Water management is one of the most effective tools for *regulating plant growth.* Evaporative cooling of an inter–mittent mist system can help control the propagation house microenvironment and reduce the heat load on cuttings, thereby permitting utilization of high light conditions to increase photosynthesis and encourage subsequent root development. A solid support medium, such as peat-perlite, is not always necessary to propagate plants; peach cuttings can be rooted under **aeroponic systems,** while woody and herbaceous ornamentals can be rooted in modified, **aero-hydroponic systems** without relying on overhead mist (*136*). Tissue culture explants are often grown in a liquid phase rather than on a solid agar media.

While leaf **water potential** (Ψ_{leaf})[3] is an important parameter for measuring water status of seedlings and cuttings, and influences rooting of cuttings—**turgor** (Ψ_{ρ}) is physiologically more important for growth processes. The water status of seedlings and cuttings is a balance between transpirational losses and uptake of water. Later in this chapter, methods to control water loss of leaves of cuttings, seedlings, and containerized grafted plants are discussed.

[3]**Water potential** (Ψ_{water}) refers to the difference between the activity of water molecules in pure distilled water and the activity of water molecules in any other system in the plant. Pure water has a water potential of zero. Since the activity of water in a cell is usually less than that of pure water, the water potential in a cell is usually a negative number. The magnitude of water potential is expressed in megapascals [1 megapascal (MPa) = 10 bars = 9.87 atmospheres]. Propagators can determine water potential by using a pressure chamber (pressure bomb) manufactured by PMS (Corvallis, OR) or Soil Moisture Corp. (Santa Barbara, CA). A psychrometer with a microvolt meter (LiCor, Lincoln, NE) can also be used. Estimation of **turgor** (Ψ_{ρ}) requires measurement of water potential (Ψ_{water}) minus the **osmotic potential** (Ψ_{π}), which is based on the formula $\Psi_{water} = \Psi_{\rho} - \Psi_{\pi}$. Osmotic potential can also be determined by either a pressure chamber or a psychrometer. The matrix potential (Ψ_{m}) is generally insignificant in determining Ψ_{water}, but is important in seed germination.

Temperature

Temperature affects plant propagation in many ways. Seed dormancy is broken in some woody species by cool-moist stratification conditions that allow the germination process to proceed. Temperature of the propagation medium can be suboptimal for seed germination or rooting due to seasonally related ambient air temperature or the cooling effect of mist. In grafting, heating devices are sometimes placed in the graft union area to speed up graft union formation, while the rest of the **rootstock** is kept dormant under cooler conditions (Figure 13–37).

It is often more satisfactory and cost-effective to manipulate temperature by heating at the propagation bench level, rather than heating the entire propagation house. The use of heating and cooling systems in propagation structures is discussed further in this chapter (see Chapter 11 for heating equipment and sensors).

Gases and Gas Exchange

High respiration rates occur with seed germination and plug development, and during adventitious root formation at the base of a cutting. These aerobic processes require that O_2 be consumed and CO_2 be given off by the propagule. Seed germination is impeded when a hard seed coat restricts gas exchange. Likewise, gas exchange at the site of root initiation and subsequent rooting are reduced when cuttings are stuck in highly water-saturated propagation media with small air pore spaces. In leaves of droughted propagules, stomata are closed, gas exchange limited, and suboptimal rates of photosynthesis occur. During propagation in enclosed greenhouses, ambient CO_2 levels can drop to suboptimal levels, limiting photosynthesis and propagule development. The buildup of ethylene gas (C_2H_2) can be deleterious to propagules during storage, shipping, and propagation conditions.

Mineral Nutrition

To avoid stress and poor development during propagation, it is important that the stock plants be maintained under optimal nutrition—prior to harvesting propagules. During propagation, nutrients are generally applied to seedlings and plugs by **fertigation** (soluble fertilizers added to irrigation water) or with slow-release fertilizers that are either preincorporated into the propagation medium or broadcast (top dressed) across the medium surface. Cuttings are normally fertilized with a slow-release fertilizer preincorporated into the propagation medium (which is discussed later in this chapter and in Chapter 11) or with soluble fertilizer applied after roots are initiated. The development of intermittent mist revolutionized propagation, but mist can severely leach cuttings of nutrients. This is a particular problem with cuttings of difficult-to-root species which have long propagation periods.

MANAGING THE PROPAGATION ENVIRONMENT

Propagation Structures

Facilities required for propagating plants by seed, cuttings, and grafting, and other methods include two basic units. One is a structure with temperature control and ample light, such as a greenhouse, modified quonset house, or hotbed—where seeds can be germinated, or cuttings rooted, or tissue culture microplants rooted and acclimatized. The second unit is a structure into which the young, tender plants (liners) can be moved for hardening, preparatory to transplanting out-of-doors. Cold frames, low polyethylene tunnels or sun tunnels covered by Saran, and lathhouses are useful for this purpose. Any of these structures may, at certain times of the year and for certain species, serve as a propagation and acclimation structure.

Aseptic Micropropagation Facilities

These are described in Chapter 18.

Greenhouses[4]

Greenhouses have a long history of use by horticulturists as a means of forcing more rapid growth of plants (*46,98*). Most of the greenhouse area in the United States is used for the wholesale propagation and production of floricultural crops, such as pot plants, foliage plants, bedding plants, and cut flowers. Lesser amounts are used for nursery stock and vegetable crops (*130*).

Greenhouse structures vary from elementary, home-constructed to elaborate commercial installations. Commercial greenhouses are usually independent structures of even-span, gable-roof construction, proportioned so that the space is well utilized for convenient walkways and propagating benches (*9,66*). In larger propagation operations, several single greenhouse units are often attached side by side, eliminating the cost of covering the adjoining walls with glass or polyethylene (see Figure 3–4). These gutter-connected houses, while more expensive to construct than independent ground-to-ground structures, allow easy access between houses and decrease the square footage (meters) of land needed for propagation houses. Heating and cooling equipment is more economical to install and operate, since a large growing area can share the same equipment (*78*). Retractable roof greenhouses reduce energy costs and are being used in cuttings and seed propagation and seedling plug production. Since the liner seedlings are partly produced under full sun conditions, they are better acclimatized for the consumer (*17*).

Quonset-type construction is very popular. Such houses are inexpensive to build, usually consisting of a framework of piping, and are easily covered with one or two layers of polyethylene (see Figure 3–5).

Arrangement of benches in greenhouses varies considerably. Some propagation installations do not have permanently attached benches, their placement varying according to the type of equipment, such as lift trucks or electric carts, used to move flats and plants (*166*). The correct bench system can increase production efficiency and reduce labor

[4]For sources of commercial greenhouses, contact the National Greenhouse Manufacturers Association, P.O. Box 567, Pana, Ill. 62557. A number of trade journals such as *GrowerTalks, Greenhouse Manager Pro,* and *Greenhouse Grower* list commercial greenhouse manufacturers and suppliers that include greenhouse structures, shade and heat retention systems, cooling and ventilation, environmental control computers, bench systems, and internal transport systems in greenhouses.

FIGURE 3–4 Gutter-connected propagation greenhouses. *Top left:* a series of gutter-connected propagation houses. *Top right:* the basic types of gutter-connected propagation greenhouses: bow or truss. Bows are less expensive, offer less structural strength. Trusses make for a stronger house, while giving propagators the ability to hang plants and equipment, such as monorails, curtain systems, and irrigation booms (*78*). *Bottom left:* non-load carrying bow propagation house. *Bottom right:* load-bearing, gutter connected truss house.

costs (*167*). An innovation that can reduce aisle space and increase the usable space by 30 percent in a propagation greenhouse is the use of rolling benches. These are pushed together until one needs to get between them, and then rolled apart (Figure 3–6).

Several new types of greenhouses are being designed and constructed to eliminate the use of any benches—plants are produced with an automated floor watering and fertility system or **floor ebb and flood system (flood floor)**. There are below-ground floor-heating pipes and irrigation lines, a system of runoff capturing tanks with filters, and computer-controlled return of appropriate levels of irrigation water mixed with soluble fertilizer to the floor growing area (*18*). While this has received limited use in the propagation of plants, it does have application in liner production of seedling plugs,

FIGURE 3–5 Versatility of a quonset-polyethylene house. *Above:* Rooting media floor beds are prepared and sterilized with methyl bromide. *Center:* Cuttings inserted and rooted under mist. *Below:* Rooted liners are pruned and maintained in beds until transplanting the following spring.

rooted cuttings, tissue culture produced plantlets, etc. (Figure 3–7). Flood floor systems are more efficient than conventional bench greenhouses. They are highly automated, require less labor, and are environmentally friendly—since irrigation runoff, including nutrients and pesticides, is recaptured and recycled. The drawback of these benchless systems is the potential for rapid disease spread.

Greenhouse construction begins with a metal framework, to which are fastened metal sash bars to support either panes of glass embedded in putty or some type of plastic material. Gutter-connected greenhouses can be constructed as bow-style houses, which are less expensive and offer less structural strength, or as load-bearing truss-style houses, which give propagators the ability to hang mist and irrigation booms, install ceiling curtains for temperature and light control, etc. (Figure 3–4). All-metal prefabricated greenhouses with prewelded or prebolted trusses also are widely used and are available from several manufacturers.

In any type of greenhouse or bench construction using wood, the wood should be pressure-treated with a preservative such as chromatid copper arsenate (CCA), which will add many years to its life (*11*). The two most common structural materials for greenhouses are steel and aluminum. Most greenhouses are made from galvanized steel, which is cheaper, stronger, lighter, and smaller than an aluminum member of equal strength. Aluminum has rust and corrosion resistance, and can be painted or anodized in various colors (*78*). With the high cost of lumber, fewer greenhouses are constructed with wood, and traditional wooden benches are being replaced by rigid plastics, metal benches, and other synthetic materials.

Greenhouse Heating and Cooling Systems

Ventilation, to provide air movement and air exchange with the outside, is necessary in all greenhouses to aid in controlling temperature and humidity. A mechanism for manual opening of panels at the ridge and sides or with passive ventilation can be used in smaller greenhouses, but most larger installations use a forced-air fan and pad-cooling ventilation system regulated by thermostats or computer controlled (*47*). Propagation greenhouses with retractable roofs and benches that can be rolled from the greenhouse structure to the outdoors have reduced many energy costs (Figure 3–8).

Traditionally, greenhouses have been heated by steam or hot water from a central boiler through banks of pipes (some finned to increase radiation surface) suitably located in the greenhouse. Unit heaters for each house, with fans for improved air circulation, are also used. If oil or gas heaters are used, they must be vented to the outside because the combustion products are toxic to plants (and

FIGURE 3–6 For more efficient use of costly greenhouse propagation space, movable benches on rollers have been installed to reduce aisle space. *Top left and right:* movable benches for seedling plug production. *Middle left and right:* movable mist benches with flexible attached hose. *Bottom left and right:* Instead of a movable bench, propagation flats are placed on rollers; notice how all trays on rollers slant towards the middle of the propagation house for easier movement of materials.

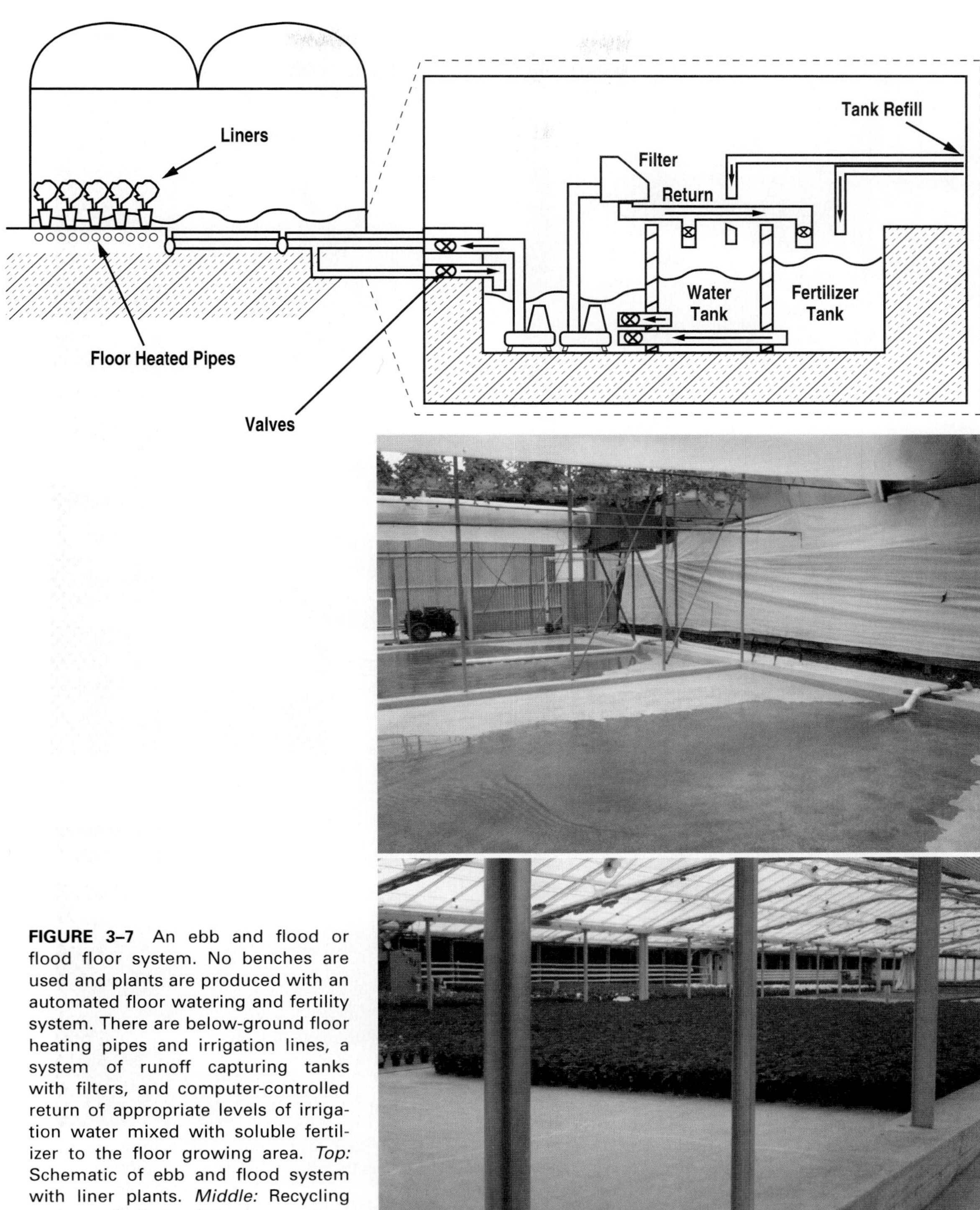

FIGURE 3–7 An ebb and flood or flood floor system. No benches are used and plants are produced with an automated floor watering and fertility system. There are below-ground floor heating pipes and irrigation lines, a system of runoff capturing tanks with filters, and computer-controlled return of appropriate levels of irrigation water mixed with soluble fertilizer to the floor growing area. *Top:* Schematic of ebb and flood system with liner plants. *Middle:* Recycling system. *Bottom:* Greenhouse crop production.

FIGURE 3–8 Propagation houses with retractable roofs and benches, which can be rolled from the greenhouse structure to the outdoors, have reduced many energy costs. *Top:* Inside of house with double-tiered benches that can be brought in at night and during inclement weather. *Bottom:* Benches slide through opening of greenhouse and can be left outside under full sun conditions.

people!) and ethylene gas generated can adversely affect plant growth. In large greenhouses, heated air is often blown into large—30 to 60 cm (12 to 24 in.)—4-mil convection polyethylene tubes hung overhead. These extend the length of the greenhouse. Small—5 to 7.5 cm (2 to 3 in.)—holes spaced throughout the length of these tubes allow the hot air to escape, thus giving uniform heating throughout the house. These same convection tubes can be used for forced air ventilation in summer, eliminating the need for manual side and top vents.

Gas-fired infrared. These vacuum-operated radiant heaters are sometimes installed in the ridges of greenhouses with the concept of heating the plants but not the air mass. Infrared heaters consist of several lines of radiant tubing running the length

FIGURE 3–9 Propagation house heating systems. *Top:* Gas-fired infrared or vacuum-operated radiant heaters (arrow). *Middle:* Forced hot air heating system. *Bottom:* heating below the bench for control of root zone temperature.

of the house with reflective shielding above the tubes, installed at a height of 1.8 to 3.7 m (6 to 12 ft) above the plants (Figure 3–9). The principal advantage of infrared heating systems in greenhouses is lower energy use (*177*). Cultural practices may need to be changed with infrared heating due to heating the plant but not the soil underneath.

Root zone heating. In contrast to infrared heating, root zone heating is done by placing pipes on or below the soil surface in the floor of the greenhouse, or on the benches, with recirculating hot water—controlled by a thermostat, circulating through the pipes. This places the heat below the plants, which hastens the germination of seeds, rooting of cuttings, or growth of liner plants. This popular system has been very satisfactory in many installations, heating the plant's roots and tops, but not the entire air mass in the greenhouse, giving substantial fuel savings. It is also excellent for controlling foliage diseases. As of the mid 1990s, the vast majority of propagation (seed germination, rooted cuttings, and **plug** growing) is done with some form of root zone heat (*9*) (Figures 3–9 and 3–10).

Solar heating. Conservation of energy in the greenhouse is important (*108*). In greenhouses, solar heating occurs naturally. The cost of fossil fuels has evoked considerable interest in methods of conserving daytime solar heat for night heating (*8,59,82,106,132,158*). Conservation methods need to be developed and utilized, otherwise high heating costs may eventually make winter use of greenhouses in colder regions economically unfeasible—relegating greenhouse operations to areas with relatively mild winters (*42,160*).

Most heat loss in greenhouses occurs through the roof. One method of reducing heat loss in winter is to install sealed polyethylene sheeting outside over the glass, or fiberglass, or to use two layers of polyethylene sheeting, as in a quonset house. This double-poly method of insulation is very effective. The two layers are kept separate by an air cushion from a low-pressure blower. Energy savings from the use of this system are substantial—more than 50 percent reduction in fuel compared to conventional glass greenhouses—but the greatly lowered light intensity with the double-layer plastic cover can lower yields of many greenhouse crops (*14,15,57*).

Another device that reduces heat loss dramatically is a *movable thermal curtain* (Figure 3–11), which, at night, is placed between the crop and the propagation house roof and walls (*77, 148,152,157,161*). Winter heating bills are reduced as much as 30 percent, since the peak of the propagation house is not heated (*86*). During summer, automated curtains also reduce heat stress on propagules and workers, and less energy is needed to run fans for cooling. Modified curtains can be used for light reduction during the day and "black clothing" for light exclusion during photoperiod manipulation of plants. Curtains range from 20 percent shade reduction to complete blackout curtains—ULS Obscura A + B (*86*). Curtain fibers are available in white, black, with aluminum coated fibers, and/or with strips of aluminum sewn in. Black shade cloth reduces light to the plants but absorbs heat and emits heat back into the propagation house. Aluminum coated curtain fabrics are good reflectors of light, but poor absorbers of heat. Some curtain materials come with an upside for reflecting heat and reducing condensation—and with a bottom side for heat retention. Insulating the north wall reduces heat loss without appreciably lowering the available light.

Efficient air movement can greatly assist in effective greenhouse heating and cooling. A low-cost horizontal air-flow system has been used effectively to maintain a constantly circulating air movement in the greenhouse. A series of blade-style fans [41 cm (16-in.) diameter] spaced 10.7 to 15 m (35 to 50 ft) apart down one side of the greenhouse and up the other sets up a circular air movement that uniformly distributes the heated or cooled air very well. It is also helpful for distributing CO_2-enriched air (*12,23*).

Greenhouses can be cooled mechanically in the summer by the use of large evaporative cooling units, as shown in Figure 3–12. The "pad and fan" system, in which a wet pad of some material, such as a special honeycombed cellulose, aluminum mesh, or plastic fiber, is installed at one side (or end) of a greenhouse with large exhaust fans at the other, has proved to be the best method of cooling greenhouses, especially in low-humidity climates (*7*). Fog can be used to cool greenhouses, but is more expensive than conventional pad and fan systems, and is inefficient in climates with high relative humidity, e.g., the Texas Gulf Coast.

Greenhouses are often sprayed on the outside at the onset of warm spring weather with a thin layer of whitewash or a white, cold-water paint. This coating reflects much of the heat from the sun, thus preventing excessively high temperatures in the greenhouse during summer. The whitewash is removed in the fall. Too heavy a coating of whitewash, however, can reduce the light irradiance to undesirably low levels.

FIGURE 3–10 Hot water, root zone heating of propagation flats. *Top left:* Gas-fired boiler for recirculating hot water to a Biotherm heating system. *Top right:* Biotherm tubing heating root zone of the plug tray; notice the probe (arrow) for regulating temperature. *Bottom left:* The flexible hot water tubing is hooked into larger PVC pipes at set distances to assure more uniform heating. *Bottom right:* Cuttings in propagation flats placed over hot water tubing; in more mild climates, the ground hot water tubing may be all that is used to control root zone temperature and the air temperature of the propagation house.

FIGURE 3–11 Movable thermal curtains, which at night, are placed between the crop and the propagation roof house and walls can dramatically reduce winter heating bills. *Top:* A propagation house with high density supplementary lighting underneath the thermal curtains. *Bottom left and right:* Even in Texas, thermal curtains reduce heating bills, and shade curtains can also be drawn to reduce the light irradiance and greenhouse cooling expenses during summer months.

FIGURE 3–12 Completely automated heating and cooling systems installed in fiberglass-covered greenhouse. *Upper left:* Hot air from hot-water heaters (at top) is blown into polyethylene distribution tubes. *Upper right:* Distribution tubes, which have outlet holes spaced to disperse heated air uniformly, extend the length of the house. *Lower left and right:* The opposite end wall of the greenhouse has an insert of a wettable pad *(right)* through which air is pulled by exhaust fans for cooling. Automatic closure panels *(left)* shut off outside air movement into the house through this pad when heating is required. All components of both heating and cooling systems are thermostatically controlled.

Environmental Controls[5]

Controls are needed for greenhouse heating and evaporative cooling systems. Although varying with the plant species, a minimum night temperature of 13 to 15.5° C (55 to 60° F) is common. Thermostats for evaporative cooling are generally set to start the fans at about 24° C (75° F). In the early days of greenhouse operation, light, temperature, and humidity were about the only environmental controls attempted. Spraying the greenhouse with whitewash in summer and opening and closing side and ridge vents with a crank to control temperatures, along with turning on steam valves at night to prevent freezing, constituted environmental control. Humidity was increased by spraying the walks and benches by hand at least once a day. Later, it was found that thermostats, operating solenoid valves, could activate electric motors to raise and lower vents, and to open and close steam and water valves, thus giving some degree of automatic control.

[5]Available from Priva Computers, Inc., P.O. Box 110, Vineland Station, Ontario, Canada LOR 2EO; Oglevee, 151 Oglevee Lane, Connellsville, Pa. 15425; Wadsworth Control Systems, 5541 Marshall Street, Arvada, Colo. 80002; Q-Com Corp, 17782 Cowan Ave., Irvine, Cal. 92714.

Analog environmental controls. Analog controls (i.e. Wadsworth Step 500) have evolved for controlling the greenhouse environment. They use proportioning thermostats or electronic sensors to gather temperature information. This information drives amplifiers and electronic logic (i.e., decision making) circuitry (*9*). Analog controls cost more than thermostats, but are more versatile and offer better performance.

Computerized environmental controls. The advent of computer technology has replaced the amplifiers and logic circuits of an analog control with a microprocessor "computer on a chip." Computer controls are quicker and more precise in combining information from a variety of sensors (temperature, relative humidity, light intensity, wind direction) to make complex judgments about how to control the propagation environment.

Although more costly than thermostats or analogs, computer controls offer significant energy and labor savings and improved production efficiency in propagation. Not only temperature, ventilation, and humidity can be controlled but many other factors, such as propagating bed temperatures, application of liquid fertilizers through the irrigation system, daylength lighting, light-intensity regulation with mechanically operated shade cloth (and thermal sheets), operation of a mist or fog system, CO_2 enrichment—all of which can be varied for different times of the day and night and for different banks of propagation units (*9,16, 52,65,150,164*). Alarms can be triggered by the computer, if deviations from preset levels occur, such as a heating failure on a cold winter night, or a mist system failure on cuttings on a hot summer day. Some of these operations are shown in Figures 3–3, 3–13, 3–14. Most importantly, the computer can provide data sheets on all factors being controlled for review to determine if changes are needed. This makes it easier for the propagator to make management decisions based on factual information (*47*).

Greenhouse Covering Materials

(64,129,155)

Glass

Glass-covered greenhouses are expensive, but for a permanent long-term installation under low-light winter conditions, may be more satisfactory than the popular, low-cost polyethylene (poly)-covered houses. Due to economics and the revolution in greenhouse-covering materials from polyethylene to polycarbonates, glass greenhouses are no longer as dominant as they were. Glass is still used, due in part to its superior light transmitting properties and less excessive relative humidity problems. Glass "breathes" (the glass laps between panes allow air to enter), whereas polyethylene, acrylic, and polycarbonate-structured sheet houses are airtight, which can result in excessive humidity and undesirable water drip on the plants if it is not properly controlled. This problem can be overcome, however, by maintaining adequate ventilation and heating. Some of the newer greenhouse covering materials are designed to channel condensation to gutters, avoiding water dripping onto plant foliage. Control of high relative humidity is a key cultural technique to manage plant pathogens, since water can both disseminate pathogens and encourage plant infection. See the section on cultural controls in propagation under integrated pest management, later in the chapter.

Flexible Covering Materials

Polyethylene (polythene, poly). More than 50 percent of the greenhouse area in the United States is covered with low-cost polyethylene (poly), most with inflated double layers, giving good insulating properties. Poly is the most popular covering for propagation houses. Several types of plastic are available, but most propagators use either single or double layered polyethylene. Poly materials are lightweight and relatively inexpensive compared to glass. Their light weight also permits a less expensive supporting framework than is required for glass. Polyethylene has a relatively short life. It breaks down in sunlight and must be replaced after one or two years, generally in the fall in preparation for winter. The new polys of the 1990s, with ultraviolet (UV) inhibitors, can last three to four years—but in the southern U.S., with higher UV levels, poly deteriorates more quickly and propagation houses need to be recovered more frequently (*9*).

A thickness of 4 to 6 mils (1 mil = 0.001 in.) is recommended. For better insulation and lowered winter heating costs, a double layer of UV-inhibited copolymer material is used with a 2.5-cm (1-in.) air gap between layers, kept separated by air pressure from a small blower.

Single layer polyethylene-covered greenhouses lose more heat at night or in winter than a glass-covered house, since polyethylene allows passage of heat energy from the soil and plants inside the greenhouse much more readily than glass. There are some newer infrared reflective polys, which save fuel, but have lower light penetration than regular poly(*9*). Glass traps most infrared radiation, whereas polyethylene is transparent to it. However, double poly covered greenhouses retain more heat than glass since the houses are more air-tight and less infrared radiation escapes.

FIGURE 3–13 Manipulating the propagation environment. *Upper left:* Greenhouse sensors that are connected to an analog or computer-controlled environmental system. *Upper right:* Analog type controller. *Middle left and right:* high density lighting for propagating plants during seasonal low-light conditions. *Bottom:* Use of incandescent lights to extend photoperiod, which encourages these rooted Japanese maple cuttings to avoid dormancy.

FIGURE 3–14 Environmental sensors for propagation. *Upper left and right:* A propagation house with an attached sensor for detecting light intensity, wind speed, and direction; this helps regulate temperature control and the fog propagation system. *Middle left and right:* The measurement of solar light is configured to conventional mist time clock system for better mist control. *Bottom left and right:* Relative humidity sensors are critical for fog propagation control.

Only materials especially prepared for greenhouse covering should be used. Many installations, especially in windy areas, use a supporting material, usually welded wire mesh, for the polyethylene film. Occasionally, other supporting materials, such as Saran cloth, are used.

Polyethylene transmits about 85 percent of the sun's light, which is low compared to glass, but it passes all wavelengths of light required for the growth of plants. A tough, white, opaque film consisting of a mixture of polyethylene and vinyl plastic is available. This film stays more flexible under low winter temperatures than does clear polyethylene, but is more expensive. Because temperature fluctuates less under opaque film than under clear plastic, it is suitable for winter storage of container-grown plants. Polyethylene permits the passage of oxygen and carbon dioxide, necessary for the growth processes of plants, while reducing the passage of water vapor.

For covering lath and shade structures, there is a number of satisfactory plastic materials prepared for the horticultural industry. Some commercially available materials include UV-treated cross-woven polyethylene fabric that resists ripping and tearing, and knitted high-density UV polyethylene shade cloth that is strong and has greater longevity.

Rigid Covering (Structured Sheet) Materials

Acrylic (Plexiglass, Lucite, Exolite). This is highly weather resistant, does not yellow with age, has excellent light transmission properties, retains twice the heat of glass, and is very resistant to impact, but is brittle. It is somewhat more expensive and nearly as combustible as fiberglass. It is available in twin-wall construction which gives good insulation properties, and has a no-drip construction that channels condensation to run down to the gutters, rather than dripping on plants.

Polycarbonate (Polygal, Lexan, Cyroflex, Dynaglas). Polycarbonate is probably the most widely used structured sheet material today (*9*). This material is similar to acrylic in heat retention properties, with about 90 percent the light transmission of glass. It has high impact strength, about 200 times that of glass. It is lightweight, about one-sixth that of glass, making it easy to install. Polycarbonate's textured surface diffuses light and reduces condensation drip. It is available in twin-wall construction which gives good insulation properties. Polycarbonate can be cut, sawn, drilled, or nailed, and is much more user-friendly than acrylics, which can shatter if nails or screws are driven into it. It is UV stabilized and will resist long outdoor exposure (some polycarbonates are guaranteed for ten years), but will eventually yellow with age (*9,113*).

Fiberglass. Rigid panels, corrugated or flat, of polyester resin reinforced with fiberglass have been widely used for greenhouse construction. This material is strong, long-lasting, lightweight, and easily applied, and comes in a variety of widths, lengths, and thicknesses but is not as permanent as glass. Only the clear material—especially made for greenhouses and in a thickness of 0.096 cm (0.038 in.) or more and weighing 4 to 5 oz per square foot—should be used. New material transmits about 80 to 90 percent of the available light, but light transmission decreases over the years due to yellowing, which is a serious problem. Since fiberglass burns rapidly, an entire greenhouse may quickly be consumed by fire, so insurance costs can be higher. Fiberglass is more expensive than polyethylene, and is not as widely used as it was in the 1970s (*9*).

The economics of using these greenhouse covering materials must be considered carefully before a decision is made. New materials are continually coming onto the market.

Hot Frames (Hotbeds) and Heated Sun Tunnels—(Closed-case propagation)

The hotbed is a small, low structure, which is used for many of the same purposes as a propagation house. Traditionally, the hotbed is a large wooden box or frame with a sloping, tight-fitting lid made of window sash. Hotbeds can be used throughout the year, except in areas with severe winters, where their use may be restricted to spring, summer, and fall. Another form of a hotbed is a **heated, low polyethylene tunnel or sun tunnel** that is made from hooped metal tubing or bent pvc pipe, which is covered with polyethylene (sometimes a white poly material is used to avoid the higher temperature buildup and temperature fluctuations of clear poly) (Figure 3–15).

Traditionally, the size of the frame conforms to the size of the glass sash available—a standard size is 0.9 by 1.8 m (3 by 6 ft). If polyethylene is used as the covering, any convenient dimensions can be used. The frame can be easily built with 3 cm (1-in.) or 6 cm (2-in.) lumber nailed to 4-by-4 corner posts set in the ground. Decay-resistant wood such as redwood, cypress, or cedar should be used, and preferably pressure-treated with wood preservatives, such as chromatid copper arsenate (CCA). This com-

FIGURE 3–15 Low polyethylene tunnel or sun tunnel that is covered with polyethylene. *Top:* Sometimes a white poly material is used to avoid the higher temperature buildup and temperature fluctuation of clear poly. Propagation flats are placed on top of hot-water tubing or electric heating cables. *Middle:* Saran shade cloth can be used to cover the poly to reduce the heat load. *Bottom:* Winterization of sun tunnels with Saran and clear poly or opaque poly.

pound retards decay for many years and does not give off fumes toxic to plants. Creosote must not be used on wood structures in which plants will be grown, since the fumes released, particularly on hot days, are toxic to plants. Publications are available giving in detail the construction of such equipment (*147*).

Plastic or pvc tubing with recirculating hot water are quite satisfactory for providing bottom heat in hotbeds. The hotbed is filled with 10 to 15 cm (4 to 6 in.) of a rooting or seed-germinating medium over the hot-water tubing. Alternatively, community propagation flats or flats with liner pots containing the medium can be used. These are placed directly on a thin layer of sand covering the hot-water tubing.

Seedlings can be started and leafy cuttings rooted in hotbeds early in the season. As in the greenhouse, close attention must be paid in hotbeds to shading and ventilation, as well as to temperature and humidity control. For small propagation operations, hotbed structures are suitable for producing many thousands of nursery plants without the higher construction expenditure for larger, walk-in propagation houses (*75*).

Cold Frames and Unheated Sun Tunnels (Closed-Case Propagation) (*1,147*)

A primary use of cold frames is in conditioning or hardening rooted cuttings or young seedlings (liners) preceding field, nursery-row, or container planting. Cold frames and unheated sun tunnels can be used for starting new plants in late spring, summer, or fall when no external supply of heat is necessary (*169*). Today, cold frames include not only low polyethylene-covered wood frames or unheated sun tunnels that people can not walk within (Figure 3–15), but also low cost, poly-covered hoop houses (Figure 3–16). The covered frames should fit tightly in order to retain heat and obtain high humidity. Cold frames should be placed in locations protected from winds, with the sash cover sloping down from north to south (south to north in the southern hemisphere).

Low-cost cold frame construction (Figure 3–16) is the same as for hotbeds, except that no provision is made for supplying bottom heat. With older style cold frames, sometimes a lath covering with open spaces between the lath boards is used to cover the cold frame. This does not prevent freezing temperatures from occurring, but does reduce high and low temperature fluctuations.

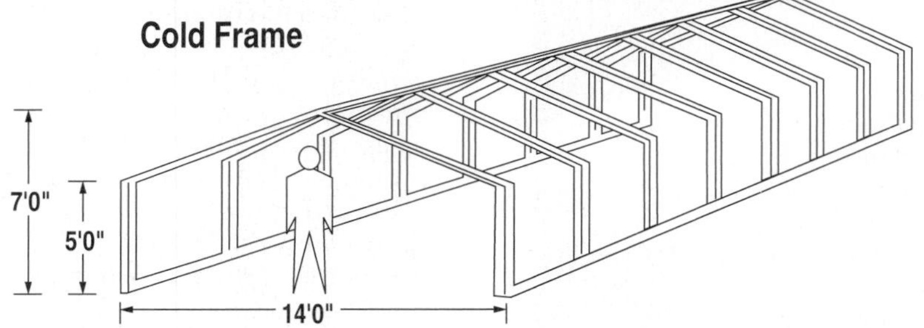

FIGURE 3–16 Cold frames that were used for propagating tender plants. Frames are opened, after protection is no longer required. *Upper left:* Older commercial use of glass-covered cold frames in propagating ground cover plants by cuttings. *Upper right:* Wood sash used for liner production in a cold frame. Glass and lath coverings are rarely used due to the high labor costs in moving the heavy sash. Plastic coverings are more suitable. *Middle:* A cold frame for propagation with poly removed. *Bottom:* Today a cold frame is most commonly a very low cost, budget, poly-covered hoop or galvanized steel bow house.

In these structures, only the heat of the sun, retained by the transparent or opaque, white polyethylene coverings, is utilized. Close attention to ventilation, shading, watering, and winter protection is necessary for success with cold frames. When young, tender plants are first placed in a cold frame, the coverings are generally kept tightly closed to maintain a high humidity, but as the plants become adjusted, the sash frames are gradually raised, or ends of the hoop house or sun tunnels opened, to permit more ventilation and drier conditions.

The installation of a mist line or frequent irrigation of plants in a cold frame is essential to maintain humid conditions. During sunny days temperatures can build up to excessively high levels in closed frames unless ventilation and shading are provided. Spaced lath, Saran or poly shade cloth covered frames, or reed mats are useful to lay over the sash to provide protection from the sun. In areas where extremely low temperatures occur, plants being overwintered in cold frames may require additional protective coverings.

Lathhouses

Lathhouses or shade houses (Figure 3–17) provide outdoor shade and protect container-grown plants from high summer temperatures and high light irradiance. They reduce moisture stress and decrease the water requirements of plants. Lathhouses have many uses in propagation, particularly in conjunction with the hardening-off and acclimation of liner plants prior to transplanting, and with maintenance of shade-requiring or tender plants. At times a lathhouse is used by nurseries simply to hold plants for sale. In mild climates, they are used for propagation, along with a mist facility, and can also be used as an overwintering structure for liner plants. Snow load can cause problems in higher latitude regions.

Lathhouse construction varies widely. Aluminum prefabricated lathhouses are available but may be more costly than wood structures. More commonly, pipe or wood supports are used, set in concrete with the necessary supporting cross-members. Today, most lathhouses are covered with high density, woven, plastic materials, such as Saran, polypropylene fabric, and UV-treated polyethylene shade cloth, which come in varying shade percentages and colors. These materials are available in different densities, thus allowing lower irradiance of light, such as 50 percent sunlight, to the plants. They are lightweight and can be attached to heavy wire fastened to supporting posts. The shade cloth is resistant to ripping, and has an optimum life of

FIGURE 3–17 Lathhouses are often covered by Saran or polypropylene shade cloth supported by wires stretched between metal upright poles. These may extend to cover many acres. *Upper:* Single layer of Saran covering a lathhouse. *Middle:* Saran-covered cold frames in a lathhouse; in winter the Saran from the lathhouse will be removed to avoid the destructive weight of ice storms. *Bottom:* Use of a retractable Saran shade cloth system in the shipping dock of a nursery.

10–15 years, depending on climate and quality of material. For winterization in less temperate areas, producers will cover the shade cloth with polyethylene. Sometimes shade is provided by thin wood strips about 5 cm (2 in.) wide, placed to give one-third to two-thirds cover, depending on the need. Both sides and the top are usually covered. Rolls of snow fencing attached to a supporting framework can be utilized for inexpensive construction.

Miscellaneous Closed-Case Propagation Systems

There are a number of closed-case propagation systems that are used in the rooting of cuttings, acclimatization and rooting of tissue culture microcuttings, and propagation of seedlings. Besides the sun tunnels or cold frames previously described, closed-case propagation systems include nonmisted enclosures in glasshouses or polyhouses (shading, tent and contact polyethylene systems, wet tents, inverted glass jars).

Propagating Frames

Even in a greenhouse, humidity is not always high enough to permit satisfactory rooting of certain kinds of leafy cuttings. Enclosed frames covered with poly or glass may be necessary for successful rooting (see Figures 3–18, 11–36). There are many variations of such devices. Small ones were called Wardian cases in earlier days (Figure 1–1). Such enclosed frames are also useful for graft union formation of small potted nursery stock, since they retain high humidity.

Sometimes in cool summer climates (as far south as Virginia in the U.S.), when fall semi-hardwood cuttings are taken, a layer of very thin (1 or 2 mils) polyethylene laid directly on top of a bed of newly prepared leafy cuttings in a greenhouse or lathhouse will provide a sufficient increase in relative humidity to give good rooting. This is sometimes referred to as a **contact polyethylene system** (Figure 11–36). Good shade control to reduce light irradiance is essential for this system.

On a more limited scale, bell jars (large inverted glass jars) can be set over a container of unrooted cuttings or freshly grafted containerized plants to speed up graft union formation (Figure 13–38). Humidity is kept high in such devices, but some shading is necessary to control temperature. As shown in Figure 3–19, polyethylene plastic bags can be placed over a simple wire framework set in the rooting container to provide an inexpensive protective cover that maintains high humidity when rooting a few cuttings.

In using all such structures, care is necessary to avoid the buildup of pathogenic organisms. The warm, humid conditions, combined with lack of air movement and relatively low light intensity, provide excellent conditions for the growth of various pathogenic fungi and bacteria. Cleanliness of all materials placed in such units is important; however, use of fungicides is sometimes necessary (see the section on **integrated pest management**).

Enclosed Poly Sweat Tent—Hydroponic System

An Australian producer of chrysanthemums uses a modified nutrient film technique (NFT) for growing greenhouse stock plants and propagating cuttings (*73*). Unrooted cuttings are stuck in Oasis root cubes, and placed in mist propagation benches containing a reservoir of water, maintained with a float valve. The system is initially enclosed in a clear poly sweat tent. Once root initiation takes place, the mist is turned off and the poly tent lifted. Cuttings are then supplied with nutrient solution in the NFT system on the propagation bench and later transplanted with the roots intact and undisturbed in the root cube. Stock plants are also maintained in the NFT system and supported in root cubes, thus allowing more precise nutritional control and reduction in environmental stress to the stock plant.

CONTAINERS FOR PROPAGATING AND GROWING YOUNG LINER PLANTS

New types of containers for propagating and growing young liner plants are continually being developed, usually with a goal of reducing handling costs. Direct sticking of unrooted cuttings into small liner containers, as opposed to sticking into conventional propagation trays, saves a production step and later avoids root disturbance of cuttings, which can lead to transplant shock (*37*) (Figure 3–27).

Flats

Flats are shallow plastic, styrofoam, wooden or metal trays, with drainage holes in the bottom. They are useful for germinating seeds or rooting cuttings, since they permit young plants to be moved easily. In the past, durable kinds of wood, such as cypress, cedar, or redwood, were preferred for flats. Galvanized-iron flats were once used, but zinc released from the galvanized flats could cause

FIGURE 3–18 Polyethylene-covered beds used in a greenhouse to maintain high humidity surrounding the cuttings during rooting. *Upper:* Propagation flats can be placed on beds. *Bottom:* Cuttings may also be stuck directly into the mist beds and covered with poly; in foreground, the poly has been removed from the rooted liners.

toxicity to plants. Today, the most popular flats are made of plastic (polyethylene, polystyrene), which come in all shapes and sizes. The 28 × 53 cm (11 × 21 in.) *1020* plastic flats are the industry standard. The number of cells or compartments per tray may range from 1 cell for a community rooting flat or seed germination tray, to 18 or more cells for a rooted liner tray, to 100–400 cells for a seedling plug tray. Trays can also be fitted with removable sheet inserts containing the cells. Plastic flats will nest, and thus require relatively little storage space. The cost of producing plastic, for flats and containers and of disposing used plastic has led to increased plastic recycling programs in horticulture.

Clay Pots

The familiar red clay flower pots, long used for growing young plants, are heavy and porous and lose moisture readily. They are easily broken, and their round shape is not economical of space. After continued use, toxic salt accumulations build up, requiring soaking in water before reuse. Clay pots are rarely used today in commercial propagation, except with specialized crops.

FIGURE 3–19 A small unit for rooting a few cuttings. A polyethylene bag has been placed over a framework made from two wire coat hangers. The bag is folded under the pot to give a tight seal. This unit should be set in a fairly light place but never in direct sunlight, which would cause overheating within the bag.

Plastic Pots

Plastic containers, round and square, have numerous advantages: they are nonporous, reusable, lightweight, and use little storage space, since they will nest. Some types are fragile, however, and require careful handling, although other types, made from polyethylene, are flexible and quite sturdy. Small liner pots for direct rooting of cuttings, seedling propagation, and tissue culture plantlet acclimatization and production have gained considerable popularity.

Many of these small containers have riblike structures to redirect root growth and prevent girdling (Figure 3–20). In forestry seedling production, ribbed book or sleeve containers are used, which consist of two matched sections of molded plastic that fit together to form a row of rectangular cells (Figure 3–20). The inner walls of small propagation containers and liner pots can also be treated with chemical root pruning agents, such as copper hydroxide ($CuOH_2$) which chemically prune liner roots at the root-wall interface (*92*). The chemically

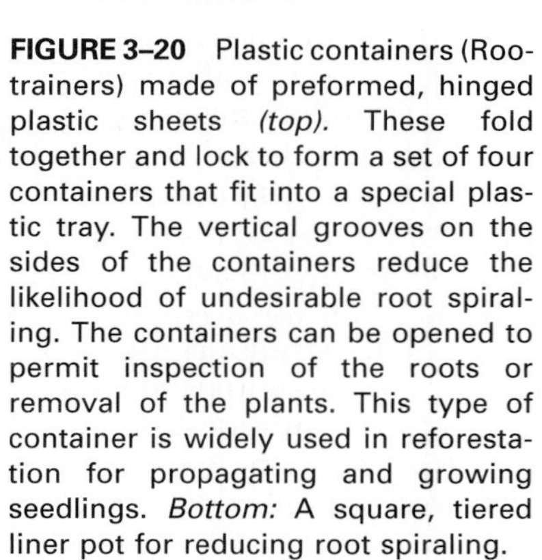

FIGURE 3–20 Plastic containers (Rootrainers) made of preformed, hinged plastic sheets *(top)*. These fold together and lock to form a set of four containers that fit into a special plastic tray. The vertical grooves on the sides of the containers reduce the likelihood of undesirable root spiraling. The containers can be opened to permit inspection of the roots or removal of the plants. This type of container is widely used in reforestation for propagating and growing seedlings. *Bottom:* A square, tiered liner pot for reducing root spiraling.

pruned lateral roots become suberized but will begin to grow again after transplanting, which results in a well-distributed root system that helps minimize transplant shock (*92*) (Figure 3–21).

Square pots are also made up into "packs" of 8 or 12 for easier handling. Plastic pots (and flats) cannot be steam sterilized, but some of the more common plant pathogens can be controlled by a hot water dip, 70° C (158° F), for three minutes followed by a rinse in a dilute bleach solution, i.e., Clorox, Purex, etc. Ultraviolet light inhibitors are sometimes incorporated in the plastic resin to prevent UV degradation of plastic pots under full sun conditions.

Fiber Pots

Containers of various sizes, round or square, are pressed into shape from peat plus wood fiber, with fertilizer added. Dry, they will keep indefinitely. Since these pots are biodegradable, they are set in the soil along with the plants. Peat pots find their best use where plants are to be held for a relatively short time and then put in a larger container or in the field. During outplanting in the field, any portion of the fiber pot transplanted above the surface of the soil will act like a wick and quickly dry out the transplant.

During production, small peat pots with plants growing in them eventually deteriorate because of constant moisture, and may fall apart when moved. On the other hand, unless the pots are kept moist, roots will fail to penetrate the walls of the pot and will grow into an undesirable spiral pattern. Units of 6 or 12 square peat pots fastened together are available. When large numbers of plants are involved, time and labor are saved in handling by the use of these units.

Paper Pots

Paper pots or paper tube pots are more popular with seedling propagation of ornamental and forestry species. They allow for greater mechanization with pot-filling machines, automatic seeders, and wire benches that allow air pruning of the root system. Typically, paper pots consist of a series on interconnected paper cells that are arranged in a honeycomb pattern that can be separated before outplanting (*92*). An advantage of the paper pot system is that pots are biodegradable, and the seedling plug can be planted intact into a larger container or into the ground without disturbing the root system. Some papier mâché pots (paper, wax, asphalt) come treated with copper hydroxide, which enhances root development and retards deterioration of the pot.

In Europe, paper tube pots with predictable degradation rates are produced by machine (*44*). The propagation medium is formed into a continuous cylinder and wrapped with a length of paper or cellulose skin which is glued and heat sealed.

Peat, Fiber, Expanded Foam, and Rockwool Blocks

Blocks of solid material, sometimes with a prepunched hole (Figure 3–22), have become popular as a germinating medium for seeds and as a rooting medium for cuttings, especially for such plants as chrysanthemums and poinsettias. Fertilizers are sometimes incorporated into the material. One type (Figure 3–23)[6] is made of highly compressed peat which, when water is added, swells to its usable size and is soft enough for the cutting or seed to be inserted. Such blocks become a part of the plant unit and are set in the soil along with the plant. These blocks replace not only the pot but also the propagating mix.

Synthetic rooting blocks (oasis, rockwool) are becoming more widely used in the nursery industry (and forestry industry for seed propagation), and are well adapted to automation. Other advantages are their light weight, consistent quality, reproducibility, and clean condition. Watering must be carefully controlled to provide constant moisture, while maintaining adequate aeration.

Plastic and Metal Growing Containers for Post-Liner Production

Many millions of nursery plants are grown and marketed each year in 3.8-liter (1-gal) and—to a lesser extent—11-liter (3-gal), 19-liter (5-gal), and larger containers. They are tapered for nesting and all have drainage holes. However, heavy-wall, injection-molded plastic containers have replaced the older-type metal containers. Machine planters have been developed utilizing containers, in which rooted cuttings or seedlings can be transplanted as rapidly as 10,000 or more a day. See the horticulture and forestry nursery production flow diagrams (Figure 3–27). Plants are easily removed from tapered containers by inverting and tapping. Some plastic containers are made of preformed, hinged plastic sheets which can be separated for easy removal of the liner (Figure 3–20). Untapered

[6]Available from Jiffy Products of America, West Chicago, Il. 60185.

FIGURE 3–21 Chemical root pruning involves treating the interior container wall with a growth-inhibiting chemical, such as copper hydroxide. This causes the lateral roots to be chemically pruned at the container wall. A well-branched root system occurs which enhances transplant establishment. *Upper:* Schematic of non-pruned versus chemically pruned seedling container roots. *Bottom left:* copper hydroxide treated container. *Bottom right:* Copper hydroxide treated *Acalpha hispida* (*left,* without visible surface roots). (Photo courtesy of M. Arnold.)

FIGURE 3–22 Solid blocks of compressed fibers with a prepunched hole are sometimes used for cuttings of easily rooted plants. After rooting, the block plus the rooted cutting is planted into a growing container or the field nursery.

metal containers must be cut down each side with can shears or tin snips to permit removal of the plant, which creates an extra production step and is potentially hazardous because of the sharp edges of the sheared can.

In areas with high summer temperatures, use of light-colored (white or silver) containers may improve root growth by reducing heat damage to the roots, which is often encountered in dark-colored containers that absorb considerable heat when exposed to the sun. However, light-colored containers show dirt marks (as opposed to black or dark green containers) and are not as attractive to the consumer. A pot-in-pot system, in which a containerized plant is inserted into a hole in the ground lined with a plastic sleeve pot, helps moderate both high and low rootball temperatures (Figure 3–39).

Polyethylene Bags and Plant Rolls

Polyethylene bags are widely used in Europe, Australia, New Zealand, and in less developed countries in the tropics, but rarely in North America for growing-on rooted cuttings or seedling liners to a salable size. They are considerably less expensive than rigid plastic containers and seem to be satisfactory (see Figure 3–24), but some types deteriorate rapidly. They are usually black, but some are black on the inside and light-colored on the outside. The lighter color reflects heat and lowers the root temperature (*159*). Polybags do not prohibit root spiraling or allow air pruning, which is a drawback to their use in propagation and liner production; however, poly tubes are open-ended, which reduces girdling problems. After planting, they cannot be stacked as easily as the rigid containers for truck transportation—the poly bags often break, and the root system of the plant is more easily damaged.

FIGURE 3–23 Use of solid block rooting medium. *Left:* Compressed sphagnum peat discs, encased in a plastic netting and containing some added mineral nutrients. *Right:* Adding water causes peat to swell to size shown. Chrysanthemum cuttings inserted into full-sized pellets rooted rapidly and are ready to be planted in soil medium.

FIGURE 3–24 Polyethylene bags used as plant-growing containers for rhododendrons in New Zealand.

A low-cost method of propagating some easy to root species is with a polyethylene plant roll (Figure 11–49). The basal ends of the cuttings are inserted in damp peat moss or sphagnum and rolled into the doubled-over plastic sheeting. The roll of cuttings is then set upright in a humid location for rooting. Polyethylene starter pouches with an absorbent paper inserted in the pouch are used for germinating selected seed lots.

Wood Containers

Large wood containers are used for growing large specimen trees and shrubs to provide "instant" landscaping for the customer. The plant material may be kept in such containers for several years, or the plants are initially field grown and then "boxed" for six months to a year to let the root system acclimatize before being sold. Heavy moving equipment is required for handling such large nursery stock (Figure 3–25).

MANAGEMENT OF EDAPHIC FACTORS IN PROPAGATION AND LINER PRODUCTION

Media[7] and Mixes for Propagating and Growing Young Liner Plants

Various materials and mixtures of materials are used for germinating seeds and rooting cuttings. For good results the following characteristics of the medium are required (*60*):

- The medium must be sufficiently firm and dense to hold the cuttings or seeds in place during rooting or germination. Its volume must be fairly constant when either wet or dry; excessive shrinkage after drying is undesirable.
- It should be highly decomposed (preferably with a 20C/1N ratio) to prevent N immobilization, and excessive shrinkage during production.
- It must be easy to wet (not too hydrophobic) and retain enough moisture to reduce frequent watering.
- It must be sufficiently porous so that excess water drains away, permitting adequate penetration of oxygen to the roots—all containers produce a perched water table, which creates a zone of saturated growing medium at the bottom of the container.
- It must be free from weed seeds, nematodes, and various pathogens.
- It must not have a high salinity level.
- It should be capable of being steam pasteurized or chemically treated without harmful effects.
- It should have a high cation exchange capacity (CEC) for retention of nutrients that may be applied preincorporated and/or in a supplementary soluble and/or slow-release fertilizer program.
- It should be of consistent quality from batch to batch, and reproducible.
- It should be readily available and of acceptable cost.

Propagation media used in horticulture and forestry consist of a mixture of organic and inorganic components which have different but complementary properties. The **organic component** includes: peat, softwood and hardwood barks, or sphagnum moss. Sawdust, leaf mulch, and rice hulls have been used, but they oxidize readily and compact easily, which decreases pore space and aeration, and they have a high C/N ratio, which can result in nutritional problems for the propagule. A **coarse mineral component** is used to improve drainage and aeration by increasing the proportion of large, air-filled pores. A variety of mineral components include: sand (avoid fine particle sands), grit, pumice, scoria, expanded shale, perlite, vermiculite, polystyrene, clay granules, and rockwool.

There is no single, ideal mix. An appropriate propagation medium depends on the species, propagule type, season, and propagation system (i.e., with fog a waterlogged medium is less of a problem than with mist); cost and availability of the medium components are other considerations. The following media components can be used in propagation systems.

Soil

A mineral soil is composed of materials in the solid, liquid, and gaseous states. For satisfactory plant growth these materials must exist in the proper proportions. The solid portion of a soil is comprised of both inorganic and organic components. The inorganic part consists of the residue from parent rock after decomposition resulting from the chemical and physical process of weathering. Such inorganic components vary in size from gravel down to

[7] *Media* is plural—*medium* is singular.

FIGURE 3–25 Wood containers are used for large nursery shrubs and trees. *Left top and center:* Wood containerized tree and heavy equipment required to lift it. *Left bottom:* Large rigid plastic container with bottom slots designed to accommodate a fork lift. *Right top and bottom:* A large specimen tree that is field grown, then dug and a wooden box constructed around its side and bottoms. The enormous weight of the rootball will require a crane for lifting; the root system will be "cured" above-ground in the box, for up to a year, before transplanting to the landscape site.

extremely minute colloidal particles of clay, the texture of the soil being determined by the relative proportions of these particle sizes. The coarser particles serve mainly as a supporting framework for the remainder of the soil, whereas the colloidal clay fractions of the soil serve as storehouses for nutrients that are released and absorbed by plants. The organic portion of the soil consists of both living and dead organisms. Insects, worms, fungi, bacteria, and plant roots generally constitute the living organic matter, whereas the remains of such animal and plant life in various stages of decay make up the dead organic material. The residue from such decay (termed **humus**) is largely colloidal and assists in holding water and plant nutrients.

The liquid part of the soil, the soil solution, is made up of water containing dissolved salts in various quantities, as well as dissolved oxygen and carbon dioxide. Mineral elements, water, and some carbon dioxide enter the plant from the soil solution.

The gaseous portion of the soil is important to good plant growth. In poorly drained, waterlogged soils, water replaces the air, thus depriving plant roots as well as certain desirable aerobic microorganisms of the oxygen necessary for their existence.

The **texture** of a mineral soil depends upon the relative proportions of *sand* (0.05 to 2 mm particle diameter), *silt* (0.05 to 0.002 mm particle diameter), and *clay* (less than 0.002 mm particle diameter). In contrast to soil *texture,* which refers to the proportions of individual soil particles, soil **structure** refers to the arrangement of those particles in the entire soil mass. These individual soil grains are held together in aggregates of various sizes and shapes.

Propagation in commercial horticulture is generally done with flats, containers and/or pot systems with "soilless" media. Some exceptions to this are field budding and grafting systems, stooling and layering systems, field propagation of hardwood cuttings without intermittent mist (Figure 11–4), and outdoor seedbeds. With the greater reliance on containerized systems for propagation, mineral soils are either unsuitable or must be amended with other components to improve aeration and prevent the compaction that occurs with the structural changes of mineral soils in a container.

Sand

Sand consists of small rock particles, 0.05 to 2.0 mm in diameter, formed as the result of the weathering of various rocks, its mineral composition depending upon the type of rock. Quartz sand, consisting chiefly of a silica complex, is generally used for propagation purposes. Sand is the heaviest of all rooting media used, a cubic foot of dry sand weighing about 45 kg (100 lb). It should preferably be fumigated or steam-pasteurized before use, as it may contain weed seeds and various harmful pathogens. Sand contains virtually no mineral nutrients and has no buffering capacity, or cation exchange capacity (CEC). It is used mostly in combination with organic materials. Sand collected near the ocean (beach sand) may be too high in salts. Calcareous sand will raise media pH and should be tested prior to mixing by using vinegar or a dilute acid.

Peat

Peat consists of the remains of aquatic, marsh, bog, or swamp vegetation which has been preserved under water in a partially decomposed state. The lack of oxygen in the bog slows bacterial and chemical decomposition of the plant material. Composition of different peat deposits varies widely, depending upon the vegetation from which it originated, state of decomposition, mineral content, and degree of acidity (*97,107,163*).

There are three types of peat as classified by the U.S. Bureau of Mines: moss peat, reed sedge, and peat humus. *Moss peat* (usually referred to in the market as **peat moss**) is the least decomposed of the three types and is derived from sphagnum or other mosses. It varies in color from light tan to dark brown. It has a high moisture-holding capacity (15 times its dry weight), has a high acidity (pH of 3.2 to 4.5), and contains a small amount of nitrogen (about 1 percent) but little or no phosphorus or potassium. This type of peat generally comes from Canada, Ireland, or Germany, although some is produced in the northern United States. Peat moss is the most commonly used peat in horticulture, the coarse grade being the best (Figure 3–26).

Reed sedge peat consists of the remains of grasses, reeds, sedges, and other swamp plants (e.g., Florida peat). This type of peat varies considerably in composition and in color, ranging from reddish brown to almost black. It should not used for propagation purposes.

Peat humus is in such an advanced state of decomposition that the original plant remains can not be identified. It can originate from either hypnum moss or reed sedge peat. It is dark brown to black in color with a low moisture-holding capacity but with 2.0 to 3.5 percent nitrogen.

When peat moss is to be used in mixes, it should be broken apart and moistened before

FIGURE 3–26 Propagation medium. *Upper left:* A front-end loader used to incorporate sand (foreground) with shredded, pulverized bark. *Upper right:* A specialized azalea propagation mix composed of peat, bark, perlite. *Left center:* Stockpiled bags of peat. *Left bottom:* Pulverized, shredded bark, which is being used in propagation to partially replace the more expensive peat component. *Bottom right:* Discarded container plants (without the containers) and soil, which will be shredded and *recycled* as compost and/or mixed with new container media.

adding to the mix. Continued addition of coarse organic materials such as peat moss or sphagnum moss to greenhouse media can initially cause a decrease in wettability. Water will not penetrate easily, and many of the peat particles will remain dry even after watering. There is no good method for preventing this nonwettability, although the repeated use of commercial wetting agents, such as Aqua-Gro, can improve water penetration (*19*).

Peat is not a uniform product and can be a source of weed seed, insects, and disease inoculum. Peat should be pasteurized along with the other media components (*20,32*). Peat moss is relatively expensive so it is used less in nursery propagation and production mixes. It is gradually being replaced by other components, such as pulverized or shredded bark. However, peat is still the main organic ingredient in propagation and greenhouse mixes.

Sphagnum Moss Peat

Commercial sphagnum moss peat or sphagnum peat is the dehydrated young residue or living portions of acid-bog plants in the genus Sphagnum, such as *S. papillosum, S. capillaceum,* and *S. palustre.* It is the most desirable peat for horticultural purposes, but its high cost limits its commercial use. It is relatively pathogen-free, light in weight, and has a very high water-holding capacity, being able to absorb 10 to 20 times its weight of water. This material is generally shredded, either by hand or mechanically, before it is used in a propagating or growing media. It contains small amounts of minerals, but plants grown in it for any length of time require added nutrients. Sphagnum moss has a pH of about 3.5 to 4.0. It may contain specific fungistatic substances, including a strain of *Streptomyces* bacteria, which can inhibit damping-off of seedlings (*5,81*).

Vermiculite

Vermiculite is a micaceous mineral that expands markedly when heated. Extensive deposits are found in Montana, North Carolina, and South Africa. Chemically, it is a hydrated magnesium-aluminum-iron silicate. When expanded, vermiculite is very light in weight [90 to 150 kg per cubic meter (6 to 10 lb per cubic foot)], neutral in reaction with good buffering properties, and insoluble in water. It is able to absorb large quantities of water—40 to 54 liters per cubic meter (3 to 4 gal per cubic foot). Vermiculite has a relatively high cation exchange capacity and thus can hold nutrients in reserve for later release. It contains magnesium and potassium, but supplementary amounts are needed from other fertilizer sources.

In crude vermiculite ore, the particles consist of many thin, separate layers with microscopic quantities of water trapped between them. When run through furnaces at temperatures near 1090° C (2000° F), the water turns to steam, popping the layers apart, forming small, porous, spongelike kernels. Heating to this temperature provides complete sterilization. Horticultural vermiculite is graded to four sizes: No. 1 has particles from 5 to 8 mm in diameter; No. 2, the regular horticultural grade, from 2 to 3 mm; No. 3, from 1 to 2 mm; No. 4, which is most useful as a seed-germinating medium, from 0.75 to 1 mm. Expanded vermiculite should not be compacted when wet, as pressing destroys its desirable porous structure. Do not use nonhorticultural (construction grade) vermiculite, as it is treated with chemicals toxic to plant tissues.

Perlite

Perlite, a gray-white silicaceous material, is of volcanic origin, mined from lava flows. The crude ore is crushed and screened, then heated in furnaces to about 760° C (1400° F), at which temperature the small amount of moisture in the particles changes to steam, expanding the particles to small, spongelike kernels that are very light, weighing only 80 to 100 kg per cubic meter (5 to 8 lb per cubic foot). The high processing temperature provides a sterile product. Usually, a particle size of 1.6 to 3.0 mm (1/16 to 1/8 in.) in diameter is used in horticultural applications (Figure 3–26). Perlite holds three to four times its weight of water. It is essentially neutral with a pH of 6.0 to 8.0 but with no buffering capacity. Unlike vermiculite, it has no cation exchange capacity and contains no mineral nutrients. Perlite presents some problems with fluoride-sensitive plants—but fluoride can be leached-out by watering heavily. It is most useful in increasing aeration in a mix. Perlite, in combination with peat moss, is a very popular rooting medium for cuttings (*31,110*).

Calcined Clay and Other Aggregates

Stable aggregates can be produced when minerals such as clay, shales, pulverized fuel ash, etc. are heated (calcined) at high temperatures. They have no fertilizer value, are porous, resistant to breakdown, and absorb water. The main purpose of these materials is to change the physical characteristics of a propagation or liner potting mix. Examples of commercial materials made from clay include: Leca,

Terragreen, and Turfice. Haydite is a combination of clay and shale, while Hortag in the UK is made from pulverized fuel ash (*25*). Clay-type kitty litter is also a calcined clay, but contains perfumes which are not desirable for propagation.

Pumice

Chemically, pumice is mostly silicon dioxide and aluminum oxide, with small amounts of iron, calcium, magnesium, and sodium in the oxide form. It is of volcanic origin and is mined in several regions in the western states of the United States. Pumice is screened to different size grades but is not heat-treated. It increases aeration and drainage in a propagation mix and can be used alone or mixed with peat moss (*80*).

Rockwool (Mineral Wool)

This material is used as a rooting and growing medium in Europe, Australia, and the United States (see Figure 11–25). It is prepared from various rock sources, such as basalt rock, melted at a temperature of about 1600° C, then, as it cools, is spun into fibers, and pressed into blocks with a binder added. Horticultural rockwool is available in several forms—shredded, prills (pellets), slabs, blocks, cubes, or combined with peat moss as a mixture. Rockwool will hold a considerable amount of water, yet retains good oxygen levels. With the addition of fertilizers it can be used in place of the Peat-Lite mixes. Before switching from more traditional media mixes, it is best to initially conduct small-scale propagation trials with rockwool and other new media components as they become commercially available (*58*).

Shredded Bark and Wood Shavings

Shredded or pulverized *softwood bark* from redwood, cedar, fir, pine, hemlock, or various *hardwood bark* species, such as oaks and maples, can be used as a component in growing and propagating mixes, serving much the same purposes as peat moss and at a lower cost (*114,118,128,139,170*). Before it is used as a growing medium, pine bark is *hammer milled* into smaller component pieces, stockpiled in the open, and often composted by turning the piles and watering as needed. Fresh barks may contain materials toxic to plants, such as phenols, resins, terpenes, and tannins—so composting for 10 to 14 weeks before using reduces phenolic levels in bark, improves its wettability as media—and the higher bark pile temperatures help reduce insect and pathogen levels (*22,25*). Because of their relatively low cost, light weight, and availability, barks are very popular and widely used in mixes for propagation and container grown plants (Figure 3–26). Wetting agents and gels increase available water content in pine bark, and may play a greater role in helping propagators reduce irrigation frequency or the volume of water required during each irrigation (*19*).

Synthetic Plastic Aggregates

These materials were used, especially in Europe and some parts of the United States, as substitutes for sand or perlite. *Expanded polystyrene flakes* improve drainage and aeration and decrease bulk density. They are chemically neutral, do not absorb water, and do not decay, but they can be difficult to incorporate uniformly in the media. The environmental problems in the production and disposal of these materials limits their commercial usage as propagation media.

Compost

In some countries, compost is synonymous with container media for propagation and plant growth; however, we define compost (composting) as a product produced from the biological decomposition of bulk organic wastes under controlled conditions, which takes place in piles or bins. The process occurs in three steps: (a) an initial stage lasting a few days in which decomposition of easily degradable soluble materials occurs; (b) a second stage, lasting several months, during which high temperatures occur and cellulose compounds are broken down; and (c) a final stabilization stage when decomposition decreases, temperatures lower, and microorganisms recolonize the material. Microorganisms include bacteria, fungi, and nematodes; larger organisms, such as millipedes, soil mites, beetles, springtails, earthworms, earwigs, slugs, and sowbugs, can often be found in compost piles in great numbers. Compost prepared largely from leaves may have a high soluble salt content, which will inhibit plant growth, but salinity can be lowered by leaching with water before use (*126*).

In the future, with dwindling land fill sites and environmental pressures to recycle organic scrapage materials, the use of composted yard wastes, chicken and cow manure, organic sludge from municipal sewage treatment plants, etc. will play a greater role as media components in the propagation and production of small liner plants. Many nurseries recycle culled, containerized plants, and shred the plant and soil as compost or as a medium component to be mixed with fresh container medium (Figure 3–26). Composted sewage sludge not only provides organic matter, but nearly all the essential trace elements, and a large percentage of major elements needed by plants in a slowly available form (*62*). Mixes should always be analyzed for heavy metals and soluble salt levels. The usual recommended rate is that compost not comprise more than 30 percent of the volume of the mix (*25*).

Suggested Mixes—Media and Preplant Granular Fertilizers for Container Growing During Propagation and Liner Production

In propagation procedures, young seedlings, rooted cuttings, or acclimated tissue culture plantlets (liners) are sometimes planted directly in the field, but frequently are started in a blended, soilless mix in some type of container. Container growing of young seedlings and rooted cuttings has become an important alternative for field growers. In the southern and western U.S., more than 80 percent of nursery plants are container produced (*39*). For this purpose special growing mixes are needed (*67,125,154,168,174*). It is sometimes more economical for a propagator to buy bags or bulk forms of premixed media. Typically, they are composed of a peat or peat-vermiculite, peat-perlite, hamermilled and composted bark, rockwool, and other combinations. **Preplant amendments** in these mixes normally include dolomitic limestone, wetting agents (surfactants) to improve water retention and drainage of the peat or bark, starter fertilizers, trace elements, and sometimes gypsum and a pH buffer.

In preparing container mixes, the media should be screened for uniformity to eliminate excessively large particles. If the materials are very dry, they should be moistened slightly; this applies particularly to peat and bark, which if mixed when dry, absorbs moisture very slowly. In mixing, the various ingredients may be arranged in layers in a pile and turned with a shovel. A power-driven cement mixer, soil shredder, or front-end loader is used in large-scale operations (Figures 3–26 and 3–31). Most nurseries omit mineral soil from their mixes. The majority of container mixes for propagation and liner production use an organic component such as a bark or peat, which solely or in combination is mixed with mineral components such as sand, vermiculite, or pumice, depending on their availability and cost.

Preparation of the mixture should preferably take place at least a day prior to use. During the ensuing 24 hours, the moisture tends to become equalized throughout the mixture. The mixture should be just slightly moist at the time of use so that it does not crumble; on the other hand, it should not be sufficiently wet to form a ball when squeezed in the hand (*49*). With barks and other organic matter, supplementary components, particularly rice hulls and sugar cane begasse, etc., it is necessary to compost the material for a period of months, before using it as a container medium component.

Container mixes require fertilizer supplements and continued feeding of the plants until they become established in their permanent locations (*174*). For example, one successful mix for small seedlings, rooted cuttings, and bedding plants consists of one part each of shredded fir or hammermilled pine bark, peat moss, perlite, and sand. To this mixture is added **preplant fertilizers**—gypsum, dolomitic limestone, and microelements. **Postplant fertilizers**—soluble forms of nitrogen, phosphorus, and potassium—are later added to the irrigation water (*fertigation*), or as a top dressing of slow-release fertilizer, such as Osmocote, Nutricote.

In summary, nurseries have changed from loam-based growing media, as exemplified by the John Innes composts (*3*) developed in England in the 1930s, to mixes incorporating such materials as finely shredded bark, peat, sand, perlite, vermiculite, and pumice in varying proportions. The trend away from loam-based mixes is due to a lack of suitable uniform soils, the added costs of having to pasteurize soil mixes, and the costs of handling and shipping the heavier soils compared to the lighter materials. Much experimentation takes place in trying to develop other low-cost, readily available bulk material to be used as a component of growing mixes such as spent mushroom compost, papermill sludge (*28,34*), composted sewage sludge (*62*), etc.

The Cornell Peat-Lite Mixes

The Cornell Peat-Lite mixes, like the earlier University of California (UC) potting mixes, are artificial "soils." First developed in the mid-1960s, they are used primarily for seed germination and for container-growing of bedding plants, annuals, and flowering potted plants (*21*). The components are lightweight, uniform, readily available, and have chemical and physical characteristics suitable for the growth of plants. Excellent results have been obtained with these mixes. It may be desirable, however, to pasteurize the peat moss before use to eliminate any disease inoculum or other plant pests. Finely shredded bark is often substituted for the peat moss.

The term *"peat-lite"* refers to peat-based media containing perlite or vermiculite.

Peat-Lite Mix A: To Make 0.76 m^3 (1 cubic yard)

- 0.39 m^3(0.5 yd^3) shredded German or Canadian sphagnum peat moss
- 0.39 m^3 (0.5 yd^3) horticultural grade vermiculite (No. 2 or 4)
- 2.25 kg (5 lb) ground limestone (preferably dolomitic), finely ground
- 0.45 to 0.9 kg (1 to 2 lb) single superphosphate (20 percent), preferably powdered
- 0.45 kg (1 lb) calcium nitrate 84 g (3 oz) fritted trace elements (FTE 555)
- 56 g (2 oz) iron sequestrene 330
- 84 g (3 oz) wetting agent

Peat-Lite Mix B (same as A, except that horticultural perlite is substituted for the vermiculite)

Peat-Lite Mix C (for germinating seeds): To Make 0.76 m^3 (1 cubic yard)

- 0.035 m^3 (1.2 ft^3) shredded German or Canadian sphagnum peat moss
- 0.035 m (1.2 ft^3) horticultural grade vermiculite No. 4 (fine)
- 42 g (1.5 oz)—4 level tbsp. ammonium nitrate
- 42 g (1.5 oz)—2 level tbsp. superphosphate (20 percent), powdered
- 210 g (7.5 oz)—10 level tbsp. finely ground dolomitic limestone

The materials should be mixed thoroughly, with special attention to wetting the peat moss during mixing. Adding a nonionic wetting agent, such as Aqua-Gro [28 g (1 oz) per 23 liter (6 gal) of water] usually aids in wetting the peat moss.

Many commercial ready-mixed preparations[8], based on the original Cornell peat-lite mixes, are available in bulk or bags and are widely used by propagators and home gardeners. Some mixes, already filled into cell packs, seed trays, or pots, are available, and ready to be planted. Some soilless proprietary mixes are very sophisticated, containing peat moss, vermiculite, and perlite, plus a nutrient charge of nitrogen, potassium, phosphorus, dolomitic limestone, micronutrients, and a wetting agent, with the pH adjusted to about 6.5.

Proprietary micronutrient materials, such as Esmigran, FTE 503, or Micromax, consisting of combinations of minor elements, are available for adding to growing media. Adding a slow-release fertilizer such as Osmocote, MagAmp, Nutriform, Nutricote, Polyon, etc. to the basic Peat-Lite mix is useful if the plants are to be grown in it for an extended period of time.

[8]Examples of suppliers in the United States are: Vaughn Seed Co., 5300 Katrine Avenue, Downers Grove, Ill. 60515; Burpee Seed Co., Burpee Building, Warminster, Pa. 18974; Premier Brands, Inc., 145 Huguenot Street, New Rochelle, N.Y. 10801; Ball Seed Co., P.O. Box 335, West Chicago, Il. 60185, Scotts (Grace-Sierra), Marysville, Oh. 43041.

Managing Plant Nutrition With Postplant Fertilization During and After the Propagation Cycle

Developing an *efficient fertilizer program* for container plants for the *21st century* depends on: (1) minimizing the loss of fertilizer from the production area, and (2) increasing the amount of fertilizer utilized or taken up by the plant (*175*). In the previous section on container media for propagation and small liner production, suggested levels of *preincorporated* (*preplant*) granular fertilizers were discussed. This section discusses some general fertilization practices for management of plant nutrition during **propagation, liner production,** and **post-liner** production (Figure 3–27). Both soluble and slow-release fertilizers are utilized.

Liquid fertilizers. For large-scale operations, it is more practical to prepare a liquid concentrate and inject it into the regular watering or irrigating system by the use of a proportioner (*13*)—"fertigation." The most economical source of fertilizers to be applied through the irrigation water is from dealers who manufacture soluble liquid fertilizer for field crops (*173*). It is possible to mix up your own nutrient concentrate formula, such as the following below, using soluble chemical formulations. It is no longer recommended to use superphosphate in soilless mixes with outdoor container production because of the phosphorus leaching that occurs. Hence more expensive, but efficient, soluble forms of phosphorus are used, such as phosphoric acid or ammonium phosphate in liquid feed programs. Potassium is typically applied as potassium chloride, or potassium nitrate, and nitrogen as Uran 30 (15 percent urea, 15 percent NH_4NO_3) or ammonium nitrate in the liquid concentrate.

An example of a liquid fertilizer system for production of containerized plants is the Virginia Tech System (VTS). With the VTS, all nutrients are supplied to the container by injecting liquid fer-

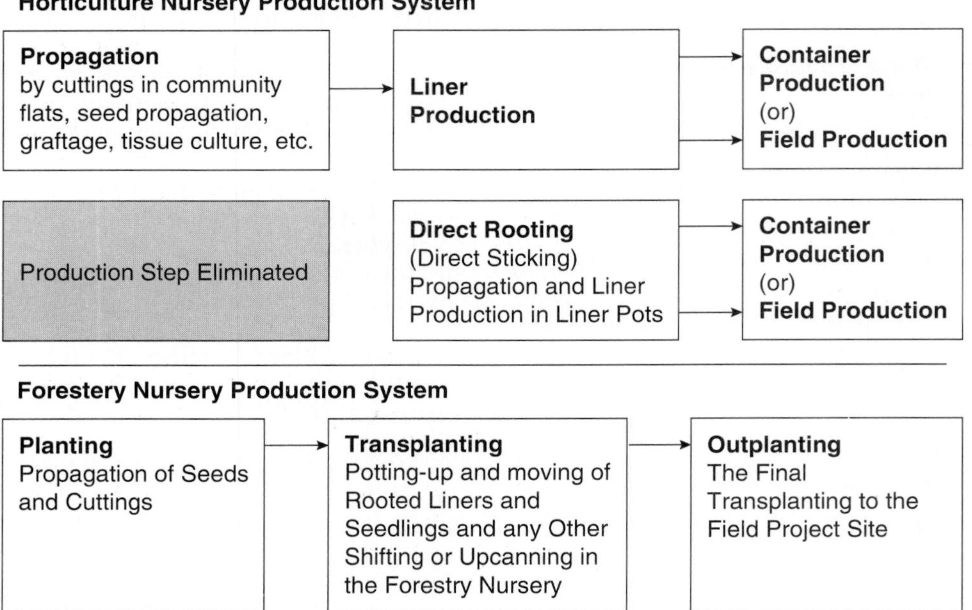

FIGURE 3–27 Flow diagram of a *Horticulture Nursery Production System* starting with propagation by rooted cuttings, seedlings, graftage or tissue culture produced plantlets—followed by transplanting into liner pots and final transplanting into larger containers or into nursery field production. *Direct rooting* (direct sticking) eliminates a production step, since both propagation and liner production occur in the same liner pot. A *Forestry Nursery Production System* of planting, transplanting, and outplanting is also described.

tilizers into the irrigation water (*173,174*). A $10N\text{-}4P_2O_5\text{-}6K_2O$ analysis liquid fertilizer is applied five times per week, 1.3 cm (0.5 in.) each irrigation at an application rate of *100 to 80 ppm N, 15 to 10 ppm P, and 50 to 40 ppm K.* Sometimes higher nitrogen levels are applied (200–300 ppm N), depending on the time of the year, plant growth conditions, or plant species. It is critical to regularly monitor soluble salt levels of the medium *prior* to fertigation. Supplemental micronutrients are also applied in a liquid form, but from separate tanks and with separate injectors to prevent fertilizer precipitation. It is best to monitor soluble salt levels of the irrigation water with a conductivity meter, i.e., to apply 100 ppm N, the injector is set so that the conductivity of the irrigation water (minus the conductivity of the water before the fertilizer was injected) reads 0.55 mS/cm {milliSiemens per cm or dS per m—they are the same units of measure} (*173*).

Controlled-Release Fertilizers. Controlled-release (slow-release) fertilizers provide nutrients to the plants gradually over a long period and reduce the possibility of injury from excessive applications (*165*). They are used chiefly on high-value, container-grown plants. Though expensive on a kg per kg (lb per lb) basis, they may be one of the most cost-effective ways to fertilize plants, since fertilizer is applied directly to the pots. In contrast, overhead fertigation with rainbird sprinkler-type systems is only about 30 percent efficient, and greater fertilizer runoff occurs from the container production area. Slow-release fertilizers include Osmocote, Phycote, Nutricote, Polyon, etc., and some are available with micronutrients incorporated in the pellets. As previously described, for both cutting and seed propagation, a low concentration of macro and micro slow-release fertilizers can be included in the propagation mix, so the newly formed roots can have nutrients available for absorption. This is particularly important with mist propagation where nutrients can be leached out from both the plant and the medium.

There are three types of slow-release fertilizers: (a) *coated water-soluble pellets* or *granules;* (b) *inorganic materials* that are slowly soluble; and (c) *organic materials* of low solubility that gradually decompose by biological breakdown or by chemical hydrolysis.

Examples of the resin-coated-type pellets are: (a) Osmocote (*124*), whose release rate depends on the thickness of the coating, and (b) Nutricote (*131*), whose release rate depends on a release agent in the coating. After a period of time the fertilizer will have completely diffused out of the pellets (*171*). Another kind of controlled-release fertilizer is the sulfur-coated urea granules, consisting of urea coated with a sulfur wax mixture so that the final product is made up of about 82 percent urea, 13 percent sulfur, 2 percent wax, 2 percent diatomaceous earth, and 1 percent clay conditioner.

An example of the slowly soluble, inorganic type is MagAmp (magnesium ammonium phosphate), an inorganic material of low water solubility. Added to the soilless mix, it supplies nutrients slowly for up to two years. MagAmp may be incorporated into media prior to steam pasteurization without toxic effects. On the other hand, steam pasteurization and sand abrasion in the preparation of mixes containing resin-coated, slow-release fertilizers, such as Osmocote, can lead to premature breakdown of the pellets and high soluble salt toxicity.

An example of the organic, low-solubility type is urea-formaldehyde (UF), which will supply nitrogen slowly over a long period of time. Another organic slow-release fertilizer is isobutylidene diurea (IBDU), which is a condensation product between urea and isobutylaldehyde, having 31 percent nitrogen.

Fertilizer Systems for Propagation. Commercial propagators often apply moderate levels of controlled slow-release macro and micro elements to the propagation media—*preincorporated* into the media—prior to sticking cuttings and starting seed germination and seedling plug production. During propagation, supplemental fertilizer is added by *top dressing (broadcasting)* with slow-release fertilizer, or injection of gradually increasing concentrations of liquid fertilizer *(fertigation)*. In the rooting of cuttings, these supplementary nutrients do not promote root initiation (*85*), but rather improve root development after root primordia initiation has occurred. Hence, supplementary fertilization is generally delayed until cuttings have begun to root. Propagation turnover occurs more quickly and plant growth is maintained by producing rooted liners and plugs that are more nutritionally fit.

Some recommended levels of slow release fertilizers for propagation are:

- 3.6 kg/m^3 (6 lbs/yd^3) 18-6-12 Osmocote (or comparable product)
- 0.6 kg/m^3 (1 lbs/yd^3) Micromax or other trace element mixtures—Perk, Esmigran, FTE 503, etc.
- For unrooted cuttings, fast germinating seeds and tissue culture plantlets, slow release fertilizers are preincorporated in the propagation media. For slower rooting or seed germinating species, Osmocote 153 g/m^2 (0.5 oz/ft^2),
- Nutricote, etc. are top dressed on the media *after* rooting or seed germination starts to occur. Determining optimum levels of fertilization for propagation depends on the propagule system, and needs to be determined on a species basis (*36*).

Fertilizer Systems for Liner Production. Soilless mixes must have added fertilizers (*135,174*). Irrigation water and the container medium should be thoroughly analyzed for soluble salts, pH, macro- and microelements, etc., before a fertility program can be established. It is always wise to conduct small trials before initiating large-scale fertility programs during propagation and liner production.

A satisfactory feeding program for growing liner plants is to combine a slowly available dry, granular fertilizer (preplant) in the original mix, with a (postplant) liquid fertilizer applied at frequent intervals during the growing season or with controlled-release fertilizers added as top dressings, as needed (*56*).

Of the three major elements—nitrogen, phosphorus, and potassium—nitrogen has the most control on the amount of vegetative shoot growth. Phosphorus is very important, too, for root development, plant energy reactions, and photosynthesis. Potassium is important for plant water relations and enhanced drought resistance (*45*).

Nitrogen and potassium are usually supplied by slow-release or solution fertilizers—*100 to 80 ppm nitrogen* and *50 to 40 ppm potassium* are optional container medium levels when the Virginia Tech Extraction Method (VTEM) is used (*176*).

Negatively charged ions, such as phosphorus, leach from soilless media, so small amounts of phosphorus must be added to the media frequently. Past research indicates that *15 to 10 ppm phosphorus* should be maintained in container medium as determined by the saturated paste or VTEM (*172,173*). Phosphorus from superphosphate leaches rapidly; so in order to maintain 10 ppm in the medium, slow-release fertilizer is used or small amounts of phosphorus in soluble form are applied. Plants respond comparably to rock phosphorus, other insoluble phosphorus carriers, or superphosphate amendments (*176*).

Calcium and magnesium are supplied as a preplant amendment in dolomitic limestone and may naturally be supplied by irrigation water. Limestone is added primarily to adjust the pH of the media. It is important to have the irrigation water checked to determine the level of dolomitic limestone needed, if any. VTEM levels of *40 ppm calcium* and *20 ppm magnesium* in the container medium are adequate.

Fertilizer Systems for Post-Liner Production. Some containerized nurseries use only slow-release fertilizers, while others use slow release in combination with liquid fertilizers, or liquid fertilizers only.

Some recommendations of slow-release fertilizers when shifting liners to 3.75 liter (one gallon containers):

- 5.3 to 7.1 kg/m^3 (9–12 lbs/yd^3) 18-6-12 Osmocote, Nutricote, Woodace, Prokote, or comparable product;
- 0.9 kg/m^3 (1.5 lbs/yd^3) Micromax, or comparable trace element mix, i.e. Perk, Esmigran, etc.

Slow-release fertilizer can be mixed (preincorporated) with the container medium, or broadcast (top dressed) on the container medium surface. Nutrients are slowly released and made available during the crop cycle. Under warmer propagation and production conditions in the southern U.S., it normally lasts for five to six months as opposed to eight to nine months in cooler climates. A nine-month formulation for container production in the southern U.S. includes Osmocote 17-7-12, which is applied at 7.1 to 9.5 kg/m^3 (12 to 16 lbs/yd^3). As with any fertility program, it is important to conduct small tests to determine what works best for your particular propagation and production system.

Salinity In Soil Mixes

Excessive salts in the propagating or growing mixes or in irrigation water [EC (electrical conductivity) over 2 mS/cm (dS per m)] can reduce plant growth, burn the foliage, or even kill the plants (*48,50*). The required fertilization programs also contribute to salt accumulation. Overfertilization causes rapid and severe salinity symptoms, starting with foliage wilting and tip and marginal leaf burning. These symptoms may be accompanied by a white surface accumulation of salts on the medium. To prevent salt buildup, the containers or flats should be leached with water periodically. Fertilizers should not be used that tend to contribute to excess salinity; for example, use potassium nitrate rather than potassium chloride.

There is considerable variability among the various plant species in their tolerance to salts in the growing medium (*50,133*). Plants such as araucaria, bougainvillea, callistemon, carnation, gladiolus, olive, petunia, hibiscus, poinsettia, and portulaca, have high tolerance, while others, such as azalea, blackberry, gardenia, mahonia, photinia, pittosporum, and strawberry have low tolerance. If only water with high salinity is available, only plants from the first group, or others of similar tolerance, should be selected for production.

MANAGEMENT OF MICROCLIMATIC FACTORS IN PROPAGATION AND LINER PRODUCTION

Water

Quality (Salinity) of Irrigation Water

Good water quality is essential for propagating quality plants (*84,151*). The salt tolerance of unrooted cuttings, germinating seeds, and tissue culture explants is much lower than that of established plants, which can be grown under minor irrigation salinity by modifying cultural conditions. Water quality[9] for propagation is considered good when the electrical conductivity (EC) reading is <0.75 mS (mili-siemens) per cm or dS (desi-siemens) per m (less than 525 ppm total soluble salts in ppm), and the sodium absorption ratio (SAR) is < 5. Except for the most salt tolerant of plants, irrigation water with total soluble salts in excess of 1,400 ppm (approximately 2 mS/cm) (ocean water averages about 35,000 ppm) would be unsuitable for propagation. The salts are combinations of such cations as sodium, calcium, and magnesium with such anions as sulfate, chloride, and bicarbonate. Water containing a high proportion of sodium to calcium and magnesium can adversely affect the physical properties and water-absorption rates of soils and should not be used for irrigation purposes. It is prudent to have nursery irrigation water tested at least twice a year by a reputable laboratory that is prepared to evaluate all the elements in the water affecting plant growth. Most producers regularly monitor EC and pH of their irrigation water and container mix with inexpensive instru-

[9]Salinity levels from irrigation water, and from water extracts of growing media (saturation-extract method) can be measured by electrical conductivity (EC) using a **Solubridge.** Various portable meters for testing salinity, as well as soil and water testing kits, etc. are available through commercial greenhouse supply companies.

ments. Some producers test and monitor their own container media nutrients, whereas plant leaf tissue is generally sent off to plant laboratories for nutrient analysis.

Although not itself detrimental to plant tissue, so-called "hard" water contains relatively high amounts of calcium and magnesium (as bicarbonates and sulfates) and can be a problem in mist-propagating units or in evaporative water cooling systems as deposits build up wherever evaporation occurs. This reduces the photosynthetic levels of cuttings, seedlings, tissue culture plantlets, etc. When "hard" water is run through a water softener, some types of exchange units replace the calcium and magnesium in the water with sodium ions. Misting and irrigating with such "soft," high-sodium water can injure plant tissue.

A better, but more costly, method of improving water quality is the **deionization** process. Calcium, magnesium, and sodium are removed by substituting hydrogen ions for them. Water passes over an absorptive medium charged with hydrogen ions, which absorbs calcium and other ions in exchange for hydrogen. For further deionization, the water is passed through a second filter charged with hydroxyl (OH) ions, which replace carbonates, sulfates, and chlorides. The proper nutritive ions may then be added back in suitable amounts.

Boron salts are not removed by deionization units and if present in water in excess of 1 ppm, can cause plant injury. There is no satisfactory method for removing excess boron from water. The best solution is to acquire another water source, and to use customized non-boron blended fertilizers.

Another good but expensive method for improving water quality is **reverse osmosis** (Figure 3–28), a process in which pressure is applied to a solvent to force it through a semipermeable membrane from a more concentrated solution to a less concentrated solution eliminating unwanted salts from an otherwise good water source.

FIGURE 3–28 Good quality water is imperative for propagation. A reverse osmosis system is shown for removing salts in commercial propagation.

Nurseries using **recycled irrigation water** (Figure 3–29) should treat the water before use. A good procedure is to:

- Initially utilize aquatic plants in runoff, catchment ponds to reduce pollutants and sediments reentering the recycling system *(141)*.
- Add chlorine or bromine to suppress algae and plant pathogens as water is pumped from the catchment pond.
- Use strainers to remove large debris. Then run the water through sand or mechanical filters with automatic back flushing. This removes coarse particles and weed seed.
- Consider running the water through an activated charcoal tank to remove soluble herbicides and other residual chemicals.
- If the water has a high salt content, it can be improved by the use of deionization or reverse osmosis, but this is very expensive.
- Water can then be treated with ultraviolet irradiation to reduce pathogens. Generally all precipitate down to at least 20 μm is filtered out for UV light to be effective.
- Recycled water is then acidified (to lower the pH, if necessary) and repumped into holding ponds with plastic liners and weed-free perimeters.
- *Fresh well water* is then pumped into the holding pond and mixed with the *recycled water.* This allows for pumping from wells during the night to meet daily irrigation needs and dilutes soluble salts of recycled water.
- This water can then be used for field watering of container nursery plants and slow-release fertilizer incorporated into containers or soluble fertilizer injected into the irrigation system.

Municipal treatment of water supplies with chlorine (0.1 to 0.6 ppm) is not sufficiently high to cause plant injury. However, the addition of fluoride

FIGURE 3–29 Systems for capturing, treating, and recycling irrigation water in commercial nurseries. *Upper left:* All irrigation water either drains into or is pumped into a holding pond. *Upper right:* irrigation water is treated with chlorine (sodium hypochlorite solution is one of the safest forms) as it is pumped from the holding pond to the water treatment and pumping facility. *Middle left:* filtration tanks for removing weed seed and particulate-suspended matter down to 20 μm (this is important if irrigation water is to be treated with ultraviolet light); some nurseries use tanks of activated charcoal to trap soluble herbicides and other undesirable chemicals. *Middle right:* Ultraviolet treatment of irrigation water with a UVS Ultra Pure model 5000 (Scoresby, Victoria, Australia). *Bottom left:* Bromination of water; some nurseries will inject acid at this point to lower water pH, if needed. *Bottom right:* Monitoring water leaving the water treatment facility for pH *(left)* and soluble salts or electrical conductivity *(right).*

to water supplies at 1 ppm can cause leaf damage to a few tropical foliage plant species.

When the water source is a pond, well, lake, or river, contamination by weed seeds, mosses, or algae can be a problem. Chemical contamination from drainage into the water source from herbicides applied to adjoining fields or from excess fertilizers on crop fields can also be damaging to nursery plants (*149*). Recycled water, which is discussed in the section **Best Management Practices,** is used in nursery and greenhouse production, and is being evaluated for general propagation in some nurseries.

The pH of Irrigation Water and Soil

The pH is a measure of the concentration of hydrogen ions, and can affect the rooting of cuttings, germination of seeds, and micropropagation of explants. Liner production is also affected by pH influence on nutrient availability and activity of beneficial microorganisms in the container medium. A pH range of 5.5 to 7.0 is best for growth of most plants (7.0 is neutral—below this level is acid and above is basic or alkaline). A method for lowering the pH and controlling carbonate problems in the nursery is to inject sulfuric or phosphoric acid into the irrigation water supplies. Softwood bark and peat-based container mixes are acid and will lower irrigation water pH. Dolomitic limestone raises soil pH and is the primary source of Ca and Mg in many propagation and liner mixes.

Water-Humidity Control

Good water management is important to limiting plant stress. Care must be exercised to avoid over-misting and over-irrigation. Too much water can be just as stressful as too little water. Root rots and damping-off organisms are favored by standing water and poor media drainage conditions.

Maintaining proper atmospheric humidity in the propagation house/beds is important, since low humidity can increase transpiration and subject unrooted cuttings and seedlings to water stress. Adequate humidity allows optimum growth, whereas extreme humidity promotes fungal pathogen, moss, and liverwort pests. Air always contains some water vapor, but at any given temperature, it can hold only a finite amount. When the physical limit is reached, the air is **saturated,** and when it is exceeded, **condensation** occurs (*93*). The unique physical properties of water affect the propagation environment. When water is converted from a liquid to a gas (water vapor), a large amount of thermal energy (540 cal/g) is required. The cooling effect of mist irrigation results as heat is absorbed and the increased relative humidity minimizes plant transpiration. A heavy mist, which condenses and forms droplets of water, should be avoided because it leaches foliage of nutrients, saturates propagation media, and can promote disease problems.

Current systems used to control water loss of plant leaves (*95,96*) are:

1. **Enclosure Systems**—outdoor propagation under low tunnels or cold frames, or non-misted enclosures in a glasshouse or polyhouse (shading, tent and contact polyethylene systems, wet tents).
2. **Intermittent Mist**—open and enclosed mist systems.
3. **Fogging Systems**

The effect of these systems on propagation environmental conditions and water relations of the propagule is discussed in greater detail in Chapter 10.

Temperature Control

As indicated in earlier sections, temperature is modified by environmental controls in the propagation structure and the type of propagation system that is used (see Chapter 10 for greater details). There is no environmental factor more critical than optimum temperature control for propagation. Optimum seed germination, rooting of cuttings, development of tissue culture plantlets, graft union formation, and specialized structure development are all temperature-driven plant responses. Hot air convection, infrared radiation, and hot water distribution systems are the three most viable ways to heat plants (Figure 3–9). Of the three, hot water is the most flexible and commonly used system in propagation houses (*122*). It allows efficient root zone heating in the form of bottom heat. Some examples include Biotherm tubing, Delta tubes, etc., which are used to maintain optimum propagation temperatures. A mist system accelerates root development of cuttings under *high* light irradiance, by evaporative cooling, which reduces the heat load on plant foliage. In fog systems, the fog particles remain suspended and *reduce* the light intensity, while a zero-transpiration environment is maintained, without the overwetting (condensation) that can occur with mist. Since only minimal condensation occurs, leaf and media temperatures are warmer with fog than mist. In liner production, DIF systems (cooler days versus night temperature) produce more compact plants. This works well for seedling plugs, bedding plants, and greenhouse crops under controlled environmental conditions (*9*).

Light Manipulation

The importance of light manipulation in propagation (irradiance, photoperiod, quality) was discussed earlier in the chapter and is covered in greater detail in later chapters on seed and cutting propagation, micropropagation, and specialized structure development and propagation. Light quality (which is commercially manipulated through greenhouse spectral filters, greenhouse coverings, and varying supplementary light sources) plays an important role in seed germination, and shoot development in macro- and micropropagation. Photoperiod can be manipulated to delay bud dormancy and extend accelerated plant growth. Photoperiod can be utilized not only to enhance root initiation, but also to increase carbohydrate reserves of deciduous, rooted cuttings (liners) for better winter survival and subsequent vigorous spring growth (*101*) (Figure 3–13).

Supplemental Photosynthetic Lighting in the Propagation House. Plant growth in the winter in propagation houses can be slow due to the lack of sufficient light for photosynthesis, especially in the higher latitudes (*27*). This is due to several reasons:

- Low number of daily light hours
- Low angle of the sun, resulting in more of the earth's atmosphere the sun's rays must penetrate
- Many cloudy and overcast days in the winter
- Shading by the greenhouse structure itself and dirt accumulating on the poly or glass or other covering materials

To overcome the problem of low natural winter light and reduced plant growth, supplemental light can be used over the plants (Figure 3–13). The best light source for greenhouse lighting is high-pressure sodium vapor lamps. Most of the radiation from these lamps is in the red and yellow wavelengths and is very deficient in the blue. However, when used in conjunction with the natural daylight radiation these lamps are quite satisfactory.

The high-pressure sodium vapor lamps emit more photosynthetically active radiation (PAR) for each input watt of electricity than any other lamp that is commercially available. Sodium vapor lamps are long-lasting and degrade very slowly. They emit a considerable amount of heat that can be a benefit in the greenhouse in winter. They use a smaller fixture than fluorescent lamps, thus avoiding the substantial shading effect from the fluorescent lamp fixture itself.

The installation should provide a *minimum* of about 65 $\mu mol\ m^{-2}s^{-1}$ or 13 $W\ m^{-2}$ PAR at the plant level with a 16-hour photoperiod. For large greenhouses the services of a lighting consultant should be used in designing the installation.

Photosynthetic lighting with high intensity discharge lights (HID) in more overcast climates has greatly expanded the production window for cuttings and seed propagation. Supplementary lighting is an important component in accelerated growth techniques (AGT) in propagating plants (Figure 3–30).

Carbon Dioxide (CO2) Enrichment in the Propagation House

Carbon dioxide is one of the required ingredients for the basic photosynthetic process that accounts for the dry-weight materials produced by the plant (*74,87,111,119*).

$$6CO_2 + 12H_2O \xrightarrow[\text{green plant cell}]{\text{light energy}} C_6H_{12}O_6 + 6O_2 + 6H_2O$$

Ambient carbon dioxide exists in the atmosphere at about 360 ppm. Sometimes the concentration in winter in closed greenhouses may drop to 200 ppm, or lower, during the sunlight hours, owing to its use by the plants. Under adequate light and temperature, but when low CO_2 concentration limits photosynthesis, an increase in CO_2 concentration from below ambient to 1,000–2,400 ppm can result in an 200 percent increase in photosynthesis. To take full advantage of this potential increase in dry-weight production, plant spacing must avoid shading of overcrowded leaves. When supplementary CO_2 is used during periods of sunny weather, the temperature in the greenhouse should be kept relatively high. Adding CO_2 at night is of no value. However, CO_2 generators can be turned on before dawn to increase photosynthesis early in the day. A tightly closed greenhouse is necessary to be able to increase the ambient CO_2.

Sources of CO_2 for greenhouses are either burners using kerosene, propane, or natural gas, or liquid CO_2 (*2*). Liquid CO_2 is expensive but almost

ACCELERATED GROWTH TECHNIQUES (AGT)

Selected rooted liners, seedling plugs, tissue culture plantlets (genetically superior)

↓

Efficient, container growing methods

↓

Protective culture (glasshouses, etc.)

↓

Programmed growth *control* throughout the year:

↓

Light
Temperature
Mineral nutrients
Water-Humidity
Carbon dioxide
Growth regulators
Mycorrhiza/other beneficial soil microbes
Growing media
Competition
Pests

↓

Production of *large* rooted cuttings, seedling plugs, TC plantlets in months *rather* than years

↓

Acclimation to natural conditions and planting

↓

Genetic selection, hybridization, propagation of superior cultivars, and release to nurseries

FIGURE 3–30 Components of accelerated growth techniques used in speeding-up vegetative and seed propagation in the production of marketable liners.

risk-free. Kerosene burners must use high-quality, low-sulfur kerosene, or SO_2 pollution can occur. With propane or natural gas, incomplete combustion is possible—byproducts include carbon monoxide (dangerous to humans) and ethylene (harmful to plants). The flames should be a solid blue color. Monitoring of the CO_2 level in the greenhouse is very important. Accurate sensors are available and should be used. With the newer computer technology, sensors in different parts of the greenhouse can give excellent control of the CO_2 levels. Excessively high levels of CO_2 in the greenhouse (over 5,000 ppm) can be dangerous to humans.

New tissue culture systems are utilizing high CO_2 enrichment and high light levels for *autotrophic micropropagation* (*83*). The plantlets are cultured without sugar in the culture medium as an energy and carbon source, and are stimulated by enriched atmospheric CO_2 and elevated light irradiance to photosynthesize and become autotrophic. The CO_2 is supplied either directly to the tissue culture vessel or indirectly via increased ambient CO_2 to permeable culture vessels. Autotrophic micropropagation improves plantlet growth and development, simplifies procedures, reduces contamination, and lowers production costs (see Chapter 18).

Accelerated Growth Techniques (AGT)

The forestry industry has developed accelerated growth systems to speed up the production of liners from cutting and seed propagation. Woody perennial plants undergo cyclic (episodic) growth, and many tree species experience dormancy. Liners are grown in protective culture facilities where photoperiod is extended and water, temperature, carbon dioxide, nutrition, mycorrhizal fungi, and growing media are optimized for each woody species and for each different phase of growth (Figure 3–30).

This concept is also being used in propagation of horticultural crops where supplementary lighting with high-pressure sodium vapor lamps and injection of CO_2 gas into mist water are used to enhance seed germination, plug development, acclimation of tissue culture plantlets, rooting of cuttings, etc. The promotive effects of AGT on rooting of *Ilex aquifolium* cuttings has been attributed, in part, to enhanced photosynthesis.

Modeling in plant propagation. Closely linked to AGT is the **modeling** of propagation environments to determine optimal light, temperature, water, CO_2, and nutritional regimes (*35*). Computer technology allows the propagator to monitor and program the propagation environment and adjust environmental conditions as needed through automated environmental control systems (Figures 3–3, 3–13, 3–14, 11–40 and 11–41).

BIOTIC FACTORS—PATHOGEN AND PEST MANAGEMENT IN PLANT PROPAGATION

Pathogen and pest management begins *prior to propagation* with the proper manipulation of stock plants or the container plants from which the propagules are harvested, as well as management of propagation beds and media preparation. If pathogens and pests are not checked during propagation, an inferior plant is produced and later production phases for finishing and selling the crop will be delayed, causing profit losses.

Pests are broadly defined as all biological organisms (bacteria, viruses, viroids, phytoplasma, fungi, insects, mites, nematodes, weeds, parasitic higher plants, birds and mammals) that interfere with plant production (*68*). *Insect pests,* such as aphids, mealy bugs, thrips, white flies, and fire ants actively seek out the plant host by migrating (flying, walking). When an infection can be spread from plant to plant, it is referred to as an infectious disease. *Infectious plant diseases* are caused by different *pathogens (infectious agents),* including pathogenic fungi, bacteria, viruses, viroids, and phytoplasma. Specific pathogens may infect only certain plant species or cultivars, or specific organs or tissue, which varies with the stage of development of the plant.

The pathogenic fungi most likely to cause disease development during propagation are species of: *Pythium, Phytophthora, Fusarium, Cylindrocladium, Thielaviopsis, Sclerotinia, Rhizoctontia,* and *Botrytis.* These are all soil-borne or aerial organisms *(Botrytis)* that infect plant roots, stems, crowns, or foliage. The so-called "damping-off" commonly encountered in seedbeds is caused by soil fungi, such as species of *Pythium, Phytophthora, Rhizoctonia, Fusarium.* Surpressing pathogens in propagation water is critical—*Phytophthora, Pythium* and *Rhizoctonia* are readily disseminated in surface water.

Conversely, intermittent mist can wash off germinating fungal spores. Mist inhibits the spore germination of powdery mildew *(Sphaerotheca pannosa)* on leaves of cuttings, and it may be that other disease organisms are held in check in the same manner. However, mist propagation is highly conducive to diseases such as aerial *Rhizoctonia* blight, *Cylindiocladium,* bacterial soft rots, etc.

A goal in propagation is to keep stock plants and propagules as clean and pest-free as possible, and to suppress pathogenic fungi, viruses, nematodes, weed seed, etc. from the propagation media. Optimum pest management depends on a thorough knowledge of the pest life cycle, environmental conditions, cultural practices, and minimizing host plant stress—the rooting of a cutting and germination of a seed are vulnerable periods of plant growth. A stressed propagule is much more susceptible to pest problems. The management of pests through integrated pest management (IPM) is discussed in this section.

Preventative Measures

Cultivar Resistance

Avoid producing crops that are susceptible to certain diseases and pests. A susceptible crop means more time, chemicals, and money spent to control the problem. In addition, the problem is passed on from the propagator to the consumer (*10*). By choosing a resistant cultivar, efforts are concentrated on propagating and producing the plant, rather than trying to control the pest, i.e., propagate disease-resistant crab apple cultivars, rather than disease susceptible *Malus* cultivars such as 'Hopa' and 'Mary Potter.' In the southern U.S., Helleri hollies (*Ilex* 'Helleri') are plagued by southern red mites, root-knot nematodes, and black root rot—so why propagate them when other holly cultivars are more resistant (*10*)?

Scouting System

All propagators should practice pest scouting. Early detection provides more effective pest and pathogen control with less reliance on pesticides. Propagation houses should be scouted on a regular basis, and all propagation employees trained to recognize and report disease and insect pests. Workers are the ones in daily contact with plants and are an invaluable resource for early detection. Some large nurseries have detailed pest management programs with crews supervised by trained plant pathologists and entomologists (*30*). Such programs involve the proactive prevention of plant diseases and the avoidance of insects, mites, and weed problems. Serological tests kits ELISA (enzyme-linked immunosorbent assays) are commercially available to propagators for the early detection of certain pathogens and viruses (*112,120*). The user-friendly Alert Diagnostic Kits can rapidly identify the *damping-off* organisms—*Pythium, Rhizoctonia, Phytophthora, Botrytis* (*5*). (Keeping viruses in check is considered in Chap. 9).

IPM—Integrated Pest Management

With propagation and liner production systems, IPM is divided into three management areas: chemical, biological, and cultural control. Total elimination of a pest is not always feasible—nor is it biologically desirable if the process is environmentally damaging or leads to new, more resistant pests and eliminates beneficial fungi and insects. In the production of clean stock plants and propagules, pest-free plants may be a requirement, but this should be accomplished by using a variety of pest management methods without an overdependence on just one method (i.e., solely using chemical control). Pest control differs from pest management in that an individual pest control technique is used in isolation to eliminate a pest and all pest-related damage (*68*). Conversely, IPM uses as many *management* (control) methods as possible in a systematic program of suppressing pests (not necessarily annihilating) to a commercially acceptable level, which is a more ecologically sound system.

Chemical Control in IPM

IPM does not imply that no chemicals are used in the control of pests and pathogens. Rather, better targeted control with less chemical usage occurs because of the integration of additional biological and cultural management measures (*137*). IPM in propagation can no longer be ignored. It is the intelligent selection and use of actions that will ensure favorable economic, ecological, and sociological consequences (*117*).

In the treatment of seeds, bulbs, corms, tubers, and roots, pesticides are sometime used in combination with cultural techniques, such as hot water soaks [43 to 57° C (110 to 135° F)]. The hot water temperature and duration is dependent on the species and propagule type being treated. For many ornamental plants, to control decay and damping-off, seeds are treated with fungicidal slurries or dusts of thiram, zineb, etc. Seeds of California poppy, and *Strelitzia* (bird of paradise) are given hot water soaks to control pathogenic fungi, while *Delphinium* (larkspur) and *Digitalis* (foxglove) seeds are given a hot water soak and then dusted with thiram to control anthracnose. Bulbs and corms of many species are treated for nematodes and pathogenic fungi with hot water soaks and/or chemical treatment.

When using pesticides, it is important that propagators follow the **Worker Protection Standard (WPS)** rules and regulations, to reduce pesticide-related illnesses and injuries (*53*). The WPS can complicate many jobs in propagation. Scheduling will become more critical so pesticide restricted-reentry intervals do not interfere with normal propagation assignments of workers. The U.S. Environmental Protection Agency (EPA) has a monthly updated bulletin that details WPS implementation information on re-entry rules and times. In the U.S., call (703)305-7666.

Fumigation with Chemicals. Chemical fumigation kills organisms in the propagating mixes without disrupting the physical and chemical characteristics of the mixes, to the extent occurring with heat treatments.[10] The mixes should be moist (between 40 and 80 percent of field capacity) and at temperatures of 18 to 24° C (65 to 75° F) for satisfactory results. Before using the mixture after chemical fumigation, allow a waiting period of two days to two weeks, depending upon the material, for dissipation of the fumes. A problem with chemically sterile media is that there are no competing microorganisms to limit the rapid recolonization of fungi and bacterium, which may create media aeration and pest problems.

Methyl Bromide

Methyl bromide is an odorless material, very volatile and quite **toxic to animals, including humans.** Its use has been banned in Holland and England because it contributes to the reduction of the earth's ozone layer. The U.S. Environmental Protection Agency has ordered a complete ban on the fumigant after January 1, 2001. It should be mixed with other materials and applied only by those trained in its use. Most nematodes, insects, weed seeds, and some fungi are killed by methyl

> An example of IPM is the *cultural control* of **aphids** in propagation by installing microscreening that covers vents and doorways of a propagation house, thereby reducing the movement of insects and the need for insecticides *(55)*. Early detection of winged aphids with yellow sticky cards that are hung in the propagation house can alert personnel to monitor plants near cards for the presence of wingless females. The option to use *biological control* is possible with an efficient scouting system that detects controllable, low aphid levels. A beneficial midge, *Aphidoletes aphidimyza,* has been used to biologically control aphid colonies. If the aphid colony is small, other *biorational products* can be used such as insecticidal soap (M-Pede), horticultural oils (UltraFine SunSpray spray oil), *botanical insecticides* such as neem (Azatin and Margosan-O), and natural pyrethrums *(55)*. *Insect growth regulators* such as kinoprene (Enstar II) and methoprene give safe, effective control of immature aphids. For large populations of aphids that were not detected early enough, *more toxic pesticides* are sometimes used, such as diazinon, bendiocarb, methiocarb, acephate; or the synthetic pyrethroids, such as fluvalinate (Mavrik), bifenthrin (Talstar), and fenpropathrin (Tame) *(55)*.

[10]Recommendations on pesticide labels must be followed to conform to permitted usages.

FIGURE 3–31 Chemical and heat treatment of propagation mixes. *Upper left and right:* Methyl bromide being applied to propagation medium. *Middle left:* Methyl bromide is extremely toxic; during soil treatment, it is important to use warning signs and restrict the movement of personnel. *Middle right:* Heat pasteurization with aerated steam. *Bottom left:* Converted cement truck mixers that contain aerated steam for pasteurizing. *Bottom right:* The media in the mixers is allowed to cool and then put in a hopper conveyer for filling propagation flats.

bromide. When properly applied with appropriate exposure duration, methyl bromide can control Verticillium, even with its relatively resistant sclerotia. Methyl bromide is most often used by injecting the material from pressurized containers into an open vessel placed under a plastic sheet which covers the soil to be treated (Figure 3–31). The cover is sealed around the edges with soil, and should be kept in place for 48 hours. Penetration is very good, and its effect extends to a depth of about 30 cm (12 in.). For treating bulk soil, methyl bromide at 333 ml per cubic meter (10 ml per cubic foot) or 0.6 kg per cubic meter (4 lb per 100 ft^3) can be used.

Methyl Bromide and Chloropicrin Mixtures

Proprietary materials are available containing both methyl bromide and chloropicrin. Such combinations are more effective than either material alone in controlling weeds, insects, nematodes, and soil-borne pathogens. The addition of chloropicrin (tear gas) to methyl bromide was primarily so that humans could detect gas leaks and evacuate before being poisoned by methyl bromide. Aeration for 10 to 14 days is required following applications of methyl bromide-chloropicrin mixtures.

> Chloropicrin and methyl bromide are hazardous materials to use, especially in confined areas. They should be applied only by persons trained in their use, who must take the necessary precautions as stated in the instructions on the containers or in the accompanying literature.

Fungicidal Soil Drenches. Fungicidal soil drenches can be applied to the container media in which young plants are growing or are to be grown to suppress growth of many soil-borne fungi. These materials may be applied either to media or to the plants. Preferably, a wetting agent should be added to the chemicals before application. It is very important in using such chemicals to read and follow the manufacturers' directions and dilutions carefully, and to try the chemicals on a limited number of plants first before going to large-scale applications. As with insect pests, pathogens can build up resistance to fungicides, so it is important to rotate fungicides and use mixtures with good residual action (*81*). Examples of fungicidal drench materials are:

Quintozene (PCNB, Terraclor) controls *Rhizoctonia, Sclerotina,* and *Sclerotium.* Benomyl (Benlate) is a fungicide that inhibits growth of such soil pathogens as *Rhizoctonia.* It is ineffective against the water molds, *Pythium* and *Phytophthora.* Etridiazole *(Terrazole, Truban)* incorporated into the propagating medium, surpresses *Pythium* and *Phytophthora.* Until 1991, Benomyl was the most widely used fungicide for propagation and ornamental use in the U.S. Unfortunately, the U.S. manufacturer no longer produces it, and future trends are for less chemicals labeled for horticultural useage.[11] Some commercial substitutes for Benomyl include Topsin M, Domain, Cleary 3336, and SysTec 1998, all of which have the systemic activity of thiophanate methyl (*37*). Banrot is a broad-spectrum fungicide that surpresses the damping-off organisms of *Pythium, Phytophthora, Rhizoctonia,* as well as *Fusarium* and *Thielaviopsis.*

Most commercial propagators use a preventive fungicidal spray program with seedlings and cuttings, while pesticides are applied when insect populations warrant their use.

Insecticidal sprays and drenches. An example of insecticidal spray and drench usage is in the control of fire ants, which is a major pest in the southern U.S. The USDA implemented the Imported Fire Ant Quarantine and Imported Fire Ant Free Nursery program in 1958 to prevent the spread of fire ants, which infest 11 southern states and Puerto Rico. The ants are easily spread by accidentally shipping them with nursery stock and small liner plants. The ants don't directly harm plants and propagules (they will tend plants with aphids, and harvest the honey dew of the aphids from the plant's leaves)—but they do damage land, livestock, have killed people, and are a nuisance to propagation workers and the public. For short-term, small-container crops, such as liners, producers will drench plant containers with Dursban, Talstar 10WP, and Diazinon (in certain states). For propagation mixes and larger container crops, producers

[11]Chemicals used in propagation and horticulture are considered minor use, as opposed to pesticides used for large commodity crops such as cotton, soybean, corn, etc. The cost for chemical companies to develop new or reregister specialty or minor-use chemicals is often prohibitive. Hence, over 1,000 minor uses of agricultural chemicals are currently at risk, and another 2,600 newly sought minor uses may never come to fruition because of the 1988 **Federal Insecticide, Fungicide and Rodenticide Act (FIFRA)** (*43*).

use soil-incorporated granular insecticide, such as Talstar and Fireban, etc. (*29*).

Biological Control in IPM

More and more insect pests and pathogens are being managed by biological methods. This is due in part to increased mite and insect resistance to pesticides, the fact that biological control can be cheaper and more effective than chemical control (i.e., two-spotted mite is effectively controlled by the Chilean predatory mite), increasing concern for environmental issues (contamination of groundwater, etc.), and worker safety, i.e., re-entry times of workers after pesticide application, etc. The goal of the Dutch government is a 50 percent reduction in pesticide usage by the year 2000 (*61*). In the U.S., there is the National Association of Biocontrol Producers for the production and utilization of beneficial insects and organisms.

In propagation, the bacterium *Bacillus thuringiensis* (BT) infects and controls most caterpillars and fungal gnat larvae but has little effect on other insects or the environment. Strains of this naturally occurring bacterium have been formulated into the biological control insecticides: Dipel, Thuricide, Bactospeine, etc.

Biofungicides are preventative, rather than curative, and must be applied or incorporated before disease onset to work properly. For example, the beneficial fungus *Gliocladium virens* is an alternative to the chemical fungicide benomyl. It is currently being marketed by Scotts Co. as Soil-Gard, and comes in an easy-to-apply granular form which is added to the propagation media. It has been cleared by the EPA for biological control of *Rhizoctonia colani* and *Pythium ultimum,* which are two of the principal pathogens causing damping-off diseases (*37*). Mycostop, a strain of *Streptomyces* bacteria isolated from Finnish peat, is being used in propagation as a drench, dip for transplants, seeds and cuttings, or as a foliar spray. It controls *Fusarium, Alternaria,* and *Phomopsis,* and supresses *Botrytis, Pythium,* and *Phytophthora* (*5,81*).

As higher plants have evolved, so have beneficial below-ground organisms interacting with the plant root system (the **plant rhizosphere**). Examples of this include symbiotic nitrogen-fixing bacterium, which are important for leguminous plants, and selected nematodes that control fungal gnats, i.e. X-Gnat from Biosys—the nematodes come in water-dispersable granules, are applied with overhead irrigation equipment, and attack gnats in the larval stage in the container medium. It is well-known that beneficial mycorrhizal fungi (which naturally colonize the root systems of <u>most</u> major horticulture, forestry, and agronomic plants) can increase plant disease resistance, and help alleviate plant stress by enhancing the host plant water and nutrient uptake (*33*). Mycorrhizae can also benefit propagation of cuttings, seedlings, and transplanting of liner plants (*38,140*).

The use of *biocontrol agents*[12] (beneficial bacteria, actinomycetes or fungi living and functioning on or near root in the rhizosphere soil) to control plant pathogens in propagation is still in its infancy (*94*). These beneficial microorganisms suppress fungal root pathogens by antibiosis (production of antibiotic chemicals), by parasitism (direct attack and killing of pathogen hyphae or spores), or by competing with the pathogen for space or nutrients, sometimes by producing chemicals such as siderophores which bind nutrients (such as iron) needed by the pathogen for its disease-causing activities. It is of interest that the inhibitory capacity of these biocontrol antagonists increases in the presence of mycorrhizal fungi, and in the absence of plant pathogens there is a stimulation of plant growth by bacterial antagonists; somehow these bacteria stimulate plant growth, but the mechanism is not known. Perhaps in the future, plant protection during propagation will be done by inoculation of bacteria or combinations of bacteria with mycorrhizal fungi, which come closest to simulating natural conditions of the plant rhizosphere (*94*).

Cultural Control in IPM

Cultural management continues to become more important in modern propagation systems with the loss of minor-use chemicals. In propagation, cultural control begins with the preplant treatment of soil mixes to suppress pathogens and pests. Other cultural control techniques include: sanitizing of propagation facilities, suppressing pathogens and insect pests of stock blocks, harvesting cuttings from stock blocks or containerized plants that are nutritionally fit and not drought stressed, providing good water drainage to reduce the potential of *Phytophthora* root rot and other damping-off organisms, reducing humidity to control *Botrytis,* minimizing the spread of pathogens by quickly disposing of diseased plants from the propagation area, and hardening-off established propagules (*121*).

[12]Recently, *Trichoderma* fungal species, which have plant growth enhancing effects, independent of their biocontrol of root pathogens, have been reported to enhance the rooting of chrysanthemum cuttings, possibly by producing growth-regulating substances *(99)*.

Suppressing pathogens in propagation water is critical, since *Phytophthora, Pythium,* and *Rhizoctonia* are readily disseminated in surface water. Checking pathogens starts with the initial removal of suspended silt and solids, which can tie up chemicals being used to treat the water supply. Sand filters are commonly used for this. Ultraviolet light irradiation is a nonchemical method of controlling pathogens, but water needs to be free of impurities which will shield some of the pathogens from the UV (Figure 3–29). The most commonly used chemical treatments of irrigation water are with chlorination or bromination; one Australian nursery aims for a 4 ppm residual chlorine at the discharge of the irrigation water. They use a swimming pool chlorine test kit (*24*).

Preplanting Treatments of Mixes—Heat Treatment of Propagation and Liner Media

Pasteurization of propagation media. Propagation mixes, such as bark, sand, and peat moss (*20,32*) can contain pathogens, and should be pasteurized. The containers (bins, flats, pots) for such pasteurized mixes should, of course, have been treated to eliminate pathogens. Never put pasteurized mixes into dirty containers. New materials, such as vermiculite, perlite, pumice, and rockwool, which have been heat-treated during their manufacture, need not be pasteurized unless they are reused.

Although the term soil "sterilization" has been commonly used, a more desirable process is **pasteurization,** since the recommended heating processes do not kill all organisms (Figure 3–31). True sterilization would require heating the propagation media to a minimum temperature of 100° C (212° F) for a sufficient period to kill all pests and pathogenic organisms; all beneficial rhizosphere organisms are also killed by the process. Pasteurization of propagation media at lower temperatures with aerated steam is generally preferable to fumigation with chemicals.

After treatment with steam, the medium can be used much sooner. Steam is nonselective for pests, whereas chemicals may be selective. Aerated steam, when properly used, is much less dangerous to use than fumigant chemicals, to both plants and the operator. Chemicals do not vaporize well at low temperatures, but steam pasteurization can be used for cold, wet media.

Moist heat can be injected directly into the soil in covered bins or benches from perforated pipes placed 15 to 20 cm (6 to 8 in.) below the surface. In heating the soil, which should be moist but not wet, a temperature of 82° C (180° F) for 30 minutes has been a standard recommendation, since this procedure kills most harmful bacteria and fungi as well as nematodes, insects, and most weed seeds, as indicated in Figure 3–32. However, a lower temperature, such as 60° C (140° F) for 30 minutes, is more desirable, since it kills pathogens but leaves many beneficial organisms that prevent explosive growth of harmful organisms if recontamination occurs. The lower temperature also tends to avoid toxicity problems, such as the release of excess ammonia and nitrite, as well as manganese injury, which can occur at high steam temperatures.

Electric heat pasteurizers are in use for amounts of soil up to 0.4 m^3 (0.5 yd^3). *Microwave ovens* can be used effectively for small quantities of soil. They do not have the undesirable drying effect of oven heating and will kill insects, disease organisms, weed seed, and nematodes.

Sanitation in Propagation

In recent years, the importance of sanitation during propagation and growing has become widely recognized as an essential part of nursery operations. During propagation, losses of young seedlings, rooted cuttings, tissue-cultured rooted plants, and grafted nursery plants to various pathogens and insect pests can sometimes be devastating, especially under the warm, humid conditions found in propagation houses (*51,89,103,109,115*). Ideally, sanitation strategies should be considered even in the construction phase of propagation structures (*116*).

Harmful pathogens and other pests are best managed by dealing with the three situations where they can enter and become a problem during propagation procedures:

- The propagation facilities—propagating room, containers, pots, flats, knives, shears, working surfaces, hoses, greenhouse benches, and the like.
- The propagation media—rooting and growing mixes for cuttings, seedlings, tissue culture plantlets.
- The stock plant material—seeds, cutting material, scion, and stock material for grafting, tissue culture, etc.

If pathogens and other pests are surpressed in each of these areas, it is likely that the young plants can be propagated and grown to a salable size with minimal disease, insect, or mite infestations. Pathogenic fungi can best be controlled by using soilless

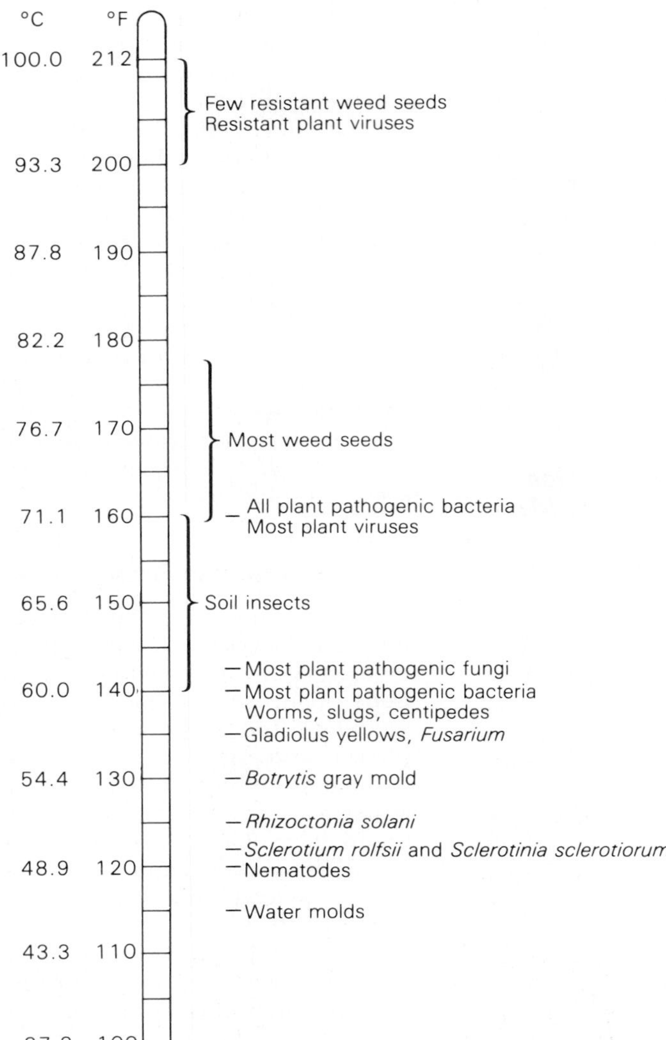

FIGURE 3–32 Soil temperatures required to kill weed seeds, insects, and various plant pathogens. Temperatures given are for 30 minutes under moist conditions.

mixes, pasteurizing propagation and growing mixes, considering general hygiene of the plants and facilities, avoiding overwatering, assuring good drainage of excess water, and using fungicides properly (*104,153*).

Sanitation of Physical Propagation Facilities

The space where the actual propagation (making cuttings, planting seeds, grafting) takes place should be a light, very clean, cool room, completely separated from areas where the soil mixing, pot and flat storage, growing, and other operations take place. Traffic and visitors in this room should be kept to a minimum. At the end of each working day all plant debris and soil should be cleaned out, the floors hosed down, and working surfaces washed with disinfectant solutions of sodium hypochlorite solution (Clorox), Consan CTA 20, benzylkonium chloride (Physan 20, Greenshield), or pine disinfectant—diluted according to directions. Benzylkonium chloride is long-lasting and can be used for several days. Dilute household vinegar gives good control of algae and moss along walkways (Figure 3–33).

Flats and pots coming into this room should have been washed thoroughly and, if used previ-

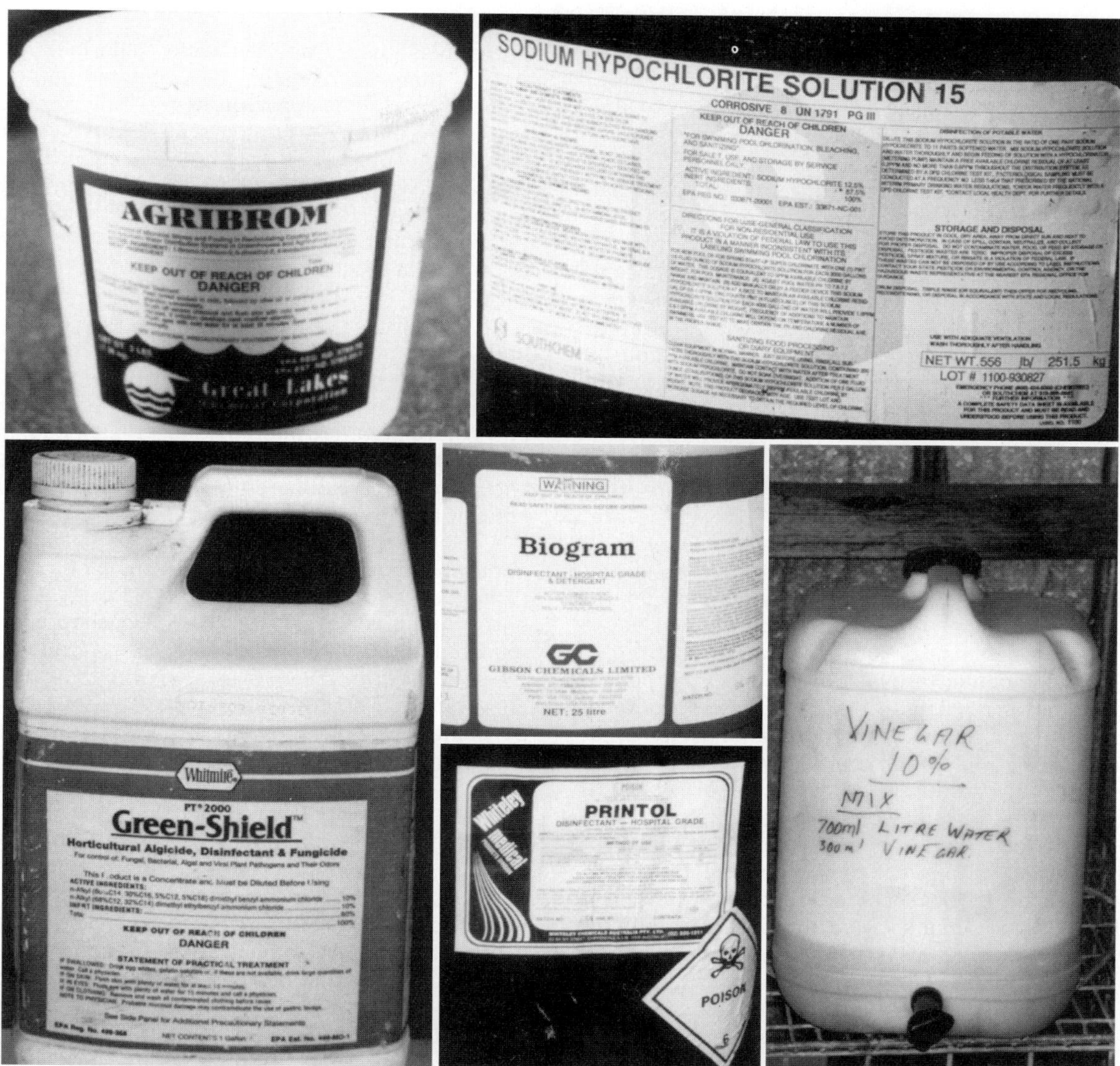

FIGURE 3–33 Some common chemicals for sanitation of propagation facilities and propagules. *Top left and right:* Diluted Agribrom and sodium hypochlorite solution can be used to treat both propagules and facilities. *Middle:* Greenshield, Biogram and Printol are used as surface disinfectants. *Bottom right:* Dilute household vinegar can control algae and moss along walkways. Always follow directions and first try small trials.

ously, should be heat-treated or disinfected with chemicals, i.e., a 30-minute soak in sodium hypochlorite (Clorox) diluted one to nine. No dirty flats or pots should be allowed in the propagation area. Knives, shears, and other equipment used in propagation should be sterilized periodically during the day by dipping in a disinfectant such as Physan.

Mist propagating and growing areas in greenhouses, cold frames, and lathhouses should be kept clean, and diseased and dead plant debris should be removed daily. Water to be used for misting should be free of pathogens. Water from ponds or reservoirs to be used for propagation purposes should be chlorinated to kill algae and pathogens. Proper

Chlorine can be used as a **sterilant,** which destroys all organisms, and as a **disinfectant,** which selectively destroys pests (*91*). When chlorine is used as a pesticide, it prevents pests from entering the propagation environment and minimizes the need for more toxic pesticides. Pest reduction or elimination is a cornerstone of IPM programs. Chlorine, in the form of laundry bleach (Clorox, etc.), is one of the most affordable and readily available chemicals *(41).* Chlorine is used to sterilize greenhouse benches, floors, and other surfaces in the propagation area. Chlorination is being increasingly used in recycled irrigation water for controlling pathogenic fungi, algae, and other pests.

Chlorine is available as: (1) a gas (Cl_2), which is liquefied in pressurized metal containers and bubbled as a gas into water, but Cl_2 gas is very toxic and its corrosive nature makes it very hazardous to handle; (2) calcium hypochlorite [$Ca(OCl)_2$] is used for domestic water treatment and is commercially available as granulated powder, large tablets, or liquid solutions; and (3) sodium hypochlorite (NaOCl), which is the active ingredient of household bleach, is the most common form of chlorine used in propagation. When a continuous supply of chlorinated water is needed, concentrated solutions of sodium or calcium hypochlorite are injected. Chlorine injectors must be installed with an approved check-valve arrangement to prevent back flow into the fresh water system *(91).* Bleach solutions are generally calculated as percent bleach or percent sodium hypochlorite; but these are not the same, since a **10 percent bleach solution** (which contains one part bleach: nine parts water) is 10 percent of 5.25 percent sodium hypochlorite or equivalent to **0.52 percent sodium hypochlorite.** Household bleach is commonly used as a disinfectant by diluting one part bleach to nine parts water.

Many chemicals, as well as organic residue from plants and propagation medium, react with chlorine and reduce its effectiveness. Enough chlorine must be added to produce an effective concentration of "free residual" chlorine (see Figure 3–34). Factors affecting chlorine activity include: (1) concentration—water treatment requires around 1 ppm, and the bleaching of propagation benches and containers a 10 percent bleach solution or 5,250 ppm and, (2) exposure time, (3) pH—around 6.5 is most effective, (4) organic matter—contaminated water with residual from soaking propagation containers or dipping propagules uses up available chlorine more rapidly than clean water, (5) water temperature, and (6) pathogen growth stage—chlorine kills fungal mycelium on contact, but is **not systemic** so fungal spores and pathogens embedded in roots and walls of styrofoam containers are much more difficult to kill; soaking materials before treating with bleach allows spores to germinate and mycelia to grow, making pathogens easier to kill (*91*).

For successful chlorination: clean container, bed, and propagule materials prior to chlorinating, monitor the chlorine solution, and ventilate the work area. Dilute chlorine solutions irritate skin and chlorine vaporization irritates eyes, nose, and throat. It is important that propagation managers know the legal exposure limits (OSHA) that workers can be exposed to chlorine.

There are some environmental concerns about the use of bleach as a disinfectant to surface disinfect cuttings and for the sterilization of tools and propagation work surfaces. The hypochlorite ion from bleach attaches to organic compounds in the soil and forms very stable chlorinated organic compounds. These compounds can be taken up by plant roots, get into the food chain, and may bioaccumulate in the body fat of animals and humans (*103*). An alternative disinfectant for propagation is hydrogen peroxide (H_2O_2). It can be used as a sterilant for both fungi and bacteria, has no toxic byproducts (it breaks down to water and oxygen), and it has no residual effect in water or soil. Hydrogen peroxide can be purchased in bulk form (35 percent concentration rate). A recommended rate for surface disinfestation of plant material is 1 part H_2O_2 (35%) to 100 parts water (*103*). Clorox (bleach) was found to be superior to hydrogen peroxide, Agrimycin 17 (agricultural streptomycin) or rubbing alcohol (isopropyl) in preventing transmission of fire blight bacteria in pear trees (*145*).

Chlorine will continue to be used as an important disinfectant in propagation. Bleach is considerably cheaper than hydrogen peroxide, and with the dilute bleach solutions typically used in propagation, there should be little if any chlorine residual in tank solutions that are allowed to sit for several days (*91*). To be environmentally safe before discharging spent chlorinated water, test kits should be used to monitor residual chlorine levels, and local water quality officials can also be contacted.

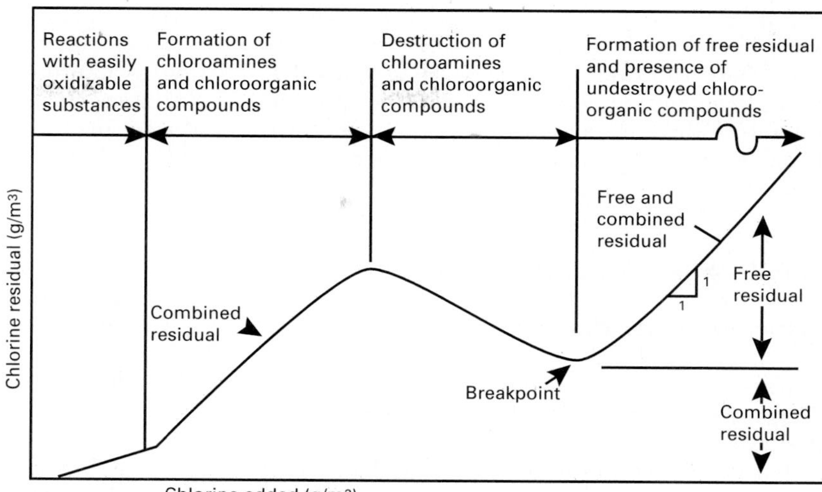

FIGURE 3–34 Many things combine with chlorine to reduce its activity in solution, and so enough must be added to produce an effective concentration of "free residual" chlorine (*91*).

chlorination will control *Phytophthora* and *Pythium* in irrigation water and can help reduce the cost of preventative fungicide programs (*40*).

Maintaining Clean Plant Material

In selecting propagating material, use only seed and those source plants that are disease- and insect-free. Some nurseries maintain stock plant blocks, which are kept meticulously "clean." However, stock plants of particularly disease-prone plants, such as euonymus, might well be sprayed with a suitable fungicide several days before cuttings are taken. Drenches of fungicides and/or Agribrom (oxidizing biocide) are sometimes applied to stock plants in the greenhouse prior to selecting explants for tissue culture.

It is best to select cutting material from the upper portion of stock plants rather than from near the ground where the plant tissue could possibly be contaminated with soil pathogens. As cutting material is being collected, it should be placed in new plastic bags.

After the cuttings have been made and before they are stuck in the flats they can be dipped in a dilute bleach solution—Agribrom, Physan 20, or various fungicides for broad-spectrum control of damping-off organisms—before any auxin treatment. One Oregon nursery disinfects Rhododendron cuttings with Consan, followed by washing in chlorinated water (*54*). Agri-strep (agricultural streptomycin) helps suppress bacterial problems, and one biological control, *Agrobacterium spp.*, helps prevent crown gall of hardwood rose cuttings (*37*). However, once a cutting or seedling *becomes infected with a bacterium, there is no effective control* other than rouging and destroying the plant propagule.

BMP—Best Management Practices

To a very limited degree, through some improper pesticide usage and inefficient irrigation and fertility systems, the nursery and greenhouse industries have been nonpoint source polluters of the environment. As a whole, the horticultural industries are good stewards of the environment. The environmentally friendly plants they produce are critical to the well-being, nutrition, and welfare of people, and are vital to enhancing the environment (reduced air and noise pollution, reduced heat loads around houses and urban areas, which lower utility cooling bills, adding O_2 to the air and contributing to the abatement of current high global CO_2 conditions, etc.).

With the increased environmental regulations facing plant propagators and as an offshoot of integrated pest management programs, the development of BMP or Best Management Practices is occurring (*76*). To help preserve the environment and head off additional state and federal regulation, BMP are being developed by the nursery industry, governmental agencies, and universities. Plans are for the nursery and greenhouse industries to help regulate themselves by adapting BMP, which many propagators have already been practicing for years. The list of the ten best management practices applies to nursery propagation systems. To date, recycled water is generally not used to propagate plants (liners and container plants are irrigated with recycled water mixed with purer well or surface collected water), but in the future with the scarcity of irrigation water and increased urban population pressure to use limited water supplies, more nurseries will have to develop propagation systems that utilize recycled water. Recycled water can present considerable challenges, since high salinity, trace levels of herbicides, pesticides, and pathogens such as *Phytophthora,* etc. can occur (Figure 3–29).

POST-PROPAGATION CARE OF LINERS

Hardening-off Liner Plants

Hardening-off or acclimatizing rooted propagules, seedlings, tissue culture plantlets, etc. is critical for plant survival and growth. In commercial production, it assures a smooth transition and efficient turnover of plant product from propagation to liner production (Figure 3–27) to finished plants in protected culture (greenhouses, etc.) or containerization and field production. This smooth transition and turnover of plant production units is essential in the marketing, sales, and profitability of plant manufacturing companies.

It is important to wean rooted cuttings from the mist system as quickly as possible. Reduction of irrigation and fertility in seedlings and plugs is done several weeks prior to shipping and/or transplanting to harden-off and ensure survival of the crop. Likewise, with acclimation of tissue-culture produced plantlets, light irradiance is increased and relative humidity gradually reduced to stimulate the plantlet to increase photosynthetic rates and have better stomatal control. All of these ensure plant survival and a speedy transition when the acclimatized plant is shifted up and finished-off as a container or field crop. Acclimatization of liners is discussed in greater detail (see Chapters 11, 18).

NURSERY BEST MANAGEMENT PRACTICES (*175*)

- Collect runoff water when injecting fertilizer.
- Apply fertilizer only when a growth response will be obtained.
- Do not broadcast fertilizer on spaced containers.
- Do not top-dress fertilizer on containers prone to blow over.
- Water and fertilize according to plant needs.
- Group plants in a nursery according to water and fertilizer needs to minimize runoff.
- Monitor quantity of irrigation applied to prevent overwatering.
- Maintain minimal spacing between containers receiving overhead irrigation.
- Use low-volume irrigation for containers larger than 26 liter (7-gallon).
- Recycle runoff water.

Handling Container Grown Plants

Irrigation. Watering of container nursery stock is a major expense and environmental concern. Hand watering of individual containers with a large-volume, low-pressure applicator on a hose several times a week is expensive, and is used only for small-scale operations. In most operations, overhead sprinklers (i.e., Rainbird-type impact sprinklers) are often used, although much runoff waste occurs. Watering of container plants by trickle or drip irrigation results in less waste (*156*), and is becoming more widely used, particularly with plants produced in larger containers (Figure 3–35). The recent development of solid-state soil tensiometers for the computer control of irrigation systems of containerized plants may help to increase water use efficiencies and decrease off-site pollution from runoff (*26*).

As part of best management practices (BMP), many nurseries are switching to computer controlled **cyclic** or **interval irrigation (pulse irrigation)** with impact sprinklers. Rather than manually turning on valves to run irrigation for 40 minutes, an environmental-control computer is programmed to precisely run the irrigation system four times daily at five minutes per cycle (*162*). Since most water is absorbed by the containers within the first five minutes, cyclic or pulse irrigation uses less water, greatly reduces water and fertility runoff, and the amount of fertilizer needed in the fertigation system is lowered.

Setting the containers on damp mats or beds of fine sand over plastic with water moving upward into the soil by capillarity (Figure 3–36) is another method of supplying water to plants (*6,123*). There have been concerns of the spread of pathogens such as *Phytophthora* from container to container with this system. In England, however, nurseries grow containerized plants on capillary sand-bed systems, which have conserved irrigation water, reduced

FIGURE 3–35 Automatic watering system for container-grown plants. *Top:* A small "spaghetti" tube carries water from the plastic feeder tubes to each plant. *Bottom:* Trickle irrigation can efficiently irrigate container plants with less water than overhead sprinkler irrigation systems.

FIGURE 3–36 Capillary irrigation. *Top:* Setting containers on damp mats or beds of fine sand over plastic with water moving upward into the soil by capillarity is another method of supplying water to plants. *Bottom:* Two main irrigation lines soak the capillary beds and water is drawn up through the bottom of the containers.

nutrient leaching, automated irrigation to reduce labor costs, reduced weed problems, and allowed for better drainage which improves the aeration of the container media (*54*). Flood floor systems for producing containerized plants was discussed earlier in this chapter (Figure 3–7). Modified flood floor systems are being used in the central U.S. to grow container plants outdoors, with the container plants sitting on top of a plastic liner (rather than a capillary sand bed).

Fertilization. Fertilizer solutions are usually injected into the irrigation system (fertigation) in commercial nurseries. Fertilizer may be supplied solely in slow-release forms (Osmocote, Phycote, Nutricote, Polyon, etc.), or used in combination with fertigation (see the section earlier in the chapter on nutrient control). After the container stock leaves the wholesale nursery, the retailer should maintain the stock by adequate irrigation and fertilization until the plants have been purchased by the consumer. Slow-release fertilizers added to containers leave a residual fertilizer supply (most retailers do not add supplementary fertilizer), and help maintain the plants until they are purchased by the consumer, and planted in the ground.

Winter protection. In areas with severe winters, attention must be given to the problems of winter injury (*71,127,134*). The amount of injury varies with the species. The chances of cold injury are lessened if the plants are well established in containers before the onset of winter (see Chapter 11 for post-propagation conditioning of liner plants for winter). Setting the plants close together in large groups (jamming or bunching) tends to prevent damage from rapid temperature fluctuation. Wrapping plastic to reduce wind penetration or using Styrofoam-type sheets around the outer rows of cans is helpful. In cold winter regions, however, some additional form of winter protection must be used; for example, placing mulch covering, such as straw or hay (can introduce weed seed), over the tops of the containers or placing the containers inside a protective structure, such as a cold frame or plastic-covered greenhouse is effective. A reliable type of winter protection, as shown in Figure 3–37, is a temporary frame constructed over the plants and covered with opaque polyethylene sheeting, 4 to 6 mils in thickness. Often two layers of polyethylene, separated by a 2.5 to 5 cm (1 to 2 in.) air space, are used.

Another method of winter protection is to cover the block of container plants, either upright or on their sides, with soft, pliable microfoam sheets, which are then covered with white polyethylene, and sealed airtight to the ground. Rodents can be a problem with this system (*79*).

Overhead sprinkling is also an effective method of protecting tender plants from subfreezing temperatures for a limited period of time, i.e., 24 to 48 hours. Water must be applied continually, changing to ice as long as the subfreezing temperatures occur. When liquid water changes to ice approximately 144 BTU of heat is released per pound of water (*72,138*).

In most woody plant species the roots do not develop as much winter hardiness as the tops, nor do containerized roots have the benefits of the insulation of a large mass of mineral soil, as with in-ground field stock. Hence, low-temperature damage of container stock occurs primarily to the root system. Winter-hardiness of roots varies with the species (*70,142*).

Root development in containerized plants. When trees and shrubs from seedlings or rooted cuttings are grown in containers, roots often begin to circle on the outside of the rootball against the slick, smooth plastic container walls. If not mechanically disrupted when the trees or shrubs are transplanted, circling roots may enlarge to the point of stressing or killing the plant by girdling (*4*). Internal walls of containers can be coated with copper compounds {such as Spin Out—which is a latex based paint containing copper hydroxide and a special formulated carrier (Figure 3–21), which enhances root absorption of copper and temporarily inhibits root elongation (*144*)}, or containers can have special wall modifications as a means to reduce or prevent root circling during liner production and

FIGURE 3–37 Winter protection of nursery container stock. *Top:* In more temperate zones, houses covered with white plastic and minimal heat to prevent freezing may be used. Plants are placed close together in beds and covered in late fall with 6-mil white polyethylene on pipe framing. The smaller-hooped sun houses *(foreground)* will rely on passive heating from the soil. *Middle:* In more mild climates, winterization may be sufficient with an open opaque poly-covered house that restricts wind movement. *Bottom:* A microfoam-insulated sheeting covered with white poly is a low cost way of winterizing container plants that have been turned on their sides.

FIGURE 3–38 One disadvantage of growing trees in containers is the possibility of producing poorly shaped root systems. Here a defective, twisted root system resulted from holding the young nursery tree too long in a container before transplanting. Such spiraling roots retain this shape after planting and are unable to anchor the tree firmly in the ground.

later container production. As shown in Figure 3–38, plants kept in containers too long will form an undesirable constricted root system from which they may never recover when planted in their permanent location (*58*). The plants should be shifted to larger containers before such "root spiraling" occurs.

The Ohio Production System **(OPS),** a system for rapidly producing container-grown shade trees (whips) in one year, compared to three years, also relies on copper-treated containers to control root growth. This eliminates or greatly reduces the need to root prune when plants are upcanned to larger containers (*143*).

Using bottomless propagation and liner pots to **"air prune"** roots, judicious root pruning, early transplanting, and careful potting during the early transplanting stages can do much to encourage the development of a good root system by the time the young plant is ready for transfer to its permanent location (*69*). Plastic containers with vertical grooves along the sides tend to prevent horizontal spiraling of the roots (Figure 3–20).

Alternatives to traditional production systems. Several in-ground alternatives to container production in the field and conventional field production of bare-root and B & B (balled-in-burlap) trees and shrubs have been developed, including: (1) the **pot-in-pot system,** (2) **in-ground plastic containers,** and (3) **in-ground fabric containers (grow bags)** (Figure 3–39). Each of these methods can influence directional root development (*4*). The pot-in-pot, in-ground system involves sinking an outer or sleeve pot into the ground and inserting a second pot, which is the production pot that is harvested with the plant. The production container may have vertical ribs, or the interior walls are treated with copper to reduce root circling. The in-ground container system is a single container (unlike the pot-in-pot system) with rows of small holes along the container sides and bottom to enhance drainage.

In-ground fabric containers or grow bags are flexible, synthetic bags, which are filled with mineral soil and placed in predug holes in the field. The synthetic woven material of the bags limits most root penetration, and directs root growth to occur within the bag [over 90 percent of the root system of conventional bare-root and balled and burlapped (B & B) plants are lost during digging]. Since the bag is placed in the ground, there is greater insulation of the root system against high and low temperatures (versus above-ground containerized crops), and the bag can be pulled out of the field, potentially reducing labor cost of traditional field techniques. This system does not work with all species, but has merits.

REFERENCES

1. Aichele, J.H. 1988. High humidity propagation using sweat box methods. *Comb. Proc. Intl. Plant Prop. Soc.* 37:497–99.
2. Aimone, T. 1986. Carbon dioxide injection (report of a talk given by J. O. Donovan and J. Tsujita at Grower Expo 66). *Grower Talks* 49(12):88–97.
3. Alvey, N.G. 1961. Soil for John Innes composts. *J. Hort. Sci.* 36:228–40.
4. Appleton, B.L. 1994. Elimination of circling tree roots during nursery production. In *The landscape below ground: proceedings of an international workshop on tree root development in urban soils,* Watson, G.A., and D. Neely, eds. Savoy, IL: Intern. Soc. Arboricul.
5. Armstrong, M. 1992. Integrated disease management—research and development using new tech-

FIGURE 3–39 Alternatives to traditional field production. *Top left:* In-ground fabric containers or grow bags. *Top center:* The pot-in-pot (P&P) system with individual pot, drip irrigation. *Top right:* Copper-treated wall of outside sleeve containers to prevent root penetration from the inner pots. *Bottom left:* P&P containers. *Bottom center and right:* The roots of the inside containers are very susceptible to heat stress when they are removed from the field. Here they are being wrapped with an insulating packing fabric prior to shipping.

niques and bioremediation at Vans pines. *Comb. Proc. Intl. Plant Prop. Soc.* 42:503–506.

6. Auger, E., C. Zafonte, and J.J. McGuire. 1977. Capillary irrigation of container plants. *Comb. Proc. Intl. Plant Prop. Soc.* 27:467–73.
7. Augsburger, N.D., H.R. Bohanon, and J.L. Calhoun. 1978. *The greenhouse climate control handbook.* Muskogee, Okla.: Acme Eng. & Mfg. Co.
8. Baird, C.D., and W.E. Waters. 1979. Solar energy and greenhouse heating. *HortScience* 14(2):147–51.
9. Ball, V., ed. 1991. *The Ball red book* (15th ed.). West Chicago, IL: Geo. J. Ball Publ. Co.
10. Baker, J.R. and R.K. Jones. 1995. Pests of the south. *Amer. Nurser.* 181(11):78–89.
11. Bartok, J.W., Jr. 1987. Chemical pressure treatment helps wood last longer. *Greenhouse Manag.* 6(2):151–52.
12. ———. 1988. Horizontal air flow. *Greenhouse Manag.* 6(10):197–212.
13. ———. 1988. Feed your plants with a fertilizer injector. *Greenhouse Manag.* 6(12):117–20.
14. Bauerle, W.L., and T.H. Short. 1977. Conserving heat in glass greenhouses with surface-mounted air-inflated plastic. *Ohio Agr. Res. and Develop. Center Spec. Circ.* 101.
15. ———. 1978. Greenhouse energy conservation and effects on plant response. *Ohio Rpt. on Res. and Develop.* 63(5):74–76.
16. Bayles, E. 1988. Computers. *Greenhouse Manag.* 7(2):90–96.
17. Beytes, C. 1994. The sky's the limit with retractable roof greenhouses. *Grower Talks* 58(4):22–26.
18. ———. 1995. Flood floor construction in 12 steps. *Grower Talks.* 58(11):40–46.
19. Bilderback, T. 1993. Wetting agents and gels—where they have a purpose. *Comb. Proc. Intl. Plant Prop. Soc.* 43:421–423.
20. Bluhm, W.L. 1978. Peat, pests, and propagation. *Comb. Proc. Intl. Plant Prop. Soc.* 28:66–70.
21. Boodley, J.W., and R. Sheldrake, Jr. 1964. Cornell "Peat-Lite" mixes for container growing. Ithaca, N.Y.: Cornell Univ. Dept. Flor. and Orn. Hort., mimeo. rpt.
22. Branson, R.L., J.P. Martin, and W.A. Dost. 1977. Decomposition rate of various organic materials in soil. *Comb. Proc. Intl. Plant Prop. Soc.* 27:94–96.
23. Brugger, M.F. 1987. Horizontal air flow improves crop growth. *Amer. Veg. Grower* 35(10):32–34.
24. Bunker, E. 1992. Water quality in propagation. *Comb. Proc. Intl. Plant Prop. Soc.* 42:81–84.
25. Bunt, A.C. 1988. *Media and mixes for container-grown plants.* (2nd Ed. of *Modern potting composts*). London: G. Allen & Unwin.
26. Burger, D.W. 1992. Water conserving irrigation systems. *Comb. Proc. Intl. Plant Prop. Soc.* 42:260–266.
27. Cathey, H.M., and L.E. Campbell. 1980. Light and lighting systems for horticultural plants. *Hort. Rev.* 2:491–537.
28. Chong, C., R.A. Cline, and D.L. Rinker. 1988. Spent mushroom compost and papermill sludge as soil amendments for containerized nursery crops. *Comb. Proc. Intl. Plant Prop. Soc.* 37:347–53.
29. Collins, H. 1995. Fighting fire ants. *GrowerTalks.* 59(2):48.
30. Connor, D. 1977. Propagation at Monrovia Nursery Company: Sanitation. *Comb. Proc. Intl. Plant Prop. Soc.* 27:102–6.
31. Cooke, C.D., and B.L. Dunsby. 1978. Perlite for propagation. *Comb. Proc. Intl. Plant Prop. Soc.* 28: 224–28.
32. Coyier, D.L. 1978. Pathogens associated with peat moss used for propagation. *Comb. Proc. Intl. Plant Prop. Soc.* 28:70–72.
33. Cuny, H. 1995. Fungi lend a hand: rooting out mycorrhizae's place in the nursery. *Nuser. Manag.* 10(4):45–49.
34. Dallon, J., Jr. 1988. Effects of spent mushroom compost on the production of greenhouse-grown crops. *Comb. Proc. Intl. Plant Prop. Soc.* 37: 323–29.
35. Davies, F.T. Jr. 1985. Plant modeling: developing an approach. *Comb. Proc. Intl. Plant Prop. Soc.* 35:770–776.
36. ———. 1988. The influence of nutrition and carbohydrates on rooting of cuttings. *Comb. Proc. Intl. Plant Prop. Soc.* 38:432–437.
37. ———. 1991. Back to the basics in propagation. *Comb. Proc. Intl. Plant Prop. Soc.* 41:338–342.
38. Davies, F.T., Jr. and C.A. Call. 1990. Mycorrhizae, survival and growth of selected woody plant species in lignite overburden in Texas. *Agric. Ecosystems Environ.* 31:243–252.
39. Davies, F.T., Jr., D.E. Kester, and T.D. Davis. 1994. Commercial importance of adventitious rooting: horticulture. In *Biology of adventitious root formation,* T.D. Davis and B.E. Haissig, eds. New York: Plenum Press.
40. Daughtry, B. 1988. Control of *Phytophthora* and *Pythium* by chlorination of irrigation water. *Comb. Proc. Intl. Plant Prop. Soc.* 38:420–422.
41. De Fossard, R.A. 1992. Treatment of plants with hypochlorite solutions. *Comb. Proc. Intl. Plant Prop. Soc.* 42:65–67.
42. Duncan, G.A., and J.N. Walker. 1980. How to save energy in the greenhouse. *Amer. Nurs.* 512(10):13, 90–112.
43. Dysart, J. 1994. Minor use chemical loss likely—but not inevitable. *GrowerTalks* 58(4):94–100.
44. Edmonds, J. 1994. Europeans embrace new container and tunnel technologies. *Amer. Nurs.* 179(12):23–24.

45. Egilla, J.N. and F.T. Davies, Jr. 1995. Response of *Hibiscus rosa-sinensis* L. to varying degrees of potassium fertilization: growth, gas exchange and mineral element concentration. *J. Plant Nutrit.* 18:1765–1783.
46. Falland, J. 1987. Greenhouses in plant propagation: A historical perspective. *Comb. Proc. Intl. Plant Prop. Soc.* 36:158–64.
47. Fessler, T.R. 1995. The propagation environment with a computer. *Comb. Proc. Intl. Plant Prop. Soc.* 45: In press.
48. Fireman, M., and R.L. Branson. 1963. Salinity in greenhouse soils. *Calif. Agr. Ext. Ser. OSA 68* (rev.).
49. Fonteno, W.C. 1988. Know your media: The air, water, and container connection. *GrowerTalks* 51(11):110–11.
50. Francois, L.E. 1980. Salt injury to ornamental shrubs and ground covers. *USDA SEA Home and Garden Bul.* 231.
51. Geard, I. D. 1979. Fungal diseases in plant propagation. *Comb. Proc. Intl. Plant Prop. Soc.* 29:589–94.
52. Germing, G. H., ed. 1986. Symposium on greenhouse climate and its control. *Acta. Hort.* 174:1–563.
53. Geistlinger, L. 1994. The Worker Protection Standard. *Amer. Nurs.* 179(1):46–57.
54. George, C. 1993. Rhododendron propagation—methods and techniques carried out in the Pacific northwest of the USA. *Comb. Proc. Intl. Plant Prop. Soc.* 43:178–182.
55. Gill, S. 1995. Early detection key to aphid control. *Grower Talks* 59(2):94.
56. Gilliam, C.H., and E.M. Smith. 1980. How and when to fertilize container nursery stock. *Amer. Nurs.* 151(2):7,117–27.
57. Goldsberry, K. L. 1979. Greenhouse heat conservation and the effect of wind on heat loss. *HortScience* 14(2):152–55.
58. ———. 1986. Future looks bright for rockwool use. *Greenhouse Manag.* 5(6):103–7.
59. Gordon, I. 1988. Structures used in Australia for plant propagation. *Comb. Proc. Intl. Plant Prop. Soc.* 37:482–89.
60. ———. 1992. A review of materials for propagation media. *Comb. Proc. Intl. Plant Prop. Soc.* 42:85–90.
61. Gough, N. 1992. Prospects for IPM in greenhouse ornamentals in Australia. *Comb. Proc. Intl. Plant Prop. Soc.* 42:103–107.
62. Gouin, F.R. 1989. Composted sewage sludge: an aid in plant propagation. *Comb. Proc. Intl. Plant Prop. Soc.* 39:489–493.
63. Graham, J.H. 1988. Interactions of mycorrhizal fungi with soilborne plant pathogens and other organisms. *An Introduction to Plant Pathology.* Vol. 78, No. 3.
64. Hamrick, D. 1988. The covering choice. *GrowerTalks* 51(12):64–74.
65. Hanan, J. 1987. A climate control system for greenhouse research. *HortScience* 22(5):704–8.
66. Hanan, J., W.D. Holley, and K.L. Goldsberry. 1978. Greenhouse construction. Chapter 3 in *Greenhouse manag.*. New York: Springer Verlag.
67. Handreck, K.A. 1985. Potting mixes, and the care of plants growing in them. East Melbourne, Australia: CSIRO.
68. Harden, J. 1992. Plant protection—management of pest control techniques. *Comb. Proc. Intl. Plant Prop. Soc.* 42:99–102.
69. Harris, R.W., W.B. Davis, N.W. Stice, and D. Long. 1971. Effects of root pruning and time of transplanting in nursery liner production. *Calif. Agr.* 25(12):8–10.
70. Havis, J.T. 1974. Tolerance of plant roots in winter storage. *Amer. Nurs.* 139(1):10.
71. Havis, J.T., and R.D. Fitzgerald. 1976. Winter storage of nursery plants. *Mass. Coop. Ext. Ser. Pubt.* 125.
72. Hendershott, C.H. 1979. Cold protection of low growing plants. *Comb. Proc. Intl. Plant Prop. Soc.* 29:533–36.
73. Herve, A.J. 1989. Production of spray chrysanthemums in a hydroponic system. *Comb. Proc. Intl. Plant Prop. Soc.* 39:66–70.
74. Hicklenton, P.R., and A.M. Armitage. 1988. *CO_2 enrichment in the greenhouse.* Portland, Oreg.: Timber Press.
75. Hildebrandt, C.A. 1987. Economical propagation structures for the small grower. *Comb. Proc. Intl. Plant Prop. Soc.* 36:506–10.
76. Hottovy, S.A. 1994. Monrovia nursery's response to new environmental restrictions. *Comb. Proc. Intl. Plant Prop. Soc.* 44:161–164.
77. Huang, K.T., and J.J. Hanan. 1976. Theoretical analysis of internal and external covers for greenhouse heat conservation. *HortScience* 11(6):582–83.
78. Humphrey, C. 1995. Basics of greenhouse design and construction. *GrowerTalks Mag.* 58(11):60–64.
79. Hutt, G.M. 1984. Microfoam use for winter protection—your fifth option. *Comb. Proc. Intl. Plant Prop. Soc.* 34:418–24.
80. Inose, K. 1971. Pumice as a rooting medium. *Comb. Proc. Intl. Plant Prop. Soc.* 21:82–83.
81. James, B.L. 1993. Update on fungicides. *Comb. Proc. Intl. Plant Prop. Soc.* 43:373–375.
82. Jensen, M.H. 1977. Energy alternative and conservation for greenhouses. *HortScience* 12(1):14–24.
83. Jeong, B.R., K. Fujiwara, and T. Kozai. 1993. Carbon dioxide enrichment in autotophic micropropagation: methods and advantages. *HortTechnology* 3(3):332–334.
84. Johnson, C.R. 1977. Some water quality problems faced by horticulturists. *Comb. Proc. Intl. Plant Prop. Soc.* 27:202–6.

85. Johnson, C.R., and D.F. Hamilton. 1977. Effects of media and controlled-release fertilizers on rooting and leaf nutrient composition of *Juniperus conferta* and *Ligustrum japonicum* cuttings. *J. Amer. Soc. Hort. Sci.* 102:320–22.

86. Kelly, M.H. 1995. Get a grip on the greenhouse climate with automated curtains. *Grower Talks.* 59(2):76–83.

87. Krizek, D.T., W.A. Bailey, H.H. Klueter, and H.M. Cathey. 1968. Controlled environments for seedling production. *Comb. Proc. Intl. Plant Prop. Soc.* 18:273–80.

88. LaCroix, L.J., D.T. Canvin, and J. Walker. 1966. An evaluation of three fluorescent lamps as sources of light for plant growth. *Proc. Amer. Soc. Hort. Sci.* 89:714–22.

89. Lambe, R.C., and W.H. Wills. 1979. The major diseases of holly in the nursery. *Comb. Proc. Intl. Plant Prop. Soc.* 29:536–44.

90. Landis, T.D. 1993. Using "limiting factors" to design and manage propagation environments. *Comb. Proc. Intl. Plant Prop. Soc.* 43:213–218.

91. ———. 1994. Using chlorine to prevent nursery diseases. *Forestry Nursery Notes.* For. Serv., USDA, Pacific NW Region, R6-CP-TP-08-94.

92. Landis, T.D., R.W. Tinus, S.E. Mcdonald, and J.P. Barnett 1990. *Containers and growing media.* The Container Tree Nursery Manual, Vol. 2. Agri. Handbk. 674. Washington, DC: For. Serv., USDA.

93. ———. 1992. *Atmospheric Environment.* The Container Tree Nursery Manual, Vol 3. Agri. Handbk. 674. Washington, DC: For. Serv., USDA.

94. Linderman, R.G. 1993. Effects of biocontrol agents on plant growth. *Comb. Proc. Intl. Plant Prop. Soc.* 43:249–252.

95. Loach, K. 1987. Mist and fruitfulness. *Horticulture Week,* April 10, 1987, pp. 28–29.

96. ———. 1989. Controlling environmental conditions to improve adventitious rooting. In *Adventitious root formation in cuttings,* T.D. Davis, B.E. Haissig, and N. Sankhla, eds. Portland, Oreg.: Dioscorides Press.

97. Lucas, R.E., P.E. Riecke, and R.S. Farnham. 1971. Peats for soil improvement and soil mixes. *Mich. Coop. Ext. Ser. Bul.* E-516.

98. Macdonald, A.B. 1986. Propagation facilities——— past and present. *Comb. Proc. Intl. Plant Prop. Soc.* 35:170–75.

99. MacKenzie, A.J., T.W. Starmann, and M.T. Windham. 1995. Enhanced root and shoot growth of chrysanthemum cuttings propagated with the fungus *Trichoderma harzianum. HortScience* 30(3):496–498.

100. Marcenaro, S., C. Voyiatzi, and B. Lercari. 1994. Photocontrol of in vitro bud regeneration: a comparative study of the interaction between light and IAA in a wild type and an *aurea* mutant of *Lycopersion esculentum. Physiol. Plant.* 91:329–333.

101. Maynard, B.K. 1993. Basics of propagation by cuttings: light. *Comb. Proc. Intl. Plant Prop. Soc.* 43:445–449.

102. McCain, A.H. 1977. Sanitation in plant propagation. *Comb. Proc. Intl. Plant Prop. Soc.* 27:91–93.

103. McClelland, M.T. and M.A.L. Smith. 1993. Alternative methods for sterilization and cutting disinfestation. *Comb. Proc. Intl. Plant Prop. Soc.* 43:526–530.

104. McCully, A.J., and M.B. Thomas. 1977. Soilborne diseases and their role in propagation. *Comb. Proc. Intl. Plant Prop. Soc.* 27:339–50.

105. McMahon, M.J., J.W. Kelly, D.R. Decoteau, R.E. Young, and R.K. Pollock. 1991. Growth of *Dendranthema x grandiflorum* (Ramat.) Kitamura under various spectral filters. *J. Amer. Soc. Hort. Sci.* 116(6):950–954.

106. Mears, D.R., ed. 1979. *Proc. 4th ann. conf. on solar energy for heating of greenhouses.* New Brunswick, N.J.: Rutgers Univ. Dept. Biol. and Agr. Eng.

107. Miller, N. 1981. Bogs, bales, and BTU'S: A primer on peat. *Horticulture* 49(4):38–45.

108. Monk, G.J., and J.M. Molnar. 1987. Energy efficient greenhouses. In *Horticultural reviews,* Vol. 9, J. Janick, ed. Westport, Conn.: AVI Publ. Co., pp. 1–52.

109. Moody, E.H., Sr. 1983. Sanitation: A deliberate, essential exercise in plant disease control. *Comb. Proc. Intl. Plant Prop. Soc.* 33:608–13.

110. Moore, G. 1988. Perlite: Start to finish. *Comb. Proc. Intl. Plant Prop. Soc.* 37:48–52.

111. Mortensen, L.M. 1987. Review: CO_2 enrichment in greenhouses—crop responses. *Scientia Hort.* 33:1–25.

112. Munro, D. 1989. Virus testing of perennial propagating stock. *Comb. Proc. Intl. Plant Prop. Soc.* 39:43–47.

113. O'Donnell, K. 1988. Polycarbonates gain as growers seek high performance coverings. *GrowerTalks* 51(12):60–62.

114. Ogdon, R.J., F.A. Pokorny, and M.G. Dunavent. 1987. Elemental status of pine bark-based potting media. In *Horticultural reviews,* Vol. 9, J. Janick, ed. Westport, Conn.: AVI Publ. Co., pp. 103–31.

115. Ormrod, D.J. 1975. Fungicides and their spectra. *Comb. Proc. Intl. Plant Prop. Soc.* 25:112–15.

116. Orndorff, C. 1983. Constructing and maintaining disease-free propagation structures. *Comb. Proc. Intl. Plant Prop. Soc.* 32:599–605.

117. Parrella M.P. 1992. An overview of integrated pest management for plant propagation. *Comb. Proc. Intl. Plant Prop. Soc.* 42:242–245.

118. Platt, G.C. 1984. Use of *Pinus radiata* bark: A four year experience. *Comb. Proc. Intl. Plant Prop. Soc.* 33:320–23.

119. Porter, M.A., and B. Grodzinski. 1985. CO_2 enrichment of protected crops. In *Horticultural*

reviews, Vol. 7, J. Janick, ed. Westport, Conn.: AVI Publ. Co.

120. Pscheidt, J.W. 1991. Diagnosis of *Phytophthora* using ELISA test kits. *Comb. Proc. Intl. Plant Prop. Soc.* 41:251–254.
121. Pyle, K. 1991. Fungus control in the post-benlate era. 1991. *GrowerTalks* 55(7):19–21.
122. Reardon, J. 1994. Motivating plant growth with your heating system. *Comb. Proc. Intl. Plant Prop. Soc.* 44:364–366.
123. Richards, M. 1978. Capillary watering of container plants. *Comb. Proc. Intl. Plant Prop. Soc.* 28:411–13.
124. Rutten, Th. 1980. Osmocote controlled release fertilizer. *Acta. Hort.* 99:187–88.
125. Sabalka, D. 1987. Propagation media for flats and for direct sticking: What works? *Comb. Proc. Intl. Plant Prop. Soc.* 36:409–13.
126. Sawhney, B.L. 1976. Leaf compost for container-grown plants. *HortScience* 11(1):34–35.
127. Self, R. 1977. Winter protection of nursery plants. *Comb. Proc. Intl. Plant Prop. Soc.* 77:303–7.
128. ———. 1978. Pine bark in potting mixes: Grades and age, disease and fertility problems. *Comb. Proc. Intl. Plant Prop. Soc.* 28:363–68.
129. Sherry, W.J. 1986. Greenhouse covering materials: Optical, thermal, and physical properties. *Grower-Talks* 49(12):53–58.
130. ———. 1988. Technology in the greenhouse: State of the industry, a survey report. *GrowerTalks* 51(9):48–57.
131. Shibata, A., T. Fujita, and S. Maeda. 1980. Nutricote coated fertilizers processed with polyolefin resins. *Acta Hort.* 99:179–86.
132. Short, T.H., and Bauerie, W.L. 1980. Greenhouse production with lower fuel costs. In *USDA 1980 yearbook of agriculture,* J. Hayes, ed. Washington, D.C.: U.S. Govt. Printing Office.
133. Skimina, C.A. 1980. Salt tolerance of ornamentals. *Comb. Proc. Intl. Plant Prop. Soc.* 30:113–18.
134. Smith, E.M., ed. 1977. *Proc. woody ornamentals winter storage symposium.* Columbus, Ohio: Ohio State Univ., Coop. Ext.
135. ———. 1980. How and when to fertilize container nursery stock. *Amer. Nurs.,* Jan. 15, pp. 365–68.
136. Soffer, H. and D.W. Burger. 1989. Plant propagation using an aero-hydroponics system. *HortScience* 24(1):154.
137. Spooner-Hart, R.N. 1988. Integrated pest management with reference to plant propagation. *Proc. Inter. Plant Prop. Soc.* 38:119–125.
138. Steavenson, H. 1976. Using water to provide cold weather protection and conserve energy. *Amer. Nurs.* 413(8):10, 65–67.
139. Stewart, N. 1986. Production of bark for composts. *Comb. Proc. Intl. Plant Prop. Soc.* 35:454–58.
140. St. John, T.V. and J.M. Evans. 1990. Mycorrhizal inoculation of container plants. *Comb. Proc. Intl. Plant Prop. Soc.* 40:222–232.
141. Street, C. 1994. Propagation of wetland species. *Comb. Proc. Intl. Plant Prop. Soc.* 44:468–473.
142. Studer, E.J., P.L. Steponkus, G.L. Good, and S.C. Wiest. 1978. Root hardiness of container-grown ornamentals. *HortScience* 11(1):34–35.
143. Struve, D.K. 1990. Container production of hard-to-find or hard-to-transplant species. *Comb. Proc. Intl. Plant Prop. Soc.* 40:608–612.
144. Struve, D.K., M.A. Arnold, R. Beeson, Jr., J.M. Ruter, S. Svenson, and W.T. Witte. 1994. The copper connection: the benefits of growing woody ornamentals in copper-treated containers. *Amer. Nurser.* 179(4):52–61.
145. Teviotdale, B.L., M.F. Wiley, and D.H. Harper. 1991. How disinfectants compare in preventing transmission of fire blight. *Calif. Agri.* 45(4):21–23.
146. Thimijan, R.W., and R.D. Heins. 1983. Photometric, radiometric, and quantum units of measure: a review of procedures for interconversion. *Hort-Science* 18:818–832.
147. U.S. Dept. of Agriculture. 1965. Hotbed and propagating frames. *USDA Misc. Publ.* 986.
148. U.S. Dept. of Energy. 1978. *Energy audit for growers: A self-inspection guide to reduce energy costs.* Alexandria, Va.: Soc. Amer. Florists.
149. Vance, B.F. 1975. Water quality and plant growth. *Comb. Proc. Intl. Plant Prop. Soc.* 25:136–41.
150. Van der Borg, H.H., ed. 1979. Symposium on computers in greenhouse climate control. *Acta Hort.* 106:1–209.
151. Vetanovetz, R.P., and J.F. Knauss. 1988. Water quality. *Greenhouse Manag.* 6(12):64–72.
152. Vollebregt, R. 1990. Analysis of greenhouse curtain systems for shading, cooling and heat retention. *Comb. Proc. Intl. Plant Prop. Soc.* 40:166–176.
153. Ward, J. 1980. An approach to the control of *Phytophthora cinnamomi. Comb. Proc. Intl. Plant Prop. Soc.* 30:230–37.
154. Ward, J., N.C. Bragg, and B.J. Chambers. 1987. Peat-based composts: Their properties defined and modified to your needs. *Comb. Proc. Intl. Plant Prop. Soc.* 36:288–92.
155. Ware, R., and 1. S. Frankhauser. 1988. What should we cover the greenhouse with? *Comb. Proc. Intl. Plant Prop. Soc.* 37:161–65.
156. Weatherspoon, D.M., and C.C. Harrell. 1980. Evaluation of drip irrigation for container production of woody landscape plants. *HortScience* 15(4): 488–89.
157. Weiler, T.C. 1977. Survival strategies of Northern Europe's greenhouse industry. *HortScience* 12(1):30–32.

158. Whitcomb, C.E. 1978. A self-contained solar-heated greenhouse. *HortScience* 13(1):30–32.

159. ———. 1979. Growing plants in poly bags. *Amer. Nurs.* 149(12):10–11, 97–98.

160. White, J.W. 1979. Energy efficient growing structures for controlled environment agriculture. In *Horticultural reviews,* Vol. 1, J. Janick, ed. Westport, Conn: AVI Publ. Co.

161. White, J.B., and R.A. Aldrich. 1980. *Greenhouse energy conservation.* University Park, Pa.: Pennsylvania State Univ.

162. Whitten, M. 1995. Watertight irrigation. *Nurser. Manag.* 10(6):45–47.

163. Whittle, J. 1987. Physical and chemical properties of peat. *Comb. Proc. Intl. Plant Prop. Soc.* 36:284–87.

164. Wildschut, H.C. 1984. Computer controls for greenhouse environments. *Comb. Proc. Intl. Plant Prop. Soc.* 33:72–78.

165. Williams, D.J. 1980. How slow-release fertilizers work. *Amer. Nurs.* 151(6):90–97.

166. Williams, T.J. 1978. *How to build and use greenhouses.* San Francisco: Ortho Books, Chevron Chemical Co.

167. Wilkerson, D.C. 1993. Improve your bottom line with innovative bench systems. *Grower Talks* 57(1):23–31.

168. Wilson, G.C.S., ed. 1980. Symposium on substrates in horticulture other than soils in situ. *Acta Hort.* 99:1–248.

169. Wood, J.S. 1985. Sun frame propagation. *Comb. Proc. Intl. Plant Prop. Soc.* 34:306–11.

170. Worrall, R. 1976. The use of sawdust in potting mixes. *Comb. Proc. Intl. Plant Prop. Soc.* 26:379–81.

171. ———. 1982. High temperature release characteristics of resin-coated slow release fertilizers. *Comb. Proc. Intl. Plant Prop. Soc.* 31:176–81.

172. Wright, R.D. 1986. The pour-through nutrient extraction procedure. *HortScience* 21:227–229

173. ———. 1987. *The Virginia Tech liquid fertilizer system for container-grown plants.* Inform. Ser. #86-5. College of Agriculture and Life Sciences, Virginia Tech University. Blacksburg, Va.

174. Wright, R.D., and A.X. Niemiera. 1987. Nutrition of container-grown woody nursery crops. In *Horticultural reviews,* Vol. 9, J. Janick, ed. Westport, Conn., AVI Publ. Co., pp. 75–101.

175. Yeager, T., D. Fare, C. Gilliam and A. Niemiera. 1994. The best ways to head off onerous environmental regulations. *Nurser. Manag.* 11(11):77–78.

176. Yeager, T.H. 1989. Developing an efficient fertilizer program for container plants. *Woody Ornamentalist* 14(11):1–3.

177. Youngsman, J.E. 1983. Ten points for infrared. *Florists' Review,* May.

SUPPLEMENTARY READING

Aldrich, R.A. and J.W. Bartok, Jr. 1990. *Greenhouse Engineering.* Northeast Regional Agricultural Engineering Service, Cooperative Extension. NRAES No. 33.

Baker, K.F. ed. 1957. The U.C. system for producing healthy container-grown plants. *Calif. Agr. Exp. Sta. Man.* 23.

Ball, V., ed. 1991. *The Ball red book* (15th ed.). West Chicago, IL.: Geo. J. Ball Publ. Co.

Bobmont, B.L. 1990. The Standard Pesticide User's Guide. ISBN 0-13 840802-5. Englewood Cliffs, N.J.: Prentice Hall.

Boodley, J.W. 1981. *The commercial greenhouse handbook.* New York: Van Nostrand Reinhold.

Bunt, A.C. 1988. *Media and mixes for container-grown plants.* (2nd Ed. of *Modern potting composts*). London: G. Allen & Unwin.

Cathey, H.M., and L.E. Campbell. 1980. Light and lighting systems for horticultural plants. *Hort. Rev.* 2:491–537.

Correll, P.G., and J.G. Pepper, eds. 1977. *Energy conservation in greenhouses.* Newark, Del.: Univ. of Delaware.

Davidson, H., R. Mecklenburg, and C. Peterson. 1994. *Nursery management: Administration and culture* (3rd ed.). Englewood Cliffs, N.J.: Prentice-Hall.

Greenhouse Manager, published monthly. Fort Worth, Tex.: Branch-Smith Publishing.

GrowerTalks, published monthly. West Chicago, Ill.: Geo. J. Ball, Inc.

Holcomb, E.J., ed. 1994. *Bedding plants IV* (4th ed.). University Park, Pa.: Pennsylvania Flower Growers.

Jones, J.B., B. Wolf, and H.A. Mills. 1991. *Plant Analysis Handbook.* Athens, Ga.: Micro-Macro Publishing, Inc.

Landis, T.D., R.W. Tinus, S.E. McDonald, and J.P. Barnett. 1989. *Seedling nutrition and irrigation.* The Container Tree Nursery Manual, Vol. 4. Agri. Handbk. 674. Washington, DC: For. Serv., USDA.

Landis, T.D., R.W. Tinus, S.E. McDonald, and J.P. Barnett. 1990. *The biological component: nursery pests and mycorrhizae.* The Container Tree Nursery Manual, Vol 5. Agri. Handbk. 674. Washington, DC: For. Serv., USDA.

LANDIS, T.D., R.W. TINUS, S.E. MCDONALD, AND J.P. BARNETT. 1990. *Containers and growing media.* The Container Tree Nursery Manual, Vol. 2. Agri. Handbk. 674. Washington, DC: For. Serv., USDA.

LANDIS, T.D., R.W. TINUS, S.E. MCDONALD, AND J.P. BARNETT. 1992. *Atmospheric Environment.* The Container Tree Nursery Manual, Vol. 3. Agri. Handbk. 674. Washington, DC: For. Serv., USDA.

LANDIS, T.D., R.W. TINUS, S.E. MCDONALD, AND J.P. BARNETT. 1995. *Nursery planning, development and management,* Vol. 1. Agri. Handbk. 674. Washington, DC: For. Serv., USDA.

LARSON, R.A., ed. 1992. *Introduction to floriculture.* (2nd ed.) New York: Academic Press.

MACDONALD, B. 1986. *Practical woody plant propagation for nursery growers,* Vol. 1. Portland, OR. Timber Press.

NAU, J. 1993. Ball culture guide: the encyclopedia of seed germination (2nd ed.). West Chicago, Ill.: Geo. J. Ball Publ. Co.

NELSON, P.V. 1991. *Greenhouse operation and management* (4th ed.). Englewood Cliffs, NJ: Prentice-Hall.

POWELL, C.C. AND R.K. LINDQUIST. 1992. *Ball pest & disease manual.* Chicago, Ill.: Geo. J. Ball Publ. Co.

PREECE, J.E., AND P.E. READ. 1993. *The biology of horticulture: an introductory textbook.* New York, NY: John Wiley & Sons, Inc.

Nursery Manager, published monthly. Fort Worth, Tex.: Branch-Smith Publishing.

REED, D.W., ed. 1996. *Water, media, and nutrition—a grower's guide.* Batavia, Ill.: Ball Publ. Co.

VERMA B.P., ed. 1983. *Greenhouse and nursery mechanization: A compilation of published papers.* St. Joseph, Mich.: Amer. Soc. Agr. Eng.

WATSON, G.A. AND D. NEELY, eds., 1994. *The landscape below ground: proceedings of an international workshop on tree root development in urban soils.* Savoy, IL.: Intern. Soc. Arboricul.

4

Principles and Practices of Seed Selection

Introduction. This chapter deals with the management of genetic variability in seedling populations and cultivars of plants. Seeds are the genetic storehouse of nature. The genetic diversity found in seeds provides the genes from which plant life covers most of the land surface in all of its environmental variations. Seed selection has enabled humans to domesticate specific kinds of plants of particular value for food, fiber, and medicine. These domesticated plants differed in important respects from their wild relatives, often growing nearby, and produced a kind of symbiotic relationship between people and plants (*33*).

During much of human existence, these special kinds of crops, sometimes referred to as **landraces,** were maintained by farmers, keeping a portion of each year's seed to produce the crops for the following year. These landraces retained local names and represent some of our most important agricultural crops coming from specific areas of diversity: southwest Asia (wheat, barley, oats, rye), Africa (rice, sorghum, watermelon), Asia (rice, millet, soybean, many vegetables), Americas (corn, squash, beans, pepper, potato, sunflower, cotton, tobacco) (*33, 64, 66*).

Over time, these groups of plants were subjected to selection and developed into specific landrace cultivars (*66*). For many years, generations of farmers and housewives in Europe, the U.S., and elsewhere maintained gardens to provide vegetables and sometimes flowers for themselves or for sale at local markets. Today these are referred to as "heirloom varieties" and efforts exist to maintain and distribute them (*4, 73*).

Modern agriculture and horticulture are more complex. Seed production, maintenance of seed stocks, and the distribution of seeds combine many large and specialized businesses and professions (*13, 28*). Plant breeders use principles and practices that have emerged over the past century of genetic research to breed new seedling cultivars (*9, 36, 57, 66*). The seed industry produces millions of kilograms (pounds) of seeds in bulk, which are stored for varying periods, transported, and provided to seed propagators worldwide (Chapter 6).

Maintenance of seed cultivars requires that genetic variability be controlled during production or its value is likely to be lost. Characteristics selected as important in agriculture, horticulture, and forestry may not be consistently perpetuated into the next seedling generation unless appropriate principles and procedures are followed. Genetically pure seed has the following attributes:

(a) trueness-to-name,
(b) trueness-to-type,
(c) freedom from pathogens, mixtures of other crop seeds, and weeds

USES OF SEEDS IN PROPAGATION

Mass-propagation by seeds is an efficient and economic method of plant production. Most agricultural crops (cereals, forages, grasses, fiber, and oil seeds) are reproduced by seed as are many vegetables, garden, and florist plants (*35*). The bedding plant industry depends upon large volumes of seeds for the plug production of flower and vegetable plants to be sold to landscapers, home gardeners, and vegetable growers (*6, 42, 51, 69, 74*).

Although many woody plant cultivars used in horticulture are propagated vegetatively, seedlings play an important role. Rootstocks for grafting fruit and nut tree cultivars include many seedlings (*60, 72*). Ornamental shrub and tree seedlings are used in landscaping (*18, 50, 53*). During the past several centuries, interest in controlled reforestation has developed with the need to replace vast areas depleted of natural forests. Billions of seeds must be produced yearly to fill this need (*62, 63, 77*). A revolution in forestry propagation is taking place in some forest species involving a combination of seedling selection and clonal propagation (*1*). A relatively recent area of propagation has developed through the interest in restoring natural areas, wetlands, and wildlife habitats. These efforts are a major component of reproducing the natural range of species (*54, 55, 76*).

SEED SELECTION IN HERBACEOUS PLANT SPECIES

Breeding Systems Controlling Genetic Variability

Genetic variability of seed-propagated populations can be **heterogeneous** (different from each other) or **homogeneous** (similar to each other) but individual genotypes are **homozygous** or **heterozygous** (see Chapter 2 for definitions) (*66*). These characteristics are determined by the **breeding system** and the management under which seed populations are grown. Three breeding systems are **self-pollination (inbreeding), cross-pollination (outbreeding),** and **apomixis** (*24*).

Self-Pollination

Self-pollination and self-fertilization occurs when the pollen germinates on the stigma and the pollen tube grows down the style to effect fertilization of the same flower or a flower of the same plant (or of the same clone). Self-pollination is a natural condition in some species because of flower structure. For example, the reproductive parts of the bean flower (*9*) are completely enclosed within the petals. (Figure 4–1). Self-pollination may vary among species from those that are highly self-pollinated, i.e., less than 4 percent cross-pollinated (barley, oats, wheat, rice, peanut, soybean, lespedeza, field pea, garden bean, cowpea, flax, and some grasses) to those that are somewhat more that 4 percent cross-pollinated (upland and Egyptian cotton, pepper, and tomato).

Homozygosity in a self-pollinated cultivar is "fixed" by consecutive generations of self-fertilizations (*23, 57, 58, 66*) (Table 4–1). If one assumes a more or less homogeneous population with individuals possessing homozygous traits, the population and the individuals will remain homogeneous and homozygous. If a mutation occurs in one of the alleles (see page 12) and is recessive, the genotype for the trait becomes heterozygous. The next generation will then produce homozygous and heterozygous pairs of the mutant allele (see Chapter 2). In consecutive generations, the proportion of homozygous individuals with the two traits will increase, while the proportion with heterozygous genotype will decrease by a factor of one-half each generation. The group of descendants of the original parent will segregate into a heterogeneous mixture of more or less true-breeding **lines.** To produce a "true-breeding" homogeneous and homozygous cultivar, plant breeders will start with a single plant and then eliminate the off-type plants each generation for a period of six to ten generations.

Cross-Pollination

In nature, many, if not most, species are naturally **cross-pollinated,** a trait which seems to be desirable. Not only does the increased heterozygosity

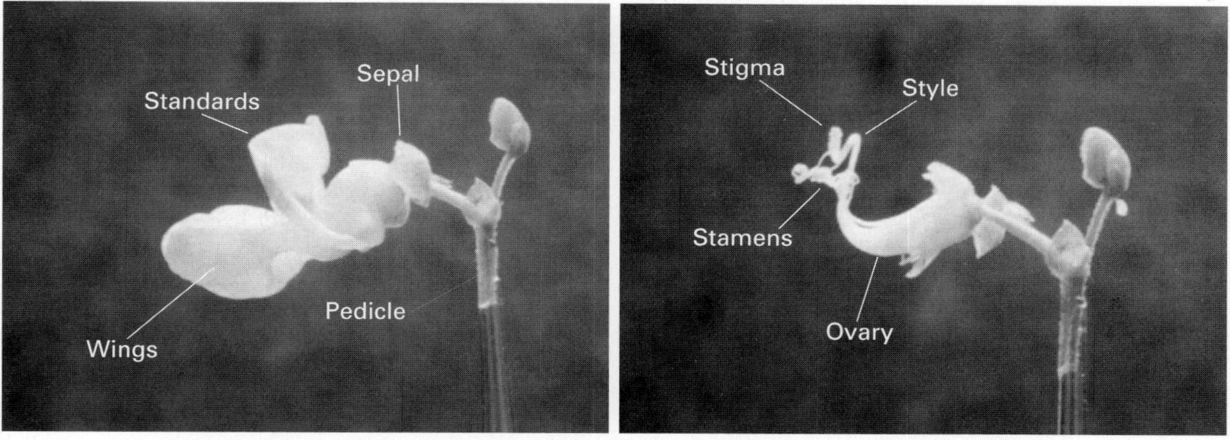

FIGURE 4–1 Self-pollination is enforced when the petals of the flower completely enclose the reproductive structures. *Left.* Flower of common bean *(Phaseolus)* with petal intact enclosing the reproductive parts. *Right.* Petals removed to show the reproductive structures. (Reproduced by permission from Bliss [10].)

provide the opportunity for evolutionary adaptation within the population if environments change but vigor tends to be enhanced. Enforced self-pollination of naturally cross-pollinated plants through consecutive generations may result in homozygous plants and a homogeneous population, but reduced vigor, size, and productivity **(inbreeding depression)** may develop. If, however, two inbred lines are crossed, the vigor of the plants of the resulting population may not only be restored but may show more size and vigor than either parent, a phenomenon known as **hybrid vigor**. In this case, the individual plants may be heterozygous but the population is likely to be homogeneous and have uniform characteristics.

Flowers are described as **perfect** when functional male and female parts are present in the same flower. Some species have evolved specific mechanisms to prevent self-pollination. These mechanisms include the following types:

TABLE 4–1

Effect of self-pollination and roguing following crossing of a tall (DD) pea and dwarf (dd) pea (Figure 2–6). "Fixing" of the two parental phenotypes can be observed in succeeding generations in the proportion of tall and dwarf plants. Continuous roguing for the recessive trait never quite eliminates its segregation from residual heterozygous individuals.

	A. Continuing self-pollination				*B. Roguing of all dwarfed plants*		
	Proportions			*Percent*			
	DD	*Dd*	*dd*	*Homozygous*	*Tall*	*Dwarf*	*%dd*
P1	1		1	100			
F1		1		0	all		
F2	1	2	1	50	3	1	25
F3	3	2	3	75	14	1	7.1
F4	7	2	7	87.5	35	1	2.8
F5	15	2	15	93.75	143	1	0.7
F6	31	2	31	96.88	535	1	0.2
F7	126	2	126	98.44	2143	1	0.05

Pollen sterility (*45*). In a number of crop species, e.g., corn and onions, specific genes have been identified that prevent normal formation of the male (pollen) reproductive structures. These traits may be bred into parental lines of specific cultivars to produce hybrid seed.

Incompatibility (*16*). Some plants (lily, cabbage, petunia) contain incompatibility alleles which prevent the pollen tube from growing down the style of a plant of the same genotype even though the pollen is viable. Several types of self-incompatibility exist and depend upon the individual species (see Figure 4–2).

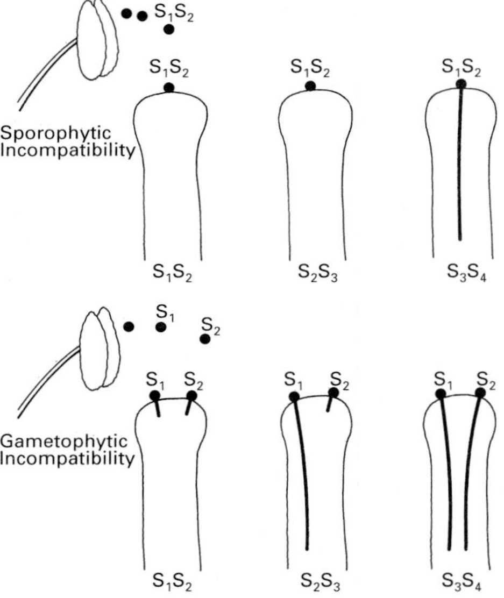

FIGURE 4–2 **Incompatibility** mechanisms prevent self-pollination in some species. *Top (cabbage):* Sporophytic incompatibility. Each pollen contains genes of both S_1 and S_2 alleles, and pollen tube will only grow down a style with a different genotype. *Bottom (clover):* Gametophytic incompatibility. Each pollen grain has a single S allele. Pollen tube will not grow down a style where that allele is represented. (Redrawn with permission from Stoskopf, et al. *Plant Breeding Theory and Practice.* Westview Press: Boulder, CO.)

Dioecy. Pistillate (female) and staminate (male) flowers are present in separate plants (buffalo grass, asparagus) (Figure 4–3).

Moneocy. Pistillate (female) and staminate (male) flowers are present on the same plant (corn, cucurbits) (Figure 4–4).

Dichogamy. Plants shed their pollen at a different time than when the pistil is receptive.

Cross-pollination of most cultivars is carried out by either wind or insects but examples exist of pollination by bats, birds, and water (*58*). **Insect pollination** is the rule for plants with white, or brightly colored, fragrant, and otherwise conspicuous flowers that attract insects. The honeybee is one of the most important pollinating insects, although wild bees, butterflies, moths, and flies also obtain pollen and nectar from the flower. Generally, pollen is heavy, sticky, and adheres to the body of the insect. Some important seed crops in this category are the following: alfalfa, birdsfoot trefoil, red clover, white clover, sweet clover, millet, onion, and watermelon. In addition, many flower and vegetable crops are insect pollinated as are many fruit plants, ornamental plants and deciduous and broad-leaved evergreens used in the landscape.

Wind pollination is the rule for many plants having inconspicuous flowers or those with monoecious, dioecious, or dichogamous flowers. Examples are grasses, corn, conifers, olive, and catkin-bearing trees such as the walnut, oak, alder, and cottonwood. The pollen produced from such plants is generally light and dry and in some cases is carried long distances in wind currents.

Apomixis (17, 32)

Apomixis occurs when an embryo is produced from a single cell of the sporophyte and does not develop from fertilization of two gametes. This new "vegetative" embryo may arise by several mechanisms which were described in Chapter 2 (pp. 31). In each case, the effect is that seed production becomes asexual such that reproduction of a clone by seed can result. In some species, both apomictic and sexual seeds are produced, sometimes within the same ovule **(facultative);** bluegrass (*Poa pratensis* L.), for instance, falls into this category. Other species are essentially 100 percent apomictic **(obligate),** for example, bahia grass (*Paspalum notatum*) and buffelgrass (*Pennisetum ciliare*). Apomixis has been reported to occur in 35 families and 300 species (*30, 57*).

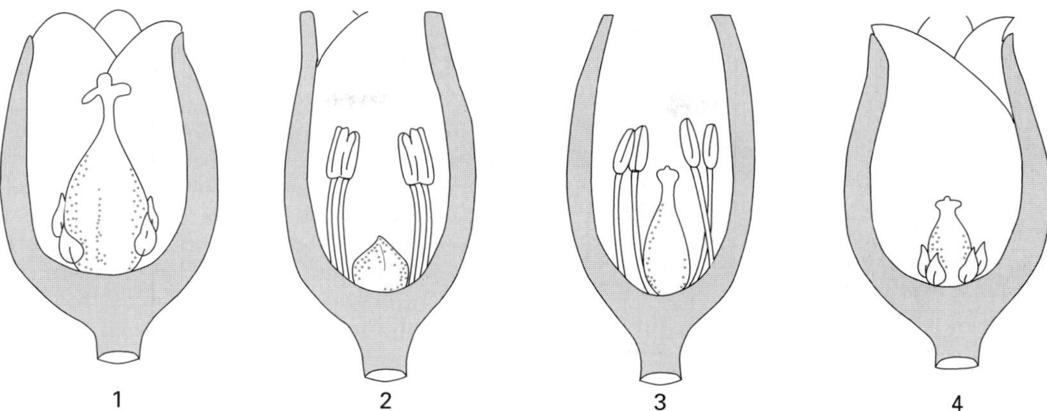

FIGURE 4–3 Range of flower structure types expressed in different asparagus flowers of individual plants.
Type 1. Completely female. Dioecious. Flowers contain only the pistil; stamens (male) are reduced and nonfunctioning.
Type 2. Completely male. Dioecious. Flowers only contain stamens. The pistil is reduced and nonfunctioning.
Type 3. Both male and female structures are functioning. Perfect.
Type 4. Both male and female structures are nonfunctioning. Sterile.
Commercial seed production of asparagus results from growing Type 1 and Type 2 plants together to enforce cross-pollination and produce the desirable hybrid plants. (Courtesy Bryan Benson.)

Breeding of apomictic cultivars requires that a genetic source for apomictic reproduction be found within that species. This trait is not identifiable by visual inspection of the parent plant but only by genetic performance, i.e., the unexpected uniformity of its progeny from normally variable populations. Apomixis has been most important in the breeding of grasses, forage crops, and sorghum. A number of cultivars have been introduced such as 'King Ranch' bluestem, 'Argentine' bahia grass and 'Tucson' sideoats grama (*9*), 'Bonnyblue' and 'Adelphi' Kentucky bluegrass (*25, 26*) and buffelgrass (*32*). Apomixis is apparently controlled by a relatively few genes and breeding systems have been described to incorporate this trait into particular species (*17*).

FIGURE 4–4 **Monoecious** plants produce male and female flowers at different locations of the same plant to favor cross-pollination. *Left.* A cucurbit shows the enlarged ovary, which will become the fruit. *Right.* The male flower does not have the ovary and is shed after flowering. (Courtesy V. Rubatsky.)

In apomixis the seedling population is immediately fixed as a "true-breeding" line without seedling variation. Such plants exhibit the apomictic cycle and express typical juvenile traits of the seedling population. Consequently, apomixis is particularly appropriate for plants whose value lies in their vegetative characteristics, as occurs in forages and grasses, rather than in plants whose value depends upon fruiting characteristics.

Synthetic seed. This term applies to seeds that have been produced from embryos developed from somatic cells in tissue culture and cell suspension systems. The regeneration process, which is known as **somatic embryogenesis,** is described in Chapters 17 and 18. The phenomenon is essentially the same as apomixis and is the basis of clonal somatic seed production.

Control of Genetic Variability During Seed Production

Isolation

Isolation is necessary (a) to prevent mechanical mixing of the seed during harvest, and (b) to prevent contamination by unwanted cross-pollination with a different but related cultivar. Isolation is achieved primarily through distance, but it can also be attained by enclosing plants or groups of plants in cages, enclosing individual flowers, or removing male flower parts and then employing artificial pollination.

Self-pollinated cultivars need to be separated only to prevent mechanical mixing of seed of different cultivars during harvest. The minimum distance usually specified between plots is 3 m (10 ft). Careful cleaning of the harvesting equipment is required when a change is made from one cultivar to another. Sacks and other containers used to hold the seed must be cleaned carefully to remove any seed that has remained from previous lots.

More isolation is needed to separate cultivars cross-pollinated by wind or insects. The minimum distance depends upon a number of factors, such as the degree of natural cross-pollination, the relative number of pollen-shedding plants, the number of insects present, and/or the direction of prevailing winds. The minimum distance for insect-pollinated plants is 0.4 km (¼ mi) to 1.6 km (1 mi). The distance for wind-pollinated plants is 0.2 km (⅛ mi) to 3.2 km (2 mi), depending on species.

Effective cross-pollination usually can take place between cultivars of the same species; it may also occur between cultivars of a different species but in the same genus; rarely will it occur between cultivars belonging to another genus. Since the horticultural classification may not indicate taxonomic relationships, seed producers should be familiar with the botanical relationships among the cultivars they grow.

Roguing

The removal of **off-type** plants, plants of other cultivars, and weeds in the seed production field is known as **roguing** (*44*). During the development of a seed-propagated cultivar, positive selection is practiced to retain a small portion of desirable plants and to maximize the frequency of desirable alleles in the population. During seed production, roguing following visual inspection exerts selection by eliminating the relatively small population that is not "true to type," thus keeping the cultivar genetically pure.

Off-type characteristics, i.e., those that do not conform to the cultivar description, arise because recessive genes may be present in a heterozygous condition even in homozygous cultivars. Recessive genes arising by mutation may not be immediately observed in the plant in which they occur. The plant becomes heterozygous for that gene, and in a later generation the gene segregates and the character appears in the offspring. Some cultivars have mutable genes that continuously produce specific off-type individuals (*56*). Off-type individual plants should be rogued out of the seed production fields before pollination occurs. Regular inspection of the seed-producing fields by trained personnel is required.

Other sources of off-type plants include contamination by unwanted pollen due to inadequate isolation, volunteer plants arising from accidentally planted seed or from seed produced by earlier crops. Seed production fields of a particular cultivar should not have grown a potentially contaminating cultivar for a number of preceding years.

Weeds are plant species that have been associated with agriculture as a consequence of their ability to exploit disturbed land areas when cultivation occurs (*33*). Some weed species have evolved seed types that closely resemble crop seeds and are difficult to screen out during seed production.

Seedling Progeny Tests

Planting representative seeds in a test plot or garden may be necessary to test for trueness to type. This procedure is used in the development of a cultivar to test its adaptability to various environments. The same device may become necessary to test whether changes in the frequency of particular genes or gene combinations may occur during seed increase generations. These changes can result from selection pressure exerted by management practices or environmental interaction. For example, intensive roguing may result in a genetic shift due to changes in the frequency of particular genes or gene combinations (*23, 44*). Shifts may also occur due to environmental exposure in the growing area which is different from the initial selection area. Seedlings of particular genotypes may survive better than others and contribute more to the next generation. If sufficiently extensive, this so-called **genetic drift** could result in populations of progeny plants that differ somewhat from those of the same cultivar grown by other producers or from the original breeder's seed.

Problems can result if seed crops of particular perennial cultivars are grown in one environment (such as a mild winter area) to produce seed to be used in a different and more severe environment (such as an area requiring cold-hardiness). This situation has occurred, for example, with alfalfa (*27*) where rules for production of forage crop seed in a mild winter area can only specify one seedling generation of increase.

Categories of Seed-Propagated Cultivars

Landraces

Historically, farmers throughout the world have maintained seed propagated plants by saving selected portions of the crop to be used to produce the next cycle. These populations, called **landraces,** evolved along with human societies and still are found throughout the world in some areas (*33*). These populations are variable but identifiable and have local names. This practice has resulted in genetic populations adapted to a localized environment which has preserved a great deal of genetic diversity. Their variability provides a buffer against environmental catastrophe.

Changes in cropping patterns have occurred during the twentieth century, particularly since about 1960. Many of the older crops' populations are being replaced by modern cultivars, which tend to be uniform and high yielding (in conjunction with high irrigation and fertility inputs). Sometimes, however, they lack adaptation to local environments. The result has been an increase in the world supply of essential food crops, but concerns have been raised that a parallel loss of genetic diversity and germ plasm has occurred. Exploration and conservation efforts have expanded to maintain these important raw materials for future use (*21*).

Heirloom Varieties

During early American and European history, many vegetable and flower varieties were maintained by families in their "kitchen gardens." These also were maintained by generations of gardeners and local farmers. The preservation and distribution of this material has been an objective of the Seed Savers Exchange, Inc., Decorah, Iowa (*4, 73*).

Lines

This term is used to designate a population of seedling plants whose genetic composition is maintained relatively intact during consecutive generations. It is one of the major categories of cultivars recognized by the International Code of Nomenclature for Cultivated Crops. The key characteristic is that genetic variability is controlled and uniformity is maintained to a standard appropriate for that cultivar. Three types of lines are recognized: (a) self-pollinated lines, (b) cross-pollinated lines and, (c) hybrid lines.

Self-pollinated cultivars are primarily **lines.** A **pure line** refers to a population in which all individuals have descended from a single homozygous plant. Sometimes a number of pure lines which are similar in appearance but actually include differing traits of importance, such as disease resistance, are combined. In herbaceous crops these are called **multilines.**

Cross-pollinated cultivars (lines) generally require a degree of heterozygosity to maintain good vigor. Inforced self-pollination with species in this group tends to result in reduced vigor. **Inbred lines** are produced mainly for later production of F_1 hybrids. Other categories include **mixtures,** which combine separate lines similar in specific traits, e.g., flower color, and **synthetic cultivars** (*66*) which combine genetically distinct but phenotypically similar seedlings in one population. To maintain the cultivar, parental lines are maintained separately as lines or clones. Some flower crops, such as petunia, pansy, and snapdragon, are made up of F_2 populations from such mixtures.

Hybrid lines are the F_1 hybrid seedling populations of two or more inbred lines. Inbred lines of cross-pollinated plants tend to lose vigor and show inbreeding depression. When crossed with another inbred line, the result is a population of uniform, heterozygous plants. Often these populations exhibit greater vigor than the parents **(hybrid vigor),** depending upon the combining ability of the parents. Hybridization is a means of "fixing" the genotype populations similar to that described for self-pollinated lines. This concept was first applied to corn (*38, 65*) but has since been applied to many other agronomic, vegetable, and flower crops (*11, 39, 42, 66*).

Hybrids may be produced between two inbred lines **(single-cross),** two single-crosses **(double-cross),** an inbred line and an open pollinated cultivar **(top-cross),** or between a single-cross and an inbred line **(three-way cross)** (*65*). Seeds from the hybrid population are normally not used for propagation because of wide range of variability in size, vigor, and other characteristics that may appear in subsequent generations. Hybrid lines may be used by seed companies as a method of maintaining control of the cultivar.

Seed Production of True-to-Type Herbaceous Cultivars

Commercial Crop Seed

Traditional methods of seed collection utilize a portion of the seed from a crop to use for planting a crop for the next year. This system would be satisfactory for self-pollinized cultivars that are easy to maintain genetically. For cross-pollinated cultivars, a knowledge of the requirements of individual crops is needed and specific conditions practiced depending upon the plant (*4*).

Pedigreed Stock System (5, 66)

Commercial seed production of most self-pollinated and cross-pollinated lines starts with a small population of seeds *(breeder's seed)* which is maintained under stringent conditions of isolation and maintenance as the primary reference material for the cultivar. In the case of a new cultivar, selection of the breeder's seed is the endpoint of the development process. Subsequent distribution to commercial growers is carried out through a **pedigree system** of production designed to maintain genetic purity by a limited sequence of seed generations produced under decreasing stringency of management standards.

The pedigree system utilizes three steps (Figure 4–5). The first is maintenance of a relatively small population of seed in a **foundation block** as the primary source of seed and grown under high standards of isolation, inspection, and roguing. To reduce the expense of production, the amount of seed is multiplied in a second generation **(increase block)** under somewhat lesser levels of management. The third generation is used to produce the **commercial seed** which provides the seed of commerce. A foundation planting originates only from breeder's seed or another foundation planting. An increase block originates only from foundation seed or from another increase block. A seed production block is planted only with seed of an increase block or a foundation block. This entire process will be carried out either by large commercial firms or groups of independent growers combined within a Crop Improvement Association which produces **certified seed** (see p. 120).

Hybrid Seed Production (66) (Figure 4–6)

Hybrid cultivars are the F_1 progeny of two of more parental lines. Parent plants are maintained either as inbred lines (corn, onion) or as vegetatively propagated clones (asparagus). The same standards of isolation as for nonhybrid seed production may be required.

Parental lines are lines are maintained under the control of the parent commercial company, which can then control the production and distribution of the seed. In order to mass-produce hybrids,

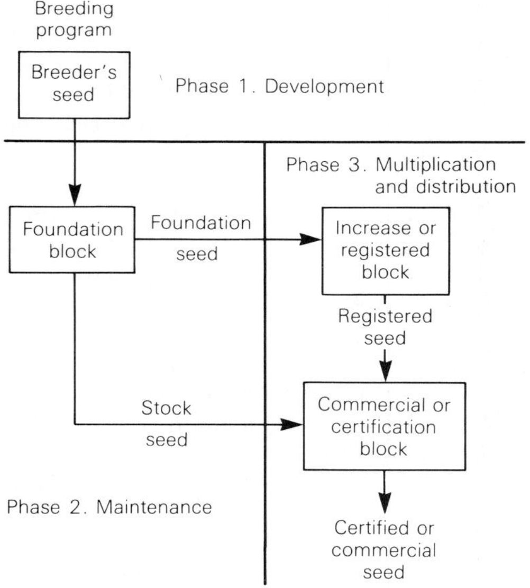

FIGURE 4–5 Pedigree system for seed production. See text for discussion.

> The first crop where hybrid seed production was used on a large scale was corn. Here, the natural breeding system was monoecy in which the female flower is the corn silk and the male flower is the tassel. Physical removal of the male flowers (detasseling) in the seed-producing inbred lines enforces cross-pollination from pollen being shed on plants of a different inbred line (*65*). Onion was the second crop where mass production of hybrid seed was accomplished. A male sterile gene was introduced into special breeding lines to prevent the production of pollen (*39*). Since then, breeders have utilized various genetic systems to produce hybrid crops, including self-incompatibility (cabbage, broccoli, brussels sprouts, *Ageratum, Bellis,* and some grasses), dioecy (spinach, asparagus, some grasses) and monoecy (cucurbits). See p. 108.

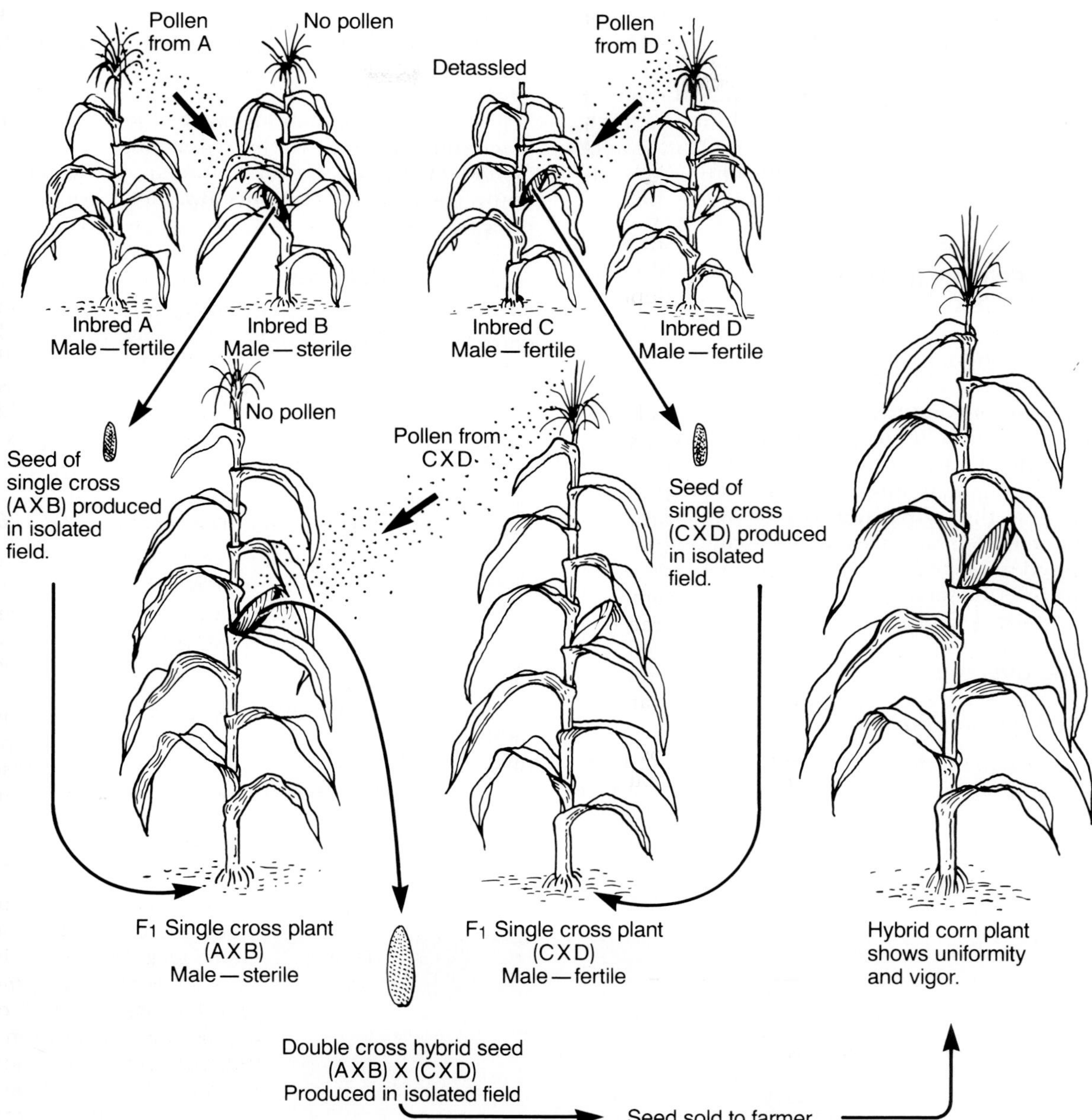

FIGURE 4–6 Hybrid seed corn production. Four inbred lines are produced to be used as parents for cross-pollination utilizing either detasseling (removal of male flower) or a pollen-sterile parent. The resulting F_1 plants are used as parents of the next (F_2) generation which are then sold to be used for commercial crop production. The individual progeny plants are highly heterozygous but the population is highly homogeneous, showing high vigor and production. (Redrawn from *USDA Yearbook of Agriculture 1937,* Washington, D.C.: U.S. Government Printing Office.)

some system must be used to prevent self-pollination and to enforce cross-pollination (*29*).

Hand pollination is sometimes practiced to produce seed in crops or situations in which the production of seed per flower is very high and/or the high value of the seed justifies the expense. Hand pollination is used to produce some hybrid flower seeds and in breeding new cultivars (*29*) (Figure 4–7).

SEED SELECTION IN WOODY PLANT SPECIES

Seed propagation is an important nursery procedure to grow many tree and shrub species for the landscape, to provide rootstocks for fruit and shade tree cultivars, and to mass-propagate forest trees. Since woody plants tend to be cross-pollinated and heterozygous, maintaining genetic purity provides challenges to the propagator. Woody plant seed may be obtained from various sources including professional seed collectors (*46*), commercial seed companies (*18*), nursery personnel, and arboreta and botanic gardens (*50*).

Breeding Systems

Self-Pollination

Self-pollination is not regularly found in woody plant species in nature but may occur if a single plant specimen is growing in the landscape, in an arboretum, or in a clonal planting. Self-pollination is likely to result in poorly developed seed, variability in the progeny from heterozygosity, and weakness in the seedlings from inbreeding depression (*20, 67, 71, 75*). Biological systems that enforce cross-pollination (see p. 106) are common, including **dioecy** (holly, pistachio, date palm), **monoecy** (walnut, pecan, conifers), **dichogamy** (walnut, pecan), and **self-incompatibility** (many fruit trees).

Cross-Pollination

Most trees and shrub species are both heterozygous and cross-pollinated, such that considerable potential for genetic variability exists among the seedling progeny. Selection of seed source plants must take into account not only the characteristics of the plant itself but the potential for cross-pollination with other species in the surrounding population. For example, presence of off-type individuals from imported seed of eucalyptus from Australia (*67*), and pear species from the Orient (*34*) could be traced to hybridization with other species in the local area.

Wind pollination is characteristic of many woody plant species with inconspicuous flowers, including such important groups as conifers (pines, spruce, etc.), olive, and catkin-bearing families (oak, walnut, alders, etc.) (*62*) (Figure 4–8). Insect pollination is the rule primarily with families that have large conspicuous flowers, including most fruits (apple, pear, peach, etc.) and some nut (almond) species (*72*).

FIGURE 4–7 F_1 hybrid cultivars of many flowering annuals have been created. Seed is produced following large scale hand pollination in a controlled environment. This system is used with some flower crops. The flowering plants shown are geranium. Seed yield per capsule is high and the seed is highly valuable. (Courtesy Goldschmidt Seeds, Inc. San Martin, CA.)

FIGURE 4–8 Some nut producing tree species have pollination systems to ensure cross-pollination. Pistachio *(Pistacio vera)* has male staminate flowers on one clone and female pistillate flowers on another. Dioecious male and female clones are propagated. *Left.* Pistillate flowers (female) of 'Red Aleppo'. *Right.* Staminate flowers (male) of 'Peters' cultivar. (Courtesy Louise Ferguson.)

Apomixis

Apomictic woody plant species and cultivars include many citrus (*12*), mangos, and some apple species (*61*). Although apomixis produces genetically uniform seedlings, it is not necessarily useful for reproducing specific fruit cultivars because of strong undesirable juvenile tendencies, such as, thorniness, excess vigor, and delayed fruiting. On the other hand, these characteristics make apomictic seedlings useful as rootstocks, characteristics exploited extensively in citrus (see p. 639).

Controlling Genetic Variation

Local Seed (*62*)

"Local seed" means seed from a natural area subjected to a restricted range of climatic and soil influences. This usually means that the collection site should be within 160 km (100 mi) of the planting site and within 305 m (1,000 ft) of its elevation. In the absence of these requirements, seed could be used from a region having as nearly as possible the same climatic characteristics. The reason for the early historical emphasis on local seed is the phenomenon known as **seed origin** or **provenance** (*31*). Natural plant populations growing within a given geographical area over a long period of time tend to evolve so that they become adapted to the environmental conditions at that site (Figure 4–9). This phenomenon is described further in the following sections of this chapter.

Seeds of a given species collected in one locality may produce plants that are completely inappropriate to another locality. For example, seeds collected from trees in warm climates or at low altitudes are likely to produce seedlings that will not stop growing sufficiently early in the fall to

FIGURE 4–9 Variation in progeny performance among different seed sources of Douglas-Fir after two years in the nursery. *Left:* Two fast-growing green strains from Washington and British Columbia. *Center:* Four slow-growing strains from Montana and Wyoming, mostly gray-green in color. *Right:* Fast-growing strain from Arizona, deep blue in color. Seedlings from separate sources varied in height from 5 to 10 cm (2 to 4 in.) up to 25 to 43 cm (10 to 16 in.) (Courtesy C.E. Heit.)

escape freezing when grown in colder regions. The reverse situation—collecting seed from colder areas for growth in warmer regions—might be more satisfactory but it also could result in a net reduction in growth resulting from the inability of the trees to utilize the growing season fully because of differences in response to photoperiod (*62*). The use of local seed is particularly important in the effort to restore any native ecosystem where the use of exotic species would be inappropriate (*54, 55*).

Pure Stands

This term refers to a group of phenotypically similar seedling plants of the same kind. This concept could apply both to plants growing in a natural environment or in a planting such as a wood lot. These populations are useful in seed collection because cross-pollination would likely occur from among this group of plants and one can judge both the female and the male parents. Although the individuals are likely to be heterozygous, they should produce good seeds and vigorous seedlings. The population should be homogeneous and reproduce the parental characteristics.

Phenotypic Tree Selection

This phrase refers to selection of a seed source *through visual inspection of the source plant*(s). The basis of this procedure is that many important traits such as stem form, branching habit, growth rate, resistance to diseases and insects, presence of surface defects, and other qualities are inherited quantitatively. Geneticists refer to this relationship as high **phenotypic correlation** between parents and offspring. In practical terms, this means that the parental performance can be a good indicator of the performance of the offspring.

When individual trees in native stands show a superior phenotype, foresters call them "plus" or "elite" trees and sometimes leave them for natural reseeding or as seed sources. Such dominant seed trees may contribute the bulk of natural reseeding in a given area (*59*).

Genotypic Selection

This term refers to selection of a seed source *based upon the performance of a* **seedling progeny test.** Seeds may be produced by open pollination (OP) where only one parent is known. However if the test is made from a controlled cross, where both parents are known, their contributions can be evaluated. A progeny test establishes the "breeding value" of a particular seed source (*7*) because genetic potential is based upon actual performance of the progeny. A representative sample of seeds is collected, planted under test conditions, and the progeny observed over a period of time. A high correlation between the average phenotypic traits of the parent(s) and the average phenotypic response of the offspring is referred to as "high additive heritability" and justifies using the "best" trees for seed sources of the next generation (*49, 59*). A low statistical correlation between parent and progeny characteristics is referred to as "low additive heritability" in that the desired traits of the progeny cannot be predicted from inspection of the parents.

Progeny testing is useful in verifying the suitability of individual seed sources for future seed collecting. The procedure is an important component to improvement of woody plants whether in forestry or horticulture.

Nursery Row Selection

Sometimes phenotypically unique individuals appear in seed populations planted in the nursery row and can be identified visually. Identification requires that the character be distinctive in vigor, appearance, or both (Figure 4–10). For example, Paradox hybrid walnut seedlings (*Juglans regia x J. hindsii*) used as a rootstock in California are produced by planting seeds of specific seed tree sources of black walnut (*Juglans hindsii*). Leaf characteristics, bark color, and greater vigor of the hybrid seedlings allow visual separation from the black walnut seedlings which are rogued out of the nursery row leaving the desired hybrid seedlings (*68*). Previous experience indicates which walnut tree sources produce the highest levels of hybrids, presumably from natural crossing with surrounding Persian Walnut (*J. regia*). "Blue" seedlings of the Colorado spruce (*Picea pungens*) appear among seedling populations having the usual green form. Differences in fall coloring among seedlings of *Liquidambar* and *Pistacia chinensis* necessitate fall selection of individual trees for landscaping or vegetative propagation as clones.

Apomixis

Apomictic cultivars produce uniform seed progeny depending upon whether they are obligate or facultative. Many citrus seedlings, for instance, are facultative but the sexual seedlings tend to be weak and do not survive or may differ from the parent morphologically.

FIGURE 4–10 Nursery seedling selection of Paradox hybrid walnuts *(Juglans hindsii* x *J. regia).* Hybrid seedlings *(left)* appear spontaneously in the nursery when particular seed sources of black walnut *(J. hindsii) (middle)* are planted. These valuable plants are identified by leaf morphology, bark color and vigor in the nursery. *(Above) Juglans regia. (middle)* Paradox. *(Below). J. hindsii*

CATEGORIES OF SEED-PROPAGATED WOODY PLANTS

Natural Wild Populations

In nature, most species can be recognized as a more or less phenotypically (and genotypically) uniform seedling population that has evolved over time through consecutive generations to be adapted to the environment at a particular site. If the species covers a wide area, local variation in environment can result in different populations adapted to that area even though the plants may be phenotypically similar. Morphologically recognizable differences may be designated as **botanical varieties.** On the other hand, subgroups of a particular species that are morphologically similar but specifically adapted to a particular environmental niche are known as **ecotypes.** Variations that occur continuously between locations are known as **clines** (*43, 53*).

The climatic and geographical locality where tree seed is produced is referred to as **seed origin** or **provenance** (*31, 43, 47, 62*). Variation can occur among groups of plants associated with latitude, longitude, and elevation. Differences may be shown by morphology, physiology, adaptation to climate and to soil, and in resistance to diseases and insects.

Distinct ecotypes have been identified by means of seedling progeny tests in many native forest tree species, including Douglas-Fir, ponderosa pine, lodgepole pine, eastern white pine, slash pine, loblolly pine, shortleaf pine, and white spruce (*62*). Other examples include the Baltic race of Scotch pine, the Hartz Mountain source of Norway spruce, the Sudeten (Germany) strain of European larch, the Burmese race of teak, Douglas-Fir from the Palmer area in Oregon, ponderosa pine from the Lolo Mountains in Montana, and white spruce from the Pembroke, Ontario (Canada) area (*18, 22, 37, 58*). Douglas-Fir has at least three recognized races, *viridis, caesia,* and *glauca,* with various geographical strains within them (*31*) that show different adaptations. For instance, progeny tests showed that a *viridis* strain from the U.S. West Coast was not winter-hardy in New York but was well suited to Western Europe. Those from Montana and Wyoming were very slow growing. Trees of the *glauca* (blue) strain from the Rocky Mountain region were winter-hardy but varied in growth rate and appearance. Strains collected further inland were winter-hardy and vigorous; similar differences occurred in Scotch pine *(Pinus sylvestris),* mugho pine *(P. mugho),* Norway spruce *(Picea abies),* and others.

Improved Seed Sources

Nursery propagation by woody plant seed can be upgraded significantly by the selection and development of improved seed sources. This applies to production of rootstocks for fruit and nut trees (*60, 72*), shade trees (*18, 50*) and trees in the landscape. Foresters (*78, 79*) have been engaged in recent years in "domesticating" and upgrading forest tree production over that of "local" seed (*47, 48, 78*).

Elite trees. Foresters refer to single seed-source trees with a superior genotype, as demonstrated by a progeny test, as **elite** trees (*7*). Nursery tests can identify and characterize specific seed sources for landscape and Christmas tree uses (*37*).

Clonal seed sources. Superior and elite seed source trees can be maintained as clones in seed orchards to preserve the genotype of the parent. Seeds from this source are then used to produce seedling trees in the nursery. This procedure is commonly used for rootstock seed production for fruit and nut cultivars. For example, 'Nemaguard' is a peach hybrid whose seedling progeny are nematode resistant and which is used for almond, peach, plum, and apricot trees in central California (*60*). Named cultivars of ornamental trees have been identified as producing uniform, superior seedling progeny (*18*).

Commercial plantings. Seeds may sometimes be collected successfully from commercial fruit or nut plantings providing that the specific cultivar or origin is known. Fruit tree seeds (apple, pear, peach) have traditionally been collected from canneries and dry yards where specific commercial cultivars are utilized. Local landscape trees may be used. In several forest-tree species, seed may be collected from above-average trees in commercial plantations. For example, in New Zealand, seed from such trees is designated "cs" ("climbing selects") and rated higher than the seed from the remainder of the trees but below that of seed orchards or from well-tested families.

Family selection. Genetic improvement of forest tree species has been brought about by family selection utilizing controlled crossing or selection from single trees (open-pollinated). Superior (elite) trees are identified to be source trees and either seedlings from them are planted together or the original parent trees are established by grafting into a seed orchard. A minimum number of individual genotypes are selected—usually around 25—to avoid the dangers of inbreeding and a limited genetic range. Seed is collected from individual trees, either after open pollination or following controlled crosses. Progeny of individual trees are planted in progeny test sites and evaluated for various forestry characteristics. Over time, inferior sources are identified and eliminated and superior sources are preserved to produce another generation of new families of improved seed genotypes.

Hybrid seed sources. F_1 hybrids of two species usually produce uniform populations of plants in the same manner as hybrid seeds of corn and other inbred lines (see p. 112). Hybridization has been valuable in producing vigorous almond x peach hybrids for almond and peach rootstocks (*40, 41*), Paradox hybrids (*Juglans hindsii* x *J. regia*) for walnuts (*68*). Forest tree hybrids, as *Pinus rigida* x *P. taeda* in Korea and *Larix decidua* x *L. leptolepsis* in Europe (*79*), are not necessarily uniform, however.

Because of expense and uncertainty of production, seeds of F_1 hybrids of the forest trees may be used to produce F_2 seedling populations within which the more vigorous plants dominate (*79*).

SEED SELECTION AND PRODUCTION SYSTEMS

Seed-Collection Zone

A **seed-collection zone** (*62*) is an area with defined boundaries and altitudinal limits in which soil and climate are sufficiently uniform to indicate high probability of reproducing a single ecotype. Seed-collection zones, designating particular climatic and geographical areas, have been established in most of the forest tree growing areas in the world (*2, 62*). California, for example, is divided into 6 physiographic and climatic regions, 32 subregions, and 85 seed-collection zones (Figure 4–11) (*63*). Similar zones are established in Washington and Oregon and in the central region of the United States.

Seed-Production Areas (50, 62)

Areas containing a group of trees can be identified and set aside specifically as a seed source. Seed trees within the area are selected for their desirable characteristics. The value of the area can be improved by removing off-type trees, those that do not meet desired standards, and other trees or shrub species that would interfere with the operations. Competing trees can be eliminated to provide adequate space for tree development and seed production. An isolation zone at least 120 m (400 ft) wide from which off-type trees are removed should be established around the area. Trees may or may not have been progeny-tested prior to establishment of the area.

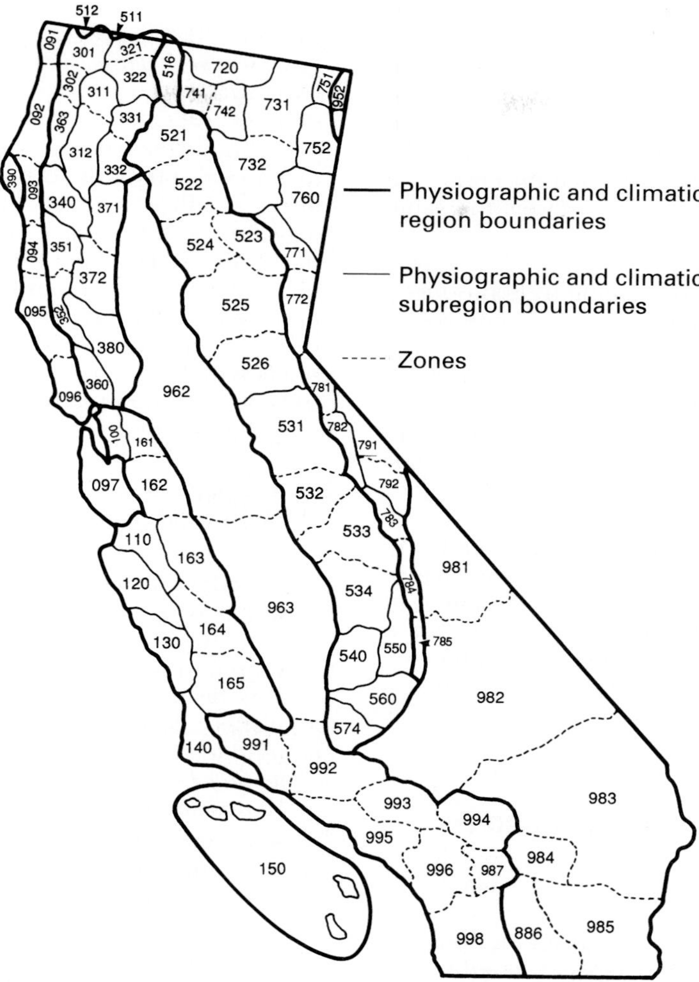

FIGURE 4–11 Seed collection zones in California. The 85 zones are identified by a three-digit number. The first gives 1 of the 6 major physiographic and climatic regions, the second gives 1 of the subregions, of which there are 32, and the third gives the individual zone.

0 = Coast
1 = South coast
3 = North coast
5 = Mountains of Sierra Nevada and Cascade ranges
7 = Northeast interior
9 = Valley and desert areas, further divided into the central valley (6), southern California (9), and the desert areas (7). (Redrawn from G. H. Schubert and R. S. Adams. 1971. *Reforestation practices for conifers in California.* Sacramento: Division of Forestry.)

Seed Orchards

Seed orchards are established to produce seeds of a particular origin or source. Fruit tree nurseries maintain seed orchards to produce seeds of specific rootstock cultivars under conditions that will prevent cross-pollination and spread of pollen-borne viruses. A clonal cultivar such as 'Nemaguard' peach is budded to a rootstock, planted in isolation to avoid chance cross-pollination by virus-infected commercial cultivars, and grown specifically for rootstock seed production. Fruit-tree rootstock clones that are self-pollinated and planted in solid blocks. An isolation zone 120 m (400 ft) wide should be established around the orchard to reduce pollen contamination from other sources. The size of this zone can be reduced if a buffer area of the same kind of tree is present around the orchard. Hybrid seed production involves planting both the parental clones in adjoining rows under the same condition.

Three general types of seed orchards are used for forest trees (*49, 62*): (a) seedling trees produced from selected parents through natural or controlled pollination; (b) clonal seed orchards in which selected clones are propagated by grafting, budding, or rooting cuttings; and (c) seedling-clonal seed orchards in which certain clones are grafted onto branches of some of the trees. The choice depends upon the particular strategy used in the seed improvement program.

A site should be selected for good seed production. For forest trees and most other native species a range of genotypes should be included in a suitable arrangement to ensure cross-pollination and to decrease effects of inbreeding depression. Seven to thirty unrelated genotypes have been recommended to avoid this problem in the purely clonal orchard.

LEGAL CONTROLS TO GENETIC PURITY

Seed Certification

Seed certification (*5, 14, 52*) is a program designed to control the production of specific seed-propagated plant cultivars to ensure the maintenance of seed purity (see Chapter 6). The system was established in the U.S. and Canada during the early 1920s to regulate the commercial production of new cultivars of agricultural crops then being introduced in large numbers by state and federal plant breeders of the U.S. The principles and accompanying regulations of seed selection maintenance were established through the cooperative efforts of public research, extension, regulatory agencies, and a seed-certifying agency known as a Crop Improvement Association that included commercial seed producers as members. These agencies were designated by law through the Federal Seed Act (1939) to conduct research, establish production standards, and certify seeds that are produced under these standards. Individual state organizations are coordinated through the Association of Official Seed Certifying Agencies (AOSCA) (*5*) in the United States and Canada. Similar programs exist at the international level where certification is regulated through the Organization for Economic Cooperation and Development (OECD).

The principal objective of seed certification is to provide standards to preserve the genetic qualities of a cultivar. Other requirements of seed quality may also be enforced as well as eligibility of individual cultivars. The seed-certifying agency may determine production standards for isolation, maximum percentage of off-type plants, and quality of harvested seed; make regular inspections of the production fields to see that the standards are maintained; and monitor seed processing. Agricultural crop classes are defined as follows:

Breeder's seed: that which originates with the sponsoring plant breeder or institution and provides the initial source of all the certified classes.

Foundation seed: progeny of breeder's seed which is so handled as to maintain the highest standard of genetic identity and purity. It is the source of all other certified seed classes but can also be used to produce additional foundation plants. (**Select seed** is a comparable seed class used in Canada.) Foundation seed is labeled with a *white* tag or a certified seed tag with the word "foundation."

Registered seed: progeny of foundation seed (or sometimes of breeder's seed or other registered seed) produced under specified standards approved and certified by the certifying agency and designed to maintain satisfactory genetic identity and purity. Bags of registered seed are labeled with a *purple* tag or with a *blue* tag marked with the word "registered."

Certified seed: progeny of registered seed (or sometimes of breeder's, foundation, or other certified seed) that is produced in the largest volume and sold to the crop producer. It is produced under specified standards designed to maintain a satisfactory level of genetic identity and purity and is approved and certified by the certifying agency. Bags of certified seed have a *blue* tag distributed by the seed-certifying agency.

The international OECD scheme includes similar classes but uses different terms. These include: **basic** (equivalent to either foundation or registered) seed, **certified first-generation** (blue tag) and **second-generation** (red tag) seed.

Tree Certification Classes

Certification of forest tree seeds is available in some states and European countries similar to that for crop seed (*14, 19, 62*). Recommended minimum standards are given by the Association of Official Seed Certifying Agencies (*5*). Since forest tree seed is in early stages of domestication, terms and conditions have less stringent standards than agricultural seeds.

Source-identified: tree seed collected from natural stands where the geographic origin (source and elevation) is known and specified or from seed orchards or plantations of known provenance, specified by seed-certifying agencies. These seeds carry a *yellow* tag.

Selected: tree seed collected from trees that have been selected for promising phenotypic characteristics but have not yet been adequately progeny-tested (phenotype selection). The source and elevation must be stated. These seeds are given a *green* label.

Certified: tree seed collected from trees of proven genetic superiority, i.e., specific set of traits defined by a certifying agency, and produced under conditions that assure genetic identity (genotypic selection). These could come from trees in a seed orchard, or from superior trees in natural stands with controlled pollination. These seeds carry a *blue* tag.

Plant Variety Protection

Breeders of a new seed-reproduced plant variety (cultivar) in the U.S. may retain exclusive propagation rights through the Plant Variety Protection Act, established in 1970 (*3, 8, 15*), and revised in 1994 (see p. 35) (*3, 70*).

The breeder applies to the U.S. Department of Agriculture for a Plant Variety Protection Certificate. For one to be granted, the cultivar must be "novel." It must differ from all known cultivars by one or more morphological, physiological, or other characteristics. It must be uniform; any variation must be describable, predictable, and acceptable. It must be stable, i.e., essential characteristics must remain unchanged during seed propagation. A certificate is good for 20 years. The applicant may designate that the cultivar be certified and that reproduction continue only for a given number of seed generations from the breeder or foundation stock. If designated that the cultivar be certified, it becomes unlawful under the Federal Seed Act to market seed by cultivar name unless it is certified. The passage of this law has greatly stimulated commercial cultivar development.

SUMMARY

This chapter focuses upon the genetic principles and practices that must be followed if seedling cultivars are to remain true-to-type after propagation. Seed cultivars are genetic populations that have been created by changing the gene frequency within the population by specific breeding techniques. Control of the genetic purity of the cultivar is the responsibility of the seed producer. Various systems, both biological and legal, have been devised for use with specific kinds of plants, whether they be short-lived herbaceous plants, such as annuals, biennials, and perennials, or long-lived woody trees and shrubs.

REFERENCES

1. Ahuja, M.R. and W.L. Libby. 1993. *Clonal Forestry: I. Genetics and Biotechnology. II. Conservation and application*. Springer-Verlag. New York.
2. Aldous, J.R. 1975. *Nursery practice.* Forestry Commission Bull. 43, London.
3. Anonymous. 1994. Congressional passage of new PVP law a triumph for seed industry. *Diversity.*
4. Ashworth, S. 1991. *Seed to Seed.* Seed Saver Publications. RR 3, Box 239. Decorah, Iowa 52101.
5. Assoc. Off. Seed Cert. Agencies. 1971. *AOSCA certification handbook.* Publ. 23.
6. Ball, V., ed. 1991. *Ball red book. Greenhouse growing* (15th ed.). West Chicago, IL. Geo. J. Ball Publ. Co.
7. Barker, S.C. 1964. Progeny testing forest trees for seed certification programs. *Ann. Rpt. Inter. Crop Imp. Assn.* 46:83–87.
8. Barton, J.H. 1982. The international breeder's rights system and crop plant innovation. *Science* 216:1071–75.
9. Bassett, M.S., ed. 1986. *Breeding vegetable crops.* Westport, Conn.: AVI Publ. Co.

10. Bliss, F.A. 1980. Common bean. In: *Hybridization of crop plants*. W.R. Fehr and H.W. Hadley, ed. Madison, WI: Amer Soc. Agron. and Crop Sci. Soc. Amer. pp. 273–284.
11. Burton, G.W. 1983. Utilization of hybrid vigor. In *Crop breeding,* D.R. Wood, K.M. Rawal, and M.N. Wood, eds. Madison, Wis.: Amer. Soc. Agron. and Crop Sci. Soc. Amer., pp. 89–107.
12. Cameron, J.W., R.K. Soost, and H.B. Frost. 1957. The horticultural significance of nucellar embryony. In *Citrus virus diseases,* J.M. Wallace, ed., Berkeley: Univ. Calif. Div. Agr. Sci., pp. 191–96.
13. Copeland, L.O., and M.B. McDonald. 1985. *Principles of seed science and technology* (2nd ed.). New York: Macmillan.
14. Cowan, J.R. 1972. Seed certification. In *Seed biology,* Vol. 3, T.T. Kozlowski, ed. New York: Academic Press, pp. 371–97.
15. Davis, D.W., and J.J. Luby. 1988. Some current options in the use of plant variety protection in horticulture. *HortScience* 23:15–18.
16. De Nettancourt, D. 1993. Self- and cross-incompatibility systems. In Hayward, M.D., N.O. Bosemark and I. Romagosa, ed. *Plant Breeding. Principles and Prospects*. London: Chapman and Hall.
17. den Nijs, A.P.M. and G.E. van Dijk. 1993. Apomixis. In: Hayward, M.D., N.O. Bosemark and I. Romagosa, ed. *Plant Breeding. Principles and Prospects*. London: Chapman and Hall.
18. Dirr, M.A., and C.W. Heuser, Jr. 1987. *The reference manual of woody plant propagation*. Athens, Ga.: Varsity Press.
19. Edwards, D.G.W., and F.T. Portlock. 1986. Expansion of Canadian tree seed certification. *For. Chron.* 62:461–66.
20. Ehrenberg, C., A. Gustafsson, G.P. Forshell, and M. Simak. 1955. Seed quality and the principles of forest genetics. *Heredity* 41:291–366.
21. Esquinas-Akacazar, J.T. Plant genetic resources. In Hayward, M.D., N.O. Bosemark and I. Romagosa, ed. *Plant Breeding. Principles and Prospects*. London: Chapman and Hall.
22. Flint, H. 1970. Importance of seed source to propagation. *Proc. Inter. Plant Prop. Soc.* 20:171–78.
23. Frey, K.J. 1983. Plant population management and breeding. In *Crop breeding,* D.R. Wood, K.M. Rawal, and M.N. Wood, eds. Madison, Wis.: Amer. Soc. Agron. and Crop Sci. Soc. Amer., pp. 55–88.
24. Fryxell, P.A. 1957. Mode of reproduction of higher plants. *Bot. Rev.* 23:135–233.
25. Funk, C.R., R.E. Engel, G.W. Pepin, A.M. Radko and R.J. Peterson. 1973. Registration of Bonnieblue Kentucky Bluegrass. *Crop Science* 14:906.
26. Funk, C.R., R.E. Engel, G.W. Pepin and R.J. Russell. 1973. Registration of Adelphi Kentucky Bluegrass. *Crop Science* 13:580.
27. Garrison, C.S., and R.J. Bula. 1961. Growing seeds of forages outside their regions of use. In *Seed Yearbook of agriculture*. Washington, D.C.: U.S. Govt. Printing Office, pp. 401–6.
28. George, R.A.T. 1985. *Vegetable seed production*. London: Longman.
29. Goldsmith, G.A. 1976. The creative search for new F1 hybrid flowers. *Proc. Inter. Plant Prop. Soc.* 26:100–103.
30. Gustafsson, A. 1946–1947. Apomixis in higher plants. Parts I–III. *Lunds. Univ. Arsskrift*. N.F. Avid. 2 Bd 42, Nr. 3:42(2); 43(2); 43(12).
31. Haddock, P.G. 1968. The importance of provenance in forestry. *Proc. Inter. Plant prop. Soc.* 17:91–98.
32. Hanna, W.W. and E.C. Bashaw. 1987. Apomixis: its identification and use in plant breeding. *Crop Science* 27:1136–1139.
33. Harlan, J.R. 1992. *Crops and Man,* 2nd Edition. Amer. Soc. of Agron., Crop Science Soc. Amer. Madison, WI.
34. Hartman, H. 1961. Historical facts pertaining to root and trunkstocks for pear trees, *Oreg. State Univ. Agr. Exp. Sta. Misc. Paper* 109, 1–38.
35. Hartmann, H.T., A.M. Kofranek, V. Rubatzky, and W.J. Flocker. 1988. *Plant science: Growth, development, and utilization of cultivated plants* (2nd ed.). Englewood Cliffs, N.J.: Prentice-Hall.
36. Hayward, M.D., N.O. Bosemark and I. Romagosa. 1993. *Plant Breeding Principles and Prospects*. Chapman & Hall, London.
37. Heit, C.E. 1964. The importance of quality, germinative characteristics and source for successful seed propagation and plant production. *Proc. Inter. Plant Prop. Soc.* 14:74–85.
38. Jones, D.F. 1918. *The effects of inbreeding and crossbreeding upon development*. Connecticut Agricultural Experiment Station Bulletin 207. U.S. Dept. of Agriculture, Washington, D.C.
39. Jones, H.A. and A.E. Clarke. 1943. Inheritance of male sterility in the onion and the production of hybrid seed. *Proc. Amer. Soc. Hort. Sci.* 43:189–194.
40. Jones, R.W. 1969. Selection of intercompatible almond and root knot nematode resistant peach rootstocks as parents for production of hybrid rootstock seed. *J. Amer. Soc. Hort. Sci.* 94:89–91.
41. Kester, Dale E. and C. Grasselly. 1987. Almond. In: R.C. Rom and R. Carlson, ed. *Rootstocks for fruit trees*. New York: Wiley.
42. Kitz, J. 1994. New varieties for today's markets. *Proc. Intern. Plant Prop. Soc.* 44:289–291.
43. Langlet, O. 1962. Ecological variability and taxonomy of forest trees. In *Tree growth,* T.T. Kozlowski, ed. New York: Ronald Press, pp. 357–69.
44. Laverack, G.K. and M.R. Turner. 1995. Roguing seed crops for genetic purity: a review. *Plant Varieties and Seeds* 8:29–45.

45. Lasa, L.M. and N.O. Bosemark. 1993. Male Sterility. In: Hayward, M.D., N.O. Bosemark and I. Romagosa, ed. *Plant Breeding. Principles and Prospects.* Chapman and Hall. London.

46. Lee, D. 1987. Seed collection and cleaning. *Comb. Proc. Intern. Plant Prop. Soc.* 37:61–65

47. Libby, W.J. 1987. Genetic resources and variation in forest trees. In *Improving vegetatively propagated plants,* A.J. Abbott and R.K. Atkin, eds. New York: Academic Press, pp. 199–209.

48. ———. 1987. Testing and deployment of genetically engineered trees. In *Tissue culture of forest trees,* J.M. Bonga, and D.J. Durzan, eds. Amsterdam: Elsevier, pp. 167–97.

49. Libby, W.J., and R.M. Rauter. 1984. Advantages of clonal forestry. *For. Chron.,* pp. 145–49.

50. Macdonald, B. 1986. *Practical woody plant propagation for nursery growers,* Vol. 1. Portland, Oreg.: Timber Press.

51. Mastalerz, J.W. 1976. Bedding plants: *A manual on the culture of bedding plants as a greenhouse crop.* University Park, Pa.: Pennsylvania Flower Growers.

52. McDonald, M.B., Jr., and W.D. Pardee, eds. 1985. *The role of seed certification in the seed industry.* CSSA Spec. pub. 10. Madison, Wis.: Crop Sci. Soc. Amer., ASA.

53. McMillan-Browse, P.D.A. 1979. *Hardy, woody plants from seed.* London: Growers Books.

54. Millar, C.I. and W.J. Libby. 1989. Disneyland or native ecosystem: genetics and the restorationist. *Restoration & Management Notes* 7:1.

55. Millar, C.I. and W.J. Libby. 1991. Strategies for conserving clinal, ecotypic, and disjunct population diversity in widespread species. In: *Genetics and conservation of rare plants.* D.A. Falk and K.E. Holsinger, ed. Oxford: Oxford University Press.

56. Pearson, O.H. 1968. Unstable gene systems in vegetable crops and implications for selection. *HortScience* 3(4):271–74.

57. Poehlman, J.M. 1995. *Breeding field crops* (4th ed.). Westport, Conn.: AVI Publ. Co.

58. Raven, P.H., R.F. Evert and S.E. Eichhorn. 1992. *Biology of plants,* 5th edition. New York: Worth Publ.

59. Ritchie, G.A., 1994. Commercial application of adventitious rooting in forestry. In T.D. Davis and B.E. Hassig, ed. *Biology of adventitious root formation.* Plenum Press. New York and London.

60. Rom, R.C., and R.F. Carlson, eds. 1987. *Rootstocks for fruit crops.* New York: John Wiley.

61. Sax, K. 1949. The use of Malus species for apple rootstocks. *Proc. Amer. Soc. Hort. Sci.* 53:219–20.

62. Schopmeyer, C.S., ed. 1974. *Seeds of woody plants in the United States.* U.S. Dept. Agr. Handbook 450. Washington, D.C.: U.S. Govt. Printing Office.

63. Schubert, G.H., and R.S. Adams, 1971. *Reforestation practices for conifers in California.* Sacramento: Calif. State Div. of Forestry.

64. Simmonds, N.W. 1979. *Principles of crop improvement.* New York: John Wiley.

65. Sprague, G.F. 1950. Production of hybrid corn. *Iowa Agr. Exp. Sta. Bul.* P48, pp. 556–82.

66. Stoskopf, N.C., D.T. Tomes and B.R. Christie. 1993. *Plant Breeding Theory and Practice.* Westview Press. Boulder, CO.

67. Stoutemyer, V.T. 1960. Seed propagation as a nursery technique. *Proc. Plant Prop. Soc.* 10:251–55.

68. Stuke, W. 1960. Seed and seed handling techniques in production of walnut seedlings. *Proc. Plant Prop. Soc.* 10:274–77.

69. Thomson, J.R. 1979. *An introduction to seed technology.* New York: John Wiley.

70. U.S. Congress. 1994. Plant Variety Protection Act Amendments of 1994. *Congressional Record.* 140:10.

71. Westwood, M.N. 1966. Arboretums—a note of caution on their use in agriculture. *HortScience* 1:85– 86.

72. ———. 1994. *Temperate zone pomology* (3rd ed.). Portland, Oreg.: Timber Press.

73. Whealy, Kent. 1992. *Garden Seed Inventory,* 3rd edition. Seed Saver Publications. RR 3, Box 239. Decorah, IA.

74. Wilson, R. 1994. Plugs and automation—"the future is here!" *Proc. Intern. Plant Prop. Soc.* 44:287–288.

75. Wyman, D. 1953. Seeds of woody plants. *Arnoldia* 13:41–60.

76. Young, J.A. and C.G. Young. 1986. *Seeds of Wild Land Plants.* Timber Press, Portland, OR.

77. Young, J.A. and C.G. Young. 1992. *Seeds of woody plants in North America.* Dioscorides Press. Portland OR.

78. Zobel, B.J., and J. Talbert. 1984. *Applied tree improvement.* New York: John Wiley.

79. Zobel, B.J., G. van Wyk, and P. Stahl. 1987. *Growing exotic forests.* New York: John Wiley.

SUPPLEMENTARY READING

Bassett, M.S., ed. 1986. *Breeding vegetable crops.* Westport, Conn.: AVI Publ. Co.

Copeland, L.O., and M.B. McDonald. 1985. *Principles of seed science and technology* (2nd ed.). New York: Macmillan.

Hartmann, H.T., A.M. Kofranek, V.E. Rubatsky, and W.F. Flocker. 1988. *Plant Science,* second edition. Englewood Cliffs, N.J.: Prentice-Hall.

Poehlman, J.M. 1987. *Breeding field crops* (3rd ed.). Westport, Conn.: AVI Publ. Co.

SCHOPMEYER, C.S., ed. 1974. *Seeds of woody plants in the United States.* U.S. Dept. Agr. Handbook 450. Washington, D.C.: U.S. Govt. Printing Office.

STOSKOPF, N.C., D.T. Tomes and B.R. Christie. 1993. *Plant Breeding Theory and Practice.* Westview Press. Boulder, CO.

WOOD, D.R., K.M. Rawal, and M.N. Wood, eds. 1983. *Crop breeding.* Madison, Wis.: Amer. Soc. Agron. and Crop Sci. Soc. Amer.

U.S. DEPT. OF AGRICULTURE. 1961. *Seeds: Yearbook of agriculture.* Washington, D.C.: U.S. Govt. Printing Office.

ZOBEL, B.J., and J. Talbert. 1984. *Applied tree improvement.* New York: John Wiley.

5

The Development of Seeds

INTRODUCTION

Four hundred million years ago, plants moved out of the oceans to colonize land. Two major adaptations made this possible. The first was the evolution of the root. The **root** not only anchored the plant in soil, but also allowed the plant to obtain water and minerals no longer brought to the plant by ocean water. A second adaptation that increased a plant's success on land was the development of a **vascular system.** This allowed transportation of materials obtained by the root system to be transported to the leafy photosynthetic parts of the plant. However, the price of these adaptations to land habitation was relative immobility. The first vascular plants (e.g., ferns) evolved spores to spread the result of sexual reproduction. The next advance in evolution produced the marvelous diversity found in seeds and their accompanying fruit structures. Seeds allowed plants, like cycads, gymnosperms, and ultimately angiosperms, the means to spread their progeny to colonize new environments. There is an incredible diversity in the mechanisms species have developed to produce and disseminate seeds. This chapter will review the basic principles, that underlie the practices used for commercial seed production.

Propagation by seeds is the major method by which plants reproduce in nature and one of the most efficient and widely used propagation methods for cultivated crops. The plants produced are referred to as **seedlings.** This term is significant in horticulture and it refers to the life cycle of a plant grown from a seed in contrast to a plant started vegetatively from a cutting or grafted plant.

The planting of the seed is the physical beginning of seedling propagation. The seed itself, however, is the end product of a process of growth and development within the parent plant that is described in this chapter. It may arise either from the fusion of male and female gametes to form a single cell (the **zygote**) within the ovule of the flower or from vegetative or unfertilized reproductive cells within the ovule (see **apomixis,** Chapter 2 and pp. 134). The zygote has the property of **totipotency,** that is, it has all of the genetic information needed to reproduce the full-sized plant and to initiate the seedling cycle of the next generation.

WHAT IS A SEED?

Botanically, a seed is a matured ovule containing an **embryo** that is usually the result of sexual fertilization (*18, 19*). Every seed consists of a **protective outer covering, storage tissue, and an embryo.** The protective layer can be either the **seed coat** or part of the fruit covering **(pericarp).** Storage tissue for dicots is contained in the cotyledons and to varying degrees, endosperm tissue. Storage tissue for monocots is the starchy endosperm, and for gymnosperms, the storage tissue is haploid gametophyte tissue (*3*). Seeds and fruits of different species vary greatly in appearance, size, shape, and location and structure of the embryo in relation to storage tissues (Figure 5–1). These points are not only useful for identification (*25, 43*) but affect germination requirements. From the standpoint of seed handling, it is not always possible to separate the fruit and seed, since they are sometimes joined in a single unit. Horticulturally, the fruit itself may be treated as the "seed," as in lettuce or corn.

The following classification of seeds is based upon morphology of embryo and seed coverings. It includes, as examples, families of herbaceous plants [after Atwater (*1*)].

I. Seeds with dominant endosperm (or perisperm) as seed storage organs (endospermic).
   A. Rudimentary embryo. Embryo is very small and undeveloped but undergoes further increase at germination (see Figure 5–9A, Magnolia).
      RANUNCULACEAE *(Aquilegia, Delphinium),* PAPAVERACEAE *(Eschscholtzia, Papaver),* FUMARIACEAE *(Dicentra),* ARALIACEAE *(Fatsia),* MAGNOLIACEAE *(Magnolia),* AQUIFOLIACEAE *(Ilex)*
   B. Linear embryo. Embryo is more developed than those in A and enlarges further at germination.
      UMBELLIFERAE *(Daucus),* ERICACEAE *(Calluna, Rhododendron),* PRIMULACEAE *(Cyclamen, Primula),* GENTIANACEAE *(Gentiana),* SOLANACEAE *(Datura, Solanum),* OLEACEAE *(Fraxinus)*
   C. Miniature embryo. Embryo fills more than half the seed.
      CRASSULACEAE *(Sedum, Heuchera, Hypericum),* BEGONIACEAE *(Begonia),* SOLANACEAE *(Nicotiana, Petunia, Salpiglossis),* SCROPHULARIACEAE *(Antirrhinum, Linaria, Mimulus, Nemesia, Penstemon),* LOBELIACEAE *(Lobelia)*
   D. Peripheral embryo. Embryo encloses endosperm or perisperm tissue.
      POLYGONACEAE *(Eriogonum),* CHENOPODIACEAE *(Kochia),* AMARANTHACEAE *(Amaranthus, Celosia, Gomphrena),* NYCTAGINACEAE *(Abronia, Mirabilis)*

II. Seeds with embryo dominant (nonendospermic); classified according to the type of seed covering.
   A. Hard seed coats restricting water entry.
      LEGUMINOSAE, GERANIACEAE *(Pelargonium),* ANACARDIACEAE *(Rhus),* RHAMNACEAE *(Ceanothus),* MALVACEAE *(Abutilon, Altea),* CONVOLVULACEAE *(Convolvulus)*
   B. Thin seed coats with mucilaginous layer.
      CRUCIFERAE *(Arabis, Iberis, Lobularia, Mathiola),* LINACEAE *(Linum),* VIOLACEAE *(Viola),* LABIATEAE *(Lavandula)*
   C. Woody outer seed coats with inner semipermeable layer (see Figure 5–1E).
      ROSACEAE *(Geum, Potentilla),* ZYGOPHYLLACEAE *(Larrea),* BALSAMINACEAE *(Impatiens),* CISTACEAE *(Cistus, Helianthemum),* ONAGRACEAE *(Clarkia, Oenothera),* PLUMBAGINACEAE *(Armeria),* Apocynaceae, Polemoniaceae *(Phlox),* HYDROPHYLLACEAE *(Nemophila, Phacelia),* BORAGINACEAE *(Anchusa),* VERBENACEAE (Lantana, Verbena), LABIATEAE *(Coleus, Moluccela),* DIPSACAEAE *(Dipsacus, Scabiosa)*
   D. Fibrous outer seed coat with more or less semipermeable membranous layer, including endosperm remnant (see Figure 5–1F).
      COMPOSITEAE (many species)

III. Unclassified
   A. Rudimentary embryo with no food storage.
      ORCHIDACEAE (orchids, in general)
   B. Modified miniature embryo located on periphery of seed.
      GRAMINEAE (grasses) (see Figure 5–1G)
   C. Axillary miniature embryo surrounded by gametophyte tissue (see Figure 5–1H)
      Gymnosperms, in particular, conifers.

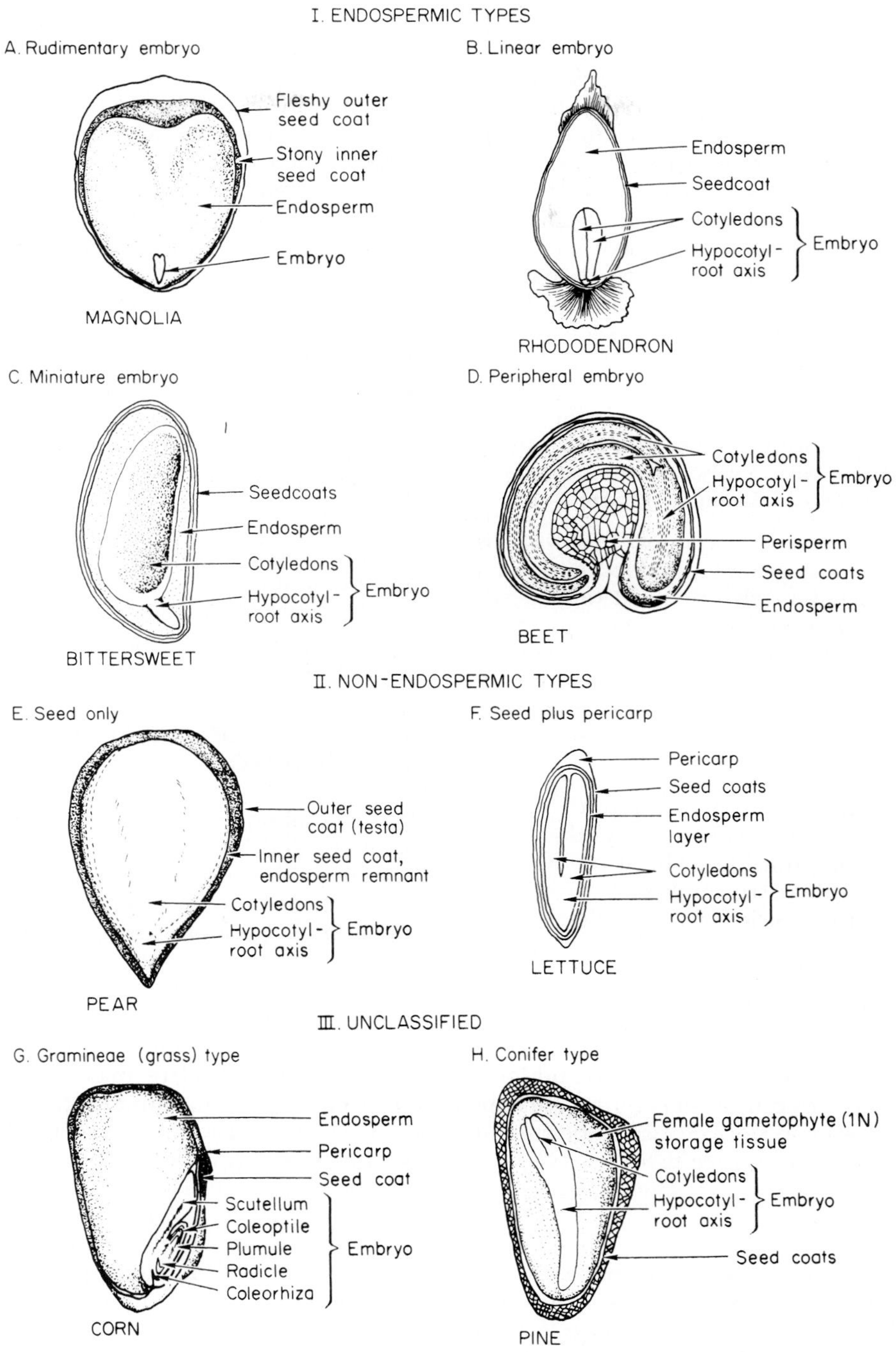

I. ENDOSPERMIC TYPES

A. Rudimentary embryo

Fleshy outer seed coat
Stony inner seed coat
Endosperm
Embryo
MAGNOLIA

B. Linear embryo

Endosperm
Seedcoat
Cotyledons
Hypocotyl-root axis
Embryo
RHODODENDRON

C. Miniature embryo

Seedcoats
Endosperm
Cotyledons
Hypocotyl-root axis
Embryo
BITTERSWEET

D. Peripheral embryo

Cotyledons
Hypocotyl-root axis
Embryo
Perisperm
Seed coats
Endosperm
BEET

II. NON-ENDOSPERMIC TYPES

E. Seed only

Outer seed coat (testa)
Inner seed coat, endosperm remnant
Cotyledons
Hypocotyl-root axis
Embryo
PEAR

F. Seed plus pericarp

Pericarp
Seed coats
Endosperm layer
Cotyledons
Hypocotyl-root axis
Embryo
LETTUCE

III. UNCLASSIFIED

G. Gramineae (grass) type

Endosperm
Pericarp
Seed coat
Scutellum
Coleoptile
Plumule
Radicle
Coleorhiza
Embryo
CORN

H. Conifer type

Female gametophyte (1N) storage tissue
Cotyledons
Hypocotyl-root axis
Embryo
Seed coats
PINE

FIGURE 5–1 Morphological types of seeds as described in the text. (Rhododendron and bittersweet redrawn from (*57*).

Embryo and Storage Tissue

The embryo is a new plant resulting from the union of a male and female gamete during **fertilization.** Its basic structure is an embryo axis with growing points at each end, one for the shoot and one for the root, and one or more seed leaves **(cotyledons)** attached to the embryo axis. Plants are classified by the number of cotyledons. Monocotyledonous plants (such as coconut palm or grasses) have a single cotyledon, dicotyledonous plants (such as bean or peach) have two, and gymnosperms (such as pine or ginkgo) may have as many as 15.

Embryos differ in size in relation to the storage tissue, reflecting the extent to which the embryo has developed within the seed. Seeds can be separated into three basic types: **nonendospermic, endospermic or unclassified seeds,** (Figure 5–1).

Endospermic type. Cotyledon growth is arrested at different stages of development, such that the embryo may be only one-third to one-half full size at the time of seed ripening. The remainder of the seed cavity contains large amounts of endosperm or **nucellus (perisperm)** depending on the species. Four basic types of endospermic seeds are recognized as characteristic of plant families (Figure 5–1).

Nonendospermic type. Rapid growth by cell division and enlargement occurs in the endosperm, digesting the enclosing nucellus. This is followed by expansion of the embryo through cell division at the periphery of the cotyledons that digests the endosperm. In these seeds, the endosperm and/or the nucellus is reduced to a remnant between the embryo and the **integuments (seed coat).**

Unclassified. This includes seeds of monocots with their modified embryo structure and location in the seed, seeds of gymnosperms that use **haploid gametophyte tissue** as storage tissue, and orchid seeds that fail to produce storage tissue.

Seed Coverings

Seed coverings may consist of the **seed coat, the remains of the nucellus and endosperm, and sometimes parts of the fruit.** The seed coat, also termed the **testa,** are derived from the integuments of the ovule. During development, the seed coat becomes modified so that at maturity they present an appearance often characteristic of the plant family (*14*). Usually, the outer layer of the seed coat becomes dry, somewhat hardened and thickened, and brownish in color. In particular families it becomes hard and impervious to water. On the other hand, the inner seed coat layers are usually thin, transparent, and membranous. Remnants of the endosperm and nucellus are sometimes found within the inner seed coat, sometimes making a distinct, continuous layer around the embryo.

In some plants, parts of the fruit remain attached to the seed so that the fruit and seed are commonly handled together as the "seed." In fruits such as achenes, caryopsis, samaras, and schizocarps, the pericarp and seed coat layers are contiguous. In others, such as the acorn, the pericarp and seed coverings separate but the fruit covering is indehiscent. In still others, such as the "pit" of stone fruits or the shell of walnuts, the covering is a hardened portion of the pericarp, but it is dehiscent (splits along an existing suture line) and can usually be removed without much difficulty.

Seed coverings provide mechanical protection for the embryo, making it possible to handle seeds without injury, and thus permitting transportation for long distances and storage for long periods of time. The seed coverings can also control germination, as discussed in Chapter 7.

FORMATION OF THE FRUIT, SEED, AND EMBRYO

The sexual cycle includes development of the male (pollen) and female (embryo sac) structures of the flower, as shown for angiospermous flowers in Figure 5–2 (*18, 19*). In this part of the cycle reduction division of chromosomes occurs to produce the haploid (n) chromosome number. Figure 5–2 shows the basic structure of the flower and Figure 5–3 shows more details of the pollen, ovary, and ovule structure. Pollen grains and an eight-celled embryo sac contain haploid (1n) male and female gametes, respectively. During flowering, pollen is transferred from the anther to the stigma **(pollination),** where it germinates. A pollen tube grows down the style into the ovary until it reaches the embryo sac within the ovule (Figure 5–3). Two male gametes from the pollen tube are discharged into the embryo sac—one to unite with a female gamete **(fertilization)** to produce the zygote and another to unite with the two polar nuclei to produce the endosperm. This is termed **double fertilization** and is unique to flowering plants (angiosperms).

With many angiosperms, the zygote is **diploid** (2n) and divides to become the embryo; the endosperm is **triploid** (3n) and develops into nutritive tissue for the developing embryo [note: not all plant species are diploid. Several important crop

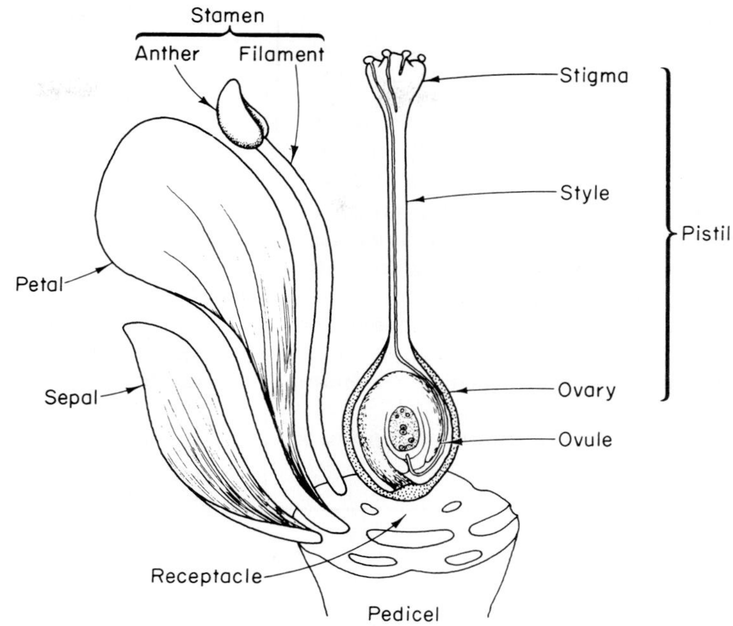

FIGURE 5–2 Flower structure of a typical angiosperm plant.

plants like potato are **tetraploid** or even **octaploid** like strawberry. However, the product of normal meiosis is still to produce gametes with half the original number of chromosomes]. Both structures are enclosed within the **nucellus** (inside the ovule) that functions initially as a nurse tissue for the developing embryo and endosperm. In some species, the nucellus develops into a food storage tissue in the seed.

Failure of the endosperm to develop properly results in retardation or arrest of embryo development, and embryo abortion can result. This phenomenon is called **somatoplastic sterility** and commonly occurs when two genetically different individuals are hybridized, either from different species (*11, 12, 13*) or from two individuals of different ploidy constitution. It can be a barrier to hybridization in angiosperms but not in gymnosperms (*59*), since the "endosperm" in these plants is haploid female gametophytic tissue. Embryos from such crosses can be "rescued" by isolating these embryos and placing them in tissue cul-

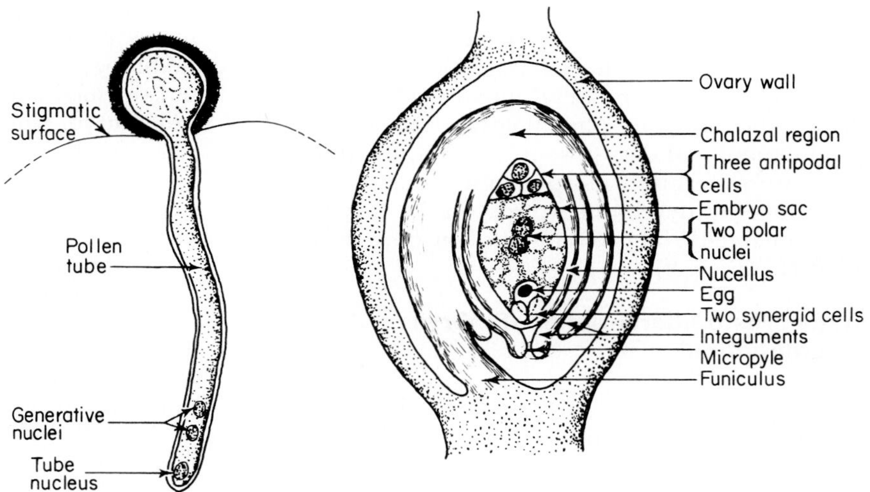

FIGURE 5–3 Sexual reproduction in an angiosperm. *Left:* Haploid pollen grain (male) germinates on the stigma, a pollen tube grows down into the style. Two generative nuclei (male sperm cells) are produced. *Right:* The ovary is the female structure (pericarp) that encloses the ovule, which consists of an embryo sac surrounded by nucellus tissue, all of which are genetically the same as the mother plant. The mature embryo sac develops eight haploid nuclei, including an egg nucleus (female sex cell) and two haploid polar nuclei, all of which take part in fertilization. The generative nuclei from the pollen tube enter through the micropyle and one nucleus fertilizes the egg, beginning the formation of the zygote (embryo), while the other generative nucleus fuses with the two polar nuclei to begin the formation of the endosperm.

ture. Details on embryo rescue are found in Chapter 17.

> The relationships between angiosperm flower structure and the parts of the fruit and seed are as follows:
>
> **ovary (pericarp) > fruit (sometimes composed of more than one ovary plus additional tissues)**
> **ovule > seed (sometimes coalesces with fruit)**
> **integuments > testa (seed coats)**
> **nucellus > perisperm (usually absent or reduced; sometimes storage tissue)**
> **2 polar nuclei + sperm nucleus > endosperm (3n)**
> **egg nucleus + sperm nucleus > zygote = embryo (2n)**

Fertilization in gymnosperms differs from angiosperms because they do not produce elaborate flower parts. There is no true stigma in gymnosperms. Rather, there is either a stigmatic surface on the opening of the ovule or a sugary pollination drop exudes from the ovule to collect wind-borne pollen (*59*). In some species, like *Ginkgo,* the male gametes are motile, swimming to fertilize the egg contained in the ovule. Double fertilization does not occur in gymnosperms. Only angiosperms produce a true triploid (3n) endosperm. In gymnosperms, haploid female gametophyte tissue surrounds the developing embryo and performs the function of the endosperm.

Stages of Seed Development

Three physiological stages of development are recognized in most seeds (Figure 5–4). These include **histodifferentiation; cell expansion** (food reserve deposits); and **maturation drying.** Figure 5–5 shows the relative growth and development in lettuce seed (fruit) showing the physiological stages of seed development and days post-pollination. In addition, there are recognized morphological stages of both seed and fruit development that can be very useful to commercial fruit growers to follow crop development (Figure 5–6).

Stage I Histodifferentiation (Embryo Differentiation)

Stage I is characterized by the differentiation of the embryo and endosperm mostly due to cell division. In stage I, the embryo reaches the beginning of the cotyledon stage of development. There is rapid increase in both fresh and dry weight. There are characteristic stages of embryogenesis that occur during stage I and these are distinct for dicots, monocots, and gymnosperms.

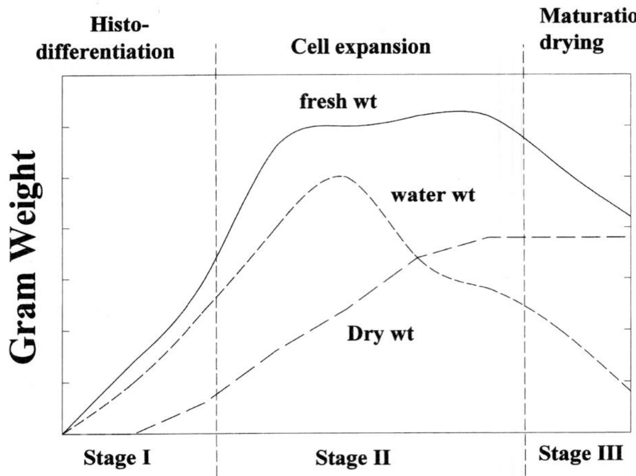

FIGURE 5–4 The stages of seed development. The stages include histodifferentiation (rapid increase in seed size due predominantly to cell division), cell expansion (largest increase in seed size for deposition of food reserves), and maturation drying (dramatic loss in seed fresh weight due to water loss). (Redrawn from Bewley and Black 1994.)

Embryo differentiation in dicots. Although there are several variations on the types of angiosperm embryogenesis (*4, 45, 53, 64*), embryo formation in **Shepard's purse** (*Capsella bursa-pastoris*) serves as a good model for dicot embryogenesis. Embryogenesis in dicots proceeds through characteristic stages of development. These include the **proembryo, globular, heart, torpedo, and cotyledon stages** (Figure 5–7, 5–8).

Following fertilization of the egg and sperm nuclei, a **proembryo** is formed by a transverse cell division to form an **apical** and **basal cell** (Figure 5–7a). The basal cell forms the suspensor, while the apical cell forms the embryo. The suspensor in dicots is usually a column of single or multiple cells. The suspensor functions to push the proembryo into the embryo sac cavity and to absorb and transmit nutrients to the proembryo. The embryo is supplied with nutrients for growth via the suspensor until later stages of embryo development when the

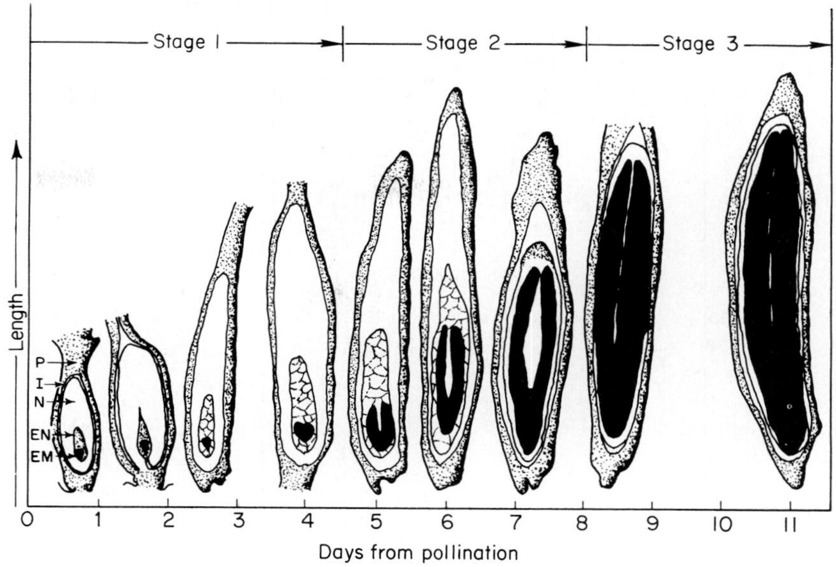

FIGURE 5–5 Growth and development of the fruit and seed in lettuce. P, pericarp; I, integuments; N, nucellus; EN, endosperm; EM, embryo. (Redrawn from Jones 1927.)

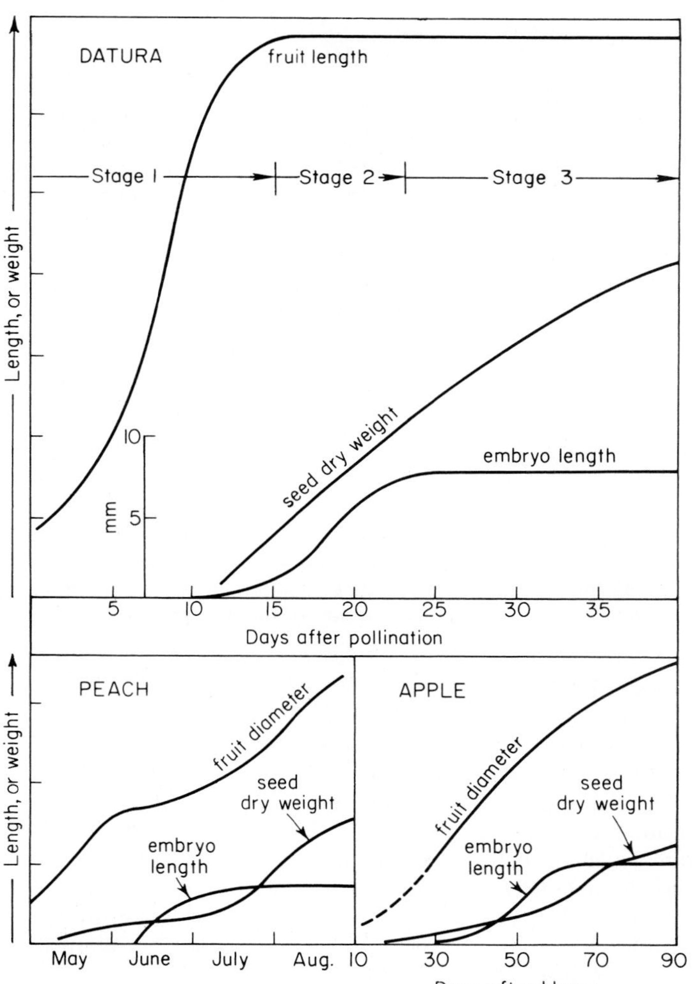

FIGURE 5–6 Comparative development of three different fruit types: Datura (dry capsule), peach (fleshy drupe), and apple (fleshy pome). Changes in the seed dry weight and embryo are the same among the species represented. Differences are apparent in fruit growth. (Data from Rietsema et al. 1955, Luckwill 1948; Hesse and Kester 1955.)

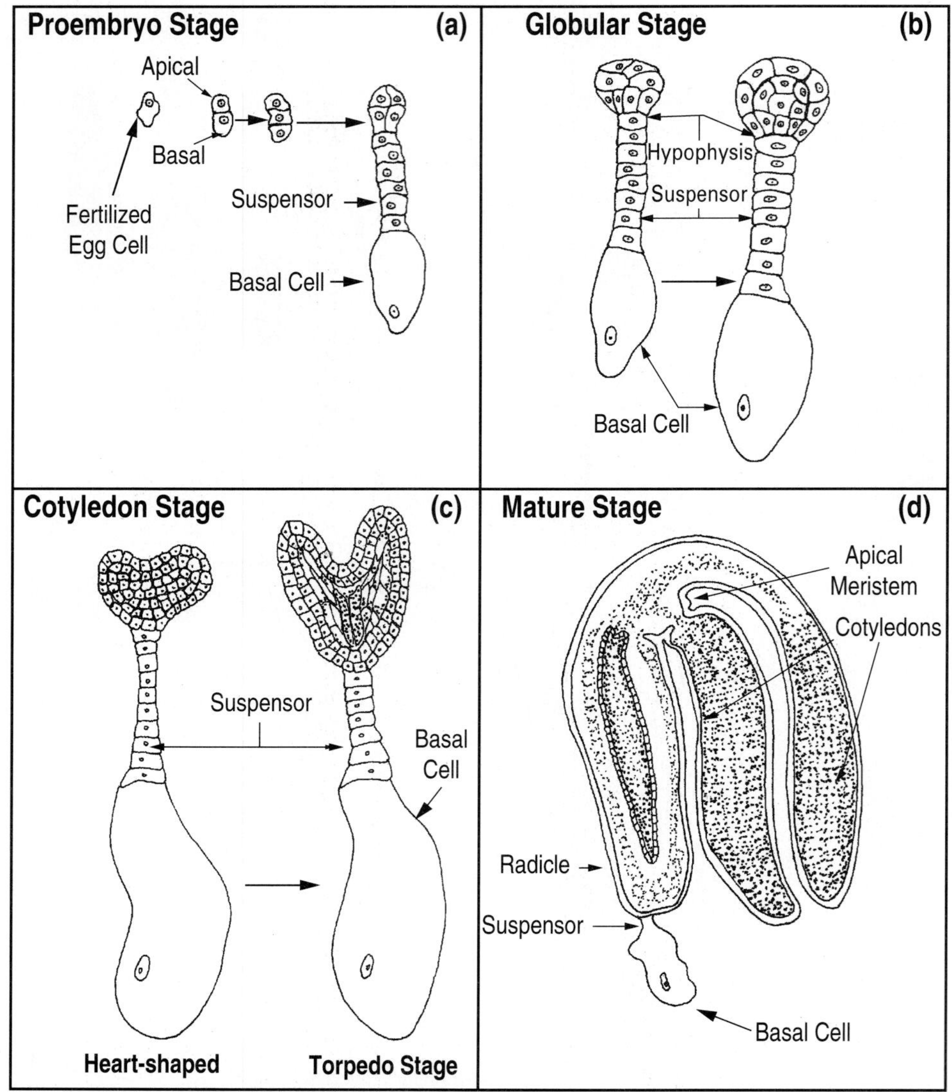

FIGURE 5–7 Diagrammatic representation of embryo development in a typical dicot (shepard's purse). See text for description of figures.

embryo is nourished by material from the endosperm. In Shepard's purse, basal cell derivatives in the globular embryo form the **hypophysis** that goes on to develop into the **radicle** (Figure 5–7b). Tissue differentiation becomes evident in the 16-celled **globular** embryo (Figures 5–7b, 5–8b and c). An outer layer of cells **(protoderm)** will develop into epidermal cells of the embryo. The inner cell layers will develop into the procambium and ground meristem.

Cotyledon primordium are evident in the **heart-shaped** stage of embryogenesis (Figures 5–7c, 5–8d). These primordia elongate to give a typical **torpedo** stage embryo (Figure 5–7e). In the torpedo stage, the embryo has organized to form an apical meristem, radicle, cotyledons, and hypocotyl.

Embryo differentiation in monocots. Monocots have a more complex embryo structure in the mature seed compared to dicots, but early embryo development is similar to dicots (*54*). The stages of embryogenesis in monocots include the **proembryo, globular, scutellar, and coleoptilar** stages (Figure 5–9). Following fertilization, an **apical** and **basal cell**

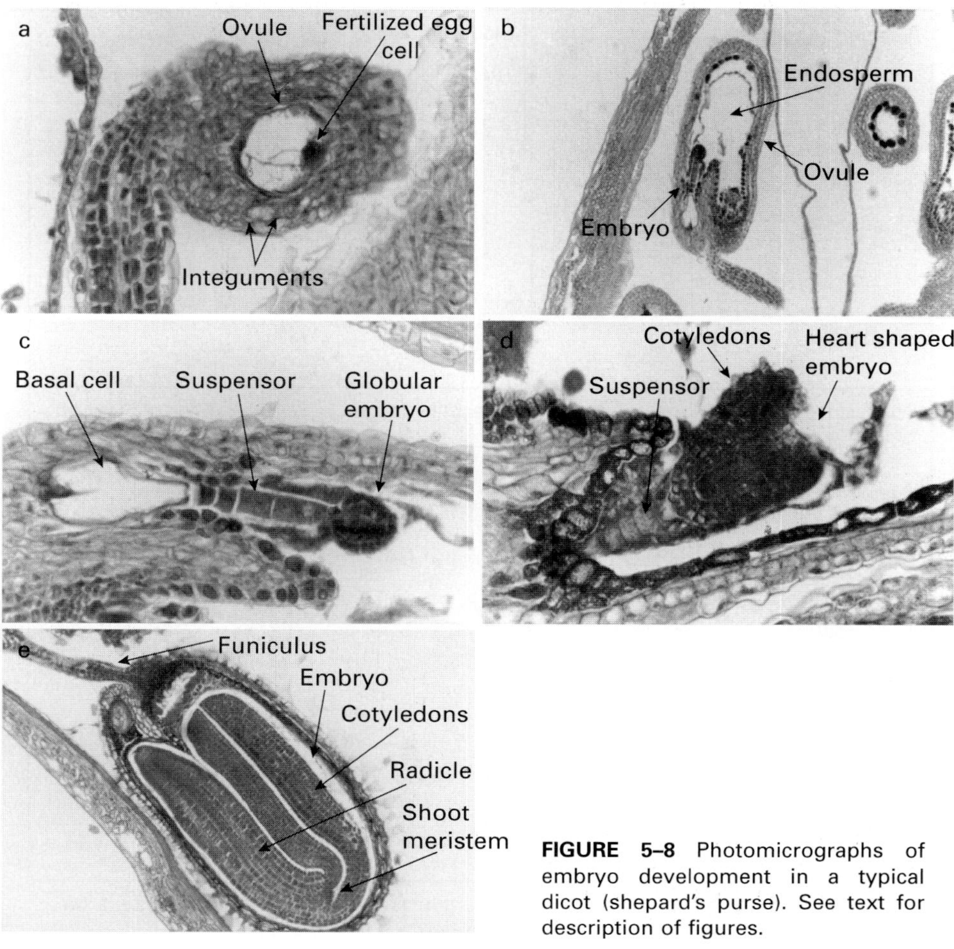

FIGURE 5–8 Photomicrographs of embryo development in a typical dicot (shepard's purse). See text for description of figures.

are visible in corn (*Zea mays*) (Figure 5–9a). The **proembryo** in the **globular** stage is similar to dicots, except that the **suspensor** is not a single or double row of cells and is less differentiated (Figure 5–9b). In the late globular stage, the outer epidermal layer is evident and a group of cells on one side of the proembryo divides more rapidly. These will give rise to the embryo axis. The remnant of the cotyledon can be seen in the **scutellar** stage of development. Monocots have reduced the pair of cotyledons represented in dicot embryos to a single modified cotyledon termed the **scutellum** (Figure 5–9c). The scutellum acts as conductive tissue between the endosperm and embryo axis. (Figure 5–10). The embryo axis differentiates into the **plumule** (shoot) and **radicle** (Figure 5–9d). In monocots, the embryo axis also has a specialized tissue surrounding the shoot and root tissue to aid in emergence during germination. These are the **coleoptile** and **coleorhiza,** respectively (Figure 5–9d, 5–10).

Embryo differentiation in gymnosperms(59). Compared to the more evolutionarily advanced angiosperms, embryo formation in gymnosperms differs in several important ways (Figure 5–11). Most conspicuous is that seeds of gymnosperms are not contained within a carpel (fruit). The term *gymnosperm* means **"naked seeded."** There is also no true triploid endosperm in gymnosperms. Rather, the developing embryo is nourished by **haploid female gametophyte tissue** (Figure 5–11f). After fertilization, several embryos begin development within a single gymnosperm seed but rarely does more than one of these embryos mature. Of commercial interest, the long embryo gestation period can slow breeding efforts especially in conifers (Figure 5–12). Some angiosperms can mature seeds within a few weeks, while gymnosperm cones can take two years to shed developed seeds.

In pine (*Pinus sp.*), the fertilized egg cell divides to form a free nuclear stage without cell

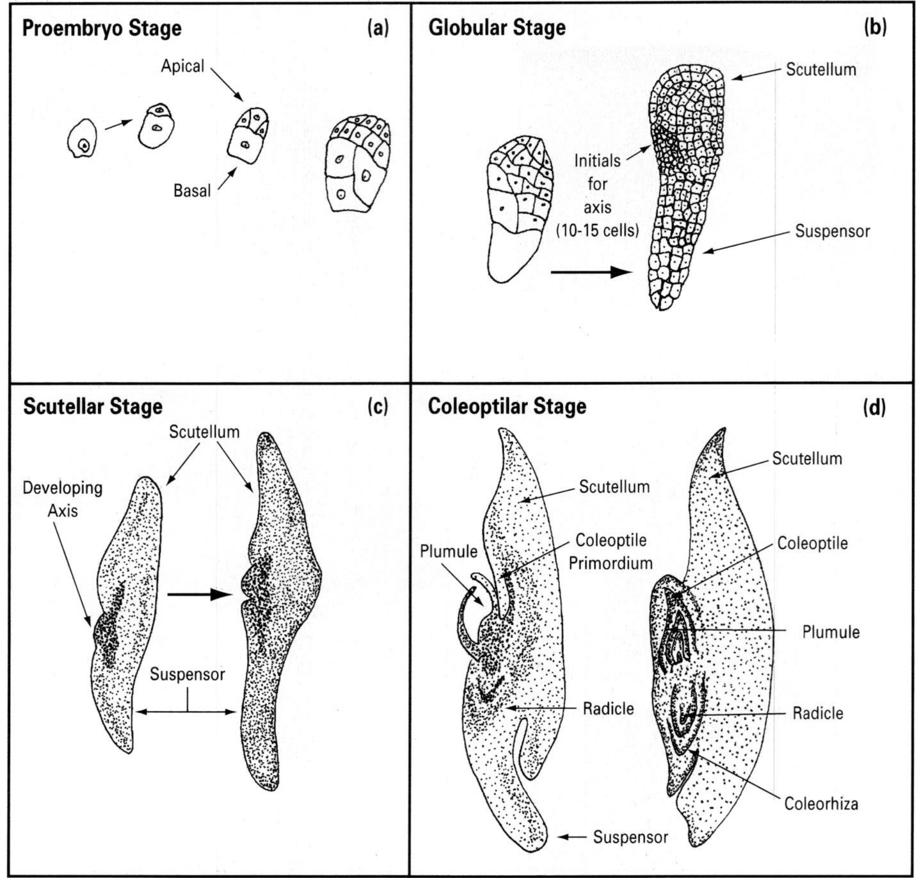

FIGURE 5–9 Embryo development in a typical monocot (corn). See text for description of figure.

walls between nuclei (Figure 5–11a). Following cell wall formation, cells organize to form an embryo tier of cells and a **suspensor tier** (Figure 5–11b). The **suspensor** differentiates into a set of primary suspensor cells **(rosette cells)** and **embryonal suspensor tubes.** The suspensor cells elongate and there are several cleavage events to give multiple embryos **(polyembryos)** inside a single seed (Figure 5–11c and e). Usually, only one of these embryos continues to develop. The proembryo differentiates an epidermal layer (Figure 5–11c) prior to the cotyledon primordium becoming evident (Figure 5–11d and f).

POLYEMBRYONY AND APOMIXIS

These two phenomena represent variations from the normal pattern of zygote formation and embryogenesis. Although related they are not the same phenomenon. **Polyembryony** means that more than one embryo develops within a single seed, sometimes many (Figure 5–13). Their occurrence can develop from several distinct causes.

Polyembryony

Adventitious Embryony (Nucellar Embryony or Nucellar Budding)

Specific cells in the **nucellus** or (sometimes) in the **integuments** have embryogenic potential and undergo embryogenesis (Figure 5–14). Genetically, these embryos have the same genotype as the parental plant and are also apomictic (see below). **Adventitious embryony** occurs in many plant species but is most prominent in some important subtropical and tropical tree species, such as *Citrus* and mango (*39*). In these species, both zygotic and apomictic embryos are produced and the stimulus of fertilization is required. In other species (e.g., *Opuntia*), no pollination or fertilization is needed.

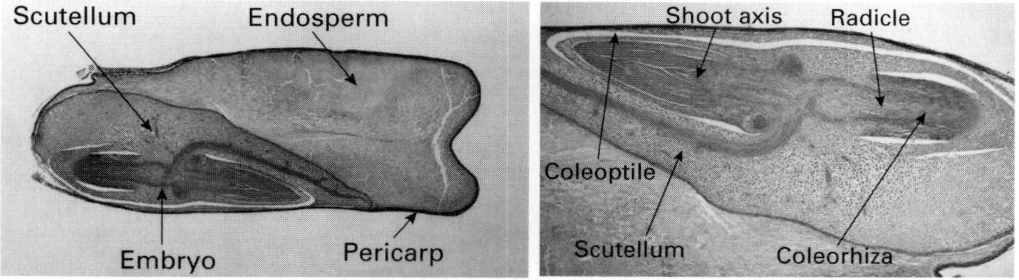

FIGURE 5–10 Cross-section of a mature seed of corn showing basic anatomical features.

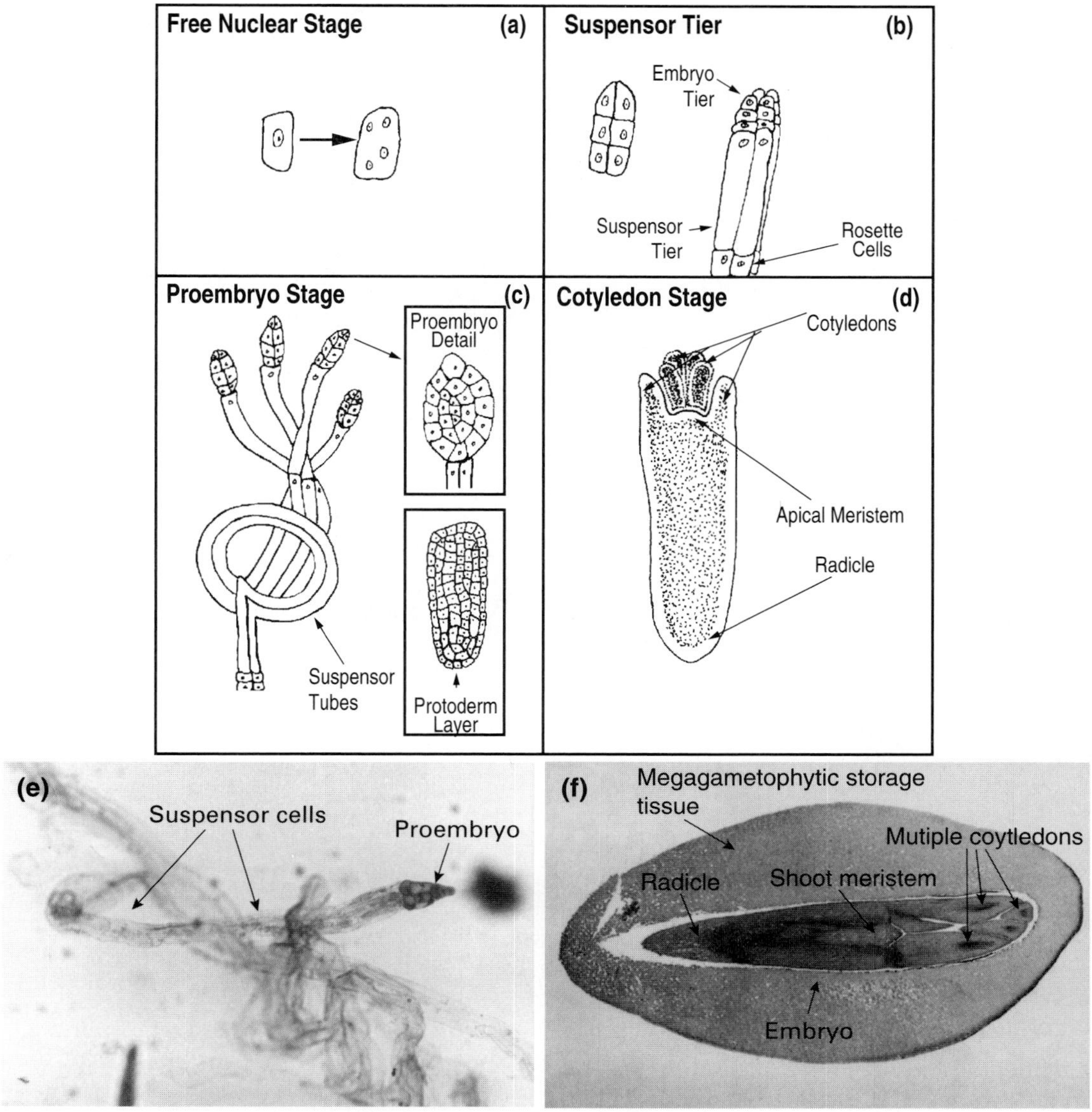

FIGURE 5–11 (a-d) Embryo development in a typical gymnosperm (pine). See text for description of figure. (e) Photomicrograph of a pine proembryo (f). Cross-section of a mature pine seed.

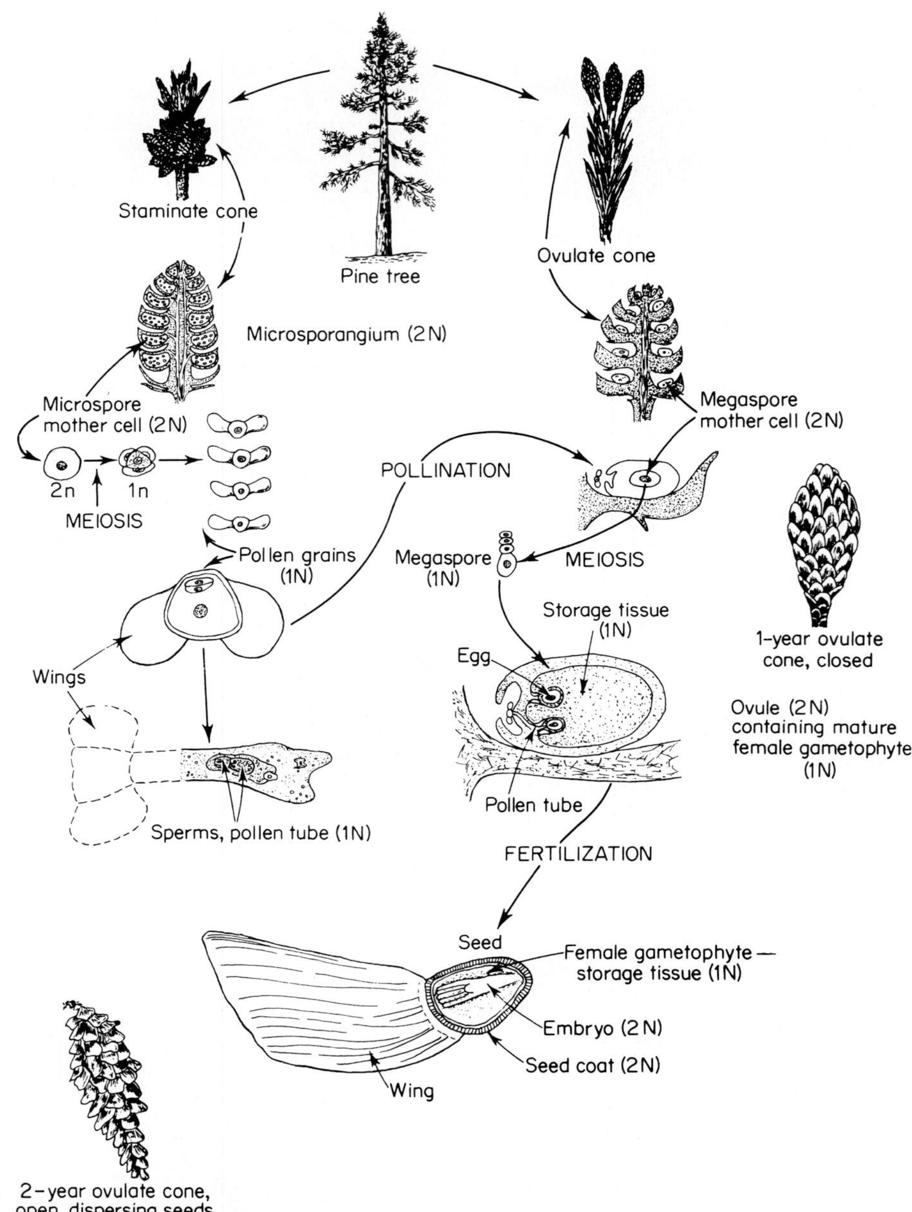

FIGURE 5–12 Diagrammatical representation of the sexual cycle in a gymnosperm (pine) showing meiosis and fertilization. It differs from that in the angiosperms by the formation of a "naked" seed, and the development of a haploid female gametophyte that comprises the storage tissue of the seed.

Polyembryogenesis

This phenomenon results with the division of the proembryo at a very early stage of development involving the initial **embryonal-suspensor mass** (ESM), which is highly embryogenic (*17*). Multiplication results from "cleavage," "splitting," or "budding." Each embryo is a duplicate of the genotype of the others and represents the new sexual generation. Polyembryogenesis is most common in gymnosperms, particularly conifers (*42, 59*).

FIGURE 5–13 Polyembryony in trifoliate orange *(Poncirus trifoliata)* seeds as shown by the several seedlings arising from each seed. One seedling, usually the weakest, may be sexual; the others arise apomictically from cells in the nucellus and are diploid copies of the mother plant.

False Polyembryogenesis

In this case, multiple embryos arise through the fertilization of separate egg nuclei within the same embryo sac apparatus. These have separate genotypes but are related, since each has the same seed parent but may have different pollen parents.

Apomixis

Apomixis (*47, 48, 53*) results from the production of an embryo that bypasses the usual process of **meiosis** and fertilization. The genotype of the embryo and resulting plant will be the same as the seed parent. **Seed production is asexual.** Such clonal seedling plants are known as **apomicts.** Some species or individuals produce only apomictic embryos and are known as **obligate apomicts;** others that produce both apomictic and sexual embryos are known as **facultative apomicts** (*4, 26, 42*).

Recurrent Apomixis

An **embryo sac** (female gametophyte) develops from the egg mother cell (or from some adjoining cell, if the egg mother cell disintegrates), but **complete meiosis does not occur.** Consequently, the egg has the normal diploid number of chromosomes, the same as the mother plant. The embryo subsequently develops directly from the egg nucleus **without fertilization.**

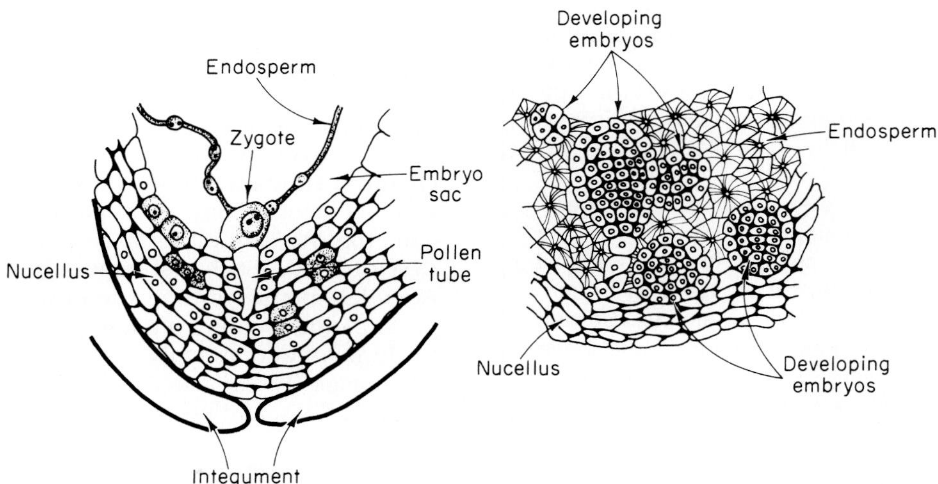

FIGURE 5–14 The development of nucellar embryos in *Citrus. Left:* Stage of development just after fertilization showing zygote and remains of pollen tube. Note individual active cells (shaded) of the nucellus, which are in the initial stages of nucellar embryony. *Right:* A later stage showing developing nucellar embryos. The large one may be the sexual embryo. (Redrawn from Gustafsson 1946.)

SIGNIFICANCE OF POLYEMBRYONY AND APOMIXIS

Polyembryony is significant because it demonstrates that embryogenic potential is not limited to the zygote but is possessed by various other somatic (nonreproductive) cells. This is the basis of in vitro somatic embryogenesis. (See Chapter 18) (*24, 56*).

Apomixis is significant in agriculture and horticulture because the seedling plants have the same genotype. This asexual process automatically eliminates variability and "fixes" the characteristics of any cultivar immediately. Uses in cultivar selection are described in Chapter 4. This apomictic life cycle as described in Chapter 4 has the same juvenile, transitional, and mature phases as does a seedling cycle. It differs in that no fertilization occurred to produce the embryo.

The value of apomictic cultivars depends on the use of the particular cultivar. Citrus is a noted example of apomixis. For cultivar propagation, apomictic seeds are not used because juvenile apomictic seedlings have many undesirable characteristics for fruit production, such as excessive vigor, slowness to come into fruit bearing, thorniness, and poor fruit quality. On the other hand, viruses are screened out during the process of embryogenesis and apomictic seedlings are used to develop "clean" versions of citrus cultivars. The most important use of apomictic seedlings is for rootstocks because of their vigor, lack of viruses, and uniformity (*8, 9*).

Apomixis is beneficial in plants where vegetative growth predominates. For example, a number of apomictic species and cultivars of grasses exist. Kentucky bluegrass *(Poa pratensis)* plants are facultative apomicts. Certain apomictic pasture grass cultivars have been developed, including 'King Ranch' bluestem *(Andropogun)*, 'Argentine; Bahia grass *(Paspalum notatum)*, and 'Tucson' side oats gama *(Bouteloua curtipendula)* (7).

This series of events is known to occur in some species of *Crepis, Taraxacum* (dandelion), *Poa* (bluegrass), and *Allium* (onion) without the stimulus of pollination; in others [e.g., species of *Parthenium* (guayule), *Rubus* (raspberry), *Malus* (apple), some *Poa* species, and *Rudbeckia*], pollination appears to be necessary, either to stimulate embryo development or to produce a viable endosperm.

Nonrecurrent Apomixis

In **nonrecurrent apomixis,** meiosis does occur and an embryo arises directly from the **egg nucleus** without fertilization. Since the egg is haploid, **the resulting embryo will also be haploid.** This case is rare and primarily of genetic interest. It does not consistently occur in any particular kind of plant as do recurrent apomixis and adventitious embryony.

Vegetative Apomixis

The term *apomixis* has been used in the past for any form of vegetative propagation. Today, its usage has been restricted to asexual production of an embryo within the ovule of flowering plants. However, some texts still include the term *vegetative apomixis* to describe the production of other structures besides an embryo. In some cases, vegetative buds or **bulbils** are produced in the inflorescence in place of flowers. This occurs in *Poa bulbosa* and some *Allium, Agave,* and grass species.

Stage II Cell Expansion

Stage II is a period of rapid cell enlargement due to the accumulation of **food reserves** (Figure 5–4). Accumulation of complex storage products including carbohydrates, fats, oils, proteins, and other biochemical substances into the storage organs of the seed (i.e., cotyledons, endosperm, nucellus, and/or gametophyte tissue) is an essential part of seed development. Such substances not only provide essential energy substrates to ensure survival of the germinating seedling but also provide essential food for humans and animals.

Stage II is an active period with large increases in DNA, RNA, and protein synthesis in the seed (*3*). The major food reserves include **carbohydrates (starch), storage proteins, and lipids (oils or fats).** Although different species may predominantly store a particular food reserve (i.e., cereal grains store starch; legumes store protein; and sunflower stores oil), most seeds contain all three types of food reserves (Table 5–1).

BIOTECHNOLOGY OF SEED RESERVES

The food reserves in seeds make up a major part of the world's diet both for human and livestock consumption. The nutritional quality of seeds can be improved by understanding the molecular genetics responsible for food reserve production. There are efforts through genetic engineering to improve the amino acid content of storage proteins in seeds (*38, 60*).

Cereals and legumes are important to the diet of most of the world's cultures and their yield and nutrition have been improved significantly over years of conventional breeding. However, most cereal proteins are nutritionally low in the essential lysine containing amino acids, and legume seeds produce storage proteins low in essential sulfur amino acids. The amino acid profile of these seeds can be improved using transformation technology (see Chapter 2) to insert new genes into crop plants to produce storage proteins high in lysine or sulfur. Exciting new germ plasm is being produced that will increase the nutritional yield of some of our major crop plants.

Plants that store oils in seeds are also the target of increased efforts to produce novel oils that can be used for detergents, lubricants, and cooking oil that is healthier by producing lipids low in unsaturated fat (*29*). Canola (rape seed) and soybeans are major crops being bio-engineered to produce novel oils.

These food reserves are manufactured in the developing seed from photosynthate being **"loaded"** into the seed from the mother plant. The process of seed reserve accumulation requires the translocation of small molecular weight compounds, such as sucrose, asparagine, glutamine, and minerals, into the seed. In dicot seeds, there is a direct vascular connection (phloem, xylem) between the mother plant and the seed through the funiculus (Figure 5–15). A vascular strand runs through the funiculus and usually down one side of the seed coat allowing transfer of photosynthate and water into the developing seed (*27*). There is no direct vascular connection from the seed coat to the nucellus, endosperm, or embryo and assimilates must reach the embryo by diffusion (*67*). At the same time, most viruses and large complex molecules are effectively screened from the embryo in this process but may accumulate in the outer layers of the seed. There is no vascular connection between the mother plant and developing seed in monocots. Rather, there is a group of cells at the seed and mother plant interface called **transfer cells** that facilitate the passage of photosynthate into the endosperm (*58*).

Specific mRNAs are required for the synthesis of storage compounds (*3, 23*). The pattern of mRNA for storage protein accumulation is similar for a number of proteins and mRNA's including phaseolin, legumin, and vicilin in legumes; cruciferin in rape seed; zein and hordein in cereals. A typical pattern for storage protein accumulation is illustrated in Figure 5–16a for broad bean (*Vicia faba*). This increase in storage protein is coincident with the increase in dry weight accumulation in Stage II embryos. Figure 5–16b shows the increase in mRNA that precedes the accumulation of the storage protein, cruciferin, in rape seed (*Brassica napus*) (*21*).

TABLE 5–1

Food reserves found in various plant species

	Average percent composition			
Species	***Protein***	***Oils***	***Starch***	***Major storage organ***
Cereals	10–13%	2–8%	66–80%	endosperm
Oil palm	9%	49%	28%	endosperm
Legumes	23–37%	1–48%	12–56%	cotyledons
Rape seed	21%	48%	19%	cotyledons
Pine	35%	48%	6%	female gametophyte

From (3, 15, 66)

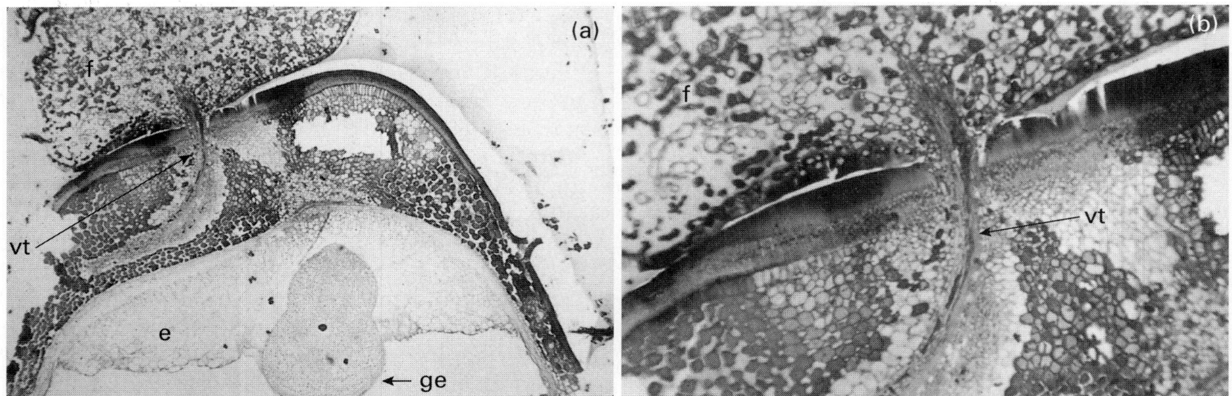

FIGURE 5–15 (a) Longitudinal section through a developing ovule of eastern redbud (*Cercis canadensis*) about 57 days post-anthesis (pollen shedding) showing the vascular connection between the funiculus and the ovule. (b) Close-up of the vascular trace. Note typical xylem cells in the vascular trace. Abbreviations: e, endosperm; f, funiculus; ge, globular embryo; vt, vascular trace. (Taken from Jones and Geneve 1995.)

Very specific genes are "turned on" during this stage of embryo growth (*23, 62*). These genes are only expressed during the embryogenesis stage of a plant's life cycle. The mRNA for storage proteins are no longer translated after maturation drying and can not be detected in germinating seeds (Figure 5–17).

Stage III Maturation Drying

Seeds in Stage III have reached **physiological maturity.** Physiological maturity is the time prior to maturation drying where the seed (embryo) has reached **maximum dry weight.** Seeds at physiologi-

Stages of Seed Development

Histo-differentiation | Cell expansion | Maturation drying

Vicia faba

Vicilin

Legumin

Protein content

Stage I | Stage II | Stage III

Days of Development

(a)

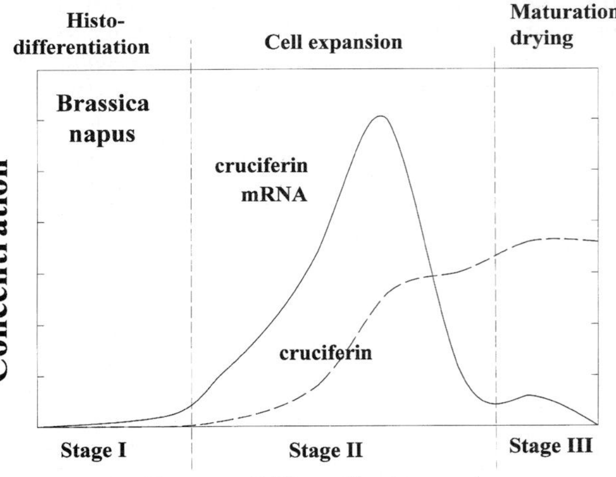

(b)

FIGURE 5–16 Accumulation of storage proteins related to the stages of seed development. (a) Pattern of protein accumulation in broad bean *(Vicia faba)* for vicilin and legumin, two major seed storage proteins in beans. (b) Accumulation of cruciferin protein and its mRNA in rape seed *(Brassica napus).* Note how the mRNA for the protein is only expressed at high levels during Stage II of seed development and is not detectable following maturation drying. (Redrawn form Finklestein and Crouch 1987.)

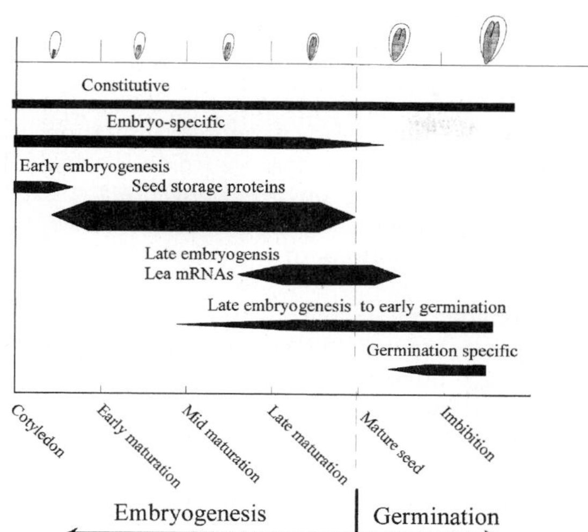

FIGURE 5–17 The relative abundance of mRNA types during seed development and germination. (Redrawn from Goldberg et al. 1989.)

cal maturity that are experimentally removed from the fruit show high germination potential as measured by seed viability and vigor (*46*). Stage III is characterized by rapid water loss. There is no longer a vascular connection with the mother plant through the funiculus. The area of the seed coat where the funiculus was attached is the **hilum.** The hilum appears to function as a valve that allows water to leave the seed completing desiccation (*30*).

Maturation drying can be considered a **"switch,"** ending the seed's developmental program and preparing the seed for germination (Figure 5–4) (*34, 37*). Synthesis of developmental proteins stops prior to drying and a new set of proteins is synthesized. A major set of these proteins is called **Lea (late embryogenesis abundant) proteins** (*16*). Lea proteins are synthesized in response to water loss in the seed. Lea proteins are very stable and hydrophilic (attracts water) and possibly function as desiccation protectants by stabilizing membranes and proteins as the seed dries. There are many ecological advantages to the production of a dry seed for seed dissemination and seed survival. However, there are few living organisms that can survive drying to 8 to 15 percent moisture (*49*). Lea proteins appear to help the seed adjust to a dry condition. In addition, the seed is also protected during desiccation by an increase in certain sugars and oligosaccharides that also provide stabilization to proteins and membranes (*5*).

Seeds acquire the ability to germinate prior to maturation drying. Usually, this potential to germinate is not expressed unless the fruit is removed from the plant and the seeds gradually dried (*35, 36*). Seeds that germinate prematurely on the plant without desiccation drying are called viviparous (see sidebar on vivipary). During normal seed development, the seed does not germinate prior to maturation drying because of high ABA content and high osmolarity in the seed. As maturation drying begins, a new set of late embryogenesis/early germination mRNA's are produced (*16, 62*). These are called **conserved mRNA** because they are stored in the dry seed. These conserved messages are expressed early in germination. Although conserved mRNA's are lost in the first few hours of germination, they allow the seed to produce proteins essential for germination before the embryo regains the capacity to synthesize new mRNA's.

Following maturation drying, the seed can be considered in a **quiescent or dormant condition.** Quiescent seeds fail to germinate because they are dry. Exposing quiescent seeds to a favorable environment will induce them to germinate. Dormant seeds fail to germinate even under favorable environmental conditions. There are several ecological advantages to seed dormancy and it is a common feature of many seeds. Over years of selection, dormancy has been bred out of most economically important crop species and these are considered as quiescent. Seed dormancy is discussed in detail in Chapter 7.

PLANT HORMONES AND SEED DEVELOPMENT

In general, concentrations of plant hormones are high in seeds compared to other parts of the plant (*3*). Seeds were the first tissue where several of the plant hormones were discovered and studied in detail. All the major hormones have been associated with seed development (*51, 52*). Plant hormones are involved in seed development in several ways. These include: (1) growth and differentiation of the embryo; (2) accumulation of food reserves; (3) storage for use during germination and early seedling growth; and (4) growth and development of fruit tissue.

Auxin. **Free and conjugated forms** of indoleacetic acid (IAA) are abundant in developing seeds. Free IAA is high during development (Stages I and II), but is reduced in mature seeds. Conjugated forms of IAA (see Chapter 2) are abundant in mature seeds and during germination. Free IAA is released from the conjugated forms for utilization during early seedling growth. There is evi-

VIVIPARY

Vivipary is the phenomenon in which seeds precociously germinate without maturation drying. These seeds germinate in the fruit while still attached to the plant. Vivipary occurs naturally in some species like mangrove *(Rhizophora mangle)*. In mangrove, vivipary is an adaptation to growing in a wet (swampy) environment. Embryos germinate directly on the tree to produce seedlings with a long javelin-shaped root (Figure 5–18). The seedling eventually falls and becomes embedded in the mud below (*61*).

For most plant species, however, vivipary is undesirable. Premature seed sprouting occurs in many species including cereal grains (wheat, corn), fleshy fruits (citrus and tomato), and nuts (pecan). Vivipary is considered a genetic mutation, but occurrence of precocious germination can be modified by the environment (*63*). Expect increased precocious germination in susceptible species during periods of wet weather (*3*).

The genetics of viviparous mutants in corn has been most extensively studied (*44*). Up to nine genes have been associated with vivipary in corn. The common feature in viviparous mutants is reduced production or insensitivity to abscisic acid (ABA). This supports the role for ABA in maintaining the embryo in the developmental mode through maturation drying.

dence that auxin from the developing seed signals the fruit to continue to develop. Fruits usually abscise if seeds abort or are unfertilized. Auxin applied to tomato or strawberry can induce parthenocarpic fruit development (see sidebar on parthenocarpy).

Gibberellins. Various forms of gibberellins are abundant during seed development (Stages I and II). Active forms decline at seed maturity and are replaced by conjugated forms of gibberellins. Like auxin, these conjugated forms are utilized during germination. Gibberellins may not play a major role in seed development. Gibberellin-deficient mutants of tomato and arabidopsis show normal seed development. Like auxin, gibberellins produced from the seed may also signal fruit development (*50*). Pea pods containing aborted seeds can continue development following application of gibberellic acid. Gibberellins can also induce parthenocarpic fruit development in crops like grapes (see sidebar on parthenocarpy).

FIGURE 5–18 Precocious (viviparous) germination in mangrove *(Rhizophora mangle)*. Note the protrusion of the radicle from the fruit while it is still attached to the plant. After sufficient radicle growth the fruit will fall from the plant and embed in the soft marshy soil around the mother plant.

Cytokinins. Several free and conjugated forms of cytokinins are high in developing seeds. The highest concentration of cytokinins are found during the cell division stages of embryogenesis (Stage I and early Stage II). Cytokinins appear to be supplied by the suspensor. Cytokinins play a role in controlling cell division in plants and appear to be important in the differentiation phase of Stage I embryos.

Abscisic acid(ABA). ABA levels are high in developing seeds (Stage II). ABA has been shown to induce the production of a number of important storage proteins in seeds. ABA appears to facilitate the accumulation of storage reserves in Stage II seeds. ABA is also a potent germination inhibitor. ABA prevents premature germination prior to maturation drying. ABA-deficient mutants fail to complete seed formation and germinate in the fruit. This phenomenon is called precocious germination or **vivipary** (see sidebar on vivipary).

PARTHENOCARPY

For many plant species, pollination is the stimulus for the beginning of fruit development. Continued fruit growth depends on seed formation. The number of seeds within a fruit strongly affects fruit size in species like apple and strawberry. Fruit that develop without seed formation (seedless) are called parthenocarpic fruit. Two types of parthenocarpy are recognized in plants (*65*). **Vegetative parthenocarpy** takes place in species like pear or fig, where the fruit develops even without pollination. **Stimulative parthenocarpy** takes place only after pollination but does not require fertilization or seed set for continued fruit growth. Grapes can form seedless fruit by stimulative parthenocarpy.

A number of species have been bred to naturally form parthenocarpic fruit. For example, parthenocarpy is essential for greenhouse cucumber fruit production because there are no reliable insect pollinators in the greenhouse. Other species (tomato, grape, some tree fruits) will form parthenocarpic fruit if sprayed with auxin or gibberellin. Interestingly, some species will only form parthenocarpic fruit if treated with auxin (tomato), while others require gibberellin (grape). The developing seed is the normal source for auxin and gibberellin for fruit growth. Both gibberellin and auxin are factors critical to fruit growth and interact during normal fruit development.

Ethylene. Significant amounts of ethylene are produced throughout seed development in *Brassica* species (*31*). Although the role of ethylene during seed development has not been extensively studied, it is interesting that ethylene production was high in developing *Brassica* embryos when embryos begin to "degreen" during maturation drying. In most seeds, embryos contain chlorophyll and are green during Stages I and II of development. There is a dramatic loss in chlorophyll during maturation drying while embryos "degreen" and appear yellow. Ethylene has a documented role in leaf senescence and could support embryo "degreening." Ethylene probably plays only a minor role during seed development. Ethylene mutants of several species produce apparently normal seeds.

RIPENING AND DISSEMINATION

Specific physical and chemical changes that take place during maturation and ripening of the fruit lead to fruit senescence and dissemination of the seed. One of the most obvious changes is the drying of the pericarp tissues. In certain species, this leads to dehiscence and the discharge of the seeds from the fruit. Changes may take place in the color of the fruit and the seed coats, and softening of the fruit may occur.

RECALCITRANT SEEDS

After developing seeds reach physiological maturity, they either proceed to desiccate (orthodox seeds), germinate on the plant (vivipary), or bypass complete desiccation (recalcitrant seeds). By definition, a recalcitrant seed loses viability after drying, while orthodox seeds tolerate drying (*3*). Germination in recalcitrant seeds must proceed soon after maturity or the seeds must be stored under conditions that prevent drying. Examples of storage life for some recalcitrant seeds stored at high humidity include: coffee *(Coffee arabica)* for 10 months; coconut *(Cocus nucifera)* for 16 months; and oak *(Quercus)* for 20 months. This compares to decades or years for many orthodox seeds. Recalcitrant seeds present challenges for propagators and limit germ plasm conservation because of their inability to store.

The biological basis for this inability in recalcitrant seeds to tolerate drying is not well understood *(2)*. *Arabidopsis* is an orthodox seeded species and its mutants have been a very useful tool for physiologists trying to study a variety of processes in plants. *Arabidopsis* mutants have been found with reduced levels of ABA, Lea proteins, and carbohydrates and these mutants are impaired for tolerance to drying. These substances are thought to be critical for survival in orthodox seeds during desiccation drying. It would be suspected that recalcitrant seeds would also show reduced ABA levels or be impaired for the production of Lea proteins or some carbohydrates. However, most recalcitrant species studied produce these substances at almost normal levels. The true nature of recalcitrance to drying remains to be found for this interesting group of seeds *(20)*.

Seeds of most species dehydrate at ripening and prior to dissemination. Moisture content drops to 30 percent or less on the plant. The seed dries further during harvest, usually to about 4 to 6 percent for storage. Germination cannot take place at this level of dryness and is an important basis for maintaining viability and controlling germination.

In certain other species, seeds must not dry below about 30 to 50 percent or they will lose their ability to germinate (*10*). These plants include (a) species whose fruits ripen early in summer, drop to the ground, and contain seeds that germinate immediately (some maples, poplar, elm); (b) species whose seeds mature in autumn and remain in moist soil over winter (oak); and (c) species from warm, humid tropics (citrus). These are called **recalcitrant** seeds (see sidebar on recalcitrant seeds) and produce special problems in handling.

Seeds of species with fleshy fruits may become dry but are enclosed with soft flesh which can decay and cause injury. In most species this fleshy tissue should be removed to prevent damage from spontaneous heating or an inhibiting substance. In some species, however (e.g., *Mahonia* and *Berberis*), the fruits and seeds may be dried together (*41*).

Seed dispersal is accomplished by many agents. Fish, birds, rodents, and bats consume and carry seeds in their digestive tracts (*22*). Fruits with spines or hooks become attached to the fur of animals and are often moved considerable distances. Wind dispersal of seed is facilitated in many plant groups by "wings" on dry fruits; tumbleweeds can move long distances by rolling in the wind. Seeds carried by moving water, streams, or irrigation canals can be taken great distances and often become a source of weeds in cultivated fields. Some plants (e.g., *Impatiens* and *Oxalis*) have mechanisms for short-distance dispersal, such as explosive liberation of seeds. Human activities in purposeful shipment of seed lots all over the world are, of course, effective in seed dispersal.

REFERENCES

1. Atwater, B.R. 1980. Germination, dormancy and morphology of the seeds of herbaceous ornamental plants. *Seed Sci. and Tech.* 8:523–73.
2. Berjak, P.L., J.M. Farrant and N.W. Pammenter. 1989. The basis of recalcitrant seed behavior. In *Recent Advances in the Development and Germination of Seeds.* R.B. Taylorson (ed.). pp. 89–108.
3. Bewley, J.D., and M. Black. 1994. *Seeds: Physiology of development and germination.* New York: Plenum Press.
4. Bhatnager, S.P., and B.M. Johri. 1972. Development of angiosperm seeds. In *Seed biology,* Vol. 1, T.T. Kozlowski, (ed.). New York: Academic Press, pp. 77–149.
5. Blackman, S.A., R.L. Obendorf and A.C. Leopold. 1992. Maturation proteins and sugars in desiccation tolerance of developing soybean seeds. *Plant Physiol.* 100:225–30.
6. Bohart, G.E. and T.W. Koeber. 1972. Insects and seed production. In *Seed Biology,* Vol. 3, T.T. Kozlowski, (ed.). Academic Press, New York. pp. 1–54.
7. Burton, G.W. 1983. Utilization of hybrid vigor. In *Crop Breeding,* D.R. Wood, K.M. Rawal, and M.N. Wood, (eds.). Amer. Soc. Agron. and Crop Sci. Amer., Madison WI. pp. 89–107.
8. Cameron, J.W., R.K. Soost, and H.B. Frost. 1959. The horticultural significance of nucellar embryony in critus. In *Citrus Virus Diseases,* J. Wallace, (ed.). Berkeley: Univ. Calif. Div. Agr. Sci., pp. 191–96.
9. Campbell, A.J., and D. Wilson. 1962. Apomictic seedling rootstocks for apples: Progress report, III. Ann. *Rpt. Long Ashton Hort. Res. Sta.* (1961): 68–70.
10. Chin, H.F. and E.H. Roberts. 1980. *Recalcitrant Crop Seeds.* Tropical Press, Kuala Lumpur.
11. Collins, G.B., and J.W. Grosser. 1984. Culture of embryos. In *Cell Culture and Somatic Cell Genetics of Plants.* Vol. 1, I.K. Vail, (ed.). New York: Academic Press, pp. 241–57.
12. Cooper, D.C., and R.A. Brink. 1940. Somatoplastic sterility as a cause of seed failure after interspecific hybridization. *Genetics* 25:593–617.
13. ———. 1945. Seed collapse following matings between diploid and tetraploid races of *Lycopersicon pimpinellifollium. Genetics* 30:375–401.
14. Corner, E.J.H. 1976. *The Seeds of Dicotyledons.* Cambridge Univ. Press. Cambridge.
15. Crocker, W. and L.V. Barton. 1957. *Physiology of Seeds.* Chronica Botanica Co., Waltham, Mass.
16. Dure, L.S. 1993. Lea proteins in higher plants. In *Control of Gene Expression.* D.P.S. Verma (ed.). CRC Press, Boca Raton, FL. pp. 325–335.

17. Durzan, D.J. 1988. Process control in somatic polyembryogenesis. In *Molecular genetics of forest trees,* J. Hällgren, ed. Umea: Swedish Agri. Univ.

18. Esau, K. 1977. *Anatomy of Seed Plants.* John Wiley and Sons. New York: Wiley.

19. Fahn, A. 1982. *Plant Anatomy.* Pergamon Press, New York: Pergamon Press.

20. Finch-Savage, W.E., S.K. Pramanik and J.D. Bewley. 1994. The expression of dehydrin proteins in desiccation-sensitive (recalcitrant) seeds of temperate trees. *Planta* 193:478–85.

21. Finkelstein, R.R., and M.L. Crouch. 1987. Hormonal and osmotic effects on developmental potential of maturing rapeseed. *HortScience* 22(5):797–800.

22. Fordham, A.J. 1984. Seed dispersal as it concerns the propagator. *Comb. Proc. Inter. Plant Prop. Soc.* 34:531–34.

23. Goldberg, R.B., S.J. Barker, and L. Perez-Grau. 1989. Regulation of gene expression during plant embryogenesis. *Cell* 56:149–60.

24. Gray, D. and A. Purohit. 1991. Somatic embryogenesis and development of synthetic seed technology. *Critical Rev. Plant Sci.* 10:33–61.

25. Gunn, C.R. 1972. Seed collecting and identification. In *Seed Biology,* Vol. 3, T.T. Kozlowski, (ed.). Academic Press, New York. pp. 55–144.

26. Gustafsson, A. 1946–1947. Apomixis in higher plants, Parts I–III. *Lunds Univ. Arsskrift,* N.F. Avid. 2 Bd 42, Nr. 3:42(2); 43(2); 43(12).

27. Hardham, A.R. 1976. Structural aspects of the pathways of nutrient flow to the developing embryo and cotyledons of *Pisum sativum* L. *Aust. Jour. Bot.* 24:711–21.

28. Hesse, C.O., and D.E. Kester. 1955. Germination of embryos of *Prunus* related to degree of embryo development and method of handling. *Proc. Amer. Soc. Hort. Sci.* 65:251–64.

29. Hills, M.J. and D.J. Murphy. 1991. Biotechnology of oil seeds. *Biotechnol. Genet. Eng. Rev.* 9:1–46.

30. Hyde, E.O. 1954. The function of the hilum in some Papilionaceae in relation to the ripening of the seed and the permeability of the testa. *Ann. Bot.* 18:241–256.

31. Johnson-Flanagan, A.M. and M.S. Spencer. 1994. Ethylene production during development of mustard *(Brassica juncea)* and canola *(Brassica napus)* seed. *Plant Physiol.* 106:601–606.

32. Jones, H.A. 1927. Pollination and life history studies of the lettuce (*Lactuca sativa* L.). *Hilgardia* 2:425–79.

33. Jones, R.E. and R.L. Geneve. 1995. Seed coat structure related to germination in eastern redbud (*Cercis canadensis* L.). *Jour. Amer. Soc. Hort. Sci.* 129:123–27.

34. Kermode, A.R. 1990. Regulatory mechanisms involved in the transition from seed development to germination. *Critical Rev. Plant Sci.* 9:155–195.

35. ———, and J.D. Bewley. 1985. The role of maturation drying in the transition from seed to germination. I. Acquisition of desiccation-tolerance and germinability during development of *Ricinus communis* L. seeds. *Jour. Exp. Bot.* 36:1906–15.

36. ———, and J.D. Bewley. 1985. The role of maturation drying in the transition from seed to germination. II. Post-germinative enzyme production and soluble protein synthetic pattern changes within the endosperm of *Ricinus communis* L seeds. *Jour. Exp. Bot.* 36:1916–27.

37. ———, J.D. Bewley, J. Dasgupta, and S. Misra. 1986. The transition from seed development to germination: A key role for desiccation? *HortScience* 21(5): 1113–18.

38. Larkins, B.A., C.R. Lending and J.C. Wallace. 1993. Modification of maize-seed-protein quality. *Am. Jour. Clin. Nutr.* 58:264S–269S.

39. Litz, R.E., R.L. Jarret, and M.P. Asokan. 1986. Tropical and subtropical fruits and vegetables. In *Tissue Culture as a Plant Production System for Horticultural Crops,* R.H. Zimmerman, R.J. Griesbach, F.A. Hammerschlag, and R.H. Lawson, (eds.). Dordrecht: Martinus Nijhoff Publishers, pp. 237–52.

40. Luckwill, J. 1948. The hormone content of the seed in relation to endosperm development and fruit drop in apple. *Jour. Hort. Sci.* 24:32–44.

41. Macdonald, B. 1986. *Practical Woody Plant Propagation for Nursery Growers,* Vol. 1. Portland, Or.: Timber Press.

42. Maheshwari, P., and R.C. Sachar. 1963. Polyembryony. In *Recent Advances in the Embryology of Angiosperm.* P. Maheshwari, (ed.). Delhi, India: Univ. of Delhi, Inter. Soc. of Plant Morph., pp. 265–96.

43. Martin, A.C. 1946. The comparative internal morphology of seeds. *Amer. Midland Nat.* 36: 5126–60.

44. McCarty, D.R. and C.B. Carlson. 1991. The molecular genetics of seed maturation in maize. *Physiol. Plant.* 81:267–72.

45. Meinke, D.W. 1991. Perspectives on genetic analysis of plant embryogenesis. *Plant Cell* 3:857–66.

46. Miles, D.F., D.M. TeKrony and D.B. Egli. 1988. Changes in viability, germination, and respiration of freshly harvested soybean seed. *Crop Sci.* 28: 700–04.

47. Naumova, T.N. 1993. Apomixis in Angiosperms: Nucellar and Integumentary Embryony. Boca Raton, FL.: CRC Press.

48. Nygren, A. 1954. Apomixis in the angiosperms II. *Bot. Rev.* 20:577–649.

49. Oliver, M.J. and J.D. Bewley. 1992. Desiccation tolerance in plants. In *Water and Life.* G.N. Somero, C.B. Osmond, and C.L. Bolis (eds.). Berlin. Springer-Verlag, pp. 141–60.

50. Ozaga, J.A., M.L. Brenner and O.M. Reinecke. 1992. Seed effects on gibberellin metabolism in pea pericarp. *Plant Physiol.* 100:88–94.

51. Quatrano, R.S. 1987. The role of hormones during seed development. In *Plant Hormones and Their Role in Plant Growth and Development.* P.J. Davies (ed.). Boston. Marinus Nijhoff Pub., pp. 494–514.

52. Radley, M. 1979. The role of gibberellin, abscisic acid, and auxin in the regulation of developing wheat grains. *Jour. Exp. Bot.* 30:381–89.

53. Raghaven, V. 1986. *Embryogenesis in Angiosperms.* Cambridge: Cambridge Univ. Press.

54. Randolf, L.F. 1936. Developmental morphology of caryopsis of maize. *Jour. Agric. Res.* 53:881–97.

55. Rietsema, J., S. Satina, and A.F. Blakeslee. 1955. Studies on ovule and embryo growth in *Datura.* 1. Growth analysis. *Amer. Jour. Bot.* 42:449–54.

56. Rowe, W.J. 1986. New technologies in plant tissue culture. In *Tissue Culture as a Plant Production System for Horticultural Crops.* R.H. Zimmerman, R.J. Griesbach, F.A. Hammerschlag, and R.H. Lawson, (eds.). Dordrecht: Martinus Nijhoff Publishers, pp. 35–51.

57. Schopmeyer, C.S. ed. 1974. *Seeds of Woody Plants in the United States.* U.S. Dept. Agr. Handbook 450. Washington, D.C.: U.S. Govt. Printing Office.

58. Shannon, J.C. 1972. Movement of ^{14}C-labeled assimilates into kernels of Zea mays L.I. pattern and rate of sugar movement. *Plant Physiol.* 49:198–202.

59. Singh, H., and B.M. Johri. 1972. Development of gymnosperm seeds. In *Seed biology,* Vol. 1, T.T. Kozlowski, ed. New York: Academic Press, pp. 22–77.

60. Sommerville, C.R. 1993. Future prospects for genetic modification of the composition of edible oils from higher plants. *Am. Jour. Clin. Nutr.* 58:270S–275S.

61. Stephens, W. 1969. The mangrove. *Oceans* 2(5):51–55.

62. Thomas, T.L. 1993. Gene expression during plant embryogenesis and germination: An overview. *Plant Cell* 5:1401–10.

63. Wellington, P.S., and V.W. Durham. 1958. Varietal differences in the tendency of wheat to sprout in the ear. *Empire Jour. Exp. Agr.* 26:47–54.

64. West, M.A.L. and J.J. Harada. 1993. Embryogenesis in higher plants: An overview. *Plant Cell* 5:1361–69.

65. Westwood, M.N. 1993. Temperate-zone Pomology. Physiology and Culture. Portland, OR.: Timber Press.

66. Winton, A.L. and K.B. Winton. 1932. *The Structure and Composition of Foods.* Vol. I. New York: Wiley.

67. Wolswinkel. P. 1992. Transport of nutrients into developing seeds: a review of physiological mechanisms. *Seed Sci. Res.* 2:59–73.

SUPPLEMENTARY READING

Asker, S.E. and L. Jerling. 1992. *Apomixis in Plants.* Boca Raton: CRC Press.

Bewley, J.D., and M. Black. 1994. *Seeds: Physiology of development and germination.* New York: Plenum Press.

Kozlowski, T.T., ed. 1972. *Seed biology,* Vol. I, *Importance, development and germination.* New York: Academic Press.

Martin, A.C., and W.D. Barkley. 1961. *Seed identification manual.* Berkeley: Univ. of Calif. Press.

Raghavan, V. 1986. *Embryogenesis in angiosperms.* Cambridge: Cambridge Univ. Press.

Rubenstein, I., R.L. Phillips, C.E. Green, and G. Gengenbach, eds. 1979. *The plant seed: Development, preservation and germination.* New York: Academic Press.

Special Review issue on plant reproduction. *Plant Cell* 5(10):1139–1251.

6

Techniques of Seed Production and Handling

More plants are propagated for food, fiber, and ornamentals from seeds than any other method of propagation. Seed propagation is the cornerstone for producing agronomic, vegetable, forestry, and many ornamental plants. The production of high-quality seeds is of prime importance to propagators. In the production of any crop, the cost of the seed is usually minor compared to other production costs. Yet, no single factor is as important in determining the success of the operation. A scheme for producing quality seed is included in Figure 6–1.

Historically, seed for next season's crop was collected as a byproduct of production. Although occasionally some seeds may be still produced in this manner (e.g., some Third World production), modern seed production has become a very specialized industry (*33*). Reasons for the emergence of specialized seed companies include:

- the increase in **improved cultivars and varieties** (many of these are hybrids; see Chapter 4) introduced by university, USDA, and private breeders.
- the need for **genetic purity** in seed lots.
- the expectation from growers for **high quality seed** with improved germination characteristics.
- the development of specialized **seed cleaning equipment** and handling (seed conditioning) that provide higher quality seeds by eliminating weed seeds, undersized or damaged seeds, and reducing seed-borne pathogens.
- development of preplanting **seed treatments** (see p. 162).

SOURCES OF SEED

Commercial Seed Sources

Agricultural, Vegetable, and Flower Seed

Commercial seed production is a specialized intensive industry with its own technology geared to the requirements of individual species (Figure 6–2). Some agricultural seeds, such as corn, wheat, small grains, and grasses, are produced in the area where the crops are grown. The advantages for producing

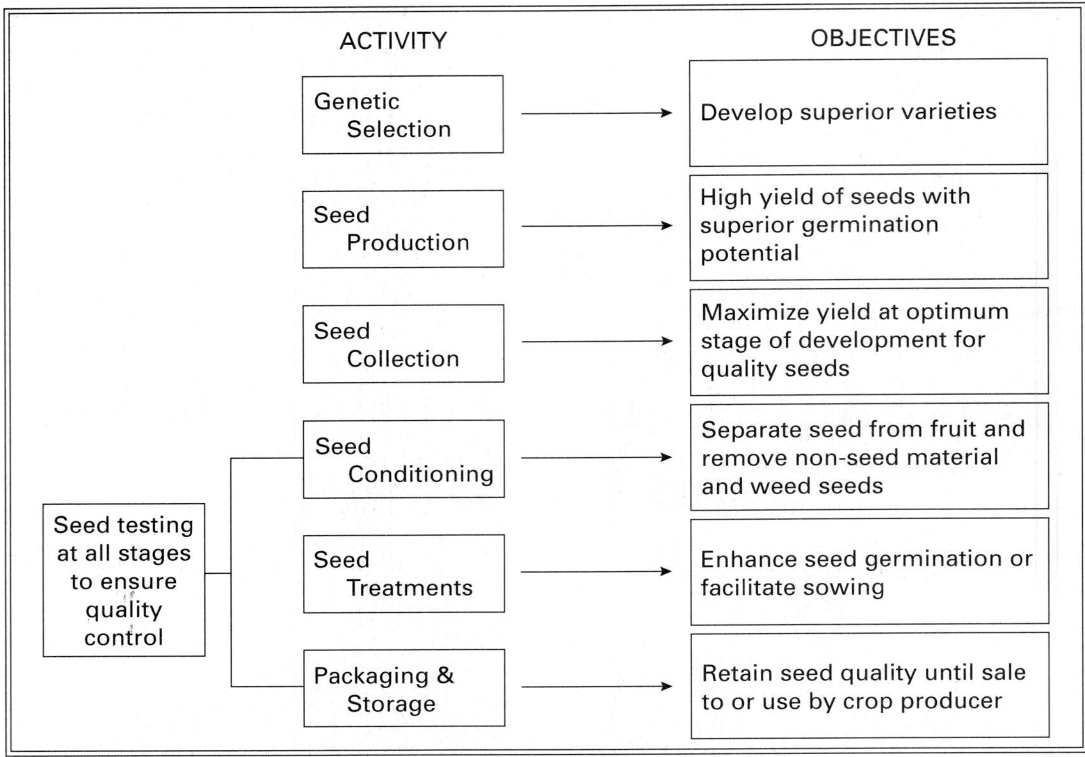

FIGURE 6–1 Procedures for producing and handling a commercial seed lot.

FIGURE 6–2 A majority of important agronomic, horticultural, and forestry crops are propagated by seeds. This includes a large diversity in seed size, shape, and requirements for germination.

seed in the production area include reduced transportation and handling costs and reduced potential for genetic shift (see Chapter 4). These are important considerations for agronomic crops where large amounts of seed are required to produce a crop. However, crop production areas may not provide the best conditions for producing high-quality, disease-free seed. Therefore, large amounts of high-value seeds such as forage, vegetable, and flowers are produced in specialized growing areas.

The major considerations for selecting areas to produce seed are environmental conditions and the cost of production (*36, 126*). Large quantities of grass, vegetable, and flower seeds are produced in areas characterized by low summer rainfall, low humidity, and limited rainfall during the seed harvest season (*12, 126*). These conditions provide good seed yields and reduce disease problems, especially during harvest when the seed must dry before being handled. There are also crops that require special environmental conditions to flower and set seeds. These include the biennial vegetable and flower corps that require **vernalization** (a period of cold temperature) to flower (*126*). Examples are onion and carrot seed production. One-year-old biennial plants used for seed production have been called **stecklings** (*61*). Plants may be chilled by overwintering in the field, or in some cases, stecklings are brought into a cooler (5° C; 40° F) to satisfy vernalization requirements.

Major production areas for high-value seed production in the United States that meet these important environmental conditions include grass and forage seed production in the Pacific Northwest and vegetable and flower seed production in the central, coastal valleys of California. Increasingly, seed production has become an international industry. High-value seeds are also being produced in South and Central America as well as Australia. Advantages of producing seed in the southern hemisphere are reduced cost of production and seed production in the season prior to planting in northern crop production areas, thus reducing storage time and cost. Hybrid seed production that requires hand pollination has moved to areas of the world with reduced labor costs. These include Central and South America, Southeast Asia, and India (*126*).

Regardless of the country where seeds are produced, there are several important considerations that must be satisfied when selecting specific sites for **seed production** (*126*). These include:

Soil type and fertility for good seed yields.

Cropping history to avoid disease or herbicide carryover.

Soil moisture or availability of supplemental irrigation.

A **dry environment** during seed harvest.

Ability to **isolate open or cross-pollinated crops.** For example, self-pollinated tomato plants require only 50 feet separation between varieties, while some insect or wind pollinated crops require up to a mile of separation between varieties to avoid unwanted cross-pollination (*61*).

Additional requirements for high quality seed production are the selection of planting density, pest control, and availability of insect pollinators (*126*). In many cases, conditions for seed production and crop production are very similar.

Woody Plant Seed

A number of commercial and professional seed-collecting firms exist which collect and sell seed of certain species. Lists of such producers are available (*37, 83*). Such seed should be properly labeled as to their origin or provenance (see Chapter 4). Some tree seed can be obtained as certified.

Seed Exchanges

Many arboreta and plant societies have seed exchanges or will provide small amounts of specialty seed.

Seed Collecting

Tree and shrub seed may be collected by the individual nursery or propagator (*104, 114, 130*). These may be collected from specific seed collection zones or from seed production areas (see Chapter 4). Seed may be collected from standing trees, trees felled for logging, or from squirrel caches. They might be collected from parks, roadways, streets, or wood lots. Seed collecting has the advantage of being under the control of the propagator but requires intimate knowledge of each species and the proper method of handling. Most important, the collector should be aware of the importance of the selection principles described in Chapter 4.

Seed Orchards

Seed orchards or plantations are used to maintain seed source trees of particularly valuable species. They are extensively used by nurseries in the production of rootstock seeds of certain species or cultivars and for forest tree improvement. The major advantage to a seed orchard is that it is a consistent source of seeds from a known (often genetically superior) parentage (*82*). They also allow the seed producer to maximize seed harvest by reducing loss due to environmental conditions or animals. Such seed orchards are described in Chapter 4.

Fruit-Processing Industries

Historically, much fruit tree rootstock seed has been obtained as byproducts of fruit-processing industries such as canneries, cider presses, and dry yards. Examples include 'Lovell' or 'Halford' peach or 'Royal' apricot in California as well as pears in the Pacific Northwest. The procedure is satisfactory if the correct cultivar is used (Chapter 4). Some seed-borne viruses might be present in certain seed sources.

HARVESTING AND PROCESSING SEEDS

Maturity and Ripening

Each crop and plant species undergoes characteristic changes leading to seed ripening which must be known to establish the best time to harvest (*1, 33, 83, 104, 129*). A seed is mature when it can be removed from the plant without impairing the seed's germination. Usually, the seed crop is harvested after the seed has reached **physiological maturity** (see Chapter 5). This is the stage on the plant when no further increase in dry weight will occur (*8*). During production the seed crop is sampled often to determine the stage of physiological maturity and the percent moisture of the seed (see sidebar on testing seed moisture). If the fruit ripens too soon or the seed is harvested when the embryo is insufficiently developed, the seed is apt to be thin, light in weight, shriveled, poor in quality, and short-lived (*36*).

TESTING SEED MOISTURE

Moisture content is found by the loss of weight when a sample is dried under standardized conditions (*120*). Oven drying at 130° C (266° F) for 1 to 4 hours is used for many kinds of seeds. For oily seeds 103° C (217° F) for 17 hours is used (*120*) and for some seeds which lose oil at that temperature (e.g., fir, cedar, beech, spruce, pine, hemlock) a toluene distillation method is used. Various kinds of electronic meters can be used for quick moisture tests (*24*).

Early seed harvest may be desirable for seeds of some species of woody plants that produce a hard seed covering in addition to a dormant embryo. If seeds become dry and the seed coats harden, the seeds may not germinate until the second spring (*127*), whereas they would have germinated the first spring. On the other hand, if harvesting is delayed too long, the fruit may **dehisce** ("split open" or "shatter"), drop to the ground, or be eaten or carried off by birds or animals. Thus, a balance must be made between late and early harvest to obtain the maximum number of high-quality seed.

Plants can be divided into three types (Figure 6–3) according to the way their fruits ripen (*96*):

Type 1 Includes plants with **dry fruits** composed of seed and fruit covers. These do not dehisce and are not disseminated immediately upon maturity. Type 1 includes most of the agricultural crops, such as corn, wheat, and other grains. Many of these have undergone considerable selection during domestication for ease of harvest and handling.

Type 2 Plants producing **dry seeds** from fruits that **dehisce readily at maturity.** This type includes seeds in follicles, pods, capsules, siliques, and cones. Many species of type 2 are either ornamental plants or are used for purposes other than food production.

Type 3 Plants with **fleshy fruits.** This includes important fruit and vegetable species used for food (such as berries, pomes, and drupes) as well as many related tree and shrub species used in landscaping or forestry.

Harvesting and Handling Procedures

Type 1. Field-grown crops (cereals, grasses, corn) can be harvested by a combine, a machine that cuts and threshes the standing plant in a single operation. Plants that tend to fall over or "lodge" are cut, piled, or windrowed for drying and curing. Low humidity is important during harvest. Rain damage results in seeds that show low vigor. The force required to dislodge the seeds may result in mechanical damage and can reduce viability and result in abnormal seedlings. Some of these injuries are internal and not noticeable, but they result in low viability after storage (*3, 62, 96*). Damage is most likely to occur if seed moisture is too high or (most frequently) too low and if the machinery is not properly adjusted. Usually less injury occurs if the seeds are somewhat moist at harvest (around 12 to 15 percent).

Type 2. Fruits of this type dehisce to release the dry seeds. Some crops of this group include many annual or biennial flowers (delphinium, pansy, petunia), various vegetables (onion, cabbage, other cruciferous crops, and okra) which must be harvested before they are fully mature, then dried or cured before extraction. Consequently, some seeds will be underdeveloped and immature. Many tree and shrub plants also have fruits which fall into this group and are handled with similar procedures. The steps are as follows:

1. **Drying.** Cut plants (sometimes by hand) or collect dry fruits and place on a canvas, tray, or screen, or lay on the floor to dry for one to three weeks. If only a few plants, cut and hang upside down in a paper bag to dry.
2. **Extraction.** Commercial seeds may be harvested by special machines that both extract by beating, flailing, or rolling the dry fruit

FIGURE 6–3 Seed conditioning begins by removing seed from the fruit. The extraction method depends on the type of fruit. These include dry fruits like the caryopsis of grains that do not dehisce at maturity; dry fruits that shed seeds at maturity like the cones of conifers; and fleshy fruits common to many vegetable crops like tomato and squash.

and separate the seeds from the fruit parts, dirt, and other debris. Tree and shrub seed may be extracted with special macerators (*104*) also suitable for fleshy fruits or seeds may be beaten, flailed, or screened by hand. Lightweight seed may be removed during this screening operation.

3. **Seed conditioning (cleaning).** Further cleaning may be required to eliminate all dirt, debris, weed, and other crop seed. Commercial seed conditioning utilizes various kinds of specialized equipment (Figure 6–4), such as screens of different sizes, air blasts, and **gravity separators** (*83, 123, 125*).

Type 3. Vegetable crops include tomato, pepper, eggplant, and various cucurbits. These are produced in commercial fields and may utilize special

FIGURE 6–4 Seed conditioning (cleaning) has become highly specialized with equipment developed specifically to clean different types of seed. Seed conditioning not only includes seed extraction from the fruit, but also includes separating the desired seeds from unwanted weed seeds, debris, and even poor quality (undersized or unfilled) seeds of the crop being conditioned. Gravity separators like the one illustrated take advantage of seed density and seed shape (roundness) to separate the desired seed from the bulk seed harvest.

macerating machinery. Cucumber and other vine crops are handled with specially developed machines (*113*).

Fleshy fruits include berries (grape), drupes (peach, plum), pomes (apple, pear), aggregate fruits (raspberry, strawberry), and multiple fruits (mulberry). In general, fleshy fruits are easiest to handle if ripe or overripe. However, fruits in the wild are subject to predation by birds (*44*).

1. **Extraction.** For small lots, fruits may be cut open and the seeds scooped out, treaded in tubs, rubbed through screens, or washed with water from a high-pressure spray machine in a wire basket. For larger lots separation is by **fermentation,** mechanical means, or washing through screens. A macerator can be used that crushes the fruits and mixes the pulverized mass with water that is diverted into a tank.
2. **Fermentation.** The macerated fruits are placed in large barrels or vats and allowed to ferment for about four days at about 21° C (70° F), with occasional stirring. If the process is continued too long, sprouting of the seeds may result. Higher temperatures during fermentation shorten the required time. As the pulp releases the seeds, the heavy, sound seeds sink to the bottom of the vat, and the pulp remains at the surface. Following extraction, the seeds are washed and dried either in the sun or in dehydrators. Additional cleaning is sometimes necessary to remove dried pieces of pulp and other materials. Extraction by fermentation is particularly desirable for tomato seed, because it can help control bacterial canker.
3. **Flotation.** A flotation process involves placing the seeds and pulp in water so that the heavy, sound seeds will sink to the bottom and the lighter pulp, empty seeds, and other extraneous materials will float to the top. This procedure can also be used to remove lightweight, unfilled seeds and other materials from dry fruits, such as acorn fruits infested with weevils, but sometimes both good and bad seed will float. In the production of rootstock seedlings in oranges, separation of the seed from the surrounding fruit pulp is facilitated by the addition of a commercial pectinase enzyme (*14*). The small berries of some species, such as Cotoneaster, Juniperus, and Viburnum, are somewhat difficult to process because of small size and the difficulty in separating the seeds from the pulp. One way to handle such seeds is to crush the berries with a rolling pin, soak them in water for several days, and then remove the pulp by flotation.
4. **Blender.** Another device that removes seeds from small-seeded fleshy fruits is an electric mixer or blender (*109*). To avoid injuring the seeds, the metal blade of the latter machine can be replaced with a piece of rubber (or Tygon tubing), 1½ in. square, cut from a tire casing. It is fastened at right angles to the revolving axis of the machine (*129*). A mixture of fruits and water is placed in the mixer and stirred for about two minutes. When the pulp has separated from the seed, the pulp is removed by flotation. This procedure is satisfactory for fruits of *Amelanchier* (serviceberry), *Berberis* (barberry), *Crataegus* (hawthorn), *Fragaria* (strawberry), *Gaylussacia* (huckleberry), *Juniperus* (juniper), *Rosa* (rose), and others (*109*).
5. **Drying.** Seeds are thoroughly washed to remove any fleshy remnants and dried except that seeds of recalcitrant species must not be

allowed to dry out (see p. 167). If left in bulk for even a few hours, seeds that have more than 20 percent moisture will heat; this impairs viability. Drying may occur either naturally in open air if the humidity is low, or artificially with heat or other devices (*58*). Drying temperatures should not exceed 43° C (110° F); if the seeds are quite wet, 32° C (90° F) is better. Too-rapid drying can cause shrinkage and cracking and can sometimes produce hard seed coats. The minimum safe moisture content for storage of most seeds is in the range of 6 to 15 percent.

Cones

Conifer cones require special procedures (*104*).

1. **Drying.** Cones of some species will open if they are dried in the open air for 2 to 12 weeks. Others must be force-dried at higher temperatures in special heating kilns. Under such conditions, the cones will open within several hours or at most two days. The temperature of artificial drying should be 46 to 60 °C (115 to 140° F) depending upon the species, although a few require even higher temperatures. Jack pine (*Pinus banksiana*) and red pine (*P. resinosa*), for example, need temperatures of 77° C (170° F) for five to six hours. Caution must be used with high temperatures; overexposure will damage seeds. After the cones have been dried, the scales open, exposing the seeds.
2. **Extraction.** The cones are shaken by tumbling or raking to remove the seeds. A revolving wire tumbler or a metal drum is used when large numbers of seeds are to be extracted. The seeds should be removed immediately upon drying, since the cones may close.
3. **Dewinging.** Conifer seeds have wings, which are removed except in species whose seed coats are easily injured, such as incense cedar (*Calocedrus*). Fir (*Abies*) seed is easily injured, but the wings can be removed if the operation is done gently. Redwood (*Sequoia and Sequoiadendron*) seeds have wings that are an inseparable part of the seed. For small lots of seed, dewinging can be done by rubbing the seeds between moistened hands or trampling or beating the seeds packed loosely in sacks. For larger lots of seeds, special dewinging machines are used.
4. **Cleaning.** The seeds are cleaned after extraction to remove the wings and other light chaff. As a final step, separation of heavy, filled seed from light seed is accomplished by gravity or pneumatic separators.

SEED TESTING

In the United States, state laws regulate the shipment and sale of agricultural and vegetable seeds within that state. Seeds entering interstate commerce or those sent from abroad are subject to the **Federal Seed Act** adopted in 1939. Such regulations require labeling (Figure 6–5) by the shipper of commercially produced seeds as to: name and cultivar; origin; germination percentage; and the percentage of pure seed, other crop seed, weed seed, and inert material. The regulations set minimum standards of quality, germination percentage, and freedom from weed seeds. Special attention must be paid to designated noxious weeds for a particular growing region. Shipment and sale of tree seed is regulated by law in some states (*102*) and in most European countries.

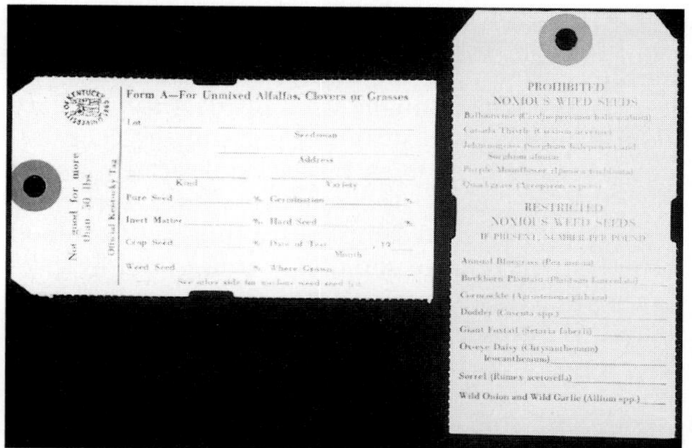

FIGURE 6–5 State and federal seed laws require testing seed lots prior to sale. Information for a seed lot includes standard germination percentage according to accepted seed testing rules, purity of the seed lot (percentage of seeds that are the desired crop and its trueness to type), percentage of weed seeds, and the amount of noxious weed seeds in the seed lot. Noxious weeds are designated as weeds that are particularly undesirable and tolerances may differ for a crop or region of the country.

Seed testing provides information to meet legal standards, determines seed quality, and establishes the rate of sowing for a given stand of seedlings. It is desirable to retest seeds that have been in storage for a prolonged period.

Procedures for testing agriculture and vegetable seed in reference to the **Federal Seed Act** are given by the U.S. Department of Agriculture (*122*). The **Association of Official Seed Analysts** also publishes procedures for testing these seeds in addition to procedures for testing seeds of many flower, tree, and shrub species (*4, 5*). International rules for testing seeds of many tree, shrub, agricultural, and vegetable species are published by the **International Seed Testing Association** (*69*). The **Western Forest Tree Seed Council** has also published testing procedures for tree seed (*128*).

A high-quality seed lot is a function of the following characteristics that are routinely tested by seed companies or state seed labs (*101, 126*):

1. Germination (viability)
2. Purity
3. Vigor
4. Seed health
5. Noxious weed seed contamination

Sampling

The first step in seed testing is to obtain a uniform sample representing the entire lot under consideration (Figure 6–6). Equally sized **primary** samples are taken from evenly distributed parts of the seed lot, such as a sample from each of several sacks in lots of less than five sacks or from every fifth sack with larger lots. The seed samples are thoroughly mixed to make a **composite** sample. A representative portion is used as a **submitted sample** for testing. This sample is further divided into smaller lots to produce a **working sample** (i.e., the sample upon which the test is actually to be run). The amount of seed required for the working sample varies with the kind of seed and is specified in the Rules for Seed Testing (*5*).

Viability Determination

Viability can be determined by several tests, the **standard germination, excised embryo,** and **tetrazolium** tests being the most important. In the standard germination test the **germination percentage** is determined by the percent of normal seedlings produced by the pure seed (the kind under consideration). To produce a good test, it is desirable to use at least 400 seeds picked at random and divided into lots of 100 each. If any two of these lots differ by more than 10 percent, a retest should be carried out. Otherwise, the average of the four tests becomes the germination percentage.

FIGURE 6–6 A sample from each seed lot must be tested prior to sale usually by a state-certified seed lab. The seed analyst uses a seed sorter to randomly select a seed sample for testing from the submitted seed lot. A portion of the seed lot will be tested for purity, while an additional subsample will be evaluated for standard germination.

Standard Germination Tests

In a standard germination test the seeds are placed under optimum environmental conditions of light and temperature to induce germination (Figure 6–7). The conditions required to meet legal standards are specified in the rules for seed testing, which may include type of test, environmental conditions, and length of test.

Various techniques are used for germinating seeds in seed-testing laboratories. Small seeds are placed on germination trays (not galvanized steel, which contains toxic zinc salts). Plastic boxes, paraffined cardboard boxes, or covered glass petri dishes also are useful containers. Absorbent paper is cut into small pieces (*blotters*) and small seeds are placed on top or between two layers. Other media are absorbent cotton, paper towels (five thicknesses), filter paper (five layers), and for large seeds, a sand, vermiculite, perlite, or soil (16 mm; 5/8 in.) layer. Containers are placed in germinators in which temperature, moisture, and light are controlled (Figure 6–8). To discourage the growth of microorganisms, all materials and equipment should be kept scrupulously clean, sterilized when

FIGURE 6–7 A standard germination test is required for all seed lots prior to sale. The two most common test procedures include the rolled towel and the petri dish tests. The tests and the procedures used for standard germination are detailed in accepted publications like the rules for testing seeds from the Association of Official Seed Analysts (*4, 5*). Included in these rules will be the preferred test (i.e., petri dish or rolled towel); the environment for the test (i.e., 20/30, this indicates daily cycles of 16 hours at 20° C followed by 30° C for 8 hours); whether light is required during the test; any seed pretreatments for dormant seeds (e.g., treatment with gibberellin or potassium nitrate); and the number of days for the first and last evaluation (counts).

FIGURE 6–8 Commercial germination chambers have light, temperature, and humidity control for conducting standard germination and vigor tests. (Courtesy C.E. Heit.)

possible, and the water amount carefully regulated. No water film should form around the seeds; neither should the germination medium be so wet that a film of water appears when that medium is pressed with a finger. Relative humidity in the germinator should be 90 percent or more to prevent drying. Containers with sand should be kept tightly closed. Water should not be added during the test.

The **rolled towel** test (Figure 6–9) is commonly used for testing cereal grains. Several layers of moist paper toweling, about 2.8 by 3.6 cm (11 by 14 in.) in size, are folded over the seeds, then rolled into cylinders and placed vertically in a germinator.

A germination test usually runs from one to four weeks but could continue for three months for some slow-germinating tree seeds with dormancy. A *first count* may be taken at one week and germinated seeds discarded with a final count taken later. At the end of the test, seeds are divided into (a) normal seedlings, (b) hard seeds, (c) dormant seeds, (d) abnormal seedlings, and (e) dead or decaying seeds. A normal seedling should have a well-developed root and shoot, although the criteria for a "normal seedling" varies with different kinds of seeds (Figure 6–10). "Abnormal seedlings" can be caused by age of seed or poor storage conditions; insect, disease, or mechanical injury; overdoses of fungicides; frost damage; mineral deficiencies (manganese and boron in peas and beans); or toxic materials sometimes present in metal germination trays, substrata, or tap water. Any ungerminated seed should be examined to determine the possible reason. "Hard seeds" have not absorbed water. Dormant seeds are those that are firm, swollen, and free from molds, but show erratic sprouting, or none.

Under seed testing rules certain environmental requirements to overcome dormancy may be specified routinely for many kinds of agricultural, vegetable, and flower seeds (*4, 69*). Tree and shrub seeds often require special pregermination treatments before tests are run (Table 6–1).

Excised-Embryo Test

The excised-embryo test is used to test the seed viability of woody shrubs and trees whose dormant embryos require long periods of after-ripening before true germination will take place (*4, 43, 63*). In this test the embryo is excised from the seed and germinated alone (see Figure 6–11).

FIGURE 6–9 Commercial seed labs process a large number of seed samples. They must keep accurate records of each seed lot and must be efficient to process samples in a timely manner, while maintaining high reproducibility from seed lot to seed lot. Seed analysts use a template board under vacuum pressure to place a standard number of seeds in precise locations on the germination paper for the rolled towel or petri dish tests. A slight tap with the mallet ensures all the seeds have been dislodged from the template. The germination media used for these tests are germination paper or blotters that have been produced to minimize any inhibitors that may be on the papers that may impact germination.

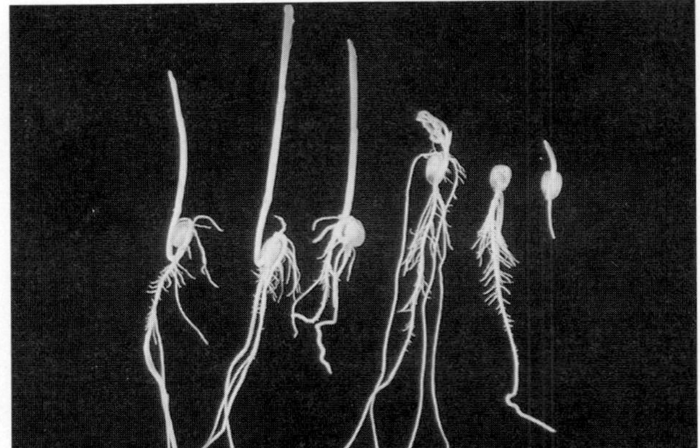

FIGURE 6–10 The criteria for germination must also be standardized from testing lab to testing lab. These criteria are published with illustrations to help aid the seed analyst in determining "normal" germination. The three corn seeds on the left are "normal" and counted as germinated in a standard rolled towel test. The three corn seeds on the right are "abnormal" because either the shoot or root has not developed normally after the final count for this seed test.

The seeds are soaked for one to four days by one of the following methods until they are completely swollen: (a) in slowly running water, (b) in standing water below 15° C (59° F), or (c) in standing water at about 25° C (68° F), with at least two changes of water daily.

Storing seeds in moist peat for three days to two weeks at cool temperatures is also satisfactory in preparing seeds for excision. The excision must be done carefully to avoid injury to the embryo. Any hard, stony seed coverings, such as the endocarp of stone fruit seeds, must first be removed.

The moistened seed coats are cut with a sharp scalpel, razor blade, or knife, under clean but nonsterile conditions with sterilized instruments, preferably under a sheet of glass. The embryo is carefully removed. If a large endosperm is present, the seed coats may be slit and the seeds covered with water, and after about a half-hour the embryo will float out or can easily be removed.

Procedures for germinating excised embryos are similar to those for germinating intact seeds. Petri dishes with a moist substratum, such as blotting or filter paper, are used. The embryos are placed on the filter paper so that they do not touch. The dishes are kept in the light at a temperature of 18 to 22° C (64 to 74° F). At higher temperatures, molds may develop and interfere with the test. The time required for the test varies from three days to three weeks.

Nonviable embryos become soft, turn brown, and decay within two to ten days; viable embryos remain firm and show some indication of viability, depending upon the species. Types of response that occur include spreading of the cotyledons, development of chlorophyll, and growth of the radicle and plumule. The rapidity and degree of development gives some indication of the vigor of the seed.

Tetrazolium Test

The **tetrazolium test** is a biochemical method in which viability is determined by the red color appearing when the seeds are soaked in a 2,3,5-triphenyltetrazolium chloride (TTC) solution (Figure 6–12). Living tissue changes the TTC to an insoluble red compound (chemically known as formazan); in nonliving tissue the TTC remains uncolored. The test is positive in the presence of dehydrogenase enzymes. This test was developed in Germany by Lakon (*81*), who referred to it as a topographical test, since loss in embryo viability begins to appear at the extremity of the radicle, epi-

TABLE 6–1

Types of germination requirements for tree and shrub seeds (with examples) when tested in the laboratory

Group 1: seeds that germinate within a wide temperature range and without light exposure

Beefwood (*Casuarina glauca)*
Italian cypress *(Cupressus sempervirens)*
Many species of eucalyptus
Honeylocust *(Gleditia triacanthos)*[a]
Some spruce species *(Picea abies, P. asperata, P. polita)*
Chinese and Siberian elm *(Ulmus parvifolia, U. pumila)*

Group 2: seeds that have specific temperature requirements but do not require light

20 to 30° C (68 to 86° F) diurnally alternating:
Catalpa
Ailanthus
Red pine (Pinus resinosa) or 25° C (77° F) constant
10 to 30° C (50 to 86° F) diurnally alternating:
Mountain mahogany *(Cercocarpus ledifolius)*
Cliffrose *(Cowania stansburiana)*
Antelope bitterbush *(Purshia tridentata)*

20° C (68° F) constant:
Several pine species *(Pinus cembroides P. halepensis, P. pinea)*
Lilac *(Syringa vulgaris)*
Arborvitae *(Thuja orientalis)*

Group 3: seeds that germinate in 7 to 12 days within a wide temperature range if exposed to artificial light

Several spruce species *(Picea engelmannii, P. mariana, P. omerika)*
Several pine species *(Pinus banksiana, P. nigra, P. mugo var. mughus, P. rigida, P. sylvestris, P. ponderosa scopulorum)*

Group 4: seeds that germinate in 14 to 28 days if exposed to artificial light and to warm alternating temperatures of 20 to 30° C (68 to 86° F); seeds may also respond to moist-chilling

Birch *(Betula)*
Elm *(Ulmus americana)*
Larch *(Larix sibirica)*
Mulberry *(Morus alba, M. nigra)*
Liquidambar styraciflua
Some spruce series *(Picea glauca, P. orientalis, P. rubens, P. sitchensis)*
Some pine species *(Pinus densiflora, P. echinata, P. elliotti, P. taeda, P. thunbergii, P. virginiana)*
Rhododendron
Sequoia and *Sequoiadendron*
Thuja plicata, T. occidentalis

Group 5: seeds that require 3 to 4 weeks moist-chilling at 3° C (37° F) before germination at alternating 20 to 30° C (68 to 86° F) (except as noted) temperatures in light for 2 to 4 weeks

Fir species *(Abies balsamea, A. fraseri, A. grandis, A. homolepsis, A. procera)*
Cedrus species 20° C (68° F)
Some pines species *(Pinus flexilis, P. glabra, P. leucodermis, P. strobus)*
Some sources of Douglas-Fir *(Pseudotsuga menziesii)*
Rosa multiflora 10 to 30° C (50 to 86° F)
Eastern hemlock *(Tsuga canadensis)* 15° C (59° F)
Sumac *(Rhus aromatica)*[a]

Group 6: seeds that require 2 to 6 months moist-chilling as a minimum requirement prior to germination; some also have other dormancy problems; embryo excision or a tetrazolium test may be useful in determining germinative capacity

Sensitive to high germination temperature 20° C (68° F):
Maple (*Acer* spp.)
Apple (*Malus* spp.)
Pear (*Pyrus* spp.)
Peach, cherry, etc. (*Prunus* spp.)
Yew (*Taxus* spp.)

Not sensitive to high germination temperature:
Some pine species *(Pinus cembra, P. lambertiana, P. monticola, P. peuce)*

Source: C.E. Heit

[a]Must be treated also for hard seed coats.

FIGURE 6–11 The excised embryo test is a quick evaluation method used for dormant seed.

cotyl, and cotyledon tips. The reaction takes place equally well in dormant and nondormant seed. Results can be obtained within 24 hours, sometimes in 2 or 3 hours (see sidebar on tetrazolium testing). TTC is soluble in water, making a colorless solution. Although the solution deteriorates with exposure to light, it will remain in good condition for several months if stored in a dark bottle. The solution should be discarded if it becomes yellow. A 0.1 to 1.0 percent concentration is commonly used. The pH should be 6 or 7. The TTC test is used primarily to obtain rapid results for both nondormant and dormant seeds.

The test distinguishes between living and dead tissues within a single seed and can indicate weakness before germination is actually impaired. Necrotic areas may be attacked by pathogenic organisms, and seeds with such dead tissues may decay during stratification or give reduced germination under unfavorable soil conditions. In the hands of a skilled technologist this test can be used for seed-quality evaluation and as a tool in seed research (*84, 88*).

On the other hand, this test may not adequately measure certain types of injury which could lead to seedling abnormality, for instance, an overdose of chemicals, seed-borne diseases, frost or heat injury. Standardized procedures and skills are required for evaluating results.

FIGURE 6–12 Tetrazolium chloride is used to test seed viability.

X-ray Analysis

X-ray photographs of seeds (*74, 108*) can be used as a rapid test for seed soundness (*2*). X-ray photographs do not normally measure seed viability but provide an examination of the inner structure for mechanical disturbance, absence of vital tissues, such as embryo or endosperm, insect infestation, cracked or broken seed coats, and shrinkage of interior tissues (possibly a sign of age; see Figure 6–13).

Standard X-ray equipment is used. A convenient Plexiglass seed holder with 100 seed compartments can be made by drilling 100 holes 1mm in diameter in a 5-mm-thick Plexiglass plate with a 0.1-mm-thick Mylar sheet glued to the underside. Place one dry seed in each compartment and expose for ½ to 3 minutes at 15 to 20 kilovolt tube potential. The plate can be processed immediately and the status of seeds determined. Seed with dimensions less than 2 mm are too small to show details. Since X-

FIGURE 6–13 X-radiograph of *Abies procera* seed. Seeds have been segregated to illustrate empty seeds (top two rows), normally filled seeds with distinct embryos (center two rows), and seeds infested with chalcid fly larvae (bottom two rows). (Courtesy Jay Allison (*2*.)

TETRAZOLIUM TESTING

Details vary for different seeds, however, general procedures include (*4, 69, 88*):

1. Any hard covering such as an endocarp, wing, or scale must be removed. Tips of dry seeds of some plants, such as *Cedrus,* should be clipped.
2. Seeds should first be soaked in water in the dark; moistening activates enzymes and facilitates cutting or removal of seed coverings. Seeds with fragile coverings, such as snap beans or citrus, must be softened slowly on a moist medium to avoid fracturing.
3. Most seeds require preparation for TTC absorption. Embryos with large cotyledons, such as *Prunus,* apple, and pear, often comprise the entire seed, requiring only seed coat removal. Other kinds of seed are cut longitudinally to expose the embryo (corn and large seeded grasses, larch, some conifers); or transversely one-fourth to one-third at the end away from the radicle (small-seeded grasses, juniper, *Carpinus, Cotoneaster, Crataegus, Rosa, Sorbus, Taxus*). Seed coats can be removed, leaving the large endosperm intact (some pines, Tilia). Some seeds (legumes, timothy) require no alteration prior to the tests.
4. Seeds are soaked in the TTC solution for 2 to 24 hours. Cut seeds require a shorter time; those with exposed embryos somewhat longer; intact seeds 24 hours or more.
5. Interpretation of results depends upon the kind of seed and its morphological structure. Completely colored embryos indicate good seed. Conifers must have both the megagametophyte and embryo stained. In grass and grain seeds only the embryo itself colors, not the endosperm. Seeds with declining viability may have uncolored spots, or be unstained at the radicle tip and the extremities of the cotyledons. Nonviability depends upon the amount and location of necrotic areas, and correct interpretation depends upon standards worked out for specific seeds (*84*).
6. If the test continues too long, even tissues of known dead seeds become red due to respiration activities of infecting fungi and bacteria. The solution itself can become red because of such contamination.

rays do not injure the seed, further tests for viability can be conducted on the same batch (*2*).

Other tests for seed viability have been developed utilizing contrast agents, such as solutions of certain salts or heavy metals (*108*), followed by X-ray photography. In one test, for example, seeds are soaked in water for 16 hours, then transferred to a concentrated (20 to 30 percent) barium chloride solution for 1 to 2 hours, washed to remove all excess material, and dried. The salts will not enter living cells because of their semipermeability but can penetrate the dead cells of damaged seeds because of membrane destruction and thus produce an exposure on the film. This penetration differentiates between living and dead seeds, or portions of seeds. Fairly consistent and reliable correlations with germination percentages can be obtained with fresh conifer seeds, although consistent results may not occur with stored seed (*39*).

Since X-ray analysis is a nondestructive means for evaluating seeds, prototype machines that provide fast, automatic, on-line sorting have been proposed (*124*). These procedures have the potential to remove nonviable seeds as well as seeds with morphological characteristics that are linked to poor vigor.

Purity Determination

Purity is the percentage by weight of the "pure seed" present in a sample. The determination of purity requires a trained seed analyst usually from a state or private seed lab. The **Society of Commercial Seed Technologists** provides training and testing to certify **Registered Seed Technologists** (*101*). There are two aspects to pure seed; a physical and a genetic component (*6, 101*). Pure seed must be separated from other physical contaminants such as soil particles, plant debris, other inert material, and weed seeds (Figure 6–14). Special care must be taken to document the occurrence of noxious weeds in a sample. Noxious weeds are identified as being particularly bad weeds for a region of the country and can vary by state. Occurrence of a single seed of

FIGURE 6–14. Purity of seeds is determined by visual examination of individual seeds in a weighed seed sample taken from the larger lot in question. Impurities may include other crop seed, weed seed, and inert, extraneous material. (Courtesy E.L. Erickson Products, Brookings, S.D.).

TESTS FOR GENETIC PURITY

Details for cultivar identification are published in the Association of Official Seed Analysts' handbook for purity testing (*6*). These can include:

Chemical tests. There are a number of chemical treatments used to separate cultivars of specific species (*32*). Examples include a fluorescence test for fescue, hydrochloric acid for oat, and peroxidase for soybean. The chemical reaction usually gives a characteristic color to identify the seed. Chemical tests are usually used in association with other tests, like seed shape and color to help determine purity.

Protein electrophoresis (Figure 6–15). A more sophisticated evaluation for cultivar identification uses differences that exist in seed proteins or enzymes. Some plant enzymes are present in different forms (isozymes) that can be separated by electrophoresis to give a pattern that is characteristic of a cultivar. Electrophoresis is a form of chromatography that uses an electrical current to separate proteins on a gel. Isozymes migrate to different locations on the gel to form a pattern identifying the cultivar.

DNA fingerprinting (Figure 6–15). This technique also uses the basic principle of electrophoresis but separates fragments of DNA called RFLP's (random fragment length polymorphisms) and RAPD's (random amplified polymorphic DNA) rather than proteins (**87**). Since this technique uses DNA it is more accurate and can identify a larger number of cultivars than isozyme analysis. DNA fingerprinting is the same process being used by law enforcement to identify suspects in criminal cases.

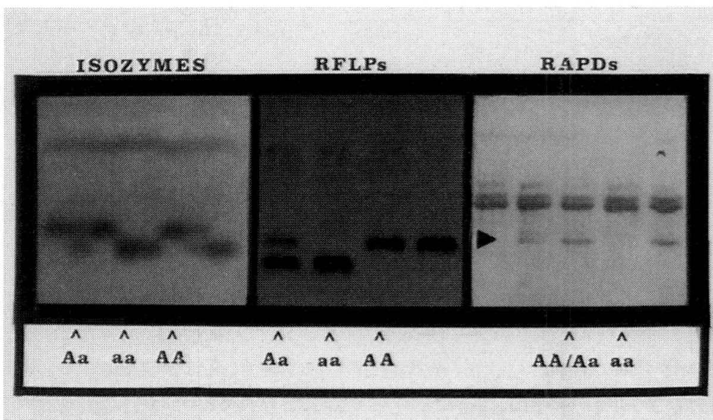

FIGURE 6–15 Isozymes and DNA fingerprinting are additional purity tests that can discriminate between genotypes of a particular crop species. The three molecular marker systems included in this figure illustrate banding phenotypes (homozygous AA, and aa, and heterozygous Aa), which can be used in cultivar discrimination. While isozymes detect genetic variation at the protein level, RFLP's and RAPD's are DNA based tests. (Courtesy Jack Staub, USDA-ARS, Univ. Wisconsin.)

some noxious weed species in a sample can render an entire seed lot unacceptable for public sale.

Purity testing also identifies how genetically pure a seed lot. is. The seed analyst determines if the sample is the proper cultivar and the percentage of the seeds that are either other contaminating cultivars or inbreds in a hybrid seed lot (see Chapter 4). Genetic purity can be difficult to determine and relies on an assortment of tests that include seed color, seed and seedling morphology, chemical tests, isozyme (characteristic seed proteins) separation by electrophoresis (*6, 101*), and DNA fingerprinting (*87*) (see sidebar on genetic purity tests).

Vigor Testing

Although state and federal seed laws currently require only purity and standard germination tests for seed lots, seed companies and many crop producers are performing vigor tests prior to sale or use. The Assocation of Official Seed Analysts states that "seed vigor comprises those seed properties which determine the potential for rapid, uniform emergence, and development of normal seedlings under a wide range of field conditions" (*7*). Standard germination tests do not always adequately predict seedling emergence under field conditions. Seed vigor tests can provide a grower with additional information that can help predict germination where conditions may not be ideal. For many vegetable crops, there is a positive relationship between seed vigor and crop yields (*118*). Various vigor tests have been developed and certain tests are applied to different species. Vigor tests include accelerated aging, controlled deterioration, cold test, cool test, electrolyte leakage, seedling growth rate, and seedling grow out tests (*4, 54, 69, 101*) (see sidebar for seed vigor tests).

Seed Health

Seed companies usually have the personnel and facilities to evaluate the health of a seed lot. Seed health comprises the occurrence of diseases, insects, or nematodes in the seed lot (*66*). Detection of these organisms requires specialized equipment and

SEED VIGOR TESTS

Details for procedures used to conduct vigor tests can be found in the Association of Official Seed Analysts' handbook on seed vigor testing (7). The more commonly conducted vigor tests include:

Accelerated aging. This test is commonly used for agronomic and vegetable seeds. Prior to a standard germination test, seeds are subjected to high temperatures (40 to 45° C) and high relative humidity (near 100 percent) for two to five days. This partially hydrates the seed without permitting radicle emergence. Higher vigor seeds tolerate this stress better than low vigor seeds as shown by higher normal germination percentages in the standard germination test.

Cold test. This is the preferred vigor test for corn seed. Seeds are planted in boxes, trays, or rolled towels that contain field soil and held at 10° C for seven days before being moved to 25° C. The number of seeds that emerge are counted after four days.

Cool test. This is a vigor test that uses procedures identical to the standard germination test except the temperature is lowered to 18° C. A similar tool being used to evaluate vegetable and flower seed vigor is the thermal gradient table (Figure 6–16). This provides a range of temperatures by circulating warm and cold water to the table. This determines the range of germination for a seed lot (*60*). Higher vigor seeds germinate better at the extreme temperatures on the table.

Electrolyte leakage. All seeds "leak" when imbibed, but the amount of electrolyte leakage increases as seeds deteriorate. Electrical conductivity can be measured by using a conductivity meter. Conductivity measurements have been correlated with field emergence, especially in large seeded crops like peas and corn (*86*).

Seedling growth rate. This test is an extension of the standard germination test for percentage germination (Figure 6–17). After a period of time at a controlled temperature (this varies between species), shoot and root length or seedling weight is determined. This permits a determination of strong versus weak seedlings in a seed lot. Recently, Ball Seed Company (West Chicago, IL) introduced the Ball vigor index that employs computer analysis of video images of seedlings in plug trays after a predetermined number of days. The index is suggestive of seedling greenhouse performance.

FIGURE 6–16 A thermal gradient table provides numerous temperatures to simultaneously test germination of a single seed lot. This is useful for determining seed vigor by evaluating germination at minimal and maximal temperatures. Breeders also use thermal gradient tables to evaluate a genotype's tendency for producing seed susceptible to thermodormancy (like lettuce).

trained personnel. Seed health is integral to the performance of the seed lot. It has also become increasingly important as international trading agreements (like the World Trade Organization and the North American Free Trade Agreement) require clean seed be made available for international sale.

Specific procedures to standardize seed health tests are available (*119*). There are three types of tests for seed health:

FIGURE 6–17 For many horticultural crops, standard germination and seedling vigor is evaluated in a seedling grow-out test. The environment for this test is standard greenhouse conditions where the crop will be commercially grown.

1. **Visual evaluation** of a seed sample for characteristic structures like spores or sclerotia of pathogens or the presence of insects.
2. **Incubation of seed** on moist germination paper or agar for growth of diseases.
3. **Biochemical tests,** such as ELISA tests, which detect the presence of specific disease organisms.

Seed Treatments

Presowing seed treatments have become a common practice in the seed industry. Seed treatments may be applied by the seed producers or on the farm. The objective of seed treatments is to either enhance the potential for germination and seedling emergence or to facilitate mechanical sowing of seed (*106, 71*). Types of seed treatments include:

Seed protectants
Germination enhancers
Inoculation with nitrogen-fixing bacteria
Facilitation of sowing

Facilities that treat seeds must consider the following aspects for quality treatments (*53*).

- Seeds must be treated **uniformly.**
- The material must continue to **adhere to the surface** of the seed during sowing.
- The treatment should **not reduce seed quality.** Any physical damage due to high temperature or mechanical injury must be minimized and monitored by seed testing.
- The treatment should be **safely applied** and allow for safe handling by the seed consumer.
- Treatments to **facilitate mechanical sowing** must produce a **uniform size** and **shape** for each seed.
- All seed treated with a pesticide must be **colored** to avoid accidental ingestion by humans or animals. Treatments can also enhance the appearance of the seed.

Modern seed treatments require specialized equipment and facilities (*31, 53*). The equipment varies depending on the type of seed treatment. The first type of seed treatment used simple powders. These are still used today, especially for on-site farm application because they require the least specialized equipment. However, powders and the dust from them present a problem for safe handling. Most commercial treatment of seeds are from liquid slurries. These are preferred because they treat seeds more uniformly, are safer to apply and handle, and are relatively cheap.

Recently, film coatings have become a popular seed treatment because the pesticide can be incorporated into the polymer that is applied in a thin uniform coat or film. The advantages of film coatings are the ability to incorporate chemical or biological materials into the coating for safe handling (this material does not rub off when handled), uniform coating size, and an attractive appearance. The cost has been prohibitive for general use with many large volume agronomic crops, but film-coated seeds have become more widely available on high-value flower and vegetable seed.

Seed Protectants

Seed protectants can be grouped as: (1) chemical treatments against pathogens, insects, and animals; (2) heat treatments against pathogens and insects; (3) inoculation with beneficial microbes against harmful fungi; and (4) safners to reduce herbicide injury (*21, 106*).

Chemical Treatment

A seed stores food reserves to provide energy and carbon for seedling growth. This makes seeds a primary food source for humankind. However, insects, pathogens, and animals also target seeds as a food source. Strategies to protect seeds probably date to man's earliest use of seeds as a food crop (*70*). Chemical treatments for seeds can be seen in the 1800s with the use of copper sulfate against a variety of cereal diseases (*106*). In the 1900s, mercury compounds were very effective against seed and seedling pathogens. These were banned in most parts of the world in the 1980s because of potential health risks. The 1940s and 1950s saw the introduction of the first broad-spectrum fungicides like captan and thiram starting the modern use of seed protectants for diseases.

The most common and important seed treatments are the chemical and physical treatments against seed-borne pathogens (*22*). It is important to understand that these treatments will not improve germination in seeds with a genetically low potential for germination or in mechanically injured seeds. These treatments are especially beneficial where germination is delayed due to poor environmental conditions like excessive water in the field, or cool soils. Under these conditions, leakage from seeds stimulate fungal spore germination and growth. A chemical seed treatment can protect the seed until the seedling emerges.

Seed treatment may be designed to protect seed from soil-borne pathogens, disinfest the seed from pathogens on the seed surface, or eliminate pathogens inside the seed (*22*). Chemical seed protectants can be applied as either **powders, liquids, slurries** or incorporated into a **film coating** (*71*).

Biocontrol

Although chemical treatments dominate industry seed treatments, the novel use of treating seeds with **beneficial microbes** presents an interesting alternative to chemical treatments (*98, 103*). Various biocontrol agents provide protection to seeds by the production of antibiotic substances; competition for space and nutrients; and parasitism (*56*). Common biocontrol agents include bacterial strains like *Eterobacter, Pseudomonas, Serratia,* and fungal strains like *Gliocladium* and *Trichoderma.* Several studies show disease prevention with biologicals to be as effective as chemical treatment with fungicides (*27, 57, 117*).

Heat Treatment (Thermotherapy)

The use of high temperature to control seed-borne diseases has been in use since 1907 (*70*). Dry seeds are immersed in hot water (49 to 57° C; 120 to 135° F) for 15 to 30 minutes, depending upon the species (*11, 12, 13*). After treatment, the seeds are cooled and spread out in a thin layer to dry. To prevent injury to the seeds, temperature and timing must be regulated precisely; a seed protectant should subsequently be used, and old, weak seeds should not be treated. Hot water is effective for specific seed-borne diseases of vegetables and cereals, such as *Alternaria* blight in broccoli and onion, and loose smut of wheat and barley.

Microwave and UV radiation can also be used to disinfect seeds (*107*). Aerated steam (see Chapter 3) is also an alternate method that is less expensive, easier to handle, and less likely to injure seeds compared to hot water (*13*). Seeds are treated in special machines in which steam and air are mixed and drawn through the seed mass to raise the temperature of the seeds rapidly (about two minutes) to the desired temperature. The treatment temperature and time vary with the organism to be controlled and kind of seed. Usually the treatment is 30 minutes, but may be as little as 10 or 15 minutes. Temperatures range from 46 to 57° C (105 to 143°

F). At the end of treatment, temperatures must be lowered rapidly to 32° C (88° F) by evaporative cooling continuing until dry. Holding seeds in moisture-saturated air at room temperature for one to three days prior to treatment will improve effectiveness.

Hot water is also used to kill insects in seeds. For example, oak (*Quercus*) seed is soaked in water at 49° C (120° F) for 30 minutes to eliminate weevils commonly found in the acorns.

Seed Coating

Seed coating uses the same technology and equipment used by the pharmaceutical industry to make medical pills (*76*). Seed coatings include pelletized or film-coated seed. Pelletized seeds are tumbled in a pan while inert powders (like diatomaceous earth) and binders form around the seed to provide a uniform round shape. The objective of this type of seed coating is to provide a round uniform shape and size to small or unevenly shaped seeds to facilitate precision mechanical sowing (Figure 6–18). There are two types of pellets distinguished by either "splitting" or "melting" when the coating is wetted. Many ornamental flower seeds are commonly pelletized for precision sowing one seed per cell in a plug flat (see Chapter 7). An increasing number of direct seeded vegetable crops are also being pelletized. Most of the lettuce seed sown in Florida and California is pelletized to provide uniform spacing and sowing depth which reduces the need to hand-thin the crop.

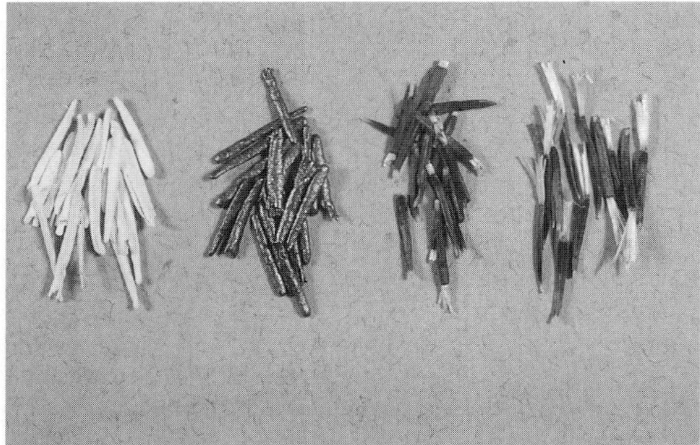

FIGURE 6–18 Coating seeds to enhance germination or facilitate sowing seeds with mechanical seeders has become an important tool for the commercial grower. Illustrated are several presowing treatments for marigold seeds. Marigold seeds are extremely difficult for the grower to sow mechanically because of its nonround shape. The seeds on the *right* are "raw" seeds that have not been pretreated; *center-right* are "de-tailed" seeds with the flower remnant removed; *center-left* are film-coated seeds called "silver bullets"; and *left* are pelletized marigold seeds.

Film coating seeds (Figure 6–19; see also jacket illustration) uses a thin polymer film to cover the seed (*76*). Film coating only adds 1 to 5 percent to the weight of a seed compared to over 1,000 percent for pelletized seed, but this can still aid in precision sowing. Fungicides and beneficial microbes can be added to both pellets and film coatings (see seed treatments, p. 162) and is the major benefit to film coating.

Germination Enhancement (77)

Two common commercial practices that provide germination enhancement are seed sizing and priming. Seed lots sold as "elite" seed have been sized to provide larger seed. This can provide seeds with a higher potential for germination viability and vigor.

Seed priming uses **osmotic** (osmoconditioning) or **matric** forces (matriconditioning or solid matrix priming) to hold seeds in the lag stage of germination for an extended time before being dried back to near the original dry weight of the seed (for procedures see sidebar on seed priming). These seeds can be handled as normal raw seeds or pelletized prior to sowing (*76*). Growth substances (*28*) or biologicals (termed *biopriming; 22, 27*) can also be included in the priming solution for added seed enhancement. Seeds that have been primed will usually have higher seed vigor compared to raw

FIGURE 6–19 Seeds tumble in this seed coating machine while a film coating is uniformly sprayed on the seeds.

TECHNIQUES FOR SEED PRIMING

Seed priming is a seed pretreatment that can significantly enhance germination efficiency in a diverse group of plants including agronomic, vegetable, and ornamental crops (*25*). Seed priming is accomplished by imbibing seeds in either osmotic solutions (osmoconditioning) or moist solid carrier materials that display appropriate matric forces (matriconditioning or solid matrix priming).

Osmotic seed priming. Commercial companies provide primed seed for sale or the crop producer can prime small seed lots prior to use. Osmotic solutions are made using various inorganic salts or more commonly polyethylene glycol (PEG) (*67, 77*). Osmotic potential of the solution, temperature during priming, and duration of priming vary for different species (*25*). Common ranges are between –5 and –15 bars (–0.5 and –1.5 MPa); 50 to 78 °F (10 to 25° C); for 1 to 15 days. Small seed lots can be primed in aerated solutions (Figure 6–20; *35*). Large commercial units use either a bubble column or stirred bioreactor (*51, 52, 90*). Following priming, seeds must be dried using forced air, fluidized beds (*85*), or centripetal dryers.

Matrix seed priming: Problems with aeration, large solution volumes, and disposal of PEG has prompted the use of matrix priming as an alternative to osmotic priming, especially in large seeded species like beans (*76, 116*). Matrix priming uses similar water potential, temperature, and durations as osmotic priming but uses materials like moistened vermiculite, Leonardite shale, diatomaceous silica, or calcined clay to prime seeds. Materials are mixed with seeds at a ratio of 0.2 to 1.5 g of material to 1 g seed and 60 to 300 percent water (based on dry weight of solid material), depending on the matrix material. The material is usually removed prior to sowing but may be left on the seed.

FIGURE 6–20 The requirements for priming include an apparatus for holding the PEG solution, controlling temperature, and providing aeration. Illustrated is a simple research treatment for a small seed lot. The seeds are contained in a cheese-cloth bag suspended in an Erlenmyer flask fitted for a glass tube for bubbling air through the PEG solution.

seed (*25*). (The physiological basis for seed priming is discussed in Chapter 7.) This can provide faster, more uniform seedling emergence for field and greenhouse crops, especially when environmental conditions for germination are not ideal (Figure 6–21). The grower must weigh the additional cost of primed seed with this potential for improved seedling emergence. It is common to prime crops like lettuce (*92*) and pansy (*29*) to overcome problems of reduced germination due to conditions of high temperature (thermodormacy, see Chapter 7).

Pregermination. The goal of each grower is to establish a **"stand"** (seedling emergence) of 100 percent (*49*). This provides a plant at each appropriate field spacing or greenhouse plug cell (see Chapter 8). This can be accomplished by using transplants or sowing more seeds than is required and thinning seedlings to the appropriate spacing. An additional treatment to improve stand establishment is **pregermination** of seeds. In concept, pregermination can take place under optimum conditions and any seeds showing radicle emergence are sown, providing near 100 percent stand. Two types of pregermination sowing techniques have been used. These include **fluid drilling** to sow germinated seeds in a gel to protect emerged radicles and **pregerminated seeds** that use a technique to dry seeds after the radicle emerges.

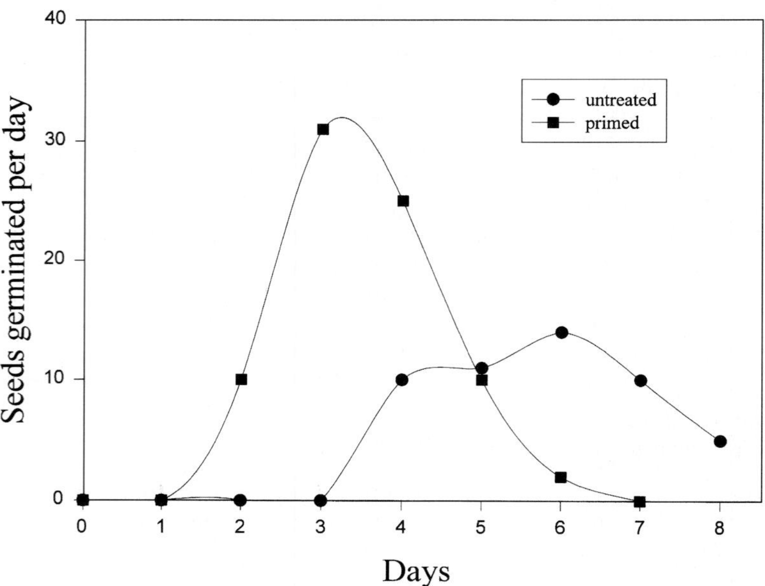

FIGURE 6–21 A major advantage for primed seeds is faster, more uniform germination. This is illustrated for a primed and control seed lot of purple coneflower (*Echinacea purpurea*) that germinated at the same percentage. (Geneve, Wartidiningsih and Kester. 1991. IPPS 41:376–379.)

Fluid Drilling. Fluid drilling (*50, 93*) is a system involving the treatment and pregermination of seeds followed by their sowing suspended in a gel. Seeds are pregerminated under conditions of aeration, light, and optimum temperatures for the species (Figure 6–22). Among the procedures that can be used are (a) germinating seeds in trays on absorbent blotters covered with paper, or (b) placing seeds in water in glass jars or plastic columns through which air is continuously bubbled and fresh water continuously supplied. Growth regulators, fungicides, and other chemicals (*47*) can potentially be incorporated into the system. Chilling (10° C; 50° F) of thermodormant celery seeds for 14 days has produced short, uniform radicle emergence without injury (*46*). Pregerminated seeds of various vegetables have been stored for 7 to 15 days at temperatures of 1 to 5° C (34 to 41° F) in air or aerated water. Separating out germinated seeds by density separation has improved the uniformity and increased overall stand (*115*).

Various kinds of gels are commercially available. Among the materials used are sodium alginate, hydrolyzed starch-polyacrylonitrile, guar gum, synthetic clay, and others. Special machines are needed to deposit the seeds and gel into the seed bed.

Pregerminated seeds. Pregerminated seeds were introduced commercially in 1995 for bedding plant species (impatiens), but the concept was introduced as early as 1897 by L. H. Bailey. A quote from Bailey's *The Nursery Book* (*9*) demonstrates that "new" is truly a relative term as he describes **"regermination."** "It is a common statement that seeds can never revive if allowed to become thoroughly dry after they have begun to sprout. This is an error. Wheat, oats, buckwheat, maize, pea, onion, radish and other seeds have been experi-

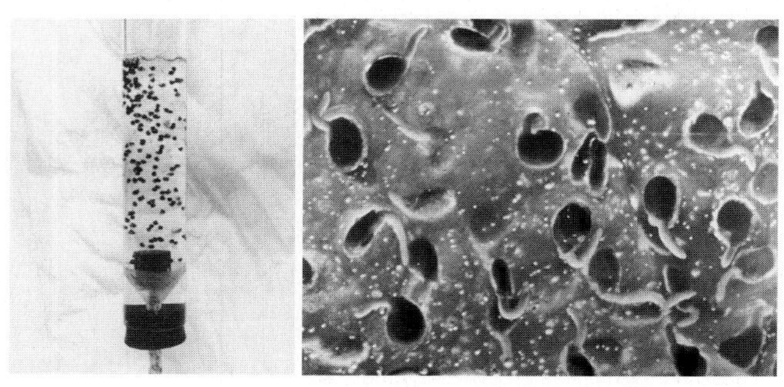

FIGURE 6–22 Pregerminated seeds are mixed in a gel prior to fluid drilling. (Courtesy of Wallace Pill, Univ. Delaware.)

mented upon in this direction, and they are found to regerminate readily, even if allowed to become thoroughly dry and brittle after sprouting is well progressed. They will even regerminate several times."

Pregermination involves germination of seeds under controlled conditions to synchronize germination after the radicle has emerged about one-sixteenth of an inch. Germinated seeds are separated from nongerminated seeds and then seeds are dried slowly to near their original dry weight (*95*). The advantages of using pregerminated seeds include: production of 95 percent or better usable seedlings; fast, uniform germination; and because the seeds are dry, mechanical seeders can be used to sow seeds. The disadvantages of using pregerminated seeds are the increased costs of the seed (up to 25 percent); seeds have a shorter shelf life (around 35 days at 5° C or 40° F); and growers must have optimized seedling growing conditions to take advantage of the benefits of pregermination.

SEED STORAGE

Seeds are usually stored for varying lengths of time after harvest. Viability at the end of storage depends on (a) the initial viability at harvest, as determined by factors of production and methods of handling; and (b) the rate at which deterioration takes place. This rate of physiological change, or aging (*97*), varies with the kind of seed and the environmental conditions of storage, primarily temperature and humidity.

Recalcitrant or Short-Lived Seeds

This group is represented by species whose seeds normally retain viability for as little as a few days, months, or at most a year following harvest (see Chapter 5). However, with proper handling and storage, seed longevity may be maintained for significant periods. A list of species with short-lived seeds has been compiled by King and Roberts (*79*). The group includes:

1. Certain spring-ripening, temperate-zone trees such as poplar *(Populus)*, maple *(Acer)* species, willow *(Salix)*, and elm *(Ulmus)*. Their seeds drop to the ground and normally germinate immediately.
2. Many tropical plants grown under conditions of high temperature and humidity; these include such plants as sugarcane, rubber, jackfruit, macadamia, avocado, loquat, citrus, many palms, litchi, mango, tea, choyote, cocoa, coffee, tung, and kola.
3. Many aquatic plants of the temperate zones, such as wild rice *(Zizania aquatics)*, pondweeds, arrowheads, and rushes.
4. Many tree nut and similar species with large fleshy cotyledons, such as hickories and pecan *(Carya)*, birch *(Betula)*, hornbeam *(Carpinus)*, hazel and filbert *(Corylus)*, chestnut *(Castanea)*, beech *(Fagus)*, oak *(Quercus)*, walnut *(Juglans)*, and buckeye *(Aesculus)*.

Orthodox Seeds

Medium-Lived Seeds

Medium-lived seeds remain viable for periods of 2 or 3 up to perhaps 15 years, providing that seeds are stored at low humidity and, preferably, at low temperatures. Seeds of most conifers, fruit trees, and commercially grown vegetables, flowers, and grains fall into this group. Crop species can be grouped according to the ability of the seed to survive under favorable ambient storage conditions (Table 6–2). The **Relative Storability Index** (*73*) indicates the storage time where 50 percent or more of seeds can be expected to germinate. Seed longevity would be considerably longer under controlled low temperature and humidity storage.

Long-Lived Seeds

These seeds generally have hard seed coats that are impermeable to water. Plant families that produce seeds with hard seed coats include the legume, geranium, and morning glory families. If the hard seed coat remains undamaged, such seeds should remain viable for at least 15 to 20 years. The maximum life can be as long as 75 to 100 years and perhaps more. Records exist of seeds being kept in museum cupboards for 150 to 200 years while still retaining viability (*100*). There are a number of claims of seeds from ancient tombs germinating after thousands of years. However, these lack definitive scientific support (*100*). Indian lotus *(Nelumbo nucifera)* seeds that had been buried in a Manchurian peat bog were originally estimated to be over 1,000 years old and germinated perfectly when the impermeable seed coats were cracked (*23*). However, recent carbon-14 dating of these and other lotus seeds estimate the age of these seeds to be only 100 to 430 years old (*100*).

TABLE 6–2

Relative storability index[a]

Crop	Category 1 (1 to 2 yr.)	Category 2 (3 to 5 yr.)	Category 3 (>5 yr.)
Agronomic			
	Bermuda grass	Barley	Alfalfa
	Cotton	Ky. Bluegrass	Sugar beet
	Field corn	Fescue	Clover
	Millet	Oats	Vetch
	Peanut	Rape seed	
	Soybean	Rice	
	Sunflower	Wheat	
Vegetable			
	Green bean	Broccoli, cabbage, cauliflower	Beet
	Lettuce	Sweet corn	Tomato
	Onion	Cucumber	
	Pepper	Melon	
		Pea	
		Spinach	
Flower			
	Begonia	Alyssum	Hollyhock
	Coreopsis	Carnation	Morning glory
	Pansy	Coleus	Salpiglossis
	Primrose	Cyclamen	Shasta daisy
	Statice	Marigold	Stocks
	Vinca	Petunia	Zinnia

[a]The relative storability index is the expected 50 percent germination in a seed lot stored under favorable ambient conditions. Storage life would be longer under controlled low temperature conditions.

Adapted from Justice and Bass 1979

A systematic study was initiated by Beal in 1879 at Michigan State University to study long-term survival of buried seed. This study is still ongoing and in 1981 (*80*) three species continued to show germination after 100 years. These species were *Malva rotundifolia, Verbascum blattaria,* and *Verbascum thapsus.* Some weed seeds retain viability for many years (50 to 70 years or more) while buried in the soil, even though they have imbibed moisture (*99*). Longevity seems related to dormancy induced in the seeds by environmental conditions deep in the soil.

Storage Factors Affecting Seed Germination

As seeds **deteriorate,** they first lose **vigor,** then the capacity for **normal germination,** and finally **viability**. Storage conditions that reduce seed deterioration are those that slow respiration and other metabolic processes without injuring the embryo. The most important conditions are low moisture content of the seed, low storage temperature, and modification of the storage atmosphere. Of these, the moisture-temperature relationships have the most practical significance. Harrington (*59*) introduced a "rule of thumb" that indicated that seed lose half their storage life for every 1 percent increase in seed moisture between 5 and 14 percent. Also, seeds lose half their storage life for every 5° C increase in storage temperature between 0 and 50° C. This is of course a generalized theory that varies between species. More accurate mathematical models have been developed to predict seed longevity at various temperature and moisture contents (*42*).

Moisture Content

Control of the moisture content of the seed is probably the most important factor in seed longevity and storage. Most crop species have orthodox seeds (medium- to long-lived) where dehydration is not only their natural state at maturity but a nonfluctuating low moisture content must also be maintained for long-term storage (*42*).

Seeds of orthodox species are desiccation-tolerant and not only can withstand dryness but must have a low moisture content for long-term storage. Seeds of many species dry naturally while still on the plant. However, there are many crop species that must be aided in seed drying by either (a) cutting the crop, windrowing in the field, and being sun dried; or (b) placing harvested seed in forced air driers. Forced air driers differ in size and design depending on the crop (*73*). A 4 to 6 percent moisture content is favorable for prolonged storage (*34*), although a somewhat higher moisture level is allowable if the temperature is reduced (*121*). For example, for tomato seed stored at 4.5 to 10° C (40 to 50° F), the percent of moisture content should be no more than 13 percent; if 21° C (70° F), 11 percent, and if 26.5° C (80° F), 9 percent.

Various storage problems arise with increasing seed moisture (*59*). At 8 or 9 percent or more, insects are active and reproduce; above 12 to 14 percent (65 percent relative humidity or more), fungi are active; above 18 to 20 percent, heating may occur; and above 40 to 60 percent, germination occurs.

If the moisture content of the seed is too low (1 to 2 percent), loss in viability and reduced germination rate can occur in some kinds of seeds (*20*). For seeds stored at these low moisture levels, it would be best to rehydrate with saturated water vapor to avoid injury to seed (*91*). Moisture in seeds is in equilibrium with the ambient relative humidity of the storage container and increases if the relative humidity increases and decreases if it is reduced (*59*). Thus moisture percentage varies with the kind of seed, which is affected by the kind of storage reserves within the seed (*15*). Longevity of seed is maximum if stored at a relative humidity range of 20 to 25 percent (*100*).

Since fluctuations in seed moisture during storage reduce seed longevity (*16*), the ability to store seeds exposed to the open atmosphere varies greatly in different climatic areas. Dry climates are conducive to increased longevity; areas with high relative humidity result in shorter seed life. Seed viability is particularly difficult to maintain in open storage in tropical areas.

Storage in hermetically sealed, moisture-resistant containers is advantageous for long storage, but the moisture content of the seed must be low at the time of sealing (*18*). Seed moisture content of 10 to 12 percent (in contrast to 4 to 6 percent) in a sealed container is worse than storage in an unsealed container (*34, 100*).

Recalcitrant seeds owe their short life primarily to their sensitivity to low moisture content (desiccation sensitivity—see Chapter 5). For instance, in silver maple *(Acer saccharinum)* seeds, the moisture content was 58 percent in the spring when the seeds matured. Viability was lost when the moisture content dropped below 30 to 34 percent (*72*). Citrus seeds can withstand only slight drying (*17, 30*) without loss of viability. The same is true for seeds of some water plants, such as wild rice, which can be stored directly in water at low temperatures (*89*). The large fleshy seeds of oaks *(Quercus),* hickories *(Carya),* and walnut *(Juglans)* lose viability if allowed to dry after ripening (*104*).

Viability of recalcitrant seeds of the temperate zone can be preserved for a period of time if kept in a moist environment and the temperature is lowered to just above freezing (*23*). Under these conditions many kinds of seeds can be kept for a year or more. Seeds of some tropical species (e.g., cacao, coffee), however, show chilling injury below 10° C (50° F).

Temperature

Reduced temperature invariably lengthens the storage life of seeds, and in general, can offset the adverse effect of a high moisture content. Subfreezing temperatures, at least down to −18° C (0° F), will increase storage life of most kinds of seeds but moisture content should be in equilibrium with 70 percent relative humidity or lower, or the free water in the seeds may freeze and cause injury (*59, 100*). Such storage is particularly useful for conifer seeds (*104, 105*). Refrigerated storage should be combined with dehumidification or with sealing dried seeds in moisture-proof containers.

Cryopreservation

Survival of seeds exposed to ultra-low temperatures has been known since 1987 (*26*). There is renewed interest in storage of seeds by cryopreservation because it is potentially a cost-effective way to preserve germ plasm for long periods of time with minimal loss of genetic information due to chromosomal mutations that accompany seed deterioration (*111*). Seeds are cryopreserved by immersion and storage in liquid nitrogen at −196° C (Figure 6–23). Seed moisture must be low (<15 percent) for survival, and gradual cooling and warming rates limit damage to the seed like cracks in the seed coat (*100*).

Cryopreservation of seeds has not replaced standard long-term storage at −18° C because long-term effects on seed survival have yet to be determined (*100*). However numerous species have been stored for short periods of time in liquid nitrogen with promising results (*110, 112*). Research is continuing, especially at the National Seed Storage Lab (see sidebar on conserving genetic resources) to

FIGURE 6–23 Seeds, shoot tips, and tissue culture explants can be stored in liquid nitrogen at low temperature for germ plasm preservation. (Courtesy Barbra Reed, USDA-ARS, National Clonal *Germplasm* Repository, (Corvalis, OR)

make cryopreservation an important tool for seed preservation.

Cryopreservation technology is also being applied to other tissue like pollen and dormant buds for possible preservation of germ plasm (*10, 38, 75*).

Types of Seed Storage

Open Storage without Moisture or Temperature Control

Many kinds of orthodox seeds only need to be stored from harvest until the next planting season. Under these conditions, seed longevity depends on the relative humidity and temperature of the storage atmosphere, the kind of seed, and its condition at the beginning of storage. Basic features (*73*) of the storage structures include (a) protection from water, (b) avoidance of mixture with other seeds, and (c) protection from rodents, insects, fungi, and fire. Retention of viability varies with the climatic factors of the area in which storage occurs. Poorest conditions are found in warm, humid climates; best storage conditions occur in dry, cold regions. Fumigation or insecticidal treatments may be necessary to control insect infestations.

Open storage can be used for many kinds of commercial seeds for at least a year (i.e., to hold seeds from one season to the next). Seeds of many species, including most agricultural, vegetable, and flower seeds, will retain viability for longer periods up to four to five years (*19, 23, 48*), except under the most adverse conditions.

Sealed Containers

Packaging dry seeds in hermetically sealed, moisture-proof containers is an important method of handling and/or merchandising seeds. Containers made of different materials vary in durability and strength, cost, protective capacity against rodents and insects, and ability to retain or transmit moisture (*33, 58, 100*). Those completely resistant to moisture transmission include tin or aluminum cans (if properly sealed), hermetically sealed glass jars, and aluminum pouches. Those almost as good (80 to 90 percent effective) are polyethylene (3 mil or thicker) and various types of aluminum-paper laminated bags. Somewhat less desirable, in regard to moisture transmission, are asphalt and polyethylene-laminated paper bags and friction-top tin cans. Paper and cloth bags give no protection against moisture change (*45*). Small quantities of seeds can be stored satisfactorily in small moisture-proof containers like mason jars or plastic food containers.

Seed may be protected against moisture uptake by mixing with a desiccant (*33, 58, 73*). A useful desiccant is silica gel treated with cobalt chloride. Silica gel (one part to ten parts seed, by weight) can absorb water up to 40 percent of its weight. Cobalt chloride turns from blue to pink at 45 percent RH and can act as a useful indicator of excess moisture. Seeds should not be stored in contact with the desiccant. Seeds in sealed containers are more sensitive to excess moisture than when subjected to fluctuating moisture content in open storage. A seed moisture content of 5 to 8 percent or less is desirable, depending upon the species.

Conditioned Storage (100)

Conditioned storage includes use of dehumidified and/or refrigerated facilities to reduce temperature and relative humidity. Such facilities are expensive but are justified where particularly valuable seeds are stored, as for research, breeding stocks, and germ plasm (see sidebar on conserving genetic resources). Also in some climatic areas, such

CONSERVING GENETIC RESOURCES

Crop cultivars produced for food, fiber, and ornamentals represent only a small proportion of the worldwide gene pool that could have economic benefit in the future. This is a genetic resource that is most easily and economically preserved by storing seed from diverse populations of crop plants. Facilities that provide long-term storage of seeds or other plant parts are called "**gene banks**" (*94*).

The **International Board for Plant Genetic Resources** (*68*) was established in 1974 to promote an international network of gene banks to conserve genetic resources mainly as long-term storage of seeds (*55*). This organization provides handbooks and describes the criteria for facilities that store seed germ plasm (*40, 41, 55*). Facilities are described for either **long-term or medium-term storage**. Long-term storage facilities provide an environment and testing regime to maintain seed viability and plant recovery for 10 to >20 years. Medium-term storage facilities are designed to preserve seeds for 5 to 10 years before having to regrow the crop to produce fresh seed. In 1984, over 100 storage facilities (55 with long-term storage) had been established worldwide (*55*).

The major facility in the United States for preserving germ plasm resources is the **National Seed Storage Laboratory** established in 1958 on the Colorado State University campus (*100*). Seeds are actively acquired from public agencies, seed companies, and individuals engaged in plant breeding or seed research. Descriptive material is recorded for each new accession on the **Germplasm Resources Information Network**. Seed samples are tested for viability, dried to approximately 6 percent moisture and stored at −18° C (0° F) in moisture-proof bags. Seed lot sizes vary for storage from between 3,000 to 4,000 seeds for cross-pollinated species and 1,500 to 3,000 seeds for pure lines. Seed lots are tested every five or ten years for germination. Seeds can be made available to breeders and researchers on request. This facility also conducts seed storage research and is one of the leading centers for research on cryopreservation of seeds. Information on germ plasm can be obtained on-line from the World Wide Web at (http://www.ars-grin.gov).

as in the highly humid tropics, orthodox seeds cannot be maintained from one harvest season to the next planting season.

Cold storage of tree and shrub seed used in nursery production is generally advisable if the seeds are to be held for longer than one year (*37, 64, 65, 104, 105*). Seed storage is useful in forestry because of the uncertainty of good seed-crop years. Seeds of many species are best stored under cold, dry conditions (*130*). Ambient relative humidity in conditioned storage should not be higher than 65 to 70 percent RH (for fungus control) and no lower than 20 to 25 percent.

It is important to control humidity in refrigerated storage, since the relative humidity increases with a decrease in temperature and moisture will condense on the seed. At 15° C (59° F) this equilibrium moisture may be too high for proper seed storage. Although the moisture content may not be harmful at those low temperatures, rapid deterioration will occur when the seeds are removed from storage and returned to ambient uncontrolled temperatures. Consequently, refrigeration should be combined with dehumidification or sealing in moisture-proof containers (*59*).

Low humidity in storage can be obtained by judicious ventilation, moisture proofing, and dehumidification as well as by the use of sealed moisture containers, or the use of desiccants, as described above. Dehumidifiers utilize desiccants (silica gel) or saturated salt solutions. The most effective storage is to dry seeds to 3 to 8 percent moisture, place in sealed containers, and store at temperatures of 1 to 5° C (41° F). Below-freezing temperatures can be even more effective if the value of the seed justifies the cost.

Moist, Cool Storage

Many recalcitrant seeds that cannot be dried can be mixed with a moisture-retaining medium, placed in a polyethylene bag or other container, and refrigerated at 0 to 10° C (32 to 50° F). The relative humidity in storage should be 80 to 90 percent. Examples of species whose seeds require this storage treatment are: Silver maple (*Acer saccharinum*), buckeye *(Aesculus spp.),* American hornbeam *(Carpinus caroliniana),* hickory *(Carya spp.),* chestnut *(Castanea spp.),* filbert *(Corylus spp.),* citrus *(Citrus spp.),* loquat *(Eriobotrya japonica),* beech *(Fagus spp.),* walnut *(Juglans spp.),* litchi, tupelo *(Nyssa silvatica),* avacado *(Persea spp.),* and oak *(Quercus spp.).* The procedure is similar to moist-chilling (stratification). Acorns and large nuts can be dipped in paraffin or sprayed with latex paint before storage to preserve their moisture content (*65*).

REFERENCES

1. Aldous, J.R. 1972. *Nursery practice.* For. Comm. Bul. 43. London: Her Majesty's Stationery Office.
2. Allison, C.J. 1980. X-ray determination of horticultural seed quality. *Comb. Proc. Inter. Plant Prop. Soc.* 30:78–86.
3. Asgrow. 1959. *A study of mechanical injury to seed beans.* Asgrow Monograph 1. New Haven, Conn.: Associated Seed Growers.
4. Assocation of Official Seed Analysts. 1978. Rules for testing seeds. *Jour. Seed Tech* 3:1–126.
5. Association of Official Seed Analysts. 1993. Rules for testing seeds. *Jour. Seed Tech.* 16:1–113.
6. Association of Official Seed Analysts. 1991. Purity testing handbook. *Assn. Offic. Seed Anal. Handbk.*
7. Association of Official Seed Analysts. 1983. Seed vigor testing handbook. *Assn. Offic. Seed Anal. Handbk.* 32.
8. Austin, R.B. 1972. Effects of environment before harvesting on viability. In *Viability of seeds,* E.H. Roberts, ed. Syracuse. N.Y.: Syracuse Univ. Press.
9. Bailey, L.H. 1897. *The Nursery Book.* New York: The MacMillan Co.
10. Bajaj, Y.P.S. 1979. Establishment of germplasm banks through freeze storage of plant tissue culture and their implications in agriculture. In *Plant cell and tissue culture principles and applications,* W.R. Sharp et al., (eds.). Columbus: Ohio State Univ. Press, pp. 745–74.
11. Baker, K.F. 1972. Seed pathology. In *Seed Biology,* Vol. 2, T.T. Kozlowski, (ed.). New York: Academic Press.
12. ———. 1980. Pathology of flower seeds. *Seed Sci. and Tech.* 8:575–89.
13. ———, and P.A. Chandler. 1957. Development and maintenance of healthy planting stock. In *The U.C. system for producing healthy container-grown plants,* K.F. Baker (ed.). Calif. Agr. Exp. Sta. Man. 23, pp. 217–36.
14. Barmore, C.R., and W.S. Castle. 1979. Separation of citrus seed from fruit pulp for rootstock propagation using a pectolytic enzyme. *HortScience* 14: 526–27.
15. Barton, L.V. 1941. Relation of certain air temperatures and humidities to viability of seeds. *Contrib. Boyce Thomp. Inst.* 12:85–102.
16. ———. 1943. Effect of moisture fluctuations on the viability of seeds in storage. *Contrib. Boyce Thomp. Inst.* 13:35–45.
17. Bass, L.N. 1943. The storage of some citrus seeds. *Contrib. Boyce Thomp. Inst.* 13:4–55.
18. ———. 1953. Seed storage and viability. *Contrib. Boyce Thomp. Inst.* 17:87–103.
19. ———. 1980 Flower seed storage. *Seed Sci. and Tech.* 8:591–99.
20. ———. 1980. Seed viability during long term storage. *Hort. Rev.* 2:117–41.
21. Bazin, M, J.F. Morin, J.P. Vergneau. 1989. New technologies in seed protection. *Acta Hort.* 253:268–69.
22. Bennett, M.A., V.A. Fritz, and N.W. Callan. 1992. Impact of seed treatments on crop stand establishment. *HortTechnology* 2:345–49.
23. Bewley, J.D., and M. Black. 1994. *Seeds: Physiology of Development and Germination.* New York: Plenum Press.
24. Bonner, F.T. 1974. Seed testing. In *Seeds of woody plants in the United States,* C.S. Schopmeyer, (ed.). U.S. Dept. Agr. Handbook 450, Washington, D.C.: U.S. Govt. Printing Office, pp. 136–52.
25. Bradford, K.J. 1986. Manipulations of seed water relations via osmotic priming to improve germination under stress conditions. *HortScience* 21: 1105–12.
26. Brown, H.T. and F. Escombe. 1879. Note on the influence of very low temperatures on the germinative power of seeds. *Proc. Roy. Soc. London* 62:160–65.
27. Callan, N.W., D.E. Mathre, and J.B. Miller. 1990. Biopriming seed treatment for biological control of *Pythium ultimum* preemergence damping-off in sh2 sweet corn. *Plant Dis.* 74:368–72.
28. Cantliffe, D.J. 1991. Benzyladenine in the priming solution reduces thermodormancy of lettuce seeds. *HortTechnology* 1:95–99.
29. Carpenter, W.J. and J.F. Boucher. 1991. Priming improves high-temperature germination of pansy seed. *HortScience* 26:541–44.
30. Childs, J.F.L., and G. Hrnciar. 1948. A method of maintaining viability of citrus seeds in storage. *Proc. Fla. State Hort. Soc.* 64:69.

31. Clayton, P.B. 1993. Seed treatment. In *Application technology for crop protection.* G.A. Matthews and E.C. Hislop, (eds.). CAB International, pp. 329–49.

32. Cooke, R.J. 1995. Variety identification: Modern techniques and applications. In *Seed Quality: Basic Mechanisms and Agricultural Implications.* A.S. Basra (ed.). Food Products Press, New York. pp. 279–318.

33. Copeland, L.O., and M.B. McDonald. 1995. *Principles of Seed Science and Technology* (3rd ed.). New York: Chapman and Hall.

34. Crocker, W., and L.V. Barton. 1953. *Physiology of Seeds.* Waltham, Mass.: Chronica Botanica.

35. Darby, R.J. and P.J. Salter. 1976. A technique for osmotically pre-treating and germinating small quantities of seeds. *Ann. Applied Biology* 83:313–1.

36. Delouche, J.C. 1980. Environmental effects on seed development and seed quality. *HortScience* 15:775–80.

37. Dirr, M.A., and C.W. Heuser, Jr. 1987. *The Reference Manual of Woody Plant Propagation.* Athens, Ga.: Varsity Press.

38. Dougall, D.K. 1978. Preservation of germ plasm. In *Propagation of higher plants through tissue culture: A bridge between research and applications,* K.W. Highes, R. Henke, and M. Constantin, (eds.). Springfield, Va.: U.S. Dept. Energy Tech. Inform. Center.

39. Eden, C.J. 1965. Use of X-ray technique for determining sound seed. *U.S. Forest Service, Tree Planters' Notes* 72:25–27.

40. Ellis, R.H., T.D. Hong and E.H. Roberts. 1985. Handbooks for seed technology for genebanks. Vol. I. *Principles and methodology. Handbooks for genebanks: No. 2. Intern. Board for Plant Genetic Resources,* IBPGR Secretariat, Rome.

41. ———, T.D. Hong and E.H. Roberts. 1985. Handbooks for seed technology for genebanks. Vol. II. *Principles and methodology. Handbooks for genebanks: No. 3. Intern. Board for Plant Genetic Resources,* IBPGR Secretariat, Rome.

42. ———, and E.H. Roberts. 1980. Improved equations for the prediction of seed longevity. *Ann. Bot.* 45:13–30.

43. Flemion, F. 1938. A rapid method for determining the viability of dormant seeds. *Contrib. Boyce Thomp. Inst.* 9:339–51.

44. Fordham, A.J. 1984. Seed dispersal as it concerns the propagator. *Comb. Proc. Inter. Plant Prop. Soc.* 34:531–34.

45. Freire, M.S. and P.M. Mumford. 1989. The efficiency of a range of containers in maintaining seed viability during storage. *Seed Sci. and Tech.* 14: 371–81.

46. Furatani, S.C., B.H. Zandstra, and H.C. Price. 1985. Low temperature germination of celery seeds for fluid drilling. *Jour. Amer. Soc. Hort. Sci.* 110:149–53.

47. Ghate, S.R., S.C. Phatak, and K.M. Batal, 1984. Pepper yields from fluid drilling with additives and transplanting. *HortScience* 19:281–83.

48. Goss, W.L. 1937. Germination of flower seeds stored for ten years in the California state seed laboratory. *Calif. Dept. Agr. Bul.* 26:326–33.

49. Gray, D. 1978. The role of seedling establishment in precision cropping. *Acta Hort.* 83:309–15.

50. ———. 1981. Fluid drilling of vegetable seeds. *Hort. Rev.* 3:1–27.

51. ———. 1994. Large-scale seed priming techniques and their integration with crop protection treatments. In *Seed treatment: progress and prospects.* BCPC Monograph No. 57. pp. 353–62.

52. ———, H.R. Rowse. W.E. Finch-Savage, W. Bujalski, and A.W. Nienow. 1992. Priming of seeds—scaling up for commercial use. *Fourth International Workshop on Seeds. Basic and Applied Aspects of Seed Biology.* Angers, France.

53. Halmer, P. 1994. The development of quality seed treatments in commercial practice—objectives and achievements. In *Seed treatment: progress and prospects.* BCPC Monograph No. 57. pp. 363–74.

54. Hampton, J.G. 1995. Methods of viability and vigor testing: A critical apraisal. In *Seed Quality: Basic Mechanisms and Agricultural Implications.* A.S. Basra (ed.). Food Products Press, New York. pp. 81–118.

55. Hanson, J. 1985. Procedures for handling seeds in genebanks. Practical manuals for genebanks: No. 1. *Intern. Board for Plant Genetic Resources,* IBPGR Secretariat, Rome.

56. Harman, G.E. and E.B. Nelson. 1994. Mechanisms of protection of seed and seedlings by biological seed treatments: implications for practical disease control. In *Seed treatment: progress and prospects.* BCPC Monograph No. 57. pp. 283–92.

57. ———, A.G. Taylor, and T.E. Stasz. 1989. Combining effective strains of *Trichoderma harzianum* and solid matrix priming to improve biological seed treatments. *Plant Dis.* 73:631–37.

58. Harrington, J.F. 1963. The value of moisture resistant containers in vegetable seed packaging. *Calif. Agr. Exp. Sta. Bul.* 792, pp. 1–23.

59. ———. 1972. Seed storage and longevity. In *Seed biology,* T.T. Kozlowski, ed. New York: Academic Press, pp. 145–245.

60. Hassell, R. 1991. Vigor testing proves to be of value. *International Floriculture Industry Plug Symposium.* The Ohio State University, pp. 24–25.

61. Hawthorn, L.R. 1961. Growing vegetable seeds for sale. In *Seeds: Yearbook of Agriculture,* A. Stefferud ed. Washington, D.C.: U.S. Govt. Printing Office. pp. 208–15.

62. ———, and L.H. Pollard. 1954. *Vegetable and Flower Seed Production.* New York: Blakiston Co.

63. Heit, C.E. 1955. The excised embryo method for testing germination quality of dormant seed. *Proc. Assn. Off. Seed Anal.* 45:108–17.

64. ———. 1967. Propagation from seed. 10. Storage methods for conifer seed. *Amer. Nurs.* 126(20): 14–15.

65. ———. 1967. Propagation from seed. 11. Storage of deciduous tree and shrub seed. *Amer. Nurs.* 126(21): 12–13, 86–94.

66. Hewett, P.D. and W.J. Rennie. 1986. Biological tests for seeds. In *Seed treatment.* K.A. Jeffs, (ed.). BCPC publications, Surrey U.K. pp. 51–82.

67. Heydecker, W. and P. Coolbear. 1977. Seed treatment for improved performance—survey and attempted prognosis. *Seed Sci. and Tech.* 5:353–25.

68. International Board for Plant Genetic Resources. 1979. *A review of policies and activities 1974–1978 and of the prospects for the future.* IBPGR Secretariat, Rome.

69. International Seed Testing Association. 1993. International rules for seed testing. *Seed Sci. and Tech.* suppl. 21:1–289.

70. Jeffs, K.A. 1986. A brief history of seed treatment. In *Seed treatment.* K.A. Jeffs, (ed.). BCPC publications, Surrey U.K. pp. 1–5.

71. ———, and R.J. Tuppen. 1986. Requirements for efficient treatment of seeds. In *Seed treatment.* K.A. Jeffs, (ed.). BCPC publications, Surrey U.K. pp. 17–50.

72. Jones, H.A. 1920. Physiological study of maple seeds. *Bot. Gaz.* 69:127–52.

73. Justice, O.L., and L.N. Bass. 1979. *Principles and Practices of Seed Storage.* London: Castle House Pub.

74. Kamra, S.K. 1964. The use of x-rays in seed testing. *Proc. Inter. Seed Testing Assn.* 29:71–79.

75. Kartha, K.K. 1985. *Cryopreservation of Plant Cells and Organs.* Boca Raton, Fla.: CRC Press.

76. Kaufman, G. 1991. Seed coating: a tool for stand establishment; a stimulus to seed quality. *HortTechnology* 1:96–102.

77. Khan, A.A. 1992. Preplant physiological seed conditioning. *Hort. Rev.* 13:131–81.

78. ———, H. Miura, J. Prusinski, and S. Ilyas. 1990. Matriconditioning of seeds to improve emergence. *Proc. Natl. Symp. on Stand Establishment for Hort. Crops.* pp. 86–96.

79. King, M.W. and E.H. Roberts. 1980. Maintenance of recalcitrant seeds in storage. In H.F. Chin and E.H. Roberts (eds.). *Recalcitrant Crop Seeds.* Tropical Press SDN. BHD. Kuala Lumpur, Malaysia. pp. 53–89.

80. Kivilaan, A. and R.S. Bandurski. 1981. The one hundred-year period for Dr. Beal's seed viability experiment. *Am. Jour. Bot.* 68:1290–92.

81. Lakon, G. 1949. The topographical tetrazolium method for determining the germinating capacity of seeds. *Plant Phys.* 24:389–94.

82. Lovelace R. 1993. Establishing and maintaining a seed orchard. *Cont. Proc. Intern. Plant Prop. Soc.* 495–96.

83. Macdonald, B. 1986. *Practical Woody Plant Propagation for Nursery Growers,* Vol. 1. Portland, Oreg.: Timber Press.

84. MacKay, D.B. 1972. The measurement of viability. In *Viability of Seeds,* E.H. Roberts, (ed.). Syracuse, N.Y.: Syracuse Univ. Press.

85. Maude, R.B., D.L. Suett, P.H. Springer, A.W. Nienow, W. Bujalski, A. Maroglou, and G.M. Petch. 1988. A prototype fluidised bed film coating seed treater. *Application to seeds and soil.* BCPC Monograph No. 39. pp. 403.

86. McDonald, M.B., Jr. 1980. Assessment of seed quality. *HortScience* 15:784–88.

87. ———, L.J. Elliot and P.M. Sweeney. 1994. DNA extraction from dry seeds for RAPD analyses in varietal identification studies. *Seed Sci. and Tech.* 22:171–176.

88. Moore, R.P. 1973. Tetrazolium staining for assessing seed quality. In *Seed ecology,* W. Heydecker (ed.). London: Butterworth, pp. 347–66.

89. Muenscher, W.C. 1936. Storage and germination of seeds of aquatic plants. *New York (Cornell Univ.) Agr. Exp. Sta. Bul.* 652, pp. 1–17.

90. Nienow, A.W. and P.A. Brocklehurst. 1987. Seed preparation for rapid germination—engineering studies. *International Conference on Bioreactors and Biotransformation.* Gleneagles, Scotland, U.K. pp. 9–12.

91. Nutile, G.E. 1964. Effect of desiccation on viability of seeds. *Crop Sci.* 4:325–28.

92. Perkins-Veazie, P. and D.J. Cantiliffe. 1984. Need for high quality seed for priming to effectively overcome thermodormancy in lettuce. *Jour. Amer. Soc. Hort. Sci.* 109:368–72.

93. Pill, W.G. 1991. Advances in fluid drilling. *HortTechnology* 1:59–64.

94. Plucknett, D.L., N.J.H. Smith, J.T. Williams and N.M. Anishetty. 1987. *Gene banks and the world's food.* Princeton Univ. Press, Princeton, N.J.

95. Polking, G. and D. Koranski. 1991. Pregerminated seeds. International Floriculture Industry Plug Symposium. The Ohio State University, pp. 58–60.

96. Pollock, B.M., and E.E. Roos. 1972. Seed and seedling vigor. In *Seed Biology,* Vol. 1. T.T. Kozlowski, (ed.). New York: Academic Press.

97. Priestley, D.A. 1986. *Seed Aging.* Ithaca, N.Y.: Cornell Univ. Press.

98. Rhodes, D.J. and K.A. Powell. 1994. Biological seed treatments—the development process. In *Seed treatment progress and prospects.* BCPC Monograph No. 57. pp. 303–10.

99. Rhodes, E.H. 1972. Dormancy: A factor affecting seed survival in the soil. In *Viability of seeds,* E. H. Roberts, (ed.). London: Chapman & Hall, pp. 32–59.

100. Roos, E.E. 1989. Long-term seed storage. In J. Janick (ed). *Plant Breeding Rev.* Timber Press, Or. pp. 129–158.

101. ———, and L.E. Wiesner. 1991. Seed testing and quality assurance. *HortTechnology* 1:65–69.

102. Rudolf, P.O. 1965. State tree seed legislation. *U.S. Forest Service, Tree Planters' Notes* 72:1–2.

103. Scheffer, R.J. 1994. The seed industry's view on biological seed treatments. In *Seed Treatment: progress and prospects.* BCPC Monograph No. 57. pp. 311–14.

104. Schopmeyer, C.S., ed. 1974. *Seeds of woody plants in the United States.* U.S. Dept. Agr. Handbook 450. Washington, D.C.: U.S. Govt. Printing Office.

105. Schubert, G.H., and R.S. Adams. 1971. *Reforestation practices for conifers in California.* Sacramento: Calif. State Div. of Forestry.

106. Schwinn, F.J. 1994. Seed treatment—A panacea for plant protection? In *Seed treatment: progress and prospects.* BCPC Monograph No. 57. pp. 3–15.

107. Sherf, A.F. and A.A. MacNab. 1986. *Vegetable Diseases and Their Control.* 2nd ed. Wiley, New York.

108. Simak, M. 1957. The X-ray contrast method for seed testing. *Medd. F. Stat. skogsforssr. Inst.* 47(4):1–22.

109. Smith, B.C. 1950. Cleaning and processing seeds. *Amer. Nurs.* 92(11):13–14, 33–35.

110. Stanwood, P.C. 1985. Cryopreservation of seed germplasm for genetic conservation. In: K.K. Kartha, (ed.). *Cryopreservation of Plant Cells and Organs.* Boca Raton, FL. CRC Press.

111. ———, and L.N. Bass. 1981. Seed germplasm preservation using liquid nitrogen. *Seed Sci. and Tech.* 9:423–37.

112. ———, and E.E. Roos. 1979. Seed storage of several horticultural species in liquid nitrogen (−196° C). *HortScience* 14:628–30.

113. Steiner, J.J., and B.F. Letizia. 1986. A seed-cleaning sluice for fleshy-fruited vegetables from small plots. *HortScience* 21:1066–67.

114. Struve, D.K., J.B. Jett, D.L. Bramlett. 1987. Production and harvest influences on woody plant seed germination. *Acta Hort.* 202:9–21.

115. Taylor, A.G. and T.J. Kenny. 1985. Improvement of germinated seed quality by density separation. *Jour. Amer. Soc. Hort. Sci.* 110:347–49.

116. ———, D.E. Klein, and T.H. Whitlow. 1988. SMP: Solid matrix priming of seeds. *Scientia Hort.* 37:1–11.

117. ———, and G.E. Harman. 1990. Concepts and technologies of selected seed treatments. *Annu. Rev. Phytopath.* 28:321–39.

118. TeKrony, D.M. and D.B. Egli. 1991. Relationship of seed vigor to crop yield: a review. *Crop Sci.* 31:816–22.

119. Tempe, J. de, and J. Binnerts. 1979. Introduction to methods of seed health testing. *Seed Sci. and Tech.* 7:601–36.

120. Thomson, J.R. 1979. *An introduction to seed technology.* New York: John Wiley.

121. Toole, E.H. 1958. Storage of vegetable seeds. *USDA Leaflet* 220 (rev.).

122. U.S. Department of Agriculture. 1952. *Manual for testing agriculture and vegetable seeds.* U.S. Dept. Agr. Handbook 30. Washington, D.C.: U.S. Govt. Printing Office.

123. Van der Berg, H.H., and R. Hendricks. 1980. Cleaning flower seeds. *Seed Sci. and Tech.* 8:505–22.

124. Van der Burg, W.J., H. Jalink, R.A. van Zwol, J.W. Aartse and R.J. Bino. 1994. Nondestructive seed evaluation with impact measurements and x-ray analysis. *Acta Hort.* 362:149–57.

125. Vaughn, C.E., B.R. Gregg, and J.C. Delouche. 1968. *Seed processing and handling.* State College, Miss.: Miss. State Univ. Seed Technology Library.

126. Watkins, J.T. 1992. The effect of environment and culture on vegetable seed quality. *HortTechnology* 2:333–34.

127. Wells, J.S. 1985. *Plant Propagation Practices* (2nd ed.). Chicago: American Nurseryman Publ. Co.

128. Western Forest Tree Seed Council. 1966. *Sampling and service testing western conifer seeds.* Portland, Oreg.: Western Forestry and Conservation Association, 36 pp.

129. Wyman, D. 1953. Seeds of woody plants. *Arnoldia* 13:41–60.

130. Young, J.A., and C.G. Young. 1986. *Collecting, processing and germinating seeds of wildland plants.* Portland, Oreg.: Timber Press.

SUPPLEMENTARY READING

BASRA, A.S. 1995. *Seed Quality: basic mechanisms and agricultural implications.* Food Products Press, New York.

CANTLIPFFE, D.J., ed. 1979. Symposium on seed quality: An overview of its relationship to horticulturists and physiologists. *HortScience* 15:764–89.

COPELAND, L.O., and M.B. MCDONALD. 1995. *Principles of Seed Science and Technology* (3rd ed.). New York: Chapman and Hall.

GORDON, A.G., and D.C.F. ROWE. 1982. *Seed manual for ornamental trees and shrubs.* Bull. 59. London: Forestry Commission.

HAWTHRON, L.R., and L.H. POLLARD. 1954. *Vegetable and flower seed production.* New York: Blakiston Co.

JEFFS, K.A. *Seed treatment.* BCPC publications, Surrey U.K.

JUSTICE, O.L., and L.N. BASS. 1979. *Principles and Practices of Seed Storage.* London: Castle House Pub.

KOZLOWSKI, T.T., ed. 1972. *Seed biology,* Vols. 1, 2, 3. New York: Academic Press.

LEE, D., 1987. Seed collection and handling. *Proc. Inter. Plant Prop. Soc.* 37:61–65.

ROBERTS, E.H., ed. 1972. *Viability of seeds.* London: Chapman and Hall.

SEED WORLD, published monthly.

SCHOPMEYER C.S., ed. 1974. Seeds of woody plants in the United States. U.S. Dept. Agr. Handbook 450. Washington, D.C.: U.S. Govt. Printing Office.

THOMSON, J.R. 1979. *An introduction to seed technology.* New York: John Wiley.

U.S. DEPT. OF AGRICULTURE. 1961. *Seeds: Yearbook of agriculture.* Washington, D.C.: U.S. Govt. Printing Office.

7
Principles of Propagation by Seed

A seed is a ripened ovule. At the time of separation from the parent plant it consists of an **embryo** and **stored food supply,** both of which are encased in a protective **covering** (Figure 7–1). The activation of the metabolic machinery of the embryo leading to the emergence of a new seedling plant is known as **germination**. This chapter describes the various conditions that determine the success of germination and initial growth of the seedling.

THE GERMINATION PROCESS

For germination to be initiated, three conditions must be fulfilled (*32, 78*): First, the seed must be **viable,** that is, the embryo must be alive and capable of germination. Second, the seed must be subjected to the appropriate environmental conditions: available **water,** proper **temperature** regimes, a supply of **oxygen,** and sometimes **light**.

Third, any **primary dormancy** condition present within the seed (*36*) must be overcome. Internal processes leading to removal of primary dormancy are collectively known as **after-ripening** and result from the interaction of the environment with the specific primary dormancy condition. After-ripening requires a period of time and sometimes specific methods of seed handling. Even in the absence of primary dormancy and/or if the seeds are subjected to adverse environmental conditions, a **secondary dormancy** can develop and further delay the period when germination takes place (*26, 82, 88*).

Transition from Seed Development to Germination

Most seeds desiccate during the maturation drying stage of seed development (Figure 7–2). These seeds are either dormant or nondormant at the time they are shed from the plant. However, some seeds either do not enter the maturation stage of seed development and germinate prior to being shed from the plant (vivipary) or can only tolerate a small degree of desiccation (recalcitrant seeds). Figure 7–2 illustrates the fate of various seeds as they approach the end of seed development. Viviparous and recalcitrant seeds are discussed in detail in

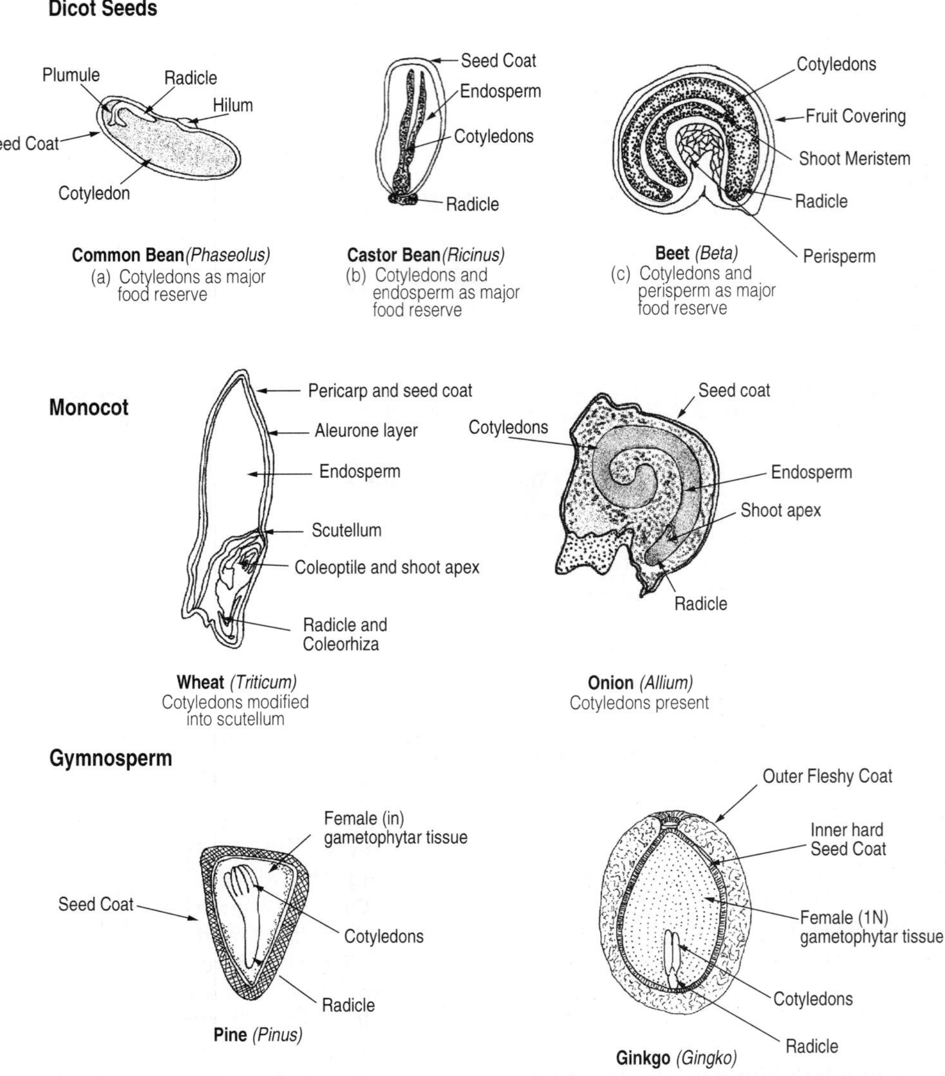

FIGURE 7–1 Seed morphology in a typical monocot, dicot, and gymnosperm.

Chapters 5 and 6. The discussion of seed germination in this chapter will focus on the basic process of seed germination in orthodox seeds that complete maturation drying and are dormant or nondormant after separation from the mother plant.

Phases of Early Germination

Early seed germination begins with imbibition of water by the seed and can be described by a triphasic increase in seed fresh weight due to increasing water uptake (Figure 7–3). These phases are characterized by a rapid increase in water uptake during the **imbibition** phase followed by a period where there is little water uptake called the **lag** phase. The fresh weight of the seed begins to increase again as water uptake drives the **emergence of the radicle.** This process relies on the water potential of the cells in the seed and embryo (see sidebar on water potential).

Water Uptake by Imbibition (Phase I)

Most seeds are dry (<15 percent moisture) after completing seed development. This results in a very low water potential in dry seeds of near −100

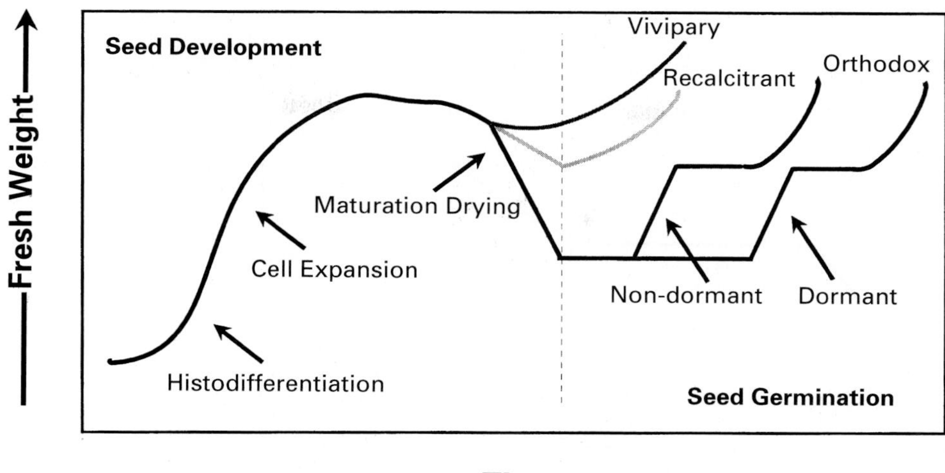

FIGURE 7–2 The transition from seed development to seed germination. Seeds may end seed development and display viviparous, recalcitrant, or orthodox seed behavior. Viviparous and recalcitrant seeds germinate before completing the maturation drying stage of development. Orthodox seeds continue to dry to about 10 percent moisture and can be either nondormant (sometimes termed *quiescent*) or dormant.

WATER POTENTIAL AND SEED GERMINATION

Water potential is described as:

Water Potential (ψ_{cell}) = Matric potential (ψ_m) + Osmotic potential (ψ_π) + Pressure potential (ψ_ρ)

Matric potential is the major force responsible for water uptake during imbibition. Matric forces are due to the hydration of dry components of the seed including cell walls and macromolecules like starch and proteins. Water uptake due to matric forces during imbibition is usually rapid as might be expected because the seed is very dry (<15 percent) moisture at the end of seed development; see Chapter 5.

Osmotic potential and pressure potential determine water uptake during the radicle emergence phase of seed germination. The initial stage of radicle emergence is due to enlargement of the cells in the radicle corresponding to increased water uptake. Osmotic potential is a measure of the osmotically active solutes in a cell including molecules like organic or amino acids, sugars, and inorganic ions. Osmotic potential is expressed as a negative value. As the number of osmotically active solutes increases in a cell, the osmotic potential becomes more negative (i.e., from −0.5 MPa to −1.0 MPa). This can result in more water moving into the cell.

On the other hand, pressure potential is an opposing force and is expressed as a positive value. The pressure potential is the turgor force due to water in the cell pressing against the cell wall. It is also an expression of the ability of the cell wall to expand. Cell wall loosening in the radicle is determined by the physical properties of the cell wall and the counter-pressure exerted by the seed tissues covering the radicle (Figure 7–4). A combination of increasing osmotic potential (more negative) and/or change in the pressure potential can result in cell enlargement and initiate radicle elongation (Figure 7–5). This is termed *growth potential* (*14*). Thus changes in osmotic potential of radicle cells, and cell wall loosening in radicle or seed covering cells are essential components controlling radicle growth and germination. An understanding of this concept is essential to understanding aspects of seed dormancy, effects of hormones on germination, and treatments like seed priming.

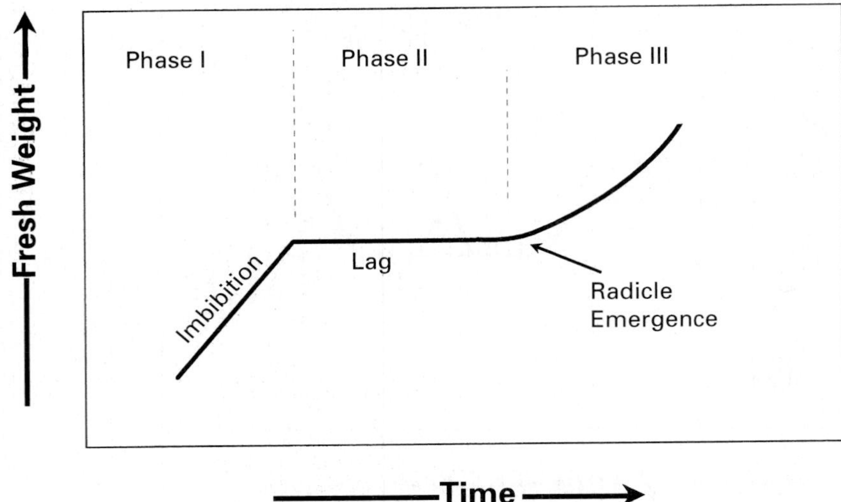

FIGURE 7–3 Phases of water uptake during germination.

MPa (*132, 133*). Imbibition is a physical process related to matric forces and occurs in dry seeds with water-permeable seed coats whether they are alive or dead, dormant or nondormant. There are two stages to imbibition (Figure 7–6) (*134, 113*). Initially, water uptake is very rapid over the first 10 to 30 minutes. This is followed by a slower wetting stage that is linear for up to an hour for small seeds or several hours (5 to 10) for large seeds. Water uptake eventually ends as the seed enters the lag phase of germination.

The seed does not wet uniformly during imbibition. There is a "wetting front" that develops as the outer portions of the seed hydrates while inner tissues are still dry. The volume of the seed also increases during imbibition in a reverse of the seed size reduction that takes place during late seed development. Another characteristic of seeds dur-

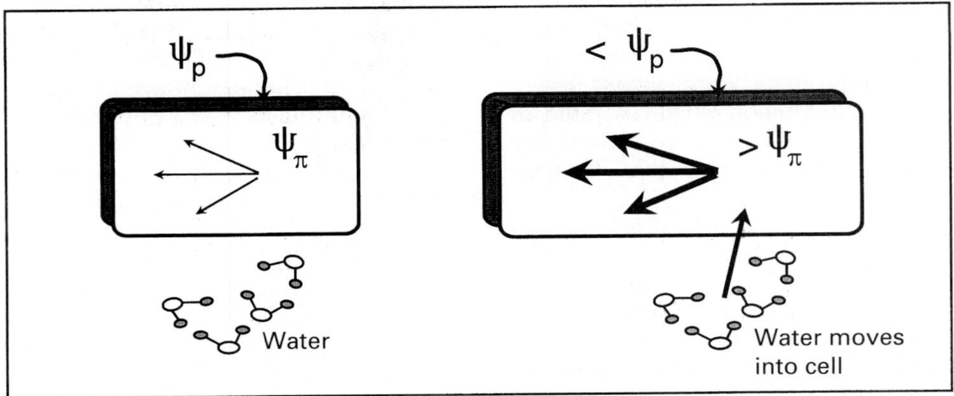

A change in the elastic properties of the cell wall or increase in the osmotic potential of the cell can lead to cell enlargement.

FIGURE 7–4 Schematic representation of water uptake in a cell. The opposing forces of osmotic potential (ψ_π) and pressure potential (ψ_p) determine water uptake by the cell and the cell's ability to expand.

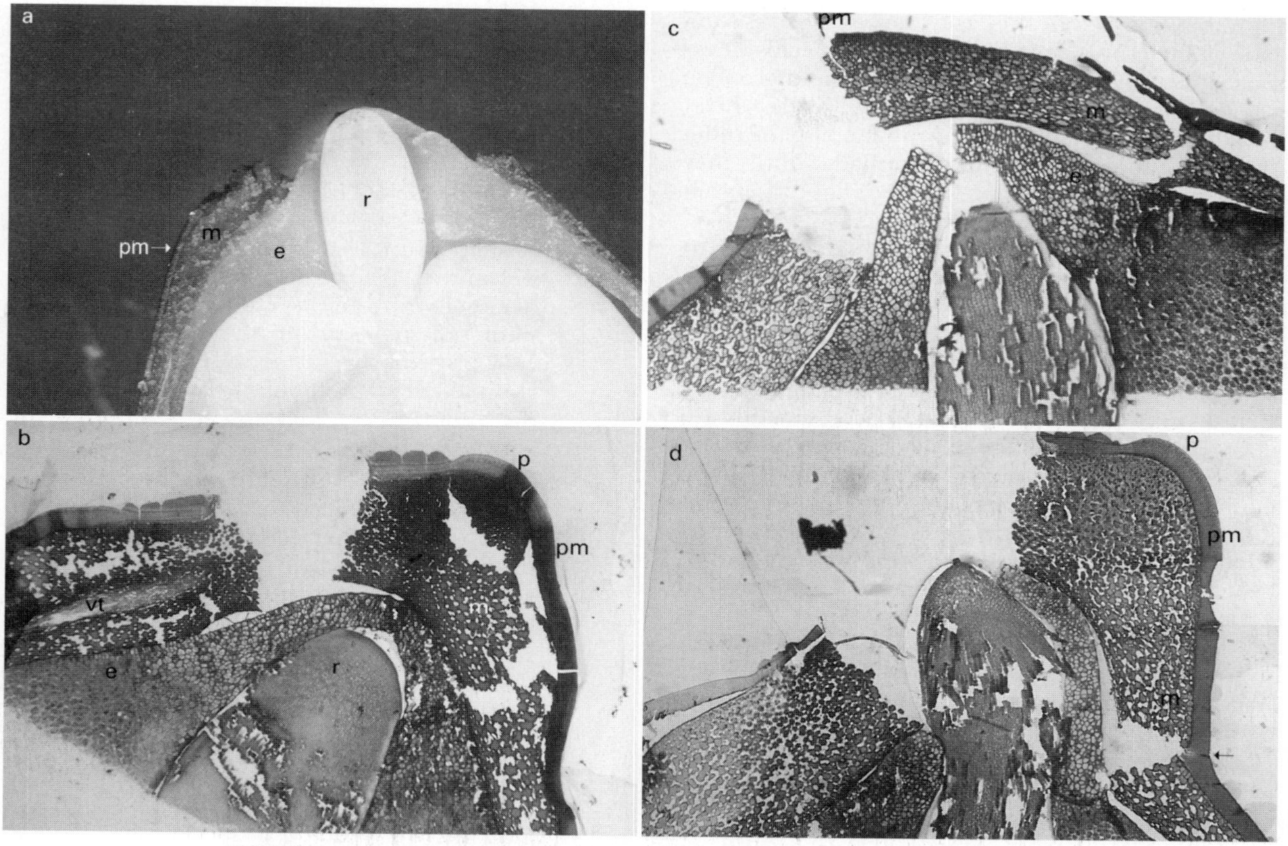

FIGURE 7–5 Longitudinal section of the hilar end of a germinating redbud *(Cercis canadensis)* seed showing the radicle elongating to rupture the seed coat. Abbreviations are radicle (r), endosperm (e), mesophyll layer of the seed coat (m), and palisade layer of the seed coat (pm).

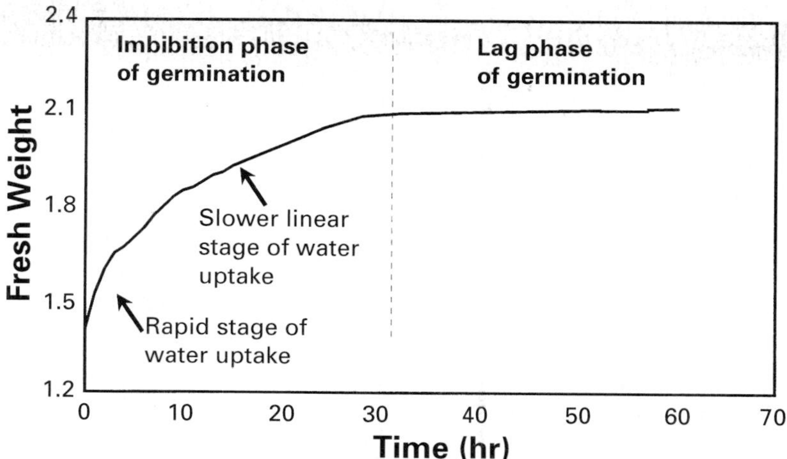

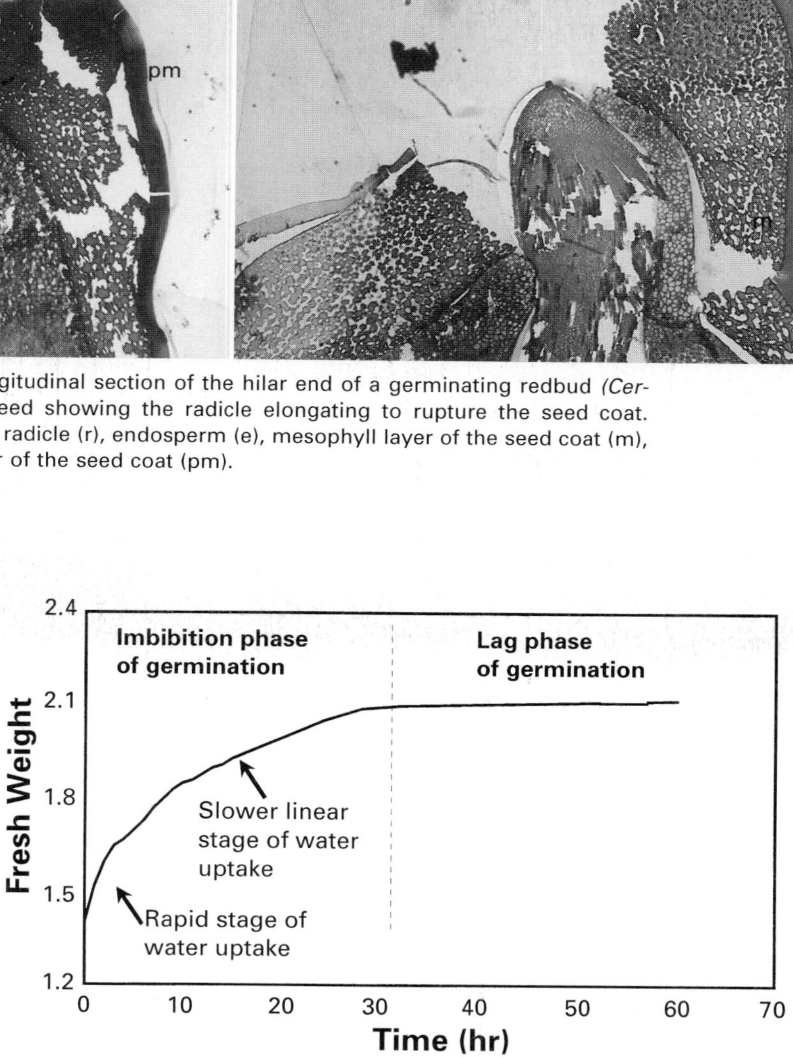

FIGURE 7–6 Water uptake during imbibition of gas plant *(Dictamnus albus)* seed. A rapid and a slower linear stage of water uptake are typical for most seeds.

ing imbibition is that they are **"leaky."** Several compounds including amino acids, organic acids, inorganic ions, sugars, phenolics, and proteins can be detected as they leak from imbibing seeds (*134*). "Leakiness" is probably due to the inability of cellular membranes to function normally until they assume their normal lipid bilayer configuration that was lost during desiccation drying (*18*).

The quantity of leaked solutes is diagnostic for seed quality and is the basis for the electrolyte leakage assay for seed vigor testing (see pp. 161). Solute leakage is also important because it influences detection of the seed by insects and pathogens during germination. Seeds that are slow to germinate or leak excessively due to poor seed quality are more susceptible to attack by insects and diseases.

Seeds can be physically damaged during imbibition. Large seeded crops like soybean can be physically damaged by the inrush of water during imbibition. It is common to raise the moisture content of soybean seeds to 17 percent prior to sowing to reduce imbibitional injury. This type of injury is more common for seeds damaged during harvest. In addition, some tropical and subtropical species (like cotton, corn, lima bean) are sensitive to chilling injury when imbibed in cold water (*73*).

Lag Phase of Germination (Phase II)

Although the lag phase is characterized as a period of reduced or no water uptake following imbibition, it is a highly active period physiologically (Figure 7–3) (*18, 134*). Cellular activities critical to normal germination during the lag phase include:

1. Mitochondria "maturation." Mitochondria are present in the dry seed and these must be rehydrated and membranes within the mitochondria must become enzymatically active. Within hours of imbibition, mitochondria appear more normal when viewed by electron microscopy and both respiration and ATP synthesis increase substantially.
2. Protein synthesis. Although mRNA is present within the dry seed (see sidebar on mRNA in dry seeds), protein synthesis does not occur until polysomes form after seed hydration. New proteins are formed within hours of the completion of imbibition. New protein synthesis during the lag period is required for germination.
3. Metabolism of storage reserve material begins. This can lead to a change in the water potential of the cells within the embryo in preparation of radicle emergence.
4. Specific enzymes including those responsible for cell wall loosening in the embryo or tissues surrounding the embryo can be produced. There is also some evidence that there is ATP-activated acidification of cell walls that could contribute to cell wall loosening in preparation for radicle elongation.

Radicle Emergence (Phase III)

The first visible evidence of germination is emergence of the radicle. This is initially the result of cell enlargement rather than cell division (*64*).

PROTEIN SYNTHESIS AND mRNA IN SEEDS

In the dry seed, there is a complement of mRNA made during the final stages of seed development (see Chapter 5). There are two types of stored mRNA in dry seeds. These are **residual** and **conserved** mRNA (*18*). Residual mRNA are messages left over from seed development. They persist in dry seeds but are rapidly degraded after imbibition and are not involved in germination. Conserved mRNA are produced (transcribed) during late seed development, stored in the dry seed, and translated into proteins during the lag phase of germination. Translation of conserved mRNA are an important step in the germination process. Most conserved mRNA are utilized during the first few hours following imbibition and then quickly degraded. Conserved mRNA code for both "housekeeping" genes necessary for normal cellular activities and germination-specific proteins like "germin" (*94*). Germin is an oxylate oxidase that may function to release calcium from calcium oxylate. Changes in cellular calcium have been shown to be important during germination (*125*). Most conserved mRNA are degraded within several hours of imbibition and a new mRNA population must be made (transcribed) for germination to be completed. All the components for new mRNA synthesis (DNA and RNA polymerases and ribonucleotide triphosphate precursors) are present in dry seed and new mRNA can be detected in the lag phase.

However, soon after radicle elongation begins, cell division can be detected in the radicle tip (*118*).

The initiation of radicle elongation occurs when either (a) the osmotic potential of the cells in the radicle becomes more negative due to metabolism of storage reserves; (b) cell walls in the radicle become more flexible to allow for cell expansion; or (c) cells in the seed tissues surrounding the radicle weaken to allow cell expansion in the radicle (*18, 110*). A combination of these factors may be involved to control germination, depending on the species.

There are a number of studies which indicate that seed coverings (especially the endosperm) can be a significant barrier to germination in some species, particularly under conditions that reduce germination, like low temperature conditions in pepper (*152*) or that inhibit germination like dormancy in iris (*21*), redbud (*57*), and lilac (*81*). In tomato (*61, 110*), hydrolytic cell wall enzymes (like endo-β-mannanase) act to degrade endosperm cell walls, permitting germination by reducing the force of the seed coverings restricting radicle elongation. There is also evidence to indicate that promotion of germination by gibberellin and the prevention of germination by abscisic acid may be mediated by the induction or inhibition of hydrolytic enzymes (Figure 7–7) (*110*).

In species like *Brassica* (rape seed), where endosperm is lacking and the seed coat is thin, changes in the osmotic potential of the cells in the radicle and cell-wall flexibility are responsible for radicle elongation (*127*). In this case, the activity of gibberellin may be to promote germination by a change in osmotic potential, while the action of abscisic acid is to inhibit germination by preventing a change in osmotic potential.

Seedling Emergence

The seedling plant begins cell division at the two ends of the embryo axis, followed by expansion of the seedling structures. The initiation of cell division in the growing points appears to be independent of the initiation of cell elongation (*16, 64*).

The embryo consists of an **axis** bearing one or more **seed leaves,** or **cotyledons**. The growing point of the root is the **radicle** and it emerges from the base of the embryo axis. The **plumule** is the growing point of the shoot and is at the upper end of the embryo axis, above the cotyledons. The seedling stem is divided into the **hypocotyl** and the **epicotyl.** The hypocotyl is the section below the cotyledons, while the epicotyl is the section above the cotyledons.

Once growth begins from the embryo axis, fresh weight and dry weight of the new seedling plant increases, but total weight of storage tissue decreases. The respiration rate, as measured by oxygen uptake, increases steadily with advance in growth. Storage tissues of the seed eventually cease to be involved in metabolic activities except in plants where the cotyledons emerge from the

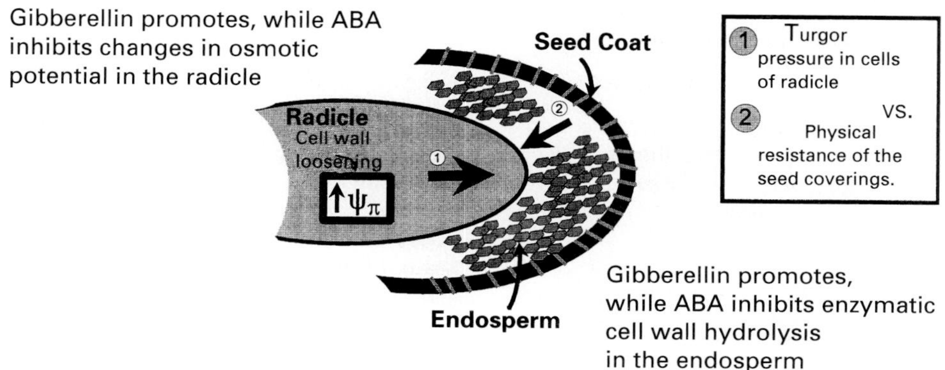

FIGURE 7–7 The balance of forces involved in germination. In many seeds, the seed coverings provide a physical resistance to radicle emergence. The ability of the radicle to penetrate the seed coverings determines the speed of germination and can be an important mechanism for controlling germination in dormant seeds.

ground and become active in photosynthesis. Water absorption increases steadily as new roots explore the germination medium and the fresh weight of the seedling plant increases. The initial growth of the seedling follows one of two patterns. In **epigeous** germination, the hypocotyl elongates and raises the cotyledons above the ground. **Hypogeous** germination is the other pattern of germination and is characterized by a lack of expansion of the hypocotyl so only the epicotyl emerges (Figure 7–8).

Storage Reserve Utilization

Initially, new growth of the embryo axis is dependent on the reserves manufactured during seed development and stored in the endosperm and cotyledons. The major storage reserves are proteins, carbohydrates (starch), and lipids (oils). These are converted to amino acids or sugars to fuel early embryo growth (Figure 7–9). The embryo is dependent on the energy and structural materials from stored reserves until the seedling emerges into the light and can begin photosynthesis.

Use of Storage Proteins

Storage proteins are stored in specialized structures called protein bodies. **Protein bodies** are located in cotyledons and endosperm of the seed (Figure 7–10). Enzymes (proteinases) are required to catabolize storage proteins into amino acids which in turn can be used by the developing embryo for new protein synthesis. These enzymes can be present in stored forms in the dry seed, but the majority of proteinases are synthesized as new enzymes following imbibition.

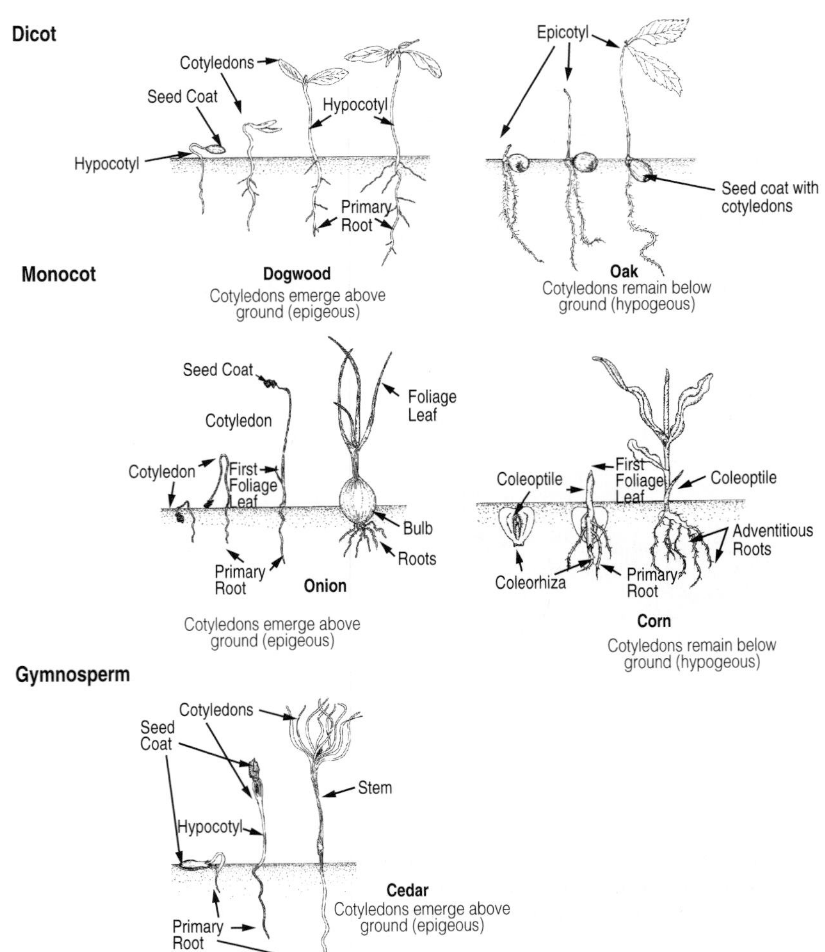

FIGURE 7–8 Typical patterns of germination for an epigeous and hypogeous dicot, a monocot, and a gymnosperm.

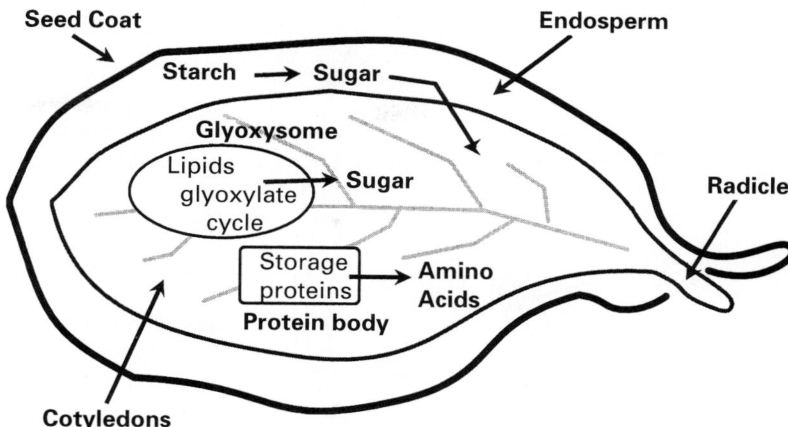

FIGURE 7–9 The general pattern of seed reserve mobilization leading to germination.

Use of Storage Carbohydrates (Starch)

Starch is the major storage material in seeds and is mostly stored in the endosperm. Catabolism of starch has been studied extensively in cereal grains (barley, wheat, and corn) and shows a coordinated system for starch mobilization (*54, 66*). Following imbibition, gibberellin in the embryo axis and the **scutellum** is translocated to the cells of the **aleurone layer** (Figure 7–11). The aleurone is a layer of secretory cells that surrounds the starchy endosperm. Gibberellin initiates the synthesis of numerous enzymes in the aleurone that are secreted into the endosperm (Table 7–1).

The major enzyme degrading starch is α-amylase (Figure 7–12). It hydrolyses starch in the **starch grains** of the endosperm to simple glucose and maltose sugar units that are eventually synthesized into sucrose for transport to the embryo axis. Enzymes break down the cell walls of the endosperm to allow movement of sucrose to the scutellum for transport to the growing axis. Gibberellin-initiated synthesis of α-amylase has been extensively studied in the cereal aleurone system and has significantly improved our understanding of the molecular mechanisms for hormone-regulated gene expression in plants (*18*).

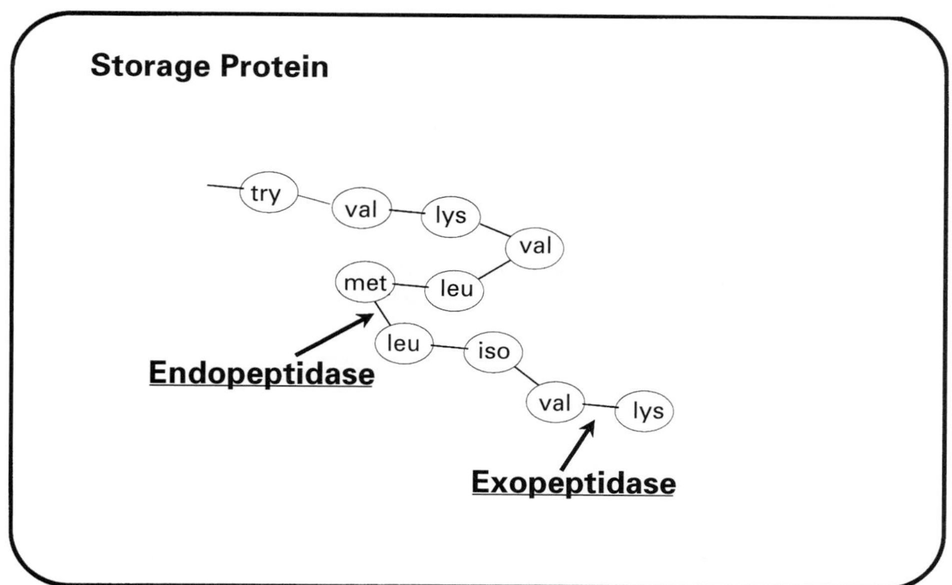

FIGURE 7–10 Enzymes degrade storage proteins into units of amino acids. Endopeptidase enzymes work on internal amino acid bonds, while exopeptidase enzymes work on the outer bonds. Amino acids are used by the developing embryo to make new proteins required for germination.

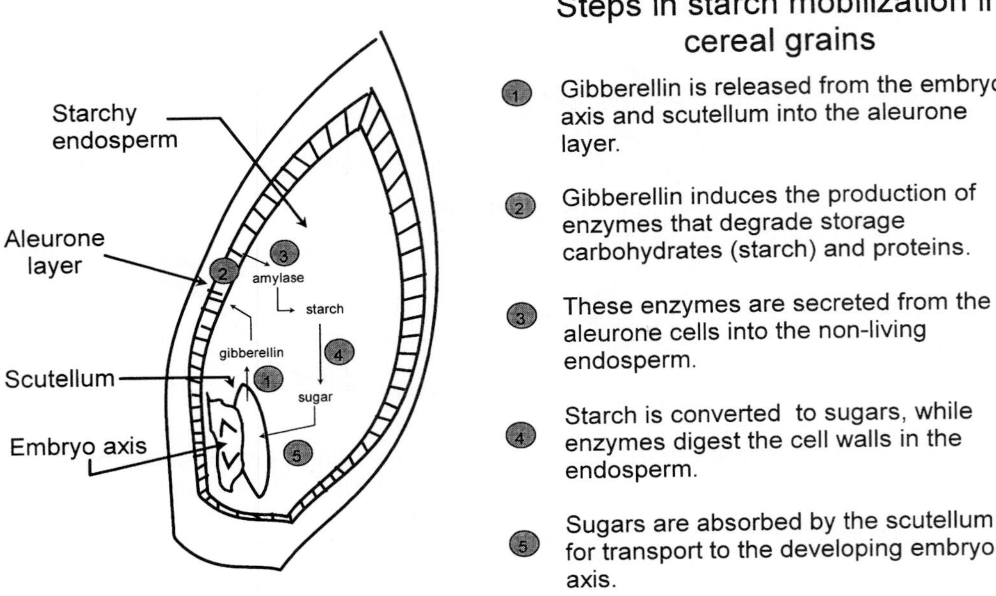

FIGURE 7–11 The cereal grain model for starch mobilization in seeds.

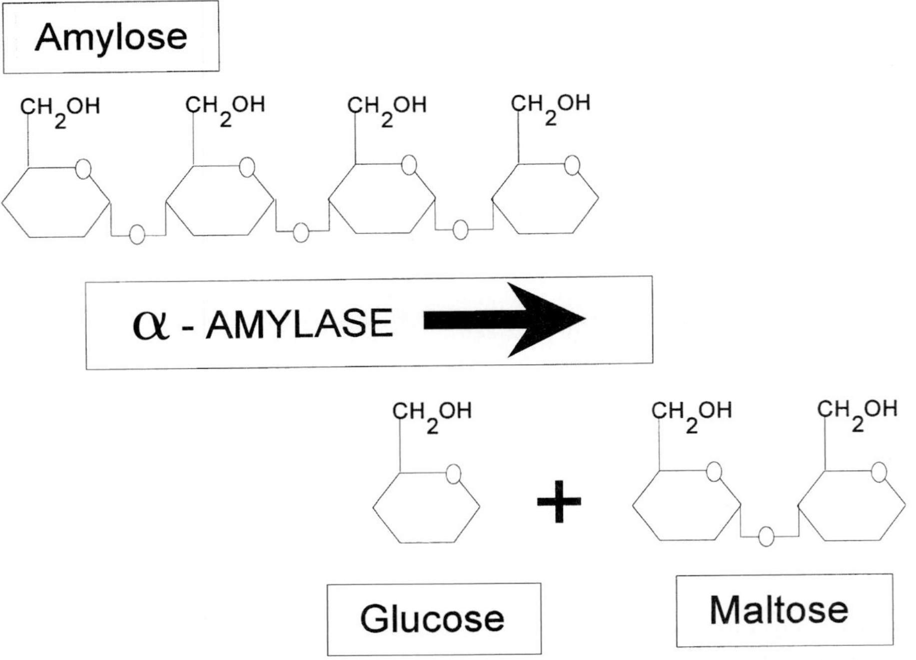

FIGURE 7–12 An important step in the degradation of starch involves the conversion of amylose to glucose and maltose sugar units.

TABLE 7–1

Selected enzymes synthesized during germination of cereals

Amylase	Starch hydrolysis to sugar
Proteases	Protein hydrolysis to amino acids
Glucanases	Cell wall degrading enzymes

Use of Storage Lipids (Oils)

Lipids are stored in specialized structures called **oil bodies** located in the endosperm and cotyledons of seeds. The main storage forms of lipid in the oil body are **triacylglycerides.** In the oil body, triacylglycerides are catabolized to glycerol and free fatty acids (Figure 7–13). Free fatty acids are moved to the **glyoxysome.** Glyoxysomes are specialized structures only present in oil-storing seeds. They function to convert free fatty acids to the organic acids, malate, and succinate using enzymes in the glyoxylate cycle. Glyoxysomes and the glyoxylate cycle are unique to germinating seeds and are not found in any other part of the plant. The end result of this biochemical process is the production of sucrose from storage lipids for use by the developing embryo (Figure 7–13).

Measures of Germination

If one measures the time sequence of germination of a given lot of seeds, or the emergence of seedlings from a seed bed, one usually finds a pattern like the germination curve shown in Figure 7–14. There is an initial delay in the start of germination, then a rapid increase in the number of seeds that germinate, followed by a decrease in their rate of appearance. When viability is less than 100 percent, the end point may not be exact.

Germination is measured on two parameters: the **germination percentage** and the **germination rate.** Vigor may be indicated by these measurements, but seedling growth rate and morphological appearance also must be considered.

Sometimes abnormally growing seedlings result from low seed quality (*74*). Statements of germination percentage should involve a time element, indicating the number of seedlings produced within a specified length of time. Germination rate can be measured by several methods. One method determines the number of days required to produce a given germination percentage like the T_{50} value. T_{50} calculates the number of days required to achieve 50 percent germination in a seed lot. Another method calculates the average number of days required for radicle or plumule emergence as follows:

$$\text{Mean days} = \frac{N_1T_1 + N_2\,T_2 + \ldots + N_xT_x}{\text{total number of seeds germinating}}$$

where *N* values are the numbers of seeds germinating within consecutive intervals of time; *T* values indicate the times between the beginning of the test and the end of the particular interval of measurement. Kotowski (*92*) has used the reciprocal of this formula multiplied by 100 to determine a **coefficient of velocity**. Gordon (*60*) has suggested the

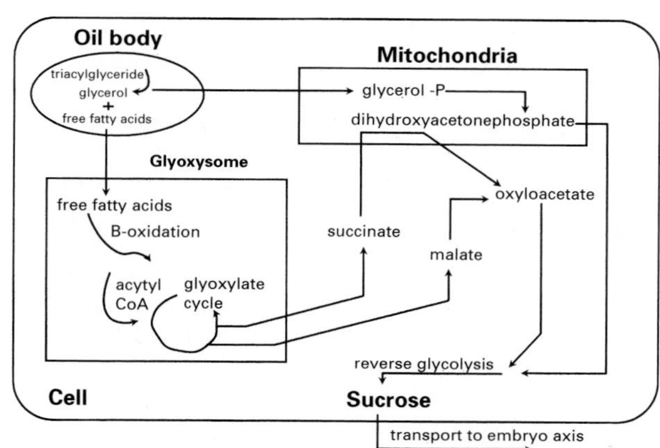

FIGURE 7–13 Lipid conversion to starch is a complex set of enzymatic reactions coordinated between the oil body, mitochondria, and glyoxysome. Key steps include the conversion of triacylglycerides (the storage form for oils in the seed) to glycerol and free fatty acids, the production of succinate and malate in the glyoxylate cycle in the glyoxysome, and then reverse glycolysis (sometimes termed *gluconeogenesis*) to produce sucrose for use by the embryo.

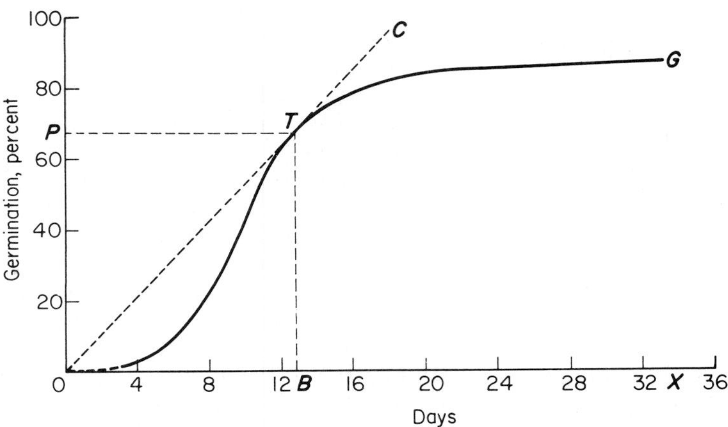

FIGURE 7–14 Typical germination curve of a sample of germinating seeds. After an initial delay, the number of seeds germinating increases, then decreases. Such a curve can be used to measure germination value. (Redrawn from Czabator 1962.)

term **germination resistance** as the time (hours or days) to average germination, based on seeds that germinate.

Czabator (*39*) has suggested another measurement for seeds of woody perennials in which germination may be slow: the **germination value** (*GV*). It includes both the germination rate and percentage. To calculate *GV*, a germination curve, as shown in Figure 7–14, must be obtained by periodic counts of radicle or plumule emergence. The important values on the curve are **T**—the point at which the germination rate begins to slow down, and **G**—the final germination percentage. These points divide the curve into two parts: a rapid phase and a slow phase. Peak value *(PV)* is the germination percentage at *T*, divided by the days to reach that point. Mean daily germination *(MDG)* is the final germination percentage divided by number of days to reach final germination. For example:

$$GV = PV \times MDG$$

$$\frac{68}{13} \times \frac{85}{34} = 5.2 \times 2.5 = 13.0$$

Environmental Factors Influencing Germination

Water

For many seeds without dormancy conditions, the availability of water is the only factor limiting germination at suitable temperatures. The mechanism for water uptake by seeds has been discussed in detail as it relates to the phases of germination (pp. 178). However, seeds within a seed lot or between seed lots can differ in their ability to germinate at any given water potential. This has been termed the base or **threshold water potential** needed to initiate radicle emergence (*24*). The time required for a seed to germinate is related to the difference in the seed's threshold water potential and the seed's water content. This difference is the basis for the **hydrotime** model (*23, 63*) that predicts the time required for radicle emergence in a seed population.

The rate of water movement into the seed is also dependent on similar properties of the germination medium. The matric potential measures the ability of water to move by capillarity through the pores of the soil or other germination medium to the seed. Rate of movement depends upon (a) the pore structure (texture) of the germination medium, (b) the soil packing, and (c) the closeness and distribution of the soil-seed contact. As moisture is removed from the soil by the imbibing seed the area nearest the seed becomes dry and must be replenished by water in pores farther away. Consequently, a firm, fine-textured seed bed closely compacted to the seed is important in maintaining a uniform moisture supply.

Osmotic potential in the soil solution depends upon the presence of solutes (salts). Excess soluble salts (high salinity) in the germination medium may exert strong negative pressure (exosmosis) and counterbalance the osmotic pressure in the seeds. Salts may also produce specific toxic effects. These may inhibit germination and reduce seedling stands (*6*). Such salts originate in the soil and other materials used in the germination medium, the irrigation water, or excessive fertilization. Since the effects of salinity become more acute when the moisture supply is low and the concentration of salts is thereby increased, it is particularly important to maintain a high moisture supply in the seed bed, where the possibility of high salinity exists. Surface evaporation from subirrigated beds can result in the accumulation of salts at the soil surface even under conditions in which salinity would not be expected. Planting seeds several inches below the top edge of a sloping seed bed can minimize this hazard (*17*).

Water stress can reduce the germination percentage (*43, 68*). Most kinds of seeds germinate equally well over the range of available soil moisture, from field capacity (FC) to permanent wilting percentage (PWP) (*6*). Germination of some seeds, particularly those with dormancy problems (e.g., beet, lettuce, endive, or celery), is inhibited as moisture levels are decreased. Such seeds apparently contain inhibitors and require leaching. Seeds of other species (e.g., spinach), when exposed to excess water, produce extensive mucilage that restricts oxygen supply to the embryo, an inhibitor may also be present (*5*). In these cases, germination improves with less moisture. Moisture content of the medium can also impact germination percentages during plug production of flower crops (*30, 31*).

Moisture stress strongly reduces the rate of seedling emergence from a seed bed. This decline in emergence rate occurs as the available moisture decreases to a level approximately halfway through the range from FC to PWP (Figure 7–15) (*6, 43, 68*). Once the seed germinates and the radicle emerges, the water supply to the seedling depends on the ability of the root system to grow into the germination medium and the ability of the new roots to absorb water.

Seed Priming

In experiments conducted in 1855, Charles Darwin hinted at the possibilities for osmotic seed priming (*2*). Darwin submerged seeds in salt water to show that they could move across the sea between land masses as a means to explain geographic distribution of plant species. Not only did seeds survive emersion in cold salt water for several weeks, but some species like cress and lettuce showed accelerated germination. The potential significance of this observation to agriculture was not recognized in Darwin's time. However, in 1963, Ells (*47*) treated tomato seed with a nutrient solution and observed improved germination. At the same time, it was observed that seeds dried following various times of imbibition showed quicker germination after subsequent rehydration (*71, 104*). This was termed "imbibitional drying" (*69*). Heydecker et al. (*75*) used polyethylene glycol to pretreat seeds and this prompted interest in "priming seeds" (*76*) that has led to a commercially significant practice for the seed industry.

Priming is a seed treatment using low water potential seed hydration. This permits the early metabolic events of germination to proceed while remaining in the lag phase of germination (Figure 7–16). Radicle emergence is prevented by the water potential of the imbibitional medium. Primed seeds are dried back to nearly their original water content before the radicle emerges (*89*). Various techniques have been used to control hydration of seeds while not permitting radicle emergence (*89*). These include (1) osmoconditioning (synonym osmotic

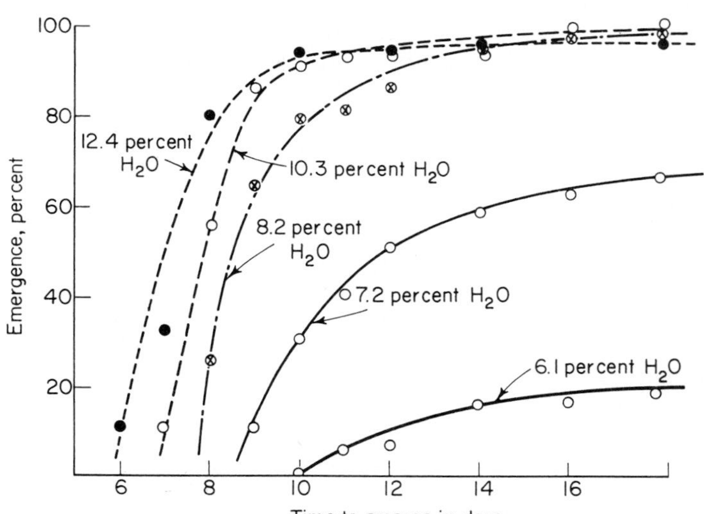

FIGURE 7–15 Effect of different amounts of available soil moisture on the germination (emergence) of 'Sweet Spanish' onion seed in Panchappa fine sandy loam. (From Ayres 1952.)

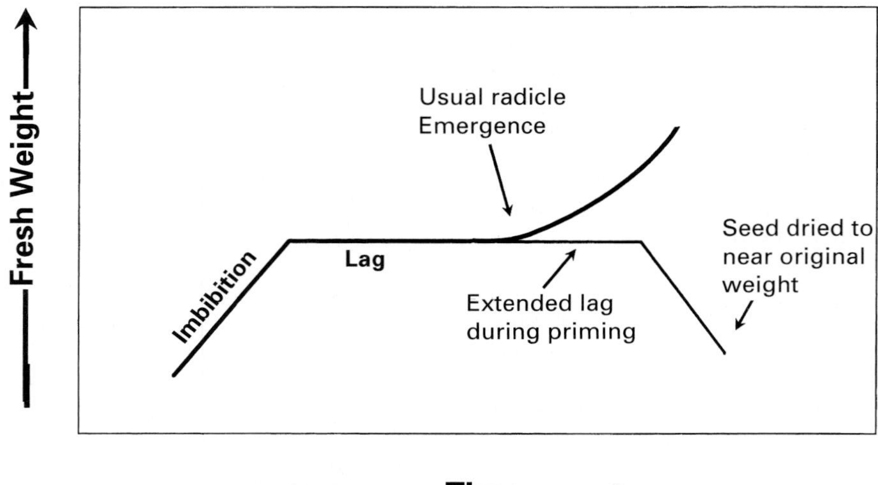

FIGURE 7–16 Phases of germination related to water uptake modified to describe seed behavior during seed priming. Seed priming extends the time the seed remains in the lag phase of germination. Primed seeds are dried to near their original weight prior to radicle emergence.

priming; osmopriming) that uses aerated solutions of PEG or salts; (2) matriconditioning (synonym solid matrix priming; moisturizing) that uses a solid carrier like vermiculite or calcined clay; and (3) large-scale drum priming (*62*).

Osmoconditioning uses the osmotic potential of the treatment solution (usually between −0.5 to −1.5 MPa) and often reduced temperatures (10 to 15° C) to keep seeds in the lag phase of germination (Figure 7–16) for an extended time (up to 20 days). Matriconditioning uses the matric properties of the solid carriers to control seed hydration (*138*). Drum priming hydrates seeds in a revolving drum using only water (*62*). It has received commercial interest in Europe because it avoids use of large amounts of PEG. All three methods wet seeds to about the same water potential and the effects on enhanced seed germination have been comparable.

Seed priming has become a commercially important seed treatment especially for high-value seeds where uniform germination is required, for example, plug production of bedding plants (see Chapter 8). It has also become important for crops that experience thermodormancy including summer seeded lettuce (*28, 145*) and summer greenhouse-sown pansy (*30*).

Several biochemical changes occur during priming (*87, 89*). There is very little increase in DNA synthesis during priming. This is expected because during priming seeds remain in the lag phase of germination prior to the onset of cell division. In contrast to DNA synthesis, RNA synthesis increases during priming. However, seeds primed in the presence of RNA synthesis inhibitors indicate that RNA synthesis is not required for the observed priming effect on seeds. One characteristic of primed seeds is that they resume RNA synthesis quicker than nonprimed seeds during germination. It is not clear if this is a cause or an effect of the priming process.

Protein synthesis increases substantially during and following priming. This includes both the quantity and the type of proteins being made. Inhibiting protein synthesis during priming prevents enhanced germination indicating that protein synthesis is an important part of the priming process. Metabolic enzymes involved in storage reserve mobilization have been shown to increase, including α-amylase, malate dehydrogenase, and isocitrate lyase. This implies that one mechanism for priming is a change in the osmotic potential of cells in the embryo due to the increase in osmotically active solutes like sugars and amino acids mobilized from starch and proteins. However, there is also some evidence to suggest that cell-wall properties of the seed coverings also change during priming.

Temperature

Temperature is, perhaps, the most important environmental factor that regulates the timing of germination, partly due to dormancy control and/or release and partly due to climate adaptation. Temperature control is also essential in subsequent seedling growth. Dry, unimbibed seeds can withstand extremes of temperature. For disease control, seeds can be placed in boiling water for short periods without killing them. In nature, brush fires are often effective in overcoming dormancy without damaging the seeds. Seeds show prolonged storage life when stored at low temperatures, even below freezing for dry seeds (see Chapter 8).

Temperature Effects on Germination

Temperature affects both germination percentage and germination rate (*92*). Germination rate is invariably low at low temperatures but increases gradually as temperature rises, similar to a chemical rate-reaction curve (*91*). Above an optimum level, where the rate is most rapid, a decline occurs as the temperatures approach a lethal limit where the seed is injured. Germination percentage, unlike the germination rate, may remain relatively constant, at least over the middle part of the temperature range, if sufficient time is allowed for germination to occur.

Three temperature points (minimum, optimum, and maximum), varying with the species, are usually designated for seed germination (*45*). **Minimum** is the lowest temperature for effective germination. **Maximum** is the highest temperature at which germination occurs. Above this the seed either is injured or goes dormant. **Optimum** temperatures for seed germination fall within the range at which the largest percentage of seedlings are produced at the highest rate. The optimum for nondormant seeds of most plants is between 25 and 30° C (77 and 86° F), but can be as low as 15° C (60° F).

Seeds of different species, whether cultivated or native, can be categorized into temperature-requirement groups. These are related to their climatic origin:

Cool-temperature tolerant. Seeds of many kinds of plants, mostly native to temperate zones, will germinate over a wide temperature range from about 4.5° C (40° F) (or sometimes near freezing) up to the lethal limit—from 30° C (86° F) to about 40° C (104° F). The optimum germination temperature is usually about 24 to 30° C (77 to 86° F). Examples include broccoli, carrot, cabbage, alyssum, and others.

Cool-temperature requiring. Seeds of some cool-season species and cultivars adapted to a Mediterranean climate require low temperatures and fail to germinate at temperatures higher than about 25° C (77° F). Species of this group tend to be winter annuals in which germination is prevented in the hot summer, but takes place in the cool fall when winter rains commence. Examples include various vegetables, such as celery, lettuce, and onion, as well as some flower seed—coleus, cyclamen, freesia, primula, delphinium, and others (*5*).

Warm-temperature requiring. Seeds of another broad group fail to germinate below about 10° C (50° F) (asparagus, sweet corn, and tomato) or 15° C (60° F) (beans, eggplant, pepper, and cucurbits). These species originated primarily in subtropical or tropical regions. Other species, such as lima bean, cotton, soybean, and sorghum, are also susceptible to "chilling injury" when exposed to temperatures of 10 to 15° C (50 to 60° F) during initial imbibition. Planting in a cold soil can injure the embryo axis and result in abnormal seedlings (*73, 114*).

Alternating temperatures. The fluctuation of day/night temperatures often gives better results than constant temperatures for both seed germination and seedling growth. Use of fluctuating temperatures is a standard practice in seed-testing laboratories, even for seeds not requiring it. The alternation should be a 10° C (18° F) difference (*144*). This requirement is particularly important with dormant, freshly harvested seeds (*4*). Seeds of a few species will not germinate at all at constant temperatures. It has been suggested that one of the reasons imbibed seeds deep in the soil do not germinate is that soil temperature fluctuations disappear with increasing soil depth (*117*).

Water and Temperature Interactions on Germination

Seedling emergence is primarily a function of the moisture available to the seed and temperature (*52*). At constant moisture levels (water potential), germination can be described by a **thermal time** model (*56, 143*). Accumulated thermal time (hours after imbibition above a critical temperature value) necessary to initiate germination varies for different species and even for seeds within a seed lot. Under field conditions, where moisture and temperature vary, the time to radicle emergence can be predicted using a **thermohydrotime** model (*23, 52, 63*).

These models predict the thermal time required following imbibition for germination. Faster germinating seeds require less accumulated thermal time to germinate than slower germinating seeds in the seed lot. At the same time, these seeds have a threshold water potential that permits radicle emergence that can also vary within a seed lot (see pp. 188). If the water potential in the soil or germination medium falls below the threshold water potential for that seed, then germination is delayed.

This helps to conceptualize the impact of moisture stress on germination in a seed bed. For example, if moisture is available following sowing, quicker germinating seeds are able to germinate and emerge (*120*). As the seed bed dries out (water potential falls below threshold water potential), the remaining seeds in the population are unable to germinate. These seeds germinate only after subsequent irrigation or rainfall. This has led Finch-Savage et al. (*53*) to be able to schedule irrigation at critical stages of germination predicted by thermal time to optimize seedling emergence for various vegetable crops.

Temperature Effects on Seedling Growth

The optimum temperature may shift after germination begins, since seedling growth tends to have different temperature requirements than seed germination. In the nursery or laboratory, the usual practice is to shift the seedlings to a somewhat lower temperature regime following germination in order to prepare the plants for transplanting and to reduce disease problems in the seed bed. If seeds are germinated and the seedlings grown at high temperatures, it is important that other environmental conditions be favorable. Plants should have increased light, preferably long photoperiods, adequate fertilization, and sterile conditions to eliminate disease pathogens. Increased carbon dioxide can also be a useful component of this system.

Aeration Effects on Germination

Exchange of gases between the germination medium and the embryo is essential for rapid and uniform germination. Oxygen (O_2) is essential for the respiratory processes in the germinating seed. Oxygen uptake can be measured shortly after imbibition of water begins. Rate of oxygen uptake is an indicator of germination progress and has been suggested as a measure of seed vigor. In general, O_2 uptake is proportional to the amount of metabolic activity taking place. Oxygen supply is limited where there is excessive water in the soil medium. Poorly drained outdoor seed beds, particularly after heavy rains or irrigation, can have the pore spaces of the soil so filled with water that little oxygen is available to the seeds. The amount of oxygen in the germination medium is affected by its low solubility in water and its slow diffusability into the medium. Thus, gaseous exchange between the germination medium and the atmosphere, where the O_2 concentration is 20 percent, is reduced significantly by soil depth and, in particular, by a hard crust on the surface, which can limit oxygen diffusion (*12, 68*).

Carbon dioxide (CO_2) is a product of respiration and under conditions of poor aeration can accumulate in the soil. At lower soil depths increased CO_2 may inhibit germination to some extent but probably plays a minor role, if any, in maintaining dormancy. In fact, high levels of CO_2 can be effective in overcoming dormancy in some seeds (*91*).

Seeds of different species vary in their ability to germinate at very low oxygen levels, as occurs under water (*18, 107*). Seeds of some water plants germinate readily under water, with germination inhibited in air. Rice seeds can germinate in a shallow layer of water. At low oxygen levels, rice seedlings, however, develop differently than those of other monocots. Shoot development is stimulated and the plumule grows to extend up through the water into the air; root growth is suppressed and poor anchorage results unless the water layer is drained away (*29*).

Light Effects on Germination

Light has been recognized since the mid-nineteenth century as a germination-controlling factor (*38*). Recent research demonstrates that light acts in both dormancy induction and release and is a mechanism that adapts plants to specific niches in the environment, often interacting with temperature. Light can involve both **quality** (wavelength) and **photoperiod** (duration). See Chapter 3 for a detailed description of light.

Light is recognized to be a factor in the following situations:

1. Certain epiphytic plants, such as mistletoe *(Viscum album)* and strangling fig *(Ficus aurea)*, have an absolute requirement for light and lose viability in a few weeks without it.
2. Most of the light-sensitive species fall into the category of physiological dormancy (including most grasses, various herbaceous vegetable and flower species, and various weed and native species). Light-sensitive seeds are characterized by being small in size, in which a shallow depth of planting would be an important factor favoring survival. Otherwise, if covered too deeply, the epicotyl may not penetrate the soil. Some important flower crops requiring light for germination include

alyssum, begonia, calceolaria, coleus, *Kalanchoe,* primrose, and *Saintpaulia* (*10*).

3. Many conifer seeds with intermediate dormancy have light sensitivity (see Table 6–1 pp. 157).
4. Germination is inhibited by light in a small number of species, such as *Phacelia, Nigelia, Allium, Amaranthus,* and *Phlox.* Some of these are desert plants where survival would be enhanced if the seeds were located at greater depths, where adequate moisture might be assured. Some flower crops are listed as dark-requiring, including calendula, delphinium, pansy, annual phlox, and annual verbena (*10*).
5. Photoperiodism affects seeds of some woody plants such as eastern hemlock *(Tsuga canadensis)* (*136*) and birch (*18*).

Additional aspects of light on germination are discussed under photodormancy (pp. 199).

Light and Seedling Growth

Light of a relatively high intensity is desirable to produce sturdy, vigorous plants, particularly if transplanting is involved. Low light intensity results in etiolation and reduced photosynthesis and poor seedling survival if transplanted.

High light intensity, on the other hand, often results in high temperatures that produce heat injury to the seedling, particularly at the soil level, in a manner resembling "damping-off" fungi attacks. Shading is desirable for many kinds of plants during their early seedling growth out-of-doors to avoid heat injury (Figure 7–17). Use of supplementary artificial light is described in Chapter 3.

FIGURE 7-17 Shade protection of one-year-old conifer seedlings grown in outdoor seed beds.

Disease Control during Seed Germination

The control of disease during seed germination is one of the most important tasks of the propagator. The most universally destructive pathogens are those resulting in "damping-off," which may cause serious loss of seeds, seedlings, and young plants (*8*). In addition, there are a number of fungal, viral, and bacterial diseases that are seed-borne and may infect certain plants (*7*). In such cases, specific methods of control are required during propagation. (See the discussion of sanitation in Chapter 3.)

Damping-Off

Damping-off is a term long used to describe the death of small seedlings resulting from attacks by certain fungi, primarily *Pythium ultimum* and *Rhizoctonia solani,* although other fungi—for example, *Botrytis cinerea* and *Phytophthora* spp.—may also be involved. Mycelia from these organisms occur in soil, in infected plant tissues, or on seeds, from which they contaminate clean soil and infect clean plants. *Pythium* and *Phytophthora* produce spores that are moved about in water.

The environmental conditions prevailing during the germination period will affect the growth rate of both the attacking fungi and the seedling. For instance, the optimum temperature for the growth of *Pythium ultimum* and *Rhizoctonia solani* is between approximately 20 to 30° C (68 and 86° F), with a decrease in activity at both higher and lower temperatures. Seeds that have a high minimum temperature for germination (warm-season plants) are particularly susceptible to damping-off, because at lower or intermediate temperatures (less than 23° C or 75° F) their growth rate is low at a time when the activity of the fungi is high. At high temperatures, not only do the seeds germinate faster, but also the activity of the fungi is less. Field planting of such seeds should be delayed until the soil is warm. On the other hand, seeds of cool-season plants germinate (although slowly) at temperatures of less than 13° C (55° F), but since there is little or no activity of the fungi, they can escape the effects of damping-off. As the temperature increases, their susceptibility increases, because the activity of the fungi is relatively greater than that of the seedling.

The control of damping-off involves two separate procedures: (a) the complete elimination of the pathogens during propagation, and (b) the control of plant growth and environmental conditions, which will minimize the effects of damping-off or give temporary control until the seedlings have passed their initial vulnerable stages of growth.

If damping-off begins after seedlings are growing, it may sometimes be controlled by treating that area of the medium with a fungicide. The ability to control attacks depends on their severity and on the modifying environmental conditions (see Chapter 3).

Symptoms resembling damping-off are also produced by certain unfavorable environmental conditions in the seed bed. Drying, high soil temperatures, or high concentrations of salts in the upper layers of the germination medium can cause injuries to the tender stems of the seedlings near the ground level. The collapsed stem tissues have the appearance of being "burned off." These symptoms may be confused with those caused by pathogens. Damping-off fungi can grow in concentrations of soil solutes high enough to inhibit the growth of seedlings. Where salts accumulate in the germination medium, damping-off can thus be particularly serious.

DORMANCY: REGULATION OF GERMINATION

In many cases, seeds may only be **quiescent** when they are separated from the plant. Quiescent seeds need only be imbibed at permissive temperatures to initiate germination. In other cases, seeds display **primary dormancy. Dormancy is a condition where seeds will not germinate even when the environmental conditions (water, temperature, and aeration) are permissive for germination.** This not only prevents immediate germination but also regulates the time, conditions, and place that germination will occur. In nature, different kinds of primary dormancy have evolved to aid the survival of the species (*91, 111, 112, 140, 142*) by programming the time of germination for particularly favorable times in the annual seasonal cycle.

Secondary dormancy is a further survival mechanism that can be induced under unfavorable environmental conditions and may further delay the time that germination occurs. Knowledge of the ecological characteristics of the natural habitat of a species can aid in establishing treatments to induce germination (*117, 159*).

In cultivation, domestication of seed-propagated cultivars of many crop plants, such as grains and vegetables, has undoubtedly included selection for sufficient primary dormancy to prevent immediate germination of freshly harvested seed but not enough to cause problems in propagation. Dormancy facilitates storage, transport, and handling. After-ripening changes take place with normal dry storage handling of most agricultural, vegetable, and flower seeds to allow germination to proceed whenever the seeds are subjected to normal germinating conditions. Problems can occur when seed testing is attempted on freshly harvested seeds. Seeds of some species are sensitive to high-temperature and light conditions related to seed dormancy (see pp. 157). Many weed seeds persist in soil due to either primary or secondary dormancy and provide "seed banks" that produce extensive weed seed germination whenever the soil is disturbed (*18*). Practical problems occur with nursery propagation of seeds of many tree and shrub species. These require specific treatments to overcome dormancy by satisfying the requirements needed to bring about germination (see Chapters 19 and 20).

Kinds of Primary Seed Dormancy

Propagators of cultivated plants have long recognized these germination-delaying phenomena and have learned to manipulate different kinds of seed dormancy through the adoption of appropriate pregermination and handling procedures discovered by trial and error (see Chapter 8). Much scientific thought has gone into defining a uniform terminology for different kinds of seed dormancy. An historically early system for seeds was formulated by Crocker in 1916 (*36, 37*) who described seven kinds based primarily on treatments to overcome them. Subsequently, Nikolaeva (*112*) has defined a system based predominantly upon physiological controls of dormancy. Atwater (*5*) has shown that morphological characteristics, including both seed morphology and types of seed covering, characteristic of taxonomic plant families could be associated with dormancy categories particularly significant in seed testing. More recently, a universal terminology for dormancy was proposed (*95, 96*). It uses the terms eco- , para- , and endo-dormancy to refer to dormancy factors related to the environment *(eco),* physical or biochemical signals originating external to the affected structure *(para),* and physiological factors inside the affected structure *(endo).*

Dormancy will be discussed in this chapter using a system adapted from Crocker (*36*) and Nikolaeva (*112*). Major categories include: **exogenous; endogenous; double;** and **secondary dormancy** (Table 7–2). Exogenous dormancy is imposed by factors outside the embryo. These include maternal tissues in the seed or fruit (like the seed coat or pericarp); or mechanical resistance imposed on the radicle from the endosperm.

TABLE 7–2

Categories of seed dormancy

Types of Dormancy	*Indicators of Dormancy*	*Conditions to Break Dormancy*
1. **Exogenous dormancy**	Imposed by factors outside the embryo	
Physical	Impermeable seed coat	Scarification
Mechanical	Seed coverings restrict radicle growth	Removal of seed covering; warm or cold stratification
Chemical	Inhibitors in seed coverings	Removal of seed coverings (fruits) or leaching seeds
2. **Endogenous dormancy**	Imposed by embryo factors	
Morphological	Underdeveloped embryo	Warm stratification
Physiological	Factors within embryo inhibits germination	
Nondeep	After-ripening	A period of dry storage
	Photodormant	Exposure to red light
Intermediate	Embryo will germinate if separated from the seed coat	Moderate periods (up to 8 weeks) of cold stratification
Deep	Embryo will not germinate when removed from seed coat or will form a physiological dwarf	Long periods (>8 weeks) of cold stratification
Epicotyl	Radicle is nondormant and growth begins when temperature and water permit, but epicotyl is dormant	Warm followed by cold stratification
3. **Double (combinational) dormancy**	Combinations of dormancy conditions that must be satisfied sequentially	
Morphophysiological	Combination of underdeveloped embryo and physiological dormancy	Cycles of warm and cold stratification
Exo-endodormancy	Combinations of exogenous and endogenous dormancy conditions Example: physical (hard seed coat) plus intermediate physiological dormancy.	Sequential combinations of dormancy-releasing treatments. Example: scarification followed by cold stratification.
4. **Secondary dormancy**	Germination conditions induce dormancy in seeds	
Thermodormancy	High temperature induces dormancy	Growth regulators or cold stratification
Conditional dormancy	Change in ability to germinate related to time of the year	Chilling or warm stratification

Endogenous dormancy is related to dormancy factors within the embryo itself. Double dormancy includes any combination of exogenous and endogenous dormancy. These dormancy factors must be relieved sequentially to allow germination. The first three categories are examples of primary dormancy. Secondary dormancy occurs in certain seeds when the germination environment does not permit germination and induces dormancy in seeds that were previously nondormant.

Exogenous Dormancy

The tissues enclosing the embryo can impact germination by (1) inhibiting water uptake; (2) providing mechanical restraint to embryo expansion and radicle emergence; (3) modifying gas exchange (i.e., limit oxygen to the embryo); (4) preventing leaching of inhibitors from the embryo; and (5) supplying inhibitors to the embryo (*18*).

Physical Dormancy (Seed Coat Dormancy)

Modification of **seed coverings** primarily affects the outer integument layer of the seed, which may become hard, fibrous, or mucilaginous during dehydration and ripening (Figure 7–18). In addition, layers of the fleshy fruit may dry and become part of the seed covering as in Cotoneaster or hawthorn *(Crataegus)*. In some drupe fruits, as the olive or *Prunus* species, these layers become the hardened endocarp (pit or stone). In other seeds, such as walnut, they become the surrounding shell. In the caryopsis or achenes of grains or grasses, the fruit covering becomes fibrous and coalesces with the seed.

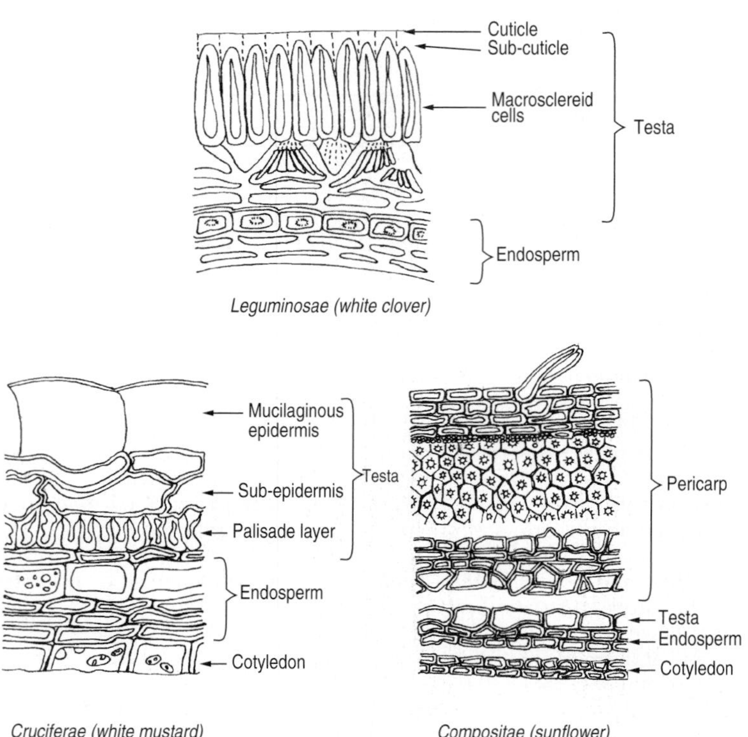

FIGURE 7–18 Types of seed coats that affect seed dormancy. (1) white clover *(Melilotus alba)* Leguminosae: the outer layer of cells becomes impervious to water uptake because of vertically oriented macrosclereid cells that are covered by a layer of cuticle. (2) white mustard *(Brassica hirta)* Cruciferae: outer seed coats develop a mucilaginous layer with inhibitors when soaked in water. (3) Sunflower *(Helianthus annus)* Compositae: pericarp hardens upon drying into a fibrous layer coalescing with the seed coats. Endosperm layer is thin, more or less membranous, and functions in dormancy control. (Courtesy Bewley and Black 1994.)

In various plant families, such as Leguminosae, the outer seed coats harden and become suberized and impervious to water (Figure 7–18, 7–19). Cells of the outer integument become rearranged, coalesce, incorporate suberin deposits, and develop external cutin coverings. These cells are called macrosclereid cells (*119*).

In other species, such as white mustard and spinach, mucilaginous layers inside and outside the seed coats are produced particularly under high moisture conditions, which also function to restrict gaseous exchange (Figure 7–18). In most families, the inner seed coat becomes membranous but remains alive and semipermeable. In the Compositae, for instance, this layer coalesces with the remnant layers of the endosperm. These layers of integument and remnants of the endosperm and nucellus remain physiologically active during ripening and for a period of time after the seed is separated from the plant (Figure 7–18). Such physiologically active layers play a role in maintaining primary dormancy, mainly because this semipermeable nature restricts aeration and inhibitor movement.

Physical exogenous dormancy is produced by seed coverings that are impervious to water. This type can preserve the dry seed for many years, even at warm temperatures. Germination can be induced by any method that can soften or scarify the covering (Figures 7–19, 7–20) (see Chapter 8). Physical dormancy is a genetic characteristic of certain plant families, including *Leguminosae, Malvaceae, Cannaceae, Geraniaceae,* and *Convolvulaceae.* Among cultivated crops, hardseededness is found chiefly in the herbaceous legumes, including clover and alfalfa, as well as many woody legumes (*Robinia, Acacia, Sophora,* etc.). Hardseededness is also increased by environ-

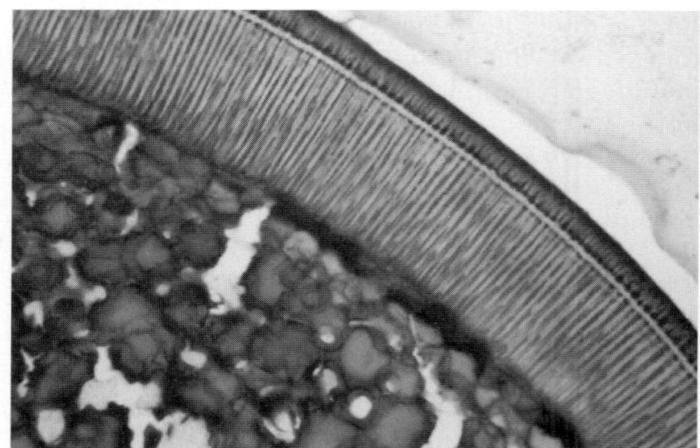

FIGURE 7–19 Cross-section of a redbud *(Cercis canadensis)* seed showing the typical macrosclereid layer in the seed coat.

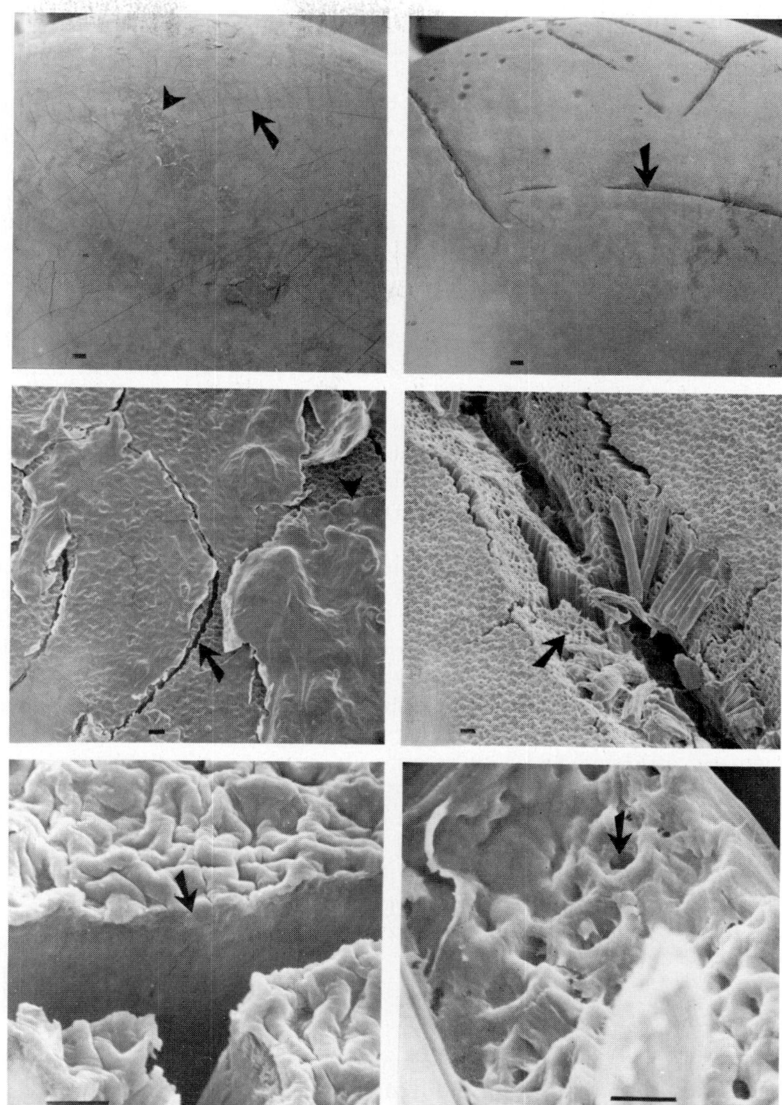

FIGURE 7–20 Scanning electron micrographs of Kentucky coffeetree *(Gymnocladus dioicus)* seed coat after different treatments to allow water uptake. (A) *Left top, middle, bottom:* Untreated seed (Bar length = 15x, 300x, and 3000x, respectively). *Right top, middle, bottom:* 150 minutes of acid scarification at the same range of magnification showing destruction of macrosclereid cells and exposing the lumen *(arrow).* [Courtesy Liu, Khatamian and Fretz (101).]

mental (dry) conditions during seed maturation, and environmental conditions during seed storage. Drying at high temperatures during ripening will increase hardseededness. Harvesting slightly immature seeds and preventing them from drying can reduce or overcome hardseededness in some cases.

Impermeability of the seed coat is due to a layer of palisadelike **macrosclereid** cells, especially thick-walled on their outer surfaces and coated with a layer of waxy, cuticular substances (*46*) (see Figures 7–18, 7–19, 7–20). Disintegration of the "caps" of such cells, or mechanical stress separating the cells, may allow water to enter and produce germination (*25, 101*). In some legume species, the point of attachment **(hilum)** of the seed acts as a one-way valve during ripening to allow water to escape in a dry atmosphere but closing in a moist atmosphere to prevent water uptake (*77*). In *Albizia lophantha* (*41*), a small opening **(strophiole)** near the hilum is sealed with a corklike plug, which can be dislodged with vigorous shaking or impaction (*67*) or by exposure to dry heat as in a fire (*41*).

In nature, impervious seed coats are softened by action of microorganisms in the soil during warm periods of the season or by passage through the digestive tracts of birds and mammals (*37*). They may be broken through mechanical abrasion, alternate freezing and thawing, and in some species by fire. In cultivation, any method to break, soften, abrade, or remove the seed coverings is immediately effective (see Chapter 8).

Mechanical Dormancy

Some seed-enclosing structures, such as shells of walnut (*37*), pits of stone fruits (*112*), and stones of olive (*35*), are too strong to allow the dormant embryo to expand during germination. Water may be absorbed but the difficulty arises in the cementing material that holds the dehiscent layers together, as shown in walnut. Softening primarily comes from soil microorganisms which are favored by nonsterile media and warm temperatures (*37*).

For a number of species, the embryo can be removed from the seed coat and will germinate normally. In these cases, the seed coverings are the primary barrier to germination. The physical strength of the endosperm and seed coverings have been shown to restrict germination in both herbaceous (lettuce, pepper, and tomato) and woody (redbud and lilac) plants. Dormancy in these species is overcome by the seed coverings weakening, by the embryo increasing in growth potential (see endogenous dormancy and pp. 183), or by a combination of seed covering and embryo effects. It has been clearly demonstrated in tomato that conditions that break dormancy are related to a weakening in the strength of the endosperm cells surrounding the radicle (110). This is accompanied by an increase in enzyme activity of cell-wall degrading enzymes in the endosperm.

Chemical Dormancy

Chemicals that accumulate in fruit and seed-covering tissues during development and remain with the seed after harvest can be shown to act as **germination inhibitors** (*50*). Proving their function as germination controls does not necessarily follow, however. Nevertheless, germination can sometimes be improved by prolonged leaching with water, removing the seed coverings, or both (*44, 112*). Some examples are the following:

1. Fleshy fruits, or juices from them, can strongly inhibit seed germination. This occurs in citrus, cucurbits, stone fruits, apples, pears, grapes, and tomatoes. Likewise, dry fruits and fruit coverings, such as the hulls of guayule, *Pennisetum ciliare,* wheat, as well as the capsules of mustard *(Brassica),* can inhibit germination. Some of the substances associated with inhibition are various phenols, coumarin, and abscisic acid.
2. Specific seed germination inhibitors play a role in the ecology of certain desert plants (*91, 154, 155*). Inhibitors are leached out of the seeds by heavy soaking rains that also provide sufficient soil moisture to ensure survival of the seedlings. Since a light rain shower is insufficient to cause leaching, such inhibiting substances have been referred to as "chemical rain gauges."
3. Dormancy in iris seeds is due to a water and ether-soluble germination inhibitor in the endosperm, which can be leached from the seeds with water or avoided by embryo excision (*3*).

Inhibitors have been found in the seeds of such families as Polygonaceae, Chenopodiaceae *(Atriplex),* Portulaceae *(Portulaca),* and other species in which the embryo is peripherally located. Likewise, seeds of a group of such families as Cruciferae (mustard), Linaceae (flax), Violaceae (violet), and Labiteae *(Lavendula)* have a thin seed coat with a mucilaginous inner layer that contains inhibitors (*5*).

Endogenous Dormancy

Morphological (Rudimentary and Linear Embryo) Dormancy

Dormancy occurs in some seeds in which the embryo is not fully developed at the time of seed dissemination. Enlargement of the embryo occurs after the seeds have imbibed water and before germination begins. The process of embryo enlargement is usually favored by a period of warm temperatures.

Atwater (*5*) has distinguished between two groups of embryos that are found in herbaceous flower crops. Seeds of some species have **rudimentary** embryos with little more than a proembryo embedded in a massive endosperm. These are found in various families, such as Ranunculaceae (anemone, Ranunculus), Papaveraceae (poppy, *Romneya*), and Araliaceae (ginseng, fatsia). Germination-inhibiting chemicals also occur in the endosperm and become active at high temperatures. Effective aids for inducing germination include (a) exposure to temperatures of 15° C (59° F) or below, (b) exposure to alternating temperatures, and (c) treatment with chemical additives such as potassium nitrate or gibberellic acid.

A second category includes seeds with **linear** embryos that are torpedo shaped and up to one-half the size of the seed cavity. Important families and species in this category include *Umbelliferae* (carrot), *Ericaceae* (rhododendron, heather), *Primulaceae* (cyclamen, primula), and *Gentianaceae* (gentian). Other conditions, such as semipermeability of the inner seed coats and internal germination inhibitors, may be involved. A temperature of about 20° C (68° F) favors germination, as does gibberellic acid treatment.

Certain temperate zone woody plants such as holly *(Ilex)* and snowberry *(Symphoricarpus)* have rudimentary embryos but, in addition, have other types of dormancy (i.e., double dormancy), such as hard seed coats and dormant embryos, which must be overcome (*112*) for germination to occur.

In seeds of some species, subsequent chilling is also required for germination after the warm embryo development period. Various temperate zone trees fall into this category, including *Fraxinus* (ash) and *Euonymous* species (*112*).

Various tropical species, many of which are monocots, have seeds with underdeveloped embryos that require an extended period at warm temperatures for germination to take place. For example, seeds of various palm species ordinarily require storage for several years to germinate, but this period can be reduced to three months by holding the seeds at temperatures of 38 to 40° C (100 to 104° F), or to 24 hours by excising the embryos and germinating them aseptically. Gibberellic acid (1,000 ppm) has accelerated germination in palm seed, but a seed coat treatment is needed to assure penetration of the chemical (*109*). Other examples include *Actinidia,* whose seed requires two months warmth, and *Annona squamosa* seed, which requires three months of warmth (*112*).

Orchids have undifferentiated embryos when the seed is shed from the mother plant. They are not considered dormant in the same sense as others in this category and are prepared for germination by special aseptic methods as discussed in Chapter 18.

Physiological Dormancy

Nondeep Physiological Dormancy

After-ripening is the time required for seeds in dry storage to lose dormancy. It is the general type of primary dormancy that exists in many, if not most, freshly harvested seeds of herbaceous plants (*4, 18, 105, 112, 139*). This type of dormancy is often transitory and disappears during dry storage, so that it generally is gone before the grower sows the seeds. Consequently, it is primarily a problem with seed-testing laboratories that need immediate germination. In seed testing laboratories such seeds respond to various short-term treatments, including short periods of chilling, alternating temperatures, and treatment with potassium nitrate and gibberellic acid (see pp. 205).

For most cultivated cereals, grasses, vegetables, and flower crops, nondeep physiological dormancy may last for one to six months and disappears with dry storage during normal handling procedures. For many noncultivated plants, nondeep physiological dormancy may not only last longer but can develop into secondary dormancy, particularly if moist seeds are buried in the ground.

Cucumber displays nondeep physiological dormancy and is typical of many crops. Cultivated cucumber *(Cucumis sativus* var. *sativus)* has been selected over many years of cultivation for a short dormancy period. It loses dormancy in dry storage at room temperature after several weeks (15 to 30 days). The hardwickii cucumber *(Cucumis sativus* var. *hardwickii)* is considered a wild progenitor species of the cultivated cucumber and it can display dormancy for 60 to 270 days (*157*). The release from dormancy for hardwickii seeds in dry storage at various temperatures is presented in Figure 7–21. The shorter storage time required to satisfy dormancy at warmer temperatures is typical of seeds with nondeep physiological dormancy.

Photodormancy

Seeds that either require light or dark conditions to germinate are termed **photodormant.** Photodormancy can be considered another type of nondeep physiological dormancy. The basic mechanism of light sensitivity in seeds involves a photochemically reactive pigment called **phytochrome,** widely present in plants (*18, 40, 139, 146*). Exposure of the imbibed seed to **red light** (660 to 760 nm) causes the phytochrome to change to the farred form of phytochrome, (or $\mathbf{P}_{fr}$), which stimulates germination. Exposure of the seed to **far-red** light (760 to 800 nm, or darkness causes a change to the alternate red form of phytochrome $\mathbf{P}_r$), which inhibits germination. These changes can be repeated indefinitely, the last treatment being the one that is effective (Figure 7–22). This was first demonstrated in the classic studies by Borthwick and co-workers at the USDA in Beltsville MD using lettuce seeds. This established the concept of photoreversibility and

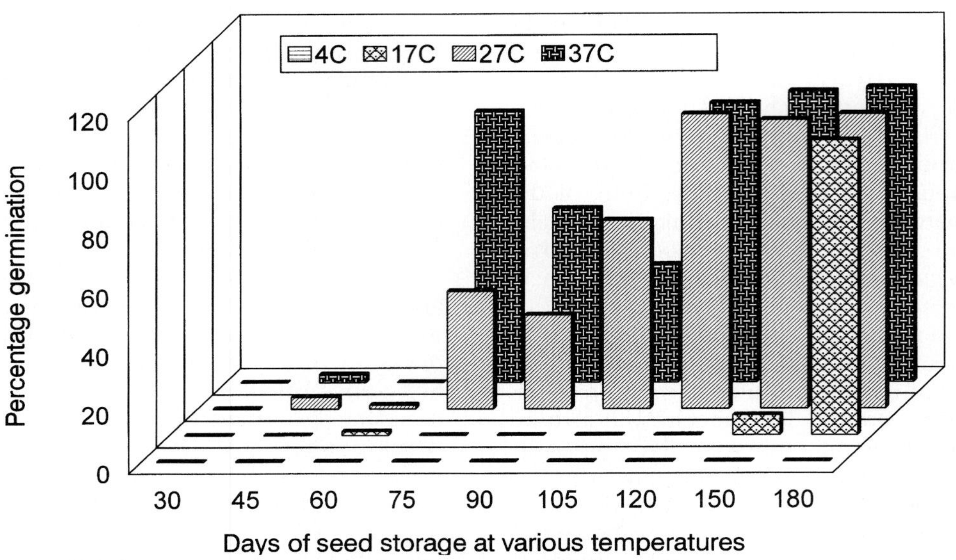

FIGURE 7–21 Release from dormancy in hardwickii cucumber *(Cucumis sativus* var. *hardwickii)* stored dry at various temperatures. The period required to after-ripen seeds and relieve dormancy is shorter at higher temperatures (*157*).

eventually the discovery of the two forms of phytochrome.

For some seeds, there is a distinct light and temperature interaction regarding dormancy and germination. A light requirement can be offset by cool temperatures and sometimes by alternating temperatures. Lettuce seeds generally require light to germinate; however, they lose their light requirement and can germinate in darkness if the temperature is below 23° C (73° F).

The membranes of the seed coat and/or the endosperm appear to act as the light sensors; if removed, the light control disappears. However, direct measurement of phytochrome levels in the seed reveals the greatest concentration of phytochrome in the embryo axis itself. Phytochrome

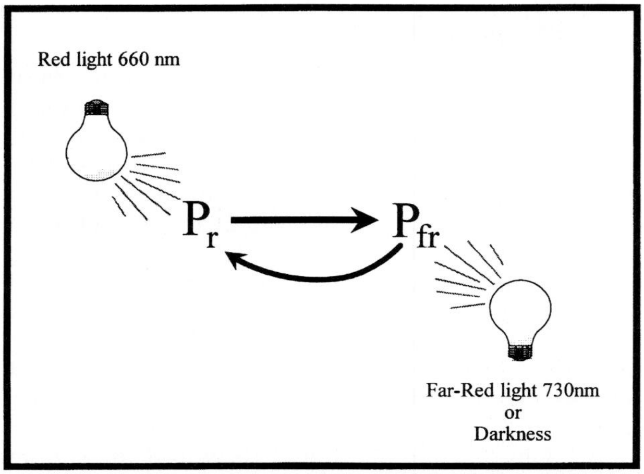

Red light 660 nm

$P_r \rightarrow P_{fr}$

Far-Red light 730nm
or
Darkness

For lettuce

Phytochrome is in the Pr form for dormant seeds.

Phytochrome must be converted to Pfr to release seeds from dormancy.

Light exposure	Seed response
Far-red	Dormant
Far-red then Red	Nondormant
Far-red then Red then Far-red	Dormant
Far-red then Red then Far-red then Red	Nondormant

FIGURE 7–22 Phytochrome controls the dormancy condition of photodormant seeds. Lettuce seeds are the model to study the photoreversibility of phytochrome. The last quality of light the seeds are exposed to determines the dormancy state. Far-red light or darkness keeps seeds dormant, while red light (natural sunlight) will relieve dormancy.

can also be active in both dry and imbibed seeds. Treatments with hormones can offset the light effect, as illustrated in Figure 7–23.

In nature, the light quality reaching the seed can have an impact even during development. Green fruits transmit light which can induce dormancy in the seeds as they mature (*18*). Experiments have shown that seeds of *Arabidopsis* are dormant if the plant is exposed to far-red light before harvest and nondormant if exposed to red light.

Furthermore, seeds of some plants (*Chenopodium album*) are dormant if plants are exposed to long days and nondormant if exposed to short days (*18*). In natural sunlight, red wavelengths dominate over far-red at a ratio of 2:1, so that the phytochrome tends to remain in the active $\mathbf{P}_{fr}$ form. Under a foliage canopy, far-red is dominant and the red/far-red ratio may be as low as 0.12:1.00 to 0.70:1.00, which can inhibit seed germination (*117*). Red light penetrates less deeply into the soil than far-red, so that the red/far-red ratio becomes lower with depth until eventually darkness is complete (*140*). Imbibed light-sensitive seeds buried in the soil will remain dormant until such time as the soil is cultivated or disturbed so as to expose them to light. Similarly, seedling survival is not favored if the seed germinates in close proximity to other plants, where there would be intense competition for light, nutrients, and water by the already established plant population. Light sensitivity can be induced in secondary dormancy by exposing imbibed nonsensitive seeds to conditions inhibiting germination, such as high temperature, high osmotic pressure, or germination-inhibiting gases (156).

Intermediate and Deep Physiological Dormancy

Seeds with intermediate and deep physiological dormancy are characterized by a requirement for a period of one to three (sometimes more) months of chilling while in an imbibed and aerated state. This

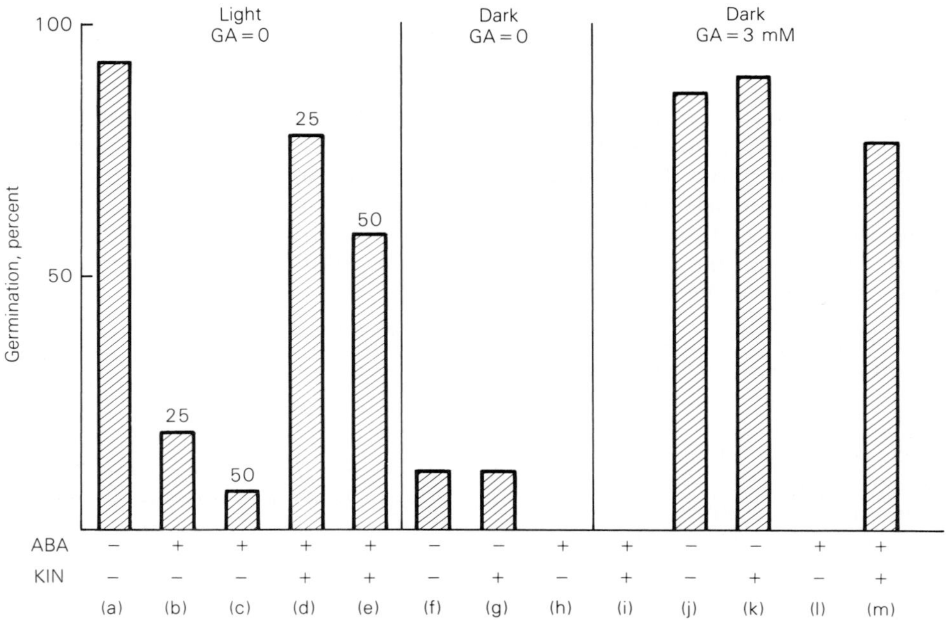

FIGURE 7–23 Interaction of light and three applied hormones on germination of 'Grand Rapids,' a light- and temperature-sensitive cultivar of lettuce with physiological dormancy. In the light (a), seeds germinate but ABA inhibits germination in the light at 25 (b) and 50 (c) mM. Kinetin overcomes the ABA inhibition (d, e). Germination is inhibited in the dark (f) and kinetin does not overcome the dark inhibition in lettuce seeds (g). ABA completely inhibits germination in the dark (h, i). Gibberellic acid overcomes dark-imposed dormancy with (j) or without (k) kinetin. ABA negates the promotive effect of gibberellic acid on germination in the dark (l) and kinetin counteracts this ABA effect and permits gibberellic acid to act. (Redrawn from Khan et al. 1971.)

is most common in seeds of trees and shrubs and some herbaceous plants of the temperate zone (*36, 128, 161*). Seeds of this type ripen in the fall, over-winter in the moist leaf litter on the ground, and germinate in the spring. Nursery propagators have known since early times that such seeds required **moist-chilling** (*18, 33, 97, 147*). This requirement led to the horticultural practice of **stratification,** in which seeds are placed between layers of moist sand or soil in boxes (or in the ground) and exposed to chilling temperatures, either out-of-doors or in refrigerators (see Chapter 8). **Successful stratification requires seeds to be stored in a *moist, aerated* medium at *chilling temperatures* for a certain period of *time*.**

Moisture. The dry dormant seed absorbs moisture by imbibition to around 50 percent (*18, 97*). In some seeds, a hard bony endocarp enclosing the seed reduces water uptake, provides inhibitors, and delays release from dormancy (*44*). Sometimes mechanically removing the covers, subjecting the seeds to warm moist nonsterile conditions prior to germination, leaching, and early harvest without drying prior to stratification reduces stratification time. The seed moisture content should remain relatively constant during stratification. Dehydration stops the stratification process (*70*) and the seeds may revert to secondary dormancy (*148*). When the end of the chilling period is reached, the seeds absorb water rapidly (*97*), the seed coats "crack," and the radicle eventually emerges, sometimes even at low temperatures. Drying at this stage can cause injury to the seed.

Aeration. The amount of oxygen needed during stratification is related to temperature (*34*). At high temperatures the moist seed coverings of dormant, imbibed seeds restrict oxygen uptake because of (a) low oxygen solubility in water, and (b) oxygen fixation by phenolic substances in the seed coats. At chilling temperatures, however, the embryos' oxygen requirement is low and oxygen is generally adequate.

Temperature. Temperature is the single most important factor controlling stratification. The most effective temperature regimes for moist-chilling are similar to those during the winter and early spring of the natural environment of the species. Temperatures somewhat above freezing [2 to 7° C (35 to 45° F)] are generally most effective with more time required at higher and lower temperatures with a minimum at -5° C (23° F) (*129*). Following release from dormancy there is a particular maximum temperature, known as the **compensation temperature,** where secondary dormancy can develop (*1, 130, 147*). For apple, this point has been determined to be 17° C (62° (*1*), but it apparently varies with individual species (*130*) and different stages of after-ripening (*137*). Toward the end of the stratification period, the maximum temperature for germination gradually increases and the minimum temperature gradually decreases. This period has been called **postdormancy** (*106*).

Time. The time required to stratify seeds results from the interaction of (a) the genetic characteristics of the seed population (*84, 85, 129, 158*), (b) the conditions during seed development (*147*), (c) the environment of the seed bed, and (d) the management of seed handling. There is a correlation between the seed-chilling requirements and the bud-chilling requirements of the plants from which the seeds were taken (*116*). In studies with almond, a high quantitative correlation was shown between the mean of the seedling populations and the mean of both the seed and pollen parents (*84*), but a low correlation between the individual seed and the buds of the new plant coming from the embryo (*85*). This difference suggests that the dormancy involves both a genetic component within the embryo and a maternal component from the seed parent. As a result, a great deal of variability in individual seed germination time can occur within a given seed lot and between different seed lots of the same species collected in different years and different locations.

Intermediate Physiological Dormancy

Seeds displaying intermediate physiological dormancy usually require chilling stratification to release the seeds from dormancy (*111, 112*). These seeds are distinguished from those with deep physiological dormancy by two key factors: (1) Embryos isolated from the surrounding seed coverings of seeds with intermediate physiological dormancy germinate readily. (2) The length of time required at chilling temperatures to satisfy dormancy is considerably shorter than for seeds with deep physiological dormancy (see Chapter 8 for techniques for stratification of seeds).

Seed coverings appear to present a significant barrier to germination in seeds with intermediate physiological dormancy because embryos can germinate and grow if isolated from the seed coverings. However, the major change that occurs in the seed is a change in the growth potential of the embryo (Figure 7–7). Growth potential is an indirect measure of the force used by the radicle to penetrate the seed coverings (see pp. 183). One way to measure growth potential is to germinate isolated embryos on solutions containing increasing amounts of polyethylene glycol. This provides a gradient of increasing (more negative) water potential. This can be illustrated using embryos isolated from eastern redbud *(Cercis canadensis)* seeds (Figure 7–24). These embryos show a greater ability to germinate as measured by radicle length at all water potentials after the seeds have been treated with 60 days of moist-chilling stratification. This change in the force generated by the radicle was the major mechanism responsible for dormancy release in eastern redbud seeds (*57*).

Possible mechanisms for dormancy release during stratification are changes in membrane fluidity at chilling temperatures (<15° C, 59° F) and differential enzyme activity (*18*). Protease and lipase enzymes have been shown to increase during chilling stratification and one lipase shows a temperature optima of 4° C (39° F) for activity (*98*). In general, there is a decrease in storage lipids, and an increase in sugars and amino acids from storage reserves during chilling stratification (*98*). This increase in osmotically active solutes could, in part, explain the increase in growth potential seen in embryos following chilling stratification and the subsequent release from dormancy.

Deep Physiological Dormancy

Seeds exhibiting deep physiological dormancy usually require a relatively long (>8 weeks) period of moist-chilling stratification to relieve dormancy. Excised embryos from seeds displaying deep physiological dormancy usually will not germinate normally and the seedlings produced may be abnormal. The relative response is the basis for the "excised embryo" viability test (*36, 112*) (see pp. 155). Responses include enlargement and greening of the cotyledons; short thickened radicle growth, no epicotyl development, or lack of normal root systems. Typically, unchilled excised embryos develop into **physiological dwarfs** (*36, 55*) (Figure 7–25).

Physiological dwarfing in excised embryos from nonchilled seeds have been shown to result from exposure of the apical meristem to warm germination temperatures (*115*). In peaches, temperatures of 23 to 27° C (73 to 81° F) and higher pro-

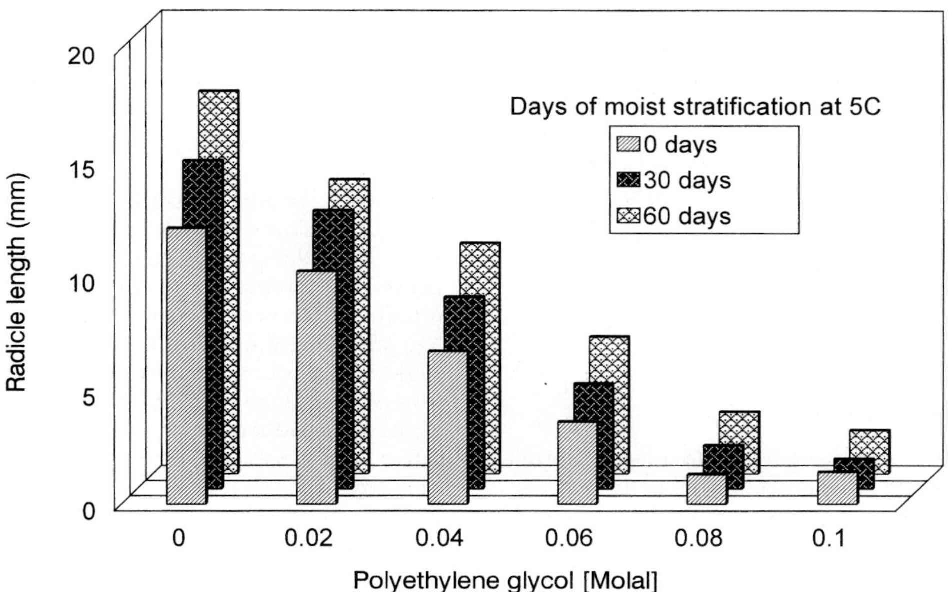

FIGURE 7–24 Embryo radicle growth in eastern redbud *(Cercis canadensis)* on polyethylene glycol solutions with increasing water potential. Embryos from seeds exposed to chilling stratification grow more than embryos from dormant seed on all polyethylene glycol solutions. This shows that one change that occurs during chilling stratification is an increase in the germination force generated by the radicle leading to release from dormancy.

FIGURE 7–25 Physiological dwarfing of seedlings from almond. Seedlings on the left have been exposed to chilling stratification, while seedlings on the right were grown from embryos isolated from dormant seeds that were never exposed to chilling temperatures.

duced symptoms of physiological dwarfing, but at lower temperatures the seedlings grew relatively normally. In almonds, exposing incompletely stratified seed to high temperatures subsequently induced physiological dwarfing in the seedling.

Pinching out the apex can circumvent dwarfing by forcing lateral growth from nondwarfed lower nodes. Dwarfing has also been offset by exposing seedlings to long photoperiods or continuous light (*55, 93*), provided that this action is taken before the apical meristem becomes fully dormant. Repeated application of gibberellic acid has also overcome dwarfing (*13, 55*). Some experiments have shown that systematic removal of the cotyledons from the dormant embryo can induce germination and overcome physiological dwarfing, suggesting the existence of endogenous inhibitors present within the cotyledons (*18*).

Epicotyl Dormancy

Some seeds have separate dormancy conditions for the radicle, hypocotyl, and epicotyl (*13, 15, 36, 112*). These species fall into two subgroups.

1. Seeds that initially germinate during a warm period of one to three months to produce root and hypocotyl growth but then require one to three months of chilling to enable the epicotyl to grow. This group includes various lily *(Lilium)* species, Viburnum spp., peony *(Paeonia)*, black cohosh *(Cimicifuga racemosa)*, and *Hepatica acutiloba*.
2. Seeds that require a chilling period, followed by a warm period for the root to grow, then a second cold period to stimulate shoot growth. In nature, such seeds require two full growing seasons to complete germination. Examples include *Trillium* and certain other native perennials of the temperate zone (see Chapter 21).

Double (Combinational) Dormancy

Double or combinational dormancy combines two (or more) kinds of dormancy, such as exo-endodormancy that combines seed coat dormancy and intermediate physiological dormancy (some tree legumes), or morphophysiological dormancy where there is an underdeveloped (rudimentary) embryo and physiological dormancy *(Ilex)*. To induce germination, all blocking conditions must be eliminated in proper sequence. For example, seed coats must be modified to allow water to penetrate to the embryo, then chilling stratification can release the seed from physiological dormancy. This type of dormancy is characteristic of species of trees and shrubs in families having seeds with hard seed coats but whose plants grow in cold winter areas. In nature, various agents of the environment—those that affect physical dormancy—soften the seed coat when the seed falls to the ground, then the seeds are chilled as they overwinter.

Secondary Dormancy

In nature, primary dormancy is an adaptation to control the time and conditions for seed germination. Secondary dormancy is a further adaptation to prevent germination of an imbibed seed if other environmental conditions are not favorable (*18, 36, 82, 88*). These conditions can include unfavorably high temperatures, temperatures too low, prolonged darkness **(skotodormancy)**, prolonged white light **(photodormancy)**, prolonged far-red light, water stress, and anoxia. These conditions are particularly involved in the seasonal rhythms and prolonged survival of weed seeds in soil (*18*).

Induction of secondary dormancy is illustrated by experiments with freshly harvested seeds of lettuce (*88*). If germinated at 25° C (77° F) the seeds require light, but if imbibed with water for two days in the dark, excised embryos germinate immediately, illustrating that only primary dormancy was present. If imbibition continues for as long as eight days, however, excised embryos will not germinate, since they have then developed secondary dormancy. Release from secondary dormancy can be induced by chilling, sometimes by light, and in various cases, treatment with germination-stimulating hormones, particularly gibberellic acid.

Secondary dormancy can come into play in some instances in cultivated crops, but prolonged dry storage may prevent its occurrence. Seeds with a dormant embryo undergoing moist-chilling may be affected if shifted to high temperatures too quickly (see pp. 190). The term could apply to hardseededness that could develop in the storage of seeds of some species, such as beans and other legumes.

Thermodormancy

For some species (like lettuce, celery, and pansy), germination at high temperatures (>25° C, 77° F) can induce **thermodormancy.** This should not be confused with the **thermal inhibition** most seeds experience when the temperature exceeds the maximum temperature for germination (see pp. 190). Seeds experiencing thermodormancy will not germinate when the temperature returns to near optimum temperatures, while thermal inhibited seeds will germinate when temperatures are lowered. Figure 7–26 shows the interaction of hormones and thermodormancy (*131*). Commercially important crops that are prone to thermodormancy (like summer-sown lettuce) are either primed prior to sowing or grown from transplants (see Chapter 8).

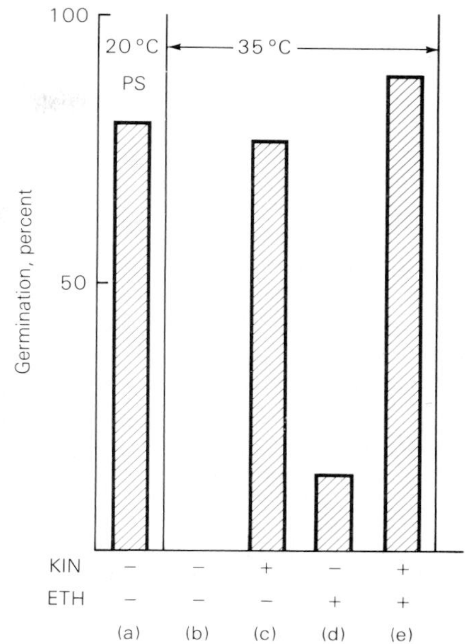

FIGURE 7–26 Hormonal effects on thermodormancy in 'Grand Rapids' lettuce seed. Seed pre-soaked for 24 hours at 20° C overcomes thermodormancy to allow germination to proceed during subsequent exposure to 35° C (a). Seeds not pre-soaked remained dormant at high temperature (b), but kinetin removed most of the dormancy (c). Ethylene had limited but significant effect on removal of thermodormancy with (d) or without (e) kinetin. (Redrawn from Sharples 1968.)

Control of Dormancy and Germination

The Involvement of Hormones in Seed Dormancy

Much experimental evidence supports the concept that specific endogenous growth promoting and inhibiting compounds are involved directly in the control of seed development, dormancy, and germination (*18, 20, 86, 97, 116*). Evidence for hormone involvement comes from correlations of hormone concentration with specific developmental stages, effects of applied hormones, and the relationship of hormones to metabolic activities.

Gibberellins. Gibberellins (GA) comprise the class of hormones most directly implicated in the control and promotion of seed germination. While there are many molecular variations of gibberellin, the one most widely used experimentally and commercially is gibberellic acid (GA_3), but GA_{4+7} is also active. These compounds occur at relatively high concentrations in developing seeds but usually drop to a lower level in mature dormant seeds, particularly in dicotyledonous plants. Gibberellins appear to play a role in two different stages of germination. One occurs in the initial enzyme induction and the second is in the activation of reserve food-mobilizing systems (see pp. 184). Applied gibberellins can relieve certain types of dormancy, including physiological dormancy, photodormancy, and thermodormancy.

Abscisic acid (ABA). This naturally occurring compound is an important growth-regulating compound not only in seed germination but in plant growth in general (*149, 150*). ABA appears to play a role in preventing "precocious germination" of the developing embryo in the ovule (see pp. 142). ABA tends to increase with maturation of the fruit and may prevent vivipary and induce primary dormancy. It has been isolated from the seed coats of dormant peach, walnut, apple, rose, and plum, but it usually decreases during stratification.

Application of ABA can inhibit germination of nondormant seeds and offset the effects of applied gibberellic acid. In general, inhibition is temporary and disappears when seeds are shifted to an ABA-free solution.

Cytokinins. These naturally occurring compounds (*141*) have the basic chemical structure of an N^6-substituted adenine (see Chapter 2). Cytokinin activity tends to be high in developing fruits and seeds, but decreases and becomes difficult to detect as the seeds mature. In seed germination, cytokinin is believed to offset the effect of inhibitors, notably ABA. It has been described, therefore, as playing a "permissive" role in germination in allowing gibberellic acid to function (*86, 90*). It is believed to be active, therefore, at a different germination stage than gibberellins. Applied cytokinin can also be effective in overcoming thermodormancy (*135*).

Ethylene. Ethylene gas is an important, naturally occurring hormone involved in many aspects of plant growth. Ethylene has been effective in overcoming seed dormancy in snowberry *(Symphoricarpos)*, honeysuckle *(Lonicera)*, and similar species, as well as seeds of corn and other cereals. Ethylene production from germinating bean and pea seed was shown in 1935. Later work demonstrated that ethylene is a natural germination-promoting agent for certain kinds of seeds. Ethylene apparently has a limited role in seed germination but has been shown to stimulate germination in subterranean clover *(Trifollium subterranean)* (*49*), Virginia-type peanut *(Arachis hypogaea)*, and witchweed *(Striga asiatica)* (*48*).

Other compounds. Certain other compounds are known to stimulate seed germination, but their role is not clear. Use of **potassium nitrate** has been an important seed treatment in seed-testing laboratories for many years without a good explanation for its action. **Thiourea** overcomes certain types of dormancy, such as the seed-coat inhibiting effect of deep embryo-dormant Prunus seeds as well as the high-temperature inhibition of lettuce seeds (*140*). The effect of thiourea may be due to its cytokinin activity in overcoming inhibition. Two other naturally occurring substances, *fusicoccin* and *cotylenin* (*87*), have been reported to mimic the combination of GA plus cytokinin.

Interactions of Hormones and Dormancy

Dormancy may involve interactions among specific hormones. In general, gibberellins promote germination in dormant seeds, while ABA inhibits germination. The following examples are typical of experimental evidence for hormonal action during dormancy release in seeds.

The interaction of exogenous hormones with light (photodormancy) is complex (see Figure 7–23) (*86, 88*). GA promotes germination, ABA inhibits it, and cytokinin counteracts the effect of the ABA. The relationship is an example of a "permissive" effect of cytokinin to allow GA stimulation of germination by offsetting ABA inhibition. Figure 7–27 illustrates similar hormone interactions for the polymorphic seeds in cocklebur (*88*).

It appears that the role ABA plays in dormancy is on the onset of dormancy during seed development. Hormone mutants have helped clarify the roles hormones play during dormancy and germination. ABA-deficient and ABA response

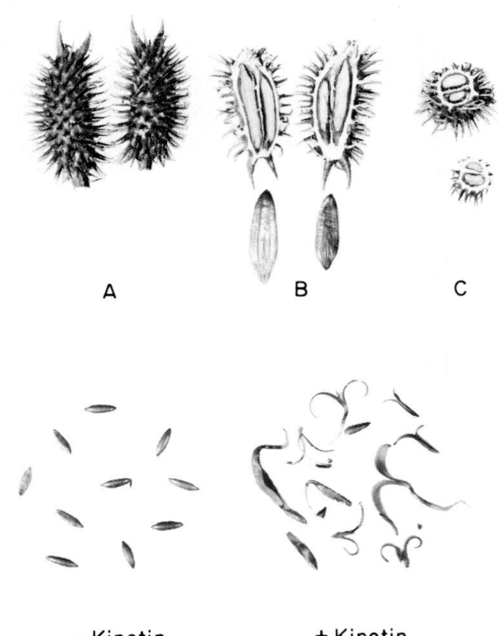

FIGURE 7–27 Cocklebur *(Xanthium)* fruits each have two seeds. The smaller of the two seeds is dormant. This is an example of polymorphic seed production. Early experiments showed that low permeability to gases was a dormancy-inducing factor. Later experiments *(151)* showed that the smaller (dormant) seed contained two water-soluble inhibitors that prevent germination. If these are removed by leaching, or if the seed is subjected to high oxygen pressure, then germination will occur. The two seeds also differ in seed coat strength and the germination forces required to rupture them. Treatment with kinetin or ethylene *(87)* will stimulate germination of both seeds, while abscisic acid will inhibit it. (From Khan et al. 1971.)

mutants in *Arabidopsis* (*83*), sunflower (*51*), and tomato (*61*) indicate that ABA must be present during seed development to induce dormancy. Recently, ABA-induced dormancy genes have been found in wheat (*108*) and *Bromus* (*59*). These genes are only expressed in dormant but not nondormant seeds. Genes associated with dormancy and nondormancy have also been found in oats (*79*). The identity and function of the proteins coded for by these genes has not yet been discovered, but suggests a molecular control of dormancy responsive to hormones.

Support for a requirement for gibberellin in the release from dormancy comes from gibberellin-deficient mutants in *Arabidopsis* and tomato (*18*). In both cases, treatments that normally relieve dormancy (after-ripening or chilling stratification) fail to stimulate germination in mutant seeds. Treatment with gibberellin initiates germination. It appears that ABA produced during seed development stimulates factors (like dormancy genes) that initiate dormancy. Endogenous ABA may be reduced during treatments to relieve dormancy, but this is not a strict requirement. Treatments that relieve dormancy show a strong correlation with gibberellin synthesis and mutants incapable of making gibberellin remain dormant unless treated with gibberellin.

Hormone Changes during Chilling Stratification

The hormonal changes that occur during the two to three months of chilling of seeds with dormant embryos have received much attention. A triphasic change in endogenous hormones is typical for many seeds during treatments to overcome dormancy (Figure 7–28) (*88*). First a reduction of ABA is seen.

ADVANTAGES OF SEED DORMANCY

Seed dormancy is an evolutionary adaptation to delay germination after the seed has been shed from the plant. There are numerous advantages to a delay in germination. These include:

1. Permitting germination only when environmental conditions favor **seedling survival.** This includes temperate species that require a period of moist, chilling conditions (i.e., winter conditions) before germination in the spring; desert species that germinate only after rainfall; weed species that require light; and even species that require extremely high temperatures prior to germination to become the primary species in an area following a forest fire.
2. Creation of a **"seed bank."** In nature, a seed bank ensures that not all the seeds for a species germinate in a single year. This is insurance against years where flowering or fruiting may not occur for some catastrophic environmental reason. Some seeds remain dormant in a seed bank for decades. Although this is a wonderful ecological adaptation, it is also the basis for persistent weed problems in agricultural fields. Some species take this concept one step further. They produce **polymorphic** seeds. In this case, seeds produced on the same plant or different plants in a population have different degrees of dormancy. Often these seeds have a different physical appearance. A classic example is found in cocklebur *(Xanthium pennsylvanicum).* The fruit of cocklebur produces two seeds of different sizes (Figure 7–27). One seed is dormant and the other seed is nondormant. One seed is for immediate germination and one seed for the seed bank and future germination.
3. Dormancy can also **synchronize germination** to a particular time of the year. This ensures that spring germinating seedlings have the entire growing season to grow and develop or for summer germinating seedlings to be at a proper stage of development entering the winter. Although environmental cues signal flowering for most crops, synchronizing germination also ensures a population of plants at the same stage of development to facilitate genetic outcrossing when all plants flower at the same time.
4. **Seed dispersal** can be facilitated by specialized dormancy conditions. An excellent example is seen in species with seeds that germinate only after the seed coverings have been modified by being processed through the digestive tract of a bird or other animal.

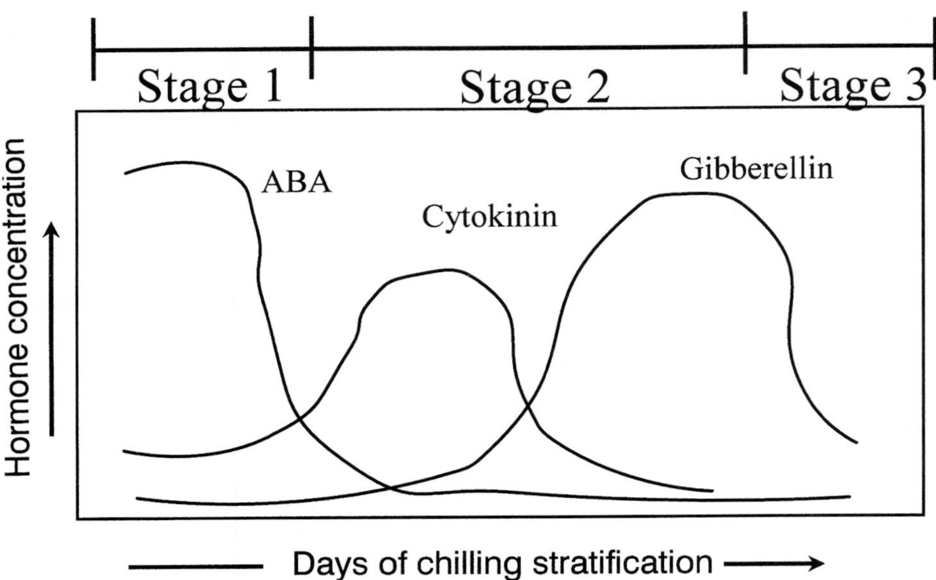

FIGURE 7–28 Triphasic changes in hormone production during chilling stratification.

The second stage is characterized by increased synthesis of cytokinin and gibberellin. The third stage can show a reduction in hormone synthesis in preparation for germination. There is evidence that conjugated stored forms of gibberellin are present in stage 3 to be converted to active forms of gibberellin during germination (*65*).

However, there are numerous exceptions to this triphasic concept of hormone change during chilling stratification. In some cases, ABA levels can remain high even though seeds are nondormant (*9, 126*). In others, there is no increase in gibberellins during stratification (*153*), and ABA levels drop in dormant seeds imbibed at nonchilling temperatures that do not relieve dormancy.

Changes in hormone levels during chilling stratification have been studied extensively in peach and apple. These studies show the dynamic changes that occur during stratification and some of the difficulties of interpreting these types of results.

In the peach (*Prunus persica*), two separate aspects of germination are affected: (a) initiation of radicle elongation, and (b) epicotyl elongation (*162, 163*). Figure 7–29 combines data from a number of experiments to show a three-phase pattern of seed response to chilling at 5° C (41° F). During stage I of stratification (0 to 30 days), no germination response occurs in intact seeds during chilling, after transfer to 25° C (77° F) at weekly intervals, or if presoaked with GA. In stage 2 (30 to 45 days), seeds being chilled do not germinate but seeds transferred to 25° C (77° F) show increasing germination response, and GA presoaking about doubles the rate of germination. In stage 3 (45 to 75 days), the seeds begin to germinate at chilling temperatures at a rapid rate (*58*) and seedlings show normal growth.

Hormone concentrations are correlated to the responses described. Freshly harvested peach seed (*42, 100*), as well as other species, including walnut (*102*), plum (*99*), apple (*11*), and hazelnut (*160*), have a high concentration of ABA in both the seed coat and a lesser amount in the cotyledons. ABA concentration drops to near zero during stage 1. Treatment of excised embryos with ABA prevents germination. Physical contact of the testa with the embryo axis has been sufficient to inhibit germination and/or to produce dwarfing (*72, 99*). Experimental treatment of intact peach seeds with a cytokinin benzylaminopurine (BAP) has overcome the inhibiting effect of the testa and allowed germination to occur (*124*).

The concentration of gibberellinlike compounds is low in stage 1 in intact seeds held at chilling temperatures (first 30 days) but shows a sharp increase in stage 2, indicating that the ability to synthesize gibberellins is either present (*58*) or there is a change from an inactive form to a free form [as shown in apple (*19*)]. Introduction of an inhibitor of gibberellin synthesis (paclobutrazol) at the beginning of stage 2 decreased gibberellin synthesis

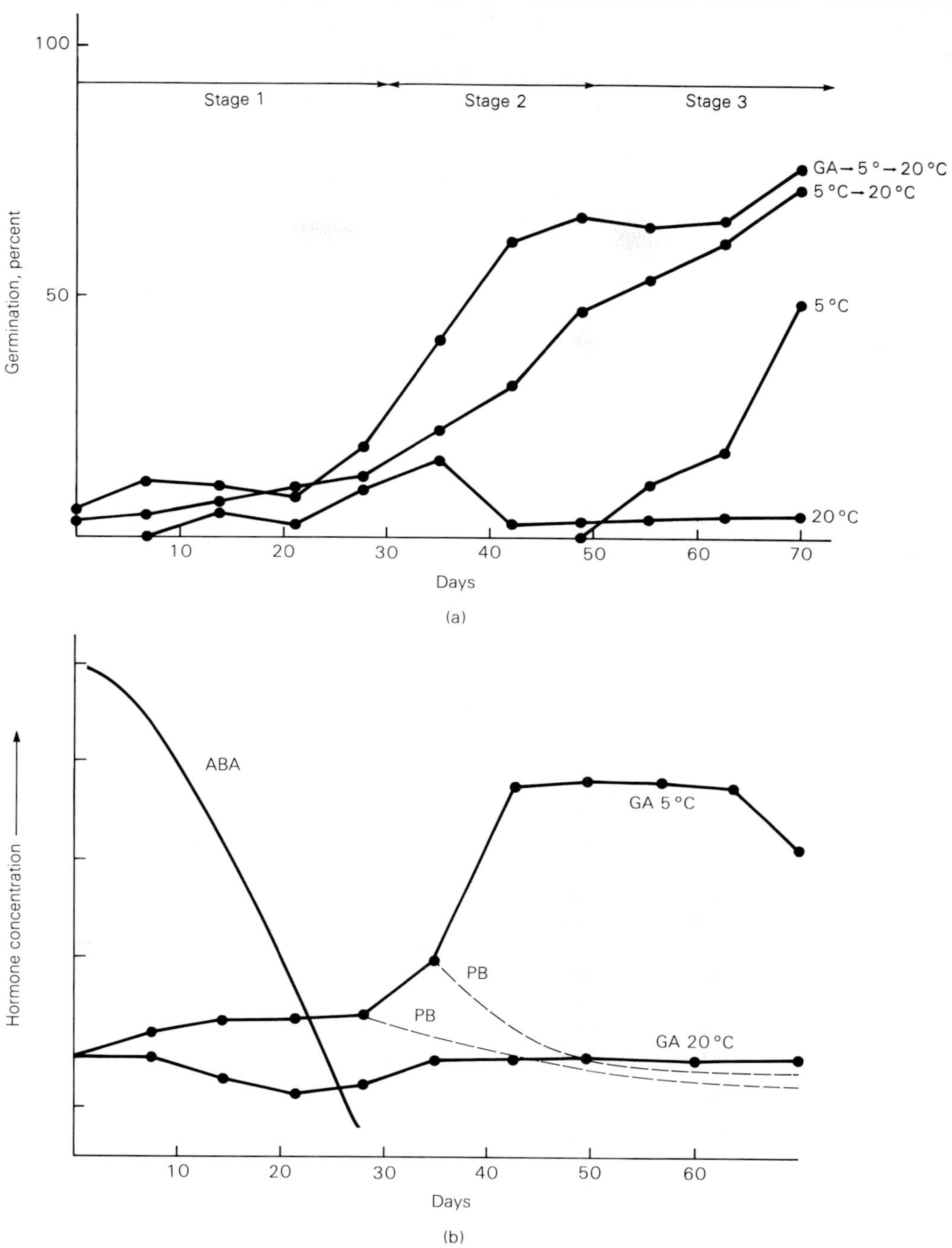

FIGURE 7–29 (A) Germination trends during stratification of peach (Gianfagna 1986). Germination was low in imbibed seed held continuously at warm temperatures (like 20° C, 77° F), but germination began to occur rapidly after 55 days in seed held continuously at cool temperatures. When stratified seeds were shifted to warm temperatures, sprouting began in about 30 days with increasing amounts with stratification time. An initial soak increased the germination rate. Paclobutrazol, an inhibitor of gibberellin synthesis, applied to seeds at 28 days decreased final germination percentages by 36, 25, and 27 percent is shifted to the warm temperature after 7, 14, and 28 days, respectively, of additional chilling. Elongation growth of the seedling was severely curtailed by paclobutrazol. (B) Correlations to hormone concentration in peach seeds during stratification. Many separate studies show that ABA is initially high in the dormant seeds, but levels drop sharply within the first 30 days. Data from Lin and Boe *(67)* and Lipe and Crane *(100),* and others is superimposed on the graph for peach *(58).* GA concentration was low during the first 30 days but increased dramatically after 30 to 45 days of stratification. Paclobutrazol sharply reduced GA concentration when applied at 28 and 35 days. (Redrawn from Gianfagna and Rachmiel 1976.)

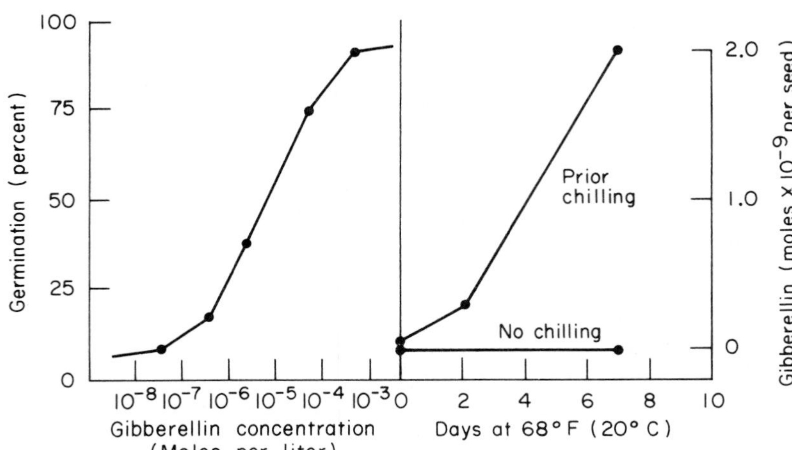

FIGURE 7–30 Interaction of gibberellin, stratification, and germination in filbert seeds. (Reproduced by permission from A.W. Galston and P.S. Davies, *Control mechanisms in plant development,* Prentice-Hall, Englewood Cliffs, N.J. 1970.)

sharply but only slightly decreased the germination percentage. Epicotyl and seedling elongation was strongly inhibited, indicating a separation between the radical and epicotyl response during chilling.

These results support the concept that inhibitors (undoubtedly ABA) are present in the testa as well as the cotyledons in the dormant seed. These disappear during the early stages of dormancy (or are neutralized by cytokinins). Gibberellins are either synthesized at the chilling temperatures or are converted to an available (or unbound) form, allowing radicle emergence (germination) to take place at warmer temperatures (*27, 103*). Epicotyl elongation is a more localized phenomenon which either has a higher threshold for gibberellin or involves a different control system.

Research with filbert *(Corylus avellana)* seeds has illustrated separation of growth-inhibiting and growth-promoting hormonal systems in control of germination. At the time of ripening, the intact seed is dormant but the embryo is quiescent. A significant amount of abscisic acid can be detected in the seed covering (*160*) as well as a detectable amount of gibberellin in the embryo (*121, 123*). When the seed is dried following harvest, the embryo becomes dormant, and gibberellin levels decrease significantly (*122*). Stratification for several months is required for germination. The gibberellin level remains low during this chilling period but increases after the seeds are placed at warm temperatures when germination begins (Figure 7–30, *right*). Gibberellic acid applied to the dormant seed (*22*) can replace the chilling requirement (Figure 7–30, *left*). ABA applied with gibberellin offsets the effect of GA and prevents germination (*123*).

REFERENCES

1. Abbott, D.L. 1955. Temperature and the dormancy of apple seeds. *Rpt. 14th Intern. Hort. Cong.* 1:746: 53.
2. Allan, M. 1977. *Darwin and his flowers.* The key to natural selection. New York: Taplinger Pub.
3. Arditti, J., and P.R. Pray. 1969. Dormancy factors in iris (Iridaceae) seeds. *Amer. Jour. Bot.* 56(3):254–59.
4. Association of Official Seed Analysts. 1993. Rules for testing seeds. *Jour. Seed Tech.* 16:1–113.
5. Atwater, B.R. 1980. Germination, dormancy and morphology of the seeds of herbaceous ornamental plants. *Seed Sci. and Tech.* 8:523–73.
6. Ayers, A.D. 1952. Seed germination as affected by soil moisture and salinity. *Agron. Jour.* 44: 82–84.
7. Baker, K.F. 1972. Seed pathology. In *Seed biology,* Vol. 2, T.T. Kozlowski, ed. New York: Academic Press.
8. ———., and P.A. Chandler. 1957. Development and maintenance of healthy planting stock. In *The U. C. system for producing healthy container-grown plants,* K.F. Baker, ed. Calif. Agr. Exp. Sta. Man. 23, pp. 217–36.
9. Balboa-Zavala, O. and F.G. Dennis. 1977. Abscisic acid and apple seed dormancy. *Jour. Amer. Soc. Hort. Sci.* 102:633–37.
10. Ball, V., 15th ed. 1991. *Ball red book. Greenhouse growing* (14th ed.). Reston, Va.: Reston Publ. Co.
11. Barthe, P., and C. Bulard. 1983. Anaerobiosis and release from dormancy in apple embryos. *Plant Physiol.* 72:1005–10.

12. Barton, L.V. 1950. Relation of different gases to the soaking injury of seeds. *Contrib. Boyce Thomp. Inst.* 16(2):55–71.

13. ———. and C. Chandler. 1957. Physiological and morphological effects of gibberellic acid on epicotyl dormancy of tree peony. *Contrib. Boyce Thomp. Inst.* 19:201–14.

14. Baskin, J.M. and C.C. Baskin. 1971. Effect of chilling and gibberellic acid on growth potential of excised embryos of *Ruellia humilis. Planta* 100: 365–69.

15. ———. and C.C. Baskin. 1985. Epicotyl dormancy in seeds of *Cimicifuga racemosa* and *Hepatica acutiloba. Bull. Torrey Bot. Club* 112:253–57.

16. Berlyn, G.P. 1972. Seed germination and morphogenesis. In *Seed biology,* Vol. 3, T. T. Kozlowski, ed. New York: Academic Press.

17. Berstein, L., A.J. MacKenzie, and B.A. Krantz. 1955. The interaction of salinity and planting practice on the germination of irrigated row crops. *Proc. Soil Science Soc. Amer.* 19:240–43.

18. Bewley, J.D., and M. Black. 1994. *Seeds: Physiology of development and germination.* New York: Plenum Press.

19. Bianco, J., S. Lassechere, and C. Bulard. 1984. Gibberellins in dormant embryos of *Pyrus malus* L. cv Golden Delicious. *Plant Physiol.* 116:185–88.

20. Black, M. 1980/1981. The role of endogenous hormones in germination and dormancy. Israel *Jour. Bot.* 29:181–92.

21. Blumenthal, A., H.R. Lerner, E. Werker, and A. Poljakoff-Mayber. 1986. Germination preventing mechanisms in *Iris seeds. Ann. Bot.* 58:551–61.

22. Bradbeer, J.W., and N.J. Pinfield. 1967. Studies in seed dormancy. III. The effects of gibberellin on dormant seeds of *Corylus avellana* L. *New Phytol.* 66:515–23.

23. Bradford, K.F. 1990. A water relations analysis of seed germination rates. *Plant Physiol.* 94:840–49.

24. Bradford, K.J. and O.A. Somasco. 1994. Water relations of lettuce seed thermoinhibition. 1. Priming and endosperm effects on base water potential. *Seed Science Res.* 4:1–10.

25. Brant, R.E., G.W. McKee, and R.W. Cleveland. 1971. Effect of chemical and physical treatment on hard seed of Penngift crown vetch. *Crop Sci.* 11. 1–6.

26. Brown, R. Germination. 1972. In *Plant physiology,* Vol. 7C, F. C. Steward, ed. New York: Academic Press.

27. Bulard, C. 1985. Intervention by gibberellin and cytokinin in the release of apple embryos from dormancy: A reappraisal. *New Phytol.* 101:241–49.

28. Cantliffe, D.J. 1991. Benzyladenine in the priming solution reduces thermodormancy of lettuce seeds. *HortTechnoloy* 1:95–99.

29. Chapman, A.L., and M.L. Peterson. 1962. The seedling establishment of rice under water in relation to temperature and dissolved oxygen. *Crop. Sci.* 2:391–95.

30. Carpenter, W.J. and S.W. Williams. 1993. Keys to successful seeding. *Grower Talks* 11:34–43.

31. Carpenter, W.J. and S. Maekawa. 1991. Substrate moisture level governs the germination of verbena seed. *HortScience* 26:786–88.

32. Ching, Te May. 1972. Metabolism of germinating seeds. In *Seed biology,* Vol. 2, T. T. Kozlowski, ed. New York: Academic Press.

33. Comé, D. 1980/1981. Problems of embryonal dormancy as exemplified by apple embryo. *Israel Jour. Bot.* 29:145–57.

34. Comé, D., and T. Tissaoui. 1973. Interrelated effects of imbibition, temperature, and oxygen on seed germination. In *Seed ecology,* W. Heydecker, ed. University Park, Pa.: Pennsylvania State Univ. Press.

35. Crisosto, C., and E.G. Sutter. 1985. Role of the endocarp in 'Manzanillo' olive seed germination. *Jour. Amer. Soc. Hort. Sci.* 110(1):50–52.

36. Crocker, W. 1916. Mechanics of dormancy in seeds. *Amer. Jour. Bot.* 3:99–120.

37. ———. 1948. *Growth of plants.* New York: Reinhold.

38. ———. 1930. Effect of the visible spectrum upon the germination of seeds and fruits. In *Biological effects of radiation.* New York: McGraw-Hill, pp. 791–828.

39. Czabator, F. 1962. Germination value: An index combining speed and completeness of pine seed germination. *For. Sci.* 8:386–96.

40. De Greef, J.A., H. Fredricq, R. Rethy, A. Dedonder, E.D. Petter and L. Van Wiemeersch. 1989. Factors eliciting germination of photobalstic kalanchoe seeds. In *Recent Advances in the Development and Germination of Seeds.* R.B. Taylorson (ed.). New York: Plenum Press, pp. 241–60.

41. Dell, B. 1980. Structure and function of the strophiolar plug in seed of *Albizia lophantha. Amer. Jour. Bot.* 67(4):556–61.

42. Diaz, D.H., and G.C. Martin. 1972. Peach seed dormancy in relation to endogenous inhibitors and applied growth substances. *Jour. Amer. Soc. Hort. Sci.* 97(5):651–54.

43. Doneen, L.D., and J.H. MacGillivray. 1943. Germination (emergence) of vegetable seed as affected by different soil conditions. *Plant Physiol.* 18:524–29.

44. du Toit, H.J., G. Jacobs, and D.K. Strydom. 1979. Role of the various seed parts in peach seed dormancy and initial seedling growth. *Jour. Amer. Soc. Hort. Sci.* 104(4):490–92.

45. Edwards, T.J. 1932. Temperature relations of seed germination. *Quart Rev. Biol.* 7:428–43.

46. Egley, G.H. 1989. Water-impermeable seed coverings as barriers to germination. In *Recent Advances in the Development and Germination of Seeds.* R.B. Taylorson (ed.). New York: Plenum Press, pp. 207–24.

47. Ells, J.E. 1963. The influence of treating tomato seed with nutrient solution on emergence rate and seedling growth. *Proc. Amer. Soc. Hort. Sci.* 83:684–87.

48. Eplee, R.E. 1975. Ethylene, a witchweed seed germination stimulant. *Weed Sci.* 23(5):433–36.

49. Esashi, Y., and A.C. Leopold. 1969. Dormancy regulation in subterranean clover seeds by ethylene. *Plant Phys.* 44:1470–72.

50. Evenari, M. 1949. Germination inhibitors. *Bot. Rev.* 15:153–94.

51. Fambrini, M., C. Pugliesi, P. Vernieri, G. Giuliano and S. Baroncelli. 1993. Characterization of a sunflower (Helianthus annus L.) mutant, deficient in caratenoid synthesis and abscisic-acid content, induced by in-vitro tissue culture. *Theor. Appl. Genet.* 87:65–9.

52. Finch-Savage, W.E., K. Phelps and J.R.A. Steckel. 1994. Seedling emergence modeling and the timing of irrigation during vegetable crop establishment. In *Fourth National Symposium on Stand Establishment of Horticultural Crops.* Davis, CA. pp. 103–8.

53. ———. 1994. Influence of seed quality on crop establishment, growth and yield. In A.S. Basra (ed.). *Seed Quality. Basic Mechanisms and Agricultural Implications.* New York: Food Products Press, pp. 361–84.

54. Fincher, G.B. 1989. Molecular and cellular biology associated with endosperm mobilization in germinating cereal grains. *Ann. Rev. Plant Physiol. Plant Mol. Biol.* 40:305–46.

55. Flemion, F., and E. Waterbury. 1945. Further studies with dwarf seedlings of non-after-ripened peach seeds. *Contrib. Boyce Thomp. Inst.* 13:415–22.

56. Garcia-Huidobro, J., J.L. Monteith and G.R. Squire. 1982. Time, temperature and germination of pearl millet (*Pennisetum typhoides* S and H). *Jour. Expt. Bot.* 33:288–96.

57. Geneve, RL. 1991. Seed dormancy in Eastern redbud (*Cercis canadensis* L.) *Jour. Amer. Soc. Hort. Sci.* 116:85–8.

58. Gianfagna, T.J., and S. Rachmiel. 1986. Changes in gibberellin-like substances of peach seed during stratification. *Physiol. Plant.* 66:154–58.

59. Goldmark, P.J., J. Curry, C.G. Morris and M.K. Walker-Simmons. 1992. Cloning and expression of an embryo-specific mRNA up-regulated in hydrated dormant seeds. *Plant Mol. Biol.* 19:433–41.

60. Gordon, A.G. 1973. The rate of germination. In *Seed ecology,* W. Heydecker, ed. University Park, Pa.: Pennsylvania State Univ. Press, pp. 391–410.

61. Groot, S.P.C. and C.M. Karssen. 1992. Dormancy and germination of abscisic acid-deficient tomato seeds. Studies with the *sitiens* mutant. *Plant Physiol.* 99:952–58.

62. Gray, D. 1994. Large-scale seed priming techniques and their integration with crop protection treatments. In *Seed treatment: progress and prospects.* BCPC Monograph No. 57. pp. 353–62.

63. Gummerson, R.J. 1986. The effect of constant temperatures and osmotic potentials on the germination of sugar beet. *Jour. Expt. Bot.* 37:729–41.

64. Haber, A.H., and H.J. Luippold. 1960. Separation of mechanisms initiating cell division and cell expansion in lettuce seed germination. *Plant Physiol.* 35:168–73.

65. Halinska, A. and S. Lewak. 1987. Free and conjugated gibberellins in dormancy and germination of apple seeds. *Physiol. Plant.* 69:523–30.

66. Halmer, P. 1985. The mobilization of storage carbohydrates in germinated seeds. *Physiol. Veg.* 23: 107–25.

67. Hamly, D.H. 1932. Softening the seeds of *Melilotus alba*. *Bot. Gaz.* 93:345–75.

68. Hanks, R.S. and F.C. Thorp. 1956. Seedling emergence of wheat as related to soil moisture content, bulk density, oxygen diffusion rate and crust strength. *Proc. Soil Sci. Soc. Amer.* 20:307–10.

69. Hanson, A.S. 1973. The effects of imbibition drying treatments on wheat seeds. *New Phytol.* 72:1063–73.

70. Haut, I.C. 1932. The influence of drying on after-ripening and germination of fruit tree seeds. *Proc. Amer. Soc. Hort. Sci.* 29:371–74.

71. Henckel, P.A. 1964. Physiology of plants under drought. *Annu. Rev. Plant Physiol.* 15:363–86.

72. Hepher, A., and J.A. Roberts. 1985. The control of seed germination in *Trollius tedebouri;* The breaking of dormancy. *Planta* 166:314–20.

73. Herner, R.C. 1986. Germination under cold soil conditions. *HortScience* 21(5):1118–22.

74. Heydecker, W. 1972. Vigour. In *Viability of seeds,* E.H. Roberts, ed. Syracuse, N.Y.: Syracuse Univ. Press.

75. ———, J. Higgins, and R.L. Gulliver. 1973. Accelerated germination by osmotic seed treatment. *Nature* (London) 246:42–4.

76. ———, J. Higgins, and Y.J. Turner. 1975. Invigoration of seeds? *Seed Sci. and Tech.* 3:881–88.

77. Hyde, E. O.C. 1956. The function of some Papilionaceae in relation to the ripening of the seed and permeability of the testa. *Ann. Bot.* 18:241–56.

78. Jann, R.C., and R.D. Amen. 1977. What is germination? In *The physiology and biochemistry of seed dormancy and germination,* A.A. Khan, ed. Amsterdam: North-Holland Publishing Co., pp. 7–28.

79. Johnson, R.R., H.J. Cranston, M.E. Chaverra and W.E. Dyer. 1995. Characterization of cDNA clones for differentially expressed genes in embryos of dormant and nondormant *Avena fatua* L. caryopses. *Plant Mol. Biol.* 28:113–22.

80. Jones, R.O. and R.L. Geneve. 1994. Seed coat structure related to germination in eastern redbud (*Cercis canadensis* L.). *Jour. Amer. Soc. Hort. Sci.* 120:123–27.

81. Junttila, O. 1973. The mechanism for low temperature dormancy in mature seeds of *Syringa* species. *Physiol. Plant.* 29:256–63.
82. Karssen, C.M. 1980/1981. Environmental conditions and endogenous mechanisms involved in secondary dormancy of seeds. *Israel Jour. Bot.* 29: 45–64.
83. ———, D.L.C. Brinkhorst-van der Swan, A.E. Breekland, and M. Koornneef. 1983. Induction of dormancy during seed development by endogenous abscisic acid deficient genotypes of *Arabidopsis thaliana* (L.) Heynh. *Planta* 157:158–65.
84. Kester, D.E. 1969. Pollen effects on chilling requirements of almond and almond hybrid seeds. *Jour. Amer. Soc. Hort. Sci.* 94:318–21.
85. ———, P. Raddi, and R. Asay. 1977. Correlations of chilling requirements for germination, blooming and leafing within and among seedling populations of almond. *Jour. Amer. Soc. Hort. Sci.* 102(2): 145–48.
86. Khan, A.A. 1971. Cytokinins: Permissive role in seed germination. *Science* 171:853–59.
87. ———. 1977. Preconditioning, germination and performance of seeds. In *The physiology and biochemistry of seed dormancy and germination,* A. A. Khan, ed. Amsterdam: North-Holland Publishing Co.
88. ———. 1980/1981. Hormonal regulation of primary and secondary dormancy. *Israel Jour. Bot.* 29:207–24.
89. ———. 1992. Preplant physiological seed conditioning. *Hort. Rev.* 13: 131–81.
90. ———, C.E. Heit, E.C. Waters, C.C. Anojulu, and L. Anderson. 1971. Discovery of a new role for cytokinins in seed dormancy and germination. In *Search.* New York Agr. Exp. Sta. (Geneva) 1(9):1–12.
91. Koller, D. 1972. Environmental control of seed germination. In *Seed biology,* Vol. 2, T. T. Kozlowski, ed. New York: Academic Press.
92. Kotowski, F. 1926. Temperature relations to germination of vegetable seeds. *Proc. Amer. Soc. Hort. Sci.* 23:176–84.
93. Lammerts, W.E. 1943. Effect of photoperiod and temperatures on growth of embryo cultured peach seedlings. *Amer. Jour. Bot.* 30:707–11.
94. Lane, B., J.M. Dumwell, J. A. Ray, M.R. Schmitt, and A.C. Cumming. 1993. Germin, a protein marker of early plant development, is an oxalate oxidase. *J. Biol. Chem.* 268:12239–42.
95. Lang, G.A. 1987. Dormancy: A new universal terminology. *HortScience* 22(5):817–20.
96. ———, J.D. Early, G.C. Martin, and R.L. Darnell. 1987. Endo-, para-, and ecodormancy: Physiological terminology and classification for dormancy research. *HortScience* 22(3):371–77.
97. Lewak, S. 1985. Hormones in seed dormancy and germination. In *Hormonal regulation of plant growth and development,* S.S. Purohit, ed. Dordrecht: Martinus Nishoff, pp. 95–144.
98. ——— and R.M. Rudnicki. 1977. After-ripening in cold-requiring seeds. In: *The physiology and biochemistry of seed dormancy and germination.* A.A. Khan (ed.). Amsterdam: North Holland Pub. pp. 193–218.
99. Lin, C.F., and A.A. Boe. 1972. Effects of some endogenous and exogenous growth regulators on plum seed dormancy. *Jour. Amer. Soc. Hort. Sci.* 97:41–4.
100. Lipe, W., and J.C. Crane. 1966. Dormancy regulation in peach seeds. *Science* 153:541–42.
101. Liu, N.Y., H. Khatamian, and T.A. Fretz. 1981. Seed coat structure of three woody legume species after chemical and physical treatments to increase seed germination. *Jour. Amer. Soc. Hort. Sci.* 106(5): 691–94.
102. Martin, G.C., H. Forde, and M. Mason. 1969. Changes in endogenous growth substances in the embryo of *Juglans regia* during stratification. *Jour. Amer. Soc. Hort. Sci.* 94:13–7.
103. Mathur, D.D., G.A. Couvillon, H.M. Vines, and C.H. Hendershott. 1971. Stratification effects of endogenous gibberellic acid (GA) in peach seeds. *HortScience* 6:538–39.
104. May, L.H., E.J. Milthorpe, and F.L. Milthorpe. 1962. Pre-sowing hardening of plants to drought. An appraisal of the contributions by P.A. Gekel. *Field Crop Abst.* 15:93–8.
105. McDonald, M.B., Sr. 1980. Assessment of seed quality. *HortScience* 15:784–88.
106. Meyer, M.M., Jr. 1987. Rest, postdormancy, and woody plant seed germination. Comb. *Proc. Inter. Plant Prop. Soc.* 37:330–35.
107. Morinaga, T. 1926. Germination of seeds under water. *Amer. Jour. Bot.* 13: 126–31.
108. Morris, C.F., R.J. Anderberg, P.J. Goldmark and M.K. Walker-Simmons. 1991. Molecular cloning and expression of abscisic acid-responsive genes in embryos of dormant wheat seeds. *Plant Physiol.* 95:814–21.
109. Nagao, M.A., K. Kanegawa, and W.S. Sakai. 1980. Accelerating palm seed germination with gibberellic acid, scarification, and bottom heat. *HortScience* 15(2):200–01.
110. Ni, B.R. and K.J. Bradford. 1993. Germination and dormancy of abscisic acid and gibberellin-deficient mutant tomato seeds. Sensitivity of germination to abscisic acid, gibberellin and water potential. *Plant Physiol.* 101:607–17.
111. Nikolaeva, M.G. 1969. *Physiology of deep dormancy in seeds.* CPST Press, Jerusalem, Israel.
112. Nikolaeva, M.G. 1977. Factors affecting the seed dormancy pattern. In *The physiology and biochemistry of seed dormancy and germination,* A.A. Khan,

ed. Amsterdam: North-Holland Publishing Co., pp. 51–76.

113. Parish, D.J. and A.C. Leopold. 1977. Transient changes during soybean imbibition. *Plant Physiol.* 59:1111–15.

114. Pollock, B.M., and E.E. Roos. 1972. Seed and seedling vigor. In *Seed biology,* Vol. 3, T. T. Kozlowski, ed. New York: Academic Press.

115. ———. 1962. Temperature control of physiological dwarfing in peach seedlings. *Plant Physiol.* 37: 190–97.

116. Powell, L.E. 1987. Hormonal aspects of bud and seed dormancy temperature-zone woody plants. *HortScience* 22(5):845–50.

117. Roberts, E.H. 1972. Dormancy: A factor affecting seed survival in the soil. In *Viability of seeds,* E.H. Roberts, ed. Syracuse, N.Y.: Syracuse Univ. Press.

118. Rogan, P.G. and E.W. Simon. 1975. Root growth and the onset of mitosis in germinating *Vicia faba. New Phytol.* 74:273–75.

119. Rolston, M.P. 1978. Water impermeable seed dormancy. *Bot. Rev.* 44:365–96.

120. Ross, H.A. and T.A. Hegarty. 1979. Sensitivity of seed germination and seedling radicle growth to moisture stress in some vegetable crop species. *Ann. Bot.* 43:241–43.

121. Ross, J.D. 1971. Studies in seed dormancy. V. The concentrations of endogenous gibberellins in seeds of *Corylus avellana* L. *Planta* (Berl.) 100:288–302.

122. ——— and J. W. Bradbeer. 1968. Concentrations of gibberellin in chilled hazel seeds. *Nature* 220:85–86.

123. ———. 1971. Studies in seed dormancy. VI. The effects of growth retardants on the gibberellin content and germination of chilled seeds of *Corylus avellana* L. *Planta* (Berl.) 100:303–8.

124. Rouskas, D., J. Hugard, R. Jenard, and P. Villemur. 1980. Physiologie végétale. Contribution à l'étude de la germination des graines de Pecher (*Prunus persica* Batsch.) cultivar INRA GF 305: effets de la benzyl-amino-purine (BAP) et les gibberellines GA_3 et GA_{4+7} sur la levee de dormance embryonnaire et l'absence des anomalies foliaires observées sur les plantes issues de graines non stratifiées. C. R. *Acad. Sci. Paris Ser* D 291:861–64.

125. Roux, S.J. 1989. Calcium-regulated metabolism in seed germination. In *Recent Advances in the Development and Germination of Seeds.* R.B. Taylorson (ed.). New York: Plenum Press, pp. 127–38.

126. Rudnicki, R. 1969. Studies on abscisic acid in apple seeds. *Planta* 86:63–8.

127. Schopfer, P. And C. Plachy. 1985. Control of seed germination by abscisic acid. III. Effect of embryo growth potential (minimum turgor potential) and growth coefficient (cell wall extensibility) in *Brassica napus* L. *Plant Physiol.* 77:676–86.

128. Schopmeyer, C.S., ed. 1974. *Seeds of woody plants in the United States.* U.S. Dept. Agr. Handbook 450. Washington, D.C.: U.S. Govt. Printing Office.

129. Seeley, S.D., and H. Damavandy. 1985. Response of seed of seven deciduous fruits to stratification temperatures and implications for modeling. *Jour. Amer. Soc. Hort. Sci.* 110(5):726–29.

130. Semeniuk, P., and R.N. Stewart. 1962. Temperature reversal of after-ripening of rose seeds. *Proc. Amer. Soc. Hort. Sci.* 80:615–21.

131. Sharples, G.C. 1973. Stimulation of lettuce seed germination at high temperatures by ethephon and kinetin. *Jour. Amer. Soc. Hort. Sci.* 98(2):207–9.

132. Shaykewich, C.F. 1973. Proposed methods for measuring swelling pressure of seeds prior to germination. *Jour. Exp. Bot.* 24:1056–61.

133. Shull, C.A. 1913. Semipermeability of seed coats. *Bot. Gaz.* 61:169–99.

134. Simon, E.W. 1984. Early events in germination. In *Seed Physiology Vol. 2.* New York: Academic Press. pp. 77–115.

135. Smith, O.E., W.W. Yen, and J.M. Lyons. 1968. The effects of kinetin in overcoming high-temperature dormancy in lettuce seeds. *Proc. Amer. Soc. Hort Sci.* 93:444–53.

136. Stearns, F., and J. Olson. 1958. Interactions of photoperiod and temperature affecting seed germination in *Tsuga canadensis. Amer. Jour. Bot.* 45:53–8.

137. Stewart, R. N., and P. Semeniuk. 1965. The effect of the interaction of temperature with after-ripening requirement and compensating temperature on germination of seed of five species of *Rosa. Amer. Jour. Bot.* 52:755–60.

138. Taylor, A.G., D.I. Klein and T.H. Whitlow. 1988. SMP: solid matrix priming of seeds. *Scientia Hort.* 37:1–11.

139. Taylorson, R.B., and S.B. Hendricks. 1977. Dormancy in seeds. *Ann. Rev. Plant Phys. 28:* 331–54.

140. Thomas, H. 1972. Control mechanisms in the resting seed. In *Viability of seeds,* E. H. Roberts, ed. Syracuse, N.Y.: Syracuse Univ. Press.

141. ———. 1977. Cytokinins, cytokinin active compounds and seed germination. In *The physiology and biochemistry of seed dormancy and germination,* A.A. Khan, ed. Amsterdam: North-Holland Publishing Co., pp. 111–24.

142. Thompson, P.A. 1973. Geographical adaptation of seeds. In *Seed ecology,* W. Heydecker, ed. University Park, Pa.: Pennsylvania State Univ. Press.

143. Thornley, H.M. and I.R. Johnson. 1990. *Plant crop modelling: a mathematical approach to plant and crop physiology.* Claredon Press. Oxford. pp. 141–44.

144. USDA. 1952. *Manual for testing agricultural and vegetable seeds.* U.S. Dept. Agr. Handbook 30. Washington, D.C.: U.S. Govt. Printing Office.

145. Valdes, V., K.J. Bradford, and K.S. Mayberry. 1985. Alleviation of thermodormancy in coated lettuce seeds by seed priming. *HortScience* 20:1112–14.

146. Van derWoude, W. 1989. Phytochrome and sensitization in germination control. In *Recent Advances in*

the Development and Germination of Seeds. R.B. Taylorson (ed.). New York: Plenum Press pp. 181–90.

147. Villiers, T.A. 1972. Seed dormancy. In *Seed biology,* Vol. 2, T.T. Kozlowski, ed. New York: Academic Press, pp. 220–82.

148. Visser, T. 1956. Some observations on respiration and secondary dormancy in apple seeds. *Proc. Koninkl. Akad. van Wetens,* Series C 59:314–24.

149. Walton, D.C. 1980. Biochemistry and physiology of abscisic acid. *Ann. Rev. Plant Phys.* 31: 453–89.

150. ———. 1980/1981. Does ABA play a role in seed germination? *Israel Jour. Bot.* 29: 168–80.

151. Wareing, P.F., and H.A. Foda. 1957. Growth inhibitors and dormancy in *Xanthium* seed. *Physiol. Plant.* 10(2):266–80.

152. Watkins, J.T. and D.J. Cantliffe. 1983. Mechanical resistance of the seed coat and endosperm during germination of *Capsicum annuum* at low temperature. *Plant Physiol.* 72:146–50.

153. Webb, D.P., J. Van Stadden and P.F. Wareing. Seed dormancy in *Acer.* Changes in endogenous germination inhibitors, cytokinins, and gibberellins during the breaking of dormancy in *Acer pseudoplatanus.* 1973. *Jour. Exp. Bot.* 24:741–50.

154. Went, F. W. 1949. Ecology of desert plants. 11. The effect of rain and temperature on germination and growth. *Ecology* 30:1–13.

155. ——— and M. Westergaard. 1949. Ecology of desert plants. III. Development of plants in the Death Valley National Monument, California. *Ecology* 30:26–38.

156. Wesson, G., and P. F. Wareing. 1969. The induction of light sensitivity in weed seeds by burial. *Jour. Exp. Bot.* 20(63):414–25.

157. Weston, L.A., R.L. Geneve and J.E. Staub. 1992. Seed dormancy in *Cucumis sativus* var. *hardwickii* (Royle) Alef. *Scientia Hort.* 50:35–46.

158. Westwood, M.N., and H.O. Bjornstad. 1948. Chilling requirement of dormant seeds of fourteen pear species as related to their climatic adaptation. *Proc. Amer. Soc. Hort. Sci.* 92:141–49.

159. Willemsen, R.W. 1975. Effect of stratification temperature and germination temperature on germination and the induction of secondary dormancy in common ragweed seeds. *Amer. Jour. Bot.* 62(1):1–5.

160. Williams, P.M., J.D. Ross, and J.W. Bradbeer. 1973. Studies in seed dormancy. VII. The abscisic acid content of the seeds and fruits of *Corylus avellana* L. *Planta (Berl.)* 110:303–10.

161. Young, J.A. and C.G. Young. 1992. *Seeds of woody plants in North America.* Revised edition. Dioscorides Press. Portland, Oregon.

162. Zigas, R.P., and B.G. Coombe. 1977. Seedling development in peach, *Prunus persica* (L.) Batsch. I. Effects of testas and temperature. *Aust. Jour. Plant Physiol.* 4:349–58.

163. ———, and B.G. Coombe. 1977. Seedling development in peach, *Prunus persica* (L.) Batsch. II. Effects of plant growth regulators and their possible role. *Aust. Jour. Plant Physiol.* 4:359–69.

SUPPLEMENTARY READING

AMEN, R. D. 1968. A model of seed dormancy. *Bot. Rev.* 34:1–31.

BARTON, L. V. 1967. *Bibliography of seeds.* Columbia Univ. Press. New York.

BEWLEY J. D., and M. BLACK. 1994. *Seeds: Physiology of development and germination.* New York: Plenum Press.

CROCKER, W., and L. V. BARTON. 1953. *Physiology of seeds.* Waltham, Mass.: Chronica Botanica.

KHAN, A. A., ed. 1977. *The physiology and biochemistry of seed dormancy and germination.* Amsterdam: North Holland Publishing Co.

KIGIL, J. and G. GALILI. 1995. *Seed development and germination.* Marce. Dekker, Inc. New York.

KOZLOWSKI, T. T., ed. 1972. *Seed biology,* Vols. 1, 2, 3. New York: Academic Press.

MAYER, A. M. 1980/1981. Control mechanisms in seed germination. *Israel Jour. Bot.* 29:1–4.

MAYER, A. M., and A. POLJAKOFF-MAYBER. 1975. *The germination of seeds* (2nd ed.). New York: Macmillan.

Proceedings of symposium. 1986. Seed germination under environmental stress. *HortScience* 21(5):1103–28.

Proceedings of symposium. 1987. Mechanisms of rest and dormancy. *HortScience* 22(5):815–50.

ROBERTS, E. H., ed. 1972. *Viability of seeds.* Syracuse, N.Y.: Syracuse Univ. press.

RUBENSTEIN, I., R. L. PHILLIPS, G. E. GREEN, and B. G. GENGENBACH. 1979. *The plant seed: Development, preservation and germination.* New York: Academic Press.

SCHOPMEYER, C. S., ed. 1974. *Seeds of woody plants in the United States.* U.S. Dept. Agr. Handbook 450. Washington, D.C.: U.S. Govt. Printing Office.

TAYLORSON, R. B., and S. B. HENDRICKS. 1977. Dormancy in seeds. *Ann. Rev. Plant Phys.* 28:331–54.

YOUNG, J.A. and C. G. YOUNG. 1992. *Seeds of woody plants in North America.* Revised edition. Portland, Oregon: Dioscorides Press.

8
Techniques of Propagation by Seed

Seed propagation involves careful management of germination conditions and facilities and a knowledge of the requirements of individual kinds of seeds. Success depends upon fulfilling the following conditions:

1. *Using seed of proper genetic characteristics to produce the cultivar, species, or provenance desired.* This can be accomplished by obtaining seed from a reliable dealer, buying certified seed, or—if producing one's own—following the principles of seed selection described in Chapter 4.
2. *Using good-quality seed.* Seeds should germinate rapidly and vigorously to withstand possible adverse conditions in the seed bed and provide a high percentage of usable seedlings.
3. *Manipulating seed dormancy.* This is accomplished by applying pregermination treatments or proper timing of planting. In the absence of specific knowledge of seed requirements, the propagator should try to duplicate the natural environmental conditions associated with germination of seed of this particular kind of plant.
4. *Supplying proper environment to the seeds and resulting seedlings.* This includes supplying sufficient water, proper temperature, adequate oxygen, and either light or darkness (depending upon the kind of seed) to the seeds and resulting seedlings until they are well established. A proper environment also includes control of diseases and insects and prevention of excess salinity.

SEED PROPAGATION SYSTEMS

Seed propagation is carried out using three basic systems: (a) **field seeding** in the location where the plant is to remain, (b) planting in **field nurseries** and transplanting to a permanent location, and (c) planting in **protected conditions,** as in a greenhouse, cold frame, or similar structure, and then transplanting to the permanent location. For commercial production of horticultural crops, the most common seed production practices are to produce vegetables from direct field seeding or transplants; annual bedding plants and herbaceous perennials from transplants (usually started as plugs); and woody seedlings from field transplant beds to produce bare root liners.

Field Seeding

Direct field seeding is used for commercial field planting of agronomic crops (grains, legumes, forages, fiber crops, oil crops), lawn grasses, many vegetable crops, and some woody perennials (Figure 8–1). The method may be used for home vegetable and flower gardeners and hobbyists. Compared to transplants, directly seeded plants properly handled are less expensive and can grow continuously without the check in growth produced by bare-root transplanting (*53*). Frequently, direct-field-seeded vegetable and other crops are precontracted for processing, whereas the more expensive transplants are targeted as an early fresh market crop. On the other hand, there are many potential field problems that must be overcome to provide the proper environmental conditions for good uniform germination. Likewise, cold weather may decrease growth. Seeding rates are critical to provide proper plant spacing for optimum development of the crop. If final plant density is too low, yields will be reduced because the number of plants per unit area is low; if too high, the size and quality of the finished plants may be reduced by competition among plants for available space, sunlight, water, and nutrients.

To maximize direct-seeding success, the following is required:

FIGURE 8–1 Lettuce production in Salinas, California.

1. A proper seed bed.
2. High quality seed.
3. The correct planting time.
4. Pretreatments to facilitate sowing or to relieve dormancy.
5. The proper mechanical seeder.
6. The proper sowing depth.
7. The proper sowing rate.
8. The proper post-sowing care.

1. Prepare the seed bed. A good seed bed should have a loose but fine physical texture that produces close contact between seed and soil so that moisture can be supplied continuously to the seed. Such a soil should provide good aeration but not too much, or it dries too rapidly. The surface soil should be free of clods and of a texture that will not form a **crust.** Soil impedance due to crusting from an improperly prepared seed bed or adverse environmental conditions during seeding can substantially reduce seedling emergence (*60*). Several materials including organic polymers and phosphorus-containing compounds have been developed to reduce soil crusting and can aid in seedling emergence (*53, 54*). The subsoil should be permeable to air and water with good drainage and aeration. Adequate soil moisture should be available to carry the seeds through the germination and early seedling growth stages, but the soil should not be waterlogged or anaerobic (without oxygen). A medium loam texture, not too sandy and not too fine, is best. A good seed bed is one in which three-fourths of the soil particles (aggregates) range from 1 to 12 mm in diameter (*35*).

Seed-bed preparation requires special machinery for field operations and spading and raking or rototilling equipment for small plots. Adding organic or soil amendments may be helpful, but these should be thoroughly incorporated and have time to decompose. Seed-bed preparation may include fumigation and other soil treatments to control harmful insects, nematodes, disease organisms, and weed seeds (see Chapter 3).

2. Select high-quality seed. Quality is based on seed testing data. A low rate of sowing requires seeds of high quality that produce not only high germination rates but also vigorous, uniform, healthy seedlings.

3. Choose correct planting time. Planting time is determined by the germination temperature requirements of the seed and the need to meet production schedules. These are determined according to the individual crop and vary with the particular kind of seed. Early season sowing of seeds requiring warm soil temperatures can result in slow and uneven germination, disease problems, and **"chilling"** injury to seedlings of some species, causing growth abnormalities. High soil temperatures can result in excessive drying, injury, or death to seedlings, or induction of thermodormancy in the case of heat-sensitive seeds such as lettuce, celery, and various flower seeds (see Chapter 21).

4. Pretreat seeds. It is often desirable to use seeds that have been pretreated for protection with a **fungicide** or enhanced for germination by a **seed coating** or **priming** treatment (see Chapter 6). These treatments can speed up germination, increase uniformity, and offset some environmental hazards in the seed bed. Coated seed with improved flowability and uniform size can improve the seeding precision of mechanical planters.

For many species (especially woody and herbaceous native plants), seeds must be pretreated to overcome dormancy conditions of the seed (see Chapter 7). The two most common treatments used by commercial propagators include **scarification** for species with hard seed coats and **stratification** for species that require periods of warm or chilling conditions to alleviate dormancy.

Scarification is any process of breaking, scratching, mechanically altering, or softening the seed coverings to make them permeable to water and gases. Typical species with hard **seed coats** include members of the legume, geranium, morning glory, and linden families. Three types of treatments are commonly used as scarification treatments. These include: **mechanical, chemical** and **hot water** treatments.

Mechanical scarification. Mechanical scarification is simple and effective with seeds of many species if suitable equipment is available (Figure 8–2). These seeds are dry after such treatment and may be stored or planted immediately by mechanical seeders. Scarified seeds are more susceptible to injury from pathogenic organisms, however, and will not store as well as comparable nonscarified seeds.

Chipping hard seed coats by rubbing with sandpaper, cutting with a file, or cracking with a hammer or a vise are simple methods useful for small amounts of relatively large seed. For large-scale mechanical operations, special scarifiers are used. Small seeds of legumes, such as alfalfa and clover, are often treated in this manner to increase germination (*10*). Seeds may be tumbled in drums lined with sandpaper or in concrete mixers containing coarse sand or gravel (*62*) (Figure 8–2). The sand or gravel should be of a different size than the seed to facilitate subsequent separation.

Scarification should not proceed to the point at which the seeds are injured. To determine the optimum time, a test lot can be germinated, the seeds may be soaked to observe swelling, or the seed coats may be examined with a hand lens. The seed coats generally should be dull but not so deeply pitted or cracked as to expose the inner parts of the seed.

Acid scarification. Dry seeds are placed in containers and covered with concentrated **sulfuric acid** (specific gravity 1.84) in a ratio of about one part seed to two parts acid. The amount of seed treated at any one time should be restricted to no more than about 10 kg (22 lb) (*48*) to avoid uncontrollable heating. Containers should be glass, earthenware, or wood—not metal or plastic. The mixture should be stirred cautiously at intervals during the treatment to produce uniform results and to prevent accumulation of the dark, resinous material from the seed coats that is sometimes present. Since stirring tends to raise the temperature, vigorous agitation of the mixture should be avoided to prevent injury to the seeds. The time of treatment may vary from as little as ten minutes for some species to six hours or more for other species. Since treatment time may vary with different seed lots, making a preliminary test on a small lot is recommended prior to treating large lots (*32,48*). With thick-coated seeds that require long periods, the progress of the acid treatment may be followed by drawing out samples at intervals and checking the thickness of the seed coat. When it becomes paper thin, the treatment should be terminated immediately.

FIGURE 8–2 Seed scarifier for ornamental seeds. Used to scarify seeds with hard seed coats.

At the end of the treatment period the acid is poured off, and the seeds are washed to remove the acid. Glass funnels are useful in removing the acid from small lots of seed. Placing seeds in a large amount of water with a small amount of baking soda (sodium bicarbonate) will neutralize any adhering acid; or the seeds can be washed for ten minutes in running water. The acid-treated seeds can either be planted immediately when wet, or dried and stored for later planting.

Large seeds of most legume species respond to the simple sulfuric acid treatment, but variations are required for some species (*48*). Some roseaceaous seeds *(Cotoneaster, Rosa)* have hard pericarps that are best treated partially with acid and then given warm stratification. A third group, such as *Hamamelis* and *Tilia,* have very "tough" pericarps that may first need to be treated with nitric acid and then with sulfuric acid.

Always use proper safety precautions while using acids for scarification. This includes personal safety equipment like gloves, face shield, eye protection, and lab coat. An eye wash and a source of running water must be available in case of an accident. Request the **MSDS safety sheet** from your chemical supplier for additional safety precautions.

Hot water scarification. Drop the seeds into four to five times their volume of hot water 77 to 100° C (170 to 212° F). The heat source is immediately removed, and the seeds soaked in the gradually cooling water for 12 to 24 hours. Following this, the unswollen seeds can be separated from the swollen ones by suitable screens and either retreated or subjected to some other treatment. The seeds should usually be planted immediately after the hot-water treatment; some kinds of seed have been dried and stored for later planting without impairing the germination percentage, although the germination rate was reduced.

Stratification. Stratification is a method of handling dormant seeds in which the imbibed seeds are subjected to a period of chilling to after-ripen the embryo. The term originated because nurseries placed seeds in stratified layers interspersed with a moist medium, such as soil or sand, in out-of-doors pits during winter. The term **moist-chilling** has been used as a synonym for stratification. However, with temperate species displaying epicotyl dormancy (like fringetree) or underdeveloped embryos (like hollies), a **warm-moist stratification** of several months followed by a moist-chilling stratification is required to satisfy dormancy conditions. This may require more than one season to achieve under natural conditions. Several tropical and semi-tropical species (like palms) require a period of **warm stratification** prior to germination to allow the embryo to continue development after fruit drop.

Outdoor planting for stratification. Seeds requiring a cold treatment may be planted out-of-doors directly in the seed bed, cold frame, or nursery row at a time of the year when the natural environment provides the necessary conditions to relieve dormancy. This is the most common treatment for seeds with endogenous physiological dormancy. Several different categories of seeds can be handled in this way with good germination in the spring following planting. Some wildflower seeds fall into this category (*5*).

Seeds must be planted early enough in the fall to allow them to become imbibed with water and to get the full benefit of the winter chilling period. The seeds generally germinate promptly in the spring when the soil begins to warm up, but while the soil temperature is still low enough to inhibit damping-off organisms and to avoid high-temperature inhibition.

Seeds with a hard endocarp, such as *Prunus* species (the stone fruits, including cherries, plums, and peaches), show increased germination if planted early enough in the summer or fall to provide one to two months of warm temperatures prior to the onset of chilling (*41*). Seeds that require high temperatures followed by chilling can thus be planted in late summer to fulfill their warm-temperature requirements, followed by the subsequent winter period that satisfies the chilling requirement.

For other seeds having hard coverings (such as juniper and some *Magnolia* species), germination can be facilitated if the fruit is harvested when ripe and the seeds planted immediately without drying (*49, 70*). Once the seeds become dry, the seed coats harden and germination may be delayed, perhaps until the second spring. Seeds that ripen early in the growing season and lose viability rapidly (recalcitrant seeds like maple and elm; see pp. 143) should be collected and planted in spring or summer as soon as they mature. Where effective treatments are not known, the propagator should attempt to reproduce the natural seeding habits of the plant and provide the germinating conditions of its natural environment. When seeds remain for a long period

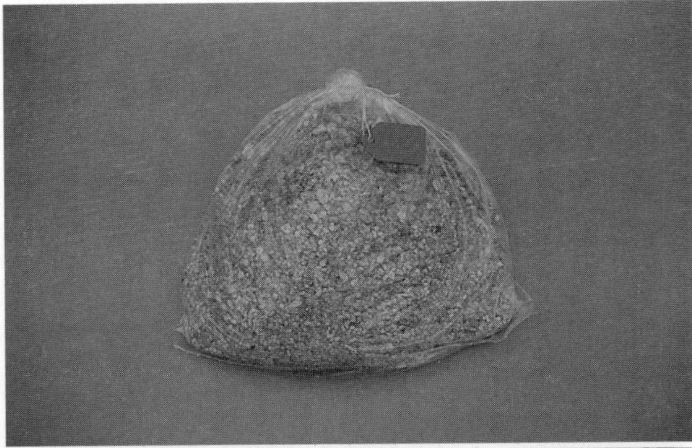

FIGURE 8–3 *Top:* Stratification of small lots of seed can be accomplished by mixing seeds with vermiculite and placed in polyethylene bags. *Bottom:* Stratification of hazelnut seed in large containers.

in an outdoor nursery or seed bed prior to germination, they must be protected from drying, adverse weather conditions, rodents, birds, diseases, and competition from weeds. Herbicides may be used for weed control.

Refrigerated stratification. An alternative to outdoor field planting is refrigerated stratification. This is useful for small seed lots or valuable seeds that require special handling. Dry seeds should be fully imbibed with water prior to refrigerated stratification. Twelve to 24 hours of soaking at warm temperatures may be sufficient for seeds without hard seed coats or coverings. Longer periods are required, with aeration, for seeds enclosed in hard endocarp or pericarp; soaking for three days to a week or more may be necessary. Leaching seeds in running water also can be used, or alternately soaking seeds for 12 hours, then draining for 12 hours (*41*).

After soaking, seeds are usually mixed with a moisture-retaining medium for the stratification period. Almost any medium that holds moisture, provides aeration, and contains no toxic substances is suitable. These include well-washed sand, peat moss, chopped or screened [0.6 to 1.0 cm (1/4 to 3/8 in.)] with smaller parts discarded sphagnum moss, vermiculite, and composted sawdust. Fresh sawdust may contain toxic substances. A good medium is a mixture of one part coarse sand to one part peat, or one part perlite to one part peat, moistened and allowed to stand 24 hours before use. Any medium used should be moist but not so wet that water can be squeezed out.

Seeds are mixed with one to three times their volume of the medium or they may be stratified in layers, alternating with similarly sized layers of the medium. Suitable containers are boxes, cans, glass jars with perforated lids, or other containers that provide aeration, prevent drying, and protect against rodents. Polyethylene bags are excellent containers either with or without media (Figure 8–3). Stratification of seeds in a plastic bag without a surrounding medium has been called *naked chilling* (*14*). A fungicide may be added as a seed protectant. Seeds may also benefit from surface disinfection prior to imbibition and stratification with a 10 percent bleach solution for 10 to 15 minutes followed by multiple rinses with water to remove the bleach.

The usual stratification temperature is 0 to 10° C (32 to 50° F). At higher temperatures seeds often sprout prematurely. Lower temperatures (just above freezing) delay sprouting.

The time required for stratification depends on the kind of seed, and sometimes upon the individual lot of seed as well (see Chapter 7). For seeds of most species, one to four months is sufficient for low-temperature stratification. During this time the seeds should be examined periodically; if they are dry, the medium should be remoistened. When sprouting begins, the seeds should be planted or moved to lower storage temperatures. The seeds to be planted are removed from the containers and separated from the medium, using care to prevent injury to the moist seeds. A good method is to use a screen that allows the medium to pass through while retaining the seeds. The seeds are usually planted without drying to avoid injury and reversion to secondary dormancy. Some success has been reported for partially drying previously stratified seeds, holding them for a time at low temperatures, then planting them "dry" without injury or loss of after-ripened condition.

FIGURE 8–4 An old-fashioned outdoor stratification box used for yew *(Taxus)* seeds at Zelenka Nursery in Michigan.

Beech and mahaleb cherry seeds were successfully dried to 10 percent, then held near freezing (*67*). Similarly, stratified fir *(Abies)* seed has been dried to 20 to 35 percent, then stored for a year at low temperatures after stratification (*18*).

Controlled outdoor stratification. Where refrigerated storage is not available, stratification may be done by storing outdoors, either in pits several feet deep or in raised beds enclosed in wooden frames (Figure 8–4) (*1, 62*). Essentially the same seed preparation procedures are used but outdoor winter chilling and natural rainfall provide the required chilling temperature and moisture. However, the seeds need to be protected against freezing, drying, and rodents (*66*).

When seeds are stratifying in a trench, a layer of wire netting should be placed on the bottom, sides, and top of the trench to protect the seeds from rodents. A layer of clean sharp sand or gravel (10 cm; 4 in.) is placed on the bottom, then the seeds are mixed with the medium or placed in alternating layers of seed and medium. The trench is then filled and covered with the netting.

5. Choosing the proper mechanical seeder for outdoor planting. The first mechanical seed drill was developed by Henry Smith in 1850 (*33*). Today, all field-sown vegetable and agronomic crops are seeded mechanically. Selection of a seeder is determined by the size and shape of the seed, soil characteristics, total acreage to be planted, and the need for precision placement of the seed in the row.

Mechanical seeders contain three basic components: a **seed hopper** for holding seeds and a **metering system** to deliver seeds to the **drill.** A drill opens the furrow for planting the seed. The drill controls seeding depth and must provide good seed-to-soil contact while minimizing soil compaction that might impede seedling emergence. The most common type of drill is a simple "Coulter" drill that places seeds into an open furrow. "Dibber" drills that punch individual holes to place seed have also shown good seeding performance (*25, 19*). In some cases, a press wheel may be used to help cover the seed and attachments to the seeder may supply fertilizer, pesticides, or anticrusting agents before or after depositing the seed.

Mechanical seeders (Figure 8–5) are available as either **random** or **precision seeders** (*53*). A **random seeder** meters seed in the row at no exact spacing. Random seeders are less complex than precision seeders and are useful where spacing between plants in the row is not critical and thinning is not applied to achieve final plant stand as in many agronomic crops. Random seeders use gravity to drop

FIGURE 8–5 Examples of field seeding machinery. *Left:* a single-row Planet Junior. *Right:* a multirow seeder for vegetable or agronomic crops. (Courtesy S. Shearer, Univ. of Kentucky)

seeds through holes located at the bottom of the hopper. The size of these holes and tractor speed determine seeding rate.

Precision seeders selectively meter seed from the hopper to maintain a preset spacing in the row and can greatly reduce the number of seeds required to seed an acre compared to random seeding. For example, to achieve the same stand for California lettuce, seeding rates were reduced by 84 percent using precision, compared to random seeding (*36*). Precision seeders use a separate power take-off on the tractor drive to power the planter and control seeding rate. Several types of precision seeders are available. These include: **belt, plate, wheel,** and **vacuum** seeders.

The **belt seeder** uses a continuously cycling belt that moves under the seed supply. Holes in the belt at specified intervals determine seed spacing. When operating correctly, one seed will move by gravity to occupy one hole on the belt and be released as it passes over the furrow.

The **plate seeder** also uses gravity to fill holes in a metal plate rotating horizontally through the seed hopper. The number of holes in the plate and the speed of plate rotation determine seed spacing.

The **wheel seeder** employs a rotating wheel oriented in a vertical position at a right angle to the bottom of the seed hopper. Seed fills the opening at the top of the wheel (bottom of the hopper) by gravity and is carried 180 degrees where it is deposited into the furrow opening.

Vacuum seeders are replacing gravity seeders in the vegetable industry because they can more precisely deliver single seeds at a specified row spacing, especially small seeds (like tomato), irregularly shaped seeds (like lettuce), or uncoated seeds (*37*). The vacuum seeder utilizes a vertical rotating plate in the hopper with cells under vacuum that pick up a single seed. Seeds are released into the planting furrow by removing the vacuum in the cell as it rotates above the seed drop tube or planting shoe. A *"singulator device"* helps displace extra seeds prior to planting. Some vacuum seeders use a burst of air to clean the cell after the seed has been dropped to avoid skips from a clogged seed hole.

For all precision seeders, different size holes in belts or plates can be used for seeds of different species with different sizes. In many cases, uniform seed size or pelleted seeds improve precision of in-row spacing. However, because of seed quality, environmental factors, insect, disease, or animal predation, seeds are usually spaced at a higher density than is optimum for a final stand and the grower must physically thin seedlings to the desired plant density following emergence. Most direct seeded vegetable crops are planted with precision seeders.

In addition to conventional seeders, **gel seeders** have been developed to deliver pregerminated seeds (see pregermination pp. 166). Pregerminated seeds are incorporated into a gel and extruded or fed into the furrow using a pumping system or by having the seed tank under pressure using compressed air (*53*). Although this method can improve seedling emergence (especially under adverse environmental conditions), gel seeding is still only a minor planting system compared to conventional seeding of dry seeds because of cost, and complexity of the operation (*37*).

6. Plant at proper depth. Depth of planting is a critical factor that determines the rate of emergence and perhaps stand density. If too shallow, the seed may be in the upper surface that dries out rapidly; if too deep, emergence of the seedling is delayed. Depth varies with the kind and size of seed and, to some extent, the condition of the seed bed and the environment at the time of planting. When exposure to light is necessary, seeds should be planted shallowly. *A rule of thumb is to plant seeds to a depth that approximates three to four times their diameter.*

7. Determine proper rate of sowing. The rate of sowing is critical in direct sowing in order to produce a desired plant density. It can be estimated by the following formula:

$$\text{Weight of seeds per unit area} = \frac{\text{density (plants/ unit area) desired}}{\text{number of seeds/ unit weight (seed count)} \times \text{germination percent*} \times \text{purity percent*}}$$

*Expresses as decimal

This rate is a minimum and should be adjusted to account for expected losses in the seed bed, determined by previous experience at that site. Many seed companies will help producers in setting up spacing requirements for direct-seeding precision planting equipment.

Rates will vary with the spacing pattern. Field crops or lawn seeds may be *broadcast* (i.e., spaced randomly over the entire area) or *drilled* at given spaces. Other field crops, particularly vegetables, are *row planted,* so that the rate per linear distance in the row must be determined. Crops may be grown in rows on *raised beds,* particularly in areas of low rainfall where irrigation is practiced and excess soluble salts may accumulate to toxic levels through evaporation. Overhead sprinkling and planting seed below the crest of sloping seed beds may eliminate or reduce this problem.

*8. **Provide post-planting care.*** Adequate moisture must be supplied to the seed once the germination process has begun. In many areas, there is adequate natural rainfall to support seed germination. In areas with irregular rainfall, supplemental irrigation is usually supplied by overhead sprinklers, subsurface furrow flooding to raised seed beds, or by trickle irrigation. The soil should also be kept from drying out and developing a crust. This is primarily a function of seed bed preparation but may be avoided by light sprinkling, shading, and covering with light mulch. With row planting, excess seed is planted, then the plants are thinned to the desired spacing. Thinning is expensive and time-consuming and can be reduced by precision planting. Competition from weeds must also be controlled by herbicide, tillage, or mulching to ensure a vigorous seedling stand.

Specialty Systems for Direct-Seed Crops

Direct-Seeded Nursery Row Production

Planting directly in separate nursery rows (Figure 8–6) is one of the primary methods used to propagate rootstocks of most fruit and nut tree species (*26, 59, 66*). Cultivars are budded or grafted to the seedlings in place (see Chapters 13 and 14). The method is also used to propagate shade trees and ornamental shrubs, either as seedlings or on rootstocks as budded selected cultivars.

Deciduous fruit, nut, and shade tree propagation usually begins by planting seeds or liners in nursery rows. Where plants are to be budded or grafted in place, the width between rows is about 1.2 m (4 ft) and the seeds are planted 7.6 to 10 cm (3 to 4 in.) apart in the row (see Figure 8–6). Seeds known to have low germinability must be planted closer together to get the desired stand of seedlings. Large seed (walnut) can be planted 10 to 15 cm (4 to 6 in.) deep, medium-sized seed (apricot, almond, peach, and pecan) about 7.6 cm (3 in.), and small seed (myrobalan plum), about 3.8 cm (1½ in.). Spacing may vary with soil type. If germination percentage is low and a poor stand results, the surviving trees, because of the wide spacing, may grow too large to be suitable for budding. Plants to be grown to a salable size as seedlings without budding could be spaced at shorter intervals and in rows closer together (Figure 8–6).

Fall planting of fruit and nut tree seeds is commonly used in mild-winter areas such as California (*26*). Seeds are planted 2.5 to 3.6 cm (1 to 1½ in.) deep and 10 to 15 cm (4 to 6 in.) apart depending on size, and then covered with a ridge of soil 15 to 20 cm (6 to 8 in.) deep, in which the seeds remain to stratify during winter. The soil ridge is removed in the spring just before seedling emergence. Herbicide control of weeds and protection of the seeds from rodents become important considerations during these procedures.

FIGURE 8–6 Direct seeding in the nursery row, illustrated by peach seedlings grown for rootstocks. These seedling plants are being budded with peach cultivars.

Field Seeding for Reforestation or Naturalizing

Field seeding of forest trees is accomplished in reforestation either through natural seed dissemination or planting. Costs and labor requirements of direct seeding are lower than those for transplanting seedlings, provided soil and site conditions favor the operation (*13*). The major difficulty is the very heavy losses of seeds and young plants that result from predation by insects, birds, and animals and from drying, hot weather, and disease (*62, 63*). A proper seed bed is essential and an open mineral soil with competing vegetation removed is best. The soil may be prepared by burning, disking, or furrowing. Seeds may be broadcast by hand or by special planters, or drilled with special seeders. Seeds should be coated with a bird and rodent repellent.

In certain situations, trees and shrubs can be directly seeded into a natural setting for landscape purposes (*8*). Dig a 4- to 5-in. hole, carefully pulverizing soil if it is compacted. On a rocky slope, place seeds in a crevice or pocket of soil. If on a slope, make a slight back slope to avoid erosion and to prevent the seed hole from being covered with loose soil from above. Place one gram of slow-release fertilizer in the bottom of the hole. Replace soil, leaving a slight depression for the seed. Plant 2 to 20 seeds, and cover with pulverized soil to 3.2 to 12.7 mm (⅛ to ½ in.) deep, depending upon size. Contact herbicides or mulches around the seed site are essential to remove weed competition. Mulches can be coarse organic materials (wood chips, coarse bark) or sheet materials (uncoated and polyethylene-coated mulching paper, black polyethylene film, asphalt roofing paper, or kraft building paper). With the coarse organic mulches use a tin can (size 2½) milk carton, or asphalt or kraft paper collar around the seed site. Cut a hole in the center of the sheet mulches to allow the seedlings to emerge. Small wire cages may be needed to cover the site for rodent and bird protection.

Wildflower seed mixtures can be naturalized to provide landscape color for public or private lands at a low cost (Figure 8–7). In many locations, wildflower establishment has become an alternative to mowing on highway right-of-ways. Seed germination and seedling establishment are improved by tillage for seed-bed preparation and a straw mulch covering for seeds (*11*). For highly erodible sites, a *"nurse"* grass crop plus wildflower seed mixture can improve wildflower establishment (*12*). Weed competition is a serious problem for wildflower plantings that must be managed to ensure a successful stand. Successful strategies include use of herbicides, tillage, fumigation, and solarization (covering soil with plastic to trap solar radiation and allow heat to pasteurize soil) (*11*).

FIGURE 8–7 Wildflowers sown for a typical highway right-of-way planting.

FIELD NURSERIES FOR TRANSPLANT PRODUCTION

Outdoor field nurseries where seeds are planted closely together in beds are used extensively for growing transplants of conifers and deciduous plants for forestry (*1, 62*), for ornamentals (*7, 15, 24, 42*), to provide rootstock liners for some fruit and nut tree species (*26, 41, 59*) (Figure 8–8), and vegetable transplants (*22*). The conditions for optimum seed germination and seedling emergence are very similar to those previously described for field seeding. However, field transplant nurseries produce seedlings at a close spacing using smaller acreage and more controlled management. It is more common to produce woody plant seedlings in transplant nurseries than direct sowing them to a permanent location. Additional practices for successful production in a transplant nursery include:

1. Prepare the site and soil. Nursery production requires a fertile, well-drained soil of medium to light texture. Preparation for planting may include rotation with other crops and incorporation of a green manure crop or animal manure (*64*). Preplant fumigation and weed control are essential aspects of most nursery operations.

Weed control can be facilitated by careful seed-bed preparation, cultivation, and chemical sprays (*39*). Three types of chemical controls are available. **Preplant** fumigation is effective and also kills disease organisms and nematodes. **Pre-emergence** herbicides are applied to the soil before the

FIGURE 8–8 Field seeding for woody plant liner production is used for production of difficult-to-root species and for understock production for budding and grafting. *Left:* seed-bed preparation prior to seeding. Note burlap used to cover seeds to enhance germination and overhead sprinkler irrigation. *Center:* Five-row liner bed for pines being cultivated for weed control. *Right:* One-year-old liner beds of deciduous plants.

weed seeds emerge. **Post-emergence** herbicides can be applied as soon as the weed seedlings have emerged. A wide range of selective and nonselective commercial products are available. Such materials should be used with caution, however, since improper use can cause injury to the young nursery plants. Not only should directions of the manufacturer be followed, but preliminary trials should be made before large-scale use.

2. Prepare the seed bed. A common size of seed bed is 1.1 to 1.2 m (3½ to 4 ft) wide with the length varying according to the size of the operation. Beds may be raised to ensure good drainage, and in some cases, sideboards are added after sowing to maintain the shape of the bed and to provide support for glass frames or lath shade. Beds are separated by walkways 0.45 to 0.6 m (1½ to 2 ft) wide. North-south orientation gives more even exposure to light than east-west.

Seed may be either broadcast over the surface of the bed or drilled into closely spaced rows with seed planters. For economy, seeds should be planted as closely together as feasible without overcrowding, which increases damping-off and reduces vigor and size of the seedlings (*31*), resulting in thin, spindly plants and small root systems. Seedlings with these characteristics do not transplant well (*30*).

3. Plant the seeds. Seeds are planted in the nursery in the fall, spring, or summer, depending on the dormancy conditions of the seed, the temperature requirements for germination, the management practices at the nursery, and the location of the nursery (in a cold-winter or a mild-winter area). Planting time varies for several general categories of seed (*42, 49, 63*):

a) Certain species (e.g., apple, pear, *Prunus,* yew) require moist-chilling **(stratification)** and are **adversely affected by high germination temperatures,** that produce **secondary dormancy.** Germination temperatures of 10 to 17° C (50 to 62° F) are optimum. Seeds of these species could be planted in the fall with germination taking place in late winter or early spring. This procedure is particularly desirable in mild-winter areas. Or seeds can be stratified during winter storage and planted as early in the spring as permitted by climate. This procedure is more likely to be used in cold-winter areas where fall-planted germinating seed might be injured by very low temperatures.

   A modification of the fall planting procedure is useful for seeds that have hard seed coats plus dormant embryos. Seeds are planted in summer or early fall to allow six to eight weeks of warm stratification in the seed bed prior to the winter chilling (*41*). Another modification is collecting seeds before they are fully ripe and planting without allowing them to dry.

b) Many kinds of seeds—including most conifers (pine, fir, spruce) and many deciduous hardwood species that benefit from moist-chilling—**do not germinate until soil temperatures have warmed up,** and are not inhibited by high soil temperatures. Optimum germination temperatures are 20 to 30° C (68 to 86° F). Seeds of these species would not germinate in the cool fall season or in early spring. Such seeds can be handled either by fall planting or by spring planting following the required chilling treatments (stratification).

c) Seeds of some species (e.g., black locust, Colorado and Norway spruce, Chinese and

Siberian elm, European larch, bristlecone pine, and Douglas-Fir) are able to **germinate over a wide range of temperatures** [15 to 32° C (60 to 90° F)]. Seeds of some of these species require moist-chilling or have hard coats requiring special treatment. If fall-planted in cold areas, seeds of this group may germinate prematurely when the seedlings can be injured by winter cold. Spring planting is recommended, although the actual time is not as critical as that for the other two groups.

d) Seeds of some species (maple, poplar, aspen, elm, cottonwood) ripen in spring or early summer. Such seeds should be planted immediately, as **their viability declines rapidly** (see recalcitrant seeds, pp. 143).

e) Several **vegetable species** including tomato, pepper, cabbage, broccoli, and onion can be produced from transplants produced in field nurseries. This is an alternative to direct seeding and is less expensive than container-grown transplants. Warm-season crops are usually seeded in spring. Plastic or fabric (floating) row covers can be used to prevent frost injury. Cool-season crops are seeded in early spring or summer for a fall harvest.

4. Determine seeding rates. The optimum seed density depends primarily on the species but also on the nursery objectives. If a high percentage of the seedlings are to reach a desired size for field planting, low densities might be desired; but if the seedlings are to be transplanted into other beds for additional growth, higher densities (with smaller seedlings) might be more practical. Once the actual density is determined, the necessary rate of sowing can be calculated from data obtained from a germination test and from experience at that particular nursery.

The following formula is useful in calculating the rate of seed sowing (*21, 42, 67*):

$$\text{Weight of seeds to sow per area} = \frac{\text{density (plants/unit area) desired}}{\text{purity percent}^{*} \times \text{germination percent}^{*} \times \text{field factor}^{*} \times \text{number of seeds/unit weight (seed count)}}$$

*Expressed as a decimal.

Field factor is a correction term applied based on the expected losses which experience at that nursery indicates will occur with that species. It is a percentage expressed as a decimal.

Seeds can be planted by (a) broadcasting by hand or seeders, (b) hand spacing (larger seeds), or (c) drilling by hand with push drills, or drilling with tractor-drawn precision drills. Seeds of a particular lot should be thoroughly mixed before planting to ensure that the density in the seed bed will be uniform. Treatment with a fungicide for control of damping-off is desirable. Small conifer seeds may be pelleted for protection against disease, insects, birds, and rodents. Depth of planting varies with the kind and size of the seed. In general, a depth of three to four times the diameter of the seed is satisfactory. Seeds can be covered by soil, coarse sand, or by various mulches.

Soil firming may be done to increase contact of seed and soil. It is used for California lettuce, for example, and carried out with a tamper, hand roller, or tractor-drawn roller either before sowing or immediately afterward. Rodent and bird protection may be necessary.

5. Provide after-care. During the first year in the seed bed, the seedlings should be kept growing continuously without any check in development. A continuous moisture supply, cultivation or herbicides to control weeds, and proper disease and insect control contribute to successful seedling growth. Fertilization with nitrogen is usually necessary, particularly when a mulch has been applied, since decomposition of organic material can produce nitrogen deficiency. In the case of tender plants, glass frames can be placed over the beds, although for most species a lath shade is sufficient. With some species, shade is necessary throughout the first season; with others, shade is necessary only during the first part of the season. Sprinkling with water to reduce ground temperature during the hot part of the day is sometimes useful (*17*).

6. Harvesting transplants. Vegetable transplants can be harvested after six to ten weeks in the seed bed. These are usually *"pulled,"* bundled, and used as bare-root transplants (Figure 8–9). In the U.S., vegetable transplant beds are either located on the producer's farm or shipped to northern growing areas from southern transplant nurseries. There is an increasing amount of vegetable and tobacco transplants being produced in plug systems and *"float beds"* (see pp. 233) that is replacing the more traditional field-nursery produced transplant.

FIGURE 8–9 *Left:* Field transplant nursery for cabbage and broccoli plants being grown on the farm site for fall production. *Right:* Floating row covers provide protection for early seeded transplant beds.

In contrast, woody plants can remain in the *"liner"* bed for one or more years before being transplanted to a permanent location (see Chapter 3). For some species, the plants may be shifted to a transplant bed after one year and then grown for a period of time at wider spacing. This basic procedure is used to propagate millions of forest tree seedlings, both conifer and deciduous species.

Liners produced in a seed-bed nursery are often designated by numbers to indicate the length of time in a seed bed and the length of time in a transplant bed. For instance, a designation of 1-2 means a seedling grown one year in a seed bed and two years in a transplant bed or field. Similarly, a designation of 2-0 means a seedling produced in two years in a seed bed and no time in a transplant bed (Figure 8–10).

PRODUCTION OF TRANSPLANTS UNDER PROTECTED CONDITIONS

Seedling production is used extensively to produce flowers and vegetables for outdoor transplanting. Historically, this method has been used to extend the growing season by producing plantlets under protection for transplanting to the field as soon as the danger of spring frosts is over or by placing seedlings under individual protectors to avoid freezing. This procedure also avoids some of the environmental hazards of germination and allows plants to be placed directly into a final spacing. Optimum germination conditions are provided in greenhouses, cold frames, or other structures to ensure good seedling survival and uniformity of plants.

FIGURE 8–10 *Left:* Pine (2-0), and *Right:* ash seedlings being mechanically undercut prior to harvesting.

Seedling growing has become an extensive **bedding plant industry** to produce small ornamental plants for home, park, and building landscaping, as well as vegetable plants for home gardening (*3, 34, 45*). Commercial vegetable growing also relies heavily on the production of transplants, involving highly mechanized operations beginning with seed germination and ending with transplanting machines which place individual plants into the field.

Production systems for transplants. Traditionally, bedding plants and vegetables have been produced by germinating seeds in flats and transplanting the seedlings to larger containers prior to field or landscape planting. However, modern greenhouse producers have adopted **plug production** as the preferred method for transplant production (*2, 34*). There are numerous advantages to plug production over conventional flat seeding, and specialized plug growers produce acres of plugs under glass each spring. Many bedding plant growers find that they can purchase plugs from specialized plug producers more economically than producing seedlings themselves. In either case, seedlings are moved to larger cell packs by the bedding plant grower for *"finishing"* prior to sale to the consumer. The advantages of plug production include:

1. Optimizes the number of plants produced per unit of greenhouse space.
2. Specialization in plug production allows growers to invest in equipment to control environmental conditions during germination.
3. Fast production (most plugs are sold within four to six weeks of seeding) allows growers to seed multiple crops per season, permits accurate crop scheduling, and allows plugs to be shipped easily to the end user.
4. Because plugs are transplanted to larger-size containers with the roots and original medium intact, plugs transplant easily with a high degree of uniformity. Plugs do not experience the same *"transplant shock"* and check in growth as seedlings removed from seedling flats.

Flat production. Traditional bedding plant production relied on flat production of seedlings. Seeds are planted in a germination flat or container and later the germinated seedlings are *"pricked out"* and transplanted to develop either in a transplant flat at a wider spacing or in individual containers, where they remain until transplanted out-of-doors (Figure 8–11). This method is still utilized by small bedding plant producers, but has largely been replaced by mechanized plug production.

Plug production. Millions of bedding plants are produced annually in greenhouses under carefully controlled environmental conditions for optimizing germination and plant growth. This has become possible mainly through the development of the **plug system** (*3, 9, 16, 29, 69*). A **plug** is a seedling produced in a small volume of medium contained in a small cell, of which between 220 up to 800 are contained on a single sheet of polystyrene, Styrofoam, or other suitable material. Sizes may range down to 1 × 1 cm (3/8 × 3/8 in.). Plug flats are filled mechanically with a growing medium and seeds are sown mechanically into each cell. (Figure 8–12).

Seed plug trays (see Chapter 3) can be placed in greenhouses under mist or fog or in

FIGURE 8–11 Traditional seedling production in community flats. *Left:* Wooden seedling flat. *Right:* Dibble board for transplanting seedlings into flats.

FIGURE 8–12 Plug production. *Top:* Plugs growing on movable benches. *Top middle:* Seeded plug trays covered with vermiculite and a porous plastic material to control moisture levels. *Bottom middle:* Good seedling emergence and uniformity. However, notice the number of plug cells without a seedling. This could be due to poor seed germination or poor mechanical seeding. *Bottom:* Typical plug-grown Begonia.

special germinators (Figure 8–13) at optimum temperature and moisture conditions. It is important to have high-quality seed. Pelleting and seed priming (see Chapter 6) is useful to improve germination rate, uniformity, and mechanical handling.

Plug growth stages. Four morphological stages of seedling growth have been recognized (*52*):

Stage 1: radicle emergence
Stage 2: cotyledon spread
Stage 3: unfolding of three or four leaves
Stage 4: more than four leaves

Providing precise environmental control for each of the stages is essential in plug production. Adequate moisture must be maintained, but too much reduces the oxygen available to the seeds. Fine misting or fog is excellent. Warm temperature

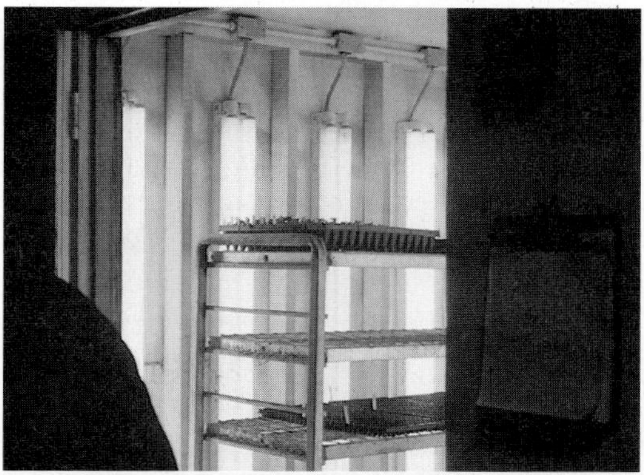

FIGURE 8–13 Specialized germinators are used by large plug growers to provide optimum germination conditions. Germinators control temperature, light, and humidity.

and consistent moisture are essential for stage 1 but can be reduced in stage 2 and in later stages (see Table 8–1). Light and supplemental fertilization is important in stages 1 and 2. Fertilization (N, K, P) is particularly important in stage 3 but must be monitored carefully.

Transplanting. Stage 4 is a seedling almost ready to transplant. Temperatures should be reduced compared to stage 3 to harden seedlings. Ammonium fertilization should be avoided at this stage. It is better to use calcium or potassium nitrate. A plug dislodger is used to remove the plugs two to three hours after watering. Place into dibbled holes in packs or pots prepared with growing medium. For large producers, mechanical transplanters are replacing hand labor for moving plugs to larger-sized containers.

Production of woody plant seedlings in containers. Production of seedling trees and shrubs in containers involves two main activities: (a) the production of landscape or fruit or nut tree plants, and (b) the mass propagation of small seedlings for reforestation. Landscape and fruit tree seedlings may be started in germination flats. Some nurseries have directly seeded into plug-tray systems (*50, 55*), avoiding the conventional propagation flats, then transplanted to either a transplant flat or a small pot. Later they are moved to slightly larger containers, or transplanted directly into the containers where they will remain until transplanted out-of-doors (Figure 8–14).

With this system, root pruning is essential to induce a desirable root system. Root pruning should be done at the first transplanting, soon after the roots reach the bottom of the flat (*27, 28*). Further root pruning should take place when the liner flat is transplanted, which should be done when the roots protrude through the pot about an inch. These roots are removed. Any delay in transplanting at either stage, or omission of pruning, will increase the incidence of poor root systems. Copper or plastic screen-bottomed flats (*23*) or supports stimulate formation of branch roots.

A second activity involving woody plant production is the mass propagation of small seedlings for direct planting in reforestation (*43, 68*), for planting into outdoor beds for an additional one or two years' growth (*61*), and for ornamental shrub production (*49*). Seeds are invariably planted directly into various kinds of small individualized containers—usually narrow and relatively deep. These may be plastic containers from which the seedling plug is removed and planted or they may be containers made of substances such as peat or fiber blocks that are planted with the seedling (see Chapter 3).

To maximize production for indoor transplant production, the following should be considered:

1. Facilities. Indoor seedling production occurs in several types of structures, including greenhouses, cold frames, and hotbeds, as described in Chapter 3. Some bedding plant operations have

TABLE 8–1

Requirements for seed germination during plug propagation of three popular bedding plants

	Petunia	*Pansy*	*Impatiens*
	Stage 1		
Temperature	80°F (27°C)	77°F (25°C)	80°F (27°C)
Moisture	100%, 1.5 vpd[a]	100%, 0.5 vpd	100%, 1.0 vpd
Light	14.3 Wm^{-2}[b]	12.6 Wm^{-2}	14.3 Wm^{-2}
Fertilizer	25–75 ppm KNO 3 1 application (1–3 days)	25–50 ppm KNO 3 (1–7 days)	None
	Stage 2		
Temperature	75°F (24°C)	66°F (18°C)	75°F (24°C)
Moisture	85%, 5.5 vpd	75%, 6.0 vpd	75%, 5.5 vpd
Light	14.3 Wm^{-2}	14.3 Wm^{-2}	14.3 Wm^{-2}
Fertilizer	50 ppm 20–10–20 (3–7 days)	None	None

Source: Ref. 52

[a]vpd = vapor pressure deficit

[b]Wm^{-2} = Watts/square meter, converted from fc assuming lights are fluorescent cool-white lamps.

FIGURE 8–14 An alternative to field production of conifers is container production. *Top:* Spruce seedlings grown in containers for future use as understocks. *Middle and bottom:* Pine seedlings being grown in containers in the greenhouse under accelerated growth conditions (see Chapter 3).

special **growth rooms** (Figure 8–13) where seed flats are placed on carts or shelves in an enclosed area and subjected to controlled environments for germination prior to being moved to the greenhouse (*57*).

Growth rooms need controlled lighting (day length and irradiance) and temperature (duration, level) and may include control of relative humidity, carbon dioxide, irrigation, and fertilization (*57*). Under proper conditions a grower may produce not only rapid uniform germination but also healthy seedlings with excellent ability to transplant without any check in plant growth.

*2. **Media.*** Germination media for herbaceous bedding plants must retain moisture, supply nutrients, permit gas exchange, and provide support for the seedling (*20*). The pH should be neutral and soluble salts should be low. Some nutrients must be present, particularly phosphorus and calcium. Regular soil mixes, for example, equal parts soil, sand, and peat, might be used. However, these have been supplanted to a large extent by special nonsoil mixes made up of such components as peat moss, perlite, ground or shredded bark, and vermiculite, fortified by mineral nutrients or slow-release fertilizers (see Chapter 3). These mixes are available commercially and are used in large quantities by bedding plant producers. The mix should have excellent drainage to inhibit damping-off and reduce crusting. A proper air and water content should be maintained for good germination and seedling growth (*51*). Small seeds should have a finer and more compact medium than is used for larger seeds. Seed flats and media should be pasteurized, and seeds should be treated for disease organisms (see Chapter 3).

*3. **Sowing seeds.*** Seeds may be broadcast over the surface of the transplant flats or planted in rows. Advantages of row planting are reduced damping-off, better aeration, easier transplanting, and less drying out. Planting at too high a density encourages damping-off, makes transplanting more difficult, and produces weaker, nonuniform seedlings. Suggested rates are 1,000 to 1,200 seeds per 29 × 54 cm (11 × 22 in.) flat for small-seeded species (e.g., petunia) and 750 to 1,000 for larger seeds. Small seeds are dusted on the surface; medium seeds are covered lightly to about the diameter of the seed. Larger seeds may be planted at a depth of two to three times their minimum diameter.

Plug production requires the use of a mechanical seeder (*4*). The choice of a seeder depends on several factors including cost, seeding speed, volume of flats to be seeded, and the need for flexibility to sow a variety of size and shaped seeds. Three types of seeders are commonly available to plug growers. These are **template, needle** and **drum** seeders (Figure 8–15).

The **template seeder** is the least expensive type of seeder. It uses a template with holes that match the location of cells in the plug flat. Template seeders use a vacuum to attach seeds to the template. Releasing the vacuum drops the seeds either directly into the plug flat or into a drop tube to precisely locate seeds in each cell of the plug flat. Templates with different size holes are available to handle different size and shape seeds. A differently sized template is also require for each plug flat size. It is a relatively fast seeder because it sows an entire flat at once. However, this is the least mechanized of the commercially available planters. It requires the operator to fill the template with seeds, remove the excess, and then move the template to the flat for sowing. Template seeders work best for round, semi-round, or pelleted seeds.

FIGURE 8–15 Mechanical seeders for greenhouse plug production include template, needle and drum seeders. (a). Template seeders sow an entire flat of seeds at once. (b). Needle seeder removing seeds for a seed hopper. (c). Needle seeder dispensing seeds into a plug flat. (d). Drum seeders have a revolving drum to quickly sow seeds. (e). Pelleted seeds work very well with drum seeders.

The **needle seeder** is a moderately priced planter. It is fully mechanical, requiring little input from the operator. Individual needles or pickup tips, under vacuum pressure, lift single seeds from a seed tray and deposit one seed per plug cell. Seeds are directly placed in plug cells or into drop tubes for more accurate seeding. A burst of air can be used to deposit seeds and clean tips of unwanted debris. The needle seeder can seed a variety of size and shaped seeds including odd-shaped seeds like marigold, dahlia, and zinnia. Although slower than the drum seeder, it is still relatively fast, sowing up to 100,000 seeds per hour.

The **drum or cylinder seeder** has a rotating drum or cylinder that picks up seeds using vacuum from a seed tray and drops one seed per plug cell. This is the fastest, most precise, and most costly of the commercial seeders. It is fully mechanical. Most drum seeders require a different drum for each plug flat, but newer models have several hole sizes per drum that can be selectively put under vacuum pressure. Drum seeders work best with round, semi-round, or pelleted seeds. Large plug growers choose drum seeders because they seed a high volume of seeds quickly. Drum seeders can sow up to 800,000 seeds per hour. Large plug growers must have the capacity to sow millions of plugs per year of over 100 different types of bedding plants (*65*). Sophisticated drum seeders "eject" seeds from the drum using an air or water stream for precise seeding location in the flat.

When evaluating a seeder, growers must consider the machine's ability to deliver seeds at the desired speed without skipping cells due to poor seed pickup or delivery, sowing multiple seeds per cell, and sow seeds without seed "bounce" that can reduce the precise location of the seed in each plug cell.

4. Moisture. The moisture content of the growing medium can be critical to germination success (*6, 38*). Species like coleus, begonia, and alyssum require a wet medium (saturated); impatiens, petunia, and pansy require a moist medium (wet but not saturated); while asters, verbena, and zinnia prefer a drier medium (watered only prior to sowing) for good germination.

For smaller growers, seed flats may be held under polyethylene tents (see Chapter 3), or in small operations, covered with glass or plastic to keep the surface from drying out. Covered flats should not be exposed directly to sunlight, as excessive heat buildup injures the seedling.

Several systems for delivering water to seed flats are available (Figure 8–16). These include **overhead** and **subirrigation systems** (*40*). Overhead irrigation can be as simple as a hose with a fine holed "rose" irrigation nozzle or a timed mist system. Automated **boom sprayers** provide fine control of overhead irrigation. The boom travels the length of the greenhouse, providing a spray of water to the flats. The speed of the boom and irrigation timing can be computer-controlled.

Subirrigation systems have the advantage of providing even moisture while reducing water runoff. **Capillary mat systems** (Figure 8–17) deliver water from a reservoir to the mat where the growing medium "pulls" water into the flat by capillary action. **Ebb and flood systems** use a sealed bench that is flooded periodically during the day and then the fertilizer solution is allowed to passively drain back into a holding tank (*56*). A variation on these systems is **"float bed"** production (Figure 8–18). A Styrofoam flat is floated in a water bed containing a nutrient solution (*44*). Float-bed tobacco transplant

FIGURE 8–16 Types of irrigation used in the greenhouse. *Top:* Mechanical traveling boom sprayer and grower supplementing irrigation with a hose utilizing a rose. *Bottom:* Mist nozzles on PVC risers.

FIGURE 8–17 Capillary mats used for subirrigation. *Top:* Seeded plug flats placed on capillary mats for germination. *Bottom:* Capillary mat folded on one edge to show fabric material used to move water by capillary action.

FIGURE 8–18 "Float bed" production of transplants. Simple frames lined with a plastic pool liner can be used for on-farm transplant production. Styrofoam flats are used to float on the nutrient solution in the float bed.

production has proven to be commercially viable and several vegetable crops are being evaluated for float production (*22*). Regardless of the system being used, water quality must be monitored during production (*40*).

In most cases, seeds are sown on the surface of the medium by mechanical seeders. Seeds can be covered with vermiculite or porous plastic sheets to maintain even moisture until seedlings emerge. In modern palletized greenhouse operations, germination occurs in specialized germination rooms or in greenhouse sections designed to optimize germination conditions, then the entire movable bench is moved to additional greenhouse sections designed for seedling growth.

*5. **Temperature.*** Temperature requirements for germination vary depending on the plant species being grown. In general, most bedding plants can be germinated in one of three temperature regimes: 26 to 30° C (78 to 80° F), 21 to 22° C (70 to 72° F), or 18 to 19° C (64 to 66° F). High temperature during germination can lead to thermodormancy in some crops.

*6. **Transplanting.*** The time required for seedling emergence can vary from about one to three weeks. When the first true leaves have appeared, the seedlings should be transplanted. Direct seeded flats should be shifted to a lower temperature for the seedlings of most kinds of plants to harden properly.

Transplanting flats or containers are filled with a growing medium and handled in the same manner as the seed flats. Holes are made in the medium at the correct spacing with a small **dibble.** The roots of each small seedling are inserted into a hole, and the medium is pressed around them to provide good contact. Dibble boards are often used to punch holes for an entire flat at once (Figure 8–11). As soon as the flat is filled, it is thoroughly watered.

Seedling growing. Once seeds have germinated and the seedlings have been transplanted, the principal objectives of production are to prevent damping-off and to develop stocky, vigorous plants capable of further transplanting with little check in growth. The usual procedure in production is to shift the flats to lower temperatures (10° C or less) and to expose them to good light. High temperatures and low light tend to produce spindly, elongated plants that will not survive transplanting. Such growth is termed "stretching." Quality plant production may require the use of growth regulators or strict environmental control (like DIF) as described in Chapter 3.

Once root systems have developed sufficiently to grow into the medium, irrigation can be scheduled to keep the medium somewhat dry on the surface but moist underneath. Such irrigation helps prevent damping-off and produces sturdy seedlings.

Transplanting Seedling Material to Permanent Locations

The final step in seedling production is transplanting to a permanent location (*47, 58*). Seedlings may be transplanted either **bare root** (vegetable transplants or deciduous fruit, nut, and shade trees), or in cells or **modular containers** (bedding plants, vegetables, forest trees), or **balled** and **burlapped** (evergreen trees), or **containerized** (ornamental shrubs and trees).

Bare-root transplanting invariably results in some root damage and transplant shock, both of which check growth. With vegetable plants these may result in premature seed-stalk formation, increased susceptibility to disease, and reduced yield potential. Handling prior to transplanting should involve **hardening-off,** which involves a controlled growth cessation. This causes accumulation of carbohydrates, making the plant better able to withstand adverse environmental conditions. This effect can be achieved by temporarily withholding moisture, reducing temperature, and gradually shifting from protected to outdoor conditions over a period of a week to ten days.

Ornamental and Vegetable Bedding Plants

During the transition to the new site, deterioration must be prevented if the plants are bare-root. Following planting, conditions must be provided for rapid root regeneration. Planting should be done as soon as possible. If not, bare-root plants can be kept (no more than seven to ten days) in moist, cool (10° C; 50° F) storage. Maintain high humidity but avoid direct watering to prevent disease.

Modular transplants (plugs or other containers) should be kept under conditions to prevent drying of the roots. Expose to light. Keep temperatures at 18 to 20° C (65 to 70° F).

Field beds should be moderately well pulverized although not necessarily finely prepared, well watered but not saturated (*58*). Transplanting is done in the field by hand or by machine. Afterward, a good amount of irrigation should be applied to increase moisture to the roots and settle the soil but not saturate it. A starter solution containing fertilizers should be applied, but if the soil is dry, should best be diluted. Temporary shade may be used for the first few days.

Trees and Shrubs

Transplanting of bare-root evergreen forest trees follows principles similar to those described. Seedling plants should be dug in the nursery in the fall after proper physiological "hardening-off" (*63*). Seedlings are packed into moisture-retaining material (vermiculite, peat moss, sawdust, shingletow) and kept in low-temperature (30 to 35° F), humid (at least 90 percent RH) storage. Polyethylene bags without moisture material are satisfactory. Some kinds of sawdust can be toxic, particularly if fresh. Bare-root nursery stock of deciduous plants and container-grown stock are handled as described for rooted cuttings in Chapter 11.

REFERENCES

1. Aldous, J.R. 1972. *Nursery practice.* For. Comm. Bul. 43. London: Her Majesty's Stationery Office.
2. Armitage, A.M. 1994. *Ornamental bedding plants.* In Ornamental Production Science in Horticulture 2. Wallingford, U.K.: CAB International.
3. Ball, V., 15th ed. 1991. *Ball red book. Greenhouse growing* (14th ed.). Reston, Va.: Reston Publ. Co.
4. Bartok, J. Jr. 1994. Facilities planning and mechanization. In *Bedding Plants IV.* J. Holcomb (ed.). Batavia, IL.: Ball Pub. pp. 233–44.
5. Brumback, W.E. 1985. Propagation of wildflowers. *Comb. Proc. Inter. Plant Prop. Soc.* 35:542–47.
6. Carpenter, W.J. and S. Maekawa. 1991. Substrate moisture level governs the germination of verbena seed. *HortScience* 26:786–88.
7. Carville, L. 1978. Seed bed production in Rhode Island. *Comb. Proc Inter. Plant Prop. Soc.* 28: 114–17.
8. Chan, F.J., R.W. Harris, and A.T. Leiser. 1971. Direct seeding of woody plants in the landscape. *Univ. of Calif. Ext. Ser.* AXT–n27.
9. Cooley, J. 1985. Vegetable plant raising using Speedling transplants. *Comb. Proc. Inter. Plant Prop. Soc.* 35:468–71.

10. Copeland, L.O., and M.B. McDonald. 1995. *Principles of seed science and technology.* (3rd ed.) New York: Chapman and Hall.

11. Corley, W.L. 1991. Seedbed preparation alternatives for establishment of wildflower meadows and beauty spots. *Southern Nurserymen's Assoc. Res. Conf.* 36:278–79.

12. ——— and J.E. Dean. 1991. Establishment and maintenance of wildflowers on erodible sites. *Southern Nurserymen's Assoc. Res. Conf.* 36:280–81.

13. Deer, H.J., and W.F. Mann, Jr. 1971. *Direct seedling pines in the South.* U.S. Dept. Agr. Handbook 391. Washington, D.C.: U.S. Govt. Printing Office.

14. Delong, S.K. 1985. Custom seed preparation for optimum conifer production. *Comb. Proc. Inter. Plant Prop. Soc.* 35:259–63.

15. Dirr, M.A., and C.W. Heuser, Jr. 1987. *The reference manual of woody plant propagation: From seed to tissue culture.* Athens, Ga.: Varsity Press.

16. Eastburn, D.P. 1984. The plug potential. *Florists Rev.* 174:6–29.

17. Eden, C.J. 1962. Conifer seed—from cone to seed bed. *Comb. Proc. Plant Pro. Soc.* 12:208–14.

18. Edwards, D.G.W. 1986. Special prechilling techniques for tree seeds. *Jour. Seed Tech. 10(2):* 151–71.

19. Finch-Savage, W.E., M. Rayment and F.R. Brown. 1991. The combined effects of a newly designed dibber drill, irrigation and seed covering treatments on lettuce and calabrese establishment. *Ann. Applied Biology* 118:453–60.

20. Fonteno, W.C. 1994. Growing media. In *Bedding Plants IV.* J. Holcomb (ed.). Batavia, IL: Ball pub. pp. 127–38.

21. Fordham, D. 1976. Production of plants from seed. *Comb. Proc. Inter. Plant Prop. Soc.* 26:139–45.

22. Frantz, J.M. and G.E. Welbaum. 1995. A comparison of four cabbage transplant production systems. In *Proceedings of the Fourth National Symposium on Stand Establishment.* Department of Vegetable Crops, University of California, Davis, CA. pp. 169–74.

23. Frolich, E.F. 1971. The use of screen bottom flats for seedling production. *Comb. Proc. Inter Plant Prop. Soc.* 21:79–80.

24. Gordon, A.G., and D.C.F. Rowe. 1982. *Seed manual for ornamental trees and shrubs.* Bull. 59. London: Forestry Commission.

25. Gray, D. and J. Reed. 1995. Use of a dibber drill and coulter drill with press wheel to improve seedling emergence in onion and lettuce. In *Proceedings of the Fourth National Symposium on Stand Establishment.* Department of Vegetable Crops, University of California, Davis, CA. pp. 125–32.

26. Hall, T. 1975. Propagation of walnuts, almonds and pistachios in California. *Comb. Proc. Inter. Plant Prop. Soc.* 25:53–7.

27. Harris, R.W., W.B. Davis, N.W. Stice, and D. Long. 1971. Root pruning improves nursery tree quality. *Jour. Amer. Soc. Hort. Sci.* 96:105–9.

28. ———. 1971. Influence of transplanting time in nursery production. *Jour. Amer. Soc. Hort. Sci.* 96:109–10.

29. Hartnett, G. P. 1985. New ideas in the use of plug systems. *Comb. Proc. Inter. Plant Prop. Soc.* 35:263–68.

30. Heit, C.E. 1964. The importance of quality, germinative characteristics and source for successful seed propagation and plant production. *Comb. Proc. Inter. Plant Prop. Soc.* 14:74–85.

31. ———. 1967. Propagation from seed. 5. Control of seedling density. *Amer. Nurs.* 125(8):14–15, 56–9.

32. ———. 1967. Propagation from seed. 6. Hardseededness, a critical factor. *Amer. Nurs.* 125(10): 10–12, 88–96.

33. Hendrick, U.P. 1933. *A history of agriculture in the state of New York.* New York: Hill and Wang.

34. Holcomb, E.J. 1995. *Bedding Plants IV.* Batavia IL: Ball Pub.

35. Hoyle, B.J., H. Yamada, and T.D. Hoyle. 1972. Aggresizing—to eliminate objectionable soil clods. *Calif. Agr.* 26(11):3–5.

36. Inman, J.W. 1967. Precision planting—a reality for vegetables. *Paper No. PC–67–12. Amer. Soc. Agricultural Engineers.*

37. ———. 1995. New developments in planting and transplanting equipment. In *Proceedings of the Fourth National Symposium on Stand Establishment.* Department of Vegetable Crops, University of California, Davis, CA. pp. 19–22.

38. Koranski, D. and R. Kessler. 1991. Seed germination: the proper conditions. *International Floriculture Industry Plug Symposium.* Ohio State Univ. Columbus, OH. pp. 53–7.

39. Kuhns, L. 1994. Weeds. In *Bedding Plants IV.* J. Holcomb (ed.). Batavia, IL: Ball Pub. pp. 273–84.

40. Langhans, R.W. and E.T. Paparozzi. 1994. Irrigation. In *Bedding Plants IV.* J. Holcomb (ed.). Batavia, IL: Ball Pub. pp. 139–50.

41. Lawyer, E.M. 1978. Seed germination of stone fruits. *Comb. Proc. Inter. Plant Prop. Soc.* 28:106–9.

42. MacDonald, B. 1986. *Practical woody plant propagation for nursery growers, Vol. 1.* Portland, Oreg.: Timber Press.

43. Maclean, N.M. 1968. Propagation of trees by tube technique. *Comb. Proc. Inter. Plant Prop. Soc.* 18:303–9.

44. Maglianti, C.G. 1987. Speedling float growing tobacco transplants on water. *Amer. Soc. Ag. Eng. Fiche no. 87–1573, 3p.*

45. Mastalerz, J.W. 1976. *Bedding plants. University Park, Pa.:* Pennsylvania Flower Growers.

46. Maude, R.B. 1978. Seed treatments for pest and disease control. *Acta Hort.* 83:205–12.

47. Mckee, J.M.T. 1981. Physiological aspects of transplanting vegetables and other crops. II. Methods used to improve transplant establishment. *Hort. Abstracts* 51(6):355–68.

48. McMillan-Browse, P.D.A. 1978. Stratification—a detail of technique. *Comb. Proc. Inter. Plant Prop. Soc.* 28:191–92.

49. ———. 1979. *Hardy woody plants from seed.* London: Grower Books.

50. Menzies, M.I. and J.T. Arnott. 1992. Comparisons of different plant production methods for forest trees. In *Transplant Production Systems.* K. Kurata and T. Kozai (eds.). Dordrecht: Kluwer Academic Pub. pp. 21–44.

51. Milks, R.R., W.C. Fonteno, and R.A. Larson. 1989. Hydrology of horticultural substrates: III Predicting air and water content in limited-volume plug cells. *Jour. Amer. Soc. Hort. Sci.* 114:57–61.

52. Ohio State University. 1987. *Tips on growing bedding plants.* Ohio Coop. Ext. Ser. and Ohio State Univ. MM 265, AGDEX 200/15. Columbus, Ohio: Ohio State Univ.

53. Orzolek, M.D. and D.R. Daum. 1984. Effect of planting equipment and techniques on seed germination and emergence: a review. *Jour. Seed Technol.* 9:99–113.

54. Page, F.R. and M.J. Quick. 1979. A comparison of the effectiveness of organic polymers as soil anti-crusting agents. *Jour. Sci. Food Agric.* 30:112–18.

55. Pinney, T. S., Jr. 1986. Update of GROPLUG® system. *Comb. Proc. Inter. Plant Prop. Soc.* 36: 577–81.

56. Poole, R.T. and C.A. Conover. 1992. Fertilizer levels and medium affect foliage plant growth in an ebb and flow irrigation system. *Jour. Environ. Hort.* 10:81–86.

57. Poynter, M.J. 1978. Building and using a growing room for seed germination of bedding plants. *Comb. Proc. Inter. Plant Prop. Soc.* 28:109–14.

58. Price, H.C., and B.H. Zandstra. 1988. Maximize transplant performance. *Amer. Veg. Grower* 36(4):10–16.

59. Rom, R.C., and R.F. Carlson, eds. 1987. *Rootstocks for fruit crops.* New York: John Wiley.

60. Royle, S. M. and T.M. Hegarty. 1978. Soil impedance and its effect on calabrese emergence. *Acta Hort.* 72:259–66.

61. Schalla, S. L., and K. Doughton. 1978. Transplanting the Douglas-Fir plug. *Comb. Proc. Inter. Plant Prop. Soc.* 28:177–84.

62. Schopmeyer, C.S., ed. 1974. *Seeds of woody plants in the United States.* U.S. For. Ser. Agr. Handbook 450. Washington, D.C.: U.S. Govt. Printing Office.

63. Schubert, G.H., and R.S. Adams. 1971. *Reforestation practices for conifers in California.* Sacramento: Calif. State Div. of Forestry.

64. Steavenson, H. 1979. Maximizing seedling growth under midwest conditions. *Comb. Proc. Intern. Plant Prop. Soc.* 29:66–71.

65. Stelk, B. 1993. Seed sowing success starts with the right equipment. *Grower's Talks* 67:33–7.

66. Stuke, W. 1960. Seed and seed handling techniques in production of walnut seedlings. *Comb. Proc Plant Prop. Soc.* 10:274–77.

67. Suszka, B. 1978. Germination of tree seed stored in a partially after-ripened condition. *Acta Hort.* 83: 181–88.

68. Thomson, J.R. 1979. *An introduction to seed technology.* New York: John Wiley.

69. Tinus, R.W., and S.E. McDonald. 1979. How to grow tree seedlings in containers in greenhouses. *USDA For. Ser. Gen. Tech. Rpt. RM–60.*

70. Vanstone, D.E. 1978. Basswood (*Tilia americana* L.) seed germination. *Comb. Proc. Inter. Plant Prop. Soc.* 28:566–69.

SUPPLEMENTARY READING

ALDOUS, J. R. 1972. *Nursery practice.* For. Comm. Bul. 43. London: Her Majesty's Stationery Office.

BALL, V., 15th ed. 1991. *Ball red book: Greenhouse growing* (15th ed.) Reston, Va.: Reston Publ. Co.

GRAY, D. 1981. Fluid drilling of vegetable seeds. *Hort. Rev.* 3:1–27.

HOLCOMB, E. J. 1995. *Bedding Plants IV.* Batavia, IL: Ball Pub.

INTERNATIONAL PLANT PROPAGATORS' SOCIETY. Proceedings of annual meetings.

LORENZ, O. A., and D. N. MAYNARD. 1988. *Knott's handbook for vegetable growers.* New York: John Wiley.

MASTALERZ, J. W. 1976. *Bedding plants.* University Park, Pa.: Pennsylvania Flower Growers.

MCMILLAN-BROWSE, P. D. A. 1979. *Hardy woody plants from seed.* London: Grower Books.

NAU, J. 1993. *Ball culture guide.* The encyclopedia of seed germination. Batavia, IL: Ball Pub.

SCHOPMEYER, C. S., ed. 1974. *Seeds of woody plants of the United States.* U.S. For. Ser. Agr. Handbook 450. Washington, D.C.: U.S. Govt. Printing Office.

STOECKELER, J. H., and P. E. SLABAUGH. 1965. *Conifer nursery practice in the prairie states.* U.S. Dept. Agr. Handbook 279. Washington, D.C.: U.S. Govt. Printing Office.

YOUNG, J.A. and C. G. YOUNG. 1992. *Seeds of woody plants in North America.* Revised edition. Portland, Or.: Dioscorides Press.

PART III—VEGETATIVE PROPAGATION

9

Selection and Management of Clones in Vegetative Propagation

Introduction The goal of vegetative propagation is to reproduce progeny plants identical in genotype to a single source plant. The biological process is known as **cloning** and the resulting population of plants is called **a clone**. The biological significance of this process is described in Chapter 2. Cloning can occur in nature by special vegetative structures (bulbs, tip layers, rhizomes, and runners, which are described in Chapters 14 and 15). These structures provide a special advantage for colonizing a specific site, but generally, cloning is not favored in nature because the process does not provide opportunity for the variation and evolutionary advancement that result from sexual reproduction. On the other hand, vegetative propagation provides a powerful tool in the selection of clones grown in cultivation (*1, 2*). Immediate selection of a single superior plant genotype is possible and can be followed by more or less indefinite multiplication.

The importance of clones to horticulture and other aspects of agriculture can hardly be overemphasized. This is not only because of benefits but also because of the problems that must be controlled to make the process successful. The first part of this chapter describes the nature of clonal propagation and the sources of variation. The second part describes propagation systems for maximizing clonal advantages and for managing potential clonal problems by maintaining genetically uniform and pathogen-free source material.

HISTORY

Early domestication of vegetatively propagated species, such as, potato *(Solanum tuberosum),* yam (*Dioscorea* sp.), sweet potato *(Ipomoea batatus)* bamboos (various genera), sugar (*Saccharum* sp.), and banana (*Musa* sp.), was due to their natural reproduction through vegetative structures with which they reproduced (*49, 98*). Clones of grapes, olives, and figs also could be easily multiplied simply by inserting pieces of stems (cuttings) into the ground to produce roots and a new plant (*133*). Similarly, poplar and willow trees were readily propagated and came into use as hedges and fences to enclose fields and mark boundaries. Tree fruit plants, such as apple, pear, cherry, plum, (southwest

Asia), peaches, plums, apricots, (eastern Asia), and citrus fruits (southeast Asia), were less easily propagated by cuttings, and the maintenance of selected individuals was accomplished by the discovery (or invention) of grafting methods. In the tropics where humidity was high, layering methods were invented which reproduced clones of tropical fruit species (litchi, breadfruit, mango, durian, rambutan, longan) (see Chapter 15).

Great strides have been made during the present century to improve the technology of rooting cuttings (*24*) through utilization of structures (greenhouses, cold frames, hotbeds), environmental controls (misting, bottom heat), media development, hormone application, and other aids (Chapters 10, 11). The application to ornamental horticulture has produced the most spectacular advances, including cultivar selection, propagation of floricultural crops and houseplants (roses, carnations, chrysanthemums, poinsettia, foliage plants) (*8*), and landscaping materials (deciduous and evergreen shrubs, conifers, landscape trees) (*28, 70*).

Traditionally agronomy (*61*) and forestry (*91*) have relied upon seedling populations except in a few unique situations. For example, Chinese fir *(Cunninghamia lanceolata)* and sugi *(Cryptomeria japonica)* have been propagated vegetatively for 1,000 years in China and Japan respectively (*91*). During the past 25 years the concepts and practices of "clonal forestry" have become a major strategy for production of specific tree crops (*2, 64*), including eucalyptus in South America, North Africa, and Europe, Monterey pine in New Zealand, and poplars and willow in the U.S. and Europe. Likewise, the technology of producing seed by vegetatively multiplying embryos in culture ("synthetic seeds") has reached a level which promises to support direct use in propagating various agronomic and vegetable crops (see Chapters 17, 18).

REASONS FOR USING CLONAL CULTIVARS

The merits of vegetative clonal propagation include:

a) fixing superior genotypes
b) uniformity of populations
c) facilitating propagation, i.e., sometimes only means to propagate
d) shorten time to flower
e) combining more than one genotype into a single plant
f) controlling phases of development

Fixing Genotypes

Clonal selection and propagation is important in plant improvement because large genetic advances can be made in a single selection step (*1, 2, 77, 108*). Clonal selection is analogous to "fixing" genetic variation in a self-pollinated population or producing an F_1 population in hybrid seed production (Chapter 4). Most cultivars selected as clones are highly heterozygous and their uniformity and unique characteristics would immediately be lost in the next seed-propagated generation. Some important cultivars are seedless and vegetative propagation is necessary to reproduce them (banana, some fig cultivars, seedling grapes, persimmons, citrus).

Monocultures. Experience in growing clonal cultivars has revealed that there are liabilities in their use. A monoculture is the mass production of plants with a single genotype. Plants in a monoculture are vulnerable to (1) changing environmental hazards, (2) particular pests, where there may not be resistant germ plasm, (3) loss of genetic diversity *(2, 30, 132)*. Environmental hazards, pests, and disease problems of commercial monoculture crops are generally controllable but may add to production costs. Much effort has been expended to offset these problems in agriculture and horticulture during the past century by developing strategies of pest and disease control. Although dependent in many cases upon chemical controls, these systems now include an array of biological and integrated pest management tools that expand the options in crop protection (see Chapter 3). Emphasis is also placed on resistant cultivars and on expanding the range of genotypes available in a given crop. A major hazard is the co-evolution of specific clones and systemic pathogens, such as viruses. Special strategies for their control are described in this chapter.

There is a temptation for commercial producers to concentrate on the most productive clones and to discard others with less immediate value. To offset these concerns, efforts need to be expended on the preservation of genetic diversity through germ-plasm collections, arboreta, natural areas, seed storage, and other systems.

Uniformity of Populations

The uniformity of individual plants within a clone population is a major advantage of clonal cultivars in commercial production (Figure 9–1). Uniformity of plant size, growth rate, time of flowering, time of harvesting, type of product, and other phenotypic characteristics make economic industrial production of fruit and nut crops possible (*126*). This characteristic applies similarly to plants in the landscape and to florists' production from greenhouses. Uniformity and the elimination of inferior individuals in the population is the major effect on increasing yield. Economic studies of clonal forests of eucalyptus (South America, Europe, Africa), Monterey pine (New Zealand), and poplars, aspens, and willows (northern Europe and North America) showed that the major genetic gain in yield (which could be as much as 3×) was due to uniformity and elimination of the less productive individuals found in a seedling mixture (*2*).

Facilate Propagation

Some cultivars may be vegetatively propagated relatively simply, e.g., by hardwood cuttings, and consistent and economical nursery production when feasible (see Chapters 10 and 11). With forest trees, vegetative propagation has been used to multiply ("bulk up") valuable seedling populations produced by breeding (*91, 92*). Tree seed production may be unpredictable in some years, and methods of seed handling may be cumbersome, whereas vegetative propagation may be more reliable.

FIGURE 9–1 Uniformity in clonally propagated trees of *Eucalyptus.* Three year old plants were established from rooted cuttings. (Courtesy Dr. Bruce M. Zobel, Zobel Forestry Associates, Raleigh, N.C.)

As a general statement, vegetative propagation is more expensive per plant than seedling production of the same species. Consequently, the primary economic benefit is to produce plants with high individual value. Vegetative propagation is feasible for many horticultural crops because their value is based on the *individual plant.* In contrast, the value of most agronomic and vegetable crops, where seeds are the method of choice, is based upon the *population.* Likewise, forest tree production, particularly in natural stands, has been based upon populations of seedlings.

Shorten Time to Flower

Plant cultivars grown by vegetative propagation invariably come into flowering at an earlier age than comparable plants grown from seed. The reason is that plant cultivars grown as clones are in their mature phase of development, as discussed further in this chapter.

Combine More than One Genotype into a Single Plant

Grafting makes possible the combination of more than one genotype in the same plant. One can choose separate genotypes for the root system, the interstem system, and the fruiting part of the plant, as discussed at length in Chapter 13. For example, tree roses require a specific interstock for the main stem with the flowering cultivar grafted on top (Figure 12–5). A weeping cultivar can be grafted on top of a larger plant (Figure 12-6). Other examples are given in Chapter 13.

In other cases, different cultivars can be grafted to different branches of the same plant. This practice can achieve a range of effects, such as combining early to late ripening in a back-yard fruit tree or adding a pollinating cultivar as a single limb in a self-incompatible cultivar.

Control Phases of Development

In Chapter 2, the seedling life cycle was described as exhibiting phase changes which involved a juvenile, transitional, and mature (adult) phase. Important traits affected include age of flowering and seed production, morphology, physiology, and regeneration competence, particularly in ability to initiate roots. Vegetative propagation can be used to maintain, enhance, or reverse specific phases.

This phenomenon will be described further in this chapter.

GENETIC BASIS OF CLONES

Clonally propagated cultivars maintained and propagated by nurseries are the basis of many horticultural industries. Other sources of clones are arboreta, plant collections, and germ-plasm repositories. Some ancient clones have been grown in cultivation for centuries. Some have originated through plant explorations and introduction, such as individual plants in orchards, landscapes, or uncultivated areas near orchards, or from controlled breeding programs.

Cultivars have originated primarily through seedling populations or as mutations. The recent emergence of biotechnology provides new tools for producing clones. In each case, clonal multiplication is used to increase the number of individual plants of the same genotype. Each plant is represented by a clonal life cycle as was described in Chapter 2.

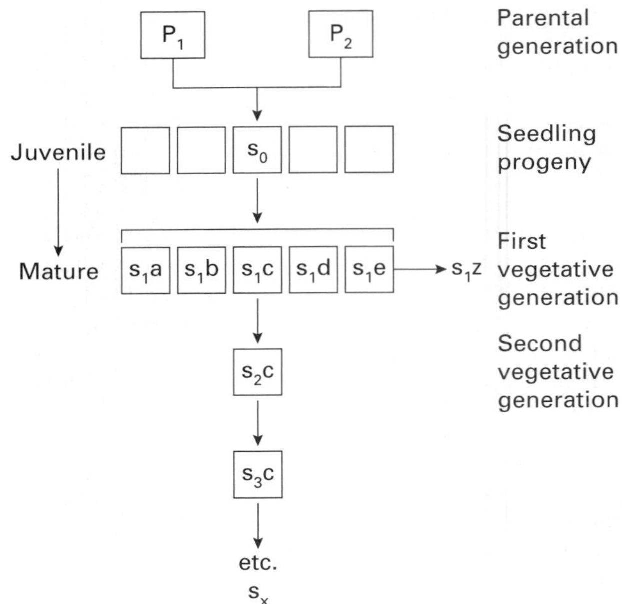

FIGURE 9–2 Model illustrating selection and subsequent cloning of a single plant from a seedling population. The original seedling plant is designated as S_0. The vegetative progeny population propagated from this plant is designated as S_1 (first generation from the seedling). Individuals of this population are designated by lowercase letters. S_{1a} S_{1b}, etc. Subsequent consecutive generations are designated as S_2, S_3 to S_x. Foresters refer to the S_0 as the ortet and subsequent generations as ramets. Two kinds of non-genetic variation are important to these early stages, (a). phase change and trueness-to-type phenotype, and (b). environment x genotype interactions. See text for explanation.

Origin of Clonal Cultivars

Seedling selection. Most cultivars grown as clones originate from a single plant in a seedling population. This original population may result from chance in which one or both of the parents are unknown or a product of a carefully planned breeding program (*1, 76, 77, 108*). The essential fact is that the superior qualities of the plant be recognized by an observant person.

Clonal multiplication from a single seedling plant takes place in two separate patterns: (a) multiple propagation from the same plant, and (b) consecutive generations of progeny from the same plant (Figure 9–2). It is convenient to refer to the seedling plant as the S_0 (originating as a seedling) generation. This plant exhibits a typical seedling life cycle as described in Chapter 2 (p. 27). The S_0 plant has been called the **ortet** and the vegetatively propagated generations as populations of **ramets.** Plants propagated vegetatively from different parts of the seedling plant can be designated as S_{1a}, S_{1b}, S_{1c}, etc.). Additional consecutive propagations from the same plant would be designated as S_2, S_3, etc.

These designations can be used to describe two specific kinds of variation that may affect the "trueness-to-type" of individual plants of the clone. One is **environmental.** The introduction of a seedling selection as a future cultivar is invariably preceded by a series of field tests in which plots of vegetatively propagated individuals are grown in different sites. One objective is to determine how much the *phenotypic* characteristics of the original seedling plant are due to a) its genetic traits and how much to the effects of b) the environment in which it was originally growing. A second objective is to determine how widely adapted plants of the cultivar are to individual sites in which they are grown. Traits that vary in different locations are referred to as **genotype × environment interactions.**

Historically, some fruit cultivars have been propagated for hundreds and sometimes thousands of years (*73, 76*). 'Cabernet Sauvignon' and 'Sultana' (now known as 'Thompson Seedless') have been grown horticulturally for about 2,000 years, as have some fig cultivars. The 'Bartlett' pear (also known as 'Williams Bon Chretien') originated as a seedling in England in 1770. 'Delicious' apple originated about 1870 in Jesse Hiatt's orchard near Peru, Iowa. 'Gros Michel' banana, which is seedless, is of ancient origin and has been the banana of commerce through much of the world.

More recently, woody ornamental shrub and tree clonal selections have become cultivars for landscaping purposes (*28, 70*). Many cultivars exist in arboreta, plant collections, and germ-plasm repositories. Some clones have originated through plant explorations and introduction. Others have been discovered as individual plants in landscapes, or uncultivated areas near orchards. Increasingly, new cultivars have been originated and introduced through controlled breeding programs.

Mutations. Chance mutations within established clones are the second major source of new cultivars. A mutation is a genetic change involving some part of the DNA molecule. Genetic mutations result from structural changes in the nuclear DNA of the chromosome in the nucleus. DNA also occurs in mitochondria and chloroplasts and defects here can produce genetic change. Chromosome changes may be due to chance rearrangement of the four bases in the DNA molecule **(point mutations)**, rearrangements of different parts of the chromosome **(deletions, duplications, translocations,** and **inversions)**, addition or subtraction of individual chromosomes **(aneuploidy)**, or the multiplication of entire set sets of chromosomes **(polyploidy)**. Changes in ploidy in grapes has resulted in giant, vigorous, low-producing "sports" (*31*). Defective plastids result in loss of chlorophyll, which may be expressed as albino or variegated plants with sectors of albino and green areas (*102, 103*).

Mutations occur as single-step genetic changes within cells of a clone. In general, these are spontaneous, taking place in most organisms at regular but rare intervals. The rate of mutation can be increased by treatment with specific mutagenic agents, e.g., X-rays, gamma rays, and certain chemicals (*15, 17, 77, 120*). These treatments provide the basis for mutation breeding.

No matter whether a mutation is by chance or induced, the sequence of discovery is the same. Figure 9–3 identifies a plant within a clone in which this change occurs as V_0 with succeeding vegetative generations as V_1, V_2, etc. The genetic change is permanent but the ability of a given mutation to produce a major change depends upon a number of conditions. *First,* the allele for the new mutant must be dominant in order for the trait to be immediately expressed (see Chapter 2). If recessive, the allele is not expressed and will only become significant after segregating in the next seedling generation, as described in Chapter 4. *Secondly,* the cell in which the mutant appears must not only survive but must divide sufficiently to occupy a significant part of the growing point of the shoot in which it appears. *Thirdly,* the trait must be sufficiently conspicuous such that the new trait can be visually identified in the plant. When such mutants meet these requirements and suddenly appear as a change in the branch of a plant, they are commonly referred to as **bud-sports** or **bud-mutations** because they appear to have originated within a single bud (Figure 9–4).

Detection of a new mutant within a clone may require a series of vegetative propagated generations and multiple propagations from many buds of the same plant. Pruning severely can accomplish

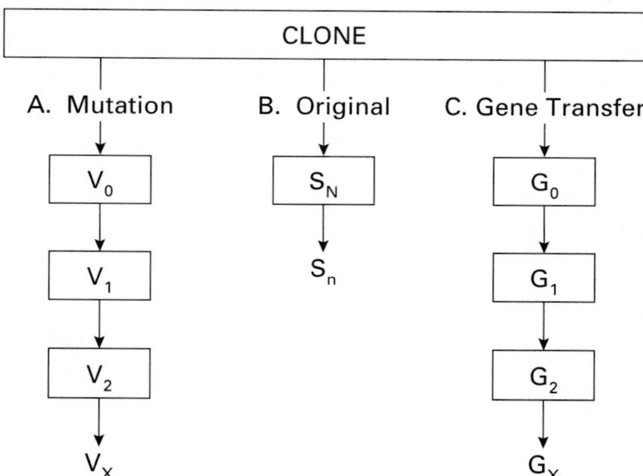

FIGURE 9–3 Model illustrating the effect of a genetic change within an individual plant on the clone. Under specific conditions (see text), a change can result in a new genotype which can be vegetatively propagated to start a new clone as a potential cultivar. The model shows two methods of how these changes occur. One is by mutation (a), either spontaneous or induced, which would initially produce a chimera (see text). Consecutive progeny of these plants are listed as V_0, V_1, etc., during which the chimera may be stabilized. Propagation sequence (b) remains the original clone. Phase change is not involved in this phenomenon because the plants are in a mature phase. The other method is (c) through genetic engineering techniques, including DNA recombination, which may also require stabilization. The generation sequence is designated here as G_0, G_1, etc.

FIGURE 9–4 Almond tree *(Prunus dulcis)* in which one branch shows conspicuously later bloom. This represents a "bud-sport" where a mutation has occurred in a gene that controls time of bloom.

the same purpose. Both increase the number of growing points that are available to show mutated sectors. In Figure 9–3, this sequence is shown as V_1, V_2, etc.

Many "sports" have become commercial cultivars. Mutations may affect fruit (color, shape, time of maturity), tree structure (spur-type), time of bloom, and a host of other traits. Sometimes these mutants are highly useful horticulturally and have given rise to important new cultivars ('Ruby Red' grapefruit, red-colored sports of apple and pear). On the other hand, mutations may be undesirable and give rise to misshapen fruit (*23, 32, 96*), low production, and susceptibility to disease.

Biotechnology

Modern science has produced a number of biotechnological strategies that can be used to alter the genotype of a clone which creates potential cultivars (*46, 86*). Cloning DNA fragments of one organism and transfering to another clone was described under the term **recombinant DNA technology** (Chapter 2). One of the potential horticultural advantages claimed for this technique is that a single trait from one organism can be inserted into an otherwise established cultivar that lacks that trait. For example, inserting a gene for increasing rooting ability of specific cultivars is possible (*95*). Figure 9–3 includes a genetically engineered sequence in which the target plant is designated as G_0 and subsequent vegetative generations as G_1, G_2, etc. These series of propagation have the same function as V_0, V_1, V_2, etc. following a mutation (same diagram) by isolating and stabilizing plants with the new genotype. Other biotechnology procedures for altering or identifying unique genotypes in clonal populations are described in Chapter 17.

Kinds of Genetic Variation within Clones

Genetic variation within a clone has three basic forms—**chimeras, transposons,** and **somaclones.**

Chimeras. When a mutation occurs within a single cell of a clone it initially produces an "island" of mutant cells within a growing point of a stem. If successful, the plant then becomes a mixture (or mosaic) of two different genotypes. This structural arrangement is known as a **chimera** and is the most important kind of genetic variant within clones typified by various kinds of **variegation** (see box) (Figure 9–5). The name *chimera* was historically given to certain unique clonal variants now known to have arisen as **graft chimeras** (see box).

Most spontaneous "bud-sports" are chimeras as are variations produced by artificially induced mutations (*62*). Chimeras develop because of the unique architecture of the apical meristem and the strategic location of the mutation in a dividing cell near the

FIGURE 9–5 Variegated pink and green 'Eureka' lemon, a type of chimera. (From Hartmann, et al., *Plant science,* Prentice-Hall, Englewood Cliffs, N.J., 1981.)

Graft chimeras. One of the early botanical mysteries was the nature of certain unusual plants that developed after certain graft combinations were made. In 1644, a strange plant called a "bizzaria" orange was produced following the grafting of a sour orange *(Citrus aurantium)* on a citron *(C. medica)* rootstock *(113, 115)*. Similarly, unique plants resulted when a medlar *(Mespilus germanica)* was grafted onto a hawthorn *(Crataegus monogyna)*, known as a hawmedlar *(9, 41)*, Laburnum, and Camellia. *(104)*. Early in this century a dispute raged as to whether or not these represented genetic hybrids produced by grafting. The question was settled by Winkler *(127, 128)* who was able to synthesize a true graft chimera of tomato *(Lycopersicon esculentum)* on black nightshade *(Solanum nigram)* and vice versa. In the procedure the scion of a young grafted plant was cut severely nearly to the callus area that arises during the healing of a wound (Figure 9–6). Adventitious shoots from this callus, turned out to be a chimera of tomato and nightshade tissue growing together and not a hybrid. Winkler gave such mixed shoots the name "chimera" after the mythological monster that was part lion and part dragon. These combinations are rare but a graft chimera of Camellia has been more recently discovered *(104)*.

apex of the apical meristem (*102, 115*), as will be described in a subsequent section.

The three important types of chimeras are based upon the distribution pattern of mutated and nonmutated cells (Figure 9–7):

1. **Periclinal:** The mutated tissue occupies a layer of cells that completely surrounds an inner core of nonmutated tissue. For example, in many red-colored fruit cultivars in apple, the red pigment is located only in the epidermal layer, whereas the cells of the inner tissue has alleles for green or yellow color. Similarly, some blackberry *(Rubus)* cultivars are thornless, because the cells making up the epidermis do not have this allele. Cells in the stem below the epidermis have the thorny allele but do not express it.

   Periclinal chimeras are relatively stable as long as reproduction includes a complete shoot apex. This type is not stable if propagated by any method that involves production of adventitious buds, as in root cuttings.
2. **Mericlinal:** This combination is similar to the periclinal except that the cells carrying the mutant gene occupy only a part of the outer cell layer. In the case of a red mutant on a yellow delicious fruit, the surface of the fruit may have longitudinal streaks or sectors of red on an otherwise yellow surface. This type is unstable and tends to change into a periclinal chimera, revert to the nonmutated form, or continue to produce mericlinal shoots.
3. **Sectorial:** The mutated cells in this combination occupy an entire sector of the stem including all layers of the shoot apex. Sectorial chimeras appear if the mutation occurs in roots and very early stages of embryos where the cells of the growing point do not occur in layers. In general, this type is unstable and tends to revert to mericlinal and periclinal chimeras as illustrated in Figure 9–7. Ferns and some conifers are exceptions as described later.

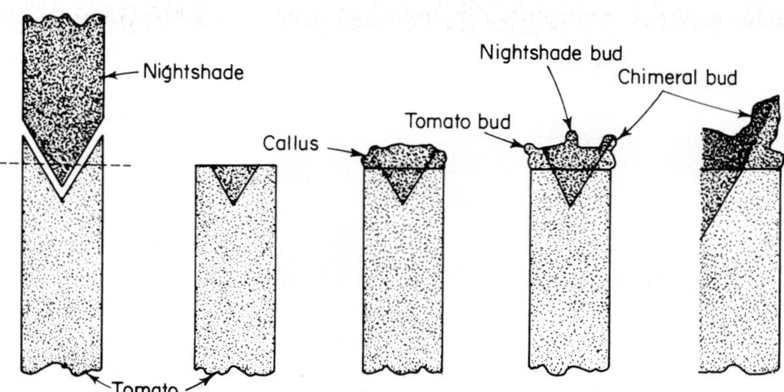

FIGURE 9–6 Method of producing a graft chimera between nightshade and tomato. (Adapted from W. N. Jones, Plant chimeras and graft hybrids. Courtesy Methuen & Co., London, Publishers.)

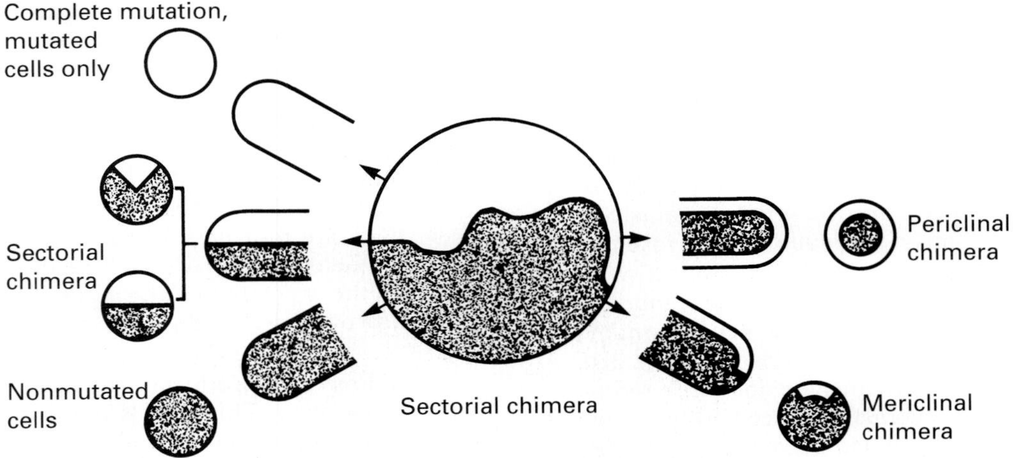

FIGURE 9–7 Buds arising at different positions on a sectorial chimera may produce shoots consisting of different patterns of mutated and nonmutated cells: entirely of mutated cells; entirely of nonmutated cells; or the shoots may be a sectorial (rarely), mericlinal, or periclinal chimera. (Adapted from W. N. Jones. *Plant Chimeras and Graft Hybrids.* Courtesy Methuen & Co., London, Publishers.)

Transposons (transposable elements) (81). This unique type of genetic variation is characteristic of specific plants. The variation typically involves uniform patterns of colored streaks and dots of different sizes and arrangements. Examples are the variegated seeds of Indian corn, color instabilities in flowers of some snapdragon and petunia cultivars as well as the variegation produced in 'Jingle Bells' poinsettia (*66, 67*). Studies of Dr. Barbara McClintock, who received the Nobel prize for this work (*69*), and others have shown these patterns to be caused by unique mobile genetic factors which have the capacity to move to different positions of the chromosome and turn genes on and off. They have been referred to as "jumping genes." These unique genotypes are inherited and the characteristics are transmitted to seedlings.

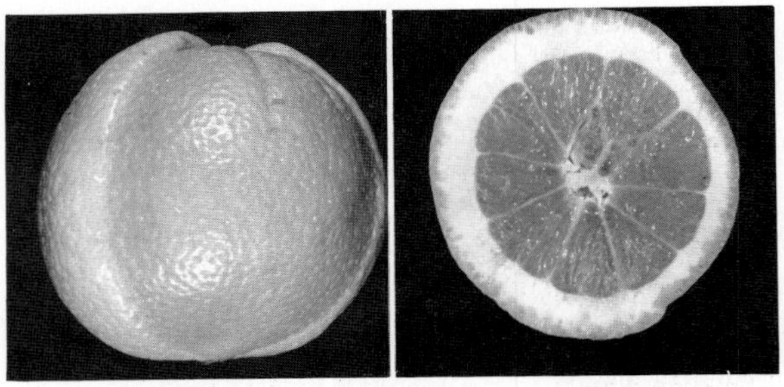

Somaclonal variation. This type of genetic variation within populations of clonal material was discovered when plant cells of certain species, for example, tomato, were grown in culture and new plants regenerated (*33*). Theoretically, all cells of these populations should have the same genotype. Instead, considerable variation may appear as illustrated by the two tobacco plants in Figure 9–9. Similarly, protoplasts of potato grown in culture regenerated plants with significant variation (*97*). These variations apparently occur in cells in the deeper areas of the growing point below the meristem, which often become polyploid (*21*). The variation is of value to plant breeders but could be a hazard in propagation by micropropagation.

FIGURE 9–8 Mericlinal chimera in 'Washington Navel' orange. Left portion of the fruit has developed from the mutation, producing a thicker rind that is yellow instead of orange in color. This example is an undesirable mutation.

Origin of Chimeras (20, 59, 66, 115)

A chimera originates with a mutation in a growing point of an individual plant within a clone (Figure 9–10). Specific conditions were previously described that must occur for a recognizable chimera to develop (p. 243). The apical meristem of plant shoots and roots consists of meristematic cells (capable of dividing) which in most flowering plants is arranged in three to four independent layers referred to as **histogens.** These overlay a "core" of inner cells to produce the arrangement known to

FIGURE 9–9 Somaclonal variation in tobacco. Plants originated by regeneration from single cells in culture. *Left:* Normal plant. *Right:* Variant plant (Courtesy T. Murashige.)

botanists as the **tunica-corpus meristem.** These layers are usually identified as L-I (outer), L-II (middle), L-III (inner), etc. The outer L-I cells divide primarily by **anticlinal** (perpendicular to surface) divisions and the daughter cells eventually become the epidermis or outer layers of the plant. L-II cells predominantly divide **anticlinally** but may undergo periclinal divisions (parallel to surface) particularly where their location is farther from the tip. Periclinal division can shift **(replace)** individual cells into the L-I layer (and vice versa). A mutation in a single cell of the L-I layer is shown. However, mutations can occur in the L-II and L-III layers.

As the growing point enlarges, various cells give rise to distinct parts of the larger stem (Figure 9–11). In general, the outer layers become the epidermal layers and, when mutant, will produce the periclinal chimera. L-II layers make up the cortex of the stem and some part of the vascular bundles. The reproductive structures are produced by this layer. This explains why a mutant in a periclinal chimera will not be seed-transmitted. The L-III and inner tissue makes up the pith and some parts of the vascular cylinder. However, there is considerable fluidity in which parts are derived from each layer.

Ferns and apparently gymnosperms differ in the origin of the chimera because they tend to have an unlayered growing point in which only a few main cells serve as the initial cells. In this case, a sectorial chimera will develop (Figure 9–10). The same pattern occurs in a root apex so that chimeras developing in roots are sectorial.

A mutation in the L-I or L-II of a structured or layered growing point normally results in a mericlinal chimera (Figure 9–10). For example, if the mutation produces red pigment on a yellow apple, it would initially appear as a red streak on the fruit but only skin-deep. The stem on which the fruit is produced should have corresponding longitudinal streaks of cells with the yellow or red gene. Lateral buds produced on the stem could produce one of three kinds of fruit: those that produce all yellow fruit (non-mutated), those that produce all red fruit (periclinal chimera), and those that produce streaked fruit (mericlinal chimera).

The patterns of chloroplast-caused variegation on leaves are produced in response to the same basic principle shown by fruits or stems. Typically, leaves are formed from three layers of cells, i.e., L-I, L-II, and L-III. The epidermal layer has no chloroplasts, so is colorless; the middle part of the leaf blade is L-II, and the central part including the midrib is derived from L-III. Thus a variegated green-albino chimera (GWG, for instance), has leaves with a white outer layer surrounded by a cen-

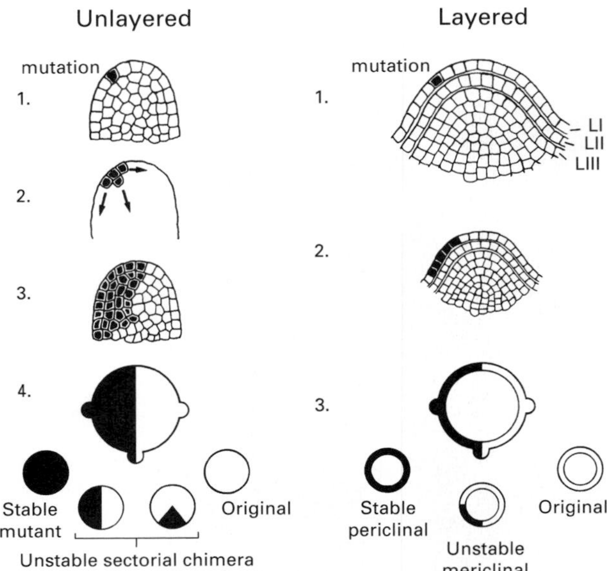

FIGURE 9–10 Origin of chimeras. *Right:* Periclinal and mericlinal chimeras develop because the mutant appears in separate layers in layered apical meristem. A mutation has occurred in a single cell of L-I layer which eventually develops into a chimera. *Left:* Sectorial chimeras develop when a mutation occurs in an unlayered apical meristem where no distinct layers occur. Unlayered growing points occur in most gymnosperms, ferns, and in roots or young embryos of other plants.

tral green. A GGG or WGG would be all green. A WGW would show a green outer layer with a central white sector.

Deviations from this basic pattern in different species can occur. The relative proportions of the leaf that originates from the separate layers are not constant and may vary in the pattern and the number of cells produced by each layer. Thus the pattern may differ and various shades of green may be present.

Modifications can result from "displacement" in which a cell of one layer, for instance, a cell of the outer G layer, can invade the tissue of the middle Y layer and suddenly produce an island of green. In Hosta, a monocot, a typical pattern is a white (W) margin (L-II) surrounding a broad green (G) center (L-III).

Genetic Variability and Stability in Propagation

Maintaining stability of chimeras. The value of many cultivars in horticulture depends upon their chimeral structure. Maintaining the chimera during propagation is sometimes a problem, depending upon the nature of the chimera and the method of propagation. Periclinal chimeras are relatively stable and maintain their integrity as long as the growing point of the propagule has continuity with the shoot system of the source plant, as true for stem cuttings, scions, layers, or division. Propagation that depends upon adventitious shoot formation, as in root cuttings, some kinds of leaf

Variegation. This general term is used to describe various types of color variation in leaves and flowers *(26, 66, 115)* (Figure 9–5). These may be combination of green-white, the white being due to defective plastids and absence of chlorophyll (albino). Other coloration patterns are yellow-green, red, purple, and various shades of each related to specific pigments such as anthocyanins and carotenes. Typically, variegation results from mericlinal variegation. Variegation can originate in seedlings, reproduce by breeding, and develop as chimeras. Chlorophyll-deficient mutants result from defective genes in the cytoplasm, as in some geranium species *(52).* Patterns of expression are related to the distribution of defective plastids in the cells of leaves.

Thornless cultivars of blackberry (*Rubus* species) exist in which the epidermis has a gene for thornlessness overlaying an inner core with the gene for thorns. Roots develop naturally from genetically "thorny" endogenous tissue. Plants grown as root cuttings or which develop as suckers on a plant show reversion to the thorny genotype. The reason is that new shoots are "suckers" which develop spontaneously from adventitious shoots originating from inner tissues. Similarly, breeding of thronlessness is difficult because the gametes originate from L-II tissue. Recovery of a completely nonchimeral thornless blackberry has been produced in tissue culture by regenerating new plants from cells derived from the outer layer (*44*).

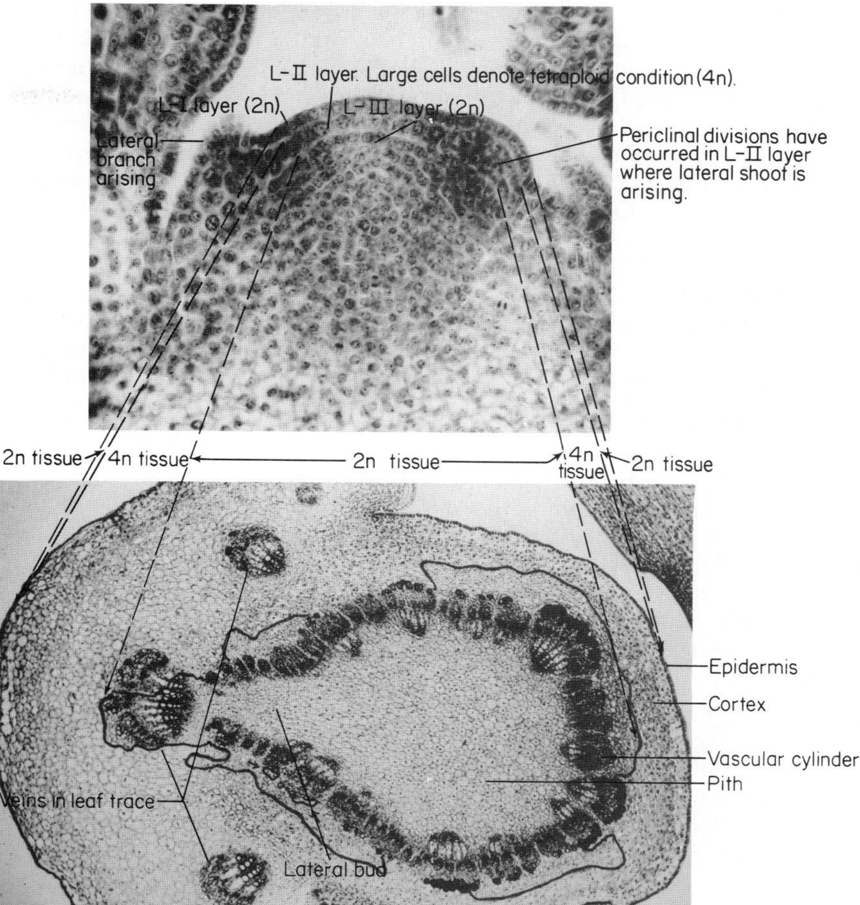

FIGURE 9–11 Relationship between cells of the histogenic layers of the apical meristem of a peach and cells of the stem that develops from them. L-I is diploid and produces the epidermis of the stem. L-II is identified in this example because the cells are tetraploid (4x) and larger in size. Progeny from this layer makes up the cortex and some of the vascular bundles. The L-III layer is diploid and becomes the inner part of the stem, including the pith most of the vascular cylinder. From Dermen (26).

cuttings, and some types of tissue culture will result in a reversion to the genotype of the inner (L-II and/or L-III) tissues. Figure 9–12 shows an albino plantlet developing from a leaf cutting of African violet in culture. It evidently arose because the adventitious shoot originated from a single cell within the inner tissue of the stem which was albino. This represents a method for developing a solid mutant plant from a chimera.

Reversions of periclinal chimeras may also occur by the displacement of a cell between apical cell layers, for example, cells of L-II into L-I. The visual result would be that individual shoots or branches of reverted phenotype appear randomly. This problem is acute in particular color "sports" of apple and pear. Or sometimes one can observe a completely solid green shoot in an otherwise variegated plant.

FIGURE 9–12 Albino plantlet of African violet originated during micropropagation. Origin is likely to have been adventitious from a single cell: hence a solid mutant resulted.

Mericlinal and sectorial (except ferns, some conifers) chimeras of plants tend to be unstable, resulting in branches which may be completely periclinal (solid mutant), completely nonmutant, and mericlinal (mixtures). Continuous selection toward the mutant type during consecutive propagations would be necessary to "fix" the mutated (or nonmutated) form, whichever is desired.

Micropropagation (see Chapters 17, 18) of chimeras follows the same rule. Shoot tip and axillary shoot cultures (p. 564) can reproduce the chimera. Adventitious shoot initiation, as used with some explants, or by accident in others result in "chimeral breakdown."

Detection of undesirable mutations. Spontaneous mutations occur at regular but very low rates in vegetatively propagated material. Nevertheless, where propagation of commercial cultivars involves millions of plants, the probability for a mutation to occur somewhere in the clone is high. The possibility also exists that exposure to specific agents in the environment could increase this risk.

Our knowledge of chimeras indicates that a mutation is initially latent, survives as an unstable chimera, and is usually not detected until after the mutated cells occupy a sufficiently large area of the stem to be observed. Consequently the propagator will usually not observe the mutation in the plant in which it actually occurs. Identification first comes in the next generation of vegetative propagation, often after the customer receives the plant.

One of the protocols of mutation breeders is to prune treated plants (M_0) to force as many lateral shoots as possible to grow continuing this treatment through additional M_1 and M_2 generations (*62*). This procedure occurs with mass propagation by budding of fruit and nut tree cultivars. Thus, although the probability of any one mutation to occur may be low, the probability of finding one is high. This explains why mutants are usually detected first in a commercial orchard rather than in the orchard where the propagation material was obtained. If the character affects flowering or fruiting, the mutant may be undetected for a number of years after the plant is in its permanent location.

NONGENETIC VARIATION WITHIN CLONES

The prior section described genetic changes within plants of the clone which lead to permanent changes in the plants. These changes can be beneficial and may lead to important new cultivars. It is just as likely, however, that these changes may be undesirable and lead to the propagation of inferior or "off-type" plants.

In this section other kinds of variation among plants of a clone will be described that do not lead to permanent change but can be manipulated often to the advantage of the propagator. This is done through the management and use of particular stock plants.

Environmental Effects Upon Phenotypic Variation

By definition, all plant members of a clone have the same genotype. Consequently, plant-to-plant variation should be environmentally related. In fact, horticulturists often note that when plants of a single clone are grown in different sites and under separate management conditions, the phenotypic expression of some traits may vary. Productivity may depend upon location and site. Plants may grow more vigorously under some management conditions than others. Stresses (drought, cold, heat, insects, diseases, mineral deficiencies) may change the appearance or performance. Some cultivars are relatively stable in a range of environments. Others tend to vary depending upon location. The latter characteristics are referred to as **genotype × environmental interactions** (see p. 242).

The initial selection of the S_0 plant described on p. 242 was based upon **phenotype** expressed as the interaction of **genotype × environment**. To determine the environmental influence on the appearance and performance of a clone, populations of plants need to be propagated and grown at different environmental sites. This was described as part of the test of a new cultivar. The purpose was to ascertain, first, if the desirable characteristics of the original plant are reproduced following vegetative propagation and, second, if the phenotypic characteristics of the clone remain the same in different environments.

This sequence is an example of phenotypic and genotypic selection. The original selection was based on **phenotypic selection,** i.e., *visual inspection of the source tree.* Actual selection is based on **genotypic selection,** i.e., the *visual inspection and performance of the progeny plants.* This provides a true test of the genetic potential of a particular source and can be a useful tool in the selection of specific sources for propagation.

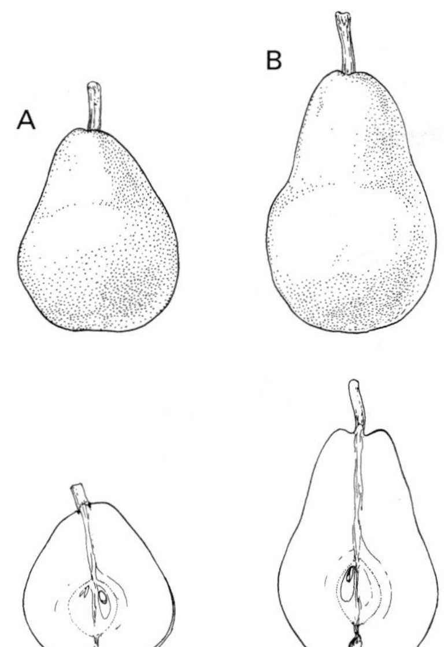

FIGURE 9–13 *Top.* "Bartlett" pear fruit from different sources (A, B) at same site suggest genetic differences. *Below left.* Typical pear fruit from California. *Below right.* Typical pear fruit from Pacific northwest. Based on (118).

Some kinds of environmental variation are subtle. 'Bartlett' pear fruits produced in Washington and Oregon tend to be longer and more "pear-shaped" than those produced in California, apparently due to climatic differences (*118*) (Figure 9–13). Fruit size, shape, and appearance may differ depending on the vigor of the plant, the size of the crop, and the age of the plant (*55*). Continued exposure to inadequate winter chilling in strawberries (*14*) or an unfavorable postbloom vegetative period can lead to deterioration of plants of many bulb species in a warm winter environment (see Chapter 15). Some plants produce leaves of a different appearance in the shade than in the sun.

A propagator needs to be familiar with the plants grown not only when propagated but also when grown in their ultimate site. In some cases, a variant may appear where it is not possible to make a judgment by visual inspection alone. In such a case, a vegetative progeny test might be made.

Phase Variation and Change in Maturation

Phase variation is an important type of nongenetic variation. Its application can have important benefits to the propagator, depending upon species and cultivar. Chapter 2 described the three kinds of life cycles of individual plants, the *seedling cycle,* the *clonal cycle,* and the *apomictic cycle.* In Figure 9–2, the S_0 generation is the only plant of the clone that has a seedling (or apomictic) life cycle as all others are vegetatively propagated. This seedling plant exhibits phases of *juvenile, transitional,* and *mature (adult)* to various degrees in different parts of the plant in a gradient from base to top (Figure 9–14). These juve-

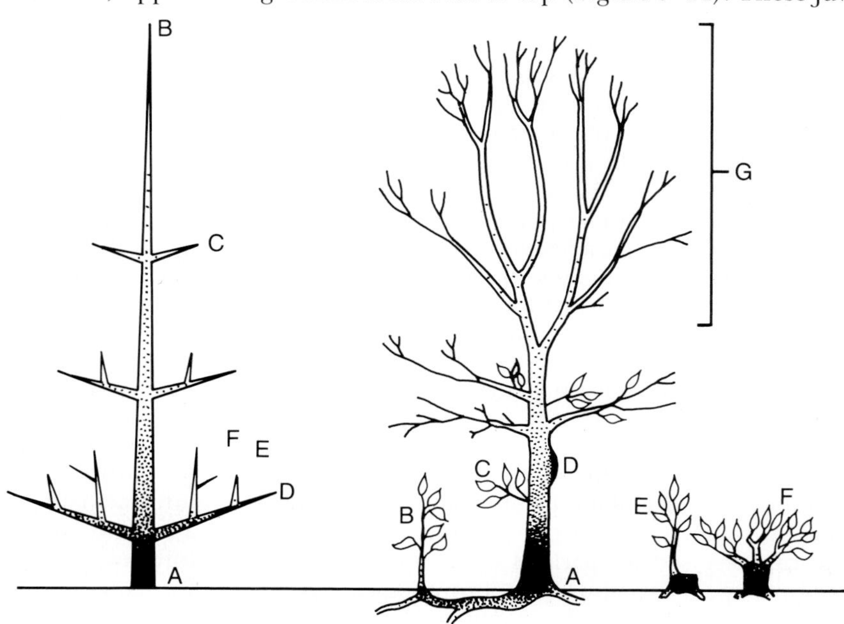

FIGURE 9–14 Juvenile-mature gradients in seedling trees represent a "cone of juvenility" near the base of the tree. *Left:* Seedling conifer. Juvenile root-shoot junction (A); mature apex (B) of the tree flowers first. Other letters refer to shoots at different locations. Gradient in juvenility in A > B > E > D > C > B results from different ontogenetic age. *Right:* Seedling deciduous tree. Juvenile root-shoot junction (A); mature flower part of tree is in apical part (G). Juvenile structures arising from "cone of juvenility" near base of tree include adventitious root "sucker" (B), watersprout (epicormic) (C), and sphaeroblast (D). A stump sprout from pruning (E) and a hedged tree (F) are also shown. (Redrawn with permission from Bonga (*11*).)

nile-mature variations are expressed in time of flowering, trueness-to-type vegetative characteristics, and rooting ability (see p. 28).

Plants differ in the pattern of age-related characteristics. Plant species where seedling development occurs with little obvious change in phenotype with age are called **homoblastic.** Those with distinct variation in specific traits which occur with age or in different parts of the plants, whether abrupt or gradual, are called **heteroblastic.**

Maturation can be considered to be biologically equivalent in animals, humans, and plants in that embryonic, juvenile, adolescent, and adult phases are present in each. The difference is in the way animals and plants grow and age. Cells in animal (and human) development are more or less determinate, i.e., cells in all parts of the body mature and age more or less together. Plants, on the other hand, grow from consecutive cell divisions in the meristem of apical and lateral shoots laid down in sequence over time. Juvenile to mature change occurs in the apical meristem as the shoot grows. This process is an epigenetic change and continues either in alternating cycles of growth and dormancy (perennials) or terminates in reproductive structures and the plant dies (annual, biennials).

"Age" has a special meaning in plant development. It is best designated by the number of nodes of development that have occurred in the apical meristem. Lateral growing points laid down at the same node at the same time retain the maturation level at which they originated (Figure 9–15). When lateral buds grow into shoots, their own independent growth pattern is maintained and the apical meristem of these shoots changes as the juvenile-to-mature shift follows its own unique rate and pattern. Eventually, however, all parts of the seedling plant attain the mature (adult) phase and produce flowers. The result is a paradox in terminology in that the part of the plant nearest the base is "oldest" in terms of chronological age but actually "youngest" (more juvenile) in terms of maturity (ontogenetic) age. Likewise, the outer periphery of the stems and branches is the "oldest" in maturity but "youngest" in chronological age (*10, 56*).

The stage of maturation in different locations of the plant is determined by the pattern of bud development in the plant. Juvenile and mature characteristics, such as flowering, leaf shape, or thorniness, can be used as markers of maturation, their location depending upon the chronological age at which ontogenetic maturity is attained in particular parts of the plant (Figure 9–16).

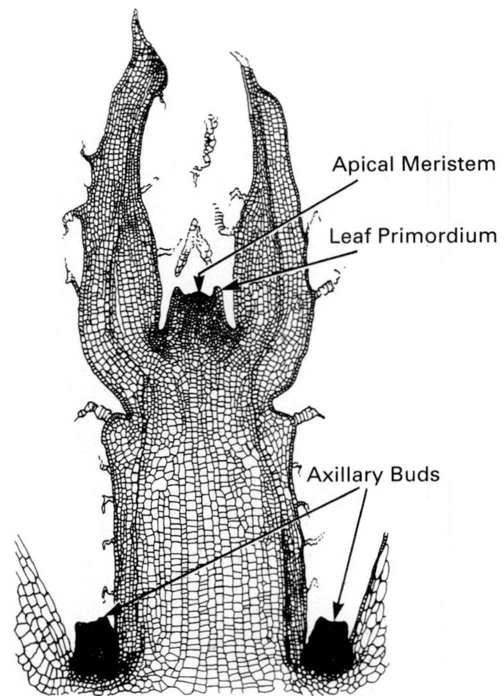

FIGURE 9–15 Apical buds on a seedling plant undergo shifts in their "ontogenetic age" as cells in their apical meristem divide and lay down new tissue. Lateral buds (axillary buds) typically are dormant due to apical dominance and retain the ontogenetic age at origin. Once lateral buds start to grow they then undergo change in maturation in the same manner as the apical bud, but in their own pattern.

Phase Variation and Vegetative Propagation

Lateral buds removed for vegetative propagation perpetuate their juvenile/maturity phase and the shift from juvenile to mature continues in subsequent growth. Consequently, vegetative propagation from mature branches of a seedling plant may not reproduce the seedling phenotype in the S_1 generation (Figure 9–2). Propagules taken from the base of the plant tend to produce biologically juvenile plants; those taken from the top of the plant tend to produce plants that are biologically mature; the entire range may be produced if plants from intermediate locations are included (see Chapter 2). The recognition of this phenomenon by foresters led to the coining of the terms **ortet** (the S_0 plant) and **ramet** (S_1 plants) (*109*). Sometimes the phenological characteristics are not the same in the ramets as the ortet, such as in earlier bloom date

FIGURE 9–16 Foliage variation gradients within a seedling tree can be markers of juvenile and mature phenotype. *Top left:* Eucalyptus seedling tree shows the large, broad, sessile (no petiole) leaves at the base of the tree; mature leaves from the upper part of the tree are narrow. *Top right:* Gradient in leaf shape is shown by a single shoot. *Lower left: Acacia melanozylon.* In this species the bipinnate leaves at the base are characteristic of the juvenile phase. The leaves at the top of the seedling are expanded petioles (phyllodia) and are characteristic of the adult form. *Lower right:* Needles of the basal juvenile part of many conifers tend to be acicular (sharp pointed), whereas needles on the upper, more mature part tend to be scalelike.

in walnut (*71*) or in a loss of rooting potential (*79*). These variations are not due to genetic changes but to *epigenetic variation* in maturation that persists during vegetative propagation (p. 19).

Juvenile Versus Adult Phenotypes

Consecutive generations of vegetative propagation leads to the stabilization (i.e., "fixing") of the mature phase in individual plants of the clone and the disappearance of seedling phase characteristics. Most clonal cultivars grown for their fruits, seeds, or flowers exhibit the mature growth phase (clonal cycle) as the typical "true-to-type" form. Individual plants come into flower and fruit reasonably quickly after a vegetative phase of development and undesirable traits associated with plants in the juvenile phase disappear. It is not appropriate to refer to the vegetative cycle of the stabilized mature clone as "juvenile."

a) Cultivars of fruit and nut crops, such as apples, pears, citrus, and walnuts, are grafted not only for genetic uniformity but also to reproduce the earlier flowering, more desirable mature form. Similarly, a flowering ornamental plant grown from seed may require many years to flower. For example, seedling Wisteria and Magnolia may take 7 and 15 years respectively (*70*). Early seed production is important for seed orchard production of specific genetic sources in forestry (*91, 132*).
b) Orchids, bulb species, and other herbaceous perennials require five to ten years of annual development from seed before flowering occurs. Once the mature (adult) stage is reached, annual flowering will occur with alternate phases of vegetative growth (bulbing) and reproduction (flowering) annually (see Chapter 16).

Invigoration versus reversion. As individual plants age chronologically, whether seedling or clonal, growth inevitably slows down, plants become primarily reproductive, and growth declines. In most cases, this type of "aging" is normal, physiological in nature, and a response to stresses relating to the complexity of the growth system, overproduction, weakened roots, or moisture and nutrient stress. This condition can normally be overcome and the plant invigorated by horticultural practices such as pruning, crop reduction, and fertilization, or by repropagation.

In some species, the juvenile (seedling) phase may be the preferred "true-to-type" phenotype. Many examples of the importance of the juvenile phase in cultivation occur in horticulture as well as in forestry.

a) Some ornamental vines, e.g., *Hedera* sp. or strangling fig (*Ficus* sp.), are primarily used in the juvenile phase, whereas the mature flowering phase is treelike (Figure 2–18). Both may be useful but for different purposes. Some ornamental conifer cultivars have juvenile foliage (needle or awl-like) distinctly different from the adult foliage (flat, rounded, or scalelike) (*70*). For instance, some (*Chamaecyperis pisifera* 'Boulevard' and *Cryptomeria japonica* 'Elegans') are more or less "fixed." Others (*Pinus sylvestris, Cryptomeria japonica,* junipers, and Lawson cypress (*Chamaecyperis lawsonia*) may develop both types of foliage.
b) The juvenile phase of most plants inherently has a higher rooting potential than the mature form. The easy-to-root juvenile vine form of Ivy (*Hedera*) has become a model system for studying root initiation (*42, 43*). The juvenile phases of the conifers listed in the previous paragraph invariably root easily and can be maintained as cultivars. The rooting potential of many if not most tree species is correlated closely with seedling age and the attainment of the adult form. One of the main reasons cultivars of fruit and nut crops are propagated by grafting is the difficulty in rooting. For clonal rootstocks, selection and maintenance of the juvenile phase are important strategies for propagation in order to improve rooting potential.
c) The uniform "true-to-type" juvenile apomictic seedlings of citrus are particularly desirable rootstock traits (*16, 38, 117*).
d) "Clonal forestry" depends on the ability to mass-propagate populations of juvenile propagules that not only exhibit easy rooting but also produce the desirable juvenile "seedling" phenotype. Figure 9–17 shows the

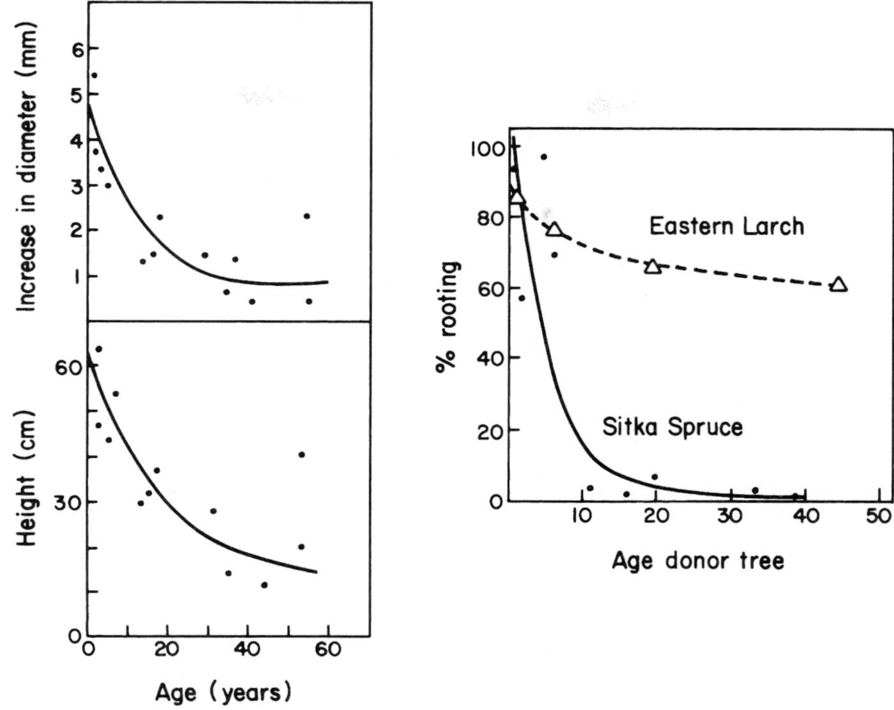

FIGURE 9–17 Graph showing the effect of age on donor tree relating to key tree characteristics that are important in clonal forestry. *Left:* Phenotypic trueness-to-type is illustrated by size of stem diameter and shoot length of grafts after multiple years. *Right:* rooting percentages of cuttings from donor trees of different ages. (From Greenwood, et al., by permission in *Clonal Forestry II,* 1993.)

relationship between rooting percentages and tree age (*2*). The desirable "true-to-type" form of forest trees grown for lumber and wood is that of the seedling or juvenile form with its straight tapered trunk, reduced branching, lateness to flower, and other characteristics. Vegetatively propagated plants of radiata pine, loblolly pine, giant sequoia, and eastern larch taken from more mature locations in the tree showed reduced height and diameter, and more branching (*40*) (Figure 9–16). Success in "clonal forestry" depends upon the ability to control maturation and to select the proper "true-to-type" stage (*2, 91*).

Promoting Shifts from Juvenile to Mature Phases

The long period of time required for seedling plants of many woody species to come into flower (Table 2–2) has important applications in horticulture. Breeding programs of woody plants primarily use seedlings but progress can be hampered by their long juvenile period (*47*). Breeding can result in shorter breeding and production cycles, since the juvenile period is controlled by genes (*123*). Growth and development can also be managed by environmental control. The most important concept is to keep the plant in *continuous growth* in order to *grow through* the juvenile phases as quickly as possible. For example, birch seedlings grown continuously under long days in the greenhouse during winter passed through the juvenile into the adult phase much faster than under normal short-day winter conditions (*93*). Crab apple (*Malus hupehensis*) seedlings grown continuously in a greenhouse to a certain height (2.5 to 3.0 m) started to flower in 13 months from germination instead of four years (*131*).

Restricting growth, girdling, bending, growing on dwarfing rootstocks, and other horticultural practices is often cited to induce flowering (*126*).

However, these are not useful until the plant approaches the transitional or mature phase. Prior to that phase, growth should be encouraged to allow the plant to grow through the juvenile stage.

Selection and Maintenance of the Juvenile Phase

Retention of the juvenile phase may be important, for instance, in increasing rooting potential. The following are methods to select, delay, or preserve propagation sources in the juvenile phase:

1. Collect material and propagate from plant parts showing juvenile characteristics. This may mean collecting from the base of a plant in the case of a seedling. Some cultivars are more or less "fixed" as juvenile, e.g., various conifer cultivars with juvenile foliage.
2. Collect material from root sprouts (suckers) or from watersprouts (epicormic shoots) arising from near the base of the plant. These arise from latent lateral buds which are located in the "cone of juvenility." The shoots may be forced by cutting back the plant to the ground to produce "stump sprouts" (*35*).
3. Some plant species produce **sphaeroblasts,** which are masses of miniature shoots from adventitious buds. Examples include sequoia, redwood (*12*), some quince, and some apple cultivars (*110*). Shoots developing from these structures invariably are juvenile, show characteristic morphology, and are easy to root.
4. Establish "hedge rows" of stock plants produced by propagating juvenile material (see Chapter 10, 11). This is perhaps the most important method to provide large populations of juvenile cutting material (*65*). Plants are severely pruned annually to a constant height to produce a continuous supply of new shoots. The original material may be seedling populations (*92*) or they may be rooted cuttings established from very juvenile material.
5. Juvenile material can also be maintained by consecutive ("serial") propagations of several generations from seedling material (*2*). Serial propagation is practiced in the annual propagation of new generations of cuttings—collected from asexually propagated container-produced woody ornamentals.
6. Establish "stoolshoot" beds by cutting shoots nearly to the ground (*70*). This is perhaps one of the reasons for success in rooting in mound and trench layering (see Chapter 15).

Reversions from Mature to Juvenile Phase (Rejuvenation)

Reversions from the mature to the juvenile phase takes place during seed reproduction whether sexual or apomictic (*57*). One of the objectives of research has been to achieve reversion from the mature to the juvenile phase in vegetative material. Examples include the following:

1. Adventitious shoots produced on roots (*42, 56*) or sphaeroblasts (*110*) tend to be juvenile. This method could produce material that could then be maintained under maintenance conditions, such as hedging.
2. Consecutive generations of grafting to seedling rootstocks (*12, 35*) has produced reversion from mature to juvenile phases in some plants (Figure 9–18). The most extensive work has been done with sequoia and with eucalyptus (*99*).
3. Consecutive subculturing of apical meristems in micropropagation has resulted in the reversion of plant shoots to the juvenile phase (see Chapter 17). Mullins (*78*) found that cultured somatic embryos from the ancient clone 'Sauvignon Blanc' produced seedlings with typical juvenile foliage. Similarly, micropropagation of cultivars over a period of many recultures has resulted in increased rooting in vitro as well as from tissue and culture-produced plantlets established as stock plants (see Chapter 17).
4. Some hormone treatments have resulted in reversions under experimental conditions. For example, applications of gibberellic acid have induced juvenile growth in ivy (*Hedera*) which could be reversed with abscisic acid (*42*). The effect is not universal, since GA can also speed up maturation in conifers.

Topophysis

Topophysis is defined as the effect of the position of the propagule on the plant upon the type of vegetative growth subsequently shown by the vegetative progeny (*74, 85*). Plants of certain species produced by cuttings taken from upright shoots **(orthotopic)** will produce plants in which the shoots grow vertically. Plants produced from cuttings taken from lateral shoots **(plagiotropic)** will

FIGURE 9–18 Rejuvenation produced by consecutive (serial) grafting from a ten-year-old mature Eucalyptus tree. *Left:* Plant produced by cleft graft of scions from the ten-year-old tree onto a seedling rootstock showed mature leaves. *Middle:* Plant produced from scions after three serial grafts. One of the shoots shows mature characteristics, the other juvenile. *Right:* Plant produced after grafting with scions after six serial grafts. All of the shoots produced juvenile (large) leaves. Cuttings from this plant rooted successfully, whereas rooting from the ten-year-old original mature tree was not successful. (From Siniscalco and Pavolettoni (99) with permission.)

grow horizontally. Figure 9–19 illustrates this phenomenon in coffee. The phenomenon can occur in many conifer species primarily as a function of the adult form (*2*). Horizontal and prostrate forms may be obtained by taking cuttings from horizontally growing shoots and may be desirable in growing specific ornamentals but is undesirable in forestry. Norfolk Island pine *(Araucaria)* is an example of topophysis which produces flat growing plants in pot culture. In some species, such as *Sequoia*, pruning can restore the orthotropic growth habit (*12*).

FIGURE 9–19 *Left:* Coffee plant propagated from a vertically oriented cutting maintains an upright growth pattern (orthotropic). *Right:* Coffee plant propagated from a lateral shoot produces a horizontal growth habit (plagiotropic).

Variable Gene Expression

Patterned genes. Many genotypes produce variation because of the variable expression of individual genes which produce specific patterns in both time and location within the plant *(66, 67, 121)*. This typical **pattern** is displayed throughout the clone and expressed by each plant. Some color variations result from patterns produced by single genes, for example, some variegated coleus cultivars and bicolored petunias. This type of variegation is inherited in seedlings, which is a general test for nonchimeral variation.

Inherited disorders. Certain fruit and nut species have specific disorders that produce unique source selection problems when vegetatively propagated. *Noninfectious bud-failure* is an inherited disorder in almond *(Prunus dulcis)* in which repeated propagation at elevated summer temperatures results in a condition known as "crazy-top" (*55, 58*). *Crinkle* and *deep-suture* in cherry *(Prunus avium)* (*80, 119*) and *yellows* in strawberry are similar problems. Control is primarily through selection of sources with low potential for the problem which are identified by vegetative progeny tests (see later in chapter).

> **"Ornamental viruses."** The striking variability produced by some plant cultivars are due to specific viruses. This is the case of the so-called "color breaks" in tulip flowers. Also, the unique variegation in certain *Abutilon* species is caused by *abutilon mosaic (66).*

PATHOGENS AND PLANT PROPAGATION

The presence of disease can be a limiting factor in both plant production and plant propagation. In this section, the unique relationships between specific plant pathogens and vegetative propagation will be examined. In nature, plants have coexisted with many other organisms, including insects, mites, fungi, bacteria, and nematodes, to evolve more or less compatible interdependent communities. In some cases, such as mycorrhizal fungi, plants and microorganisms form mutually beneficial associations (*88, 105, 106, 122*). When humans began to cultivate plants the balances changed. Humankind began to propagate specific kinds of crop plants and adopt monocultural systems, making them more vulnerable to attack by specific parasites. As the civilized world expanded, individual plant species were transported to new areas of the world (see Chapter 1) where they were introduced to new environments and different pathogens and parasites. Pests and pathogens existing only in one area "hitchhiked" along with plant introductions to new areas and became established there. A classic example of this problem was the introduction of the "root louse" phylloxera from the U.S. to Europe in infested grapevines. The high susceptibility of European grape cultivars, which were grown largely as rooted cuttings, devastated the European wine industry. The industry was saved by the subsequent introduction of resistant grape rootstocks from America to which the European cultivars could be grafted. Another example was the introduction of infected potato tubers to Ireland where local cultivars succumbed to *Phytophthora* fungus (*72, 94*).

Plant Pathogen Complexes

Two major developments occurred during the middle 1800s which, among others, have dominated plant culture ever since. One was the demonstration that plant diseases were caused by pathogens, as Louis Pasteur and others had shown earlier with human diseases. Since then, generations of plant pathologists have identified the *causal agents*, determined their *etiology*, and devised *control strategies* for numerous diseases that caused short-term disasters and long-term declines.

A second major discovery was that plant diseases and pathogens could be controlled by *chemical methods*, first demonstrated by the application of the famous Bordeaux Mixture (copper sulfate, and lime) which controlled downy mildew on grapes (*72, 94*). Since that time, the "chemical revolution" has extended across the agricultural spectrum. Concerns about the environmental impact of chemicals as well as the destabilization of the natural ecosystem of plant and pathogen have led to additional concepts of Integrated Pest Management (IPM) and alternate methods of control described in Chapter 3.

Systemic Pathogens and Clones

Propagation is a vulnerable phase of plant production, since an environment of high moisture and warm temperatures is also optimal for pathogens. Some propagation practices were designed to suppress disease rather than controlling the causes of disease.

Vegetatively propagated clones were associated with many production problems. Horticulturists and orchardists of the 1600s and 1700s (*60, 74, 116*) believed that fruit cultivars tended to "run out," caused by aging and repeated propagation; hence cultivars needed to be replaced periodically with new seedling-produced scions (*60*). Some decline problems were recognized as being induced by diseases and a concept of viruses emerged. Adolf Mayer in 1886 demonstrated the transmission of *tobacco mosaic* with juice from a diseased plant to a healthy plant. Erwin Smith in 1888, a USDA scientist, showed that *peach yellows*, a disease that was devastating peach orchards of the eastern U.S., could be transmitted from a diseased to a healthy plant by budding. Since that time, plant virologists have demonstrated that a number of very small, transmissible organisms were the cause of numerous plant disorders and the primary reason for clonal deterioration (*116*).

a. Viruses. These small submicroscopic organisms are composed of nucleic acids (RNA or DNA) and encased in an outer sheath of a protein (*63*). Viruses replicate within plant cells or in the bodies of insect vectors as obligate parasites, multiplying and growing in collaboration with chromosomes. Viruses are very small but can be observed with the aid of an electron microscope. The etiology of many plant virus diseases is now known and control procedures are a major aspect of propagation systems.

Cultivars and species differ in their tolerance to viruses (*20, 80, 119*). The effect may be sufficiently severe to kill the plant or make it stunted and unproductive. Others affect fruit and flower quality or produce symptoms of chlorosis, albinism, mosaic, and/or variegation patterns. Some are essentially symptomless and have little obvious effect on plant appearance, production, or growth (*72*). Opinions differ as to the seriousness of such latent viruses (*20, 45*). However, if tolerant themselves, infected plants are reservoirs of infection to other potentially susceptible cultivars.

b. Phytoplasma (mycoplasma-like organisms). Phytoplasmas are extremely small one-celled parasitic organisms. They live in the phloem of susceptible plants and in specific phloem-feeding leafhoppers (*Homoptera*). These organisms multiply rapidly, utilizing host cells for their development and survival. They disrupt translocation in the phloem, causing wilting and yellowing. Hormonal disturbances can occur that result in proliferation of shoots (commonly named "witches brooms") and other malformations, such as conversion of flower parts to leafy structures.

Plant diseases caused by phytoplasma include *aster yellows* in strawberry, some vegetables, and ornamentals, blueberry *stunt disease,* and *X-disease* in stone fruit trees. Differential susceptibility to phytoplasmas, i.e., between scion cultivar and rootstock, causes *pear decline.* Phytoplasmas move among plants by insect vectors, usually leafhoppers or psylla, and are carried in propagating materials. A similar organism, known as *Spiroplasma citri* Saglio et al., causes *stubborn disease* in citrus.

c. Fastidious bacteria. These single-celled bacterialike organisms do not have a nucleus, are difficult to culture, and occur in the xylem and phloem of susceptible plants and in the bodies of leafhopper vectors (*34*). They are carried in vegetative propagating material. *Club leaf* in clover and *greening* in citrus are examples of phloem-limited fastidious bacteria. *Pierce's disease* in grape, *almond leaf scorch, phony peach, plum leaf scald,* and *leaf scorch* of various trees are caused by xylem-limited fastidious bacteria. Symptoms include leaf necrosis, wilting, small fruit, and decline. Control can be achieved by removing diseased orchard trees and wild hosts and using tolerant cultivars and rootstocks. To free infected bud-wood, dormant cuttings can be treated with hot water (see p. 263).

d. Viroids. These are extremely small agents, about one-tenth the size of the smallest virus, that cause such infectious diseases as *potato spindle-tuber, citrus exochordis, chrysanthemum mottle, chrysanthemum stunt, avocado sunblotch,* and *cadang-cadang* in coconut (*27, 112*).

Their structure is a circular, single-stranded RNA molecule of low molecular weight. Some, such as *citrus exocortis,* are transmitted mechanically on pruning shears and during grafting. Viroids are carried in vegetative propagules as well as in some seeds. Some multiply most rapidly at high temperatures.

Symptoms are sometimes slow to develop and difficult to identify. Methods of detection involve the extraction and purification of nucleic acids followed by electrophoresis to detect the unique RNA viroid (*34, 63*) and by dot-blot hybridization assays using nucleic acid probes.

MANAGEMENT OF SOURCES FOR VEGETATIVE PROPAGATION

Characteristics of Source Material

One of the keys to successful vegetative propagation is the proper source of plant material. In chapters to follow, the influence of the source plants on the ability to make cuttings, to successfully graft or bud, to micropropagate, and to do other procedures will be considered. In this chapter, we are dealing with a more fundamental question: How can we be sure that the plants produced are the correct cultivar **(true-to-name),** have not deviated from the standard for that cultivar **(true-to-type),** and are not infected with serious pathogens **(pathogen-free)**? The existence of any of the three problems can frustrate propagators, enrage customers, and sometimes cause economic losses.

Trueness-to-Name

Plants have names (Chapter 2) and a successful nursery industry requires that plants being produced and sold be correctly labeled either by scientific name, cultivar name, or both. The process starts by identifying the source of plant material at the time of collection and is continued by maintaining records during propagation and completed by properly labeling the final product. Nevertheless, mistakes can happen, a switch in labels or plants and/or change in name can be deliberately made.

Identification of species and cultivar is usually made by visual inspection based on personal knowledge or written descriptions, but the plant may need to be grown under appropriate conditions to counteract any genotype × environment interaction. Plants must be examined at the proper stage of development to ascertain its essential phenotypic characteristics. For example, fruiting or flowering cultivars should have their fruit or flowers inspected.

Fingerprint methods to distinguish among cultivars of specific crops are laboratory biochemical tests which produce characteristic patterns of lines or dots on a gel which can be correlated to a specific cultivar. These tests vary in their reliability and applicability to particular groups of plants.

The most widely used biochemical tests are described in the box.

Trueness-to-Type

The propagation source must not only be the correct cultivar, but the characteristics of the progeny plants must conform to an accepted norm for that cultivar when grown in its ultimate location, i.e., be *true-to-type*. Trueness-to-type deviations among clones can result from several causes as described in this chapter including epigenetic changes, i.e., juvenile-mature phases, and genetic changes. Some problems may be unique to specific plant species, cultivars, or individual propagation sources and the

Isozyme analysis (4, 25, 75, 114). Enzymes are proteins that may occur in more than one molecular form (isozymes) which differ in their mobility in an electric field. This test is carried out by extracting proteins from plant tissue (active leaf, pollen, embryo), applying the protein extract to one end of a slab of starch or polyacrylamide matrix held in an electrophoretic apparatus, and subjecting it to an electric current. Over a period of hours the isozymes separate into bands that can be made visible by specific stains (see Figure 9–20).

Each plant genotype has a characteristic pattern of isozymes which in principle can provide a "fingerprint" for that cultivar. To be useful, there must be a range of variation among different genotypes such that an array of standard patterns is available. A sample from an unknown genotype is compared against this standard.

RFLP (Restriction Fragment Length Polymorphisms) (75, 88). This is a DNA-based test. DNA is extracted from the organism, and digested with **restriction enzymes,** which cut the DNA molecule at specific locations. The resulting DNA fragments are subjected to electrophoresis in an agarose gel much like the isozyme procedure. The fragments have different mobilities based on their size and produce a characteristic banding pattern. The separated bands are transferred to a nylon membrane (Southern Blotting) and recognized by hybridization with specific labeled probes. Labeling of the bands is carried out by a complex laboratory technique, which utilizes radioactive or chemiluminescent procedures to reveal the specific banding pattern. The technique is difficult and expensive but it results in an almost limitless series of "markers" which are highly specific to individual genotypes.

PCR (Polymerase Chain Reaction) (130) and RAPD (Randomly Amplified Polymorphic DNA). These DNA-based procedures provide a method by which a given segment of DNA (target DNA) is experimentally amplified in the laboratory. The extracted target DNA is heated to separate the two complementary strands. The extract is then cooled and combined with two **primers** at cooler temperatures. Primers are specially prepared complex molecules known as **oligonucleotides** which provide bases which make up the genetic code. Each primer binds to one of the DNA strands and, in the presence of the enzyme **thermostable DNA polymerase** (see Chapter 2), primes the DNA to replicate according to the genetic code present in the DNA. Through repeated cycles, the DNA can be amplified up to a million times within a few hours. The RAPD test is similar but utilizes only a single primer at a time and the annealing temperature is less stringent. The material is electrophoresed on an agarose gel and stained to produce characteristic bands that identify different patterns of DNA (see Figure 9–20).

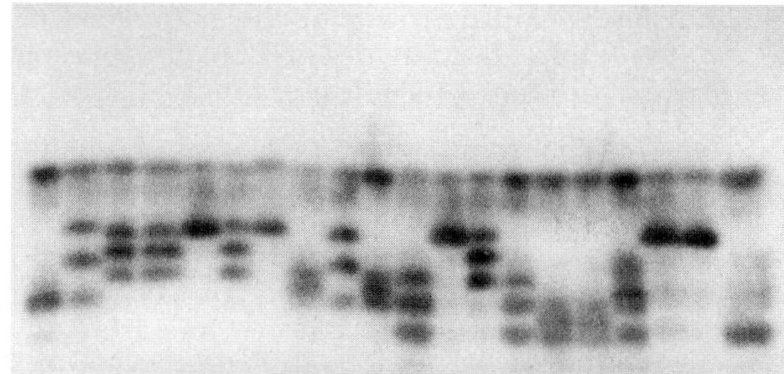

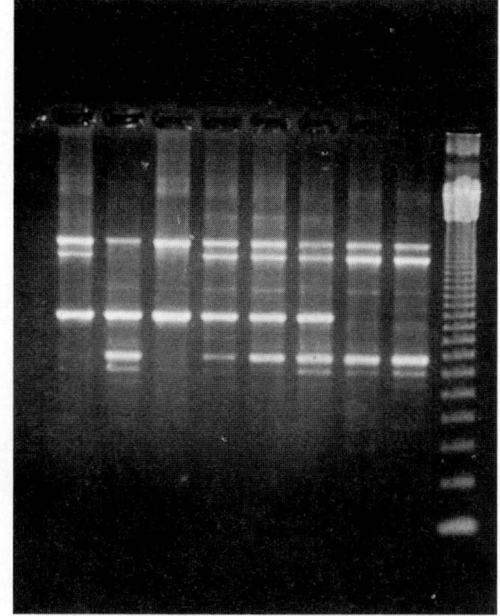

FIGURE 9–20 Fingerprinting for cultivar identification. *Above:* Isozyme patterns of *glucose phosphate isomerase* for twenty grape *(Vitis vinifera)* cultivars. *Below:* RAPD profiles for eight apricot cultivars. (Courtesy A. Arulsekar.)

procedures used for evaluation are species- or cultivar-based.

Although objective fingerprinting systems are useful, they have not normally been sufficiently sensitive to distinguish most differences in trueness-to-type. Evaluating for trueness-to-type is mainly by **visual inspection** by individuals knowledgeable about a particular plant species or cultivar.

True-to-type evaluation is comprised of two steps: (a) **phenotypic selection,** i.e., visual inspection and evaluation of the source plants, and (b) **genotypic selection,** i.e., visual inspection and evaluation of the progeny plants (vegetative progeny testing). In either case, one must take into account possible environmental interactions. These may provide a "false" reading by indicating that a plant is off-type when it is merely expressing environmental variation (Figure 9–12, for instance). Stock plants that are pruned severely for cuttings or scions may produce vigorous vegetative shoots which may produce atypical fruit or nut phenotypes (*58*).

The limitations of visual inspection for phenotypic selection should be recognized. Some variants (either genetic or nongenetic) may be latent in the source plant and not expressed. Mutant sectors may not be sufficiently visible to be immediately detected. Potential for genetic disorders may not be high enough to be expressed until after a period of growth at high temperatures (see p. 257). In these cases expression of the variant does not appear until the next vegetative generation.

Freedom from Pathogens

Viruses and related organisms such as phytoplasmas, viroids, and fastidious bacteria have plagued clonally propagated cultivars. Once a plant is infected, the pathogen may become systemic, spreading to unaffected parts of the plant, crossing graft unions, and becoming a source of inoculum in bud-wood, seed, or pollen. The infected source plant serves as a reservoir for disease dissemination by insect vector or pollen transfer.

During the 1950s, the concept of **pathogen-tested** source trees used for propagation was developed. Precautions taken included the absence of targeted pathogens that could affect the source plants and the prevention of infections during propagation (*6, 7, 107*). Part of this work dealt with control of fungi and bacteria during greenhouse propagation. Similarly, nematodes were serious problems but were controlled by eliminating these pathogens in the nursery soil, mostly by fumigation (see Chapter 3).

Methods to Detect Pathogens in Propagation Sources

a. Visual inspection. Pathogens alter the biochemical processes in susceptible plants which may express characteristic symptoms. In highly susceptible plants, the pathogen may be self-eliminating by producing such drastic effects that the host plant does not survive. The absence of disease symptoms, however, does not indicate absence of a pathogen. Some viruses, such as *Prunus necrotic ring spot* virus may produce an initial shock after which the plant recovers and shows no further visual symptoms (*82*). Commonly, virus symptoms are expressed more at cool temperatures (22° C or less; 70° F); higher temperatures apparently inhibit their activity (*37*). Multiple infections by different viruses may produce a severe reaction, whereas each alone produces little effect.

Visual inspection for specific symptoms is an important skill for the propagator in detecting problems among stock plants. It is not a reliable indicator, however, in the source plant itself. Rather visual inspection becomes more important in the inspection of grafted sensitive indicator plants.

*b. **Culture indexing.*** This procedure is used to detect specific fungi and bacteria in the production of ornamentals, such as chrysanthemum, carnations, and pelargonium (Figure 9–21). Pieces of surface sterilized plant parts are cultured on a medium favoring the growth of the pathogens (*7, 51, 84, 90*). Pathogens are identified by morphological features or biochemical tests.

*c. **Virus indexing.*** The primary method for detecting viruses has been to transmit them by grafting or budding to sensitive indicator plants, which then develop identifiable symptoms (*63, 80, 119*) (Figure 9–22). One widely used procedure for indexing woody plants is to place a bud from the plant to be tested beneath a bud of the virus indicator plant onto the same rootstock growing in a container. The indicator bud is forced to grow. If the test bud contains virus, it moves into the rootstock and up into the growing indicator shoot, where symptoms appear (*36*).

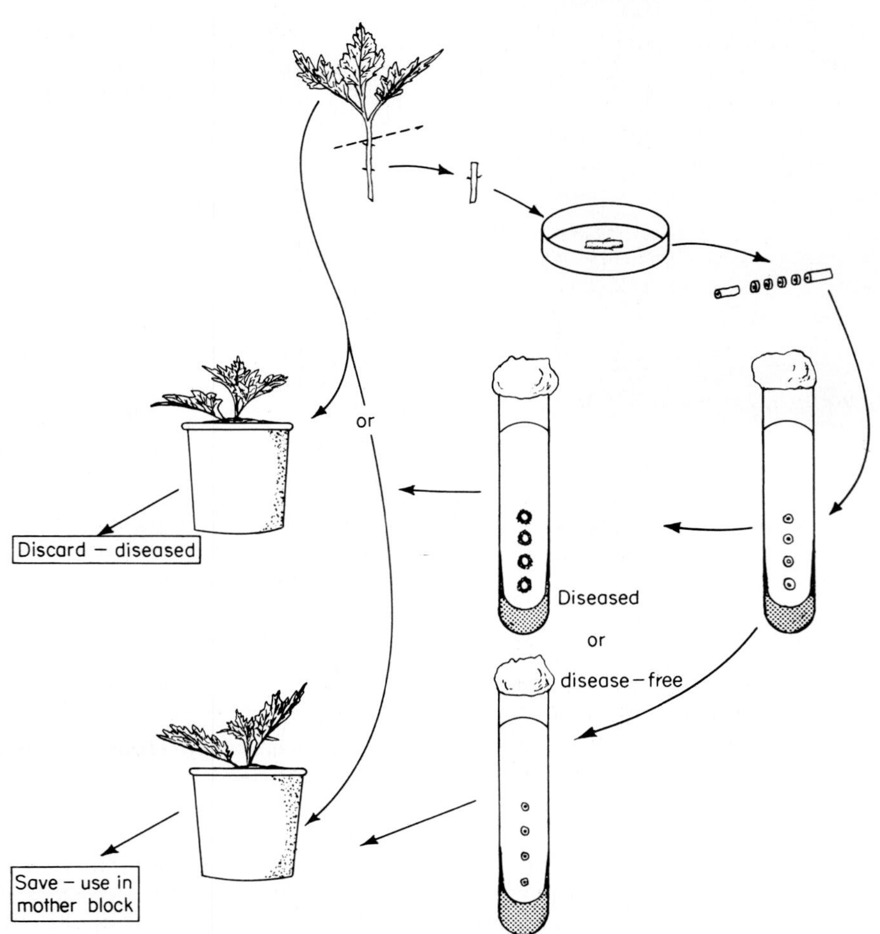

FIGURE 9–21 Culture indexing of chrysanthemum for *Verticillium* wilt pathogen. Stem sections are removed from surface-sterilized cuttings and placed on culture media. Only healthy plant sources are used to mass propagate the clone or cultivar. (Redrawn from Baker and Chandler, California Division of Agricultural Sciences Manual 23.)

Leaf grafting is used in strawberry (*13*). Among *Prunus* spp., the necrotic reaction of highly sensitive tissue of 'Shirofugen' flowering cherry to the *Prunus* necrotic rings spot virus (*80, 119*) (Figure 9–22) is a widely used "quick" test that requires about a month's incubation to complete. In contrast, two years is required to detect many virus diseases. Certain viruses can be detected in herbaceous hosts by sap transmission.

d. Serology. Serology identifies unique proteins associated with a particular pathogen. The ELISA (enzyme-linked immuno-sorbent assay) can be used to detect essentially any pathogen or virus (Figure 9–23). Antibodies are produced by a rabbit against specific proteins characteristic of a particular virus or virus strain. After a series of antigen injections, blood is drawn and the serum fraction containing antibodies is separated from red blood cells. Whole serum or purified antibodies are employed in the serology tests with ELISA (*19*). Commercial kits can be obtained to screen hundreds of samples to produce immediate reactions.

e. Biochemical methods. Methods used to identify viral nucleic acids, such as **polyacrylamide gel electrophoresis,** are used to detect viroids and the replicated form (double-stranded RNA) of RNA plant viruses (*53, 63*).

Procedures to Eliminate Pathogens from Plant Parts

a. Selection of uninfected parts. Some parts of a plant may be infected, while others are not. Some soil-borne pathogens, such as *Phytophthora* spp., can be avoided by taking tip cuttings from uninfected portions of the plant (*Fusarium,* and *Verticillium*) (*7*).

b. Shoot apex culture and tip-grafting. The small apical dome of a growing point (Figure 17–14) including the meristem and a few subtending leaf primordia is often free of virus and other pathogens even in systemically infected plants. Excision and aseptic culture of this small segment can be the start of a foundation clone (see Chapter 18 for culture procedures). The smaller the explant, the more effective is the elimination of pathogens. For example, meristem tips of 0.10 to 0.15 mm have given 100 percent virus elimination, but the percentage decreases as the size increased to 1 mm. On the other hand, the smaller the explant, the more difficult it is to establish and to survive. A compromise is to use pieces 0.25 to 1.00 mm long and confirm virus status by indexing. A combination of thermotherapy and meristem excision can increase the probability of virus exclusion (Figure 9–24).

Some examples where shoot apex culture has been used to eliminate viruses (*76, 89, 111*) include the flower crops (amaryllis, carnation, chrysanthemum, dahlia, freesia, geranium, gladiolus, iris, lily, narcissus, nerine, orchid), fruit crops (apple, banana, citrus, gooseberry, grape, pineapple, raspberry, stone fruits, strawberry), vegetables (brussels sprouts, cauliflower, garlic, potato, rhubarb, sweet potato), and miscellaneous (cassava, hops, sugarcane, taro).

Micrografting is a method used for woody plants where tip cultures will not survive. Shoot tips can be grown by placing on a shoot apex of an already rooted seedling plant in culture. It has an added advantage, for instance in Citrus, in that the procedure is done with buds of mature clones which avoids the juvenile phenotype of nucellar seedlings (see Chapter 17).

c. Heat treatments. Sufficient heat to kill the pathogen but not injure the plant can be used to treat plants, plant parts, bulbs, and seeds for fungi, bacteria, and nematodes (*5, 7*). Treatments vary with different plant species from 43.5 to 57° C (110 to 135° F) ½ to 4 hours. Methods include hot water soaking, exposure to hot air or to aerated steam. Hardening-off vegetative material or reducing moisture content in seeds prior to heat treatment generally improves plant survival. Damage to heat-treated seeds can be alleviated by submersion in polyethylene glycol 6000 (*5*).

d. Thermotherapy. This procedure eliminates viruses within plants (*54, 63, 83, 89, 111*). Plants in containers are preconditioned at high temperatures and then placed at 37 to 38° C (98 to 100° F) for two to four weeks or longer (Figure 9–25). After treatment, individual buds are removed and budded to clean rootstocks. Each budded progeny must be tested to determine if the new plant is indeed "virus-free." The combination of thermotherapy and shoot apex culture is used to improve the procedures where only one of the procedures is not adequate. The plant is first heat-treated and then apical shoot tips about 0.33 mm long are removed and grown *in vitro* in test tubes. For plants that do not root easily, shoot tip grafting may be used.

e. Growing seedlings. Most viruses either are not transmitted through seeds or the percentage of infected seedlings is low. This characteristic is appar-

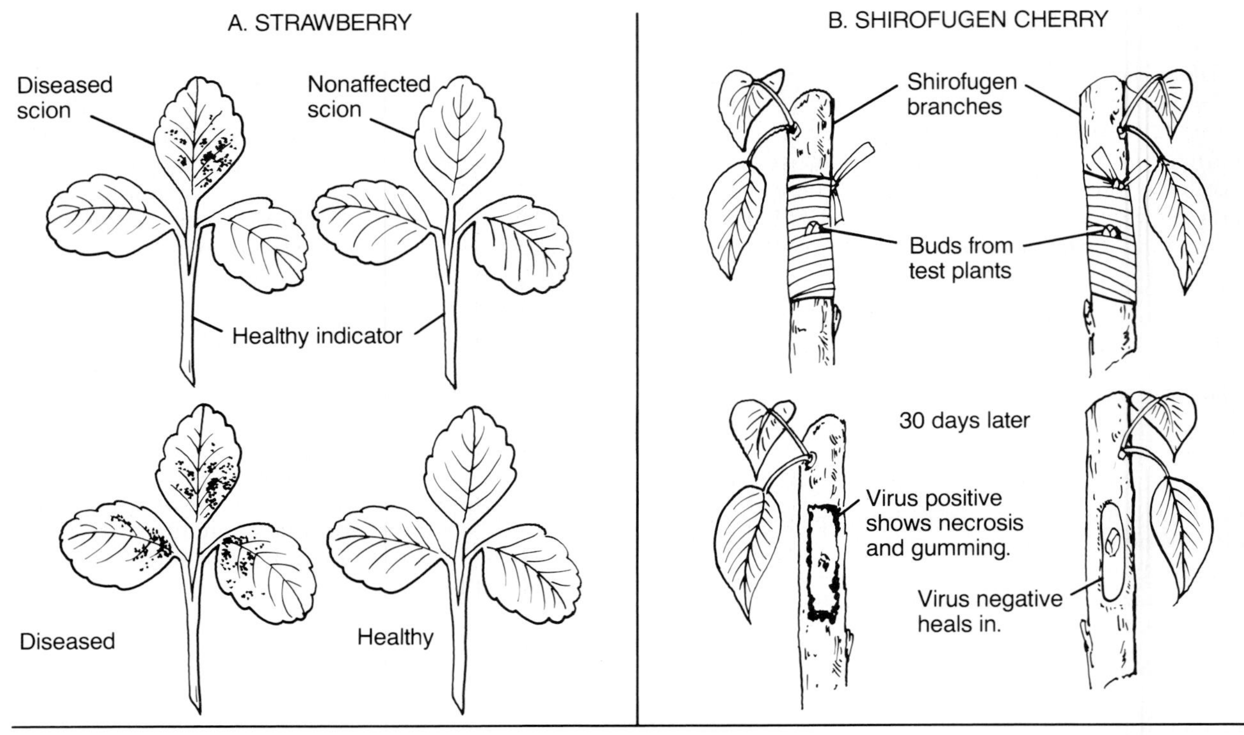

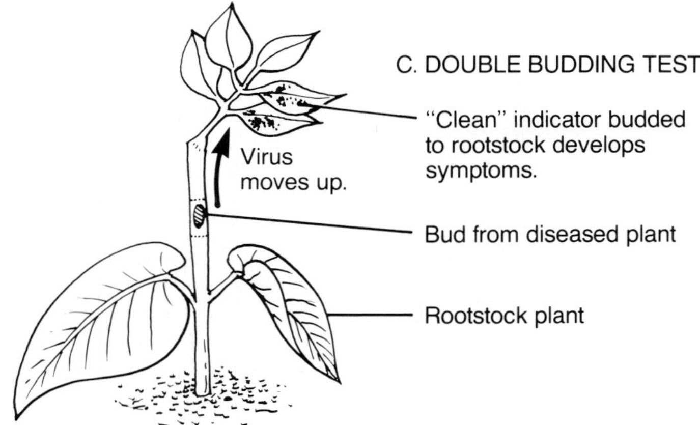

FIGURE 9–22 Indexing for virus transmission. *Top left:* Excised leaf method for transmitting a virus from a symptomless plant to a sensitive indicator. A terminal leaf from the test plant is removed, a petiole of the indicator plant is split, and the excised test leaf is inserted. The graft union is wrapped with latex tape. Leaves on the indicator stock plant shows symptoms. *Top right:* Shirofugen cherry indexing. Buds are removed from a test plant and inserted in sequence (usually three per test) in a 'Shirofugen' flowering cherry, which is very sensitive to certain fruit tree viruses, such as the prunus ring spot complex. If the test plant is infected, the bud will die; gum and necrotic tissue will appear around the bud shield within a month. *Below:* Double budding test for fruit trees. A rootstock plant is grown in a container in the greenhouse. A bud from the test plant is placed into the base of the plant and allowed to heal. At the same time a bud from a sensitive indicator plant is budded into the rootstock plant. If the test plant is diseased, the virus will move across the graft union, move through the conducting system of the rootstock, and infect the indicator, which should develop specific symptoms.

FIGURE 9–23 Typical serological results produced by the ELISA procedure. Each cell is a separate test whose results are shown by the intensity of the color staining. *Left panel:* Top four rows were made on a virus positive sample in which the concentration of the sample has been continuously diluted from left to right. Bottom four rows show results when all samples were virus-negative. *Right panel:* Results of ELISA testing of a random series of samples whose virus content is unknown. Note random distribution of staining results. (Courtesy of Adib Rowhani).

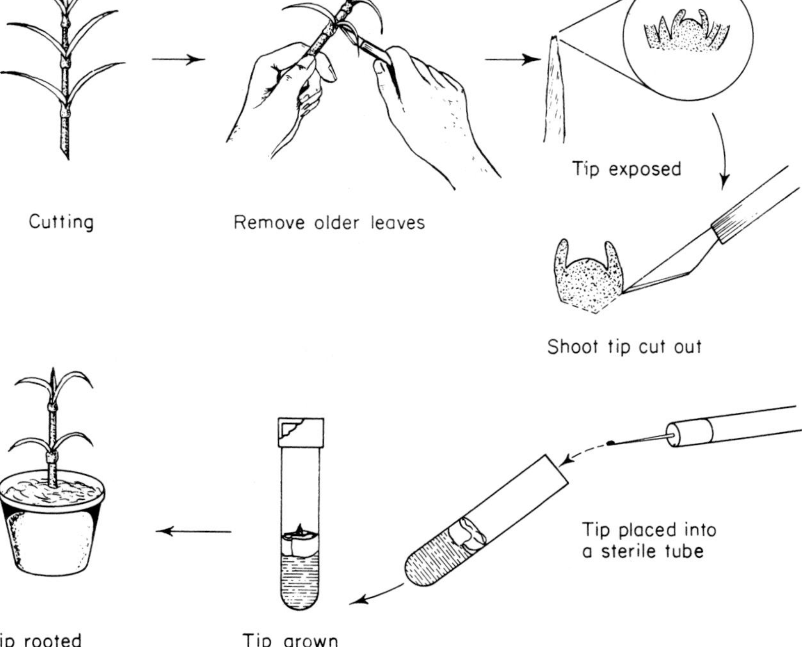

FIGURE 9–24 Meristem-tip culture of carnation. Following the arrows: the carnation cutting is obtained, the larger leaves are stripped away, and then the small, enclosing leaves at the extreme tip are removed to expose the growing point. The tip and the next subtending leaf primordia are removed with a scalpel and placed on the surface of a paper wick in a test tube with nutrient media. The shoot tip grows in the tube until large enough to be transplanted to a container. A full-size plant is shown at bottom left. (Redrawn from Holley, W.O. and R. Baker. 1963. *Carnation Production.* Dubuque, IA: William C. Brown.)

FIGURE 9–25 Heat chamber for conducting thermotherapy of plant material to eliminate viruses. Plants are grown at approximately 38° C (100° F) for four to six weeks. At the conclusion of this period, buds are removed to be indexed for the continued presence of viruses.

ently a major biological advantage of seed reproduction in nature, since most viruses are screened out in the sexual reproductive process. Horticulturally, this procedure has important uses. A new cultivar originating as a seedling generally starts its vegetative sequence (Figure 9–2) as a "clean" individual. If protected from infection during the early testing process, the cultivar can be released as a "virus-free" cultivar.

Citrus provides an important application of the seedling principle in that essentially all of the very serious viruses in "old-line" citrus cultivars can be eliminated by growing nucellar seedlings. To make these very juvenile plants useful horticulturally, nucellar plants are grown into their S_1 or S_2 generation to pass through their juvenile phase *(16, 38).* "True Potato Seed" (TPS) is also being used to grow potato crops in some countries instead of the so-called "seed" potato tuber based material. The reason is lower cost and freedom from viruses.

PROPAGATION SOURCES AND THEIR MANAGEMENT

Propagation of cultivars in specific species requires that a supply of propagules be available either continually (some florists' crops, such as carnations, chrysanthemum, foliage plants) or at specific times of the year (fruit and nut crops, woody ornamentals). To do so requires that **propagation source plants** be available to meet these needs. How to select, maintain, and manage propagation sources is one of the most basic decisions that a propagator or nursery has to make. The source material must be the correct cultivar, produce true-to-type plants, and generally be free of serious pathogens. Management of the material must be such that sufficient propagating material is available in an optimum condition at a time when it is to be used. The system must be economical and fit into the overall nursery operation. The situation of the individual nursery and requirements of the species and cultivar determine the method of source maintenance.

Traditionally, nurseries managed their own source procurement systems as part of their overall operations. As technological requirements to control disease and maintain quality have increased, specialized nursery operations, both private and governmental, have appeared that maintain the source material, supervise its distribution, and sometimes provide material, such as "liners," either rooted or unrooted, to produce commercial plants directly or to provide propagating material to establish source plants.

In general, nurseries use one of the following systems:

Commercial plantings. This option is economical, requires no stock blocks, and depends on commercial performance as the criterion of selection. Individual plants are visually inspected prior to use and any "suspicious" off-type plants can be rejected whether or not the cause is known. For example, commercial orchards have been the traditional source of bud-wood for fruit and nut cultivars in California. The source trees are grown for crop and not bud-wood. Propagators use limited amounts of propagating material from each of many trees. Maintaining identity from individual source plant to commercial material is virtually impossible. Bud-wood "lines" may evolve from consecutive generations from different nursery sources. As growth in older trees declines, fewer new shoots are available and the source must be replaced by younger trees with greater vigor. This system utilizes a combination of propagation from many trees combined with consecutive generations of propagation. Problems associated with any one tree may be diluted in the total population, but the variant will continue to persist and the proportion will sometimes increase (*58*).

European vineyardists, who have grown specific wine grape cultivars for centuries, practice "clonal selection" from selected vines in their commercial planting to establish new vineyards (*129*). Differences have apparently evolved over time to produce "strains" or "clones" of established cultivars which may not differ phenotypically but are claimed to have different product uses.

Careful field observations can result in the discovery of useful mutants and many new cultivars of apples, pears, and peaches have been discovered. Undesirable mutants may have been avoided. Historically, however, this system has resulted in the increasing contamination by viruses and other pathogens of major cultivars of many if not most intensively clonally propagated crops. In some crops, the viability of the industry is threatened. In other crops, increasing incidence of inherited disorders and nonproductive mutants have resulted. For these reasons, Foundation Clone systems have become a major aspect of production of many crops.

Production material within the nursery. Collecting cuttings or buds from nursery material is comparable to field selection except that the commercial nursery crop is the source of propagation material for the next cycle of propagation. For example, the prunings of ornamental shrubs being grown in containers may be used as a source of cutting material. Plants may be set aside and pruned severely to force growth. Surplus material may be utilized (*70*). Unsold budded fruit and nut trees may be maintained in the nursery row for several years.

The advantages are efficiency, convenience, and economy. However, the same problems described for field selection potentially may develop.

Stock blocks. A stock block includes source plants maintained in a more or less permanent location separate from commercial propagation and managed to produce cuttings, bud-wood, scion-wood, or divisions. These blocks may be planted either in the ground in a permanent site, or in containers, in the open or under protection (*70*). Hedge rows can be maintained to keep plants in a juvenile state for forest tree production (*2, 65*). Similarly, hedge rows are useful for the production of fruit tree rootstocks (*50*). Management of these blocks is geared to production of propagation source material except that a few plants might be maintained for observation of "trueness-to-type."

Fruit tree nurseries establish scion or bud-wood orchards where trees are planted relatively closely together and pruned for the production of propagation material rather than for fruit (Figure 9–26). Nurseries sometimes maintain some trees or a branch of each tree as a check for verification of cultivar and type.

Individual plants in stock blocks need to be examined for trueness-to-cultivar and trueness-to-type prior to their use. This operation may require that the plants be allowed to fruit or flower or that a vegetative progeny test be conducted. Severe pruning in the stock block may change the growth habit, general appearance, and fruit or nut quality such that either a false reading of off-type may be made or specific variants may fail to be observed.

Maintaining stock blocks is expensive, time-consuming, and may require much space. Once identity and trueness-to-type is verified, however, this system provides a safeguard for good quality and a convenient supply of propagation material. Unless these blocks are protected against infection and/or tested for disease periodically, the same disease contamination problems described for field and nursery commercial material may develop (Figure 9–27).

Foundation Clone System (Pedigreed Production)

This system is based on (a) identification of a single plant or a small group of source plants (sometimes called *nuclear*) which meet specific standards of pathogen freedom and genetic quality and, (b) subsequent multiplication in large volume for limited consecutive generations under controlled conditions. The system restricts consecutive propaga-

FIGURE 9–26 Nursery stock plants for providing scions and buds for fruit tree production. (Photograph from Robert Geneve.)

FIGURE 9–27 Tag is attached to stock blocks to indicate that the trees have been tested as "virus-free." (Photograph from Robert Geneve.)

tion generations and expands propagation from the same plant. (Figure 9–28).

Two versions are described: (a) Foundation Clonal System, based upon selection of a single plant as the initial source; and (b) Mass Clonal System, based upon the selection of a group of plants as the initial stock which are renewed at regular intervals. The initial stimulus to develop these systems was to control virus diseases in major vegetatively propagated crops. The concepts of "pedigree production" and "nuclear stock" are similar to Registration and Certification systems of agricultural crops (Chapter 4).

Historically, these programs began with certification of pathogen-free potato stocks. They then were developed for other major crops, such as grape, strawberry, blueberry, apple, citrus, and stone fruits, as serious virus problems began to appear in these intensively cultivated monocultures (*25, 29, 119, 125, 129*) Many major fruit and nut industries currently utilize some form of controlled program. Foundation clonal systems are effective in selecting and maintaining low-potential sources of inherited disorders, such as *noninfectious bud-failure* in almond, and solving trueness-to-type problems (*58*). For the latter disorders, however, selection of the Foundation clones requires genotypic selection.

A Foundation Clonal System has three basic phases:

1. *Initial selection or development.* The minimum requirements are true-to-name (correct cultivar), true-to-type, and free of prescribed pathogens. Other requirements might be imposed for individual crops or become part of the description for that material. This material may be referred to as "Foundation clones" (often "clones") or as "elite plants."
2. *Maintenance in a Foundation block or planting.* A limited number of Foundation plants are established in a **foundation block** (or orchard) under close supervision and appropriate management, depending upon the needs of the crop. In Registration and Certification programs, individual plants are registered with a registration agency, usually governmental (*36, 39*).
3. *Multiplication and distribution.* **Foundation source** material is used to establish stock blocks (mother block, scion orchard, or nursery increase block) which provide sufficient source material (cuttings, buds, scions) for production of commercial nursery material (*48*).

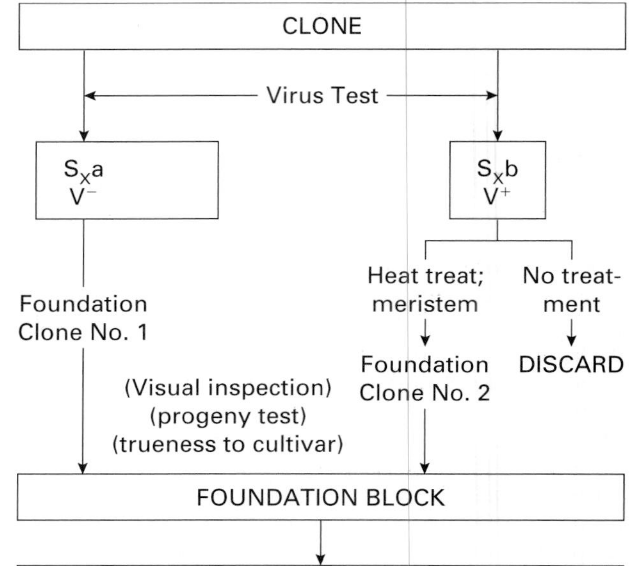

FIGURE 9–28 Diagram to illustrate the infection of a virus within a clone and the steps needed to identify "virus free" Foundation Clones to meet requirements of registration and certification programs.

Repositories, Botanical Gardens, and Plant Collections

Plant collections that exist around the world can be a source of clones. Usually these sources provide only enough material to establish a source block. The standards of quality already described should be followed in obtaining the material, but this may depend upon the type of collection. Botanical gardens maintained by private, public, or governmental agencies in general emphasize the genetic aspects of species and cultivars. These may be a part of a breeding and clonal selection program where propagation material may be controlled by patents and other protective protocols (see Chapter 2).

Clonal repositories are of two types. One maintains clonal germ plasm, as in the U.S. Clonal Repository System. The second includes repositories to maintain and distribute Foundation clones for virus control programs. The USDA IR-2 Repository at Prosser, Washington is of this type (*36*). Many states and countries in the world maintain governmental repositories that provide virus-tested stocks of various crops.

In California, strawberry, deciduous fruit, nut, vine crops, and citrus can be produced under a system of Registration and Certification administered by the California Department of Agriculture under regulations enacted by the California state legislature *(3, 68)*. Foundation clones used in these programs originated as single plants from research and development programs of private nurseries, government research groups, or breeders. These meet virus detection standards that are specified in the regulations. Foundation plants of each foundation clone are established in a Foundation orchard or vineyard maintained either by a private nursery or Foundation Plant Materials Service (FPMS) at the University of California, Davis. Individual trees are registered when trueness-to-cultivar and trueness-to-type is verified by visual inspection in bearing plant (phenotypic selection). Genotypic selection by progeny testing is not required but may be carried out by the originator (particularly for inherited disorders) and become part of the description of each foundation clone. Foundation trees are virus-tested at specified intervals. Trees may be maintained either in open blocks with specific isolation requirements or kept in screenhouses with insect-proof mesh. Programs exist both for scion material and rootstocks.

Commercial propagators obtain Registered Foundation propagating material and establish mother blocks, scion orchards, or increase blocks in nursery rows and each tree is registered. Each type has separate requirements for isolation, indexing tests, and management. Buds from these nursery blocks are used to propagate commercial certified nursery plants (Figure 9–29).

Mass Clonal Systems are used with herbaceous plant species, such as strawberry, carnation, chrysanthemums, and geraniums, where single-source foundation clones are not practical. A system used for producing pathogen-free, true-to-type geranium cuttings (CVI) is described *(18, 84)*. *Stage I:* Cuttings are produced from source plants that have been visually inspected for trueness-to-type and culture-indexed for Verticillium and Xanthomonas. oted cuttings are then subjected to thermotherapy for 4 weeks. *Stage II:* A population of meristem-tip cultured plants from Stage I plants is produced, indexed, and multiplied in culture for 12 weeks. Virus negative plants provide an elite mother block. *Stage III:* These plants are reindexed and cuttings are propagated. Visual inspection identifies poorly performing cuttings which are discarded along with their mother plants. *Stage IV:* A new group of true-to-type and indexed plants is propagated to produce the *nucleus* (foundation) block (3 months), multiplied in an *increase* block to produce *stock plants* that are used to provide the cuttings which are harvested and sold to greenhouse growers to produce plants.

Quarantines and Movement of Vegetatively Propagated Materials (34)

Propagation of vegetatively propagated material may include shipment from country to country or from state to state in the United States. Because of the potential risk of transporting dangerous pests on the material or systemically infected clones, particularly with viruses, specific regulations and quarantines are in place to control such movement (*30,*

Sweet potato is an example of a clonal crop whose commercial success depends upon production through Foundation Clonal Systems *(22)*. Sweet potatoes *(Ipomoea battatus)* and yams *(Dioscorea spp.)* are not only susceptible to a range of viral, fungal and bacterial diseases but tend to be genetically unstable and produce "off-type" root tubers which may be chimeral. Propagation is accomplished by adventitious shoots (suckers) emerging from the fleshy edible root and can result in the production of all three types of chimeras, including sectorial.

124). Soil on the plant is usually not allowed because of nematodes and other pests. Plants are inspected before entry and sometimes treated. Post-entry quarantines and virus indexing may be required for specific plants.

Propagators should be familiar with any national, state, or local regulation that affects distribution of their products.

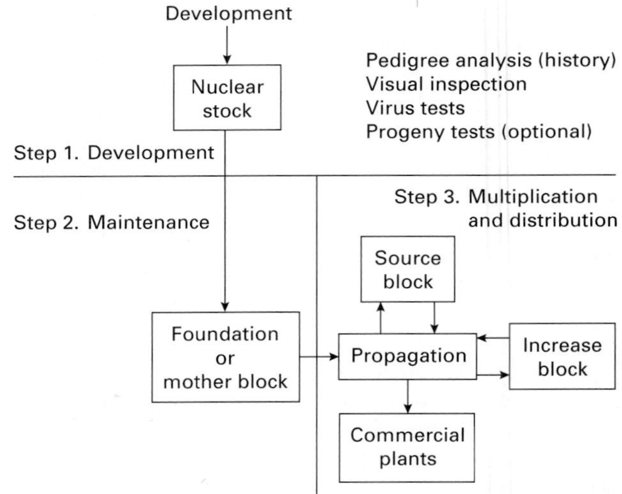

Includes:
Visual inspecion of source plants (pheontypic selection) at each step
Indexing (as required at each step)
Visual inspection of progeny (genotypic selection) (Optional)
Registration of individual source trees

FIGURE 9–29 Relationship among different steps in establishing a Foundation clonal stock system for selection, maintenance, and distribution of true-to type, virus tested propagation material.

SUMMARY

This chapter has described the special benefits and problems that are associated with clones and vegetative propagation. Clones have been particularly important in the evolution of fruits, nuts, some vegetables, and certain ornamentals. The application to landscape plants and street trees is more recent but intensive. Clonal forestry is becoming important.

The importance of clones is primarily due to their uniformity and capacity to capture major genetic gains in a single step. Clones grown on a large scale as practiced in modern horticulture have some important drawbacks. Because they are monocultures, clones are vulnerable to specific environmental hazards, particularly new pests and diseases or environmental conditions. Despite the fact that clones are selected for their genetic uniformity, they are subject to variation within the clone that can cause problems both to the propagator and to the customer, particularly if the problem does not appear until after the plant leaves the nursery.

The three main selection problems in clonal propagation emphasized in this chapter include: (a) having the correct cultivar, (b) maintaining trueness-to-type, and (c) producing pathogen-free material. Selection, maintenance, and distribution programs are described to control these three problems and ensure the highest quality material to consumers.

REFERENCES

1. Abbott, A.J., and P.K. Atkin, eds. 1987. *Improving vegetatively propagated crops.* New York: Academic Press.
2. Ahuja, M.R. and W.J. Libby. 1993. *Clonal Forestry.* Vol. I and II. Springer-Verlag. Heidelberg, Berlin.
3. Anonymous. 1990. *Regulations.* Calif Dept. of Food and Agriculture: Sacramento, CA.
4. Arulsekar, S. and D.E. Parfitt. 1986. Isozyme analysis procedures for stone fruits, almond, grape, walnut and fig. *HortScience* 31:928–933.

5. Baker, K.F. 1962. Thermotherapy of planting material. *Phytopathology* 52:1244–55.

6. Baker, K.F. 1984. Development of nursery techniques. *Proc. Inter. Plant Prop. Soc.* 34:152–64.

7. Baker, K.F., and P.A. Chandler. 1957. Development and maintenance of healthy planting stock. In *The U.C. system for producing healthy container-grown plants,* K.F. Baker, ed. Calif. Agr. Exp. Sta. Man. 23, pp. 217–36.

8. Ball, V. 1991. *Ball Red Book: greenhouse growing, 6th edition.* Reston, VA: Reston Publishing Co.

9. Baur, E. 1909. Pfropfbastarde periclinal Chimäeren and Hyperchimäeren. *Ber Deuts. Bot. Ges.* 27:603–5.

10. Bonga, J.M. 1982. Vegetative propagation in relation to juvenility, maturity, and rejuvenation. In *Tissue culture of forest trees.* J.M. Bonga and D.J. Durzan, eds. Amsterdam: Elsevier.

11. ———. 1987. Clonal propagation of mature trees: Problems and possible solutions. In *Tissue culture of forest trees.* J.M. Bonga and D.J. Durzan, eds. Amsterdam: Elsevier, pp. 249–71.

12. Boulay, M. 1987. Conifer micropropagation: Applied research and commercial aspects. In *Tissue culture of forest trees,* J.M. Bonga and D.J. Durzan, eds. Amsterdam: Elsevier, pp. 185–206.

13. Bringhurst, R.S., and V. Voth. 1956. Strawberry virus transmission by grafting excised leaves. *Plant Disease Rpt.* 40(7):596–600.

14. Bringhurst, R.S., V. Voth, and D. van Hook. 1960. Relationship of root starch content and chilling history to performance of California strawberries. *Proc. Amer. Soc. Hort. Sci.* 75:373–81.

15. Broerties, C., and A.M. Van Harten. 1987. Application of mutation breeding methods. In *Improving vegetatively propagated crops,* A.J. Abbott and R.K. Atkin, eds. New York: Academic Press, pp. 335–48.

16. Cameron, J.W., R.K. Soost, and H.B. Frost. 1957. The horticultural significance of nucellar embryony in citrus. In *Citrus virus diseases,* J.M. Wallace, ed. Berkeley: Univ. Calif. Div. Agr. Sci., pp. 191–96.

17. Campbell, A.I., ed. 1986. Workshop on clonal selection in tree fruit. *Acta Hort.* 180:1–131.

18. Cassels, A.C. 1992. Micropropagation of commercial Pelargonium species and hybrids (Glasshouse Geraniums). In *Biotechnology in Agriculture and Forestry.* Vol. 20. Y.P.S. Bajaj, ed. Springer-Verlag. Berlin Heidelberg. pp. 286–306.

19. Clark, M.F., and A.N. Adams. 1977. Characteristics of the microplate method of enzymelinked immunosorbent assay for the detection of plant viruses. *Jour. Gen. Virol.* 34:475–83.

20. Converse, R.H. 1985. Latent viruses: Harmful or harmless? *HortScience* 20(5):845–48.

21. D'Amato, F. 1977. Cytogenetics of differentiation in tissue and cell cultures. In *Plant cell, tissue and organ cultures,* J. Reinert and Y.P.S. Bajaj, eds. Berlin: Springer-Verlag. pp. 343–93.

22. Danglar, J.M. 1994. Compendium. Sweet potato foundations programs. *HortTechnology* 4(3):224–238.

23. Darrow, G.M., R.A. Gibson, W.E. Toenjes, and H. Dermen. The nature of giant apple sports. *Jour. Hered.* 39:45–51.

24. Davies, F.T. Jr., T.D. Davis and D.E. Kester. 1994. Commercial importance of adventitious rooting to horticulture. In *Biology of adventitious root formation,* T.D. Davis and B.E. Haissig, eds. New York and London: Plenum Press. pp. 53–60.

25. DeGrandcamp, M.V. 1993. Differentiating plant clones in culture and maintaining "virus tested" blueberry clones. *Proc. Intern. Plant Prop. Soc.* 43 513–516.

26. Dermen, H. 1960. Nature of plant sports. *Amer. Hort. Mag.* 39:123–73.

27. Diener, T.O. 1979. *Viroids and viroid diseases.* New York: John Wiley.

28. Dirr, M.A. and C.W. Heiser, Jr. 1987. *The reference manual of woody plant propagation.* Athens GA: Varsity Press.

29. Durling, D.L. 1990. International citrus nursery production as it relates to phytophythora, viruses, and growing media. *Proc. Intern. Plant Prop. Soc.* 40:182–187.

30. Duvick, D.N. 1978. Risks of monoculture via clonal propagation. In *Propagation of higher plants through tissue culture; bridge between research and application,* K.W. Hughes et al., eds. Springfield, Va.: U.S. Dept. of Energy Tech. Infor. Center, pp. 73–84.

31. Einset, J., and C. Pratt. 1954. "Giant" sports of grapes. *Proc. Amer. Soc. Hort. Sci.* 63:251–56.

32. ———. 1959. Spontaneous and induced apple sports with misshapen fruit. *Proc. Amer. Soc. Hort. Sci.* 73:1–8.

33. Evans, D.A., W.R. Sharp and H.P. Medina-Filho. 1984. Somaclonal and gametoclonal variation. *Amer. J. Bot.* 71:759–774.

34. Foster, J.A. 1988. Regulatory actions to exclude pests during the international exchange of plant germplasm. *HortScience* 23(1):60–66.

35. Franclet, A., M. Boulay, F. Bekkaoui, Y. Fouret, B. Verschoore-Martouzet, and N. Walker. 1987. Rejuvenation. In *Tissue culture of forest trees,* J.M. Bonga and D.J. Durzan, eds. Amsterdam: Elsevier, pp. 232–48.

36. Fridlund, P.R. 1980. Maintenance and distribution of virus-free fruit trees. In *Proc. conf. on nursery production of fruit plants through tissue culture,* R.H. Zimmerman, ed. U.S. Dept. Agr. Sci. and Education Administration, ARR-NE-11, pp. 11–22.

37. ———. 1970. Temperature effects on virus disease symptoms in some *Prunus, Malus,* and *Pyrus* cultivars. *Wash. Agr. Exp. Sta. Bul.* 726.

38. Frost, H.B. 1938. Nucellar embryony and juvenile characters in clonal varieties of citrus. *Jour. Hered.* 29:423–32.
39. Goff, L.M., 1986. Certification of horticultural crops: a state perspective. In *Tissue culture as a plant production system for horticultural crops,* R. H. Zimmerman, R. J. Griesbach, F. A. Hammerschlag, and R.H. Lawson, eds. Dordrecht: Martinus Nijhoff Publishers, pp. 139–46.
40. Greenwood, M.S. and K.W. Hutchison. 1993. Maturation as a developmental process. In *Clonal Forestry* Vol 1. M.R. Ajuga and W. Libby, eds. Berlin Heidelberg: Springer-Verlag. pp. 14–33.
41. Haberlandt, G. 1930. *Das Wesen der Crataegomespili Sitzber.* Preuss. Akad. Wiss. 20:374–94.
42. Hackett, W.P. 1985. Juvenility, maturation and rejuvenation in woody plants. In *Horticultural Reviews,* Vol. 7, J. Janick, ed. Westport, Conn.: AVI Publ. Co., pp. 109–55.
43. Hackett, W.P. and J.R. Murray. (in press). Approaches to understanding maturation or phase change. In *Biotechnology of ornamental crops.* R. Geneve, and J. Preece eds. Wallingford, UK: C.A.B. International.
44. Hall, H.K., R.M. Skirvin, and W.F. Braam. 1986. Germplasm release of 'Lincoln Logan,' a tissue culture derived genetic thornless 'Loganberry.' *Fruit Var. Jour.* 40:134–135.
45. Hamilton, R.1. 1985. Using plant viruses for disease control. *HortScience* 20(5):848–52.
46. Hammerslag, F.A. and R.E. Litz. eds. 1992. *Biotechnology of perennial fruit crops.* Wallingford, UK: C.A.B. International.
47. Hansche, P. and Wm. Beres. 1980. Genetic remodeling of fruit and nut trees to facilitate cultivar improvement. *HortScience* 15:710–715.
48. Hansen, A.J. 1985. An end to the dilemma. Virus-free all the way. *HortScience* 20(5):852–59.
49. Harlan, J.R. 1992. *Crops and Man,* 2nd edition. Wisconsin: Madison. Amer. Soc. of Agron. Inc.
50. Howard, B.H. 1994. Manipulating rooting potential in stockplants before collecting cuttings. In *Biology of adventitious root formation.* T.M. Davis and B.E. Haussig, eds. New York and London: Plenum Press. pp. 123–142.
51. Joiner, N., ed. 1981. *Foliage plant production.* Englewood Cliffs, N.J.: Prentice-Hall.
52. Jones, W.N. 1969. *Plant chimeras and graft hybrids.* 2nd edition. London: Methuen & Co.
53. Jordan, R.L. 1986. Diagnosis of plant viruses using double-stranded RNA. In *Tissue culture as a plant production system for horticultural crops,* R. H. Zimmerman, R.J. Griesbach, F.A. Hammerschlag, and R. H. Lawson, eds. Dordrecht: Martinus Nijhoff Publishers, pp. 125–34.
54. Kartha, K.K. 1986. Production and indexing of disease-free plants. In *Plant tissue culture and its agricultural applications,* L.A. Withers and P.G. Alderson. London: Butterworth, pp. 219–38.
55. Kester, D.E. 1976. Noninfectious bud failure in almond. In *Virus diseases and noninfectious disorders of stone fruits in North America.* U.S. Dept. Agr. Handbook 437. Washington, D.C.: U.S. Govt. Printing Office, pp. 278–82.
56. ———. 1976. The relationship of juvenility to plant propagation. *Proc. Inter. Plant Prop. Soc.* 26:71–84.
57. ———. 1983. The clone in horticulture. *HortScience* 18(6):831–37.
58. Kester D.E. and T. Gradziel. 1996. Genetic disorders. In *Almond production manual,* W.C. Micke, ed. Univ. of Calif. Publ. 3364.
59. Klekowski, E.J. Jr., 1988. *Mutation, developmental selection and plant evolution.* New York: Columbia Univ. Press.
60. Knight, T.A. 1795. Observations on the grafting of trees. *Phil. Trans. Roy. Soc.,* London 85:290.
61. Kovar, J.L. and R.O. Kuckenbuch. 1994. Commercial importance of adventitious rooting to agronomy. In *Biology of adventitious root formation.* T.D. Davis and B.E. Haissig, eds. New York and London: Plenum Press.
62. Lacey, C.N.D., and A.I. Campbell. 1987. Selection, stability and propagation of mutant apples. In *Improving vegetatively propagated crops,* A.J. Abbott and R.K. Atkin, eds. New York: Academic Press, pp. 349–62.
63. Lawson, R.H. 1986. Pathogen detection and elimination. In *Tissue culture as a plant production system for horticultural crops,* R.H. Zimmerman, R.J. Griesbach, F.A. Hammerschlag, and R.A. Lawson, eds. Dordrecht: Martinus Nijhoff Publishers, pp. 97–118.
64. Libby, W.J. and M.R. Ahuja. 1993. The genetics of clones. In *Clonal Forestry,* I. M.R. Ahuja and W.J. Libby eds. Berlin, Heidelberg: Springer-Verlag.
65. Libby, W.J., and J.V. Hood. 1976. Juvenility in hedged radiata pine. *Acta Hort.* 56:91–98.
66. Marcotrigiano, M. 1990. Genetic mosaics and chimeras: Implications in biotechnology. In *Biotechnology in agriculture and forestry.* Vol 11. Y.P.S. Bajaj, ed. Berlin, Heidelberg: Springer-Verlag.
67. Marcotrigiano, M. 1991. Understanding foliar variegation as it relates to propagation. *Proc. Intern. Plant Prop. Soc.* 41:410–415.
68. Mather, S.M. 1961. Nursery stock registration and certification in California. *Bull. Dept. of Agr., State of Calif.* 1:173–184.
69. McClintock, B. 1984. The significance of responses of the genome to challenge. *Science* 226:792–801.

70. McDonald, B. 1986. *Practical woody plant propagation for nursery growers.* Portland, OR: Timber Press.

71. McGranahan, G., and H.I. Forde. 1985. Relationship between clone age and selection trait expression in mature walnuts. *Jour. Amer. Soc. Hort. Sci.* 110(5):692–95.

72. Melhus, I.E. and G.C. Kent. 1939. *Elements of Plant Pathology.* New York: Macmillan.

73. Miller, E.V. 1954. The natural origins of some popular varieties of fruit. *Econ. Bot.* 8:337–48.

74. Molisch, H. 1938. *The longevity of plants* (1928). Lancaster, Pa.: E. Fulling (English translation).

75. Moore, G.A. and R.E. Durham. 1992. Molecular markers. In Hammerschlag, F.A. and R.E. Litz. *Biotechnology of perennial fruit crops.* Wallingford, Eng. C.A.B. International.

76. Moore, J.M., and J. Janick, eds. 1996. *Advances in fruit breeding.* Vol. I-IV. West Lafayette, Ind.: Purdue Univ. Press.

77. Moore, J.M. and J. Janick, eds. 1983. *Methods of fruit breeding.* West Lafayette, Ind.: Purdue Univ. Press.

78. Mullins, M.G., Y. Nair, and P. Sampet. 1979. Rejuvenation *in vitro:* Induction of juvenile characters in an adult clone of *Vitis vinifera. Ann. Bot.* 44:623–27.

79. Nelson, S.H. 1977. Loss of productivity in clonal apple rootstocks. *Proc. Inter. Plant. Prop. Soc.* 27:350–55.

80. Nemeth, M. 1986. *Virus, mycoplasma and rickettsia diseases of fruit trees.* Dordrecht: Martinus Nijhoff Publishers.

81. Nevers, P., N.S. Shepherd, and H. Saedler. 1986. Plant transposable elements. In *Advances in botanical research.* J. A. Callow, ed. Orlando, Fla.: Academic Press, pp. 103–203.

82. Nyland, G., R.M. Gilmer, and J.D. Moore. 1976. "Prunus" ring spot group. In *Virus diseases and noninfectious disorders of stone fruits in North America.* U.S. Dept. Agr. Handbook 437. Washington, D.C.: U.S. Govt. Printing Office, pp. 104–32.

83. Nyland G. and A.C. Goheen. 1969. Heat therapy of virus disease of perennial plants. *Ann. Rev. Phytopath.* 7:331–54.

84. Oglevee-O'Donovan, W. 1986. Production of culture virus-indexed geraniums. In *Tissue culture as a plant production system for horticultural crops,* R.H. Zimmerman, R.J. Griesbach, F.A. Hammerschlag, and R.H. Lawson, eds. Dordrecht: Martinus Nijhoff Publishers, pp. 119–24.

85. Oleson, P O. 1978. On cyclophysis and topophysis. *Silvae Genetica* 27:173–78.

86. Potrylus, I. 1993. Gene transfer to plants: approaches and available techniques. In *Plant Breeding: principles and prospects.* M.D. Hayward, N.O. Bosemark and I. Romagosa. London: Chapman & Hall. pp. 126–137.

87. Perez de la Vega, M. 1993. Biochemical characterization of populations. In *Plant breeding: principles and prospects,* M.D. Hayward, N.O. Bosemark and I. Romagosa. pp. 184–200.

88. Powell, C.L., and D.J. Bagyaraj, eds. 1984. *Mycorrhizae.* Boca Raton, Fla.: CRC Press.

89. Quak, F. 1977. Meristem culture and virus-free plants. In *Applied and fundamental aspects of plant cell, tissue, and organ culture,* J. Reinert and Y. P. S. Bajaj, eds. Berlin, Springer-Verlag, pp. 598–615.

90. Raju, B.C., and J.C. Trolinger. 1986. Pathogen indexing in large-scale propagation of florist crops. In *Tissue culture as a plant production system for horticultural crops.* R.H. Zimmerman, R.J. Griesbach, F.A. Hammerschlag, and R.H. Lawson, eds. Dordrecht: Martinus Nijhoff Publishers, pp. 135–38.

91. Ritchie, G.A. Commercial application to adventitious rooting in forestry. In *Biology of adventitious root formation,* T.D. Davis and B.E. Haissig, eds. New York and London: Plenum Press. pp. 37–52.

92. Ritchie, G.A. 1993. Production of Douglas-Fir, *Pseudotsuga menziesii* (Mirb.) Franco, rooted cuttings for reforestation by Weyerhaeuser Company. *Proc. Intern. Plant Prop. Soc.* 43:284–288.

93. Robinson, L.W. and P.F. Wareing. 1969. Experiments on the juvenile-adult phase change in some woody species. *New Phytol.* 68:67–78.

94. Robinson, R.A. 1996. *Return to resistance.* Davis, CA: AgAccess.

95. Rugini, E., A. Pellegrineschi, A. Mencuccini, M & D. Maruitti. 1991. Increase of rooting ability in the woody species kiwi (*Actinidia deliciosa* a chev) by transformation with *Agrobacterium rhizogenes* rol genes. *Plant Cell Reports* 10:191–195.

96. Shamel, A.D., C.S. Pomeroy, and R.E. Caryl. 1929. Bud selection in the Washington Navel orange; progeny tests of limb variations. *USDA Tech. Bul.* 123.

97. Shepherd, J.F., D. Bidney and E. Shakin. 1980. Potato protoplasts in crop improvement. *Science* 208: 17–24.

98. Simmonds, N.W., ed. 1979. *Evolution of cultivated crops.* London: Longman.

99. Siniscalco, C. and L. Pavolettoni, 1988. Rejuvenation of *Eucalyptus x trabuti* by successive grafting. *Acta Hort.* 227:98–100.

100. Soost, R.K., J.W. Cameron, W.P. Bitters, and R.G. Platt. 1961. Citrus bud variation. *Calif. Citrograph* 46:176, 188–93.

101. Stace-Smith, R. 1985. Role of plant breeders in dissemination of virus diseases. *HortScience* 20(5):834–37.

102. Stewart, R.N. 1978. Ontogeny of the primary body in chimeral forms of higher plants. In *The clonal*

basis of development, S. Subtelny et al., eds. New York: Academic Press, pp. 131–60.

103. ———. 1965. The origin and transmission of a series of plastogene mutants in *Dianthus and Euphorbia. Genetics* 52:925–47.

104. Stewart, R.N., F.G. Meyers, and H. Derman. 1972. *Camellia* + 'Daisy Eagleson', a graft chimera of *Camellia sasanqua* and *C. japonica. Amer. J. Bot.* 59:515–524.

105. St. John, T. 1994 Propagation of mycorrhizal plants for restoration. *Proc. Intern. Plant Prop. Soc.* 44:344–347.

106. St. John, T. and J.M. Evans. 1990. Mycorrhizal inoculation of container plants. *Proc. Intern. Plant Prop. Soc.* 40:222–232.

107. Stone, O.M. 1978. The production and propagation of disease free plants. In *Propagation of higher plants through tissue culture: A bridge between research and application,* K.W. Hughes et al., eds. Springfield, Va.: U.S. Dept. of Energy Tech. Inform. Center, pp. 25–34.

108. Stoskopf, N.C., D.T. Tomes and B.R. Christie. 1993. *Plant breeding theory and practice.* Boulder, CO: Westview Press.

109. Stout, A.B. 1940. The nomenclature of cultivated plants. *Amer. Jour. Bot.* 27:339–47.

110. Stoutemyer, V.T. 1937. Regeneration in various types of apple wood. *Iowa Agr. Exp. Sta. Res. Bul.* 220:308–52.

111. Styer, D.J., and C.K. Chin. 1983 Meristem and shoot-tip culture for propagation, pathogen elimination, and germplasm preservation. In *Horticultural reviews,* Vol. 5, J. Janick, ed. Westport, Conn.: AVI Publ. Co., pp. 221–27.

112. Symons, R.H. 1980. Rapid indexing of sunblotch viroid in avocados and of exocortis viroid in citrus. *Proc. Inter. Plant Prop. Soc.* 30:578–83.

113. Tanaka, T. 1927. Bizzarria—a clear case of periclinal chimera. *Jour. Gen.* 18:77–85.

114. Tanksley, S.D. and T.J. Orton. 1983. *Isozymes in plant breeding and genetics.* Elsevier: Amsterdam.

115. Tilney-Bassett, R.A.E. 1986. *Plant chimeras.* London: Edward Arnold.

116. Tincker, M.A.H. 1945. Propagation, degeneration, and vigor of growth. *Jour. Roy. Hort. Soc.* 70:333–37.

117. Tolley, 1. S. 1975. A technique for the accelerated production of commercially acceptable citrus clones from seed. *Proc Inter. Plant Prop. Soc.* 25:294–97.

118. Tufts, W.P., and C.J. Hansen. 1931. Variation in shape of Bartlett pears. *Proc. Amer. Soc. Hort. Sci.* 28:627–33.

119. U.S. Dept. Agr. 1976. *Virus and other disorders with virus-like symptoms of stone fruits in North America.* U.S. Dept. Agr. Handbook 10. Washington, D.C.: U.S. Govt. Printing Office.

120. van Oosten, H.J., and H.H. van der Borg, eds. 1977. Symposium on clonal variation in apple and pear. *Acta Hor.* 75:1–185.

121. Vaughn, K. 1983. Chimeras: Problems in propagation. *HortScience* 18(6):845–48.

122. Verkade, S.D. 1986. Mycorrhizal inoculation during plant propagation. *Proc. Inter. Plant Prop. Soc.* 36:613–18.

123. Visser, T., H.H. Vergaegh and D.P. de Vries. 1976. A comparison of apple and pear seedlings with reference to the juvenile period. I. Seedling growth and yield. *Acta Hort.* 56:205–14.

124. Waterworth, H.E. 1993. Processing foreign plant germplasm at the National Plant Germplasm Quarantine Center. *Plant Dis.* 77:854–60.

125. Weeks, J.W. 1991. Producing virus-indexed liner plants for small fruit production. *Proc. Intern. Plant Prop. Soc.* 41:211–213.

126. Westwood, M.N. 1994. Temperate zone pomology. *Physiology and culture,* 3rd ed. Portland, OR: Timber Press.

127. Winkler, H. 1907. Uber Pfropfbastarde und pflanzliche Chimiiren. *Ber. Deuts. Bot. Ges.* 25:568–76.

128. ———. 1910. Uber die Nachkommenschaft der Solanum Pfropfbastarde und die Chromosomenzahlen ihrer Keimzellen. *Zetis. f Bot.* 2:1–38.

129. Wolbert, J., M.A. Walker, E. Weber and D. Roberts. 1995. *Clonal selection.* Proc. of Natl Symp. Davis CA.: Soc. Enol. and Vit.

130. Xoconostle-Cazares, B., E. Lozoya-Gloria and L. Herrera-Estrella. 1993. Gene cloning and identification. In *Plant breeding: principles and prospects.* M.D. Hayward, N.O. Bosemark, and I. Romagosa, eds. London: Chapman & Co.

131. Zimmerman, R.H. 1971. Flowering in crabapple seedlings: Methods of shortening the juvenile phase. *Jour. Amer. Soc. Hort. Sci.* 96(4):404–11.

132. Zobel, B.J., G. van Wyk, and P. Stahl. 1987. *Growing exotic forests.* New York: John Wiley.

133. Zohary, D., and P. Spiegel-Roy. 1975. Beginnings of fruit growing in the old world. *Science* 187(4174):319–27.

SUPPLEMENTARY READING

ABBOTT, A. J., and P. K. ATKIN, eds. 1987. *Improving vegetatively propagated crops.* New York: Academic Press.

BAKER, K. F., ed. 1957. *The U. C. system for producing healthy container-grown plants.* Calif. Agr. Exp. Sta. Man. 23.

BROERTJES C., and A. M. VAN HARTEN. 1987. Application of mutation breeding methods. In *Improving vegetatively propagated crops,* A. J. Abbott and R. K. Atkin, eds. New York: Academic Press, pp. 335–48.

CAMPBELL, A. I., ed. 1986. Workshop on clonal selection in tree fruit. *Acta Hort.* 180:1–131.

CONVERSE, R. H., ed. 1987. *Virus diseases of small fruits.* U.S. Dept. Agr. Handbook 631. Washington, D.C.: U.S. Govt. Printing Office.

FRAZIER, N. W., ed. 1970. *Virus diseases of small fruits and grapevines.* Berkeley: Univ. Calif. Div. Agr. Sci.

MARSTON, M. E. 1955. The history of vegetative propagation. *Rept. 4th Inter. Hort. Cong.,* Vol. 2, pp. 1157–64.

NEMETH, M. 1986. *Virus, mycoplasma and rickettsia diseases of fruit trees.* Dordrecht: Martinus Nijhoff Publishers.

POSNETTE, A. F., ed. 1963. *Virus diseases of apples and pear.* Tech. Comm. 30. Bucks, England: Commonwealth Agricultural Bureaux, Farnham Royal.

Proceedings of symposium. 1983. The clone in horticulture. *HortScience* 18(6):829–48.

Proceedings of symposium. 1988. Genetic considerations in the collection and maintenance of germplasm. *HortScience* 23(1):78–97.

Proceedings of symposium. 1988. Temperate fruit crops: History and management of genetic resources. *HortScience 23*(1):49–75.

Proceedings of symposium. 1985. Virus diseases: A dilemma for plant breeders. *HortScience* 20(5):831–59.

SIMMONDS, N. W., ed. 1979. *Evolution of cultivated crops.* Longman: London.

10

The Biology of Propagation by Cuttings

The main focus of this chapter is on adventitious root formation, since it is the primary regenerative process required in most cutting propagation. Adventitious bud and shoot development, events important in the regeneration of leaf and root cuttings, are also discussed.

Vegetative or clonal reproduction is the most important propagation method used for the commercial production of many, if not most, horticultural crops (ornamentals, fruits, nuts, and vegetables). Adventitious root formation is a prerequisite to successful cutting propagation. In forestry, cutting propagation has been around for hundreds of years. Vegetative propagation of forest planting stock through adventitious rooting is one of the most exciting emerging technologies in forestry. Yet, many economically important woody plants have a low genetic and physiological capacity for adventitious root formation, which limits their commercial production. Furthermore, rooting and acclimatization of tissue-culture produced plants will need to be improved if biotechnology (manipulating genes for new flower color, disease resistance, fruit yield, etc.) is to be incorporated into the propagation and production of genetically transformed woody plant species. Labor costs contribute as much as 80 percent of propagation expenditures, so there is considerable financial pressure to streamline propagation techniques and improve rooting success.

Commercial propagators have developed technologies that successfully manipulate environmental conditions to maximize rooting, i.e., intermittent mist and fog systems, temperature and light manipulation. What has lagged behind is the knowledge of the biochemistry, the genetic and molecular manipulation of rooting. *After years of very intensive research, the fundamental biology of what triggers adventitious root formation remains largely unknown.* The new tools of biotechnology offer exciting opportunities to understand the molecular keys to rooting, and to enable propagators to develop cultivars that can be commercially rooted.

The first section of this chapter explores the *scientific approaches utilized to understand the regenerative process of adventitious root and bud formation.* This includes:

- the descriptive—observations made of rooting and bud formation,
- correlative effects—studies on how hormonal control affects rooting and bud formation,
- the biochemical basis for adventitious root formation, and
- molecular—the biotechnological advances in asexual propagation.

The second section of the chapter deals with *factors affecting rooting and bud formation.* This includes:

- the management of stock plants to maximize cutting propagation,
- treatment of cuttings, and
- environmental manipulation of cuttings.

FIGURE 10–1 The ultimate in adventitious root production is shown on this banyan tree *(Ficus benghalensis)* at Hilo, Hawaii. Such roots arise from branches, extend downward, and grow into the soil.

DESCRIPTIVE OBSERVATIONS OF ADVENTITIOUS ROOT AND BUD FORMATION

Propagation by **stem** and **leaf-bud cuttings (single-eye cuttings)** requires only that a new adventitious root system be formed, because a potential shoot system (a bud) is already present. **Root cuttings** and **leaf cuttings** must initiate both a new shoot system—from an adventitious bud—as well as new adventitious roots.

The formation of adventitious roots and buds is dependent on plant cells to **dedifferentiate** and develop into either a root or shoot system. The process of **dedifferentiation** is the capability of previously developed, differentiated cells to initiate cell divisions and form a new meristematic growing point. Since this characteristic is more pronounced in some cells and plant parts than in others, the propagator must do some manipulation to provide the proper conditions for plant regeneration. A sound understanding of the underlying biology of regeneration is very helpful in this regard.

Adventitious Root Formation

Adventitious roots form naturally on various plants. Corn, pandanus, and other monocots develop "brace" roots, which arise from the intercalary regions at the base of internodes. Banyan trees (Figure 10–1) produce long aerial roots from their shoots that grow to and into the ground. Plants that are regenerated from rhizomes, bulbs, and other such structures also develop adventitious roots.

Adventitious roots are of two types: **preformed roots** and **wound-induced roots** (Figures 10–2, 10–3, 10–4, and 10–5). Preformed root initials and primordia develop naturally on stems while they are still attached to the parent plant and roots may or may not emerge prior to severing the stem piece.

Preformed, or **latent root initials** (*192*) generally lie dormant until the stems are made into cuttings and placed under environmental conditions favorable for further development and emergence of the primordia as adventitious roots. In *Populus* × *robusta,* root initials form in stems in midsummer and then emerge from cuttings made the following spring (*266*). In some spiecies, primordia develop into aerial roots on the intact plant and become quite prominent (Figures 10–4 and 10–5). Such preformed root initials occur in a number of easily rooted genera, such as willow (*Salix*), hydrangea (*Hydrangea*), poplar (*Populus*), coleus, jasmine (*Jasminum*), currant (*Ribes*), citron (*Citrus medic*), and others (*100*). The position of origin of these preformed root initials is similar to *de novo* adventitious root formation (Table 10–1) (*192*). In some of the clonal apple rootstocks and in old trees of some apple and quince cultivars, these preformed latent roots cause swellings, called **burr knots.**[1] Species with preformed root initials generally root rapidly

[1]Burr knots are not desirable and are selected against in modern apple rootstock breeding programs. Though rooting of cuttings is easier, clusters of burr knots can later girdle the stem.

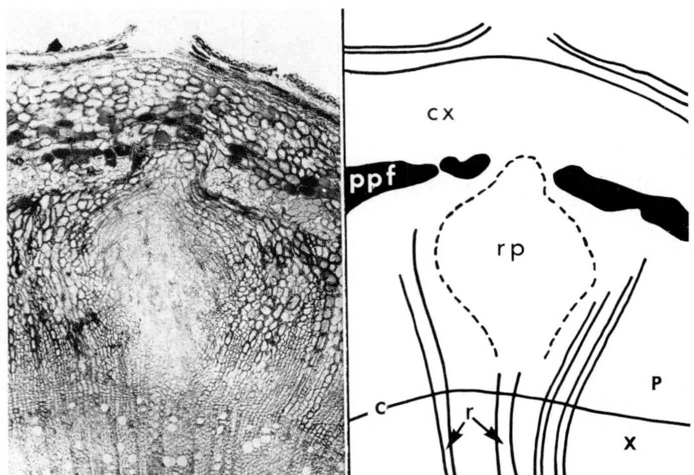

FIGURE 10–2 Tissues involved in adventitious root formation in 'Brompton' plum hardwood cuttings. cx, cortex; ppf, primary phloem fibers; rp, root primordium; p, phloem; r, rays; c, cambium; x, xylem. (Courtesy A. Beryl Beakbane, East Malling Research Station (*13*).)

and easily, but cuttings of many species without such root initials root just as easily.

In willow, latent root primordia can remain dormant, embedded in the inner bark for years if the stems remain on the tree (*40*). Their location can be observed by peeling off the bark and noting the protuberances on the wood, with corresponding indentations on the inside of the bark that was removed.

Wound-induced roots, on the other hand, develop only after the cutting is made, in response to wounding in preparing the cutting. In effect, they are considered to be formed **de novo** (anew) (*58*).

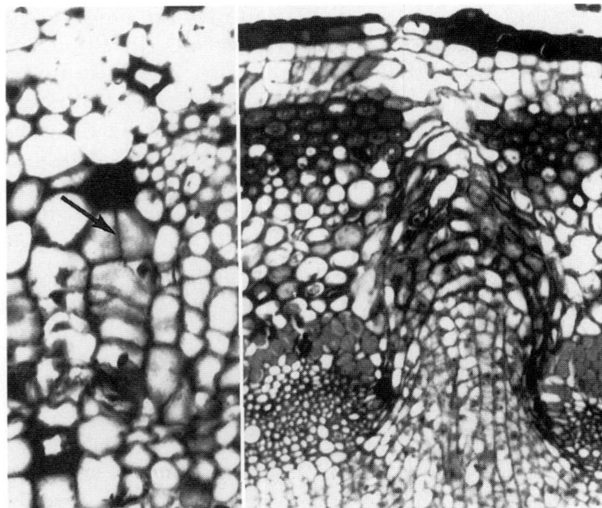

FIGURE 10–3 Developmental stage of rooting in mature cutting of *Ficus pumila.* First anticlinal division of a phloem ray cell occurs during dedifferentiation *(left)* (see arrow), and a young root primordium elongates through the cortex *(right)* (*58*).

Any time living cells at the cut surfaces are injured and exposed, a **response to wounding** begins (*45*). The subsequent wound response and root regeneration process includes three steps:

1. The outer injured cells die, a necrotic plate forms, the wound is sealed with a corky material **(suberin)**, and the xylem plugs with gum. This plate protects the cut surfaces from desiccation and pathogens.
2. Living cells behind this plate begin to divide after a few days and a layer of parenchyma cells **(callus)** forms a wound periderm.
3. Certain cells in the vicinity of the vascular cambium and phloem begin to divide and initiate *de novo* adventitious roots.

The developmental changes that occur in *de novo* adventitious root formation of wounded roots can generally be divided into four stages:

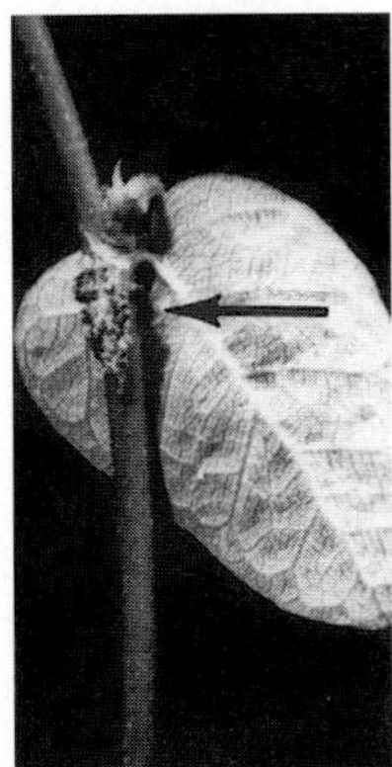

FIGURE 10–4 Preformed aerial roots at node of *Ficus pumila.*

FIGURE 10–5 Clusters of adventitious roots developing from preformed root initials at the base of shoots of the 'Colt' cherry rootstock. (Photographed at East Malling Research Station, England.)

1. **Dedifferentiation** of specific differentiated cells.
2. **Formation of root initials** from certain cells near vascular bundles, or vascular tissue, which have become meristematic by dedifferentiation.
3. Subsequent development of root initials into organized **root primordia.**
4. Growth and **emergence** of the root primordia outward through other stem tissue plus the formation of vascular (conducting) tissue between the root primordia and the vascular tissues of the cutting.

Alternatively, other scientists have divided the process of adventitious root formation in Monterrey pine (Pinus radiata) into three stages (264, 265), or five distinct stages for hypocotyl cuttings of bean (Vigna) (258).

The precise location inside the stem where adventitious roots originate has intrigued plant anatomists for centuries. Probably the first study of this phenomenon was made in 1758, by a French dendrologist, Duhamel du Monceau (*70*). A great many subsequent studies have covered a wide range of plant species (*7, 100, 192*).

Adventitious roots usually originate on **herbaceous plants** just outside and between the vascular bundles (*229*), but the tissues involved at the site of origin can vary widely depending upon the kind of plant (Figure 10–6) and propagation technique (*1*). In tomato, pumpkin (*223*), and mung bean (*22*), adventitious roots arise in the phloem parenchyma; in *Crassula* they arise in the epidermis (*201*); and in coleus they originate from the pericycle (*39*).

Adventitious roots in castor bean (*Ricinus communis*) cuttings arise between the vascular bundles (Figure 10–7). Root initials in carnation cuttings arise in a layer of parenchymatous cells inside a fiber sheath; the developing root tips, upon reaching this band of impenetrable fiber cells, do not push through it, but turn downward, emerging from the base of the cutting (*269*).

TABLE 10–1

Origin of preformed root initials (primordia, burr knots, and/or rootgerms) in stems of woody plants

Origin	*Genera*
Rays	
Wide rays	*Populus*
Medullary rays, associated with buds	*Ribes*
Nodal and connected with wide radial bands of parenchyma	*Salix*
Internodal medullary rays	*Salix*
Medullary ray	*Citrus*
Phloem ray parenchyma	*Hydrangea*
Cambium	
Cambial ring in branch and leaf gap; 1° and 2° medullary rays	*Malus*
Cambial region of an abnormally broad ray	*Acer, Chamaecyparis, Fagus, Fraxinus, Juniperus, Populus, Salix, Taxus, Thuja, Ulmus*
Leaf and Bud Gaps	
Bud gap	*Cotoneaster*
Median and lateral leaf trace gaps at node	*Lonicera*
Parenchymatous cells in divided bud gap	*Cotoneaster*

Source: M. B. Jackson (ed.). 1986 *New root formation in plants and cuttings.* Martinus Nijhoff Publishers. Dordrecht, The Netherlands.

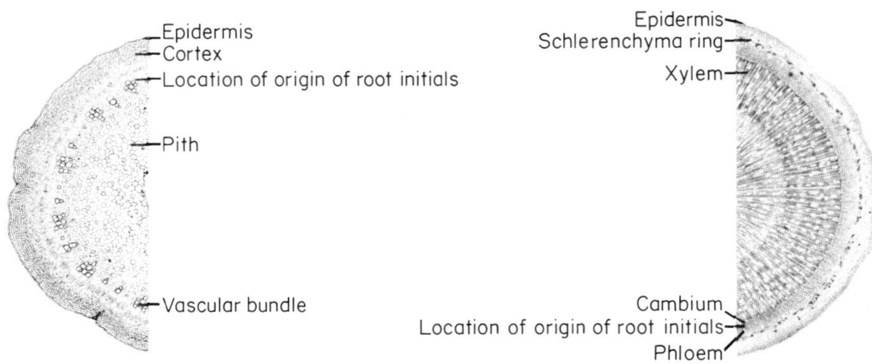

FIGURE 10–6 Stem cross sections showing the usual location of origin of adventitious roots. *Left:* Young, herbaceous, dicotyledonous plant. *Right:* Young, woody plant.

Adventitious roots in stem cuttings of **woody perennial plants** usually originate from living parenchyma cells, in the young, secondary phloem (Figure 10–2), but sometimes in vascular rays, cambium, phloem, callus, or lenticels (*99, 129, 192*) (Table 10–2). Two patterns of adventitious root formation emerge: **direct root formation** of cells in close proximity to the vascular system (i.e., generally more easy-to-root species); and **indirect root formation,** where nondirected cell divisions, including callus formation, occur for an interim period before cells divide in an organized pattern to initiate adventitious root primordium (i.e., generally more difficult-to-root species). See the flow diagram of adventitious root formation (Figure 10–8) (*97, 192*).

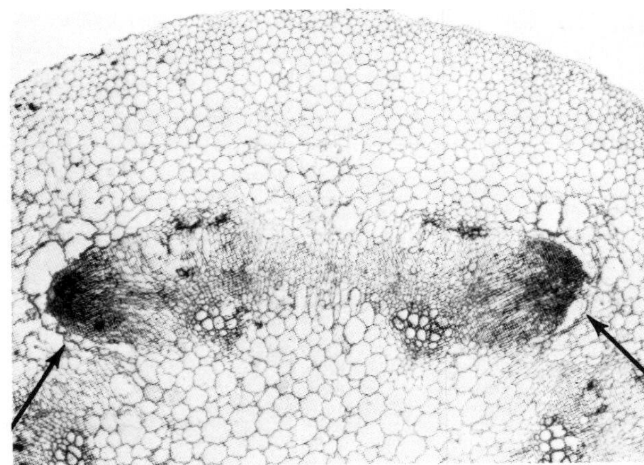

FIGURE 10–7 Adventitious root primordia (arrows) arising laterally and adjacent to vascular bundles in the castor bean *(Ricinus communis),* a herbaceous plant. Vascular connections will develop between the adventitious roots and the plant's vascular bundles. Epidermis is at top, pith at bottom. (From Priestley and Swingle (*229*).)

Generally, the origin and development of *de novo* adventitious roots takes place next to and just outside the central core of vascular tissue. Many easy-to-root woody plant species develop adventitious roots from phloem ray parenchyma cells. Figure 10–3 depicts the first *anticlinal division* of a phloem ray cell during dedifferentiation and a young root primordium elongating through the cortex. Upon emergence from the stem (Figure 10–9), the adventitious roots have already developed a root cap as well as a complete vascular connection with the originating stem.

The time for root initials to develop after cuttings are placed in the propagating bed varies widely. In one study (*269*), they were first observed microscopically after three days in chrysanthemum, five days in carnation (*Dianthus caryophyllus*), and seven days in rose (*Rosa*). Visible roots emerged from the cuttings after ten days for the chrysanthemum, but three weeks were required for the carnation and rose.

Phloem ray parenchyma cells in juvenile (easy-to-root) cuttings of *Ficus pumila* undergo early anticlinal cell division and root primordia formation more quickly than mature (difficult-to-root) plants under optimal auxin treatments (Table 10–3). Once primordia are formed there is a comparable time period (seven to eight days) between root primordia elongation (emergence) and maximum rooting in both the easy- and difficult-to-root plants (*58*). This also was reported with *Agathis australis,* where primordia formation was variable in cuttings from different-aged stock plants, but once root primordia formed, root emergence consistently occurred within a three-to-four-week period (*192, 313, 314*).

Stem Structure and Rooting

There have been attempts to correlate stem structure with rooting ability of cuttings. A continuous **sclerenchyma ring** (Figure 10–6) between the

TABLE 10–2

Origin of wound-induced *de novo* adventitious roots in stems of woody plants

Origin	Genera
Cambial and ray	
Cambial and phloem portions of ray tissues	*Acanthopanax, Chamaecyparis, Cryptomeria, Cunninghamia, Cupressus, Metasequoia*
Medullary rays	*Vitis*
Cambium	*Acanthus, Lonicera*
Fascicular cambium	*Clematis*
Phloem ray parenchyma	*Ficus, Hedera*
Secondary phloem in association with a ray	*Malus* (Malling stocks), *Camellia,* 'Brompton' plum
Phloem area close to the cambium	*Pistacia*
Cambium and inner phloem ray also in leaf gap	*Griselinia*
Bud and leaf gaps	
Outside the cambium in small groups	*Rosa, Cotoneaster, Pinus, Cephalotaxus, Larix, Sciadopitys, Malus, Acanthus*
Pericycle	
Callus, internal	
Irregularly arranged parenchymatous tissues	*Abies, Juniperus, Picea, Sequoia*
Callus, external	
Callus tissues (external)	*Abies, Cedrus, Cryptomeria, Ginkgo, Larix, Pinus, Podocarpus, Sequoia, Sciadopitys, Taxodium, Pinus*
Bark and basal callus	*Citrus*
Within callus at base of cutting	*Pseudotsuga*
Other	
Hyperhydric outgrowth of the lenticels	*Tamarix*
Margin of differentiating resin duct or parenchyma within the inner cortex	*Pinus*

Source: M. B. Jackson (ed.). 1986. *New root formation in plants and cuttings.* Martinus Nijhoff Publishers. Dordrecht, The Netherlands.

phloem and cortex, exterior to the point of origin of adventitious roots, occurs as the stem matures and gets older. Such a ring occurs in difficult-to-root species such as olive stem cuttings (*44*), mature *Hedera helix* (*102*), and *Ficus pumila* (*58*), while easy-to-root types were characterized by discontinuity or fewer cell layers of this scelerenchyma ring (*13*).

With most difficult-to-root species, stem structure does not influence rooting potential. While a sheath of lignified tissue in stems may in some cases act as a *mechanical barrier* to root emergence, there are so many exceptions that this *is not* the primary cause of rooting difficulty. Moreover, auxin treatments and rooting under mist (*191*) cause considerable cell expansion and proliferation in the cortex, phloem, and cambium, resulting in breaks in continuous sclerenchyma rings—yet in difficult-to-root cultivars, even with wounding, there is still no formation of root initials.

Easily rooted carnation cultivars have a band of sclerenchyma present in the stems, yet the developing root primordia emerge from the cuttings by growing downward and out through the base (*269*). In other plants, in which an impenetrable ring of sclerenchyma could block root emergence, this same rooting pattern can occur. Rooting is related to the genetic potential and physiological conditions for root initials to form, rather than to the mechanical restriction of a sclerenchyma ring barring root emergence (*58, 251, 312*).

Callus

Callus is an irregular mass of parenchyma cells in various stages of lignification that commonly develops at the basal end of a cutting placed under environmental conditions favorable for rooting. Callus growth proliferates from cells at the base of the cutting, primarily from the vascular cambium, although cells of the cortex and pith may also contribute to its formation (Table 10–2).

Roots frequently emerge through the callus, leading to the belief that callus formation is essential for rooting. In easy-to-root species, the formation of callus and the formation of roots are independent of each other, even though both involve cell division. Their simultaneous occurrence is due to their dependence upon similar internal and environmental conditions.

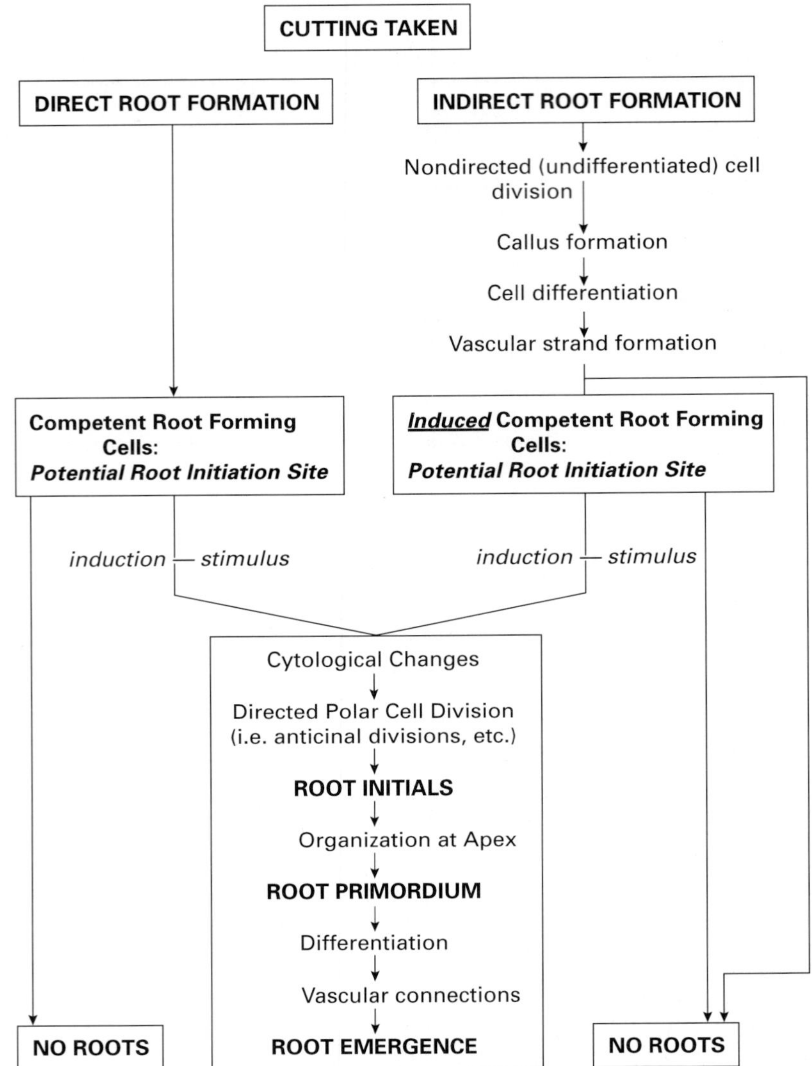

FIGURE 10–8 Flow diagram of adventitious root formation through **direct** (cells in close proximity to vascular system—i.e., generally more easy-to-root species) and **indirect model** (interim period of undifferentiated cell division—i.e., generally more difficult-to-root species). When a potential root initiation site is already present the initial cell divisions lead to root production *in situ.* When a site is not present alternative routes leading to the creation of a site are shown. Rooting does not always occur. (Modified from Lovell and White *(192)* and Geneve (*97*).)

Callus formation is a precursor of adventitious root formation in some species, however. Origin of adventitious roots from callus tissue has been associated with difficult-to-root species (Table 10–2) (*58, 149*), such as *Pinus radiata* (*38*), *Sedum* (*321*), and the mature phase of *Hedera helix* (*101*). Adventitious roots originate in the callus tissue formed at the base of the cutting and from "tracheary nests," i.e., in callus of *Ficus pumila* (Figure 10–10). It is possible to have adventitious roots originating from different tissues on the same cutting—epicotyl stem cuttings of *Pinus sylvestris* can form roots from resin duct wound (callus) tissue, central and basal wound (callus) tissue, and vascular tissue (Figure 10–11) (*92*).

Leaf Cuttings

Many plant species, including both monocots and dicots, can be propagated by leaf cuttings (*115*). The origin of new shoots and new roots in leaf cuttings is quite varied, and develops from primary or secondary meristems.

Preformed primary meristems are groups of cells directly descended from embryonic cells that

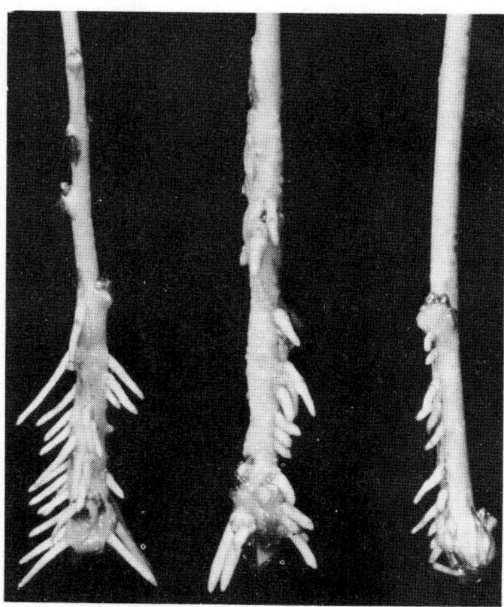

FIGURE 10–9 Emergence of adventitious roots in plum stem cuttings. Observe the tendency of the roots to form in longitudinal rows, which appear directly below buds.

have never ceased to be involved in meristematic activity.

Wound-induced, secondary meristems are groups of cells that have differentiated and functioned in some previously differentiated tissue system and then dedifferentiate into new meristematic zones *(de novo)*, resulting in the regeneration of new plant organs. This is the most common type of meristems in leaf cuttings.

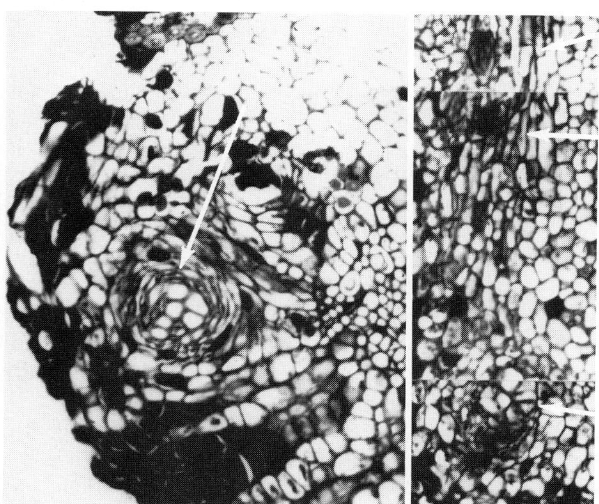

FIGURE 10–10 *Left:* Root primordia (see arrow) in callus originating in the vicinity of differentiating tracheary elements which have been described as "callus xylem" or "tracheary nests." *Right:* Callus primordia distinguished by "tracheary nests" (see arrows) connected to the main vascular system of *Ficus pumila* (*58*).

TABLE 10–3

Time of adventitious root formation in juvenile and mature leaf-bud cuttings of *Ficus pumila* treated with IBA

	Juvenile	***Mature***
Anticlinal cell divisions of ray parenchyma	Day 4	Day 6
Primordia	Day 6	Day 10
First rooting[a]	Day 7	Day 20
Maximum rooting[b]	Day 14	Day 28

Source: Ref. *58*.

[a]Based on 25 percent or more cuttings with roots protruding from stem.

[b]Based on 100 percent rooting and maximum root number.

Leaf Cuttings with Preformed, Primary Meristems

Detached leaves of *Bryophyllum* produce small plantlets from notches around the leaf margin (see Figure 11–16). These small plants originate from so-called foliar "embryos," formed in the early stages of leaf development from small groups of vegetative cells at the edges of the leaf. As the leaf expands, a foliar embryo develops until it consists of two rudimentary leaves with a stem tip between them, two root primordia, and a "foot" that extends toward a vein (*142, 320*). As the leaf matures, cell division in the foliar embryo ceases, and it remains dormant. If the leaf is detached and placed in close contact with a moist rooting medium, the young plants rapidly break through the leaf epidermis and become visible in a few days. Roots extend downward, and after several weeks many new independent plants form while the original leaf dies. The new plants thus develop from latent primary meristems—from cells that have not fully differentiated. Production of new plants from leaf cuttings by the renewed activity of primary meristems is found in species such as the piggyback plant (*Tolmiea*) and walking fern (*Camptosorus*).

Leaf Cuttings with Wound-Induced, Secondary Meristems

In leaf cuttings of *Begonia rex, Sedum,* African violet (*Saintpaulia*), snake plant (*Sansevieria*), *Crassula,* and lily, new plants may develop from secondary meristems arising from differentiated cells

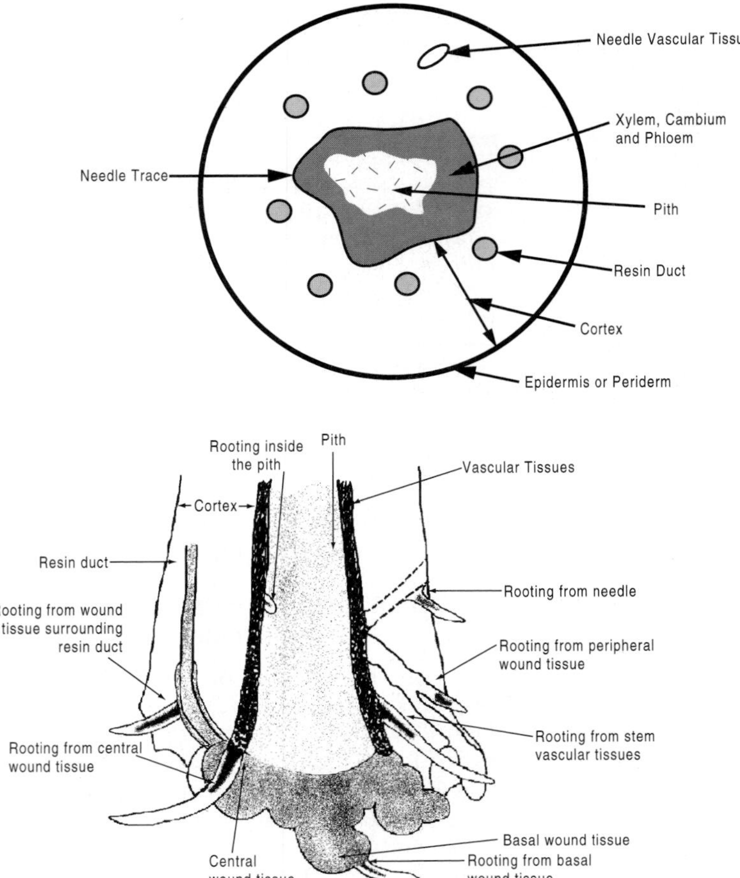

FIGURE 10–11 It is possible to have adventitious rooting originating from different tissues on the same cutting. Top: Tissue map of transverse section of epicotyl stem of one-year-old *Pinus sylvestris.* Bottom: Schematic longitudinal section showing examples of rooting occurring from resin duct wound (callus) tissue. No single cutting developed roots from all potential tissues. (Redrawn from Flygh, et al., (92).)

at the base of the leaf blade or petiole as a result of wounding.

In African violet, new roots and shoots arise *de novo* by the formation of meristematic cells from previously differentiated cells in the leaves. The roots are produced from thin-walled cells lying between the vascular bundles. The new shoots arise from cells of the subepidermis and the cortex immediately below the epidermis. Roots first emerge, form branch roots, and continue to grow for several weeks before adventitious buds and their subsequent development into adventitious shoots occurs. Root initiation and development are independent of adventitious bud and shoot formation (*302*). The same process occurs with many begonia species (Figure 10–12). Although the original leaf supplies nutrient materials to the young plant, it does not become a part of the new plant (*214*).

In *Lilium longiflorum* and *L. candidum,* the bud primordium originates in parenchyma cells in the upper side of the bulb scale, whereas the root primordium arises from parenchyma cells just below the bud primordium. Although the original scale serves as a source of food for the developing plant, the vascular system of the young bulblet is independent of that of the parent scale, which eventually shrivels and disappears (*305*).

In several species, for example, sweet potato, *Peperomia,* and *Sedum,* new roots and new shoots on leaf cuttings arise in callus tissue which develops over the cut surface through activity of secondary meristems. The petiole of *Sedum* leaf cuttings forms a considerable pad of callus within a few days after the cuttings are made. Root primordia are organized within the callus tissue, and shortly thereafter four or five roots develop from the parent leaf. Following this, bud primordia arise on a lateral sur-

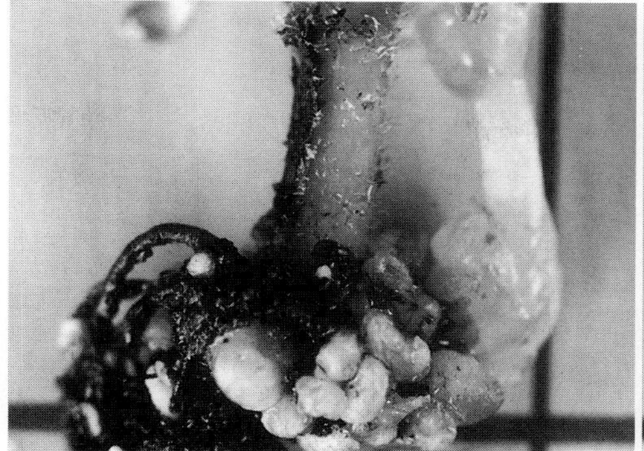

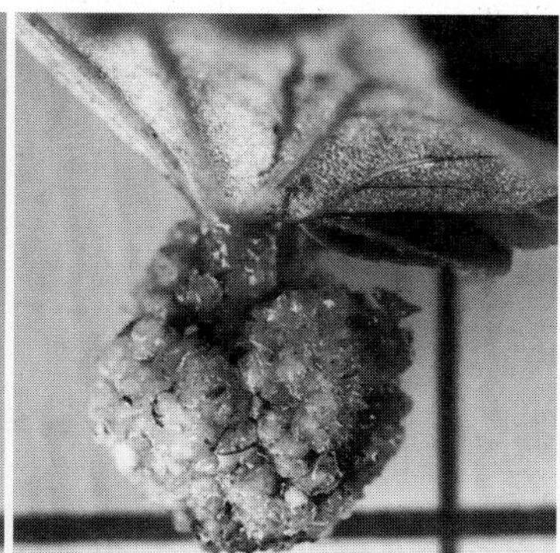

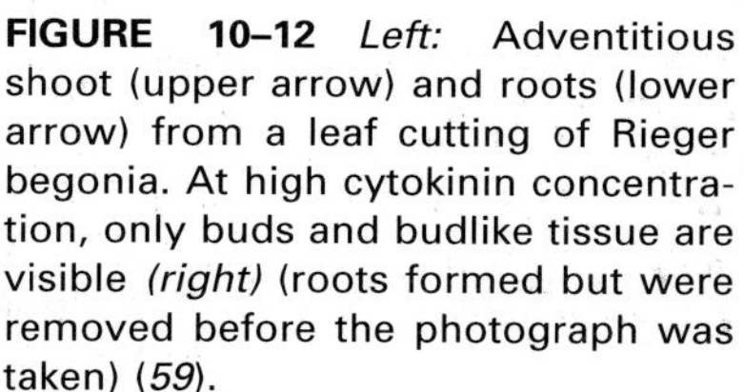

FIGURE 10–12 *Left:* Adventitious shoot (upper arrow) and roots (lower arrow) from a leaf cutting of Rieger begonia. At high cytokinin concentration, only buds and budlike tissue are visible *(right)* (roots formed but were removed before the photograph was taken) *(59)*.

face of the callus pad and develop into new shoots (*321*).

The limiting factor in leaf cutting propagation is generally the formation of adventitious buds, not adventitious roots. Adventitious roots form on leaves much more readily than do adventitious buds. In some plants, such as the India rubber fig (*Ficus elastica*), the cutting must include a portion of the old stem containing an axillary bud (a leaf-bud cutting) because although adventitious roots may develop at the base of the leaf, an adventitious shoot is not likely to form. In fact, rooted leaf cuttings of some species will survive for years without producing an adventitious shoot.

Root Cuttings

Development of adventitious shoots, and in many cases adventitious roots, must take place if new plants are to be regenerated from root pieces (root cuttings) (*243*). Regeneration of new plants from root cuttings takes place in different ways, depending upon the species. Commonly, the root cutting first produces an adventitious shoot, and later produces roots, often from the base of the new shoot rather than from the original root piece itself. With root cutting propagation of apples, and the storage roots of sweet potato—these adventitious shoots can be removed and rooted as stem cuttings when treated with auxin (*244*). In other plants, a well-developed root system has formed by the time the first shoots appear.

In some species, adventitious buds form readily on roots of intact plants, producing **suckers.** When roots of such species are dug, removed, and cut into pieces, buds are even more likely to form. In young roots, such buds may arise in the pericycle near the vascular cambium (*253*). The developing buds first appear as groups of thin-walled cells having a prominent nucleus and a dense cytoplasm (*80, 300*). In old roots, buds may arise in a callus-like growth from the phellogen; or they may appear in a callus-like proliferation from vascular ray tissue. Bud primordia may also develop from wound callus tissue that proliferates from the cut ends of injured surfaces of the roots (*229*), or they may arise at random from cortex parenchyma (*244*).

Sometimes regeneration of new root meristems on root cuttings is more difficult than the production of adventitious buds (*29*). New roots may not always be adventitious and can develop from latent lateral root initials contained in the root piece or

ROOT CUTTING PROPAGATION OF CHIMERAL PLANTS

One of the chief advantages claimed for asexual propagation is the exact reproduction of all characteristics of the parent plant. With root cuttings, however, this generalization does not always hold true. In periclinal chimeras, in which the cells of the outer layer are of a different genetic makeup from those of the inner tissues, the production of a new plant by root cuttings (derived from nonmutated inner tissues) results in a plant that is different in appearance from the parent. This is well illustrated in the thornless boysenberry, and the 'Thornless Evergreen' trailing blackberry, in which stem or leaf-bud cuttings produce plants that retain the (mutated) thornless condition, but root cuttings develop into (normal, nonmutated) thorny plants.

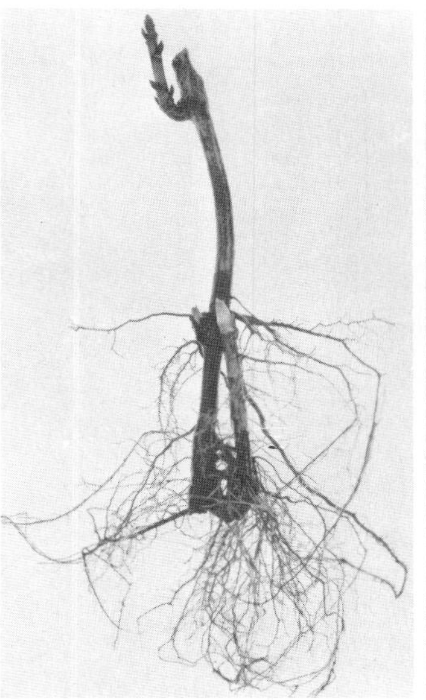

FIGURE 10–13 Results of planting a cutting of red currant *(Ribes sativum)* upside-down (reversed polarity). *Left:* Several months after starting. *Right:* One year later. The shoot from the center bud with new roots at its base has become the main plant and is growing with correct polarity. The shoot from the top bud, while still alive, has failed to develop normally.

attached lateral roots (*256*). Generally, such branch roots arise from differentiated cells of the pericycle adjacent to the central vascular cylinder (*19*). Adventitious root initials have been observed to arise in the region of the vascular cambium in roots.

A list of plants commonly propagated by root cuttings is given in Table 11–2.

Polarity

The polarity inherent in shoots and roots is shown dramatically in the rooting of cuttings (Figures 10–13 and 10–14). Polarity is the quality or condition inherent in a cutting that exhibits different properties in opposite parts, i.e., stem cuttings form shoots at the **distal** end (nearest the shoot tip), and roots form at the **proximal** end (nearest the *crown,* which is the junction of the shoot and root system). Root cuttings of many species form roots at the distal end and shoots at the proximal end. Changing the position of the cuttings with respect to gravity does not alter this tendency (*26*) (Figure 10–14). Polarity is also observed in leaf cuttings even though roots and shoots arise at the same position, usually the base of the cutting (see Figure 10–12).

In 1878, Vöchting (*304*) advanced the theory that polarity could be attributed to individual cellular components, since no matter how small the piece, regeneration was consistently polar. A general explanation of polarity is that when tissue segments are cut, the physiological unity is disturbed. This must cause a redistribution of some substance, probably auxin, thus accounting for the different growth responses. The correlation of polarity of root initiation with auxin movement has been noted in several instances (*122, 195, 245, 261, 307*). It is also known that the polarity in auxin transport varies in intensity among different tissues. The polar movement of auxins is an active transport process, mediated by a membrane transport carrier, which occurs in phloem parenchyma cells (*184*).

FIGURE 10–14 Polarity of root regeneration in grape hardwood cuttings. Cuttings at left were placed for rooting in an inverted position, but roots still developed from the morphologically basal (proximal) end. Cuttings on right were placed for rooting in the normal, upright orientation with roots forming at the basal end.

CORRELATIVE EFFECTS: HOW HORMONAL CONTROL AFFECTS ADVENTITIOUS ROOT AND BUD FORMATION

The Effects of Buds and Leaves

In 1758, Duhamel du Monceau (*70*) explained the formation of adventitious roots in stems on the basis of the downward movement of sap. In extending this concept, Sachs, the great German plant physiologist, in 1882 postulated the existence of a specific root-forming substance manufactured in the leaves, which moves downward to the base of the stem, where it promotes root formation (*250*). It was shown by van der Lek in 1925 that strongly sprouting buds promote the development of roots just below the buds in cuttings of such plants as willow, poplar, currant, and grape (*183*). It was assumed that hormone-like substances were formed in the developing buds and transported through the phloem to the base of the cutting, where they stimulated root formation.

The existence of a specific root-forming factor was first determined by Went in 1929 when he found that if leaf extracts from the *Acalypha* plant were applied back to *Acalypha* or to *Carica* tissue, they would induce root formation (*311*). Bouillenne and Went found substances in cotyledons, leaves, and buds that stimulated rooting of cuttings; they called this material **"rhizocaline"** (*30, 311*)

Bouillenne's and Went's (31) conclusion, in 1933, about the existence of the root-forming **rhizocaline** has been misinterpreted (even today). Use of the term implies that rhizocaline or evidence for it was discovered; on the contrary, Bouillenne and Went should be *credited* with having obtained extensive *physiological evidence* for and only for the root-promoting properties of *endogenous auxin*—an unidentified substance at that time *(122)*. In 1933, IAA had not been chemically discovered, nor was it known that IAA was an auxin, nor was auxin associated with rooting. Bouillenne and Went *(31)* did *postulate* that the rooting stimulus (rhizocaline) was a *complex of substances,* including sugar and oxygen, based on their experiments—more factors were to be added to the hypothetical complex later (see p. 293-94).

In 1938, Cooper *(46)* reached similar conclusions to Bouillenne and Went that a substance other than just auxin was required for rooting, and referred to this substance as rhizocaline. He concluded that basal treatment of cuttings with auxin caused a rapid basipetal movement of rhizocaline, which was produced in leaves and upper stems.

Bud Effects on Rooting

In Went's 1934 pea test for root-forming activity of various substances, it is significant that the presence of at least one bud on the pea cutting was essential for root production (*311*). A budless cutting would not form roots even when treated with an auxin-rich preparation. This finding indicated again that a factor other than auxin, presumably one produced by the bud, was needed for root formation. In 1938, Went (who was a colleague of W.C. Cooper at the California Institute of Technology) postulated that specific factors other than auxin were manufactured in the leaves and were necessary for root formation. Thus, rhizocaline was more than just auxin. Later studies (*82, 203*) with pea cuttings confirmed this theory.

For root initiation, the presence of an actively growing shoot tip (or a lateral bud) is necessary during the first three or four days after the cuttings are made (*122*). But after the fourth day the shoot terminal and axillary buds can be removed without interfering with subsequent root formation.

Bud removal from cuttings in certain species will stop root formation, especially in species without preformed root initials (*183*). In some plants, if the tissues exterior to the xylem are removed, just below a bud, root formation is reduced, indicating that some root-promoting compound(s) travels through the phloem from the bud to the base of the cutting. It has been shown that if hardwood, deciduous cuttings are taken in midwinter, when the buds are in the rest period,[2] they have either no effect or can inhibit rooting (*90, 153*). But if the cuttings are made in early fall or in the spring, when the buds are active and not at rest, they show a strong root-promoting effect.

With cuttings of apple and plum rootstocks, the capacity of shoots to regenerate roots increases during the winter, reaching a high point just before

[2]The "rest period" is a physiological condition of the buds of many woody perennial species beginning shortly after the buds are formed. While in this condition, they will not expand into flowers or leafy shoots even under suitable growing conditions. After exposure to sufficient chilling hours (0 to 5° C), however, the "rest" influence is broken, and the buds will develop normally with the advent of favorable growing temperatures.

FIGURE 10–15 Effect of leaves on cuttings of 'Lisbon' lemon. Both groups were rooted under intermittent mist and were treated with indolebutyric acid at 4,000 ppm by the concentrated-solution-dip (quick-dip) method.

bud-break in the spring; this is believed to be associated with a decreasing level of bud dormancy following winter chilling (*152*). Studies with Douglas-fir cuttings showed a pronounced relationship between bud activity and rooting of cuttings—cuttings taken in early fall (September to October in the U.S.), root the poorest (*242*).

Leaf Effects on Rooting

It has long been known that the presence of leaves on cuttings exerts a strong stimulating influence on rooting (see Figures 10–15 and 10–16). The stimulatory effect of leaves on rooting in stem cuttings is nicely shown by studies (*237*) with avocado. Cuttings of difficult-to-root cultivars under mist soon shed their leaves and die, whereas leaves on the cuttings of cultivars that have rooted are retained as long as nine months. While the presence of leaves can be important in rooting, leaf retention is more a consequence of rooting than a direct cause of rooting. After five weeks in the rooting bed, there was five times more starch in the base of the easy-to-root avocado cuttings than there was at the beginning of the tests. In hibiscus, rooting is also enhanced when leaves are retained on the cuttings (*249, 297*).

Carbohydrates translocated from the leaves are important for root development. However, the strong root-promoting effects of leaves and buds are probably due to other, more direct factors (*35*). Leaves and buds produce auxin, and the effects of the polar apex-to-basal (basipetal) transport of auxins are observed on rooting at the base of the cutting.

Plant Growth Substances

All classes of growth regulators—auxins, cytokinins, gibberellins, ethylene, and abscisic acid, as well as ancillary compounds such as growth retardants/inhibitors, polyamines, phenolics, etc.—influence root initiation either directly or indirectly (*61, 180*). However, auxins have the greatest effect on root formation in cuttings. See Chapter 2 for a more complete description of plant hormones, plant growth regulators, and how they function.

Auxins

In the mid-1930s and later, studies of the physiology of auxin action showed that auxin was involved in such varied plant activities as stem growth, adventitious root formation (*122, 288, 290, 311*), lateral bud inhibition, abscission of leaves and fruits, and activation of cambial cells.

Indole-3-acetic acid (IAA) was identified as a naturally occurring compound having considerable auxin activity (*177*). Synthetic indoleacetic acid

FIGURE 10–16 Effect of leaves, buds, and applied auxin on adventitious root formation in leafy 'Old Home' pear cuttings. *Top:* Cuttings treated with auxin (indolebutyric acid at 4,000 ppm for five seconds). *Bottom:* Untreated cuttings. *Left to right:* with leaves; leaves removed; buds removed; one-fourth natural leaf area. (Courtesy W. Chantarotwong.)

(IAA) (see Figure 2–14) was subsequently tested for its activity in promoting roots on stem segments, and in 1935, investigators demonstrated the practical use of this material in stimulating root formation on cuttings (*288*). About the same time it was shown (*324*) that two synthetic materials, **indolebutyric acid (IBA)** and **napthaleneacetic acid (NAA)** (see Figure 2–14), were even more effective than the naturally occurring or synthetic IAA for rooting. Today, IBA and NAA are still the most widely used auxins for rooting stem cuttings and for rooting tissue culture-produced microcuttings. It has been repeatedly confirmed that auxin is required for initiation of adventitious roots on stems, and indeed, it has been shown that divisions of the first root initial cells are dependent upon either applied or endogenous auxin (*96, 116, 196, 279, 289, 292*).

Indolebutyric acid, although less abundant that IAA, is also a naturally occurring substance in plants (*193*). In *Arabidopsis,* endogenously formed IAA is more readily transported than endogenously formed IBA. IAA also conjugates via amide bonds, while IBA conjugates form ester bonds. In *Malus* (apple), when IBA is applied to stem cuttings or microcuttings to stimulate rooting, it is, in part, converted to IAA (*295*). IBA may also enhance rooting via increased internal-free IBA, or may synergistically modify the action of IAA or endogenous synthesis of IAA; IBA can enhance tissue sensitivity for IAA and increase rooting (*296*). In avocado microcuttings, IBA increased endogenous IAA and indole-3-acetyl-aspartic acid (IAAsp) before root differentiation occurred, and as root formation proceeded (*95*).

In mung bean cuttings, IBA applied to the cutting base was transported to the upper part of the cuttings to a greater extent than IAA, and rapidly metabolized into IBA conjugates. These IBA conjugates were reported to be superior to free IBA in serving as an auxin source during later stages of rooting (*316*).

There are advantages to integrating biochemical, physiological, and developmental anatomical approaches in rooting studies. Adventitious root formation is a developmental process marked by a sequence of histological events. Each developmental stage has different requirements for growth substances, such as auxins, cytokinins, and gibberellic acid. Thus, establishing the time intervals of adventitious root initiation and development has made possible the correlation of sequential physiological and histological events in rooting (*28, 57, 85, 98, 264, 265*). With pea cuttings, the role of auxins in the intricate developmental processes of rooting occurred in two basic stages (*82, 83, 204*):

1. **A root initiation .stage** in which root meristems were formed (including dedifferentiation, root-initial and root-primordia formation). This stage could be further divided into:
   a) **An auxin-active stage,** lasting about four days, during which auxin had to be supplied continuously for roots to form, coming either from terminal or lateral buds, or from applied auxin (if the cutting has been decapitated) (*83, 204*).
   b) **An auxin-inactive stage** occurred next. Withholding auxin during this stage (which lasts about four days) did not adversely affect root formation.
2. **Elongation of root primordia stage,** during which the root tip grew outward through the cortex, finally emerging from the epidermis of the stem (see Figure 10–9). A vascular system was developed in the new root and became connected to adjacent vascular bundles of the stem. At this stage there was no further response to applied auxin.

Cytokinins

Cytokinins have the greatest effect on initiating buds and shoots from leaf cuttings and in tissue culture systems (*27, 59, 78, 246, 299*). Natural and synthetic cytokinins include: **zeatin, zeatin riboside, kinetin, isopentenyladenine (2iP), thidiazuron (TDZ), and benzyladenine (BA** or **BAP).** See Chapter 2. Generally, a high auxin/low cytokinin ratio favors adventitious root formation and a low auxin/high cytokinin ratio favors adventitious bud formation (*32, 141*) (Figure 10–17). Cuttings of species with high natural cytokinin levels have been more difficult to root than those with low cytokinin levels (*216*). Applied synthetic cytokinins normally inhibit root initiation in stem cuttings (*224, 254*). However, cytokinins at *very low* concentrations, when applied to decapitated pea cuttings at an early developmental stage (*84*), or to begonia leaf cuttings (*141*), promote root initiation, while higher concentrations inhibit root initiation. Application to pea cuttings at a later stage in root initiation does not show such inhibition; the influence of cytokinins in root initiation may thus depend upon the particular stage of initiation and the concentration (*28, 57, 69, 265*). To date, the quantitative determination of endogenous cytokinins at various stages of rooting has yet to be determined (*299*).

It has been suggested that the few cases of rooting success using exogenous applications indicate that cytokinins may play an indirect rather than

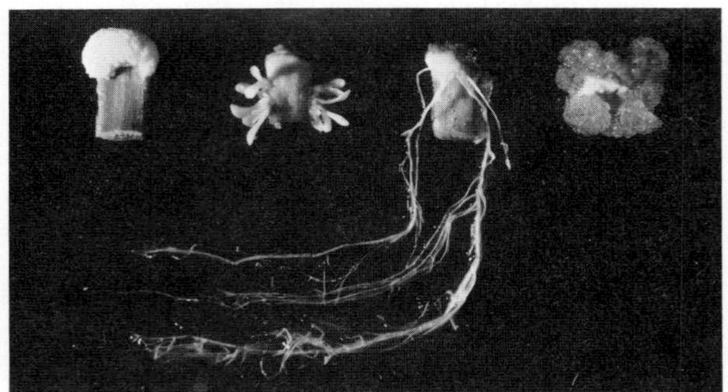

FIGURE 10–17 Effects of adenine sulfate (a cytokinin precursor) and indoleacetic acid (auxin) on growth and organ formation in tobacco stem segments. *Far left:* Control. *Central left:* Adenine sulfate, 40 mg per liter. Bud formation with decrease in root formation. *Center right:* Indoleacetic acid, 0.02 mg per liter. Root formation with prevention of bud formation. *Far right:* Adenine sulfate, 40 mg per liter plus indoleacetic acid, 0.02 mg per liter. Growth stimulation but without organ formation. (Courtesy Folke Skoog.)

a direct role on rooting (*299*). Cytokinins may also be indirectly involved in rooting through effects on rejuvenation and accumulation of carbohydrates at the cutting base (i.e., carbohydrate loading) (*299*).

Leaf cuttings provide good test material for studying auxin-cytokinin relationships, since such cuttings must initiate both roots and shoots. Cytokinin application at relatively high concentrations promoted bud formation and inhibited root formation of *Begonia* and *Bryophyllum* (*51, 142*) leaf cuttings, while auxins, at high concentrations, stimulated roots and inhibited buds. Too high a cytokinin concentration applied to leaf cuttings maximizes adventitious bud formation, but reduces the quality of new shoots (Figures 10–12 and 10–18); from a *horticultural standpoint,* adventitious shoot quality, not just adventitious bud formation, is an important criterion in regenerating new plants from leaf cuttings (*59*).

The considerable seasonal changes in the regenerative ability of *Begonia* leaf cuttings are due to a complex interaction of environmental cues: temperature, photoperiod, and irradiance, which affect the levels of endogenous cytokinins, auxins, and other growth regulators (*144*).

Gibberellins (GA)

The gibberellins (Figure 2–14) are a group of closely related, naturally occurring compounds first isolated in Japan in 1939 and known principally for their effects in promoting stem elongation. At relatively high concentrations (i.e., $10^{-3}M$) they have consistently inhibited adventitious root formation (*259*). There is evidence that this inhibition is a direct local effect that prevents the early cell divisions involved in transformation of differentiated stem tissues to reach a meristematic condition (*36*). Gibberellins have a function in regulating nucleic acid and protein synthesis and may suppress root initiation by interfering with these processes, particularly transcription (*127*). At lower concentrations (10^{-11} to $10^{-7}M$), however, gibberellin has promoted root initiation in pea cuttings, especially when the stock plants were grown at low light levels (*126*).

In *Begonia* leaf cuttings, gibberellic acid (*145*) inhibited both adventitious bud and root formation, probably by blocking the organized cell divisions that initiate formation of bud and root primordia. Inhibition of root formation by gibberellin depends on the developmental stage of rooting. With herbaceous materials, inhibition is usually greatest when

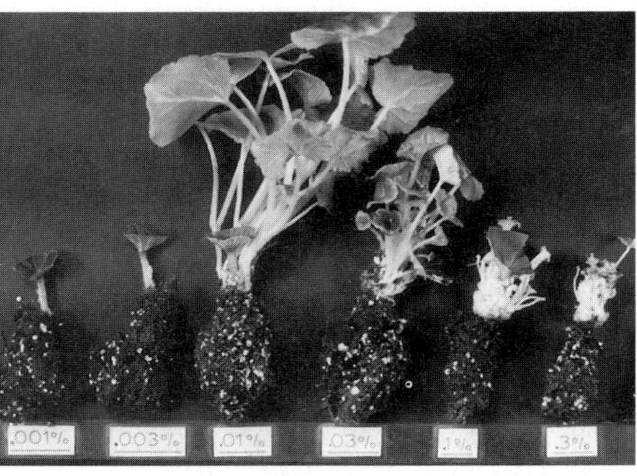

FIGURE 10–18 *Left:* Application of the cytokinin PBA mixed with talc to leaf cutting petiole base of Rieger begonias. *Right:* Concentrations of talc-applied cytokinin ranged from 0 to 1 percent with 0.01 percent (100 mg/l PBA) producing the optimum shoot number. At higher concentrations more adventitious buds are produced, but they fail to develop a commercially acceptable shoot system (*59*).

GA is applied three to four days after cutting excision (*127*). However, woody plant species such as *Salix* (*116*) and *Ficus* (*57*) were not adversely affected by GA during root initiation, but were inhibited if GA was applied after root primordia were initiated. GA caused reduction in cell numbers in older established primordia which was deleterious to root formation. The biochemical and physiological mechanisms by which applied gibberellins inhibit adventitious rooting remains unknown (*122*).

Ethylene (C_2H_4)

Ethylene can enhance, reduce, or have no effect on adventitious root formation (*260*). In 1933, Zimmerman and Hitchcock (*322*) showed that applied ethylene at about 10 ppm causes root formation on stem and leaf tissue as well as the development of pre-existing latent roots on stems. They and other scientists (*324*) also showed that auxin applications can regulate ethylene production, and suggested that auxin-induced ethylene may account for the ability of auxin to cause root initiation. Centrifuging *Salix* cuttings in water, or just soaking them in hot or cold water, stimulates ethylene production in the tissues as well as root development, suggesting a possible causal relationship between ethylene production and subsequent root development (*169, 170*).

Ethylene promotion of rooting occurs more frequently in intact plants than cuttings, herbaceous rather than woody plants, and plants having preformed root initials. The effects of ethylene on rooting are not as predictable or consistent as those of auxin (*122*). A large body of evidence suggests that endogenous ethylene is not directly involved in auxin-induced rooting of cuttings (*210*).

Abscisic Acid (ABA)

Reports on the effect of abscisic acid (ABA) on adventitious root formation are contradictory (*12, 143, 233*), apparently depending upon the concentration and upon the environmental and nutritional status of the stock plants from which the cuttings are taken. ABA is important to rooting, since it (1) antagonizes the effects of gibberellins and cytokinins, both of which can inhibit rooting, and (2) influences the ability of cuttings to withstand water stress during propagation. If the role of ABA in rooting is to be understood, than endogenous ABA levels will need to be determined at the site of root initiation, during the developmental stages of rooting (*61*).

Ancillary Compounds

There are ancillary compounds that modify main hormone effects on rooting, and adventitious bud and shoot formation. These compounds include growth retardants/inhibitors, polyamines, and phenolics. Salicylate, which is classified as a phytohormone by some plant physiologists, has been reported to enhance rooting in combination with auxin (*61, 232*).

Growth retardants/inhibitors. Growth retardants, generally applied to reduce shoot growth, have been used to enhance rooting based on the rationale that they (a) antagonize GA biosynthesis or activity (GA is normally inhibitory to rooting), or (b) reduce shoot growth, resulting in less competition and consequently more assimilates are available for rooting at cutting bases (*65*). Synthetic antigibberellins and inhibitors of GA biosynthesis include chlormequat chloride (CCC), paclobutrazol (PP333, Bonzi), uniconazole (a triazole growth retardant related to PP333), morphactins, ancymidol (Arest), gonadotropins, and daminozide (SADH, Alar) (*169, 234*). Growth retardants frequently promote rooting (generally in combination with exogenous auxin) (*65*). However, the mode of action of how these compounds enhance rooting is not well understood. Hence, rooting enhancement by GA biosynthesis inhibitors has been inconsistent, and none are commercially used for rooting (*61*).

The polyamines. The effect of polyamines on rooting of woody plant species is quite variable. Putrescine, spermidine, and spermine in combination with IBA improved rooting of hazel microshoots (*238*). The rooting of olive microshoots increased by using polyamines along with NAA, but rooting of almond, pistachio, chestnut, jojoba, apricot, and walnut did not increase (*247*). In NAA-treated *Hedera helix* cuttings, there were increases in endogenous polyamines, particularly putrescine (*98*). Polyamines may serve as secondary messengers for rooting. To date, polyamine enhancement of rooting appears to occur only in the presence of auxin.

In summary, the principal plant growth regulator used to enhance rooting of cuttings is auxin, while cytokinins are used to stimulate adventitious bud formation. The other plant growth regulators and ancillary compounds can influence organogenesis, but not consistently enough to merit their commercial use in cutting propagation.

> Much of the research dealing with hormones and rooting has been based upon exogenous treatments *(122)*. In contrast, little work has critically tested the roles of endogenous hormones and their interactions with applied hormones. Particularly lacking is research aimed at determining how hormones might regulate gene expression and thereby influence rooting, directly and indirectly. Hence, it is difficult to distinguish between possible controlling roles of hormones on rooting, and indirect hormonal effects on other physiological processes of cuttings (*122*).

Classification of Plant Rooting Response to Growth Regulators

Plants can be divided into three classes with regard to growth regulator effects on rooting:

1. Those which have all the essential endogenous substances (root morphogens) plus auxin. When cuttings are made and placed under proper environmental conditions, rapid root formation occurs.
2. Those in which the naturally occurring root morphogen(s) are present in ample amounts, but auxin is limiting. With the application of auxin, rooting is greatly increased.
3. Those that lack a rooting morphogen(s) and/or lack the cell sensitivity to respond to the morphogen(s), even though natural auxin may or may not be present in abundance. External application of auxin gives little or no rooting response.

THE BIOCHEMICAL BASIS FOR ADVENTITIOUS ROOT FORMATION

The fundamental basis of adventitious root formation remains one of the least understood of plant processes (*62, 122*). Despite years of active research, the primary chemical stimulus for dedifferentiation and root initial formation (the critical steps of adventitious root formation) and the subsequent organization of meristemoids (meristematic centers of cells actively dividing into root primordia) remain unknown. The following is a brief history of post-World War II research on the biochemistry of rooting.

The Revival of the Hypothetical Rooting Stimulus—Rhizocaline

After World War II, Bouillenne and Bouillenne-Walrand (*30*) revived the concept of **rhizocaline.** They proposed (without supporting experimental evidence) that rhizocaline be considered as a complex of three components: (a) a specific factor (possibly an orthodihydroxy phenol), which is translocated from the leaves; (b) a nonspecific factor (auxin), which is translocated from the leaves and is found in biologically low concentrations; and (c) a specific enzyme located in cells of certain tissues (pericycle, phloem, cambium), which is possibly a polyphenol-oxidase enzyme. They further proposed that the ortho-dihydroxy phenol reacts with auxin wherever the required enzyme is present, giving rise to the complex "rhizocaline," which may be considered one step in a chain of reactions leading to root initiation.

Endogenous Rooting Inhibitors

In the early 1950s, endogenous chemical inhibitors were reported to retard rooting in selected plant species, as indicated below. This was found to be the case with selected grape cultivars (*268*); leaching cuttings with water enhanced the quantity and quality of roots. Difficult-to-root hardwood cuttings of wax flower (*Chamaelaucium uncinatum*) have a cinnamic acid derivative which inhibits rooting, while no detectable levels of this phenolic compound were found in easy-to-root softwood cuttings (*50*). Cuttings of difficult-to-root mature eucalyptus (*49, 222*), chestnut (*303*), and dahlia cultivars (*16, 17*) also had higher rooting inhibitors than easy-to-root forms.

In summary, inhibitors are not universally found in difficult-to-root species, which strongly suggests that rooting ability is controlled by other biochemical mechanisms and/or molecular factors.

Rooting Co-factors (Auxin Synergists)

Various *model rooting bioassay systems* have been used to test adventitious root formation. During the 1950s and 1960s, the easy-to-root mung bean (*Phaseolus aureus*) was used by Hess as a rooting bioassay to screen biochemical effects on rooting. Hess was not able to demonstrate any difference in rooting inhibitors between the juvenile easy-to-root, and mature difficult-to-root forms of English ivy (*Hedera helix*). Instead, he determined that the juvenile, easy-to-root forms of English ivy, and easy-to-root cultivars of chrysanthemum and *Hibiscus rosa-sinensis* contained greater nonauxin rooting stimuli than their difficult-to-root forms (*147, 148*). He termed these nonauxin rooting stimuli **rooting co-factors,** which was a modification of the rhizocaline theory that biochemical factors, other than just auxin, were controlling rooting. These rooting co-factors were naturally occurring substances that appeared to act synergistically with indoleacetic acid in promoting rooting.

One of rooting co-factors (No. 4) of English ivy represented a group of active substances, tentatively characterized as oxygenated terpenoids. Another (No. 3) was identified as isochlorogenic acid (*148*). In testing the biological activity of compounds structurally related to co-factor No. 3, Hess (*147*) reported that the phenolic compound catechol reacts synergistically with indoleacetic acid in root production in the mung bean bioassay. Catechol is readily oxidized to a quinone, and since the mung bean itself is a good source of phenolase, it was theorized that oxidation of an *ortho-dihydroxy phenol* (o-diphenol) is one of the first steps leading to root initiation, as postulated in the rhizocaline theory of Bouillenne and Bouillenne-Walrand (*30*). In addition, various other compounds were found to react synergistically with auxin in promoting rooting (*104*) (see Figure 10–19).

Rooting co-factors have since been found in maple (*Acer*) species (*176*). Fadl and Hartmann (*89, 90*) isolated an endogenous root-promoting factor from basal sections of hardwood cuttings of an easily rooted pear cultivar ('Old Home'). Extracts from basal segments of similar cuttings of a difficult-to-root cultivar ('Bartlett'), treated with IBA, did not show this root-promoting factor.

The action of these phenolic compounds in root promotion was theorized to be in protecting the root-inducing, naturally occurring auxin—indoleacetic acid—from destruction by the enzyme, indoleacetic acid oxidase (*68, 88*). This possibility was suggested also in rooting (*111*) juvenile English ivy, in which the phenol, catechol, showed remarkable synergism with IAA in promoting rooting. However, when naphthaleneacetic acid (NAA), which is not affected by IAA oxidase, was used as the auxin, NAA alone was as effective as NAA plus catechol. This finding implied that catechol was protecting IAA from destruction by IAA oxidase.

Jarvis (*164*) attempted to integrate the biochemical and developmental anatomy of adventitious root formation by examining the four developmental stages of rooting (Figure 10–20). His premise was: (a) that the initial high concentrations of auxin needed in early rooting events are later inhibitory to organization of the primordium and its subsequent growth—hence the importance of regulating endogenous auxin concentration with the IAA oxidase/peroxidase enzyme complex playing a central role (i.e., IAA oxidase metabolizes or breaks down auxin); and (b) IAA oxidase activity is controlled by phenolics (*o*-diphenols are inhibitory to IAA oxidase), while borate complexes with *o*-diphenols result in greater IAA oxidase activity—and hence a reduction of IAA to levels that are optimal for the later organizational stages of rooting.

In the *in vitro* rooting of poplar (*Populus*) shoots, endogenous free IAA activity is highest dur-

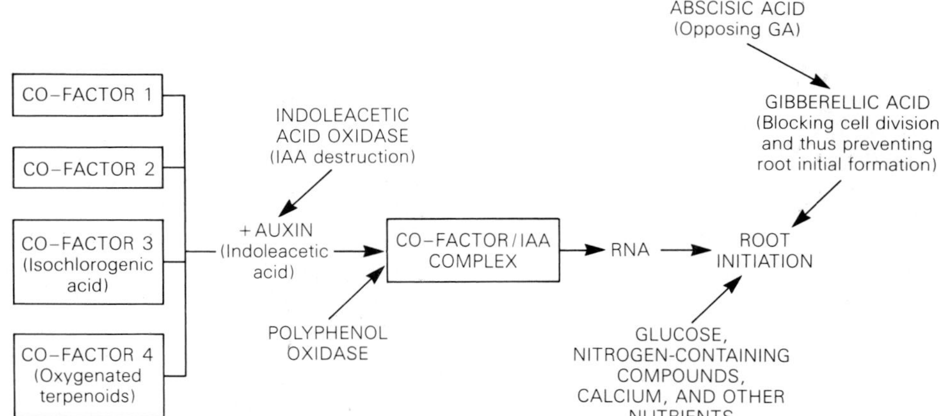

FIGURE 10–19 Hypothetical relationships of various components leading to adventitious root initiation. In addition, specific root-inhibiting factors may be present, which interfere with root development.

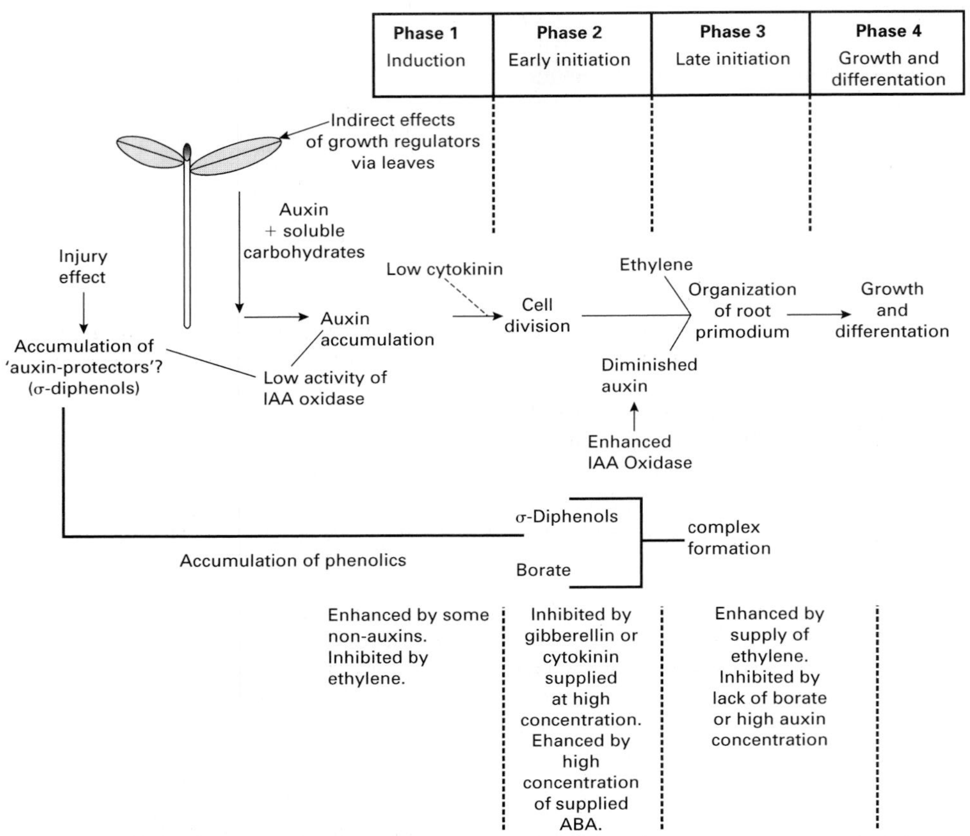

FIGURE 10–20 Hypothesized scheme of Jarvis *(164)* which proposes the role of phenolics, IAA oxidase/peroxidase, borate, and phytohormones in the four developmental stages of adventitious root production. (From M. B. Jackson (ed.). 1986. *New root formation in plants and cuttings.* Martinus Nijhoff Publishers. Dordrecht, The Netherlands.)

ing root induction, followed by a peak of soluble peroxidase activity and subsequent decrease in free IAA—preceding root emergence (*139*). These events correspond to the initiative phase of rooting suggested by Jarvis (*164*).

In summary, rooting co-factors and inhibitor studies have intrigued scientists trying to understand rooting phenomena. While rooting co-factors have been correlated to rooting responses, there has been no cause and effect relationship demonstrated. Application of rooting co-factors to difficult-to-root forms has not increased rooting response. It has been proposed that predisposition of cells to initiate root primordia (**cell competence,** see Figure 10–21) may be dependent on active enzymes (*15*) and/or substrate, which difficult-to-root cuttings may lack for synthesis of auxin-phenolic conjugates. However, *controversy* exists since auxin-phenol-enzyme complexes have not been found *in vivo* (in plant tissue), and promoter-inhibitor systems of rooting have not been universally observed in plants (*10*).

Biochemical Changes in the Development of Adventitious Roots

Once adventitious roots have been initiated in cuttings, considerable metabolic activity occurs as new root tissues are developed and the roots grow through and out of the surrounding stem tissue. Protein synthesis and RNA production were both shown indirectly to be involved in adventitious root development in etiolated stem segments of *Salix tetrasperma* (*163*). To date, it is not clear to what extent RNA metabolism is altered within that small pool of cells actually involved in root initiation (*165*). More definitive studies need to include microautoradiographic and histochemical approaches.

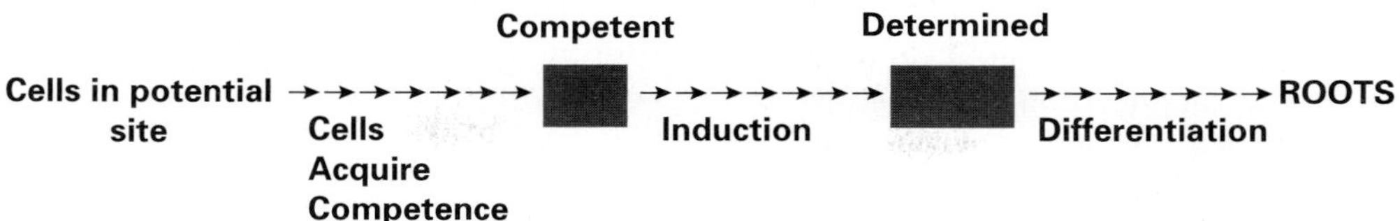

FIGURE 10–21 Phases in the organ-ogenic process of rooting. (Adapted from Christianson and Warnick (*43*).)

Some significant studies in hydrangea (*207, 208*) have been made of the biochemical changes taking place during the *development* of preformed root initials into emerging roots. These studies followed the change in DNA and enzyme levels as the roots developed. The total protein content of the root initials increased over 100 percent in the first four days after the cuttings were made, but there was no pronounced increase in the DNA content of the cells until the sixth day. Apparently, then, considerable protein synthesis occurs within existing cells, prior to large-scale replication of nuclear DNA and cell division.

Initially, the enzymes peroxidase, cytochrome oxidase, succinic dehydrogenase, and starch-hydrolyzing enzymes increased in the phloem and xylem ray cells of the vascular bundles. During subsequent root development, enzyme activity shifted from the vascular tissues to the periphery of the vascular bundles. These increases in enzyme activity occurred two to three days after the cuttings were made.

During rooting, starch is converted to soluble carbohydrate. In hydrangea, starch disappeared from the endodermis, phloem and xylem rays, and pith—in tissues adjacent to the developing root primordia—and was converted to soluble carbohydrate. Similarly, in the development of adventitious roots on IBA-treated plum cuttings, as soon as callus and roots started forming, pronounced carbohydrate increases of sucrose, glucose, fructose, and sorbitol—and starch losses—occurred at the base of the cuttings where rooting occurs (*34*). While soluble carbohydrates are not the cause of rooting, the developing callus and roots at the cutting base act as

The following synopsis summarizes some current thinking on the role of rhizocaline, inhibitors, and rooting co-factors in rooting:

In recent years, the role of phenolic compounds in the control of rooting has been seriously challenged. Wilson and Van Staden *(318)* argue that the concept of rhizocaline, inhibitors, and promoters (including rooting co-factors) represents a traditional approach to understanding rooting. The concept is founded on a bioassay principle, in which plant extracts or known compounds promote or inhibit rooting when supplied to cuttings. It is sometimes assumed that activity in a rooting bioassay reflects *in vivo* activity, and hence has physiological significance. They argue that this assumption is not warranted because even though a large diversity of known and unknown compounds has been found to promote or inhibit rooting in rooting bioassays, no well-substantiated mode of action has been established. Furthermore, there is no good evidence that rhizocaline consists of an auxin-phenolic conjugate, and other explanations for the actions of phenolics are not well substantiated. They suggest that the action of rooting promoters and inhibitors is mediated by chemical injury (see the later discussion on wounding in this chapter). Irrespective of their chemical identity, low concentrations promote rooting, while higher concentrations are inhibitory *(318)*.

Wilson (*317*) further proposed that a rooting morphogen (arbitrarily defined as a nonauxin endogenous substance(s) promoting rooting) can be assumed to induce roots in woody stem cuttings. Whereas auxins promote rooting of most herbaceous cuttings, they may have little effect on more difficult-to-root woody cuttings. The interaction between a rooting morphogen(s) of vascular origin and potential sites for root initiation are likely to be dynamic and variable. Potential rooting sites are not equally sensitive to the rooting morphogen, since each cell has a unique lineage, ontogeny, and position, i.e., the competency of cells varies, which affects their ability to respond to the morphogen and root. Hence, he concludes that no simply defined morphogen can be said to limit rooting, and the chemical identity of the morphogen (if definable) is probably unimportant.

Physiological and biochemical studies have addressed post-translational rooting processes, including the influences of plant growth regulators on the biochemistry of rooting (Figure 10–22) (*123*). Such studies have *not* led to an understanding of the control of rooting, *nor* enhanced our ability to manipulate rooting. Indeed, past studies may have led us to believe that rooting is hopelessly complex. The sought-for controls of rooting may be discovered by searching for and studying direct genetic effects, using current molecular genetic technologies, nonrooting mutants, and rootless plants (*123*). Integrating molecular genetic technologies at the transcriptional and translational levels with biochemical, post-translational processes may eventually reveal the controls of rooting.

a "sink" for the movement of soluble carbohydrates from the top of the cutting.

MOLECULAR/BIOTECHNOLOGICAL ADVANCES IN ASEXUAL PROPAGATION

Development of synthetic seed from vegetative (somatic) tissue in somatic seed technology (see Chapter 17), artificially inducing roots by nonpathogenic *Agrobacterium,* and the potential transformation of cells using a disarmed **plasmid**[3] from a root-inducing bacterium or from an auxin-inducing fragment of the **T-DNA**[4] may play important roles in the vegetative propagation of plants (see Chapter 2). Applying biotechnology studies at the transcriptional and translational levels can reveal the controls of rooting, adventitious bud formation, tuberization, and other developmental processes important to vegetative propagation. As an example, if an enzyme negatively affected rooting, then antisense DNA or RNA could be used to turn off the gene that produced the enzyme. Initially, the genetically transformed plant would be micropropagated, and then once established *ex vitro,* conventional propagation techniques would be used to mass-produce the genetically transformed plant (*54*).

Which genes or gene groups affect rooting have yet to be fully determined (*114*) (see Figure 10–23). Changes in gene expression were observed during the formation of adventitious root primordia of sunflower (*Helianthus annus*) hypocotyl cuttings (*217*), and in the rooting of juvenile and mature English ivy (*252*).

Today, difficulties in rooting *in vitro* and *ex vitro* (outside the test tube), developing successful tissue culture multiplication systems, and transformation systems for rooting limit the production of transgenic woody plants, i.e., commercially important plants for the production of fruits, nuts, wood, paper, and landscape ornamentals (*55, 239, 255*). Some difficult-to-root woody species have been genetically "transformed" to easy-rooters. Rooting of kiwi (*Actinidia deliciosa*) cuttings was improved by introducing genes from the root-inducing bacterium *Agrobacterium rhizogenes* (*247, 248*) (see Chapter 2).

There is progress being made by using this root-inducing bacterium to enhance root regeneration of bare-root almond stock (*277, 278*) and *in vitro* rooting of difficult-to-root apple (*220*). *Agrobacterium rhizogenes* has been used as an effective rooting agent in hazelnut (*Corylus avellana*) cuttings (*8*), and

Cell Competency-to-Root. The formation of new centers of cell divisions—called *de novo meristems,* that differentiate into adventitious roots—requires that a cell or group of cells (i.e., phloem ray parenchyma cells or callus cells) embark upon a new developmental program (*205*). Do all cells have the potential (competency) for adventitious organ formation? What is the molecular mechanism that controls adventitious organ formation? What is the molecular basis for the plasticity that allows differentiated cells (phloem ray parenchyma) to start new developmental programs? How many different signals are needed for root induction? Why is there a decline or loss of competence for the formation of adventitious roots in physiologically mature-phase shoot tissue, as compared to physiologically juvenile-phase tissue? Competence to root can be assessed by determining whether or not tissue is capable of responding in a specific way to inductive treatments (*212*). A model of the events in the organogenic process of rooting is given in Figure 10–21, as is a model of the molecular events in the developmental pathway of root initiation in Figure 10–23.

[3]Plasmids are small molecules of extrachromosomal DNA that carry only a few genes and occur in the cytoplasm of a bacterium.

[4]T-DNA is the portion of the root-inducing (Ri) plasmid (i.e., from *Agrobacterium rhizogenes*) which is inserted into the plant genome (i.e., of a difficult-to-root species) and stabilized; hence this normally difficult-to-root species is potentially "transformed" to an easy-to-root clone. See Chapter 2 for molecular biology terminology.

INVESTIGATION OF ROOTING BY PROCESS

EXPERIMENTAL TREATMENTS	MEDIATING PROCESSES	OBSERVATION OF EFFECTS
Light Temperature Oxygen Carbon Dioxide Water Auxins Gibberellins Cytokinins Ethylene Minerals Organic Nitrogen Phenolics Carbohydrates Amino acids Polyamines Histones Acid / Bases Nucleotides Nucleosides Nucleic Acids Root Symbionts Various Inhibitors Many Others	TRANSCRIPTIONAL	ASSESSMENTS NEEDED
	TRANSLATIONAL	ASSESSMENTS NEEDED
	POST-TRANSLATIONAL	**MOST PAST ASSESSMENTS HERE**

INVESTIGATION OF ROOTING BY DISCIPLINE

Morphology	Cytology Anatomy	Physiological Biochemistry (Most past assessments here)	Molecular Genetics

FIGURE 10–22 Some environmental and chemical factors (in the left column) that have been implicated in rooting. Investigation of rooting research is by *process* (upper figure) and investigation by *discipline* (lower figure). In past research, effects of experimental treatments may have been at any or all process levels, but were usually assessed only post-translationally, in physiological and/or biochemical studies. (From Haissig, et al. (*123*).)

with *in vitro* and *ex vitro* rooting of pine (Pinus) and larch (Larix). How the bacterium enhances rooting is not well understood. It may be modifying the root environment by secreting hormones or other compounds, or by transformation of plant cells (*200*).

In studies of tobacco plants transformed with root-inducing (Ri T-DNA) of *Agrobacterium rhizogenes,* rooting the transformed tobacco explants was due to *genes* which *increased auxin sensitivity* of the tissue. Rooting of transformed plants was not due to genes which regulated auxin production, or to a substantially altered balance of auxin to cytokinin ratio (*267*). In other studies with nonrooting tobacco mutants, sensitivity to auxin was due to general alteration of the cellular response to auxin and was not due to increased rate of conjugation of auxins by these tissues, or by disruption of auxin transport (*37*). Thus, there are implications that the lack of *cell competency* in difficult-to-root species may be due to lack of cell sensitivity to auxin rather than to a suboptimal level of endogenous auxin.

Vascularization and Rooting. Direct and circumstantial evidence suggests that the control and progression of adventitious root formation is related to vascularization. Vascularization is the normal development of the xylem and phloem tissue systems, subsequent natural growth, and hardening of structural support cells with the deposition of lignin, i.e., fibers, xylem vessels, and sieve tubes. The timing and the degree of lignification may be involved in determining the potential for root formation.

Phenylpropanoid metabolism is of interest because some phenylpropanoids are lignin precursors and, therefore, related to vascularization. Genes that encode the enzymes of phenylpropanoid metabolism have been cloned. Thus, the use of molecular biology techniques to study vascularization may provide strong inferences about the controls of adventitious rooting (*122, 212*).

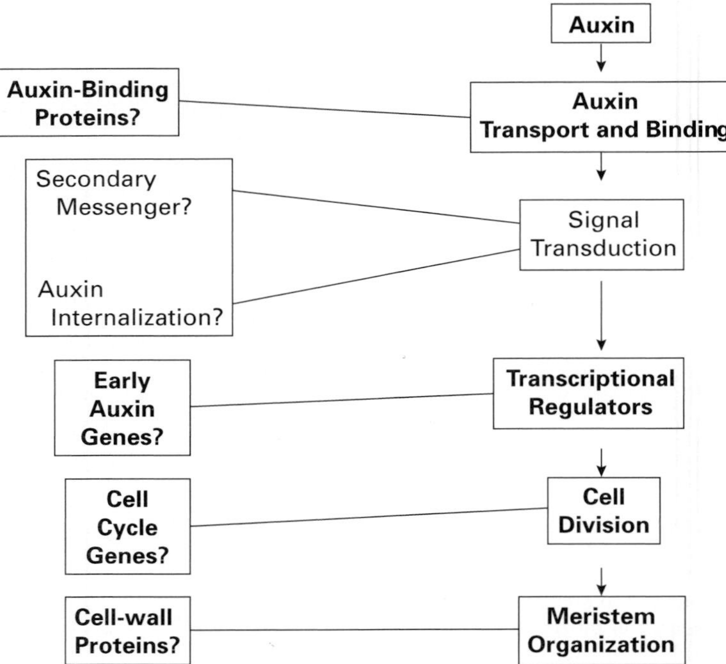

FIGURE 10–23 A model for the molecular events in the developmental pathway of root initiation. On the right are proposed steps, and on the left are potential molecules to be tested for their specific roles. (Courtesy B. Goldfarb, M. Greenwood, and K. Hutchison.)

FACTORS AFFECTING REGENERATION OF PLANTS FROM CUTTINGS

Great differences in the rooting ability of cuttings exist among species and cultivars. Stem cuttings of some cultivars root so readily that the simplest facilities and care give high rooting percentages. On the other hand, cuttings of many cultivars or species have yet to be rooted. Cuttings of some "difficult" cultivars can be rooted only if specific influencing factors are taken into consideration and if the cultivars are maintained at the optimum condition. With most species, the careful selection of cutting material from stock plants or containerized plants, management of cuttings, and control of environmental conditions during rooting are the difference between commercial success or failure. These influencing factors include:

Management of Stock Plants to Maximize Cutting Propagation

1. Selection and maintenance of source material that is easy-to-root (juvenile); rejuvenation of stock plant material
2. Manipulating the environmental conditions and physiological status of the stock plant in relation to: (a) water status, (b) temperature, (c) light (irradiance, photoperiod, quality), (d) stock plant etiolation, (e) girdling (f) CO_2 enrichment, (g) carbohydrates, and (h) managing carbohydrate/nitrogen levels
3. Type of wood selected
4. Seasonal timing

Treatment of Cuttings

1. Storage of cuttings
2. Auxins
3. Mineral nutrition of cuttings
4. Leaching of nutrients
5. Wounding

Environmental Manipulation of Cuttings

1. Water relations—humidity control
2. Temperature
3. Light (irradiance/photon flux, photoperiod, quality)
4. Accelerated growth techniques (AGT)
5. Photosynthesis of cuttings

The remainder of this chapter discusses these factors in detail.

Selection and Maintenance of Stock Plants for Cutting Propagation

Management of stock plants (or containerized plants) to maximize rooting begins with *selection* of source material that is easy-to-root (juvenile), *maintenance* of stock plants in the juvenile/transition phase to maximize rooting, and *rejuvenation* of stock plant material (reversal from the mature to a juvenile/transition phase) to reestablish high rooting potential.

For new cultivars to be commercially successful, they must be easy to propagate and suitable for existing propagation and production systems. New cultivars are, in part, selected for their ease-of-rooting (juvenile characteristics). Despite how desirable the form, flower color, ornamental characteristics, or yield (fruit crops), it is not economically feasible to use cutting propagation with a new cultivar that has less than 50 percent rooting. (Ways of overcoming low rooting percentage by utilizing more expensive systems of graftage and budding, mounding and stooling, micropropagation are discussed elsewhere.) Nurseries continually select for plants that are easy-to-root through the annual harvesting and rooting of cuttings from previously rooted containerized plants in production blocks or stock plants. This *serial propagation* of new generations of rooted cuttings helps *maintain* easy-to-root characteristics of a cultivar.

There are other horticultural and forestry practices that can maintain stock plants in a physiologically juvenile or transition phase and improve rooting success (*57, 155, 157, 175*). The development of systems for obtaining whole populations of juvenile and partially juvenile/transition cuttings is revolutionizing clonal forestry. For example, growing seedling populations of elite germ plasm of Monterrey pine, loblolly pine, Douglas-Fir rather than taking cuttings from this elite material and using hedging and pruning systems and serial-cutting generations to maintain a high rooting potential—has exciting opportunities for increasing timber yield. The hedging or shearing treatments given Monterrey pine (*Pinus radiata*) trees (Figure 10–24) is quite effective in maintaining juvenility and increasing the rooting potential of cuttings taken from them, compared to nonhedged trees (*185*).

Rejuvenation of Stock Plants

In difficult-to-root woody plant species, ease of adventitious root formation declines with age of parent stock, resulting in a propagation enigma, since desirable characteristics are frequently not expressed until after a plant has reached maturity. The transition from the *juvenile* to the *mature* phase has been referred to as *phase change, ontogenetic aging,* or *meristem aging.* There are progressive changes in such morphological and developmental characteristics as leaf shape, branching pattern, shoot growth, vigor, and the ability to form adventitious buds and roots (*108, 112, 113, 209*) (see Chapter 9). Experiments with apple, pear, eucalyptus, live oak, and Douglas-Fir have shown that the ability of cuttings to form adventitious roots decreased with increasing age of the plants from seed, in other words, when the stock plant changed from the juvenile to the adult phase (*136*). With many woody species, it is the **physiological or ontogenetic age,** not **chronological age,** of the cutting that is most important in rooting success (see Chapter 9).

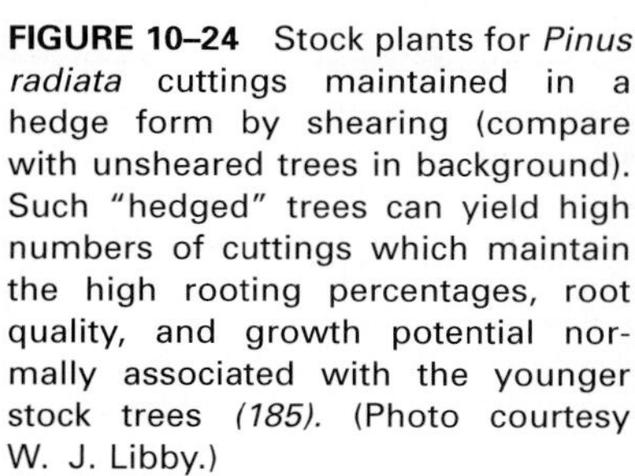

FIGURE 10–24 Stock plants for *Pinus radiata* cuttings maintained in a hedge form by shearing (compare with unsheared trees in background). Such "hedged" trees can yield high numbers of cuttings which maintain the high rooting percentages, root quality, and growth potential normally associated with the younger stock trees *(185)*. (Photo courtesy W. J. Libby.)

CLONAL MULTIPLICATION OF ELITE FORESTRY GERM PLASM

- Elite germ plasm is selected for superior growth rates, form, disease resistance, etc.

↓

- **Seed taken from the elite germ plasm is planted and the seedling stock plants (ortets) are established.** The ortets are continually maintained in a juvenile/transition condition by hedging and shearing treatments.

↓

- **Cuttings (ramets) are taken from the ortets and rooted.**

↓

- **Rooted ramets are transplanted to timber production sites and further evaluated for desirable growth characteristics.**

↓

- **Ortets of superior ramets are maintained and clonally multiplied to bulk-up ramets needed for commercial production.** Ortets of less desirable ramets are culled from the selection program.

↓

- **Clonally regenerated plants are established in a mosaic of clonal blocks with other clonal material to avoid monoculture production and reinforce genetic diversity in the timber plantations.**

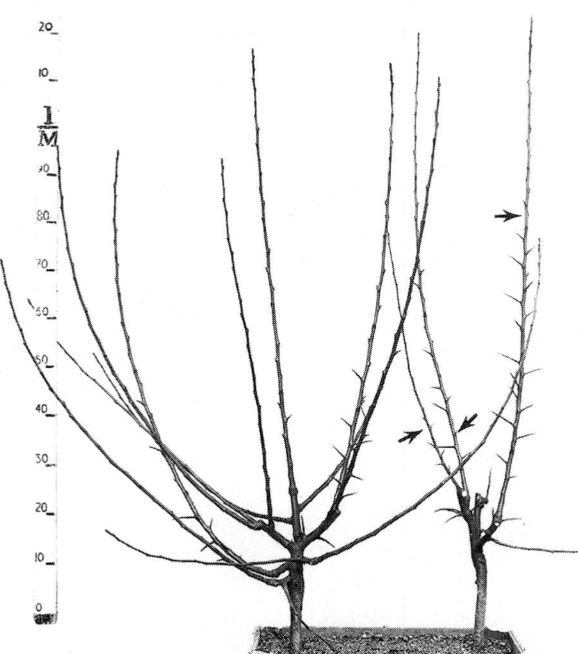

FIGURE 10–25 Optimum rooting of hardwood cuttings for 'M-26' apple rootstock occurs from the subordinate, thinner shoots that develop in the shoot hierarchy (framework) of severely pruned stock plant hedges. *Right:* The thinner, subdominant cuttings have been collected, while only the dominant, spiny shoots (arrows) remain to provide the framework for next year's generation of shoots (cuttings). *Left:* An unpruned stock plant with subordinate and dominant, spiny shoots. (Courtesy B. H. Howard, Horticulture Research International, East Malling, England.)

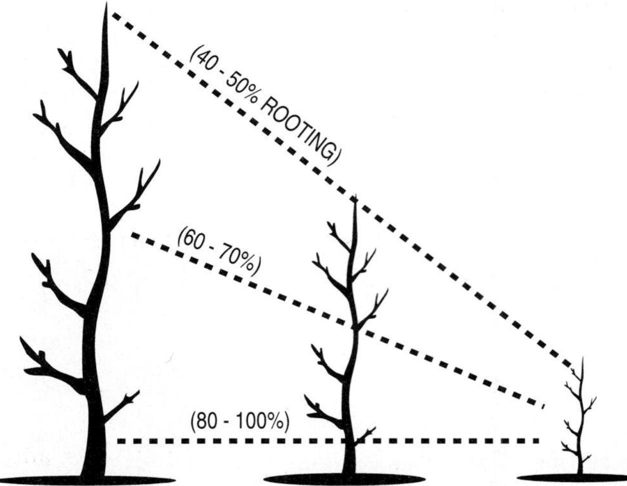

FIGURE 10–26 Rooting potential (typical values in brackets) of hardwood cuttings in a hard-pruned (severely cut back) hedge is more influenced by the relative position of the shoots than by their absolute position in terms of distance between themselves or from the root system. (Redrawn from B. H. Howard, Horticulture Research International, East Malling, England (*154*).)

Stock plants for cuttings of selected fruit tree rootstocks and woody ornamentals are maintained as hedges rather than allowed to grow to a tree form. Proper hedge management (pruning) of permanent stock plants can maintain large numbers of cuttings in an apparent juvenile stage of development. See Chapter 11 for a discussion on different types of pruning and girdling systems for stock plants.

Severe or hard pruning gives rise to many shoots suitable for cuttings, but their higher rooting potential is not necessarily due to greater vigor (as has long been supposed), but rather to the less vigorous (thinner-diameter), subordinate shoots that root better than the more vigorous ones, e.g., *Prunus, Rhododendron, Syringa,* etc. *(154, 155).*

Enhanced rooting potential, with relatively thin shoots of both hardwood and softwood cuttings, is only achieved with an improved propagation environment. Thin cuttings are more susceptible to basal rotting, so good drainage and mist management are critical; when planting hardwood cuttings directly in the field, a compromise needs to be made between the thinner shoots that root more quickly and larger stem diameter cuttings that survive longer in the poor conditions often present in field soil during winter *(154).*

Competence to root appears to be controlled independently in individual shoots, and is indirectly related to shoot thickness, which favors the subordinate (subdominant) shoots that develop in the shoot hierarchy of the severely pruned hedges (Figure 10–25). Rooting potential among shoots in a hedge is then more dependent on their relative position, rather than their proximity to the ground (Figure 10–26). The most vigorous shoots are the poorest rooters, but make better hardwood cuttings *(155).* Thin-stemmed shoots are better propagated as softwood cuttings. They have a higher leaf-to-stem ratio, and greater accumulation of dry matter at the cutting base before the first roots emerge (*158*).

In some species, such as apple, English ivy, olive, eucalyptus, *Acacia koa,* differences in certain morphological characteristics, such as leaf size and shape, make it easy to distinguish between the adult and the lower, juvenile portions of the plant. Cuttings should be taken from the latter type for best rooting (*226*). In some kinds of deciduous trees, such as oak and beech, leaf retention late into the fall occurs on the basal parts of the tree and indicates the part **(cone of juvenility)** still in the physiologically juvenile stage. Cuttings should be taken from this type of wood.

In rooting cuttings of difficult species it would be useful to be able to induce rejuvenation to the easily rooted juvenile or transition stage from plants in the mature form. This has been done in several instances by the following methods:

- Rejuvenation of apple can be done with mature trees by causing adventitious buds/shoots to develop from root pieces, which are then made into softwood stem cuttings and rooted (*273*).
- By removing terminal and lateral buds and spraying stock plants of *Pinus sylvestris* with a mixture of cytokinin, tri-iodobenzoic acid, and Alar (daminozide), many fascicular buds can be forced out. With proper subsequent treatment, high percentages of these shoots can be rooted (*315*).
- Chemical manipulation with gibberellin sprays on English ivy plants in the growth phase has caused substantial stimulation of growth and reversion of some of the branches to the juvenile stage (*240*), as well as improved rooting of ivy cuttings taken from the sprayed plants (*275*).
- In some plants juvenile wood can be obtained from mature plants by forcing juvenile growth from **sphaeroblasts**—wartlike protuberances containing meristematic and conductive tissues sometimes found on trunks or branches (*310*). These are induced to develop by disbudding and heavily cutting back stock plants. Using the mound-layering (stooling) method on these rooted sphaeroblast cuttings produces rooted shoots that continue to possess juvenile characteristics.
- Grafting mature forms of ivy onto juvenile forms has induced a change of the mature to the juvenile stage, provided that the plants are held at fairly high temperatures (*69, 275*)*;* such transmission of the juvenile rooting ability from seedlings to mature forms by grafting has also been accomplished in rubber trees (*Hevea brasilensis*) (*213*), and with serial graftage of mature difficult-to-root scions onto seedling rootstock of *Eucalyptus* × *trabutii* (Figure 10–27).
- Ready-rooting cuttings can be produced by utilizing stock plants that were originally produced by micropropagation. Micropropagated stock plants exhibit certain juvenile characteristics and produce an increased number of higher-rooting thin-stemmed cuttings that conventionally produced stock plants. The tissue

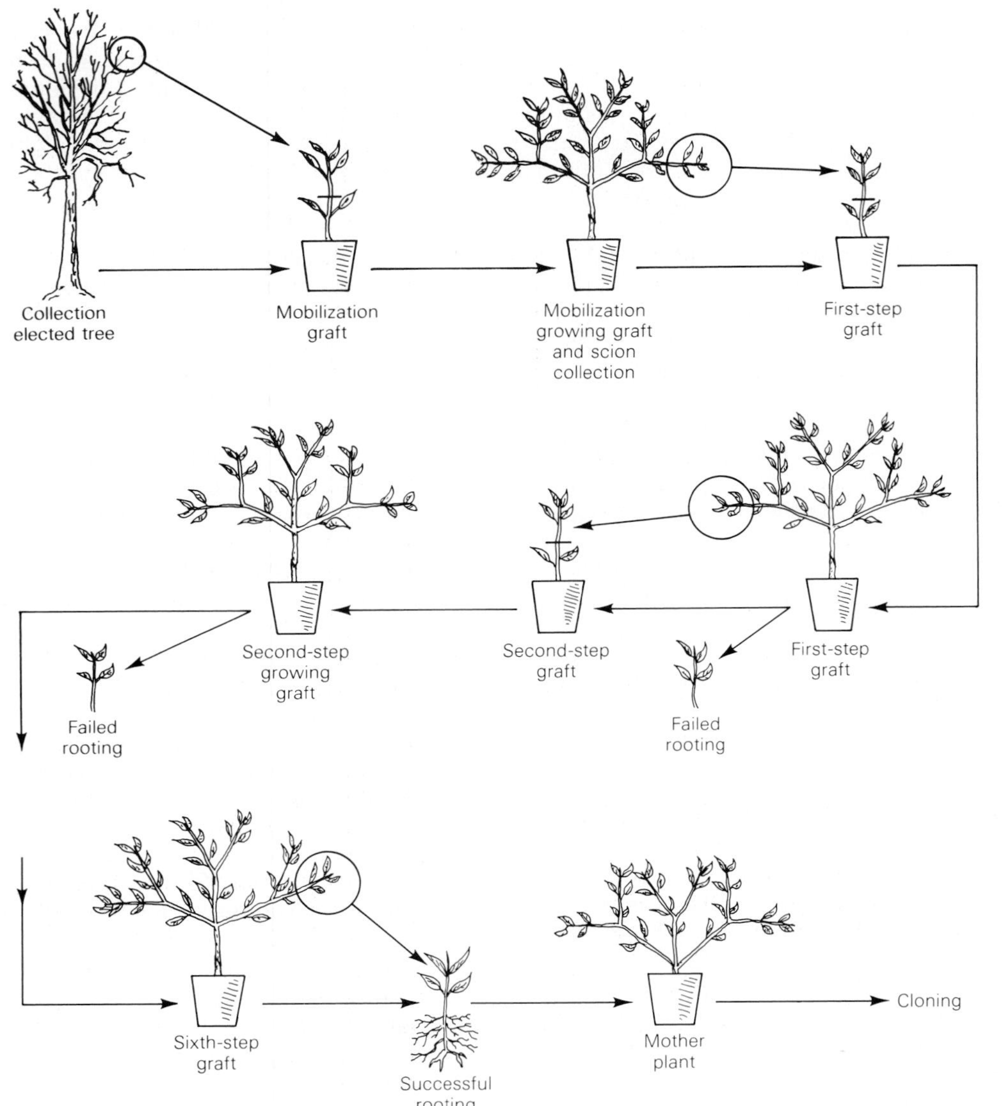

FIGURE 10–27 Scheme for rejuvenation techniques used in serial graftage of ten-year-old *Eucalyptus* × *trabutii* onto juvenile seedling understock. Six serial grafts were needed before mature grafted scions could be used as cuttings and rooted (*257*).

culture effect is long-lasting and appears to be maintained by severe hedge pruning (*160, 161*).

- Rejuvenation of tissue *in vitro* has tremendous potential to increase rooting of tissue-culture-derived cuttings (*3, 110, 161, 175, 225, 282*). Likewise, recent advances with transgenic plants, somatic embryogenesis, and synthetic seed technology may be used to restore the juvenile condition from physiologically mature plants (*72, 215*).

Manipulating the Environmental Conditions and Physiological Status of the Stock Plant

Water status. The physiological condition of stock plants is a function of genotype (species, cultivar) and environmental conditions (water, temperature, light, CO_2, and nutrition). To avoid water stress, plant propagators often emphasize the desirability of taking cuttings early in the morning when the plant material is in a turgid condition. There is experimental evidence to support this view. Studies with both cacao (*87*) and pea (*231*) cuttings showed reduced rooting when the cuttings were taken from stock plants having a water deficit. Unrooted cuttings are particularly vulnerable to water stress, since rehydration of the tissue is very difficult without a root system. Furthermore, droughted cuttings are more prone to disease and pest problems.

Temperature. Information on temperature interactions with stock plant water relations, irradiance, and CO_2 is limited. Research has shown that there is a complex interaction of temperature and stock plant photoperiod on the level of endogenous auxins and other hormones (*144*). With deciduous woody species (apple, plum), higher air temperatures can produce more rapid growth of stock plants and the production of higher-rooting, thinner-stemmed cuttings (*154*). *In general,* the air temperature of stock plants (12 to 27° C; 54 to 81° F) appears to play only a minor role in ease of rooting of cuttings (*202*).

Light. *Light duration* (photoperiod), *irradiance* [($W \cdot m^{-2}$) or photon flux ($\mu mol\ m^{-2} \cdot s^{-1}$)], and *spectral quality* (wavelength) influence the stock plant condition and subsequent rooting of cuttings (*202*). For instance, sufficient irradiance to stock plants is needed to maintain minimal endogenous auxin for rooting chrysanthemum cuttings. Conversely, too high an irradiance can cause photo destruction of auxin or adversely affect stock plant water relations.

There is some evidence that the **photoperiod** under which the stock plants are grown exerts an influence on the rooting of cuttings taken from them (*42, 106, 168, 202*). This could be a photosynthetic or morphogenic effect. If the photoperiod influences photosynthesis, it may be related to carbohydrate accumulation, with best rooting obtained under photoperiods promoting increased carbohydrates. If manipulation of photoperiod favors vegetative growth (rooting) and suppresses reproductive growth (flowering), then the effect is photomorphogenic (*128, 271*). Long-day conditions (sufficient hours of light to satisfy the critical photoperiod) have been used with some short-day flowering cultivars of chrysanthemum; where flowering is antagonistic to rooting, the long-day conditions promote vegetative growth and enhance rooting (*91*). Likewise, with some woody perennials where the onset of dormancy shuts down vegetative growth and/or reduces rooting, propagators can manipulate stock plants by extending the photoperiod with low irradiance from an artificial light source.

The conflicting reports on the influence of **light quality** on stock plants and subsequent rooting of cuttings is attributed to the effect of red and far-red light on rooting (i.e., via the phytochrome system) (*202*). In one experiment, when stock plants were exposed for six weeks to different light quality sources, cuttings from plants exposed to blue light rooted most readily (*276*).

Stock plant etiolation. Reducing irradiance levels of stock plants can sometimes enhance rooting of difficult-to-root species. By definition, **etiolation**[5] is the total exclusion of light; however, plant propagators also use this term when forcing new stock plant shoot growth under conditions of heavy shade. Softwood cuttings are then taken from new growth and often root more readily. **Banding** is a localized light exclusion pretreatment which excludes light from that portion of a stem that will be used as the cutting base (*11, 197*) Banding can be applied to etiolated shoots or applied to light-grown shoots which are still in the softwood stage. In the latter case, a band of Velcro or black adhesive tape is said to **blanch** the underlying tissues, since the stock plant shoot accomplished its initial growth in light prior to banding. **Shading** refers to any stock-plant growth under reduced light conditions (*168*). For techniques of banding, blanching, etiolation, and shading, see Chapter 13, Table 11–3, Figures 11–23, 11–24.

Anatomical and physiological changes can occur in etiolated stem tissue that enhance rooting. Etiolation of chestnut (*Castanea*) (i.e., covering the stool bed with soil) caused a greater accumulation of starch grains—but no significant change in stem anatomy; however, girdling and then etiolating the shoots increased parenchyma and storage cells above the girdle, reduced sclerenchyma formation—and was the only treatment that rooted (*18*). Exclusion of light by etiolation, stem banding, or shading greatly enhances a stem's sensitivity to auxin (*198, 199*). Translocatable factors produced distal to (above) an etiolated segment also enhanced the etiolation effect (*11*). Etiolation may

[5]Etiolation is the development of plants or plant parts in the absence of light. This results in such characteristics as small unexpanded leaves, elongated shoots, and lack of chlorophyll, resulting in a yellowish or whitish color.

FIGURE 10–28 *Left:* Etiolation frames in place over stock plant hedges of *Syringa vulgaris*. *Right:* Improved rooting following etiolation of *S. vulgaris* 'Madame Lemoine' (far left) and *S. vulgaris* 'Charles Joly' (second from right). Cuttings from non-etiolated stock plants have poor rooting (second from left and far right). (Courtesy B. H. Howard, Horticulture Research International, East Malling, England (*156*).)

also reduce the production of lignin [for structural support cells (sclereids, fibers)]; thus, instead of forming lignin, phenolic metabolites may be channeled to enhance root initiation (*56, 81*).

Rooting in cuttings of *Syringa vulgaris* 'Madame Lemoine' was enhanced when stock plants were grown in the dark for a short period after bud break. Cuttings grown initially in the dark were found to have relatively thin stems, resulting in a higher leaf-to-stem ratio than normal light-grown ones. This was associated with a net accumulation of dry matter at the cutting base before the first roots emerged (Figure 10–28) (*158*). A cutting must produce and/or rely on stored carbohydrates in excess of its maintenance requirement for successful rooting to occur, which is why stock plant manipulation (etiolation, hedging) and the rooting environment of the cutting are so critical for successful rooting. Besides carbohydrates, there are other biochemical changes occurring during etiolation and hedging that enhance rooting.

Girdling. Girdling, or otherwise constricting the stem, blocks the downward translocation of carbohydrates, hormones, and other possible root-promoting factors and can result in an increase in root initiation. Using this technique on shoots prior to their removal for use as cuttings can improve rooting. This practice has been remarkably successful in some instances. For example, rooting of citrus and hibiscus cuttings was stimulated by girdling or binding the base of the shoots with wire several weeks before taking the cuttings (*272*).

In cuttings from mature trees of the water oak (*Quercus nigra*), a threefold improvement in rooting was obtained when cuttings were taken from shoots that had been girdled six weeks previously, especially if a talc powder consisting of auxin, growth retardant, carbohydrate, and a fungicide was rubbed into the girdling cuts (*130*). Enhanced rooting of cuttings taken from girdled stock plants has also been obtained with sweet gum, slash pine, and sycamore. Girdling just below a previously etiolated stem section was particularly effective in promoting rooting in apple cuttings (*66*).

Carbon dioxide enrichment. With many species, carbon dioxide enrichment of the stock plant environment has increased the number of cuttings that can be harvested from a given stock plant, but there is considerable variation of rooting response among species. Principal reasons for increased cutting yields are increased photosynthesis, higher relative growth rate, and greater lateral branching of stock plants (*202*). Any benefits of CO_2 enrichment have been limited to greenhouse-grown stock plants during conditions when greenhouse vents are closed and ambient CO_2 becomes a limiting factor to photosynthesis (i.e., October-March in central Europe). Without adequate light (supplementary greenhouse lighting during low-light-irradiance months), CO_2 enrichment is of minimal benefit (*206*) (see Chapter 3).

Carbohydrates. The relationship between carbohydrates and adventitious root formation remains controversial. Since Krause and Kraybill (*178*) *hypothesized* the importance of carbohydrate-to-nitrogen (C/N) ratio in plant growth and development, rooting ability of cuttings has been discussed in relation to carbohydrate content. The carbohydrate pools of sugars (soluble carbohydrates) and storage carbohydrates (starches or insoluble carbohydrates) are important to rooting as building blocks of complex macromolecules, structural elements, and energy sources (*109, 121, 280*).

Although stock plant carbohydrate content and rooting may sometimes be positively correlated (*124, 146*), *carbohydrates do not have a regulatory role in rooting.* A positive correlation between carbohydrate content and rooting may reveal that the supply of current photosynthate is insufficient for supporting optimal rooting (*301*). High C/N ratios in cutting tissue promote rooting but do not accurately predict the degree of rooting response (*280*). Cuttings use stored carbohydrates in root regeneration, but only in small amounts. Differences in C/N ratios are due mainly to nitrogen rather than carbohydrate content. Nitrogen has been negatively correlated to rooting (*124*), which suggests that the correlation between high C/N ratios and rooting may be due to low N levels. Rooting can be benefited by controlling nitrogen fertility of stock plants, such that cutting shoot development is not stimulated by high N levels (*236, 309*). This avoids the disadvantage of adventitious rooting competing with rapidly developing shoots for carbohydrates, mineral nutrients, and hormones (*121*).

Managing carbohydrate/nitrogen levels of stock plants. Generally, maintaining stock plants under a high carbohydrate/high nitrogen level is optimal for rooting cuttings under mist, and a high carbohydrate/low-to-moderate nitrogen ratio is optimal for rooting dormant hardwood cuttings. Cuttings of *Hypericium, Ilex, Rosa,* and *Rhododendron* rooted best when stock plants were suboptimally fertilized, resulting in less than maximal shoot growth (*236*). Very low nitrogen leads to reduced vigor, whereas high nitrogen caused excess vigor; either extreme is unfavorable for rooting. Adequate nitrogen is necessary for nucleic acid and protein synthesis.

It is important to distinguish between the role of carbohydrates in enabling a cutting to *survive* (*until it roots*), and the role of carbohydrates *in rooting* itself. In species where unrooted hardwood cuttings were propagated directly in the field without mist, survival is necessary before rooting occurs. Hence, the need to compromise between thin-rooting cuttings (which root better, but have poorer field survival) and larger diameter, carbohydrate-rich cuttings that survive better in the field, but have lower rooting capacity.

To maintain high carbohydrate/low-to-moderate nitrogen ratios of stock plants for optimal rooting of hardwood cuttings, producers can manipulate stock plants as follows:

- Reducing nitrogen fertilization, thus reducing shoot growth and allowing for carbohydrate accumulation.
- Selecting cutting material from lateral shoots, which have slower growth rates and higher carbohydrate storage than fast-growing terminal shoots. [But for plants showing a plagiotropic growth pattern (see Figure 10-29), use of lateral shoots should be avoided.]
- For maintenance of adequate carbohydrate levels, photosynthate production of greenhouse-grown stock plants can be controlled by increasing light irradiance of supplementary high-pressure sodium-vapor lights.

Type of Wood Selected from Stock Plants

In woody perennials, types of materials to use range from softwood terminal shoots of current growth to dormant hardwood cuttings. No one type of cutting material is best for all plants. What may be ideal for one plant would be a failure for another. Procedures for certain species or cultivars, however, often may be extended to related species or cultivars (see Chapters 19 and 20).

> Leafy softwood cuttings may be the best way to propagate certain difficult-to-root species [e.g., maple (*Acer*), crabapple (*Malus*), redbud (*Cercis*)]. Softwood cuttings would tend to have higher auxin, and lower endogenous carbohydrate. They have a moderate light requirement, since some photosynthesis enhances their rooting. They are propagated only with more intensive (critical) water management, using mist or fog. Whereas dormant hardwood cuttings have low auxin and high carbohydrate storage, photosynthesis is initially not needed for rooting, and they can be propagated under lower light, without mist or under less critical mist regimes.

Differences between lateral and terminal shoots. In general, with exceptions, softwood cuttings root better from terminal shoots, and the more lignified, semi-hardwood cuttings root better from lateral shoots. In rooting different types of *softwood* plum cuttings taken in the spring, there was a marked superiority in rooting of lateral shoots, compared with terminal shoots. Similarly, lateral branches of Fraser fir (*Abies*), white pine, Norway and Sitka spruce gave consistently higher percentages of rooted cuttings than did terminal shoots (*23, 270*). In rhododendrons, too, thin cuttings made from lateral shoots consistently give higher rooting percentages than those taken from vigorous, strong terminal shoots. In certain species, however, plants propagated from cuttings taken from lateral branches may have an undesirable growth habit: They tend to become plagiotropic and have a horizontal branchlike growth after rooting, whereas cuttings taken from primary axes grow upright (orthotropically) and produce symmetrical trees (e.g., *Taxus cuspidata,* coffee, Norfolk Island pine) (see Figure 9–19 and 10–29). This is referred to as *topophysis.*

Differences between various parts of the shoot. With some woody plants, *hardwood cuttings* are made by sectioning shoots a meter long and obtaining four to eight cuttings from a single shoot. Marked differences are known to exist in the chemical composition of such shoots from base to tip (*293*). Variations in root production on cuttings taken from different portions of the shoot are often observed, with the highest rooting, in many cases, found in cuttings taken from the basal portions of the shoot. Cuttings prepared from shoots of three cultivars of the highbush blueberry (*Vaccinium corymbosum*) have greater rooting if taken from the basal portions of the shoot rather than from terminal portions (*219*). Exceptions to this are found in rose (*124*) and other species. The number of preformed root initials in woody stems (in some species at least) distinctly decreases from the base to the tip of the shoot (*116*). Consequently, the rooting capacity of basal portions of such shoots would be considerably higher than that of the apical parts. This factor is of little importance, however, in cuttings of easily rooted species which root readily regardless of the position of the cutting on the shoot.

Flowering or vegetative wood. With most plants, cuttings can be made from shoots that are in either a flowering or a vegetative condition. Again, with easily rooted species it makes little difference which is used, but in difficult-to-root species this can be an important factor. For example, in blueberry (*Vaccinium atrococcum*), hardwood cuttings from shoots bearing flower buds did not root as well as cuttings with only vegetative buds (*218*). Herbaceous dahlia cuttings bearing flower buds are more difficult to root than cuttings having only vegetative buds (*17*).

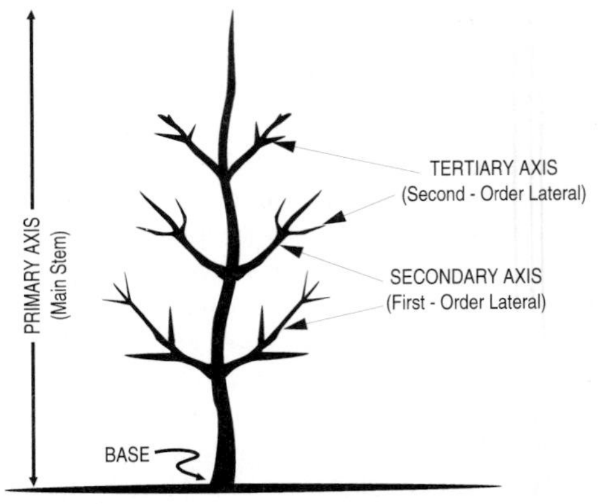

FIGURE 10–29 The line drawing shows the location where cuttings were taken on stock plants of Frasier fir *(Abies fraseri). Top:* A schematic of the branch order. *Bottom:* Demonstrates multiple terminal used as cuttings. Cuttings from lateral branches root readily, but have an undesirable horizontal growth habit (plagiotropic) after rooting. Cuttings taken from the tips of primary axes (main stem) produce symmetrical, upright (orthotropic) trees. (Redrawn from Blazich and Hinesley (*23*).)

Flowering is a complex phenomenon and can serve as a **competing sink** to the detriment of rooting. Removal of flower buds increased rooting in rhododendron by eliminating the strong competing sink of flower buds for metabolites necessary for rooting (*166*). With many ornamental species (e.g., *Abelia, Ligustrum, Ilex,* etc.) it is commercially desirable to remove flower buds from cuttings for more rapid root development, earlier vegetative growth, and more efficient liner production (*171*).

Seasonal Timing

Seasonal timing, or the period of the year in which cuttings are taken, can play an important role in rooting (*9, 132*) With many species there is an optimal period of the year for rooting (*2*). Propagators strive to *maintain the plants' momentum* by rooting during these optimal periods to maximize the rooting process and speed up the production of liners. Climate permitting, it is possible to make cuttings of easy-to-root species throughout the year.

In propagating **deciduous** species, hardwood cuttings can be taken during the dormant season (from leaf fall, when buds are dormant, and before buds start to force out in the spring). Leafy softwood or semi-hardwood (greenwood) cuttings could be prepared during the growing season, using succulent and partially matured wood, respectively. The narrow- and broad-leaved evergreen species have one or more flushes of growth during the year, and cuttings can be obtained year-round in relation to these flushes of growth.

Certain species, such as privet, can be rooted readily if cuttings are taken almost any time during the year; on the other hand, excellent rooting of leafy olive cuttings under mist can be obtained during late spring and summer, whereas rooting drops almost to zero with similar cuttings taken in midwinter. Seasonal changes influenced rooting of both juvenile and mature (difficult-to-root) *Ficus pumila* cuttings; however, treating juvenile (easy-to-root) cuttings with IBA overcame the seasonal fluctuation in rooting (Figure 10–30). Shoot RNA was found to be an index of bud activity and subsequent seasonal rooting differences. Highest shoot RNA levels and increased vascular cambial activity occurred during peak rooting periods in both the easy- and difficult-to-root forms (*52*).

Softwood cuttings of many deciduous woody species [e.g., cherries (*135*), lilac (*Syringa*)] taken during spring or summer usually root more readily than hardwood cuttings procured in the winter. The Chinese fringe tree (*Chionanthus retusus*) is notoriously difficult to root, but by taking cuttings during a short period in midspring, high rooting percentages can be obtained (*274*).

The effect of timing is also strikingly shown by difficult-to-root deciduous azalea cuttings. These root readily if the cuttings are taken from succulent growth in early spring; by late spring, however, the rooting percentages decline rapidly. For any given species, small experiments are required to determine the optimum time to take cuttings, which is more related to the physiological condition of the plant than to any given calendar date.

Methods to document the most advantageous time to collect cuttings have included calendar days, days from bud-break, use of indicator plants, number of hours of sunlight, degree-day chilling units, and the morphological condition of the stock plant. In a novel approach, a degree-day (heat-unit) system was utilized to predict successful rooting in difficult-to-root adult Chinese pistache (*Pistacia*) (*71*). Maxi-

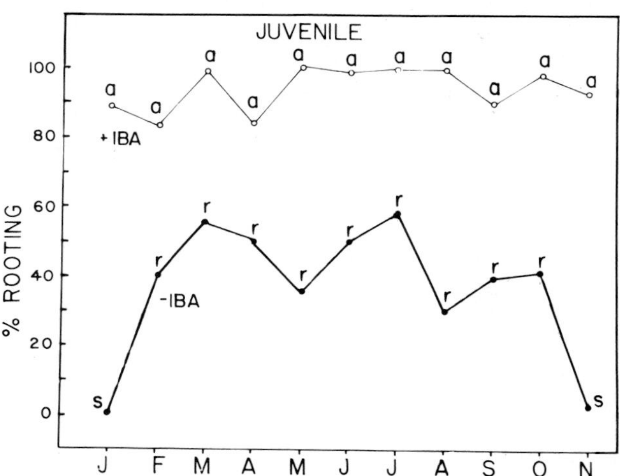

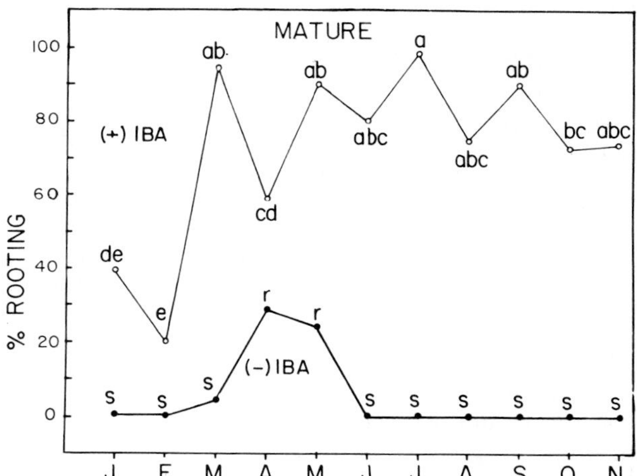

FIGURE 10–30 Seasonal fluctuation in percent rooting of juvenile (easy-to-root) and mature (difficult-to-root) *Ficus pumila* cuttings. Mature control (IBA) plants root only from March to May, while juvenile control roots poorly—from November to January. When treated with IBA (+IBA), the juvenile cuttings overcome the seasonal fluctuation in rooting, whereas mature cuttings still show poor rooting in January and February (*52*).

mum rooting occurred when cuttings were collected from stock plants with green softwood stems, which had 380 degree days (using a threshold temperature of 7.2° C/45° F) after bud-break.

Often the effects of timing are merely a reflection of the response of the cuttings to environmental conditions at different times of the year. When hardwood cuttings of deciduous species are taken and planted in the nursery in early spring, after the rest period of the buds has been broken by winter chilling, the results are quite often a complete failure, since the buds quickly open with the onset of warm days. The newly developing leaves will start transpiring and remove the moisture from the cuttings before they have the opportunity to form roots; thus they soon die. Newly expanding buds and shoots are also competing sinks for metabolites and phytohormones, to the detriment of rooting. This has been shown with *Rosa multiflora* under an intermittent mist system where water stress was not a factor (*124*). If cuttings can be taken and planted in the fall while the buds are still in the rest period, roots may form and be well established by the time the buds open in the spring.

Broad-leaved evergreens usually root most readily if the cuttings are taken after a flush of growth has been completed and the wood is partially hardened-off or lignifed. This occurs, depending upon the species, from spring to late fall. In rooting cuttings of *narrow-leaved evergreens,* best results may be expected if the cuttings are taken during the period from late fall to late winter (*181*). With junipers, and yew (*Taxus*), rooting was lowest during the season of active vegetative growth and highest during the dormant period. Furthermore, the low temperatures occurring at the time when such coniferous evergreens root best apparently is not a requirement, since juniper stock plants held in a warm greenhouse from early fall to midwinter produced cuttings that rooted better than outdoor-grown stock plants exposed to seasonal conditions (*182*).

In many containerized ornamental nurseries, cuttings from difficult-to-root species are taken early in the propagation season, whereas cuttings of easy-to-root species are taken later in the summer. This seasonal scheduling of propagation also more efficiently utilizes propagation facilities and personnel.

TREATMENT OF CUTTINGS

Only high quality cuttings should be collected for propagation. As the wise instructions to employees in a commercial propagation department go—"A cutting that is barely good enough is *never* good enough, so don't put it in the bunch!" Quality control of cuttings begins with stock plant quality control. Since many containerized ornamental nurseries no longer use stock plants, it is essential to maintain quality control of all production container plants from which propagules are taken. It is important that propagules be collected from stock plants free of viruses, bacteria, fungi, and other pathogenic organisms. Propagation is the foundation on which production horticulture hinges. Marginal quality propagules delay product turnover and create cultural and quality problems throughout the production cycle (*5*).

Storage of Cutting Material

As noted earlier, many propagators prefer to collect propagules from stock plants early in the day when cuttings are still turgid. If the cuttings cannot be stuck immediately, they are misted to reduce transpiration and held overnight in refrigeration facilities (Figure 11–49) at 4 to 8° C (*40 to 48° F*) and generally stuck the next day.

Cuttings of some temperate-zone woody species have been stored at low temperatures for extended periods without any deleterious effects on subsequent root formation and leaf retention. Storage of *Rhododendron catawbiense* cuttings in moist burlap bags at either 21 or 2° C (70 or 36° F) for 21 days did not reduce rooting (*64*), although carbohydrate concentrations in the bases of cuttings changed with time and storage temperature. However, cuttings of Foster's holly root poorly even after the shortest cold storage.

Cuttings of many tropical foliage plants are imported from Central America to be rooted and finished in the U.S. or Europe. It can take 4 to 14 days to deliver cuttings from Central America to U.S. nurseries. Duration in transit may affect cutting quality due to excess respiration, light exclusion, moisture loss, pathogen invasion, and/or ethylene buildup. Croton (*Codiaeum variegatum*) cuttings had excellent quality when stored 5 to 10 days at 15 to 30° C (59 to 86° F) or 15 days at 15 to 20° C (59 to 68° F) (*306*).

Unrooted cuttings of chrysanthemum, poinsettia, and carnation are routinely shipped by air transportation. Storage life of unrooted cuttings of geranium (*Pelargonium* × *hortorum* Bailey) was improved by high-humidity storage in polyethylene bags at 4° C (39 ° F) and low-irradiance illumination. Prestorage application of antitranspirants was detrimental, but soaking cutting bases in 2 to 5 percent sucrose for 24 hours prior to storage improved rooting (*221*). The ethylene inhibitor, silver nitrate, was more effective in maintaining storage life than silver thiosulfate, which reduced rooting (*221*). Abscisic acid will reduce transpiration in geranium cuttings, which may be of practical value in the shipment and storage of geranium cuttings (*4*).

In general, successful storage of unrooted cuttings depends on storage conditions, state of the cuttings, and species. It is important that dry matter losses and pathogens be minimized. Within the storage unit, it is best to maintain nearly 100 percent humidity, and temperature should be as low as the hardiness of the given species can tolerate (*14*). Reduced oxygen and ethylene levels and high CO_2 [controlled atmospheric storage (CA)] help to maintain rooting capacity (*14*). Storage duration can vary from days to several months depending on cutting carbohydrate reserves, frost hardiness, and degree of lignification (woodiness of the material) (see the discussion in Chapter 11).

Auxins

Before the use of synthetic root-promoting growth regulators (auxins) in rooting stem cuttings, many chemicals were tried with limited success (*173*). The discovery that natural auxins, such as **indoleacetic acid** (IAA) and synthetic ones such as **indolebutyric acid** (IBA) and **naphthaleneacetic acid** (NAA), stimulated the production of adventitious roots in stem and leaf cuttings, and was a milestone in propagation history (*324*). The response, however, is not universal; cuttings of some difficult-to-root species still root poorly after treatment with auxin. Hence auxin is not always the limiting chemical component in rooting, as discussed earlier in this chapter.

An ancient practice of some Middle Eastern and European gardeners in early days was to embed grain seeds into the split ends of cuttings to promote rooting. This seemingly odd procedure had a sound physiological basis, for it is now known that germinating seeds are good sources of auxin which aids root formation in cuttings.

Mixtures of root-promoting substances are sometimes more effective than either component alone. For example, equal parts of indolebutyric acid (IBA) and naphthaleneacetic acid (NAA), when used on a number of widely diverse species, were found to induce a higher percentage of cuttings to root and more roots per cutting than either auxin alone (*150*). Species are also known to react differently when treated with equal amounts of NAA or IBA; NAA was more effective than IBA in stimulating rooting of Douglas-Fir (*245*).

Adding a small percentage of certain phenoxy compounds to either IBA or NAA increased rooting and produced root systems better than those obtained with phenoxy compounds alone (*61, 151*). Amino acid conjugates of IAA sometimes stimulate better rooting than IAA alone. It has been suggested that activity of IAA in rooting may depend on its covalent bonding to low molecular weight phenolic compounds, i.e., chemical linkage with sugars, sugar alcohols, etc. (*61, 117*). Likewise, other tests (*79*) have shown that a combination of IBA and NAA, gave much better rooting than the use of any of these auxins alone.

The acid form of auxin is relatively insoluble in water, but can be dissolved in a few drops of alcohol or ammonium hydroxide before adding to water. Salts of some auxins may be more desirable than the acid form in some instances, because of their comparable activity and greater solubility in water (*323*). Also, solvents used to dissolve the acid formulations at higher concentrations, such as alcohol, NaOH etc. may be toxic to cuttings.

The aryl esters of both IAA and IBA, and amides of IBA [Phenyl-IAA **(P-IAA),** Phenyl-IBA **(P-IBA),** phenyl thioester **(P-ITB),** and phenyl amide **(NP-IBA)** (see Figure 2–14)], have been reported to be more effective than the acid forms in promoting root initiation (*61, 118, 119, 281*). The physiologically active phenyl-modified auxins are probably enzymatically hydrolyzed after cellular uptake, yielding the free parent acid (i.e., IAA or IBA) and phenolic moiety or portion (*61*). Again, this is species-dependent. It may also be that these new formulations are less toxic to plant material than the acid form.

For general use in rooting stem cuttings of the majority of plant species, IBA and/or NAA are recommended (*61*). To determine the best auxin and optimum concentration for rooting any particular species under a given set of conditions, small trials are necessary and should be repeated over several occasions, since repeated experiments can give conflicting results.

Application of synthetic auxins to stem cuttings at high concentrations can inhibit bud development, sometimes to the point at which no shoot growth will take place even though root formation has been adequate. Application of auxins to root cuttings may also inhibit the initiation and development of shoots from such root pieces. Basally applied IBA increased rooting, but inhibited bud-break of single-node rose stem cuttings. IBA was translocated to the upper part of the cutting, where it inhibited bud-break and increased ethylene synthesis of the cuttings (*285*).

Early bud-break and shoot growth of newly rooted cuttings are important in the overwinter survival of *Acer, Cornus, Hamamelis, Magnolia, Prunus, Rhododendron,* etc. These species need to put on a growth flush (after rooting, but prior to winter dormancy) so that sufficient levels of carbohydrates are stored in the root system to ensure winter survival. Hence, there is concern about auxin-suppressing bud-break and growth of rooted cuttings—and reduced winter survival. See Chapter 11 for discussion on post-propagation care of rooted cuttings.

There is often a question of how long the various root-promoting preparations will keep without losing their activity. Bacterial destruction of IAA occurs readily in unsterilized solutions. A widely distributed species of *Acetobacter* destroys IAA, but the same organism has no effect on IBA. Uncontaminated solutions of NAA and 2,4-D maintained their strength for as long as a year. Of course, alcohol solutions of auxin will depress any microbial activity.

IAA is sensitive to light and is readily inactivated. Concentrated IBA solutions in 50 percent isopropyl alcohol are quite stable and can be stored up to six months at room temperature in clear glass bottles under low light conditions without loss in activity (*241*). Both NAA and 2,4-D seem to be light-stable. Indoleacetic acid oxidase in plant tissue will break down IAA but has no apparent effect on IBA or NAA.

In stem tissue, auxin generally moves in a basipetal direction (apex to base). In early work, synthetic auxin applications were made to cuttings at the apical end to conform to the natural downward flow. As a practical matter, however, it was soon found that basal applications gave better results. Sufficient movement apparently carried the applied auxin into parts of the cutting where it stimulated root production. In tests using radioactive IAA for rooting leafy plum cuttings, IAA was absorbed and distributed throughout leafy cuttings in 24 hours, whether application was at the apex or base (*283*). However, with basal application, *most* of the radioactivity remained in the basal portion of the cuttings. Leafless cuttings absorbed the same amount of IAA as leafy cuttings, indicating that transpiration "pull" was not the chief cause of absorption and translocation.

> *In summary,* new synthetic auxins are needed to more effectively promote rooting rather than new formulations (i.e., carriers such as DMSO, polyethylene glycol, and other solvents), because rooting has not greatly increased by changing auxin solvents and carriers (*61*). The development of more efficient synthetic auxins and identification of other chemicals that act with auxin in promoting rooting (e.g., auxin synergists, rooting morphogens, etc.) would almost certainly lead to the formulation of better commercial chemical aids for rooting of cuttings than are presently available (*21*).
>
> For application methods of root-promoting growth regulators and listing of commercial auxin sources, solvents, and carriers see Chapter 11.

Mineral Nutrition of Cuttings During Rooting

It has been difficult to quantify the effect of nutrition on root primordia initiation versus root primordia elongation (*20, 286*). It is important that stock plants be maintained under optimum nutrition prior to the collection of cuttings. Mobilization studies have been conducted to examine the movement of mineral nutrients into the base of cuttings during root initiation (*335*). The redistribution of nitrogen in stem cuttings during rooting was accelerated by auxin treatment of plum (*284*). However, N was not mobilized, nor was any redistribution of P, K, Ca, and Mg detected during root initiation in stem cuttings of Japanese holly (*24, 25*).

There are conflicting reports on mobilization. During root initiation in chrysanthemum cuttings P, but not N, K, or Ca, was mobilized (*103*). Although considered immobile, redistribution of Ca was reported during rooting of Japanese holly. Apparently, Ca was redistributed to support tissue development in the upper cutting sections and not for root growth and development.

The importance of N in root initiation is supported by nutrition studies on rooting of cuttings and the importance of N in nucleic acid and protein synthesis (*20*). The influence of N on root initiation and development also relates to such factors as carbohydrate availability, C/N ratio, and hormonal interactions.

Zinc can promote the formation of the auxin precursor, tryptophan, and the formation of auxin (IAA) from tryptophan. Conversely, Mn acts as an activator of the IAA-oxidase enzyme system (*291*) and B may enhance IAA-oxidase activity, thus regulating endogenous auxin levels (Figure 10–20) (*164, 308*). Higher endogenous auxin levels are required for early root initiation than for later root development (see Figure 10–20 and discussion). If root initiation is related to the relative activity of IAA and

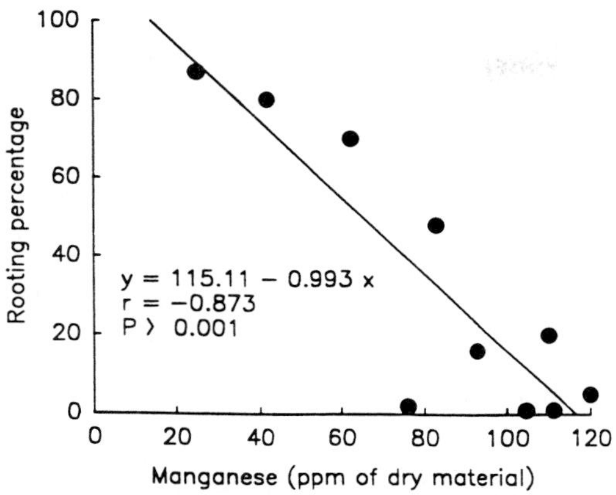

FIGURE 10–31 Correlation between manganese content of leaves of different avocado clones and rooting percentage. (From data of Reuveni and Raviv (*237*).)

IAA-oxidase, then rooting may be correlated with changes in relative Zn, Mn, and B concentration at the site of root initiation during the developmental stages of *de novo* rooting.

In a study with poinsettia where mineral element concentration was analyzed during the developmental stages of rooting, Fe, Cu, and Mo increased in the basal portion of stem cuttings during early root initiation, while P, K, Ca, and Mg decreased (*286*). During root primordia elongation and root emergence, Fe, Cu, Mo, Mg, Mn, B, and Zn concentration continued to increase at the cutting bases, but P and K concentrations remained low compared to when cuttings were initially inserted into the propagation medium.

High levels of Mn were found (*237*) in leaves of cuttings taken from difficult-to-root avocado cultivars, whereas cuttings from easy-to-root cultivars had a much lower manganese level (Figure 10–31). The negative correlation with rooting may be linked to manganese's activating the IAA-oxidase system and lowering endogenous IAA levels (*291*).

Leaching of Nutrients

The development of intermittent mist revolutionized propagation, but mist can severely leach cuttings of nutrients. This is a particular problem with cuttings of difficult-to-root species which take a longer time to root under mist. Mineral nutrients such as N, P, K, Ca, and Mg are leached from cuttings while under mist (*25, 103*).

Nitrogen and Mn are easily leached; Ca, Mg, S, and K are moderately leached; and Fe, Zn, P, and Cl are leached with difficulty (*294*). Both leaching and mineral nutrient mobilization contribute to foliar deficiencies of cuttings (*20*). The amount of leaching depends on the growth stage of the cutting material; leafy hardwood cuttings are reported to be more susceptible than softwood or herbaceous cuttings (*103*). Apparently, young growing tissues more quickly tie up nutrients by using them in the synthesis of cell walls and other cell components. Greater leaching occurs with leafy hardwood cuttings, since a greater portion of nutrients is in exchangeable forms (*103*).

As a whole, mist application of nutrients has not been a viable technique to maintain cutting nutrition. Nutrient mist application can inhibit rooting (*172*) and stimulate algae growth, which causes sanitation and media aeration problems (*319*).

A commercial technique is to apply moderate levels of controlled slow-release macro and micro elements to the propagation media either preincorporated into the media prior to sticking cuttings or by top dressing (broadcast) during propagation. These supplementary nutrients do not promote root initiation (*167*), but rather improve root development after root primordia initiation has occurred. Hence, turnover of rooted cuttings occurs more quickly and plant growth is maintained by producing rooted liners that are more nutritionally fit. Optimum levels of fertilization for rooting need to be determined on a species basis (see Chapter 11).

Wounding

Basal wounding is beneficial in rooting cuttings of certain species, such as rhododendrons and junipers, especially cuttings with older wood at the base. Following wounding, callus production and root development frequently are heavier along the margins of the wound. Wounded tissues are stimulated into cell division and production of root primordia (*194*) (Figures 10–32, 10–33). This is due to a natural accumulation of auxins and carbohydrates in the wounded area and to an increase in the respiration rate in the creation of a new "sink area." In addition, injured tissues from wounding produce ethylene, which may indirectly promote adventitious root formation (*179, 322*).

It has been proposed that wounding a cutting initiates a chemical signal that induces changes in the metabolism of affected cells (*318*). A listing of metabolic responses to wounding is given in Table 10–4. Potentially, cells at the base of the cutting influenced by wounding have enhanced receptivity to respond to auxin and other morphogens

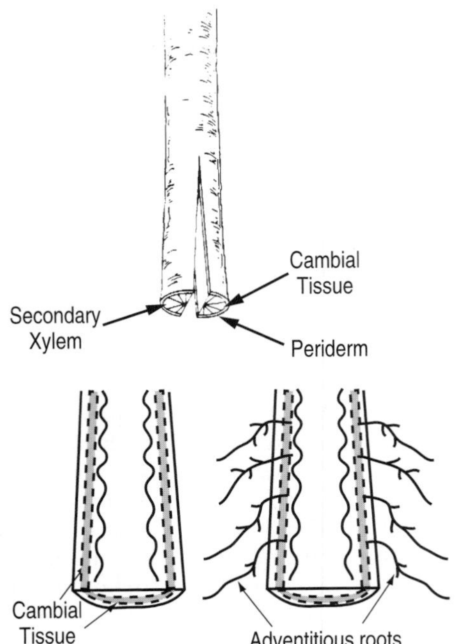

FIGURE 10–32 *Top:* Split-base treatment to enhance root initiation in leafless, dormant apple rootstock cuttings. *Bottom left:* The inner surface of a hardwood cutting wounded by splitting. When split longitudinally, much more cambium is exposed than with a normal cut across the stem base; cambial cells are able to regenerate in response to auxin treatment and produce cambial callus. *Bottom right:* Roots emerging from cambial callus. (Redrawn from MacKenzie, et al. (*194*) and Howard, Horticulture Research International, East Malling, England (*158*).)

(nonauxin endogenous compounds) essential to rooting (*317, 318*).

Wounding cuttings may also permit greater absorption of applied growth regulators by the tissues at the base of the cuttings. In stem tissue of some species there is a sclerenchymatic ring of tough fiber cells in the cortex external to the point of origin of adventitious roots. There is evidence in a few species (*13, 44*) that newly formed roots may have difficulty penetrating this band of cells. In those species, a shallow wound would cut through these cells and enhance the emergence of the developing roots.

ENVIRONMENTAL MANIPULATION OF CUTTINGS

Water Relations—Humidity Control

The loss of water from leaves may reduce the water content of the cuttings to such a low level that they do not survive. Propagation systems are designed:

- to maintain an atmosphere with low evaporative demand, so minimizing transpirational water losses from cuttings and thereby avoiding substantial tissue water deficits (cuttings without roots lack effective organs to replace transpired water lost, and cells must maintain adequate **turgor** for initiation and development of roots);
- to maintain acceptable temperatures for the regeneration metabolism needed at the cutting base, while avoiding heat stress of leaves; and

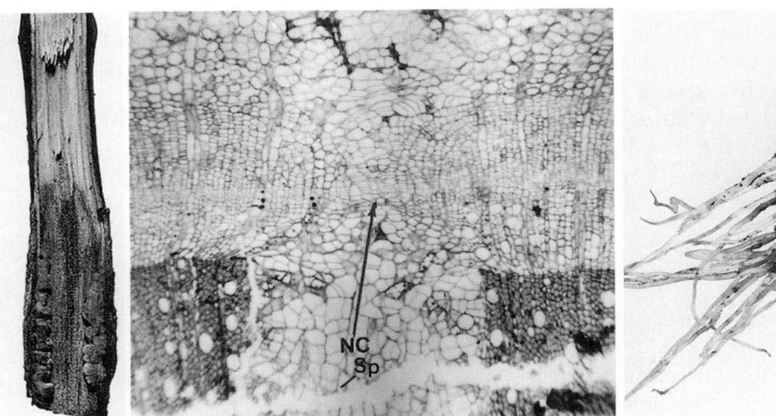

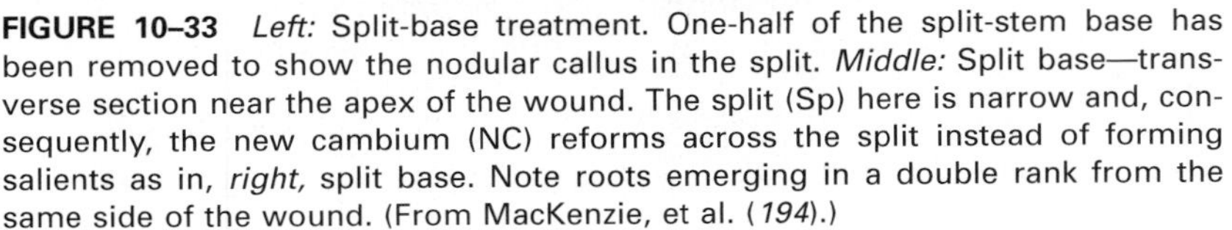
FIGURE 10–33 *Left:* Split-base treatment. One-half of the split-stem base has been removed to show the nodular callus in the split. *Middle:* Split base—transverse section near the apex of the wound. The split (Sp) here is narrow and, consequently, the new cambium (NC) reforms across the split instead of forming salients as in, *right,* split base. Note roots emerging in a double rank from the same side of the wound. (From MacKenzie, et al. (*194*).)

TABLE 10–4

Some plant metabolic responses to wounding

Increase in ascorbic acid	Increase in phenolics
Increase in fatty acids	Evolution of ethylene
Increase in lipids	Increase in terpenoids
Systemic chemical signal	Systemic electrical signal
New membrane synthesis	Peroxidation of membranes
Weakened cell membranes	Induction of cyanide insensitive pathway
Ion influx into cells	Increased capacity for protein synthesis

Source: Wilson and van Staden *(318).*

- to maintain light levels suitable for photosynthesis and carbohydrate production for the maintenance of the cuttings and for use, once root initiation has occurred, without causing water stress (*189*). See Chapter 3 for the discussion on environmental management and the water relations of propagules.

The water status of cuttings is a balance between transpirational losses and uptake of water. Water absorption through the leaves is *not* the major contributor to water balance in most species. Rather, the cutting base and any foliage immersed in the propagation media are main entry points for water (*190*). Water uptake of cuttings is directly proportional to volumetric water content of the propagation media, with wetter media improving water uptake (see Figure 10–34) (*105, 235*). However, excess water reduces media aeration (*86*) and can cause disease problems.

Water uptake in cuttings declines after they are initially inserted into propagation media. This decline in hydraulic conductivity of cuttings is apparently caused by blockage of xylem vessels and/or collapse of tracheids, which is similar to problems observed with cut flowers (*162*). Another advantage in wounding cuttings is to increase the contact area between the cutting base and propagation medium, thus improving water uptake of cuttings (*105, 190*). With the formation of functional adventitious roots, new vascular connections occur between the roots and stem. Thus, the hydraulic contact between the propagation medium water and the cutting is maintained.

Degree of stomatal opening can be a useful indicator to determine if a given propagation system is maintaining adequate turgor of cutting leaves. A simpler and more useful system is to measure evaporation rates directly with evaporimeters (Figures 10–35 and 11–40) (*132, 187, 190*). When water deficits cause stomata to close, CO_2 diffusion into the leaf is restricted, limiting photosynthesis and any subsequent carbohydrate gain in the cuttings. Carbon gain due to photosynthesis is probably more important *after* root initiation has occurred to promote rapid development of roots. It has been reported that translocation of photosynthate from

FIGURE 10–34 Water uptake by cuttings is directly proportional to the volumetric water content of the rooting medium. Here, softwood cuttings of *Escallonia × exoniensis* are inserted in a peat-pumice mix containing 15, 20, 40 and 60 percent by volume of water and in water *(left to right).* The degree of wilting relates to the water content. (Courtesy K. Loach.)

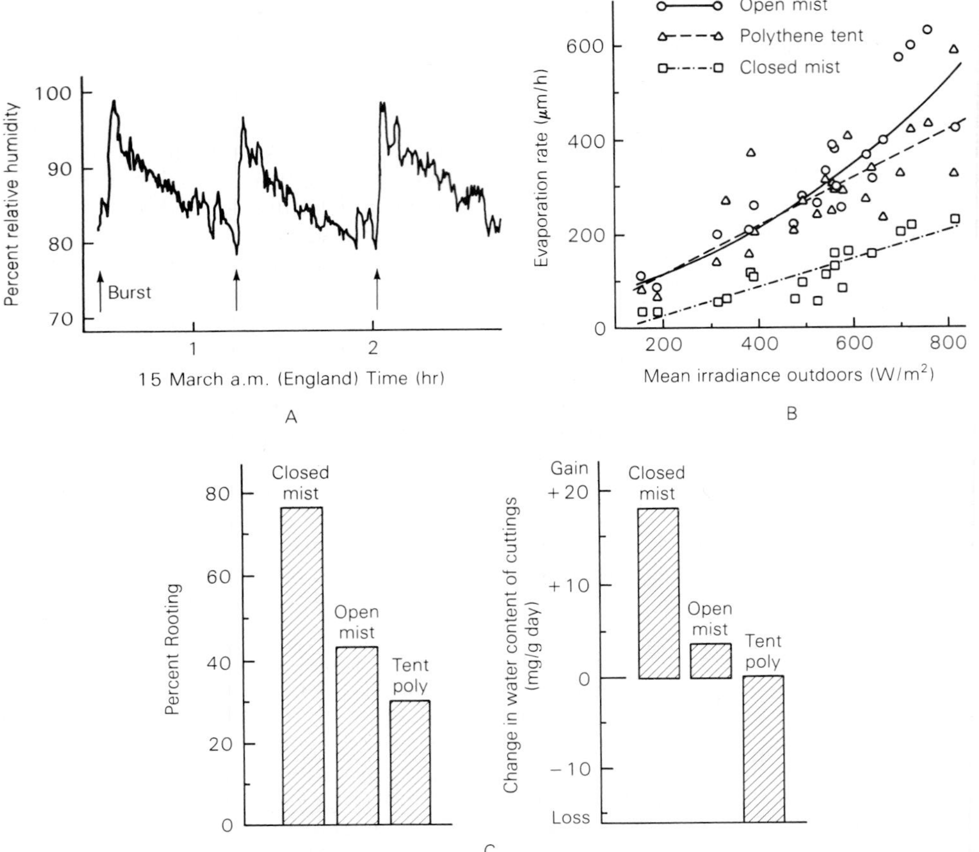

FIGURE 10–35 *Top left:* Cyclic changes in relative humidity under open-bench mist. *Top right:* Evaporation rates measured in three propagation systems: open mist, closed mist, polyethylene closed tent. *Below:* Rooting and water loss of six woody species in the three propagation systems (*188, 189*).

leaves of intact plants continues under moderate or severe stress (*33*).

The driving force that determines the rate at which cuttings lose water is the difference in pressure between water vapor in the leaves (V_{leaf}) and that in the surrounding air (V_{air}). Commercial propagation systems aim to minimize this difference either by decreasing V_{leaf} through reducing leaf temperature (e.g., with intermittent mist) and/or by increasing the V_{air} by preventing the escape of water vapor (i.e., with an enclosed poly tent). Enclosed systems use **humidification** (*94*), since only the V_{air} is increased. Intermittent mist affects primarily V_{leaf}, but also provides a modest increase in V_{air}. Methods used to control water loss of leaves (*189*) are:

1. **Enclosures**—outdoor propagation under low tunnels or cold frames, or nonmisted enclosures in a glasshouse or polyhouse (shading, tent, and contact polyethylene systems, wet tents)
2. **Intermittent mist**—open and enclosed mist systems
3. **Fogging systems**

Major advantages of **enclosed systems** are their simplicity and low cost. The main disadvantage is that they trap heat if light irradiance is high. The trapped heat reduces the relative humidity of the air, and leaf temperature rises to increase the leaf-to-air vapor pressure gradient and consequently leaves lose water. Shading must be used with these systems. Polyethylene films have a low permeability to water vapor loss, but allow gas exchange. They are used to cover outdoor propagation structures as well as for closed mist systems in greenhouses. Modified polyethylene films are now available with additives of vinyl acetate, aluminum, or magnesium silicates which increase their opacity to long-wave radiation (i.e., reduce heat buildup). Polyethylene covered structures have replaced much of the traditional glass-covered cold frames (see Figures 3–4, 3–5, 3–15, 3–18).

Nonmisted enclosures in a glasshouse or polyhouse can be used for difficult-to-root species and have the advantage of avoiding nutrient-leaching problems of intermittent mist, yet affording greater environmental control than outdoor propagation. The shading system entails applying shading compounds to greenhouse roofs and/or utilizing automatically operating light-regulated shading curtains (see Figure 3–11). Shading systems are integrated with temperature control by ventilated fogging or pad-and-fan cooling and heaters.

Contact systems entail laying polyethylene sheets *directly on cuttings* that are watered-in (Figure 11–36). When the irradiance and air temperature can be controlled, leaves tend to be cooler under contact polyethylene because they are in direct contact with the polyethylene and are moistened by condensation forming under the cover. Thus, there is the dual benefit that some evaporative cooling can occur, and that water loss from the foliage is reduced, since the condensation contributes to relative humidity of the air, rather than solely internal water from the leaf tissue. Hence, there is less internal water stress than with drier leaves in an **indoor polytent system.** Well-managed, enclosed nonmist systems offer a low-tech, cost-effective alternative to mist and fog systems, and may be superior to mist when irradiance and temperature levels are relatively low (*211*).

Intermittent mist has been used in propagation since the 1940s and 1950s. Mist systems minimize V_{leaf}, which lowers the leaf-to-air vapor pressure gradient and slows down leaf transpiration. Mist also lowers ambient air temperature and the cooler air consequently lowers leaf temperature by **advection,** in addition to cooling occurring through evaporation of the applied film of water (*189*). Advective cooling occurs only minimally in enclosed nonmist systems. Since intermittent mist lowers medium temperature, suboptimal temperatures can occur which reduce rooting. A common commercial technique to control the rooting medium temperature is to use bottom (basal) heat both with indoor and outdoor mist systems (see Figures 3–10 and 11–37). **Enclosed mist** utilizes polyethylene-covered structures in glasshouses that reduce the fluctuation in ambient humidity that are common to open-bench mist (Figure 10–35). Enclosed mist also ensures more uniform wetting of foliage, since air currents are reduced. There are advantages in using enclosed mist with difficult-to-root species compared to **open mist** or the polytent system without mist (Figures 11–36 and 11–37).

Fog systems maximize V_{air} by raising the ambient humidity. Fog generators produce very fine water droplets that average 15 µm diameter which remain suspended in the air for long periods to maximize evaporation (see Figure 11–42). Their surface/volume ratio is high (compared to larger water droplets produced from a deflector-type mist nozzle) so that the finely divided mist particle has a larger surface which also increases evaporation (*189*). Since water passes into the air as a vapor, rather than directly wetting leaf surfaces as occurs with mist, foliar leaching and overwetting of media are avoided. The propagation media are not cooled to the same degree as with mist, thus avoiding suboptimal rooting temperatures, and less basal heat is needed. A disadvantage of the fog system is the higher initial cost and maintenance.

There are advantages to using fog over either open or enclosed mist systems, particularly with difficult-to-root plants (*131, 133, 134*) and with the acclimation and *ex vitro* rooting needs of tissue-culture-produced liners (see Chapter 18). Propagators must decide which is the most cost-effective system for their particular needs.

Temperature

Temperature of the propagation medium can be suboptimal for rooting due to the cooling effect of mist or seasonally related ambient air temperature. It is often more satisfactory and cost-effective to manipulate temperature by heating at the propagation bench level rather than heating the entire propagation house. See Chapters 3 and 11 for heating equipment systems.

The consensus regarding the optimum medium temperature for propagation is 18 to 25° C (65 to 77° F) for temperate climate species and 7° C (13° F) higher for warm-climate species (73, 174). Daytime air temperatures of about 21 to 27° C (70 to 80° F) with night temperatures about 15° C (60° F) are satisfactory for rooting cuttings of most temperate species, although some root better at lower temperatures. High air temperatures tend to promote bud elongation in advance of root initiation, and to increase water loss from the leaves. It is important that adequate moisture status be maintained by the propagation system so that cuttings gain the potential benefit of the higher basal temperature.

Root initiation in cuttings is temperature-driven but subsequent root growth is strongly dependent on available carbohydrates. This is particularly evident in leafless hardwood cuttings in which excessive root initiation and growth can so deplete stored reserves that there are insufficient available carbohydrates for satisfactory bud growth. The same principle holds true for leafy cuttings (semi-hardwood, softwood, herbaceous), where shoot growth can divert carbohydrates away from developing root initials and thereby slow root growth (Figure 10–36).

The optimum air temperature for growing a crop is probably the best for rooting cuttings (*228*). Bottom heat should be manipulated in two phases, with a higher beginning temperature for root **initiation** and a lower temperature for root **development** and growth (*73, 174*). Optimum temperature for root **initiation** in *Forsythia* and *Chrysanthemum* was 30° C (86° F), whereas root **development** (elongation of primordia and protruding of roots from the stem cutting) was optimum at lower temperatures of 22 to 25° C (72 to 77° F). Respiration is reduced at the lower temperature, which allows more optimum photosynthate accumulation for root development. See Chapter 3 for the discussion on temperature in propagation.

Light

As discussed earlier in the stock plant manipulation section and in Chapter 3, light is a contributing factor in adventitious root and bud formation of cuttings (*67, 75, 76*).

Irradiance. Cuttings of some woody plant species root best under relatively low irradiance (*167, 186*). However, cuttings of certain herbaceous plants, such as chrysanthemum, geranium, and poinsettia, rooted better in tests when the irradiance increased to 116 W/m^2 during trials in winter months. Very high irradiance (174 W/m^2) damaged leaves on the cuttings, delayed rooting, and reduced root growth.

With selected temperate species under an English propagation system, acceptable light ranges were 20 to 100 W/m^2 (*105*). Propagators need to determine irradiance levels to fit their particular production systems.

Photoperiod. In some species, the photoperiod under which the cuttings are rooted may affect root initiation; long days or continuous illumination are generally more effective than short days (*41*), although in other species photoperiod has no influence (*271*).

This situation is complex, however, since photoperiod can affect shoot development as well as root initiation. For example, in propagation by leaf cuttings there must be development of adventitious buds and roots. Using *Begonia* leaf cuttings (*140*), where the light irradiance was adjusted so that the total light energy was about the same under both long days and short days, it was found that short days and relatively low temperatures promoted adventitious bud formation on the leaf pieces, whereas short days suppressed adventitious root formation. Roots formed best under long days with relatively high temperatures.

FIGURE 10–36 Cuttings of *Forsythia* × *intermedia* 'Lynwood' rooted under open-bench mist *(left)* and in a misted, polyethylene enclosure *(right)*. Note that cuttings rooted in the warmer, more humid enclosure break bud and grow faster than those under open mist. However, for some species and circumstances, too much top growth can divert carbohydrates away from developing root initials and slow root growth. (Courtesy K. Loach.)

In rooting cuttings of 'Andorra' juniper, pronounced variations in rooting occurred during the year, but the same variations took place whether the cuttings were maintained under long days, short days, or natural daylength (*181*). A number of tests have been made of the effect of photoperiod on root formation in cuttings, but the results are conflicting; hence it is difficult to make any generalization (*6, 74, 77, 181*).

In some plants, photoperiod will control growth after the cuttings have been rooted. Certain plants cease active shoot growth in response to natural decreases in daylength. This is the case with spring cuttings of deciduous azaleas and dwarf rhododendrons which had rooted and were potted in late summer or early fall. Improved growth of such plants was obtained during the winter in the greenhouse if they were placed under continuous supplementary light, in comparison with similar plants subjected only to the normal short winter days. The latter plants, without added daylength, remained in a dormant state until the following spring (*48*).

High carbohydrate reserves are important for rooted cuttings (liners), since spring growth in a deciduous plant depends on reserves accumulated during the previous growing season. With red maple (*Acer rubrum*), a night interruption lighting period to extend the natural photoperiod to maintain high carbohydrate reserves enhances growth of rooted liners; however, this was not economically justified, since growth of natural daylength liners were comparable after two years of field culture (*262*).

Quality. Lighting that provides more red than far-red light appears to increase rooting in many greenhouse crops (*202*). It is conceivable with certain plant species that root initiation is regulated by red and far-red light through the phytochrome system. Radiation in the orange-red end of the spectrum seems to favor rooting of cuttings more than that in the blue region, but there are conflicting reports.

Accelerated Growth Techniques

Accelerated growth techniques (AGT) were developed by the forestry industry to speed up the production of liners from vegetative propagules and from seed propagation (*125*). Woody perennial plants undergo cyclic growth, and many tree species experience dormancy. Liners are grown in protective culture facilities where photoperiod is extended and water, temperature, carbon dioxide, nutrition, mycorrhizal fungi, and growing media are optimized for each woody species and for each different phase of growth. See Figure 3–30 and the discussion on AGT in Chapter 3.

This concept is also being used in propagation of horticultural crops where supplementary lighting with high-pressure sodium vapor lamps and injection of CO_2 gas into mist water are used to enhance rooting of *Ilex aquifolium* (*93*). The promotive effects on cuttings have been attributed to enhanced photosynthesis. In another study, CO_2 injection into enclosed fog tunnels enhanced root formation of *Chamelaucium* and *Correa.* This was attributed, in part, to decreased leaf transpiration and increased water potential of cuttings, implying that the higher CO_2 reduced stomatal conductance and improved water relations of the cuttings (*107*).

There is a growing trend for **modeling** propagation environments to determine optimal light, temperature, water, CO_2, and nutritional regimes (*53*). Computers can be programmed to monitor the propagation environment and adjust environmental conditions as needed through automated environmental control systems (Figures 3–3, 3–14, 11–40, and 11–41).

Photosynthesis of Cuttings

Photosynthesis by cuttings is not an absolute requirement for root formation. This has been observed in leafy cuttings forming roots when placed in the dark (*63*) and with leafless hardwood cuttings that root. Increasing light irradiance has not always promoted rooting, and net photosynthesis of unrooted cuttings is saturated at relatively low PAR (irradiance measured as photosynthetically active radiation) levels (*63*); hence high PAR does not enhance photosynthesis and could potentially lead to desiccation of cuttings. Unrooted *Acer rubrum* cuttings are much more prone to drought stress, which lowers photosynthetic rates and stomatal conductance (*263*).

It has long been thought that carbohydrate content of cuttings is important to rooting, and carbohydrates do accumulate in the base of cuttings during rooting (*120*). The amount of carbohydrates accumulated at cutting bases has been correlated with photosynthetic activity (*63*), but carbohydrates can also accumulate in the upper portion of leafy cuttings until after roots have formed (*35*).

In poinsettia cuttings, stomatal conductance and photosynthetic levels were initially low and remained low until root primordia were first microscopically observed (*287*); stomatal conductance and photosynthesis increased rapidly as root primordia began to elongate and emerge from the cuttings (Figure 10–37). Most likely, root primordia were producing phytohormones such as cytokinins which increased stomatal conductance and subsequently affected photosynthetic rates. In the rooting of cuttings, initial lower light irradiance could be used to hasten root initiation by reducing water stress (*189*), and light irradiance increased during root primordia emergence to support rapid primordia elongation and root system development.

If a generalization can be made, photosynthesis in cuttings is probably more important *after* root initiation has occurred and helps aid root development and the more rapid growth of a rooted liner (*60*) (see Chapter 11).

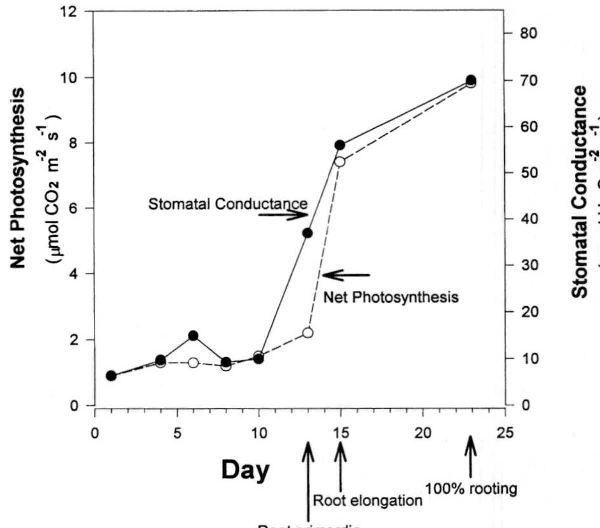

FIGURE 10–37 Influence of adventitious root formation on gas exchange of poinsettia (*Euphorbia pulcherrima* cv. Lilo) cuttings. Root primordia were microscopically observed at day 13, when stomatal conductance began to increase (*287*).

REFERENCES

1. Altamura, M.M., F. Capitani, D. Serafini-Fracassini, P. Torigiani, and G. Falasca. 1991. Root histogenesis from tobacco thin cell layers. *Protoplasma* 161: 31–42.
2. Anand, V. K., and G. T. Heberlein. 1975. Seasonal changes in the effects of auxin on rooting in stem cuttings of *Ficus infectoria*. *Physiol. Plant.* 34:330–34.
3. Arnaud, Y., A. Franclet, H. Tranvan and M. Jacques. 1993. Micropropagation and rejuvenation of *Sequoia sempervirens* (Lamb) Endl: a review. *Ann Sci. For.* 50:273–95.
4. Arteca, R. N., D. S. Tsai, and C. Schlagnhaufer. 1985. Abscisic acid effects on photosynthesis and transpiration in geranium cuttings. *HortScience* 20:370–72.
5. Baker, K. F. 1984. The obligation of the plant propagator. *Comb. Proc. Intl. Plant Prop. Soc.* 34:195–203.
6. Baker, R. L., and C. B. Link. 1963. The influence of photoperiod on the rooting of cuttings of some woody ornamental plants. *Proc. Amer. Soc. Hort. Sci.* 82:596–601.
7. Barlow, P.W. 1994. The origin, diversity and biology of shoot-borne roots. In *Biology of adventitious root formation,* T.D. Davis and B.E. Haissig, eds. New York N.Y.: Plenum Press.
8. Bassil, N.V., W.M. Proebsting, L.W. Moore, and D.A. Lightfoot. 1991. Propagation of hazelnut stem cuttings using *Agrobacterium rhizogenes*. *HortScience* 26(8):1058–1060.
9. Bassuk, N. L., and B. H. Howard. 1981. Seasonal rooting changes in apple hardwood cuttings and their implications to nurserymen. *Comb. Proc. Intl. Plant Prop. Soc.* 30:289–93.
10. Bassuk, N. L., L. D. Hunter, and B. H. Howard. 1981. The apparent involvement of polyphenol oxidase and phloridzin in the production of apple rooting co-factors. *J. Hort. Sci.* 56: 313–22.
11. Bassuk, N., and B. Maynard. 1987. Stock plant etiolation. *HortScience* 22:749–50.
12. Basu, R. N., B. N. Roy, and T. K. Bose. 1970. Interaction of abscisic acid and auxins in rooting of cuttings. *Plant and Cell Physiol.* 11:681–84.
13. Beakbane, A. B. 1969. Relationships between structure and adventitious rooting. *Comb. Proc. Intl. Plant Prop. Soc.* 19:192–201.
14. Behrens, V. 1988. Storage of unrooted cuttings. In *Adventitious root formation in cuttings,* T. D. Davis, B. E. Haissig, and N. Sankhla, eds. Portland, Oreg.: Dioscorides Press.

15. Bhattacharya, N. C. 1988. Enzyme activities during adventitious rooting. In *Adventitious rooting formation in cuttings,* T. D. Davis, B. E. Haissig, and N. Sankhla, eds. Portland, Oreg.: Dioscorides Press.

16. Biran, I., and A. H. Halevy. 1973. Endogenous levels of growth regulators and their relationship to the rooting of dahlia cuttings. *Physiol. Plant* 28:436–42.

17. ———. 1973. The relationship between rooting of dahlia cuttings and the presence and type of bud. *Physiol. Plant.* 28:244–47.

18. Biricolti, S., A. Fabbri, F. Ferrini, and P.L. Pisani. 1994. Adventitious rooting in chestnut: an anatomical investigation. *Scientia Hort.* 59:197–205.

19. Blakely, L. M., S. J. Rodaway, L. B. Hollen, and S. G. Croker. 1972. Control and kinetics of branch root formation in cultured root segments of *Haplopappus ravenii. Plant Physiol.* 50:35–49.

20. Blazich, F. A. 1988. Mineral nutrition and adventitious rooting. In *Adventitious root formation in cuttings,* T. D. Davis, B. E. Haissig, and N. Sankhla, eds. Portland, Oreg.: Dioscorides Press.

21. ———. 1988. Chemicals and formulations used to promote adventitious rooting. In *Adventitious root formation in cuttings,* T. D. Davis, B. E. Haissig, and N. Sankhla, eds. Portland, Oreg.: Dioscorides Press.

22. Blazich, F. A., and C. W. Heuser. 1979. A histological study of adventitious root initiation in mung bean cuttings. *J. Amer. Soc. Hort. Sci.* 104(1):63–67.

23. Blazich, F.A., and L.E. Hinesley. 1995. Fraser fir. *Amer. Nurser. Mag.* 181(5):54–67.

24. Blazich, F. A., and R. D. Wright. 1979. Nonmobilization of nutrients during rooting of *Ilex crenata* cv. Convexa stem cuttings. *HortScience* 14:242.

25. Blazich, F. A., R. D. Wright, and H. E. Schaffer. 1983. Mineral nutrient status of 'Convexa' holly cuttings during intermittent mist propagation as influenced by exogenous auxin application. *J. Amer. Soc. Hort. Sci.* 108:425–29.

26. Bloch, R. 1943. Polarity in plants. *Bot. Rev.* 9:261–310.

27. Boe, A. A., R. B. Steward, and T. J. Banko. 1972. Effects of growth regulators on root and shoot development of *Sedum* leaf cuttings. *HortScience* 74(4):404–5.

28. Bollmark, M., and L. Eliasson. 1986. Effects of exogenous cytokinins on root formation in pea cuttings. *Physiol. Plant.* 68:662–66.

29. Bonnett, H. T., Jr., and J. G. Torrey. 1965. Chemical control of organ formation in root segments of *Convolvulus* cultured in vitro. *Plant Physiol.* 40:1228–36.

30. Bouillenne, R., and M. Bouillenne-Walrand. 1955. Auxines et bouturage. *Rpt. 14th Inter. Hort. Cong.* 1:231–38.

31. Bouillenne, R. and F. Went. 1933. Recherches expérimentales sur la néoformation des racines dans lex plantules et les boutures des plantes supérieures. *Ann. Jard. Bot. Buitenzorg* 43:25.

32. Bouza, L., M. Jacques, B. Sotta, and E. Miginiac. 1994. Relations between auxin and cytokinin contents and in vitro rooting of tree Peony (*Paeonia suffruticosa* Andr.) *Plant Growth Regul.* 15:69–73.

33. Boyer, J. S. 1976. Photosynthesis at low water potentials. *Phil. Trans. Royal Soc. (London), Ser.* B, 273:501–12.

34. Breen, P. J., and T. Muraoka. 1973. Effect of indolebutyric acid on distribution of ^{14}C photosynthate in softwood cuttings of Marianna 2624 plum. *J. Amer. Soc. Hort. Sci.* 98(5):436–39.

35. ———. 1974. Effect of leaves and carbohydrate content and movement of ^{14}C-assimilate in plum cuttings. *J. Amer. Soc. Hort. Sci.* 99(4):326–32.

36. Brian, P. W., H. G. Hemming, and D. Lowe. 1960. Inhibition of rooting of cuttings by gibberellic acid. *Ann. Bot.* 24:407–9.

37. Caboche, M., J. F. Muller, F. Chanut, G. Aranda, and S. Cirakoglu. 1987. Comparison of the growth promoting activities and toxicities of various auxin analogs on cells derived from wild type and a nonrooting mutant tobacco. *Plant Physiol.* 83:795–800.

38. Cameron, R. J., and G. V. Thomson. 1969. The vegetative propagation of *Pinus radiata:* Root initiation in cuttings. *Bot. Gaz.* 130(4): 242–51.

39. Carlson, M. C. 1929. Origin of adventitious roots in *Coleus* cuttings. *Bot. Gaz.* 87:119–26.

40. ———. 1950. Nodal adventitious roots in willow stems of different ages. *Amer. J. Bot.* 37:555–61.

41. Carpenter, W. J., G. R. Beck, and G. A. Anderson. 1973. High intensity supplementary lighting during rooting of herbaceous cuttings. *HortScience* 8(4): 338–40.

42. Christiansen, M. V., E. N. Eriksen, and A. S. Andersen. 1980. Interaction of stock plant irradiance and auxin in the propagation of apple rootstocks by cuttings. *Scientia Hort.* 12:11–17.

43. Christianson, M.L., and D.A. Warnick. 1985. Temporal requirement for phytohormone balance in the control of organogenesis in vitro. *Dev. Biol.* 112: 494–97.

44. Ciampi, C., and R. Gellini. 1958. Anatomical study on the relationship between structure and rooting capacity in olive cuttings. *Nuovo Giorn. Bot. Ital.* 65:417–24.

45. Cline, M. N., and D. Neely. 1983. The histology and histochemistry of the wound healing process in geranium cuttings. *J. Amer. Soc. Hort. Sci.* 108: 452–96.

46. Cooper, W.C. 1938. Hormones and root formation. *Bot. Gaz.* 99:599.

47. ———. 1944. The concentrated-solution-dip method of treating cuttings with growth substances. *Proc. Amer. Soc. Hort. Sci.* 44:533–41.

48. Crossley, J. H. 1965. Light and temperature trials with seedlings and cuttings of *Rhododendron molle. Comb. Proc. Intl. Plant Prop. Soc.* 15:327–34.

49. Crow, W. D., W. Nicholls, and M. Sterns. 1971. Root inhibitors in *Eucalyptus grandis:* Naturally occurring derivatives of the 2,3-dioxabicyclo (4,4,0) decane system. *Tetrahedron Letters* 18. London: Pergamon Press, pp. 1353–56.
50. Cuir, P., S. Sulis, F. Mariani, C.F. Van Sumere, A. Marchesini, and M. Dolci. 1993. Influence of endogenous phenols on rootability of *Chamaelaucium uncinatum* Schauer stem cuttings. *Scientia Hort.* 55:303–314.
51. Danckwardt-Lilliestrõm, C. 1957. Kinetin-induced shoot formation from isolated roots of *Isatis tinctoria. Physiol. Plant* 10:794–97.
52. Davies, F. T., Jr. 1984. Shoot RNA, cambial activity and indolebutyric acid effectively in seasonal rooting of juvenile and mature *Ficus pumila* cuttings. *Physiol. Plant.* 62:571–75.
53. ———. 1985. Plant modeling: developing an approach. *Comb. Proc. Intl. Plant Prop. Soc.* 35:770–776.
54. ———. 1993. What's new in the biology of adventitious root formation. *Comb. Proc. Intl. Plant Prop. Soc.* 43:382–84.
55. Davies, F. T., Jr., T.D. Davis, and D.E. Kester. 1994. Commercial importance of adventitious rooting to horticulture. In *Biology of adventitious root formation,* T.D. Davis and B.E. Haissig, eds. New York, N.Y.: Plenum Press.
56. Davies, F. T., Jr., and H. T. Hartmann. 1988. The physiological basis of adventitious root formation. *Acta Hort.* 227:113–20.
57. Davies, F. T., Jr., and J. N. Joiner. 1980. Growth regulator effects on adventitious root formation in leaf bud cuttings of juvenile and mature *Ficus pumila. J. Amer. Soc. Hort. Sci.* 100:643–46.
58. Davies, F. T., Jr., J. E. Lazarte, and J. N. Joiner. 1982. Initiation and development of roots in juvenile and mature leafbud cuttings of *Ficus pumila* L. *Amer. J. Bot.* 69:804–11.
59. Davies, F. T., Jr., and B. C. Moser. 1980. Stimulation of bud and shoot development of Rieger begonia leaf cuttings with cytokinins. *J. Amer. Soc. Hort. Sci.* 105(1):27–30.
60. Davis, T. D. 1988. Photosynthesis during adventitious rooting. In *Adventitious root formation in cuttings,* T. D. Davis, B. E. Haissig, and N. Sankhla, eds. Portland, Oreg.: Dioscorides Press.
61. Davis, T.D., and B.E. Haissig. 1990. Chemical control of adventitious root formation in cuttings. *Plant Growth Reg. Soc. Amer. Quart.* 18(1):1–17.
62. ———. 1994. *Biology of adventitious root formation.* New York, N.Y.: Plenum Press.
63. Davis, T. D. and J. R. Potter. 1981. Current photosynthate as a limiting factor in adventitious root formation in leafy pea cuttings. *J. Amer. Soc. Hort. Sci.* 106:278–82.
64. ———. 1985. Carbohydrates, water potential and subsequent rooting of stored rhododendron cuttings. *HortScience* 20:292–93.
65. Davis, T. D., and N. Sankhla. 1989. Effect of shoot growth retardants and inhibitors on adventitious rooting. In *Adventitious root formation in cuttings,* T. D. Davis, B. E. Haissig, and N. Sankhla, eds. Portland, Oreg.: Dioscorides Press.
66. Delargy, J. A., and C. E. Wright. 1978. Root formation in cuttings of apple (cv. Bramley's Seedling) in relation to ringbarking and to etiolation. *New Phytol.* 81:117–27.
67. ———. 1979. Root formation in cuttings of apple in relation to auxin application and to etiolation. *New Phytol.* 82:341–47.
68. Donoho, C. W., A. E. Mitchell, and H. N. Sell. 1962. Enzymatic destruction of C^{14} labelled indoleacetic acid and naphthaleneacetic acid by developing apple and peach seeds. *Proc. Amer. Soc. Hort. Sci.* 80:43–49.
69. Doorenbos, J. 1954. Rejuvenation of *Hedera helix* in graft combinations. *Proc. Kon. Ned. Akad. Wet.* Series C 57:99–102.
70. Duhamel du Monceau, H. L. 1758. *La physique des arbres,* Vols. 1 and 2. Paris: Guerin and Delatour.
71. Dunn, D.E., J.C. Cole, and M.W. Smith. 1996. Calendar date, degree days, and morphology influence rooting of *Pistacia chinensis. J. Amer. Soc. Hort. Sci.* In press.
72. Durzan, D. J. 1988. Rooting in woody perennials: problems and opportunities with somatic embryos and artificial seeds. *Acta Hort.* 227: 121–25.
73. Dykeman, B. 1976. Temperature relationship in root initiation and development of cuttings. *Comb. Proc. Intl. Plant Prop. Soc.* 26:201–7.
74. Economou, A. S., and P. E. Read. 1987. Light treatments to improve efficiency of in vitro propagation systems. *HortScience* 22:751–54.
75. Eliasson, L. 1971. Growth regulators in *Populus tremula.* II. Effect of light on inhibitor content in root suckers. *Physiol. Plant.* 24:205–8.
76. ———. 1980. Interaction of light and auxin in regulation of rooting in pea stem cuttings. *Physiol. Plant.* 48:78–82.
77. Eliasson, L. and L. Brunes. 1980. Light effects on root formation in aspen and willow cuttings. *Physiol. Plant.* 48:261–65.
78. Ellis, D.D., H. Barczynsha, B.H. McCown and N. Nelson. 1991. A comparison of BA, zeatin and thidiazuron for adventitious bud formation from *Picea glauca* embryos and epicotyl explants. *Plant Cell, Tis. Organ Cult.* 27:281–287.
79. Ellyard, R. K. 1981. Effect of auxin combinations on the rooting of *Persoonia chameapitys* and *P. pinifolia* cuttings. *Comb. Proc. Intl. Plant Prop. Soc.* 31:251–55.

80. Emery, A. E. H. 1955. The formation of buds on roots of *Chamaenerion angustifolium* (L.) *Scop. Phytomorphology* 5:139–45.
81. Englert, J.M., B.K. Maynard, and N.L. Bassuk. 1991. Correlation of phenolics with etiolated and light-grown shoots of *Carpinus betulus* stock plants. *Comb. Proc. Intl. Plant Prop. Soc.* 41:290–295.
82. Eriksen, E. N. 1973. Root formation in pea cuttings. I. Effects of decapitation and disbudding at different development stages. *Physiol. Plant.* 28:503–6.
83. ———. 1974. Root formation in pea cuttings. II. The influence of indole-3-acetic acid at different development stages. *Physiol. Plant.* 30:158–62.
84. ———. 1974. Root formation in pea cuttings. III. The influence of cytokinin at different development stages. *Physiol. Plant.* 30(2):163–67.
85. Eriksen, E. N., and S. Mohammed. 1974. Root formation in pea cuttings. II. Influence of indole-3-acetic acid at different developmental stages. *Physiol. Plant.* 32:158–62.
86. Erstad, J.L.F., and H.R. Gislerød. 1994. Water uptake of cuttings and stem pieces as affected by different anaerobic conditions in the rooting medium. *Scientia Hort.* 58:151–160.
87. Evans, H. 1952. Physiological aspects of the propagation of cacao from cuttings. *Proc. 13th Inter. Hort. Cong.* 2:1179–90.
88. Fadl, M. S., A. S. El-Deen, and M. A. El-Mahady. 1979. Physiological and chemical factors controlling adventitious root initiation in carob *(Ceratonia siliqua)* stem cuttings. *Egyptian J. Hort.* 6(1):55–68.
89. Fadl, M. S., and H. T. Hartmann. 1967. Isolation, purification, and characterization of an endogenous root-promoting factor obtained from the basal sections of pear hardwood cuttings. *Plant Physiol.* 42:541–49.
90. ———. 1967. Relationship between seasonal changes in endogenous promoters and inhibitors in pear buds and cutting bases and the rooting of pear hardwood cuttings. *Proc. Amer. Soc. Hort. Sci.* 91:96–112.
91. Fischer, P., and J. Hansen. 1977. Rooting of chrysanthemum cuttings: Influence of irradiance during stock plant growth and of decapitation and disbudding of cuttings. *Scientia Hort.* 7:171–78.
92. Flygh, G., R. Grönroos, L. Gulin and S. Von Arnold. 1993. Early and late root formation in epicotyl cuttings of *Pinus sylvestris* after auxin treatment. *Tree Physiol.* 12:81–92.
93. French, C. J., and W. C. Lin. 1984. Seasonal variations in the effect of CO_2 mist and supplementary lighting from high pressure sodium lamps on rooting of English holly cuttings. *HortScience* 19:519–21.
94. Gaffney, J. J. 1978. Humidity: Basic principles and measurement techniques. *HortScience* 13(5): 551–55.
95. Garcia-Gomez, C. Sanchez-Romero, A. Barcelo-Munoz, A. Heredia and F. Pliego-Alfaro. 1994. Levels of endogenous indole-3-acetic acid and indole-3-acetyl-aspartic acid during adventitious rooting in avocado microcuttings. *J. Exper. Bot.* 45(275): 865–870.
96. Gasper, T., and M. Hofinger. 1989. Auxin metabolism during rooting. In *Adventitious root formation in cuttings,* T. D. Davis, B. E. Haissig, and N. Sankhla, eds. Portland, Oreg.: Dioscorides Press.
97. Geneve, R.L. 1991. Patterns of adventitious root formation in English Ivy. *J. Plant Growth Regul.* 10:215–220.
98. Geneve, R.L. and S.T. Kester. 1991. Polyamines and adventitious root formation in the juvenile and mature phase of English ivy. *J. Exp. Bot.* 42(234):71–75.
99. Ginzburg, C. 1967. Organization of the adventitious root apex in *Tamarix aphylla Amer. J. Bot.* 54:4–8.
100. Girouard, R. M. 1967. Anatomy of adventitious root formation in stem cuttings. *Comb. Proc. Intl. Plant Prop. Soc.* 17:289–302.
101. ———. 1967. Initiation and development of adventitious roots in stem cuttings of *Hedera helix. Can. J. Bot.* 45:1883–86.
102. ———. 1969. Physiological and biochemical studies of adventitious root formation. Extractible rooting co-factors from *Hedera helix. Can. J. Bot.* 47(5): 687–99.
103. Good, G. L., and H. B. Tukey, Jr. 1964. Leaching of nutrients from cuttings under mist. *Comb. Proc. Intl. Plant Prop. Soc.* 14:138–42.
104. Gorter, C. J. 1969. Auxin-synergists in the rooting of cuttings. *Physiol. Plant.* 22:497–502.
105. Grange, R. I., and K. Loach. 1983. The water economy of unrooted leafy cuttings. *J. Hort. Sci.* 58:9–17.
106. ———. 1985. The effect of light on the rooting of leafy cuttings. *Scientia Hort.* 27:105–11.
107. Grant, W.J.R., H.M. Fan, W.J.S. Downton, and B.R. Loveys. 1992. Effects of CO_2 enrichment on the physiology and propagation of two Australian ornamental plants, *Chamelaucium unicanatum* (Schauer) × Chamelaucium *floriferum* (MS) and *Correa schlechtendalii* (Behr). *Scientia Hort.* 52:337–342.
108. Greenwood, M. S. 1987. Rejuvenation in forest trees. *J. Plant Growth Regul.* 6:1–12.
109. Greenwood, M. S., and G. P. Berlyn. 1973. Sucrose: Indoleacetic acid interactions on root regeneration by *Pinus lambertiana* embryo cuttings. *Amer. J. Bot.* 60:42–47.
110. Gupta, P. K., and D. J. Durzan. 1987. Micropropagation and phase specificity in mature, elite Douglas fir. *J. Amer. Soc. Hort. Sci.* 112:969–71.
111. Hackett, W. P. 1970. The influence of auxin, catechol, and methanolic tissue extracts on root initiation in aseptically cultured shoot apices of the juve-

nile and adult forms of *Hedera helix J. Amer. Soc. Hort. Sci.* 95:398–402.

112. ———. 1985. Juvenility, maturation and rejuvenation in woody plants. *Hort. Rev.* 7:109–15.

113. ———. 1989. Donor plant maturation and adventitious root formation. In *Adventitious root formation in cuttings,* T. D. Davis, B. E. Haissig, and N. Sankhla, eds. Portland, Oreg.: Dioscorides Press.

114. Hackett, W. P. 1994. Research on adventitious root formation: where do we go from here? In *Biology of adventitious root formation,* T.D. Davis and B.E. Haissig, eds. New York, N.Y.: Plenum Press.

115. Hagemann, A. 1932. Untersuchungen an Blattstecklingen. *Gartenbauwiss* 6:69–202.

116. Haissig, B. E. 1965. 1972. Meristematic activity during adventitious root primordium development. Influences of endogenous auxin and applied gibberellic acid. *Plant Physiol.* 49:886–92.

117. ———. 1974. Influence of auxins and auxin synergists on adventitious root primordium initiation and development. *N. Zealand For. Sci.* 4:299–310.

118. ———. 1979. Influence of aryl esters of indole-3-acetic and indole-3-butyric acids on adventitious root primordium initiation and development. *Physiol. Plant.* 47:29–33.

119. ———. 1983. N-phenyl indolyl-3-butyramide and phenyl indole-3-thiolobutyrate enhance adventitious root primordium development. *Physiol. Plant.* 57:435–40.

120. ———. 1984. Carbohydrate accumulation and partitioning in *Pinus banksiana* seedlings and seedling cuttings. *Physiol. Plant.* 61:13–19.

121. ———. 1986. Metabolic processes in adventitious rooting of cuttings. In *New root formation in plants and cuttings,* M. B. Jackson, ed. Dordrecht: Martinus Nijhoff Publishers.

122. Haissig, B.E. and T.D. Davis. 1994. A historical evaluation of adventitious rooting research to 1993. In *Biology of adventitious root formation,* T.D. Davis and B.E. Haissig, eds. New York N.Y.: Plenum Press.

123. Haissig, B.E., T.D. Davis, and D.E. Riemenschneider. 1992. Researching the controls of adventitious root formation. *Physiol. Plant.* 84:310–317.

124. Hambrick C. E., F. T. Davies, Jr., and H. B. Pemberton. 1991. Seasonal changes in carbohydrate/nitrogen levels during field rooting of *Rosa multiflora* 'Brooks 56' hardwood cuttings. *Scientia. Hort.* 46:137–146.

125. Hanover, J.W. 1976. Accelerated-optimal-growth: a new concept in tree production. *Amer. Nurser. Mag.* 144(10):12–11, 58, 60, 64–65.

126. Hansen, J. 1976. Adventitious root formation induced by gibberellic acid and regulated by irradiance to the stock plants. *Physiol. Plant.* 36:77–81.

127. ———. 1988. Influence of gibberellins on adventitious root formation. In *Adventitious root formation in cuttings,* T. D. Davis, B. E. Haissig, and N. Sankhla, eds. Portland, Oreg.: Dioscorides Press.

128. Hansen, J., L. H. Strömquist, and A. Ericsson. 1978. Influence of the irradiance on carbohydrate content and rooting of cuttings of pine seedlings *(Pinus sylvestris L.). Plant Physiol.* 61:975–79.

129. Harbage, J.F., D.P. Stimart, and R.F. Evert. 1994. Anatomy of adventitious root formation in microcuttings of *Malus domestica* Borkh. 'Gala'. *J. Amer. Soc. Hort. Sci.* 118:680–688.

130. Hare, R. C. 1977. Rooting of cuttings from mature water oak *(Quercus nigra). Southern J. Appl. For.* 1(2):24–25.

131. ———. 1981. Improved rooting powder for chrysanthemums. *HortScience* 16:90–91.

132. Harrison-Murray, R. S. 1991. A leaf-model evaporimeter for estimating potential transpiration in propagation environments. *J. Hort. Sci.* 66:131–139.

133. Harrison-Murray, R. S., B. H. Howard, and R. Thompson. 1988. Potential for improved propagation of cuttings through the use of fog. *Acta Hort.* 227:205–10.

134. Harrison-Murray, R. S., and R. Thompson. 1988. In pursuit of a minimum stress environment for rooting leafy cuttings: Comparison of mist and fog. *Acta Hort.* 227:211–16.

135. Hartmann, H. T., and R. M. Brooks. 1958. Propagation of Stockton Morello cherry rootstock by softwood cuttings under mist sprays. *Proc. Amer. Soc. Hort. Sci.* 71:127–34.

136. Hartmann, H. T., and C. J. Hansen. 1958. Rooting pear and plum rootstocks. *Calif. Agr.* 12(10):4, 14, 15.

137. Hartmann, H. T., W. H. Griggs, and C. J. Hansen. 1963. Propagation of own-rooted Old Home and Bartlett pears to produce trees resistant to pear decline. *Proc. Amer. Soc. Hort. Sci.* 82:92–102.

138. Hartmann, H. T., and F. Loreti. 1965. Seasonal variation in the rooting of olive cuttings. *Proc. Amer. Soc. Hort. Sci.* 87:194–98.

139. Hausman, J.F. 1993. Changes in peroxidase activity, auxin level and ethylene production during root formation by poplar shoots raised in vitro. *J. Plant Growth Reg.* 13:263–268.

140. Heide, O. M. 1965. Photoperiodic effects on the regeneration ability of *Begonia* leaf cuttings. *Physiol. Plant.* 18:185–90.

141. ———. 1965. Interaction of temperature, auxin, and kinins in the regeneration ability of *Begonia* leaf cuttings. *Physiol. Plant.* 18:891–920.

142. ———. 1965. Effects of 6-benzylamino-purine and 1-naphthaleneacetic acid on the epiphyllous bud formation in *Bryophyllum. Planta* 67:281–96.

143. ———. 1968. Stimulation of adventitious bud formation in *Begonia* leaves by abscisic acid. *Nature* 219(5157):960–61.

144. ———. 1968. Auxin level and regeneration of *Begonia* leaves. *Planta* 81:153–59.
145. ———. 1969. Non-reversibility of gibberellin-induced inhibition of regeneration in *Begonia* leaves. *Physiol. Plant.* 22:671–79.
146. Henry, P.H., F.A. Blazich, and L.E. Hinesley. 1992. Nitrogen nutrition of containerized eastern redcedar. II Influence of stock plant fertility on adventitious rooting of stem cuttings. *J. Amer. Soc. Hort. Sci.* 117:568–570.
147. Hess, C. E. 1962. Characterization of the rooting cofactors extracted from *Hedera helix* L. and *Hibiscus rosa-sinensis* L. *Proc. 16th Inter. Hort. Cong.*, pp. 382–88.
148. ———. 1968. Internal and external factors regulating root initiation. In *Root growth: Proc. 15th Easter School in Agricultural Science, University of Nottingham.* London: Butterworth.
149. Hiller, Charlotte. 1951. A study of the origin and development of callus and root primordia of *Taxus cuspidata* with reference to the effects of growth regulators. Master's thesis, Cornell Univ., Ithaca, N.Y.
150. Hitchcock, A. E., and P. W. Zimmerman. 1940. Effects obtained with mixtures of root-inducing and other substances. *Contrib. Boyce Thomp. Inst.* 11:143–60.
151. ———. 1942. Root inducing activity of phenoxy compounds in relation to their structure. *Contrib. Boyce. Thomp. Inst.* 12:497–507.
152. Howard, B. H. 1965. Increase during winter in capacity for root regeneration in detached shoots of fruit tree rootstocks. *Nature* 208:912–13.
153. ———. 1968. Effects of bud removal and wounding on rooting of hardwood cuttings. *Nature* 220 (5164):262–264.
154. ———. 1991. Stock plant manipulation for better rooting and growth from cuttings. *Comb. Proc. Intl. Plant Prop. Soc.* 41:127–130.
155. ———. 1994. Manipulating rooting potential in stockplants before collecting cuttings. In *Biology of adventitious root formation,* T.D. Davis and B.E. Haissig, eds. New York N.Y.: Plenum Press.
156. Howard, B. H. and R. S. Harrison-Murray. 1995. Responses of dark-preconditioned and normal light-grown cuttings of *Syringa vulgaris* 'Madame Lemoine' to light and wetness gradients in the propagation environment. *J. Hort. Sci.* 70:989–1191.
157. Howard, B. H., R. S. Harrison-Murray, J. Vesek, and O. P. Jones. 1988. Techniques to enhance rooting potential before cutting collection. *Acta Hort.* 227:176–86.
158. Howard, B.H., and M.S. Ridout. 1992. A mechanism to explain increased rooting in leafy cuttings of *Syringa vulgaris* 'Madame Lemoine' following dark-treatment of the stockplant. *J. Hort. Sci.* 59:131–139.
159. ———. 1994. Partitioning sources of rooting potential in plum hardwood cuttings. *J. Hort. Sci.* 69:735–745.
160. Howard, B.H., O.P. Jones, and J. Vasek. 1989. Long-term improvement in the rooting of plum cuttings following apparent rejuvenation. *J. Hort. Sci.* 64:147–156.
161. ———. 1989. Growth characteristics of apparently rejuvenated plum shoots *J. Hort Sci.* 64:157–162.
162. Ikeda, T., and T. Suzaki. 1985. Influence of hydraulic conductance of xylem on water status in cuttings. *Can. J. For. Res.* 16:98–102.
163. Jain, M. K., and K. K. Nanda. 1972. Effect of temperature and some antimetabolites on the interaction effects of auxin and nutrition in rooting etiolated stem segments of *Salix tetrasperma. Physiol. Plant.* 27:169–72.
164. Jarvis, B. C. 1986. Endogenous control of adventitious rooting in non-woody species. In *New root formation in plants and cuttings,* M. B. Jackson, ed. Dordrecht: Martinus Nijhoff Publishers.
165. Jarvis, B. C., S. Yasmin, and M. T. Coleman. 1985. RNA and protein metabolism during adventitious root formation in stem cuttings of *Phaseolus aureus. Physiol. Plant.* 64:53–59.
166. Johnson, C. R. 1970. The nature of flower bud influence on root regeneration in the *Rhododendron* shoot. Ph.D. Dissertation. Oreg. State Univ., Corvallis, Oreg.
167. Johnson, C. R., and D. F. Hamilton. 1977. Effects of media and controlled-release fertilizers on rooting and leaf nutrient composition of *Juniperus conferta* and *Ligustrum japonicum* cuttings. *J. Amer. Soc. Hort. Sci.* 102:320–22.
168. Johnson, C. R., and A. N. Roberts. 1971. The effect of shading rhododendron stock plants on flowering and rooting. *J. Amer. Soc. Hort. Sci.* 96:166–68.
169. Kawase, M. 1964. Centrifugation, rhizocaline, and rooting in *Salix alba. Physiol. Plant.* 17:855–65.
170. ———. 1971. Causes of centrifugal root promotion. *Physiol Plant.* 25:64–70.
171. Keever, G. J., G. S. Cobb, and D. R. Mills. 1987. Propagation of four woody ornamentals from vegetative and reproductive stem cuttings. *Ornamentals Res. Rep. 5,* Alabama Agr. Exp. Sta., Auburn University, Montgomery.
172. Keever, G. I., and H. B. Tukey, Jr. 1979. Effect of nutrient mist on the propagation of azaleas. *HortScience* 14:755–56.
173. Kefford, N. P. 1973. Effect of a hormone antagonist on the rooting of shoot cuttings. *Plant. Physiol.* 51:214–16.
174. Kester, D. E. 1970. Temperature and plant propagation. *Comb. Proc. Intl. Plant Prop. Soc.* 20:153–63.
175. Kester, D. E. 1983. The clone in horticulture. *HortScience* 18:831–37.

176. Kling, G. J., M. M. Meyer, Jr., and D. Seigler. 1988. Rooting co-factors in five *Acer* species. *J. Amer. Soc. Hort. Sci.* 113:252–57.
177. Kögl, F., A. J. Haagen-Smit, and H. Erxleben, 1934. Uber ein neues Auxin ("Heteroauxin") aus Harn, XI. *Mitteilung. Z. Physiol. Chem.* 228:90–103.
178. Kraus, E. J., and H. R. Kraybill. 1918. Vegetation and reproduction with special reference to the tomato. *Oreg. Agr. Exp. Sta. Bul.* 149.
179. Krishnamoorthy, H. N. 1970. Promotion of rooting in mung bean hypocotyl cuttings with Ethrel, an ethylene releasing compound. *Plant Cell Physiol.* 11:979–82.
180. Krul, W. R. 1968. Increased root initiation in Pinto bean hypocotyls with 2,4-dinitrophenol. *Plant Physiol.* 43:439–41.
181. Lanphear, F. O., and R. P. Meahl. 1961. The effect of various photoperiods on rooting and subsequent growth of selected woody ornamental plants. *Proc. Amer. Soc. Hort. Sci.* 77:620–34.
182. ———. 1966. Influence of the stock plant environment on the rooting of *Juniperus horizontalis* 'Plumosa'. *Proc. Amer. Soc. Hort. Sci.* 89:666–71.
183. Lek, H. A. A., van der. 1925. Root development in woody cuttings. *Meded. Landbouwhoogesch. Wageningen* 38(1).
184. Leopold, A. C. 1964. The polarity of auxin transport in *Meristems* and *Differentiation,* Brookhaven Symposia in Biology, Rpt. 16. Upton, N.Y.: Brookhaven Natl. Lab., pp. 218–34.
185. Libby, W. J., A. G. Brown, and J. M. Fielding. 1972. Effect of hedging radiata pine on production, rooting, and early growth of cuttings. *New Zealand J. For. Sci.* 2:263–83.
186. Loach, K. 1979. Mist propagation: Past, present, future. *Comb. Proc. Intl. Plant Prop. Soc.* 29:216–29.
187. ———. 1983. Propagation systems in New Zealand: A means of comparing their effectiveness. *Comb. Proc. Intl. Plant Prop. Soc.* 33:291–94.
188. ———. 1987. Mist and fruitfulness. *Horticulture Week,* April 10, 1987, pp. 28–29.
189. ———. 1988. Controlling environmental conditions to improve adventitious rooting. In *Adventitious root formation in cuttings,* T. D. Davis, B. E. Haissig, and N. Sankhla, eds. Portland, Oreg.: Dioscorides Press.
190. ———. 1988. Water relations and adventitious rooting. In *Adventitious root formation in cuttings,* T. D. Davis, B. E. Haissig, and N. Sankhla, eds. Portland, Oreg.: Dioscorides Press.
191. Long, W. G., D. V. Sweet, and H. B. Tukey. 1956. The loss of nutrients from plant foliage by leaching as indicated by radioisotopes. *Science* 123: 1039–40.
192. Lovell, P. H., and J. White. 1986. Anatomical changes during adventitious root formation. In *New root formation in plants and cuttings,* M. B. Jackson, ed. Dordrecht: Martinus Nijhoff Publishers.
193. Ludwig-Muller, J., and E. Epstein. 1994. Indole-3-butyric acid in *Arabidopsis thaliana* III. In vivo biosynthesis. *J. Plant Growth Regul.* 14:7–14.
194. MacKenzie, K. A. D., B. H. Howard, and R. S. Harrison-Murray. 1986. The anatomical relationship between cambial regeneration and root initiation in wounded winter cuttings of the apple rootstock M.26. *Annal. Bot.* 58:649–61.
195. Maini, J. S. 1968. The relationship between the origin of adventitious buds and the orientation of *Populus tremuloides* root cuttings. *Bul. Ecol. Soc. Amer.* 49:81–82.
196. Maldiney, R., F. Pelese, G. Pilate, L. Sossountzov, and E. Miginiac. 1986. Endogenous levels of abscisic acid, indole-3-acetic acid, zeatin and zeatin-riboside during the course of adventitious root formation in cuttings of *Craigella* and *Craigella* lateral suppressor tomatoes. *Physiol. Plant.* 68:426–30.
197. Maynard, B. K., and N. L. Bassuk. 1987. Stock plant etiolation and blanching of woody plants prior to cutting propagation. *J. Amer. Soc. Hort. Sci.* 112: 273–76.
198. ——— 1991. Stock plant etiolation and stem banding effect on the auxin dose-response of rooting in stem cuttings of *Carpinus betulus* L. 'Fastigiata.' *J. Plant Growth Regul.* 10:305–311.
199. ———. 1992. Stock plant etiolation, shading, and banding effects on cutting propagation of *Carpinus betulus.* 1992. *J. Amer. Soc. Hort. Sci.* 117:740–744.
200. McAfee, B.J., E.E. White, L.E. Pelcher, and M.S. Lapp. 1993. Root induction in pine *(Pinus)* and larch *(Larix)* spp. using *Agrobacterium rhizogenes. Plant Cell, Tissue Org. Cult.* 34:53–62.
201. McVeigh, I. 1938. Regeneration in *Crassula multicava. Amer. J. Bot.* 25:7–11.
202. Moe, R., and A. S. Andersen. 1988. Stock plant environment and subsequent adventitious rooting. In *Adventitious root formation in cuttings,* T. D. Davis, B. E. Haissig, and N. Sankhla, eds. Portland, Oreg.: Dioscorides Press.
203. Mohammed, S. 1975. Further investigations on the effects of decapitation and disbudding at different development stages of rooting in pea cuttings. *J. Hort. Sci.* 50:271–73.
204. Mohammed, S., and E. N. Eriksen. 1974. Root formation in pea cuttings. IV. Further studies on the influence of indole-3-acetic acid at different development stages. *Physiol. Plant.* 32:94–96.
205. Mohnen, D. 1994. Novel experimental systems for determining cellular competence and determination. In *Biology of adventitious root formation,* T.D. Davis and B.E. Haissig, eds. New York, N.Y.: Plenum Press.

206. Molitor, H. D., and W. U. von Hentig. 1987. Effect of carbon dioxide enrichment during stock plant cultivation. *HortScience* 22:741–46.

207. Molnar, J. M., and L. J. LaCroix. 1972. Studies of the rooting of cuttings of *Hydrangea macrophylla:* Enzyme changes. *Can. J Bot.* 50:315–22.

208. ———. 1972. Studies of the rooting of cuttings of *Hydrangea macrophylla:* DNA and protein changes. *Can. J. Bot.* 50:387–92.

209. Morgan, D. L., E. L. McWilliams, and W. C. Parr. 1980. Maintaining juvenility in live oak. *HortScience* 15:493–94.

210. Mudge, K. W. 1988. Effect of ethylene on rooting. In *Adventitious root formation in cuttings,* T. D. Davis, B. E. Haissig, and N. Sankhla, eds. Portland, Oreg.: Dioscorides Press.

211. Mudge, K.W., V.N. Mwaja, F.M. Itulya, and J. Ochieng. 1995. Comparison of four moisture management systems for cutting propagation of bougainvillea, hibiscus and kei apple. *J. Amer. Soc. Hort. Sci.* 120:366–373.

212. Murray, J.R., M. C. Sanchez, A.G. Smith and W.P. Hackett. 1993. Differential competence for adventitious root formation in histologically similar cell types. In *Biology of adventitious root formation,* T.D. Davis and B.E. Haissig, eds. New York N.Y.: Plenum Press.

213. Muzik T. J., and H. J. Cruzado. 1958. Transmission of juvenile rooting ability from seedlings to adults of *Hevea brasiliensis. Nature* 181:1288.

214. Naylor, E. E., and B. Johnson. 1937. A histological study of vegetative reproduction in *Saint-paulia ionantha. Amer. J. Bot.* 24:673–78.

215. Nørgaard, J.V. 1992. Artificial seeds in micropropagation. *Comb. Proc. Intl. Plant Prop. Soc.* 42:182–184.

216. Okoro, O. O., and J. Grace. 1978. The physiology of rooting *Populus* cuttings. II. Cytokinin activity in leafless hardwood cuttings. *Physiol. Plant.* 44:167–70.

217. Oliver, M.J., I. Mukherjee and D.M. Reid. 1994. Alteration in gene expression in hypocotyls of sunflower *(Helianthus annuus)* seedlings associated with derooting and formation of adventitious root primordia. *Physiol. Plant.* 90:481–489.

218. O'Rourke, F. L. 1940. The influence of blossom buds on rooting of hardwood cuttings of blueberry. *Proc. Amer. Soc. Hort. Sci.* 40:332–34.

219. ———. 1944. Wood type and original position on shoot with reference to rooting in hardwood cuttings of blueberry. *Proc. Amer. Soc. Hort. Sci.* 45:195–97.

220. Patena, L., E. G. Sutter, and A. M. Dandekar. 1988. Root induction by *Agrobacterium rhizogenes* in a difficult-to-root woody species. *Acta Hort.* 227:324–29.

221. Paton, F., and W. W. Schwabe. 1987. Storage of cuttings of *Pelargonium × hortorum Bailey. J. Hort. Sci.* 62:79–87.

222. Paton, D. M., R. R. Willing, W. Nichols, and L. D. Pryor. 1970. Rooting of stem cuttings of eucalyptus: A rooting inhibitor in adult tissue. *Austral. J. Bot.* 18:175–83.

223. Petri, P. S., S. Mazzi, and P. Strigoli. 1960. Considerazióne sulla formazióne delle radici avventìzie con particolare riguardo a: *Cucurbita pepo, Nerium oleander, Menyanthes trifoliatae, Solanum lycopersicum. Nuovo Giorn. Bot. Ital.* 67:131–75.

224. Pierik, R. L. M., and H. H. M. Steegmans. 1975. Analysis of adventitious root formation in isolated stem explants of *Rhododendron. Scientia Hort.* 3:1–20.

225. Pliego-Alfaro, F., and T. Murashige. 1987. Possible rejuvenation of adult avocado by graftage onto juvenile rootstocks in vitro. *HortScience* 22:1321–24.

226. Porlingis, I. C., and I. Therios. 1976. Rooting response of juvenile and adult leafy olive cuttings to various factors. *J. Hort. Sci.* 51:31–39.

227. Pounders, C.T., and G.S. Foster. 1992. Multiple propagation effects on genetic estimates of rooting for western hemlock. *J. Amer. Soc. Hort. Sci.* 117:651–655.

228. Preece, J.E. 1993. Basics of propagation by cuttings—temperature. *Comb. Proc. Intl. Plant Prop. Soc.* 43:441–443.

229. Priestley, J. H., and C. F. Swingle. 1929. Vegetative propagation from the standpoint of plant anatomy. *USDA Tech. Bul.* 151.

230. Proebsting, W. M. 1984. Rooting of Douglas-fir stem cuttings: Relative activity of IBA and NAA. *HortScience* 19:854–56.

231. Rajagopal, V., and A. S. Andersen. 1980. Water stress and root formation in pea cuttings. *Physiol. Plant.* 48:144–49.

232. Raskin, I. 1992. Salicylate—a new plant hormone. *Plant Physiol.* 99:799–803.

233. Rasmussen, S., and A. S. Andersen. 1980. Water stress and root formation in pea cuttings. II. Effect of abscisic acid treatment of cuttings from stock plants grown under two levels of irradiance. *Physiol. Plant.* 48:150–54.

234. Read, P. E., and V. C. Hoysler. 1969. Stimulation and retardation of adventitious root formation by application of B-Nine and Cycocel. *J. Amer. Soc. Hort. Sci.* 94:314–16.

235. Rein, W.H., R.D. Wright, and J.R. Seiler. 1991. Propagation medium moisture level influences adventitious rooting of woody stem cuttings. *J. Amer. Soc. Hort. Sci.* 116:632–636.

236. Rein, W.H., R.D. Wright, and D.D. Wolf 1991. Stock plant nutrition influences the adventitious rooting of 'Rotundifolia' holly stem cuttings. *J. Environ. Hort.* 9:83–85.

237. Reuveni, O., and M. Raviv. 1981. Importance of leaf retention to rooting avocado cuttings. *J. Amer. Soc. Hort. Sci.* 106:127–30.

238. Rey, M., C. Díaz-Sala, and R. Rodríguez. 1994. Exogenous polyamines improve rooting of hazel microshoots. *Plant Cell, Tissue Org. Cult.* 36:303–308.

239. Ritchie, G.A. 1994. Commercial applications of adventitious rooting to forestry. In *Biology of adventitious root formation,* T.D. Davis and B.E. Haissig, eds. New York, N.Y.: Plenum Press.

240. Robbins, W. J. 1960. Further observations on juvenile and adult *Hedera. Amer. J. Bot.* 47:485–91.

241. Robbins, J. A., M. J. Campidonica, and D. W. Burger. 1988. Chemical and biological stability of indole-3-butyric acid (IBA) after long-term storage at selected temperatures and light regimes. *J. Environ. Hort.* 6:33–38.

242. Roberts, A. N., and L. H. Fuchigami. 1973. Seasonal changes in auxin effect on rooting of Douglas-fir stem cuttings as related to bud activity. *Physiol. Plant.* 28:215–21.

243. Robinson, J. C. 1975. The regeneration of plants from root cuttings with special reference to the apple. *Hort. Abst.* 45(6):305–15.

244. Robinson, J. C., and W. W. Schwabe. 1977. Studies on the regeneration of apple cultivars from root cuttings. I. Propagation aspects. *J. Hort. Sci.* 52:205–20.

245. ———. 1977. Studies on the regeneration of apple cultivars from root cuttings. II. Carbohydrate and auxin relations. *J. Hort. Sci.* 52: 221–33.

246. Rowland, L.J. and E.L. Ogden. 1992. Use of a cytokinin conjugate for efficient shoot regeneration from leaf sections of highbush blueberry. *HortScience* 27:1127–1129.

247. Rugini, E. 1992. Involvement of polyamines in auxin and *Agrobacterium rhizogenes*-induced rooting of fruit trees in vitro. *J. Amer. Soc. Hort. Sci.* 117:532–536.

248. Rugini, E., A. Pellegrineschi, M. Mencuccini and D. Mariotti. 1991. Increase of rooting ability in the woody species kiwi (*Actinidia deliciosa* A. Chev.) by transformation with *Agrobacterium rhizogenes* rol genes. *Plant Cell Rept.* 10:291–295.

249. Ryan, G. F., E. F. Frolich, and T. P. Kinsella. 1958. Some factors influencing rooting of grafted cuttings. *Proc. Amer. Soc. Hort. Sci.* 72:454–61.

250. Sachs, J. 1880 and 1882. Stoff und Form der Pflanzenorgane. I and II. *Arb. Bot. Inst. Würzburg* 2:452–88; 4:689–718.

251. Sachs, R. M., F. Loreti, and J. DeBie. 1964. Plant rooting studies indicate sclerenchyma tissue is not a restricting factor. *Calif. Agr.* 18(9): 4–5.

252. Sanchez, M.C., A.G. Smith, and W.P. Hackett. 1995. Localized expression of a proline-rich protein gene in juvenile and mature ivy petioles in relation to rooting competence. *Physiol. Plant.* 93:207–216.

253. Schier, G. A. 1973. Origin and development of aspen root suckers. *Can. J. For. Res.* 3:39–44.

254. Schraudolf, H., and J. Reinert. 1959. Interaction of plant growth regulators in regeneration processes. *Nature* 184:465–66.

255. Schuerman, P.L. and A.M. Dandekar. 1993. Transformation of temperate woody crops: progress and potentials. *Scientia Hort.* 55:101–124.

256. Siegler, E. A., and J. J. Bowman. 1939. Anatomical studies of root and shoot primordia in 1-year apple roots. *J. Agr. Res.* 58:795–803.

257. Siniscalco, C., and L. Pavolettoni. 1988. Rejuvenation of *Eucalyptus* × *trabutii* by successive grafting. *Acta Hort.* 227:98–100.

258. Sircar, P. K., and S. K. Chatterjee. 1973. Physiological and biochemical control of meristemation and adventitious root formation in *Vigna* hypocotyl cuttings. *Plant Propagator* 19(1):1.

259. ———. 1974. Physiological and biochemical changes associated with adventitious root formation in *Vigna* hypocotyl cuttings. II. Gibberellin effects. *Plant Propagator* 20(2):15–22.

260. ———. 1980. Effect of foliar applications of kinetin and ethrel on adventitious root formation at the base of *Vigna* hypocotyl cuttings. *Plant Propagator* 26(4):3–5.

261. Skoog, F., and C. Tsui. 1948. Chemical control of growth and bud formation in tobacco stem and callus. *Amer. J. Bot.* 35:782–87.

262. Smally, T. J., and M. A. Dirr. 1988. Effect of night interruption photoperiod treatment on subsequent growth of *Acer rubrum* cuttings. *HortScience* 23: 172–74.

263. Smally, T. J., M.A. Dirr, A.M. Armitage, B.W. Wood, R.O. Teskey and R.F. Severson. 1991. Photosynthesis, leaf water, carbohydrate, and hormone status during rooting of stem cuttings of *Acer rubrum. J. Amer. Soc. Hort. Sci.* 116:1052–1057.

264. Smith, D. R., and T. A. Thorpe. 1975. Root initiation in cuttings of *Pinus radiata* seedlings. I. Developmental sequence. *J. Exp. Bot.* 26: 184–92.

265. ———. 1975. Root initiation in cuttings of *Pinus radiata* seedlings. II. Growth regulator interactions. *J. Exp. Bot.* 26:193–202.

266. Smith, N. G., and P. F. Wareing. 1972. The distribution of latent root primordia in stems of *Populus* × *robusta* and factors affecting emergence of preformed roots from cuttings. *J. Forestry* 45:197–210.

267. Spano, L., D. Mariotti, M. Cardarelli, C. Branra, and P. Costantino. 1988. Morphogenesis and auxin sensitivity of transgenic tobacco with different complements of Ri T-DNA. *Plant. Physiol.* 87:476–83.

268. Spiegel, P. 1954. Auxins and inhibitors in canes of *Vitis. Bul. Res. Coun., Israel* 4:176–83.

269. Stangler, B. B. 1949. An anatomical study of the origin and development of adventitious roots in stem cuttings of *Chrysanthemum morifolium* Bailey, *Dianthus caryophyllus* L., and *Rosa dilecta* Rehd. Ph.D. dissertation, Cornell Univ., Ithaca, N.Y.

270. Steele, M.J., M.M. Yeoman and M.P. Coutts. 1990. Developmental changes in Sitka spruce as indices of physiological age. II. Rooting of cuttings and callusing of needle explants. *New Phytol.* 114:11–120.

271. Steponkus, P. L., and L. Hogan. 1967. Some effects of photoperiod on the rooting of *Abelia grandiflora* Rehd. 'Prostrata' cuttings. *Proc. Amer. Soc. Hort. Sci.* 91:706–15.

272. Stoltz, L. P., and C. E. Hess. 1966. The effect of girdling upon root initiation. Carbohydrates and amino acids. *Proc. Amer. Soc. Hort. Sci.* 89:734–43.

273. Stoutemyer, V. T. 1937. Regeneration in various types of apple wood. *Iowa Agr. Exp. Sta. Res. Bul.* 220:309–52.

274. ———. 1942. The propagation of *Chionanthus retusus* by cuttings. *Nat. Hort. Mag.* 21(4):175–78.

275. Stoutemyer, V. T., O. K. Britt, and J. R. Goodin. 1961. The influence of chemical treatments, understocks, and environment on growth phase changes and propagation of *Hedera canariensis*. *Proc. Amer. Soc. Hort. Sci.* 77:552–57.

276. Stoutemyer, V. T., and A. W. Close. 1947. Changes of rooting response in cuttings following exposure of the stock plants to light of different qualities. *Proc. Amer. Soc. Hort. Sci.* 49:392–94.

277. Strobel, G. A., and A. Nachmias. 1985. *Agrobacterium rhizogenes* promotes the initial growth of bare rootstock almond. *J. Gen. Microbiol.* 131:1245–49.

278. ———. 1988. *Agrobacterium rhizogenes:* A root inducing bacterium. In *Adventitious root formation in cuttings,* T. D. Davis, B. E. Haissig, and N. Sankhla, eds. Portland, Oreg.: Dioscorides Press.

279. Strömquist, L., and J. Hansen. 1980. Effects of auxin and irradiance on the rooting of cuttings of *Pinus sylvestris*. *Physiol. Plant.* 49:346–50.

280. Struve, D. K. 1981. The relationship between carbohydrates, nitrogen and rooting of stem cuttings. *Plant Propagator* 27:6–7.

281. Struve, D. K., and M. A. Arnold. 1986. Aryl esters of IBA increase rooted cutting quality of red maple 'Red Sunset' softwood cuttings. *HortScience* 21: 1392–93.

282. Struve, D.K. and R.D. Lineberger. 1988. Restoration of high adventitious root regeneration potential in mature *Betula papyrifera* Marsh. softwood stem cuttings. *Can. J. For. Res.* 18:265–269.

283. Strydom, D. K., and H. T. Hartmann. 1960. Absorption, distribution, and destruction of indoleacetic acid in plum stem cuttings. *Plant. Physiol.* 35:435–42.

284. ———. 1960. Effect of indolebutyric acid on respiration and nitrogen metabolism in Marianna 2624 plum softwood stem cuttings. *Proc. Amer. Soc. Hort. Sci.* 76:124–33.

285. Sun, W.Q., and N.L. Bassuk. 1993. Auxin-induced ethylene synthesis during rooting and inhibition of budbreak of 'Royalty' rose cuttings. *J. Amer. Soc. Hort. Sci.* 118:638–643.

286. Svenson, S.E. and F.T. Davies, Jr. 1995. Change in tissue elemental concentration during root initiation and development of poinsettia cuttings. *HortScience* 30:617–619.

287. Svenson, S.E., F.T. Davies, Jr., and S.A. Duray. 1995. Gas exchange, water relations, and dry weight partitioning during root initiation and development of poinsettia cuttings. *J. Amer. Soc. Hort. Sci.* 120: 454–459.

288. Thimann, K. V., and J. B. Koepfli. 1935. Identity of the growth-promoting and root-forming substances of plants. *Nature* 135:101–2.

289. Thimann, K. V., and E. F. Poutasse. 1941. Factors affecting root formation of *Phaseolus vulgaris*. *Plant. Physiol.* 16:585–98.

290. Thimann, K. V., and F. W. Went. 1934. On the chemical nature of the root-forming hormone. *Proc. Kon. Ned. Akad. Wet.* 37:456–59.

291. Thomaszewski, M., and K. V. Thimann. 1966. Interactions of phenolic acids, metallic ions, and chelating agents on auxin-induced growth. *Plant. Physiol.* 41:1443–54.

292. Tillburg. E. 1974. Levels of indole-3-acetic acid and acid inhibitors in green and etiolated bean seedlings *(Phaseolus vulgaris)*. *Physiol. Plant.* 31: 106–11.

293. Tukey, H. B., and E. L. Green. 1934. Gradient composition of rose shoots from tip to base. *Plant. Physiol.* 9:157–63.

294. Tukey, H. B., Jr., H. B. Tukey, and S. H. Wittwer. 1958. Loss of nutrients by foliar leaching as determined by radioisotopes. *Proc. Amer. Soc. Hort. Sci.* 71:496–506.

295. Vander Krieken, W.M., H. Breteler, and M.H.M. Visser. 1992. The effect of the conversion of indolebutyric acid into indoleacetic acid on root formation on microcuttings of *Malus*. *Plant Cell Physiol.* 33: 709–713.

296. Vander Krieken, W.M., H. Breteler, M.H.M. Visser, and D. Mavridou. 1993. The role of the conversion of IBA into IAA on root regeneration in apple: introduction of a test system. *Plant Cell Reports.* 12:203–206.

297. Van Overbeek J., S. A. Gordon, and L. E. Gregory. 1946. An analysis of the function of the leaf in the process of root formation in cuttings. *Amer. J. Bot.* 33:100–107.

298. Van Staden, J., and F. E. Drewes. 1994. The effect of benzyladenine and its glucosides on adventitious bud formation on Lachenalia leaf sections. *S. Afri. J. Bot.* 60(3):191–192.

299. Van Staden, J., and A. R. Harty. 1988. Cytokinins and adventitious root formation. In *Adventitious root formation in cuttings,* T. D. Davis, B. E. Haissig, and N. Sankhla, eds. Portland, Oreg.: Dioscorides Press.

300. Vasilevskaya, V. K. 1957. The anatomy of bud formation on the roots of some woody plants. *Russian Vest. Leningr. Univ., Ser. Bio. Bul.* 1:3–21.

301. Veierskov, B. 1988. Relations between carbohydrates and adventitious root formation. In *Adventitious root formation in cuttings,* T. D. Davis, B. E. Haissig, and N. Sankhla, eds. Portland, Oreg.: Dioscorides Press.

302. Venverloo, G. J. 1976. The formation of adventitious organs. III. A comparison of root and shoot formation on *Nautilocalyx* explants. *Z. Pflanzenphysiol.* 80:310–22.

303. Vieitez, J., D. G. I. Kingston, A. Ballester, and E. Vieitez. 1987. Identification of two compounds correlated with lack of rooting capacity of chestnut cuttings. *Tree Physiol.* 3:247–55.

304. Vöchting, H. 1878. *Uber Organbildung in Pflanzenreich.* Bonn: Verlag Max Cohen, pp. 1–258.

305. Walker, R. I. 1940. Regeneration in the scale leaf of Lilium *candidum* and *L. longiflorum. Amer. J. Bot.* 27:114–17.

306. Wang, Y. T. 1987. Effect of temperature, duration and light during simulated shipping on quality and rooting of croton cuttings. *HortScience* 22:1301–2.

307. Warmke, H. E., and G. L. Warmke. 1950. The role of auxin in the differentiation of root and shoot primordia from root cuttings of *Taraxacum* and *Cichorium, Amer. J. Bot.* 37:272–80.

308. Weiser, C. J., and L. T. Blaney. 1967. The nature of boron stimulation to root initiation and development in beans. *Proc. Amer. Soc. Hort. Sci.* 90:191–99.

309. Welander, M. 1995. Influence of environment, fertilizer and genotype on shoot morphology and subsequent rooting of birch cuttings. *Tree Physiol.* 15:11–18.

310. Wellensiek, S. J. 1952. Rejuvenation of woody plants by formation of sphaeroblasts. *Proc. Kon. Ned. Akad. Wet.* 55:567–73.

311. Went, F. W. 1934. A test method for rhizocaline, the root-forming substance. *Proc. Kon. Ned. Akad. Wet.* 37:445–55.

312. White, J., and P. H. Lovell. 1984. The anatomy of root initiation in cuttings of *Griselinia littoralis* and *Griselinia lucida. Ann Bot.* 54:7–20.

313. ———. 1984. Anatomical changes which occur in cuttings of *Agathis australis* (D. Don) Lindl. 1. Wounding responses. *Ann. Bot.* 54:621–32.

314. ———. 1984. Anatomical changes which occur in cuttings of *Agathis australis* (D. Don) Lindl. 2. The initiation of root primordia and early root development. *Ann Bot.* 54:633–45.

315. Whitehill, S. J., and W. W. Schwabe. 1975. Vegetative propagation of *Pinus sylvestris. Physiol. Plant.* 35:66–71.

316. Wiesman, Z., J. Riov, and E. Epstein. 1989. Characterization and rooting ability of indole-3-butyric acid conjugates formed during rooting of mung bean cuttings. *Plant Physiol.* 91:1080–1084.

317. Wilson, P.J. 1994. The concept of a limiting rooting morphogen in woody stem cuttings. *J. Hort. Sci.* 69: 591–600.

318. Wilson, P.J. and J. Van Staden. 1990. Rhizocaline, rooting co-factors, and the concept of promoters and inhibitors of adventitious rooting—a review. *Annals Bot.* 66:476–490.

319. Wott, J. A., and H. B. Tukey, Jr. 1967. Influence of nutrient mist on the propagation of cuttings. *Proc. Amer. Soc. Hort. Sci.* 90:454–61.

320. Yarborough, J. A. 1932. Anatomical and developmental studies of the foliar embryos of *Bryophyllum calycinum. Amer. J. Bot.* 19:443–53.

321. ———. 1936. Regeneration in the foliage leaf of *Sedum. Amer. J. Bot.* 23:303–7.

322. Zimmerman, P. W. 1933. Initiation and stimulation of adventitious roots caused by unsaturated hydrocarbon gases. *Contrib. Boyce Thomp. Inst.* 5:351–69.

323. ———. 1937. Comparative effectiveness of acids, esters, and salts as growth substances and methods of evaluating them. *Contrib. Boyce Thomp. Inst.* 8:337–50.

324. Zimmerman, P. W., and F. Wilcoxon. 1935. Several chemical growth substances which cause initiation of roots and other responses in plants. *Contrib. Boyce Thomp. Inst.* 7:209–29.

325. Zucconi, F., and A. Pera. 1978. The influence of nutrients and pH effects on rooting as shown by mung bean cuttings. *Acta Hort.* 79:57–62.

SUPPLEMENTARY READING

Brouwber, R., O. Gasparikova, J. Kolek, and B. C. Loughman, eds. 1981. *Structure and function of plant roots.* Dordrecht: Martinus Nijhoff Publishers.

Davis, T. D., B. E. Haissig, and N. Sankhla, eds. 1988. *Adventitious root formation in cuttings.* Portland, Oreg.: Dioscorides Press.

Davis, T.D. and B.E. Haissig. 1994. *Biology of adventitious root formation.* New York, N.Y.: Plenum Press.

International Plant Propagators' Society. Proceedings of annual meetings.

Jackson, M. B., ed. 1986. *New root formation in plants and cuttings.* Dordrecht: Martinus Nijhoff Publishers.

Jackson, M. B., and A. D. Stead. 1983. *Growth regulators in root development.* Monograph 10. Wantage, England: British Plant Growth Regulator Group.

Loreti, F., ed. 1988. Special issue on the International Symposium on Vegetative Propagation of Woody Species. Pisa, Italy. (Sept. 3–5, 1987). *Acta Hort.* 227:1–516.

Special issue on vegetative propagation of conifers and hardwoods. 1974. *N. Zealand Jour. For. Sci.* 4(2).

Sutton, R. F., and R. W. Tinus. 1983. *Root and root system terminology.* Forest Science Monograph. Washington, D.C.: Soc. Amer. Foresters.

Torrey, J. G., and D. T. Clarkson, eds. 1975. *The development and function of roots.* London: Academic Press.

11 Techniques of Propagation by Cuttings

This chapter covers different types of cuttings, and the techniques and procedures utilized to maximize cutting propagation in commercial production systems. In cutting propagation, a portion of stem, root, or leaf is cut from the parent or stock plant and induced to form roots and shoots by chemical, mechanical, and/or environmental manipulation. In most cases the new independent plant produced is a clone, which is identical to the parent plant.

THE IMPORTANCE AND ADVANTAGES OF PROPAGATION BY CUTTINGS

Today's propagation systems are more market-driven than product-driven (*111*). This means that propagators must first analyze market demands, and then select and develop cultivars, utilizing optimum propagation techniques to produce plants for their customers. Cuttings are still the most important means of propagating ornamental shrubs—deciduous species as well as the broad- and narrow-leaved types of evergreens. Cuttings are also used widely in commercial greenhouse propagation of many florists' crops—poinsettias, chrysanthemums, geraniums, in propagating foliage crops, certain fruit crops, and some vegetables (horseradish, chicory, artichoke, sweet potato), and forestry species (*27, 93*).

Generally, vegetative propagation is more costly (per unit propagule) than sexual (seedling) propagation. For instance, the production costs of propagating broad-leaved evergreens and leafy cuttings can require protected culture (glass or polyethylene-covered structures), bottom-heated rooting structures, intermittent mist and/or fog systems. For many species, the superiority of the clonally produced cultivars justifies the higher cash value that is necessary to offset the added propagation costs that invariably are associated with this process.

For species that can be propagated easily by cuttings, this method has numerous advantages. Many new plants can be started in a limited space from a few stock plants. It is inexpensive (compared to other asexual methods), rapid, and simple, and does not require the special techniques necessary in grafting, budding, or micropropagation. Cutting propagation avoids the graftage problems of incompatibility with rootstocks and poor graft union formation. Greater uniformity is obtained by absence of the variation which sometimes appears as a result of the variable seedling rootstocks of grafted plants. The parent plant is usually reproduced exactly, with no genetic change.

In forestry, cutting propagation has been used for hundreds of years and is routinely done in commercial production of Cryptomeria or "Sugi" *(Cryptomeria japonica)*, poplars, *(Populus spp)*, willows *(Salix spp)*, chinese fir *(Cunninghamia lanceolata)*, hybrid eucalyptus *(Eucalyptus)*, Monterrey pine *(Pinus radiata)*, and Norway maple, *(Acer platanoides)* (*93*). In timber production, most conifer species are seed propagated, however, vegetative propagation of conifers for reforestation accounts for an annual production of more that 65 million plants worldwide (*93*). Rooted cuttings from elite Douglas-Fir *(Pseudotsuga menziesii)* selections can show superior growth characteristics to seed-propagated plants (*94*).

It is not always desirable, however, to produce plants on their own roots by cuttings even if it is possible to do so. It is often advantageous or necessary to use a rootstock resistant to some adverse soil condition or soil-borne organism, or to utilize available dwarfing or invigorating rootstocks. Cutting propagation, like any other asexual technique, may potentially increase disease and insect susceptibility, since the clonal plants may lack the genetic diversity of seedling-produced plants.

TYPES OF CUTTINGS

Cuttings are made from the vegetative portions of the plant, such as **stems, modified stems** (rhizomes, tubers, corms, and bulbs), **leaves,** or **roots**. Cuttings can be classified according to the part of the plant from which they are obtained:

- Stem cuttings
    - Hardwood
        - Deciduous
        - Narrow-leaved evergreen
    - Semi-hardwood
    - Softwood
    - Herbaceous
- Leaf cuttings
- Leaf-bud cuttings (single-eye or single-node cuttings)
- Root cuttings

Many plants can be propagated by several different types of cuttings with satisfactory results. The preferred type depends on individual circumstances; the least expensive and easiest method is usually selected.

For easy-to-root woody perennial plants, hardwood stem cuttings in an outdoor nursery are frequently used because of the simplicity and low cost. For more tender herbaceous species, or for those more difficult to propagate, it is necessary to resort to the more expensive and more elaborate facilities required for rooting the leafy types of cuttings. In today's containerized nurseries, a larger portion of easy- and difficult-to-root species are propagated with mist, fog, or contact polyethylene sheet systems (Figure 11–36). Root cuttings of some species are also satisfactory, but cutting material may be difficult to obtain in large quantities.

In selecting cutting material it is important to use stock plants that are free from diseases, moderately vigorous, and of known identity. Propagators should avoid stock plants that have been injured by frost or drought, defoliated by insects, stunted by excessive flowering or fruiting, or by lack of soil moisture or proper nutrition—and plants that have overly vigorous growth. A marginal cutting slows down the whole production process, creates cultural problems, and produces an inferior quality plant.

A desirable practice for the propagator is the establishment of stock blocks as a source of propagating material, where uniform, true-to-type, pathogen-free mother plants can be maintained and held under the proper nutritive condition for the best rooting of cuttings taken from them. Another advantage of permanent stock plants is that they can be manipulated to enhance rooting (layering, stooling, hedging-severe pruning), whereas this is not usually feasible with containerized stock plants. However, in most container nurseries no stock plants are maintained. Rather, propagules are collected from the container plants in production—hence the need for good cultural controls, and the maintenance of cultivar records/identity of *all* plants in production.

Stem Cuttings

Stem cuttings can be divided into four groups, according to the nature of the wood used: **hardwood, semi-hardwood, softwood,** and **herbaceous**. In propagation by stem cuttings, segments of shoots containing lateral or terminal buds are obtained with the expectation that under the proper conditions adventitious roots will develop and thus produce independent plants.

The type of wood, the stage of growth used in making the cuttings, and the time of year when the cuttings are taken are some of the factors important in satisfactory rooting of plants. Information concerning these factors is given in Chapters 10 and 20.

It is imperative that propagators keep thorough records of procedures and seasonal condition of plant materials, and conduct small tests to achieve optimum success for their particular propagation system (*39, 42*). The type of cutting utilized and cultural manipulation during propagation is dependent on the market the plant is being targeted for e.g., ornamental trees with multistemmed flowering stems versus high-branched trees with strong, straight central leaders. Other considerations are the plant species, environmental conditions, propagation system utilized, propagation facilities, available personnel—and ultimately what is cost-effective for the producer. See Table 11–1 for a comparison of different types of cuttings.

Hardwood Cuttings (Deciduous Species)

Hardwood cuttings are those made of matured, dormant firm wood after leaves have abscised. The use of hardwood cuttings is one of the least expensive and easiest methods of vegetative propagation. Hardwood cuttings are easy to prepare, are not readily perishable, may be shipped safely over long distances if necessary, and require little or no special equipment during rooting. Hardwood cuttings are easily transplanted after rooting, and some producers report that liners produced from hardwood cuttings are larger than those from softwood cuttings (*45*).

The low cost of hardwood cutting propagation makes feasible high-density meadow orchards, consisting of precocious dwarfed fruit trees. Some peach cultivars, for example, can be propagated on a large scale from rooted hardwood cuttings (*41, 51*).

Hardwood cuttings are prepared during the dormant season—late fall, winter, or early spring—usually from wood of the previous season's growth, although with a few species—fig, olive, and certain plum cultivars—two-year-old or older wood can be used. Hardwood cuttings are most often used in propagation of deciduous woody plants, although some broad-leaved evergreens, such as the olive, can be propagated by leafless hardwood cuttings. Many deciduous ornamental shrubs are started readily by this type of cutting. Some common ones are privet, forsythia, wisteria, honeysuckle, willow *(Salix)*, poplar *(Populus)*, dogwood *(Cornus)*, *Pontentilla*, *Sambucus*, crape myrtle, and spiraea. Rose rootstocks such as *Rosa multiflora* are propagated in great quantities by hardwood cuttings. A few fruit species are propagated commercially by this method, for example, fig, quince, olive, mulberry, grape, currant, pear, gooseberry, pomegranate, and some plums.

The propagating material for hardwood cuttings should be taken from healthy, moderately vigorous stock plants growing in full sunlight. The wood selected should not be from extremely rank growth with abnormally long internodes, or from small, weakly growing interior shoots. Generally, hardwood cutting material is ready when you can remove the leaves without tearing the bark. Wood of moderate size and vigor is the most desirable. The cuttings should have an ample supply of stored carbohydrates to nourish the developing roots and shoots until the new plant becomes self-sustaining. Tip portions of a shoot, which are usually low in stored carbohydrates, are discarded. Central and basal parts generally make the best cuttings, but there are exceptions (*50*).

Hardwood cuttings vary in length from 10 to 76 cm (4 to 30 in.). Long cuttings, when they are to be used as rootstocks for fruit trees, permit the insertion of the cultivar bud into the original cutting following rooting, rather than into a smaller new shoot arising from the original cutting.

At least two nodes are included in the cutting; the basal cut is usually just below a node and the top cut 1.3 to 2.5 cm (1/2 to 1 in.) above a node. When hardwood cuttings of *Rosa multiflora* are to be field-planted in raised soil beds, it is common procedure to "de-eye" or remove all lower basal axillary buds prior to sticking cuttings to prevent suckering from the base of the cutting.

Hardwood cuttings will desiccate, so it is important that they not dry out during handling and storage. After cuttings are cut with a band saw (Figure 11–3), some producers will take the bundled cuttings and dip the tops (apex) in wax. The wax helps reduce desiccation and indicates the orientation of the cuttings for late fall or spring field propagation.

The diameter of the cuttings may range from 0.6 to 2.5 or even 5 cm (¼ in. to 1 or 2 in.), depending upon the species. Three different types of cuttings are shown in Figure 11–1: the **mallet,** the **heel,** and the **straight** cutting. The mallet includes a short section of stem of the older wood, whereas the heel cutting includes only a small piece of the older wood. The straight cutting, not including any of the

TABLE 11–1

Propagation systems with different types of cuttings

Cutting Type	*Hardwood (Deciduous)*	*Hardwood (Evergreen)*	*Semi-hardwood*	*Softwood*	*Herbaceous*	*Leaf*	*Leaf-bud (Single eye or node)*	*Root*
Description	Mature, dormant, or quiescent hardwood stems; woody species.	Mature hardwood stems; woody species.	Partially mature wood on current season's growth; woody species.	New, soft succulent growth; woody species.	Succulent stems from nonwoody plants.	Leaf blade or leaf blade and petiole; generally from non-woody, herbaceous plants.	Leaf blade + petiole, + short piece of stem with attached axillary bud; woody and herbaceous plants.	Root pieces from thin to fleshy roots; woody and herbaceous plants.
Season Propagated	Dormant season: late fall to early spring.	Dormant season: late fall to late winter.	Late spring to late summer.	Late winter to early summer.	Year-round—with greenhouse and/or tropical field production.	Year-round—as long as leaves are available.	Generally during growing season; year-round for tropical plants.	Take in late winter or early spring when roots contain stored carbohydrates—but before new shoot growth.
Propagation System	Field propagated; also greenhouse propagated with light intermittent mist, fog, humidity tent, or contact polyethylene.	Light intermittent mist, fog, humidity tent, or contact polyethylene.	Intermittent mist, fog, humidity tent, or contact polyethylene.	Intermittent mist, fog, or humidity tent.	Intermittent mist, fog, or humidity tent.	Intermittent mist, fog, or humidity tent.	Intermittent mist, fog, humidity tent, or contact polyethylene.	Depending on species, directly planted into field, or direct planted in flats, and covered with contact polyethylene, or stuck in containers and held in dormant storage.
Cutting Length	10–76 cm (4–30 in.); normally at least two nodes with basal cut just below the node; mallet, heel, and straight (most common) cuttings used; longer cuttings for rootstocks.	10–20 cm (4–8 in.)	7.5–15 cm (3–6 in.)	7.5–12.5 cm (3–5 in.)	7.5–12.5 cm (3–5 in.)	Varies with species, and leaf size, e.g., *Sansevieria* 7.5–10 cm (3–4 in.); other species just use section of leaf.	2.0–7.5 cm (1–3 in.); bud may sometimes be placed 1.3–2.5 cm (0.5–1.0 in.) below the surface.	*Small, delicate roots*—2.5–5.0 cm (1–2 in.); *somewhat fleshy*—5.0–7.5 cm (2–3 in.); *large roots*—5–15 cm (2–6 in.).

Chemical Treatment	IBA or NAA at 2,500–5,000 ppm; generally, 10,000 ppm maximum with difficult-to-root.	IBA or NAA at 2,000 ppm or slightly higher. Difficult-to-root; 5,000–10,000 ppm.	IBA or NAA at 1000–3000 ppm; generally, 5,000 ppm maximum.	IBA or NAA at 500 to 1,250 ppm; generally, 3,000 ppm maximum.	Auxin not needed, or may apply IBA or NAA from 500–1,250 ppm.	Apply cytokinin for adventitious buds, e.g., BA at 100 ppm for begonia, African violets (spray application).	IBA or NAA at 1,000 to 3,000 ppm with *Ficus.*	Usually no hormones applied.
Examples of Plants	Privet, Forsythia, Landscape Roses, Willow, Crape Myrtle, *Euonymous,* Dogwood, Fig, Quince, Apple, Pear, Plum rootstock	Juniper, Yew, Spruce, Firs *(Abies).*	Holly, Pittosporum, Rhododendron, Citrus, Olive.	Lilac, Forsythia, Maple, Magnolia, Weigelia, Apple, Peach, Pear, Plum.	Geraniums, Poinsettia, Dieffenbachia, Chrysanthemum; most Floral Crops; Pineapple (slips and suckers); Sweet Potato (slips).	Begonia, African violet, Sansevieria.	Black Raspberry, Boysenberry, Maple, Ficus, Rhododendron.	Poppy, Aralia, Geranium. Viburnum, Horse Radish.
Other Observations	One of the least expensive; easy to ship over long distances; little equipment needed, i.e., field propagation without mist; start when you can remove leaves without tearing bark; central and basal parts generally better cuttings than apical; cuttings can be pretreated with bottom heat, then field planted; or direct rooted in field during spring or fall.	Very slow to root; bottom heat 23°– 27° C (75°–80° F) helpful; basal wounding may be helpful; important to maintain adequate moisture levels during rooting.	Leaf may be reduced to control transpiration; wounding may be helpful.	Roots relatively quickly(2–5 weeks); bottom heat helps; 23°–27° C (75°–80° F); requires more attention and equipment (mist/fog); do **not** use weak, thin, or thick, heavy cuttings; remove all flower buds; use lateral or side branches for cuttings. Very susceptible to desiccation.	Probably the fastest to root; can be susceptible to decay; bottom heat is helpful.	Periclinal chimeras will not reproduce true-to-type; with fibrous rooted begonias—cut large leaves into triangular sections containing a piece of large vein; entire leaf blade plus petiole used with African violet.	Useful when propagating materials are scarce—since twice as many plants can be produced as with stem cuttings; each node can be used as a cutting; high humidity essential and bottom heat is helpful.	To maintain polarity and avoid planting upside down, make a straight cut near the crown (proximal) end, and a slanting cut at the distal end; should be planted with the proximal end up or cutting planted horizontally; periclinal chimeras can not be propagated by root cuttings.

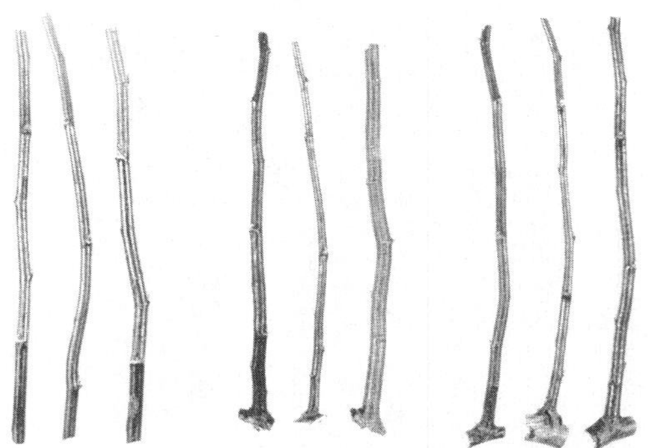

FIGURE 11–1 Types of hardwood cuttings. *Left:* Straight—the type ordinarily used. *Center:* Heel cuttings. A small piece of older wood is retained at the base. *Right:* Mallet cuttings. An entire section of the branch of older wood is retained.

older wood, is the most common and gives satisfactory results in most instances. Hardwood cuttings of quince, freshly prepared and after a summer's growth in the nursery, are shown in Figure 11–2.

Where it is difficult to distinguish between the top and base of the cuttings, it is advisable to make one of the cuts at a slant rather than at right angles. In large-scale operations, bundles of cutting material are cut to the desired lengths by band saws or other types of mechanical cutters rather than individually by hand (Figure 11–3). For large-scale commercial operations, the planting of cuttings is mechanized, using equipment as illustrated in Figure 11–4.

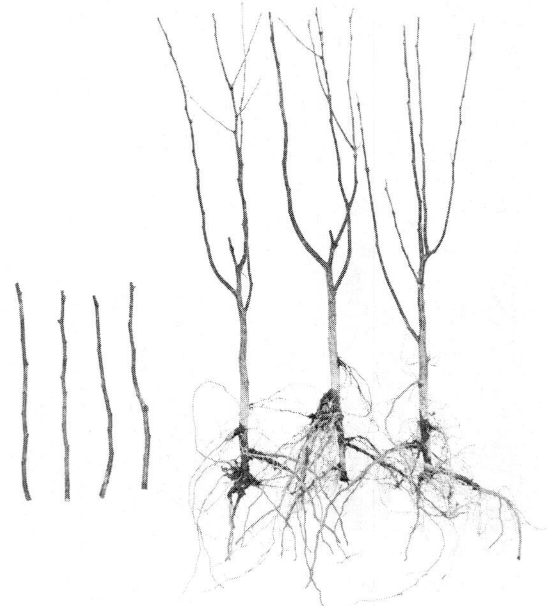

FIGURE 11–2 *Left:* Hardwood cuttings of quince *(Cydonia oblonga)* prepared and ready for planting in an outdoor nursery in early spring. *Right:* Rooted cuttings after one summer's growth.

Direct spring planting. It is often sufficient with easily rooted species to gather the cutting material during the dormant season, wrap it in newspaper or slightly damp peat moss in a polyethylene bag, and store at 0 to 4.5° C (32 to 40° F) until spring. The cutting material should not be allowed to dry out or to become excessively wet during storage. At planting time, the cuttings are made into proper lengths and planted into a field nursery propagation bed without intermittent mist or in propagation flats with a very light intermittent mist.

Stored cutting material should be examined frequently. If signs of bud development appear, lower storage temperatures should be used or the cuttings should be made and planted without delay. If the buds are forcing their way out when the cuttings are planted, leaves will form and the cuttings will die, because of water loss from the leaves and depletion of stored carbohydrate reserves prior to rooting.

Direct fall planting. In regions with mild winters or reliable snow cover, cuttings can be made in the autumn and planted immediately in the nursery. Rooting may take place during the dormant season, or the formation of roots and shoots may occur simultaneously the following spring. Hardwood cuttings will take a longer time to root in the field at the relatively low and declining soil temperatures of fall. Hardwood cuttings of peach and peach × almond hybrids have been successfully rooted in the nursery by this method, provided that they were treated prior to planting with indolebutyric acid (IBA) (*51*). Selected ornamental shrubs can be fall-propagated as far north as southern Canada without auxin pretreatment (*45*). Field rose rootstock for landscape and cut rose for production are fall-planted in Spain, California, and Texas.

Field propagated hardwood cuttings are dug after a growing season as rooted liners using an apparatus such as a modified potato digger, or with a U-blade attached to a tractor (Figure 11–52). With greater spacing, or in the case of landscape roses, rooted cuttings are left in the field for an additional season and sold as finished plants.

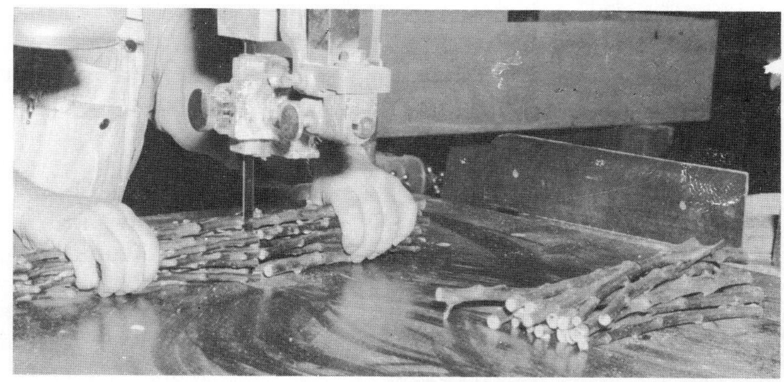

FIGURE 11–3 Sawing hardwood cuttings to length with a band saw. This method is much faster than preparing each cutting individually and gives good results for easy-to-root species.

Warm-temperature pretreatment for root initiation. Hardwood cuttings are taken in fall while the buds are in or entering the "rest period." The cuttings are treated with auxin, stored under moist conditions at relatively warm temperatures—18 to 21° C (65 to 70° F)—for three to five weeks to stimulate root initiation. The cuttings are then planted in the nursery (in mild climates) or held in cold storage (2 to 4.5° C; 35 to 40° F) until spring. This system works well with hardwood cuttings of pear which are allowed to initiate roots while the buds were under the "rest." Since the buds are dormant and not growing, they are not actively competing with the rooting process for carbohydrate reserves in the cuttings (*2*).

In dealing with deciduous hardwood cuttings (which are leafless, cannot photosynthesize, and are thus surviving on stored carbohydrate reserves), it is important to avoid temperatures that cause excessive callusing and the loss of stored reserves; otherwise, rooting and field survival are poor (Figure 11–5).

Callusing and rooting can occur simultaneously. Callusing can sometimes interfere with rooting, and—except for some difficult-to-root species that initiate roots from callus—is *not* a prerequisite to rooting.

Initiating roots with bottom heat. This method has been successful for difficult-to-root species such as some apple, pear, and plum rootstocks. Cuttings are collected in either the fall or late winter, the basal ends treated with IBA at 2,500 to 5,000 ppm. Cuttings are then bundled and placed upright for about four weeks on a 20 cm (8 in.) sand base with heating mats or circulating hot-water tubing, just below the sand surface. The bundled cuttings are packed between 6 cm (3 in.) of moist peat and maintained with bottom heat at 18 to 21° C (65 to 70° F). The top portion of the cuttings are left exposed to the cool or ambient outdoor temperatures. It is best to do this in a covered, unheated shed for protection against excessive

FIGURE 11–4 Machine for large-scale planting of hardwood cuttings. Developed at the Tree Nursery Division, P. F. R. A., Indian Head, Saskatchewan, Canada, primarily for propagation of willow and poplar for shelter belt use. This four-unit machine will plant 10,000 to 12,000 cuttings per hour. (Courtesy Canada Department of Regional Economic Expansion.)

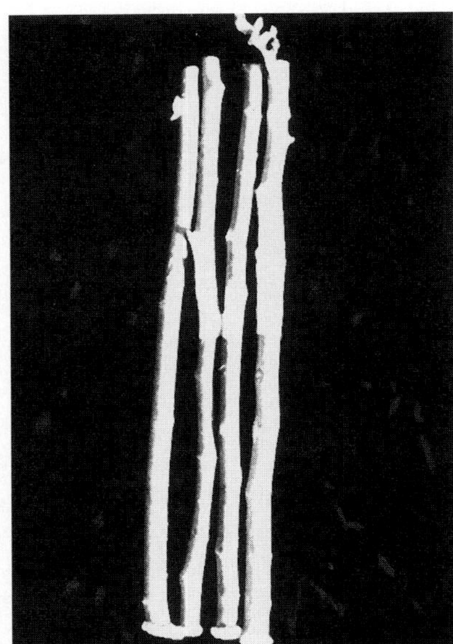

FIGURE 11–5 *Rosa multiflora* hardwood cuttings that were simultaneously chip-budded and callused for rooting. Too high a temperature 27° C (80° F) caused profuse callusing at the base, but poor rooting and field survival.

moisture from rains. The East Malling Research Station in England developed (*58, 60, 61, 62*) commercial procedures for propagating difficult species by this method. Cuttings must be transplanted before buds begin growth; this is usually done as roots first emerge. It is essential to prevent the cuttings from rotting by avoiding excessive application of water to the rooting medium. It is not necessary to await root emergence before transplanting.

This procedure is probably best suited for regions having relatively mild winters (*55, 56, 62*). When soil or weather conditions are not suitable for planting after roots become visible, it has been satisfactory to leave the cuttings undisturbed in the rooting bed, shut off the bottom heat, then plant them in the nursery when conditions become suitable (*15*).

Hardwood Cuttings (Narrow-Leaved Evergreen Species)

Narrow-leaved evergreen cuttings must be rooted under moisture conditions that will prevent excessive drying as they usually are slow to root, sometimes taking several months to a year. Some species root much more readily than others. In general, *Chamaecyparis, Thuja,* and the low-growing *Juniperus* species root easily and the yews (*Taxus* spp.) fairly well, whereas the upright junipers, the spruces (*Picea* spp.), hemlocks (*Tsuga* spp.), firs (*Abies* spp.), and pines (*Pinus* spp.) are more difficult. In addition, there is considerable variability among the different species in these genera in regard to the ease of rooting of cuttings. Cuttings taken from young seedling stock plants root much more readily than those taken from older trees because of the juvenility factor. Treatments with IBA enhance rooting.

Narrow-leaved evergreen cuttings ordinarily are best taken between late fall and late winter (Figure 11–6). Rapid handling of the cuttings after the material is taken from the stock plants is important. The cuttings are usually best rooted in a greenhouse or polyhouses with relatively high light irradiance and under conditions of high humidity or very light misting, but without heavy wetting of the leaves. A bottom heat temperature of 24 to 26.5° C (75 to 80° F) has given good results. Dipping the cuttings into a fungicide helps prevent fungal diseases. Sand alone is a satisfactory rooting medium, as is a 1:1 mixture of perlite and peat moss. Some individual cuttings take longer to root than others. The slower-rooting ones can be inserted again in the rooting medium, and eventually root.

FIGURE 11–6 Narrow-leaved evergreen cuttings. *Above:* Juniper cuttings ready for sticking in rooting medium. *Below:* Cuttings after rooting.

The type of wood to use in making the cuttings varies considerably with the particular species being rooted. As shown in Figure 11–6, the cuttings are made 10 to 20 cm (4 to 8 in.) long with all the leaves removed from the lower half. Mature terminal shoots of the previous season's growth are usually used. In some instances, as in with *Juniperus chinensis* 'Pfitzeriana,' older and heavier wood also can be used, thus resulting in a larger plant when it is rooted. On the other hand, some propagators use small tip cuttings, 5 to 8 cm (2 to 3 in.) long, placed very close together in a flat for rooting. In some species, such as *Juniperus excelsa,* older growth taken from the sides and lower portion of the stock plant roots better than the more succulent tips. Cuttings of *Taxus* root best if they are taken with a piece of old wood at the base of the cutting; such cuttings seem less subject to fungus attacks (*119*). In certain of the narrow-leaved evergreen species, some type of basal wounding is often beneficial in inducing rooting.

Semi-hardwood Cuttings

Semi-hardwood (greenwood) cuttings are those made from woody, broad-leaved evergreen species, and leafy summer and early fall cuttings of deciduous plants with partially matured wood. Cuttings of broad-leaved evergreen species are generally taken during the summer (or late spring through early fall in warmer climates) from new shoots just after a flush of growth has taken place and the wood is partially matured. Many ornamental shrubs, such as camellia, pittosporum, rhododendron, euonymus, the evergreen azaleas, and holly, are commonly propagated by semi-hardwood cuttings. A few fruit species, such as citrus and olive, can also be propagated in this manner.

Semi-hardwood cuttings are made 7.5 to 15 cm (3 to 6 in.) long with leaves retained at the upper end, as shown in Figure 11–7. If the leaves are very large, they should be trimmed to reduce the leaf surface area, which lowers transpirational water loss and allows closer spacing in the cutting bed (Figure 11–27). The shoot terminals are often used in making cuttings, but the basal parts of the stem will usually root also. The basal cut is usually just below a node. The cutting wood should be obtained in the cool, early morning hours when leaves and stems are turgid. Cuttings should be placed in large containers, which are covered with clean moist burlap to maintain a high humidity, or put in large polyethylene bags. Cuttings should be kept out of the sun until they can be stuck and propagation is initiated.

It is necessary that leafy cuttings be rooted under conditions that will keep water loss from the leaves at a minimum; commercially, they are ordinarily rooted under intermittent mist, fog, or under polyethylene sheets laid over the cuttings—(in cool temperate climates or during fall in the southern U.S. (Figure 11–36)). Bottom heat and auxin treatment are also beneficial.

Softwood Cuttings

Cuttings prepared from the soft, succulent, new spring growth of deciduous or evergreen species may properly be classed as **softwood cuttings**. The softwood condition for most woody plants ranges from two to eight weeks. Softwoods are produced during growth flushes and may occur just once per year, as with *Chionanthus virginicus* and *Ulmus parvifolia,* or several times during the year, e.g., April to August in Texas with *Lagerstroemia* and *Spirea.* August softwood cuttings in the southeastern U.S. are not physiologically the same as June softwood cuttings

FIGURE 11–7 Semi-hardwood cuttings as shown by euonymus. *Left:* Cuttings as prepared for rooting from partially matured wood in midsummer. *Right:* Cuttings after rooting.

e.g., June softwood cuttings generally survive the winter better than August cuttings. Many ornamental woody plants can be started by softwood cuttings. Typical examples are the hybrid French lilacs, forsythia, magnolia, weigella, spiraea, maples, magnolia, and flowering dogwood. Various crab apple cultivars can also be started in this manner (*18*). Other examples are shown in Figure 11–8. Although not commonly done, apple, peach, pear, plum, apricot, and cherry can be propagated by softwood cuttings.

For some difficult-to-root species, softwood cuttings may be the only commercial method to clonally regenerate cultivars. Softwood cuttings generally root easier and quicker than the other types, but require more attention and sophisticated equipment. This type of cutting is always made with leaves attached. They must, consequently, be handled carefully to prevent desiccation, and be rooted under conditions which will avoid excessive water loss from the leaves. Temperature should be maintained during rooting at 23 to 27° C (75 to 80° F) at the base of cuttings. As long as light is adequate, ambient air temperature of the mist or fog system can rise to 30 to 32° C (86 to 90° F) without detriment to rooting. Softwood cuttings produce roots in two to four or five weeks in most cases. In general, they respond well to auxin treatment.

It is important in making softwood cuttings to obtain the proper type of cutting material from the stock plant. Such material will vary greatly, however, with the species being propagated. Extremely fast-growing, soft, tender shoots are not desirable, as they are likely to deteriorate before rooting. At the other extreme, older woody stems are slow to root or may just drop their leaves and not root. The best cutting material has some degree of flexibility, but is mature enough to break when bent sharply. Weak, thin, interior shoots should be avoided as well as vigorous, abnormally thick or heavy ones. Average growth from portions of the plant in full light is the most desirable to use. Some of the best cutting material is the lateral or side branches of the stock plant. Heading back the main shoots will usually force out numerous lateral shoots from which cuttings can be made. Softwood cuttings are 7.5 to 12.5 cm (3 to 5 in.) long with two or more nodes. The basal cut is usually made just below a node.

The leaves on the lower portion of the cutting are removed, with those on the upper part retained. Large leaves can be reduced in size to lower the transpiration rate and to occupy less space in the propagating bed. But drastically cutting back leaves and the mutual shading of leaves of crowded cuttings in a flat can reduce rooting, and encourage diseases such as *Botrytis*. For difficult-to-root plants, factors which favor a *high leaf:stem ratio*—such as not trimming leaves, and selecting cuttings with relatively thin stems—favor rooting; whereas leaf-trimming and selecting thick fleshy-stemmed cuttings may result in stem rotting (Figure 11–9). All flowers or flower buds should be removed.

Softwood cuttings stress easily, so it is important to collect cutting material early in the day. Softwood cuttings should be kept moist, cool, and turgid at all times. Laying the cutting material or prepared cuttings in the sun for even a few minutes

FIGURE 11–8 Softwood cuttings of several ornamental species. *Top:* Cuttings made in late spring from young shoots. *Below:* Cuttings after rooting. *Left to right: Myrtus, Pyracantha,* Oleander, and *Hebe.*

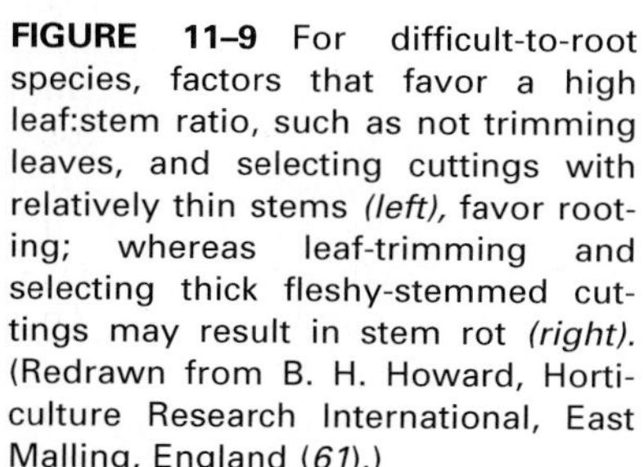

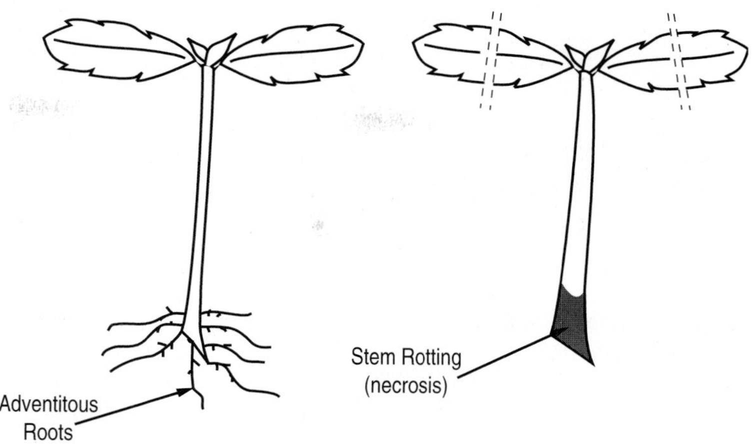

FIGURE 11–9 For difficult-to-root species, factors that favor a high leaf:stem ratio, such as not trimming leaves, and selecting cuttings with relatively thin stems *(left)*, favor rooting; whereas leaf-trimming and selecting thick fleshy-stemmed cuttings may result in stem rot *(right)*. (Redrawn from B. H. Howard, Horticulture Research International, East Malling, England (*61*).)

will cause serious damage. Soaking the cutting material or cuttings in water for prolonged periods to keep them fresh is undesirable. Refrigerated storage (4 to 8° C; 40 to 47° F) for one to two days is another option. Cuttings of some species, such as *Forsythia* × *intermedia*, can be safely stored for a month.

Some producers of ornamental shrubs will fall-propagate hardwood cuttings in polyhouses, with heated benches under mist, and then harvest softwood cuttings from the developing hardwood shoots as the various species flush in the spring (*3*). Before the end of summer, both the original crop of hardwood cuttings and the bonus crop of softwood cuttings can be lined out.

Herbaceous Cuttings

Herbaceous cuttings are made from succulent, nonwoody plants like geraniums, chrysanthemums, coleus, carnations, and many foliage crops. They are 8 to 13 cm (3 to 5 in.) long with leaves retained at the upper end, as shown in Figure 11–10, or without leaves (Figure 11–11). Most florists' crops are propagated by easily rooted herbaceous cuttings. They are rooted under the same conditions as softwood cuttings. Bottom heat is also helpful. Under proper conditions, rooting is rapid and in high percentages. Although auxins are usually not required, they are often used to gain uniformity in rooting and the development of heavier root systems. The basal ends of pineapple slips and suckers and cactus cuttings are normally callused and suberized for a week or more before propagating in rooting media. These practices tend to prevent the entrance of decay organisms.

Types of cuttings (apical versus basal) and nodal position of herbaceous cuttings can influence shoot growth and finished plant quality of rooted liners. For example, basal cuttings of *Hedera helix* and *Schefflera arboricola* develop longer shoots and more roots than apical cuttings. With golden pothos (*Epipremnum aureum*), a 3-cm (1-in.) or longer internode section below the node and a fraction of the old aerial root should be retained on cuttings for most rapid axillary shoot development (*117*). With foliage plants such as *Dracena* spp., branching

FIGURE 11–10 Typical herbaceous cuttings. *Left to right:* Chrysanthemum, begonia, and geranium. It is often necessary with large-leaved plants, such as the begonia, to trim back some of the leaves to prevent wilting and to conserve space in the propagating bench.

FIGURE 11–11 A stem cutting consisting only of a leafless stem piece; used here in propagating the monocotyledonous plant, *Dieffenbachia maculata.* Latent buds develop into shoots along with the formation of adventitious roots.

of herbaceous canes during propagation is done by cutting one-third to one-half of the way through the canes (Figure 11–12).

Leaf Cuttings

In **leaf cuttings,** the leaf blade, or leaf blade and petiole, is utilized in starting new plants. Adventitious buds, shoots, and roots form at the base of the leaf and develop into the new plant; the original leaf does not become a part of the new plant. Only a limited number of plant species can be propagated by leaf cuttings.

One type of propagation by leaf cuttings is illustrated by *Sansevieria.* The long tapering leaves are cut into sections 8 to 10 cm (3 to 4 in.) long as shown in Figure 11–13. These leaf pieces are inserted three-fourths of their length into the rooting medium, and after a period of time a new plant forms at the base of the leaf piece. The variegated form of *Sansevieria, S. trifasciata laurenti* is an example of a periclinal chimera that will not reproduce true-to-type from leaf cuttings; to retain its characteristics, it must be propagated by division of the original plant.

In starting plants with fleshy leaves, such as *Begonia rex,* by leaf cuttings, the large veins are cut on the undersurface of the mature leaf, which is then laid flat on the surface of the propagating medium. The leaf is pinned or held down in some manner, with the natural upper surface of the leaf exposed. After a period of time under humid conditions, new plants form at the point where each vein was cut. The old leaf blade gradually disintegrates.

Another method, sometimes used with fibrous-rooted begonias, is to cut large, well-developed leaves into triangular sections, each containing a piece of a large vein. The thin outer edge of the leaf is discarded. These leaf pieces are then inserted upright in sand or on a peat-perlite medium, with the pointed end down. The new plant develops from the large vein at the base of the leaf piece.

New plants arise from leaves in a variety of ways. An interesting example is illustrated in Figure 11–14, where the new plant of the piggyback plant develops at the junction of the leaf blade and petiole, even while the leaf is still growing on the mother plant.

The African violet *(Saintpaulia)* is typical of leaf cuttings that can be made of an entire leaf (leaf blade plus petiole), the leaf blade only, or just a portion of the leaf blade. The new plant forms at the base of the petiole or midrib of the leaf blade (see Figure 11–15).

An unusual type of leaf cutting is illustrated in Figure 11–16, were many new plants or **offsets** arise at the margins of the leaf. The leaf itself eventually

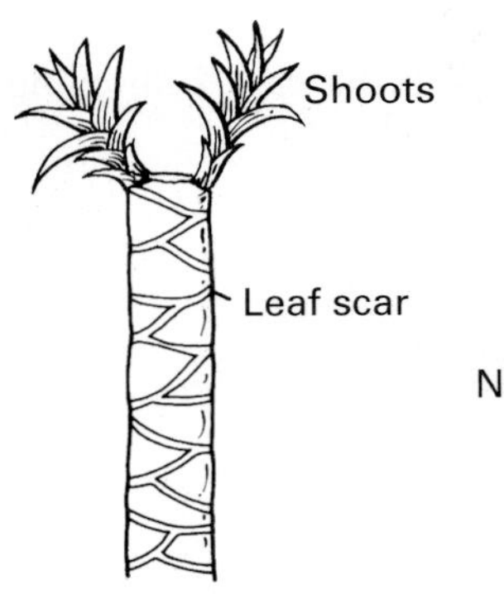

Unnotched cane without prominent axillary buds.

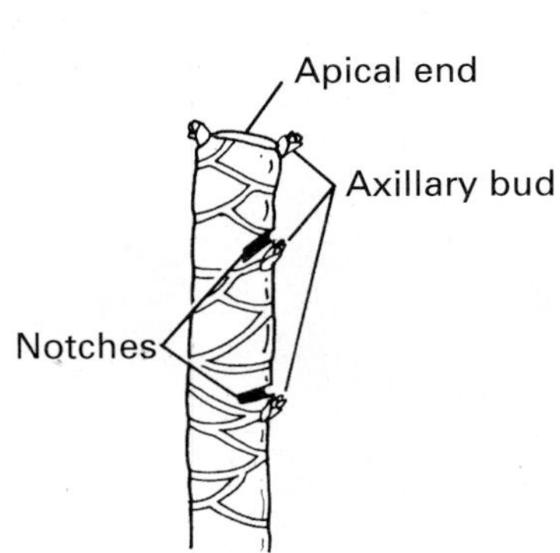

Freshly cut cane following notching.

Notched cane with developing shoots.

FIGURE 11–12 Branching of herbaceous canes of Dracena during propagation is done by cutting one-third to one-half way through the canes (*22*).

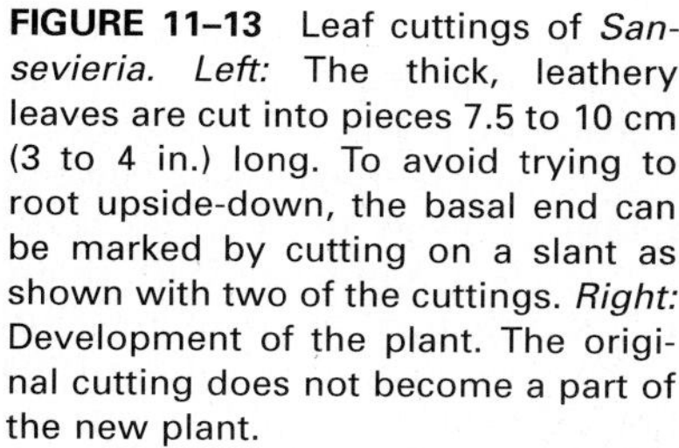

FIGURE 11–13 Leaf cuttings of *Sansevieria. Left:* The thick, leathery leaves are cut into pieces 7.5 to 10 cm (3 to 4 in.) long. To avoid trying to root upside-down, the basal end can be marked by cutting on a slant as shown with two of the cuttings. *Right:* Development of the plant. The original cutting does not become a part of the new plant.

deteriorates. Leaf cuttings should be rooted under the same conditions of high humidity as those used for softwood or herbaceous cuttings. Most leaf cuttings root readily, but the limitation to regeneration is adventitious bud and shoot development. Hence, cytokinins can be used to induce buds to form (Figure 10–18) (*29*). Methods of applying cytokinins are the same as those discussed for auxins later in the chapter.

Leaf-Bud Cuttings

A **leaf-bud cutting (single-eye** or **single-node cutting)** consists of a leaf blade, petiole, and a short piece of the stem with the attached axillary bud (Figure 11–17). They differ from leaf cuttings in that only adventitious roots need form. The axillary bud at the nodal area of the stem provides the new shoot. A number of plant species such as the black raspberry *(Rubus occidentalis)*, blackberry, boysenberry, lemon, camellia, maple, and rhododendron are readily started by leaf-bud cuttings, as well as many tropical shrubs and most herbaceous greenhouse plants usually started by stem cuttings.

Leaf-bud cuttings are particularly useful when propagating material is scarce, because they will produce at least twice as many new plants from the same amount of stock material as can be started by stem cuttings. Each node can be used as a cutting.

Treatment of the cut surfaces with auxin should stimulate root production. The cuttings are inserted in the rooting medium with the bud 1.3 to

FIGURE 11–14 Leaf cuttings of piggyback plant *(Tolmiea menziesi)*. The large parent leaf (underneath) is set in a moist rooting medium in a humid location. The new plants (arrows) arise at the junction of the leaf blade and petiole.

FIGURE 11–15 Leaf cuttings of African violet *(Saintpaulia). Left:* Each cutting consists of a leaf blade and petiole. *Right:* Leaf cuttings after adventitious bud, shoot, and root formation. The new plant that forms at the base of the petiole is then potted and finished with or without the original leaf.

FIGURE 11–16 Leaf cuttings of *Kalanchoe pinnata (Bryophyllum pinnata),* air plant. *Left:* New plants developing from foliar "embryos" in the notches at the margin of the leaf. *Right:* Leaves ready to lay flat on the rooting medium. They should be partially covered or pegged down to hold the leaf margin in close contact with the rooting medium.

2.5 cm (½ to 1 in.) below the surface. High humidity is essential, and bottom heat is desirable for rapid rooting.

Root Cuttings

Best results with **root cuttings** are likely to be attained if the root pieces are taken from young stock plants in late winter or early spring when the roots are well supplied with stored carbohydrates but before new growth starts. Taking the cuttings during the spring when the parent plant is rapidly making new shoot growth should be avoided. Root cuttings of the Oriental poppy *(Papaver orientale)* should be taken in midsummer, the dormant period for this species. Securing cutting material in quantities for root cuttings can be quite labor-intensive, unless it can be obtained by trimming roots from nursery plants as they are dug.

FIGURE 11–17 Leaf-bud or single-eye cutting used in the propagation of rhododendrons. *Left:* Cutting when made. *Center:* Root development after several weeks. *Right:* Appearance of rootball and new shoot after five months. (Courtesy H. T. Skinner.)

It is essential to maintain the correct polarity when planting root cuttings. To avoid planting them upside-down, the proximal end (nearest the crown of the plant) may be made with a straight cut and the distal end (away from the crown) with a slanting cut. The proximal end of the root piece should always be up. In planting, insert the cutting vertically so that the top is at about soil level. With many species, however, it is satisfactory to plant the cuttings horizontally 2.5 to 5 cm (1 to 2 in.) deep (Figure 11–18), avoiding the possibility of planting them upside-down.

Not all species should be propagated by root cuttings. If root cuttings are used to propagate chimeras of *Aralias* and *Pelargonium* with variegated foliage, the new plants produced lose their variegated form. Propagation by root cuttings is very simple, but the root size of the plant being propagated may determine the best procedure to follow.

Plants with Small, Delicate Roots

Root cuttings of plants with small, delicate roots should be started in propagation flats or cell packs in the greenhouse, hotbed, or heated polyhouse. The roots are cut into short lengths, 2.5 to 5 cm (1 to 2 in.) long, and scattered horizontally over the surface of the medium. They are covered with a layer of medium measuring 1 to 2 cm (½ in.) After watering, a polyethylene cover or a pane of glass should be placed over the flat to prevent drying until the plants are started. Figure 11–19 shows a rooting chamber with small root cuttings of geranium rooting in cells of propagation flats. The flats are set in a shaded place. After the plants become well formed, they can be transplanted to other flats or lined-out in nursery rows for further growth. See Table 11–2 for a list of selected species that can be propagated by root cuttings.

Plants with Somewhat Fleshy Roots

Cuttings of plants with fleshy roots are best started in a flat in the greenhouse or hotbed, e.g. the lilacs in Figure 11–19. The root pieces should be 5 to 7.5 cm (2 to 3 in.) long and planted vertically, observing correct polarity. New adventitious shoots should form rapidly, and as soon as the plants become well established with good root develop-

FIGURE 11–18 Propagation by root cuttings. *Left:* Horseradish *(Armoracia rusticana)* root pieces planted horizontally. *Right:* Apple *(Malus pumila)* root pieces set vertically. Adventitious buds arising from the root piece form the new shoot system.

FIGURE 11–19 A rooting chamber for propagating root cuttings of geranium and lilac *(top left)*. Root cuttings of geranium are maintained in the cells of propagation flats *(bottom left)*, and then finished off in pots. Rooted lilac cuttings *(right)* are transplanted to containers or lined-out in the field.

TABLE 11–2

Some species that can be propagated by root cuttings

Acanthopanax pentaphyllus (coralberry)	*Malus* spp. (apple, flowering crab apple)
Actinidia deliciosa (kiwifruit)	*Myrica pennsylvanica* (bayberry)
Aesculus parviflora (bottle-brush buckeye)	*Papaver orientale* (oriental poppy)
Ailanthus altissima (tree-of-heaven)	*Phlox* spp. (phlox)
Albizia julibrissin (silk tree)	*Plumbago* spp. (leadwort)
Anemone japonica (Japanese anemone)	*Populus alba* (white poplar)
Aralia spinosa (Devil's walking stick)	*Populus tremula* (European aspen)
Artocarpus altilis (breadfruit)	*Populus tremuloides* (quaking aspen)
Broussonetia papyrifera (paper mulberry)	*Prunus glandulosa* (dwarf flowering almond)
Campsis radicans (trumpet vine)	*Pyrus calleryana* (oriental pear)
Celastrus scandens (American bittersweet)	*Rhus copallina* (shining sumac)
Chaenomeles japonica (Japanese flowering quince)	*Rhus glabra* (smooth sumac)
Chaenomeles speciosa (flowering quince)	*Rhus typhina* (staghorn sumac)
Clerodendrum trichotomum (glory-bower)	*Robinia pseudoacacia* (black locust)
Comptonia peregrina (sweet fern)	*Robina hispida* (rose acacia)
Daphne genkwa (daphne)	*Rosa blanda* (rose)
Dicentra spp. (bleeding heart)	*Rosa nitida* (rose)
Eschscholzia californica (California poppy)	*Rosa virginiana* (rose)
Ficus carica (fig)	*Rubus* spp. (blackberry, raspberry)
Forsythia intermedia (forsythia)	*Sassafras albidum* (sassafras)
Geranium spp. (geranium)	*Sophora japonica* (Japanese pagoda tree)
Hypericum calycinum (St. Johnswort)	*Stokesia laevis* (Stokes aster)
Koelreuteria paniculata (golden-rain tree)	*Symphoricarpos hancockii* (coralberry)
Liriope spp. (liriope)	*Syringa vulgaris* (lilac)
Liquidambar styraciflua (American sweet gum)	*Ulmus carpinifolia* (smooth-leaved elm)

ment, they can be transplanted. These root pieces can also be stuck directly in containers and held in dormant storage in cool greenhouses during the winter season, then undergo a period of active spring growth followed by midsummer planting in the field.

Plants with Large Roots, Propagated Out-of-Doors

Large root cuttings are made 5 to 15 cm (2 to 6 in.) long. They are tied in bundles, care being used to keep the same ends together to avoid planting upside-down later. The cuttings are packed in boxes of damp sand, bark sawdust, or peat moss for about three weeks and held at about 4.5° C (40° F). After this they should be planted 5 to 7.5 cm (2 to 3 in.) apart in well-prepared nursery soil with the tops of the cuttings level with, or just below, the top of the soil.

Some deciduous shrubs grown from root pieces are converted to softwood summer cuttings by taking elongating shoots from the root pieces and rooting these softwood cuttings under mist. This conversion technique is reported to produce heavier, faster-growing plants and reduce production time by one year with *Aronia, Clethra, Comptonia, Euonymus, Spiraea,* and *Viburnum* species (*88*).

STOCK PLANTS: SOURCES OF CUTTING MATERIAL

In cutting propagation, the source of the cutting material is very important (*8*). The stock plants, from which the cutting material is obtained, should be:

- True-to-name and type
- Free of disease and insect pests
- In the proper physiological state so that cuttings taken from them are likely to root

Several sources are possible for obtaining cutting material:

- *From plants growing in the landscape in parks, around houses or buildings, or in the wild.* For nurseries, this can be a dangerous practice, since identification of the cultivar may be unknown.

- *Prunings from nursery plants as they are trimmed and shaped.* Many nurseries use prunings as the primary source of their cutting material. Sometimes, however, the trimming is not done at the optimum time to root the cuttings—and the unrooted cuttings must be stored. Most ornamental nurseries in the southern U.S. take cuttings from containerized plants, rather than from stock blocks. As a general rule of thumb, cuttings of easy-to-root species are taken during normal production pruning cycles. Conversely, softwood cuttings of more difficult-to-root species are taken during a brief window-of-time when rooting is optimum.
- *Stock plants can be specially maintained as a source of cutting material* (Figure 11–20). Although such plants may occupy valuable land space, this is probably the ideal source of cutting material. The history and identity of each stock plant can be determined accurately. To maintain high rooting potential, it much easier to use techniques such as hedging-back, mounding, stooling, banding (see p. 346) on stock plants than nonpermanently maintained container-produced plants. By culturally maintaining uniformity of growth in stock plants, the propagator ensures that evenly graded batches of cutting material are available during a given period. Consequently, the success rate and uniformity of rooting is that much greater.

  Field-grown stock plants are insurance for future propagules, particularly during unusually cold winters that might wipe out an outdoor container-grown crop.
- *It is becoming more common to use tissue-culture-produced liners as sources of stock plants* in the development of new cultivars and disease-indexed plants. Conventional macropropagation techniques can then be used after establishment of micropropagated stock plants (Figure 11–21).

Stock Plant Manipulation

Pruning and Girdling

Annual pruning is an important aspect of stock plant management in relation to (a) maintenance of juvenility to improve rooting, (b) plant shaping for easier and faster collection of propag-

FIGURE 11–20 Stock plants maintained as a source of cutting material. Stock plants of *Pittosporum tobira (top left),* English ivy *(top right), Liriope* spps. *(bottom left),* and poinsettia for greenhouse propagation *(bottom right).*

ules (Figures 10–24, 10–25, 11–20), (c) increased cutting production, (d) timing of flushes, and (e) reducing reproductive shoots (*97*).

Types of pruning (*97*) include:

Modified stooling. Plants are severely cut back to their base but not mounded with soil as with traditional stooling; this eliminates reproductive shoots and is beneficial for Hydrangea, Senecio.

Hard pruning. Stock plants are cut back to half their size annually; this avoids the rank growth that can occur from modified stooling and eliminates reproductive growth; used with *Forsythia, Weigela,* and heather. The advantage of hard-pruned hedges is **not** to increase the vigor of shoots or to mimic juvenile material, as has been long assumed. Thick cuttings from vigorous shoots may survive better than thinner shoots, but the thinner shoots root faster as long as propagation conditions are designed to rapidly drain away excess water, e.g., thinner cuttings are particularly prone to rotting *(61)*. The faster rooting of thinner hardwood cuttings suggests that the rooting potential among shoots in a hedge is more influenced by the relative positions of shoots than by their absolute position in terms of the distance between themselves or from the root system (see Figures 10–25, 10–26). In general, hedges should be grown to produce the maximum number of relatively thin-stemmed cuttings with a high leaf:stem ratio if the species is difficult-to-root, e.g., *Syringa vulgaris.* Conversely, large, fleshy-stemmed cuttings are perfectly acceptable for the easy-to-root *Forsythia* × *intermedia.*

Moderate pruning. Plants are cut back by one-third to one-half of the previous annual shoot each year and there is less die-back than with the foregoing two methods; used with *Viburnum,* and deciduous azaleas.

Light pruning. Light pruning implies tipping back or just normal removal of cuttings from the stock.

Hedging. The severity of pruning to maintain the hedge form is generally determined by the ease with which cuttings can be collected from the stock plant (e.g., *Berberis* and *Pyracantha* are heavily pruned, while *Eleagnus* and *Cornus* are lightly trimmed).

Double pruning. Spring pruning produces a flush of cuttings for summer softwood cut-

FIGURE 11–21 *Upper left:* Custom propagation greenhouse specializing in producing unrooted and rooted cuttings. Disease-indexed stock plants produced by tissue culture are maintained in isolation. Stock plants are replaced each one to two years. *Upper right:* Rooted poinsettia cuttings. *Lower left:* Unrooted geranium cuttings. *Lower right:* Cuttings are boxed and shipped to greenhouse producers to be finished off.

tings or semi-hardwood fall cuttings. In England, a second trimming in June delays the softwood cutting collection period until fall; cutting production is increased, but growth is weaker than during the normal summer flush.

With *Dracaena* stock plant production, **incisions** are made above axillary buds by cutting one-third to one-half through the cane (Figure 11–12). This breaks apical dominance and induces additional buds to develop in the plant without sacrificing the apical heads. This technique further promotes greater branching during field propagation (*22*).

Girdling shoots of stock plants prior to taking cuttings has been used successfully to root slash pine, sweetgum, sycamore, and 19- to 57-year-old water oak (*52, 53, 54*). The treatment consists of girdling shoots by removing 2.5 cm (1 in.) of bark, applying IBA talc, and wrapping the shoot with polyethylene film and aluminum. Once primordia become visible as small bumps in the callus, the cutting is removed from the stock plant and rooted under mist (Figure 11–22).

FIGURE 11–22 *Left:* Girdled shoot of slash pine *(Pinus caribaea)* with 2.5 cm (1 in.) of bark removed, treated with an IBA talc slurry, then wrapped with plastic film and aluminum foil. *Upper right:* Root primordia visible as small bumps on the callus. *Lower right:* The cutting is removed from the stock plant and rooted under mist. Only girdled cuttings will root when removed from the stock plant. (Courtesy R. C. Hare.)

TECHNIQUES OF SHADING, ETIOLATION, BANDING, AND BLANCHING *(74)*

- Field-grown or containerized dormant stock plants are ready for treatment when all chilling or rest requirements have been fulfilled and the buds begin to swell, usually early spring in the field or midwinter in a greenhouse.
- During the **shading** process, entire plants or several branches of a plant are then covered with an opaque material, usually black cloth or plastic; taking care to allow space for the new growth to extend. A wire or wooden frame may be used to support the covering.
- Cuts should be made in the covering material or corners left slightly open near the top of the structure to allow for ventilation. A small heat buildup under the structure is desirable, but enough ventilation should be supplied so that plants are not scorched. It is not necessary or desirable to exclude 100 percent of the light. Between 95 and 98 percent light exclusion is preferable.
- Initial growth is allowed to progress in the dark **(etiolation)** until new shoots are between 5 and 7 cm (2 to 3 in.) long, after which time the shade is gradually removed over the period of one week so as not to scorch the very tender shoots.
- On the first day of shade removal, **banding** is initiated with the placement of self-adhesive black bands at the base of each new shoot (the future cutting base). Black plastic electrical tape works well. To speed later removal, the end of the tape should be folded over to form a small loop. These bands keep the base of the shoot in an etiolated condition while the tops of the shoots are allowed to turn green in the light. (Banding mimics mound stooling and is applied to actively developing tissue of shoots above-ground.)
- The bands are approximately 2.5 cm wide × 2.5 cm (1 in.) long but can vary in length. Strips of Velcro are made up of two pieces—one a "woolly" side and the other a material with "hooks" or "nubs" that adhere to the wool when they are pressed together. The Velcro band "sandwiches" the shoot so that the hooks gently pierce the surface of the shoot when they are pressed together. In addition, auxin in a talc base is added to both parts of the Velcro before banding the shoot, thereby delivering a rooting stimulus to the shoot prior to its being made into a cutting. The hooks of the Velcro aid in wounding the stem so there is better penetration of the stem with auxin. Typically, 8,000 ppm IBA in talc is used. Both pieces of Velcro are dipped in the talc preparation, the excess powder is tapped off, and the Velcro band is then firmly pressed onto the shoot base.
- Velcro bands are generally left on the shoots for four weeks, although as short as 2 weeks and as long as 12 weeks of banding has also been successful.
- After 4 weeks, the cuttings are removed from the stock plant below (proximal to) the banded area. The bands are removed and the cutting bases are again treated with hormone before sticking them in the rooting medium. This second IBA treatment is generally a quick dip of 4,000 ppm in 50 percent ethanol.
- The area underneath the band is typically yellowish white and swollen, and occasionally root primordia can be seen already emerging.
- Cuttings are stripped of lower leaves and placed in a rooting medium consisting of peat and perlite (1:1 by volume). Bottom heat (25° C; 77° F) is supplied during the winter months. Mist is applied intermittently, beginning at seven seconds of mist every two minutes, becoming less frequent with time and amount of cloud cover. Fifty percent Saran shading is also applied over the rooting bench to keep down air temperatures around the cuttings. Greenhouse ambient temperature is maintained at 20° C (68° F), and daylength is regulated to 16 hours of daylight using 60-W incandescent bulbs spaced 1 meter (1.1 yd) apart and hung 1 meter (1.1 yd) above the bench.
- The cuttings are left in the rooting bench from 2 to 5 weeks for deciduous plants and up to 12 weeks for pines.
- After rooting, the cuttings are potted up, fertilized, and kept under long days (16 hours) to encourage growth. Depending on the time of year, they may be placed outside to continue growing during the late spring and summer (for winter-propagated plants) or they may be hardened-off and placed in a protective overwintering structure (for summer-propagated plants).

Etiolation, Shading, Blanching, and Banding

A modification of the traditional technique of etiolation and blanching, using Velcro adhesive fabric strips as the blanching material is shown in Figures 11–23 and 11–24 (*74*). This etiolation technique for softwood cutting propagation has improved the rooting success of a wide range of difficult-to-root woody species (Table 11–3) (*74, 76*). See Chapter 10 for the discussion of the physiological implications of these techniques. Rooting, subsequent bud-break, and growth of difficult-to-root apple cultivars have been promoted by a short period of banding light-grown shoots of stock plants with Velcro, before taking cuttings (*107*). Stock plant etiolation and stem banding are estimated to be 30 percent more expensive than traditional cutting propagation, but 50 percent less expensive than grafted plants, and 30 percent less or equal to that of micropropagated plants (*75*). The trade-off in using these light exclusion techniques is the greater range of plant species that can be propagated on their own roots, the increased success of plant establishment, and extending the production season, i.e., propagation occurs earlier by forcing containerized stock plants in the greenhouse, which can allow for additional top-growth of rooted liners—shortening production times and reducing costs.

FIGURE 11–23 Etiolation of *Kalmia latifolia* 'Obtusa red.' Black Velcro (arrows) with an auxin talc is wrapped around the base of the etiolated shoots which are gradually exposed to higher light irradience. After the shoots green up, they are removed and rooted under mist.

BLANCHING

1. All the previous steps on page 348 are followed except that stock plants are not initially covered, so new growth occurs in the light.
2. When the soft, green shoots are 5 to 7 cm long, banding with Velcro or electrical tape, as previously described, proceeds in exactly the same way as etiolation and banding.

ROOTING MEDIA

There is no universal or ideal rooting mix for cuttings. An appropriate propagation medium depends on the species, cutting type, season, propagation system (i.e., with fog, a high water-holding medium is less of a problem than with mist); the cost and availability of the medium components are other considerations.

The rooting medium has four functions:

1. To hold the cutting in place during the rooting period.
2. To provide moisture for the cutting.
3. To permit penetration to and exchange of air at the base of the cutting.
4. To create a dark or opaque environment by reducing light penetration to the cutting base.

Propagation media includes an **organic component:** peat, sphagnum moss, or softwood and hardwood barks. The **coarse mineral component** is used to increase the proportion of large, air-filled pores and drainage and includes perlite, vermiculite, expanded shale, coarse sand or grit, pumice, scoria, polystyrene, and rockwool (*6, 16, 87*). Rarely is mineral soil used as a propagation medium component, except in field propagation of hardwood cuttings. Most propagators use a combination of organic and mineral components (e.g., peat-perlite,

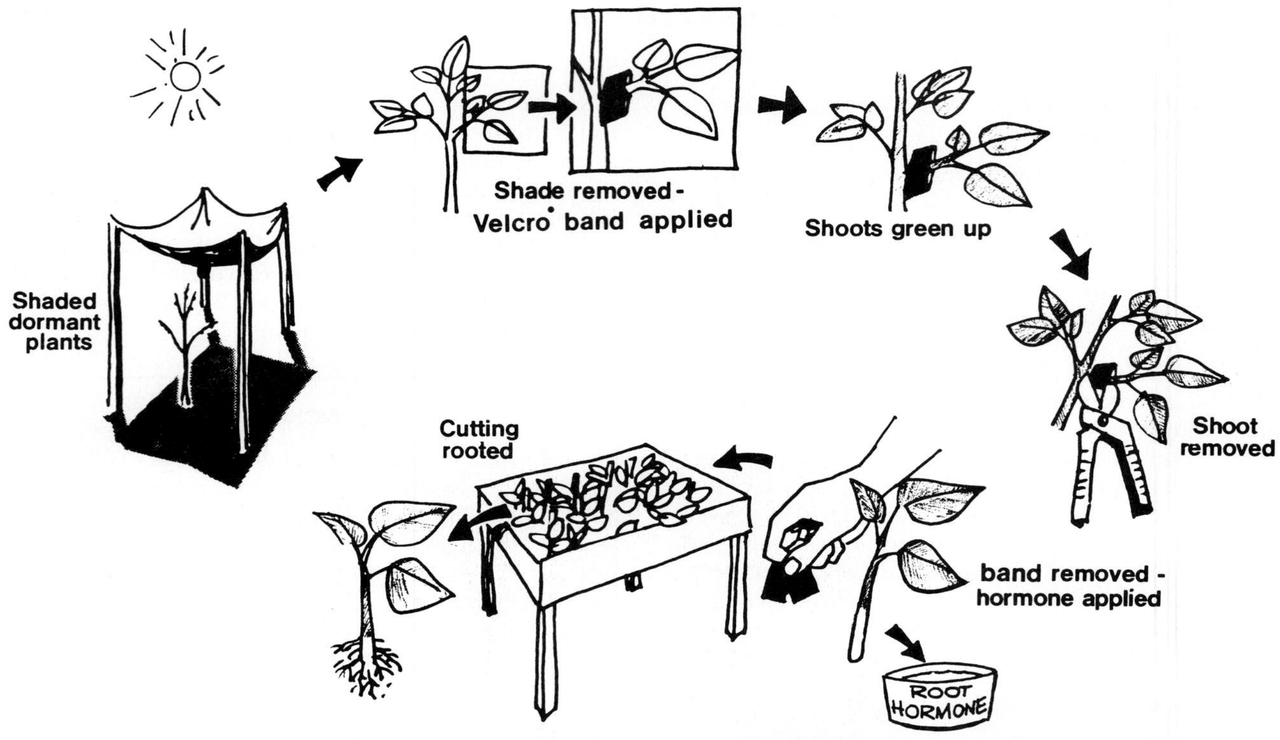

FIGURE 11–24 Scheme for etiolated softwood cutting propagation using Velcro fabric strips as the blanching material. (Courtesy B. K. Maynard and N. L. Bassuk

TABLE 11–3

Enhanced rooting of cuttings from stock plant etiolation, shading, and banding treatments

Treatment	*Species*	
Banding/blanching	*Acer platanoides*	*Platanus ocidentalis*
	Tilia cordata	*Rhododendron* cvs.
	Pinus elliottii	*Rubus idaeus*
Etiolation	*Artocarpus heterophyllus*	*Syringa vulgaris* cvs.
	Bryophyllum tubiflorum	*Malus sylvestris*
	Camphora officinarum	*Mangifera indica*
	Clematis spp.	*Persea americana*
	Corylus maxima	*Prunus domestica*
	Cotinus coggygria	*Rubus idaeus* 'Meeker'
	Polygonum baldschuanicum	*Tilia tomentosa*
Etiolation plus banding	*Acer spp.*	*Pinus strobus*
	Betula papyrifera	*Malus* × *domestica*
	Carpinus betulus	*Persea americana*
	Castanea mollissima	*Pistachia vera*
	Corylus americana	*Syringa vulgaris*
	Pinus spp.	*Carpinus betulus*
	Quercus spp.	*Tilia* spp.
	Hibiscus rosa-sinensis cvs.	
Shading	*Crassula argentea*	*Rhododendron spp.*
	Schefflera arboricola	*Rosa* spp.
	Hibiscus rosa-sinensis	*Euonymus japonicus*
	Picea sitchensis	

Source: Maynard and Bassuk *(74, 77).*

peat-expanded shale, peat-vermiculite-perlite, bark-haydite [clay and shale], peat-rockwool, etc.) (Figure 3–26) (*95*). Sometimes the mineral component is used alone (e.g., sand, rockwool, Oasis cubes, perlite) or in combination (e.g., vermiculite-perlite, sand-polystyrene). Sufficient *coarse* mineral component should be added to improve aeration.

A trend in U.S. nursery propagation is for partial replacement of expensive peat with fine bark. In container production, barks have generally replaced peat as the dominant media component. In Europe, Australia, and Israel, rockwool sheets and cubes are being used for direct sticking of cuttings (Figure 11–25). Rockwool can be handled efficiently (no containers or trays need be filled), and since cuttings are directly stuck into cubes or sheets, there is minimal disturbance of roots, which avoids transplant shock when rooted cuttings are shifted up to liner production stages (Figure 3–27).

There are a myriad of premixed commercial propagation media available, which can help reduce production steps, save time and labor, and assure

FIGURE 11–25 Various rooting media used by propagators. *Upper left:* Rockwool. *Upper right:* Polystyrene root cubes. *Center left:* Compressed peat pellet. *Center right:* Conventional cutting flat with peat-perlite mix. *Lower left:* Individualized small liner container for direct sticking. *Lower right:* Polystyrene root sheets or blocks.

better media standards [i.e., Promix BX (sphagnum peat, perlite, vermiculite), W.R. Grace Co.]. There is generally no advantage in adding a wetting agent (surfactant) to propagation media, and with some species a wetting agent may decrease rooting (*12*). However, some propagators will incorporate a wetting agent with peat-perlite-based propagation mixes.

An ideal propagation medium provides sufficient porosity to allow good aeration and has a high water-holding capacity, yet is well drained and free from pathogens. Pathogens can occur in peat (*Phythium, Penicillium,* etc.) and other organic components of media. A good commercial practice is to pasteurize with aerated steam. Some mineral components, such as vermiculite, perlite, and rockwool, are relatively sterile due to their high-temperature exposure during manufacturing, and need not be pretreated for pathogens. Integrated pest management (IPM) is used during rooting to control damping-off organisms *(Pythium, Phytophthora, Rhizoctonia Pestalotiopsis, Glomerella, Botrytis, Peronospora).* IPM includes the selective use of broad-spectrum fungicides on cuttings (see Chapter 3).

The key to successful propagation medium is **good water management.** Rarely can a rooting response be attributed to differences in aeration due to the physical properties of the various media (*110*). Apparently, gaseous diffusion proceeds relatively freely through the propagation media. Most of the aerobic requirements for rooting (*67*) are supplied by diffusion of oxygen through the aerial portion of the cutting to its base.

However, water films both within and around the base of the cutting can obstruct the free passage of oxygen to developing root initials. Thus, it is very difficult to pinpoint the relationship of rooting with physical characteristics such as volumetric air and water contents of the medium (*67*). During cool winter months and/or with a closed mist system, the medium should be sufficiently loose to allow adequate aeration even when water utilization is lower.

Chemical and physical standards of a propagation media used at a successful commercial nursery are listed in Table 11–4.

Propagation Unit Systems

Besides rockwool rooting cubes and blocks, and the standard plastic propagation flat, there is a variety of *propagation unit systems* to work with (*73*). These include Jiffy-7's (compressed peat blocks), synthetic foam media Rootcubes, Japanese paper pot containers with an expandable honeycomb form, bottomless polystyrene or plastic trays with wedge or cylindrical cells for air pruning of roots (Speedling, Gro-Plug, etc.), and various modifications of plastic flats (see Chapter 3).

Direct sticking or rooting of cuttings into small liner plastic containers for rooting, as opposed to sticking in conventional flats or rooting trays, is an important technique for utilizing personnel and materials more efficiently (*80*). Over 50 percent of cutting propagation costs are due to labor. By direct sticking, the production step of transplanting rooted cuttings and potential transplant shock due to a disturbed root system is avoided (Figure 3–27). The plant materials must be easy to root (greater than 80 percent rooting) to justify the additional propagation space required, but the labor savings

TABLE 11–4

Suggested chemical and physical standards for rooting medium

	Property	*Comments*
Chemical		
	pH	4.5-6.5; 5.5-6.5 preferred
	Buffer capacity	As high as possible
	Soluble salts	400-1,000 ppm (1 media: 2 water by volume)
	Cation exchange capacity	25 to 100 meg/liter
Physical		
	Bulk density	0.3-0.80 g/cm^3 (dry) or 0.60-1.15 g/cm^3 (wet)
	Air-filled porosity	15-40% by volume, ideally 20-25% range
	Water-holding capacity	20-60% volume after drainage
	Particle stability	Materials should resist decomposing quickly; decomposition can alter other media components

Source: Maronek, Studebaker, and Oberly (*72*).

and versatility of this system are substantial (Figure 11–26).

WOUNDING

Root production on stem cuttings can be promoted by wounding the base of the cutting. This has proved useful in a number of species, such as juniper, arborvitae, rhododendron, maple, magnolia, and holly (*120*). Wounds may be produced in cuttings of narrow-leaved evergreen species, such as arborvitae or Thuja, by **stripping** off the lower side branches of the cuttings (Figure 11–27). The benefits of stripping basal leaves is species-dependent; *Berberis* and *juniperus* cuttings benefit, whereas *Spiraea,* forsythia, and weigela do not (*71*). Stripping basal leaves of cuttings reduces the propagation bench space required for some species, allows the propagator more flexibility to work with different size propagules, may serve to improve the contact area between the cutting and media (see Chapter 10), and can potentially improve absorption of auxins.

A common method of wounding—making one to four vertical cuts with the tip of a sharp grafting knife down each side of the cutting for 2.5 to 5 cm (1 to 2 in.), penetrating through the bark and into the wood—may be enough (Figure 11–27). A more drastic wound is made with a razor blade device. This consists of four single-edged blades soldered together along their backs. Four wound cuts are then made simultaneously with this equipment. Wounding hardwood cuttings by splitting the stem base is an effective technique to allow auxin to reach the cambium and stimulate rooting (Figures 10–32, 10–33).

Larger cuttings, such as magnolias and rhododendrons, may be more effectively wounded by removing a thin slice of bark for about 2.5 cm (1 in.) from the base on two sides of the cutting, exposing the cambium but not cutting deeply into the wood (*120*). A simple carrot peeler works just as well for wounding rhododendron cuttings (Figure 11–27). For the greatest benefit, the cuttings should be treated after wounding with auxins, either in a talc or liquid quick-dip. The device shown in Figure 11–28 can be used to make rapid and uniform wounded cuttings.

One type of wounding to avoid is inadvertent crushing and damaging of basal cutting tissue with dull shears. It is from this basal tissue that both the movement of water from the propagation media must occur for cutting survival, as well as root initiation for regenerating new plants. Always use sharp,

FIGURE 11–26 Ornamental nursery propagation systems. More cuttings can be rooted per unit area in a conventional plastic root flat *(upper left),* but additional labor is needed in transplanting rooted cuttings into small liner pots, prior to producing a finished container or field-grown crop. Direct sticking (direct rooting) allows cuttings to be rooted directly into small liner pots *(upper and lower right),* or into large 3.8 liter (1-gal) containers *(lower left),* which saves labor and avoids transplant shock to the root system.

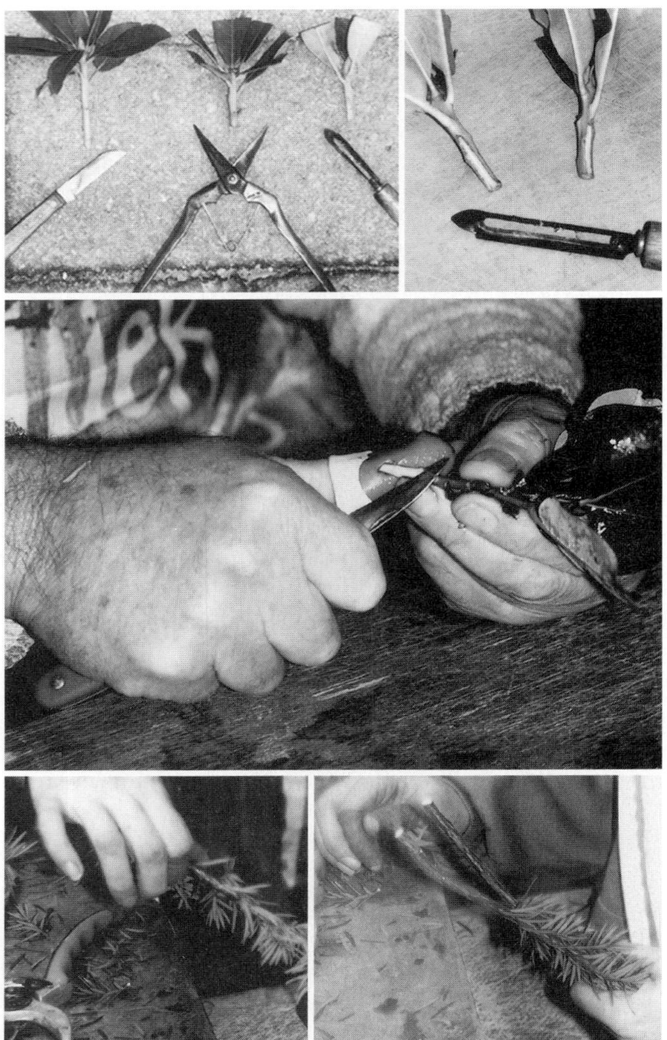

FIGURE 11–27 Wounding and stripping of cuttings. *Top left and right:* A carrot peeler is used for wounding rhododendron cuttings—part of the leaf surface area has been removed by the pruning shears. *Center:* Wounding a cutting with a knife—notice the thumb protection for the propagator. *Bottom left and right:* Preparing a *Thuja* cutting by stripping off the lower needles.

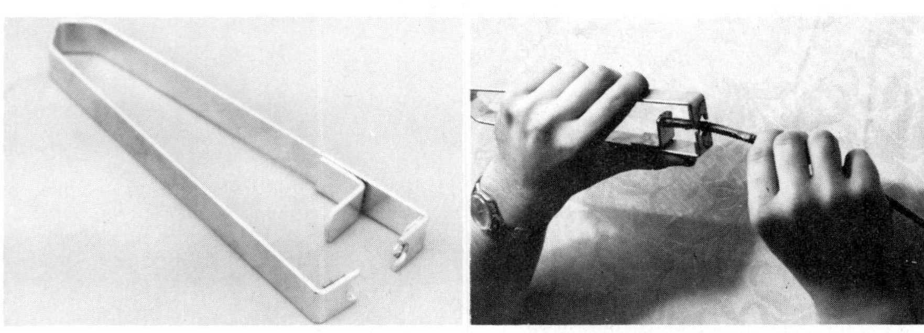

FIGURE 11–28 Tool designed for making wounding cuts in the base of cuttings to stimulate rooting. Four sharp prongs make the actual cuts as the cutting is pulled through the opening, as shown in the photo on the right.

periodically disinfected pruning shears, such as Felco No. 7 (Figure 11–33). A pruning shear system with tubing supplying disinfectant has been developed to reduce pathogen contamination when making cuttings (Figure 11–34).

TREATING CUTTINGS WITH AUXINS

The purpose of treating cuttings with auxins, which are plant growth regulators, is to increase the percentage of cuttings that form roots, to hasten root initiation, and to increase uniformity of rooting (Figure 11–29). Some difficult-to-root species will not respond to auxin treatment. Plants whose cuttings root easily may not justify the additional expense and effort of using these materials. Best use of auxins is with moderately difficult to difficult-to-root species. The value of auxins in propagation is well established. Although treatment of cuttings with auxins is useful in propagating plants, and can increase production efficiency time from propagule to rooted liner, the ultimate size and vigor of such treated plants is no greater than that obtained with untreated plants.

Rooting Chemicals, Formulations, and Carriers

The synthetic root-promoting chemicals that have been found most reliable in stimulating adventitious root production in cuttings are the auxins: **indolebutyric acid** (IBA) and **naphthaleneacetic acid** (NAA), although others can be used. IBA and NAA are often used in combination. IBA is the best auxin for general use because it is nontoxic to plants over a wide concentration range, and is effective in promoting rooting of a large number of plant species. IBA is a relatively stable compound, and the shelf-life of IBA products can be extended by darkness and refrigeration (Figure 11–30). NAA is quite stable both when mixed as a powder or a liquid. Generally, if a cutting does not respond to IBA, other root-promoting compounds will not compensate. IBA may be toxic to softwood cuttings of certain species, which leads to poor cutting regrowth and overwintering losses. These chemicals are available in commercial preparations, dispersed in talc or in concentrated liquid formulations that can be diluted with water (aqueous solution) to the proper strength (see Table 11–5).

FIGURE 11–29 Effect of indolebutyric acid at four concentrations on rooting of *Escallonia* leafy cuttings under mist.

TABLE 11–5

Partial list of commercial rooting compounds, sources, formulations, and ingredients

Trade name	*Source*	*Formulation*	*Ingredient*
Chryzopon	ACF Chemiefarma Maarssen, The Netherlands	Powder (talc)	0.1 to 8% IBA
C-mone	Coor Farm Supply Services, Inc., Smithfield, NC	Liquid (isopropyl alcohol)	1% and 2% IBA
Dip 'N Grow	Astoria-Pacific Inc. Clackamas, OR	Liquid (alcohol)	1% IBA + 0.5% NAA + boron
Hormex	Brooker Chemical Corp. North Hollywood, CA	Powder (talc) Liquid	Rooting Powder—0.1 to 4.5% IBA Hormex Concentrate—.013% IBA + 0.24% NAA + Vitamin B-1
Hormodin	E.C. Geiger, Inc. Harleysville, PA	Powder (talc)	0.1 and 0.8% IBA
Hormo-Root	Rockland Chemical Co. Newfoundland, NJ	Powder (talc)	1, 2, 3 or 4.5% IBA
Rhizopon	Hortus USA Corp. Inc. New York, NY	Powder and water-soluble tablet forms	0.5 to 1.0% IAA or 0.1 to 0.2% NAA or 0.5 to 8.0% IBA
Roots	Wilson Laboratories, Inc. Dundas, Ontario, Canada	Liquid	0.4% IBA + ethazol (fungicide)
Synergol	Silvaperl Products Ltd. Harrogate, North Yorkshire, Great Britain	Liquid	0.5% K-IBA + 0.5% K-NAA + fungicides and other additives
Woods Rooting Compound	Earth Science Products Corp., Wilsonville, OR	Liquid (ethanol)	1.03% IBA + 0.56% NAA

Note: The EPA registration number is on the finished formulation container. If no number is present the product has not been registered and/or is being sold illegally (*36*).

Until recently, it was possible for propagators to prepare their own solutions and talc formulations with technical-grade auxins. Technical-grade IBA has never undergone registration with the U.S. Environmental Protection Agency (EPA). It contains a trace amount of dioxin, to which some people are very sensitive. It is important to always wear gloves during propagation when handling formulations containing auxin and other chemicals.

In the U.S., nurserymen/propagators can only use *finished (end-use) formulations* containing auxin(s), which are registered with the EPA. Some of the EPA-approved liquid formulations include Dip N' Grow and Woods Rooting Compound, and commercial talc (powder) auxin formulations such as HormoRoot and Hormex (Table 11–5, Figure 11–30). Information is presented in Chapters 19, 20, and 21 for rooting with technical-grade auxins, since recommended auxin concentrations can still be legally made by diluting end-use formulation products. *Note the EPA registration number on the container.* If the number is not present, the product has not been registered or is not being sold legally *(36)*. See page 358 for examples of making recommended auxin concentrations with end-use formulation products for quick-dip applications.

FIGURE 11–30 Auxin formulations. *Top left:* Seradix rooting powder. *Top right:* Hormo Root powder and Woods liquid rooting compound. *Bottom left:* Dip 'n Grow liquid rooting compound. *Bottom right:* Refrigeration of liquid auxin formulations to extend their shelf life.

The **K + salt formulations** enable IBA and NAA to be dissolved in water. Otherwise, the **acid formulations** of these auxins need to be dissolved initially in alcohol (isopropyl, ethanol, methanol), acetone, or another solvent or carrier before water can be added. It is not advisable to use any carriers that have not been registered with the EPA. There may be some benefit in mixing auxins with carriers, such as polyethylene glycol (Carbowax, laboratory-grade PEG) or propylene glycol (potable water antifreeze)—particularly with extremely difficult-to-root species that require higher auxin concentrations and are sensitive to alcohol (*21, 20, 33, 35*). Windshield washer fluid (methanol based) is also an effective carrier of auxin (*21*). At auxin concentrations above 5,000 ppm there is less burning with propylene glycol than alcohol (*9*). The auxins NAA and IBA do not readily dissolve in 50 percent polyethylene glycol (readily available as antifreeze), so it is necessary to heat the solvent to 72° C/160° F, although upon cooling after mixing, auxins will not precipitate out (*9*).

Generally, ethyl alcohol (95 percent/190 proof ethanol or grain natural spirits that can be purchased at liquor stores and later diluted to a 50 percent concentration) causes less burning to plant tissue than isopropyl alcohol (rubbing alcohol). Some U.S. nurseries will use K-IBA (water-based solution)

Generally, auxin concentrations of 500 to 1,250 ppm are used to root the majority of *softwood* and *herbaceous cuttings;* 1,000 to 3,000 ppm with a maximum of 5,000 ppm for *semi-hardwood cuttings;* 1,000 to 3,000 ppm with a maximum of 10,000 ppm for *hardwood cuttings* (see Table 11–1). One Texas nursery that produces some 2,500 cultivars and species propagates over 70 percent of their cuttings with 3- to 5-second quick-dips in aqueous dilutions of Dip N'Grow at either 1,250 ppm IBA + 625 ppm NAA (i.e., 1 Dip n' Grow:7 water, v/v), or 2,000 ppm IBA + 1000 ppm NAA. No product endorsement is intended.

with cuttings during active growth stages (softwood, semi-hardwood) and use IBA with ethanol during dormant periods (hardwood) cuttings to avoid burning and dehydrating the tissue of the cuttings. See Table 11–5 for a listing of commercial rooting compounds containing K-salt formulations of auxin.

What is needed are not new solvents or carriers for auxins, but rather new, more effective auxin formulations. With some woody plant species, the **aryl esters** of IAA and IBA and the **aryl amid** of IBA are equal or more effective than the acid formulation in promoting root initiation (*34, 104*). Phenyl indole-thiolobutyrate (P-ITB) has been approved by the EPA and is *as effective* as IBA in rooting a wide number of woody species (*34, 36*).

Methods of Application

Commercial Powder Preparations

Complete directions come with the commercial materials, together with a list of plants that are likely to respond to the particular preparation. Woody, difficult-to-root species should be treated with higher auxin concentrations, whereas tender, succulent, and easily rooted species should be treated with lower-strength formulations. Fresh cuts can be made at the base of the cuttings shortly before they are dipped into the powder. The operation is faster if a bundle of cuttings is dipped at once rather than dipping individual cuttings. The inner cuttings in the bundle need to receive as much powder as those on the outside. The powder adhering to the cuttings after they are lightly tapped is sufficient. It may be beneficial to pre-wet cutting bases with water so that powder adheres better. Enhanced rooting has occurred by *predipping cuttings* in 50 percent aqueous solutions of acetone, ethanol, or methanol prior to applying IBA talc (*60*). Some propagators have used *combined treatments of auxin quick-dips* (solution concentrate) followed by *auxin mixed in talc* applied to cutting bases.

It is advisable in using powder and liquid preparations to place a small portion of the stock material into a temporary container, sufficient for the work at hand, and discard any remaining portion after use, rather than dipping the cuttings into the entire prepared chemical stock. This prevents early deterioration because of contamination with moisture, fungi, or bacteria.

Talc preparations have the advantage of being easy to use (Figure 11–31). However, uniform root-

PREPARING QUICK-DIPS WITH TECHNICAL GRADE IBA OR NAA IN 50 PERCENT ALCOHOL FOR EXPERIMENTAL ROOTING TRIALS

Concentration: 500 to 10,000 ppm
Duration of basal dip: 3 to 5 seconds

Final concentration	*Auxin (per liter of solution)*	
(ppm)	*(mg)*	*(g)*
500	500	0.5
1,000	1,000	1.0
5,000	5,000	5.0
10,000	10,000	10.0

- To make a **10,000 ppm** (10,000 mg/liter or **1 percent stock solution)** of auxin, dissolve 10 g of auxin in 15 to 20 ml of alcohol (ethyl, isopropyl, or methyl), then top to 1,000 ml (1 liter) with 50 percent alcohol. To make 1 liter (1,000 ml) of a **1,000 ppm** auxin solution from the **1 percent stock solution,** add 100 ml of the 1 percent stock solution and 900 ml of 50 percent alcohol.
- With K-IBA or K-NAA follow the same procedures, except that water is used as the solvent and no alcohol is needed.
- Avoid precipitation problems with auxin solutions by using distilled or deionized water, not tap water.
- Label solutions and color-code different solution concentrations with food dyes, which can be purchased at supermarkets.

PREPARING AN IBA:NAA QUICK-DIP WITH AN EPA-APPROVED END-USE FORMULATION PRODUCT

Concentration: 500 to 10,000 ppm
Duration of basal dip: 3 to 5 seconds
Use an EPA-approved end-use formulation, e.g., Dip 'N Grow (1% IBA + 0.5% NAA)

Final concentration	*IBA (per liter of solution)*			*NAA (per liter of solution)*			*Stock concentrate and dilution*		
	(ppm)	(mg)	(g)	(ppm)	(mg)	(g)	Dip 'N Grow concentrate	50% alcohol	Final solution
250 IBA : 125 NAA	250	250	0.25	125	125	0.125	25 ml	975 ml	1,000 ml
500 IBA : 250 NAA	500	500	0.5	250	250	0.25	50 ml	950 ml	1,000 ml
1,000 IBA : 500 NAA	1,000	1,000	1.0	500	500	0.5	100 ml	900 ml	1,000 ml
5,000 IBA : 2,500 NAA	5,000	5,000	5.0	2,500	2,500	2.50	500 ml	500 ml	1,000 ml
10,000 IBA : 5,000 NAA	10,000	10,000	10.0	5,000	5,000	5.0	1,000 ml	0 ml	1,000 ml

The Dip 'N Grow is the stock solution (1% IBA = 10,000 ppm : 0.5% NAA = 5,000 ppm); To make 1 liter (1,000 ml) solution of 250 ppm IBA: 125 ppm NAA, use the formula:

$Conc_{D'NG} \times Vol_{D'NG} = Con_{sol} \times Vol_{sol}$

10,000 ppm IBA (D'NG concentrate) × Vol. = 250 ppm × 1000 ml;

25 ml of D'NG concentrate stock solution + 975 ml 50% alcohol = 250 ppm IBA: 125 ppm NAA in one liter.

ing may be difficult to obtain, due to variability in the amount of the talc adhering to the base of cuttings, the amount of moisture at the base of the cutting, the texture of the stem (i.e., coarse or smooth), and loss of the talc during insertion of the cutting into the propagation medium. Talc formulations are generally less effective than IBA in solution at comparable concentrations (*20*).

Dilute Solution Soaking Method

In an older procedure, the basal part—2.5 cm (1 in.)—of the cuttings is soaked in a dilute solution of the material for about 24 hours just before they are inserted into the rooting medium. The concentrations used vary from about 20 ppm for easily rooted species to about 200 ppm for the more difficult species. During the soaking period, the cuttings should be held at about 20° C (68° F), but not placed in the sun.

This is generally a slow, cumbersome technique and is not commercially popular. Equipment is needed for soaking cuttings, and with the long time duration, there is variability of results, with environmental changes occurring during the soaking period (*69*). There has been success with soaking basal portions of hard-to-root cuttings (*Prunus,* conifers, evergreen, and deciduous shrubs) for a maximum of four hours at 50 to 150 ppm IBA. The cuttings are then propagated under mist (*81*).

Quick-Dip (Concentrated Solution Dip)

In the quick-dip method, a concentrated solution varying from 500 to 10,000 ppm (0.05 to 1.0 percent) of auxin in aqueous solution or 50 percent alcohol is prepared, and the basal 0.5 to 1 cm (1/5 to 2/5 in.) of the cuttings are dipped in it for a short time (usually three to five seconds, sometimes longer). The cuttings are then inserted into the rooting medium. Cuttings are most efficiently dipped as a bundle, not one by one (Figures 11–31, 11–32). There is no absolute ideal dipping depth. But deeper dipping of the cutting base can compensate for a lower auxin concentration. There is less variation with a shallow dipping depth if the auxin concentration is adequate (*59*).

Many propagators prefer the quick-dip compared to a talc application because of the consistency of results and application ease. The base of the bundled cuttings can be dipped in solution, inadvertently removing talc from the cutting base while sticking cuttings in the medium is avoided (*13, 31, 60*). Greater rooting and more consistent rooting response have been reported with quick-dips than with talc, due to more uniform coverage and reduced environmental influence on chemical uptake (*32, 69*). In talc formulation, auxins (which have low solubility in the acid forms) must first go into solution *after* cuttings are stuck in the propaga-

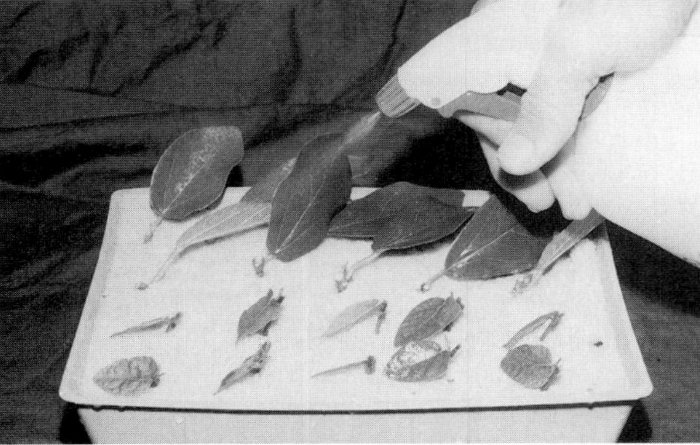

FIGURE 11–31 Application of auxin by *(top)* talc, *(center)* liquid concentrate quick-dip, or *(bottom)* spraying foliage or cutting bases.

tion media. Hence there is a time delay before auxins are absorbed through the cutting base. Only in a few limited cases with certain *Elaeagnus, Rhododendron,* and *Ilex* species does talc outperform the quick-dip method (*11*).

Quick-dip stock solutions must be tightly sealed when not in use because the evaporation of the alcohol will increase the auxin concentration. It is best to use only a portion of the material at a time, just sufficient for immediate needs, discarding it after use at the end of the day, rather than pouring it back into the stock solution. On extremely hot days, in open areas where evaporation is high, it is best to discard the old and add fresh solution several times during the day. Stock solutions that contain a high percentage of alcohol will retain their activity almost indefinitely if kept clean. No matter what the product, protect yourself and workers from undue exposure. Use rubber or plastic gloves when working with any of these rooting compounds.

Auxins used in excessive concentrations can inhibit bud development, causing yellowing and dropping of leaves, blackening of the stem, and eventual death of the cuttings. An effective, nontoxic concentration has been used if the basal portion of the stem shows some swelling, callusing, and profuse root production just above the base of the cutting.

To avoid rooting inconsistency (with <u>any</u> application method!), it is critical that propagators strictly adhere to established procedures.

It is also widely held that the most critical factor affecting the response of cuttings to hormone treatment is the concentration of auxin in the liquid or powder preparation—this is important. However, *it is the total dose of auxin received by those tissues capable of responding that determines rooting.* For quick-dip preparations, factors such as duration and depth of dipping, and the position in which the cutting is dried, affect the amount of solution taken in through the cut end of the stem. Both are as important as the auxin concentration when treating cuttings (*59*). When using powder preparations, important factors are those which affect the transfer of auxin from the powder carrier through the epidermis as well as via the cut end of the stem. These may include whether or not the cutting is predipped in organic solvent, the amount of powder which adheres to the cuttings, and whether the powder is retained or lost at planting. *Systematically following standardized procedures of a technique (method of application) is as important as the concentration of auxin applied. Propagators must set up simple procedures that will improve overall survival and rooting of cuttings, and increase the uniformity of response* (*61*).

FIGURE 11–32 Innovations in propagation. *Top left:* Rather than counting individual cuttings to be placed in a bundle of 50, the average weight of the total cuttings in bundles is measured and the number of cuttings estimated. *Top right and bottom left:* Bundles of cuttings are placed in a large bin that is flooded with preformulated auxin concentrations for a given time. The vacuum (below the bin) drains the tray and the auxin is recycled for other cuttings; this takes the guesswork out of what constitutes a 1- , 3- , or 5-second quick-dip. The auxin solution is discarded at the end of the day. *Bottom right:* Auxin preparations are stained with food dye to denote different concentrations and stored in color-coded containers.

Alternative Auxin Application Methods

Rather than quick-dipping, some propagators will **spray auxins to the point of runoff** over the bases of cuttings or on the foliage prior to sticking (Figure 11–31). Cuttings can also be stuck in trays, and the auxin solution sprayed on the leaves and stems until beads of liquid drop down into the rooting medium. (*65*). A 50 to 250 ppm IBA spray solution is used for chrysanthemum, begonia, dieffenbachia, heath, and hibiscus (*65*).

Excellent rooting has occurred with the **total immersion of whole cuttings** of herbaceous cuttings of plumbago, ivy, clematis, delphinium, *Ficus*, etc. for a few seconds—at 50–250 ppm IBA. Rooting was enhanced when *Berberis, Cotoneaster, Lavandula, Prunus, Pyracantha* and *Viburnum* cuttings were totally immersed for two minutes at 1,000 ppm—compared to powder formulations. Immersing the entire cuttings into the concentrated solution dip containing a wetting agent (surfactant) has been more effective in promoting rooting in some cases, than just dipping the base alone (*112*). There is an initial retardation of shoot growth, but this does not seem to be a disadvantage. Pretreatment of stock plants with foliar application of auxins prior to removing cuttings has also been used to promote rooting (*89*).

PREVENTIVE DISEASE CONTROL

Disease-free stock plants. As part of a preventive disease program, cuttings should be harvested from disease-free stock plants under non-stress conditions. It is best to collect turgid cuttings early in the day to assure optimum water conditions. It is important that the pruning shears used to collect cuttings be disinfected periodically (Figures 11–33, 11–34). Physan 20 (benzyl chloride), isopropyl alcohol, and monochloramine are better disinfectants than sodium hypochlorite (Clorox), which is quickly inactivated when it comes in contact with organic matter (stem material, media components). Monochloramine was found to be equal in efficacy to alcohol—less corrosive and costly, and with excellent stability under high organic contamination (*99*). See the referenced article on how a

FIGURE 11–33 Preventive disease control measures. *Upper left and right:* Collecting cuttings in buckets containing cups for periodically disinfecting knives and shears. *Upper center:* Bundling cuttings with rubber bands for processing. *Lower left:* Reducing stress on cuttings before sticking by maintaining them under low light, cool temperatures, and high relative humidity (note the mist line for syringing). *Lower center and right:* Soaking cuttings in a broad spectrum fungicide and bactericide prior to treating with rooting hormones and sticking.

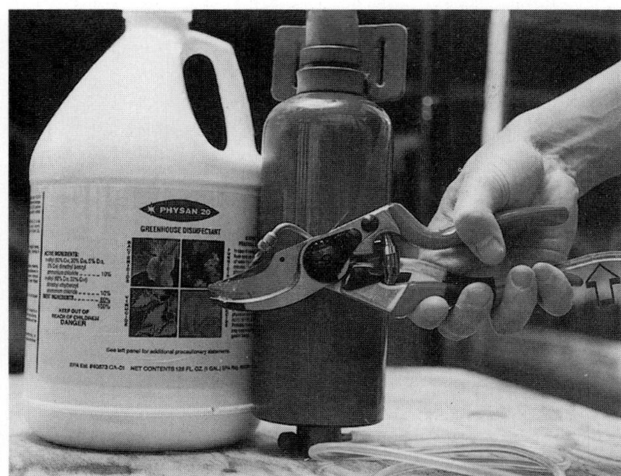

FIGURE 11–34 A pruning shear system with tubing (arrow) supplying disinfectant to the blades of the shear. This reduces pathogen contamination when making cuttings.

propagator can make monochloramine from local materials (*99*). Also, refer back to Chapter 3 for preventative measures discussed, including integrated pest management **(IPM),** current best management practices **(BMP),** and post-propagation care of rooted liner plants.

Once cuttings are collected, they should be selectively treated with broad-spectrum fungicidal dips prior to sticking, and/or chemical drenches, during propagation. Cuttings can be dipped in solutions of Argribrom, which is an oxidizing biocide. One Texas nursery dips its unrooted cuttings in a 25 ppm Agribrom bath. The nursery reports that it is more effective for disease control than dipping cuttings in baths containing fungicides for control of damping-off organisms, and agricultural streptomycin for bacterial control (Figure 11–33). The cutting bases are allowed to dry (keeping the leaves wet) and then treated with auxin. Benlate, which was the most widely used systemic fungicide for propagation, is no longer commercially available, and future trends are for less chemicals labeled for propagation use. Methyl bromide—used in fumigating and sterilizing propagation beds and media for very effective control of diseases, insects, nematodes, and weeds—will be completely fazed out in the U.S. by the year 2001 (*4*). Some possible Benlate substitutions in propagation include rotating the use of the fungicides Topsin (Atochem N.A.), Domain (Sierra-Grace), and Cleary 3396 (W.A. Cleary) (*26*). Always follow directions and conduct small tests to check for phytotoxicity, and protect yourself and workers from chemical exposure. See the discussion of fungicides in Chapter 3.

The use of **biocontrol agents** (beneficial bacteria, actinomycetes, mycorrhizal fungi, and other beneficial fungi living and functioning on or near roots in the rhizosphere soil) to control pathogens and enhance rooting in propagation is still in its infancy (*66, 82*). The beneficial fungus, *Gliocladium virens* (SoilGard 12G, W.R. Grace & Co.), may be an alternative to Benlate. It has been cleared by the EPA for biological control of *Rhizoctonia colani* and *Phythium ultimum,* which are two of the principal pathogens causing damping-off diseases. With hardwood cutting propagation of roses and cherry, crown gall *(Agrobacterium tumefaciens)* is controlled by dipping cutting bases in a special *Agrobacterium* isolate that is antagonistic to the virulent form.

Beneficial Microbes for Enhancing Rooting and Pathogen Control

As plants have evolved, so have rhizosphere organisms, some of which show great promise for propagation systems. Although industry still relies on chemical application of auxins to stimulate rooting of cuttings and application of pesticides to control pathogens and pests during propagation, utilizing beneficial microbes are novel approaches to reduce chemical treatments, control soil pathogens (*66*), and enhance rooting and cutting survival (*83*). Until recently it was thought that mycorrhizal fungi enhanced root development and cutting survival <u>after</u> colonization of adventitious roots (*116*), but new evidence indicates that mycorrhizae can enhance root initiation, prior to root colonization (*24, 37, 103*). *Trichoderma harzianum,* a fungus that controls soil-born pathogens, enhanced root and shoot growth of chrysanthemum cuttings during propagation, possibly by the production of growth-regulating substances or by chemically antagonizing or competing with pathogens. (See Chapter 3 for the discussion of biocontrol agents and IPM methods, including cultural controls.)

ENVIRONMENTAL CONDITIONS FOR ROOTING LEAFY CUTTINGS

For successful rooting of leafy cuttings, some essential environmental requirements are:

- Rooting media temperature 18 to 25° C (65 to 77° F) for temperate species and 7° C (12° F) higher for most tropical species

- Atmosphere conducive to low water loss and maintenance of turgor in leaves
- Ample, but not excessive light—100 W/m^2 with selected temperate woody species (exceptions are with species propagated under full sun irradiance in outdoor mist beds)
- Clean, moist, well-aerated, well-drained rooting medium

A wide range of equipment is satisfactory for providing these conditions—ranging from a simple glass jar or polyethylene cover placed over a few cuttings stuck in sand, easy-to-root cuttings placed in polyethylene rolls filled with media (Figure 11–35), or to very elaborate greenhouse propagation facilities having raised benches, automatic mist and fog systems—with computerized environmental control of relative humidity, temperature, photoperiod, light irradiance, and CO_2 enrichment (Figure 3–3) (*85*).

FIGURE 11–35 Polyethylene plastic sheeting can be used for starting cuttings of easily rooted species. The basal ends of the cuttings are inserted in damp sphagnum or peat moss and rolled in the polyethylene as shown here. The roll of cuttings should then be set upright in a cool, humid location for rooting.

Enclosure Systems (Closed-Case Propagation)

Rooting of cuttings can be done with simple **enclosure systems (closed-case propagation)** outdoors with **low polyethylene tunnels** (sun tunnels), or **cold or hot frames** covered with glass or polyethylene. Enclosed systems are also used inside a greenhouse with **contact poly** systems, where thin 1- to 3-mil polyethylene sheets are laid in direct contact with watered-in cuttings on a raised bench or on propagation flats placed on the floor (Figure 11–36). **Indoor polytents,** which are nonmisted polyethylene

FIGURE 11–36 *Upper left and right:* Rooting cuttings in enclosed polyethylene tents under mist. The shade cloth shown in upper left corner can be readily pulled if light irradiance becomes too high. *Below left:* Nonmisted polytent. *Below right:* Nonmisted contact polyethylene sheet system. Note condensation on the underside of the poly sheet.

tents supported by wire or wooden frames, are another low-cost way to propagate (Figure 11–36). Nonmisted enclosures in a greenhouse can be used to propagate difficult-to-root species and have the advantage of avoiding nutrient leaching problems of mist propagation—yet afford greater environmental controls than outdoor propagation. See the discussion in Chapter 10.

With enclosed systems, a reduction in the water loss from leaves is accomplished by the resulting increase in relative humidity, but enclosure also tends to trap heat. Leaf tissue is not readily cooled, since there is minimal air movement and evaporative cooling. To help reduce the heat load, light irradiance reaching the enclosed poly system is regulated by shading, and the greenhouse temperature is controlled by fan and pad cooling. Another variation of the contact poly system is to use **rooting beds on the ground, out-of-doors** in full sun and lay **Microfoam** sheets [0.63-cm (1/4 in.)] directly on cuttings in the fall, covering them with white 4-mil co-polymer film sealed to the ground by gravel or pieces of pipe (*49*). With propagation in temperate climates, the ideal cycle may be to root cuttings under contact polyethylene film in the fall and winter, and utilize intermittent mist during the spring and summer.

Intermittent Mist

Intermittent mist systems are widely used and have given propagators great flexibility in rooting softwood, semi-hardwood, hardwood, and herbaceous cuttings. Such sprays provide a film of water over the cuttings and media. A most important function of the film of water on the leaf surface is to intercept the irradiation of the light so that water is *evaporated from the leaf surface* rather than internal water from leaf tissue. Intermittent mist controls water loss from cuttings by reducing both leaf and surrounding air temperature via evaporative cooling, and raising relative humidity. To counteract the lower media temperatures caused by mist, bottom heat is frequently used in outdoor and indoor rooting structures (Figure 11–37).

Open mist systems are used in both outdoor propagation in cold frames, polyethylene tunnels, and lath and shade houses, and under full sun (Figures 11–26, 11–37, and 11–38). The open mist system is also used in glass and poly-covered greenhouses and set up on the floor area, or on, or above, the propagation bench (Figure 11–38). A very short duration (usually 3 to 15 seconds) should be used for misting. Unless mist actually wets the leaves, rooting is likely to be unsatisfactory. Besides using fixed risers containing mist nozzles (Figure 11–39), mechanized **traveling boom systems** are used to deliver mist to cuttings (Figure 3–3).

Enclosed mist systems are covered polyethylene structures inside greenhouses to reduce the fluctuation in ambient humidity and ensure more uniform coverage of mist, since air currents that disturb mist patterns are avoided (Figure 11–36). This system has been very effective in propagating difficult-to-root species, softwood cuttings of large-leaved species (e.g., *Corylus maxima*), and broad-leaved evergreens; it is not effective for conifers. The enclosed mist system has fewer disease problems than the open mist system, since there is less mist required, less media saturation, and fewer leaching problems (*68*).

Mist Nozzles

The choice of mist nozzles is based on: (a) cost, (b) maintenance, (c) convenience in operation, (d) availability from suppliers, (e) size of mist droplet [ideally, 50 to 100 μm (0.002 to 0.004 in.)], (f) amount of water used (fine orifice mist nozzles use less water but clog-up more readily), and (g) mist pattern (sufficient coverage but avoiding overwetting media).

The two main types of nozzles are: (a) **pressure jet** or **whirl-type nozzle** and (b) the **deflection** or **anvil nozzle.**

In the whirl-type, water is forced under pressure through small grooves set on angles to each other, which produces a mist when water exits the orifice. There are improved designs that operate under lower water pressures, which are nondripping and self-cleaning. Many of these nozzles have a low water output of 9 to 20 liters (2 to 5 gallons) per hour. Pressure jet nozzles have curved internal grooves, and when pressurized water is forced through the grooves, the impact at the orifice of the nozzle breaks up the water flow into mist. The Spray Systems Parasol is one such nozzle used in U.S. nurseries and has with a larger, more maintenance-free orifice. Figure 11–39.

The deflection nozzle develops a mist when pressurized water passes through the orifice and strikes a flat surface or anvil. The larger aperture used in this type reduces clogging but uses more water. Again, there are many variations among the two principal types in orifice size and water efficiency. There are also excellent hard-plastic nozzles that are less expensive and more durable than metal ones (Figure 11–39) (*106*).

FIGURE 11–37 *Upper left and right:* Ground bed heating in a glass propagation house with hot water forced through solar heating panels. *Lower left:* Outdoor propagation facilities relying on bottom heat by circulating hot water through PVC pipe embedded in scoria (note the partially exposed PVC pipe). *Lower right:* Outdoor hot-water-heated concrete bed.

Mist Controls

Intermittent mist during the daylight hours, which supplies water at intervals frequent enough to keep a film of water on the leaves but no more, gives better results than continuous mist. Since it would be impractical to turn the mist on and off by hand at short intervals throughout the day, automatic control devices are necessary. Several types are available, all operating to control a solenoid (magnetic) valve in the water line to the nozzles.

In a mist installation, the cuttings will be damaged if the leaves are allowed to become dry for very long. Even ten minutes without water on a hot, sunny day can be disastrous. In setting up the control system to provide an intermittent mist, every precaution should be taken to guard against accidental failure of the mist applications. This includes the use of a "normally open" solenoid valve, that is, one constructed so that if the electric power becomes disconnected, the valve is open and water passes through it. Application of electricity closes the valve and shuts off the water. If an accidental power failure occurs or any failure in the electrical control mechanism takes place, the mist remains on continuously, and no desiccation damage to the cuttings results.

Timers

Electrically operated timer mechanisms are available that operate the mist as desired. A successful type uses two timers acting together in series—one turns the entire system on in the morning and off at night; the second, an interval timer, operates the system during the daylight hours to produce an intermittent mist—at any desired combination of

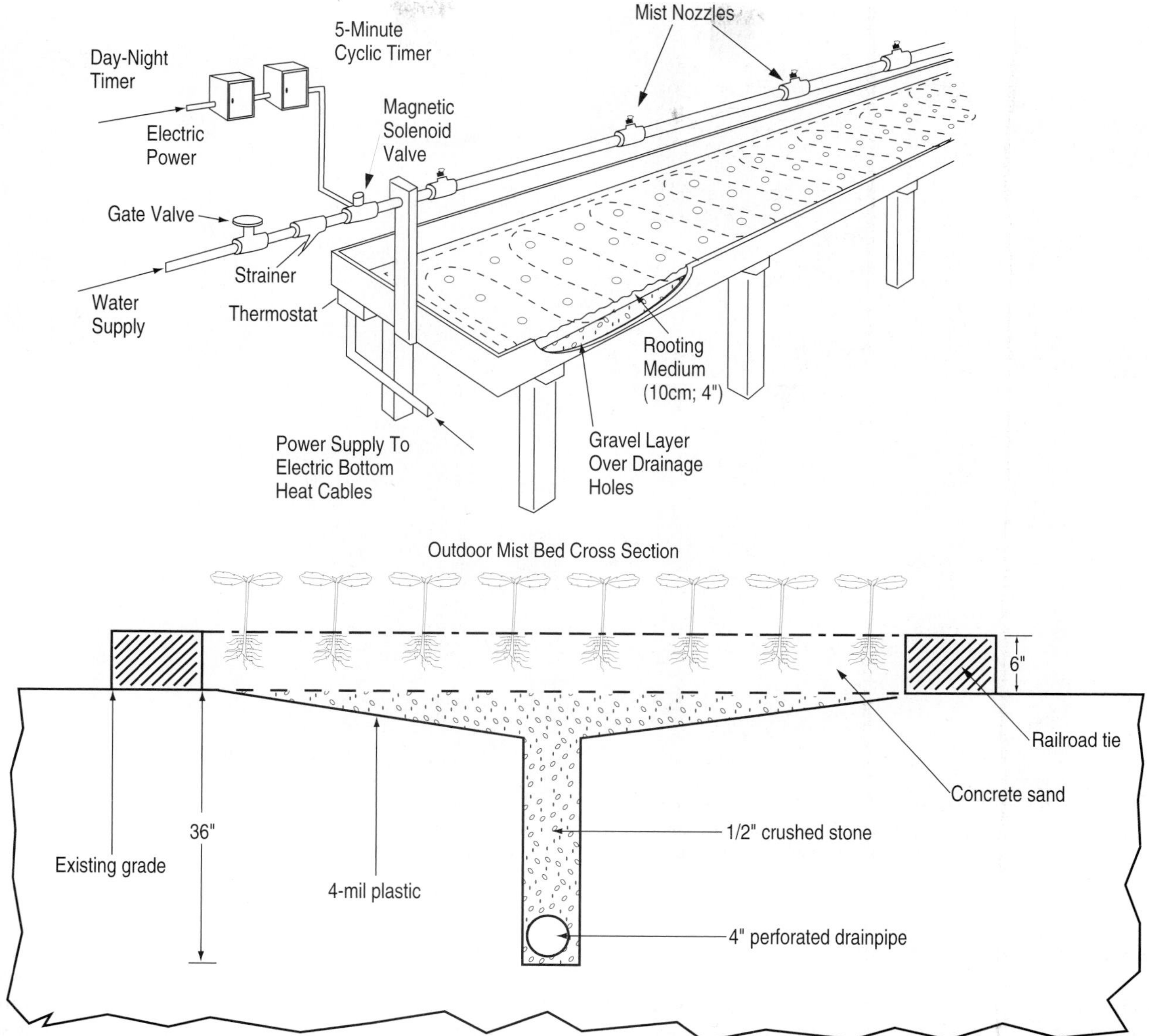

FIGURE 11–38 *Top:* Basic component parts of an open intermittent mist propagating installation with electric bottom-heat cable. One timer turns the mist system on in the morning and off at night. The second is a short interval timer to provide the intermittent mist cycles. *Bottom:* Cross section of an outdoor mist bed. Cuttings of Thuja, Taxus, and Juniperus cultivars, etc. are stuck in the concrete sand of the beds between the railroad ties (*90*).

timing intervals, such as six seconds **ON** and two minutes **OFF.** Time clocks for regulating the application of water are preferred by many propagators because they are easily installed, inexpensive, and dependable (*57*). Some electronic timers are very versatile and can operate many banks of mist nozzles in sequence (Figure 11–40). Timers have the disadvantage of not responding to daily fluctuation in light irradiance, cloud-cover, relative humidity, and temperature. Although mechanically and electronically reliable, the propagator must make daily adjustments with this equipment.

FIGURE 11–39 Deflection-type mist nozzles made of hard plastic *(top left and far left)* or metal *(top right)* and pressure jet or whirl-type nozzles *(top far right, bottom right).* A pneumatic pump tank *(bottom left)* is one method of maintaining sufficient water pressure so that mist (not coarse droplets) is produced instantaneously on demand.

The danger of electrical shock should always be kept in mind when installing and using any electrical control unit in a mist bed where considerable water is present. Low voltage systems are safer. The complete electrical installation should be done by a competent electrician.

Electric Leaf

In another type of control mechanism, the so-called "electric leaf," two carbon electrodes are set in a small circular ebonite block and placed under the mist along with the cuttings. The alternate wetting and drying of the terminals makes and breaks the low voltage circuit, which in turn, controls the solenoid valve. There are several variations in this type of control (*115*). The electric leaf is responsive to environmental changes, and theoretically maintains a film of water on the leaves of the cuttings at all times, automatically compensating for hourly and daily changes in leaf evaporation. The principal defect of the electric leaf is the gradual buildup of salt deposits between the terminals, which will conduct electricity. The leaf must therefore be cleaned periodically. The Aquamonitor overcomes salt deposit problems, since it has an air gap between the electrodes, which can be adjusted.

Screen Balance

Another type of control is based upon the weight of water. A small stainless-steel screen is attached to a lever that actuates a switch. When the

FIGURE 11-40 Mist can be controlled by *(top left)* producer-made systems comprised of a 24-hour time clock wired in series to 30- to 60-minute clocks which control banks of mist beds. More sophisticated electronic mist control systems are available *(top right)*. A light sensor *(bottom left)* wired to a computer records radiant energy. The programmed computer then determines when shade cloth will be drawn mechanically over cutting flats being propagated with a contact poly system. A screen balance system *(middle right)* more closely apporximates changing environmental conditions and leaf evaporation than time clock systems. A simple evaporimeter *(bottom right)* made of a pipette filled with water attached to filter paper and PVC tape can be used for measuring propagation environmental conditions and for determining the adequacy of the mist being applied. (Bottom right photo courtesy of K. Loach.)

mist is on, water collects on the screen until its weight trips the switch, shutting off the solenoid. When the water evaporates from the screen, it raises, closing the switch connection, which opens the solenoid, again turning on the mist. This type of control is best adapted to regions where considerable fluctuation in weather patterns may occur throughout the day, from warm and sunny to overcast, cool, and rainy; the unit compensates for changes in leaf evaporation. A Mist-A-Matic is a common screen balance unit (Figure 11–40). These units have greater maintenance requirements than time clocks, and are prone to salt deposits, algae growth, and wind currents which distort balance accuracy.

Photoelectric Cell

Controls based upon the relationship between light irradiance and transpiration rate are available. These contain a photoelectric cell which conducts current in proportion to light irradiance. In essence, they are controlled by light irradiance and convert light energy into electrical energy. The photoelectric cell activates a magnetic counter, or charges a condenser, so that after a certain period of time the solenoid valve is opened and the mist is applied. The higher the light irradiance, the more frequently the mist is applied. Between dusk and dawn, very little water is transpired. During cloudy days, less mist is used than during bright, sunny days. Such a control system would not be well suited for outdoor mist beds, where transpiration is affected by wind movement as well as by light irradiance. There may be advantages in using a time clock to override the system at night, since some photoelectric cells are prone to overwater during cloudy weather and at night. The MacPenny Solarmist utilizes such a photoelectric cell system.

Solar-activated mist-control devices can be quite suitable for greenhouse propagation, where wind velocities are negligible, rapid changes in temperature rarely occur, and light is the most important environmental parameter contributing to evapotranspiration of the cuttings. See Figure 3–14. The Weather Watcher solar-powered mist controller (Jeffery Electronics, New South Wales, Australia) makes sole use of solar energy to control mist systems. It also uses 70 percent less water than intermittent mist benches controlled by conventional time clocks. This conserves water and reduces potential leaching of nutrients from cuttings during rooting (*14*).

Computerized Controllers

There are computerized propagation controllers that can be programmed to monitor air, media, leaf cutting temperature, light irradiance and vapor pressure differences between air and leaf; the environmental information can then be coupled to frequency and duration of mist or fog needed. These systems are common in Holland, which has more than 2000 hectares (5,000 acres) of glasshouses with computer-controlled climates that regulate temperature, supplementary light, shade, and CO_2 enrichment. Likewise, larger U.S. and English propagators are using computer-controlled systems to record environmental conditions of temperature, light, and mist in order to **model** optimum conditions for rooting cuttings (*26*) Figures 3–3, 11–41).

Fog Systems

True fog is made by fog generators and atomizers that produce very fine water droplets from 2–40 microns (μm). By comparison, human hair is about 100 μm in diameter (Figure 11–42). High-pressure foggers produce *both* fog and micromist. In reality, the best atomizers used in micromist systems produce an array of droplets ranging in size from 2–100 μm in diameter (*79*). The volume median diameter of such a micromist is about 40 μm, so that half of the volume of water is in droplets larger than 40 μm. Fog remains airborne sufficiently long for evaporative cooling, and for an increased relative humidity of 93 to 100 percent to occur. Manufacturers may claim that their atomizers produce 10–20 μm droplets, but what is important is the *average* micron size of the water droplet (Figure 11–42). As a general rule of thumb in greenhouse propagation, all droplets smaller than 40 μm will stay airborne as fog, but droplets larger than 40 μm tend to settle and condense as water on leaf surfaces, especially in a high-humidity propagation environment (*79*). With true fog, water is suspended in the air as a vapor, whereas mist droplets (generally 50–100 μm) lose their suspension, fall onto the surface of leaves and media, and condense. This liquid from mist cools the leaf surface where it evaporates, but leaches nutrients from the leaf and can easily overwet the media. There are many advantages with fog systems, however, they are more costly to purchase, install, and maintain than conventional mist systems.

Fogging Equipment

Centripetal foggers (direct-pressure swirl jet atomizers) for ventilated high humidity consist of a self-contained unit incorporating a large fan that forces a stream of air through water ejected from a rapidly rotating nozzle (Figure 11–43). The water is atomized into an average 30+μm droplet, which is then forced into a cooling air stream through the propagation house by a fan attached to the rear of the unit. The water droplet size can be two times larger than those generated by impaction-pin atomizers. Centripetal foggers produce "wet fog," since they wet the leaf and humidify the air. It combines the advantages of both mist and dry fog (2–40 μm) and is very suitable for rooting large-leaved cuttings during the spring and summer. Larger mist droplets tend to fall out closer to the fan, making that area wetter, while smaller size mist is dispersed at greater distances. High-humidity propagation is successful only when incoming air passes through the fogger while operating for effective ventilation (*84*). Greenhouses must be shaded and good fan ventilation is essential. Best results are obtained with an oscillating humidifier that produces a large volume [10 to 50 gal (38 to 190 liters) per hour] of fog with

FIGURE 11–41 *Upper left and right:* "Low-tech" system of commercially rooting dracena cane cuttings in a polyethylene bag filled with sphagnum. *Lower left:* Some commercial nurseries utilize high-tech systems of plant modeling to determine optimal environmental conditions for rooting cuttings. *Lower right:* A data logger records environmental conditions for later evaluation.

20- to 30-μm droplets. One such unit is the Agritech (Broadway, N.C.) (Figure 11–43). Another variation is the Humidifan (Centre Hall, PA.), which has a single motor and is operated without a nozzle, eliminating potential blockage problems.

With **high-pressure fogging (impaction pin atomizers),** water is forced under high pressure (500 to 1,000 psi) through mist nozzles with very fine orifices. The water then hits an impact pin attached to the nozzle, which atomizes droplets to less than 20 μm in size, and subsequently forms a dense fog. Individual nozzles typically put out 5 to 8 liters (1 to 2 gal) per hour and are spaced 2 m (6 ft) apart. This is the "Cadillac" of fog systems. It is more expensive, but the most energy-efficient system for producing true fog droplets. One such system is produced by Mee (San Gabriel, Calif.) (Figure 11–43).

Pneumatic or ultrasonic humidifier nozzles (air atomizers) use compressed air and water. Water is disrupted by passage through a field of high-frequency sound waves generated by compressed air in a resonator located in front of the nozzle. In essence, water is being accelerated and atomized to fog. The nozzle orifice is much larger and less prone to blockage than are high-pressure fogging nozzles. Outputs range from 20 to 55 liters (5 to 14 gal) per hour. Air atomizing fog systems are more cost-effective for small propagation areas, but are not practical for large propagation ranges. Energy requirements for producing fog with this system are 20 times more than with impaction pin atomizers. One such unit is the Sonicore ultrasonic humidifier (Ultrasonics Ltd., West Yorkshire, England) (Figure 11–43).

Fogging Controllers

The key to successful fogging hinges on a good ventilation system to avoid heat buildup from stagnant warm air. Relative humidity sensing (Figure 3–14) is needed for fog systems. Unfortunately,

Propagation Water System	Fog	Micromist	Mist	Sprinklers (coarse mist; rain size drops)
Droplet size range	2-40 µm[1]	2-100 µm	50-100+µm	100+µm
Average droplet size	15 µm	40 µm	>50 µm	>100+µm

[1]Human hair has an average diameter of 100µm.

FIGURE 11–42 A comparison of fog, micromist, and mist systems used for propagation (*79*)

FIGURE 11–43 High-pressure fogging systems *(upper left and right)*, centripetal foggers for ventilated high humidity *(middle right)*, and pneumatic or ultrasonic humidifier nozzle systems *(bottom)* have many advantages over intermittent mist, albeit more expensive. (Bottom photo courtesy of K. Loach.)

accurate control of high humidity presents problems. Time clocks are not satisfactory for controlling the rate of fogging. Most fog controllers operate to maintain a fixed relative humidity, which is the simplest and least expensive option, albeit less than perfect (*68*).

PREPARING THE PROPAGATION BED, BENCH, ROOTING FLATS, AND CONTAINERS AND INSERTING THE CUTTINGS

The rooting frames or benches should preferably be raised or, if on the ground, equipped with drainage tile, to assure perfect drainage of excess water (Figures 3–6, 11–38). It has become popular to propagate in flats and liner containers placed on the ground of greenhouses, quonset-style houses, or outdoors on washed gravel or concrete base mist beds that have been sloped for good water drainage (Figures 3–4, 3–5, 3–10, 11–37). This avoids the high cost of propagation bench construction and utilizes space more efficiently.

The frames or flats should be deep enough so that about 10 cm (4 in.) of rooting medium can be used, and a cutting of average length—7.5 to 13 cm (3 to 5 in.)—can be inserted up to half its total length, with the end of the cutting still 2.5 cm (1 in.) or more above the bottom of the flat (Figure 11–44). The rooting medium should be watered thoroughly before the cuttings are inserted, which should be as soon as possible after they are prepared. It is very important that the cuttings be protected from drying at all stages during their preparation and insertion.

After a section of the rooting bench or flat or small liner container has been filled with cuttings, it should be well watered to settle the rooting medium around the cuttings.

PREVENTING OPERATION PROBLEMS WITH MIST PROPAGATION

Difficulties may arise in **operating mist systems.** Low water pressure can be a problem. Many propagators like to operate with a minimum mist-line water pressure of 34.5 Newtons/cm^2 (50 psi) to assure that fine mist (not coarse mist) is produced and that uniform coverage is maintained within the specification and the spacing of the nozzle used (*70*). A small, electrically powered rotary booster pump that builds up pressure into a pressurized holding tank assures that mist can be produced instantly on demand (Figure 11–39).

If there is much sand or debris in the water, filters or strainers should be installed in the supply line and cleaned periodically. Filters should always precede the solenoid valve in a mist line (Figures 3–28, 3–29, 11–38). Fog systems utilize very elaborate filtration systems (particularly high-pressure fog systems with ultra-fine nozzle orifices); chlorination is also included to remove algae. The biocide Agribrom can be injected into high-pressure fog systems (and can also be used in conventional intermittent mist systems) at 0.25 ppm to control algae, fungi, bacteria, viruses, and other microorganisms (*78*).

Algal growth often develops a gelatinous green coating on and around mist-propagation installations after an extended period of operation. This coating is not directly harmful to the cuttings but is unsightly and a hazard—because it is very slippery. This slimy material is principally blue-green (*Oscillatoria, Phormidium,* and *Arthrospira*) and green (*Stichococcus* and *Chlamydomonas*) algae (*23*). Allowing the water to remain off for a period each night so that the mist area can completely dry out will generally hold the algae in check. Algaecides such as Algimine, Algofen (dichlorophen), Algae-Go 36-20, Cyprex (dodine acetate), and Agribrom can also be used and are effective on mosses and liverworts. Diluted household vinegar and chlorox give good control of algae and moss along walkways (Figure 3–33). Algae will reduce aeration of propagation media and create cultural problems.

The **quality of water** used in mist can influence the rooting response (see also Chapter 3 for the discussion on water management and treatment systems). A complete water analysis should be done (by the local municipal water department if using municipality water; otherwise, by university soil labs or commercial nutrient and water analyses laboratories) to determine pH, total soluble salts, SAR (sodium absorption ratio), total carbonates, electrical conductivity, etc. High pH is commonly controlled by acidification (acid injection of sulfuric or phosphoric acid) into mist water. Water high in salts, such as sodium, or potassium carbonate, bicarbonates, or hydroxides, can be detrimental, especially when coupled with low calcium levels. Adding gypsum ($CaSO_4$) to the rooting media is one way to partially offset sodium problems. Sometimes a reverse osmosis system is used for removing salts from propagation mist water (Figure 3–28). Another option could be switching to a propagation system that requires less water on the foliage (e.g., a

FIGURE 11–44 *Top left, top right, and center:* Sticking cuttings in raised rooting frames. *Right and bottom left:* Different methods to line up and uniformly space cuttings using various types of templates or marking boards.

contact poly system or closed mist versus a traditional open mist with greater water demands).

It makes good sense to work with chlorinated, chloraminated, or brominated (Agribrom) mist water (Figures 3–33, 3–34), which reduces algae growth and controls some damping-off organisms (*78, 99, 100*). This is particularly appropriate, since many Australian, European and U.S. nursery propagators are being forced to capture and reuse runoff water. See the discussion in Chapter 3.

MANAGEMENT PRACTICES

Record keeping. There are many factors that go into the making of a good propagator: education, training, personal interest, a keen eye, and the ability to learn from success and failure. Good record keeping is essential in helping the propagator to hone skills and reduce failures (*39, 96*). With such a large number of different cultivars and

Daily Propagation Record Sheet

Date	Quantity Propagated	Cultivar	Description of Activity Performed	Number of Personnel	Total Personnel Hours

FIGURE 11–45 A sample record sheet that a propagator could also computerize to chart daily progress of materials propagated, units produced, activity performed, and total personnel hours—in conjunction with Figure 11–46. (Modified from (*39*).)

species, it is difficult to remember the details of propagating a particular crop. Both written records (computerized and printed copies) and pictures (color prints or Polaroid snapshots) are important. Both show new propagation personnel how to propagate plants, and what optimum results look like. Videotapes can be effective in training personnel.

It is equally important to *protect your records* with printed hard copies and back-up files on computer disks. One Texas nursery had an electrical fire which occurred at night and burned down their propagation facilities, destroying their computer and all propagation file records.

Propagation is the critical first key step in producing a finished crop (*42*). If problems occur in propagation, then scheduling and planning are for naught, and production, marketing, and sales are delayed. It is from good record keeping that the data and details can be used to develop lists of scheduling and planning in propagating and producing a finished crop. Examples of record keeping forms are shown in Figures 11–45, 11–46, and 11–47.

In order to manage a propagation facility, one must collect data on critical activities. Record keeping and evaluation is the first management step. Record keeping compels the propagator to monitor cuttings for subtle changes in callusing and rooting, from which optimum environmental conditions can be determined. The propagator recognizes what type of cuttings can be rapidly produced, and those which can not. The system also becomes an excellent method to track experimental results, leading to improved techniques and implementation of new crop propagation systems. Data from records can be used by propagation managers to make lists of plants to be propagated each month, schedule critical propagation facilities, assist in budgeting, line up supplies, schedule labor needs, and establish yearly, monthly, and daily propagation quotas (*39, 96*).

Costing variables. These need to be closely monitored (through record keeping). For projecting costs, many nurseries and greenhouse operations have some form of costing labor (*loaded labor rates*) that includes salaries, wages, benefits, maintenance costs, grounds upkeep, and depreciation (*92*). *Reasonable Expectancy* (RE) programs are used in propagation by timing propagation procedures, figuring crew averages, and developing rates per worker hour. This has led to **piece-work systems** where propagation workers receive cash incentives when their units of propagules produced exceed the daily established quotas (*86*). Once a costing basis has been established, the cost of a procedure can be determined (*92*).

The value of producing a propagation crop is generally based on the price of selling the crop to commercial customers, or in buying the crop if it were not propagated in-house. Labor is the single most expensive item in a propagation budget, so nurseries tend to base the majority of their costing on labor hours.

Addressing and changing standard operation procedures in propagation is done with the implementation of an **action plan/cost-benefit analysis** (*92*). The planning process used is called an action plan and states the goal and the step-wise procedure for producing a crop. Each step identifies the person responsible, due dates, and the final completion date. The action plan forces the individual doing the planning to outline the details. The action plan is also coupled with a cost-benefit analysis. Costs of propagating the crop can then be figured by the *Reasonable Expectancy* (RE) established and by *loaded labor rates.* The revenues can be figured and the fixed assets required can be calculated. All this allows key weaknesses in propagation to be identified. Action plans are thus written to improve each shortfall. Using this analysis, nurseries have realized that the increased labor in carrying out an additional propagation procedure may be more costly than the gain resulting from increased rooting percentages, i.e., the extra labor and time in stripping or wounding a cutting to get marginally higher rooting percentages may not be cost-effective. By forcing propagators to evaluate each crop, some producers may decide that it is more cost-effective *not* to propagate all crops. A more profitable alternative would

Record Card for Cutting Propagation

Cutting
Botanical Name:__________
Common Name:__________
Cultivar:__________
Date Propagated:__________
Date Rooted:__________
Cutting Type (i.e. semihardwood, terminal, basal, etc.):__________
Cutting Size (length or number of nodes):__________
Stock Plant Characteristics & Any Pretreatment (shading, banding, etc.):__________
Cutting Treatment (wounding, stripping cuttings, etc.):__________
Auxin(s): Formulation __________ Concentration __________ Method of Application __________
Rooting Medium:__________
Propagating System (mist, fog, contact poly sheets, etc.):__________
Environmental Requirements (bottom heat, temperature, special mist conditions, light conditions, etc. __
____
Flat, Bed or Container Size Planted & Location__________
No. of Cuttings per Flat__________ (or) No. of Direct-Stuck Liner Pots per Flat__________
Source of Cuttings__________
Propagator's Name and ID No. (to correspond with Label No. on propagation flat)__________
Date Rooted Cuttings Potted Up:__________No. of Liners__________
Area to be Placed, Customer, or Department Shipped to :__________
Results: Total Rooted________% Rooted________
Total Rooted Cuttings Shifted-up Liner Pots__________
% Rooted Cuttings Shifted-up to Liner Pots__________
Total Rooted Liner Pots Shifted-up to One-Gallon Containers__________
% Rooted Liner Pots Shifted-up to One-Gallon Containers__________
Observations & Comments:__________

FIGURE 11–46 A sample record card charting the propagation history in a production cycle of a plant cultivar from propagation through liner production. This could easily be computerized. (Modified from (*39, 92*).)

Genus	**Species**	**Cultivar**	**Patent #(if appropriate)**
Nandina	domestica	'Gulf Stream'	#5656

Company Catalog No.	**Propagation System**	**Date Propagated**
No. 4928	Direct stuck into 3P liner pots	14-Sept-97

No. Liner Pots (Cuttings) Per Flat; Location Cuttings Taken from Stock Plants

36 liner pots (cuttings) per flat; Cuttings harvested from Section D, Area 1, from 1-gallon plants

Propagator's ID # (to track who propagated the tray)

No. 18

FIGURE 11–47 Some sample propagation information to be printed on plastic labels and inserted in propagation flats or direct-stuck liner pots in liner trays.

be to buy rooted liners of selected species from outside contractors.

There is no one correct way for costing cuttings (*7*), just as there is no one correct way to propagate a species. But the importance of accurate record keeping and cost analysis helps the propagator determine: (1) which crops are profitable, (2) which crops need changes in propagation/production procedures, and (3) which crops should be dropped from production. The bottom line is to realize an acceptable profit on *all* plants produced (*92*).

Timing and scheduling. Commercial priorities determine scheduling in a nursery (*10*). When cuttings are stuck may be determined by competing heavy labor demands in the spring to help with shipping nursery product to retailers and mass merchandisers, availability of propagation space, and efficient use of personnel—rather than the optimum biological time to take cuttings. However, with some species, it is critical that cuttings be taken during a specific period if rooting is to occur. For example, *Ulmus parvifolia* cuttings must be taken six to eight weeks after bud-break. With some species, taking cuttings during the optimum time of the year is more crucial than using auxins. As a general rule of thumb, more difficult-to-root plants are stuck early in the propagation season. Easier-to-root cuttings have greater flexibility in propagation requirements; they root more quickly and tie up propagation space for shorter periods (*26, 96*).

Efficiently run nursery and greenhouse companies are constantly striving to reduce *plant wastage* (*scrapage*). That is, to reduce the percentage of plants that are propagated and later discarded because of poor quality and/or poor market demand. Plant wastage is caused by poor propagation and production techniques, scheduling problems, and poor marketing strategies. Reducing the *plant residency* period—so that the time from propagation to production to sale of *quality,* finished plants is condensed—is important in holding down production costs and maintaining profitability (*63*).

There have been some shifts in the nursery and greenhouse industries from production-led to market-led propagation systems (*111*). Problems with traditional *production-led propagation* include:

- The marketing strategy is constrained by the production process, i.e., plants for next year's spring sales are produced from late spring onwards and require additional space for overwintering.
- Cuttings for the year's production are taken at one time, requiring large amounts of propagation and production space in the greenhouse, which may be poorly utilized during other periods.
- The utilization of propagation and production facilities and systems is poor; bottlenecks occur at crucial stages, i.e., filling propagation flats with media, sticking cuttings, shifting rooted liners up to larger containers.
- Mass factory production techniques, rather than more careful individual selection, can lead to variable quality and high failure rates.

With a *market-led propagation system,* the producer negotiates with retail outlets the quantity of plants required at particular times during the sales season and can adjust the growing program accordingly to deliver fresh plants, in prime condition as required. In addition, by careful selection of plant species, and improved propagation techniques and facilities, it is possible to enhance the sales appeal of many species by producing better quality plants in flower (color!) for delivery to garden centers throughout the selling season (*111*). Figure 11–48 charts a market-driven propagation system for *Abelia* in England, where it is propagated seven different times during the year to meet market demands, instead of mass propagating in more traditional production systems during only one period in July.

> *In propagation, quality is the result of meticulous attention to detail* (Ben Davis).

CUTTING NUTRITION

It is essential that cuttings be taken from nutritionally healthy stock plants (*25, 109*). Most cuttings have sufficient tissue nutrients to allow root initiation to occur. Intermittent mist will rapidly deplete nutrients from cutting leaves. However, until a cutting initiates roots, its ability to absorb nutrients from propagation media is limited.

As a whole, mist application of nutrients has not been a feasible technique for maintaining cutting nutrition (*102, 123*). The algae formed create sanitation and media aeration problems, which inhibit rooting.

A commercial technique that works with many plant species is to supply a low level of slow-release

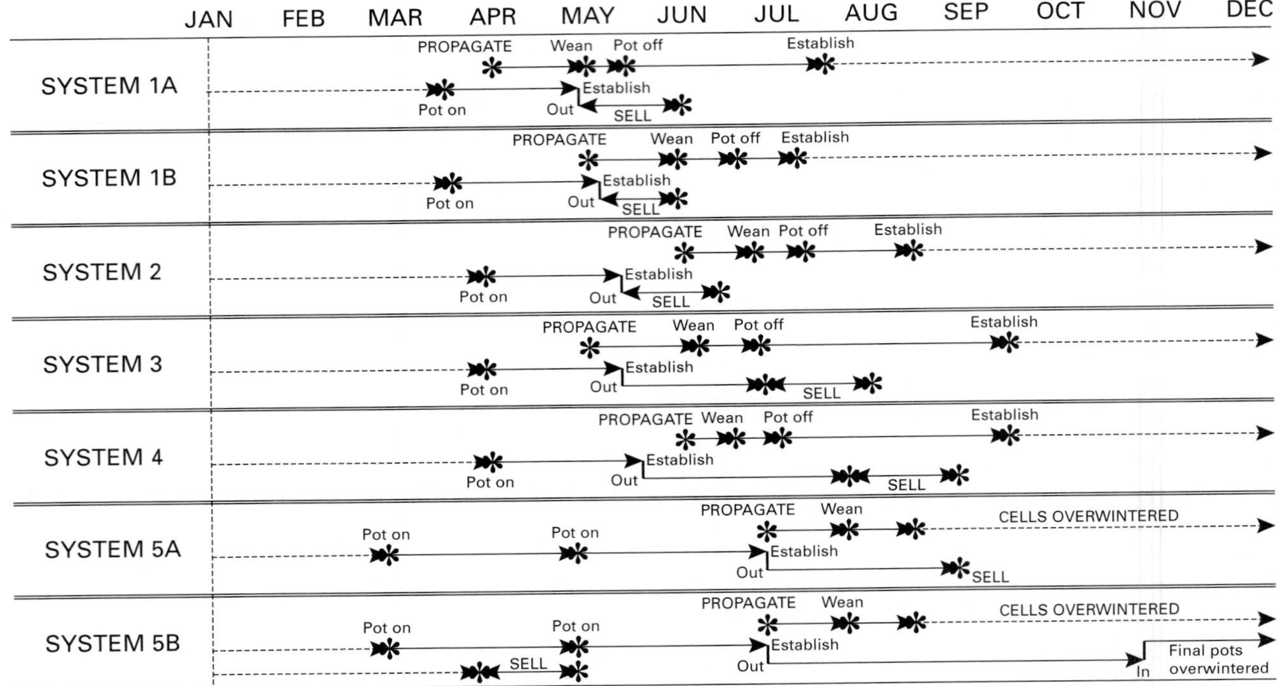

FIGURE 11–48 An example of a market-driven propagation system for *Abelia* × *grandiflora* in England that more efficiently utilizes propagation facilities, and delivers finished plants during designated selling periods. There are seven different propagation periods and seven targeted sales periods that extend the marketing period of the crop. *(111).*

fertilizer either top dressed (broadcast) on the media or preincorporated (e.g., Osmocote, Ficote, Nutricote, etc.). Top dressing Osmocote 18-6-12 at 6.8 to 13.8 g/m^2 (2.6 to 5.3 oz/ft^2) enhanced both root and shoot development of *Ligustrum japonicum* (too high a rate will delay the rooting of cuttings). These supplementary nutrients do not promote root initiation, but rather improve root development once root primordia formation and subsequent root elongation have occurred. Generally, slow-release formulations with a slow initial release rate (e.g., 12–14 months, Osmocote 17N-3.1P-8.1K) are better suited than short-release (3-month formulations) for top dressing on the surface of propagation media under mist systems (*1*).

Dilute liquid fertilizer can be applied to the propagation medium *after* roots have been initiated. Propagation turnover occurs more quickly and plant growth is maintained by producing rooted liners that are nutritionally balanced. However, with some species, such as *Thuja occidentalis,* there is no initial benefit of fertilizer application on root development and early growth of rooted liners (*19*).

CARE OF CUTTINGS DURING ROOTING

Hardwood stem cuttings or root cuttings started out-of-doors in the nursery require only the usual care given to other crop plants, such as adequate soil moisture, freedom from weed competition, and insect and disease control. Best results are obtained if the propagation facility is established in full sun, where shading and root competition from adjacent vegetation does not occur.

Environmental conditions. The temperature of leafy softwood, herbaceous, or semi-hardwood stem cuttings and leaf-bud or leaf cuttings should be carefully controlled throughout the rooting period. Glass-covered frames, particularly the tighter polyethylene-covered structures, exposed to a few hours of strong sunlight will build up excessively high and injurious temperatures. Such structures should always be shaded by saran screens, or whitewash on the poly or glass, thus reducing light irradiance.

If *bottom heat* is provided, thermometers or remote recording sensors should be inserted in the rooting medium to the level of the base of the cutting and checked at frequent intervals (Figures 11–37 and 11–41). Excessively high temperatures in the rooting medium, even for a short time, are likely to result in death of the cuttings.

It is important to *maintain humidity* as high as possible in rooting leafy cuttings to keep water loss from the leaves to a minimum. If cuttings become droughted for any length of time, the cuttings will not root, even though they are rehydrated and high-humidity conditions are reestablished. However, most nursery and greenhouse operations use intermittent mist, contact polyethylene sheets, or fog to overcome such problems.

Adequate drainage must be provided so that excess water can escape and not cause the rooting medium to become soggy and waterlogged. When peat or sphagnum moss is used as a component of the rooting medium, it is especially important to see that it does not become excessively wet.

Weed control. Weeds can be a serious problem during propagation of woody nursery crops. Weeds should be removed to prevent their seeding and competing with cuttings. Weed control in propagation begins with: (a) using weed-free, pasteurized, or gas-sterilized rooting media, (b) keeping the perimeters adjacent to the propagation area free of weeds, and (c) herbiciding the propagation area.

Herbicide use in propagation. Some preemergent herbicides, such as Ronstar and Rout, can be used for effective weed control in flats of unrooted cuttings of selected species of *Rhododendron, Euonymus, Ilex,* and *Cotoneaster* (*64*). However, granular dinitroaniline herbicides (Rout 3G, OH-2 3G, Ronstar 2G, Snapshot 2.5TG, SWGC 2.68G, etc.) can also suppress root initiation and development in stem cuttings (*46*). Sensitivity of cuttings to herbicides is dependent on the herbicide, application rate, and plant species. The rooting and subsequent liner growth of hibiscus was adversely affected by preemergence herbicides (applied after cuttings were stuck), while Asian jasmine was unaffected (*28*). Preemergence herbicides can effectively be used in propagation when applied to propagation flats, prior to sticking cuttings, but propagators should:

- choose the correct herbicide to control the particular weed species,
- follow label directions,
- *always* conduct trials to evaluate specific herbicides and the depth of sticking cuttings of individual species to be propagated,
- determine potential phytotoxicity, reduced rooting, and reduced growth of the cutting species prior to large-scale application of *any* herbicide (*46*).

Suspicions have persisted in the nursery industry that preemergence herbicides applied to container plants and stock plants cause reduced rooting of stem cuttings. Research shows that herbicides applied at normal-use rates generally have no effect on rooting of cuttings of most woody landscape species, even when stock plants are treated repeatedly over several years (*17*).

Sanitation and IPM. It is also necessary to maintain sanitary conditions in the propagating frame. Leaves that drop should be removed promptly, as should any obviously dead cuttings. Pathogens find ideal conditions in humid, closed propagating structures with low light irradiance. If not controlled, pathogens can destroy thousands of cuttings in a short time. Disease control is done on a preventive and scheduled weekly basis, selectively rotating fungicides. See the discussion on IPM in Chapter 3. Disease problems under mist or fog propagation conditions have generally not been serious. Frequent washing of leaves by aerated water can remove spores before they are able to germinate. The higher allowable light irradiance conditions with mist—a cutting must produce carbohydrates through photosynthesis in excess of its maintenance requirements for root development to proceed, and good ventilation and movement of air—also decrease the incidence of disease (*61*). Cuttings propagated with a fog system experience minimal stress conditions. The cuttings are not subjected to high irradiance, heat, or desiccation, and optimum water status is maintained without severe leaching. Hence, they are much less susceptible to disease problems.

Pests such as mites, aphids, and mealy bugs are controlled by miticides and insecticides, and by immediately roguing (discarding) infested cuttings and other IPM techniques.

Hardening-Off and Post-Propagation Care

Hardening-off. This is the process of gradually acclimating rooted cuttings from the high humidity of mist, fog, or a contact polysheet system in order to reduce humidity. This weaning process enables the rooted cutting to become more self-sufficient in absorbing nutrients and water through the root system, in photosynthesizing, and in conditioning new developing leaves and stems to better tolerate the stresses of lower relative humidity coupled with higher temperature and light irradiance.

Cuttings *deteriorate* when they are left under mist too long after they have rooted. This reduces root quality, causes premature leaf drop, and slows down the plant's momentum, which can delay the production period and produce a poorer quality plant. Cutting deterioration is one reason why flats or propagation liner containers of easy-to-root and difficult-to-root species are not mixed together in the same propagation system or house. One species would need to have the mist reduced, the plant hardened-off, and removed while the slower-rooting species is still rooting. Hardening-off encourages better root development from rooted cuttings. The *key to plant survival* is to reduce mist once roots start to develop, allow secondary root growth, but avoid excessive root development, which can be particularly detrimental to shoot development of leafless hardwood cuttings (*121*). Chemically root pruning with copper-treated propagation containers can also help root development (*105*).

There are several ways of successfully weaning rooted cuttings from the mist conditions:

- The cuttings may be left in place in the mist bed but with the duration of the misting periods gradually decreased, either by lessening the "on" periods and increasing the "off" periods or by leaving the misting intervals the same but gradually decreasing the time for which the mist is in operation each day.
- Another method is to root the cuttings in flats and move the flats after rooting to a lathhouse or cold frame, where they are "hardened-off" and then potted into containers as rooted liners. Cuttings may be left in the rooting medium until the dormant season, when they can be dug more safely, to be either lined out in the nursery row for further growth or potted and brought into the greenhouse. If the rooted cuttings are left in the rooting medium for a considerable time, it is advisable to water them at intervals with a nutrient solution, or top dress with a slow-release fertilizer.

  Some propagators **direct stick** cuttings in small containers set up in flats, or in modules preformed in plastic trays (Figure 11–26). Then, after rooting, the plants may be easily moved for transplanting without disturbing the roots. An alternative method is to root the cutting in a solid, block-type rooting medium, which, after rooting, permits transplanting without disturbing the roots. Such products are made from compressed peat and synthetic materials such as rockwool, polystryrene, etc. (Figure 11–25).
- Another method is to pot the cuttings immediately after rooting and hold them for a time in a cool, humid, shaded location (e.g., a fog chamber, closed frame, or greenhouse).
- With contact poly, slits can be made in the polyethylene with a knife; these are gradually increased in size and number over time to lower relative humidity, and increase light irradiance and ultraviolet exposure.

Avoiding overwintering problems of rooted liners. Poor winter survival of rooted cuttings can occur with certain deciduous woody plants, such as *Acer, Betula, Cornus, Hamamelis, Lilac, Magnolia, Prunus, Quercus, Rhododendron, Stewartia* and *Viburnum.* Newly rooted cuttings go dormant in the fall, but die either during the winter or after bud-break in the spring (*48, 79, 108*). With some species it is essential that after rooting a flush of growth occurs in mid- to late summer so that adequate carbohydrate reserves are produced. The rooted liner is hardened-off prior to winter. This assures winter survival. Late summer flushes of growth can be accomplished by extending the photoperiod (Figure 3–13), and manipulating fertilizer regimes (*25, 38, 48*). However, there must be sufficient time after growth flush(es) for rooted liners to harden-off before the onset of winter. In the following spring the rooted liners are then transplanted to containers or lined out in the field. See the discussion in Chapter 10.

Residual auxin applied to enhance rooting may suppress bud-break and growth flushes of rooted cuttings—and reduce winter survival. In rose cuttings, basally applied IBA increased rooting, but also increased ethylene synthesis, and subsequently inhibited bud-break of the rooted cuttings (*108*).

COLD STORAGE OF ROOTED AND UNROOTED CUTTINGS

Sometimes it may be convenient to collect cuttings when nursery plants are pruned and store them for later propagating. Most nurseries have refrigerated storage facilities (4 to 8° C; 40 to 47° F) for holding cuttings one to two days or longer before processing for propagation (Figure 11–49). Cuttings of *Rhododendron catawbiense* can be stored for 21 days in moist burlap bags at 2 to 21° C (36 to 70° F) with no reduction in rooting (*30*). Softwood cuttings of Kurume-type azaleas were taken in spring and held

COLD STORAGE SYSTEMS (*122*)

The *direct-cooled refrigeration system* (similar to the common kitchen refrigerator) is the most popular for nursery stock. Fans circulate air over cooling coils to directly cool the cuttings and the system's interior. This system is relatively stable as long as the humidity is not too high. Excessive humidity in the storage chamber will condense and freeze the coils, requiring defrosting and subsequent temperature fluctuations. It is generally not advisable to install misting systems or sprinkle water on the floor to raise humidity. This causes more condensation on the coils and increases potential mold development. Unrooted Taxus cutting wood is stored by a Michigan nursery in slotted pallet boxes to allow air circulation. Temperature is maintained at 1° to 2° C (34° to 36° F) and relative humidity is maintained with a humidifier (*98*). In general, to avoid desiccation and retain moisture of cuttings, it is best to pack cuttings in an insulation material and seal them in polyethylene bags.

The *jacketed cooler* is another popular system for storing plants, cutting wood, and understock. It is essentially a box within a box. Refrigerated air passes between the boxes, cooling the inner walls. It is a very stable cooling system, and it can maintain nearly 100 percent relative humidity. Hence, cuttings don't require packing in sealed poly bags to maintain high relative humidity. Because the jacketed cooler is a closed system, growers can convert it into a controlled atmosphere storage system. Researchers are still studying the usefulness of controlled-atmosphere storage on nursery crops.

for ten weeks in polyethylene bags at –0.5 to 4.5° C (31 to 40° F) with no adverse effect on rooting (*91*).

Many nurseries will overwinter rooted cuttings either in flats or in small liner containers protected by minimum-heat-maintained structures (e.g., quonset, polyhoop houses, greenhouses). It is possible to store rooted cuttings of certain species safely for up to five months at 1 to 4° C (34 to 39° F) in polyethylene bags (*101*). Cuttings of 31 woody ornamentals stored for six months had better survival at 0° C than at 4.5° C (32° F versus 40° F), although with some species there was no difference (*43*).

With rooted, deciduous hardwood cuttings which are lifted from propagation beds in late winter, it may be necessary to refrigerate them for several months until they can be transplanted. Figure 11–49 shows a plant roll system, where dormant, rooted deciduous cuttings are placed on long strips of plastic with moist peat moss. The plants, peat moss, and plastic are then rolled up like a jelly roll and placed in cold storage for up to two months at 1° C (34° F) until field planted (*90*).

As previously mentioned, newly rooted softwood cuttings of selected species can be difficult to overwinter, and simply cannot be transplanted to the field from their rooting beds. Storing rooted cuttings in *cold storage* allows growers to commit valuable greenhouse overwintering space to the production of other crops. It also permits earlier deliveries to warmer regions because the plants are available for shipping all winter. The planting season can also be extended because the rooted cuttings are held in a state of dormancy into the spring (*122*). Late in the year (during the slow season) is probably the safest time to harvest crops for refrigerated storage. With rooted softwood cuttings, their natural growth cycle has been disrupted, so they need not be dormant before cold storage (*122*). Cuttings should be allowed to partially dry before pulling them from the rooting beds and then put in poly bags packed with materials to act as insulation and to absorb excess water. The poly bags allow gas exchange, but retain cutting moisture to *prevent desiccation* during cold storage. Rooted softwood cuttings of seven species successfully survived up to seven months when stored at either –2° or 2° C (28° and 35° F) (*122*).

Douglas-Fir cuttings can successfully be stored at –1° C (30° F). Mold is also prevented at this lower storage temperature. Cuttings are stored at nearly 100 percent humidity by placing them in clear poly bags with a large block of water-saturated oasis material. The cold storage conditions may be satisfying the chilling requirement for bud dormancy, and therefore satisfying the need to have a sufficient cold period requirement prior to taking cuttings in early winter (*93*).

Unrooted chrysanthemum and carnation cuttings can be stored in sealed plastic bags for several weeks at –0.5° C (31° F) for subsequent rooting. In tests on the effects of storage on subsequent performance of plants, cuttings rooted after storage gave better results than those stored after rooting. Prestorage of chrysanthemum for 12 days at 10° C (50° F) enhanced rooting of cuttings compared to nonstored cuttings (Figure 11–50). Roots were initi-

FIGURE 11–49 *Left top and bottom:* Refrigerated storage for holding unrooted cuttings. High humidity is maintained by overhead sprinkler systems. *Top right:* Cold storage of rooted and unrooted cuttings. *Bottom right:* Winter-lifted rooted deciduous cuttings on long strips of plastic with moist peat moss. The plants, peat moss, and plastic are rolled up like a jelly roll and placed in cold storage. They will later be transplanted in the field in late spring (Maryland, USA). (Courtesy V. Priapi (*90*).)

ated while cuttings were in storage (*113*). This could enable producers to store cuttings for a more convenient time to propagate, and also reduce the time needed for propagating under a mist or fog system.

Storage and subsequent rooting of carnations was better if auxin was applied *after* storage temperatures of *less than* 13° C (55° F), whereas auxin was more effective if applied *prior* to cuttings being maintained at storage temperatures *above 13° C* (*114*). Again, the implications are for using storage for convenience of propagating, as well as for reducing the propagation period.

Carnation cuttings, either rooted or unrooted, store well at −0.5° C (31 to 33° F) for at least five months if placed in polyethylene-lined boxes with a small amount of moist sphagnum or peat moss. The poly film should not be sealed.

The proper storage temperature is species-dependent. Some species survive cold temperature better than others. The storage unit for unrooted cuttings should be maintained at close to 100 percent humidity—pathogens must be controlled. Topsin (Atochem N.A.), Domain (Sierra-Grace), and Cleary 3396 (W.A. Cleary), Alliette, and other systemic fungicides should be considered. Storage temperature should be as low as possible without impairing rooting of cuttings or survival of rooted liners.

HANDLING FIELD-PROPAGATED PLANTS

Rooted deciduous, hardwood cuttings[1] are dug in the nursery row during the dormant season after the leaves have dropped. With fast-growing species, the cuttings may be sufficiently large to dig after one season's growth. Slower-growing species may require two or even three years to become large enough to transplant.

Most deciduous trees and shrubs harvested bare-root in the late fall and early winter will lose more than 90 percent of their root system between the nursery digging process and the final transplanting site. Conversely, smaller sized liners may have 90 percent of their root system still intact. Many species of herbaceous perennials are also handled bare-root, though harvest time often depends on the species. By properly performing five major steps when working with bare-root plants: (1) harvesting, (2) processing, (3) storing, (4) shipping, and (5) transplanting—most nurseries should be able to avoid problems (Figure 11–51). With some species, it may be advantageous to root prune plants one year prior to digging by pruning the roots with a sharp-shooter spade, or mechanically digging, slightly lifting, and placing the rootball in its original ground location (*47, 118*). This slows growth in the nursery, but promotes a more compact, fibrous root system and reduces transplant shock.

The digging should take place on cool, cloudy days when there is no wind. If possible, digging should not be done when the soil is wet, especially if the soil has a high clay content. Most of the soil should drop readily from the roots after the plants are removed. If only a few plants are to be removed from the nursery row, they can be dug with shovels (spades), but in large-scale nursery operations some type of mechanical digger, such as is shown in Figure 11–52, is generally used. This digger "undercuts" the plants. A sharp U-shaped blade travels 30 to 60 cm (1 to 2 ft) below the soil surface under the nursery row, cutting through the roots. Sometimes a horizontal, vibrating, "lifting" blade is also attached, and travels behind the cutting blade. This blade lifts the plants out of the soil, and shakes the soil from the root system, making them easy to pull by hand.

Once a bare-root plant is dug, it is imperative that the roots not dry out. Excessive moisture loss or desiccation causes large transplant losses. Roots of bare-root plants will lose water five times faster than the stems (*40*). After the plants are dug, they should be quickly heeled-in in a convenient location, placed in cold storage, or replanted immediately in their permanent location. **Heeling-in** is placing dug, **bare-rooted** deciduous or coniferous nursery plants close together in trenches with the roots well covered. This is a temporary provision for holding the young plants until they can be set out in their permanent location (see Figure 11–53). Healing-in outdoors in a well-drained site is an adequate method of overwintering fall-harvested plants, especially when the

FIGURE 11–50 Effect of length of time at a prestorage temperature of 10° C (50° F) on rooting of *Chrysanthemum morifolium* 'Pink Boston.' Cuttings were stored 0 days *(left)* and 12 days *(right)*. Both were propagated at the same time and evaluated after 7 days. (Courtesy P. A. Van de Pol (*113*).)

[1]The procedures described in this section also apply to nursery plants propagated as seedlings, as tissue-culture-produced liners, or as budded or grafted trees.

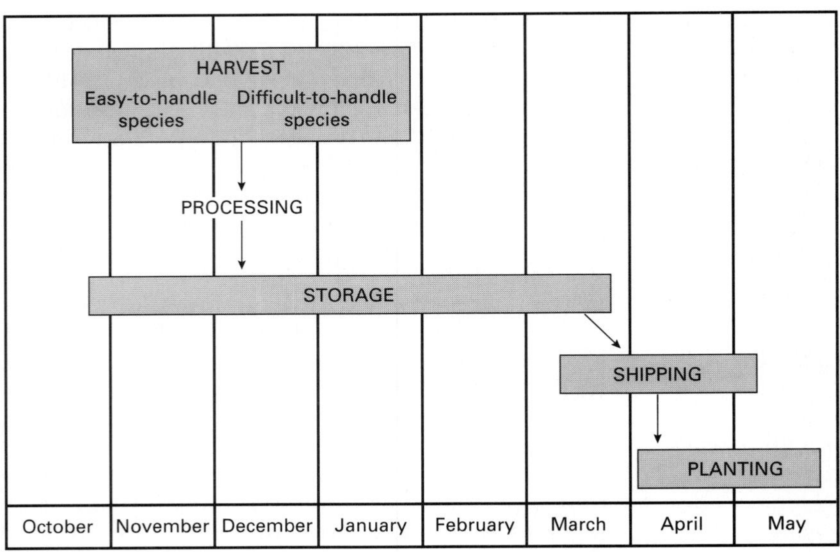

FIGURE 11–51 A general outline and schedule for handling bare-rooted trees and shrubs in Oregon (*40*).

winter is mild and the plants will be shipped in early spring (*40*).

Commercial nurseries often store quantities of deciduous plants for several months through the winter in cool, dark rooms with the roots protected by damp bark wood shavings, moist sawdust, or some similar material. Nursery stock to be kept for extended periods should be held in cold storage, ideally with jacketed cooler systems, with high relative humidity (+ 95 percent) and temperatures of 0 to 2° C (32 to 35° F). Bare-root plants are usually placed onto pallets, and the pallets are stacked on top of each other (*44*). Wooden boxes filled with moist sawdust or bark are also used to hold larger plants and seedlings until they can be field planted in the spring. If bare-root liners leaf out during stor-

FIGURE 11–52 Digging machines specially constructed for undercutting nursery stock in commercial nurseries. Photo on left shows large U-shaped blade which can be lowered to travel 0.3 to 0.6 m (1 to 2 ft) below the soil surface to cut roots. Vibrating "lifter" behind the cutting blade facilitates the movement of the blade through the soil.

FIGURE 11–53 Temporary storage of nursery stock by "heeling-in" in raised beds filled with damp wood shavings sufficiently deep to cover the roots.

age, they should be stripped of their leaves prior to transplanting in the field (*5*).

Generally, *water stress* is the chief factor limiting transplanting success. Waxing the stems of stored, dormant roses and other shrubs has been a common nursery practice. Some of the waxes used in the fresh-fruit industry, such as Shield Brite (Shield-Brite Corp., Kirkland, WA), have been somewhat effective in reducing water loss and improving survival of bare-root plants (*40*). The use of anti-transpirants and film-forming compounds or dipping roots in hydrogels is not nearly as effective as defoliating plants (*5*). The anti-transpirant, Moisturin (Burke's Protective Coatings, Washougal, WA), enhanced transplant survival and growth of bare-root Washington hawthorns and roses, which are difficult-to-transplant because of excessive desiccation during post-harvest production. However, with selected oak species, where transplanting success is limited by poor root regeneration potential, anti-desiccants are ineffective (*40*). The greatest amount of water loss during post-harvest handling comes after storage—during transport, and transplanting to the final site. In general, any treatment that reduces water loss and plant stress during handling and transplanting is beneficial.

Balled and Burlapped Stock

Unless very small, plants of broad- or narrow-leaved evergreen species usually are not handled successfully bare-root, as is done with *dormant and leafless deciduous plants*. The presence of leaves on evergreen plants requires continuous contact of the roots with soil. Therefore, large, salable plants of broad- or narrow-leaved evergreens, and occasionally deciduous plants, are either grown in containers or dug and sold "balled and burlapped" (B&B). By the latter method, the plants are removed from the soil by carefully digging a trench around each individual plant or using a digging machine such as a mechanized tree spade (Figure 11–54). The soil mass around the roots is sometimes tapered at top and bottom, resulting in a ball of soil in which the roots are embedded. It is important that the soil be at the proper moisture level—not too wet and not too dry—otherwise, it will fall apart. The ball is gently placed on a large square of burlap, which is then pulled tightly around the ball, pinned with nails or hog wings, and wrapped with twine. When done properly, the burlap ensures that roots are kept in contact with the soil, and the plant can be moved safely for considerable distances and replanted successfully (Figure 11–54). Larger specimen B&B plants are sometimes set in a wire cage to help keep the rootball intact during handling and shipping. Some field-grown B&B plants are **shifted** into larger, rigid-plastic containers or wooden boxes to ease handling and marketing, and to increase the sale value (Figure 3–25).

CONTAINER-GROWN PLANTS AND ALTERNATIVE FIELD PRODUCTION SYSTEMS

Production of plants in containers has replaced much of the traditional nursery field production methods, primarily because of handling ease, improved plant marketability, greater cultural control, and faster product turnover. However, many plant species are still field-grown because of their growth habits, and lower production costs. Field-grown plants do well in climates experiencing low temperatures, which limit survival of containerized crops without overwintering protection (see Figure 3–37).

See Chapter 3 for the discussion on best management practices, post-propagation care of liners, handling *container-grown* plants—irrigation systems, winter protection, root development in containerized plants; alternatives to traditional production systems—*pot-in-a-pot system, in-ground plastic containers,* and *in-ground fabric containers* or *grow bags* (Figure 3–39); and avoiding transplant problems.

FIGURE 11–54 Digging and handling balled and burlapped (B&B) shrubs and trees. *Top left:* Mechanical spade digger. *Top right:* B&B with burlap and twine. *Left center and bottom:* Transporting palletized B&B plants from the field to shipping dock with forklift equipment. *Bottom right:* Holding area of palletized B&B plants; overhead irrigation of 6 meter (20 ft) risers is used to reduce transpiration stress.

DIFFICULT-TO-HANDLE BARE-ROOT SPECIES (*40*)

Difficult-to-handle bare-root species include *Betula* (birch) and *Crataegus* (hawthorn). Generally, poor transplanting success and survival is due to excessive desiccation during post-production handling. However, some species of *Quercus* (oak) exhibit poor or slow root regeneration potential after transplanting.

Apparently there are physiological differences in dormancy and stress tolerance between easy-to-handle *Acer platanoides* (Norway maple) and difficult-to-handle *Crataegus phaenopyrum* (Washington hawthorn). Washington hawthorn does not achieve the same level of deep dormancy, cold hardiness, and desiccation-tolerance that Norway maple does. The desiccation tolerance of Norway maple increases substantially in November (Oregon), whereas desiccation tolerance in Washington hawthorn does not begin until late December. Thus, it is best to schedule difficult-to-transplant species for harvest as late in the season as possible to allow desiccation tolerance to increase (see Figure 11–51) (*40*).

REFERENCES

1. Ahmad, M., P.B. Lombard, and R.L. Ticknor. 1992. Effect of slow-release fertilizers on propagation medium and on rooting and growth of cuttings. *Comb. Proc. Intl. Plant Prop. Soc.* 42:238–241.
2. Ali, N., and M.N. Westwood. 1966. Rooting of pear cuttings as related to carbohydrates, nitrogen, and rest period. *Proc. Amer. Soc. Hort. Sci.* 88:145–50.
3. Alward, T.M., 1984. Softwood cuttings taken from developing hardwood cuttings. *Comb. Proc. Intl. Plant Prop. Soc.* 34:535–36.
4. Anonymous. 1994. After methyl bromide: No easy answers. *Calif. Agricul.* 48(3):7–9.
5. Askew, J.C., C.H. Gilliam, H.G. Ponder, and G.J. Keever. 1985. Transplanting leafed-out bare root dogwood liners. *HortScience* 20:219–21.
6. Avery, J.D. and C.B. Beyl. 1991. Propagation of peach cuttings using foam cubes. *HortScience* 26(9):1152–1154.
7. Badenhop, M.B. 1984. How much does it cost to produce a rooted cutting? *Amer. Nurser.* 160:104–111.
8. Baldwin, I., and J. Stanley. 1981. How to manage stock plants. *Amer. Nurs.* 153(8):16, 74–80.
9. Barnes, H.W. 1989. Propylene glycol quick-dips: practical applications. *Comb. Proc. Intl. Plant Prop. Soc.* 39:427–432.
10. Beeson, R.C. Jr. 1991. Scheduling woody plants for production and harvest. *HortTechnology.* 1(1): 30–36.
11. Berry, J. B. 1984. Rooting hormone formulations: A chance for advancement. *Comb. Proc. Intl. Plant Prop. Soc.* 34:486–91.
12. Bhat, N.R., T.L. Prince, H.K. Tayama, and S.A. Carver. 1992. Root cutting establishment in media containing a wetting agent. *HortScience* 27(1):78.
13. Bonaminio, V.P. 1983. Comparison of quick-dips with talc for rooting cuttings. *Comb. Proc. Intl. Plant Prop. Soc.* 33:565–68.
14. Burger, D.W. 1994. Intermittent mist control via solar cells. *HortTechnology* 4(3):273–274.
15. Carlson, R.F. 1966. Factors influencing root formation in hardwood cuttings of fruit trees. *Mich. Quart. Bul.* 48:449–54.
16. Carter, A. and M. Slee. 1991. Propagation media and rooting cuttings of *Eucalyptis grandis*. *Comb. Proc. Intl. Plant Prop. Soc.* 41:36–39.
17. Catanzaro, C.J., W.A. Skroch, and P.H. Henry. 1993. Rooting performance of hardwood stem cuttings from herbicide-treated nursery stock plants. *J. Environ. Hort.* 11(3):128–130.
18. Chapman, D.J. 1989. Consider softwood cuttings for tree propagation. *Amer. Nurser.* 169(12):45–50.
19. Chong, C. 1982. Rooting response of cuttings of two cotoneaster species to surface-applied Osmocote slow-release fertilizer. *Plant Propagator* 28(3): 10–12.
20. Chong, C., O.B. Allen, and H.W. Barnes. 1992. Comparative rooting of stem cuttings of selected woody landscape shrub and tree taxa to varying concentrations of IBA in talc, ethanol and glycol carriers. J. Environ. Hort. 10(4): 245–250.
21. Chong, C. and B. Hamersma. 1995. Automobile radiator antifreeze and windshield wiper fluid as IBA carriers for rooting woody cuttings. *HortScience* 30(2):363–365.
22. Cialone, J. 1984. Developments in dracaena production. Comb. Proc. Intl. Plant Prop. Soc. 34: 491–94.

23. Coorts, G. D., and C. C. Sorenson, 1968. Organisms found growing under nutrient mist propagation. *HortScience* 3(3):189–90.
24. Cuny, H. 1995. Fungi lend a hand—rooting out mycorrhizae's place in the nursery. *Nurser. Manag. Prod.* 11(4):44–50.
25. Davies, F.T., Jr. 1988. Influence of nutrition and carbohydrates on rooting of cuttings. *Comb. Proc. Intl. Plant Prop. Soc.* 38:432–437.
26. ———. 1991. Back to the basics in propagation. *Comb. Proc. Intl. Plant Prop. Soc.* 41:338–342.
27. Davies, F. T., Jr., T.D. Davis, and D.E. Kester. 1994. Commercial importance of adventitious rooting to horticulture. In *Biology of adventitious root formation,* T.D. Davis and B.E. Haissig, eds. New York N.Y.: Plenum Press.
28. Davies, F.T., Jr., and S.A. Duray. 1992. Effect of preemergent herbicide application on rooting and subsequent liner growth of selected nursery crops. *J. Environ. Hort.* 10(3):181–186.
29. Davies, F. T., Jr., and B. C. Moser. 1980. Stimulation of bud and shoot development of Rieger begonia leaf cuttings with cytokinins. *J. Amer. Soc. Hort. Sci.* 105(1):27–30.
30. Davis, T. D., and T. R. Potter. 1985. Carbohydrates, water potential and subsequent rooting of stored rhododendron cuttings. *HortScience:* 20: 292–93.
31. Dirr, M. A. 1982. What makes a good rooting compound? *Amer. Nurs.* 155(8):33–40.
32. ———. 1983. Comparative effects of selected rooting compounds on the rooting of *Photinia* × *fraseri. Comb. Proc. Intl. plant Prop.* Soc. 33:536–40.
33. ———. 1989. Rooting response of *Photinia* × *fraseri* Dress 'Birmingham' to 25 carrier and carrier plus IBA formulations. *J. Environ. Hort.* 7(4): 158–60.
34. ———. 1990. Effects of P-ITB and IBA on the rooting response of 19 landscape taxa. *J. Environ. Hort.* 8(2):83–85.
35. ———. 1992. Update on root-promoting chemicals and formulations. *Comb. Proc. Intl. Plant Prop. Soc.* 42:361–65.
36. ———. 1994. The latest status of IBA and other root-promoting chemicals. *Nurser. Manag.* 10(5): 24–25.
37. Douds, D.D. Jr., G. Bécard, P.E. Pfeffer, L.W. Doner, T.J. Dymant and W.M. Kayser. 1995. Effect of vesicular-arbuscular mycorrhizal fungi on rooting of *Sciadopitys verticillata* Sib & Zucc. cuttings. *HortScience* 30(1):133–34.
38. Drew, J.J., M.A. Dirr, and A.M. Armitage. 1993. Effects of fertilizer and night interruption on overwinter survival of rooted cuttings of *Quercus* L. *J. Environ. Hort.* 11:97–101.
39. Elliott, F.A. 1989. Benefits of good record keeping in propagation. *Comb. Proc. Intl. Plant Prop. Soc.* 39:135–138.
40. Englert, J.M., L.H. Fuchigami, and T.H.H. Chen. 1993. Bare-root basics—how to handle bare-root trees and shrubs after harvesting. *Amer. Nurser.* 177(4):56–61.
41. Erez, A., and Z. Yablowitz. 1981. Rooting of peach hardwood cuttings for the meadow orchard. *Scientia Hort.* 15:137–44.
42. Finnerty, T.L. 1994. Native woody plant propagation—three key steps part II. Planning, scheduling, and good record keeping. *N. Amer. Plant Prop.* 6(2):16–17.
43. Flint, H. L., and J. J. McGuire. 1962. Response of rooted cuttings of several woody ornamental species to overwinter storage. *Proc. Amer. Soc. Hort. Sci.* 80:625–29.
44. Foster, S. 1994. Handling bare-root tree whips at Greenleaf Nursery. *Comb. Proc. Intl. Plant Prop. Soc.* 44:478–79.
45. Fourrier, B. 1984. Hardwood cutting propagation at McKay nursery. *Comb. Proc. Intl. Plant Prop. Soc.* 34:540–43.
46. Gilliam, C.H., D.J. Eakes, and J.W. Olive. 1993. Herbicide use during propagation affects root initiation and development. *J. Environ. Hort.* 11(4):157–159.
47. Gilman, E.F., and T.H. Yeager. 1988. Root initiation in root-pruned hardwoods. *HortScience.* 23(4):775.
48. Goodman, M. A., and D. P. Stimart. 1987. Factors regulating overwinter survival of newly propagated stem tip cuttings of *Acer* palmatum Thunb. 'Bloodgood' and *Cornus florida* L. var. rubra. *HortScience* 22:1296–98.
49. Gouin, F. R. 1981. Vegetative propagation under thermoblankets. *Comb. Proc. Intl. Plant Prop. Soc.* 30:301–5.
50. Hambrick, C. E., F. T. Davies, Jr., and H. B. Pemberton. 1991. Seasonal changes in carbohydrate/nitrogen levels during field rooting of *Rosa multiflora* 'Brooks 56' hardwood cuttings. *Scientia Hort.* 46:137–146.
51. Hansen, C. J., and H. T. Hartmann. 1968. The use of indolebutyric acid and captan in the propagation of clonal peach and peach-almond hybrid rootstocks by hardwood cuttings. *Proc. Amer. Soc. Hort. Sci.* 92:135–40.
52. Hare, R. C. 1976. Rooting of American and Formosan sweetgum cuttings taken from girdled and nongirdled cuttings. *Tree Planters Notes* 27(4): 6–7.
53. ———. 1976. Girdling and applying chemicals promote rapid rooting of sycamore cuttings. *U.S. For. Serv. Res. Note* SO-202.
54. ———. 1977. Rooting of cuttings from mature water oak. *S. J. Appl. Forest.* 1(2):24–25.
55. Hartmann, H. T., W. H. Griggs, and C. J. Hansen. 1963. Propagation of ownrooted Old Home and Bartlett pears to produce trees resistant to pear decline. *Proc. Amer. Soc. Hort. Sci.* 82:92–102.

56. Hartmann, H. T., C. J. Hansen, and F. Loreti. 1965. Propagation of apple rootstocks by hardwood cuttings. *Calif. Agr.* 19(6):4–5.

57. Hess, C. E., and W. E. Snyder. 1953. A simple and inexpensive time clock for regulating mist in plant propagation procedures. *Proc. Plant Prop. Soc.* 3: 56–61.

58. Howard, B. H. 1968. The influence of indolebutyric acid and basal temperature on the rooting of apple rootstock hardwood cuttings. *J. Hort. Sci.* 43:23–31.

59. ———. 1985. The contribution to rooting in leafless winter plum cuttings of IBA applied to the epidermis. *J. Hort. Sci.* 60:153–159.

60. ———. 1985. Factors affecting the response of leafless winter cuttings of apple and plums to IBA applied in powder formulation. *J. Hort. Sci.* 60: 161–68.

61. ———. 1993. Understanding vegetative propagation. *Comb. Proc. Intl. Plant Prop. Soc.* 43:157–162.

62. Howard, B. H., and R. J. Garner. 1965. High temperature storage of hardwood cuttings as an aid to improved establishment in the nursery. *Ann. Rpt. E. Malling Res. Sta. for 1964,* pp. 83–87.

63. Johnson, D.R., T.E. Bilderback and C.E. Safely. 1986. Determining container production costs on a per plant basis. *Comb. Proc. Intl. Plant Prop. Soc.* 36:378–384.

64. Johnson, J. R., and J. A. Meade. 1986. Preemergent herbicide effect on the rooting of cuttings. *Comb. Proc. Intl. Plant Prop. Soc.* 36:567–70.

65. Kroin, J. 1992. Advances using indole-3-butyric acid (IBA) dissolved in water for root cuttings, transplanting, and grafting. *Comb. Proc. Intl. Plant Prop. Soc.* 42:489–492.

66. Linderman, R.G. 1993. Effects of biocontrol agents on plant growth. *Comb. Proc. Intl. Plant Prop. Soc.* 43:249–252.

67. Loach, K. 1985. Rooting of cuttings in relation to the propagation medium. *Comb. Proc. Intl. Plant Prop. Soc.* 35:472–85.

68. ———. 1989. Controlling environmental conditions to improve adventitious rooting. In *Adventitious root formation in cuttings,* T. D. Davis, B. E. Haissig, and N. Sankhla, eds. Portland, Oreg.: Dioscorides Press.

69. ———. 1988. Hormone applications and adventitious root formation in cuttings—a critical review. *Acta Hort.* 227:126–33.

70. Macdonald, B. 1986. *Practical woody plant propagation for nursery growers,* Vol. 1. Portland, Oreg.: Timber Press.

71. Maronek, D. M., D. Studebaker, T. McCloud, V. Black and R. St. Jean. 1983. Stripping vs. nonstripping on rooting of woody ornamental cuttings—grower results. *Comb. Proc. Intl. Plant Prop. Soc.* 33:388–97.

72. Maronek D. M., D. Studebaker, and B. Oberly. 1985. Improving media aeration in liner and container production. *Comb. Proc. Intl. Plant Prop. Soc.* 35:591–97.

73. Maunder, C. 1983. A comparison of propagation unit systems. *Comb. Proc. Intl. Plant Prop. Soc.* 33:233–38.

74. Maynard, B. K., and N. L. Bassuk. 1987. Stockplant etiolation and blanching of woody plants prior to cutting propagation. *J. Amer. Soc. Hort. Sci.* 112:273–76.

75. ———. 1990. Comparisons of stock plant etiolation with traditional propagation methods. *Comb. Proc. Intl. Plant Prop. Soc.* 40:517–523.

76. ———. 1992. Stock plant etiolation, shading and banding effects on cutting propagation of *Carpinus betulus. J. Amer. Soc. Hort. Sci.* 117(5):740–744.

77. Maynard, B.K., W.Q. Sun, and N.L. Bassuk. 1990. Encouraging bud break in newly-rooted softwood cuttings. *Comb. Proc. Intl. Plant Prop. Soc.* 40:597–602.

78. Mazalewski, R.L. 1989. Practical aspects of high pressure fog systems. *Comb. Proc. Intl. Plant Prop. Soc.* 39:101–105.

79. Mee, T.R. 1994. Understanding fog technology. *Comb. Proc. Intl. Plant Prop. Soc.* 44:250–253.

80. Merker, R. 1985. System of direct stick propagation. *Comb. Proc. Intl. Plant Prop. Soc.* 35:182–83.

81. MacDonald, B. 1986. *Practical woody plant propagation for nursery growers, Vol. 1.* Portland, Oreg.: Timber Press.

82. MacKenzie, A.J., T. W. Starmann, and M.T. Windham. 1995. Enhanced root and shoot growth of chrysanthemum cuttings propagated with the fungus *Trichoderma harzianum. HortScience* 30(3): 496–498.

83. McLean, C., A.C. Lawrie, and K.L. Blaze. 1994. The effect of soil microflora on the survival of cuttings of *Epacris impressa.* Plant Soil 166:295–297.

84. Milbocker, D. C. 1987. The use of humidifan in propagation. *Comb. Proc. Intl. Plant Prop. Soc.* 37: 513–18.

85. Molnar, J. M., and W. A. Cumming. 1968. Effect of carbon dioxide on propagation of softwood, conifer, and herbaceous cuttings. *Can. J. Plant Sci.* 48:595–99.

86. Motley, B. 1994. Incentive pay in propagation. *Comb. Proc. Intl. Plant Prop. Soc.* 44:449–453.

87. Noland, D. A., and D. J. Williams. 1980. The use of pumice and pumice-peat mixtures as propagation media. *Plant Propagator* 26(4):6–7.

88. Orndorff, C. 1987. Root pieces as a means of propagation. *Comb. Proc. Intl. Plant Prop. Soc.* 37:432–35.

89. Preece, J. E. 1987. Treatment of stock plant with plant growth regulators to improve propagation success. *HortScience* 22:754–59.

90. Priapi, V.M.. 1993. Outdoor mist propagation. *Amer. Nurser.* 178(11):30–35.

91. Pryor, R. L., and R. N. Stewart. 1963. Storage of unrooted azalea cuttings. *Proc. Amer. Soc. Hort. Sci.* 82:483–84.

92. Richey, M.L. 1989. Costing variables in propagation techniques. *Comb. Proc. Intl. Plant Prop. Soc.* 39:502–506.
93. Ritchie, G.A. 1994. Commercial applications of adventitious rooting to forestry. In *Biology of adventitious root formation,* T.D. Davis and B.E. Haissig, eds. New York, N.Y.: Plenum Press.
94. Ritchie, G.A., Y. Tanaka, and S.D. Duke. 1992. Physiology and morphology of Douglas-fir rooted cuttings compared to seedlings and transplants. *Tree Physiol.* 10:179–194.
95. Sabalka, D. 1986. Propagation media for flats and for direct sticking: What works? *Comb. Proc. Intl. Plant Prop. Soc.* 36:409–13.
96. Santana, C. 1995. Production scheduling of cuttings. *Comb. Proc. Intl. Plant Prop. Soc.* 45: In press.
97. Scott, M. A. 1987. Management of hardy nursery stock plants to achieve high yields of quality cuttings. *HortScience* 22:738–41.
98. Shugert, R. 1989. Storage of cutting wood prior to sticking. *Comb. Proc. Intl. Plant Prop. Soc.* 39:422–426.
99. Skimina, C. A. 1984. Use of monochloramine as a disinfectant for pruning shears. *Comb. Proc. Intl. Plant Prop. Soc.* 34:214–20.
100. Smith, I. 1993. Government regulations and nursery accreditation. *Comb. Proc. Intl. Plant Prop. Soc.* 43:117–121.
101. Snyder, W. E., and C. E. Hess. 1956. Low temperature storage of rooted cuttings of nursery crops. *Proc. Amer. Soc. Hort. Sci.* 67:545–48.
102. Sorenson, D. C., and G. D. Coorts. 1968. The effect of nutrient mist on propagation of selected woody ornamentals. *Proc. Amer. Soc. Hort. Sci.* 92:6 96–703.
103. Stein, A., J.A. Fortin, and G. Vallée. 1990. Enhanced rooting of *Picea mariana* cuttings by ectomycorrhizal fungi. *Can. J. Bot.* 68:468–470.
104. Struve, D. K., and M. A. Arnold. 1986. Atyl esters of IBA increase rooted cutting quality of red maple 'Red Sunset' softwood cuttings. *HortScience* 21:1392–93.
105. Struve, D.K., M.A. Arnold, R. Beeson, Jr., J.M. Ruter, S. Svenson, and W.T. Witte. 1994. The copper connection: the benefits of growing woody ornamentals in copper-treated containers. *Amer. Nurser.* 179(4): 52–61.
106. Sumner, P. E. 1987. Uniformity analysis of various types of mist propagation nozzles. *Comb. Proc. Intl. Plant Prop. Soc.* 37: 522–26.
107. ———. 1991. Stem banding enhances rooting and subsequent growth of M.9 and MM.106 apple rootstock cuttings. *HortScience* 26(11): 1368–1370.
108. ———. 1993. Auxin-induced ethylene synthesis during rooting and inhibition of budbreak of 'Royalty' rose cuttings. *J. Amer. Soc. Hort. Sci.* 118(5):638–643.
109. Svenson, S.E. and F.T. Davies, Jr. 1995. Change in tissue elemental concentration during root initiation and development of poinsettia cuttings. *HortScience* 30(3):617–619.
110. Tilt, K. M., and T. E. Bilderback. 1987. Physical properties of propagation media and their effects on rooting of three woody ornamentals. *HortScience* 22:245–47.
111. Vallis, G. 1991. The development of a market-led propagation system. *Comb. Proc. Intl. Plant Prop. Soc.* 41:134–41.
112. Van Braght, J., H. van Gelder, and R. L. M. Pierik. 1976. Rooting of shoot cuttings of ornamental shrubs after immersion in auxin-containing solutions. *Scientia Hort.* 4:91–94.
113. Van de Pol, P. A. 1988. Partial replacement of the rooting procedure of *Chrysanthemum morifolium* cuttings by pre-rooting storage in the dark. *Acta Hort.* 226: 519–524.
114. Van de Pol, P. A., and J. V. M. Vogelezang. 1983. Accelerated rooting of carnation 'Red Baron' by temperature pretreatment. *Scientia Hort.* 20: 287–94.
115. Vanstone, F. H. 1959. Equipment for mist propagation developed at the N.I.A.E. *Ann. Appl. Bio.* 47:627–31.
116. Verkade, S. D. 1986. Mycorrhizal inoculation during plant propagation. *Comb. Proc. Intl. Plant Prop. Soc.* 36:613–18.
117. Wang, Y. T., and C. A. Boogher. 1988. Effect of nodal position, cutting length, and root retention on the propagation of golden pothos. *HortScience* 23:347–49.
118. Watson, G.W., and T.D. Sydnor. 1987. The effect of root pruning on the root system of nursery trees. *J. Arboricul.* 13:126-130.
119. Wells, J. S. 1952. Pointer on propagation. Propagation of *Taxus. Amer. Nurs.* 96(11):13, 37–38, 43.
120. ———. 1962. Wounding cuttings as a commercial practice. *Comb. Proc. Intl. Plant Prop. Soc.* 12:47–55.
121. Whalley, D. N., and K. Loach. 1981. Rooting of two genera of woody ornamentals from dormant, leafless (hardwood) cuttings and their subsequent establishment in containers. *J. Hort. Sci.* 56: 131–38.
122. Wood, T., and A.C. Cameron. 1989. Cold storage—overwintering your softwood cuttings in the refrigerator releases valuable greenhouse space. *Amer. Nurser. Mag.* 170(7): 49–54.
123. Wott, J. A., and H. B. Tukey, Jr. 1967. Influence of nutrient mist on the propagation of cuttings. *Proc. Amer. Soc. Hort. Sci.* 90:454–61.

SUPPLEMENTARY READING

American Nurseryman Magazine. Chicago: American Nurseryman Publ. Co.

DIRR, M. A., and C. W. HEUSER, JR. 1987. *The reference manual of woody plant propagation.* Athens, Ga.: Univ. Press.

INTERNATIONAL PLANT PROPAGATORS' SOCIETY. Combined Proceedings of annual meetings.

JOINER, J. N. 1981. *Foliage plant production.* Englewood Cliffs, N.J.: Prentice-Hall.

LAMB, J. G. D., J. C. KELLY, and P. BOWBRICK. 1985. *Nursery stock manual.* London: Grower Books.

LANDIS, T.D., R.W. TINUS, S.E. MCDONALD, and J.P. BARNETT. 1990. *Containers and growing media.* The Container Tree Nursery Manual, Vol. 2. Agri. Handbk. 674. Washington, DC: For. Serv., USDA.

LANDIS, T.D., R.W. TINUS, S.E. MCDONALD, and J.P. BARNETT. 1990. *The biological component: nursery pests and mycorrhizae.* The Container Tree Nursery Manual, Vol 5. Agri. Handbk. 674. Washington, DC: For. Serv., USDA.

LANDIS, T.D., R.W. TINUS, S.E. MCDONALD, and J.P. BARNETT. 1992. *Atmospheric Environment.* The Container Tree Nursery Manual, Vol 3. Agri. Handbk. 674. Washington, DC: For. Serv., USDA.

LANDIS, T.D., R.W. TINUS, S.E. MCDONALD, and J.P. BARNETT. 1995. *Nursery planning, development and management.* The Container Tree Nursery Manual, Vol 1. Agri. Handbk. 674. Washington, DC: For. Serv., USDA.

MACDONALD, B. 1986. *Practical woody plant propagation for nursery growers,* Vol. 1. Portland, Oreg.: Timber Press.

12
The Biology of Grafting

Historically, fruit and nut trees have been grafted because of difficulty in propagating by cuttings, and the superiority and high value of the grafted crop. Grafting is among the most expensive propagation techniques, surpassing even micropropagation. **Budding,** which is a form of grafting, is 3-times more costly than cuttings, and 14-times more expensive than seedling propagation (*88*). Horticulture and forestry have sought to develop clonal propagation systems that avoid labor-intensive graftage. Yet, traditional and highly efficient grafting and budding systems are essential for the propagation of many woody plant species. New markets continue to require grafted and budded plants for improved plant quality, fruit yield, superior forms, and greater plant ecological ranges. In the U.S. southeast, where high temperatures and periodic flooding of soils (low soil oxygen) are the norm, cultivars of birch (*Betula*), fir (*Abies*), and oak (*Quercus*) are grafted onto adapted rootstock (*130, 131, 133*). The propagator benefits via new markets, while the consumer gains a greater variety of tolerant landscape plants.

With the greater reliance on integrated pest management, and reduced availability of pesticides and soil fumigants, disease-tolerant rootstocks will play a greater role not only with woody perennial fruit crops and ornamentals but also with grafted vegetable crops (*15, 156*).

This chapter reviews grafting and budding: its historical and present importance, man-made and natural grafting systems, graft union formation, graft incompatibility, scion-rootstock and rootstock-scion relationships, and the mechanisms of scion-rootstock and rootstock-scion effects. Chapters 13 and 14, respectively, describe the techniques of grafting and budding. Chapter 19 enumerates grafting and budding systems for selected fruit and nut trees, as does Chapter 20 for selected woody ornamental plants. A better understanding of the fundamental biology of grafting (and the causes of graft incompatibility) will enhance the development of superior cultivars and increase the ecological range of species for new markets in horticulture and forestry.

THE HISTORY OF GRAFTING

The origins of grafting can be traced to ancient times. There is evidence that the art of grafting was known to the Chinese at least as early as 1560 B.C. Aristotle (384–322 B.C.) and Theophrastus (371–287 B.C.) discussed grafting in their writings with considerable understanding. During the days of the Roman Empire, grafting was very popular, and methods were precisely described in the writings of that era. Paul the Apostle, in his Epistle to the Romans, discussed grafting between the "good" and the "wild" olive trees (Romans 11:17–24).

The Renaissance period (1350–1600 A.D.) saw a renewed interest in grafting practices. Large numbers of new plants from foreign countries were imported into European gardens and maintained by grafting. By the sixteenth century, the cleft and whip grafts were widely used in England and it was realized that the cambium layers must be matched, although the nature of this tissue was not then understood or appreciated. Propagators were handicapped by a lack of a good grafting wax; mixtures of wet clay and dung were used to cover the graft unions. In the seventeenth century, orchards in England were planted with budded and grafted trees.

Early in the eighteenth century, Stephen Hales, in his studies on the "circulation of sap" in plants, approach-grafted three trees and found that the center tree stayed alive even when severed from its roots. Duhamel studied wound healing and the uniting of woody grafts. The graft union at that time was considered to act as a type of filter that changed the composition of the sap flowing through it. Thoüin (*164*), in 1821, described 119 methods of grafting and discussed changes in growth habit resulting from grafting. Vöchting (*171*), in the late nineteenth century, continued Duhamel's earlier work on the anatomy of the graft union. Development of some of the early grafting techniques have been reviewed by Wells (*179*).

Liberty Hyde Bailey in *The Nursery Book* (*3*), published in 1891, described and illustrated the methods of grafting and budding commonly used in the United States and Europe at that time. The methods used today differ very little from those described by Bailey.

TERMINOLOGY

Grafting is the art of connecting two pieces of living plant tissue together in such a manner that they will unite and subsequently grow and develop as one *composite plant.* As any technique that will accomplish this could be considered a method of grafting, it is not surprising that innumerable procedures for grafting are described in the literature on this subject. Through the years several distinct methods have become established that enable the propagator to cope with almost any grafting problem at hand. These are described in Chapter 13 with the realization that there are many variations of each and that there are other, somewhat different, forms that could give the same results. Figure 12–1 illustrates a grafted plant and the parts involved in the graft.

Budding is a form of grafting. However, the scion (see below) is reduced in size to usually contain only one bud. An exception to this is patch bud-

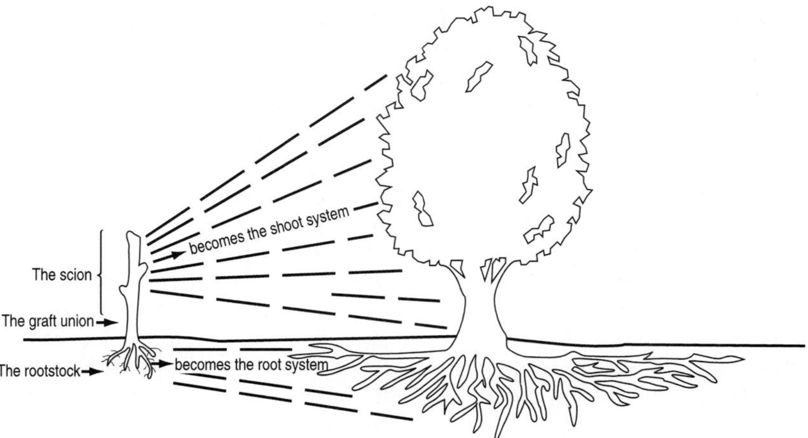

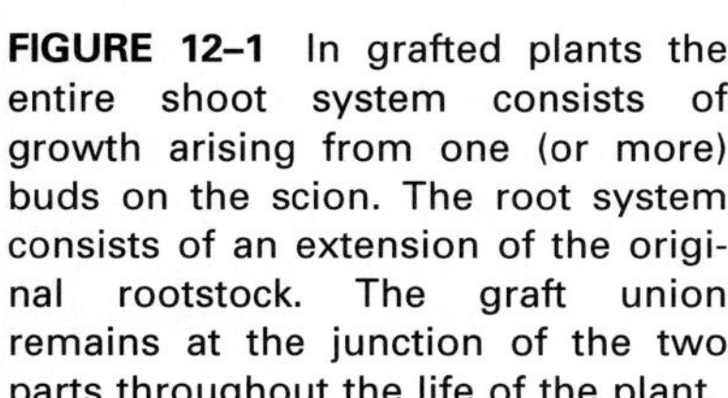

FIGURE 12–1 In grafted plants the entire shoot system consists of growth arising from one (or more) buds on the scion. The root system consists of an extension of the original rootstock. The graft union remains at the junction of the two parts throughout the life of the plant.

ding of pecan, where secondary and tertiary buds are adjacent at the same node to the primary bud. The various budding methods are described in Chapter 14.

The **scion** is the short piece of detached shoot containing several dormant buds which, when united with the rootstock, comprises the upper portion of the graft and from which will grow the stem or branches, or both, of the grafted plant. It should be of the desired cultivar and free from disease. In essence, it becomes the shoot system of the graft.

The **rootstock** (*understock, stock*) is the lower portion of the graft, which develops into the root system of the grafted plant. It may be a seedling, a rooted cutting, or a layered or micropropagated plant. If the grafting is done high in a tree, as in topworking, the rootstock may consist of the roots, trunk and scaffold branches.

The **interstock** (*intermediate stock, interstem*) is a piece of stem inserted by means of two graft unions between the *scion* and the *rootstock*. Interstocks are used to avoid an incompatibility between the rootstock and scion, to produce special tree forms, to control disease (e.g., fire-blight resistance), or to take advantage of its growth-controlling properties.

Vascular cambium is a thin tissue located between the bark (periderm, cortex, and phloem) and the wood (xylem) (Figure 12–8). Its cells are meristematic, that is, they are capable of dividing and forming new cells. For a successful graft union, it is essential that the cambium of the scion be placed in close contact with the cambium of the rootstock.

Callus is a term applied to the mass of parenchyma cells that develop from and around wounded plant tissues. It occurs at the junction of a graft union, arising from the living cells of both the

ROOTSTOCK CATEGORIES

Rootstocks can be divided into two groups: **seedling** and **clonal.**

Seedling rootstocks, those developed from germinated seeds, have certain advantages. Production of seedlings is relatively simple and economical and is well adapted to mass propagation methods (see Chapter 8). Most seedling plants do not retain viruses occurring in the parent plant (although some viruses are seed-transmitted). In some instances (e.g., plum rootstocks), the root system developed by seedlings tends to grow deeper and to be more firmly anchored than rootstocks grown from cuttings.

However, seedling rootstocks have the disadvantage of genetic variation, which may lead to variability in the growth and performance of the scion of the grafted plant. Such variation is most likely to occur if the seed is obtained from unknown, unselected sources. The seed source plants may be unusually heterozygous or may have been cross-pollinated with related species. Within the same species, some seed sources produce much better rootstock plants than others. With careful testing, individual plants in a species could be selected to be a mother-tree seed source, or the start of a clonal rootstock.

Variability among seedling rootstocks can be reduced by careful selection of the parental seed source regarding its identity, and by its protection from cross-pollination. Variation can be reduced if all nursery trees of the same age are dug from the nursery row at one time, and small or obviously off-type seedlings or budded trees are discarded. In most nurseries, the young trees are graded by size, and all those of the same grade are sold together. The practice of retaining slow-growing seedlings or budded trees for an additional year's growth is undesirable as it tends to perpetuate variability. In the United States, many fruit orchards, if grown on uniform seedling rootstocks, show no more variability resulting from the rootstock used than from unavoidable environmental differences in the orchard, principally soil variability.

Historically, **clonal rootstocks** received much attention in European countries, especially England, in regard to their development and use. Today, much of U.S. commercial apple production is on clonal rootstock because of the greater size control and fruit yield. Other fruit crops such as pear, quince, plum, cherries, and grapes are routinely produced on clonal rootstock (*180*). Each individual rootstock plant is genetically the same as all the other plants in the clone and can be expected to have identical growth characteristics in a given environment.

scion and rootstock. The production and interlocking of these parenchyma (or callus) cells constitute one of the important steps in callus bridge formation between the scion and rootstock in a successful graft.

Propagation and Utilization of Clonal Rootstock

Clonal rootstocks can be propagated vegetatively either by stool layering, rooted cuttings, or by micropropagation. Micropropagation of clonal rootstocks makes possible the production of great numbers of such plants, upon which the scion cultivar can be grafted or budded. Micropropagated apple rootstocks sometimes display undesirable juvenile-like characteristics—more vigorous shoot growth, suckering, and a delay in fruit cropping (*79, 81*).

Rootstock of citrus is produced from apomictic seed, and is genetically uniform; this is a more cost-effective method of propagating clonal rootstock than traditional asexual techniques.

Clonal rootstocks are desirable not only to produce uniformity but—equally important—to preserve their special characteristics and the specific influences they have on scion cultivars, such as disease resistance, dwarfing characteristics, growth, or flowering habit. Because of difficulty in rooting, there is currently not a good method to clonally propagate pecan and pistachio rootstock, hence they are all grafted on variable seedling rootstock. Development of successful clonal propagation techniques would facilitate the selection and multiplication of outstanding clonal rootstock that would enhance the zinc uptake of pecans or reduce biennial bearing tendencies of pistachio.

To maintain rootstock influence, deep planting of the nursery tree—which may led to **"scion rooting"**—must be avoided, as illustrated in Figure 12–2. The deeper the graft union below the soil surface, the higher the incidence of scion rooting is likely to be (*27*).

The combinations of different clonal rootstocks with different scion cultivars allow much refinement in the performance of grafted trees. Each particular scion-rootstock combination requires thorough testing, however, before its performance is established and can be predicted.

In propagating and using clonal rootstocks, it is very important in nursery propagation that only pathogen-free material be obtained, as any diseases present in such stocks are maintained and spread, along with the rootstock material. On such "clean" rootstock material, only disease-free scion material should be grafted or budded.

FIGURE 12–2 Scion roots of an 'Old Home' pear grafted on quince rootstock. (A) Original quince roots. (B) Scion roots arising from the 'Old Home' pear above the graft union. These have assumed the major support of the tree. The dwarfing influence due to the quince rootstock has been lost.

REASONS FOR GRAFTING AND BUDDING

Grafting and budding serve many different purposes:

- Perpetuating clones that cannot be readily maintained or economically propagated by cuttings, layers, division, or other asexual methods
- Obtaining the benefits of certain rootstocks
- Obtaining the benefits of certain interstocks (*double-working*)
- Changing cultivars of established plants (*top-working*)
- Hastening reproductive maturity and earlier fruit production
- Hastening plant growth rate and reducing production time
- Obtaining special forms of plant growth
- Repairing damaged parts of trees

- Study and elimination of virus diseases
- Study of plant developmental and physiological processes

Each of these reasons is discussed in detail below.

Perpetuating Clones That Cannot Be Readily Maintained or Economically Propagated by Cuttings, Layers, Division, or Other Asexual Methods

Cultivars of some groups of plants, including most fruit and nut species and many other woody plants, such as selected cultivars of fir, eucalyptus, beech, oak, and spruce, are not propagated commercially by cuttings because of poor rooting. Additional individual plants often can be started by the slow and labor-intensive techniques of layering or division. But for propagation in large quantities, it is necessary to resort to budding or grafting scions of the desired cultivar on compatible seedling rootstock plants.

Most woody plants are highly heterozygous, so seed propagation will not maintain genetic uniformity. In established cultivars, buds and scions are invariably taken from the adult growth phase to avoid juvenility delays in fruiting.

In forestry, grafting is used almost exclusively for the clonal production of genetically improved seed orchards of radiata pine (*Pinus radiata*), hoop pine (*Araucaria cunninghamii*), slash pine (*P. elliottii*), Caribbean pine (*P. caribaea*), eucalyptus (*Eucalyptus nitens*), Douglas-Fir (*Pseudotsuga menziesii*), etc. (*121*). The major advantage of using grafts is that superior germ plasm from older, elite trees can be clonally regenerated as parent trees for seed orchards. Frequently, trees selected for breeding or seed orchard purposes are so old (often greater than 15 or 20 years) that clonal production by rooted cuttings is either impossible or far more costly than grafting. Where graft incompatibility is not a serious problem, grafting scions of elite trees onto established seedling rootstock is a quick, straightforward, and cost-effective way of developing seed orchards.

Obtaining the Benefits of Certain Rootstocks

Many plant cultivars selected for their desirable fruit or ornamental qualities do not have comparably suitable root systems, but require grafting onto other roots to give satisfactory plants. For many kinds of plants, rootstocks are available that tolerate unfavorable conditions, such as heavy, wet soils (*128, 130, 131, 133*), or resist soil-borne insect or disease organisms (*97*) better than the plant's own roots. See Chapters 19 and 20 for detailed discussions of the rootstocks available for the various fruit and ornamental species. Also, for some species, size-controlling rootstocks are available that can cause the composite grafted plant to have exceptional vigor or to become dwarfed (Figure 12–3). Scions grafted on selected rootstock of some citrus, pear, and apple rootstocks produce larger size and/or better quality fruit than when grafted onto other rootstock (*180*).

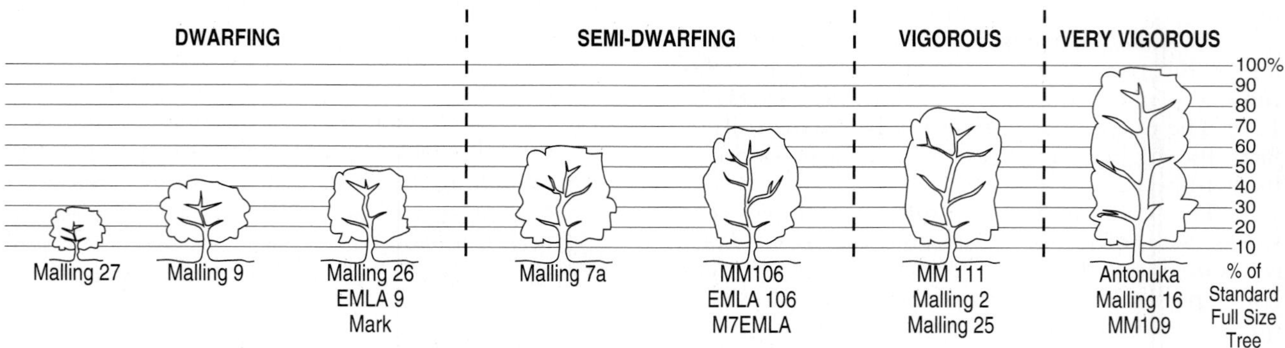

FIGURE 12–3 *Relative size* of apple trees on different rootstock. The reduction in tree size ranges from dwarfing (25 to 50 percent of a standard full size tree) to semi-dwarfing (60 to 70 percent) to vigorous to very vigorous (same size as a seedling tree). With the exception of Antonovka, all listed are clonal rootstock. The *absolute size* of the mature, composite tree is determined by soil, climate, culture, and the vigor of the scion cultivar, e.g., the scions of the vigorous cultivar 'Mutsu' are twice as large as 'Golden Delicious' on 'Malling 9' dwarfing rootstock.

Scions of many fruit crops can be established more quickly in the orchard and come into bearing more rapidly when grafted onto dwarfing rootstock as opposed to being grown as seedlings or as rooted cuttings. An exception to this is peach production in Mexico, where precocious seedlings are selected for fruit production; seedlings fruit as rapidly as grafted plants.

Some plants rooted by cuttings make such poor root systems that they are difficult or impossible to transplant, for example, the Koster spruce (*Picea pungens*) can be rooted in commercial numbers, but can not be successfully transplanted unless the root system is produced from grafted plants (*49*). Many Asiatic maples form poor root systems from cuttings, and must be grafted (*170*).

Special rootstocks for glasshouse vegetable crops are used in Europe and Asia to avoid root diseases such as Fusarium and Verticillium wilt (*70*). In the Netherlands, greenhouse cucumbers are grafted onto *Cucurbita ficifolia,* and commercial tomato cultivars are grafted onto vigorous F_1 hybrid, disease-resistant rootstocks (*15, 156*).

Obtaining the Benefits of Certain Interstocks (Double-Working)

More than two plants can be combined together in a vertical arrangement. In addition to the rootstock and scion, one may insert a third plant system between them by grafting. Such a section is termed an **interstock, intermediate stem section, interstem,** or **intermediate stock.** This is done by making two grafts (Figure 13–43), or double budding, e.g., a thin plate (minus the bud) of 'Old Home' pear interstock is budded on the quince rootstock, then a shield bud of the 'Bartlett' scion is inserted directly over the 'Old Home' plate and wrapped with a budding rubber (Figure 14–19).

There are several reasons for using double-working in propagation:

- The interstock makes it possible to avoid certain kinds of incompatibility.
- The interstock may possess a particular characteristic (such as disease resistance or cold-hardiness) not possessed by either the rootstock or the scion.
- A certain scion cultivar may be required for disease resistance in cases where the interstock characteristics are the chief consideration, such as in the control of leaf blight on *Hevea* rubber trees (*85*).
- The interstock may reduce vegetative growth and enhance reproductive growth of the tree. For example, when a stem piece of the dwarfing 'Malling 9' apple rootstock is used as an interstock and inserted between a vigorous rootstock and a vigorous scion cultivar, it reduces growth of the composite tree and stimulates flowering and fruiting in comparison with a similar tree propagated without the interstock (*134*) (see Figure 12–4).

Nurseries supplying trees on seedling or clonal rootstocks, or with a clonal interstock, should identify such stocks on the label just as they do for the scion cultivar.

Changing Cultivars of Established Plants (Topworking)

A fruit tree, or an entire orchard, may be an undesirable cultivar. It could be unproductive, or an old cultivar whose fruits are no longer in demand; it could be one with poor growth habits, or possibly one that is susceptible to prevalent diseases or insects. Topworking is done by California producers of peaches, plums, and nectarines every two to three years. This allows producers to remain competitive by introducing new cultivars and extending the production season of their orchards with different cultivars. Examples of topworking are shown in Figures 13–39, 13–40, and 13–41.

In an orchard of a single cultivar of a species requiring cross-pollination, provision for adequate cross-pollination can be obtained by topworking scattered trees throughout the orchard to a proper pollinating cultivar. A single pistillate (female) plant of a dioecious (pistillate and staminate flowers borne on separate individual plants) species, such as the hollies (*Ilex*), may be unfruitful because of the lack of a nearby staminate (male) plant to provide proper pollination. This problem can be corrected by grafting a scion taken from a staminate plant onto one branch of the pistillate plant.

The home gardener may be interested in growing several cultivars of a fruit species together on a single tree of that species by topworking each primary scaffold branch to a different cultivar. In a few cases, different species can be worked on the same tree. For example, a single citrus tree would grow oranges, lemons, grapefruit, mandarins, and limes; or plum, almond, apricot, and nectarine can be grafted on peach rootstock. Some different cultivars (or species), however, grow at different degrees of vigor, so careful pruning is required to cut back

FIGURE 12–4 Effect of interstock on size of six-year-old 'Cox's Orange Pippin' apple scion grafted on a vigorous 'MM 104' rootstock. *Top left:* Cox/'M 9' dwarfing interstock/ 'MM 104.' *Top right:* Cox/'M 27' dwarfing interstock/'MM 104.' *Lower left:* Cox/'MM 104' vigorous interstock/'MM 104.' *Lower right:* Cox/'M 20' dwarfing interstock/'MM 104.'

the most vigorous cultivar on the tree to prevent it from becoming dominant over the others.

Walnut and pistachio are difficult to transplant. Producers will plant seedling rootstock in the orchard, and then graft two years later.

Hastening Reproductive Maturity of Seedling Selections

In fruit breeding programs, if young seedling selections are grown on their own roots, they may take five to ten or more years to grow out of the juvenile phase and come into bearing. It is possible to hasten the onset of maturity by grafting terminal shoots from seedlings onto large, established trees (*183*) or onto certain dwarfing rootstocks (*169*). Such grafting takes advantage of an existing large root system of the rootstock plant to speed up maturation of the juvenile scion.

Hastening Plant Growth Rate and Reducing Nursery Production Time

In nursery production of shade trees, budded or grafted trees grow more rapidly than seedling or cutting-produced trees, e.g., *Acer platanoides* 'Crimson King' budded on a vigorous rootstock, budded *Tilia cordata* or budded *Zelkova serrata* grow more in one year than rooted cuttings will in three or four years (*49*).

Obtaining Special Forms of Plant Growth

By grafting certain combinations together it is possible to produce unusual types of plant growth, such as "tree" roses (Figure 12–5) or "weeping" cherries or birches. Often an intermediate upright growing stock must be used to obtain the height needed before grafting on the "weeping" top scion cultivar (see Figure 12–6).

FIGURE 12–5 Double-working in the production of specialty "tree" roses. The interstock *Multiflore de la Grifferaie* forms the straight trunk of the tree rose.

FIGURE 12–6 "Weeping" plant forms may be obtained by grafting. An upright growing cultivar is grafted at the top by a side graft with another cultivar having a hanging growth pattern.

Cactus is easily grafted to produce unusual plant forms, as shown in Figure 12–7.

Repairing Damaged Parts of Trees

Occasionally, the roots, trunk, or large limbs of trees are severely damaged by winter injury, cultivation implements, diseases, or rodents. By the use of bridge grafting, or inarching, such damage can be repaired and the tree saved. This is discussed in detail in Chapter 13.

Study and Elimination of Virus Diseases

Virus diseases can be transmitted from plant to plant by grafting. This characteristic makes possible testing for the presence of the virus in plants that may carry the pathogens but show few or no symptoms. By grafting scions or buds on a plant suspected of carrying the virus onto an indicator plant known to be highly susceptible, and which shows prominent symptoms, detection is easily accomplished (*98*). This procedure is known as *indexing* (see Chapter 9).

In order to detect the presence of a latent virus in a symptomless carrier, it is not necessary to use combinations which make a permanent, compatible graft union. For example, the 'Shirofugen' flowering cherry (*Prunus serrulata*) is used to detect viruses in peach, plum, almond, and apricot. Cherry does not make a compatible union with these species, but a temporary, incompatible union is a sufficient bridge for virus transfer.

Thermotherapy (heat treatment) is used to rid scion material of viruses (Figure 9–25). After the virus-free material is indexed, as indicated above, or tested with serological techniques, it can be multiplied by traditional grafting/budding techniques. Micrografting under aseptic tissue culture conditions is another technique used to clean up viruses and bacterial problems with budwood (*115*).

Study of Plant Developmental and Physiological Processes

Grafting has enabled plant biologists to study some unique physiological and developmental processes, beginning in the 1700s with Stephen Hale's studies on the circulation of plant sap. Grafting has been used successfully to study transmissible factors (*99*) in flowering (*40*), tuber initiation, the control of

FIGURE 12–7 Grafted ornamental cactus. An easily rooted cultivar is used as the rootstock and an unusual attractive type is used as the scion. These grafts are made in large quantities in Japan and shipped to wholesale nurseries in other countries for rooting, potting, and growing until ready for sale in retail outlets.

branching (*14*), and promotion of cold-hardiness between induced and noninduced organs. The use of multiple graft combinations, including reciprocal and autografting, has facilitated studies on promoters and inhibitors in adventitious rooting (*54*), root regeneration potential, and rejuvenation of mature phase plant material (*120*).

NATURAL GRAFTING

Occasionally, branches become naturally grafted together following a long period of being pressed together without disturbance. In commercial orchards, limbs of fruit trees are sometimes deliberately "braced" together and allowed to naturally graft, forming a stronger scaffold system to better support the fruit load of the tree (Figure 13–29).

Natural grafting of roots is not so obvious, but more significant and widespread, particularly in stands of forest species, such as pine, hemlock, oak, and Douglas-Fir (*56, 97*). Such grafts are common between roots of the same tree or between roots of trees of the same species. Grafts between roots of trees of different species are rare. In the forest, living stumps sometimes occur, kept alive because their roots have become grafted to those of nearby intact, living trees, allowing the exchange of nutrients, water, and metabolites (*95*).

The anatomy of natural grafting of aerial roots has been studied (*132*). Natural root grafting also permits transmission of fungi, viruses, and phytoplasmas from infected trees to their neighbors (*46*). This problem can occur in orchard and nursery plantings of trees, and urban shade tree sites where numerous root grafts may result in the slow spread of pathogens throughout the planting. Natural root grafting is a potential source of error in virus-indexing procedures where virus-free and virus-infected trees are grown in close proximity (*57*). In addition, fungal pathogens such as those causing oak wilt and Dutch elm disease can be spread by such natural root connections.

FORMATION OF THE GRAFT UNION

A number of detailed studies have been made of graft union formation, with woody (*4, 5, 33, 47, 135, 159*) and herbaceous plants (*91, 104, 108, 127, 152, 160, 165, 190*). Just as *de novo* meristems are necessary for adventitious bud and root formation, a *de novo* formed meristematic area must develop between the scion and rootstock if successful graft union formation is to occur (*190*). The parts of the graft that are originally prepared and placed in close contact do not themselves move about or grow together. Rather, the union is accomplished entirely by cells that develop *after* the actual grafting operation has been made. The graft union is initially formed by rapidly dividing callus cells, originating from the scion and rootstock, which later differentiate to form the vascular cambium (a lateral meristem) and the associated vascular system.

The development of a compatible graft is typically comprised of three major events: *adhesion* of the rootstock and scion, proliferation of callus cells at the graft interface or *callus bridge;* and *vascular differentiation* across the graft interface (*106*).

The scion will not resume its growth successfully unless a vascular connection has been established so that it may obtain water and mineral nutrients. Likewise, degeneration of the rootstock will occur if the phloem in the graft union is disrupted from sending carbohydrates and other metabolites from the scion to the root system. In addition, the scion must have a terminal meristematic region—a bud—to resume shoot growth, and eventually to supply photosynthates to the root system.

Considering in more detail the **steps involved in graft union formation** (Figures 12–8 and 12–9), the first one listed below is a preliminary step, but nevertheless, it is essential and one over which the propagator has control.

1. Lining up of vascular cambiums of the rootstock and scion. The statement is often made that for successful grafting the cambium layers of rootstock and scion must be "matched." Although this is desirable, it is unlikely that complete matching of the two cambium layers is obtained. They normally are only one to several cell layers thick. In fact, it is only necessary that the cambial regions be close enough together so that the parenchyma cells from both rootstock and scion produced in this region can become interlocked. In a mismatched rootstock and scion, where one partner has a greater diameter than the other, lining up the periderm on at least one side of the rootstock and scion generally assures that their cambiums are in close enough proximity to interconnect through the callus bridge (Figures 13–4 and 14–7). The cambium is critical for maintaining vascular connections in the callus bridge.

Two badly matched cambial layers may delay the union or, if extremely mismatched, prevent the graft union from taking place, leading to graft failure (*152*). With vanilla, which is a herbaceous, monocotyledonous plant, the cambium layer is not necessarily required for forming the graft union, since any parenchyma cells capable of dividing will produce callus tissue and lead to the formation of a union between the rootstock and scion (*114*). However, a continuous cambium layer in the graft union

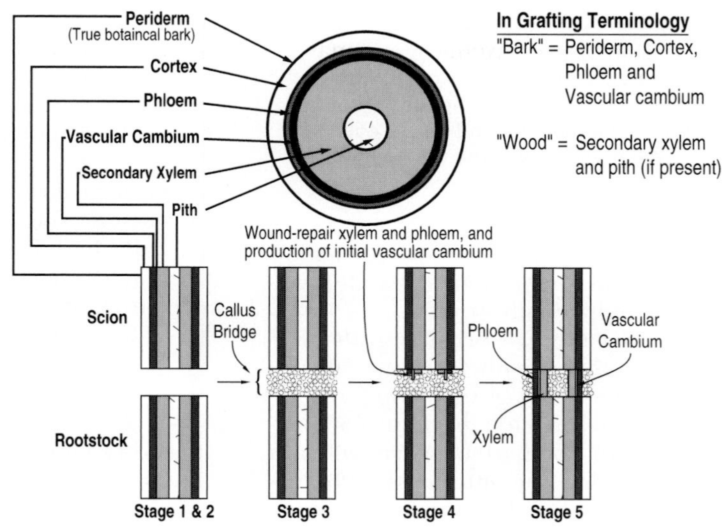

FIGURE 12–8 *Top:* Grafting terminology of the "bark" and "wood" and associated tissues with schematic drawing of a stem cross section of a young woody plant stem. *Bottom:* Schematic longitudinal section of the stages of graft union formation: (1) Lining up of vascular cambiums of the rootstock and scion, and (2) subsequent wound healing response. (3) Callus bridge formation. (4) Wound-repair xylem and phloem occur in the callus bridge just prior to initial cambium formation. (5) The vascular cambium is completed across the callus bridge and is forming secondary xylem and phloem.

GRAFT UNION FORMATION IN PEA ROOTS

Day 1-2	Initial cell divisions to compartmentalize the wounds in both the scion and rootstock.
Day 2-7	Continued proliferation of parenchymatous callus from both grafting partners in the graft union area.
Day 4	First wound-repair vascular tissue differentiation in the graft union area. The necrotic layer (cells killed by wounding) is disrupted and largely disappears.
Day 7	Callus bridge formation is finished with the complete filling of graft union area with parenchymatous callus. Wound-repair xylem (from callus cells) links the callus bridge.
Day 8	Wound-repair phloem (from callus cells) links the callus bridge.
Day 12	Wound-repair cambium (from callus cells) links the callus bridge.
After Day 12	Production of secondary xylem and phloem by reconstituted cambium in the callus bridge.

FIGURE 12–9 Graft union formation in grafted pea roots (*91, 160*). This sequence of grafting events is common to top-grafting and root grafting in many other woody and herbaceous plant species. What will vary is the time period in grafting events with different species.

is necessary for successful graft union formation with woody perennial angiosperms and gymnosperms that are commercially grafted.

It is essential that the two original graft components be held together firmly by some means, such as wrapping, tying, stapling, or nailing, or better yet, by wedging (as in the cleft graft, or machine-notched chip budding) so that the parts will not move about and dislodge the interlocking parenchyma cells after proliferation has begun.

2. Wounding response. A **necrotic layer** or **isolation layer** forms from the cell contents and cell walls of the cut scion and rootstock cells. In the actual cutting of cells of both scion and rootstock during the preparation of the graft, cells are killed at each surface at least one cell layer deep. Much of the necrotic layer material later disappears or it may remain in pockets between subsequently formed callus produced by actively dividing parenchyma cells. Undifferentiated **callus tissue** is produced from uninjured, rapidly dividing parenchyma cells (adjacent and internal to the necrotic layer). The callus tissue initially forms a **wound periderm.**

WOUNDING RESPONSE

Some literature refers to a "wound healing response" (*19*), "wound healing process" (*29*), or "healing of the graft union." A wounded area of a plant is not healed per se by the replacement of injured tissues, rather it is compartmentalized or walled-off from the rest of the plant as a defensive mechanism to eliminate invasion of pathogens, etc. (*138, 139, 150, 151*); this is all part of the *response* or *reaction* to wounding, which occurs in grafting, budding, or the propagation of a cutting. A *necrotic plate* or *isolation layer* at the graft interface is first formed, which helps adhere the grafted tissues together, especially near the vascular bundles (*165*). Wound repair occurs by meristematic activity, which results in the initial formation of a wound periderm between the necrotic layer and uninjured tissue—the wound periderm becomes suberized to further reduce pathogen entry (*29*). In graftage, with the close-proximity physical contact of scion and stock cells, and pressure exerted on the graft union area from the scion and rootstock tied or wedged together, the necrotic layer does not form a barrier to graft union formation. Rather, profuse callusing causes the majority of the necrotic layer to disappear (in most situations) (*160, 165*). Further meristematic activity occurs in graft union formation, culminating with the formation of a vascular cambium in the callus bridge area.

3. Callus bridge formation. New parenchymatous callus proliferates in one to seven days from both the rootstock and scion (*165*) (Figures 12–8, 12–9, 12–10, 12–11). The callus tissue continues to

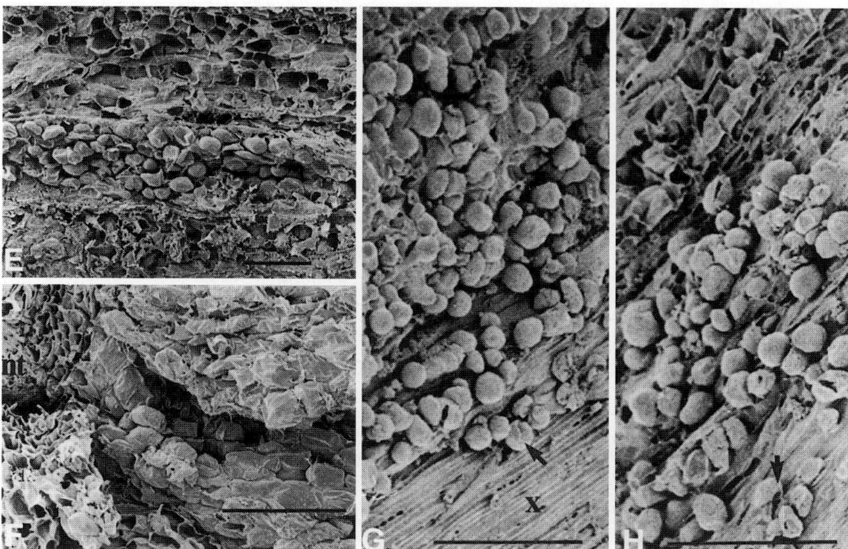

FIGURE 12–10 Early callus bridge formation in Sitka spruce *(Picea sitchensis). Top left:* scanning electron photos of a cross section of rootstock wound surface at seven days with a cluster of callus cells (arrow) formed in the cortical region. *Bottom left:* scion wound surface at seven days with callus cells (arrow) associated with the needle trace (nt) in the outer cortex. *Middle:* rootstock wound surface of a nine-day-old graft with well-established callus originating from ray cells in the xylem (x) close to the cambium (arrow). *Right:* Scion wound surface of a nine-day-old graft showing callus formation mainly in the cambium region. Callus is also produced from ray cells in the xylem (arrow), and from phloem parenchyma cells. (Courtesy of J. R. Barnett (*4*).)

form by further cell divisions of the outer layers of undamaged parenchyma cells [in the *cambial region, cortex, pith* (*160*)—or *xylem ray parenchyma* (*5*)] in the scion and rootstock. The actual cambial tissue plays a lesser role in callus formation of the wound periderm and callus bridge formation than originally supposed (*146, 160*). These new parenchyma cells produced are adjacent and internal to the necrotic layer; soon they intermingle and interlock, filling up the spaces between scion and stock (Figure 12–12).

In grafting scions on larger, established rootstocks (e.g., topworking in the field), the rootstock produces most of the callus. However, when the graft partners are of equal size, the scion forms much more callus than does the rootstock (*33, 160*). This result is explained by natural polarity, since the root-tip-facing end of any segment of root or shoot (proximal end) forms more callus than the shoot-tip-facing (distal) end (Figure 12–20) (*18*). In budding, the sizes of the cut surfaces are so different that it is difficult to distinguish which grafting partner contributes the most callusing (*23*).

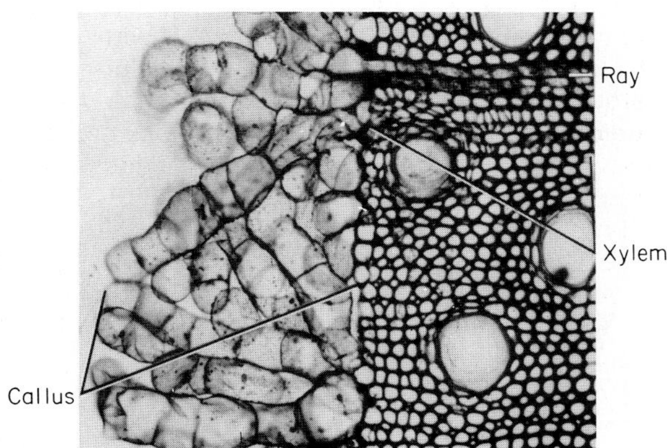

FIGURE 12–11 Callus production from incompletely differentiated xylem, exposed by excision of a strip of bark. × 120. (Reprinted with permission from K. Esau, *Plant Anatomy,* John Wiley & Sons, Inc., New York, 1953.)

Adhesion between cells of the scion and rootstock is aided by a "cement" or binding material, which projects beadlike projections from the surface of the callus cells of both grafting partners. A general fusion of the cell walls then follows (*5, 75, 160*). The beadlike projections are a mix of pectins, carbohydrates, and proteins (*96*). The cells do not need to divide before producing the cement, and the cement can bond the graft partners, regardless of the absence or presence of the necrotic layer (*160*).

It is not clear if a specific cell-to-cell recognition in grafting is required as part of adhesion and the events that follow successful graft union formation. The formation of superimposed sieve areas and sieve plates (in phloem sieve elements), pits and performation plates (in xylem elements), and the *plasmodesmata*[1] (in vascular parenchyma) may require some sort of cellular recognition or

[1]Plasmodesmata are minute cytoplasmic threads that extend through openings in cell walls and connect the protoplasts of adjacent living cells at the graft interface.

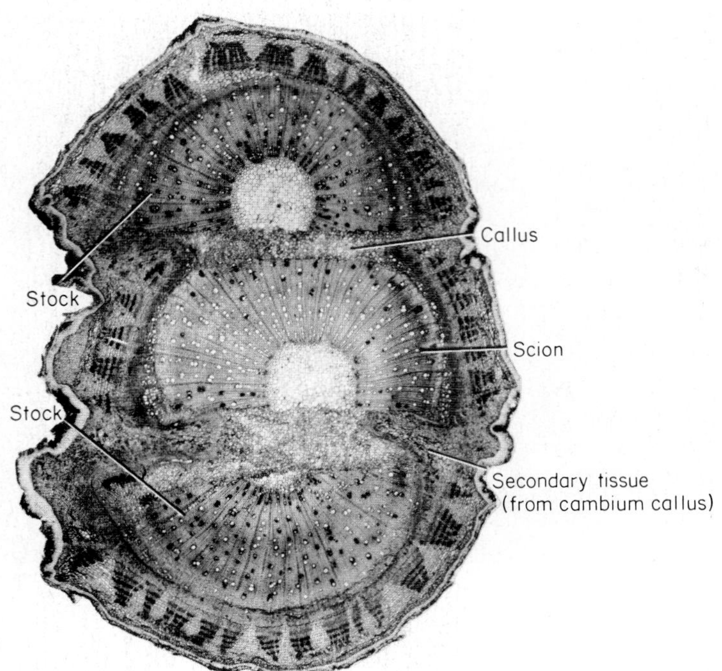

FIGURE 12–12 Cross section of a *Hibiscus* wedge graft showing the importance of callus development in the healing of a graft union. Cambial activity in the callus has resulted in the production of secondary tissues which have joined the vascular tissues of the stock and scion. × 10. (Reprinted with permission from K. Esau, *Plant Anatomy,* John Wiley & Sons, Inc., New York, 1953.)

cellular communication (*108*). For cell recognition, the pectin fragments during the adhesion process may act as signalling molecules. This is discussed later in the section on graft compatability-incompatibility.

Underneath the necrotic layer, parenchyma cells show increased cytoplasmic activity with, in some plants at least, a pronounced accumulation of *dictyosomes*[2] along the graft interfaces (Figure 12–13) (*104, 107, 108*). These dictyosomes appear to secrete materials into the cell wall space between the graft components via vesicle migration to the plasmalemma, resulting in a rapid adhesion between parenchymatous cells at the graft interface.

4. Wound-repair xylem and phloem; differentiation of vascular cambium across the callus bridge. In both woody and herbaceous plants, the initial xylem and phloem are generally differentiated prior to the bridging of vascular cambium across the callus bridge (Figures 12–8 and 12–9) (*33, 35, 53, 79, 160*). The *wound-repair xylem* (wound-type vascular elements) is generally the first *differentiated* tissue to bridge the graft union, fol-

[2]Dictyosomes are a series of flattened plates or double lamellae—one of the component parts of the Golgi apparatus.

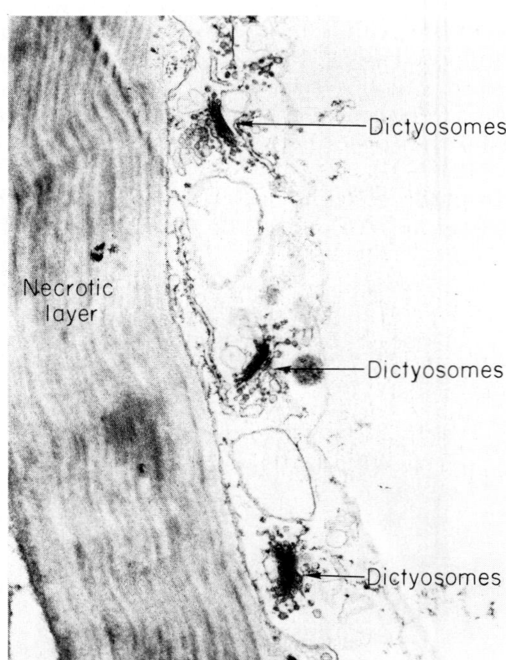

FIGURE 12–13 Accumulation of dictyosomes along the cell walls adjacent to the necrotic layer at six hours after grafting in the compatible autograft in *Sedum telephoides.* × 17,500. (Courtesy R. Moore and D. B. Walker (*108*).)

lowed by *wound-repair phloem.* Initial xylem tracheary elements and frequently initial phloem sieve tubes form directly by differentiation of callus into these vascular elements. A vascular cambium layer subsequently forms between the vascular systems of the scion and rootstock.

Exceptions to this developmental sequence are in bud graftage in citrus, apple, and rose where the wound cambium differentiates prior to the bridging of vascular tissue, and in autografts of *Sedum* (Crassulaceae) where procambial differentiation occurs before vascular differentiation (*108*). With budding, the scion is considerably smaller and normally limited to one bud and a short shoot piece; hence, any early vascular differentiation from callus cells is probably limited by lower phytohormone levels. The cambium can form independent of any xylem or phloem (*23*), or the cambium may differentiate between the wound-bridging xylem and phloem (*160*). It is important that the vascular cambium unite so that the continuity of wound-bridging xylem and phloem can be maintained and so that secondary vascular development occurs for successful graft union formation.

At the edges of the newly formed callus mass, parenchyma cells touching the cambial cells of the rootstock and scion differentiate into new cambium cells within two to three weeks after grafting. This cambial formation in the callus mass proceeds farther and farther inward from the original rootstock and scion cambium, and on through the **callus bridge,** until a continuous cambial connection forms between rootstock and scion.

5. Production of secondary xylem and phloem from the new vascular cambium in the callus bridge. The newly formed cambial layer in the callus bridge begins typical cambial activity, laying down new secondary xylem toward the inside and phloem toward the outside.

In the formation of new vascular tissues following cambial continuity, it appears that the type of cells formed by the cambium is influenced by the cells of the graft partners adjacent to the cambium. For example, xylem ray cells are formed where the cambium is in contact with xylem rays of the rootstock, and xylem elements where they are in contact with xylem elements (*125*).

Production of new xylem and phloem thus permits the vascular connection between the scion and the rootstock. It is essential that this stage be completed before much new leaf development arises from buds on the scion. Otherwise, the enlarging leaf surfaces on the scion shoots will have little or no water to offset that lost by transpiration, and the scion will quickly become desiccated and die. It is possible, however, even though vascular connections fail to occur, that enough translocation can take place through the parenchyma cells of the callus to permit survival of the scion. In grafts of *Vanilla* orchid, a monocot, scions survived and grew for two years with only union of parenchyma cells, however, the grafted plants did not survive when subjected to transpirational stress (*114*).

CORRELATIVE EFFECTS OF SCION BUDS AND LEAVES ON XYLEM AND PHLOEM FORMATION

The first vascular tissues produced in the callus bridge are wound-repair xylem and phloem. The new wound-repair xylem tissue originates from the activities of the scion tissues rather than from that of the rootstock (*147, 188*). The amount of initial graft-bridging xylem is strongly influenced by the presence of leaves and branches on the scion, and not by the presence of the rootstock (*160*). The scion buds are effective in inducing differentiation of vascular elements in the tissues onto which they are grafted. This bud influence has been shown by inserting a scion bud into a root piece of *Cichorium* rootstock. Under the influence of auxin produced by the bud, the old parenchyma cells differentiate into groups of conducting xylem elements (*56*).

Induction of vascular tissues in callus is under the control of phytohormones (principally auxins) and other metabolites originating from growing points of shoots (*167*). Auxins (IAA or NAA) will cause the induction of wound-repair xylem, while auxins and carbohydrates can induce wound-repair phloem in callus tissue (*1, 118*). Auxin can also induce cambial formation when applied to wounded vascular bundles of cactus rootstock (*152*). For successful graft union formation of *in vitro* grafted internodes, auxin is an absolute requirement, cytokinin stimulates graft development, but gibberellic acid is inhibitory (*119*). Auxins enhance grafting success in root-grafting pecan trees (*187*). In cactus grafts, auxin can also promote vascular connections (Figure 12–14) (*15*).

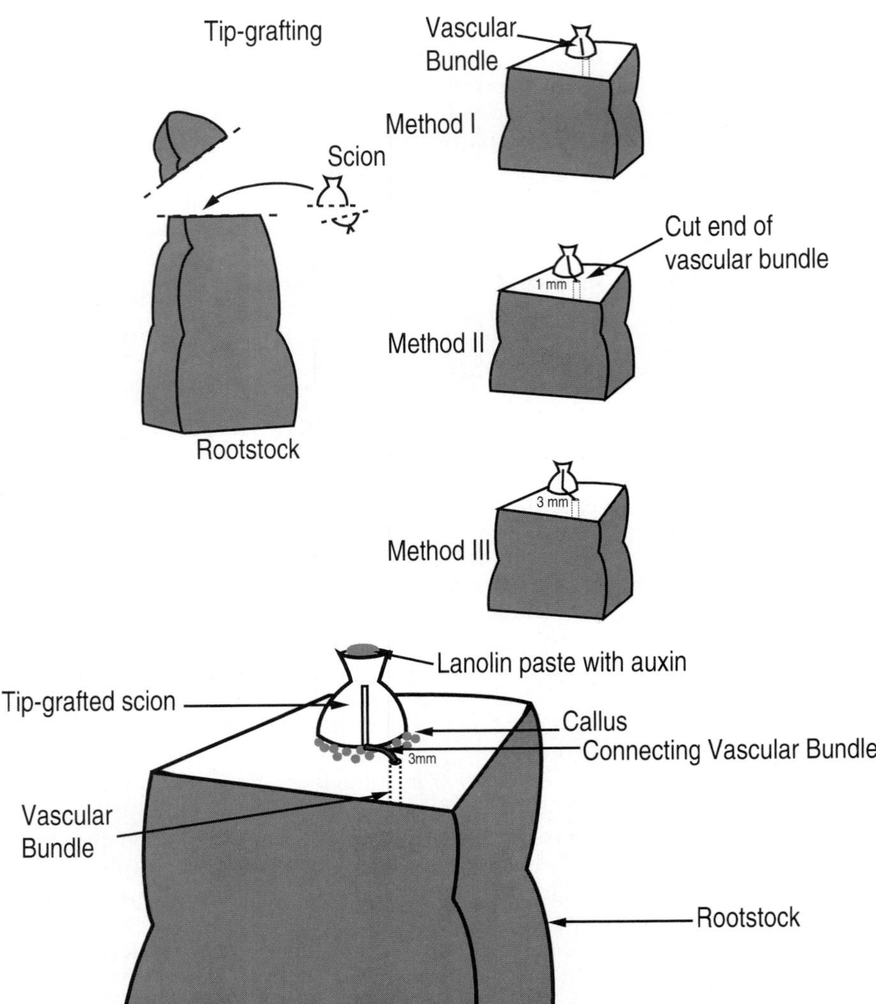

FIGURE 12–14 Schematic of tip grafting of cactus. *Top:* In Method I, the vascular bundles of the scion and rootstock were placed together, or 1 mm (Method II) or 3 mm (Method III) apart. *Bottom:* Auxin in lanolin paste promoted vascular connections between misaligned graft partners, and increased the diameter of the connecting vascular bundle. (Redrawn from Shimomura and Fuzihora (*152*).)

GRAFT UNION FORMATION IN T- AND CHIP BUDDING

In T-budding, the bud piece usually consists of the "bark" (periderm, cortex, phloem, cambium), and often some "wood" (xylem tissue). Attached externally to this is a lateral bud subtended, perhaps, by a leaf petiole. In budding, this piece of tissue is laid against the exposed xylem and cambium of the rootstock, as shown diagrammatically in Figure 12–16.

Detailed studies of the grafting process in T-budding have been made for the rose (*23*), citrus (*94*), and apple (*113*).

In the apple, when the flaps of bark on either side of the "T" incision on the rootstock are raised, separation occurs from the young xylem. The entire cambial zone remains attached to the inside of the bark flaps. Very shortly after the bud shield is inserted, a necrotic plate or layer of material develops from the cut cells. Next, after about two days,

SYMPLASTIC AND APOPLASTIC CONNECTIONS BETWEEN THE SCION AND ROOTSTOCK

There is little evidence to support the notion that cells in the callus bridge are fused per se (as in protoplast fusion), but there is evidence that parenchym cells of the graft partners are interconnected by plasmodesmata (*75, 110*); these cytoplasmic strands form continuous, *symplastic cell connections,* linking up cell membranes that form a potential pathway of communication among cells in the graft bridge. This may be important in cell recognition, and compatibility/incompatibility response, which is discussed later. *Apoplastic connections* occur during adhesion of the graft with free cell walls of both graft partners coming together and adhering by means of their extracellular pectinacous beads.

In a compatible graft, the wound response is followed by dissolution of the necrotic layer, perhaps as a prerequisite to the formation of secondary plasmodesmata between cells of the graft partners (*165*).

The secondary plasmodesmata are formed *de novo* across the fused callus walls, particularly near cut vascular strands (*83*). In the formation of *de novo* plasmodesmata, development of continuous cell connections starts with the thinning and loosening of local wall regions opening the chance of fusion of plasmalemma (cell or protoplast membrane) and endoplasmic reticulum[3] between the adjoining cells (*83*). Golgi vescicles secreting wall material are also involved in the process (Figure 12–15). Sieve elements in the connecting phloem of the grafting partners are also interconnected, further demonstrating *symplastic connections* between the graft partners (*110*).

callus parenchyma cells start developing from the rootstock xylem rays and break through the necrotic plate. Some callus parenchyma from the bud scion ruptures through the necrotic area in a similar manner. As additional callus is produced it surrounds the bud shield and holds it in place. The callus originates almost entirely from the rootstock tissue, mainly from the exposed surface of the xylem cylinder. Very little callus is produced from the sides of the bud shield (scion).

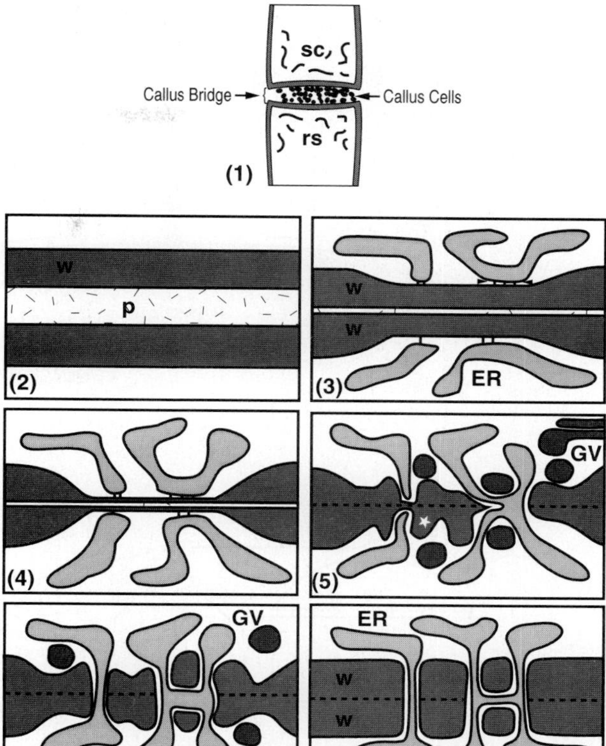

FIGURE 12–15 Schematic diagram of secondary *(de novo)* formation of plasmodesmata at the graft interface (callus bridge). **(1)** approaching callus cells of scion (sc) and rootstock (rs). Pectic material (p) between adjoining callus cell walls. Region between arrows: wall parts where secondary plasmodesmata will be formed as shown in detail. Formation of *continuous* cell connections **(2 to 7)** by plasmalemma and endoplasmic reticulum (ER) fusion of adjoining cells **(5, 6)**, within wall parts which have been thinned synchronously with both cell partners. Elongation of the branched and single strands during rebuilding the modified wall parts **(6, 7)**. w=cell wall, GV=golgi vesicles, *new deposited wall material. (Redrawn from Kollman and Glockmann (*83*).)

Cell proliferation continues rapidly for two to three weeks until all internal air pockets are filled with callus. Following this, a continuous cambium is established between the bud and the rootstock. The callus then begins to lignify, and isolated xylem tracheary elements appear. Lignification of the callus

[3]The endoplasmic reticulum (ER) is a membrane system that divides the cytoplasm into compartments and channels. Rough ER is densely coated with ribosomes, whereas smooth ER has fewer ribosomes.

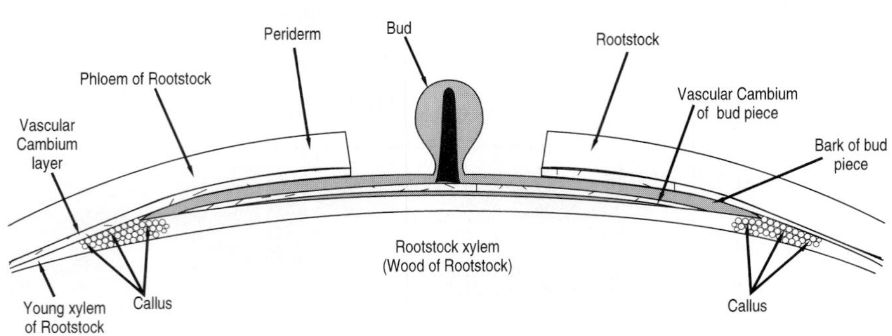

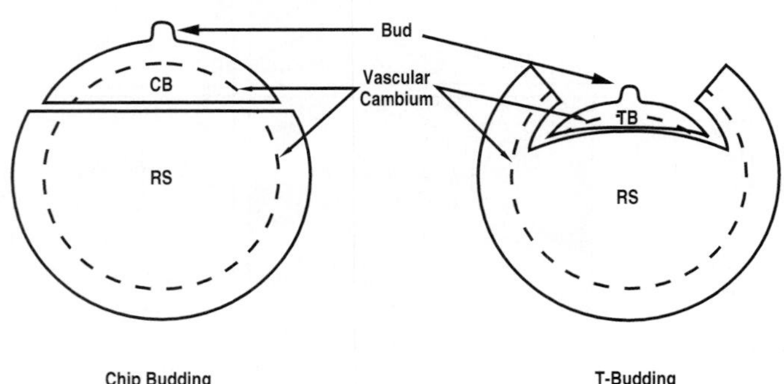

FIGURE 12–16 *Top:* Tissues involved in healing of an inserted T-bud as prepared with the "wood" (xylem) attached to the scion bud piece. Graft union formation occurs when callus cells developing from the young xylem of the rootstock intermingle with callus cells forming from exposed cambium and young xylem of the T-bud piece. As the bark is lifted on the rootstock for insertion of the bud piece it detaches by separation of the youngest xylem and cambial cells. *Bottom:* A cross section of a chip bud (CB), T-bud (TB), and rootstock (RS). Because the chip bud substitutes exactly for the part of the rootstock which is removed, the cambium of the roots and scion are placed close together, resulting in a rapid and strong union. When a T-bud (right) is slipped under the "bark," the cambium of the rootstock and scion are not adjacent, and the union formation can be weak and slow. (Redrawn from B. H. Howard, Horticulture Research International, East Malling, England (*71*).)

VARIOUS STAGES AND TIME INTERVALS INVOLVED IN GRAFT UNION FORMATION IN T-BUDDED CITRUS HAVE BEEN DETERMINED AS FOLLOWS *(94):*

Stage of Development	*Approximate Time After Budding*
1. First cell division	24 hours
2. First callus bridge	5 days
3. Differentiation of cambium	
a) In the callus of the bark flaps (rootstock)	10 days
b) In the callus of the shield bud (scion)	15 days
4. First occurrence of xylem tracheids	
a) In the callus of the bark flaps	15 days
b) In the callus of the shield	20 days
5. Lignification of the callus completed	
a) In the bark flaps	25 to 30 days
b) Under the shield	30 to 45 days

is completed about 12 weeks after budding (*113, 172*). The developmental stages and time intervals for graft union formation in T-budded citrus are listed on page 408.

More rapid union development in chip budding. Anatomical studies (*155*) have been made comparing graft union formation in T- and chip budding. Early union formation between 'Lord Lambourne' apple scion and 'Malling 26' dwarfing rootstock showed a more rapid and complete union of xylem and cambial tissues of the scion and rootstock after chip budding compared to T-budding. This is probably due to a much closer matching of the scion tissue to the rootstock stem (Figure 12–16). Also in T-budding, the cambium of the rootstock is lifted in the flap of "bark," so considerable callus in-filling and the development of a new cambium must occur. There is more flexibility in chip budding, which can be done over longer periods on either an active or dormant rootstock, unlike T-budding, which requires an active rootstock. In part this is due to less callus filling needed and because there is no requirement for an active cambium to lift the flap of rootstock bark as with T-budding.

The advantages of chip budding compared to T-budding have also been demonstrated with 'Crimson King' maple on *Acer platanoides* rootstock, 'Conference' pear on 'Quince A' rootstock, and 'Rubra' linden on *Tilia platyphyllos* rootstock.

FACTORS INFLUENCING GRAFT UNION SUCCESS

As anyone experienced in grafting or budding knows, the results are often inconsistent, an excellent percentage of "takes" occurring in some operations, whereas in others the results are disappointing. A number of factors can influence the healing of graft unions.

Incompatibility

One of the symptoms of incompatibility in grafts between distantly related plants is a complete lack, or a very low percentage, of successful unions. Incompatibility is discussed in greater detail starting on page 416. Grafts between some plants known to be incompatible, however, will initially make a satisfactory union, even though the combination eventually fails.

Plant Species and Type of Graft

Some plants are much more difficult to graft than others even when no incompatibility is involved. Difficult species include the hickories, oaks, and beeches. Nevertheless, such plants, once successfully grafted, grow very well with a perfect graft union. In grafting apples, grapes, and pears (Figure 12–17), even the simplest techniques usually give a good per-

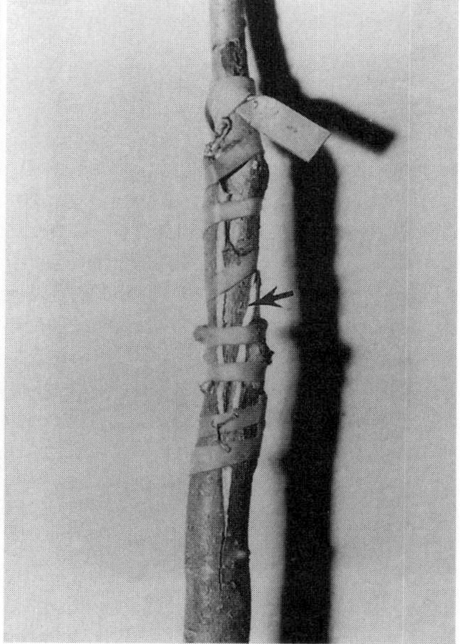

FIGURE 12–17 Some species form profuse callusing (arrows) which helps increase graft union success. *Left:* A saddle grafted grape. *Right:* Pear is easily grafted by a whip-and-tongue graft.

centage of successful unions, but grafting certain stone fruits, such as peaches and apricots, requires more care and attention to detail. Strangely enough, grafting peaches to some other compatible species, such as plums or almonds, is more successful than reworking them back to peaches. One method of grafting may give better results than another, or budding may be more successful than grafting, or vice versa. For example, gymnosperms are grafted, whereas many angiosperm cultivars tend to be budded, rather than grafted (*13*). In topworking black walnut *(Juglans hindsii)* to the Persian walnut *(Juglans regia)* in California, the bark graft method is more successful than the cleft graft. In nursery propagation of pecans, patch budding in Texas is preferred to the whip graft, which does better in climates with higher humidity, such as Mississippi.

Some species, such as mango *(Mangifera indica)* and *Camellia reticulata* are so difficult to propagate by the usual grafting or budding methods that they are *approach grafted.* Both graft partners are maintained for a time after grafting onto their own roots as containerized plants. This variation among plant species and cultivars in their grafting ability is probably related to their ability to produce callus parenchyma, and differentiate a vascular system across the callus bridge.

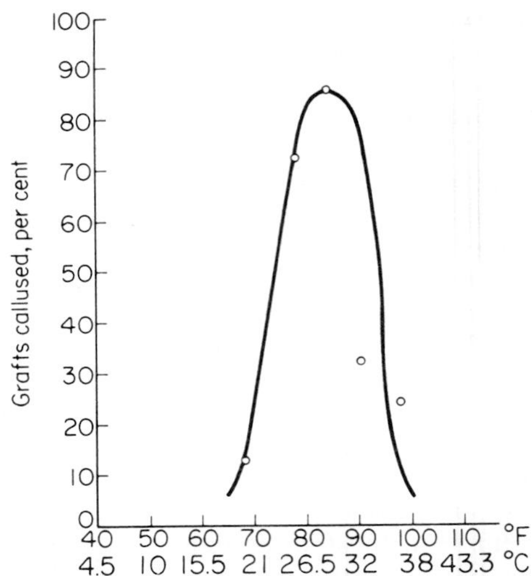

FIGURE 12–18 Influence of temperature on the callusing of walnut *(Juglans)* grafts. Callus formation is essential for the healing of the graft union. Maintaining an optimum temperature following grafting is very important for successful healing of walnut grafts. (Adapted from data of Sitton (*154*).)

Environmental Conditions During and Following Grafting

Certain environmental requirements must be met for callus tissue to develop.

Temperature

Compared to field grafting and budding, temperature levels for greenhouse containerized rootstock and bench grafting can, of course, be readily controlled, thereby permitting greater reliability of results and more flexibility of scheduling grafting and budding over a longer period of time. Temperature has a pronounced effect on the production of callus tissue (Figure 12–18). In apple grafts, little, if any, callus is formed below 0° C (32° F) or above about 40° C (104° F). At 32° C (90° F) and higher, callus production is retarded, with cell injury becoming more apparent as the temperature increases, until death of the cells occurs at 40° C. In bench grafting, callusing may be allowed to proceed slowly for several months by storing the grafts at relatively low temperatures, 7 to 10° C (45 to 50° F), or if rapid callusing is desired, they may be kept at higher temperatures for a shorter time. Too high a temperature to induce rapid callus development of bench-grafted plants can deplete needed carbohydrate reserves, which limits field survival (Figure 11–5) (*37*).

Following bench grafting of grapes, a temperature of 24 to 27° C (75 to 80° F) is about optimum; 29° C (85° F) or higher results in profuse formation of a soft type of callus tissue that is easily injured during the planting operations. At 20° C callus formation is slow, and below 15° C (60° F) it almost ceases.

Outdoor grafting operations should thus take place at a time of year when such favorable temperatures can be expected and when the plant tissues, especially the cambium, are in a naturally active state. These conditions generally occur during the spring months. Delay of outdoor grafting operations performed late in the spring, i.e., in the southern U.S. where excessively high temperatures may occur, often results in failure. Tests (*60*) of top-grafting walnut in California during hot weather in May showed that whitewashing the area of the completed graft union promoted healing of the union. Whitewash reflects the radiant energy of the sun, thus resulting in lower bark temperatures.

Moisture and Plant Water Relations

The cambium of the graft partners, and parenchyma cells comprising the important callus tissue are thin-walled and tender, with no provision for resisting desiccation. If exposed to drying air they will be killed. This was found to be the case in studies of the effect of humidity on healing of apple grafts. Air moisture levels below the saturation point inhibited callus formation. Desiccation of cells increased as the humidity dropped. *In vitro* studies (*41*) of stem pieces of ash *(Fraxinus excelsior)* have shown that callus production on the cut surfaces was markedly reduced as the water potential decreased.

Water is one of the driving forces for cell enlargement and is necessary for callus bridge formation between the stock and scion. Water must be utilized initially from scion tissue, and if below a certain water potential, insufficient water is available for callus formation. Failed grafts of well-hydrated Sitka spruce rootstocks produced no callus at the graft union, suggesting that callus formation at the cut surface is controlled or dependent on the formation of callus from the scion (*6*). Until vascular connections are formed between the rootstock and scion, the callus bridge provides the initial pathway for water, bypassing damaged xylem vessels and tracheids of the scion and rootstock. Within the first three to four days of callus bridge formation, there is a recovery of scion leaf water potential (Ψ_{leaf}) (*6*); with maturation of the connecting tracheids, (Ψ_{leaf}) and osmotic potential (Ψ_{π}) continue to increase (*10, 11*). Photosynthesis declines and does not increase until xylem connections become reestablished (*9*).

Unless the adjoining cut tissues of a completed graft union are kept at a very high humidity level, the chances of successful healing are poor. With most plants, thorough waxing of the graft union or sealing of the graft union with polyethylene grafting tape or Parafilm, helps retain the natural moisture of the tissues, which is all that is necessary. Often root grafts are not waxed but stored in a moist (not overly wet) packing material during the callusing period. Damp peat moss or wood shavings are good media for callusing, providing adequate moisture and aeration.

Growth Activity of the Rootstock

Some propagation methods, such as T-budding and bark grafting, depend upon the bark "slipping," which means that the cambium cells are actively dividing, producing young thin-walled cells on each side of the cambium. These newly formed cells separate easily from one another, so the bark "slips" (Figures 12–16, 13–17, and 13–18). Chip budding can be done on a dormant or active rootstock. Hence, there is much more flexibility in scheduling chip budding, because there is no requirement for an active cambium to lift the flap of rootstock bark, as with T-budding.

Initiation of cambial activity in the spring results from the onset of bud activity, since shortly after the buds start growth, cambial activity can be detected beneath each developing bud, with a wave of cambial activity progressing down the stems and trunk. This stimulus is due, in part, to production of auxin originating in the expanding buds (*175*). Callus proliferation—essential for a successful graft union—occurs most readily at the time of year just before and during "bud-break" in the spring. This is due to auxin gradients that diminish through the summer and into fall. Increasing callus proliferation takes place again in late winter, but this is not dependent upon breaking of bud dormancy.

In T-budding seedlings in the nursery in late summer, it is important that they have an ample supply of soil moisture just before and during the budding operation. If they should lack water during this period, active growth is checked, cell division in the cambium stops, and it becomes difficult to lift the bark flaps to insert the bud. At certain periods of high growth activity in the spring, plants exhibiting strong root pressure (such as the walnut, maple, and grape) show excessive sap flow or "bleeding" when cuts are made preparatory to budding and grafting. Grafts made with such moisture exudation around the union will not heal properly. Such "bleeding" at the graft union can be overcome by making slanting knife cuts below the graft around the tree. They should be made through the bark and into the xylem to permit such exudation to take place below the graft union. Containerized rootstock plants of *Fagus, Betula,* or *Acer* are relocated to a cool place with reduced watering until the "bleeding" stops. The plants are then grafted after the excessive root pressure subsides.

On the other hand, dormant containerized rootstocks of junipers or rhododendrons, when first brought into a warm greenhouse in winter for grafting, should be been held for several weeks at 15 to 18° C (60 to 65° F) until new roots begin to form. The rootstocks are then physiologically active enough to be successfully grafted.

When the rootstock is physiologically overactive (excessive root pressure and "bleeding"), or underactive (no root growth), some form of side graft can be used, in which the rootstock top is retained. On the other hand, top-grafting, in which the top of the rootstock is completely removed at the time the graft is made, is likely to be successful in plants in which the rootstock is neither overactive nor underactive (*42*).

The Craftmanship of Grafting

The art and craftsmanship in grafting and budding is critical for successful grafting. This is particularly true with difficult-to-graft species, such as conifers, e.g., *Picea pungens,* which callus poorly, making the lining up of the cambial layers of the rootstock and scion critical. Conversely, the grafting technique is less critical in grape or pear grafts, which callus much more profusely and have high grafting success (Figure 12–17).

Sometimes the techniques used in grafting are so poor that only a small portion of the cambial regions of the rootstock and scion are properly aligned. Graft union formation may be initiated and growth from the scion may start; but after a sizable leaf area develops and if high temperatures and high transpiration occurs, water movement through the limited conducting area is insufficient, and the scion subsequently dies. Other errors in technique resulting in graft failure include poor or delayed waxing, uneven cuts, use of desiccated scions, and girdling that occurs when polyethylene wrapping tape is not removed.

Virus Contamination, Insects, and Diseases

Some *delayed incompatibilities* are caused by viruses and phytoplasma (mycoplasma-like organisms). The cherry leaf roll virus causes blackline in walnut when it is initially spread by virus-infected pollen of the symptomless English walnut *(Juglans regia)*. The virus then travels down the scions of *J. regia* into the susceptible rootstocks—black walnut *(J. hindsii)* or Paradox walnut *(J. hindsii* × *J. regia)* (*101*). The black walnut rootstock (used for resistance to Phytophora root-rot in the soil) has a hypersensitive reaction and puts down a chemical barrier to wall-off the virus. This causes the graft to fail, and a characteristic black line forms at the graft union. Apple union necrosis and decline (AUND) (*36*) and brownline of prune (*100*) is caused by tomato ring-spot virus that is transmitted by soil-borne nematodes to the rootstock and then to the graft union. Graft unions appear to be normal until the virus has moved, either from the rootstock or the fruiting branches to the graft union. Apparently because of hypersensitivity and death of the scion cells (in prunes and apples) or rootstock cells (in walnut), the graft union deteriorates and graft failure occurs. Virus and phytoplasma-induced delayed incompatibility is probably more common than expected (*139*).

Using virus-infected propagating materials in nurseries can reduce bud "take" as well as the vigor of the resulting plant (*122*). In stone fruit propagation, bud-wood, free of ring spot virus, has consistently greater "takes" than infected bud-wood.

Top-grafting olives in California is seriously hindered in some years by attacks of the American plum borer *(Euzophera semifuneralis)*, which feeds on the soft callus tissue around the graft union, resulting in the death of the scion. In England, nurseries are often plagued with the red bud borer *(Thomasiniana oculiperda)*, which feeds on the callus beneath the bud-shield in newly inserted T-buds causing them to die (*52*).

Sometimes bacteria or fungi gain entrance at the wounds made in preparing the graft or bud unions. For example, it was found that a rash of failures in grafts of *Cornus florida* 'Rubra' on *C. florida* stock was due to the presence of the fungus *Chalaropsis thielavioides* (*31*).

Plant Growth Regulators and Graft Union Formation

Plant growth regulators, particularly auxin, applied to tree wounds or to graft unions, give variable results in wounding response and graft union formation (*93, 119, 152*). Auxin (IBA, NAA) and cytokinin (BA) enhance graft success when applied to the base of side grafted *Picea* scions, while dikegulac stimulates scion growth by retarding rootstock development (*12*). Cytokinins enhance patch budding of Persian Walnut. The eloquent work of Shimomura (*152*) in tip grafting of cactus demonstrated how auxins enhanced vascular connections of deliberately misaligned scions (Figure 12–14). TIBA, a well-known inhibitor of basipetal transport of auxin, inhibits vascular connections in the graft union; but by subsequent reapplication of auxin, the inhibitory effect of TIBA is eliminated and vascular connections occur.

However, unlike auxin usage in cutting propagation, no plant growth regulators are used in commercial grafting and budding systems. In general, plant growth regulators do not uniformly enhance grafting, nor do they overcome graft incompatibility.

Post Graftage—Bud-Forcing Methods

After graft union formation has occurred in grafting or budding, it is often necessary to force out the scion or the scion bud. In field budding of roses, two to three axillary buds of the rootstock remain distal to the scion bud. The axillary buds of the rootstock, which develop into photosynthesizing branches, are initially important for the growth of the composite plant. But they can inhibit growth of the scion through apical dominance, which is an auxin response. By *"crippling"* (cutting half-way through the rootstock shoot above the bud union and breaking the shoot over the rootstock stem), girdling, or totally removing the rootstock above the scion bud union, apical dominance is broken and the scion bud rapidly elongates (*48*) (Figure 12–19).

With budded citrus, plants on which rootstock shoots remained attached (lopping; or bending the rootstock shoot to its base and tying it in position) had the greatest gains in scion growth. This was due to the greater transfer of photosynthate from the rootstock leaves to scion shoots during growth flushes, and to roots during periods between growth flushes (*184, 185*).

POLARITY IN GRAFTING

Correct polarity is strictly observed in commercial grafting operations. As a general rule, as shown in Figure 12–20, in *top-grafting,* the morphologically *proximal* end of the scion should be inserted into the *distal* end of the rootstock. But in normal *root grafting,* the *proximal* end of the scion should be inserted into the *proximal* end of the rootstock.

Should a scion be inserted with reversed polarity—"upside-down"—in *bridge grafting* for example, it is possible for the two graft unions to be successful and the scion to stay alive for a time. But, as seen in Figure 12–21, the reversed scion does not increase from its original size, whereas the scion with correct polarity enlarges normally.

Nurse-root grafting is a *temporary* graft system to allow a difficult-to-root plant to form its own adventitious roots. The rootstock may be turned upside-down, its polarity reversed, and then grafted to the desired scion. A temporary union will form and the rootstock will supply water and mineral nutrients to the scion, but the scion is unable to supply necessary organic materials to the rootstock, which eventually dies. In nurse-root grafting, the graft union is purposely set well below the ground level, and the scion itself produces adventitious roots, which ultimately become the entire root system of the plant.

In T-budding or patch budding, the rule for observance of correct polarity is not as exacting. The buds (scion) can be inserted with reversed polarity and still make permanently successful unions. As shown in Figure 12–22, inverted T-buds start growing downward, then the shoots curve and grow upward. In the inverted bud piece, the cam-

FIGURE 12–19 Forcing or "crippling" of T-budded apples *(top)* and chip budded roses *(bottom).* Notice that the rootstock is partially severed on the same side (arrows) that the rootstock was budded. This breaks the apical dominance of the rootstock shoot system on the scion, and helps force out the scion bud. By not totally severing the rootstock top, growth of the composite plant is maximized, since the shock of total severance to the composite plant is avoided, and photosynthate is still produced by the rootstock (*184*). The rootstock shoot system will later be totally severed and the scion will fully develop into the shoot system of the composite plant.

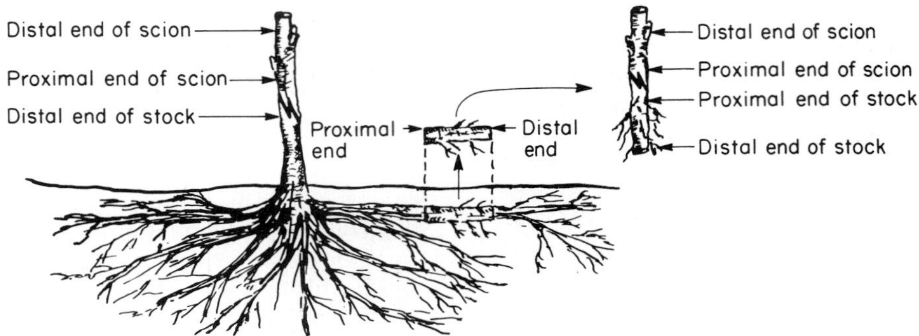

FIGURE 12–20 Polarity in grafting. In top-grafting, the proximal end of the scion is attached to the distal end of the rootstock. In root grafting, however, the proximal end of the scion is joined to the proximal end of the rootstock.

bium is capable of continued functioning and growth. In the xylem, phloem, and fibers formed from cambial activity, there is a twisting configuration which apparently allows for normal translocation and water conduction (*32*). However, it is still desirable to maintain polarity when budding.

FIGURE 12–21 Bridge graft on a pear tree five months after grafting. Center scion was inserted with reversed polarity. Although the scion is alive it has not increased from its original size. The two scions on either side have grown rapidly.

> The **proximal** end of either the shoot or the root is that nearest the stem-root junction of the plant. The **distal** end of either the shoot or the root is that farthest from the stem-root junction of the plant and nearest the tip of the shoot or root.

GENETIC LIMITS OF GRAFTING

Since one of the requirements for a successful graft union is the close matching of the callus-producing tissues near the cambium layers, grafting is generally confined to the dicotyledons in the angiosperms, and to gymnosperms. Both have a vascular cambium layer existing as a continuous tissue between the

FIGURE 12–22 Two-year-old 'Stayman Winesap' apple budded on 'McIntosh' seedling by inverted T-bud (reversing the scion bud polarity). Note the development of wide-angle crotches. (Courtesy Arnold Arboretum, Jamaica Plain, Mass.)

xylem and the phloem. Grafting is more difficult, with a low percentage of "takes" in monocotyledonous plants of the angiosperms. Monocots have vascular bundles scattered throughout the stem, rather than the continuous vascular cambium of dicots. However, there are cases of successful graft unions between monocots. By making use of the meristematic properties found in the intercalary tissues (located at the base of internodes), successful grafts have been obtained with various grass species as well as the large tropical monocotyledonous vanilla orchid (*114*).

Before a grafting operation is started, it should be determined that the plants to be combined are capable of uniting and producing a permanently successful union. There is no definite rule that can predict exactly the ultimate outcome of a particular graft combination except that *the more closely the plants are related botanically, the better the chances are for the graft union to be successful* (*74*). However, there are numerous exceptions to this rule.

Grafting within a Clone

A scion can be grafted back onto the plant from which it came, and a scion from a plant of a given clone can be grafted onto any other plant of the same clone. For example, a scion taken from an 'Elberta' peach tree could be grafted successfully to any other 'Elberta' peach tree in the world.

Grafting between Clones within a Species

In tree fruit and nut crops, different clones within a species can almost always be grafted together without difficulty and produce satisfactory trees. However, in some conifer species, notably Douglas-Fir *(Pseudotsuga menziesii),* incompatibility problems have arisen in grafting together individuals of the same species, such as selected *P. menziesii* clones onto *P. menziesii* seedling rootstock (*33*). Incompatibility is also a problem in grafting clones of deciduous species, such as red maple *(Acer rubra),* Chinese chestnut *(Castanea mollissima),* and red oak *(Quercus rubra).*

Grafting between Species within a Genus

For plants in different species but in the same genus, grafting is successful in some cases but unsuccessful in others. Grafting between most species in the genus *Citrus,* for example, is successful and widely used commercially. Almond *(Prunus amygdalus),* apricot *(Prunus armeniaca),* European plum *(Prunus domestica),* and Japanese plum *(Prunus salicina)*—all different species—are grafted commercially on peach *(Prunus persica),* a still different species, as a rootstock. But on the other hand, almond and apricot, both in the same genus, cannot be intergrafted successfully. The 'Beauty' cultivar of Japanese plum *(Prunus salicina)* makes a good union when grafted on almond, but another cultivar of *P. salicina,* 'Santa Rosa,' cannot be successfully grafted on almond. Thus, compatibility between species in the same genus depends upon the particular genotype combination of rootstock and scion.

Reciprocal interspecies grafts are not always successful. For instance, 'Marianna' plum *(Prunus cerasifera × P. munsoniana)* on peach *(Prunus persica)* roots makes an excellent graft combination, but the reverse—grafts of the peach on 'Marianna' plum roots—either soon die or fail to develop normally (*2, 90*).

Grafting between Genera within a Family

When the plants to be grafted together are in the same family but in different genera, the chances of a successful union become more remote. Cases can be found in which such grafts are successful and used commercially, but in most instances such combinations are failures. Intergeneric grafts are rarely used in conifers. However, high success rates occur between Nootka cypress *(Chamaecyparis nootkatensis)* grafted on Chinese arborvitae *(Thuja orientalis)* rootstock (*74*).

Trifoliate orange *(Poncirus trifoliata)* is used commercially as a dwarfing rootstock for the orange *(Citrus sinensis),* which is a different genus. The quince *(Cydonia oblonga)* has long been used as a dwarfing rootstock for certain pear *(Pyrus communis* and *P. pyrifolia)* cultivars. The reverse combination, quince on pear, though, is unsuccessful. The evergreen loquat *(Eriobotrya japonica)* can be grafted on deciduous and dwarfing quince rootstock *(Cydonia oblonga).* See p. 129 of Westwood (*180*) for other examples of graft compatibility between related pome genera.

Intergeneric grafts in the nightshade family, Solanaceae, are quite common. Tomato *(Lycopersicon esculentum)* can be grafted successfully on Jimson weed *(Datura stramonium),* tobacco *(Nicotiana tabacum),* potato *(Solanum tuberosum),* and black nightshade *(Solanum nigrum).*

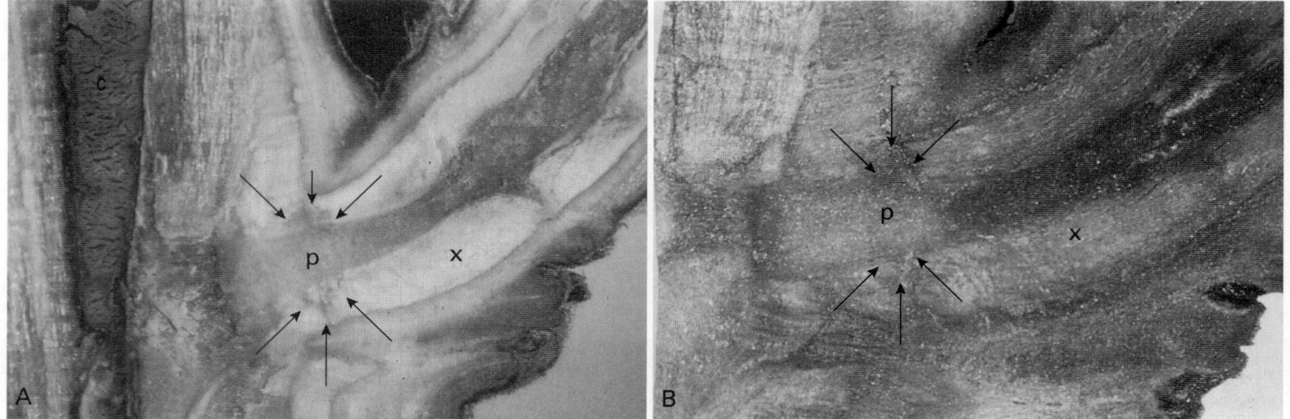

FIGURE 12–23 Graft incompatibility in 'Jonagold' apple scions budded to dwarfing 'Mark' rootstock. *Left:* Unstained section, with callus tissue (c) between the rootstock and scion. *Right:* Section stained with toluidine blue O. The xylem (x) in the graft union is interrupted by parenchyma tissue (arrows) which limits water flow and survival of the scion (Courtesy of M. R. Warmund (*176*).)

Grafting between Families

Successful grafting between plants of different botanical families is usually considered to be impossible, but there are reported instances (*191*) in which it has been accomplished. These are with short-lived, herbaceous plants, though, for which the time involved is relatively brief. Grafts, with vascular connections between the scion and rootstock, were successfully made (*117*) using white sweet clover, *Metilotus alba* (Leguminosae) the scion, and sunflower, *Helianthus annuus* (Compositae) as the rootstock. Cleft grating was used, with the scion inserted into the pith parenchyma of the stock. The scions continued growth with normal vigor for more than five months. As far as is known, however, there are no instances in which woody perennial plants belonging to different families have been successfully and permanently grafted together.

GRAFT INCOMPATIBILITY

The ability of two different plants, grafted together, to produce a successful union and to develop satisfactorily into one composite plant is termed **compatibility** (*139*). **Graft failure** can be caused by anatomical mismatching, poor craftmanship,

FIGURE 12–24 Breakage at the graft union resulting from incompatibility. *Left:* One-year-old nursery trees of apricot on almond seedling rootstock. *Right:* Fifteen-year-old 'Texas' almond tree on seedling apricot rootstock which broke off cleanly at the graft union—a case of "delayed incompatibility" symptoms.

FIGURE 12–25 Apple grafts (right portion of tree) five months after being grafted on a pear tree. This is an incompatible combination; the apple grafts eventually died although they initially grew strongly.

adverse environmental conditions, disease, and **graft incompatibility.** Graft incompatibility occurs because of: (1) adverse physiological responses between the grafting partners, (2) virus or phytoplasma transmission, and (3) anatomical abnormalities of vascular tissue in the callus bridge (Figure 12–23).

The distinction between a compatible and an incompatible graft union is not always clear-cut. Incompatible rootstock-scion combinations can completely fail to unite. Frequently they unite initially, with apparent success (*21*) (Figures 12–24 and 12–25), but gradually develop distress symptoms with time, due either to failure at the union or to the development of abnormal growth patterns (*16, 52*). A strong, well-healed union and one showing incompatibility symptoms are illustrated in Figure 12–26. Nelson (*116*) has developed an extensive survey of incompatibility in horticultural plants which should be consulted before attempting graft combinations between species whose graft reactions are unknown to the grafter.

External Symptoms of Incompatibility

Graft union malformations resulting from incompatibility can usually be correlated with certain external symptoms. The following symptoms have been associated with incompatible graft combinations:

- Failure to form a successful graft or bud union in a high percentage of cases.
- Yellowing foliage in the latter part of the growing season, followed by early defoliation. Decline in vegetative growth, appearance of shoot die-back, and general ill health of the tree.
- Premature death of the trees, which may live for only a year or two in the nursery.
- Marked differences in growth rate or vigor of scion and rootstock.

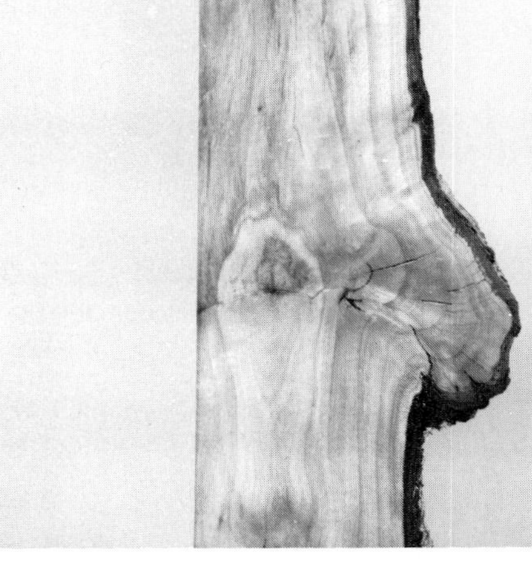

FIGURE 12–26 Radial sections through two fruit-tree graft unions. *Left:* A strong, well-knit compatible graft union. *Right:* A weak, poorly connected union, showing incompatibility symptoms.

- Differences between scion and rootstock in the time at which vegetative growth for the season begins or ends.
- Overgrowths at, above, or below the graft union.
- Suckering of rootstock.
- Graft components breaking apart cleanly at the graft union.

An isolated case of one or more of the above symptoms (except for the last) does not necessarily mean the combination is incompatible. Incompatibility is clearly indicated by *the breaking off trees at the point of union, particularly when they have been growing for some years and the break is clean and smooth, rather than rough or jagged.* This break may occur in a year or two after the union is made, for instance, in the apricot on almond roots (see Figure 12–24), or much later with conifers and oaks. Scion overgrowth at the graft union is not a reliable indicator, but is sometimes associated with incompatibility (Figure 12–27) (*2, 20*).

FIGURE 12–27 It is possible for the scion to overgrow the rootstock and yet develop into a large, strong tree. Such overgrowth is shown here for the Caucasian wingnut *(Pterocarya fraxinifolia)* grafted on *Pterocarya stenoptera.* Overgrowth or undergrowth of the scion is more related to genetic tendency for growth than to incompatibility.

Anatomical Flaws Leading to Incompatibility

With incompatible cherry *(Prunus)* grafts, the number of well-differentiated phloem sieve tubes is much lower at and below the union. There is a greater autolysis of cells, and generally a very low degree of differentiation (*149*). Poor differention of the phloem below the union may be due to a lack of hormones, carbohydrates, and other factors—the size of the sieve tubes depends on auxin, cytokinin, and sucrose levels (*149*). With incompatible apricot/plum *(Prunus)* grafts, some callus differentiation into cambium and vascular tissue does occur; however, a large portion of the callus never differentiates (Figure 12–28) (*45*). The union that occurs is mechanically weak.

With incompatible apple grafts, vascular discontinuity occurs with xylem interrupted by parenchyma tissue (Figure 12–23) (*176*). This disrupts normal xylem function leading to death of the budded scion.

Delayed incompatibility. Some apricot cultivars grafted onto myrobalan plum rootstock will not break at the graft union until the trees are full-grown and bearing crops (*44, 126*). Graft incompatibility can take as long as 20 years to occur with conifers and oaks.

Physiological and Pathogen-Induced Incompatibility

For lack of better terminology, graft incompatibility has been traditionally classified as *nontranslocatable* (localized) or *translocatable* (*112*). *Pathogen-induced* (virus, phytoplasma) *incompatibility* is a third category. It is difficult to distinguish differences between the symptoms of nontranslocatable and translocatable incompatibility. Anatomical symptoms of incompatibility can include phloem degeneration or phloem compression, and cambial or vascular discontinuity in the union area, causing mechanical weakness and subsequent breakdown of the union.

Localized (*nontranslocatable*) *incompatibility* includes graft combinations in which a *mutually compatible interstock* overcomes the incompatibility of the scion and rootstock. The interstock prevents physical contact of the rootstock and scion, and affects the physiology of the normally incompatible scion and rootstock. In some innovative research, membrane filters placed between graft partners demonstrated that physical contact is not necessary to develop compatible grafts (*105, 106*). A good exam-

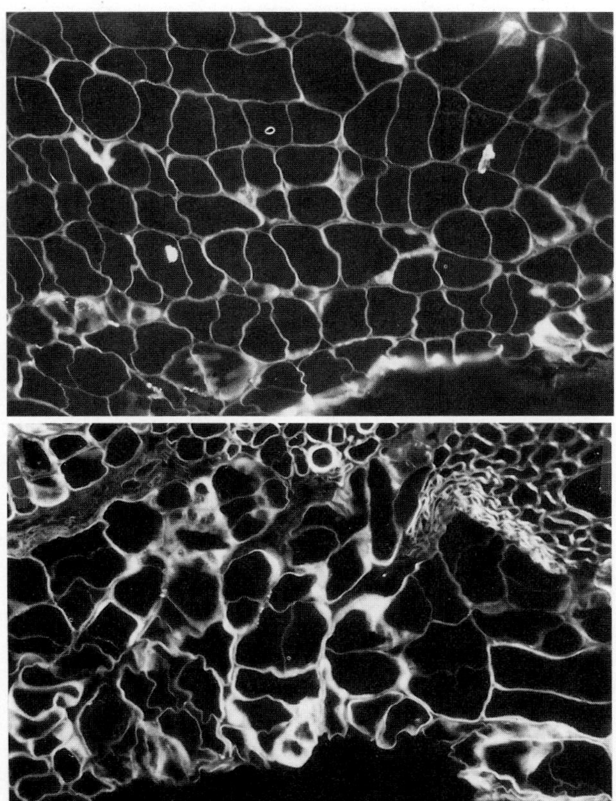

FIGURE 12–28 Graft establishment between compatible and incompatible *Prunus* spp. *Top:* Callus from a compatible graft 21 days after grafting. The cells show an orderly disposition and are uniformly stained (160× magnification). *Bottom:* Callus from incompatible graft ten days after grafting. The cells show an irregular disposition and the cell walls are thick and irregular. (Courtesy P. Errea (*45*).)

ple of nontranslocatable incompatibility is 'Bartlett' ('Williams') pear grafted directly onto dwarfing quince rootstock. When mutually compatible 'Old Home' or ('Beurré Hardy') is used as an interstock, the three-graft combination is completely compatible and satisfactory tree growth takes place (*111, 125, 126, 177*).

Translocatable incompatibility includes certain graft/rootstock combinations in which the insertion of a mutually compatible interstock does *not* overcome incompatibility. Apparently, some labile influence can move across the interstock and cause phloem degeneration. This can be recognized by the development of a brownline or necrotic area in the bark at the rootstock interface. Consequently, carbohydrate movement from the scion to the rootstock is restricted at the graft union.

'Hale's Early' peach grafted onto 'Myrobalan B' plum rootstock is an example of translocatable incompatibility. The tissues are distorted and a weak union forms. Abnormal quantities of starch accumulate at the base of the peach scion. If the mutually compatible 'Brompton' plum is used as an interstock between the 'Hale's Early' peach and the 'Myrobalan B' rootstock the incompatibility symptoms persist, with an accumulation of starch in the 'Brompton' interstock. 'Nonpareil' almond on 'Marianna 2624' plum rootstock shows complete phloem breakdown, although the xylem tissue connections are quite satisfactory. In contrast, 'Texas' almond, on 'Marianna 2624' plum rootstock produces a compatible combination. Inserting a 15 cm (6-in.) piece of 'Texas' almond as an interstock between the 'Nonpareil' almond and the 'Marianna' plum rootstock fails to overcome the incompatibility between these two components. Bark disintegration occurs at the normally compatible 'Texas' almond/'Marianna' plum graft union (*93*).

Pathogen-induced incompatibility [virus or phytoplasma (mycoplasma-like organisms)] cases are widespread and more are continually being found. These graft union failures resemble incompatibility symptoms, but are due to pathogens. In certain cases abnormalities first attributed to rootstock-scion incompatibility were later found to be due to latent virus or phytoplasma introduced by grafting from a resistant, symptomless partner to a susceptible partner (*28, 39, 95*). Figure 12–29 shows such an occurrence in apple.

Tristeza—which comes from the Spanish and Portuguese word *triste,* meaning "sad" or "wretched," is an important example of virus-induced incompatibility in citrus. Failure of sweet orange *(Citrus sinensis)* budded onto sour orange *(C. aurantium)* rootstock in South Africa (1910) and in Java (1928) was at one time blamed on incompatibility, even though this combination was a commercial success in other parts of the world. Incompatibility was believed due to production of a substance by the scion that was toxic to the rootstock (*168*). Subsequent studies involving Tristeza or "quick decline" of orange in Brazil and California made clear that the toxic substance from the sweet orange scions was instead a virus tolerated by the sweet orange, but lethal to the sour orange rootstock (*17, 178*).

Other examples of virus-induced incompatibility include: blackline in English walnut *(Juglans regia),* which infects susceptible walnut rootstock (see p. 412); apple union necrosis and decline

TYPES OF GRAFT INCOMPATIBILITY

Anatomical Flaws

- Incompatible cherry *(Prunus)* grafts with poor phloem development and/or weak unions
- incompatible apricot/plum *(Prunus)* grafts—mechanically weak unions
- Some budded apple *(Malus)* combinations—vascular discontinuity

Nontranslocatable (Localized) Incompatibility

- 'Bartlett' pear on quince roots; incompatibility overcome with 'Old Home' interstock

Translocated Incompatibility

- 'Hale's Early' peach on 'Myrobalan B' plum roots
- 'Nonpareil' almond on 'Marianna 2624' plum roots
- Peach cultivars on 'Marianna 2624' plum roots

Pathogen-induced Incompatibility (Virus, Phytoplasma)

- Citrus quick decline or Tristeza
- Pear decline
- Walnut blackline
- Apple union necrosis and decline (AUND)
- Prune brownline

(AUND) (*36*); and brownline of prune (*100*), which is caused by tomato mosaic virus that is transmitted by soil-borne nematodes to the rootstock, and then to the graft union. Pear decline is due to a phytoplasma, rather than a virus.

Causes and Mechanisms of Incompatibility

Physiological and biochemical mechanisms. Tissue compatibility or incompatibility in plants can be regarded as a physiological tolerance or intolerance, respectively, between different cells (*103, 106, 107*). Although incompatibility is clearly related to genetic differences between rootstock and scion, the mechanisms by which incompatibility is

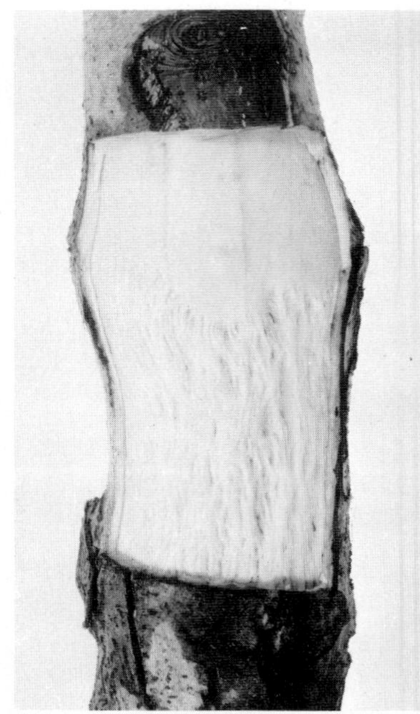

FIGURE 12–29 Latent viruses in the scion portion of graft combination may cause symptoms to appear in a susceptible rootstock following grafting. Here "stempitting" virus symptoms have developed in the sensitive 'Virginia Crab' apple rootstock. The wood of the scion cultivar—above the graft union—is unaffected. (Courtesy H. F. Winter (*183*).)

expressed is not clear. The large numbers of different genotypes that can be combined by grafting produces a wide range of different physiological, biochemical, and anatomical interactions when grafted. Several hypotheses have been advanced in attempts to explain incompatibility.

One proposed *physiological and biochemical* mechanism concerns incompatible combinations of certain pear cultivars on quince rootstock (*59*). The incompatibility is caused by a *cyanogenic glucoside, prunasin,* normally found in quince, but not in pear tissues. Prunasin is translocated from the quince into the phloem of the pear. The pear tissues break down the prunasin in the region of the graft union, with hydrocyanic acid as one of the decomposition products (Figure 12–30). The presence of the hydrocyanic acid leads to a lack of cambial activity at the graft union, with pronounced anatomical *disturbances* in the *phloem and xylem* at the resulting union. The phloem tissues are gradually destroyed at and above the graft union. Conduction of water and materials is seriously reduced in both xylem and phloem. *The presence of cyanogenic glycosides in woody*

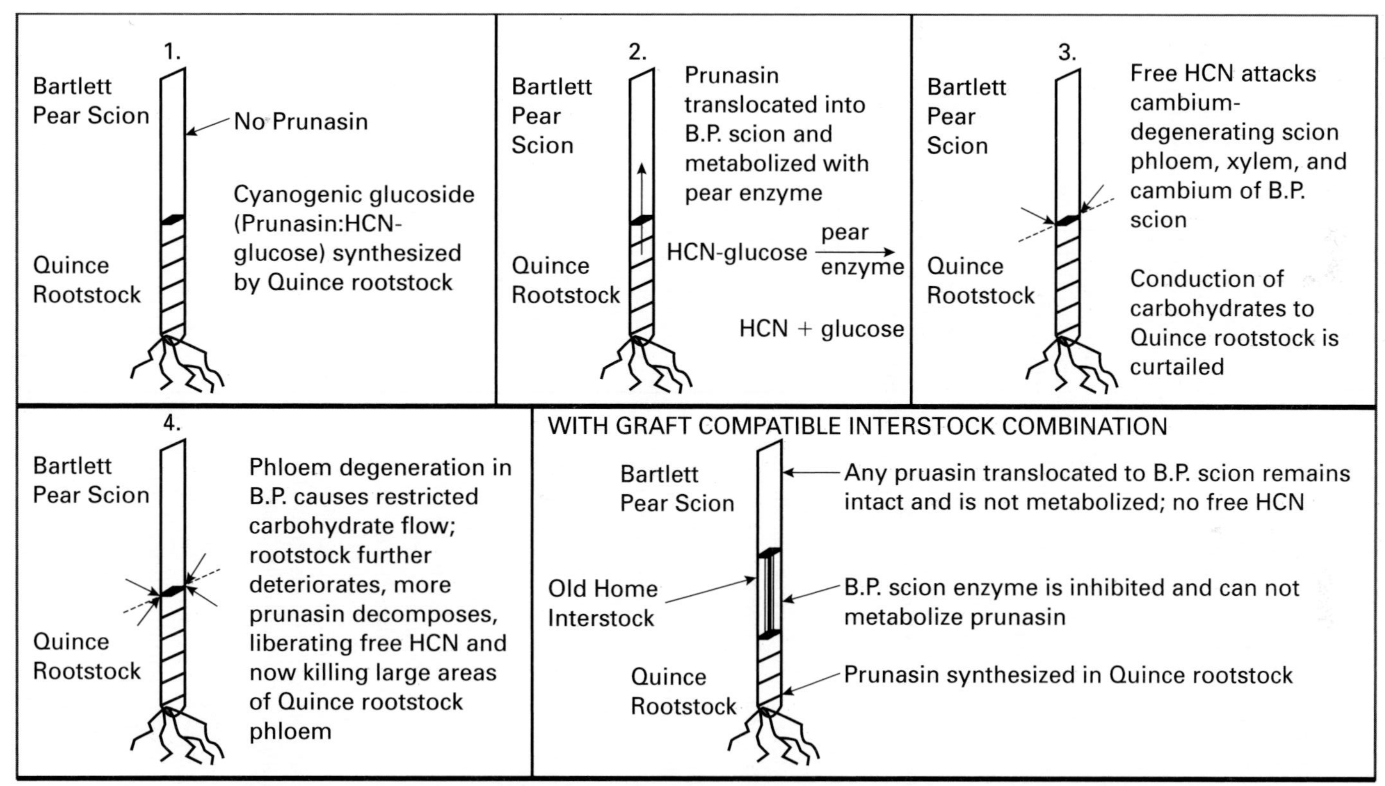

FIGURE 12–30 Nontranslocatable incompatibility of Bartlett pear scion on quince rootstock (*59*).

plants is restricted to a relatively few genera. Hence, this reaction cannot be considered a universal cause of graft incompatibility.

Modification of cells and tissue. The **lignification** processes of cell walls are important in the formation of strong unions in pear-quince grafts. Inhibition of lignin formation and the establishment of a mutual middle lamella results in weak graft unions. In compatible pear-quince graft combinations, the lignin in cell walls at the graft union is comparable to adjacent cells outside the union (*22*). Conversely, adjoining cell walls in the graft union of incompatible combinations contain no lignin, and are interlocked only by cellulose fibers.

With incompatible apricot/plum *(Prunus)* grafts, some callus differentiation into cambium and vascular tissue does occur; however, a large portion of the callus never differentiates (Figure 12–28) (*45*). The union that occurs is mechanically weak.

Cell recognition of the grafting partners. It has been postulated that the critical event deciding compatible and incompatible grafts may occur when the callus cells first touch (*189*). There may be *cellular recognition*[4] that must occur in successful graft union formation. Alternatively, the failure of procambial differentiation in incompatible grafts may be the result of a direct form of cellular communication between the graft partners (*108*).

In a compatible graft, the wound response is followed by a dissolution of the necrotic layer, perhaps as a prerequisite to the formation of secondary plasmodesmata between cells of the graft partners (*165*). There *is* direct cellular contact of plasmodesmata (minute cytoplamic threads that extend through openings in cell walls and connect the protoplasts of connecting cells) in the callus bridge that symplastically connects the grafting partners (*84*). This forms a potential communication pathway among cells in the graft bridge, which may be important in cell recognition, compatibility/incompatibility responses.

Conversely, *cellular recognition* may *not* be a factor in grafting compatibility/incompatibility. Partners of compatible and incompatible grafts adhere during the early stages of graft union formation; this passive event does not require mutual cell recognition [grafted *Sedum* will even adhere to inert wooden objects (*107, 108*)], nor is it related to compatibility (*106*). Adhesion of graft partners results from the deposition and subsequent polymerization of cell wall materials that occur in response to wounding. Callus proliferation is not related to graft compatibility-incompatibility systems, since it does not require a recognition event to occur, i.e., callus proliferation occurs in wounded cuttings, as well as in incompatible and compatible graft systems (*107, 108*).

Vascular differentiation in the callus bridge, which typically occurs from the severed vascular strands of the scion and rootstock, can occur even when the scion and rootstock are physically separated by a porous membrane filter (inserted in order to prevent direct cellular contact without impeding the flow of diffusible substances between the graft partners) (*105*)*;* this was done with autografts of *Sedum* (a herbaceous species), which may

[4]Cellular recognition is defined as the union of specific cellular groups on the surfaces of the interacting cells that results in a specific defined response, e.g., pollen-stigma compatibility-incompatibility recognition responses with glycoprotein surface receptors in flowering plants *(26)*.

CELLULAR RECOGNITION

It is currently not known if some kind of cell-to-cell recognition in grafting must occur as part of adhesion and the events that follow in successful graft union formation. Possibly, the formation of superimposed sieve areas and sieve plates (in sieve elements), pits and perforation plates (in xylem elements), and the plasmodesmata (in vascular parenchyma) require some sort of cellular recognition or cellular communication (*108*). Evidence suggests that in the graftage of *Cucumis* and *Cucurbita,* changes in protein banding may be due to polypeptides migrating symplastically across the graft union via the connecting phloem (*166*). Translocation of signaling molecules, such as polypeptides in the phloem, could be significant in cell recognition and compatibility between the graft partners. (In graft incompatibility, phloem degeneration frequently takes place at the graft union.)

Pectin fragments formed during the adhesion process of grafting may act as signaling molecules—and influence cell recognition. In Sitka spruce, the beadlike projections from callus formed during graftage are in part composed of pectins, proteins, carbohydrates, and fatty acids. These beadlike projections, besides binding or cementing cells, may serve a more active role in cell recognition and the successful merging of tissues of the grafting partners (*96*).

not be representative of graftage in woody perennial plants. Nonetheless, it is evidence that successful graft union formation: (1) can occur in the absence of direct cellular contact, and (2) does not require a positive recognition system.

The lining up of tissue [(e.g., vascular cambium of woody plants, vascular bundles of Cacti (*151*)] determines what cell types and tissue will be differentiated in the callus bridge. It has been proposed that phytohormones are released from wounded vascular bundles into the surrounding tissue where they function as morphogenic substances inducing and controlling the regeneration of cambia and vascular tissue (*1*). This hypothesis can be applied to graft union formation, with phytohormones such as auxin as potential morphogens needed for graft union formation. Auxin should not be considered as a specific recognition molecule per se because of its ubiquitous occurrence and its involvement in numerous other developmental processes (*106, 105*). Phytohormones (and carbohydrates, etc.), predominantly released from the scion, enable vascular connections to develop and join as a functional unit in the graft union, without any cellular recognition required.

A model for graft compatibility-incompatibility is presented that suggests grafts will be incompatible only if naturally occurring morphogens that promote the formation of a successful graft (e.g., auxin) are overridden by toxins [e.g., hydrocyanic acid, benzaldehyde (*64, 65*)] that elicit graft incompatibility (Figure 12–31) (*106*).

Most likely, graft compatibility-incompatibility is a combination of the morphogen-toxin scenario of Figure 12–31 and/or some chemical recognition response. To date, we have little understanding of the molecular chain of events that occurs during wounding (*182*) and graft union formation, or how those chain of events vary between compatible-incompatible graft partners. In Douglas-Fir, graft incompatibility is apparently controlled by multiple genes with additive effects (*34*).

Predicting Incompatible Combinations

To be able to predict in advance of grafting whether or not the components of the proposed scion-stock combination are compatible would be of tremendous value. An *electrophoresis test* is being used for testing *cambial peroxidase* banding patterns of the

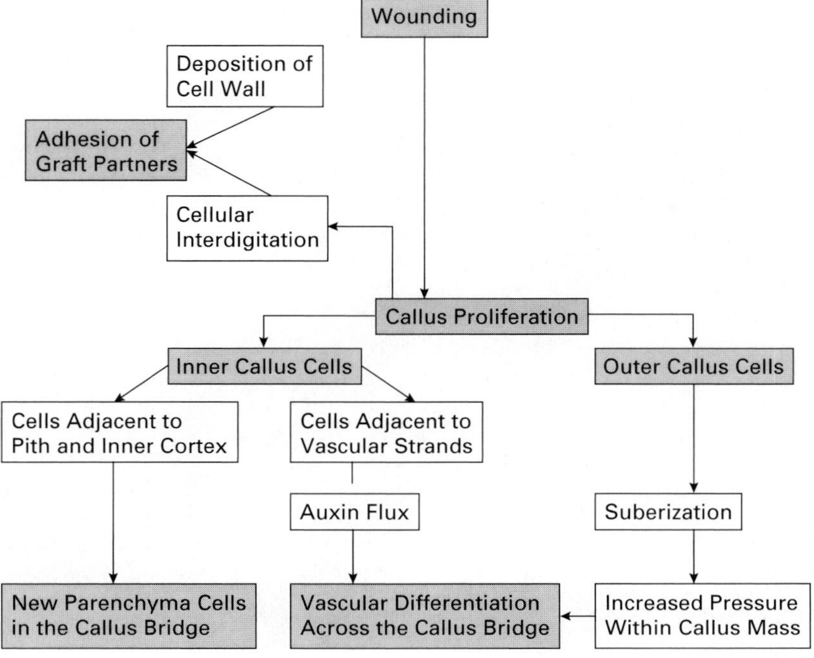

FIGURE 12–31 A model to explain the development of a compatible graft union. The stages are: adhesion of the scion and rootstock, proliferation of callus cells to form the callus bridge, and vascular differentiation across the graft interface. The outer callus cells are from the periderm and outer cortex. The pressure exerted on the graft is from the physical contact of the scion to the rootstock—and the development of a suberized periderm. Auxin is a potential morphogen, enhancing vascular dedifferentiation. In this model, incompatibility is not caused by specific cellular recognition events between the graft partners. Rather, incompatibility may occur when a toxin, such as hydrocyanic acid (HCN) or benzaldehyde, counteracts naturally occurring mophogens (e.g., auxin), thus inhibiting or degenerating vascular tissues in the graft union (*106*).

CELLULAR RESPONSES OF COMPATIBLE AND INCOMPATIBLE GRAFTS

At the cellular level, the initial stages of graft union formation were similar between the incompatible combination of *Sedum telephoides* (Crassulaceae) on *Solanum pennellii* (Solanaceae) and those occurring in a compatible autograft of *Sedum* on *Sedum.* However, after 48 hours, *Sedum* cells in the incompatible graft deposited an insulating layer of suberin along the cell wall. The cell walls later underwent lethal senescence and collapse, and formed a necrotic layer of increasing thickness (Figure 12–32) (*107, 108*). Associated with this cellular senescence in *Sedum* cells was a dramatic increase in a hydrolytic enzyme, acid phosphatase (*109*). Rather than callus cells interlocking, cambial formation and vascular connection, the thick necrotic layer prevented cellular connections, which led to scion desiccation and eventual death. Interestingly, the *Solanum* rootstock did not show the rejection response that the *Sedum* scion did.

scion and rootstock of chestnut, oak and maple (*139, 140, 141, 142, 144, 145*). Peroxidases mediate lignin production. Increased peroxidase activity occurs in incompatible grafts as compared to compatible autografts, and adjacent rootstock and scion cells must produce similar lignins and have identical peroxidase enzyme patterns to ensure the development of a functional vascular system across the graft union (*38*). With electrophoresis, if the peroxidase bands match, the combination may be compatible; if they do not, incompatibility may be predicted. This is an important step in developing diagnostic tests for graft compatibility. Perhaps serological tests for graft compatibility may be developed in the future, as are currently being utilized in disease diagnostic kits of plant pathogens.

The introduction of new *Prunus* rootstock can be difficult (and very costly!) because incompatibility can occur some years after grafting. The composite tree can grow "normally" for years, and then a breakdown occurs at the graft union area. It is now known that with incompatible apricot/plum (*Prunus*) grafts, some callus differentiation into cambium and vascular tissue does occur; however, a large portion of the callus never differentiates (Figure 12–28) (*45*). The fact that this process can be *detected histologically* within weeks after grafting facilitates early detection of graft incompatibility in fruit trees (*42*).

Magnetic resonance imaging (*MRI*) can be used to detect vascular discontinuity in bud unions of apple (*176*). A high MRI signal intensity is associated with bound water in live tissue and the establishment of vascular continuity between the rootstock and scion. MRI may be useful for detecting graft incompatibilities caused by poor vascular connections.

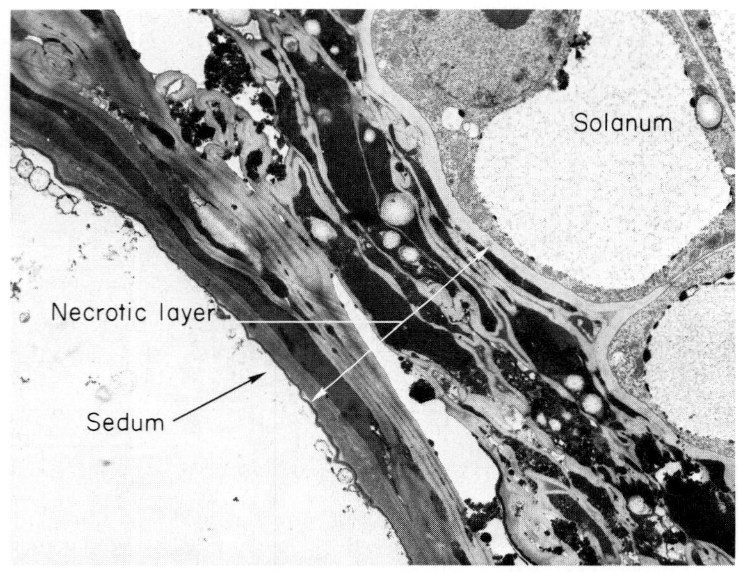

FIGURE 12–32 The graft interface of an incompatible graft between *Sedum telephoides* and *Solanum pennellii* at eight days after grafting. Lethal cellular senescence in *Sedum* has resulted in the formation of a necrotic layer of collapsed cells that separates the two graft partners. × 5,000. (Courtesy R. Moore and D. B. Walker (*115*).)

Correcting Incompatible Combinations

There is not a practical, cost-effective way to correct large-scale plantings of incompatible graft partners. Plants would normally be rogued and discarded. With perhaps some isolated, specimen trees of value, if the incompatibility were discovered before the tree died or broke off at the union, a bridge graft could be done with a mutually compatible rootstock. Another costly alternative is to inarch with seedlings of a compatible rootstock. The inarched seedlings would eventually become the main root system (see Chapter 13).

SCION-ROOTSTOCK (SHOOT-ROOT) RELATIONSHIPS

Combining two (or more, in the case of interstocks) different plants (genotypes) into one plant by grafting—one part producing the top and the other part the root system—can produce growth patterns that are different from those that would have occurred if each component part had been grown separately. Some of these effects are of major importance in horticulture and forestry, while others are detrimental and should be avoided. These altered characteristics may result from (a) specific characteristics of the graft partners not found in the other, for example, resistance to certain diseases, insects, or nematodes—or tolerance of certain adverse environmental or soil conditions; (b) interactions between the rootstock and the scion that alter size, growth, productivity, fruit quality, or other horticultural attributes; and (c) incompatibility reactions.

In practice, it may be difficult to separate which of the three kinds of influencing factors is dominant in any given graft combination growing in a particular environment.

Long-term results depend on the rootstock-scion combination, environment (climate, edaphic factors such as soil), propagation, and production management; this affects yield, quality, plant form and ornamental characteristics (if applicable), and the economics of production.

Effects of the Rootstock on the Scion Cultivar

Size and Growth Habit

Size control, sometimes accompanied by change in tree shape, is one of the most significant rootstock effects. Rootstock selection in apple has produced a complete range of tree size—from dwarfing to very vigorous—by grafting the same scion cultivar to different rootstocks (Figure 12–3).

That specific rootstocks can be used to influence the size of trees has been known since ancient times. Theophrastus—and later the Roman horticulturists—used dwarfing apple rootstocks that could be easily propagated. The name "Paradise," which refers to a Persian (Iranian) park or garden—*pairidaeza*—was applied to dwarfing apple rootstocks about the end of the fifteenth century.

A wide assortment of size-controlling rootstocks has now been developed for certain of the major tree fruit crops (*21*). Most notable is the series of clonally propagated apple rootstocks collected and developed at the East Malling Research Station in England, beginning in 1912. These were classified into four groups, according primarily to the degree of vigor imparted to the scion cultivar: *dwarfing, semi-dwarfing, vigorous,* and *very vigorous*—same size as seedling rootstock (Figure 12–3) (*61, 123*). Similarly, the size-controlling effects of the rootstock on sweet cherry *(Prunus avium)* scion cultivars has been known since the early part of the eighteenth century (*63*). Mazzard *(P. avium)* seedling rootstocks produce large, vigorous, long-lived trees, whereas *P. mahaleb* seedlings, as a rootstock, tend to produce smaller trees of shorter life. However, individual seedlings of these species, when propagated asexually and maintained as clones, can produce different, distinct rootstock effects. Rootstock effects on tree size and vigor are recognized also in citrus, pear, and other species (*51*). A discussion of specific rootstocks for the various fruit and nut crops is given in Chapter 19.

Fruiting

Fruiting precocity, fruit bud formation, fruit set, and yield of a tree can be influenced by the rootstock used. In general, fruiting precocity is associated with dwarfing rootstocks, and delay in fruiting with vigorous rootstocks. Apple rootstocks are used primarily for reducing tree size and for increasing precocity and yield efficiency.

Besides being more precocious, intensive plantings of small trees resulting from dwarfing rootstock intercept more light and have less internal shading, which is related to greater dry matter production and fruit yield. The higher ratio of fruit weight to trunk and branch weight (partitioning of photosynthate to fruit rather than wood formation) may also contribute to higher *yield effi-*

ciencies for trees growing on dwarfing rootstock than more vigorous clonal and seedling rootstock (*124, 161*).

Vigorous, strongly growing rootstocks in some cases result in a larger plant that produces a bigger crop (per individual tree) over a long period of years. On the other hand, trees on dwarfing rootstocks are more fruitful, and if closely planted, produce higher yields per hectare (acre). Because of the earlier and higher productivity of an apple crop on dwarfing rootstock the cash flow and return on investment to the producer is much improved. Furthermore, the management costs of harvesting, pruning, spraying, and general maintenance is much greater on large-sized trees.

Size, Quality, and Maturity of Fruit

There is considerable variation among plant species in regard to the effect of the rootstock on fruit characteristics of the scion cultivar. However, in a grafted tree there is no transmission of fruit traits characteristic of the rootstock to the fruit produced by the grafted scion. For example, quince, commonly used as a pear rootstock, has fruits with a pronounced tart and astringent flavor, yet this flavor does not appear in the pear fruits. The peach is often used as a rootstock for apricot, yet there is no indication that the apricot fruits have taken on any characteristics of peach fruits.

Although there is no transfer of fruit characteristics between the rootstock and the scion, certain rootstocks can affect fruit quality of the scion cultivar. A good example of this is the "black-end" defect of pears. 'Bartlett,' 'Anjou,' and some other pear cultivars on several different rootstocks often produce fruits that are abnormal at the calyx end.

In citrus, striking effects of the rootstock appear in fruit characteristics of the scion cultivar (*16*). If sour orange *(Citrus aurantium)* is used as the rootstock, fruits of sweet orange, tangerine, and grapefruit are smooth, thin-skinned, and juicy, with excellent quality, and they store well without deterioration. Sweet orange *(C. sinensis)* rootstocks also result in thin-skinned, juicy, high-quality fruits. The larger fruit size of 'Valencia' oranges are associated with the dwarfing trifoliate orange rootstock, whereas sweet orange rootstocks produce the smaller fruits. Semi-dwarfing clonal rootstock will enhance the fruit size of 'Red Delicious' and 'Granny Smith' apples, compared to seedling rootstock, while 'Gala' is unaffected by rootstocks (*72*).

Miscellaneous Effects of the Rootstock on the Scion Cultivar

Winter-hardiness. In citrus, the rootstock used can affect the cold-hardiness of the scion cultivar. Grapefruit cultivars on 'Rangpur' lime rootstock survive cold better than those on sour orange rough lemon rootstock. The rootstock can affect the rate of maturity of the scion wood as it hardens-off in the fall (*51*).

Extending scion tolerance of adverse edaphic conditions. For many kinds of plants, rootstocks are available that tolerate unfavorable conditions, such as heavy, wet soils (*128, 130, 131, 133*). In the U.S. southeast, where high temperatures and periodic flooding of soils (low soil oxygen) are the norm, cultivars of birch *(Betula),* fir *(Abies),* and oak *(Quercus)* are grafted onto rootstock that tolerate these atmospheric and edaphic environments (*130, 131, 133*). 'Whitespire' Japanese birch *(Betula populifolia)* is an excellent landscape tree for the U.S. southeast. It tolerates heat and drought, but will not tolerate poorly drained soils. The ecological niche of 'Whitespire' may be expanded by grafting it onto flood-tolerant rootstock of river birch *(B. nigra)* (*131*). Compared to many other genera of temperate woody plants, trees in the genus *Prunus* are often intolerant of poor drainage conditions. Ornamental *Prunus* cultivars can be adapted to poorly drained landscape sites by grafting onto more flood-tolerant 'Newport' plum *(Prunus hybrida)* and 'F-12/1' Mazzard cherry *(P. avium)* (*128*). Japanese Momi fir *(Abies firma)* is one of the few firs that will tolerate the heavy clay, wet soil conditions, and heat of the southeastern U.S. Consequently, it is being recommended as the rootstock for more desirable fir cultivars (*133*).

Disease resistance. It is known that some rootstocks are more tolerant than others to adverse soil pests, such as nematodes (*Meloidogyne* spp.), e.g., 'Nemaguard' peach rootstock. The growth of the scion cultivar is subsequently enhanced by the rootstock's ability to withstand the soil pest.

Effects of the Scion Cultivar on the Rootstock

Although there is a tendency to attribute all cases of dwarfing or invigoration of a grafted plant to the rootstock, the effect of the scion on the behavior of the composite plant may be as important as that of the rootstock.

Effect of the Scion on the Vigor and Development of the Rootstock

Vigor is a major effect of the scion on the stock, just as it was in the case of rootstock effect on scion cultivar. If a strongly growing scion cultivar is grafted on a weak rootstock, the growth of the rootstock will be stimulated so as to become larger than it would have been if left ungrafted. Conversely, if a weakly growing scion cultivar is grafted onto a vigorous rootstock, the growth of the rootstock will be lessened from what it might have been if left ungrafted. In citrus, for example, when the scion cultivar is less vigorous than the rootstock cultivar, it is the scion cultivar rather than the rootstock that determines the rate of growth and ultimate size of the tree (*68*).

Interstock Effects on Scion and Rootstock

The ability of certain dwarfing rootstock clones, inserted as an *interstock* between a vigorous top and vigorous root, to produce dwarfed and early bearing fruit trees has been known for centuries and is used commercially to propagate dwarfed trees. The degree of size control induced in apples by various dwarfing interstocks is shown in Figure 12–4. Dwarfing of apple trees by the use of a dwarfing interstock, such as 'Malling 9,' has been widely used commercially for many years. This method has the advantage of allowing the use of well-anchored, vigorous rootstock rather than a brittle, poorly anchored dwarfing clone. However, excessive suckering from the roots may occur due to the dwarfing interstock, even in rootstock types that normally do not sucker freely.

Possible Mechanisms for the Effects of Rootstock on Scion and Scion on Rootstock

While many of the effects of rootstock-scion relations are known, the fundamental mechanisms of control, particularly on the molecular basis, are not well understood. A better understanding of the mechanisms controlling growth and development in grafted plants would speed up the design, development, and commercialization of new composite plant systems. By understanding these mechanisms, breeders could better predict the growth responses of new potential graft partners (while they are still on the "drawing board") and/or develop more efficient screening tests—rather than relying on cumbersome trial and error processes that may take up to ten years or more in evaluating grafted, woody perennial plants.

Without question, the nature of the rootstock-scion relationship is very complex and differs among genetically different combinations. Furthermore, in a composite plant system, size control, plant form, flowering, fruiting, disease resistance, flood tolerance, etc. are not controlled by the same genes or physiological/morphological mechanisms. Theories advanced as possible explanations for the interaction between the rootstock and scion include: (a) *anatomical factors,* (b) *nutritional and carbohydrate levels,* (c) *absorption and translocation of nutrients and water,* (d) *phytohormones,* and (e) *other physiological factors.*

Anatomical Factors

The roots and stems of dwarfing apple rootstocks, which can reduce vegetative growth and increase flowering, are characterized by several anatomical features. These include: (a) a high ratio of *bark* (periderm, cortex, and phloem tissue) to *wood* (xylem tissue); (b) a large proportionate volume occupied by living cells (axial parenchyma and ray parenchyma cells) relative to functionally dead xylem cells (vessels and fibers); and (c) fewer and smaller xylem vessels (*7, 8, 30, 92, 153*).

Much of the functional wood tissue of roots of dwarfing apple stocks is composed of living cells, whereas in nondwarfing, vigorous rootstocks, the wood consists of a relatively large amount of lignified tissue without living cell contents (i.e., a larger vessel/tracheid system for more efficient water transport). At the graft interface between the scion bud and dwarfing apple rootstock, xylem vessels with smaller than normal diameter are formed, whereas semi-dwarfing rootstock produces normal xylem after a brief interruption (*157*). It has been proposed that failure of auxin to cross the bud-union interface in the case of the dwarfing rootstock leads to reduced rootstock xylem formation, and hence a reduced supply of water and minerals to the scion, thus causing the dwarfing effect (*15*). Defects in the graft union that cause a partial discontinuity of the vascular tissues may in part explain the marked depletion of solutes, nutrients, and cytokinins (produced from root apices) in the sap contents of dwarfing interstocks and rootstocks (*76*).

Conversely, with kiwifruit, the roots of flower-promoting rootstock tend to have more and larger xylem vessels, more crystalline idioblasts, and more starch grains (*173, 174*). Most likely the water supply from the rootstock to the scion in early spring determines the abundance of flower production of the kiwifruit scions.

Nutritional and Carbohydrate Levels

Dwarfing rootstock of apple tends to partition a greater proportion of carbon to reproductive areas (spurs,[5] spur leaves, fruit) and less to the tree branch and frame dry weight, compared to nondwarfing rootstocks (*161*). The greater water and nutrient uptake of the vigorous rootstock contributes to production of new vegetative growth, which is a competing sink with reproductive growth.

The rootstock affects the partitioning of dry matter between above- and below-ground tree components. Vigorous rootstocks accumulate more dry matter in the root system during the growing season than dwarfing stock (*162*). At the time apples are being harvested, the insoluble root starch supply is greater, but soluble sucrose and sorbitol are less in vigorous rootstock compared to dwarfing rootstock.

There appears to be no direct role of apple rootstock influencing mineral nutrition at the site of flower formation (*67*). Most likely, rootstock effects on flowering are due to internal control mechanisms that affect the proportion of spurs that become floral (*66*). To summarize, dwarf apple rootstocks do affect precocity and flowering, *in part,* because of differences in carbohydrate metabolism and the greater carbon partitioning to the reproductive areas. The contributing influence of hormones, which also affects carbon partitioning and flowering, is discussed in the next column.

Absorption and Translocation of Nutrients and Water

Apple rootstocks do affect Ca, Mg, Mn, and B uptake, but there is no apparent direct relationship of mineral status with rootstock vigor, productivity, or spur characteristics (*67*).

Rootstocks do differ in their ability to absorb and translocate P (*77*), but a direct role of phosphorus at the site of flower formation induced by rootstock seems unlikely (*67*). In a study of the translocation of radioactive phosphorus (^{32}P) and calcium (^{45}Ca) from the roots to the tops of one-year 'McIntosh' apple trees grown in solution culture, it was shown that over three times as much of both elements was found in the scion top when vigorous rootstock was used in comparison with the dwarfing rootstock (*25*). This may indicate a superior ability of the vigorous rootstock to absorb and translocate mineral nutrients to the scion in comparison with the dwarfing rootstock. Or it may only mean that roots of the dwarfing rootstock, with their higher percentage of living tissue, formed a greater "sink" for these materials, retaining them in the roots.

The fact that interstocks of such dwarfing apple clones as 'Malling 9' will cause a certain amount of dwarfing suggests that translocation is involved. This is due to partial blockage at the graft unions or to a reduction in movement of water or nutrient materials (or both) through the interstock piece itself. Differences among rootstocks in water translocation have been demonstrated with a steady-state, heat-balance technique that accurately measures xylem sap flow rate and sap flow accumulation over time. Under nonstress conditions, sap flow was greater in 'Granny Smith' apple scions grafted to very vigorous seedling (standard) rootstock while sap flow was similar between the dwarfing and semidwarfing rootstock (*73*). Moisture stress affects the sap flow of the vigorous seedling rootstock the least and reduced sap flow on the dwarfing rootstock the most.

The cause of the differences in the sap flow induced by the rootstock is not known, but it may be related to differences in root characteristics, xylem anatomy, or other features of the hydraulic architecture from the roots to the graft union or of the graft union itself.

Although a restricted graft union or smaller xylem elements may contribute to the dwarfing influence of certain rootstocks, the *primary* influence probably lies more in the nature of the growth characteristics of such rootstocks.

In summary, as long as mineral elements are not limiting, the greater uptake of P and Ca by the more vigorous rootstock does not adequately account for dwarfing effects (*25*). While rootstocks can influence leaf mineral nutrition, results have been inconsistent (*67*).

Phytohormones and Correlative Effects

Plants maintain a constant root/shoot ratio, and any attempt to alter this ratio results in the plant redirecting its growth pattern until the ratio

[5]The principal fruiting unit in apple are *spurs,* which may be classified as short shoots. The terminal bud of a spur may be either vegetative, containing only leaves, or reproductive. Reproductive buds of apple are mixed buds that produce both flowers and leaves.

is reestablished. This also applies to grafted plants and plants transplanted into a landscape site or orchard. Producing a composite plant by grafting onto a dwarfing rootstock is an alteration in the normal growth pattern (*87*). Growth in the composite plant will be redirected until an equilibrium is reached between the rootstock-scion system. Intimately involved in redirecting growth are the correlative effects of root (rootstock)/shoot (scion) systems, mediated by phytohormones. Auxins, which are produced predominantly in the shoot system, are basipetally translocated primarily through the phloem and into the root system, where they affect root growth. Cytokinins are produced predominantly from root apices, and are translocated primarily through the xylem, where they can influence physiological responses and growth in the scion.

Of all the phytohormones, auxin probably plays one of the most important roles in dwarfing rootstock control of apple scion growth (*78*). The dwarfing effect may be explained by reduced auxin transport through the bark of the dwarf rootstock (*87*); this could alter the hormonal balance between shoots and roots, and account for the reduced growth and vigor of the scion. Auxin affects vascular differentiation, and is important for stimulating cambial activity and xylem development (*1*) in the graft union area and the vascular system of the grafting partners. Greater reduction in cambial activity and xylem formation in the rootstock occurs at the bud union with dwarfing compared to vigorous rootstock (*157*). This may be explained by reduced capacity to support polar auxin transport (*not* auxin uptake into cells) and a reduced capacity for auxin efflux from transporting cells (*158*). Since auxin is known to stimulate its own transport (*55*), lower endogenous auxin levels in the dwarfing rootstock may limit its capacity to support polar auxin transport. A chain of events is set off with less auxin being transported, which leads to reduced cambial activity and subsequent reduced xylem formation. Reduced xylem formation limits conduction in the dwarf rootstock which concurs with the reports on lower xylem sap flow (*73*).

Auxin can indirectly affect cytokinin production. Reduced auxin transport leads to a smaller root system in the dwarf rootstock which produces less cytokinin and/or the root metabolism is sufficiently altered to affect cytokinin synthesis. There is less cytokinin translocated upward from the roots to the shoots and reduced top growth occurs, hence, the dwarfing effect. This correlative effect is mediated by auxin and cytokinin as growth in the composite plant is redirected and an equilibrium is reached between the dwarf rootstock/scion system.

Abscisic acid (ABA) and gibberellic acid (GA) may also play a role in the correlative effects of dwarfing rootstock. Root apices are an important site of ABA synthesis. In a comparison of endogenous growth-regulating factors in the bark of the dwarfing 'Malling 9' apple rootstock with those in the bark of the very vigorous 'Malling 16,' the dwarfing rootstock contained lower amounts of growth promoting materials—but *more* growth inhibitors—than did the very vigorous rootstock (*89*). ABA levels are higher in dwarfing rootstock (*186*), and in the stems of dwarfed apple trees than in more vigorous ones (*137*).

There are conflicting reports that higher GA is found in more vigorous rootstock. It is known that roots do produce gibberellin and that the amounts of gibberellin in the transpiration stream are sufficient to have a decided growth-controlling influence (*80*). To date, there is little evidence to support a role for gibberellin in vigorous compared to dwarfing rootstock (*87, 136, 137*). Furthermore, there has been no consistent relationship between dwarfism and GA content in higher plants.

In summary, with apple, auxin is directly involved in dwarfing rootstock effects, and cytokinins (which are affected by auxin-mediated root growth and subsequent cytokinin biosynthesis) are either directly or indirectly involved in plant size control. There is a strong case for ABA-mediated dwarfing effects, while there are conflicting reports on the role of GA. Most likely, there is an *interaction of factors* affecting dwarfing phenomena such as phytohormones, anatomical factors, nutrition and carbohydrate levels, and translocation of carbohydrates across the graft union.

Other Physiological Factors

A wide range of physiological characteristics have been found to affect rootstocks, scions, and their resulting interactions (*87, 129, 163*). For example, rootstocks have been found to influence transpiration rate and crop water-use efficiency in peach, leaf conductance, and osmotic potential in apple, and midday leaf water potential in citrus, peach, and apple trees. Rootstock-scion combinations can also influence net photosynthesis and growth characteristics of grafted *Prunus* species under droughted conditions (*129*). The greater tolerance to flooding found in selected rootstock of *Prunus* (*128*), and fir (*Abies*) is probably due to physiological and/or morphological mechanisms (*43*) that allow selected rootstock to handle anaerobic conditions better than other rootstock.

Leaf net photosynthesis tends to be higher with apple scions on vigorous rootstock than on dwarfing rootstock (*148*). But photosynthetic rates can not be used to explain differences in yield and yield efficiencies induced by the rootstock. Part of this complexity is because the presence of fruit increases leaf net photosynthesis by some unknown mechanism (*148*). Cytokinins are known to promote photosynthesis, and root-produced ABA—translocated in xylem sap—can reduce stomatal conductance and photosynthetic rates in the shoot system.

Very little work has been done in studying the molecular basis of rootstock-scion relations. Possibilities are that certain genes are being turned on and off, and/or that genetic information may be transmitted between the graft partners of the composite plant (*118*). Epigenetic changes occur in grafting with the speeding up of maturation on grafted versus seedling-grown plants (see discussion on epigenetic changes in Chapter 9). Conversely, micropropagated dwarfing apple rootstocks that are grafted can have more juvenilelike characteristics, which delays bearing and fruit cropping of trees (*79*).

REFERENCES

1. Aloni, R. 1987. Differentiation of vascular tissues. *Ann. Rev. Plant Physiol.* 38:179–204.
2. Amos, J., T.N. Hoblyn, R.J. Garner, and A. Witt. 1936. Studies in incompatibility of stock and scion. I. Information accumulated during twenty years of testing fruit tree rootstocks with various scion varieties at East Malling. *Ann. Rpt. E. Mailing Res. Sta for 1935,* pp. 81–99.
3. Bailey, L.H. 1891. *The nursery book.* New York: Rural Publishing Company.
4. Barnett, J.R., and H. Miller. 1994. The effect of applied heat on graft union formation in dormant *Picea sitchensis* (Bong.) Carr. *J. Exp. Bot.* 45:135–143.
5. Barnett, J.R., and I. Weatherhead. 1988. Graft formation in Sitka spruce: a scanning electron microscope study. *Ann. Bot.* 61:581–587.
6. Barnett, J.R. and I. Weatherhead. 1989. The effect of scion water potential on graft success in Sitka spruce *(Picea sitchensis). Ann. Bot.* 64:9–12.
7. Beakbane, A.B. 1953. Anatomical structure in relation to rootstock behavior. *Rpt. 13th Inter. Hort. Cong.,* Vol. 1, pp. 152–58.
8. Beakbane, A.B., and W.S. Rogers. 1956. The relative importance of stem and root in determining rootstock influence in apples. *J. Hort. Sci.* 31:99–110.
9. Beeson, R.C., Jr. and W.M. Proebsting. 1988. Photosynthate translocation during union development in *Picea* grafts. *Can. J. For. Res.* 18:986–990.
10. ———. 1988. Relationship between transpiration and water potential in grafted scions of *Picea. Physiol. Plant.* 74:481–486.
11. ———. 1988. Scion water relations during union development in Colorado blue spruce grafts. *J. Amer. Soc. Hort. Sci.* 113:427–431.
12. ———. 1989. *Picea* graft success: effects of environment, rootstock disbudding, growth regulators, and antitranspirants. *HortScience* 24(2):253–254.
13. Beeson, R.C. Jr. 1991. Scheduling woody plants for production and harvest. *HortTechnology.* 1(1):30–36.
14. Beveridge, C.A., J.J. Ross, and I.C. Murfet. 1994. Branching mutant rms-2 in *Pisum sativum:* grafting studies and indole-3-acetic acid levels. *Plant Physiol.* 104:953–959.
15. Biggs, F. and T. Biggs. 1990. Tomato grafting. *Comb. Proc. Intl. Plant Prop. Soc.* 40:97–101.
16. Bitters, W.P. 1961. Physical characters and chemical composition as affected by scions and rootstocks. Chapter 3 in *The orange: Its biochemistry and physiology,* W.B. Sinclair, ed. Berkeley: Univ. Calif. Div. Agr. Sci.
17. Bitters, W.P., and E.R. Parker. 1953. Quick decline of citrus as influenced by top-root relationships. *Calif. Agr. Exp. Sta. Bul.* 733.
18. Bloch R. 1943. Polarity in plants. *Bot. Rev.* 9:261–310.
19. Bloch, R. 1952. Wound healing in higher plants. *Bot. Rev.* 18:655–79.
20. Bradford, F.C., and B.G. Sitton. 1929. Defective graft unions in the apple and pear. *Mich. Agr. Exp. Sta. Tech. Bul. 99.*
21. Brase, K.D., and R.D. Way. 1959. Rootstocks and methods used for dwarfing fruit trees. *N. Y. State Agr. Exp. Sta Bul.* 783.
22. Buchloh, G. 1960. The lignification in stock-scion junctions and its relation to compatibility. In *Phenolics in plants in health and disease,* J.B. Pridham, ed. Long Island City, N.Y.: Pergamon Press.
23. Buck G.J. 1953. The histological development of the bud graft union in roses. *Proc. Amer. Soc. Hort. Sci.* 63:497–502.
24. Buck G.J., and B.J. Heppel. 1970. A budgraft incompatibility in *Rosa. J. Amer. Soc. Hort. Sci.* 95(4): 442–46.
25. Bukovac, M.J., S.H. Wittwer, and H.B. Tukey. 1958. Effect of stock-scion interrelationships on the trans-

port of ^{32}P and ^{45}Ca in the apple. *J. Hort. Sci.* 33:145–52.

26. Burnet, F.M. 1971. Self-recognition in colonial marine forms and flowering plants in relation to the evolution of immunity. *Nature* (London) 232:230–235.
27. Carlson, R.F. 1967. The incidence of scion-rooting of apple cultivars planted at different soil depths. *Hort. Res.* 7(2):113–15.
28. Cation, D., and R.F. Carlson. 1962. Determination of virus entities in an apple scion/rootstock test orchard. *Quart. Bul. Mich. Agr. Exp. Sta. Rpt. I,* 43(2):435–43, 1960. *Rept. II,* 45(17): 159–66.
29. Cline, M.N., and D. Neely. 1983. The histology and histochemistry of the wound healing process in geranium cuttings. *J. Amer. Soc. Hort. Sci.* 108:452–96.
30. Colby, H.L. 1935. Stock-scion chemistry and the fruiting relationships in apple trees. *Plant Physiol.* 19:483–98.
31. Collins, R.P, and S. Waxman. 1958. Dogwood graft failures. *Amer. Nurs.* 108(8):12.
32. Colquhoun, T.T. 1929. Polarity in *Casuarina paludosa. Trans. and Proc. Roy. Soc. South Australia.* 53: 353–58.
33. Copes, D. L. 1969. Graft union formation in Douglas-fir. *Amer. J. Bot.* 56(3):285–89.
34. ———. 1974. Genetics of graft rejection in Douglas-fir. *Can. J. For. Res.* 4:186–192.
35. Crafts, A.S. 1934. Phloem anatomy in two species of *Nicotiana,* with notes on the interspecific graft union. *Bot. Gaz.* 95:592–608.
36. Cummins, J.N., and D. Gonsalves. 1982. Recovery of tomato ring-spot virus from inoculated apple trees. *J. Amer. Soc. Hort. Sci.* 107:798–800.
37. Davies, F.T., Jr., Y. Fann, and J.E. Lazarte. 1980. Bench chip budding of field roses. *HortScience* 15:817–818.
38. Deloire, A. and C. Hebant. 1982. Peroxidase activity and lignification at the interface between stock and scion of compatible and incompatible grafts of *Capsicum* on *Lycopersicum. Ann Bot.* 49:887–891.
39. Dimalla, G.G., and J.A. Milbrath. 1965. The prevalence of latent viruses in Oregon apple trees. *Plant Dis. Rpt.* 49(1):15–17.
40. Dole, J.M. and H.F. Wilkins. 1991. Vegetative and reproductive characteristics of poinsettia altered by a graft-transmissible agent. *J. Amer. Soc. Hort. Sci.* 116:307–311.
41. Doley, D., and L. Leyton. 1970. Effects of growth regulating substances and water potential on the development of wound callus in *Fraxinus. New Phytol.* 69:87–102.
42. Dorsman, C. 1966. Grafting of woody plants in the glasshouse. *Proc. 17th Inter. Hort. Cong.* 1: 366.
43. Drew, M.C. and L.H. Stolzy. 1993. Growth under oxygen stress. In *Plant roots: the hidden half,* Y. Waisel, E. Amram, U. Katkafi, eds. New York, N.Y.: Marcel Dekker, Inc.
44. Eames, A.J., and L.G. Cox. 1945. A remarkable tree-fall and an unusual type of graft union failure. *Amer. J. Bot.* 32:331–35.
45. Errea, P., A. Felipe, and H. Herrero. 1994. Graft establishment between compatible and incompatible *Prunus spp. J. Exp. Bot.* 45:393–401.
46. Epstein, A.H. 1978. Root graft transmission in tree pathogens. *Ann. Rev. Phytopathol.* 16:181–92.
47. Evans, G.E., and H.P. Rasmussen. 1972. Anatomical changes in developing graft unions of *Juniperus. J. Amer. Soc. Hort. Sci.* 97(2):228–32.
48. Fann, Y.S., F.T. Davies, Jr., and D.R. Paterson. 1983. Correlative effects of bench chip budded 'Mirandy' roses. *J. Amer. Soc. Hort. Sci.* 108:180–183.
49. Flemer, W. III. 1989. Why we must still bud and graft. *Comb. Proc. Intl. Plant Prop. Soc.* 39:516–522.
50. Fujii, T., and N. Nito. 1972. Studies on the compatibility of grafting of fruit trees. I. Callus fusion between rootstock and scion. *J. Jap. Soc. Hort. Sci.* 41(1):110.
51. Gardner, F.E., and G.H. Horanic. 1963. Cold tolerance and vigor of young citrus trees on various rootstocks. *Proc. Fla. State Hort. Soc.* 76:105–10.
52. Garner, R.J., and D.H. Hammond. 1939. Studies in nursery technique. Shield budding. Treatment of inserted buds with petroleum jelly. *Ann. Rpt. E. Malling Res. Sta for 1938,* pp. 115–17.
53. Gebhardt, K. and H. Goldbach. 1988. Establishment, graft union characteristics and growth of *Prunus* micrografts. *Physiol. Plant.* 72:153–159.
54. Geneve, R.L., M. Mokhtari, and W.P. Hackett. 1991. Adventitious root initiation in reciprocally grafted leaf cuttings from the juvenile and mature phase of *Hedera helix* L. *J. Exp. Bot.* 42(234): 65–69.
55. Goldsmith, M.H.A. 1982. A site responsible for polar transport of indole-3-acetic acid in sections of maize coleoptiles. *Planta* 155:68–75.
56. Graham, B.F., Jr., and F.H. Bornmann. 1966. Natural root grafts. *Bot. Rev.* 32(3):255–92.
57. Guengerich, H.W., and D.F. Milliken. 1965. Root grafting, a potential source of error in apple indexing. *Plant Dis. Rpt.* 49:39–41.
58. Gur, A. and A. Blum. 1973. The role of cyanogenic glycoside in incompatibility between peach scions and almond rootstocks. *Hort. Res.* 13:1–10.
59. Gur, A., R.M. Samish, and E. Lifshitz. 1968. The role of the cyanogenic glycoside of the quince in the incompatibility between pear cultivars and quince rootstocks. *Hort. Res.* 8:113–34.
60. Hansen, C.J., and H.T. Hartmann. 1951. Influence of various treatments given to walnut grafts on the percentage of scions growing. *Proc. Amer. Soc. Hort. Sci.* 57:193–97.

61. Hatton, R.G. 1927. The influence of different rootstocks upon the vigor and productivity of the variety budded or grafted thereon. *J. Pom. Hort. Sci.* 6:1–28.

62. Hayward, H.E., and E.M. Long. 1942. Vegetative responses of the Elberta peach on Lovell and Shalil rootstocks to high chloride and sulfate solutions. *Proc. Amer. Soc. Hort. Sci.* 41:149–55.

63. Hesse, H. 1710. *Teutscher* Gärtner. Leipsig.

64. Heuser, C.W. 1984. Graft incompatibility in woody plants. *Comb. Proc. Intl. Plant Prop. Soc.* 34:407–12.

65. ———. 1987. Graft incompatibility: Effect of cyanogenic glycoside on almond and plum callus growth. *Comb. Proc. Intl. Plant Prop. Soc.* 37:91–97.

66. Hirst, P.M. and D.C. Ferree. 1995. Rootstock effects on the flowering of 'Delicious' apple. I. Bud development. *J. Amer. Soc. Hort. Sci.* 120:1010–1017.

67. Hirst, P.M. and D.C. Ferree. 1995. Rootstock effects on the flowering of 'Delicious' apple. II. Nutritional effects with specific reference to phosphorus. *J. Amer. Soc. Hort. Sci.* 120:1018–1024.

68. Hodgson, R.W. 1943. Some instances of scion domination in citrus. *Proc. Amer. Soc. Hort. Sci.* 43:131–38.

69. Homes, J. 1965. Histogenesis in plant grafts. In *Proc. Inter. Conf. Plant Tissue Cult.,* P.R. White and A.R. Groves, eds. Amer. Inst. Biol. Sci.

70. Honma, S. 1977. Grafting eggplants. *Scientia Hort.* 7:207–11.

71. Howard, B.H. 1993. Understanding vegetative propagation. *Comb. Proc. Intl. Plant Prop. Soc.* 43:157–162.

72. Hussein, I.A., and D.C. Slack. 1994. Fruit diameter and daily fruit growth rate of three apple cultivars on rootstock-scion combinations. *HortScience* 29(2):79–81.

73. Hussein, I.A. and M.J. McFarland. 1994. Rootstock-induced differences in sap flow of 'Granny Smith' apple. *HortScience* 29:1120–1123.

74. Jayawickrama, K.J., J.B. Brett, and S.E. McKeand. 1991. Rootstock effects in grafted conifers: A review. *New Forests* 5:157–173.

75. Jeffree, C.E., and M.M. Yeoman. 1983. Development of intercellular connections between opposing cells in a graft union. *New Phytol.* 93:491–509.

76. Jones, O.P. 1974. Xylem sap composition in apple trees: effect of the graft union. *Ann. Bot.* 38:463–467.

77. ———. 1976. Effect of dwarfing interstocks on xylem sap composition in apple trees: effect on nitrogen, potassium, phosphorus, calcium and magnesium content. *Ann. Bot.* 40:1231–1235.

78. ———. 1986. Endogenous growth regulators and rootstock/scion interaction in apple and cherry trees. *Acta Hort.* 179:177–183.

79. ———. 1989. Juvenile-like character of apple trees produced by grafting scions and rootstock produced by micropropagation. *J. Hort. Sci.* 64:395–401.

80. Jones, O.P., and H.J. Lacey. 1968. Gibberellin-like substances in the transpiration stream of apple and pear trees. *J. Exp. Bot.* 19:526–31.

81. Jones, O.P., and C.A. Webster. 1993. Nursery performance of 'Cox' apple trees with rootstocks of M.9 from either micropagation or improved conventional propagation from micropropagated plants. *J. Hort. Sci.* 68:763–766.

82. Kester, D.E., C.J. Hansen, and C. Panetsos. 1965. Effect of scion and interstock variety on incompatibility of almond on Marianna 2624 rootstock. *Proc. Amer. Soc. Hort. Sci.* 86:169–77.

83. Kollman, R., and C. Glockmann. 1991. Studies on graft unions. III. On the mechanism of secondary formation of plasmodesmata at the graft interface. *Protoplasma* 165:71–85.

84. Kollmann, R., S. Yang, and C. Glockmann. 1985. Studies on graft unions. II. Continuous and half plasmodesmata in different regions of the graft interface. *Protoplasma.* 126:19–29.

85. Langford, M.H., and C.H.T. Townsend, Jr. 1954. Control of South American leaf blight of *Hevea* rubber trees. *Plant Dis. Rpt. Suppl.* 225.

86. Lapins, K. 1959. Some symptoms of stock-scion incompatibility of apricot varieties on peach seedling rootstock. *Can. J. Plant Sci.* 39:194–203.

87. Lockard, R.G., and G.W. Schneider. 1981. Stock and scion growth relationships and the dwarfing mechanism in apple. *Hort. Rev.* 3:315–75.

88. Maynard, B.K., and N.L. Bassuk. 1990. Comparisons of stock plant etiolation with traditional propagation methods. *Comb. Proc. Intl. Plant Prop. Soc.* 40:517–523.

89. Martin, G.C., and E.A. Stahly. 1967. Endogenous growth regulating factors in bark of EM IX and XVI apple trees. *Proc. Amer. Soc. Hort. Sci.* 91:31–38.

90. McClintock J.A. 1948. A study of uncongeniality between peaches as scions and the Marianna plum as a stock. *J. Agr. Res.* 77:253–60.

91. McCully, M.E. 1983. Structural aspects of graft development. In *Vegetative compatibility responses in plants,* R. Moore, ed. Waco, Tex.: Baylor Univ. Press.

92. McKenzie, D.W. 1961. Rootstock-scion interaction in apples with special reference to root anatomy. *J. Hort. Sci.* 36:40–47.

93. McQuilkin, W.E. 1950. Effects of some growth regulators and dressings on the healing of tree wounds. *J. For.* 48(9):423–28.

94. Mendel, K. 1936. The anatomy and histology of the bud-union in citrus. *Palest. J. Bot. (R),* 1(2):13–46.

95. Milbrath, J.A., and S.M. Zeller. 1945. Latent viruses in stone fruits. *Science* 101:114–15.

96. Miller, H., and J.R. Barnett. 1993. The structure and composition of beadlike projections on Sitka spruce callus cells formed during grafting and in culture. *Ann. Bot.* 72:441–448.

97. Miller, L., and F.W. Woods. 1965. Root grafting in loblolly pine. *Bot Gaz.* 126:252–55.

98. Miller, P.W. 1952. Technique for indexing strawberries for viruses by grafting to *Fragaria vesca. Plant Dis. Rpt.* 36.

99. Millner, M.E. 1932. Natural grafting in *Hedera helix. New Phytol.* 31:2–25.

100. Mircetich, S.M. and J.W. Hoy. 1981. Brownline of prune trees, a disease associated with tomato ringspot virus infection of myrobalan and peach rootstocks. *Phytopathology* 71:31–35.

101. Mircetich, S.M.J., J. Refsguard, and M.E. Matheron. 1980. Blackline of English walnut trees traced to graft-transmitted virus. *Calif. Agr.* 34(11–12):8–10.

102. Moing, A., and J.P. Gaudillere. 1992. Carbon and nitrogen partitioning in peach/plum grafts. *Tree Physiol.* 10:81–92.

103. Moore, R. 1981. Graft compatibility and incompatibility in higher plants. 1981. *Develop. Compar. Immunol.* 5:377–389.

104. Moore, R. 1982. Graft formation in *Kalanchoe blossfeldiana. J. Exp. Bot.* 33:533–40.

105. ———. 1982. Studies of vegetative compatibility-incompatibility in higher plants. V. A morphometric analysis of the development of a compatible and an incompatible graft. *Can. J. Bot.* 60:2780–2787.

106. ———. 1984. A model for graft compatibility-incompatibility in higher plants. *Amer J. Bot.* 71:752–758.

107. Moore, R., and D. B. Walker. 1981. Graft compatibility-incompatibility in plants. *BioScience* 31(5): 389–91.

108. ———. 1981. Studies of vegetative compatibility-incompatibility in higher plants. I. A structural study of a compatible autograft in *Sedum telephoides* (Crassulaceae). II. A structural study of an incompatible heterograft between *Sedum telephoides* (Crassulaceae) and *Solanum penellii* (Solanaceae). *Amer. J. Bot.* 68(6):820–42.

109. ———. 1981. Studies of vegetative compatibility-incompatibility in higher plants. III. The involvement of acid phosphatase in the lethal cellular sensescence associated with an incompatible heterograft. *Protoplasma* 109:317–34.

110. Monzer, J. and R. Kollmann. 1986. Vascular connections in the heterograft *Lophophora williamsii* Coult. on *Trichocereus spachianus* Ricc. *J. Plant Physiol.* 123:359–372.

111. Mosse, B. 1958. Further observations on growth and union structure of double-grafted pear on quince. *J. Hort. Sci.* 33:186–93.

112. ———. 1962. *Graft-incompatibility in fruit trees.* Tech. Comm. 28. East Malling, England: Comm. Bur. Hort. and Plant. Crops.

113. Mosse, B., and M.V. Labern. 1960. The structure and development of vascular nodules in apple bud unions. *Ann. Bot.* 24:500–507.

114. Muzik, T.J. 1958. Role of parenchyma cells in graft union in *Vanilla* orchid. *Science* 127:82.

115. Navarro, L., C.N. Roistacher, and T. Murashige. 1975. Improvement of shoot tip grafting in vitro for virus-free Citrus. *J. Amer. Soc. Hort. Sci.* 100:471–479.

116. Nelson, S.H. 1968. Incompatibility survey among horticultural plants. *Comb. Proc. Intl. Plant Prop. Soc.* 18:343–407.

117. Nickell, L.G. 1946. Heteroplastic grafts. *Science* 108:389.

118. Ohta, Y. 1991. Graft-transformation, the mechanism for graft-induced genetic changes in higher plants. *Euphytica* 55:91–99.

119. Parkinson, M, and M.M. Yeoman. 1982. Graft formation in cultured explanted internodes. *New Phytol.* 91:711–19.

120. Pliego-Alfaro, F., and T. Murashige. 1987. Possible rejuvenation of adult avocado by graftage onto juvenile rootstocks in vitro. *HortiScience* 22:1321–24.

121. Porada, H. 1993. Timber species propagation. *Comb. Proc. Intl. Plant Prop. Soc.* 43:80–85.

122. Posnette, A.F. 1966. Virus diseases of fruit plants. *Proc. 17th Inter. Hort. Cong.* 3:89–93.

123. Preston, A. P 1958. Apple rootstock studies: Thirty-five years' results with Cox's Orange Pippin on clonal rootstocks. *J. Hort. Sci.* 33: 194–201.

124. ———.1967. Apple rootstock studies: Fifteen years' results with some M.IX crosses. *J. Hort Sci.* 42:41–50.

125. Proebsting, E.L. 1928. Further observations on structural defects of the graft union. *Bot. Gaz.* 86:82–92.

126. ———. 1926. Structural weaknesses in interspecific grafts of *Pyrus. Bot. Gaz.* 82:336–38.

127. Rachow-Brandt, G., and R. Kollmann. 1992. Studies on graft unions. IV. Assimilate transport and sieve element restitution in homo- and heterografts. *J. Plant Physiol.* 139:579–583.

128. Ranney, T.G. 1994. Differential tolerance of eleven *Prunus* Taxa to root zone flooding. *J. Environ. Hort.* 12(3): 138–141.

129. Ranney, T.G., N.L. Bassuk, and T.H. Whitlow. 1991. Influence of rootstock, scion, and water deficits on growth of 'Colt' and 'Meteor' cherry trees. *HortScience* 26: 1204–1207.

130. Ranney, T.G. and R.E. Birr. 1994. Comparative flood tolerance of birch rootstocks. *J. Amer. Soc. Hort. Sci.* 119:43–48.

131. Ranney, T.G. and E.P. Whitman II. 1995. Growth and survival of 'Whitespire' Japanese birch grafted on rootstocks of five species of birch. *HortScience* 30(3): 521–522.

132. Rao, A.N. 1966. Developmental anatomy of natural grafts in *Ficus globosa*. *Aust. J. Bot.* 14:269–76.
133. Raulston, J.C. 1995. New concepts in improving ornamental plant adaptability with stress tolerant rootstocks. *Comb Proc. Intl. Plant Prop. Soc.* 45: In press.
134. Roberts, A.N., and L.T. Blaney. 1967. Qualitative, quantitative, and positional aspects of interstock influence on growth and flowering of the apple. *Proc. Amer. Soc. Hort. Sci.* 91:39–50.
135. Robitaille, R.H., and R.F. Carlson. 1970. Graft union behavior of certain species of *Malus* and *Prunus*. *J. Amer. Soc. Hort Sci.* 95(2):131–34.
136. ———. 1971. Response of dwarfed apple trees to stem injections of gibberellic and abscisic acids. *HortScience* 6(6):539–40.
137. ———. 1976. Gibberellic and abscisic acid-like substances in the regulation of apple shoot extension. *J. Amer. Soc. Hort. Sci.* 101:388–392.
138. Santamour, F.S., Jr. 1979. Inheritance of wound compartmentalization in soft maples. *J. Arboricul.* 5:220–225.
139. ———. 1988. Graft compatibility in woody plants: An expanded perspective. *J. Environ. Hort.* 6(1): 27–32.
140. ———. 1988. Graft incompatibility related to cambial peroxidase isozymes in Chinese chestnut. *J. Environ. Hort.* 6(2):33–39.
141. ———. 1988. Cambial peroxidase enzymes related to graft incompatibility in red oak. *J. Environ. Hort.* 6(3):87–93.
142. ———. 1989. Cambial peroxidase enzymes related to graft incompatibility in red maple. *J. Environ. Hort.* 7:8–14.
143. ———. 1992. Predicting graft incompatibility in woody plants. *Comb Proc. Internl. Plant Prop. Soc.* 42:131–134.
144. Santamour, F.S., and M.V. Coggeshall. 1996. Cambial peroxidases as predictors of graft incompatibility in red oak. *J. Environ. Hort.* 14: In press.
145. Santamour, F.S., Jr., A.J. McArdle, and R.A. Jaynes. 1986. Cambial isoperoxidase patterns in *Castanea*. *J. Environ. Hort.* 4(1):14–16.
146. Sass, J.E. 1932. Formation of callus knots on apple grafts as related to the histology of the graft union. *Bot. Gaz.* 94:365–80.
147. Sax, K., and A. Q. Dickson. 1956. Phloem polarity in bark regeneration. *J. Arn. Arb.* 37: 173–79.
148. Schechter, I., D.C. Elfving, and J.T.A. Proctor. 1991. Apple tree canopy development and photosynthesis as affected by rootstock. *Can. J. Bot.* 69:295–300.
149. Schmid, P.P., and W. Feucht. 1981. Differentiation of sieve tubes in compatible and incompatible *Prunus* graftings. *Scientia Hort.* 15:349–354.
150. Shigo, L.A. 1993. *100 Tree Myths*. Durham, N.H.: Shigo and Trees Assoc.
151. Shigo, L.A., and H.G. Marx. 1977. Compartmentalization of decay in trees. *U.S Dept. Agric. For. Ser. Agric. Inform. Bull.* No. 405.
152. Shimomura, T., and K. Fuzihara. 1977. Physiological study of graft union formation in cactus. II. Role of auxin on vascular connection between stock and scion. *J. Japan. Soc. Hort. Sci.* 45:397–406.
153. Simmons, R.K., and M.C. Chu. 1984. Tissue development within the graft union as related to dwarfism in apple. *Acta Hort.* 146:203–210.
154. Sitton, B. G. 1931. Vegetative propagation of the black walnut. *Mich. Agr. Exp. Sta. Tech. Bul.* 119.
155. Skene, D.S., H.R. Shepard, and B.H. Howard. 1983. Characteristic anatomy of union formation in T- and chip-budded fruit and ornamental trees. *J. Hort. Sci.* 58(3):295–99.
156. Smith, J.W.M., and P. Proctor. 1965. Use of disease resistant rootstocks for tomato crops. *Exp. Hort* 12:6–20.
157. Soumelidou, K., N.H. Battey, P. John, and J.R. Barnett. 1994. The anatomy of the developing bud union and its relationship to dwarfing in apple. *Ann. Bot.* 74:605–611.
158. Soumeiidou, K., D.A. Morris, N.H. Battey, J.R. Barnett, and P. John, 1994. Auxin transport capacity in relation to the dwarfing effect of apple rootstocks. *J. Hort. Sci.* 69:719–725.
159. Soule, J. 1971. Anatomy of the bud union in mango (*Mangifera indica* L.). *J. Amer. Soc. Hort. Sci.* 96(3):380–83.
160. Stoddard, F.L., and M.E. McCully. 1980. Effects of excision of stock and scion organs on the formation of the graft union in coleus: A histological study. *Bot. Gaz.* 141:401–2.
161. Strong, D., and A. Miller-Azarenko. 1991. Dry matter partitioning in 'Starkspur supreme delicious' on nine rootstocks. *Fruit Var. J.* 45:238–241.
162. Stutte, G.W., T.A. Baugher, S.P. Walter, D.W. Leach, D.M. Glenn and T.J. Tworkoski. 1994. Rootstock and training system affect dry-matter and carbohydrate distribution in 'golden delicious' apple trees. *J. Amer. Soc. Hort. Sci.* 119:492–497.
163. Syvertsen, J.P. 1985. Integration of water stress in fruit trees. *HortScience* 20:1039–1043.
164. Thoüin, A. 1821. *Monographie des greffes, ou description technique* (in Royal Hort. Soc. Library, London).
165. Tiedemann, R. 1989. Graft union development and symplastic phloem contact in the heterograft *Cucumis sativus* on *Cucurbita ficifolia*. *J. Plant Physiol.* 134:427–440.
166. Tiedemann, R., and U. Carsens-Behrens. 1994. Influence of grafting on the phloem protein patterns in Cucurbitacea. I. additional phloem exudate proteins in *Cucumis sativus* grafted on two Cucurbita species. *J. Plant Physiol.* 143:189–194.

167. Torrey, J.G., D.E. Fosket, and P.K. Hepler. 1971. Xylem formation: A paradigm of cytodifferentiation in higher plants. *Amer. Sci.* 59:338–52.

168. Toxopeus, H.J. 1936. Stock-scion incompatibility in citrus and its cause. *J. Pom. and Hort. Sci.* 14:360–64.

169. Tydeman, H.M., and F.H. Alston. 1965. The influence of dwarfing rootstocks in shortening the juvenile phase of apple seedlings. *Ann. Rpt. E. Malling Res. Sta. for 1964,* pp. 97–98.

170. Vertrees, J.D. 1991. Understock for rare *Acer* species. *Comb. Proc. Intl. Plant Prop. Soc.* 41:272–275.

171. Vöchting, H. 1892. Veber transplantation am pflanzenköper.

172. Wagner, D.F. 1969. Ultrastructure of the bud graft union in *Malus.* Ph.D. dissertation, Iowa State Univ., Ames, Iowa.

173. Wang, Z.Y., K.S. Gould, and K.J. Patterson. 1994. Comparative root anatomy of five *Actinida* species in relation to rootstock effects on kiwifruit flowering. *Ann Bot.* 73:403–414.

174. Wang, Z.Y., K.J. Patterson, K.S. Gould, and R.G. Lowe. 1994. Rootstock effects on budburst and flowering in kiwifruit. *Scientia. Hort.* 57:187–199.

175. Wareing, P.F., C.E.A. Hanney, and J. Digby. 1964. The role of endogenous hormones in cambial activity and xylem differentiation. In *The formation of wood in forest trees,* M.H. Zimmerman, ed. New York: Academic Press.

176. Warmund, M.R., B.H. Barritt, J.M. Brown, K.L. Schaffer, and B.R. Jeong. 1993. Detection of vascular discontinuity in bud unions of 'Jonagold' apple on mark rootstock with magnetic resonance imaging. *J. Amer. Soc. Hort. Sci.* 118:92–96.

177. Waugh, F.A. 1904. The graft union. *Mass. Agr. Exp. Sta. Tech. Bul. 2.*

178. Webber, H.J. 1943. The "tristeza" disease of sour orange rootstock. *Proc. Amer. Soc. Hort. Sci.,* 43: 160–68.

179. Wells, R.B. 1986. A historical review of grafting techniques. *Comb. Proc. Inter. Plant Prop. Soc.* 35:96–101.

180. Westwood, M.N. 1993. Temperate-zone pomology: physiology and culture. (3rd Ed.) Portland, Oregon: Timber Press.

181. Wetmore, R.H., and J.P. Rier. 1963. Experimental induction of vascular tissues in callus of angiosperms. *Amer. J. Bot.* 50:418–30.

182. Wildon, D.C., J.F. Thain, P.E.H. Minchin, I.R. Gubb, A.J. Reilly, Y.D. Skipper, H.M. Doherty, P.J. O'Donnell, and D.J. Bowles. 1992. Electrical signalling and systemic proteinase inhibitor induction in the wounded plant. *Nature.* 360:62–65.

183. Winter, H.F. 1963. Prevalence of latent viruses in Ohio apple trees. *Ohio Farm and Home Res.* 48: 58–59, 63.

184. Williamson, J.G., and B.E. Maust. 1995. Growth of budded, containerized, citrus nursery plants when photosynthesis of rootstock shoots is limited. *HortScience* 30(7):1363–1365.

185. Williamson, J.G., W.S. Castle, and K.E. Koch. 1992. Growth and ^{14}C-photosynthate allocation in citrus nursery trees subjected to one or three bud-forcing methods. *J. Amer. Soc. Hort. Sci.* 117:37–40.

186. Yadava, U.L., and D.F. Dayton. 1972. The relation of endogenous abscisic acid to the dwarfing capability of East Malling apple rootstocks. *J. Amer. Soc. Hort. Sci.* 97(6):701–5.

187. Yates, I.E., and D. Sparks. 1992. Pecan cultivar conversion by grafting onto roots of 70-year-old trees. *HortScience* 27:803–807.

188. Yeager, A.F. 1944. Xylem formation from ring grafts. *Proc. Amer. Soc. Hort. Sci.* 44:221–22.

189. Yeoman, M.M. 1984. Cellular recognition systems in grafting. *Encyclopedia of Plant Physiology.* 453–72.

190. Yeoman, M.M. and R. Brown. 1976. Implications of the formation of the graft union for organization in the intact plant. *Ann. Bot.* 40:1265–76.

191. Zebrak, A.R. 1937. Intergeneric and interfamily grafting of herbaceous plants. *Timirjazey Seljskohoz Akad.* 2:115–33. *Herb. Abst.* 9:675. 1939.

SUPPLEMENTARY READING

BEAKBANE, A. B. 1956. Possible mechanisms of rootstock effect. *Ann. of App. Bio.* 44:517–21.

BRASE, K. D., and R. D. WAY. 1959. Rootstocks and methods used for dwarfing fruit trees. *N. Y. Agr. Exp. Sta. Bul.* 783.

ESAU, K. 1977. *Anatomy of seed plants* (2nd ed.). New York: John Wiley.

FEUCHT, W. 1987. Graft incompatibility of tree crops: An overview of the present scientific status. *Acta Hort.* 227:33–41.

GARNER, R.J. 1988. *The grafter's handbook* (5th ed.). London: Cassell PLC.

LOCKARD, R. G., and G. W. SCHNEIDER. 1981. Stock and scion growth relationships and the dwarfing mechanism in apple. *Hort. Rev.* 3:315–75.

MOORE, R., ed. 1983. *Vegetative compatibility responses in plants.* Waco, Tex.: Baylor Univ. Press.

MOORE, R., and D. B. WALKER. 1981. Graft compatibility-incompatibility in plants. *BioScience* 31(5): 389–91.

MOSSE, B. 1962. Graft incompatibility in fruit trees. *Commonwealth Agr. Bur. Tech. Comm. 28.*

NELSON, S. H. 1968. Incompatibility survey among horticultural plants. *Comb. Proc. Intl. Plant Prop. Soc.* 18:343–93.

ROGERS, W. S., and A. B. BEAKBANE. 1957. Stock and scion relations. *Ann. Rev. Plant Physiol.* 8:217–36.

ROM, R. C. and R. F. CARLSON, eds. 1987. *Rootstocks for fruit crops.* New York: John Wiley.

SIMONS, R. K. 1987. Compatibility and stock-scion interactions as related to dwarfing. In *Rootstocks for fruit crops,* R. C. Rom and R. F. Carlson, eds. New York: John Wiley.

WESTWOOD, M.N. 1993. *Temperate-zone pomology: physiology and culture.* (3rd Ed.) Portland, Oregon: Timber Press.

13

Techniques of Grafting

Since people first learned to graft plants, a myriad of grafting techniques have developed. In *The Grafters' Handbook,* Garner (*19*) has enumerated and described many of these—others have come and gone.

Here we describe the most important grafting methods. Among them, a person who can use a sharp knife can find one that meets any specific grafting need. Success in grafting depends, however, not only on a technically correct graft, but in preparation of the scion and rootstock for graftage. Equally critical is the optimum time for grafting, and proper aftercare.

This chapter is divided into three sections: (1) the *types of grafts,* (2) *production processes of graftage*—including the preparation, craftsmanship, and aftercare of grafted plants, and (3) *grafting systems,* including field grafting, bench grafting, and miscellaneous grafting systems—such as herbaceous graftage, cutting grafts, and micrografting.

For any successful grafting operation, producing a plant as shown in Figure 13–1 requires five important elements:

1. *The rootstock and scion must be compatible.* They must be capable of uniting. Usually, but not always, closely related plants, such as two apple cultivars, can be grafted together. Distantly related plants, such as oak and apple, cannot make a successful graft combination (see Chapter 12 for a discussion of this factor).
2. *The vascular cambium of the scion must be placed in intimate contact with that of the rootstock.* The cut surfaces should be held together tightly by wrapping, nailing, wedging, or some similar method. Rapid development of the graft union is necessary so that the scion may be supplied with water and nutrients from the rootstock by the time the buds start to open.
3. *The grafting operation must be done at a time when the rootstock and scion are in the proper physiological stage.* Usually, this means that the scion buds are dormant while at the same time, the cut tissues at the graft union are capable of producing the callus tissue neces-

FIGURE 13–1 Cultivars of the Persian (English) walnut *(Juglans regia)* grafted on *J. hindsii* rootstocks. The characteristics of these two species remain distinctly different after grafting, exactly to the junction of the graft union.

sary for healing of the graft. For deciduous plants, dormant scionwood is collected during the winter and kept inactive by storing at low temperatures. The rootstock plant may be dormant or in active growth, depending upon the grafting method used.

4. *Immediately after the grafting operation is completed, all cut surfaces must be protected from desiccation.* The graft union is covered with tape, grafting wax, Parafilm tape (American Can Co.), or the grafts are placed in moist material or a covered grafting frame.
5. *Proper care must be given the grafts for a period of time after grafting.* Shoots coming from the rootstock below the graft will often choke out the desired growth from the scion. In some cases, shoots from the scion will grow so vigorously that they break off unless staked and tied or cut back.

TYPES OF GRAFTS

Grafting may be classified according to the part of the rootstock on which the scion is placed—a root, or various places in the top of the plant. Types of grafts can be categorized as: (1) *detached scion graftage,* which includes apical, side, bark and root graftage, (2) *approach graftage,* where the root system of the scion and shoot system of the rootstock are not removed until after successful graft union formation occurs, and (3) *repair graftage* of established trees. Grafts categorized in the side box are described in greater detail in the chapter.

TYPES OF GRAFTS

1. **Detached Scion Graftage**
   - **A. Apical Graftage**
     - Whip-and-tongue graft
     - Splice graft (whip graft)
     - Cleft graft (split graft)
     - Wedge graft (saw-kerf graft)
     - Saddle graft
     - Four-flap graft (banana graft)
   - **B. Side Graftage**
     - Side-stub graft
     - Side-tongue graft
     - Side-veneer graft
   - **C. Bark Graftage**
     - Bark graft (rind graft)
     - Inlay bark graft
   - **D. Root Graftage**
     - Whole-root and piece-root grafting
     - Nurse-root grafting
2. **Approach Graftage**
   - Spliced approach graft
   - Tongued approach graft
   - Inlay approach graft
3. **Repair Graft**
   - Inarching
   - Bridge graft
   - Bracing

With high labor costs, only a few of the more efficient grafts are utilized in U.S. woody ornamental nurseries, such as the side veneer, splice (whip graft), and whip-and-tongue graft. There is only limited use of approach and repair graftage. With fruit crops, depending on the species, a number of different apical, side, and root grafts are still utilized around the world. Chip budding and T-budding, which are described in detail in Chapter 14, are two of the most common budding methods for woody ornamental and fruit crops. In automated and manual graftage of herbaceous vegetable crops, the splice graft is one of the most important.

Detached Scion Graftage—Apical Graftage

There are many variations of apical graftage. As the name suggests, the scion is inserted into the top of the severed rootstock shoot.

Whip-and-Tongue Graft

The whip-and-tongue graft, shown in Figures 13–2 and 13–3 is particularly useful for grafting relatively small material, 6 to 13 mm (¼ to ½ in. in diameter). It is highly successful if done properly because there is considerable vascular cambium contact. It heals quickly and makes a strong union. Preferably, the scion and rootstock should be of equal diameter. The scion should contain two or three buds with the graft made in the smooth internode area below the lower bud.

The cuts made at the top of the rootstock should be the same as those made at the bottom of the scion. First, a smooth, sloping cut is made, 2.5 to 6 cm (1 to 2½ in.) long. The longer cuts are made when working with large material. This first cut should preferably be made with one single stroke of the knife, so as to leave a smooth, flat surface. To do this, the knife must be razor sharp. Wavy, uneven cuts made with a dull knife will not result in a satisfactory union.

On each of these cut surfaces, a reverse cut is made. It is started downward at a point about one-third of the distance from the tip and should be about one-half the length of the first cut. To obtain a smooth-fitting graft, this second cut should not just split the grain of the wood, but should follow along under the first cut, tending to parallel it.

The rootstock and scion are then inserted into each other, with the tongues interlocking. It is extremely important that the vascular cambium layers match along at least one side, preferably along both sides. The lower tip of the scion should not overhang the stock, as there is a likelihood of the formation of large callus knots. The use of scions larger than the rootstock should be avoided for the same reason.

After the scion and rootstock are fitted together, they are securely held by tying with budding rubber strips, plastic (poly) budding/grafting tape, or raffia. Is important that the tissues in the graft union area not dry out, so either sealing the graft union with grafter's wax or Parafilm, or placing the plants under high relative humidity is essential until the graft union has formed.

In *bench graftage* (page 470) the bare-root grafted plants can be stored in a grafting box (without sealing the graft union with grafter's wax), and packed with slightly moist peat or bark. Grafted plants in liner pots can be placed in a polytent in a temperature-controlled greenhouse (Figure 13–16). If the bare-root, bench-grafted plants are to be directly planted in a field nursery, the union is placed below the soil level. Any poly budding tape will need to be removed after graft union formation to prevent girdling the stem. Grafts wrapped with budding rubbers and covered with soil should be inspected later, since the rubber decomposes very slowly below ground and may also cause a constriction at the graft union.

If the whip-and-tongue graft is used in *field grafting* (page 472), the graft union of the *topworked* (page 473) plant must be tied and also sealed with grafter's wax or Parafilm. Aftercare of grafted plants is further described in the section on Production Processes of Graftage.

Splice Graft (Whip Graft)

The splice graft is simple and easy to make (Figure 13–4). It is the same as the whip-and-tongue graft, except that the second, or "tongue," cut is not made in either the rootstock or scion. A simple slanting cut of the same length and angle is made in both the rootstock and the scion. These are placed together and wrapped or tied as described for the whip graft. If the scion is smaller than the rootstock it should be set at one side of the rootstock so that the vascular cambium layers will be certain to match along that side (Figure 13–4).

The splice graft is particularly useful in grafting plants that have a very pithy stem or that have wood that is not flexible enough to permit a tight fit when a tongue is made as in the whip-and-tongue graft. The splice graft is used in greenhouse production of vegetable crops for grafting disease-resis-

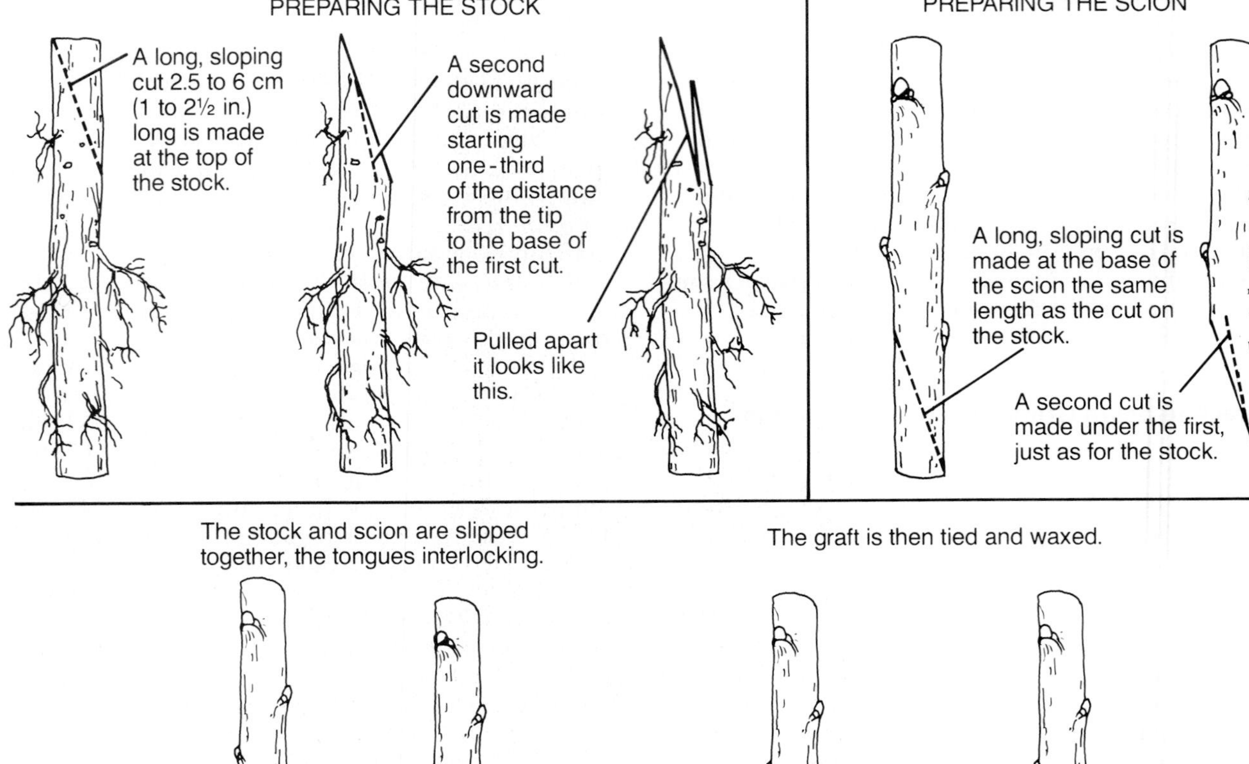

FIGURE 13–2 Whip-and-tongue graft. This method is widely used in grafting small plant material and is especially valuable in making root grafts as illustrated here.

tant rootstocks. The graft can be performed manually or with sophisticated, robotic grafting machines (Figures 13-34 and 13–35).

The rootstock and scion must be held together while tying the splice graft. In field grafting, it is not a convenient method to use at ground level, and must be performed higher up on the rootstock, where the grafter must do both the cutting and tying. A splice graft is more commonly used in bench grafting and grafting of container plants. The whip-and-tongue does not have this limitation, since the tongue holds the graft together, so that the

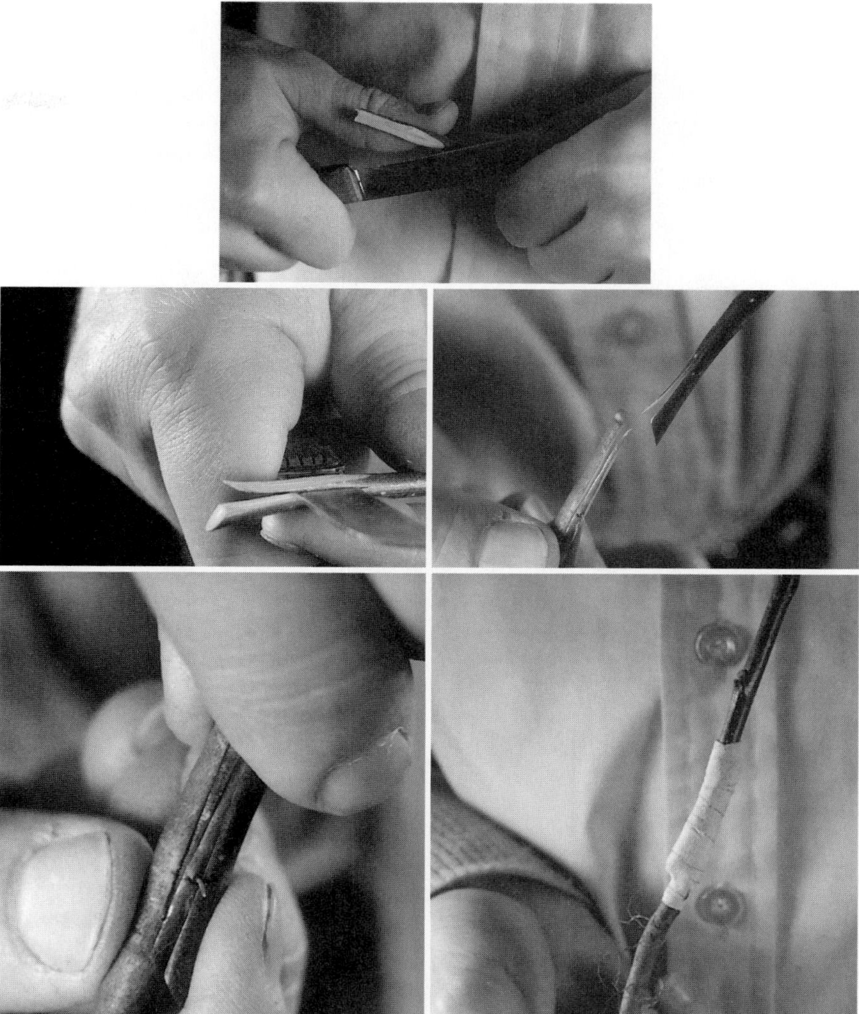

FIGURE 13–3 Procedures in making a whip-and-tongue graft. *Top:* A slice cut is made across both the rootstock and scion. *Center left:* A second cut is made to the tongue; the grafter's hands are locked together to avoid injury. *Center right* and *bottom left:* Fitting and locking the tongues of the graft partners. *Bottom right:* Wrapping with grafting tape.

grafter has both hands free for tying, or the fitted graft can be left for a helper or "tier" to tie and seal.

Cleft Graft (Split Graft)

The cleft or split graft is one of the oldest methods of field grafting. It is used to topwork trees, either in the trunk of a small tree or in the scaffold branches of a larger tree (Figures 13–5 and 13–6). Cleft grafting is used for *crown grafting* (see the Grafting Systems section) or grafting smaller plants such as established grapevines or camellias. In topworking trees, this method should be limited to stock branches about 2.5 to 10 cm (1 to 4 in.) in diameter and to species with fairly straight-grained wood that will split evenly.

In all types of grafting the scion must be inserted right side up. That is, the apical tip of the buds on the scion should be pointing upward and away from the rootstock. Unless this rule is observed, the graft will not be successful.

Although cleft grafting can be done any time during the dormant season, the chances for successful healing of the graft union are best if the work is done in early spring just when the buds of the rootstock are beginning to swell but before active growth has started. If cleft grafting is done after the

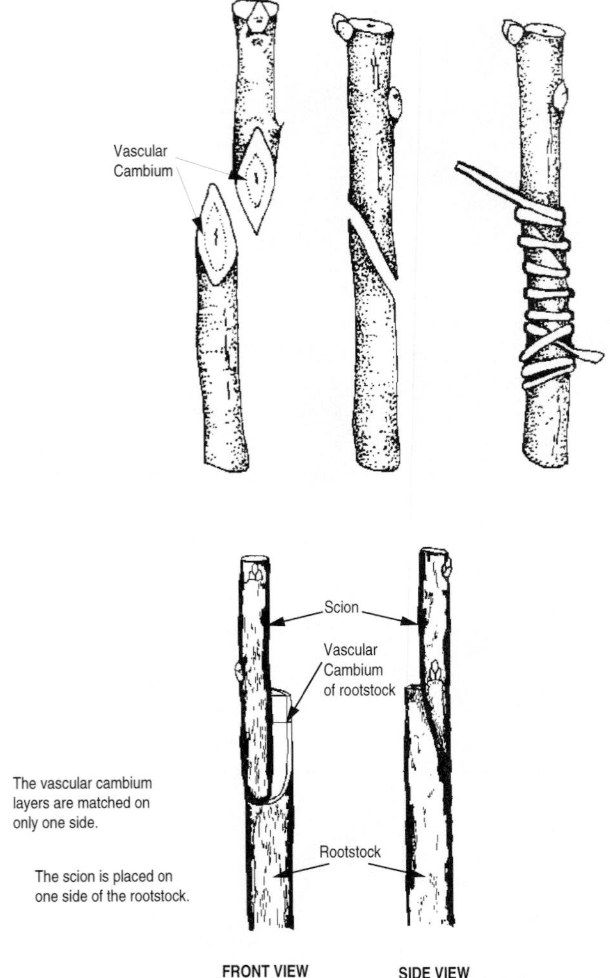

FIGURE 13–4 Splice graft (whip graft). *Top:* Procedures in making the splice graft with a slice cut that slants across the grafting partners. Ideally, the rootstock and scion are of the same caliber. *Bottom:* Method of making a splice graft when the scion is considerably smaller than the rootstock. It is important that the cambium layers be matched on one side.

tree is in active growth, the bark of the rootstock may separate from the wood, causing difficulties in obtaining a good union. When this separation occurs, the loosened bark must be firmly nailed back in place. The scions should be made from dormant, one-year-old wood. Unless the grafting is done early in the season (when the dormant scions can be collected and used immediately), the scionwood should be collected in advance and refrigerated. In sawing off the branch for this and other topworking methods, the cut should be made at right angles to the main axis of the branch.

In making the cleft graft, a heavy knife, such as a butcher knife, or one of several special cleft grafting tools, is used to make a vertical split for a distance of 5 to 8 cm (2 to 3 in.) down the center of the stub to be grafted (Figure 13–5). This split is made by pounding the knife in with a hammer or mallet. The branch is sawed off in such a position that the end of the stub which is left is smooth, straight-grained, and free of knots for at least 15 cm (6 in.). Otherwise, the split may not be straight, or the wood may split one way and the bark another. The split should be in a tangential rather than radial direction in relation to the center of the tree. This permits better placement of the scions for their subsequent growth. Sometimes the cleft is made by a longitudinal saw cut rather than by splitting. After a good, straight split is made, a screwdriver, chisel, or the wedge part of the cleft-grafting tool is driven into the top of the split to hold it open.

Two scions are inserted, one at each side of the stock where the vascular cambium layer is located. The scions should be 8 to 10 cm (3 to 4 in.) long, about 10 to 13 mm (⅜ to ½ in.) in thickness, and should have two or three buds. The basal end of each scion should be cut into a gently sloping wedge about 5 cm (2 in.) long. It is not necessary that the end of the wedge come to a point. The side of the wedge which is to go to the outer side of the rootstock should be slightly wider than the inside edge. Thus, when the scion is inserted and the tool is removed, the full pressure of the split rootstock will come to bear on the scions at the position where the vascular cambium of the rootstock touches the vascular cambium layer on the outer edge of the scion. Since the bark of the rootstock is almost always thicker than the bark of the scion, it is usually necessary for the outer surface of the scion to set slightly in from the outer surface of the rootstock in order to match the vascular cambium layers.

The long, sloping wedge cuts at the base of the scion should be smooth, made by a single cut on each side with a sharp knife. Both sides of the scion wedge should press firmly against the rootstock for their entire length. A common mistake in cutting scions for this type of graft is to make the cut on the scion too short and the slope too abrupt, so that the point of contact is only at the top. Slightly shaving the sides of the split in the stock will often permit a smoother contact.

After the scions are properly made and inserted, the tool is withdrawn, without disturbing the scions, which should be held tightly by the pres-

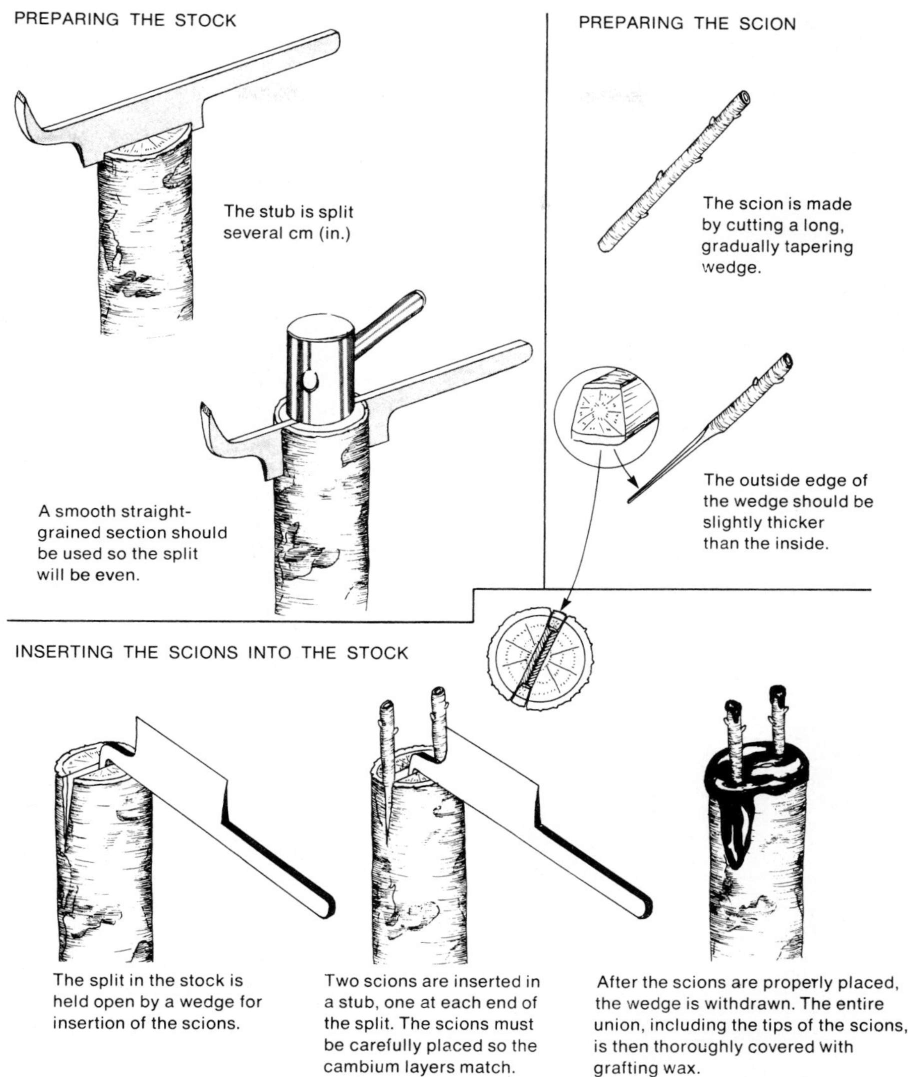

FIGURE 13–5 Steps in making the cleft graft (split graft). This method is very widely used and is quite successful if the scions are inserted so that the cambium layers of stock and scion match properly.

sure of the rootstock so that they cannot be pulled loose by hand. No further tying or nailing is needed unless very small rootstock branches have been used. In this case, the top of the rootstock can be wrapped tightly with poly grafting tape or adhesive tape to hold the scions in place more securely.

Thorough waxing of the completed graft is essential. The top surface of the stub should be entirely covered, permitting the wax to work into the split in the stock. The sides of the grafted stub should be well covered with wax as far down the stub as the length of the split. The tops of the scions should be waxed but not necessarily the bark or buds of the scion. Two or three days later all the grafts should be inspected and rewaxed where openings appear. Lack of thorough and complete waxing in this type of graft is a common cause of failure.

Wedge Graft (Saw-Kerf Graft)

Wedge grafting is illustrated in Figure 13–7. Like the cleft graft, it can be made in late winter (in mild climates) or early spring before the bark begins to slip (separates easily from the wood).

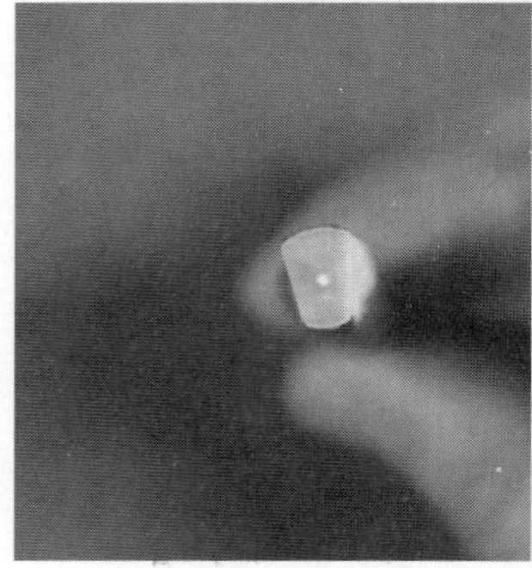

FIGURE 13–6 *Top:* Tools used in making a cleft (split) graft. *Center:* Making the cleft by splitting the rootstock top. *Bottom left:* Scionwood with the outside wedge slightly thicker than the inside. *Bottom right:* Inserting the first of two scions. The split rootstock is temporarily separated by the tool.

The diameter of the stock to be grafted is the same as for the cleft graft—5 to 10 cm (2 to 4 in.), and the scions also are the same size—10 to 13 cm (4 to 5 in.) long and 10 to 13 mm (⅜ to ½ in.) in thickness.

A sharp, heavy, short-bladed knife is used for making a V-wedge in the side of the stub, about 5 cm (2 in.) long. Two cuts are made, coming together at the bottom and as far apart at the top as the width of the scion. These cuts extend about 2 cm (¾ in.) deep into the side of the stub. After these cuts are made, a screwdriver is pounded downward behind the wedge chip from the top of the stub to knock out the chip, leaving a V opening for insertion of the scion. The base of the scion is trimmed to a wedge shape exactly the same size and shape as the opening. With the two vascular cambium layers matching, the scion is tapped downward firmly into place and slanting outward slightly at the top so that the vascular cambium layers cross. If the cut is long enough and gently tapering, the scion should be so tightly held in place that it would be difficult to dislodge.

In a stub that is 5-cm (2-in.) wide, two scions should be inserted 180° apart; in a 10-cm (4-in.) stub, three scions should be used, 120° apart. After all scions are firmly tapped into place, all cut surfaces, including the tips of the scion, should be waxed thoroughly.

Saddle Graft

The saddle graft can be bench grafted by hand or machine (Figure 13–31). The rootstock and scion should be the same size. The scion is prepared by cutting upward through the bark and into the wood on opposite sides of the scion (Figure 13–8). The knife should penetrate more deeply into the wood as the cuts are lengthened. Before the knife is withdrawn, it is turned into the middle of the scion piece, and the saddle shape is gradually formed by removing pieces of the wood. The rootstock is cut transversely and receives two upward cuts on either side. This should be done to expose the vascular cambia of the rootstock to match vascular cambia in the saddle of the scion. The apex of the rootstock is carved to fit the saddle. The graft needs to be tied, and all exposed cut surfaces sealed or stored in a grafting case until the graft union has formed. The saddle graft is used for bench grafting grape and *Rhododendron* cultivars (*19*).

Four-Flap Graft (Banana Graft)

The four-flap graft is used in topworking small-caliper trees or tree limbs up to 2.5 cm (1 in.) in diameter. This field graft is normally done manually (Figures 13–9 and 13–10), but there is a tool that aids in stripping the rootstock bark flaps from the wood (Figure 13–11). Both the scion and rootstock should be of equal diameter, and the best fit is obtained when the scion is slightly larger than the

PREPARING THE ROOTSTOCK

A heavy sharp knife is pounded into the side of the stub to make two cuts to form a V.

A screwdriver is used to flip out the V-shaped chip, leaving a space for insertion of the scion.

PREPARING THE SCION

The scion should be about 10 to 13 cm (4 to 5 in.) long, 10 to 12 mm (3/8 to 1/2 in.) thick, and with 2 or 3 healthy vegetative buds. The basal ends should be cut to a V-shaped wedge, matching the opening in the stock.

INSERTING THE SCIONS INTO THE ROOTSTOCK

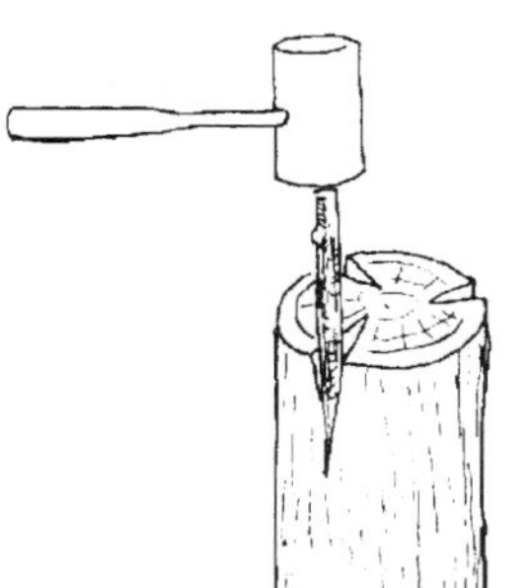

The scion is gently tapped into the V-shaped opening in the stock, matching the cambium layers at a slight angle so that the cambium of stock and scion cross.

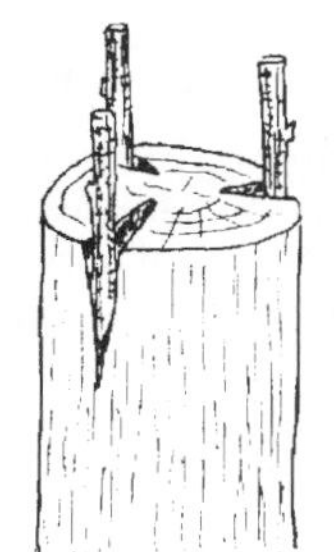

Scion should be inserted at an angle so that the cambium layers of stock and scion are closely matched, barely crossing each other.

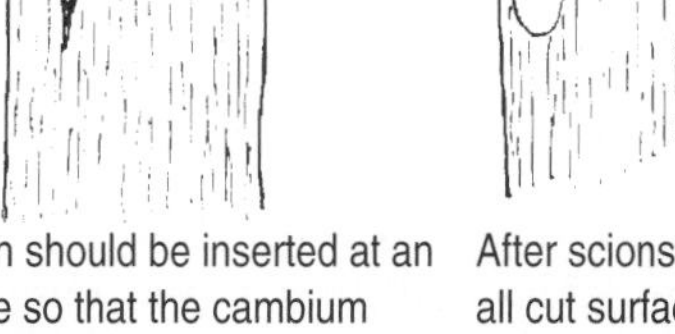

After scions are in place all cut surfaces are thoroughly covered with grafting wax.

FIGURE 13–7 Wedge graft (saw-kerf graft). Sometimes called the saw-kerf because the cuts in the side of the rootstock can be made with a saw, rather than with the sharp tool depicted.

rootstock. The four-flap graft is done with pecans in Texas from April to mid-May, when the rootstock bark is actively slipping (*37*). Scionwood, which is collected while dormant during the winter, is taken from cold storage and used immediately.

The rootstock with a primary stem or lateral limb is severed horizontally with sharp pruning shears. On the rootstock where the horizontal cut was made, four vertical, equally spaced cuts 4 cm (1.5 in.) long are made with a grafting knife which penetrates from the bark down to the interior wood. A 15 cm (6 in.) piece of scionwood with three axillary buds is cut on four sides with a knife. Cuts are made on the scion through the bark down to the wood—without removing much wood. There should be four thin slivers of bark, with the vascular cambium at the corners, which gives the prepared scion a square diameter appearance. The four flaps of bark are pulled down 4 cm (1.5 in.) on the rootstock, and the inner wood is removed with pruning shears. The scion piece is inserted upright on the rootstock and the four flaps of the rootstock are pulled up to cover the four cut surfaces of the scion. A rubber band is rolled up onto the flaps to hold them in place. The

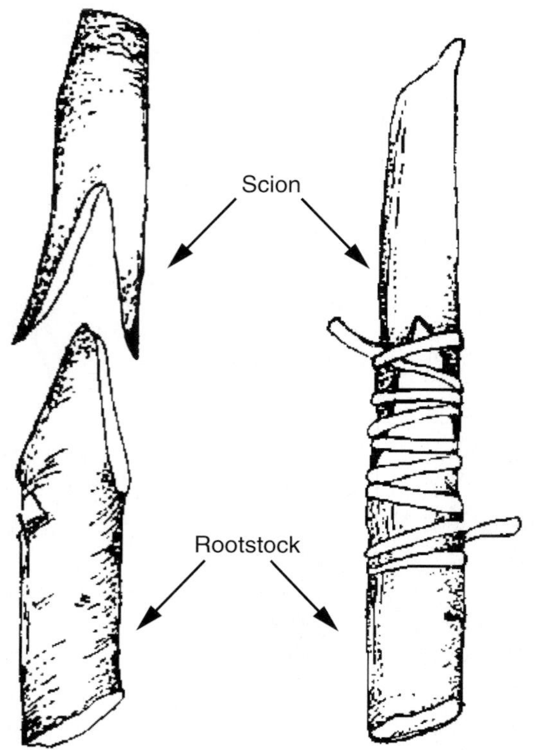

FIGURE 13–8 Steps in making the saddle graft. The scion is cut to have a saddle appearance and the understock to form a blunt point. The cambium layers are matched up and the graft tied.

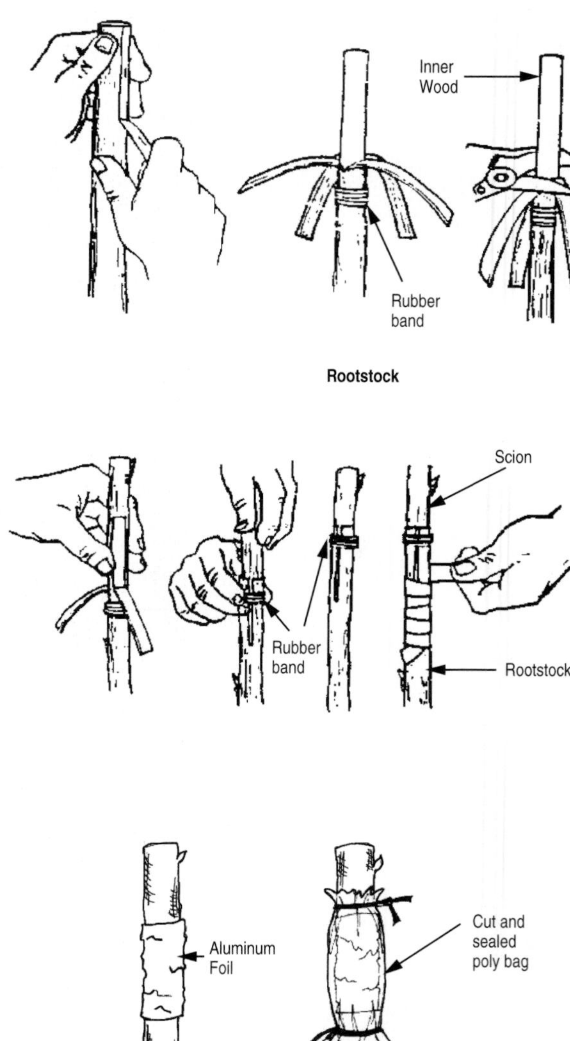

FIGURE 13–9 The four-flap or banana graft. *Top:* The top of the rootstock is cut horizontally, and the bark is cut vertically into four strips. The four bark flaps are peeled down and the inner wood removed. *Middle:* The scion bark is removed and the wood retained. The flaps of the rootstock cover the cut surfaces of the scionwood and are temporarily held by a rubber band. The graft is then tied with white grafting tape. *Bottom left:* Aluminum foil is wrapped around the graft to exclude heat from the graft. *Bottom right:* The grafted area covered with aluminum foil is wrapped with a cut poly bag, which is sealed to retain high relative humidity until the graft takes (*37*).

cut flap areas are then tied with flagging tape, green floral tape, or white budding tape. The tip of the scion is painted with tree paint or sealed with white glue to prevent it from drying out.

The taped graft area is then covered with aluminum foil to protect it from heat. A hole is made in the corner of a clear poly bag (freezer bag) and the poly slid down over the graft area so that it is just covering the aluminum foil (no poly should cover the apex of the scion, nor should it touch any exposed wood). Air is expelled so the poly fits snugly over the aluminum foil, and it is tied at both ends with stretchable plastic budding tape or rubber bands. The function of the poly bag is to maintain a high relative humidity in the graft area. In four to six weeks *after* the graft has taken, the ties, poly bag, and aluminum foil are removed.

The vegetative growth of the rootstock plant must be kept in check, since many new shoots will appear on the rootstock below the graft. Some of

FIGURE 13–10 Steps in the four-flap graft. See Figure 13–9.

these shoots are needed for maintaining tree vigor, but the rootstock shoots should not become dominant or exceed the height of the scion—the growing tips of the rootstock shoots will have to be removed several times during the growing season. After two to three years, all rootstock branches are removed below the graft and the scion becomes the dominant shoot system.

FIGURE 13–11 A tool for the four-flap graft which is slid over the rootstock and used to cut & peel the four flaps.

Detached Scion Graftage—Side Grafting

There are many types of side graftage. As the name suggests, the scion is inserted into the side of the rootstock, which is generally larger in diameter than the scion. This method has proven useful for large-scale propagation of nursery trees (*33*). Generally, the rootstock shoot is removed after the graft takes, and the scion becomes the dominant shoot system.

Side-Stub Graft

The side-stub graft is useful in grafting branches of trees that are too large for the whip-and-tongue graft yet not large enough for other methods such as the cleft or bark graft. For this type of side graft, the best rootstocks are branches about 2.5 cm (1 in.) in diameter. An oblique cut is made into the rootstock branch with a chisel or heavy knife at an angle of 20 to 30 degrees. The cut should be about 2.5 cm (1 in.) deep, and at such an angle and depth that when the branch is pulled back the cut will open slightly but will close when the pull is released.

The scion should contain two or three buds and be about 7.5 cm (3 in.) long and relatively thin. At the basal end of the scion, a wedge about 2.5 cm (1 in.) long is made. The cuts on both sides of the scion should be very smooth, each made by one sin-

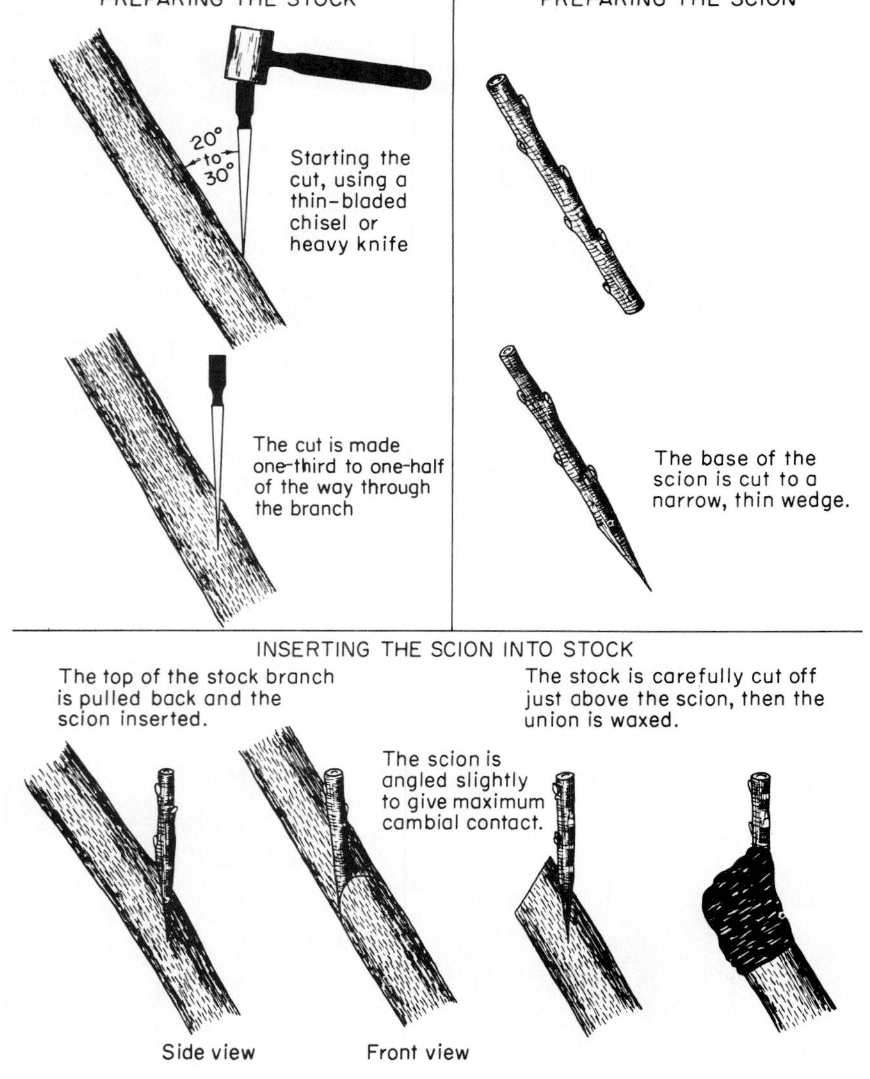

FIGURE 13–12 Steps in preparing the side-stub graft. A thin-bladed chisel, as illustrated here, is ideal for making the cut, but a heavy butcher knife could be used satisfactorily.

gle cut with a sharp knife. The scion must be inserted into the rootstock at an angle as shown in Figure 13–12 so as to obtain maximum contact of the vascular cambium layers. The grafter inserts the scion into the cut while the upper part of the rootstock is pulled backward, using care to obtain the best cambium contact. Then the rootstock is released. The pressure of the rootstock should grip the scion tightly. The scion can be further secured by driving two small flat-headed wire nails [20 gauge, 1.5 cm (⅝ in.) long] into the stock through the scion. Wrapping the rootstock and scion at the point of union with nursery tape also may be helpful. After the graft is completed the rootstock may be cut off just above the union. This must be done very carefully or the scion may become dislodged. The entire graft union must be thoroughly covered with grafting wax, sealing all openings. The tip of the scion also should be covered with wax or sealed with white glue (*50*).

Side-Tongue Graft

The side-tongue graft, shown in Figure 13–13, is useful for small plants, especially some of the broad- and narrow-leaved evergreen species. The rootstock plant should have a smooth section in the stem just above the crown of the plant. The diameter of the scion should be slightly smaller than that of the rootstock. The cuts at the base of the scion are made as for the whip-and-tongue graft. Along a smooth portion of the stem of the rootstock a thin piece of bark and wood, the same length as the cut surface of the scion, is completely removed. Then a reverse cut is made downward in the cut on the rootstock starting one-third of the distance from the top of the cut. This second cut in the rootstock should be the same length as the reverse cut in the scion. The scion is then inserted into the cut in the rootstock, the two tongues interlocking, and the vascular cambia matching. The graft is wrapped tightly, using one of the methods described for the whip-and-tongue graft.

The top of the rootstock is left intact for several weeks until the graft union has healed. Then it may be cut back above the scion gradually or all at once. This forces the buds on the scion into active growth.

Side-Veneer Graft

The side-veneer graft is widely used for grafting small potted liner plants such as seedling conifers, deciduous trees and shrubs, and fruit crops (Figures 13–14 and 13–15). A shallow downward and inward cut from 25 to 38 mm (1 to 1½ in.) long is made in a smooth area just above the crown of the rootstock. At the base of this cut, a second short inward and downward cut is made, intersecting the first cut, so as to remove the piece of wood and bark. The scion is prepared with a long cut along one side and a very short one at the base of the scion on the opposite side. These scion cuts should be the same length and width as those made in the rootstock so that the vascular cambium layers can be matched as closely as possible.

After inserting the scion, the graft is tightly wrapped with poly budding strips, budding rubbers, or with nursery adhesive tape. The graft may or may not be covered with wax, depending upon the species. A common practice in side grafting small potted plants of some woody ornamental species is to plunge the grafted plants into a slightly moist medium, such as peat moss, so that it just covers the graft union. Inserting the grafted liner plants in polytents in temperature-controlled greenhouses is commonly done (Figure 13–16). To maintain high humidity, the newly grafted plants may also be placed for healing in a mist propagating house (but the grafts are not directly placed under mist), or set in grafting cases. The latter are closed boxes with a transparent cover, which permits retention of high humidity around the grafted plant until the union has healed. The grafting cases are kept closed for a week or so after the grafts are put in, then gradually opened over a period of several weeks; finally, the cover is taken off completely.

After the union has healed, the rootstock can be cut back above the scion either in gradual steps or all at once.

Detached Scion Graftage—Bark Grafting

Bark grafting is done in topworking established plants. The rootstock must be in an active stage of growth so that the bark will slip. The scion is inserted between the bark and wood of the rootstock. Bark grafting can be performed on branches ranging from 2.5 cm (1 in.) up to 30 cm (1 ft) or more in diameter. The latter size is not recommended, as it is difficult to heal over such large stubs before decay-producing organisms attack.

Scions must be collected for deciduous species during the dormant season and held under refrigeration. For evergreen species, freshly collected scionwood can be used. In the bark graft, scions are not as securely attached to the rootstock as in some of the other methods and are more susceptible to wind breakage during the first year, even though

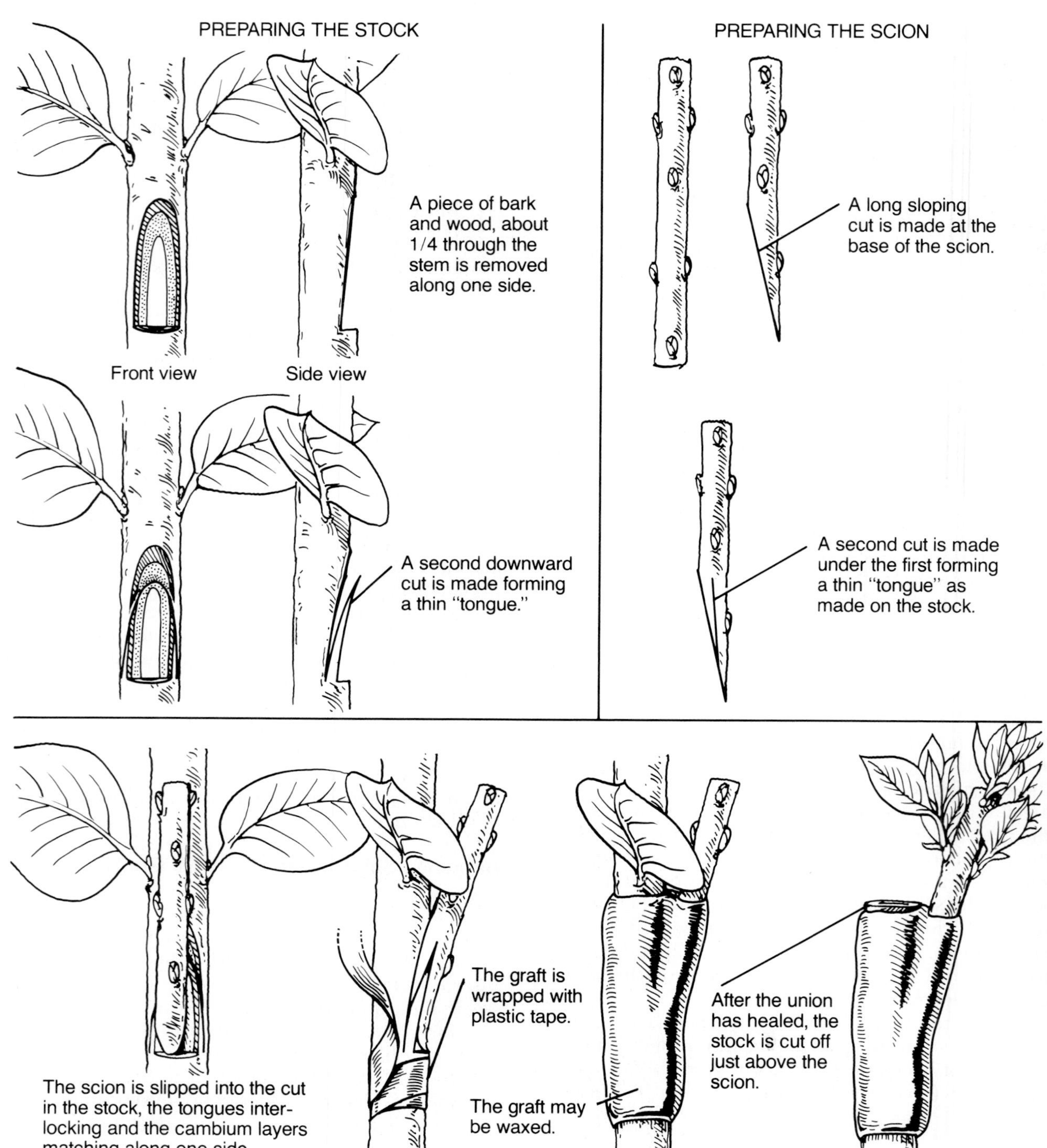

FIGURE 13–13 Side-tongue graft. This method is very useful for grafting broad-leaved evergreen plants. Final tying may be done with budding rubbers, poly tape, or waxed string. The graft may be waxed, or wrapped with a sealing tape such as Parafilm.

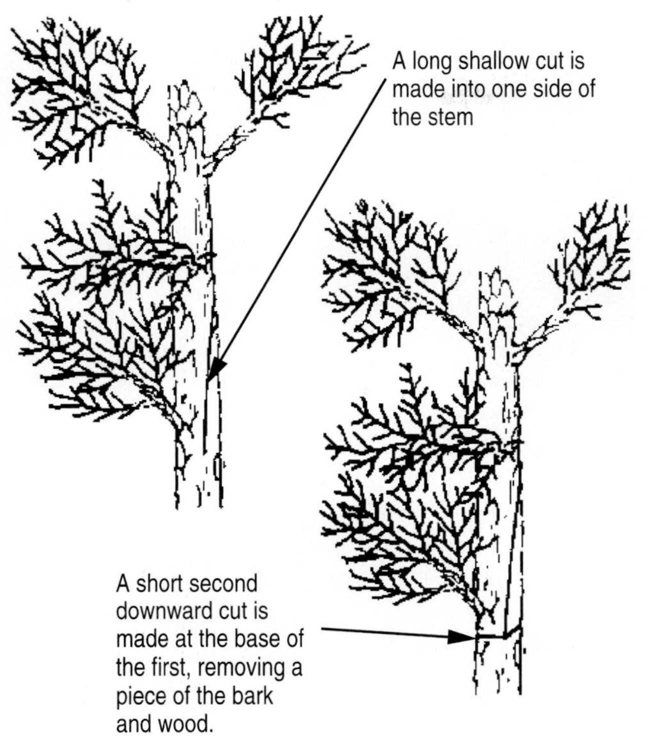

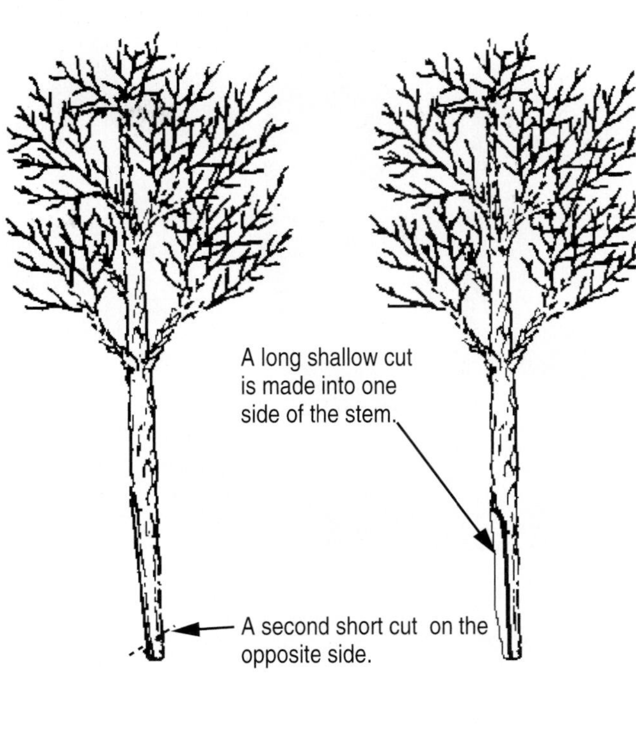

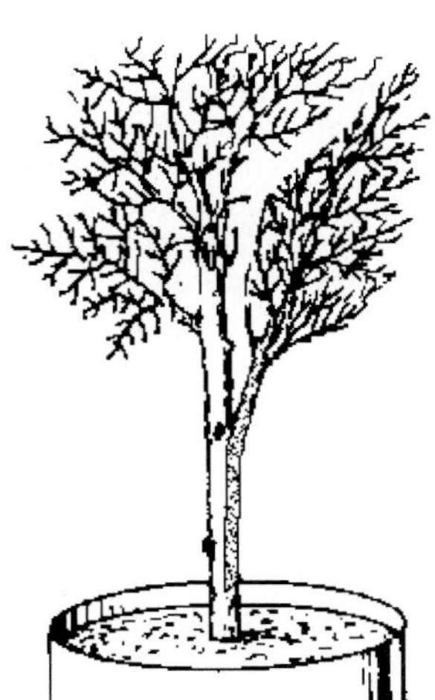

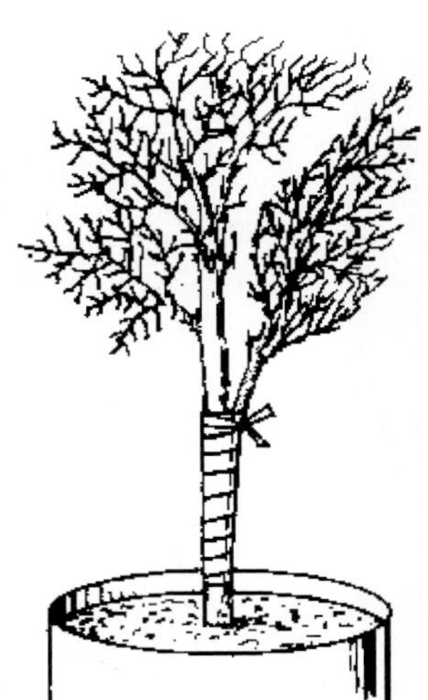

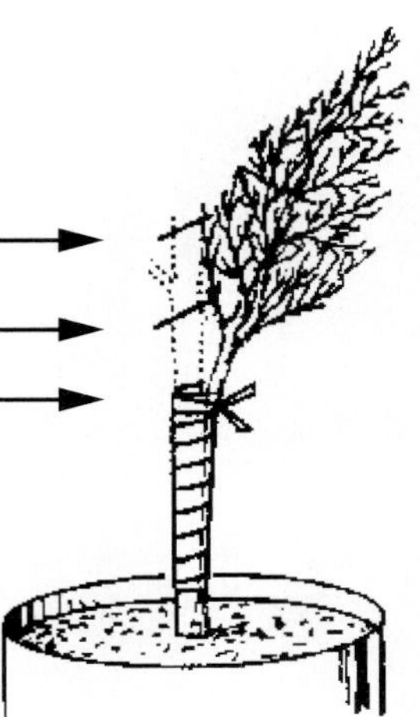

FIGURE 13–14 Steps in making the side-veneer graft. This method is one of the most popular grafts for propagating conifers and shrubs. The graft is quite versatile and can be used on a larger number of species than other grafts, such as the whip-and-tongue.

FIGURE 13–15 *Top:* In the side-veneer graft, a budding rubber is used to wrap the dormant, leafless scion to the rootstock. *Bottom:* Side-veneer-grafted Japanese maples in potted liner pots, which will be moved to a polytent area for callusing.

FIGURE 13–16 Polytent systems for maintaining grafted plants under high relative humidity. The greenhouse is temperature controlled. *Top:* The plants have callused and the above poly cover removed. *Bottom:* the poly sheet is placed on top of the grafted maple liners and along the sides; the poly has been temporarily lifted so the grafted plants could be photographed. It is important that light irradiance be controlled to prevent supraoptimal temperatures.

healing has been satisfactory. Therefore, the new shoots arising from the scions probably should be staked during the first year, or cut back to about half their length, especially in windy areas. After a few years' growth, the bark graft union is as strong as the unions formed by other methods. Two modifications of the bark graft are described next.

Bark Graft (Rind Graft)

Several scions are inserted into each rootstock stub. For each scion, a vertical knife cut 2.5 to 5 cm (1 to 2 in.) long is made at the top end of the rootstock stub through the bark to the wood. The bark is then lifted slightly along both sides of this cut, in preparation for the insertion of the scion. The dormant scions should be 10 to 13 cm (4 to 5 in.) long, containing two or three buds, and be 6 to 13 mm ($\frac{1}{4}$ to $\frac{1}{2}$ in.) in thickness. One cut—about 5 cm (2 in.) long—is made along one side at the base of the scion. With large scions, this cut extends about one-third of the way into the scion, leaving a "shoulder" at the top. The purpose of this shoulder is to reduce the thickness of the scion to minimize the separation of bark and wood after insertion in the rootstock. The scion should not be cut too thin, or it will be mechanically weak and break off at the point of attachment to the rootstock. If small scions are used, no shoulder is necessary. On the side of the scion opposite the first long cut, a second, shorter cut is made, as shown in Figure 13–17, bringing the basal end of the scion to a wedge shape. The scion is then inserted between the bark and the wood of the rootstock, centered directly under the vertical cut through the bark. The longer cut on the scion is placed against the wood, and the shoulder on the scion is brought down until it rests on top of the stub. The scion is then ready to be fastened in place. Nail the scion into the wood, using two nails per scion. Flat-headed nails 15 to 25 mm ($\frac{5}{8}$ to 1 in.) long, of 19 or 20 gauge wire, depending on the size of the scions, are satisfactory. The bark on both sides of the scion should be nailed down securely or it will tend to peel back from the wood.

Another method commonly used with soft-barked trees, such as the avocado, is to insert all the

PREPARING THE ROOTSTOCK

A vertical cut 2.5 to 5 cm (1 to 2 in.) long is made through the bark to the wood.

The bark on both sides of the cut is slightly separated from the wood.

PREPARING THE SCION

The scion is cut as shown below, a long cut with a shoulder on one side, and a shorter cut on the opposite side.

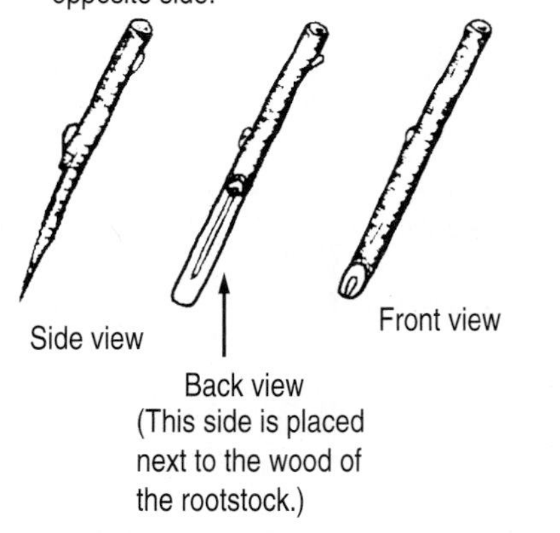

INSERTING THE SCIONS INTO THE ROOTSTOCK

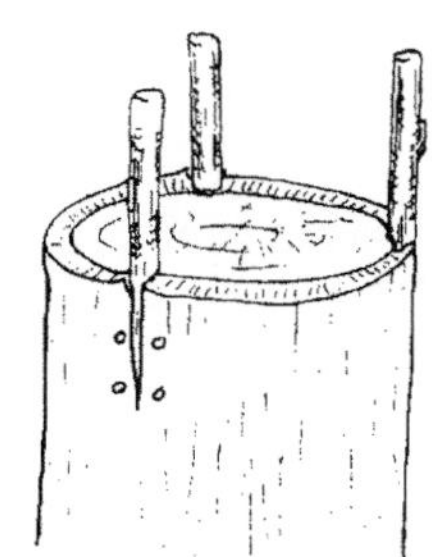

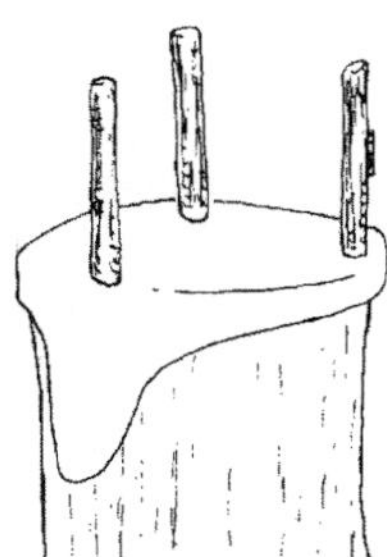

FIGURE 13–17 Steps in preparing the bark graft (rind graft). In grafting some thick-barked plants the vertical cut in the bark is unnecessary; the scion is inserted between the bark and wood of the stock.

scions in the stub and then hold them in place by wrapping waxed string, adhesive tape, or poly budding tape around the stub. This is more effective than nailing for preventing the scions from blowing out but probably does not give as tight a fit. Both nailing and wrapping are advisable for maximum strength. If a wrapping material is used, it must be checked to avoid constricting the rootstock. After the stub has been grafted and the scions fastened by nailing or tying, all cut surfaces, including the end of the scions, should be covered thoroughly with grafting wax.

Inlay Bark Graft

Two knife cuts about 5 cm (2 in.) long are made through the bark of the rootstock down to the wood, rather than just one (Figure 13–18). The distance between these two cuts should be exactly the same as the width of the scion. The piece of bark between the cuts should be lifted and the terminal two-thirds cut off. The scion is prepared with a smooth slanting cut along one side at the basal end completely through the scion. This cut should be about 5 cm (2 in.) long but *without* the shoul-

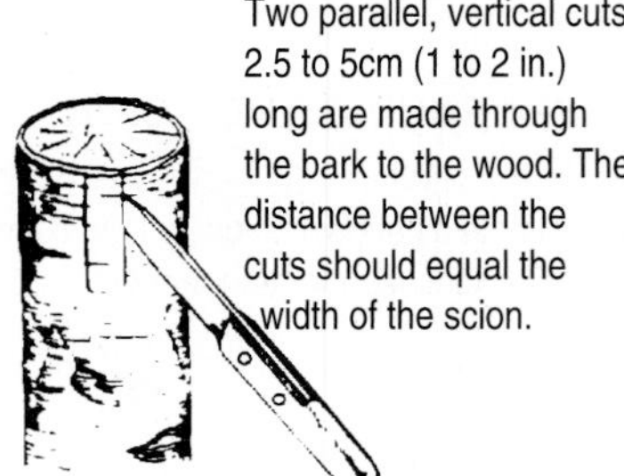

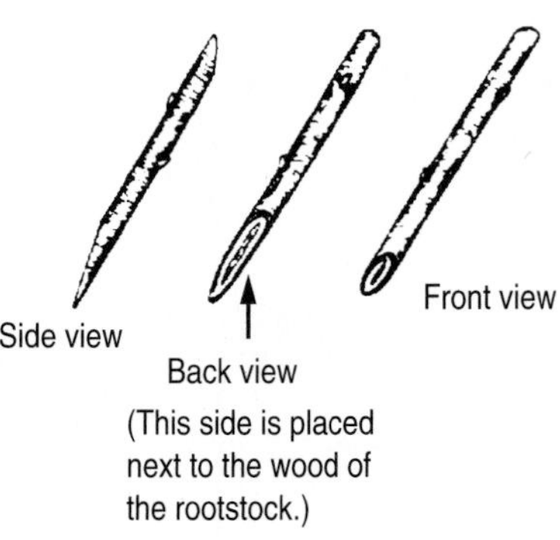

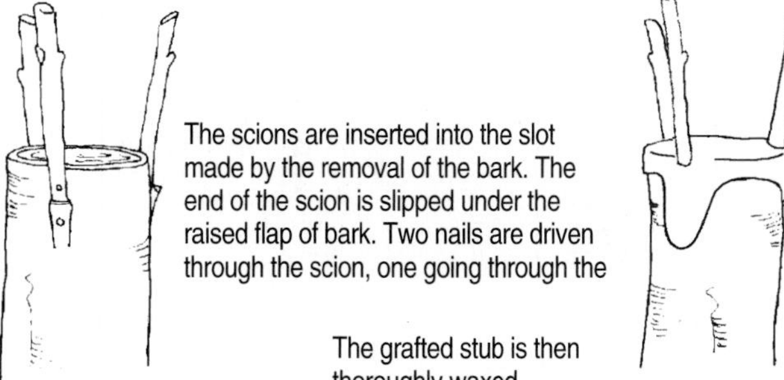

FIGURE 13–18 Inlay bark graft. With Texas pecans, the scion of the inlay bark graft is stapled or nailed, and aluminum foil and a cut poly bag are used instead of grafting wax (*36*).

der, in contrast to the bark graft. On the opposite side of the scion, a cut about 13 mm (½ in.) long is made, forming a wedge at the base of the scion. The scion should fit snugly into the opening in the bark with the longer cut inward and with the wedge at the base slipped under the flap of remaining bark.

The scion should be nailed into place with two nails, the lower nail going through the flap of bark covering the short cut on the back of the scion. If the bark along the sides of the scion should accidentally become disturbed, it must be nailed back into place. Flat-point staples in the vertical position, budding or flagging tape have all been used to secure the graft (*36*). The inlay bark graft is well adapted for use with thick-barked trees, such as walnuts and pecans, on which it is not feasible to insert the scion under the bark.

Detached Scion Graftage—Root Graftage

A number of plants are propagated commercially by root grafting—apples, pears, grapes, and many woody ornamental shrubs and trees (*16, 17*).

Root Grafting (Whole-root and Piece-root Graftage)

In root grafting, the rootstock seedling, rooted cutting, or layered plant is dug up, and the roots are used as the rootstock for the graft. The entire root system may be used (**whole-root graft**—Figures 13–19 and 13–20), or the roots may be cut up into small pieces and each piece used as a rootstock **(piece-root graft)** Figure 13–21. Both methods give satisfactory results. As the roots used are relatively small [0.6 to 1.3 cm (¼ to ½ in.) in diameter], the *whip-and-tongue* graft is frequently used. In England, *Rhododendron* cultivars are saddle grafted on roots of *R. ponticum;* the root graft is then tied and placed in a propagation case (*19*). Root grafts are usually bench grafted indoors during the late winter or early spring. The scionwood collected previously is held in storage, while the rootstock plants are also dug in the late fall and stored under cool (1.5 to 4.5° C; 35 to 40° F) and moist conditions until the grafting is done. The term *bench grafting* is given to this process, because it is performed indoors with dormant scions and rootstocks at benches by skilled grafters as a large-scale operation.

In making root grafts, the root pieces should be 7.5 to 15.0 cm (3 to 6 in.) long and the scions about the same length, containing two to four buds. After the grafts are made and properly tied, they are bundled together in groups of 50 to 100 and stored for callusing in damp sand, peat moss, or other packing material.

FIGURE 13–19 Bundles of one-year-old apple seedlings of a size suitable for whole root bench grafting.

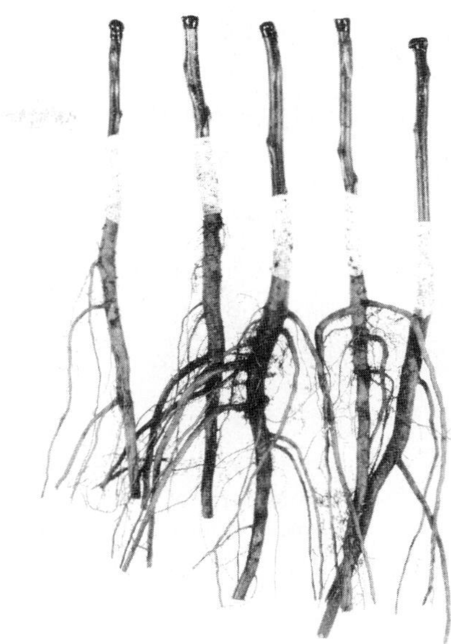

FIGURE 13–20 Whole root apple grafts made by the whip-and-tongue method, and wrapped with adhesive nursery tape. To maintain the correct polarity of the roots, cut the top part (proximal end) horizontally and the distal part (farthest from the crown of the tree) with a slanted cut.

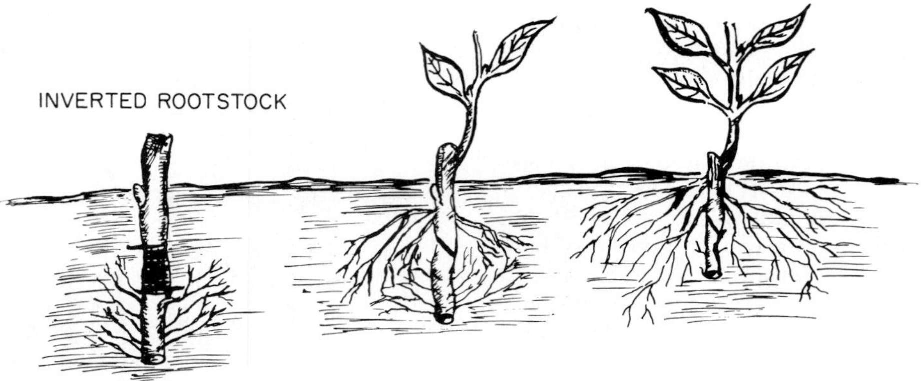

FIGURE 13–21 *Reversing the polarity* of the rootstock piece of the root graft is one method of "nurse-root" grafting. The nurse-root graft is a temporary graft used to induce the scion to develop its own roots. The nurse root sustains the plant until the scion roots form, then it dies. In the method shown, the rootstock piece is inverted, so the distal of the rootstock is temporarily joined to the proximal of the scion.

Nurse-Root Grafting

Stem cuttings of a difficult-to-root species can sometimes be induced to develop adventitious roots by making a *temporary "nurse-root" graft.* The plant to be grown on its own roots is temporarily grafted as the scion. The scion may be made longer than usual and the graft planted deeply with the major portion of the scion below ground. **Scion rooting** can be promoted by applying an auxin, such as indolebutyric acid, into several vertical cuts made through the bark at the base of the scion, just above the graft union. This is done just before planting, and the grafts are set deeply so that most of the scion is covered *(mound layered)* with soil (*28*). After one or two seasons of growth the scions have roots. The temporary nurse rootstock is then cut off and discarded. The rooted scion is replanted to grow on its own roots; it is later used as a rootstock and grafted to a desired scion fruit cultivar.

Methods of root grafting include: (a) *Reversing the polarity* of the nurse-root rootstock. The rootstock piece will eventually die if it is grafted onto the scion in an *inverted position* (Figure 13–21) (*34*). A graft union is formed—the inverted rootstock piece sustains the scion until it roots, but the rootstock fails to receive sufficient carbohydrates from the scion and eventually dies—leaving the scion on its own roots. (b) Another method is *girdling the rootstock just above the graft union* at the scion base (Figure 13–22). The rootstock is girdled with budding rubber strips (0.016 gauge) (*7*). Budding rubbers disintegrate within a month when exposed to sun and air; however, when buried in the soil, they will last as long as

FIGURE 13–22 A second nurse-root graft method is to *girdle the rootstock.* A 'Malling 9' apple scion was root grafted to an apple seedling nurse root. Just above the graft union the scion was wrapped with a budding rubber strip (see arrow). After two years in the nursery, vigorous scion roots were produced. The budding rubber has effectively constricted development of (girdled) the seedling nurse root, which can now be cut off and discarded. (Courtesy D. S. Brown.)

FIGURE 13–23 Approach grafting used in obtaining a desirable scion cultivar on a seedling camellia. *Top:* Seedling plant in container is set close to a large plant of the desired cultivar. The graft union is made and tightly wrapped with adhesive tape. *Below:* After the graft union has healed, which may take several months, the rootstock plant is cut off above the graft union, and the scion is severed from the parent plant just below the graft union. Approach grafting is sometimes necessary for plants very difficult to graft by other methods.

two years, allowing sufficient time for the scion to become rooted. (c) In a third method, an *incompatible rootstock* is used. When the graft is planted deeply, scion roots will gradually become more dominant in sustaining the plant. An example of this is an apple scion on pear rootstock.

APPROACH GRAFTAGE

The distinguishing feature of approach grafting is that two independent, self-sustaining plants are grafted together. After a union has occurred, the top of the rootstock plant is removed above the graft and the base of the scion plant is removed below the graft. Sometimes it is necessary to sever these parts gradually rather than all at once. Approach grafting provides a means of establishing a graft union between certain plants in which successful graft unions are difficult to obtain. Usually it is performed with one or both of the plants to be grafted growing in a container. Rootstock plants in containers may also be placed adjoining an established plant which is to furnish the scion part of the new, grafted plant (see Figure 13–23).

This type of grafting should be done at times of the year when growth is active and rapid healing of the graft union will take place. Three useful methods of making approach grafts are described below and illustrated in Figure 13–24.

Spliced Approach Graft

In the spliced approach graft the two stems should be approximately the same size. An exception to this is the spliced approach graft of Mango in India, where the scion is considerably smaller than the field-grown rootstock; the scion, in a pot, is hung from the branch of the larger rootstock (*19*). At the point where the union is to occur, a slice of

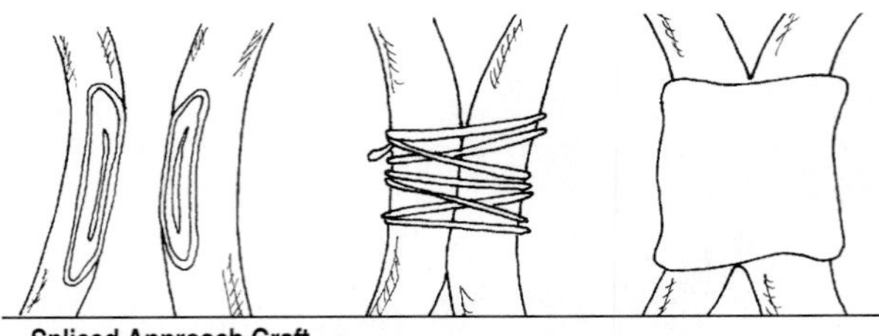
Spliced Approach Graft

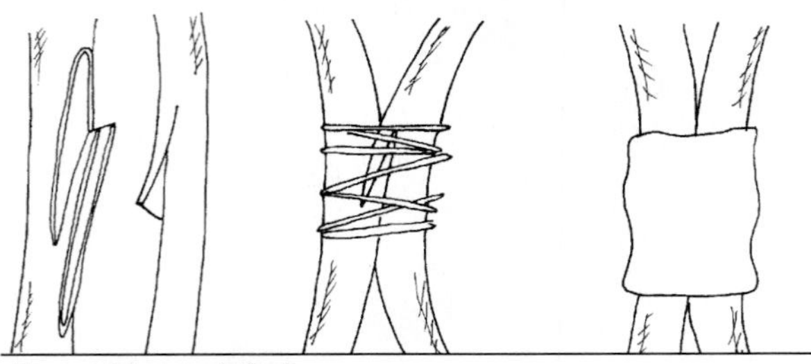
Tongued Approach Graft

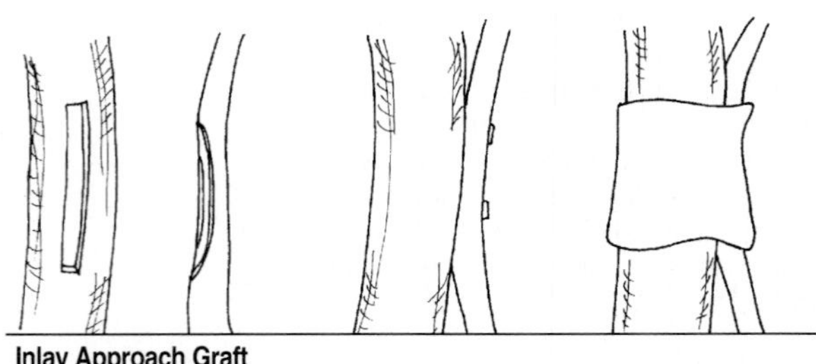
Inlay Approach Graft

FIGURE 13–24 Three methods of making an approach graft.

bark and wood 2.5 to 5 cm (1 to 2 in.) long is cut from both stems. This cut should be the same size on each so that identical cambium patterns are made. The cuts must be perfectly smooth and as nearly flat as possible so that when they are pressed together there will be close contact of the vascular cambium layers. The two cut surfaces are then bound tightly together with raffia, or poly grafting tape. The whole union should then be covered with grafting wax. After the parts are well united, which may require considerable time in some cases, the rootstock above the union and the scion below the union are cut, and the graft is then completed. It may be necessary to reduce the leaf area of the scion if it is more than the root system of the rootstock can sustain. Figure 13–23 shows the use of this method with camellias.

Tongued Approach Graft

The tongued approach graft is the same as the spliced approach graft except that after the first cut is made in each stem to be joined, a second cut—downward on the stock and upward on the scion—is made, thus providing a thin tongue on each piece. By interlocking these tongues a very tight, closely fitting graft union can be obtained.

Inlay Approach Graft

The inlay approach graft may be used if the bark of the rootstock plant is considerably thicker than that of the scion plant. A narrow slot, 7.5 to 10 cm (3 to 4 in.) long, is made in the bark of the rootstock plant by making two parallel knife cuts and removing the strip of bark between. This can be done only when the rootstock plant is actively growing and the bark "slipping." The slot should be exactly as wide as the scion to be inserted. The stem of the scion plant, at the point of union, should be given a long, shallow cut along one side, of the same length as the slot in the rootstock plant and deep enough to go through the bark and slightly into the wood. This cut surface of the scion branch should be laid into the slot cut in the rootstock plant and held there by nailing with two or more small, flat-headed wire nails. The entire union must then be thoroughly covered with grafting wax. After the union has healed, the rootstock can be cut off above the graft and the scion below the graft.

REPAIR GRAFTAGE

Inarching

Inarching is similar to approach grafting in that both rootstock and scion plants are on their own roots at the time of grafting. It differs in that the top of the new rootstock plant usually does not extend above the point of the graft union as it does in approach grafting. Inarching is used to replace roots damaged by cultivation equipment, rodents, or disease. It can be used to very good advantage in saving a valuable tree or improving its root system (Figure 13–25).

Seedlings (or rooted cuttings) planted beside the damaged tree, or suckers arising near its base, are grafted into the trunk of the tree to provide a new root system to supplant the damaged roots. The seedlings to be inarched into the tree should be spaced about 13 to 15 cm (5 to 6 in.) apart around the circumference of the tree if the damage is exten-

FIGURE 13–25 *Left:* Inarches that have just been inserted. The one on the left has been waxed. The one on the right has been nailed into place and is ready for waxing. *Right:* Inarching can be used for invigorating established trees by replacing a weak rootstock with a more vigorous one. Here a Persian walnut tree has been inarched with Paradox hybrid seedlings *(Juglans hindsii × J. regia)* seedlings.

sive. A damaged tree will usually stay alive for some time unless the injury is very severe. The procedure for inarching is to plant seedlings of a compatible species around the tree during the dormant season, and graft when active growth commences in early spring. Inarching will enhance growth of uninjured, older trees (*22*).

As illustrated in Figure 13–26, the graft is similar to an inlay bark graft. The upper end of the seedling, which should be 6 to 13 mm (¼ to ½ in.) thick is cut shallowly along the side for 10 to 15 cm (4 to 6 in.). This cut should be on the side next to the trunk of the tree and deep enough to remove some of the wood, thus exposing two strips of cambium tissue. Another, shorter cut, about 13 mm (½ in.) long, is made on the side opposite the long cut, making a sharp, wedge-shaped end on the seedling stem.

A long slot is cut in the trunk of the tree by removing a piece of bark the width of the seedling and just as long as the cut surface made on the seedling. A small flap of bark is left at the upper end of the slot, under which the wedge end of the seedling is inserted. The seedling is nailed into the slot with four or five small, flat-headed wire nails. The nail at the top of the slot should go through the flap of bark and through the end of the seedling. If the bark of the tree along the sides of the seedling should accidentally be pulled loose, nail it back in place. The entire area of the graft union should then be thoroughly waxed.

Bridge Grafting

Bridge grafting is another form of repair grafting—used when there is injury to the trunk, such as by cultivation equipment, rodents, disease, or winter injury. If the damage to the bark is extensive, the tree is almost certain to die, because the roots will be deprived of their carbohydrate supply from the top of the tree. Trees of some species, such as the elm, cherry, and pecan, can compartmentalize extensively injured areas by the development of a wound periderm of callus tissue. However, most woody species with severely damaged bark should be bridge grafted if they are to be saved, as illustrated in Figure 13–27.

Bridge grafting is best performed in early spring as active growth of the tree is beginning and the bark is slipping easily. The scions should be obtained when dormant from one-year-old growth, 6 to 13mm (¼ to ½ in.) in diameter, of the same or compatible species, and refrigerated until grafted. In an emergency, one may successfully perform bridge grafting late in the spring, using scionwood whose buds have already started to grow; the developing buds or new shoots are removed.

The first step in bridge grafting is to trim the wounded area back to healthy, undamaged tissue by removing dead or torn bark. A scion is inserted every 5 to 7.5 cm (2 to 3 in.) around the injured section and attached at both the upper and lower ends into live, undamaged bark. It is important that the

PREPARING THE TREE TO BE INARCHED

Vertical cuts about 15 cm (6 in.) long are made through the bark near the base of the tree to be inarched.

PREPARING THE SEEDLING FOR INARCHING

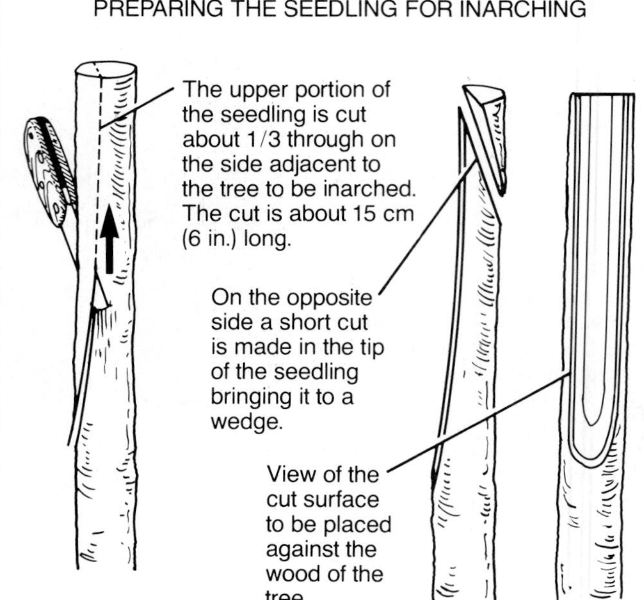

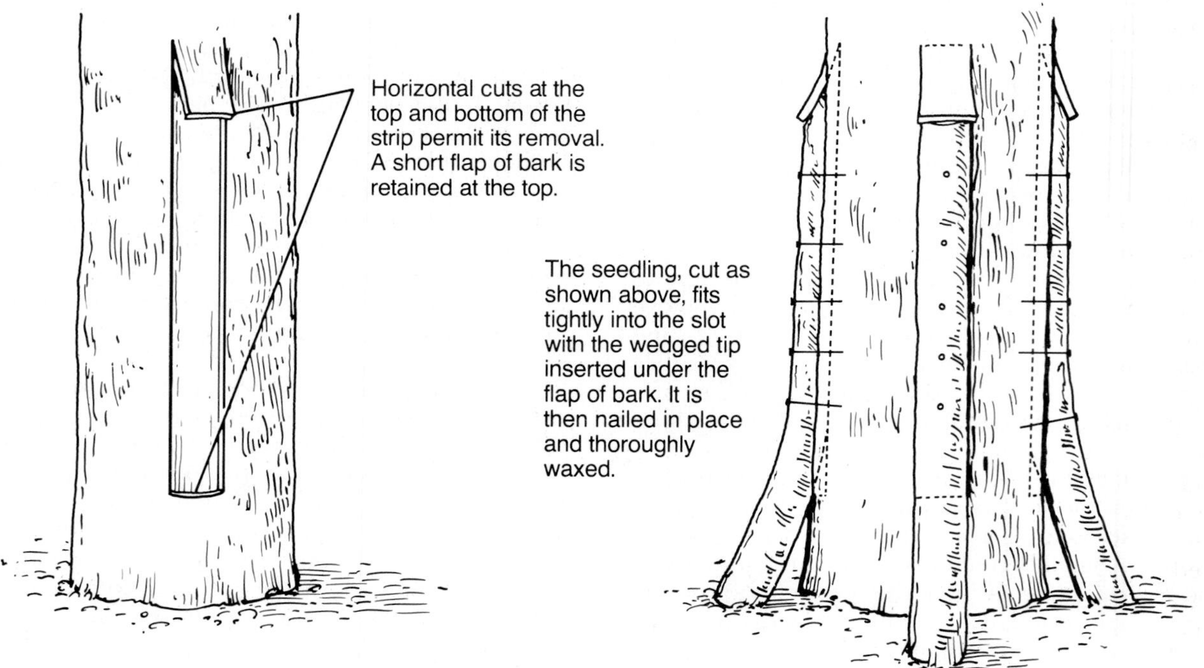

FIGURE 13–26 Steps in inarching a large plant, with smaller ones planted around its base.

FIGURE 13–27 Injured trunk of a cherry tree successfully bridge grafted by a modification of the bark graft.

scions are right side up. If reversed, a union may form, but the scions will not enlarge in diameter as they would if inserted correctly. Figure 13–28 shows the details of making a bridge graft.

After all the scions have been inserted, the cut surfaces must be thoroughly covered with grafting wax, particular care being taken to work the wax around the scions, especially at the graft unions.

Bracing

Bracing is a form of natural branch grafting that is used by fruit producers to strengthen scaffolding limbs of a tree in order to support the weight of the fruit crop. Natural grafting of roots or shoot systems occurs in species such as fig *(Ficus)*, rubber trees *(Hevea)*, birch *(Betula)*, beech *(Fagus)*, ash *(Fraxinus)*, maple *(Acer)*, pine *(Pinus)*, climbing species such as English ivy *(Hedera)*, and other species (see Chapter 12). Branches and trunks can naturally graft when they come in contact with each other during early development. The union begins with compression and constant and increasing pressure which ruptures the outer bark of the graft partners. This is followed by continued secondary growth, which leads to graft union formation and the joining of the independent vascular systems of the partners.

In the bracing of limbs, fruit producers will pull together two strong, young lateral shoots from the limbs to be braced. A rope or electrical cord is used to temporarily brace the larger limbs. The weaved smaller shoots, which will naturally graft, are tied with waxed string or poly tape to keep them together (Figure 13–29).

PRODUCTION PROCESSES OF GRAFTAGE

Preparation for Graftage
- Tools and accessories for grafting
- Grafting machines and grafting automation/robotics
- Selection, handling, and storage of scionwood

The Craftsmanship of Grafting
- Manual techniques: speed, accuracy, success rates
- Ergonomics/carpel tunnel syndrome
- Piecework
- Record keeping
- Advantages in the mechanization of grafting

Aftercare of Grafted Plants
- In bench grafting systems
- In field and nursery grafting systems

PRODUCTION PROCESSES OF GRAFTAGE

Success in grafting depends 45 percent on preparation, including the quality and preparation of the scion and rootstock material, 10 percent on craftsmanship, and 45 percent on the aftercare of the grafted plant (*26, 35*). The production goals of grafting are achieving a high success rate, or "take," and obtaining high speed and accuracy in performing the graft. Preparation for grafting begins with the proper tools and accessories, as well as the selection and handling of the scion and rootstock. Since grafting is a repetitive, labor-intensive process, graft-

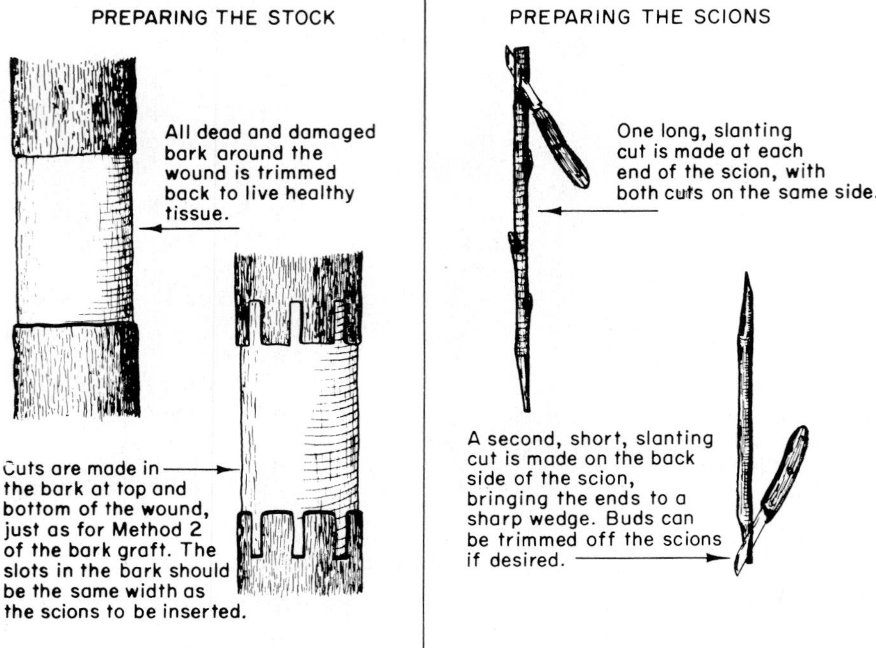

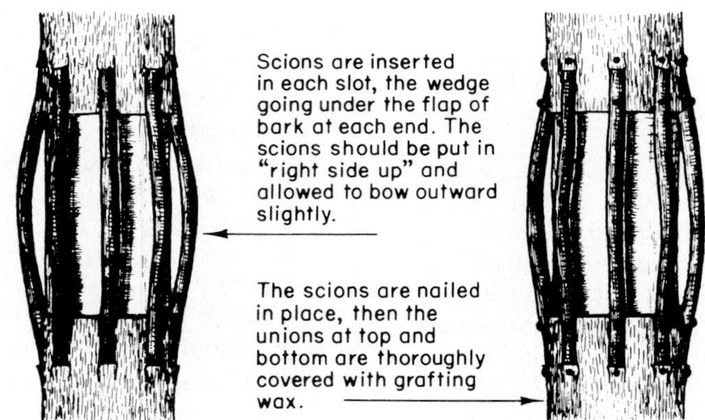

FIGURE 13–28 A satisfactory method of making a bridge graft, using a modification of the inlay bark graft.

ing machines and grafting automation, including robotics, continue to play a greater role.

Preparation for Grafting

Tools and Accessories for Grafting

Special equipment needed for any particular method of grafting has been illustrated along with the description of the method. There are some pieces of equipment, however, that are used in all types of grafting.

Knives. For propagation work the two general types of knives used are the budding knife and the grafting knife (Figure 13–30). Where a limited amount of either budding or grafting is done, the budding knife can be used satisfactorily for both operations. The knives have either a folding or a fixed blade. The fixed-blade type is stronger, and if a holder of some kind is used to protect the cutting edge, it is probably the most desirable. A well-built, sturdy knife of high-carbon steel is essential. Grafting blades are flat on one side and have a tapered

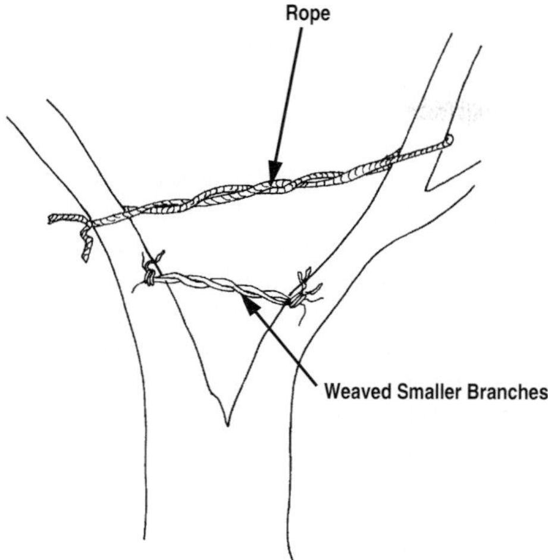

FIGURE 13–29 Bracing of fruit tree limbs by encouraging natural grafting. The tree limbs are braced with a twisted rope or electrical cord, and smaller shoots from the limbs are woven together and naturally graft as secondary growth occurs.

edge on the other to make a sharp, clean cut. Hence, grafting knives are available for either right- or left-handed people.

Tying and Wrapping Materials

Grafting methods, such as the whip-and-tongue, splice (whip), side veneer graft, and budding methods, such as chip budding and T-budding, etc., require that the graft union be held together by tying until the parts unite. A number of materials can be used for tying or wrapping grafts and budding. Some of these tying materials can also seal and help maintain a high relative humidity in the graft union area. This can help eliminate production steps for applying a hot wax sealant on top of the tying materials, or the need for maintaining the grafts in special poly chambers or "sweat boxes."

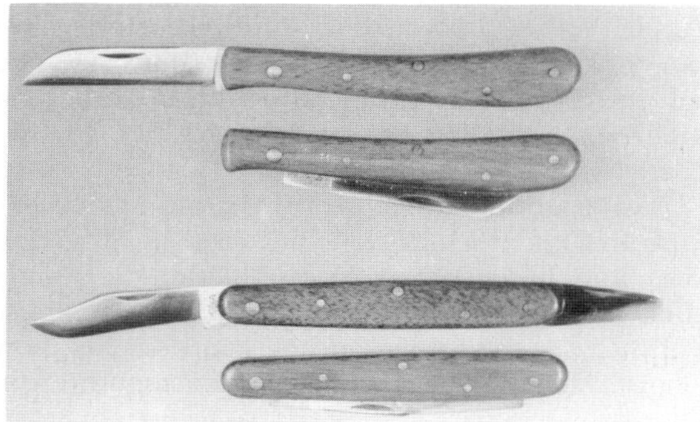

FIGURE 13–30 Types of folding knives used in plant propagation. *Top:* Grafting knife. *Below:* Budding knife. The blunt part (quill) on the right of this knife is used in T-budding to open the bark flap for insertion of the shield bud piece.

Some common tying materials for budding and grafting include:

- *budding rubbers* (also used in grafting).
- clear or colored *polyethylene* or polyvinyl chloride (*PVC*) budding and grafting strips, which are 0.5 to 1.3 cm (3⁄16 to ½ in.) wide and slightly elastic, allowing for a more secure wrap; this is also called flagging tape, green floral tape, white budding tape, or orange grafting tape.
- *raffia* (strips of palm leafstalk fiber—an older wrapping material, but still used).

These wrapping materials are not self-adhesive, so they must be tied with a half-hitch knot, which is done at the final turn of the tape by slipping it under the previous turn. With the exception of budding rubber, which deteriorates in full sunlight (but not when buried with the graft below ground), the wrapping materials must later be removed to prevent girdling the plant.

Self-adhering tying materials. A time-tested tying material is *waxed string* or *twine,* which adheres to itself and to the plant parts without tying. It should be strong enough to hold the grafted parts together yet weak enough to be broken by hand.

A special *nursery adhesive tape* is manufactured that is similar to surgical adhesive tape but lighter in weight and not sterilized. It is more convenient to use than waxed cloth tape. Adhesive tape is useful for tying and sealing whip grafts. When using any kind of tape or string for wrapping grafts, it is important not to use too many layers or the material may eventually girdle the plant unless it is cut. When this type of wrapping is covered with soil, it usually rots and breaks before damage can occur, unless it has been wrapped in too many layers. On a limited scale, adhesive tapes, such as *duct* and *electrical* tape can be used, while *masking* tape tends to unravel (*48*). Regardless of the wrapping material, it is best to cut them *after* the graft has taken to avoid girdling.

Self-Sealing tying materials include *Parafilm tape,* which has been used with successful results to wrap graft unions rapidly (*5*) and for chip-budding roses (Figure 14–7). This material is a waterproof, flexible, stretchable, thermoplastic film with a paper backing. The film is removed from the paper, wrapped around the graft union, and pressed into place by hand. Sometimes budding rubbers are used to tie a graft, which is then sealed with *Parafilm tape.* Self-sealing cure *crepe rubber sheets* are used for herbaceous grafts. *Rubber patches* up to 4 cm (1.5 in.) are fixed with a staple and used for budding.

Splints made of *toothpicks, bamboo,* or *metal skewers* are used with bench grafting of herbaceous plants such as cacti. The splints are later removed after the graft has taken.

Miscellaneous fastenings and wrapping material include: 18-gauge 2-cm (¾ in.) *nails,* 1.6 cm (⅝ in.) *flat-point staples,* as used in the inlay bark graft of pecans (*36*), and *plastic graft clips,* used in manual and machine splice grafting of vegetable crops (Figure 13–34). The combination of *aluminum foil* and *polyethylene bags* wrapped around a four-flap or inlay bark graft replaces the need for waxing the graft (which would melt and be unsuitable in spring field grafting in Texas or other warmer regions). *Metal shoot guide clips* (Figure 14–8) are used for field budded, dormant rootstock to compel upright growth from the bud.

In a novel approach for developing robotic grafting systems for vegetable crops in Japan, Chinese cabbage seedling (scions) are horizontally grafted to turnip rootstocks. In grafting Solanacea vegetables, the graft partners are joined by a chemical adhesive, followed by spraying a chemical hardener to set-up and solidify the adhesive (*46*) (Figures 13–35 and 13–36).

Whether a sealant, such as grafting wax, is to be applied to a graft depends on the type of graft, the grafting system, and the type of material used. Sealants are generally not used with budding, since tying with budding rubbers, rubber patches, or Parafilm tape is sufficient to alleviate desiccation problems. If the bench grafted plant is to be placed in a high relative-humidity graft box or temperature-controlled polytent, or immediately outplanted in the field with the union below the soil surface—waxing may be omitted.

Grafting waxes. Grafting wax has two chief purposes: (a) It seals over the graft union, thereby preventing the loss of moisture and death of the tender, exposed cells of the cut surfaces of the scion and rootstock. These cells are essential for callus production and healing of the graft union. (b) It prevents the entrance of various decay-producing organisms that rot wood.

An ideal grafting wax should adhere well to the plant surfaces, not be washed off by rains, not be so brittle as to crack and chip during cold weather, or so soft that it will melt and run off during hot days, but still be pliable enough to allow for swelling of the scion and growth enlargement of the stock without cracking. *Hot waxes* require heating, while *cold waxes* contain volatile solvents which keep the wax liquid. The cold wax solidifies when the solvents evaporate. Most nurseries develop their own hot wax, which is low-melting, soft, and flexible, so that subsequent handling of the graft does not cause cracking and flaking. Thermostatically controlled wax heaters are available to provide instant liquid wax when needed. *The wax should be hot enough to flow easily, yet not be boiling—which damages plant tissue.*

Various recipes for making hot and cold waxes are listed by Garner (*19*). For hot wax, blocks of premixed grafting wax containing the necessary ingredients (e.g., TrowBridge's grafting wax, Walter E. Clark & Son, Orange, Conn., USA) can be purchased from nursery supply houses.

Grafting Machines

Several *bench grafting* machines or devices have been developed to prepare graft and bud unions and a few have been widely used, especially in propagating grapevines (*2, 3*).

One of the most successful machines is the hand-operated device shown in Figure 13–31, originally developed in France.[1] This device makes a type of *wedge graft,* cutting out a long V-notch in the rootstock and a corresponding long, tapered cut at the base of the scion. By reversing the position of the rootstock and scion, it could also make a *saddle graft* (Figure 13–8). Although the cuts fit together very well, the operation is slow because the graft union must be either tied with a budding rubber, poly tape, or stapled together. This machine has been used successfully in propagating both grapes and fruit trees. A similar, but portable, device is manufactured in New Zealand.[2]

Another machine[3] used in grafting grapes is shown in Figure 13–32. It is powered by an electric motor and has two sets of saw blades on a single shaft for cutting notches, one in the base of the scion and one at the top end of the rootstock, as shown in Figure 13–33. Three persons work together: one

[1]Available in the United States from Heitz Wine Cellars, 500 Taplin Road, St. Helena, Calif. 94574.

[2]Available from Raggett Industries, 10 Cheesman Road, P.O. Box 1286, Gisborne, New Zealand.

[3]Available from Spink Refrigeration Co., 677 St. Helena Highway South, St. Helena, Calif. 94574 U.S.A.

FIGURE 13–31 A type of wedge grafting made by a French grafting device. *Top left:* Appearance of graft as made by the cutting blades (before wrapping with grafting tape). *Top right:* Apple graft union after one year's growth in the nursery. *Below:* Grafting device in use. Grafts can be made with this device much faster than by the whip-and-tongue graft method. A saddle graft could also be made by this machine, by reversing the cuts so the scion piece has the saddle shape (Figure 13–8).

FIGURE 13–32 Machine in operation for cutting notches in ends of grape scion and rootstocks for making the type of graft illustrated in Figure 13–33. The blades in front are making cuts in the rootstocks; those in the back are making cuts for the scions.

cutting out the scions, one cutting the rootstocks, and one assembling the grafts. After the rootstock and scion have been pushed together, the graft fit is so tight that no wrapping or stapling is necessary.

A third device, widely used for grapes, is the Pfropf-Star grafting machine[4] manufactured in Germany. It cuts through both the rootstock and scion, one laid on top the other, making an *omega-shaped cut* and leaving the two parts interlocked. The device is available with either a foot-operated or electromagnetic-operated mechanism. It is the fastest of the three machines.

[4]Manufactured by E. & H. Wahler, 7056 Weinstadt-Schnait, Buchhaldenstrasse 21, Stuttgart, Germany. Available in the United States from Tree & Vine Management Co., P.O. Box 311, Reedley, Calif. 93654 U.S.A.

Grafting Automation/Robotics

Prototypes and commercial robotic machines for grafting vegetable seedlings have been developed (Figures 13–34, 13–35, 13–36). The production of grafted vegetable crops is seldom done in the U.S. or European countries, where rotation of crops is practiced and land use is less intensive. However, in highly populated Japan, Korea, and some Asian and European countries, where the land use is highly intensive and farming areas quite small, grafted vegetable seedlings are used extensively. Grafted seedlings account for 81 percent of the commercial outdoor and greenhouse vegetable production in Korea, and 54 percent and 81 percent, respectively for Japan (*29, 32*). Vegetable rootstock used are resistant to soil-borne pathogens and nematodes, which build up under these intensive cultivation conditions (*45*). Some of the automated machines utilize horizontal grafting plates, and a chemical adhesive and hardener are applied to cement the graft (Figures 13–35 and 13–36) (*45, 46*). One of the commercialized grafting robots is reported to graft 1000 Solanaceae vegetable seedlings per hour (Figure 13–35).

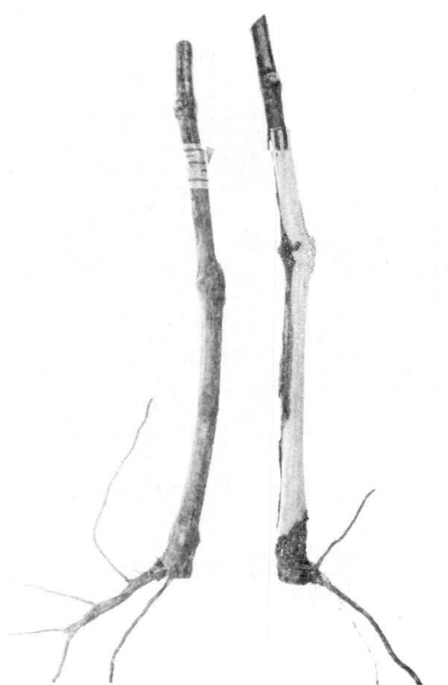

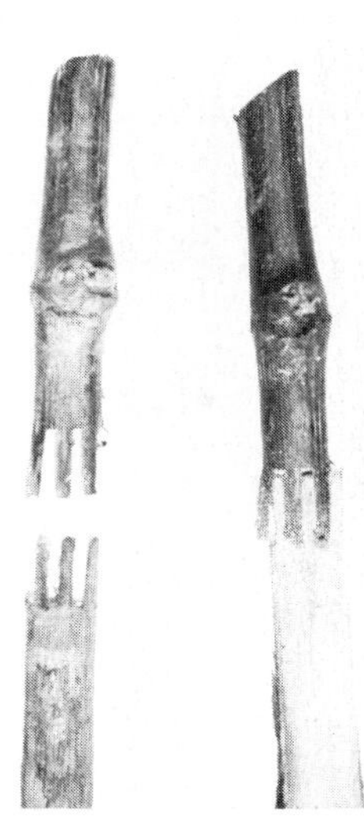

FIGURE 13–33 Grapes propagated by machine grafting. Small, one-budded scions are grafted on rooted cuttings during the dormant season. The graft union is wrapped with a budding rubber, then allowed to callus before planting. Because of the shape, the machined grafts formed are sometimes referred to as *notch* or *groove* grafts.

Selection and Handling of Scionwood

Kind of wood. Since bench or field grafting of deciduous species takes place in winter or early spring, the use of scionwood that grew the previous summer is necessary.

In selecting such scion material the following points should be observed:

- For most species, the wood should be one-year-old or less (current season's growth). Avoid including older growth, although with certain species, such as the fig or olive, two-year-old wood is satisfactory, or even preferable, if it is of the proper size.
- Healthy, well-developed vegetative buds should be present. Avoid wood with flower buds. Usually, vegetative buds are narrow and pointed, whereas flower buds are round and plump (see Figure 14–1).
- The best type of scion material is vigorous (but not overly succulent), well-matured, hardened shoots from the upper part of the tree, which have grown 60 to 90 cm (2 to 3 ft) the previous summer. Such growth develops on relatively young, well-grown, vigorous plants; high production of scion material can be promoted by pruning the plant back heavily the previous winter. Water sprouts from older trees sometimes make satisfactory scionwood, but suckers arising from the base of grafted trees should not be used, since they may consist of rootstock material. A satisfactory size is from 0.6 to 1.2 cm (¼ to ½ in.) in diameter.
- The best scions are obtained from the center portion or from the basal two-thirds of the shoots. The terminal sections, which are likely to be too succulent, pithy, and low in stored carbohydrates, should be discarded. Mature wood with short internodes should be selected.

Source of material. Scionwood should be taken from source plants of the correct cultivar known to be pathogen-tested and genetically true-to-type (see Chapter 9). Virus diseased, undesirable sports, chimeras, and virus-like genetic disorders must be avoided. Source plants may be of two basic types:

1. Plants produced in an orchard, vineyard, ornamental field or container nursery, or landscape are selected when the flowering, fruiting, and growth habits are known. It is best to take propagation material from bearing plants where production history is known. Visual inspection, however, may not reveal the true condition of the proposed source plant and, to be sure, appropriate indexing and progeny tests would be required (see Chapter 9).
2. In commercial nurseries, special scion blocks, where plants are grown particularly for propagation, may be maintained. Such plants are

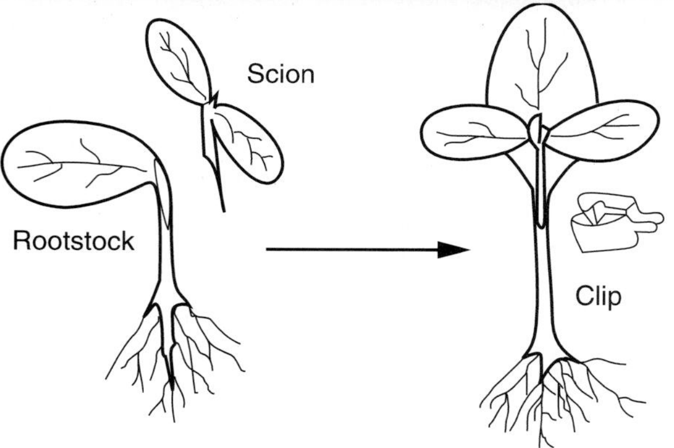

FIGURE 13–34 Prototype grafting robot for bench grafting cucurbit vegetable seedlings to pathogen- and nematode-resistant rootstock. *Top:* Cutting unit for rootstock. *Middle:* Grafting robot G892. *Bottom:* Schematic of machine graft with plastic clips used to hold the graft partners. (Photographs courtesy of M. Suzuki, Japanese Institute of Agricultural Machinery—BRAIN.)

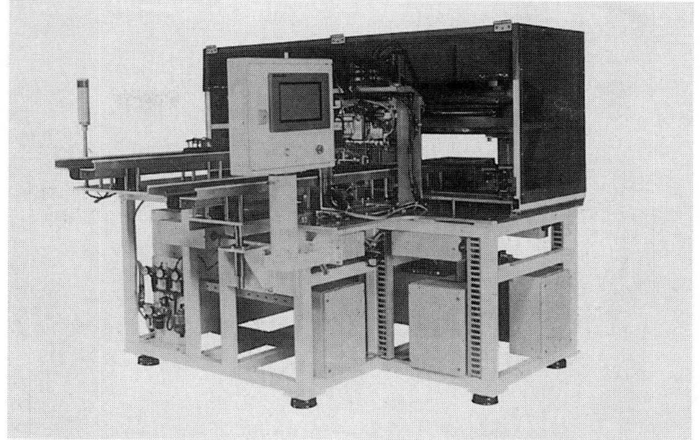

FIGURE 13–35 Commercial grafting robot for Solanaceae vegetable seedlings. The robot makes a horizontal cut across the surfaces of the graft partners, positions the scion and rootstock, and then cements the graft together with a chemical adhesive and hardener. The production output of this robot is 1000 grafted seedlings per hour. (Photograph courtesy of M. Oda (*45*).)

handled differently than they would be for producing a crop. For example, fruit trees may be pruned back each year to produce a large annual supply of long, vigorous shoots well-suited for scionwood. Such special blocks would usually be handled to conform to registration and certification programs and would be subject to isolation, indexing, and inspection requirements. In addition, it is important to maintain source identity of scion material through the entire propagation sequence, so that over a period of time proper sources of the various cultivars can be identified and maintained.

Collection and handling. For deciduous plants to be grafted in early spring, the scionwood can be collected almost any time during the winter season when the plants are fully dormant (*6*). In climates with severe winters, the wood should *not* be gathered when it is frozen. Any wood that shows freezing injury should *not* be used. Where considerable winter injury is likely, it is best to collect dormant scionwood and put it in cold storage after leaf fall but before the onset of winter.

Storage. Scionwood collected prior to grafting must be properly stored. It should be kept slightly moist and at a low enough temperature to prevent elongation of the buds. A common method is to

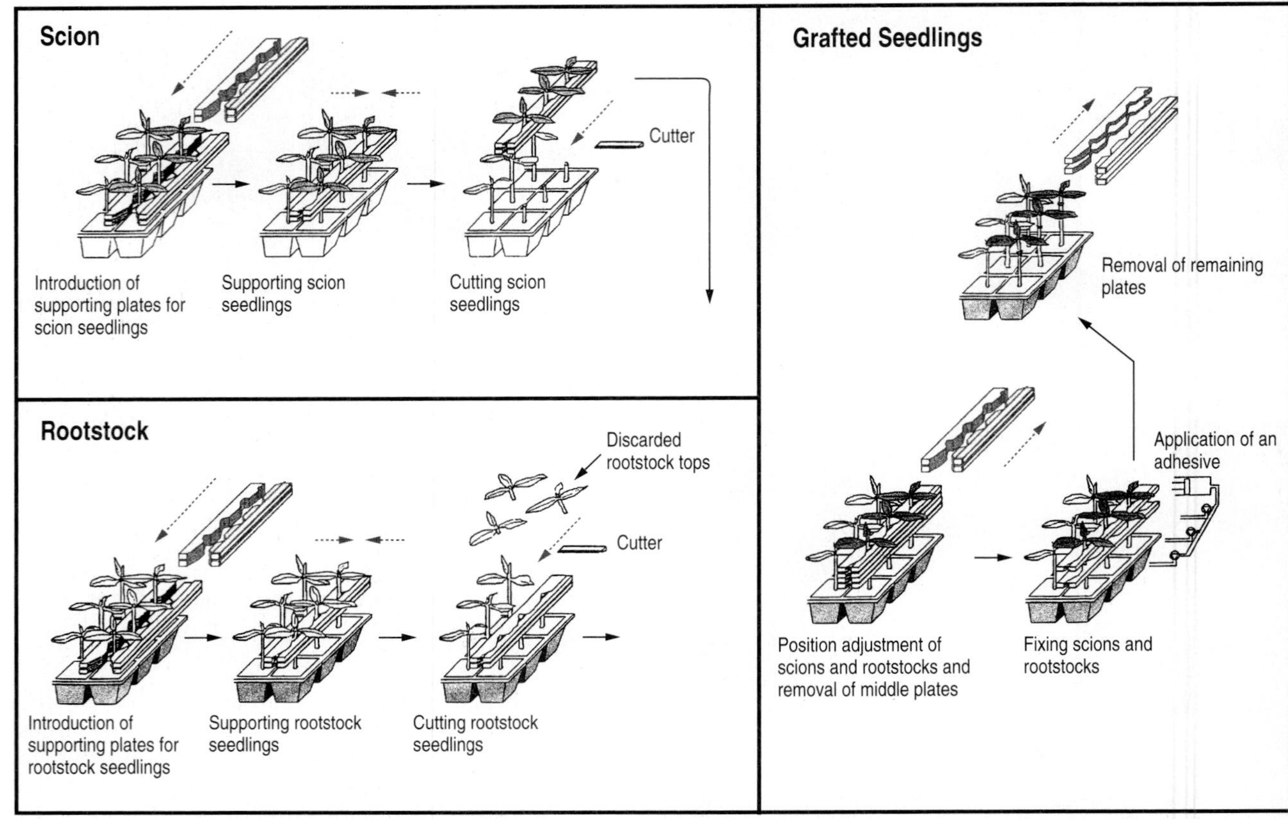

FIGURE 13–36 Schematic of the operations performed by the grafting robot viewed in Figure 13–35. (Illustration courtesy of K. Kurata (*29*).)

wrap the wood, in bundles of 25 to 100 sticks, in heavy, waterproof paper or in polyethylene sheets or bags. *It is absolutely critical that all bundles be labeled accurately.*

A small amount of some clean, *slightly moist material,* such as sawdust, wood shavings, or peat moss, should be sprinkled through the bundle. Sand should not be used, because it will adhere to the scionwood and dull the edge of the knife during the grafting operation. If the packing material is wet, various mold and fungi may develop and damage the buds, even at low storage temperatures. Packaged scionwood is more apt to be damaged by being too wet than by being too dry. If it is to be stored for a prolonged period, the bundle should be examined every few weeks to see that the wood is not becoming either dried out or too wet. When the buds show signs of swelling, the wood should either be used for grafting immediately or moved to a lower storage temperature.

Polyethylene bags are useful for storing small quantities of scionwood. They allow the passage of oxygen and carbon dioxide, which are exchanged during the respiration process of the stored wood, but retard the passage of water vapor. Sometimes the natural moisture in the wood is sufficient, so slightly moist packing material is not needed in the sealed poly bags. In commercial field rose production, the scionwood (budwood) is harvested dormant, wrapped in slightly moist newspaper and sealed in poly bags, and maintained at –1.7 to –0.6° C (29 to 31° F) for up to seven months.

The temperature at which the wood is stored is important. If it is to be kept only two or three weeks before grafting, the temperature of a home refrigerator—about 5° C (40° F)is satisfactory. If stored for a period of one to three months, scionwood should be held at about 0° C (32° F) (*5*) to keep the buds dormant. However, buds of some species, such as the almond and sweet cherry, will start growth after about three months even at this temperature. Do not store scionwood in a home freezer because the very low temperatures, about –18° C (0° F), may injure the buds.

Storage of scions should not be attempted if succulent, herbaceous plants are being grafted; such scions should be obtained at the time of grafting and used immediately. Certain broad-leaved evergreen species, such as camellias, olives, and citrus, can be grafted in the spring before much active growth starts without previous collection and storage of the scionwood. It is taken directly from the tree as needed, using the basal part of the shoots containing dormant, axillary buds. The leaves are removed at the time of collection.

Attempting to use scionwood in which the buds are starting active growth is almost certain to result in failure. In such cases, the buds quickly leaf out before the graft union has healed; consequently, the leaves withdraw water from the scions by transpiration, and cause the scions to die. In addition, the strong competing sink of a developing shoot can interfere with graft union formation.

In topworking pecans (*4*) good results are obtained by using precut scions; that is, scions cut in advance by skilled persons at a convenient time, then held in cold storage in polyethylene bags for up to nine days before inserting in the graft unions. Grafting success was reduced only slightly by the use of precut scions.

Handling of rootstock for bench grafting. Seedling rootstock of maple *(Acer)* is established in liner pots for one year, brought into a greenhouse in the fall after leaf drop, and placed in bottom-heated benches at 13 to 16° C (55 to 60° F). Bench grafting of the container rootstock with a splice graft is then done in January and February (Canada), when white roots appear along the perimeter of the potballs (*26*).

Failure is likely when bench grafting onto rootstock with inactive roots. Bench grafting is best when new, white root tips of 6 mm (¼ in.) occur or buds start to swell on the rootstock.

THE CRAFTSMANSHIP OF GRAFTING

Grafting is both an art and a skill. Successful grafting is a *repetitive* task that requires a high degree of accuracy and speed. To become skilled it is essential for the grafter to eliminate all unnecessary movements. To increase grafting efficiency, it is important to organize the workplace so that scion material, knives, and grafting tape are all within easy reach (*38*). Grafting is generally more efficient with a *team approach,* where each worker performs a certain task. This reduces inefficient motion of constantly picking up and putting down different tools.

Manual Techniques: Speed, Accuracy, Success Rates

Tips on improving grafting techniques and ergonomics include (*38*):

- Concentrate on accuracy first, and allow grafting speed to build up—aim at initially completing at least 200 bench grafts a day.
- Use a graft method that is less time-consuming (yet still successful!) that can be done with a lesser amount of skill and preparation of the rootstock, e.g., bench grafting with a splice (whip) graft, compared to a whip-and-tongue or side veneer graft (*40*); this works well with *Betula, Cornus, Fagus, Ginkgo, Quercus, Acer* (*26*).
- Grafting is best with two people: in bench grafting with the whip-and-tongue, one person does the graft and the other moves the potted liner rootstocks, and waxes the graft union. In T-budding field roses, the budder prepares the rootstock and inserts the shield bud of the scion, while the "tier" follows and ties the budding rubber around the budded graft (Figure 14–10).
- The grafting knife should always be held with a relaxed grip to improve accuracy and reduce repetitive strain injuries (e.g., *carpal tunnel syndrome*); it is necessary to restrict and control your arm movements.
- There are two basic cuts in grafting: the *slice cut,* which is made using your arm and shoulder to pull the knife, e.g., in the making of a splice *(whip)* graft; in the *cross cut,* the grafter's arm and knife are rotated using the thumb as a pivot, with both hands joined to prevent the knife from cutting the grafter—a whip-and-tongue is created when the scion and rootstock are sliced and then cut across in *four* (economical) movements.
- Tie the grafting tape around your neck so that you know where it is and keep it free of contamination.
- Speed is more important than 100 percent accuracy; increase speed by developing a rou-

tine when grafting and avoiding useless movements.

- Hold the grafting knife in your hand at all times, e.g., in one little finger.
- Make sure that everything is at a height that is easy to reach.

Record Keeping

- *Keep records—on the grafters* to determine daily quantities grafted; some operations pay on a *piecework* or bonus system (see Chapter 11). Records kept by the supervisor can be constructively used to help grafters improve their technique and efficiency (*9*).
- *Keep records—on the plant material* to determine the optimum windows of time to graft (*40*) [p. 471], and to assure having the best available material to graft. Records should be kept on the conditions of the grafting material, grafting problems encountered, and aftercare of the grafts. Should a crop failure occur, records can generally help pinpoint the cause and help managers improve efficiency (*9*).

Advantages in Mechanizing Grafting

There are a number of advantages of **bench grafting with machines.** This includes utilizing less-skilled workers who can easily become proficient in a day—and fewer injuries related to muscle strains and carpal tunnel syndrome (*47*). Larger-caliber grafts are machine grafted much easier than by hand grafting, and there is greater flexibility in handling different size graft partners, and less wasted plant material.

AFTERCARE OF GRAFTED PLANTS

In Bench Grafting Systems

A common method is to wrap the union with budding rubbers, poly or plastic tape, biodegradable cloth tape, or older materials such as raffia. After wrapping, the entire union can be covered with grafting wax.

Root grafting. The root grafts may be placed under refrigeration 7° C (45° F) for about two months. For general callusing purposes, temperatures from 7 to 21° C (45 to 70° F) are the most satisfactory. The callusing period for apples can be shortened to around 30 days if the grafts are stored at a temperature of about 21° C (70° F) and at a high humidity. To use this higher callusing temperature the material should be collected in the fall and the grafts made before any cold weather has overcome the rest period of the scion buds. After the unions are well healed, the grafts must be stored at cool temperatures—2 to 4° C (35 to 40° F)—to overcome the "rest period" of the buds and to hold them dormant until planting (*24*). The root grafts are lined-out in early spring in the nursery row directly from the low-temperature storage conditions.

Hot-pipe callusing system. With some plants, the graft union should be kept warm, 24 to 27° C (75 to 82° F), but the roots and the buds on the scion should be kept cool, about 7° C (45° F) to prevent premature growth before the graft union has callused and healed together. An ingenious system for regulating temperature was developed for whip grafting hazelnuts, which are notoriously difficult to root graft (Figure 13–37). The graft union itself is kept warm by PVC pipe containing recirculating hot water. The graft union is laid against the pipe, but the scion and roots protrude into areas having lower temperatures. This hot-pipe callusing system, when used outdoors in late winter or early spring, has increased the root grafting "takes" in difficult species, such as hazelnut. Some aeration of the callusing grafts is required, so airtight containers should not be used.

Virtually any graft used in bench grafting can be callused with the hot-pipe system, including root grafting apples, pears, peaches, and plums (*31*), and ornamentals such as *Acer, Cedrus, Corylus,* and *Fagus.* Not all species respond—the higher graft union temperature does not enhance graftage of spruce *(Picea).*

Closed case. Waxing may be omitted if the *bare-root grafts* are to be protected from drying by packing the grafts in boxes containing slightly moist peat. Some producers still dip the grafts in rose wax from the scion end to the taped union of the graft prior to boxing (*47*). The boxes are then moved to a callusing room at 21° C (70° F) for about 12 days. Once the grafts have formed sufficient callus, the boxed grafts are then held in cold storage at 2° C (35° F) until outplanted in the field (*47*). Another form of the closed case is the use of a *polytent* in a heated greenhouse which is used for callusing the grafts of *potted rootstock liners* (Figure 13–16). Provided light irradiance is controlled, glass mason jars can be used as a closed-case system for grafted plants in containers (Figure 13–38).

"Optimum Windows" of the year when selected ornamental species can be grafted in the greenhouse in Oregon, USA *(40)*

	Month											
Crop	***Jan.***	***Feb.***	***Mar.***	***Apr.***	***May***	***June***	***July***	***Aug.***	***Sept.***	***Oct.***	***Nov.***	***Dec.***
Abies	X	X										X
Acer palmatum	X	X	X	X	X	X	X	X	X	X		
Aesculus	X	X	X				X	X	X			X
Carpinus	X	X	X	X						X	X	
Cedrus	X	X	X						X			
Cercis		X	X	X			X	X				
Cornus	X	X	X						X	X		
Fagus	X	X	X	X					X	X	X	
Ginkgo	X	X	X	X				X	X			
Hamamelis	X	X	X	X				X	X	X		
Larix	X	X										
Liquidambar		X	X					X	X			
Liriodendron		X	X						X	X		
Picea	X	X					X	X				X
Pinus spp.	X	X									X	X
Wisteria	X	X	X	X								

Open case (open bench). Grafting is also done in a temperature-controlled greenhouse or unheated polyhouse (depending on the season). Waxing can be omitted in the bench grafting of *potted rootstock liners* by plunging the container and burying the graft in slightly moist peat moss or bark in a temperature-controlled greenhouse. The medium is bottom heated and kept at 18 to 21° C (65 to 70° F) for three to six weeks for callusing; ideally, the air temperature should be cooler to discourage any initial top growth.

Outplanting of bare-root grafts. As soon as the ground can be prepared in the spring, the grafts are lined-out in the nursery row 10 to 15 cm (4 to 6 in.) apart. They should be planted before growth of the buds or roots begins. If growth starts before the grafts can be planted, they should be moved to lower temperatures (−1 to 2° C; 30 to 35° F). The grafts are usually planted deep enough so that the graft union is just below the ground level, but if the roots are to arise only from the rootstock the graft should be planted with the union well above the soil level, i.e., 7 to 15 cm (3 to 6 in.). It is very important to prevent **scion rooting** where certain definite influences, such as dwarfing or disease resistance, are expected from the rootstock.

After one summer's growth, the grafts should be large enough to transplant to their permanent location. If not, the scion may be cut back to one or two buds, or headed-back somewhat to force out

FIGURE 13–37 Hot-pipe callusing system for root grafting difficult plants. It can be used for bench grafting with other kinds of grafts. The graft union is placed in a slot in a large plastic pipe. Inside the large pipe is a smaller pipe through which thermostatically controlled hot water circulates. Insulating material laid over this pipe retains the heat. The protected roots and scions protrude into areas having cooler temperatures, which retards their development. (Courtesy H. B. Lagerstedt, *(30, 31)*.)

scaffold branches, and allowed to grow a second year. With the older root system—a strong, vigorous top is obtained the second year.

Aftercare in Field and Nursery Grafting Systems

Aftercare of field- and nursery-grafted plants is described in Production Processes Of Graftage. In the section on Types of Grafts, see the descriptions for grafting and aftercare using such grafts as the whip-and-tongue, four-flap, and inlay bark graft. Chapter 12 describes heading-back and crippling techniques which are used in field and container nurseries to encourage the scions of grafted and budded plants to overcome the apical dominance of the rootstock and begin final production development.

Aftercare in topworking systems (top-grafting and top-budding). After the actual top-grafting (or top-budding) operation is finished, much important work needs to be done before the topworking is successfully completed. A good grafting job can be ruined by improper care of the grafted trees.

In three to five days after grafting, the trees should be carefully inspected and the graft unions rewaxed if cracks or holes appear in the wax.

It is essential to prevent sunburn on the trunk and large branches exposed to the sun by retaining some protective top foliage. Protection is especially important if the grafting has been done late in the season when hot weather can be expected, and when no protective nurse branches have been retained. The energy from the sun absorbed by the dark-colored bark can raise the temperature of the living cells below the bark to a lethal level.

It is sometimes advisable to whitewash the trunk, branches, and scions of the top-grafted trees. Various paints are available (Figure 13–40). The white color reflects a considerable portion of the sun's radiant energy, thus keeping the temperature of the living tissues within safe limits. Interior, water-based white house paints (both latex and acrylic) mixed with water, 1:1, prevent sunburn for one season. Exterior paints give longer protection but are more likely to cause injury to the tree (*41*).

FIGURE 13–38 A closed case system of covering containerized grafted plants with glass Mason jars. These grafted plants are maintained under shade to reduce the light irradiance and minimize heat buildup.

GRAFTING SYSTEMS

Some of the different grafting systems have been described in the sections on types of grafting, and the production processes of graftage. Grafting systems are categorized as field, bench, and miscellaneous grafting systems. See the box on page 473.

Field Grafting Systems

Crown Grafting

The *crown graft* originally referred to scions grafted onto larger rootstock. The large rootstock stem was grafted with a number of scions which sometimes were in a crownlike circle (*19*). Today, the term includes grafting onto an established rootstock with single or multiple scions, using the whip-and-tongue, cleft, wedge, side veneer, inlay bark graft, etc. The choice of the graft depends upon the species and size of the rootstock. In California, seedling walnut trees are planted in the orchard and then grafted at the crown—close to the junction of the root and shoot—of the rootstock.

GRAFTING SYSTEMS

Field Grafting Systems
- Crown graftage
- Topworking
- Frame working
- Repair graftage—see pg. 458
- Double-working (by grafting and budding)—sometimes bench grafted

Bench Grafting Systems—see page 470
- Root graftage
- Nurse-root graftage
- Grafting of plants in liner pots under protected culture
- Approach graftage—sometimes field grafted

Miscellaneous Grafting Systems
- Herbaceous grafting—including grafting of vegetable crops
- Cutting grafts—simultaneous rooting and grafting of roses, stenting, etc.
- Micrografting

Crown grafting of deciduous plants is done from late winter to late spring. In each species grafting should take place shortly before new growth starts. The scions should be prepared from mature, dormant wood of the previous season's growth.

If the graft is above the soil level, the union must be well tied (or nailed) and sealed to firmly hold the graft and prevent desiccation. However, when the operation is performed just below, at, or just above the soil level, it is possible to cover the graft union, or even the entire scion, with soil and thus eliminate the necessity for waxing or ceiling. In all cases, the union should be tied securely with tape to hold the grafted parts together until graft take occurs.

Topworking (Top-grafting or Top-budding)

Topworking is used primarily to change the cultivar of an established plant—tree, shrub, or vine—either by grafting (Figures 13–39 and 13–40) or by budding (see Chapter 14). Crown grafting is also a form of topworking. Topworking can be done with any of the apical or side graftage grafts described earlier in this chapter, depending on the plant species. This procedure may be preferred to removal and replacement of the entire plant, since a return to flowering and fruiting is faster with topworking an established plant than with transplanting a new nursery plant—particularly if the topworked plant is young, healthy, and well cared for. Plants that are old, diseased, or of a short-lived species are not satisfactory candidates for topworking.

Preparation for topworking. Top-grafting is usually done in the spring, shortly before new growth starts. The exact time depends upon the method to be used. The cleft, side, whip, and wedge grafts can be done before the bark is slipping, but the bark graft must be done when the bark is slipping, preferably just as the buds of the stock tree are starting to grow. Top-budding can be done with the T-, patch-, or chip-bud methods.

It is usually advisable to obtain an ample amount of good-quality scionwood prior to grafting and store it under the proper conditions, although for broad-leaved evergreens, such as avocado or citrus, scionwood can be collected at the time of the grafting operation. See the earlier section on the selection and handling of scionwood.

In preparing for topworking, one must decide, for each individual stock tree, how many scaffold branches, if any, should be used (usually three to five). However, no scaffold branches are retained when topworking pecans in Texas with the inlay bark method. **Frameworking** is when many grafts are made high in the tree in the small secondary scaffold branches. Frameworking allows the tree to return to bearing earlier than topworking a few larger scaffold limbs lower in the tree. However, frameworking is expensive and requires removing all new shoots from the rootstock. Hence, it is rarely used today.

The branches to be grafted should be well distributed around the tree and up and down the main trunk, avoiding branches with weak, narrow crotches. All others can be removed unless one or

FIGURE 13–39 Extensive topworking operation. Young apple trees have been top-grafted to a more profitable cultivar. Each tree has a small "nurse" branch.

FIGURE 13–40 Proper method of topworking (top-grafting) trees. *Left:* Scions, which have been inserted into fairly small branches, are starting to grow. Tree has been whitewashed to prevent sunburn injury. *Right:* Same tree about six weeks later. Stakes have been nailed to branches; the shoots developing from scions are tied to these stakes to prevent their breaking off in winds.

more nurse branches are used. Figure 13–41 shows a worker preparing the branches for topworking.

Topworking is an extremely severe pruning operation for the tree and results in a considerable imbalance between the root system and the top. However, deciduous trees soon recover and the new growth from the scions and from latent buds on the rootstock restores a balance without damage. Under some situations, the tree will be adversely affected by the heavy cutting back. It may show intense leaf burn, and may even fail to survive.

It is advisable to retain some of the foliage of the rootstock plant as well as small lower branches and even large limbs to be held temporarily as **nurse branches.** Such nurse limbs should be pruned back fairly severely, otherwise the grafted scions may not make adequate growth. It is essential that nurse limbs be left when topworking broad-leaved evergreen trees, such as citrus or olive. If they are retained on the south and west parts of the tree they will shade the grafted portions and reduce the chances of sunburn injury to the exposed branches.

Topworking is most successful when done on relatively young trees where the branches to be grafted are no larger than 7.5 to 10 cm (3 to 4 in.) in diameter and are relatively close to the ground. When attempting to topwork large old trees, it is often necessary to go high up in the trees to find branches with a diameter as small as 10 cm. If the grafting is done on such branches, the new top is inconveniently high for the various orchard operations, such as thinning and harvesting. The other alternative is cutting off the main trunk close to the ground and inserting the scions into wood 30 cm (1 ft) or more in diameter. Although many scions can be inserted around the tree by bark grafting, it is quite likely, as shown in Figure 13–42, that wood rot will develop in the center of the stub before the growth of the scions can heal it over.

Double-Working (Grafting or Budding)

A double-worked plant has three parts, all different genetically: the rootstock, the interstock, and the scion (*18*) (see Figure 13–43). Such a plant has two unions, one between the rootstock and interstock and one between the interstock and the scion. The interstock may be less than 25 mm (1 in.) in length or extensive enough to include the trunk and secondary scaffold branches of a tree.

Double-working is used for various purposes (see Chapter 12).

Examples of double-working are (a) the propagation of 'Bartlett' pears on quince as a dwarfing rootstock by using a compatible interstock such as 'Old Home' or 'Hardy' pear (Figure 13–43), and (b) the propagation of dwarfed apple trees consisting of the scion cultivar grafted onto a dwarfing 'M 9' or 'M 27' interstock that is grafted onto a more vigorous rootstock such as 'MM 106,' 'M 111,' or apple seedlings Figure 12–4 (*11*).

Several methods are used for developing double-worked nursery trees. The grafting in these techniques can be done by machine grafting devices as illustrated in Figure 13–32, or by use of the whip graft.

- Rootstock "liners," either seedlings, clonal rooted cuttings, or rooted layers, are set out in

FIGURE 13–41 Sawing off a branch in preparation for top-working (top-grafting) so that no bark-tearing occurs. *Top left:* The first cut is made in a smooth area starting on the underside of the branch and continuing about one-third of the distance through. *Top right:* The second cut is made starting from the upper part of the branch and cutting downward. The second cut should be back 2.5 or 5 cm (1 or 2 in.) on the branch from the first cut. *Below left:* After the branch breaks off and falls, the second cut is continued. *Below right:* Final smooth cut, ready for grafting.

the nursery row in early spring. These are then fall-budded with the interstock buds, growth from which, a year later, is fall-budded with the scion cultivar buds. Generally, three years are required to produce a nursery tree by this method.

- The interstock piece is bench grafted onto the rooted rootstock—either a seedling or a clonal stock—in late winter. After callusing, the grafts are lined-out in the nursery row in the spring. These are then fall-budded to the scion cultivar. By this method the nursery tree is propagated in two years.
- A variation of the previous method is to prepare, by bench grafting, two graft unions—the scion grafted to the interstock and the interstock grafted to the rooted rootstock. After callusing, the completed graft, with two unions, is lined-out in the nursery row. Depending upon growth rate, a nursery tree can be obtained in one or two years.
- The scion piece is bench grafted onto the interstock in late winter or early spring; then, after callusing, this component is grafted—as shown in Figure 13–43—onto rootstocks that have been grown in place in the nursery row. A nursery tree is obtained in one or two years by this method.
- Double-shield budding is used for double-working in one operation by budding (Figure 14–19). A nursery tree is produced in one year, or if growth is slow, two years after budding.
- The interstock shoots still on the plant can be T-budded in late summer with the scion buds inserted about 15 cm (6 in.) apart. During late winter, the budded interstock shoots are removed with the budded scion at the terminal end of each piece and bench grafted with a whip graft onto seedling rootstocks. After callusing, the completed graft—now consisting of rootstock interstock and a budded scion—is ready for planting in the nursery row (*15*).

FIGURE 13–42 An improper method of top-working trees. This walnut tree was cut off close to the ground, and a number of scions were inserted around its circumference by the bark-graft method. Two of the scions grew successfully, but the remainder failed. Most of the original tree is dead; healing of such a large cut is almost impossible. Better methods of top-grafting trees are shown in Figures 13–39 and 13–40.

Bench Grafting Systems

The term **bench grafting (bench working)** traditionally refers to any graft procedure performed on a rootstock and scion that are not initially planted. This would include *root graftage, nurse-root graftage* or any graftage performed on *bare-root rootstock.* Bench grafting also applies to *potted liner rootstock* that is grafted on a bench or table, as is commonly done with selected woody ornamental species (*40*) (Figures 13–15 and 13–16). Certain *approach grafts* are bench grafted, while others, such as the spliced approach graft of mangos in India are field grafted, with the potted rootstock grafted to the established scion in the field.

Bench grafting is performed manually, by machine or robotics (Figure 13–35).

Miscellaneous Grafting Systems

Herbaceous grafting. Grafting herbaceous types of plants is used for various purposes, such as studying virus transmission, stock-scion physiology, and grafting compatibility, as well as for the commercial greenhouse production of selected vegetable crops, particularly in Japan, Korea, and Europe (Figure 13–34). The rootstock is grafted shortly after seed germination, while the plants are quite small. Such material is generally very soft, succulent, and susceptible to injury. In one technique (see Figure 13–44), a simple splice graft is used with a diagonal cut made through the seedling rootstock, just above the cotyledons (*23*). A piece of thin-walled polyethylene tubing is slipped over the cut

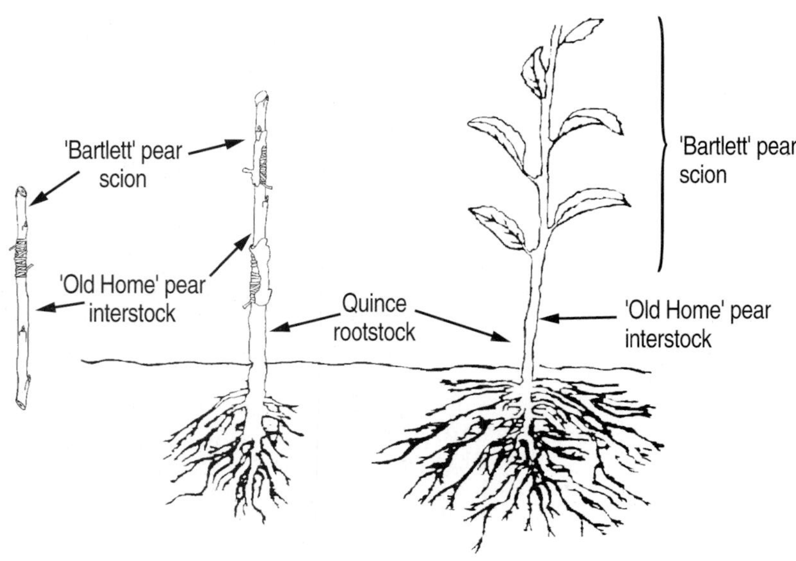

FIGURE 13–43 There are three distinct parts and two graft unions in a double-worked plant. The 'Bartlett' pear scion is grafted on an 'Old Home' pear interstock—then grafted to quince rootstock.

end of the rootstock to give a snug fit. The basal end of the scion receives a diagonal cut similar in length and angle to that given the stock. The scion is then slipped down into the plastic tubing so that the two cut surfaces make intimate contact. The tubing holds the graft in place until healing occurs—about 12 days after grafting. Then the tubing may be slipped off over the scion if there are no leaves and if the bud has not expanded; otherwise, it may be cut off with a razor blade.

Automated procedures of vegetable grafting using robotics, plastic grafting clips, or adhesives to glue the horizontally grafted partners together were described earlier in the chapter (Figures 13–34, 13–35, 13–36).

In another procedure (*12*) used in grafting older herbaceous rootstock plants, several leaves are retained below the graft union. A cleft graft is used, and the graft union is bound with raffia, budding rubbers, or adhesive latex tape. To prevent drying out, the entire plant—following grafting—is covered with a supported polyethylene bag. The grafted plant is then set in the shade until the graft has healed; then the plastic cover can be removed.

A method of establishing *Vitis vinifera* grapes on phylloxera or nematode-resistant rootstocks is by "greenwood" grafting, in which the grafting is done in the spring by placing the scion—taken from new growth—onto a new green shoot developed from the rootstock plant. A simple splice graft is made with the sloping cuts 2.5 to 4 cm (1 to 1 ½ in.) long. The rootstock and scion pieces must be the same diameter. The scion piece has two or three buds. Cuts are matched as closely as possible, and the graft union is completely covered by wrapping with a budding rubber. The graft union should be healed and the bud should be growing by two weeks after grafting. Since these grafts are quite fragile, they must be tied to a stake (*10, 21*).

Cutting-grafts. In the cutting-graft, a leafy scion is grafted onto a leafy, unrooted stem piece (which is to become the rootstock), and the combination is then placed in a rooting medium under intermittent mist for *simultaneous grafting and rooting* of the rootstock. Leaves must be retained on the rootstock piece in order for it to root. This procedure was utilized many years ago in studying stock-scion physiology in citrus (*20*). It has been used in commercial propagation of various types of citrus on clonal dwarfing rootstocks (*13*). It is also of value in propagating certain difficult-to-root conifers (*49*), rhododendrons (*14, 39*), and macadamias (*1*), as well as a number of apple, plum, and pear cultivars (*42*). It is used in the Netherlands in propagating greenhouse roses, where it is called **stenting** (Figure 13–45) (*52, 53*).

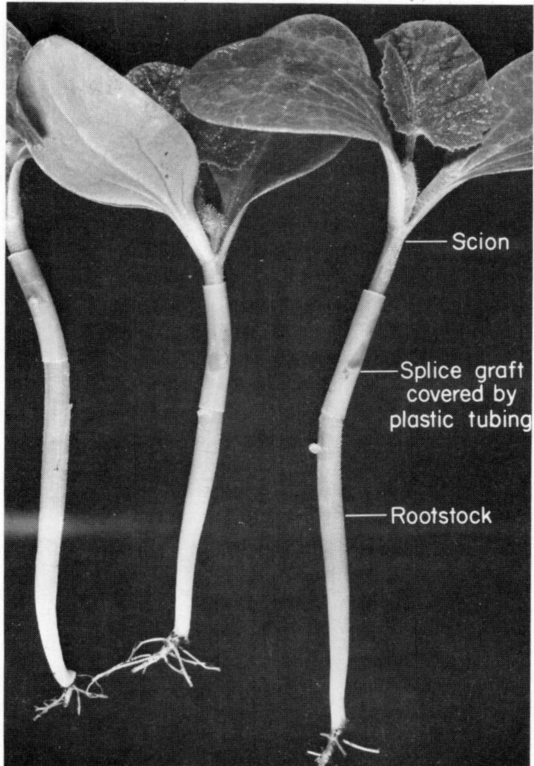

FIGURE 13–44 Grafting young, herbaceous plant material together by using a splice graft. A single, slanting cut is made in the upper part of the stock. A piece of transparent plastic tubing is slipped over this to give a snug fit. A similar slanting cut is made in the lower portion of the scion, which is then slipped into the plastic tubing so that the cut surfaces match. After the graft has healed, the tubing is cut away with a razor blade.

For citrus, a simple splice graft is used. The slope of the cut is at a 30-degree angle 1⅓ to 2 cm (½ to ¾ in.) long; the union is tied with a rubber band. The base of the rootstock is dipped into an auxin, such as indolebutyric acid, and then the grafts are placed under mist, or in a closed case, in flats of the rooting medium over bottom heat. After healing of the union and rooting of the stock, the grafts are allowed to harden by discontinuing the mist and bottom heat for about two weeks. Then the grafts are ready for planting in 3.8-liter (1-gal) containers.

FIGURE 13–45 Roses in the Netherlands being propagated by simultaneous rooting and grafting. (A) *Left:* Shoot cut apart to be used for rootstocks. Only internodes are used. *Right:* Sections cut for scions. One leaf is used per scion. (B) Saddle graft made with an Omega grafting machine. (C) Graft wrapped with tape for healing. (D) Completed graft with union healed and stock well rooted, ready for planting. In the Netherlands this process is called "stenting," a contraction of the Dutch words *stekken* ("to strike a cutting") and *enten* ("to graft"). (Courtesy P. A. Van de Pol et al. (*52, 53*).)

Micrografting. Grafting of tiny plant parts can be done aseptically using tissue culture techniques described in Chapters 17 and 18, in which the small grafts are grown in closed containers until they are large enough to be transferred to open conditions. Micrografting has been used mostly with citrus, apple, and some *Prunus* species, particularly in developing virus-free plants, where a virus-free shoot tip can be obtained but cannot be rooted. The shoot tip is grafted aseptically onto a virus-free seedling thus providing a complete virus-free plant from which other "clean" plants can then be propagated (*8, 25, 27, 44*). Tests conducted some years after the shoot-tip grafts were made and the resultant trees were fruiting showed that the fruits were normal for the cultivar, disease-free, and with no variations appearing (*43*). The procedures for micrografting are described in Chapter 17.

Micrografting can be done without tissue culture techniques. The nonaseptic propagation of very small nursery trees by grafting tiny seedlings with matchlike scions, then growing the minute grafts long enough to have a viable plant, is a promising procedure (*51*). This is useful particularly in the tropics, where nursery plants sometimes must be shipped long distances, often by air, into inaccessible regions. Quantities of such tiny plants can be transported much more readily than full-sized nursery trees.

REFERENCES

1. Ahlswede, J. 1985. Twig grafting of macadamia. *Comb. Proc. Intl. Plant Prop. Soc.* 34:211–14.
2. Alley, C.J. 1957. Mechanized grape grafting. *Calif. Agr.* 11(6):3, 12.
3. ———. 1970. Can grafting be mechanized? *Comb. Proc. Intl. Plant Prop. Soc.* 20:244–48.
4. Anonymous. 1968. Pre-cut scions. *Agr. Res.* 17(6):11.
5. Beineke, W.F. 1978. Parafilm: A new way to wrap grafts. *HortScience* 13(3):284.
6. Bhar, D.S., R.J. Hilton, and G.C. Ashton. 1966. Effect of time of cutting and storage treatment on growth and vigor of scions of *Malus pumila* cv. McIntosh. *Can. J. Plant Sci.* 46:69–72.
7. Brase, K.D. 1951. The nurse-root graft, an aid in rootstock research. *Farm Res.* 17(1):16.
8. Burger, D.W. 1985. Micrografting: A tool for the plant propagator. *Comb. Proc. Intl. Plant Prop. Soc.* 34:244–48.
9. Carpenter, E.L. 1989. How records can improve grafting. *Comb. Proc. Intl. Plant Prop. Soc.* 39:413–415.
10. Carlson, V. 1963. How to green graft grapes. *Calif. Agr. Ext. Publ. AXT 115.*
11. Cummins, J.N. 1973. Systems for producing multiple-stock fruit trees in the nursery. *Plant Prop.* 19(4):7–9.
12. Denna, D.W. 1962. A simple grafting technique for cucurbits. *Proc. Amer. Soc. Hort. Sci.* 81:369–70.
13. Dillon, D. 1967. Simultaneous grafting and rooting of citrus under mist. *Comb. Proc. Intl. Plant Prop. Soc.* 17:114–17.
14. Eichelser, J. 1967. Simultaneous grafting and rooting techniques as applied to rhododendrons. *Comb. Proc. Intl. Plant Prop. Soc.* 17:112.
15. Fisher, E. 1977. The pre-budded interstem: A new technique. *Fruit Var. J.* 31(1):14–15.
16. Flemer, W., III. 1986. New advances in bench grafting. *Comb. Proc. Intl. Plant Prop. Soc.* 36:545–49.
17. Gaggini, J.B. 1985. Bench grafting of trees under polyethylene. *Comb. Proc Intl. Plant Prop. Soc.* 34:646–47.
18. Garner, R.J. 1940. Studies in nursery technique: The production of double worked pear trees. *Ann. Rpt. E. Malling Res. Sta. for 1939,* pp. 84–86.
19. Garner, R.J. 1988. *The grafter's handbook* (5th ed.). New York: Oxford Univ. Press.
20. Halma, F.F., and E.R. Eggers. 1936. Propagating citrus by twig-grafting. *Proc. Amer. Soc. Hort. Sci.* 34:289–90.
21. Harmon, F.N., and E. Snyder. 1948. Some factors affecting the success of greenwood grafting of grapes. *Proc. Amer. Soc. Hort. Sci.* 52:294–98.
22. Hearman, J., A.B. Beakbane, R.G. Hatton, and W.A. Roach. 1936. The reinvigoration of apple trees by the inarching of vigorous rootstocks. *J. Pom. Hort. Sci.* 14:376–90.
23. Holt, J. 1958. A simple way of grafting herbaceous plants. *Gard. Chron.* 143:332.
24. Howard, G.S., and A.C. Hildreth. 1963. Induction of callus tissue on apple grafts prior to field planting and its growth effects. *Proc. Amer. Soc. Hort. Sci.* 82:11–15.
25. Huang, S., and D.F. Millikan. 1980. *In vitro* micrografting of apple shoot tips. *HortScience* 15(6):741–43.
26. Intven, W.J., and T.J. Intven. 1989. Apical grafting of *Acer palmatum* and other deciduous plants. *Comb. Proc. Intl. Plant Prop. Soc.* 39:409–412.
27. Jonard, R., J. Hugard, J. Macheix, J. Martinez, L. Mosella-Chancel, J. Luc Poessel, and P. Villemur. 1983. In vitro micrografting and its application to fruit science. *Scientia Hort.* 20:147–59.
28. Kerr, W.L. 1936. A simple method of obtaining fruit trees on their own roots. *Proc. Amer. Soc. Hort. Sci.* 33:355–57.
29. Kurata, K. 1994. Cultivation of grafted vegetables II. Development of grafting robots in Japan. *HortScience* 29:240–244.
30. Lagerstedt, H.B. 1982. A device for hot callusing graft unions of fruit and nut trees. *Comb. Proc. Intl. Plant Prop. Soc.* 31:151–59.
31. ———. 1984. Hot callusing pipe speeds up grafting. *Amer. Nurs.* 160(8):113–17.
32. Lee, J.M. 1994. Cultivation of grafted vegetables 1. Current status, grafting methods, and benefits. *HortScience* 29(4):235–239.
33. Leiss, J. 1987. Modified side graft for nursery trees. *Comb. Proc. Intl. Plant Prop. Soc.* 36:543–44.
34. Lincoln, F.B. 1938. Layering of root grafts—a ready method for obtaining self-rooted apple trees. *Proc. Amer. Soc. Hort. Sci.* 35:419–22.
35. MacDonald, B. 1986. *Practical woody plant propagation for nursery growers,* Vol 1. Portland, Oreg.: Timber Press.
36. McEachern, G.R., S. Helmers, L. Stein, J. Lipe, and L. Shreve. 1993. Texas inlay bark graft. In *Texas pecan handbook,* TAEX Hort Handbook 105., G.R. McEachern and L.A. Stein, eds. College Station, Tex.: Texas Agricultural Extension Service, Texas A&M University.
37. McEachern, G.R. and A. Stockton. 1993. The four-flap graft. In *Texas pecan handbook,* TAEX Hort Handbook 105., G.R. McEachern and L.A. Stein, eds. College Station, Tex.: Texas Agricultural Extension Service, Texas A&M University.
38. McPhee, G.R. 1992. Grafting techniques. *Comb. Proc. Intl. Plant Prop. Soc.* 42:51–53.
39. McGuire, J.J., W. Johnson, and C. Dawson, 1987. Leaf-bud or side graft nurse grafts for difficult-to-root rhododendron cultivars. *Comb. Proc. Intl. Plant Prop. Soc.* 37:447–49.
40. Meacham. G.E. 1995. Bench grafting, when is the best time? *Comb. Proc. Intl. Plant Prop. Soc.* 45: In press.

41. Micke, W.C., J.A. Beutel, and J.A. Yeager. 1966. Water base paints for sunburn protection of young fruit trees. *Calif. Agr.* 20(7):7.

42. Morini. S. 1984. The propagation of fruit trees by grafted cuttings. *J. Hort. Sci.* 59(3):287–94.

43. Nauer, E.M., C.N. Roistacher, T.L. Carson, and T. Murashige. 1983. In vitro shoot-tip grafting to eliminate citrus viruses and virus-like pathogens produces uniform bud-lines. *HortScience* 18(3):308–9.

44. Navarro, L., C.N. Roistacher, and T. Murashige. 1975. Improvement of shoot tip grafting *in vitro* for virus-free citrus. *J. Amer. Soc. Hort. Sci.* 100:471–79.

45. Oda, M., T. Nagaoka, T. Mori and M. Sei. 1994. Simultaneous grafting of young tomato plants using grafting plates. *Scientia Hort.* 58:259–264.

46. Oda, M. and T. Nakajima. 1992. Adhesive grafting of Chinese cabbage on turnip. *HortScience* 27(10):1136.

47. Patrick, B. 1992. Budding and grafting of fruit and nut trees at Stark Bothers. *Comb. Proc. Intl. Plant Prop. Soc.* 42:354–356.

48. Singha, S. 1990. Effectiveness of readily available adhesive tapes as grafting wraps. *HortScience* 25:579.

49. Teuscher, H. 1962. Speeding production of hard-to-root conifers. *Amer. Nurs Mag.* 116(7):16.

50. Upshall, W.H. 1946. The stub graft as a supplement to budding in nursery practice. *Proc. Amer. Soc. Hort. Sci.* 47:187–89.

51. Verhey, E. W.M. 1982. Minute nursery trees, a breakthrough for the tropics? *Chronica Hort.* 22(1):1–2.

52. Van de Pol, P.A., and A. Breukelaar. 1982. Stenting of roses: A method for quick propagation by simultaneously cutting and grafting. *Scientia Hort.* 7:187–96.

53. Van de Pol, M.H., A.J. Joosten, and H. Keizer, 1986. Stenting of roses, starch depletion and accumulation during the early development. *Acta Hort.* 189:51–59.

SUPPLEMENTARY READING

Chandler, W. H. 1957. Chapter 13 in *Deciduous orchards* (3rd ed.). Philadelphia: Lea & Febiger.

———. 1958. *Evergreen orchards* (2nd ed.). Philadelphia: Lea & Febiger.

Garner, R. J. 1988. *The grafter's handbook* (5th ed.). New York: Oxford Univ. Press.

Hartmann, H. T., and J. A. Beutel. 1979. Propagation of temperate zone fruit plants. *Calif. Agr. Exp. Sta. Leaflet 21103.*

International Plant Propagators' Society. Combined Proceedings of annual meetings.

Rom, R. C. and R. F. Carlson, eds. 1987. *Rootstocks for fruit crops.* New York: John Wiley.

Snyder, J. C., and R. D. Bartram. 1965. Grafting fruit trees. *Pacific Northwest Ext. Publ.* (Idaho, Oregon, and Washington).

Spangelo, L. P. S., R. Watkins, and E. J. Davies. 1968. Fruit tree propagation. *Can. Dept. Agr. Publ. 1289,* Ottawa.

Way, R. D., F. G. Dennis, and R. M. Gilmer. 1967. Propagating fruit trees in New York. *N. Y. Agr. Exp. Sta. Bul. 817.*

14
Techniques of Budding

Budding is a form of grafting. However, budding utilizes only one bud and a small section of bark with or without wood. This is in contrast to grafting, in which the scion consists of a short detached piece of stem tissue with several buds.

Chip budding and T-budding are the two most important types of budding for woody ornamentals and fruit trees. A strategy of some ornamental and fruit tree nurseries in England is to chip bud or T-bud initially in the nursery row. Grafting with a whip-and-tongue graft is used as a back-up for budded plants which do not take. There are other variations of this where different budding and grafting systems are combined in nursery field production.

Budding methods, such as T-budding and patch budding, depend upon the bark's *"slipping."* This term indicates the condition in which the *"bark"* (periderm, cortex, phloem, and cambium) can be easily separated from the *"wood"* (xylem)—see Chapter 12, and Figure 12–8. It denotes the period of the year when the plant is in active growth, when the cambium cells are actively dividing, and when newly formed tissues are easily torn as the bark is lifted from the wood. In Oregon, the bark slips during periods of late June through August (*19*). However, adverse growing conditions, such as lack of water, insect or disease problems, defoliation, or low temperatures, may reduce growth, lead to a tightening of the bark, and seriously interfere with the budding operation. If the bark has only the slightest adherence to the wood, the percentage of bud take will be severely limited. Bud take can be enhanced by irrigating the nursery rootstock a week to several days prior to budding, which should be the optimum growing temperature for the species. Of the budding types described here, only one—the chip bud—can be done when the bark is not slipping.

The budding operation, particularly chip budding and T-budding, can be performed more rapidly than the simplest method of grafting. Some rose budders insert as many as 2,000 to 3,000 or more T-buds a day with the tying done by helpers. If performed under the proper conditions, the percentage of successful unions in T-budding is very high—90 to 100 percent. Budding is widely used in producing nursery stocks of rose, fruit, and shade tree cultivars, where hundreds of thousands of individual plants are propagated each year.

The use of budding is confined generally to young plants or the smaller branches of large plants where the buds can be inserted into shoots from 6 to 26 mm (1/4 to 1 in.) in diameter. Topworking young trees by top-budding (frameworking) is quite successful (Figure 14–18). Here the buds are inserted in small, vigorously growing branches in the upper portion of the tree.

Budding may result in a stronger union, particularly during the first few years, than is obtained by some of the grafting methods, and thus the shoots are not as likely to blow out in strong winds. Budding makes more economical use of propagating wood than grafting, each bud potentially being capable of producing a new plant. This may be important if propagating wood is scarce. In addition, budding is simple and can be performed easily by an amateur.

Budding height varies on nursery rootstock. Budding is done higher on dwarfing rootstock for fruit crops (Figure 14–6) to prevent scion rooting and allow deep planting; higher budding also increases the dwarfing effect of the rootstock. Budding is done lower on ornamental shade trees (Figure 14–8), where tree shape and appearance is of commercial importance.

ROOTSTOCKS FOR BUDDING

Rootstock should have the desired characteristics of vigor, proper growth habit, and resistance to soil-borne pests, as well as being easily propagated. Rootstock may be a rooted cutting, a rooted layer, a seedling, or a micropropagated plant (see Chapter 12). The length of time before budding depends on rootstock vigor, length of growing season, and climate. As little as six months to one year's growth in the nursery row is needed to produce a rootstock plant large enough to be budded, but seedlings of slow-growing species, such as pecan, and those grown under unfavorable conditions may require two or three seasons.

To produce nursery trees free of harmful pathogens (such as viruses, fungi, or bacteria) the rootstock plant, as well as the budwood, must be free of such organisms.

TIME OF BUDDING—FALL, SPRING, OR JUNE

Budding methods are used at seasons of the year when the rootstock is actively growing—so that the cambium divides and the bark separates readily from the wood—although chip budding can be done when the bark is not slipping. Well-developed buds of the desired cultivar must be available at the same time. For spring budding, the scion buds are used from dormant, stored budwood, whereas quiescent (nonactive) buds are used from the current season's wood in June and fall budding. Budding is normally done during three periods of the year. In the northern hemisphere, these periods are late July to early September **(fall budding),** March and April **(spring budding),** and late May and early June **(June budding).** In the southern hemisphere, similar periods would be late January to early March *(fall budding),* September and October *(spring budding),* and late November and early December *("June" budding).*

The budding schedules listed (Fall, Spring, June) are for production of *nursery plants* that will later be dug from the nursery row and transplanted into an orchard or landscape site. Budding schedules can also be applied to container-produced nursery plants such as citrus, pecans, and selected woody ornamentals. Top-budding (topworking) is done with established trees and shrubs *in the field* and is scheduled during the year based on the plant species and budding method to be utilized. *Bench grafting/budding* (e.g., chip budding) can be done with dormant rootstock during wintertime, working on a table. Bench grafting/budding also includes budding onto *liner rootstocks in containers* under protected culture conditions (greenhouse, polyhouse). The budding could extend throughout the year, depending on the species and budding method. See the chart on p. 471 for optimum production windows in grafting/budding in the greenhouse (*11*).

Fall Budding

Late summer is the most important time for budding in the propagation of fruit tree nursery stock and shade trees. Fall budding is particularly important in northern areas (or southern areas in the southern hemisphere) where the growing season is short. "Fall budding" is done in *late summer, not* in autumn. The rootstock plants are large enough by late summer to accommodate the bud, and the plants are still actively growing, with the bark slipping easily. Once growth has stopped and the bark adheres tightly to the wood, a T- or patch bud can no longer be done.

The budsticks, consisting of the current season's shoots, are obtained at or near the time of fall budding. They should be moderately vigorous and contain healthy vegetative buds in the leaf axils (Fig-

FIGURE 14–1 In budding, it is important to use vegetative rather than flower buds. Vegetative buds are usually small and pointed, while flower buds are larger and more plump. Differences between vegetative and flower buds in three fruit species are illustrated here. *Left:* Almond. The shoot on the left has primarily flower buds and should not be selected for budding. The shoot on the right has vegetative buds, which are more suitable. *Center:* Peach. The shoot on the right has excellent vegetative buds, while those on the left shoot are mostly flower buds. *Right:* Pear. All the buds on the shoot at the left are flower buds. Buds on the shoot at the right are good vegetative buds, suitable for budding.

ure 14–1). Do not select shoots whose buds have been broken off. Short, slowly growing shoots on the outer portion of the tree should be avoided, because they may have mostly flower buds rather than vegetative buds. Flower buds are usually round and plump, whereas vegetative buds are smaller and pointed. Some species have mixed buds (see Chapter 12), i.e., the node contains both vegetative and flower buds. These are satisfactory for use in budding.

Every effort should be made to make sure that the trees from which the budsticks are obtained are free of any bacterial, fungal, or viral diseases. *Using infected budsticks can infect every budded nursery tree with the disease.* See Chapter 9 for sources of certified virus and disease-free scions.

As the budsticks are selected, the leaves should be cut off immediately, leaving only a short piece of the leaf petiole attached to the bud to aid in handling the bud later on. Budsticks should be kept from drying by wrapping in some material, such as clean, moist burlap, and placed in a cool, shady location until they are needed. Budsticks should be used promptly after cutting, although they can be refrigerated for a short time. It is best to collect the budsticks as they are being used, a day's supply at a time.

The best buds on the stick to use are usually on the middle and basal portions. Buds on the succulent terminal portion of the shoot should be discarded. In certain species, such as the sweet cherry, buds on the basal portion of the shoots are flower buds which, of course, should not be used.

After the buds have been inserted and tied, nothing needs to be done until the following spring. *Although the rootstock eventually is cut off above the bud, in no case should this be done immediately after the bud has been inserted.* Graft union of the bud piece to the rootstock is greatly facilitated by the normal movement of water, nutrients, and photosynthate in the stem of the rootstock. This is stopped if the top of the rootstock is cut off above the bud.

If the budding operation is done properly, the bud piece should unite with the rootstock in two to three weeks, depending upon growing conditions. Abscission of the leaf petiole next to the bud is a good indication that the bud has united, especially if the bark piece retains its normal light brown or green color and the bud stays plump. On the other hand, if the leaf petiole does not drop off cleanly but adheres tightly and starts to shrivel and darken, and the bark piece turns black—the operation has failed. If the bark of the rootstock is still slipping and budwood is still available, budding may be repeated.

Even though the union has formed, buds on most deciduous species usually do not grow or "push out" in the fall, since they are either in a physiological rest or inhibited by "apical dominance" (presence of other buds distal to it). By spring, the chilling winter temperatures have overcome the rest influence and the bud is ready to grow. Exceptions to this occur in maples, roses, honey locust, and certain other plants, where some of the buds start growth in the fall. In northern areas, such fall-forced buds usually fail to mature before cold weather starts, and are likely to be winter-killed.

In the spring, the rootstock is cut off immediately above the bud, just before new growth begins. Cutting back the rootstock breaks apical dominance of the upper axillary buds, and forces the inserted bud into growth. In citrus, the rootstock is partially cut above the bud and the top is bent or lopped over, away from the bud. This procedure is referred to as "crippling" (Figure 12–19). Leaves of the rootstock still supply the roots with photosynthate, but the partial cutting more rapidly forces the bud into growth than severely severing the rootstock top. After the new shoot from the bud has started to grow, the top is completely removed. Sometimes corn choppers or machines that can cut and mulch the rootstock tops are used to top the rootstock about 8 to 15 cm (3 to 6 in.) above the graft; hydraulic or pneumatic pruning shears are used later to do the final pruning just above the graft.

In colder northern regions, fall-inserted buds are sometimes covered with soil during the winter until danger of frost has passed and are then uncovered and topped-back in late spring.

Where strong winds occur, support for the newly developing shoot may be necessary in species where new shoots grow vigorously. Sometimes the rootstock is cut off several centimeters above the bud, and the projecting stub is used as a support to tie the tender young shoot arising from the bud. Pecan patch buds are forced in the spring by girdling the rootstock a few centimeters above the inserted bud and stripping the phloem away. The tender scion shoot growing from the patch bud can then be tied to the rootstock. This stub is removed after the shoot has become well established. With pistachio, stakes may be driven into the ground next to the rootstock and developing scion shoots tied at intervals during their growth.

Metal shoot guide clips are used to obtain upright scion growth as the bud elongates (Figure 14–8). Since they are attached to the rootstock or inserted into the soil, they also give some mechanical support to the elongating scion shoot.

Cutting back to force the scion bud also forces many latent buds on the rootstock. These must be rubbed off as soon as they appear, or they will soon choke out the inserted bud. It may be necessary to go over the budded plants several times before the developing shoot inhibits these "sprouts" from appearing. Nursery people refer to this procedure as "suckering."

The shoot arising from the inserted bud becomes the top portion of the plant (Figure 14–2). After one season's growth in the nursery, this shoot will have developed sufficiently to enable the plant to be dug and moved to its permanent location during the following dormant season. Such a tree would have a one-year-old top and a two- or perhaps three-year-old rootstock, but it is still considered a "yearling" tree. If the top makes insufficient growth the first year, it can be allowed to grow a second year, and is then known as a two-year-old tree. Abnormal, slow-growing, stunted trees should be discarded.

FIGURE 14–2 A row of fall-budded nursery trees about one year after budding. The inserted buds have grown through the spring and summer. The rootstocks have grown through two seasons. The tops of the rootstocks were cut off above the inserted buds the spring following budding.

Spring Budding

Spring budding is similar to fall budding except budding takes place the following spring as soon as active growth of the rootstock begins and the bark separates easily from the wood. The period for successful spring budding is short, and budding should be completed before the rootstocks have made much new growth.

Budsticks are chosen from the same type of shoots—in regard to vigor of growth and type of buds (Figure 14–1)—that would have been used in fall budding. However, they are collected when dormant, in late fall or winter, and placed in cold storage. Budwood collected in late winter must still be dormant, before there is any evidence of the buds swelling. Budwood is stored at about −2 to 0° C (29 to 32° F) to hold the buds dormant. The budsticks should be wrapped in bundles with *slightly* damp sphagnum, wood shavings, peat moss, moistened newspapers, or paper towels to prevent drying out.

With chip budding, the bark does not need to slip, and dormant rootstock can be worked with. In Oregon, spring chip budding of shade trees is done just before bud-break of the rootstock.

Other types of budding are done in the spring as soon as the bark on the rootstock slips easily. Then, about two to four weeks after budding, when the bud unions have formed, the top of the rootstock must be cut off above the bud to force the inserted bud into active growth. At the same time, axillary buds on the rootstock begin to grow and should be removed. In California, spring budding is used to propagate plums, apricots, and almonds. The rootstock is propagated as a hardwood cutting in the nursery row in the fall, and spring budded in April after the rooted cutting begins new growth. After the bud takes, the rootstock top is cut back and a nursery plant is produced in one growing season.

Sometimes it is helpful to permit shoots from the rootstock to develop in order to prevent sunburn and help nourish the plant. They must be held in check however, and eventually removed. In two-year field rose bush production in California, Texas, and Spain, spring, T-budded rootstock plants are not cut back until the late winter of the second and final growing season.

Although the new shoot from the inserted bud gets a later start in spring budding than in fall budding, spring buds will usually develop rapidly enough, if growing conditions are favorable, to make a satisfactory top by fall. Fall budding, however, is preferred for several reasons:

- The higher temperatures at that time (late summer) promote more extensive bud union formation.
- The budding season is longer.
- There is no necessity to store the budsticks.
- The inserted buds start growth earlier in the spring.
- The pressure of other work is usually not so great for propagators in the late summer as in the spring.

June Budding

June budding is used to obtain a budded tree in a single growing season. Both roots and tops of the budded tree develop during a single growing season. Budding is done in the early part of the growing season and the inserted bud (from the current season's growth) is forced into growth immediately. As a method of nursery propagation, June budding is confined to regions that have a relatively long growing season—in the United States this region includes the central valley of California and some of the southern states—such as Tennessee and Arkansas. In Texas, June budding is not successful due to the excessive heat during this time, so it is done in May before high temperatures become a limiting factor.

June budding is used mostly to produce stone fruits—peaches, nectarines, apricots, almonds, and plums. Peach seedlings are generally used as the rootstock because they have the necessary vigor. Budding is done by the T-bud method. If seeds are planted in the fall, or stratified seeds planted as early as possible in the spring, the seedlings usually attain sufficient size [30 cm (12 in.) high and at least 3 mm (⅛ in.) in diameter] to be budded by mid-May or early June (in the northern hemisphere). Preferably, June budding should not be done after late June, or a nursery tree of satisfactory size will not be obtained by fall. June-budded trees are not as large by the end of the growing season as those propagated by fall or spring budding, but they are of sufficient size—10 to 16 mm (⅜ to ⅝ in.) caliper and 90 to 150 cm (3 to 5 ft) tall—to produce entirely satisfactory trees (*12*).

Budwood used in June budding consists of the current season's growth, that is, of new shoots which have developed since growth started in the spring. By late May or early June, these shoots will usually have grown sufficiently to have a well-developed bud in the axil of each leaf. At this time of year these buds will not have entered the rest period, so when they are used in budding they continue their growth on through the summer, producing the top portion of the budded seedling.

For June-budded trees, handling subsequent to the actual operation of budding is somewhat more exacting than for fall- or spring-budded trees. The rootstocks are smaller and have less stored carbohydrates than those used in fall or spring budding. The object behind the following procedures—shown in Figure 14–3—is to keep the rootstock (and later the budded top) actively and continuously growing so as to allow no check in growth, while at the same time changing the seedling shoot to a budded top. The bud should be inserted high enough (about 14 cm; 5½ in.) on the stem so that a number of leaves—at least three or four—can be retained below the bud.

The method of T-budding with the "wood out," also known as a "flipped bud," should be used (see page 491). The bud union forms quickly, since temperatures are relatively high, and rapidly growing, succulent plant parts are used. By four days after budding, the bud union has started, and the top of the rootstock can be partially cut back—

FIGURE 14–3 It is important in June budding that the rootstock be cut back to the bud properly. *Far left:* The bud is inserted high enough on the rootstock so that there are several leaves below the bud. *Center left:* Three or four days after budding, the stock is partially cut back about 9 cm (3½ in.) above the bud. *Center:* Ten days to two weeks after budding, the rootstock is completely removed just above the bud. *Center right:* This forces the bud and other buds on the rootstock into growth; the latter must subsequently be removed. *Far right:* Appearance of the budded tree after the new shoot has made considerable growth.

about 9 cm (3½ in.) above the bud—leaving at least one leaf above the bud and several below it. This operation will force the inserted bud into growth and will check terminal growth of the rootstock. It will also stimulate shoot growth from basal buds of the rootstock, which will produce additional leaf area. This continuous leaf area is necessary so that there will always be enough leaves to keep manufacturing photosynthate for the small plant. Ten days to two weeks after budding, the rootstock can be cut back to the bud. If the budding rubber has not broken, it should be cut at this time. Poly budding strips would need to be removed or cut on the opposite side of the bud. Other shoots arising from the rootstock should be headed back to retard their growth. After the inserted bud grows and develops a substantial leaf area, it can supply the plant with the necessary photosynthates. By the time the shoot from the inserted bud has grown about 25 cm (10 in.) high, it should have enough leaves so that all other shoots and leaves can be removed. Later inspections should be made to remove any rootstock shoots below the budded shoot. The steps in fall, spring, and June budding are compared in Figure 14–4.

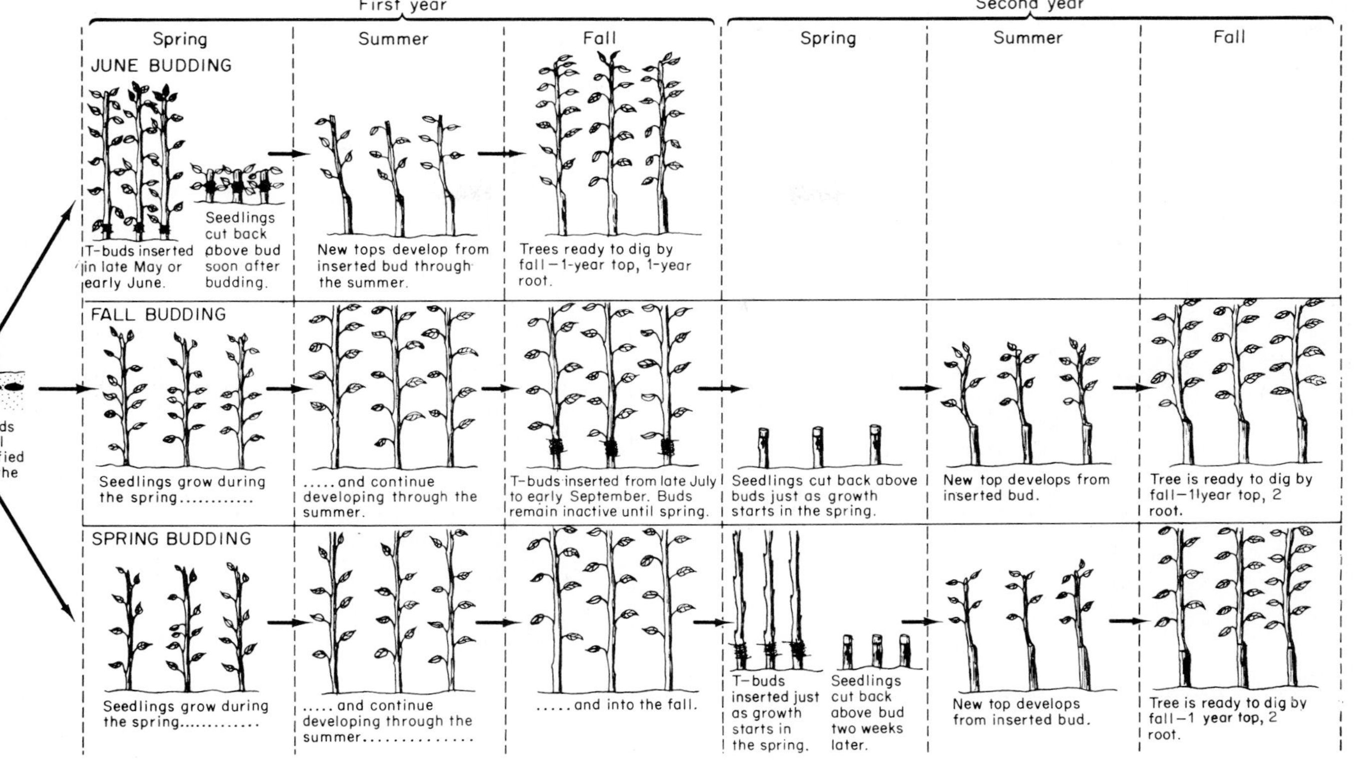

FIGURE 14–4 Comparison of the steps in June, fall, and spring budding for nursery production. The actual techniques in budding are not difficult, but it is very important that the various operations be done at the proper time. Rootstocks are propagated from fall-planted or spring-planted stratified seeds, seedling transplants, rootcuttings or layers, or micropropagated plantlets.

TYPES OF BUDDING

Chip Budding

Chip budding works well in regions with shorter growing seasons. T-budding remains an important budding system, particularly in production areas with long, hot growing seasons, i.e., for June budding. In recent years, chip budding has gradually replaced T-budding as the primary budding method for many woody ornamental trees, shrubs, and fruit trees in numerous parts of the world (*9, 13, 18*). Commercial nurseries have switched to chip budding because of better takes and straighter, more uniform tree growth (*14*). Chip budding in the fall gives excellent results in budding grape cultivars on phylloxera or nematode-resistant rootstocks (*6, 10*).

The production "window" for chip budding—or for bench grafting dormant rose rootstock (*3, 4*)—is large, since active rootstock growth is *not* required, e.g., early in the spring before growth starts, during the summer when active growth has stopped prematurely owing to lack of water, high temperatures, or other causes. More vigorous initial growth has been reported with chip buds than with T-buds (*7, 8*). Studies in England have shown that a better union is obtained with chip budding than with T-budding (*13*). See Chapter 12 for further discussion on the anatomical and physiological advantages of chip budding, compared to T-budding (*15*). Chip budding can be easily mechanized and performed with grafting machines (Fig-ure 14–7).

Chip budding is generally used with small material, 13 to 25 mm (½ to 1 in.) in diameter. It works well with late winter (Texas) grafting of small pistachio rootstocks that have too thin a bark for spring T-budding. As illustrated in Figures 14–5, 14–6, 14–7, and 14–8, a chip of bark is removed from a smooth place between nodes near the base of the rootstock and replaced by another chip of the same size and shape from the budstick

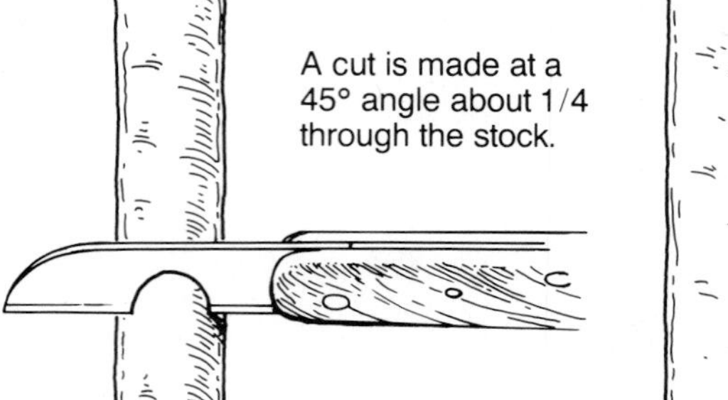

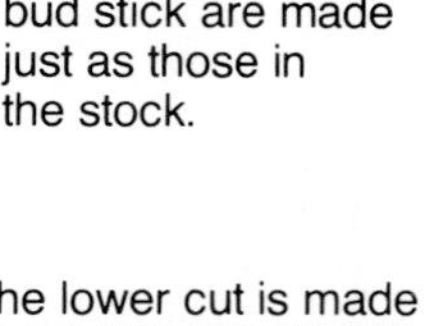

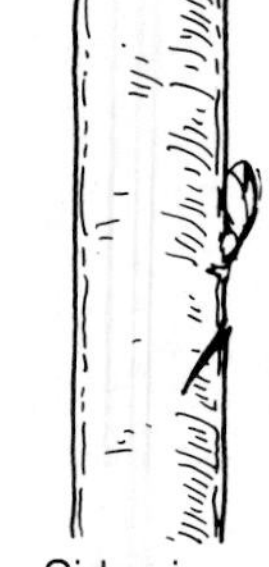

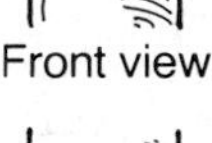

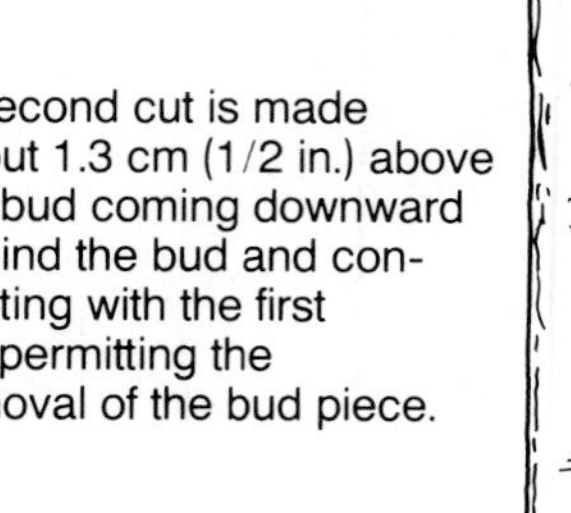

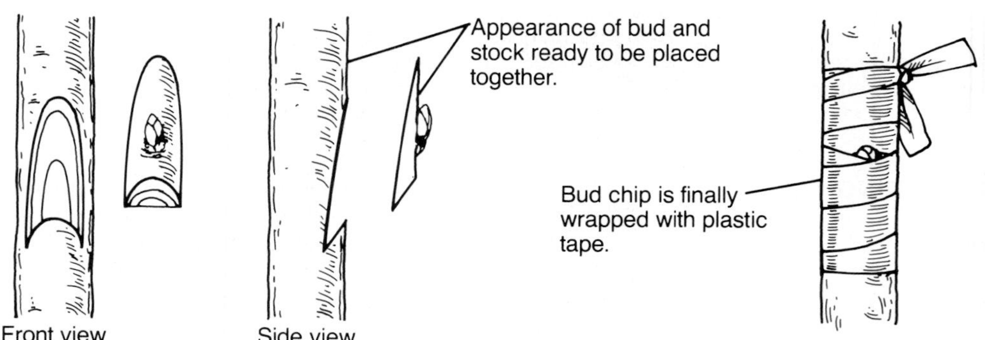

FIGURE 14–5 Chip budding is widely used in propagating woody ornamentals and fruit trees. The bud piece is cut as shown here and covered completely with poly tape.

which contains a bud of the desired cultivar. The chips in both rootstock and budstick are cut out in the same manner. In the budstick, the first cut is made just below the bud and down into the wood at an angle of 30 to 45 degrees. The second cut is started about 25 mm (1 in.) above the bud and goes inward and downward behind the bud until it intersects the first cut. (The order of making these two cuts may be reversed.) The chip is removed from the rootstock and replaced by the one from the budstick. The bark on the rootstock is generally thicker than that of the scionwood. Therefore, the chip removed from the rootstock is slightly larger than that removed from the scion. The cambium layer of the bud piece must be placed to coincide with that of the stock, preferably on both sides of the stem, but at least on one side (Figure 14–7).

FIGURE 14–6 *Left:* A block of clonal apple rootstocks that have been budded to the desired cultivar by chip budding in England. Arrows point to insertion of bud. *Right:* Close-up of two stems in each of which two chip buds have been inserted. Top buds, in both cases, have been covered by transparent plastic tape. Lower buds have been covered by opaque white tape; buds are completely covered by tape, which will be removed after healing takes place.

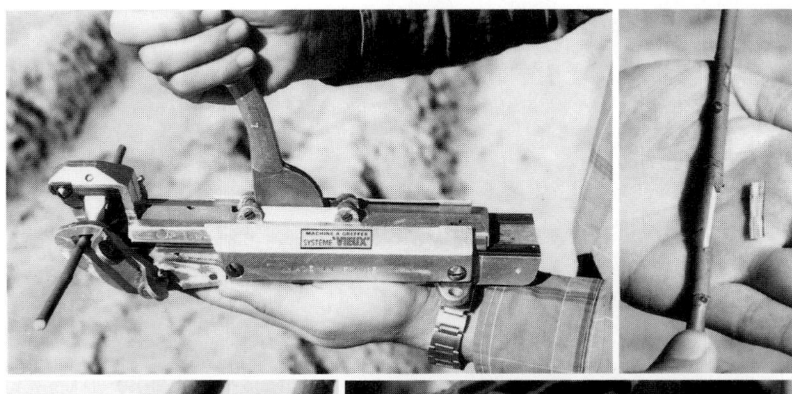

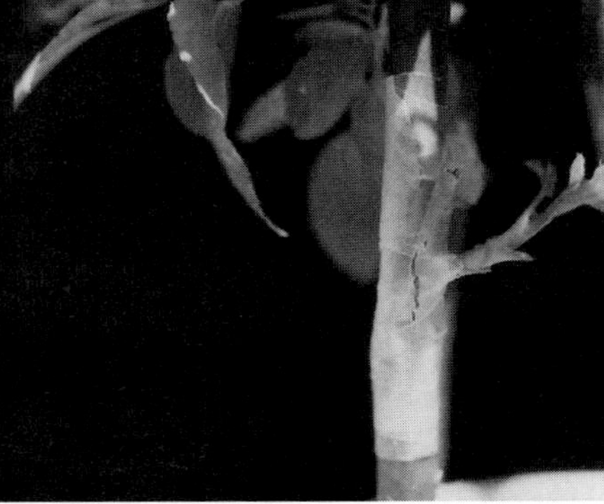

FIGURE 14–7 Chip budding of field rose bushes. *Top left:* A tool for chip budding. *Top right:* Machine-cut chip bud—the scion bud with stem section (right) is slid into the groves of the machine-notched rootstock (left). *Bottom left:* Ideally, the scion and rootstock should be of equal diameter. When the diameters are different, it is important that the cambiums of the graft partners be matched on one side. *Bottom right:* The scion bud penetrates through the Parafilm tape.

(continued)

Figure 14-7 continued

Left: Chip-budded plants form a strong graft union (see Chapter 12 for the discussion on chip budding). The rootstock top has been cut off and the scion shoot system established. *Right:* Chip-budded 'Mirandy' on *Rosa multiflora* rootstock.

FIGURE 14–8 *Top left:* Manually chip budding *Acer platanoides* with a knife. *Top right:* A chip-budded crab apple is being wrapped with poly tape, which will be removed after the graft has taken to prevent girdling the plant. *Bottom left:* Grow Straight metal shield to produce straight, upright growth from the scion bud. The top of the rootstock has been cut off to force out the scion bud. *Bottom right:* The metal shield system with 'Crimson King' maple T-budded to a seedling rootstock. (Photographs courtesy K. Warren (*19*).)

The chip bud must be wrapped to seal the cut edges and to hold the bud piece tightly into the rootstock, since there are no protective flaps of bark to prevent the bud piece from drying out. Nursery adhesive tape works well for this purpose, although white or transparent plastic tape is more often used, covering the bud. Wrapping must be done immediately to prevent drying out (*1*). When the bud starts growth, the tape must be cut.

The rootstock is not cut back above the bud until the union is complete. If the chip bud is inserted in the fall, the rootstock is cut back just as growth starts the next spring. If the budding is done in the spring, the rootstock is cut back about ten days after the bud has been inserted.

T-Budding (Shield Budding)

This method of budding is known by both names—the "T-bud" designation arises from the T-like appearance of the cut in the rootstock, whereas the "shield bud" is derived from the shield-like appearance of the bud piece when it is ready for insertion in the rootstock.

T-budding is widely used by nurserymen in propagating nursery stock of many fruit trees, shade trees, roses, and some ornamental shrubs. Its use is generally limited to rootstocks that range from 6 to 25 mm (¼ to 1 in.) in diameter, and are actively growing so that the bark will separate readily from the wood.

The bud is inserted into the rootstock 5 to 25 cm (2 to 10 in.) above the soil level, where the bark is smooth. Opinions differ on the proper side of the rootstock in which to insert the bud. If extreme weather conditions are likely to occur during the critical graft union period following budding, the bud is placed on the side of the rootstock so as to give as much protection as possible. Some believe that placing the bud on the windward side gives less chance for the young shoot to break off. Otherwise, it probably makes little difference where the bud is inserted, and the convenience of the operator and the location of the smoothest bark are controlling factors. When rows of closely planted rootstocks are budded, it is more convenient to have all the buds on the same side for later inspection and manipulation.

The cuts to be made in the rootstock plant are illustrated in Figure 14–9. Most budders prefer to make the vertical cut first, then the horizontal crosscut at the top of the T. As the horizontal cut is made, the knife is given a twist to open the flaps of bark for insertion of the bud. Neither the vertical nor horizontal cut should be made longer than necessary, because additional tying is required later to close the cuts.

After the proper cuts are made in the rootstock and the incision is ready to receive the bud, the shield piece or shield bud is cut out of the budstick.

To remove the bark shield with the bud, an upward slicing cut is started at a point on the stem about 13 mm (½ in.) below the bud, continuing under the bud to about 2.5 cm (1 in.) above. The shield piece should be thin, but thick enough to have some rigidity. A second horizontal cut is then made 1.3 to 1.9 cm (½ to ¾ in.) above the bud, thus permitting the removal of the shield piece. On many cultivars of shade trees in Oregon, professional budders use only a single cut to remove the bud. Two cuts are used on "wood-out" cultivars (see box, below).

There are two methods of preparing the shield—with the *"wood-in"* or with the *"wood-out."* This refers to the sliver of wood just under the bark of the shield piece, which remains attached if the second, or horizontal, cut is deep and goes through the bark and wood, joining the first slicing cut. Some professional budders prefer to remove this sliver of wood, but others retain it. In budding certain species, however, such as maples and walnuts, better success is usually obtained with *"wood-out"* buds. To prepare the shield with the wood-out, the second horizontal cut, above the bud, should be just deep enough to go through the bark and not the wood. If the bark is slipping, the bark shield can be snapped (slipped) loose from the wood (which still remains attached to the budstick) by pressing it against the budstick and sliding it sideways. The small core of wood comprising the vascular tissues remains with the wood. If the shield is pulled *outward* rather than *slid sideways* from the wood, the core pulls out of the bud—leaving a hole in the shield and eliminating any chances of success. June-budding fruit trees require that the shield piece be prepared with the wood-out. In most other instances, however, the wood is left in. In spring budding, using *dormant* budwood, this sliver of wood is tightly attached to the bark and cannot be removed.

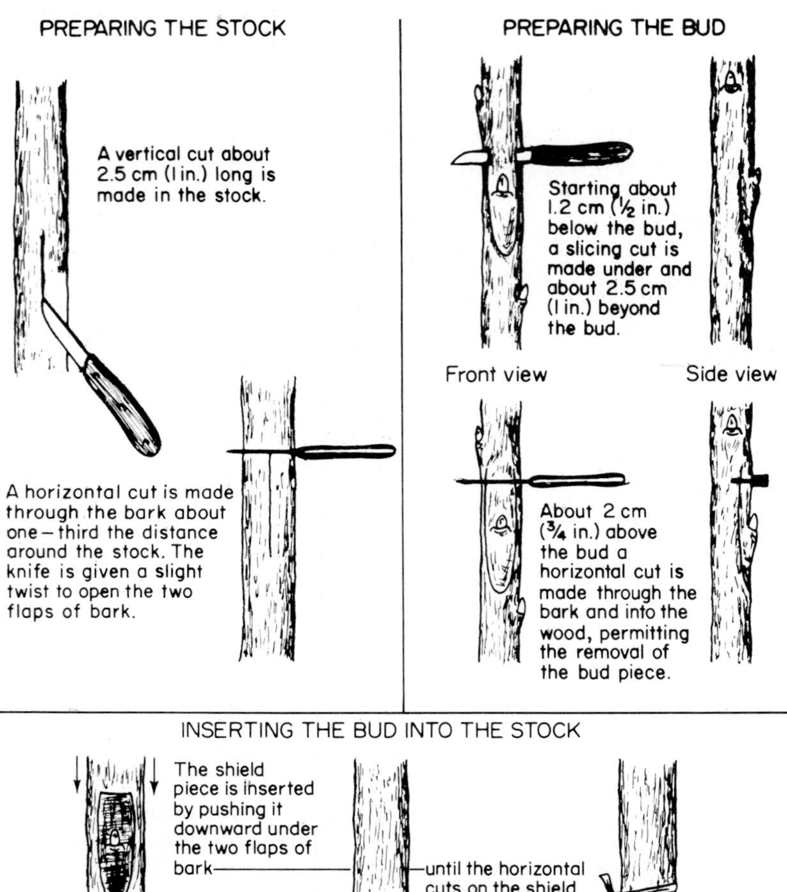

FIGURE 14–9 Basic steps in making the T-bud (shield bud).

The next step is the insertion of the shield piece containing the bud into the incision in the rootstock. The shield is pushed downward under the two raised flaps of bark until its upper, horizontal cut matches the same cut on the stock. The shield should fit snugly in place, well covered by the two flaps of bark but with the bud itself exposed (see Figure 12–16). A "budder" and "tier" in T-budding field roses is shown in Figure 14–10.

Waxing is not necessary, but the bud union must be wrapped, using tape, budding rubbers, or raffia to hold the two components firmly together until healing is completed. Parafilm tape works well for tying and sealing the bud, and the elongating shoot easily penetrates the tape (Figure 14–7) (*3, 4*). Rubber budding strips, especially made for wrapping, are widely used for this purpose (Figure 14–11). Their elasticity provides sufficient pressure to hold the bud securely in place. The rubber, being exposed to the sun and air, usually deteriorates, breaks, and drops off after several weeks, at which time the bud should be healed in place. If the budding rubber is covered with soil, the rate of deterioration will be much slower. Use of this material elim-

FIGURE 14–10 Improving the ergonomics of field budding. *Top:* To enhance the efficiency of T-budding field roses in California, the "budder" (left) and "tier" (right) are supported above the rootstock plants with a body harness, which minimizes muscle strain. The technicians work in the shade and propel the carts down the field row with their legs. *Bottom:* The budder inserts the shield bud which will be wrapped with a budding rubber by the tier. The box next to the budder's head contains the scionwood (budwood). Both the budder and tier work as a team and are paid on piecework, with an incentive bonus based on successful "takes."

inates the need to cut the wrapping ties, which can be costly if many thousands of plants have been budded. The rubber will expand as the rootstock grows, reducing the danger of constriction.

In tying the bud, the ends of the budding rubbers are held in place by inserting them under the adjacent turn. The bud itself should not be covered. The amount of tension given the budding rubber is quite important. It should not be too loose, or there will be too little pressure holding the bud in place. On the other hand, if the rubber is stretched extremely tight, it may be so thin that it will deteriorate rapidly and break too soon—before the bud union has taken place. Often the tying is done from the top down to avoid forcing the bud out through the horizontal cut. Novice budders should avoid the tendency to over-tie as if the bud were a mummy. The inner layers of the tie won't deteriorate, so girdling can occur. See Chapter 13 for the description of different tying and sealing materials for budding and grafting.

FIGURE 14–11 Steps in the development of a T-bud. *Left:* Bud after being inserted and wrapped. *Center:* Bud has healed in place, the budding rubber has dropped off, and the rootstock has been cut back above the bud. *Right:* Shoot development from the inserted bud. All buds arising from the rootstock have been rubbed off.

Inverted T-Incision of the Rootstock

In rainy localities, water running down the stem of the rootstock may enter the T-cut, soak under the bark of the rootstock and prevent the shield piece from healing into place. Under such conditions, an inverted T-incision in the rootstock gives better results, since excess water is shed—*the scion bud is inserted with normal polarity.* In citrus budding, the inverted T-incision method is widely used, even though the conventional method also gives good results. In species that bleed (excessive sap flow) during budding, such as chestnuts, the inverted T-incision allows better drainage and better healing. Proponents of both conventional and inverted T-budding can be found, and in a given locality the usage of either with a given species tends to become traditional.

In the inverted T-incision method, the rootstock has the transverse cut at the bottom rather than at the top of the vertical cut. In removing the shield piece from the budstick the knife starts above the bud and cuts downward below it. The shield is removed by making the transverse cut 13 to 19 mm (½ to ¾ in.) below the bud. The shield piece containing the bud is inserted with normal polarity into the lower part of the incision and pushed upward until the transverse cut of the shield meets that made in the rootstock.

Patch Budding

The distinguishing feature of patch budding and related methods is that a rectangular patch of bark is removed completely from the rootstock and replaced with a patch of bark of the same size containing a bud of the cultivar to be propagated.

Patch budding is somewhat slower and more difficult to perform than T-budding. It is widely and successfully used on thick-barked species, such as walnuts and pecans, in which T-budding sometimes gives poor results, presumably owing to the poor fit around the margins of the bud—particularly the top and bottom. Patch budding, or one of its modifications, is also extensively used in propagating tropical species, such as the rubber tree *(Hevea brasiliensis)*.

Patch budding requires that the bark of both the rootstock and budstick be slipping easily. It is usually done in late summer or early fall, but can be done in the spring also. In propagating nursery stock, the diameter of the rootstock and the budstick should preferably be about the same, about 13 to 26 mm (½ to 1 in.) (*17*).

Special knives (see Figure 14–12) have been devised to remove the bark pieces from the rootstock and the budstick. Some type of double-bladed knife that makes two transverse parallel cuts 25 to 35 mm (1 to 1⅜ in.) apart is necessary. These cuts, about 25 mm (1 in.) in length, are made through the bark to the wood in a smooth area of the rootstock about 10 cm (4 in.) above the ground. Then the two transverse cuts are connected at each side by vertical cuts made with a single-bladed knife.

The patch of bark containing the bud is cut from the budstick in the same manner in which the bark patch is removed from the rootstock. Using the same two-bladed knife, the budder makes two transverse cuts through the bark one above and one below the bud. Then two vertical cuts are made on each side of the bud so that the bark piece will be about 25 mm (1 in.) wide. The cut on the right side of the bud should form a 90° angle with the horizontal cut. The bark piece containing the bud is now ready to be removed. It is important that it be

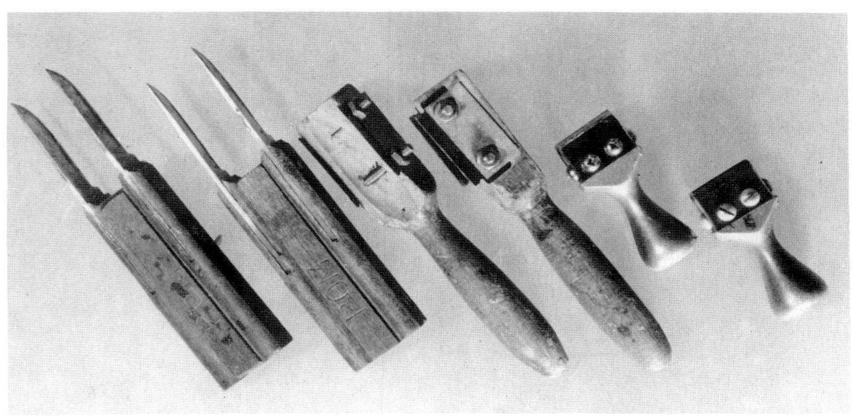

FIGURE 14–12 Tools used in patch budding and related methods. *Left:* Manufactured double-bladed knives. *Center:* Double-bladed knives made with two razor blades. *Right:* Manufactured patch-budding tools consisting of four rectangular cutting blades.

slid off sideways rather than being lifted or pulled off. There is a small core of wood, the *bud trace,* which must remain inside the bud if a successful "take" is to be obtained. By sliding the bark patch to one side, this core is broken off, and it stays in the bud. If the bud patch is lifted off, this core of wood is likely to remain attached to the wood of the budstick, leaving a hole in the bud, as shown in Figure 14–13.

After the bud patch is removed from the budstick, it must be inserted immediately on the rootstock, which should already be prepared, needing only to have the bark piece removed. The patch from the budstick should fit snugly at the top and bottom into the opening in the rootstock, since both transverse cuts were made with the same knife. It is more important that the bark piece fits tightly at top and bottom than along the sides. These procedures are illustrated Figures 14–14, 14–15, and 14–16.

The inserted patch is now ready to be wrapped. Often the bark of the rootstock will be thicker than the bark of the inserted bud patch so that upon wrapping, it is impossible for the wrapping material to hold the bud patch tightly against the rootstock. In this case, the bark of the rootstock is pared down around the bud patch so that it will be of the same thickness, or preferably slightly thinner than the bark of the bud patch. Then the wrapping material will hold the bud patch tightly in place.

The patch bud should be covered by a material that not only holds the bark tightly in place but covers all the cut surfaces. Entrance of air under the patch is prevented to avoid drying and death of the tissues. The bud itself must not be covered during wrapping. The most satisfactory material is nursery adhesive tape or ploy budding strips.

Patch buds, especially with walnuts, should not be wrapped to cause a constriction at the bud union. When the rootstock is growing rapidly, the tape is cut about ten days after budding. A single vertical knife cut on the side opposite the bud is sufficient, but care should be taken not to cut into the bark.

In California, Texas, and other areas with hot summers, patch budding is best performed in late summer (fall budding) when both the seedling rootstock and the source of budwood are growing rapidly and their bark slipping easily. The budsticks for patch budding done at this time should have the leaf blades cut off two to three weeks before the budsticks are taken from the tree. The petiole or leaf stalk is left attached to the base of the bud, but by the time the budstick is removed, this petiole has dropped off or is easily pulled off.

Patch budding can be done in the spring after new growth has started on the rootstocks and the bark is slipping. There is a problem, however, in obtaining satisfactory buds to use at this time of year, since it is necessary that the bark of the budstick separate readily from the wood. At the same time, the buds should not be starting to swell. There are two methods by which satisfactory buds can be obtained for patch budding in the spring.

In one, used in Texas for pecans, the budsticks are selected during the dormant winter period and stored at low temperatures (about 27° C; 36° F) and wrapped in *slightly moist* sphagnum or wood shavings to prevent their drying out. Then, about two or three weeks before the spring budding is to be done, they are brought into a warm room. The budsticks may be left in the moist sphagnum or set with their bases in a container of water. The increased temperature will cause the cambium layer to become active, and soon

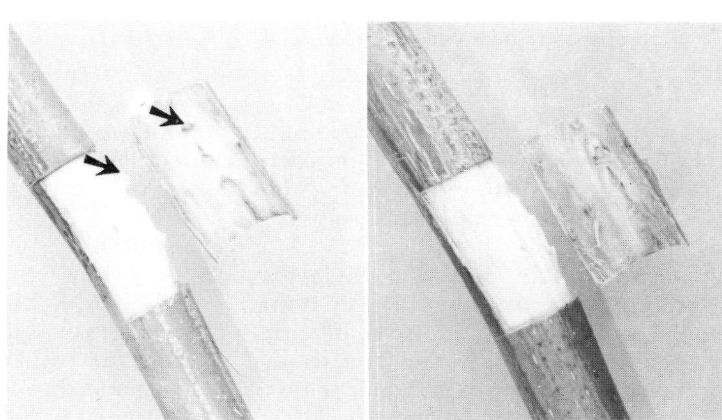

FIGURE 14–13 Removing the bud patch from the budstick in patch budding. *Left:* Incorrect. The core of wood in the bud, comprising the vascular tissues, has broken off, leaving a hole in the bud. Such a bud is not likely to grow. *Right:* Correct. The patch was pushed sideways, and the core of wood has remained inside the bud.

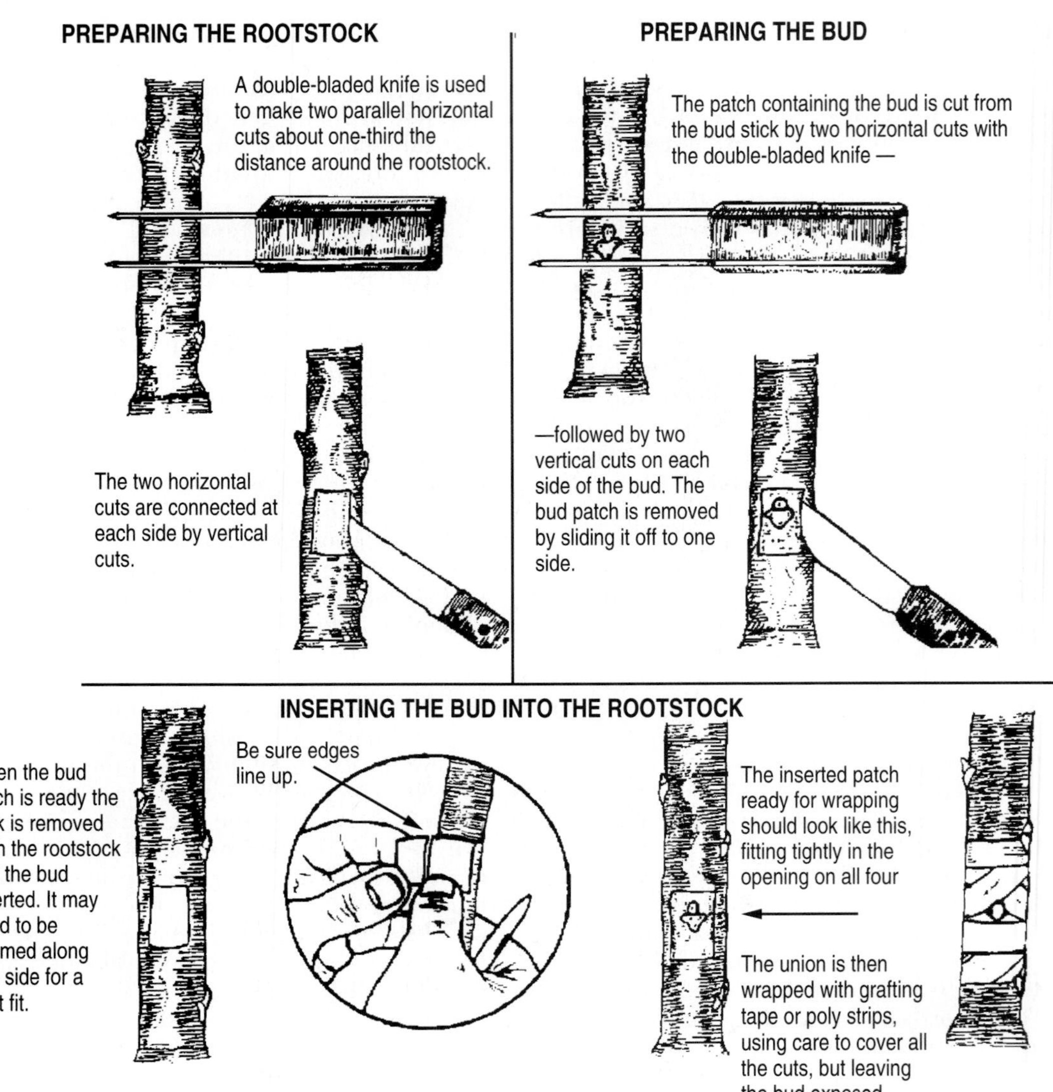

FIGURE 14–14 Steps in making the patch bud. This method is widely used for propagating thick-barked plants.

the bark will slip sufficiently for the buds to be used. Although a few of the more terminal buds on each stick may start swelling in this time and cannot be used, there should be a number of buds in satisfactory condition.

The second method of obtaining buds for spring patch budding is to take them directly from the tree that is the source of the budwood, in early spring, before the buds begin to force. The scions are then refrigerated and stored as dormant scions, and later brought to room temperature for a few hours to stimulate cambial activity, and then budded. Pecans should not be budded until at least two weeks after the rootstock begins to grow.

I-Budding

In I-budding, the bud patch is cut in the form of a rectangle or square, just as for patch budding. With the same parallel-bladed knife, two transverse cuts are made through the bark of the rootstock. These are joined at their centers by a single vertical cut to

FIGURE 14–15 Steps in the development of a patch bud. *Top left:* Patch bud after being inserted and wrapped with tape. *Top right:* After about ten days, the tape is slit along the back side to release any constricting pressure. *Below left:* The tape is completely removed after about three weeks, at which time the patch should be healed in place. *Below right:* Cutting back the rootstock above the bud forces it into growth.

produce the shape of the letter **I.** The two flaps of bark can then be raised to insert the bud patch beneath them. A better fit may occur if the side edges of the bud patch are slanted. In tying the I-bud, be sure that the bud patch does not buckle outward and leave a space between the rootstock (see Figure 14–17).

I-budding is most appropriate when the bark of the rootstock is much thicker than that of the budstick. In such cases, if the patch bud were used, considerable paring down of the bark of the rootstock around the patch would be necessary. This operation is not necessary in the I-bud method (Figure 14–17).

Flute Budding and Ring Budding

See details in Figure 14–17.

FIGURE 14–16 *Left:* Shoot development from patch-budding operation shown in Figure 14–15 after two years' growth. *Right:* Strong, smooth union developed after 14 years. *Juglans regia* 'Hartley' on Paradox rootstock (*Juglans regia* × *J. hindsii*).

TOP-BUDDING (TOPWORKING)

Young trees with an ample supply of vigorous shoots at a height of 1.2 to 1.8 in (4 to 6 ft), can be top-budded (topworked) rapidly with a high degree of success (Figure 14–18). Older trees can be top budded by cutting back severely the prior year to provide a quantity of vigorous water-sprout shoots fairly close to the ground.

Depending upon the size of the tree, 10 to 15 buds are placed in vigorously growing branches 6 to 19 mm (¼ to ¾ in.) in diameter in the upper portion of the tree—about shoulder height. A number of buds can be placed in a single branch, but usually only one will be saved to develop into secondary branches, to form the permanent new top of the tree. The T-bud or chip-bud method is used on thin-barked species, and the patch bud on those with thick bark.

Top-budding is usually done in midsummer, as soon as well-matured budwood can be obtained and while the rootstock is actively growing and the bark slipping. Orchard trees generally stop growth earlier in the season than young nursery trees; therefore the budding must be done earlier. When top-budding is done at this time of year, the buds usually remain inactive until the following spring. Just as vegetative growth starts, the rootstock branches are cut back just above the buds to force the buds into active growth. They should develop into good-sized branches by the end of the summer. At the time the shoots are cut back to the buds, all nonbudded branches should be removed at the trunk. Inspect the trees carefully through the summer and remove all shoots that arise from any but the inserted buds. See the discussion on topworking on p. 473.

DOUBLE-WORKING BY BUDDING

In propagating nursery trees, budding methods can be used in developing double-worked trees. Interstocks can be budded on the rootstocks; the following year the cultivar is budded on the interstock. Although effective, this process takes three years (Figure 12–5). Development of a double-worked tree—in one operation in one year—is possible by using the double-shield bud method (*2, 5, 7*) (Figure 14–19).

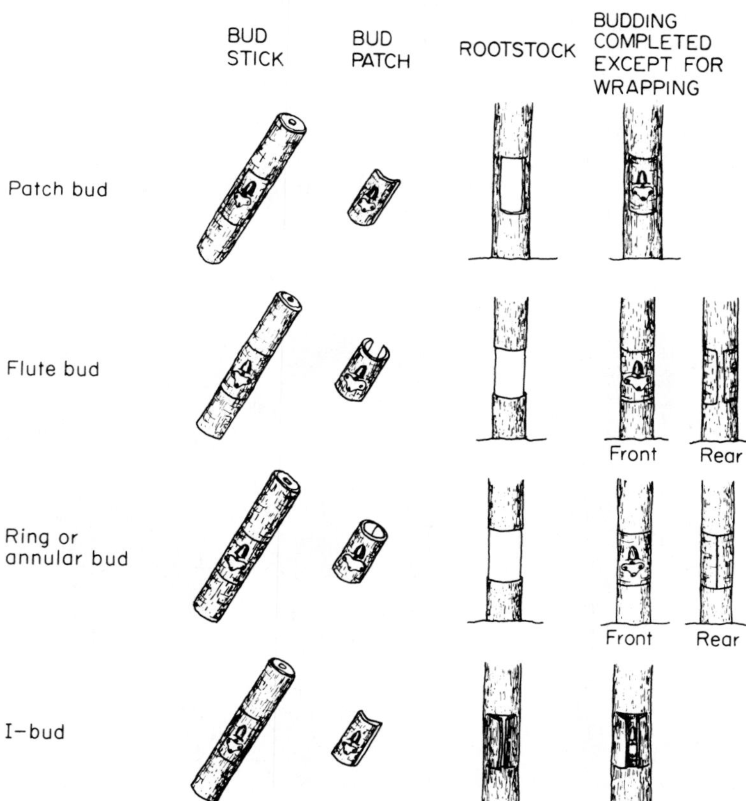

FIGURE 14–17 There are many variations of the patch bud, some of which are shown here. The naming of these types is somewhat confused; the most generally accepted names are given.

FIGURE 14–18 Topworking a young tree by top-budding. T-buds were inserted in the positions shown by arrows and have grown for one season. All other shoots have been removed. From L. H. Day, "Apple, Quince, and Pear Rootstocks in California," *Calif. Agr. Exp. Sta. Bul. 700.*

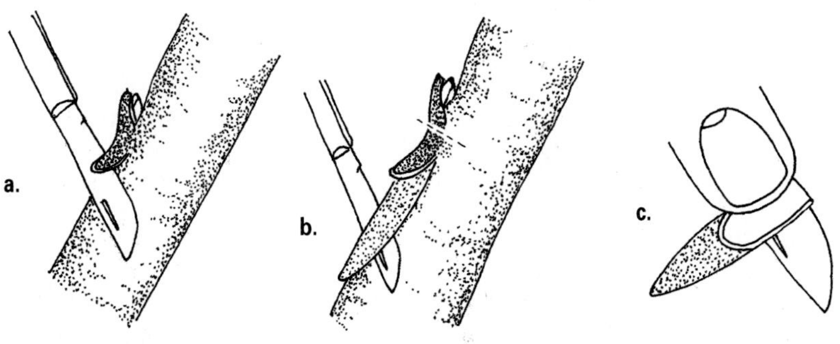

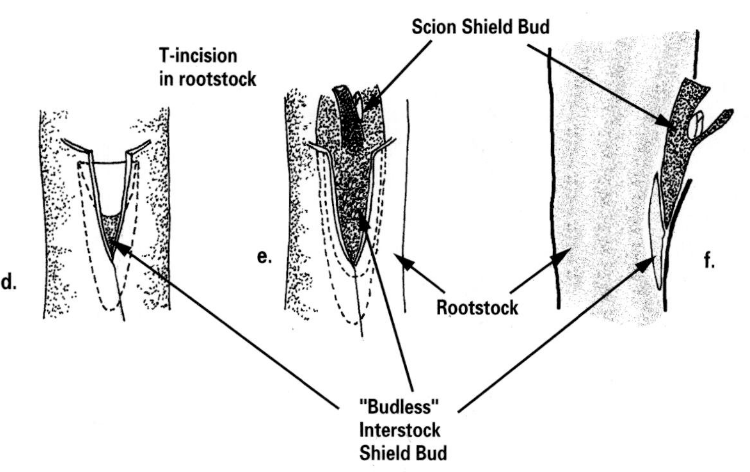

FIGURE 14–19 Double working by budding. (a) A shallow incision is made in the interstock; (b) a second incision is made to debud the interstock shield bud, which is detached (c); (d) a T-cut incision has been made in the rootstock and the budless interstock shield bud inserted—as depicted by the shadowed area; (e) the scion shield bud is inserted in the rootstock T-incision on top of the budless interstock shield bud; (f) side-view of the budding. The budless interstock will later grow and form a complete bridge between the scion and rootstock.

A thin budless shield piece of the desired interstock is first inserted in the T-incision of the rootstock. The scion shield bud is then inserted on top of the interstock. The budless interstock will then grow and form a complete bridge between the scion and rootstock.

MICROBUDDING

This type of budding is used successfully in propagating citrus trees and probably also could be utilized for other tree and shrub species. It has been of commercial importance in the citrus districts of southeastern Australia (*17*). Microbudding is similar to T-budding, except that the bud piece is reduced to a very small size. The leaf petiole is cut off just above the bud, and the bud is removed from the budstick with a razor-sharp knife. A flat cut is made just underneath the bud. Only the bud and a small piece of wood under it are used. In the rootstock an inverted T-cut is made, and the microbud is slipped into it, right side up. The entire T-cut, including the bud, is covered with thin plastic budding tape. The tape remains for 10 to 14 days for spring budding and three weeks for fall budding, after which it is removed by cutting with a knife. By this time, the buds should have healed in place; subsequent handling is the same as for conventional T-budding.

REFERENCES

1. Bremer, A.H. 1977. Chip budding on a commercial scale. *Comb. Proc. Intl. Plant Prop. Soc.* 27:366–67.
2. Bryden, J.D. 1957. Use of plastic ties in fruit tree budding. *Agr. Gaz. New S. Wales (Austral.)* 68:-87–88.
3. Davies, F.T., Jr., Y. Fann, and J.E. Lazarte. 1980. Bench chip budding of field roses. *HortScience,* 15(6):817–18.
4. Fann, Y., F.T. Davies, Jr., and D.R. Paterson. 1983. Correlative effects of bench chip-budded 'Mirandy' roses. *J. Amer. Soc. Hort Sci.* 108(2): 180–83.
5. Garner, R.J. 1953. Double-working pears at budding time. *Ann. Rpt. E. Malling Res. Sta. for 1952,* pp. 174–75.
6. Harmon, F.N., and J.H. Weinberger. 1969. The chip-bud method of propagating vinifera grape varieties on rootstocks. *USDA Leaflet 513.*
7. Howard, B.H. 1977. Chip budding fruit and ornamental trees. *Comb. Proc. Intl. Plant Prop. Soc.* 27:357–64.
8. Howard, B.H., D.S. Skene, and J.S. Coles. 1974. The effect of different grafting methods upon the development of one-year-old nursery apple trees. *J. Hort. Sci.* 49:287–95.
9. Kidd, E.L., Jr., 1987. Asexual propagation of fruit and nut trees at Stark Brothers Nurseries. *Comb. Proc. Intl. Plant Prop. Soc.* 36:427–30.
10. Lider, L.A. 1963. Field budding and the care of the budded grapevine. *Calif. Agr. Ext. Ser. Leaflet 153.*
11. Meacham, G.E. 1995. Bench grafting, when is the best time? *Comb. Proc. Intl. Plant Prop. Soc.* 45: In press.
12. Mertz, W. 1964. Deciduous June-bud fruit trees. *Comb. Proc. Intl. Plant Prop. Soc.* 14:255–59.
13. Osborne, R.H. 1987. Chip budding techniques in the nursery. *Comb. Proc. Intl. Plant Prop. Soc.* 36:550–55.
14. Patrick B. 1992. Budding and grafting of fruit and nut trees at Stark Brothers. *Comb. Proc. Intl. Plant Prop. Soc.* 42:354–356.
15. Skene, D.S., H.R. Shepard, and B.H. Howard. 1983. Characteristic anatomy of union formation in T- and chip-budded fruit and ornamental trees. *J. Hort. Sci.* 58(3):295–99.
16. Smith, N.G., R.J. Garner, and W. S. Rogers. 1962. Delayed growth of apple scions in relation to early budding, bud constriction, and some other factors. *Ann. Rpt. E. Malling Res. Sta. for 1961,* pp. 51–56.
17. Taylor, R.M. 1972. Influence of gibberellic acid on early patch budding of pecan seedlings. *J. Amer. Soc. Hort. Sci.* 97(5):677–79.
18. Tubesing, C.E. 1988. Chip budding of magnolias. *Comb. Proc. Intl. Plant Prop. Soc.* 87:377–79.
19. Warren, K. 1989. A crab apple system. *Amer. Nurser.* 170(5):31–35.
20. Wishart, R.D. A. 1961. Microbudding of citrus. *S. Austral. Dept. Agr. Leaflet 3660.*

SUPPLEMENTARY READING

CARLSON, R. F., and A. E. MITCHELL. 1971. Budding and grafting fruit trees. *Mich. Agr. Ext. Bul. 508.*

GARNER, R. J. 1988. *The grafter's handbook* (5th ed.). New York: Oxford Univ. Press.

ROM, R. C. and R. F. CARLSON, eds. 1987. *Rootstocks for fruit crops.* New York: John Wiley.

HARTMANN, H. T., and J. A. BEUTEL. 1979. Propagation of temperate zone fruit plants. *Calif. Agr. Exp. Sta. Leaflet 21103.*

HOWARD, B. H. 1993. Understanding vegetative propagation. *Comb. Proc. Intl. Plant Prop. Soc.* 43:157–162.

INTERNATIONAL PLANT PROPAGATORS' SOCIETY. Combined proceedings of annual meetings.

MACDONALD, B. 1986. *Practical woody plant propagation for nursery growers,* Vol. 1. Portland, Oreg.: Timber Press.

PLATT, R. G., and K. W. OPITZ. 1973. Propagation of citrus. In *The citrus industry,* Vol. 3, W. Reuther, ed. Berkeley: Univ. of Calif. Press.

15

Layering and Its Natural Modifications

Layering is a form of rooting cuttings in which adventitious roots are initiated on a stem while it is still attached to the plant. The rooted stem **(layer)** is then detached, transplanted, and becomes a separate plant on its own roots. Some plant species utilize layering as a natural means of reproduction, as *tip layers* of black raspberries and trailing blackberries *(Rubus)*, *runners* of strawberry *(Fragaria)*, and *stolons* of Bermuda grass (*Cynodon* spp.). Structures such as *offsets, suckers,* and *crowns* are handled essentially as rooted layers.

Layering is an ancient nursery technique which was used extensively by European nurseries from the eighteenth to early twentieth centuries for propagating woody shrub and tree species (*26, 53*). Mound and trench layering were developed to mass produce hard-to-root clonal rootstocks for apple and other fruit species in England in the early 1900s (*40, 45*). To a large extent, layering as a nursery technique has been replaced by more modern methods of rooting and container production. Nevertheless, the procedure is highly reliable for hard-to-root clones and continues to be used in the production of horticulturally important rootstocks and other plants which are sufficiently valuable to justify the higher costs and hand labor requirement associated with layering. Fruit crops propagated by layering include: filbert, Muscadine grape *(Vitus rotundfolia)*, size-controlling apple rootstocks, and some tropical fruit plants such as mango and litchi (see Chapter 19).

Layering also has specialty uses such as the production of specimen ornamental plants in the greenhouse by air layering, for example, rubber plant *(Ficus elastica)*, croton, and *Dieffenbachia* (*17*). In subtropical and tropical regions, air layering in stock-plant blocks of ornamental species is done in the field under full sun or shade conditions. Layering is valuable in enabling an amateur or specialist horticulturist to produce a relatively small number of large-sized plants of a special cultivar in an outdoor environment with a minimum of propagation facilities.

The purpose of this chapter is to describe the applications of layering as a propagation technique, the physiological conditions that underlie its success, and the various methods by which it is practiced.

PHYSIOLOGY OF REGENERATION BY LAYERING

The physical attachment of the stem to the plant during rooting allows for a continual supply of water, minerals, carbohydrates, and hormones through the intact xylem and phloem to the rooting area. Water stress associated with stem cutting propagation is avoided during the long rooting period required for hard-to-root genotypes (*39*). Leaching of nutrients and metabolites associated with mist systems is avoided.

Accumulation of photosynthates and hormones in the rooting area is an important factor in the success of root initiation observed during layering. This can be promoted by girdling, incision, or bending of the stem (*9*). Enhancing rooting by adding auxins, such as indolebutyric acid (IBA), to the girdled cuts of layers is generally as effective as with cutting propagation (*42, 48*).

Light exclusion in the rooting zone is another aspect of the success with hard-to-root clones. A distinction must be made between **blanching,** which is the covering of an intact stem *after* it has grown, and **etiolation,** which is the effect produced as the shoot *elongates* in the absence of light (*14, 23*). (See Chapters 10 and 11 for discussions on *etiolation, blanching, banding,* and *shading* on root initiation.) Some plants are able to produce roots on intact stems after blanching. Others require phloem interruption as well. The greatest stimulus to root induction during layering occurs when the growing shoots are progressively covered by the rooting medium as they develop to produce etiolation. The success of trench layering occurs because approximately 2.5 cm (in.) of the base of the layered shoots is never exposed to light (*14*).

Another condition of layered plants is the apparent reinvigoration and possible rejuvenation of the layered shoots that arise (see p. 256). Vigorous new shoots are stimulated from the base of the plants next to the root system. There is a possible parallel to the hedging methods used to "rejuvenate" stock plants for improved rooting of the cuttings that are taken from them. Stool shoots—produced by severely pruning back stock plants and using PVC black tape to cover the base of shoots—have enhanced the rooting of apple (*23*).

Root initiation and development on stems during layering is associated with seasonal patterns. In most cases, layering is started in the spring with dormant hardwood shoots. Rooting may not begin until later in the season and is associated with the accumulation of carbohydrates and other substances toward the latter part of the growth cycle.

MANAGEMENT OF PLANTS DURING LAYERING

Most layering methods are field nursery operations which may continue in place for 10 to 20 years once beds are established. Proper choice of the site for good soil, drainage, and climate are essential (*20*). Only virus-tested, true-to-type propagation sources should be used.

During the rooting period, layers are covered with soil or other rooting media not only to exclude light, but to provide continuous moisture, allow for good aeration, and help insulate the layer from temperature extremes (*32, 44, 46*). Special attention needs to be paid to weed, disease, and insect control—see Chapter 3. Without proper management, layering beds tend to decline over time.

In general, much hand labor and attention is required, such that layering procedures tend to be expensive. Nevertheless, modern nurseries have developed specific kinds of mechanical devices and machinery to facilitate the management of large layering beds (Figure 15–1) (*8, 12, 22*).

PROCEDURES IN LAYERING

The most commonly used systems to layer plants include:

- simple layering
- compound layering
- serpentine layering
- air layering
- mound layering, or stooling
- trench layering
- drop layering

Of these, the most commercially important are mound layering for fruit understocks and air layering for some tropical fruits.

Simple Layering

Simple layering consists of bending an intact shoot to the ground to cause adventitious roots to form (Figure 15–2). The method can be used to propagate a wide range of plants, indoor or outdoor, on woody shrubs that produce numerous new shoots

FIGURE 15–1 Mounding up of friable soil with a ridge plow in England. *Left:* Soil is mounded up to and between new shoots until 15 cm (6 in.) of the proximal shoot portion of apple stool is covered. During the harvest, soil is plowed away from the sides of the ridges. *Right:* The rooted apple rootstocks are cut with a tractor-mounted rotary saw. (Courtesy B. H. Howard. 1987. Propagation. In *Rootstocks for fruit crops,* R. C. Rom and R. F. Carlson, eds. © John Wiley.

annually (*26, 40, 50, 53*) or on trees that tend to produce suckers, such as filberts (Figure 15–3). European nurseries have established permanent layering beds by planting shrubs at eight to ten feet apart (*53*), and growing them in place for several years prior to use. Once the layering beds are established, the new shoots are bent to the ground annually for layering while new shoots develop for the following year. All available shoots are worked with, thus utilizing all of the area surrounding the plant.

Layering (Figure 15–4) is usually done in the early spring using flexible, dormant, one-year-old shoot branches of the plant, which can be bent easily to the ground. These shoots are bent and "pegged down" at a location 15 to 20 cm (6 to 9 in.) from the tip, forming a "U." Bending, twisting, cutting, or girdling at the bottom of the "U," stimulates rooting at that location. The base of the layer is covered, leaving the tip exposed.

Shoots layered in the spring will usually be rooted by the end of the growing season and removed either in the fall or in the next spring before growth starts. Mature shoots layered in summer should be left through the winter and

FIGURE 15–2 Propagation of two *Dieffenbachia* plants by simple layering. Leggy stems were curved and placed into containers of soil. After several months, strong root systems formed at the curved portion of stems (*right*); new plants are then severed from the mother plant for independent growth.

FIGURE 15–3 Propagation of filberts in Oregon by simple layering. Center row consists of mother plants. Two outer rows are layered plants, which will be dug at end of the growing season. Arrows point to shoots from mother plants, which have been bent over and placed under the soil for rooting in the manner shown in Figure 15-4.

either removed the next spring before growth begins or left until the end of the second growing season. When the rooted layer is removed from the parent plant, it is treated essentially as a rooted cutting (*9*).

New shoots growing from the base of the plant during the rooting year are used for layering during the next season. With this system a supply of rooted layers can be produced over a period of years by establishing a layering bed composed of stock plants far enough apart to allow room for all shoots to be layered.

Compound or French Layering (*26*)

Compound layering is similar to simple layering, except that the branch to be layered is laid horizontally to the ground and numerous shoots develop for rooting rather than just one. This older method was used extensively in Europe, but has been replaced by more modern methods of rooting cuttings and growing plants in containers.

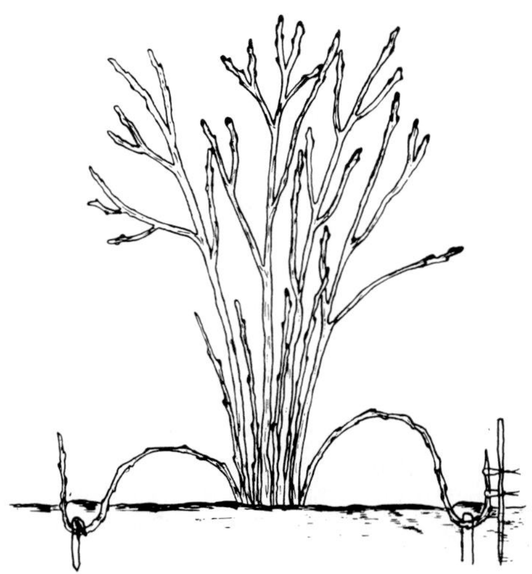

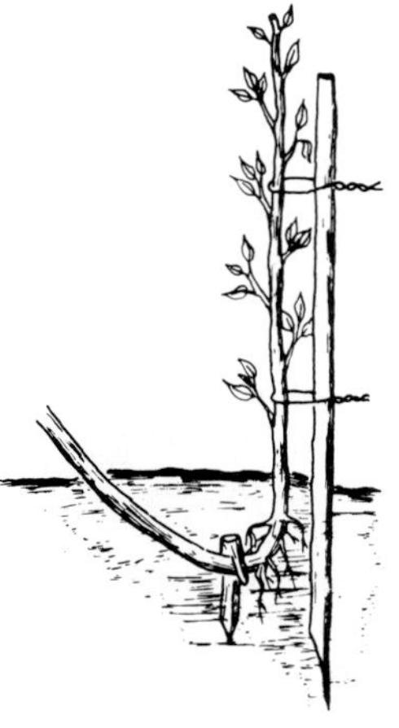

FIGURE 15–4 Steps in propagation by simple layering.

Permanent layering beds are established with plants spaced 1.8 to 3 m (6 to 10 ft) apart and grown for several years to establish a good root system. The vegetative top is then cut back to 2.5 cm (1 in.) from the ground and shoots are allowed to grow for the following season. Before the beginning of the season, long shoots are bent over horizontally to the ground and held down with "U"-shaped wire pegs, being sure that the shoots are completely level. Once new shoots grow about 10 cm (4 in.), the pegs are removed, a shallow trench is dug adjacent to the stem, and the shoots are laid in the bottom of the trench and repegged to hold them in place. Soil or other media is filled in as the shoots grow.

Serpentine Layering

This method is used for propagating plants that have long, flexible shoots, for example, the Muscadine grape and ornamental vines such as *Wisteria* and *Clematis.* The horizontal shoots are alternately covered and uncovered to produce roots at different nodes.

Air Layering (Pot Layerage, Circumposition, Marcottage, Gootee)

Air layering is an ancient method used to propagate a number of tropical and subtropical trees and shrubs (*5, 24, 34, 37*), including the litchi (*15*), longan (*54*), and the Persian lime *(Citrus aurantifolia)* (*43*). The method is useful for producing a few plants of relatively large size for special purposes. For instance, greenhouse and field production (in subtropical/tropical regions) of *Ficus* species, croton, *Monstera,* and philodendron will result in rapid production of large plants (*24, 35, 38, 49*) (Figure 15–5). Air layering has been used to root mature pines (*2, 16, 30*). By using polyethylene film and aluminum foil to wrap the layers, outdoor air layering is possible with many woody plant species (Figure 15–6) (*15, 52*).

Air layers are made in the spring or summer on stems of the previous season's growth or, in some cases, in the late summer with partially hardened shoots. Stems older than one year can be used in some cases, but rooting is less satisfactory and the larger plants produced are somewhat more difficult to handle after rooting. The presence of active leaves on the layered shoot speeds root formation. With tropical greenhouse plants, layering should be done after several leaves have developed during a period of growth.

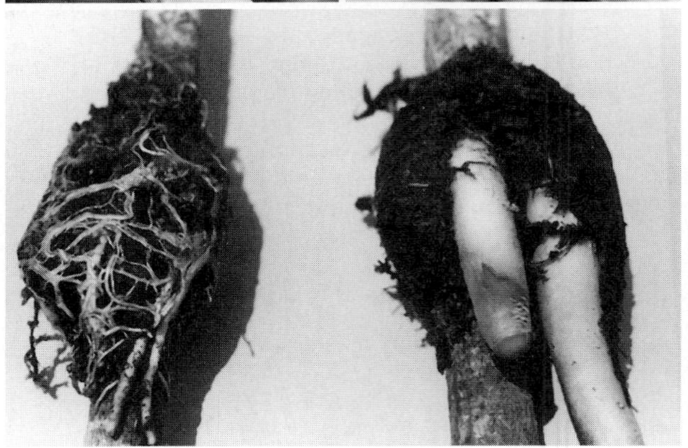

FIGURE 15–5 Air layering of *Dracaena marginata.* Fine roots are induced by a double-slit method *(upper and lower left),* while coarse roots are produced when shoots are girdled during the air-layering technique *(upper and lower right).* (Courtesy T. K. Broschat and H. Donselman (*4*).)

First, girdle the stem or remove a strip of bark 1.8 to 2.5 cm (½ to 1 in.) wide completely around the stem, depending upon the kind of plant. Width is generally three to four times wider than the branch diameter. Scrape the exposed surface to insure complete removal of the phloem and cambium to avoid premature healing. Making a slanting upward cut on one or both sides of the stem about 3 cm (1½ in.) long will keep the two surfaces apart, particularly if sphagnum or a piece of wood (toothpick) is inserted. Girdling reduces water conductivity more than a double-split technique, but need not impede rooting (*5*). Growing plants in 50 percent shade reduces water stress (*4*).

Application of IBA to the exposed wound can be beneficial. One method described for the commercial air layering of *Mahonia aquifolium* 'Com-

FIGURE 15–6 Steps in making an air layer on a *Ficus elastica* plant using polyethylene film. *Right (top):* The stem should be girdled for a distance of about 2.5 cm (1 in.) to induce adventitious root formation above the cut. A ball of slightly damp sphagnum moss is placed around the girdled section *(middle).* A wrapping of polyethylene film is placed around the sphagnum moss and tied at each end *(below). Left:* Roots on a *Ficus elastica* air layer showing through the clear plastic covering. At this stage the layer is ready to be removed from the parent plant and potted. Aluminum foil may also be used to wrap the layer.

pacta' is to insert a small amount of sphagnum moss soaked with 60 ppm IBA under the wounded flap of tissue (*51*). Increased concentrations up to 2 percent IBA in talc has increased rooting and survival in pecan air layers (*42*).

The cut area around the stem is enclosed in a medium which holds moisture and is well aerated. A suitable material is two handfuls of *slightly moistened* sphagnum moss which has excess moisture squeezed out. If the moisture content of the sphagnum moss is too high, the stem will decay. The size of the rootball can be important: too large a rootball holds excess moisture that inhibits root growth—12 × 8 cm (4.7 to 3.1 in.) is considered ideal by some propagators (*53*). Placement is also important. The top of the girdle should be in the top one-third of the rooting medium.

Polyethylene film, 20 to 25 cm (8 to 10 in.) square, wrapped carefully about the branch so that the sphagnum moss is completely covered, is an excellent covering. The ends of the sheet should be folded (as in wrapping meat) with the fold placed on the lower side. The two ends must be twisted to make sure that no water can seep inside. Aluminum foil is useful as an additional wrap and helps to maintain moderate temperatures by reflecting sunlight (*6, 44*). Foil is wrapped around the polyethylene or may be used as the sole wrapping material in climates with high relative humidity. Adhesive tape, such as electricians' waterproof tape, serves well to wrap the ends; the winding should be started above the edge of the cover to enclose the ends, particu-

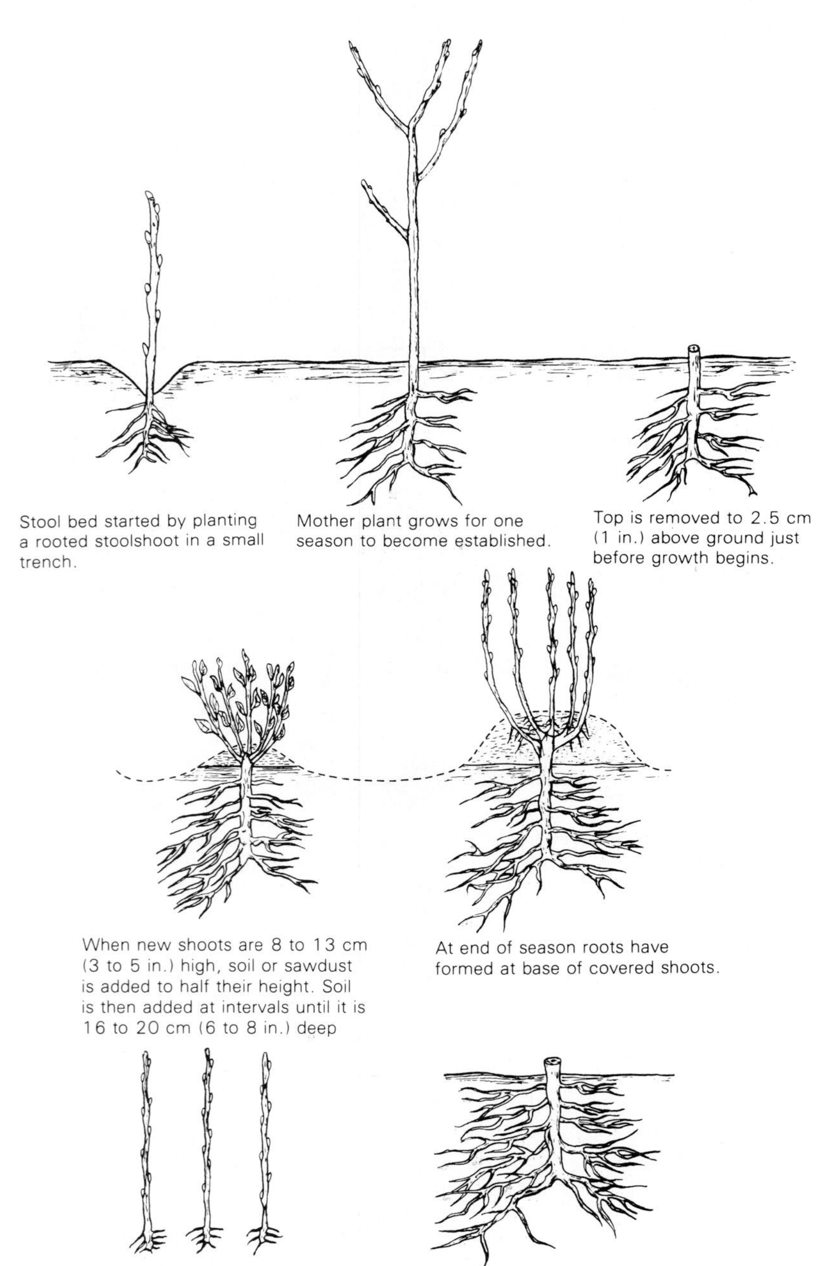

FIGURE 15–7 Steps in propagation by mound (stool) layering.

larly the upper one, securely. Budding rubbers, twist ties, and florists' ties are other materials that can be used for this purpose. Pre-filled plastic pouches and/or bags as well as Jiffy-7 pots enclosed with aluminum have been described as faster to apply (*54*). To avoid breakage during air layering, some operators will attach short canes as a "splint" across the girdled or incised section.

The layer is removed from the parent plant when roots are observed through the transparent film. The earliest adventitious roots are generally thick and corky and the propagator should wait until production of fibrous secondary roots occurs. In some plants, rooting occurs in two to three months or less. Layers made in spring or early summer are best left until the shoots become dormant in the fall, and are removed at that time. Some species, e.g., holly, lilac, azalea, and magnolia, should be left for two seasons (*9*). In general, removal of the layer for transplanting is best when growth is not active.

Pruning to reduce the top in proportion to the roots is usually advisable. Pot the rooted layer into a suitable container and place it under *very reduced* mist, fog, or under low light conditions to allow the plant to acclimate. If dry, initially place the rooted layer in water within five to ten minutes of removal. If potted in the fall, a sufficiently large root system usually develops by spring to permit successful growth under more optimal high-light conditions. Placing the rooted layers under mist for several weeks, followed by gradual hardening-off, is probably the most satisfactory procedure (*36*).

FIGURE 15–8 *Above:* Stool bed used in propagating clonal apple rootstocks. *Bottom:* Rooting medium of wood shavings and soil pulled away showing root production by end of summer.

Mound (Stool) Layering or Stooling

Mound layering is a method where the shoots are cut back to the ground and soil or rooting medium is mounded around them to stimulate roots to develop at their bases. This old nursery propagation method is was standardized and improved to mass propagate specific apple clonal rootstocks in England in the early 1900s. The procedure is used to commercially propagate millions of apple, pear, and some other fruit tree roots each year and is useful for quince, currants, and gooseberries (*3, 7, 11, 14*). Stooling produces "stool shoots," whereas other layering techniques produce "layers." (Figures 15–7, 15–8, and 15–9).

Establishing the stool bed. Plant healthy stock plants of suitable size (8 to 10 mm diameter), in loose, fertile, well-drained soil one year before propagation is to begin. The plants are set 30 to 38 cm (12 to 15 in.) apart in a row, but the spacing between rows may vary with different requirements of the nursery. Width between rows should allow for cultivation and hilling operations during spring and summer. A minimum of 2.5 m (8 ft) row spacing would usually be required to accommodate tractors (*3*). Plants are cut back to 45–60 cm (18–24 in.) and allowed to grow for one year.

Managing the bed. Before new growth starts in the spring, plants are cut back to 2.5 cm (1 in.) above ground level. Two to five new shoots usually develop from the crown the first year, more in later years. When these shoots have grown 7.6 to 12.7 cm (3 to 5 in.), loose soil, bark, sawdust, or a soil-sawdust mixture is used to cover each shoot to one-half of its height (Figure 15–7). When shoots have grown 19 to 25 cm (8 to 10 in.), a second hilling operation takes place. Additional rooting medium is added around the bases of the shoots to about half the total height. A third and final hilling operation is made in midsummer when the shoots have developed approximately 45 cm (18 in.). The bases of the

FIGURE 15–9 Clonal propagation of pecan by mound (stool) layering. *Upper left:* Five-year-old 'Stuart' pecan trees in the background and five-year-old pecan stooling beds in the foreground. *Upper right:* Stool plants prior to mounding with soil. *Lower left:* Larger rooted propagules *(lower right)* can then be lined out in the nursery.

shoots will then have been covered to a depth of 15 to 20 cm (6 to 8 in.).

Harvest. Stool shoots should have rooted sufficiently by the end of the growing season to be separated from the parent plant. Rooted layers are cut close to their bases, handled as rooted liners, including grading, packing, storing, and delivered to customers to be transplanted directly into the nursery row as "rooted liners."

Third year. After the shoots have been cut away, the mother stool beds remain exposed until new shoots have grown 7.6 to 12.7 cm (3 to 5 in.). At this time, the "hilling-up" begins for the next year.

A stool bed can be used for 15 to 20 years with proper handling, providing it is maintained in a vigorous condition and diseases, insects, and weeds are controlled. Selective and biennial harvesting has been used to invigorate declining stool beds and produce large shoots for high budding, although the method has led to increased apple mildew infection (*47*). Tipoff™-sprays, containing the auxin NAA, can eliminate small nonproductive shoots from apple stool beds (*21*).

Cutting back whole plants, then mound layering the vigorous juvenile shoots, has been described as a method of rooting six- to seven-year-old seedling cashew (*34*), seedling pecan (*29*), and other difficult-to-root plants. Mounding (stooling) pecan rootstock has been successfully done in Peru (Figure 15–9).

Girdling the bases of the shoots by wiring about six weeks after they begin to grow may stimulate rooting (*28*). The size of the root system in apple layers has been increased on shoots growing through the spaces of a galvanized screen 0.5 cm (3/16 in.) square laid in a 45-cm (18-in.) strip down the row over the top of the cut-back stumps. New shoots grow through this screen and gradually become girdled as the season progresses (*19*).

Budded plants of apple and citrus (*11, 30*) have been produced by budding the layer in place in the stool bed in the middle of the growing season. The budded layer is transplanted to the nursery in the fall for an additional season's growth. Budding rootstocks **in situ** is generally not recommended unless only virus-free scions are used, because the rootstocks may become permanently infected. A containerized stooling system for limited quantities of clonal apple rootstock is illustrated in Figure 15–10 (*33*).

Trench Layering

Trench layering is a modified layering method in which the mother plants are established in a sloping position horizontally such that shoots can be layered horizontally in the base of a trench (Figure 15–11). Soil, bark, sawdust, or other rooting material is filled in around the new shoots as they develop so as to bring about etiolation. Primarily because of the etiolation, this method is used essentially only for the most difficult-to-root clones. In fact, the procedure was developed in response to the need for propagating a bacterial canker-resistant cherry rootstock Mazzard 'F12-1.' The method can be used for other woody species which are difficult-to-root by stooling such as quince, apple, mulberry, and walnut (*25, 26*).

Establishing the layer bed (27). One-year-old nursery plants or well-rooted layers are planted 65 to 70 cm (20 to 30 in.) apart in a straight line down the row. The distance between rows depends upon equipment. Double rows may also be used (*26*). The trees are planted at an angle of 30 to 45° in order to get the required amount of growth and still allow layering. The plants are allowed to grow during the first year to establish a good root system. By the end of the growing season or during that winter

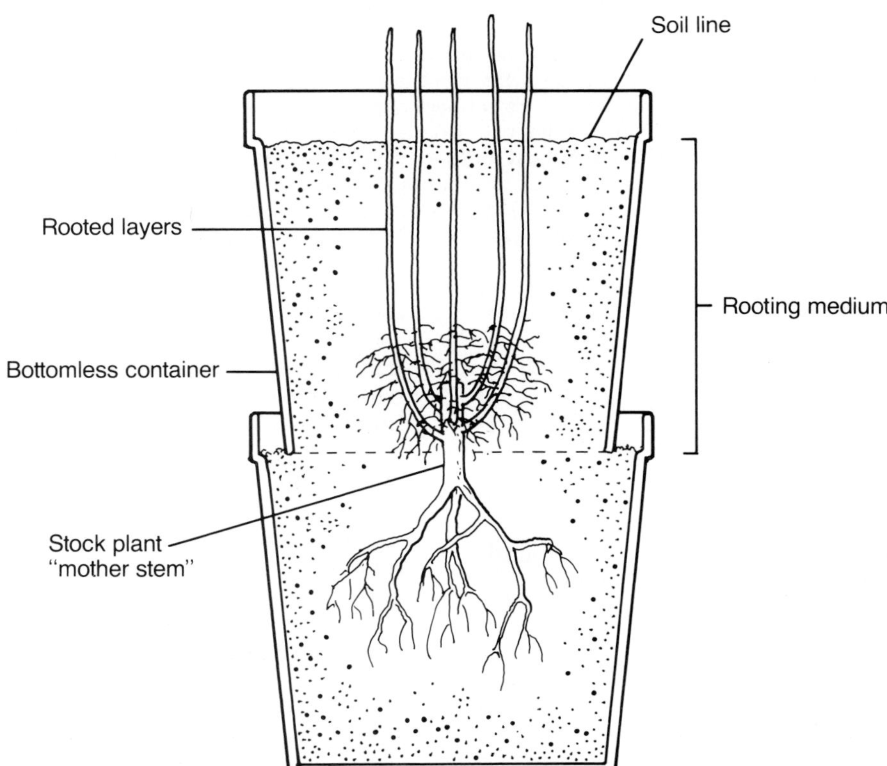

FIGURE 15–10 For limited quantities of clonal rootstock this containerized layering system can be used. It also makes a good classroom demonstration of mounding or stooling. (Redrawn from R. H. Munson (*33*).)

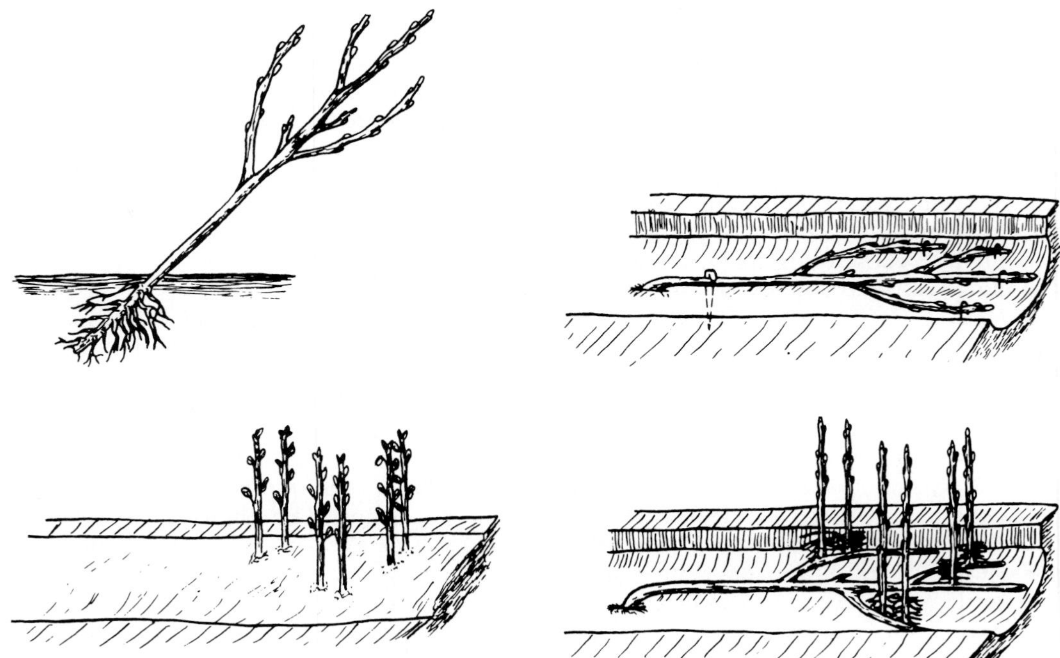

FIGURE 15–11 Steps in propagation by trench layering. *Upper left:* Mother plant after one year's growth in nursery. The trees were planted in the row at an angle of 30° to 45°. The trees are 46 to 78 cm (18 to 30 in.) apart down the row. *Upper right:* Just before growth begins, the plant is laid flat on the bottom of a trench about 5 cm (2 in.) deep. Shoots are cut back slightly and weak branches removed. Tree must be kept completely flat with wooden pegs or wire fasteners. *Lower left:* Rooting medium, as fine soil, peat moss, or sawdust is added at intervals to produce etiolation on 5 to 7.5 cm (2 to 3 in.) of the base of the developing shoots. Apply first 2.5 to 5 cm (1 to 2 in.) layer before the buds swell. Repeat as shoots emerge and before they expand. Later coverings are less frequent and only cover half of shoot. Final medium depth is 15 to 19 cm (6 to 8 in.) *Lower right:* At the end of the season the medium is removed and the rooted layers are cut off close to the parent plant. Shoots left at internodes can be layered the following season.

a shallow trench 5 × 23 cm (2 × 9 in.) is dug down the row. The plants are brought down to a horizontal level and "pegged" carefully so that they are flat on the bottom of the trench along with strong lateral branches. All of the plant must be level to produce even shoot growth.

Second year. Successful layering depends upon etiolation. Buds are covered with about a 2.5 cm (1 in.) of soil before they emerge. Subsequent applications of rooting medium such as sawdust are added periodically to etiolate 5 to 7.5 cm of the developing shoots. Final depth should be 15 to 19 cm. Rooting should take place by the end of the season.

Harvesting. At the end of the season, the sawdust is removed and rooted layers cut off close to the original branch, leaving a small stub for next year's growth. The process is repeated in subsequent years. A well-cared-for stock bed (mother bed) should last 15 to 20 years (*27*).

Trench layering is used primarily for woody species which are difficult to propagate by other methods of stooling. In some cases (walnuts, for instance), plants can be placed horizontally in the trench and the developing shoots layered the first year (*41*).

Drop Layering

This technique is a combination of crown division and layering that has been used for a limited number of shrubby species, such as dwarf ericaceous plants, some dwarf rhododendrons, some conifers, *Berberis,* and boxwood (*26*). Well-grown and well-branched plants are planted deeply in a hole or trench and covered almost completely with only the tips of the branches exposed. New growth comes from the branch tips, but the older bases of the branches are blanched and roots form. At the end of the season the entire plant is dug and divided into all of its rooted parts, each used as rooted liners.

PLANT MODIFICATIONS RESULTING IN NATURAL LAYERING

Tip Layering

Tip layering occurs naturally in trailing blackberries, dewberries, and black and purple raspberries *(Rubus)*. Biennial canes arise from the crown each spring. The canes are vegetative during the first year, fruitful the second, and then die. New canes are produced annually to produce the so-called "bramble bush." Propagation consists of removing the rooted tip layers at the end of the season and transplanting them into a permanent location. In the nursery, healthy young plants should be set aside as stock plants solely for propagation. The original plants are set 3.6 m (12 ft) apart to give room for subsequent layering. The plants are cut to within 23 cm (9 in.) of the ground as soon as they are planted. Vigorous new canes are "summer topped" by pinching off 7.6 to 9.2 cm (3 to 4 in.) of the tip after growth of 45 to 76 cm (18 to 30 in.) to encourage lateral shoot production. This operation will increase the number of potential tip layers, and also can increase next season's fruit crop.

Canes begin to arch over in late summer. Their tips assume a characteristic appearance in that the terminal ends become elongated and the leaves small and curled to give a "rat-tail" appearance. The best time for layering is when only part of the lateral tips have attained this appearance. If the operation is done too soon, the shoots may continue to grow instead of forming a terminal bud. If done too late, the root system will be small.

Tips can be layered by hand, using a spade or trowel to make a hole, with one side vertical and one sloping slightly toward the parent plant. The tip is placed in the hole along the sloping side and soil is pressed firmly against it. Placed thus, the tip stops growing and becomes "telescoped," forming an abundant root system and a vigorous vertical shoot. Rooting takes place below the tip of the current season's shoot. The shoot tip recurves upward to produce a sharp bend in the stem from which roots develop (Figure 15–12).

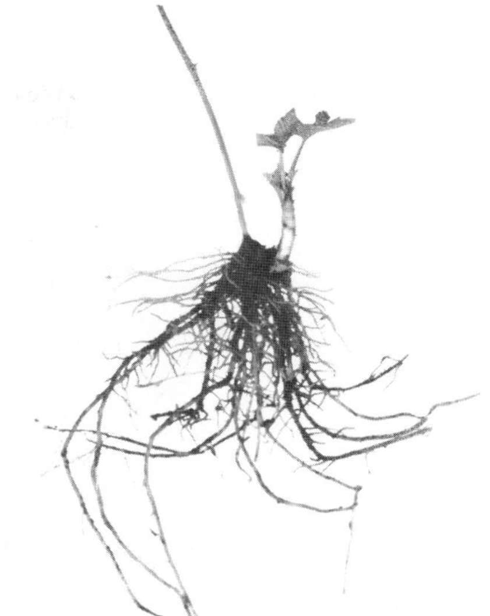

FIGURE 15–12 Tip layer of boysenberry *(Rubus)*.

The plants are ready for digging at the end of the season. The rooted tip consists of a terminal bud, a large mass of roots, and 15 to 20 cm (6 to 8 in.) of the old cane that serves as a "handle" and marks the location of the new plant. Since tip layers are tender, easily injured, and subject to drying out, digging just prior to replanting is preferable. The remainder of the layered shoot attached to the parent plant is cut back to 23 cm (9 in.) as in the first year. Economical quantities of shoots can be produced annually for as long as ten years.

Rooted tip layers are planted in the late fall or early spring. New canes develop rapidly during the first season.

Runners

A **runner** is a specialized stem that develops from the axil of a leaf at the crown of a plant, grows horizontally along the ground, and forms a new plant at one of the nodes. The strawberry is a typical plant with this growth habit (Figure 15–13). Other plants which produce runners used in propagation include bugle *(Ajuga)*, strawberry geranium *(Saxifraga sarmentosa)*, and the ground cover *Duchesnea indica*. Plants of these species grow as a rosette or crown. Some ferns, such as Boston fern *(Nephrolepsis)*, produce runnerlike branches, as do certain orchid species *(Dendrobium)*, which form small plants known as *"keikies."* Runners of the spider plant *(Chlorophytum comosum)* are shown in Figure 15–14.

FIGURE 15–13 Runners arising from the crown of a strawberry plant. New plants are produced at every second node. The daughter plants, in turn, produce additional runners and runner plants.

Most strawberry cultivars form runners in response to length of day and temperature. Runners are produced in long days of 12 to 14 hours or more with high midsummer temperatures. New plants produced at alternate nodes produce roots but remain attached to the mother plant. New runners are in turn produced by daughter plants. The connecting stems die in the late fall and winter, separating each daughter plant from the others.

In propagating by runners, daughter plants are dug when they have become well rooted, and then transplanted to the desired locations.

Stolons

Stolons are modified stems that grow horizontally to the ground and produce a prostrate or sprawling mass of stems growing along the ground, for example, in dogwood *(Cornus stolonifera).* The term also describes the horizontal stem structure occurring in Bermuda grass *(Cynodon dactylis), Ajuga,* mint *(Mentha),* and *Stachys.* Stolonlike underground stems are involved in tuberization of the potato tuber (see p. 536).

A stolon can be treated as a naturally occurring rooted layer and cut from the plant and planted.

Offsets

An **offset** is a characteristic type of lateral shoot or branch that develops from the base of the main stem in certain plants. This term is applied generally to a shortened, thickened stem of rosettelike appearance. Many bulbs reproduce by producing typical offset bulblets from their base (see Chapter

FIGURE 15–14 *Left:* Masses of runners developing from spider plants *(Chlorophytum comosum). Right:* Rooted plantlets at end of runners can be cut off and planted.

16 for details). The term *offset* (or **offshoot**) also applies to lateral branches arising on stems of monocotyledons as in date palm, pineapple, or banana (see Chapter 19) (Figure 15–15).

Offsets are cut close to the main stem with a sharp knife. If well rooted, the offset can be potted and established like a rooted cutting. If insufficient roots are present, the shoot can be placed in a favorable rooting medium and treated as a leafy stem cutting. **Slips** and **suckers,** which are two types of pineapple offsets, readily form aerial rootlets which facilitate commercial field propagation (Figure 15–16).

If offset development is meager, cutting back the main rosette may stimulate the development of more offsets from the old stem in a manner similar to the removal of the terminal bud in the stimulation of lateral shoots in any other type of plant.

Natural multiplication by offsets tends to be slow and not well suited for commercial propagation. The natural reproduction rate of many monocotyledonous plants can be increased greatly by micropropagation (see Chapters 17 and 18). Offsets can be induced in *Hosta* by BA (*13*).

Offshoots of the date palm do not root readily if separated from the parent plant. They are usually layered for a year prior to removal. Micropropagation is an alternate method.

Suckers

A **sucker** is a shoot that arises on a plant from below ground, as shown in Figure 15–17. The most precise use of this term is to designate a shoot that arises from an adventitious bud on a root. In practice, shoots that arise from the vicinity of the crown are often called "suckers," even though originating from stem tissue. Propagators generally designate any shoot produced from the rootstock below the bud union of a budded tree as a "sucker" and refer to the operation of removing them as "suckering." In contrast, a shoot arising from a latent bud of a stem several years old as, for instance, on the trunk or main branches, should be termed a **watersprout.**

Suckers are dug out and cut from the parent plant. In some cases, part of the old root may be retained, although most new roots arise from the base of the sucker. It is best to dig out the sucker rather than pull it, to avoid injury to its base. Suckers are treated as a rooted layer or cutting. They are usually dug during the dormant season.

Crown Division

The term **crown,** as generally used in horticulture, designates that part of a plant at the surface of the ground from which new shoots are produced. In trees or shrubs with a single trunk, the crown is principally a point of location near the ground surface marking the general transition zone between stem and root. In herbaceous perennials, the crown is the part of the plant from which new shoots arise annually. The crown of herbaceous perennials consists of many branches, each being the base of the current season's stem, which originated from the base of the preceding year's branch. These lateral shoots are stimulated to grow from the base of the old stem as it dies back after blooming. Adventitious

FIGURE 15–15 Removing a date offshoot with chisel and sledge hammer. (From R. W. Nixon, Date culture in the United States, *USDA Cir. 728.*)

FIGURE 15–16 *Upper left:* Cross section of a pineapple showing the crown atop the fruit and slip attached to the peduncle supporting the fruit. *Upper right:* Slips *(left)* from the peduncle are more commonly used propagules than suckers *(right). Lower left:* Preformed root primordia enable slips to root readily when mass propagated in black polyethylene mulched field beds *(lower right).* Slips, suckers, and crowns are three types of pineapple offsets that can be used in pineapple vegetative propagation.

roots develop along the base of the new shoots. These new shoots flower either the same year they are produced or the following year. As a result of annual production of new shoots and the dying back of old shoots, the crown may become extensive within a period of a relatively few years and may need to be divided every few years to prevent overcrowding.

With asparagus, which is a herbaceous perennial, the crown is composed of fleshy roots. One way to propagate asparagus is by dividing the crown. New adventitious buds are produced from the crown and develop into the shoot system of the new asparagus plant.

Multibranched woody shrubs may develop extensive crowns. Although an individual woody stem may persist for a number of years, new vigorous shoots are continuously produced from the crown, and eventually crowd out the older shoots. Crowns of such shrubs can be divided in the dormant season and treated as a large rooted cutting.

Crown division is an important method of propagation for herbaceous perennials, and to some extent for woody shrubs, because of its simplicity and reliability. Such characteristics make this method particularly useful to the amateur or professional gardener who is generally interested in only a modest increase of a particular plant.

FIGURE 15–17 Suckers arising as adventitious shoots from the roots of a red raspberry plant. After they are well rooted the suckers may be cut from the parent plant and transplanted to their permanent location.

FIGURE 15–18 Crown division of herbaceous perennials by *(top)* knife with blue fescue (*Festuca ovina* 'Glauca'), *(middle)* hand ax with feather reed grass (*Calamagrostis acutiflora* 'Stricta'), and *(bottom)* handsaw with giant reed *(Arundo donax)*. (Courtesy R. A. Simon.)

Crowns of outdoor herbaceous perennials are usually divided in the spring just before growth begins or in late summer or autumn at the end of the growing season. As a general rule, plants that bloom in the spring and summer and produce new growth after blooming should be divided in the fall. Those that bloom in summer and fall and make little or no new growth until spring should be divided in early spring. Potted plants are divided when they become too large for the particular container in which they are growing. Division is necessary to maintain the variegated form in some plants usually propagated by leaf cuttings, such as *Sanseviera* (*17*).

In crown division, plants are dug and cut into sections with a knife, hand ax, handsaw, or other sharp instrument (Figure 15–18). In herbaceous perennials, such as the Shasta daisy *(Chrysanthemum superbum)* or day lily *(Hemerocallis)* (Figure 15–19), where an abundance of new rooted offshoots are produced from the crown, each may be broken from the old crown and planted separately. The older part of the plant clump is discarded.

For commercial cultivar production, division can be very slow. Shoot production can be increased by cutting back to the crown in the early spring after new growth starts and treating with a cytokinin (*1*). The cytokinin, BA, enhances offset formation in a number of *Hosta* cultivars (*13*). Rapid multiplication is achieved in micropropagation, regenerating plants from callus produced on flower petals and sepals (*18, 31*) (see Chapters 17 and 18).

Herbaceous perennials propagated by division can generally be propagated by micropropagation (*55*). Axillary branching is greatly enhanced and multiplication can be made at high rates with a minimum number of stock plants.

FIGURE 15–19 Propagation by crown division illustrated by day lily *(Hemerocallis)* clump.

REFERENCES

1. Apps, D.A., and C.W. Heuser. 1975. Vegetative propagation of *Hemerocallis*—including tissue culture. *Comb. Proc. Intl. Plant Prop. Soc.* 25:362–67.
2. Barnes, R.D. 1974. Air-layering of grafts to overcome incompatibility problems in propagating old pine trees. *New Zealand J. For. Sci.* 4(2):120–26.
3. Brase, K.D., and R.D. Way. 1959. Rootstocks and methods used for dwarfing fruit trees. *N. Y. Agr. Exp. Sta. Bul.* 783.
4. Broschat, T.K., and H.M. Donselman. 1981. Effects of light intensity, air layering, and water stress on leaf diffusive resistance and incidence of leaf spotting in *Ficus elastica. HortScience* 16(2):211–12.
5. ———. 1983. Effect of wounding method on rooting and water conductivity in four woody species of air-layered foliage plants. *HortScience* 18:445–47.
6. Cameron, R.J. 1968. The leaching of auxin from air layers. *New Zealand J. Bot.* 6(2):237–39.
7. Carlson, R.F., and H.B. Tukey. 1955. Cultural practices in propagating dwarfing rootstocks in Michigan. *Mich. Agr. Exp. Sta. Quart. Bul.* 37: 492–97.
8. Chase, H.H. 1964. Propagation of oriental magnolias by layering. *Comb. Proc. Intl. Plant Prop. Soc.* 14:67–69.
9. Creech, J.L. 1954. Layering. *Nat. Hort. Mag.* 33:37–43.
10. Doud, S.I., and R.F. Carlson. 1977. Effects of etiolation, stem anatomy, and starch reserves on root initiation of layered *Malus* clones. *J. Amer. Soc. Hort. Sci.* 102(4):487–91.
11. Duarte, O., and C. Medina. 1971. Propagation of citrus by improved mound layering. *HortScience* 6:567.
12. Dunn, N.D. 1979. Commercial propagation of fruit tree rootstocks. *Comb. Proc. Intl. Plant Prop. Soc.* 29:187–90.
13. Garner, J.M., G.J. Keever, D.J. Eakes, and J.R. Kessler. 1995. BA-induced offset formation in hosta dependent on cultivar. *Comb. Proc. Intl. Plant Prop. Soc.* 45: In press.
14. Garner, R.J. 1988. *The grafters handbook* (5th ed.). New York: Oxford Univ. Press.
15. Grove, W.R. 1947. Wrapping air layers with rubber plastic. *Proc. Fla. State Hort. Sci.* 60:184–89.
16. Hare, R.C. 1979. Modular air-layering and chemical treatments improve rooting of loblolly pine. *Comb. Proc. Intl. Plant Prop. Soc.* 29:446–54.
17. Henley, R.W. 1979. Tropical foliage plants for propagation. *Comb. Proc Intl. Plant Prop. Soc.* 29:454–67.
18. Heuser, C.W., and J. Harker. 1976. Tissue culture propagation of daylilies. *Comb. Proc. Intl. Plant Prop. Soc.* 25:269–72.
19. Hogue, E.J., and R.L. Granger. 1969. A new method of stool bed layering. *HortScience* 4:29–30.
20. Howard, B.H. 1977. Effects of initial establishment practice on the subsequent productivity of apple stoolbeds. *J. Hort. Sci.* 52:437–46.
21. ———. 1984. The effects of NAA-based Tipoff sprays on apple shoot production in MM.106 stoolbeds. *J. Hort. Sci.* 59:303–11.
22. ———. 1987. Propagation. In *Rootstocks for fruit crops,* R.C. Rom and R.F. Carlson, eds. New York: John Wiley.
23. Howard, B.H., R.S. Harrison-Murray, and S.B. Arjyal. 1985. Responses of apple summer cuttings to severity of stockplant pruning and to stem blanching. *J. Hort. Sci.* 60:145–52.
24. Joiner, J.N., ed. 1981. *Foliage plant production.* Englewood Cliffs, N.J.: Prentice-Hall.

25. Maurer, K.J. 1950. Möglichkeiten der vegetativen Vermehrung der Walnuss. *Schweiz. Z. Obst. V. Weinb.* 59:136–37.
26. McDonald, B. 1986. *Practical woody plant propagation for nursery growers.* Portland, OR: Timber Press.
27. McKensie, R. 1993. Propagation of rootstocks by trench layering. *Comb. Proc. Intl. Plant Prop. Soc.* 43:345–347.
28. Medina, C., and O. Duarte. 1971. Propagating apples in Peru by an improved mound layering method. *Jour. Amer. Soc. Hort. Sci.* 96:150–51.
29. Medina, J.P. 1981. Studies of clonal propagation on pecans at Ica, Peru. *Plant Propagator* 27(2):10–11.
30. Mergen, F. 1955. Air layering of slash pine. *J. For.* 53:265–70.
31. Meyer, M.M. 1976. Propagation of daylilies by tissue culture. *HortScience* 11:485–87.
32. Modlibowska, I., and C.P. Field. 1942. Winter injury to fruit trees by frost in England (1939–1940). *J. Pom. Hort. Sci.* 19:197–207.
33. Munson, R.H. 1982. Containerized layering of *Malus* rootstocks. *Plant Propagator* 28(2):12–14.
34. Nagabhushanam, S., and M.A. Menon. 1980. Propagation of cashew *(Anacardium occidentale* L.*)* by etiolation, girdling and stooling. *Plant Propagator* 26:11–13.
35. Neel, P.L. 1979. Macropropagation of tropical plants as practiced in Florida. *Comb. Proc. Intl. Plant Prop. Soc.* 29:468–80.
36. Nelson, R. 1953. High humidity treatment for air layers of lychee. *Proc. Fla. State Hort. Soc.* 66: 198–99.
37. Nelson, W.L. 1987. Innovations in air layering. *Comb. Proc. Intl. Plant Prop. Soc.* 37:88–89.
38. Poole, R.T., and C.A. Conover. 1988. Vegetative propagation of foliage plants. *Comb. Proc. Intl. Plant Prop. Soc.* 37:503–7.
39. Raviv, M., and O. Reuveni. 1984. Mode of leaf shedding from avocado cuttings and the effect of its delay on rooting. *HortScience* 19:529–31.
40. Rom, R.C. and R.F. Carlson. 1987. *Rootstocks for fruit crops.* New York: John Wiley.
41. Serr, E.F. 1954. Rooting paradox walnut hybrids. *Calif. Agr.* 8(5):7.
42. Sparks, D., and J.W. Chapman. 1970. The effect of indole-3-butyric acid on rooting and survival of air-layered branches of the pecan, *Carya illinoensis* Koch, cv. 'Stuart.' *HortScience* 5(5):445–46.
43. Sutton, N.E. 1954. Marcotting of Persian limes. *Proc. Fla. State Hort. Soc.* 67:219–20.
44. Thomas, L.A. 1938. Stock and scion investigations. II. The propagation of own-rooted apple trees. *J. Counc. Sci. Industr. Res. Org., Austral.* 11:175–79.
45. Tukey, H.B. 1964. *Dwarfed fruit trees.* New York: MacMillan.
46. Tukey, H.B., and K. Brase. 1930. Granulated peat moss in field propagation of apple and quince stocks. *Proc. Amer. Soc. Hort. Sci.* 27:106–13.
47. Vasek, J., and B.H. Howard. 1984. Effects of selective and biennial harvesting on the production of apple stoolbeds. *J. Hort. Sci.* 59:477–85.
48. Vieitez, E. 1974. Vegetative propagation of chestnut. *New Zealand For. Sci.* 4(2):242–52.
49. Vinzant, K.L. 1978. Propagation of *Scheffelera arboricola* by air layering and factors affecting long term storage. *The Plant Propagator* 24(1):5–6.
50. Wells, J.S. 1985. *Plant Propagation Practices.* Chicago, IL: Amer Nurs. Publ. Co.
51. Wells, R. 1986. Air layering: an alternative method for the propagation of *Mahonia aquifolia* 'Compacta.' *Comb. Proc. Intl. Plant Prop. Soc.* 36:97–99.
52. Wyman, D. 1952. Air layering with polyethylene films. *J. Roy, Hort. Soc.* 77:135–40.
53. Wyman, D. 1952. Layering plants in Holland. *Amer. Nurs.* XCV (10).
54. Young, P.J. 1994. Commercial marcotting of fruit trees. *Comb. Proc. Intl. Plant Prop. Soc.* 44:86–89.
55. Zilis, M., D. Zwagerman, D. Lamberts, and L. Kurtz. 1979. Commercial propagation of herbaceous perennials by tissue culture. *Comb. Proc. Intl. Plant Prop. Soc.* 29:404–13.

SUPPLEMENTARY READING

GARNER, R. J. 1988. *The grafter's handbook* (5th ed.). New York: Oxford Univ. Press.

GARNER, R. J., and S. A. CHAUDHRI. 1976. *The propagation of tropical fruit trees.* Hort. Rev. 4. East Malling, Maidstone, Kent: Commonwealth Bureau of Hort. and Plant. Crops.

MCDONALD, B. 1986. *Practical woody plant propagation for nursery growers.* Portland, OR: Timber Press.

MCMILLAN-BROWSE, P. D. A. 1979. *Plant propagation.* New York: Simon and Schuster.

MINISTER OF AGRICULTURE, FISHERIES AND FOOD 1969. *Fruit tree raising-rootstocks and propagation* (5th ed.). Bul. 135. London.

ROM, R. C., and R. F. CARLSON, eds. 1987. *Rootstocks for fruit crops.* New York: John Wiley.

STILL, S. M. 1988. *Manual of herbaceous ornamental plants.* Champaign, Ill.: Stipes Publ. Co.

TUKEY, H. B. 1964. *Dwarfed fruit trees.* New York: Macmillan.

16
Propagation by Specialized Stems and Roots

Bulbs, corms, tubers, tuberous roots and stems, rhizomes, and pseudobulbs are specialized vegetative structures that function primarily in the storage of food, nutrients, and water during adverse environmental conditions. Plants possessing these modified plant parts are generally herbaceous perennials in which the shoots die down at the end of a growing season, and the plant survives in the ground as a dormant, fleshy organ that bears buds to produce new shoots the next season. Collectively, plants that survive as underground storage organs are called **geophytes** (*16, 62*). Such plants are well suited to withstand periods of adverse growing conditions in their yearly growth cycle. The two principal climatic cycles for which such performance is adapted are the warm-cold cycle of the temperate zones and the wet-dry cycle of tropical and subtropical regions (*4, 20*).

It has become common to refer to all geophytes as bulbs regardless of the morphology of the underground organ. Ornamental geophytes are usually discussed as flowering bulbs and are used extensively for cut flowers, pot plants, and for landscaping (*16, 65*). However, for propagation, these specialized organs also function in vegetative reproduction and it is very important to distinguish between the various structures (i.e., bulbs versus corms). The propagation procedure that utilizes the production of naturally detachable structures, such as the bulb and corm, is generally spoken of as **separation**. In cases in which the plant is cut into sections, as is done with the rhizome, stem tuber, and tuberous root, the process is spoken of as **division**. A summary of propagation techniques used for species with specialized organs is presented in Table 16–1.

BULBS

Definition and Structure

A **bulb** is a specialized underground organ consisting of a short, fleshy, stem axis (**basal plate**), bearing at its apex a growing point or a flower primordium enclosed by thick, fleshy scales (see Figure 16–1). Bulbs are mostly produced by monocotyledonous plants in which the usual plant structure is modified

TABLE 16–1

Propagation of representative species with specialized structures

Specialized structure	Distinguishing characteristics	Plant species	Vegetative propagation			
			Separation by offsets	Stem cuttings	Cutting or division of specialized organs	Tissue culture
Bulb	Specialized underground organ produced mainly by monocots. Consists of a short, modified stem surrounded by fleshy leaves (scales) modified for food storage.	Tulip Daffodil Onion Lily Hyacinth	X	X	X	X
Corm	Specialized underground, rounded stem consisting of compacted nodes with lateral buds. Although plants persist for years, corms are replaced each season.	Crocus Gladiolus Liatris Freesia	X		X	X
Tuber	Swollen underground stem modified for food storage with easily distinguished nodes and buds. Tubers are similar to corms except that tubers are lateral structures, while corms are produced directly under the flowering stem.	Potato Caladium Anemone		X	X	X
Tuberous Stem	Flattened, swollen stem produced by the enlargement of the hypocotyl at the root-shoot junction. Tuberous stems are perennial structures and can get quite large.	Tuberous begonia Gloxinia Cyclamen		X	X	X
Tuberous Root	Enlarged fleshy root with shoots produced at one end and roots produced at the other. Tuberous roots are biennial structures.	Dahlia Iris Sweet potato Day lily		X	X	X
Rhizome	Specialized stem that grows horizontally at or just below the soil surface.	Iris Bamboo Lily-of-the-valley			X	X
Pseudobulb	Above ground, enlarged stem with several nodes produced by orchids.	Epiphytic orchids	X			X

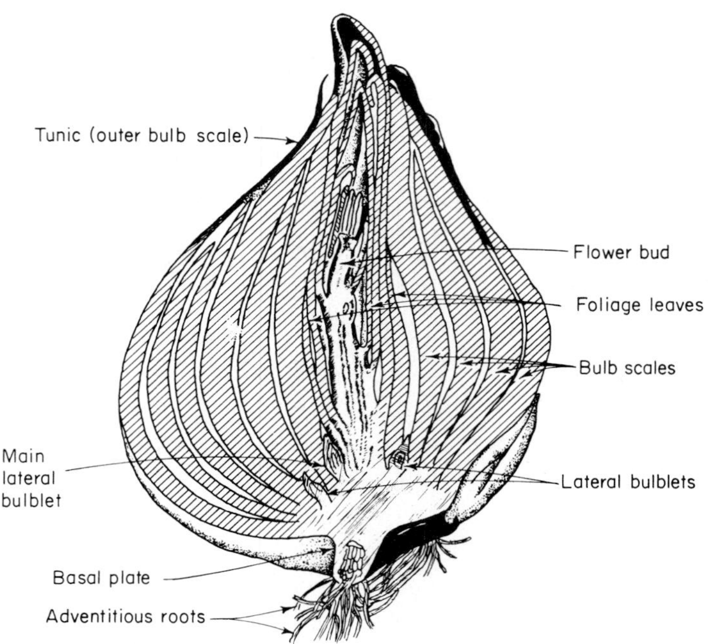

FIGURE 16–1 The structure of a tulip bulb—an example of a **tunicated laminate** bulb. Longitudinal section representing stage of development shortly after the bulb is planted in the fall. (Redrawn from Mulder and Luyten (*54*).)

for storage and reproduction. Oxalis is the one dicot genus producing bulbs (*65*).

Most of the bulb consists of **bulb scales**. The outer bulb scales are generally fleshy and contain reserve food materials, whereas the bulb scales toward the center function less as storage organs and are more leaflike. In the center of the bulb, there is either a vegetative meristem or an unexpanded flowering shoot. Meristems develop in the axil of these scales to produce miniature bulbs, known as **bulblets,** which when grown to full size are known as **offsets**. In various species of lilies, bulblets may form in the leaf axils either on the underground portion or on the aerial portion of the stem. The aerial bulblets are called **bulbils,** while the underground organs are called **stem bulblets**. There are two types of bulbs: **Tunicate** and **nontunicate** (scaly).

Tunicate Bulbs

Tunicate (laminate) bulbs are represented by the onion, garlic, daffodil, and tulip. These bulbs have outer bulb scales that are dry and membranous. This covering, or tunic, provides protection from drying and mechanical injury to the bulb. The fleshy scales are in continuous, concentric layers, or **lamina,** so that the structure is more or less solid (Figure 16–2).

There are three basic bulb structures, defined by type of scales and growth pattern. The amaryllis *(Hippeastrum)* is an example of one type, in which the expanded bases of leaves are used for food storage. These are no scale leaves in this type of bulb (Figure 16–3). Tulip is an example of the second type of bulb. This type of bulb has only true scales, with leaves produced on the flowering or vegetative shoot. A third type, which includes the *Narcissus,* has both expanded leaf bases and true scales (*63*).

Adventitious root primordia are present on the dormant stored bulb. They do not elongate until planted under proper conditions and at the proper time. They occur in a narrow band around the outside edge on the bottom of the basal plate.

Nontunicate Bulbs

Nontunicate (scaly) bulbs are represented by the lily (Figures 16–4 and 16–5). These bulbs do not possess the enveloping dry covering. The scales are separate and attached to the basal plate. In general, nontunicate bulbs are easily damaged and must be handled more carefully than the tunicate bulbs, they must be kept continuously moist because they are injured by drying. In the nontunicate lily bulb new roots are produced in midsummer or later, and persist through the following year (*20, 68*). In most

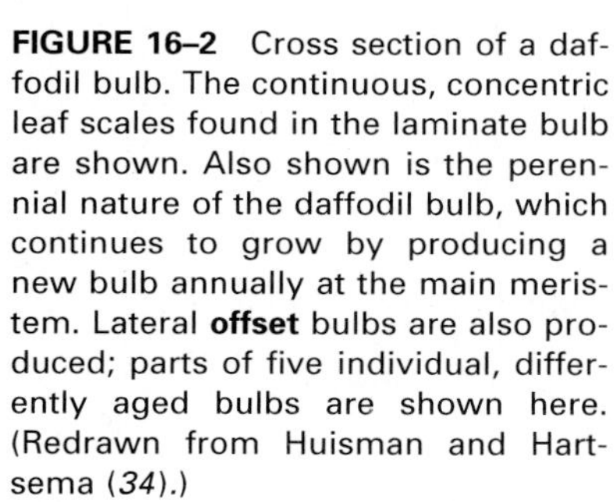

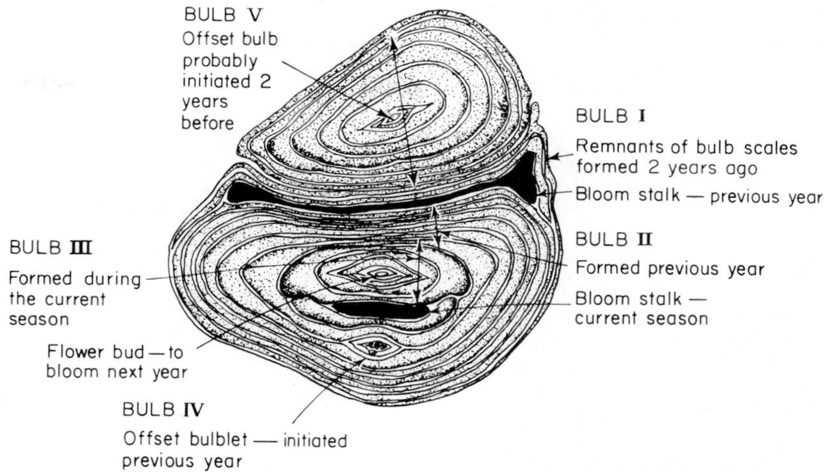

FIGURE 16–2 Cross section of a daffodil bulb. The continuous, concentric leaf scales found in the laminate bulb are shown. Also shown is the perennial nature of the daffodil bulb, which continues to grow by producing a new bulb annually at the main meristem. Lateral **offset** bulbs are also produced; parts of five individual, differently aged bulbs are shown here. (Redrawn from Huisman and Hartsema (*34*).)

lily species, roots also form on the stem above the bulb.

In many species thickened **contractile** roots (Figure 16–6) shorten and pull the bulb to a given level in the ground (*63, 84*). Tulips do not produce contractile roots but produce **droppers** (*16, 63, 65*), stolonlike structures that grow from the bulb and produce a bulb at the tip.

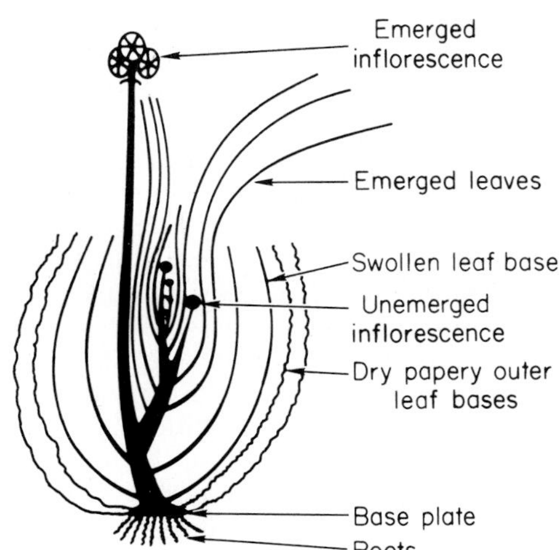

FIGURE 16–3 Diagram showing morphology and growth cycle of *Hippeastrum* bulb. New bulbs continually develop from the center in a cycle of four leaves and an inflorescence. Bases of these leaves enlarge to become the scales that contain stored food. Oldest scales disintegrate. (From A. R. Rees, *The growth of bulbs,* Academic Press, New York, 1972.)

Growth Pattern

An individual bulb goes through a characteristic cycle of development, beginning with its initiation as a meristem and terminating in flowering and seed production. This general developmental cycle is composed of two stages: (a) *vegetative* and (b) *reproductive.* In the vegetative stage, the bulblet grows to flowering size and attains its maximum weight. The subsequent reproductive stage includes the *induction* and *initiation* of flowering, *differentiation* of the floral parts, *elongation* of the flowering shoot, and finally *anthesis* (flowering). Sometimes seed production occurs. Various bulb species have specific environmental requirements for the individual phases of this cycle that determine their seasonal behavior, environmental adaptations, and methods of handling. They can be grouped into classes according to their time of bloom and method of handling.

Spring-Flowering Bulbs

Important commercial crops included in the spring-flowering group are the tulip, daffodil, hyacinth, and bulbous iris, although other kinds are grown in gardens (*16, 65*).

Bulb formation. The vegetative stage begins with the initiation of the bulblet on the basal plate in the axil of a bulb scale. In this initial period, which usually occupies a single growing season, the bulblet is insignificant in size, since it is present within another growing bulb and can be observed only if the bulb is dissected. Its subse-

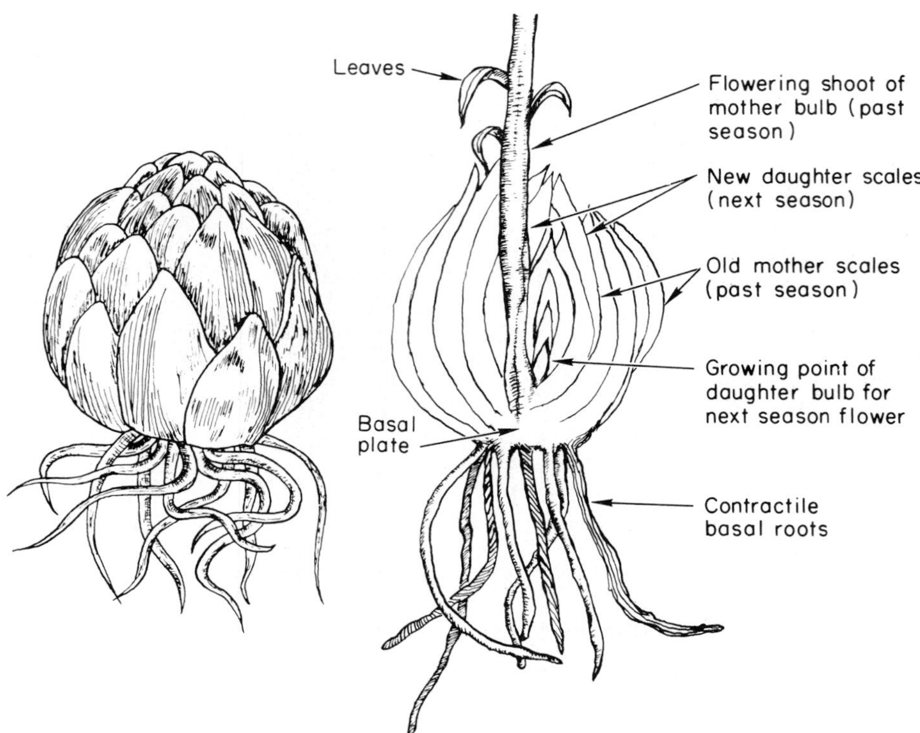

FIGURE 16–4 *Left:* Outer appearance of a scaly bulb of lily *(Lilium hollandicum). Right:* Longitudinal section of a bulb of *L. longiflorum* 'Ace,' after flowering stage, showing old mother bulb scales and new daughter bulb scales. Bulb obtained in fall near digging time (*14*).

quent pattern of development and the time required for the bulblet to attain flowering size differ for different species. The bulbs of the tulip and the bulbous iris, for instance, disintegrate upon flowering and are replaced by a cluster of new bulbs and bulblets initiated the previous season. The largest of these may have attained flowering size at this time, but smaller ones require additional years of growth (see Figures 16–1 and 16–7). The flowering bulb of the daffodil, on the other hand, continues to grow from the center, year by year, producing new offsets, which may remain attached for several years (sees Figures 16–2 and 16–7). The hyacinth bulb also continues to grow year by year, but because the number of offsets produced is limited, artificial methods of propagation are usually used.

The size and quality of the flower are directly related to the size of the bulb. A bulb must reach a certain minimum size to be capable of initiating flower primordia. Commercial value is largely based on bulb size (*2*), although the condition of the bulb and freedom from disease are also important quality factors.

Increase in size and weight of the developing bulb takes place in the period during and (mostly) after flowering, as long as the foliage remains in good condition (*12*). Cultural operations that include irrigation; weed, disease, and insect control; and fertilization encourage vegetative growth. The benefit, however, is to the next year's flower production, because larger bulbs are produced. Conversely, adverse growing conditions, removal of foliage, and premature digging of the bulb results in smaller bulbs and reduced flower production.

Moderately low temperatures tend to prolong the vegetative period, whereas higher temperatures may cause the vegetative stage to cease and the reproductive stage to begin. Thus a shift from cool to warm conditions early in the spring, as occurs in mild climates, will shorten the vegetative period, result in smaller bulbs, and consequently produce inferior blooms the following year (*63*). Commercial bulb-producing areas for hardy spring-flowering

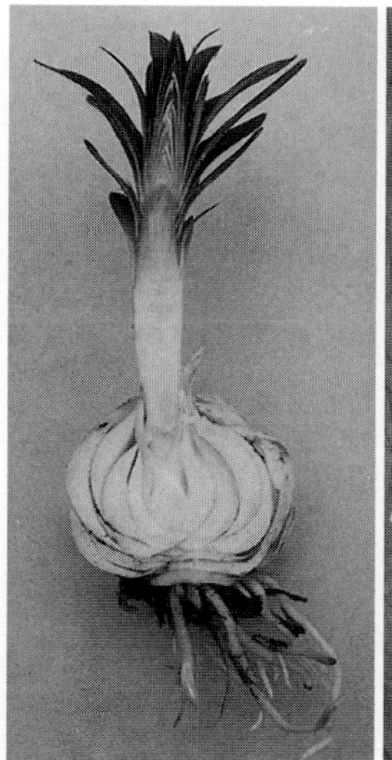

FIGURE 16–5 In the production of a lily crop, producers count leaf emergence from potted bulbs *(left)* (pot and soil removed) and use this as a gauge to speed up or slow down the production cycle, in timing flowering plants *(right)* for certain market periods. (Left photo courtesy A. E. Nightingale.)

FIGURE 16–6 Contractile roots in lily.

bulbs are largely in regions of cool springs and summers, such as the Netherlands and the Pacific Northwest of the United States.

The relative length of photoperiod apparently is not an important factor affecting bulb formation in most species. It has been shown, however, to be significant in some *Allium* species, such as onion and garlic (*33, 44*).

Flower bud formation and flowering. The beginning of the reproductive stage and the end of the vegetative stage is indicated by the drying of the foliage and the maturation of the bulb. From then on, no additional increase in size or weight of the bulb takes place. The roots disintegrate, and the bulb enters a seemingly "dormant" period. However, important internal changes take place, and in some species the vegetative growing point undergoes transition to a flowering shoot. In nature, all bulb activity takes place underground during this period; in horticultural practice, the bulbs are dug, stored, and distributed during this three- to four-month period.

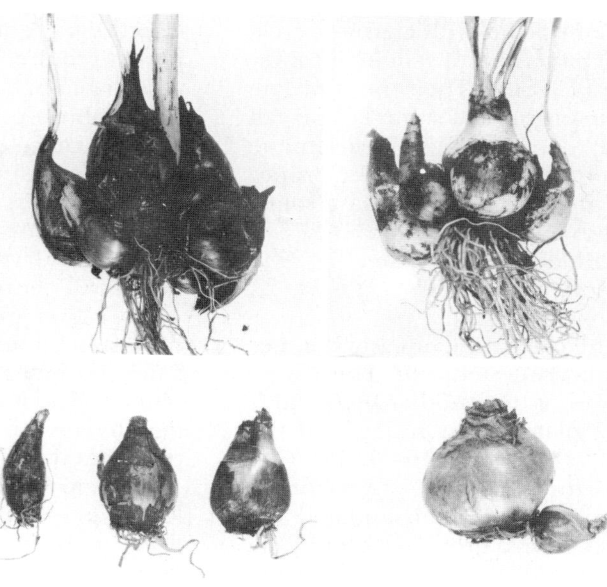

FIGURE 16–7 Propagation by offset bulbs. *Top left:* Bulbous iris. The old bulb disintegrates, leaving a cluster of bulbs. *Top right:* Daffodil. Bulb continues to grow from the inside each year, but continuously produces lateral bulbs, which eventually split away. *Below left:* Daffodil. Three types of bulbs: "split," or "slab," bulb; "round" bulb; and "double-nose" bulb. *Below right:* Hyacinth. Lateral bulblets produced, but old bulb continues to develop from inside.

Temperature controls the progression from the vegetative stage to flowering (*12, 32, 33, 80*). Differentiation of flower primordia for the spring flowering group occurs at moderately warm temperatures in late summer or early fall either in the ground or in storage. Subsequent exposure to lower but above-freezing temperatures is required to promote flower stalk elongation. As temperatures increase in the spring, the flower stalk elongates and the bulb plant subsequently flowers. Bulbous iris is an exception in that flower induction is induced by low temperatures in fall and early winter. For species in groups II (i.e., *Narcissus*) and III (i.e., tulip), induction occurs in spring after the storage period.

The optimum temperature—determined by the shortest time in which the bulb would flower—has been established for these developmental phases for the important bulb species, such as tulip (*12*), hyacinth (*11*), lily (*80*), and daffodil, depending also upon cultivar. Such information is important in establishing forcing schedules for retail florist sales (*9*) (see Figure 16–5). Holding the bulbs continuously at high temperatures (30 to 32° C; 86 to 90° F or more), or at temperatures near freezing, will inhibit or retard floral development and can be used to lengthen the period required for flowering. With a shift to favorable temperatures, flower bud development will continue. This treatment (below or above optimum) can be used when shipping bulbs from northern- to southern-hemisphere countries.

Summer Flowering Bulbs

Lilies are important plants with a growth cycle geared to the seasonal pattern of the summer-winter cycles of the temperate zone. Their nontunicate bulbs do not go "dormant" in late summer and fall as does the tunicate type of bulb, but have unique characteristics that must be understood for proper handling. Although different lily species have somewhat different methods of reproduction (*68, 85*) the pattern, as determined for the Easter lily (*Lilium longiflorum*), can serve as a model (*4, 13, 14, 52, 80*).

Lilies flower in the late spring or early summer at the apex of the leaf-bearing stem axis. The flower-producing bulb is known as the **mother bulb** and is made up of the basal plate, fleshy scales, and the flowering axis (Figures 16–3 and 16–5). Prior to flowering, a new daughter bulb(s) is developing within the mother bulb. It had been initiated the previous fall and winter from a growing point in the axil of a scale at the base of the stem axis. During spring the daughter bulb initiates new scales and leaf primordia at the growing point. Natural chemical inhibitors in the daughter scales prevent elongation of the daughter axis, which remains dormant—but it can be promoted to grow by exposure to high temperature (37.5° C; 100° F), to low temperature (4.5° C; 40° F), or by treatment with gibberellic acid (*79*).

After flowering of the mother bulb, no more scales are produced by the daughter bulb, but it increases in size (circumference) and weight until it equals the weight of the mother bulb surrounding it. Inhibitory effects of the daughter scales decrease, as does the response to dormancy-breaking treatments (*79*). Fleshy basal roots persist on the mother bulb through the fall and winter, and new adventitious roots develop in late summer or fall from the basal plate of the new bulb formed above the mother bulb. Warm temperatures promote root formation (*13*). Bulbs should be dug for transplanting after they "mature" in the fall. The top may or may not have died down. Bulbs should be handled carefully to avoid injury and to prevent drying. The commercial value of the bulb depends on size (transverse circumference) and weight at the time of digging (*42*), and on the condition of the fleshy roots and the freedom of the bulb from disease.

Transition of the meristem to a flowering shoot does not take place until the stem axis has protruded through the "nose" of the bulb and the shoot is *induced* with chilling temperatures (*39*). The critical temperatures are 15.5 to 18.5° C (60 to 65° F) or less, and the chilling effect becomes most effective at 2 to 4.5° C (35 to 40° F). The cold period requirement to induce flowering is known as **vernalization.** The shoot emerges 10 to 15 cm (4 to 6 in.) above ground after chilling and flower primordia start to *develop*. Storing bulbs at warm (21° C; 70° F or more) or low (–0.5° C; 31° F) temperatures will keep bulbs dormant and delay blooming (*75*). Moisture content of the storage medium is important; if too dry, the bulb will deteriorate, and if too wet, it will decay.

Following the flower induction stage, and with the onset of higher temperatures, the stem elongates, initiating first leaves and then flowers. The outer scales of the "old" mother bulb rapidly disintegrate early in the spring as the new mother bulb produces the flower.

Stem bulblets may develop in the axils of the leaves underground or, in some species, bulbils may develop above ground. These appear about the time of flowering.

Flowering time and the size and quality of the Easter lily bloom can be closely regulated by manipulating the temperature at various stages following the digging of bulbs in the fall (*14*). Daylength also influences development, but to a lesser extent (*81*).

Tender, Winter-Flowering Bulbs

There are a number of flowering bulbs from tropical areas whose growth cycle is related to a wet-dry, rather than a cold-warm, climatic cycle. The amaryllis *(Hippeastrum vittata)* is an example (Figure 16–3) (*32, 59*). This bulb is a perennial, growing continuously from the center with the outer scales disintegrating. New leaves are produced continuously from the center during the vegetative period extending from late winter to the following summer. In the axil of every fourth leaf (or scale) that develops, a meristem is initiated. Thus throughout the vegetative period a series of vegetative offsets is produced. By fall, the leaves mature and the bulb becomes dormant, during which time the bulb should be dry. In this period, the *fourth* growing point from the center and any external to it differentiate into flower buds, and the shoot begins to elongate slowly. After two or three months of dry storage, the bulbs can be watered, causing the flowering shoots to elongate rapidly, with flowering taking place in midwinter. Maximum foliage development and bulb growth are essential to produce a bulb large enough to form a flowering shoot.

Propagation

Offsets

Offsets are used to propagate many kinds of bulbs. This method is sufficiently rapid for the commercial production of tulip, daffodil, bulbous iris, and grape hyacinth but, in general, is too slow for the lily, hyacinth, and amaryllis.

If undisturbed, the offsets may remain attached to the mother bulb for several years. They can also be removed at the time the bulbs are dug and replanted into beds or nursery rows to grow into flowering-sized bulbs. This may require several growing seasons, depending upon the kind of bulb and size of the offset.

Tulip. Tulip bulb planting takes place in the fall (*12, 40*). Two systems of planting are used: the bed system, used extensively in Holland, and the row or field system, used mostly in the United States and England. Beds are usually 1 m (3 ft) wide and separated by 31- to 45-cm (12- to 18-in.) paths. The soil is removed to a depth of 9 cm (4 in.), the bulbs set in rows 15 cm (6 in.) apart, and the soil replaced. In the other system, single or double rows are placed wide enough apart to permit the use of machines. To improve drainage, two or three adjoining rows may be planted on a ridge. Bulbs are spaced one to two diameters apart, with small bulbs scattered along the row. A mulch may be applied after planting but removed the following spring before growth.

Planting stock consists principally of those bulbs of the minimum size for flowering, 9 to 10 cm (3.6 to 4 in.) or smaller in circumference. Since the time required to produce flowering size varies with the size of the bulb, the planting stock is graded so that all those of one size can be planted together. For instance, an 8-cm (3.2-in.) or larger bulb normally requires a single season to become flowering size; a 5- to 7-cm (2- to 3-in.) bulb, two seasons; and those 5 cm or less, three years (*12*).

During the flowering and subsequent bulb growing period of the next spring, good growing conditions should be provided so that the size and weight of the new bulbs will be at a maximum. Foliage should not be removed until it dries or matures. Important cultural operations include removal of competing weed growth, irrigation, fungicidal sprays to control *Botrytis* blight (*8, 21*), and fertilization. Beds should be inspected for disease early in the season and for trueness-to-cultivar at the time of blossoming. All diseased or off-type plants should be rogued out (*30*). It is desirable to remove the flower heads at blooming time, because they may serve as a source of *Botrytis* infection and can lower bulb weight.

Bulbs are dug in early to midsummer when the leaves have turned yellow or the outer tunic of the bulb has become dark brown in color. In the Pacific Northwest of the United States, where summer temperatures are cool and the leaves remain green for a longer period, digging may take place before the leaves dry. If the bulbs are dug too early or if warm weather causes early maturation, the bulbs may be small in size. The bulbs are dug by machine or by hand with a short-handled spade. After the loose soil is shaken from the bulbs, they are placed in trays in well-ventilated storage houses for drying, cleaning, sorting, and grading. General storage temperatures are 18 to 20° C (65 to 68° F). To force early flowering, the bulbs should be held at 20° C for three to five weeks and then placed at 9° C (48° F) for eight weeks. Later flowering can be produced by holding bulbs at 22° C (72° F) for ten weeks. For shipment from northern- to southern-hemisphere countries, the bulbs can be held at −1° C (31° F) until late December, when they are shifted to a higher temperature (25.5° C; 78° F) (*11*).

Daffodil. Daffodil bulbs are perennial and produce a new meristem growing point at the center every year (*23, 28*). **Offsets** are produced that grow in size for several years until they break away from the original bulb, although they are still attached at the basal plate. An offset bulb, when it first separates from the mother bulb, is known as a "split," "spoon," or "slab," and can be separated from the mother bulb and planted. Within a year it becomes a "round," or "single-nose," bulb containing a single flower bud. One year later a new offset should be visible, enclosed within the scales of the original bulb, indicating the presence of two flower buds. At this stage the bulb is known as a "double-nose." By the next year the offsets split away, then the bulb is known as a "mother bulb." Grading of daffodil bulbs is principally by age, that is, as splits, round, double-nose, and mother bulbs. The grades marketed commercially are the round and the double-nose bulb. The mother bulbs are used as planting stock to produce additional offsets, and only the surplus is marketed. Offsets, or splits, are replanted for additional growth.

Storage should be at 13 to 16° C (55 to 60° F) with a relative humidity of 75 percent. To force earlier flowering, they can be stored at 9° C (48° F) for eight weeks. To delay flowering, store at 22° C (72° F) for 13 to 15 weeks. For shipment from the northern to the southern hemisphere, the bulbs can be held at 30° C (86° F) until October, then stored at −1° C (31° F) until late December, and then at 25° C (77° F) (*11*).

Hot water treatment plus a fungicide for stem and bulb disease and nematode control is important (*30*). A three- or four-hour treatment at 43° C (110° F) is used, but that temperature must be carefully maintained or the bulbs may be damaged.

Lilies. Lilies increase naturally, but except for a few species this increase is slow and of limited propagation value except in home gardens (*68, 85*). Several methods of bulb increase are found among the different species. For instance, *Lilium concolor, L. hansonii, L. henryi,* and *L. regale* increase by bulb splitting. Two to four lateral bulblets are initiated about the base of the mother bulb, which disintegrates during the process, leaving a tight cluster of new bulbs. *Lilium bulbiferum, L. canadense, L. pardalinum, L. parryi, L. superbum,* and *L. tigrinum* multiply from lateral bulblets produced from the rhizomelike bulb. This process is sometimes called "budding-off."

Bulblet Formation on Stems

Underground stem bulblets are used to propagate the Easter lily *(Lilium longiflorum)* and some other lily species (Figure 16–8). In the field, flowering of the Easter lily occurs in early summer. Bulblets form and increase in size from spring throughout summer (*67*). Between mid-August and mid-September in the northern hemisphere the stems are pulled from the bulbs and stacked upright in the field. Periodic sprinkling keeps the stems and bulblets from drying out. Similarly, the base of the stem can be "heeled in" the ground at an angle of 30 to 45 degrees or laid horizontally in trays at high humidity.

About mid-October the bulblets are planted in the field 10 cm (4 in.) deep and an inch apart in double rows spaced 91 cm (36 in.) apart. Here they remain for the following season. They are dug in September as yearling bulbs and again replanted, this time 12.5 cm (6 in.) deep and 10 to 12.5 cm (4 to 6 in.) apart in single rows. At the end of the second year, they are dug and sold as commercial bulbs.

Digging is done in September after the stem is pulled. The bulbs are graded, packed in peat moss, and shipped. Commercial bulbs range in size from 17.5 to 20 cm (7 to 10 in.) in circumference. Lily bulbs must be handled carefully so they will not be injured and must be kept from drying out. The fleshy roots should also be kept in good condition. For long-term storage that will prevent flowering,

FIGURE 16–8 Two methods of lily propagation. *Left:* Bulblets are produced at the base of individual bulb scales; this method of propagation is possible for nearly all lily species. *Right:* Underground stem bulblets; these are produced only by some lily species.

the bulbs should be packed in polyethylene-lined cases with peat moss at 30 to 50 percent moisture and stored at −1° C (31° F) (*75*).

Control of viruses, fungus diseases, and nematodes during propagation is important in bulb production. Methods of control include using pathogen-free stocks for propagation (*3, 8, 67*), growing plants in pathogen-free locations with good sanitary procedures, and treating the bulbs with fungicides.

Aerial stem bulblets, commonly known as *bulbils,* are formed in the axil of the leaves of some species, such as *Lilium bulbiferum, L. sargentiae, L. sulphureum,* and *L. tigrinum.* Bulbils develop in the early part of the season and fall to the ground several weeks after the plant flowers. They are harvested shortly before they fall naturally and are then handled in essentially the same manner as underground stem bulblets. Increased bulbil production can be induced by disbudding as soon as the flower buds form. Likewise, some lily species that do not form bulbils naturally can be induced to do so by pinching out the flower buds and a week later cutting off the upper half of the stems. Species that respond to the latter procedure include *Lilium candidum, L. chalcedonicum, L. hollandicum, L. maculatum,* and *L. testaceum* (*68*).

Bulbil formation rarely occurs in Easter lily (*L. longiflorum*). Exogenous cytokinin application to foliage induced large numbers of bulbils in leaf axils on above-ground stems (*56*), which might be used for Easter lily propagation.

Stem Cuttings

Lilies may be propagated by stem cuttings. The cutting is made shortly after flowering. Instead of roots and shoots forming on the cutting, as would occur in other plants, bulblets form at the axils of the leaves and then produce roots and small shoots while still on the cutting.

Leaf-bud cuttings, made with a single leaf and a small heel of the old stem, may be used to propagate a number of lily species. A small bulblet will develop in the axil of the leaf. It is handled in the same manner as it would be according to the other methods described here.

Bulblet Formation on Scales (Scaling)

In **scaling,** individual bulb scales are separated from the mother bulb and placed in growing conditions so that adventitious bulblets form at the base of each scale (Figure 16–8). Three to five bulblets will develop from each scale. This method is particularly useful for rapidly building up stocks of a new cultivar or to establish pathogen-free stocks. Almost any lily species can be propagated by scaling (*46, 78*).

In Japan and the United States, commercial-sized Easter lily bulbs are produced after two to three growing seasons from scaling. The normal development of **de novo** bulblet formation during scale propagation is (a) callus formation followed by organized meristem formation, (b) leaf primordia formation, (c) formation of a scale bulblet which enlarges, and (d) leaf emergence from the primordia (*45*).

In the Netherlands, there is a program to produce forcible commercial bulbs after only one growing season from scaling with **preformed** bulblets attached, that is, taking scales which had been induced to form a small bulblet. Basically, there are four types of plant development during detached scale propagation of lilies with preformed bulblets attached (*78*). These types include:

> **ETP:** epigeous-type plant which produces a direct bolting bulblet with foliage leaves and no foliage scales. This is the most desirable development form.
>
> **HETP:** hypoepigeous-type plant which first forms a rosette with foliage scales and bolts at a later stage of growth.
>
> **HTP:** hypogeous-type plant which forms only a rosette with foliage scales—both HTP and HETP are susceptible to diseases and spray damage, since leaves remain close to the soil.
>
> **NLB:** non-green leaf bulblet which does not produce foliage, since the bulbs remain dormant (Figure 16–9). Bulblet formation is influenced by temperature (*78*).

Increasing the duration of bulb storage prior to propagation by bulb scales will decrease the harvest weight of newly generated bulbs (*46*). Outer or middle scales show increased bulb weight and number of forcible commercial bulbs, while innermost scales resulted in low weights and few forcible commercial bulbs.

Scaling is done soon after flowering in midsummer, although it might be done in late fall or even in midwinter. The bulbs are dug, the outer two layers of scales are removed, and the mother bulb is replanted for continued growth. It is possible to remove the scales down to the core, but this will reduce subsequent growth of the mother bulb. The scales should be kept from drying and handled so as

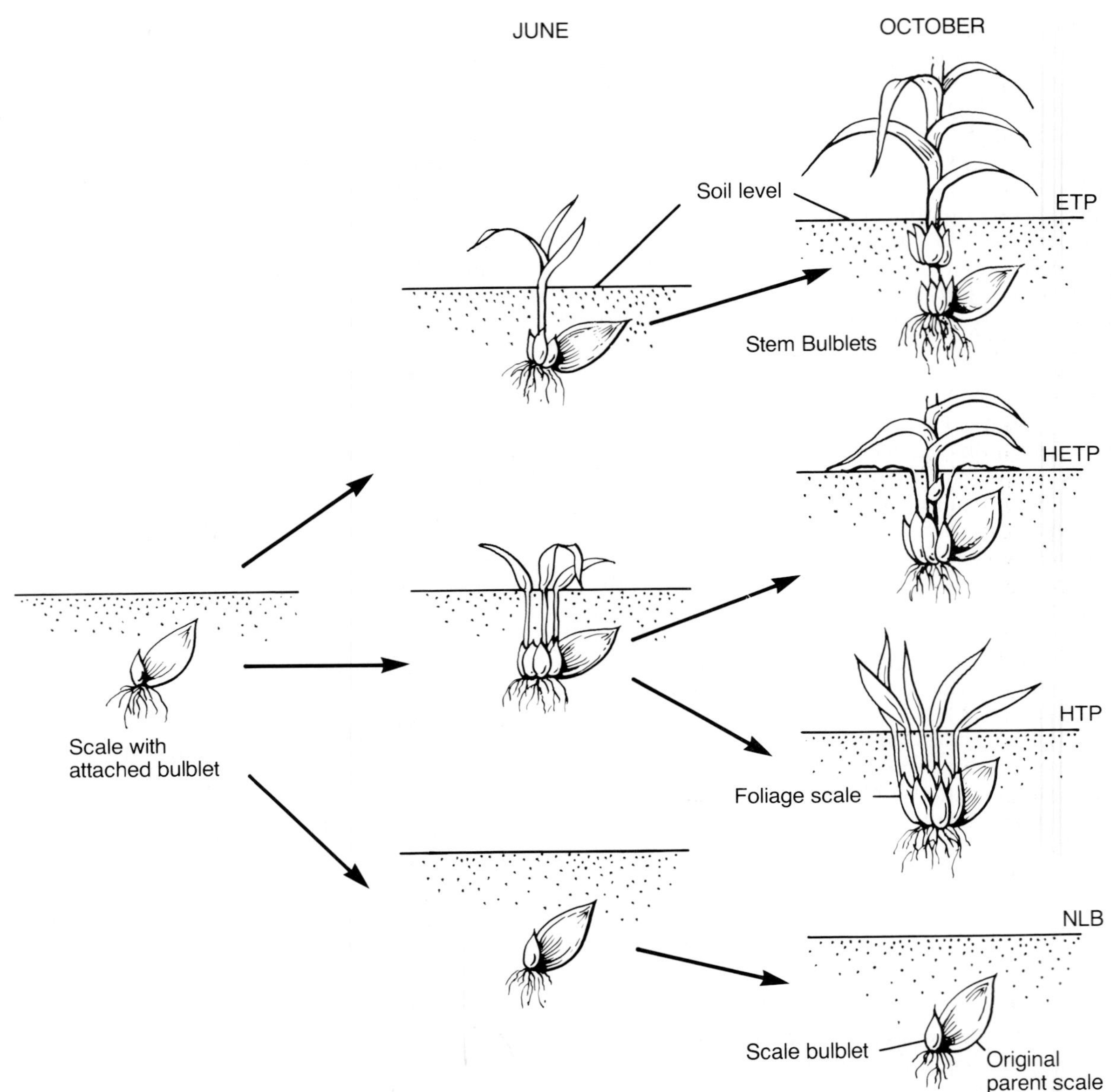

FIGURE 16–9 Types of development during scale propagation with attached bulblets. ETP, epigeous-type plant (bulblet bolts and produces the most desirable form); HETP, hypoepigeous-type plant (rosettes first, then bolts); HTP, hypogeous-type plant (only a rosette with foliage leaves forms—not commercially desirable); NLB, non-green-leaf bulblet (bulblet remains dormant and does not bolt—not desirable). (Redrawn from Van Tuyl (*78*).)

to avoid injury. Scales with evidence of decay should be discarded and the remaining ones dusted with or dipped into a fungicide. Naphthaleneacetic acid (1 ppm) will stimulate bulblet formation.

Scales may be handled by several methods. (a) They may be field planted in beds or frames no more than 6.25 cm (2½ in.) deep. Bulblets form on the scale during the first year to produce **yearlings**. These are replanted for a third year to produce **commercials**. (b) Scales may be placed in trays or flats of moist sand, peat moss, sphagnum moss, or vermiculite for six weeks at 18 to 21° C (65 to 70° F).

The scales are inserted vertically to about half their length. Small bulblets and roots should form at the base within three to six weeks. The scales are transplanted either into the open ground or into pots or flats of soil, and then planted in the field the following spring. Subsequent treatment is the same as described for underground bulblets. (c) Scales may be packed in layers in moist vermiculite in plastic-lined boxes. These are incubated for 6 to 12 weeks at 15 to 26° C (60 to 80° F). The lower temperature will encourage rooting. The boxes are then placed in cold temperature over winter and planted in rows in the spring. Two years is required to reach planting size (*49*).

A simple method of propagating lilies by scales, after removing them from the bulb, is to dust them with a fungicide, then place the scales so they are not touching in damp vermiculite in a polyethylene bag. The bag is closed and tied, then set for six to eight weeks where the temperature is fairly constant at about 21° C (70° F). After bulblets are well developed at the base of the scales, the bag with the scales still inside should be refrigerated at 2 to 4.5° C (35 to 40° F) for at least eight weeks to overcome dormancy. The small bulblets can then be potted and placed in the greenhouse or out-of-doors for further growth.

Tissue culture utilizing aseptic techniques has been used as a means of scaling, particularly to obtain, maintain, and multiply virus-free stock (see Chapters 17 and 18 and Figure 16–10).

Keep in mind that a bulb is a compressed shoot system where apical dominance limits axillary bud growth. If new bulblets are to be formed then apical dominance must be removed using the same principle as in pruning an above-ground shoot structure. An exception to this is daughter bulb formation in the bulbous *Iris* cultivar 'Ideal'. In contrast to certain other bulbous plants, apical dominance does not directly limit daughter bulb number in *Iris,* but does prevent lateral bud sprouting (*17*).

FIGURE 16–10 *Above:* Micropropagation of *Iris hollandica* 'Prof Blaauw' from single- and double-scale explants. Shoots formed on bulb-scale explants *(upper left)* are sliced longitudinally and subcultured. New shoots and bulblets are then regenerated from these shoots *(upper right). Hyacinthus oriental* 'Pink Pearl' forms bulblets *(lower left).* The leaves from the bulblets are subcultured and many plantlets *(lower right)* are regenerated. (Courtesy P. C. G. van der Linde and J. van Aartrijk.)

Basal Cuttage

The hyacinth is the principal plant propagated by this method, although others such as *Scilla* can be handled in this way. Specific methods include "scooping" and "scoring" to remove apical dominance and encourage bulblet formation (*11, 24*). Mature bulbs are used, which have been dug after the foliage has died down and are 17 to 18 cm or more in circumference. In **scooping,** the entire basal plate is scooped out with a special curve-bladed scalpel, a round-bowled spoon, or a small-bladed knife. Adventitious bulblets develop from the base of the exposed bulb scales. Depth of cutting should be enough to destroy the main shoot. In **scoring,** three straight knife cuts are made across the base of the bulb, as shown in Figure 16–11, each deep enough to go through the basal plate and the growing point. Growing points in the axils of the bulb scales grow into bulblets.

To combat decay that may develop during the later incubation period, infected bulbs should be discarded; the tools disinfected frequently with alcohol, Physan, formalin, or mild carbolic acid solution; and the cut bulbs dusted with a fungicide.

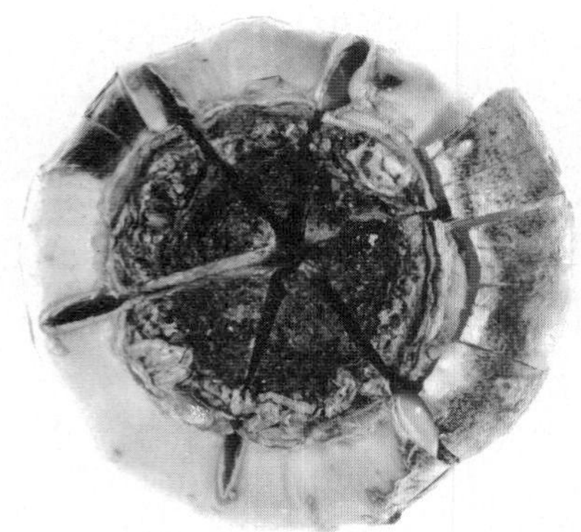

FIGURE 16–11 Basal cuttage in a *Hyacinth* bulb which has been scored. Note the bulblets starting to appear.

Most important is to callus the bulbs at about 21° C (70° F) for a few days to a few weeks in dry sand or soil or in open trays, cut side down. After callusing, the bulbs are incubated in trays or flats, in dark or diffuse light, at 21° C (70° F), which is increased to 29.5 to 32° C (85 to 90° F) over a two-week period and held at high humidity (85 percent) for 2½ to 3 months.

The mother bulbs are planted about 10 cm (4 in.) deep in nursery beds in the fall. The next spring, bulblets produce leaves profusely. Normally, the mother bulb disintegrates during the first summer. Annual digging and replanting of the graded bulblets is required until they reach flowering sizes. Bulbs for greenhouse forcing should be 17 cm (6¾ in.) or more in circumference; bulbs for bedding should be 14 to 17 cm (5½ to 6¾ in.) in circumference (*2*). On the average, a scooped bulb will produce 60 bulblets, but four to five years will be required to produce flowering sizes; and a scored bulb will produce 24, requiring three to four years (*11*).

Hot-water treatment of hyacinth bulbs used for controlling *Xanthomonas hyacinthi* has been reported to induce bulblet formation and could substitute for basal cuttage (*1*). Bulbs must be at least 1.5 cm in circumference and treated from mid-July to early September. Treatment is 43° C (110° F) for 4 days or 38° C (100° F) for 30 days at relative humidity of 60 to 70 percent.

Leaf Cuttings

This method is successful for blood lily *(Haemanthus),* grape hyacinth *(Muscari),* hyacinth, and Cape cowslip *(Lachenalia)* (*18*), although the range of species is probably wider.

Leaves are taken at a time when they are well developed and green. An entire leaf is cut from the top of the bulb and may in turn be cut into two or three pieces. Each section is placed in a rooting medium with the basal end several inches below the surface, as described for rooting cuttings. The leaves should not be allowed to dry out, and bottom heat is desirable. Within two to four weeks small bulblets form on the base of the leaf and roots develop. At this stage the bulblets are planted in soil.

Bulb Cuttings

Among plants that respond to the bulb-cutting method of propagation are the *Albuca, Chasmanthe, Cooperia, Haemanthus, Hippeastrum, Hymenocallis, Lycoris, Narcissus, Nerine, Pancratium, Scilla, Sprekelia,* and *Urceolina* (*18*).

A mature bulb is cut into a series of eight to ten vertical sections, each containing a part of the basal plate. These sections are further divided by sliding a knife down between each third or fourth pair of concentric scale rings and cutting through the basal plate. Each of these fractions makes up a bulb cutting consisting of a piece of basal plate and segments of three or four scales. This technique is also referred to as **bulb chipping** or **fractional scale-stem cuttage** (*9*).

The bulb cuttings are planted vertically in a rooting medium, such as peat moss and sand, with just their tips showing above the surface. The subsequent technique of handling is the same as for ordinary leaf cuttings. A moderately warm temperature, slightly higher than for mature bulbs of that kind, is required. New bulblets develop from the basal plate between the bulb scales within a few weeks, along with new roots. At this time they are transferred to flats of soil to continue development.

A variation of this method, called **twin scaling** (*64, 72*), involves dividing bulbs into portions, each containing a pair of bulb scales and piece of basal plate. These are kept in plastic bags and incubated with damp vermiculite three to four weeks at 21° C (70° F) or planted in compost. Bulbils then develop at the edge of the basal plate (Figure 16–12) (*29*).

Twin-scale types most likely to produce a bulbil consisting of a third-year leaf base plus a second-year bulb scale, or scales and (or) leaf bases. Types with first-year organs or a flower stalk are less productive but can still form bulbils (*27*). Twin scaling and tissue culture remain the two most important multiplication programs for *Narcissus.*

FIGURE 16–12 Twin-scale propagation of *Narcissus* (daffodil) cut from whole bulbs and kept at 23° C in moist vermiculite for two months. (From A. R. Rees, *The growth of bulbs,* Academic Press, New York, 1972.)

Micropropagation by Tissue Culture

Many bulb species are highly adapted to micropropagation techniques, utilizing enhanced axillary shoot formation, adventitious shoot formation, bulblet induction on scales, or very often, flower scapes (Table 16–2 and Figure 16–10). These methods are especially valuable to multiply new cultivars rapidly and to develop, maintain, and produce the initial stock of specific virus-tested propagation material. These methods are described in Chapters 17 and 18.

TABLE 16–2

Partial list of bulbous crops capable of tissue culture propagation and their explant source

		Explant Source				
Family	***Genus***	***Bulb Scale***	***Leaf***	***Stem***	***Bud***	***Flower Petal***
Liliaceae	*Lilium*	x	x	x	x	x
	Tulipa	x		x	x	x
	Hyacinthus	x	x	x	x	x
	Ornithogalum	x			x	
	Muscari	x	x			x
	Fritillaria	x				
	Lachenalia		x			
	Alstroemeria	*(Rhizome explants)*				
Iridaceae	*Iris*	x		x	x	
	Gladiolus			x	x	x
	Freesia			x		
	Crocus				x	x
Amaryllidaceae	*Narcissus*	x	x	x		
	Hippeastrum	x		x		
	Nerine	x		x		

Source: Compiled from J. Van Aartrijk and P. C. C. Van der Linde (*76*) and DeHertogh and Le Nard (*16*).

CORMS

Definition and Structure

A **corm** is the swollen base of a stem axis enclosed by the dry, scale-like leaves. In contrast to the bulb, which is predominantly leaf scales, a corm is a solid stem structure with distinct nodes and internodes. The bulk of the corm consists of storage tissue composed of parenchyma cells. In the mature corm, the dry leaf bases persist at each of these nodes and enclose the corm. This covering, known as the **tunic,** protects it against injury and water loss. At the apex of the corm is a terminal shoot that will develop into the leaves and the flowering shoot. Axillary buds are produced at each of the nodes. In a large corm, several of the upper buds may develop into flowering shoots, but those nearer the base of the corm are generally inhibited from growing. However, should something prevent the main buds from growing, these lateral buds would be capable of producing a shoot (see Figures 16–13 and 16–14).

Two types of roots are produced from the corm: a *fibrous* root system developing from the base of the mother corm, and enlarged, fleshy *contractile* roots developing from the base of the new corm. The latter roots apparently develop in response to the fluctuating temperatures near the soil surface and with exposure of the leaves to light. At lower soil depths temperature fluctuations decrease (*36*), and contraction ceases once the corm is at a given depth.

Growth Pattern

Gladiolus and crocus are typical cormous plants (*5, 10*). The gladiolus is semi-hardy to tender and, in areas with severe winters, the corm must be stored over winter and replanted in the spring. At the time of planting, the corm is a vegetative structure (*31, 60*). New roots develop from its base, and one or more of the buds begins to develop leaves. Floral initiation takes place within a few weeks after the shoot begins to grow. At the same time the base of the shoot axis thickens, and a new corm for the succeeding year begins to form above the old corm. Stolonlike structures bearing miniature corms or **cormels** on their tip develop from the base of the new corm.

In gladiolus there can be competition for assimilates between flower and corm development, which is controlled by photoperiod (*70*). Short-day (SD) conditions stimulate corm development, while corms are checked under long-days (LD), when competition with flower development and anthesis occurs. However, plant size is related to size of the corm, since final corm weight is greater under LD conditions (larger plants) than SD (smaller plants), which apparently is due to more total assimilates being available in the larger LD plants for corm production (*56*).

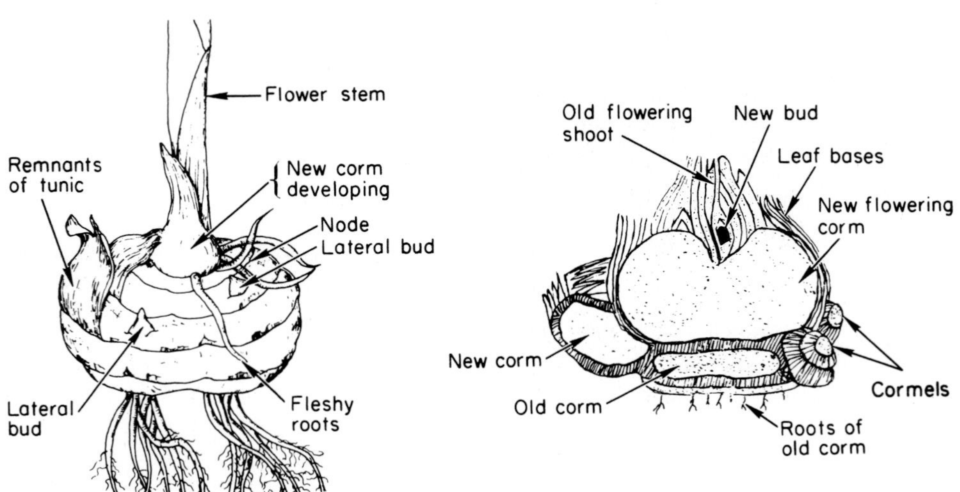

FIGURE 16–13 Gladiolus corm. *Left:* External appearance. *Right:* Longitudinal section showing solid stem structure.

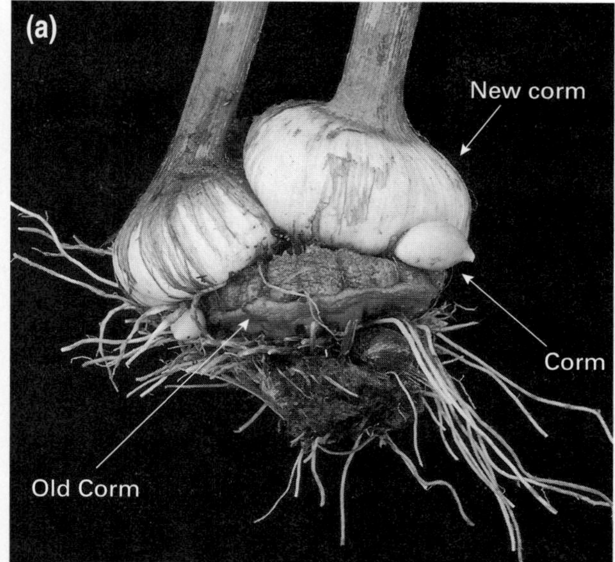

FIGURE 16–14 Stage of gladioulus corm development during the latter part of the growing season. The remnants of the originally planted corm (see arrows) are evident just below the newly formed corm. Many small white cormels also have been produced (Arrow 16-14b).

As new corms continue to enlarge, the old corm begins to shrivel and disintegrate as its contents are utilized in flower production. After flowering, the foliage continues to manufacture food materials, which are stored in the new corm. At the end of the summer, when the foliage dries, there are one or more new corms and perhaps a great number of little cormels. The corms are dug and stored over winter until they are planted the following spring.

Propagation

New Corms

Propagation of cormous plants is principally by the natural increase of new corms. Flower production in corms, as in bulbs, depends upon food materials stored in the corm the previous season, particularly during the period following bloom. In gladiolus, cool nights and long growing periods are favorable for production of very large corms. Fertilization and other good management practices during bloom have their greatest effect on the next year's flowers. Plants are left in the ground for two months following blooming, or until frost kills the tops. After digging, the plants are placed in trays with a screen or slat bottom arranged to allow air to circulate between them, and cured at about 32° C (90° F) at 80 to 85 percent relative humidity. A few hours at 35° C (95° F) may be helpful. Then the new corms, old corms, cormels, and tops can be easily separated. The corms are graded according to size, sorted to remove the diseased ones, treated with a fungicide, and returned to a 35° C (95° F) temperature for an additional week. This curing process suberizes the wounds and helps combat *Fusarium* infection. The corms are then stored at 5° C (40° F) with a relative humidity of 70 to 80 percent in well-aerated rooms to prevent excessive drying. It may also be desirable to treat them with a suitable fungicide (*43*) immediately before planting.

Cormels

Cormels are **miniature** corms that develop between the old and the new corms. One or two years' growth is required for them to reach flowering size. Shallow planting of the corms, only a few inches deep, results in greater production of cormels; increasing the depth of planting reduces cormel production.

Cormels are separated from the mother corms and stored over winter for planting in the spring. Dry cormels become very hard and may be slow to start growth the following spring, but if they are stored at about 5° C (40° F) in slightly moist peat moss, they will stay plump and in good condition. Soaking dry cormels in cool running water for one to two days and holding them moist until planting at first sign of root development will hasten the onset of growth.

Disease-free cormels can be obtained by hot-water treatments, which should be done between two and four months after digging (*8*). Holding cormels at room temperatures to keep them dormant will increase tolerance to this treatment. The cormels are soaked in water at air temperature for two days, then placed in a 1:200 dilution of commercial 37 percent formaldehyde for four hours, and then immersed in a water bath at 57° C (135° F) for 30 minutes. At the end of the treatment, the cormels are cooled quickly, dried immediately, and stored at 5° C (40° F) in a clean area with good air circulation.

The cormels are planted in the field in furrows about 5 cm (2 in.) deep in the manner of planting large seeds. Only grasslike foliage is produced the first season. The cormel does not increase in size but produces a new corm from the base of the stem axis, in the manner described for full-sized corms. At the end of the first growing season, the beds are dug and the corms separated by size. A few of the corms may attain flowering size, but most require an additional year of growth. Size grades in gladiolus are determined by diameter. There are seven grades, the smallest 0.9 to 1.2 cm (⅛ to ½ in.) in diameter, the largest 5 cm (2 in.) or more (*2*).

Division of the Corm

Large corms can be cut into sections, retaining a bud with each section. Each of these should then develop a new corm. Segments should be dusted with a fungicide because of the great likelihood of decay of the exposed surfaces (*47*).

TUBERS

Definition and Structure

A **tuber** is a special kind of swollen, modified stem structure that functions as an underground storage organ (Figure 16–15). The potato *(Solanum tuberosum)* is a notable example of a tuber-producing plant, as is the *Caladium,* grown for its striking foliage, and the Jerusalem artichoke *(Helianthus tuberosus)*.

A tuber has all the parts of a typical stem but is very much swollen. Externally the **eyes,** present in regular order over the surface, represent nodes, each consisting of one or more small buds subtended by a leaf scar. The arrangement of the nodes is a spiral, beginning with the terminal bud on the end opposite the scar resulting from the attachment to the stolon. The terminal bud is at the apical end of the tuber, oriented farthest (distally) from the crown of the plant. Consequently, tubers show the same apical dominance as any stem. Apical dominance also occurs with tuber pieces of yam *(Dioscorea alta)*. Vine production is much greater with proximal than with distal tuber sections (*61*).

FIGURE 16–15 Tubers of white (Irish) potato showing their development from stolons arising from stem tissue. Note the adventitious root system originating from the main plant stem. The tuber is attached to stolon at the tuber's morphological basal (proximal) end.

Internally, a potato tuber is composed of enlarged parenchyma-type cells containing large amounts of starch. It has the same internal structure as any stem with pith, vascular areas, and cortex.

Growth Patterns

The tuber is a storage structure that is produced in one growing season, remains dormant during the winter, and then functions to regenerate new shoots the following spring. After a new seasonal cycle begins, the shoots utilize the stored food in the old tuber, which then disintegrates (*6, 22, 73*). As the main shoot develops, adventitious roots are initiated at the base, and lateral buds grow out horizontally into the soil to produce elongated, etiolated stems *(stolons)* as shown in Figure 16–15. Continued elongation of the stolon takes place during long photoperiods and is associated with the presence of auxin and a high gibberellin level. Tuberization begins with inhibition of terminal growth and the initiation of cell enlargement and division in the subapical region of the stolons. This process is associated with short or intermediate daylengths, reduced temperatures (particularly at night), high light intensity, low mineral content, increased cytokinins and inhibitors (ABA), and a reduction in gibberellin levels in the plant (*51*).

Tuberization is caused by the production of a tuber-inducing substance which is linked to a tuberization regulatory protein that is produced in the leaves and the mother tuber (*19, 57*). It seems to be necessary for the stolon tip to have attained a particular physiological age. Continued tuber enlargement is dependent on a continuing adequate supply of photosynthate. Conditions that favor rapid and luxurious plant growth above ground, such as an abundance of nitrogen, or high temperatures, are not conducive to tuber production. In the fall, the tops of the plants die down and the tubers are dug. At this time, the buds of potato tubers are dormant for six to eight weeks. This condition must disappear before sprouting will take place.

FIGURE 16–16 "Seed" potato, which in fact is a vegetative propagule made of a diced tuber. The shoots form from eyes (axillary buds) on the tuber and the root system occurs from the newly elongating shoot. Potatoes can be commercially produced from true seed, but the production period is longer and the clone can be lost.

Propagation

Division

Propagation by tubers can be done either by planting the tubers whole or by cutting them into sections, each containing one or more axillary buds or "eyes." These small pieces of tuber to be used for propagation are commonly referred to as "seed" potatoes (Figure 16–16). The weight of the tuber piece should be 28 to 56 g (1 to 2 oz) to provide sufficient stored food for the new plant to become well established, although in some areas only the "eyes" with a very small piece of tuber are sold and planted.

Division of tubers is done by machine or with a sharp knife shortly before planting. The cut pieces should be stored at warm (20° C; 68° F) temperatures and relatively high humidities (90 percent) for two to three days prior to planting. During this time the cut surfaces heal and become **suberized,** which protects the "seed" piece against drying and decay. Treatment of potato tubers prior to cutting for the control of *Rhizoctonia* and scab may be desirable.

Caladium tubers are produced commercially in Florida (*69, 83*). The tubers are cut into sections, usually two buds per piece. These are planted 7.6 to 9 cm (3 to 4 in.) deep, 9 to 15 cm (4 to 6 in.) apart in rows 45 to 60 cm (18 to 24 in.) apart. Harvest begins in November. After harvest, the tubers are dried in open sheds for six weeks or artificially dried for 48 hours. Further storage should be at temperatures above 16° C (60° F).

Tubercles

Begonia evansiana and the cinnamon vine *(Dioscorea batatas)* produce small aerial tubers, known as **tubercles,** in the axils of the leaves. These tubercles may be removed in the fall, stored over winter, and planted in the spring (*18*). Short days induce tuberization (*57*).

TUBEROUS ROOTS AND STEMS

Definition and Structure

The tuberous root and stem class includes several types of structures with thickened tuberous growth that function as storage organs. Botanically, these differ from true tubers, although common horticultural usage sometimes utilizes the term "tuber" for all of them.

Fleshy and Tuberous Roots

Various herbaceous perennial species show massive enlargement of secondary roots. Typical examples are sweet potato *(Ipomoea batatus)* (Figures 16–17 and 16–18), cassava *(Manihot esculenta),* and *Dahlia* (Figures 16–17 and 16–19). Sweet potato has a **fleshy root** from which both adventitious buds and roots are produced, while dahlia has a **tuberous root** with a section of the attached crown containing a preformed bud for shoot development. With dahlia, fibrous roots are commonly produced on the opposite *(distal)* end, and tuberous roots closer to the crown or stem *(proximal)* end.

Tuberous Stems

Tuberous stems are produced by the enlargement of the hypocotyl section of the seedling plant, but may include the first nodes of the epicotyl and the upper section of the primary root (*25, 35, 66*). Typical plants with this structure are the tuberous begonia *(Begonia × tuberhybrida)* (*26*) and cyclamen *(Cyclamen persicum)*. These structures have a vertical orientation with one or more vegetative buds produced on the upper end or crown. Fibrous roots are produced on the basal part of the structure.

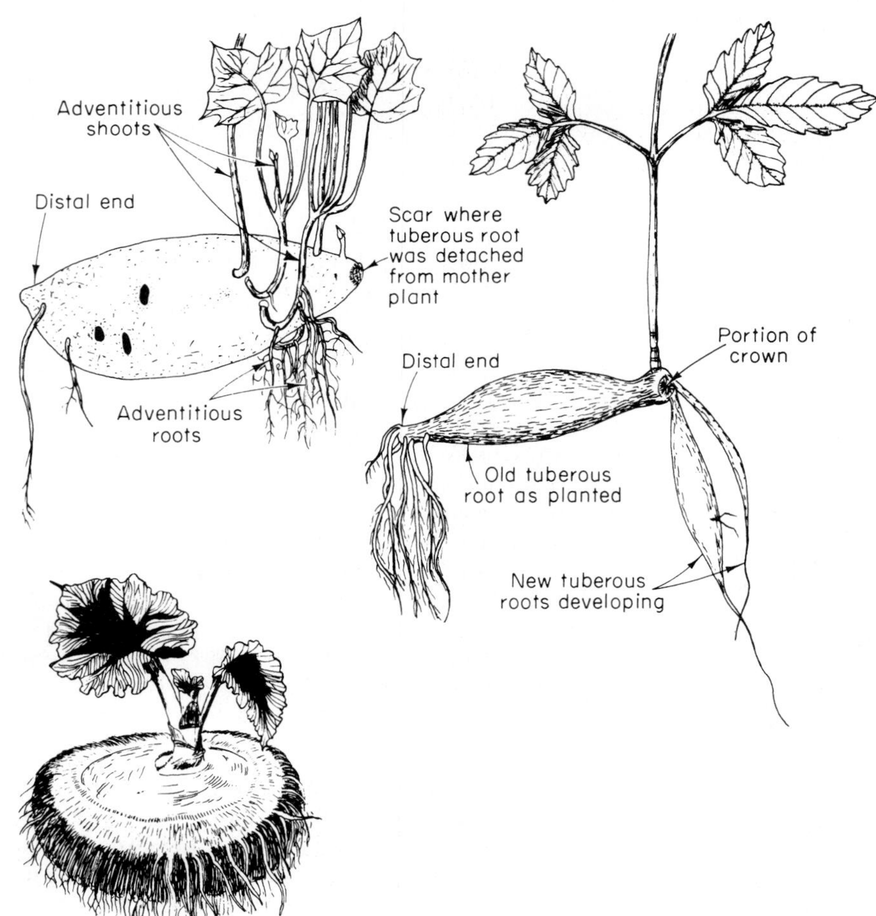

FIGURE 16–17 Types of fleshy and tuberous roots and shoots. *Top left:* Sweet potato fleshy root showing adventitious shoots. *Top right:* Dahlia during early stages of growth. The old tuberous root piece will disintegrate in the production of the new plant; the new roots can be used for propagation. *Below left:* A tuberous begonia stem, showing its vertical orientation. This type continues to enlarge each year.

FIGURE 16–18 Propagation by sweet potato fleshy roots. Adventitious shoots ("slips") develop when the mother root is placed under warm, moist conditions. *Left:* Root on the left has been subjected to 34.5° C (100° F) for 26 hours to overcome the proximal dominance shown on the unheated root on the right. (Courtesy Welch and Little (*82*).) *Right:* After slips are well rooted, they are removed and planted.

FIGURE 16–19 Propagation of dahlia. To produce a new plant each separate tuberous root must have a section of the crown bearing a shoot bud, as shown by the detached root on the left.

Growth Pattern

Tuberous roots are biennial. They are produced in one season, after which they go dormant as the herbaceous shoots die. These function as storage organs to allow the plant to survive the dormant period. In the following spring, buds from the crown produce new shoots, which utilize the food materials from the old root during their initial growth. The old root then disintegrates, and new tuberous roots are produced, which in turn maintain the plant through the following dormant period (*22, 41, 53*).

Photoperiod, not temperature, is the dominant controlling factor of tuberization in dahlia (*15*). Tuberized roots are formed under short-day conditions (five inductive cycles of an 11- to 12-hour critical photoperiod) or when a growth retardant is applied (*53*). Fibrous roots form under long-days or when gibberellic acid is applied. Apparently, conditions favoring tuberous root growth are antagonistic to vegetative (shoot) growth (*53*). Tuberous root formation in cassava is attributed to cytokinin control of meristematic activity (*50*).

Tuberous stems of tuberous begonia and cyclamen, on the other hand, are perennial and continue to enlarge laterally every year (*15, 25*). Normally, these species are propagated by seed but the "tuber" can be dug, stored, and used for annual propagation over a period of years.

Propagation

Division

The usual method for propagating tuberous roots is by dividing the crown so that each section bears a shoot bud. Dahlia, for example, is dug with its cluster of roots intact, dried for a few days, and stored at 4 to 10° C (40 to 50° F) in sawdust or vermiculite. Open storage may result in shriveling. The root cluster is divided in the late winter or spring shortly before planting. In warm, moist conditions the buds begin to grow, and the tubers can be divided with assurance that each section will have a bud.

The tuberous stem of the tuberous begonia can be divided shortly after growth starts in the spring as long as each section has a bud. To combat decay, the cut surface should be dusted with a fungicide and each section dried for several days after cutting and before placing in a moist medium.

Adventitious Shoots

The fleshy roots of a few species of plants such as sweet potato have the capacity to produce adventitious shoots if subjected to the proper conditions. The roots are laid in sand so that they do not touch one another and are covered to a depth of about 5 cm (2 in.). The bed is kept moist. The temperature should be about 27° C (80° F) at the beginning and about 21 to 24° C (70 to 75° F) after sprouting has started. As the new shoots, or **slips,** come through the covering, more sand is added so that eventually the stems will be covered for 10 to 12.5 cm (4 to 5 in.). Adventitious roots develop from the base of these adventitious shoots. After the slips are well rooted, they are pulled from the parent plant and transplanted into the field. If sweet potato roots are cut in half and the pieces are subjected to 43° C (110° F) for about 26 hours, slip production increases. This treatment overcomes the apical dominance and also controls nematodes and fungus diseases (*82*) (see Figure 16–18). This procedure can be modified in certain cultivars of the sweet potato by dividing the fleshy root into 20- to 25-g pieces, treating with a fungicide, then giving a presprouting treatment for four weeks of 26.5° C (80° F) and 90 percent relative humidity before planting (*7*).

Cyclamen can be multiplied vegetatively by cutting off the upper one-third of the tuberous stem and notching the surface into 1-cm squares. Adventitious shoots develop (12 to 13 per tuber) and can be used for propagation (*55*).

Stem Cuttings

Vegetative propagation in plants of this group, such as dahlia or tuberous begonia, is often more satisfactory with stem, leaf, or leaf-bud cuttings. The cuttings will develop tuberous roots at their base. This process can be stimulated if the stem cutting initially includes a small piece of the fleshy root or stem. Vine cuttings from established beds also can be used in sweet potato propagation.

RHIZOMES

Definition and Structure

A **rhizome** is a specialized stem structure in which the main axis of the plant grows horizontally at or just below the ground surface. A number of economically important plants, such as bamboo, sugar cane, banana, and many grasses, as well as a number of ornamentals, such as rhizomatous *Iris* and lily-of-the-valley, have rhizome structures. Most are monocotyledons, although a few dicotyledons—for example, low-bush blueberry *(Vaccinium angustifolium)*—have analogous underground stems classed as rhizomes. Many ferns and lower plant groups have rhizomes or rhizome-like structures.

Figure 16–20 shows structural features of a rhizome (*71*). The stem appears segmented because it is composed of nodes and internodes. A leaflike *sheath* is attached at each node; it encloses the stem and, in an expanded form, becomes the foliage leaves. When the leaves and sheaths disintegrate, a scar is left at the point of attachment identifying the node and giving a segmented appearance. Adventitious roots and lateral growing points develop in the vicinity of the node. Upright-growing, above-ground shoots and flowering stems *(culms)* are produced either terminally from the rhizome tip or from lateral branches.

Two general types of rhizomes are found (*48*). The first (the **pachymorph**) is illustrated by rhizomatous *Iris* in Figure 16–21 and by ginger in Figure 16–22. The rhizome is thick, fleshy, and shortened in relation to length. It appears as a many-branched

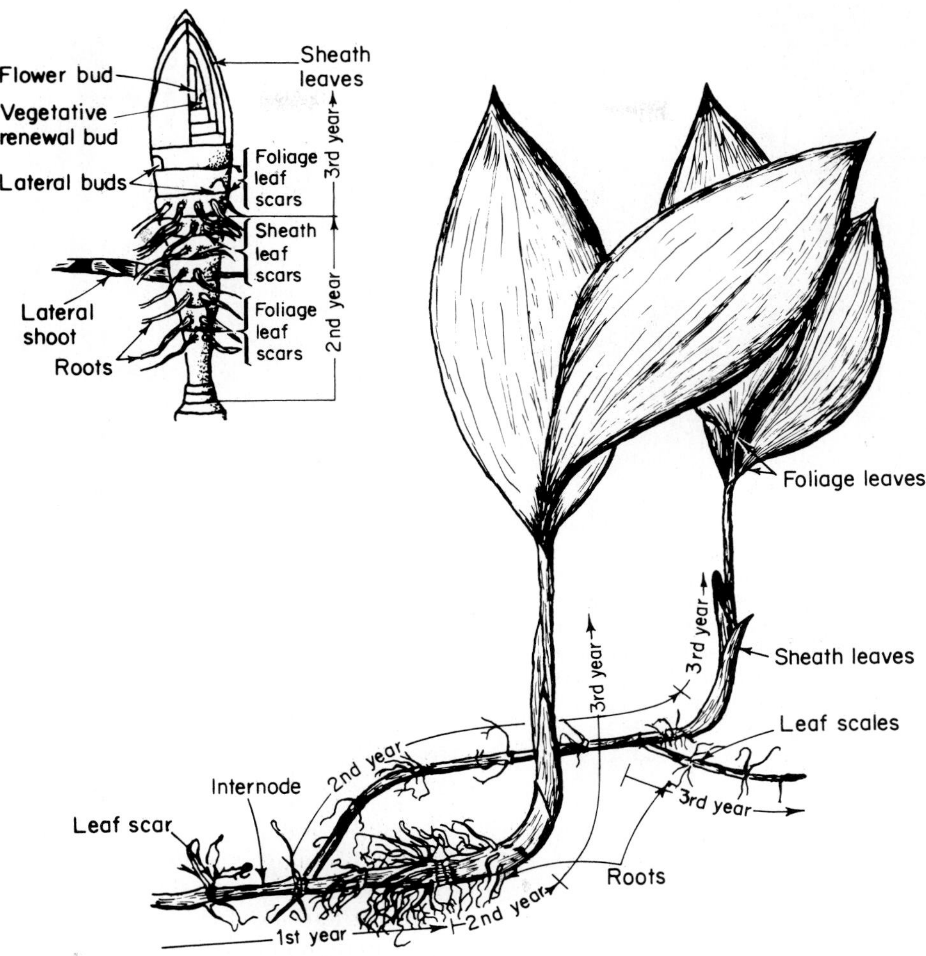

FIGURE 16–20 Structure growth cycle of lily-of-the-valley *(Convallaria majalis). Right:* Section of rhizome as it appears in late spring or early summer with one-, two-, or three-year-old branches. A new rhizome branch begins to elongate in early spring and terminates in a vegetative shoot bud by the fall. The following spring, leaves of the bud unfold; food materials manufactured in the leaves by photosynthesis are accumulated in the rhizome. Growth the second season is again vegetative. Early in the third season a flower bud begins to form, and at the same time a vegetative growing point forms in the axil of the last leaf. *Top left:* Section of the three-year-old branch showing terminal flower bud and lateral shoot bud enclosed in leaf sheaths. Such a section is sometimes known as a **pip** or **crown** and is forced for spring bloom. In the early spring the flowering shoot expands, blooms, and then dies down, and the shoot bud begins a new cycle of development. (Redrawn from Zweede (*86*).)

clump made up of short, individual sections. It is *determinate,* that is, each clump terminates in a flowering stalk, growth continuing only from lateral branches. The rhizome tends to be oriented horizontally with roots arising from the lower side.

The second type (the **leptomorph**) is illustrated by the lily-of-the-valley in Figure 16–20. The rhizome is slender with long internodes. It is *indeterminate,* that is, it grows continuously in length from the terminal apex and from lateral branch rhizomes. The stem is symmetrical and has lateral buds at most nodes, nearly all remaining dormant. This type does not produce a clump but spreads extensively over an area.

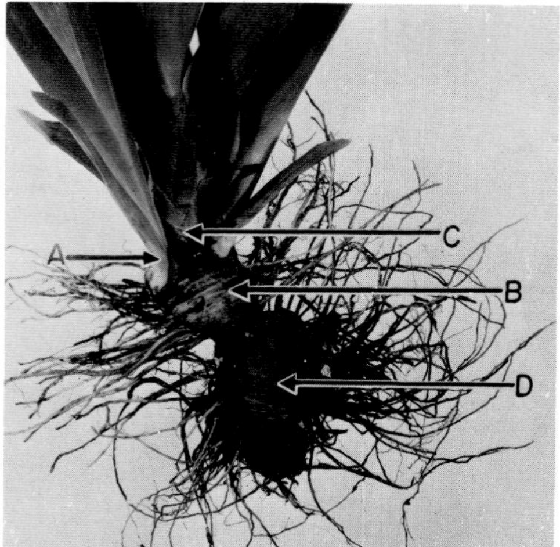

FIGURE 16–21 Structure of an iris (rhizomous type) plant as it appears about the time of flowering. A two-year-old section that had flowered the previous year and is now dying back is shown at D. The lateral branch arising from it consists of the one-year-old vegetative section (B); the current-season, lateral vegetative branches (A); and the current-season, terminal flowering shoot (C). The vegetative shoot (A) will flower the following year.

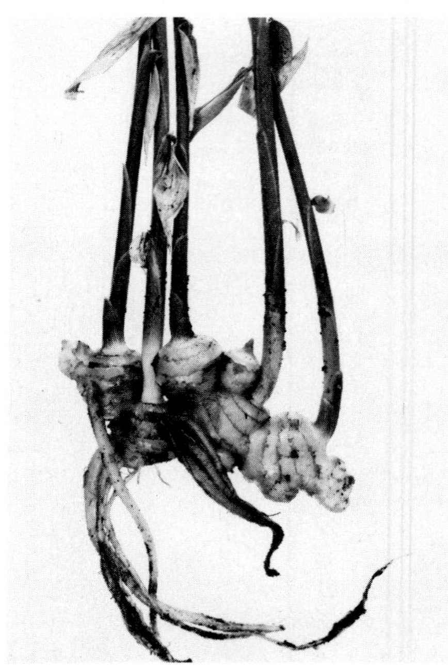

FIGURE 16–22 Rhizome of the tropical ginger plant *(Zingiber officinale)*. This is easily propagated by division of the thickened rhizome, which is the source of commercial ginger.

Intermediate forms between these two types also exist. These are called **mesomorphs** (*48*).

Growth Pattern

Rhizomes grow by elongation of the growing points produced at the terminal end and on lateral branches. Length also increases by growth in the intercalary meristems in the lower part of the internodes. As the plant continues to grow and the older part dies, the several branches arising from one plant may eventually become separated to form individual plants of a single clone.

Rhizomes exhibit consecutive vegetative and reproductive stages, but the growth cycle differs somewhat in the two types described. In the pachymorph rhizome of *Iris* (Figure 16–21), a growth cycle begins with the initiation and growth of a lateral branch on a flowering section. The flowering stalk dies, but these new lateral branches produce leaves and grow vegetatively during the remainder of that season. Continued growth of the underground stem, storage of food, and the production of a flower bud at the conclusion of the vegetative period are dependent upon photosynthesis. Consequently, foliage should not be removed during this period. A flowering stalk is produced the following spring and no further terminal growth can take place. In general, plants with this structure flower in the spring and grow vegetatively during the summer and fall.

Plants with a leptomorph habit as a general rule (with exceptions) grow vegetatively during the beginning of the growth period and flower later in the same period. The length of time during which an individual rhizome section remains vegetative varies with different kinds of plants. An individual branch in the lily-of-the-valley in Figure 16–20, for instance, is vegetative three years before a flower bud forms.

Bamboo is divided into clump growers (pachymorphs), which have constricted rhizomes, and running bamboos (leptomorphs), which spread rapidly by vigorous rhizomes that grow several feet or more (*71*). Generally speaking, the pachymorphs are more desirable ornamentally. Some bamboo species remain vegetative for many years, but then they change abruptly and the entire plant produces flowers.

In some rhizomatous plants, such as blueberry, rhizome development is increased by higher temperatures and a long photoperiod, and is correlated with vigorous above-ground growth and high-nitrogen status (*37, 74*).

Propagation

Division of Clumps and Rhizomes

Division is the usual procedure for propagating plants with a rhizome structure, but the procedure may vary somewhat with the two types. In pachymorph rhizomes, individual sections (or *culms*) are cut off at the point of attachment to the rhizome, the top is cut back, and the piece is transplanted to the new location. Leptomorph rhizomes can be handled in essentially the same way by removing a single lateral "offshoot" from the rhizome and transplanting it. The tip of the lily-of-the-valley rhizome bearing a flower bud (see Figure 16–20), called a "pip," is removed along with the rooted section below and transplanted.

The bird of paradise *(Strelitzia reginae)* has a slow rate of multiplication when propagated by rhizome division. Mechanical induction of branching to eliminate apical dominance of a branch leaf sheath (fan) attached to the rhizome encourages lateral shoot formation and rapid multiplication (Figure 16–23) (*77*). Division is usually carried out at the beginning of a growth period (as in early spring) or at or near the end of a growth period (i.e., in late summer or fall).

Propagation is carried out by cutting the rhizome into sections, being sure that each piece has at least one lateral bud, or "eye"; it is essentially a stem cutting. Bananas, for instance, are propagated in this way. This general method works well for the leptomorph rhizomes, in which a dormant lateral growing point is present at most nodes. The rhizomes are cut or broken into pieces, and adventitious roots and new shoots develop from the nodes. Rhizome-producing turf grasses, for instance, are cut up into sections and the individual "sprigs" transplanted. New plants can be established readily by this method.

Culm Cuttings

In large rhizome-bearing plants, such as bamboos, the aerial shoot, or culm, may be used as a cutting. These may be whole culm cuttings, in which the entire aerial shoot is laid horizontally in a trench. New branches arise at the nodes. A stem cutting of three- or four-node sections may be planted vertically in the ground.

PSEUDOBULBS

Definition and Structure

A **pseudobulb** (literally "false bulb") is a specialized storage structure, produced by many orchid species, consisting of an enlarged, fleshy section of the stem made up of one to several nodes (see Figure 16–24). In general, the appearance of the pseudobulb varies with the orchid species. The differences can be used to identify species.

Growth Pattern

These pseudobulbs arise during the growing season on upright growths that develop laterally or terminally from the horizontal rhizome. Leaves and flowers form either at the terminal end or at the base of the pseudobulb, depending upon the species. During the growth period, they accumulate stored food materials and water and assist the plants in surviving the subsequent dormant period.

Propagation

Offshoots

In a few orchids, such as the *Dendrobium* species, the pseudobulb is long and jointed, being made up of many nodes. Offshoots develop at these nodes. From the base of these offshoots roots develop. The rooted offshoots are then cut from the parent plant and potted.

Division

The most important commercial species of orchids, including the *Cattleya, Laelia, Miltonia,* and *Odontoglossum,* may be propagated by dividing the rhizome into sections, the exact procedure used depending upon the particular kind of orchid. Division is done during the dormant season and preferably just before the beginning of a new period of growth. With a sharp knife, the rhizome is cut back far enough from the terminal end to include four to five pseudobulbs in the new section, leaving the old rhizome section with a number of old pseudobulbs, or "back bulbs," from which the leaves have dehisced. The section is then potted, whereupon growth begins from the bases of the pseudobulbs and at the nodes. The removal of the new section of the rhizome from the old part stimulates new growth, or "back breaks," to occur from the old parts of the rhizome. These new growths grow for a season and can be removed the following year.

FIGURE 16–23 Propagation of *Strelitzia reginae* by mechanical induction of branching to eliminate apical dominance and increase multiplication rate. To remove the apex, a transverse lateral incision is made above the basal plate through the basal leaf sheath of a branch, keeping the leaves in contact with the roots *(upper left and right).* Lateral shoot formation occurs 4.5 months after excision of the apex *(lower left).* After one year, multiple clusters with roots *(middle right)* can be separated into individual plantlets *(lower right).* (Courtesy P. A. Van de Pol (77).)

FIGURE 16–24 Cattleya orchid showing rhizome structure and upright elongated psuedobulbs as basal part of shoots.

FIGURE 16–25 A "back bulb" of a *Cymbidium* orchid was removed from the parent plant and placed in a rooting medium; the offshoot shown above then developed. This offshoot is now ready for removal and potting. When this is done a second offshoot, or "break," should appear. (Courtesy A. Kofranek.)

An alternate procedure is to cut partly through the rhizome and leave it for one year. New back breaks will develop, which can be removed and potted.

Back Bulbs and Green Bulbs

Back bulbs (i.e., those without foliage) are commonly used to propagate clones of *Cymbidium.* These are removed from the plant, the cut surface is painted with a grafting compound, and they are placed in a rooting medium for new shoots to develop. When the stage shown in Figure 16–25 is reached, the shoot can be removed from the bulb and potted. This back bulb can be repropagated and a second shoot developed from it.

Green bulbs (i.e., those with leaves) also can be used in *Cymbidium* propagation. Treatment with indolebutyric acid, either by soaking or by painting with a paste, has been shown to be beneficial (*38*).

Other Modified Stem Structures

Other forms of specialized structures include runners, stolons, crowns, offsets, and suckers, which are modified stem structures. These structures are discussed in Chapter 15 under the section on plant modifications for natural layering.

REFERENCES

1. Amano, N., and K. Tsutsui. 1980. Propagation of hyacinth by hot water treatment. *Acta Hort.* 109:279–87.
2. Amer. Assoc. Nurserymen, Inc., Comm. on Hort. Stand. 1986. *American standard for nursery stock.* Washington, D.C.: Amer. Assoc. Nurs.
3. Baker, K.F., and P.A. Chandler. 1957. Development and maintenance of healthy planting stock. Sec. 13 in *Calif. Agr. Exp. Sta. Man. 23.*
4. Ball, V., ed. 1991. *The Ball red book: Greenhouse growing* (15th ed.). Reston, VA.: Reston Publ. Co.
5. Benschop, M. 1993. *Crocus.* In *The physiology of flower bulbs.* A. DeHertogh and M. Le Nard (eds.). Elsevier, Amsterdam. pp. 257–72.
6. Booth, A. 1963. The role of growth substances in the development of stolons. In *The growth of the potato,* J.D. Ivins and F.L. Milthorpe, eds. London: Butterworth, pp. 99–113.
7. Bouwkamp, J.C., and L.D. Scott. 1972. Production of sweet potatoes from root pieces. *HortScience* 7(3):271–72.

8. Byther, R.S. and G.A. Chastangner. 1993. Diseases. In *The physiology of flower bulbs.* A. DeHertogh and M. Le Nard (eds.) Amsterdam: Elsevier, pp. 71–100.

9. Christie, C.B. 1985. Propagation of amaryllids: A brief review. *Comb. Proc. Inter. Plant Prop. Soc.* 35:351–57.

10. Cohat, J. 1993. *Gladiolus.* In *The physiology of flower bulbs.* A. DeHertogh and M. Le Nard (eds.). Amsterdam: Elsevier, pp. 297–320.

11. Crossley, J.H. 1957. Hyacinth culture; narcissus culture; tulip culture. *Handbook on bulb growing and forcing.* Northwest Bulb Growers Assoc., pp. 79–84, 99–104, 139–44.

12. DeHertogh, A.A., L.H. Aung, and M. Benschop. 1983. The tulip: Botany, usage, growth and development. *Hort. Rev.* 5:45–125.

13. ———, and N. Blakely. 1972. The influence of temperature and storage time on growth of basal roots of nonprecooled and precooled bulbs of *Lilium longiflorum* Thunb. cv. 'Ace'. *HortScience* 74:409–10.

14. ———, A.N. Roberts, N.W. Stuart, R.W. Langhans, R.G. Linderman, R.H. Lawson, H.F. Wilkins, and D.C. Kiplinger. 1971. A guide to terminology for the Easter lily. *HortScience* 6:121–23.

15. ———, and M. Le Nard. 1993. *Dahlia.* In *The physiology of flower bulbs.* A. DeHertogh and M. Le Nard (eds.). Amsterdam: Elsevier, pp. 273–284.

16. ———, and ———. 1993. *The physiology of flower bulbs.* Amsterdam: Elsevier.

17. Doss, R.P. 1979. Some aspects of daughter bulb growth and development and apical dominance in bulbous iris. *Plant Cell Phys.* 20:387–94.

18. Everett, T.H. 1954. *The American gardener's book of bulbs.* New York: Random House.

19. Ewing, E.E. 1985. Cuttings as simplified models of the potato plant. In *Potato Physiology,* P.H. Li, ed. New York: Academic Press.

20. Genders, R. 1973. *Bulbs.* New York: Bobbs-Merrill.

21. Gould, C.J. 1953. Blights of lilies and tulips. *Plant diseases: USDA yearbook of agriculture.* Washington, D.C.: U.S. Govt. Printing Office, pp. 611–16.

22. Gregory, L.E. 1965. Physiology of tuberization in plants (tubers and tuberous roots). In *Encyclopedia of Plant Physiology,* Vol. 15. Berlin: Springer-Verlag, pp. 1328–54.

23. Griffiths, D. 1930. Daffodils. *USDA Circ. 122.*

24. ———. 1930. The production of hyacinth bulbs. *USDA Circ. 112.*

25. Haegeman, J. 1979. *Tuberous begonias.* Vaduz: A.R. Gantner Verlag KG.

26. ———. 1993. Begonia—Tuberous hybrids. In *The physiology of flower bulbs.* A. DeHertogh and M. Le Nard (eds.). Amsterdam: Elsevier, pp. 227–39.

27. Hanks, G.R. 1985. Factors affecting yields of adventitious bulbils during propagation of *Narcissus* by the twin-scaling technique. *Jour. Hort. Sci.* 60(4):531–43.

28. ———. 1993. *Narcissus.* In *The physiology of flower bulbs.* A. DeHertogh and M. Le Nard (eds.). Amsterdam: Elsevier, pp. 462–558.

29. ———, and A.R. Rees. 1979. Twin-scale propagation of *Narcissus:* a review. *Scientia Hort.* 10:1–14.

30. Harrison, A.D. 1964. *Bulb and corm production.* Bul. 62. London: Minist. Agr., Fish. and Foods, pp. 1–84.

31. Hartsema, A. M. 1937. Periodieke ontwikkeling van *Gladious hybridum* var. Vesuvius. *Verh. Koninkl. Ned. Akad. van Wet.* 36(3):1–34.

32. ———. 1961. Influence of temperatures on flower formation and flowering of bulbous and tuberous plants. In *Encyclopedia of Plant Physiology,* Vol. 16. Berlin: Springer-Verlag, pp. 123–67.

33. Heath, O.V., and M. Holdsworth. 1948. Morphologenic factors as exemplified by the onion plant. *Symposia for the Soc. Exp. Biol.* II:326–50.

34. Huisman, E., and A.M. Hartsema. 1933. De periodieke ontwikkeling van *Narcissus pseudonarcissus L. Meded. Landbouwhoogesch., Wageningen,* DL. 37 (Meded. No. 38, Lab v. Plantenphys. onderz., Wageningen).

35. Jacobi, E.F. 1950. *Plantikunde voor tuinbouwscholen.* Zwolle, The Netherlands: W.E.J. Tjeenk Willink.

36. Jacoby, B., and A.H. Halevy. 1970. Participation of light and temperature fluctuations in the induction of contractile roots of gladiolus. *Bot. Gaz.* 131(1): 74–7.

37. Kender, W.J. 1967. Rhizome development in the lowbush blueberry as influenced by temperature and photoperiod. *Proc. Amer. Soc. Hort. Sci.* 90:144–48.

38. Kofranek, A.M., and G. Barstow. 1955. The use of rooting substances in *Cymbidium* green bulb propagation. *Amer. Orch. Soc. Bul.* 24(11):751–53.

39. Langhans, R.W., and T.C. Weiler. 1968. Vernalization in Easter lilies. *HortScience* 3:280–82.

40. Le Nard, M. and A.A. DeHertogh. 1993. *Tulipa.* In *The physiology of flower bulbs.* A. DeHertogh and M. Le Nard (eds.). Amsterdam: Elsevier, pp. 617–82.

41. Lewis, C.A. 1951. Some effects of daylength on tuberization, flowering, and vegetative growth of tuberous-rooted begonias. *Proc. Amer. Soc. Hort. Sci.* 57:376–78.

42. Lin, P.C., and A.N. Roberts. 1970. Scale function in growth and flowering of *Lilium longiflorum,* Thunb. 'Nellie White'. *Jour. Amer. Soc. Hort. Sci.* 95(5): 559–61.

43. Magie, R.O. 1953. Some fungi that attack gladioli. In *Plant Diseases: USDA yearbook of agriculture.* Washington, D.C.: U.S. Govt. Printing Office, pp. 601–7.

44. Mann, L.K. 1952. Anatomy of the garlic bulb and factors affecting bulb development. *Hilgardia* 21:195–251.

45. Matsuo, E., T. Ohkurano, K. Arisumi, and Y. Sakata. 1986. Scale bulblet malformations seen in *Lilium longiflorum* during scale propagation. *HortScience* 21:150.

46. ———, and J.M. van Tuyl. 1986. Early scale propagation results in forcible bulbs of easter lily. *HortScience* 21:1006–7.

47. Mckay, M.E., D.E. Bythe and J.A. Tommerup. 1981. The effects of corm size and division of the mother corm in gladioli. *Aust. Jour. Exp. Agric. Animal Husbandry* 21:343–48.

48. McClure, F.A. 1966. *The bamboos: A fresh perspective.* Cambridge, Mass.: Harvard Univ. Press.

49. McRae, E.A. 1978. Commercial propagation of lilies. *Comb. Proc. Inter. Plant Prop. Soc.* 28:166–69.

50. Melis, R.J.M., and J. van Staden. 1985. Tuberization in cassava *(Manihot esculenta)*. Cytokinin and abscisic acid activity in tuberous roots. *Jour. Plant Physiol.* 118:357–66.

51. Menzel, C.M. 1985. Tuberization in potato at high temperatures: Responses to erogenous gibberellin, cytokinin and ethylene. *Potato Res.* 28:263–66.

52. Miller, W.B. 1993. *Lilium longiflorum.* In *The physiology of flower bulbs.* A. DeHertogh and M. Le Nard (eds.). Amsterdam: Elsevier, pp. 391–422.

53. Moser, B.C., and C.E. Hess. 1968. The physiology of tuberous root development in dahlia. *Proc. Amer. Soc. Hort. Sci.* 93:595–603.

54. Mulder, R., and I. Luyten. 1928. De periodieke ontwikkeling van der *Darwin* tulip. *Verh. Koninkl. Ned. Akad. van Wet.* 26:1–64.

55. Nakayama, M. 1980. Vegetative propagation of cyclamen by notching of tuber. II. Effect of scooping size and notching size on the regeneration of cyclamen tuber. *Jour. Japan. Soc. Hort. Sci.* 49(2):228–34.

56. Nightingale, A.E. 1979. Bulbil formation on *Lilium longiflorum* Thunb. cv. Nellie White by foliar applications of PBA. *HortScience* 14:67–8.

57. Nitsch, J.P. 1971. Perennation through seeds and other structures. In *Plant physiology,* Vol. 6A, F.C. Steward, ed. New York: Academic Press, pp. 413–79.

58. Nowak, J. 1993. *Hyacinthus.* In *The physiology of flower bulbs.* A. DeHertogh and M. Le Nard (eds.). Amsterdam: Elsevier, pp. 335–48.

59. Okubo, H. 1993. *Hippeastrum* (Amaryllis). In *The physiology of flower bulbs.* A. DeHertogh and M. Le Nard (eds.). Amsterdam: Elsevier, pp. 321–34.

60. Pfieffer, N.E. 1931. A morphological study of *Gladiolus. Contrib. Boyce Thomp. Inst.* 3:173–95.

61. Quamina, J.E., B.R. Phills, and W.A. Hill. 1982. Vine production from tuber pieces of various sizes and sections of yam (*Dioscorea alata* L.). *HortScience* 17:73.

62. Raunkiaer, C. 1934. *Life forms of plants and statistical plant geography.* Oxford: Clarendon Press.

63. Rees, A.R. 1972. *The growth of bulbs.* New York. Academic Press.

64. ———, and G.R. Hanks. 1980. The twin-scaling technique for narcissus propagation. *Acta Hort.* 109: 211–16.

65. ———. 1992. *Ornamental bulbs, corms and tubers.* Wallingford: CAB International.

66. Reinders, E., and R. Prakken. 1964. *Leerboek der Plantkunde.* Amsterdam: Scheltema & Holkema N. V.

67. Roberts, A.N., and L.T. Blaney. 1957. Easter lilies: Culture. In *Handbook on bulb growing and forcing.* Northwest Bulb Growers Assoc., pp. 35–43.

68. Rockwell, F.F., E.C. Grayson, and J. de Graaf. 1961. *The complete book of lilies.* Garden City, N.Y.: Doubleday.

69. Sheehan, T.J. 1955. Caladium production in Florida. *Fla. Agr. Ext. Circ. 128.*

70. Shillo, R., and A.H. Halevy. 1981. Flower and corm development in gladiolus as affected by photoperiod. *Scientia Hort.* 15:187–96.

71. Simon, R.A. 1986. A survey of bamboos: Their care, culture and propagation. *Comb. Proc. Inter. Plant Prop. Soc.* 36:528–31.

72. Skelmersdale, L. 1978. Propagation of bulbous and bulbous-like plants. *Comb. Proc. Inter. Plant Prop. Soc.* 28:209–15.

73. Slater, J.W. 1963. Mechanisms of tuber initiation. In *The growth of the potato,* J. D. Ivins and F.L. Milthorpe, eds. London: Butterworth, pp. 114–20.

74. Smagula, J.M., and P.R. Hepler. 1980. Effect of nitrogen status of dormant rooted lowbush blueberry cuttings on rhizome production. *Jour. Amer. Soc. Hort. Sci.* 105:283–85.

75. Stuart, N.W. 1954. Moisture content of packing medium, temperature and duration of storage as factors in forcing lily bulbs. *Proc. Amer. Soc. Hort. Sci.* 63:488–94.

76. Van Aartrijk, J., and P.C.G. Van der Linde. 1986. In vitro propagation of flower-bulb crops. In *Tissue culture as a plant production system for horticultural crops,* R. H. Zimmerman et. al., eds. Dordrecht: Martinus Nijhoff Publishers.

77. Van de Pol, P.A., and T.F. Van Hell. 1988. Vegetative propagation of *Streilitzia reginae. Acta Hort.* 226: 581–86.

78. Van Tuyl, J.M. 1983. Effect of temperature treatments on the scale propagation of *Lilium longiflorum* 'White Europe' and *Lilium* × 'Enchantment'. *HortScience* 18:754–56.

79. Wang, S.Y., and A.N. Roberts. 1970. Physiology of dormancy in *Lilium longiflorum* Thunb. 'Ace'. *Jour. Amer. Soc. Hort. Sci.* 95:554–58.

80. Wang, Y.T., and A.N. Roberts. 1983. Influence of air and soil temperatures on the growth and development of *Lilium longiflorum* Thunb. during different growth phases. *Jour. Amer. Soc. Hort. Sci.* 108:810–15.

81. Weiler, T.C., and R.W. Langhans. 1972. Growth and flowering responses of *Lilium longiflorum* Thunb. 'Ace' to different day lengths. *Jour. Amer. Soc. Hort. Sci.* 97:176–77.

82. Welch, N.C., and T.M. Little. 1967. Heat treatment and cutting for increased sweet potato slip production. *Calif. Agr.* 21(5):4–5.

83. Wilfret, G.J. 1995. *Caladium.* In *The physiology of flower bulbs.* A. DeHertogh and M. Le Nard (eds.). Amsterdam: Elsevier, pp. 239–48.

84. Wilson, K., and J.N. Honey. 1966. Root contraction in *Hyacinthus orientalis. Ann. Bot.* 30:47–61.

85. Woodcock, H.B.D., and H.T. Stearn. 1950. *Lilies of the world.* New York: Scribner's.

86. Zweede, A.K. 1930. De periodieke ontwikkeling van *Convallaria majalis. Verh. Koninkl. Ned. Akad. van Wet.* 27:1–72.

SUPPLEMENTARY READING

CROCKETT, J. V. 1971. *Bulbs.* New York: Time-Life Books.

DEHERTOGH, A. A., L. H. AUNG, and M. BENSCHOP. 1983. The tulip: botany, usage, growth and development. *Hort. Rev.* 5:45–125.

DEHERTOGH, A.A., and M. LE NARD. 1993. *The physiology of flower bulbs.* Amsterdam: Elsevier.

GENDERS, R. 1973. *Bulbs, a complete handbook.* New York: Bobbs-Merril.

GILES, F. A., R. MCINTOSH, and D. C. SAUPE. 1980. *Herbaceous perennials.* Reston, VA.: Reston Publ. Co.

HARRISON, A. D. 1964. *Bulb and corm production.* Bul. 62. London: Ministry of Agriculture, Fisheries, and Food, pp. 1–84.

HARTSEMA, A. M. 1961. Influence of temperatures on flower formation and flowering of bulbous and tuberous plants. In *Encyclopedia of plant physiology,* Vol. 16, V. Ruhland, ed. Berlin: Springer-Verlag, pp. 123–67.

KOHLEIN, F. 1988. *Iris.* Portland, Oreg.: Timber Press.

LI, P. H., ed. 1985. *Potato physiology.* New York: Academic Press.

N. A. GLADIOLUS COUNCIL. 1972. *The world of the gladiolus.* Edgewood, Md.: Edgewood Press.

REES, A. R. 1972. *The growth of bulbs.* London: Academic Press.

———. 1966. The physiology of ornamental bulbous plants. *Bot. Rev.* 32:1–22.

———. 1992. *Ornamental bulbs, corms and tubers.* CAB International, Wallingford.

STILL, S. M. 1988. *Herbaceous ornamental plants* (3rd ed.). Champaign, Ill.: Stipes Publ. Co.

Third international symposium on flower bulbs. 1980. *Acta Hort.* 109:1–533.

VAN AARTRIJK, J., and P. C. G. VAN DER LINDE. 1986. In vitro propagation of flower-bulb crops. In *Tissue culture as a plant production system for horticultural crops.* R. H. Zimmerman, et al., eds. Dordrecht: Martinus Nijhoff Publishers.

PART IV—METHODS OF MICROPROPAGATION

17

Principles of Tissue Culture for Micropropagation

INTRODUCTION

The ability to establish and maintain plant organs (embryos, shoots, roots, flowers), and plant tissues (cells, callus, and protoplasts) in aseptic culture and to regenerate new plants from them is a result of basic and applied research in scientific laboratories (botany, plant pathology, and genetics) since before the turn of the century (*51, 103, 196*). These separate procedures have been collectively called **tissue and organ culture, in vitro culture, micropropagation,** and most recently **biotechnology** (see Chapter 2).

Table 17–1 lists various procedures that fall within the area of biotechnology. The table is organized by the structures formed in tissue culture (Figure 17–1). These include the formation of seedlings, plantlets, callus, and somatic embryos. Of these techniques, the process of plantlet formation via micropropagation has the most direct application for plant propagation and will be emphasized in this chapter.

Micropropagation differs from traditional propagation in that the biological components of each procedure are separated into stages to produce a high degree of control over each aspect of the regeneration and development processes. Each step of the sequence can be manipulated (or programmed) by selection of explants and control of the culture environment. New concepts have had to be identified and inevitably new problems have emerged that are unique to these procedures. Insight into the biological mechanisms involved in micropropagation has also found application in conventional propagation methods.

This chapter describes the biological nature of the systems listed in Table 17–1 and defines the morphological, physiological, and genetic basis for the growth and development of plants in tissue culture.

Terminology (Also see Chapter 18, pp. 601)

Tissue culture Term used for the range of procedures used to maintain and grow plant tissues (callus, cells, protoplasts) and organs (stems, roots, embryos) in aseptic (or *in vitro*) culture. The technique is used for propagation, genotype modification (i.e., plant breeding), biomass production of biochemical secondary products, plant pathology, germplasm preservation, and scientific investigations. These procedures are included in the term **biotechnology** (see Chapter 2).

TABLE 17–1

Techniques used to regenerate plants through tissue culture

Structures Formed	*Regeneration Method*	*Explant Source*	*Uses*
Seedling Formation			
	Seed culture	Seeds	Primarily used to produce orchids. Orchid seeds lack the typical storage reserves found in other seeds and respond well to tissue culture.
	Embryo culture Embryo rescue Ovule culture Ovary culture	Embryos are isolated from the fruit and seed coverings.	Mature embryos germinate easily in tissue culture to form seedlings. Used for research, understocks for micrografting, and occasionally for propagation. Isolation of immature embryos is primarily used for breeding interspecific crosses. These crosses usually fail to set seed, but early embryo development can occur. Embryos in the precotyledonary stage are separated from the plant and complete their development in tissue culture
Plantlet Formation			
	Axillary shoot formation		
	1. Meristem culture	Shoot tip less than 1 mm in size.	Initially developed as a micropropagation system, but now mostly used for virus elimination.
	2. Shoot culture Axillary branching Nodal cultures Stool shoots Pseudocorms Minitubers	Stem with one to four nodes. May include leaves and shoot tip.	Shoot cultures are the most often used micropropagation systems.
	Adventitious shoot formation (organogenesis)		
	1. Diploid plant regeneration (full complement of chromosomes)	Leaf pieces, petioles, bulb scales, stem internodes, roots, and callus.	Often used for micropropagation, especially in monocots. Adventitious shoot regeneration is one of the key steps in obtaining plants that have been genetically transformed.
	2. Haploid plant regeneration (half the original chromosome number)	Anther (pollen mother cells)	Used in breeding to obtain haploids. Shoots or somatic embryos may be obtained.

Micropropagation Term for *in vitro* propagation of plants. There are four specific developmental stages to micropropagation. These are establishment, multiplication, rooting, and acclimatization stages (see Chapter 18; Figures 18–7 and 18–8).

Explant The piece of the plant (propagule) used to initiate the micropropagation or tissue culture process. Other propagules include *cutting, layer, scion,* or *seed.*

Organogenesis The process of developing adventitious shoots and/or roots (see Chapter

Structures Formed	*Regeneration Method*	*Explant Source*	*Uses*
	Micrografting	Small scion shoot tip usually grafted to a seedling rootstock.	Useful for virus elimination. Can also be used in special situations (like grape) as an alternative to conventional grafting.
Callus formation			
	Callus cultures (stationary)	Any vegetative tissue.	Callus cultures are used for research, breeding, and genetic transformation studies. Callus cells can be used to produce enzymes, medicines, natural flavors, and colors.
	Callus suspension cultures	Callus subcultured from stationary cultures.	Suspension cultures are shaken constantly to perpetuate callus formation. Uses are the same as stationary callus cultures.
	Protoplast cultures	Protoplasts are isolated single cells without a cell wall. The cell wall has been digested by fungal enzymes.	Protoplasts are used in plant research to study basic cell function. Protoplasts can also be used in breeding. Under the right conditions, two protoplasts can fuse to form a single cell. The nuclei in these cells can merge, combining genetic information, even in species that are not sexually compatible. New cell walls form and the resultant callus can be induced to form adventitious shoots.
Somatic (vegetative) embryo formation	Adventitious somatic embryogenesis		
	1. Type 1	Nucellus or ovule.	Used to regenerate clonal copies of the mother plant.
	2. Type 2 (polyembryogenesis)	Embryogenic suspensor mass.	Breeding and genetic transformation
	3. Type 3	Developing embryos or seedling parts.	Breeding and genetic transformation.
	Induced somatic embryogenesis	Callus and cell suspension cultures.	Breeding and genetic transformation. Most systems proposed for synthetic seeds use this procedure.

10). Changes take place in the cells that lead to the development of a unipolar structure (stem or root primordium) whose vascular system is often connected to the parent tissue (*183*).

Somatic embryogenesis The development of embryos from vegetative cells rather than from the union of male and female gametes to produce a zygote, as described in Chapter 5. In this process, a bipolar structure is produced with a root-shoot axis and a closed independent vascular system (*183*). Other naturally occurring examples of embryo development

FIGURE 17–1 Common structures formed in tissue culture: (a) shoots; (b) roots; (c) somatic embryos; and (d) callus.

within the plant are **zygotic embryogenesis** and **apomictic embryogenesis** (see Chapter 5).

Callus Production of undifferentiated cells (usually parenchyma cells).

REPRODUCTION OF SEEDLING PLANTS IN TISSUE CULTURE

In vitro culture can be effectively used in a number of germination and reproduction procedures where protection from contaminating organisms is needed and when specific genetic barriers need to be overcome.

Aseptic Seed Culture

Orchid seeds are extremely tiny, lack stored-food reserves, and in nature depend upon a symbiotic relationship with specific fungi to obtain nutrients. It is estimated that 30,000 seeds are produced in a single *Cattleya* pod. In 1922, Lewis Knudson reported the successful culture of orchids in a

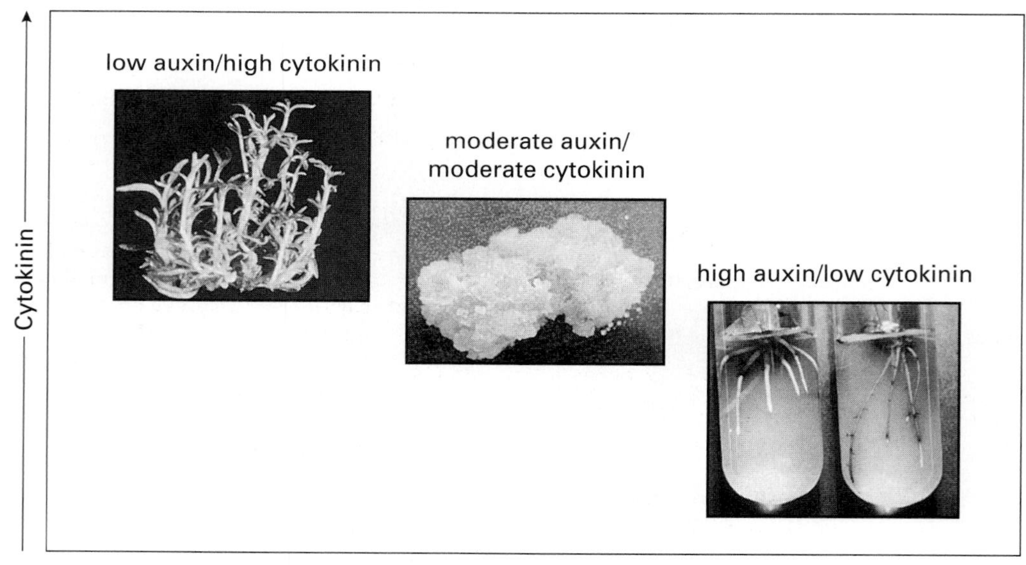

FIGURE 17–2 The relationship between the ratio of cytokinin to auxin on organogenesis in culture. A high cytokinin to auxin ratio tends to promote shoot organogenesis, while a low cytokinin to auxin ratio promotes rooting. More equal concentrations tend to promote both shoots and roots from the same culture or undifferentiated callus.

Tissue culture is based upon the principle of totipotency (see Chapter 2), first described by the German plant physiologist Haberlandt in 1902 (*66*). He predicted the indefinite culture of tissues, cells, and organs. However, it was not until 1934 that P.R. White (*196*) was able to grow tomato roots continuously on an *in vitro* medium supplemented with yeast extract. The essential ingredients that stimulated this growth turned out to be certain B vitamins, notably B_1 (thiamine). In 1939, Nobecourt and Gautheret in France and White working in the United States independently reported the indefinite culture of plant callus tissue on a synthetic medium supplemented by auxin (*196*). Thus, the first requirement for micropropagation was provided, i.e., the ability to grow plants in artificial culture. A second requirement for successful micropropagation was the regulation of shoot and root regeneration. This was elegantly presented in the classic work by Skoog and co-workers utilizing tobacco pith callus (*163*). They demonstrated that shoots, roots, and callus could be regenerated in culture depending on the auxin to cytokinin ratio (Figure 17–2).

A French plant pathologist, Morel, first observed the rapid vegetative proliferation of rooted plantlets in *Cymbidium* orchid cultures (*127*) when he tried to eliminate viruses in shoot tips (*126*). Toshio Murashige, of the University of California, Riverside, expanded this work and developed the concept of developmental stages of micropropagation leading to plantlet establishment (*131*). Micropropagation and tissue culture protocols are based on the concept of *competence* and *determination* as controlled by epigenetic processes within the plant (see Chapter 2). Explants either must contain cells that are already competent to regenerate shoots, roots, embryos, or callus, or competence must be induced during the culture process (see Chapter 10).

Application of micropropagation began in the 1960s and early 1970s (*3, 16, 31, 67, 69, 83, 112, 127, 130, 131, 158*), extended to the development of commercial **laboratory-nurseries** in the 1970s in the United States, Europe, Australia, and Asia (*16, 74, 78, 137, 138, 172, 200*), and expanded in the 1980s (*33, 47, 56, 103, 202*). Future use depends upon adaptation to industrial conditions (*102*), reduction in costs (*128, 177*), and coordination of production to markets (*90*).

Genetic engineers and plant breeders utilize *in vitro* culture techniques to carry out breeding strategies for the genetic improvement of plants (*5, 44, 60, 101*). These efforts must be integrated with propagation industries to make the products available for practical use.

nonsymbiotic culture medium that supplied minerals and sucrose. The basis of his success was his discovery that calcium hypochlorite (bleach) could surface-sterilize seeds but was not toxic to the germinating embryo. This culture procedure revolutionized orchid propagation at the time but was eventually replaced by micropropagation (see Chapter 18). However, it is still commonly used to germinate hybrid seeds (see box, p. 555).

Embryo Culture and Embryo Rescue

Embryo culture involves the excision of an embryo from a seed and germinating it in aseptic culture (*27, 81, 140, 198*). A principal use of this technique is to "rescue" embryos that would have aborted within the seed before the fruit in which they have been produced is mature. Many interspecific hybridizations are initially successful but the embryo aborts sometime during development **(somatoplastic sterility)**. Early ripening cultivars of many fruit tree species tend to mature the fruit before their embryos are fully developed. Under natural conditions these embryos can abort, but they can be excised and grown in an appropriate sterile culture (*75, 96, 146*).

A second use is to induce prompt germination of mature dormant seeds and thus shorten breeding cycles (*106*). Excising the embryo to eliminate dormancy-inducing restraints in the surrounding endosperm has been described for *Iris* (*147, 173*), *Maranta* (*73*), peach (*96, 106*), and others. Appropriate environmental sequences may be needed for embryos with internal epicotyl dormancy, as in *Paeonia* (*122*) (Figure 17–3).

One protocol for the development of clover (*Trifolium*) embryo cultures (*27*) calls for the following sequence of stages:

1. High sucrose, moderate auxin, and low cytokinin for one to two weeks. At this time the embryo stops growing and must be transferred.
2. Normal sucrose, low auxin, and moderate cytokinin. This allows for resumption of

FIGURE 17–3 Embryo culture of peony *(Paeonia)* seed with epicotyl dormancy. *Left top:* Seed (left). Seed coat removed (right) to show massive endosperm. *Left center:* Very small rudimentary embryo is located inside base of endosperm. Shows cutting away sections of endosperm to extract the embryo on probe. *Right top:* First stage of embryo germination after two months in test tube on agar medium with MS salts and 4 percent sucrose. Only radicle grows at warm temperature. *Lower right:* A subsequent cold treatment stimulates the epicotyl to grow. Shows seedling after seven weeks at 27° C (80° F), five weeks at 3° C (37° F), and four weeks at 27° C (80° F) in light. *Lower left:* Seedling transplanted to growing medium. (Courtesy M.M. Meyer, Jr. Reprinted from *Amer. Peony Soc. Bul.* 217, 1978 *(122).*)

ASEPTIC ORCHID SEED GERMINATION

Aseptic seed culture for orchids is currently used in breeding programs to produce seedlings. Approximately seven years is required from germination to flowering. Seeds can be extracted from green pods that are about 60 percent mature after externally disinfesting the pod with alcohol. The usual procedure is to remove seeds from a mature fruit when it naturally dehisces and disinfest it with calcium or sodium hypochlorite (*25*). A small amount of seed is placed in a vial or flask and covered with five to ten times its volume with disinfestant plus a drop or two of a wetting agent. Seeds are kept immersed for five minutes, shaking periodically, after which the seeds will sink. Pour off the disinfestant and then wash several times with sterile water. Pour the seeds over a sterile medium containing 1 to 1.5 percent agar, 2 percent sucrose and basic minerals solution.

embryo growth and, sometimes, shoot development.

3. Embryos with shoots are shifted to a medium with low auxin and high cytokinin to stimulate shoot proliferation. Shoots are rooted and transplanted.
4. Embryos with renewed, but disorganized, growth are transferred to a somatic embryogenesis induction medium.

Rudimentary embryos up to the heart-shaped stage of development are difficult to handle physically and sensitive to the medium used. Apparently, it is essential that the suspensor tissue be retained when the proembryo is excised at this particular stage. Embryos at this stage could best be handled by ovule culture.

Ovary and Ovule (Ovulo) Culture

This system includes the aseptic culture of the excised ovary, ovule (fertilized or unfertilized), and placenta attached within the ovule. Although this technique was first utilized to investigate problems of fruit and seed development (*10, 148*), adaptations of the procedure have uses in propagation, particularly in genetic improvement.

Unfertilized ovules have been excised, grown in culture, supplied with pollen, and subsequently fertilized *in vitro* (*54, 150*). Such procedures have been used most successfully with plants having multiple ovules. Pollen can be placed directly on the placenta inside the ovule, where pollen tubes can develop and grow immediately into the ovules without passing down the style.

EMBRYO CULTURE

Culture procedures are based on the stage of embryo development (see Chapter 5). Methods of handling are different for the three broad categories of embryo development: (a) *rudimentary embryos* from proembryo to approximately the heart stage; (b) *immature embryos* that are at approximately the heart stage up to about one-half full size; and (c) *mature embryos* whose cotyledons have reached nearly full size and are desiccation-tolerant (see pp. 140). Fruits and seeds can be easily disinfested, opened, and the embryo extracted aseptically. Agar is the preferred support, but the ingredients of the basal medium vary greatly and individual species may require experimentation. Immature embryos separated from the ovule cease embryogenesis and show precocious abnormal germination. A major breakthrough in embryo nutrition came with the discovery by Van Overbeek (*190*) in 1942 that 1 to 15 percent coconut milk (a liquid endosperm) would prevent precocious germination and allow embryo development to proceed. Since then, it has been found that success with embryos excised at this stage required high osmotic pressure and high nitrogen, either ammonium or nitrate, depending upon the species. Important media constituents include sucrose (8 to 18 percent), casein hydrolysate (5 to 30 percent), a moderate or low level of auxin and cytokinin, and/or gibberellin, depending upon the species. GA may, however, induce precocious germination, which can be overcome by abscisic acid. ABA has been shown to promote embryo development. A shift from embryonic to germination potential may be promoted by exposing the embryos to controlled desiccation (*95*) or by chilling at 2° C (36° F) (*75*). Mature or nearly mature embryos can be germinated on a basal medium of only inorganic salts and agar. Sucrose (1 or 2 percent or none) may be helpful if the embryo is immature but otherwise may be inhibitory.

Cultured ovules are useful for rescuing embryos that abort at a very young stage if not separated from the plant (*39, 57, 146, 149, 168*). This is a simpler technique than culturing isolated embryos. The dissection of such small embryos is difficult and the medium required is more complex than culturing the entire ovule. Ovules can easily be disinfested and generally will grow in a basal medium of inorganic salts, sucrose, and sometimes growth regulators.

The technique of ovary (fruit) culture is less common and varies in nutritional requirements (*147*). For most species, a basic medium of inorganic salts and sucrose is essential, but auxin may be highly stimulatory.

MICROPROPAGATION OF PLANTLETS FROM TISSUE CULTURE

The cultural systems described under this classification (Table 17–1) are based on the maintenance and multiplication of **microshoots** in culture to produce rooted **microcuttings.** A distinction is made between microshoots originating from axillary buds and those arising adventitiously directly from tissue without axillary buds (i.e., leaf or petiole) or indirectly from callus developed from the original explant. However, in either case, the stages of micropropagation are very similar.

Developmental Stages in Micropropagation

Success of micropropagation is largely due to separating different developmental aspects of culture into stages, each of which is manipulated by media modification and environmental control. Four distinct stages are recognized for most plants. The procedures for these stages are illustrated in Chapter 18 (Figures 18–7 and 18–8) and include:

Stage I. Establishment
Stage II. Multiplication
Stage III. Root formation
Stage IV. Acclimatization

Stage I Establishment and Stabilization of Explants in Culture

The objectives for Stage I are to successfully place an explant into aseptic culture by avoiding contamination and then to provide an *in vitro* environment that promotes stable shoot production (*114*).

Explant selection. The selection and management of the source plant is an important aspect of successful micropropagation. Three aspects require particular attention: (1) genetic and epigenetic characteristics of the source, (2) control of pathogens, and (3) physiological conditioning of the plant to optimize its ability to establish itself in culture.

First of all, correct genotype identification and trueness-to-type are essential because a mistake at this stage can multiply the problem many times and cause much economic loss before it is eventually discovered. Among different species and cultivars, much variation exists in their ability to regenerate and to respond to cultural conditions. Systematic research and development is generally required to provide the needed knowledge and experience for each kind of plant. A period of time may be needed to determine the requirements for any unfamiliar species or cultivar, although generalizations can be made from the information available in published literature.

In general, those plants that are easy to propagate conventionally are also easy to micropropagate. For example, most herbaceous plants can be vegetatively propagated relatively easily by either conventional propagation or micropropagation (*55, 56, 131, 140, 171*). Woody perennials have been more difficult to regenerate and new principles of culture have had to be developed (*6*).

Success in micropropagation of woody plant species is to a large extent a function of the juvenility status of the source plant (see Chapters 2 and 9). Seedling plants are invariably easier to micropropagate than cultivars of that species, and cultivars of the same species often differ in their specific requirements for micropropagation. Typically, the limiting factor in micropropagation is the initiation of roots rather than the ability to proliferate shoots. Manipulations to obtain juvenile explants include selection from the basal juvenile parts of seedlings, induction of adventitious shoots on roots, and consecutive grafting to seedling plants (described in Chapter 9).

The micropropagation sequence of subculturing has resulted in rejuvenation that ultimately leads to increased competence for rooting (*202*). Shifts in leaf morphology and other juvenility markers can result after micropropagation, and the rooted products of micropropagation invariably show some juvenile characteristics (pp. 579). A progressive increase in the ability to root, paralleled by shifts in vigor and morphology of the shoot with consecutive subcultures, has been found in several woody species. This process may be the function of the stabilization period of shoot production described in a subsequent section.

Explant disinfestation (*21, 177*). Pathogen problems include: (a) contamination by fungi, molds, yeasts, and bacteria on the surface of stems or lodged in cracks, bud scales, and elsewhere; (b) systemic viruses and viruslike organisms; and (c) internal pathogens. In some cases, pathogens can inhibit growth and rooting but remain suppressed until the explant is subcultured. Examples include *Bacillus subtilis, Erwinia,* and *Pseudomonas* (*21, 98, 99*).

External contaminants are present literally everywhere—in the air and on the surface of plants, tables, hands, and so on. Spores move by air currents carrying dust particles. One must disinfest the explants, the tools, and the working area to remove such contaminants from the surface. All work must be done in special transfer areas where contaminants have been eliminated and precautions taken to prevent recontamination (see Chapter 18 for procedures to disinfest explants).

Reduction in surface contaminants begins with the stock plants used as the source of explants. Usually, contaminants are present only on the surface of plant parts, although they may be lodged in cracks, between bud scales, and elsewhere. Internal structures, such as growing points of buds and inside seeds and fruits, tend to be relatively free of pathogens. However, if the plant is growing in a humid atmosphere, mycelia may invade plant interiors and become a persistent problem.

Reduction in the surface contaminants on stock plants aids in later disinfestation attempts (*73*). In general, stock plants growing in containers in protected environments in a greenhouse are "cleaner" sources of explants than are those growing out-of-doors. Overhead watering, sprinkling, or any activity that increases humidity around the plant should be avoided. Likewise, keep plant parts off the ground and avoid the use of roots or underground portions as explant sources, if possible. Insect and mite populations should be controlled. In one case, withholding water and keeping plants cool and dry for three weeks was necessary to reduce contaminants on *Dieffenbachia,* even inside a greenhouse (*98*). If plants are growing out-of-doors, covering shoots with plastic bags or spraying with fungicides may be useful (*31*).

Disinfestation requires the use of chemicals that are toxic to the microorganism but relatively nontoxic to plant material. Tissue culture became possible with the development of convenient and effective disinfestants, such as **calcium hypochlorite** (*103*) and **sodium hypochlorite** (now widely available as commercial household bleaches under various trade names).

One should not take for granted that the explants are free of viruses and any internal pathogen even if the extreme meristem tip of the shoot is utilized. Specific indexing or other detection procedures should be part of the system (see Chapter 9). Explant selection is emphasized because very large numbers of plants can be propagated and disseminated before a problem may be detected.

Culture indexing can be used to identify explants showing positive results for the presence of pathogens. Indexing at the initial explant stage, however, does not always identify all potential contaminants (*74*).

Physiological conditioning is the treatment of the stock plants to provide more favorable material for explanting. It may include: (a) growing plants in containers in the greenhouse with environmental controls to achieve the proper stage of shoot development in relation to seasonal patterns, (b) producing healthy plants, and (c) treatment with growth regulators (*30*).

Culture medium. The culture medium (see Chapter 18) includes a (usually) semisolid support (agar or other commercial product like Gelrite), a basal medium (BM) of inorganic elements (both major and minor), an energy source, primarily sucrose, and often some vitamin supplements. Most cultures require auxin and cytokinin at particular ratios and/or amounts for establishment and maintenance. To develop axillary shoots, lower levels are needed. The explant initially grows by elongation of the main terminal shoot, with limited proliferation of axillary shoots. A mass of microshoots is produced within a few weeks, the number depending upon the apical dominance of the particular kind of plant (*32, 140*). The culture mass is subdivided after two to four weeks and subcultured on fresh medium. Division and subculturing is repeated at intervals.

Exudation. One of the most frequently encountered problems with establishing explants (especially woody perennials) in culture is the production of exudates by the explants. These exudates are usually considered to be various phenolic compounds that oxidize to form a brown material in the medium (Figure 17–4). These substances tend to be inhibitory to development (*30, 55*). Treatments to minimize this problem include pretreating the explant with an antioxidant (citric or ascorbic acid), including an adsorbent material in the medium

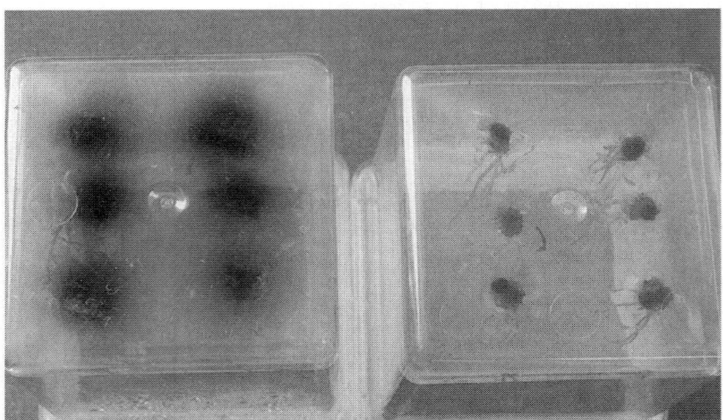

FIGURE 17–4 Exudation from explants in tissue culture. Explants on the left are grown on agar, while cultures on the right are grown on Gelrite. It is not clear from this illustration whether the gelling agent is preventing exudation or simply the oxidation of that exudation leading to the darkened coloration around each explant.

(polyvinylpyrrolidone or activated charcoal), and frequent transfers to a new medium.

Stabilization. Explants may require repeated subculturing to produce a uniform, well-growing culture. Most annuals and herbaceous plants stabilize quickly within a few subcultures. Woody plants invariably require longer periods and sometimes stabilized cultures may never be achieved (Figure 17–5). Growth characteristics change from unpredictable, often abnormal, shoot development, at first, to a uniform predictable growth pattern (*114, 116*) (Figure 17–6). Difficulty in stabilization appears to be associated with the phase state of the explant: the more juvenile, the more easily stabilized. The characteristic seasonal growth pattern of the species being cultured has been associated with ease of stabilization.

Stage II Multiplication

The purpose of Stage II is to maintain the microculture in a stabilized state and multiply the microshoots to the number required for rooting. The basic medium of Stage II is similar to Stage I and specific media formulations are provided in Chapter 18. Intensive commercial micropropagation may require the optimization of not only the inorganic elements but also of other factors of the medium. This can be done by systematically testing a range of concentrations of individual elements combined in various ways with other elements. Suggested methods have been a broad-spectrum approach using all possible combinations (*31, 32*), or a factorial approach (*4, 32, 171*), where a range of each ingredient (others held constant) is tested in consecutive subcultures. One or more subcultures should occur between each test to equalize the test material.

Growth regulators are used to support a basic level of growth but are equally important to direct the developmental response of the propagule. Shoot initiation is strongly supported by **cytokinin** concentration (Figure 17–7). In general, the minimum concentration of cytokinin that stimulates lateral shoot initiation is selected during the multiplication stage. Increased cytokinin levels promote

FIGURE 17–5 Kentucky coffee tree *(Gymnocladus dioicus)* fails to stabilize when cultured from explants collected from mature trees. Left: Short shoots are produced that fail to elongate. Right: However, explants from juvenile explants *(52)* and explants from stump sprouts of mature trees *(166)* will stabilize and form elongated shoots for micropropagation.

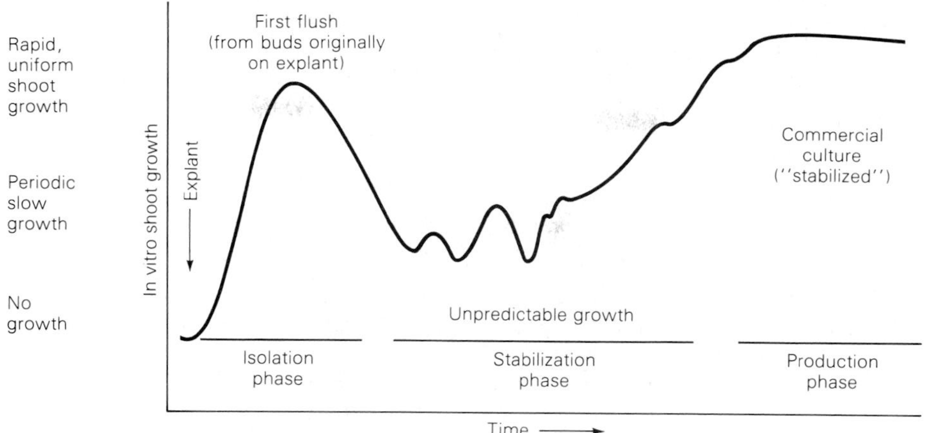

FIGURE 17–6 Generalized scheme of the three important phases in the microculture period through which a shoot must progress to be successfully microcultured. The second period is rooting and acclimatization of shoots. (Reprinted by permission from D. D. McCown and B. H. McCown. In *Cell and tissue culture in forestry.* Vol. 3. J. M. Bonga and D. J. Durzan, eds. Dordrecht: Martinus Nijhoff *(116).*)

lateral shoot proliferation, but can inhibit shoot elongation (Figure 17–8). Requirements may vary at different stages of culture such that variable growth responses may occur during consecutive subcultures (Figure 17–9). Adjustment of cytokinin concentrations may be necessary (*171*). Auxin is usually low or absent in Stage II cultures.

The **propagation ratio** (PR) is the number of new microshoots that can be produced per microculture per subculture. In general, the propagation ratio is greater from adventitious shoots than from axillary shoot cultures. However, the proportion of off-type and aberrant shoots is more apt to be greater with the adventitious origin. Consequently, the highest PR may not be the most desirable to maintain quality and trueness-to-type.

Subculturing frequency may vary from two to eight weeks. Promptness is essential to maintain shoot proliferation and to establish optimum shoot multiplication rates. Variation in routine may

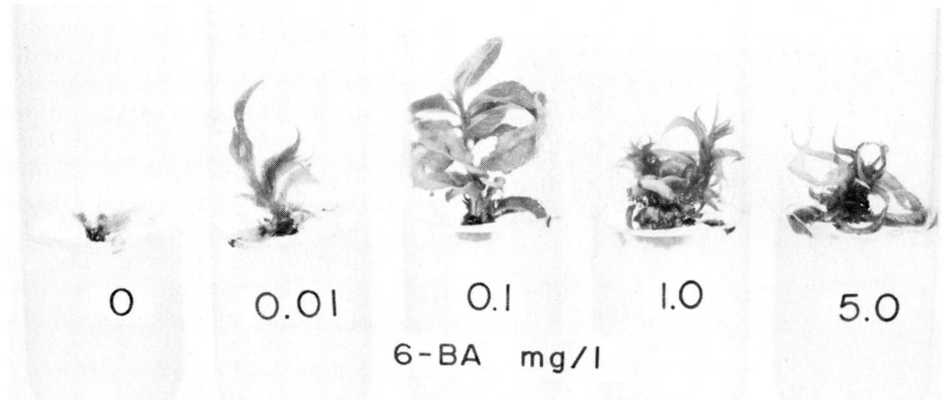

FIGURE 17–7 Cytokinin (BA) concentration changes the kind of shoot development in the multiplication stage. At 0.1 mg/l, the terminal shoot elongates; at 1.0 mg/l, lateral shoots develop and the terminal shoot is inhibited. Propagules are almond-peach hybrid.

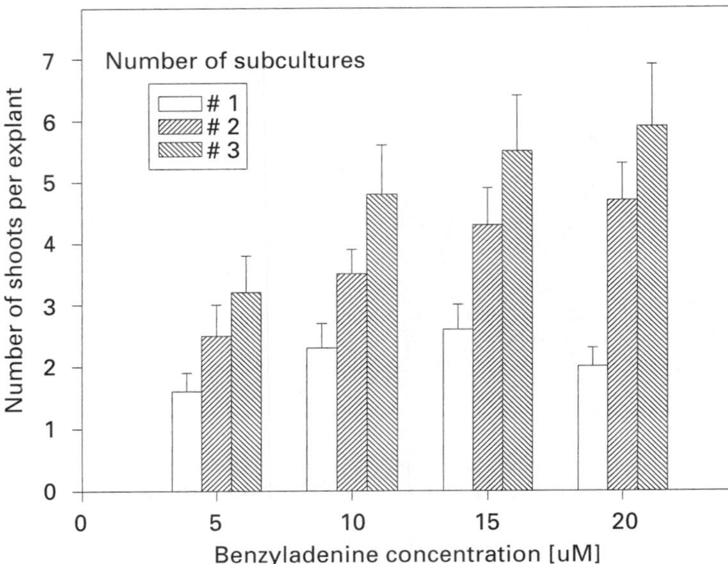

FIGURE 17–8 Relationship between cytokinin (BA) concentration and shoot initiation in shoot cultures of eastern redbud *(Cercis canadensis)* *(199)*. The number of shoots per explant increases as cytokinin level increases. This response is consistent between subcultures, but the overall number of shoots per explant increases.

adversely affect multiplication rate, yield and quality of shoots, rooting, and subsequent growth. If subculturing is delayed too long, leaf yellowing and necrosis can develop.

Subculture technique needs to be optimized for the particular kind of culture. This includes size of new explant, method of cutting, and sometimes orientation of planting. Seasonal rhythmic patterns in proliferation sometimes occur (*171*), even though subculturing is done under uniformly controlled conditions. In general, proliferation is greater in summer than in winter.

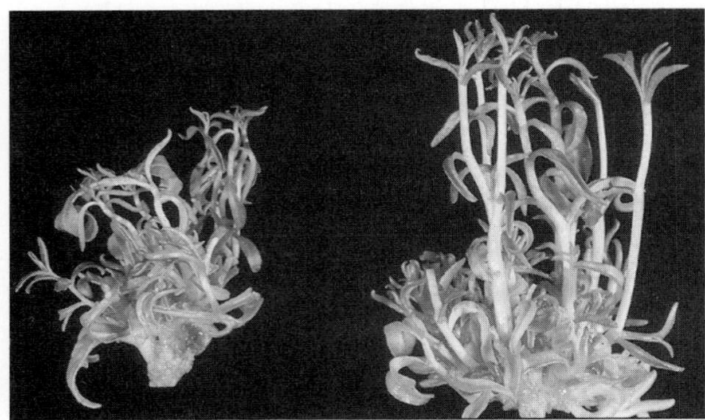

FIGURE 17–9 Although cytokinin promotes shoot formation, high concentrations also inhibit shoot elongation. The culture of gas plant *(Dictamnus albus)* on the left was treated with 20 μM benzyladenine (BA), while the culture on the right was treated with 5 μM BA *(91)*.

Stage III Root Formation

The function of this stage is to root the microcuttings and to prepare them for transplanting out of the aseptic protected environment of the test tube to the outdoor conditions of the greenhouse or transplant area (*17, 38, 140, 203*). Rooting can take place in the *in vitro* or *ex vitro* environment (Figure 17–10). Alternate sequences for rooting include the following protocols:

In vitro rooting

a. Subculture individual microcuttings to a root-inducing medium in which the growth regulator balance of the medium is changed to reduced (or no) cytokinin and increased auxin. The concentration of basal salts in the rooting medium is usually cut in half.
b. Transfer individual microcuttings into a root-inducing (often liquid) medium for a few days and then transfer to an auxin-free medium for rooting. This procedure may take place in the light or darkness and for some species, including additional factors in the medium such as phloroglucinol can be helpful (*87*).

Ex vitro rooting

Many commercial as well as experimental micropropagation systems avoid *in vitro* rooting by treating microcuttings with auxin, inserting them directly in the rooting medium, and placing them

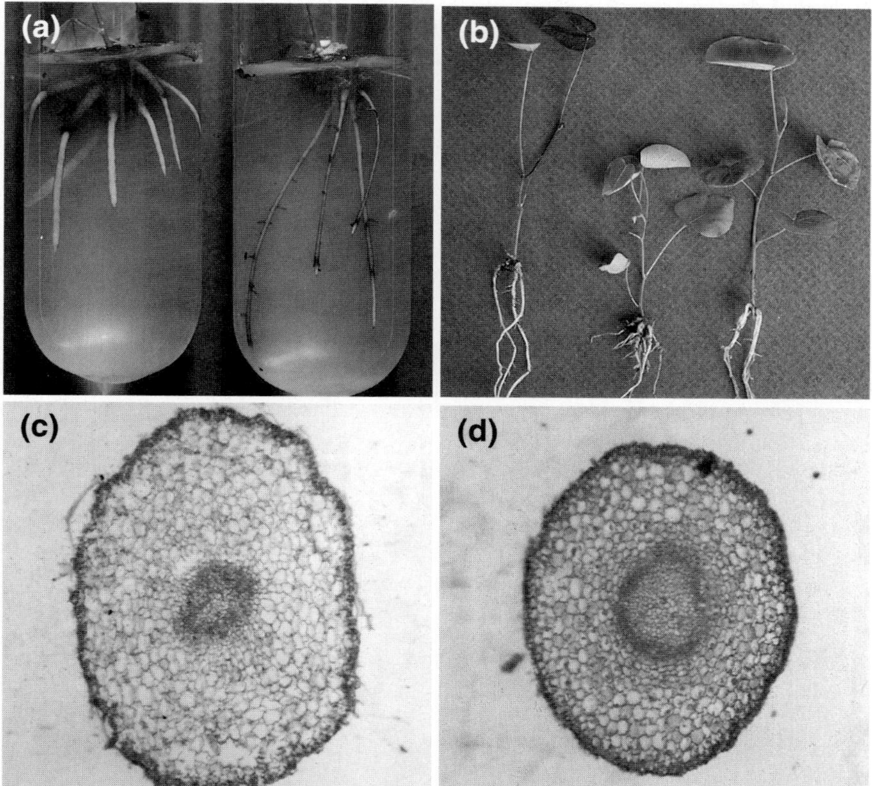

FIGURE 17–10 Root formation in microcuttings of eastern redbud *(Cercis canadensis).* (a) Root formation *in vitro.* It is common that roots on NAA-treated microcuttings (left) are shorter and thicker than roots on IBA-treated microcuttings (right). IBA is used most often to root microcuttings in a wide variety of species. (b) Three cultivars of redbud microcuttings rooted *ex vitro.* (c) Anatomy *in vitro* and (d) *ex vitro* developed roots. Observe the swollen cortical cells and the less developed vascular system on *in vitro* roots.

under mist or high humidity conditions for rooting. The procedure not only provides an excellent transition from the culture environment to the open air but saves labor requirements for handling plants in micropropagation. Protocols include either:

a. Treat microcuttings with auxin (usually IBA) as a quick-dip and stick in greenhouse medium under high humidity (see Chapter 18).
b. Treat microcuttings with auxin for 5 to 15 days as *in vitro* treatment in agar or liquid culture, then wash off the agar and place microcuttings under an *ex vitro* environment.

There is a distinct morphological difference between roots formed *in vitro* compared to *ex vitro* roots (Figure 17–10). *Ex vitro* microcuttings tend to form roots that are more "normal" than *in vitro* microcuttings. This includes less swollen cortical cells and a more well-formed vascular system (Figure 17–10). After transplanting to a greenhouse medium, roots that developed *in vitro* may not survive. In other cases, *in vitro* roots must adopt a "normal" morphology to function properly (*61, 113*).

One positive aspect of *in vitro* culture is that many plants that are normally difficult-to-root from cuttings can be rooted as microcuttings (*151, 181*). This has been shown to be related to the number of subcultures (*169*) and probably reflects the maturation state of the microcuttings. It is generally considered that microcuttings are in a more juvenile stage of development while in culture. Interestingly, this appears to be a short-term rejuvenation because micropropagated plants can flower 1 or 2 years after transplanting to the field, while seedlings of the same species can take up to 15 years to flower.

Stage IV Acclimatization (143)

This stage involves the shift from a **heterotrophic** (sugar-requiring) to an **autotrophic** (free-living) condition and the **acclimatization** of the microplant to the outdoor environment. Keeping the shoot actively growing is important because acclimatization and development of autotrophic conditions depends on new growth after transfer from the test-tube environment. Immediately upon transplanting, rooted microplants should be kept in very high humidity, gradually exposing them to outdoor conditions to prevent dehydration (*36, 37*).

Microshoots developed *in vitro* have a unique **leaf morphology.** Leaf size and the number of cell layers that comprise the leaf are much reduced compared to leaves developed outdoors. Because of the high humidity *in vitro,* leaves do not develop **epicuticular waxes** on the leaf surface in normal amounts or with the same chemical properties (Figure 17–11) (*143*). As a result, leaves on microshoots desiccate quickly after they are removed from *in vitro* culture.

Stomata are also morphologically different on leaves of *in vitro* grown plants. For example, *in vitro* sweet gum *(Liquidambar)* leaves had stomata with guard cells that were rounded and raised on the lower leaf surface, while greenhouse plants typically had elliptical and sunken guard cells (*195*). Stomata on plants grown *in vitro* usually remain open and fail to close after initially being removed from the *in vitro* environment, again leading to rapid wilting and desiccation of these leaves.

Treatments to Enhance Acclimatization

Prior to transplanting:

a. Reduce humidity (approximately 35 percent) in the *in vitro* environment prior to transplanting. This can be accomplished by placing a desiccant in the vessel or by cooling the bottom of the vessel. Uncapping vessels for five to seven days prior to transplanting has also been effective.

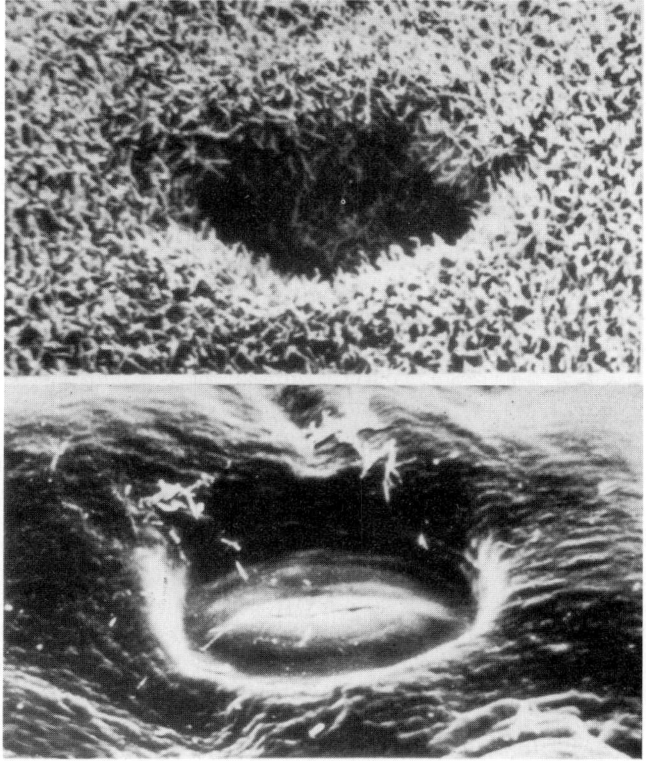

FIGURE 17–11 Scanning electron micrograph of stomata from the lower leaf surface of carnation plants grown in the greenhouse (top) or *in vitro* (bottom). Note lack of wax on in vitro grown plants. (Courtesy of E. Sutter. University of California at Davis)

b. Increasing osmotic potential of the medium (usually with additional sucrose).
c. Growth regulators (like paclobutrazol) have been used with variable success as an *in vitro* pretreatment before transplanting.

After transplanting:

a. Gradually reducing humidity is the most common treatment to acclimatize plantlets. Humidity may be maintained by using intermittent mist, fog, or enclosing plantlets in a polyethylene tent. This is a critical step in the acclimatization process.
b. Antitranspirants have been used with variable results and can be phytotoxic. They are not commonly used commercially.

TYPES OF SYSTEMS USED TO REGENERATE PLANTLETS BY MICROPROPAGATION (TABLE 17–1)

There are basically two developmental patterns for plantlet formation in tissue culture (Figure 17–12). These are described by the way shoots originate from the initial explant in Stages I and II. Once shoots have formed, the rooting and acclimatization stages are the same for the various developmental patterns. These patterns are: (a) **axillary shoot formation,** and (b) **adventitious shoot formation**. Adventitious shoots may arise directly from an explant without nodes or may arise from callus produced by the initial explants.

Axillary shoot formation

Meristem culture. This procedure utilizes the smallest part of the shoot tip as the explant (see Figure 17–13), including the meristem dome and a few subtending leaf primordia. The number of additional structures depends upon the length of the excised stem. The primary reason for this procedure is to produce a rooted microplant that is free of systemic viruses, viruslike organisms, and superficial fungi and bacteria (*80, 92, 93, 144, 164, 174*). The smaller the explant, the more effective is the elimination of pathogens (*92, 103*). Meristem tips of 0.10 to 0.15 mm have given 100 percent virus elimination with the percentage decreasing as the size increased to 1 mm. On the other hand, the smaller the explant, the more difficult it is to establish and the lower the survival rate. A general compromise is to use explants 0.25 to 1.0 mm long (*118*). Virus status must be confirmed by indexing

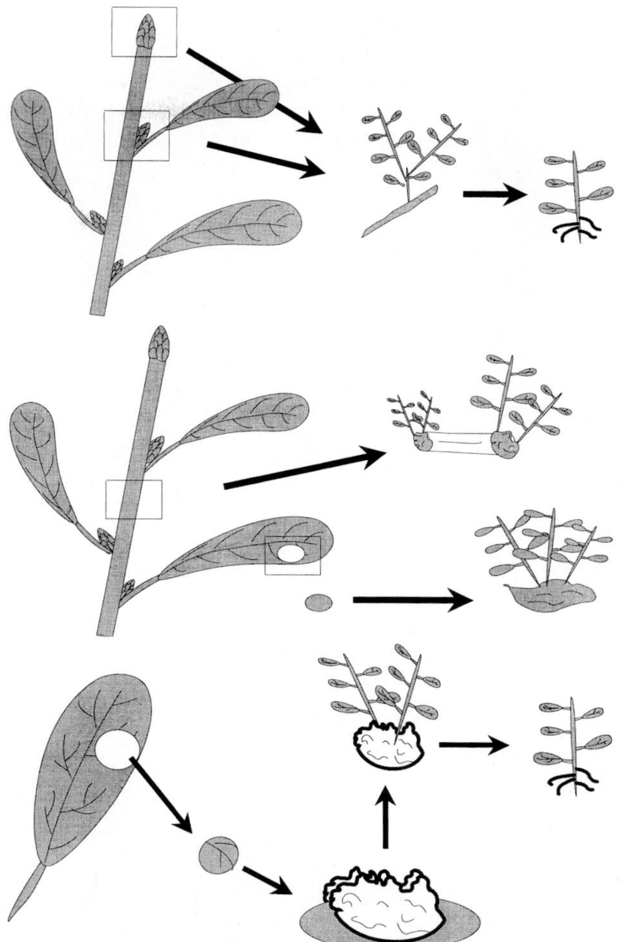

FIGURE 1–12 Patterns of plantlet development by micropropagation: (a) axillary shoot formation from nodal explants; (b) adventitious shoots forming from tissue without nodes; (c) adventitious shoots formed after callus is formed from the initial explant.

(*107*). **Heat treatment** (thermotherapy) of the test plant prior to meristem excision can increase the probability of virus removal. Some antiviral agents, such as Ribavirin (*56, 140*), added to the medium have increased the recovery of virus-free plants.

Once a rooted plant is produced, it can be used as the start of a foundation clone in a "clean stock" program for that particular species and cultivar (*145*) (see Chapter 9).

Meristem-tip culture has been successful with many important herbaceous crops. Material from this kind of program has become an essential production aspect for such commercial crops as carnation (*79*), chrysanthemum, orchid, geranium,

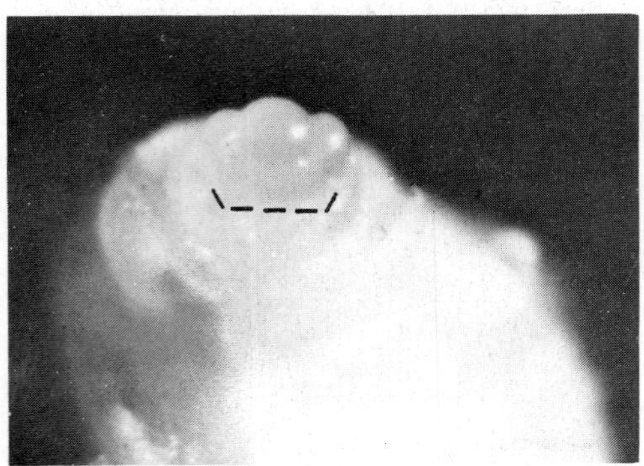

FIGURE 17–13 Shoot tip of carnation stem with outer leaves removed, showing the apical and lateral meristems (growing points). Part of shoot tip to be excised for culturing is indicated in lines. (Courtesy W. P. Hackett.)

potato, sweet potato, cassava, banana, and others. The procedure is more complex with woody plants, and micrografting is an alternative procedure. Improvements in the micropropagation of woody plants have been able to extend its usefulness (*15*).

Axillary shoot cultures. Shoot cultures are like miniaturized stem cuttings (Chapters 10, 11), divisions, or layers (Chapter 15) that grow from lateral and terminal meristems. Plantlets developed in this way tend to be reliable in the reproduction of the genotype of the source plant because the system involves extension of existing apical meristems.

Typical of biological material, different species and different clones have unique growth patterns and requirements that may differ in culture. Several different growth patterns have been described (*55*):

a. *Axillary branching.* In this system, the apical shoot tip is repressed and lateral shoots are stimulated to form a dense clump of shoots (Figure 17–14). The base of this clump may be thickened and calluslike. Cultures are subcultured by cutting the clump into sections and planting each in a separate container or by spacing out into larger containers.
b. *Nodal cultures.* This pattern involves plants whose shoots have rather strong apical dominance. Long shoots are cut into single nodes and planted vertically in the medium. Axillary buds at each node elongate and grow in length. This pattern is repeated by again cutting into nodal segments at each subculture. Cotyledonary node cultures have been very effective in micropropagating a variety of plants, especially legumes (Figure 17–15).
c. *"Stool shoots."* This pattern is analagous to that of plants grown by layering. A shoot of several nodes is laid horizontally on the surface of the medium, sometimes on a slant. Lateral growing points form a thicket of small vertical growing shoots. These "layers" may be either subdivided at each subculture or the entire unit transferred to a new culture vessel when the medium is exhausted.
d. *Proliferation of pseudocorms.* Growing points of orchids in culture produce clusters of small protuberances that resemble the pseudocorms which are initiated when an orchid seed germinates. The structure is subcultured by cutting into sections.
e. *Minitubers.* Potato plants in culture form miniature tubers at the end of small stolons extending from the lateral meristems located at each node of potato plants when treated with high cytokinin levels, particularly in the darkness (*34*). These storage organs can be removed and used in the production of virus-free planting stock. Similarly, yam *(Dioscorea)* produces root tubers at the base of stem node cuttings (*135*).

Adventitious Shoot Formation

Adventitious shoots are initiated either directly on the explant or indirectly in the callus that is produced on the explant. This type is analogous to leaf and root cuttings (Chapters 10, 11), bulb scales, and bulbils (Chapter 16). The selection of explant and the growth regulator regime determine the success of adventitious shoot initiation.

Indirect development of adventitious shoots first involves the initiation of basal callus from excised shoots in culture (*3*). Shoots arise from the periphery of the callus and are not initially connected to the vascular tissue of the explant (Figures 17–16, 17–17). Adventitious shoot formation can result in high rates of multiplication, in general, higher than rates compared to axillary shoot cultures. On the other hand, adventitious shoots can also result in increased numbers of aberrant, off-type plants resulting from conditions described in a later section.

FIGURE 17–14 Shoot-tip propagation by axillary shoot formation on poplar *(Populus). Upper left:* Vegetative bud explant after two weeks in culture (3 mm in diameter). *Upper right:* Axillary buds on explant after four to six weeks (15 mm in diameter). *Lower left:* Advanced proliferation of microshoots (5 cm in diameter). *Lower right:* Rooted microplant. (Courtesy C. B. Christie (*24*).)

FIGURE 17–15 Shoots arising from cotyledonary node explants of eastern redbud *(Cercis canadensis).*

Some kinds of explants are as follows:

Leaf pieces. Plants such as African violet *(Saintpaulia)* (*11*), *Salpiglossis* (*108*), and horseradish (*125*) easily regenerate in this manner.

Thin layer epidermal strips. Experimentally, whole plants have been regenerated from epidermal strips five to seven layers thick, usually originating from flower scape tissue (*100, 189*).

Fragmented shoot apices (*7,8*). Grape shoot apices 0.1 mm in length are cut into two to four segments and cultured in drops of medium. These small segments proliferate into small leafy structures that can develop into whole plants.

Cotyledons and hypocotyls. Excised cotyledon and hypocotyl segments are useful as starting points (Figure 17–17) for conifers from which adventitious shoots regenerate (*13, 22*).

FIGURE 17–16 Adventitious shoot formation on Douglas-Fir *(Pseudotsuga menziesii)* cotyledons. (a) Cotyledons are excised from germinated seeds, cut into segments, and placed on a medium with 1 ppm BA plus 0.001 ppm NAA to initiate adventitious buds. (b) Medium is changed by omitting hormones to allow shoots to develop. (c) Individual shoots are excised and rooted in culture with 0.001 ppm NAA, 0.5 percent sucrose, and the temperature reduced to 19° C. (d) Plantlets grown in greenhouse. (Courtesy Dr. Tsai-Ying Cheng.)

Young needle fascicles. These have been used to regenerate shoots from older conifer trees (*13*). Explants from rejuvenated plant tissues are particularly responsive (*12*).

Immature inflorescences on flower scapes. Segments from these structures are highly regenerative in some monocots, such as *Gladiolus* (*205*), *Hemerocallis* (*76*), *Iris* (*124*), *Hosta* (*123*), and *Freesia* (*142*), as well as some dicotyledons such as *Gerbera* (*141*) and *Chrysanthemum* (*156*).

Bulb scales and other storage structures. Bulbous-type monocots (*83–85*) characteristically have rings of meristematic tissue at the base of bulb scales near the basal plate (see Chapter 16). Excised scale explants in culture develop adventitious shoots directly. Initiation of adventitious shoots begins in single parenchyma cells located either in the epidermis or just below the surface of the stem; some of these cells become meristematic and develop into pockets of small, densely staining cells (*77*).

Development of Haploid Plants

Anther and pollen (microspore) culture. The discovery that pollen grains could develop into embryos was made by accident with *Datura* by Guha and Masheshwari (*62*). Development of microspores into plants was later realized in tobacco by Nitsch and co-workers (*14, 136*). Since then, whole plants have been produced in a number of species either directly from the immature pollen grain (*136, 180*) or from callus that develops from the microspore (*162, 180*).

Micrografting. Grafting very small meristem tips comparable to those described in the preceding section can be used as an alternative method to produce virus-free materials for various woody plants, such as citrus (*133*), apple (*82*) (see Figure 17–19), and Prunus (*15, 134*). For example, this procedure is important in citrus not only because it is successful but also because explants can be used from ontogenetically mature trees (see pp. 252) and avoid the juvenile phenotype of nucellar seedlings also used in virus cleanup in citrus (see pp. 263). Micrografting also shows promise for early detection of graft incompatibility relationships (*89*) (see box, p. 568).

Anther culture is used for plant breeding to produce haploid plants. Subsequent doubling of the chromosomes results in an isogenic, homozygous line (*18*). Two basic procedures have been used for tobacco (*180*). Figure 17–18 describes a simple technique that is effective with this plant. A second technique utilizes a "stress" period of the excised buds prior to culture. Flower buds are placed into a sealed container and stored in the dark for a period of time based on a time-temperature pattern, such as 7° C for two to three weeks. The temperature may be higher, but the time must be shorter. After this treatment the anthers are floated on liquid medium in a petri dish. Embryos become discernible in about 14 days. It is necessary to transfer the embryos to new medium, or new medium must be added. Procedures have also been described (*140*) for potato (*192*), *Brassica* spp. (*94*), cereals, and grasses (*193*).

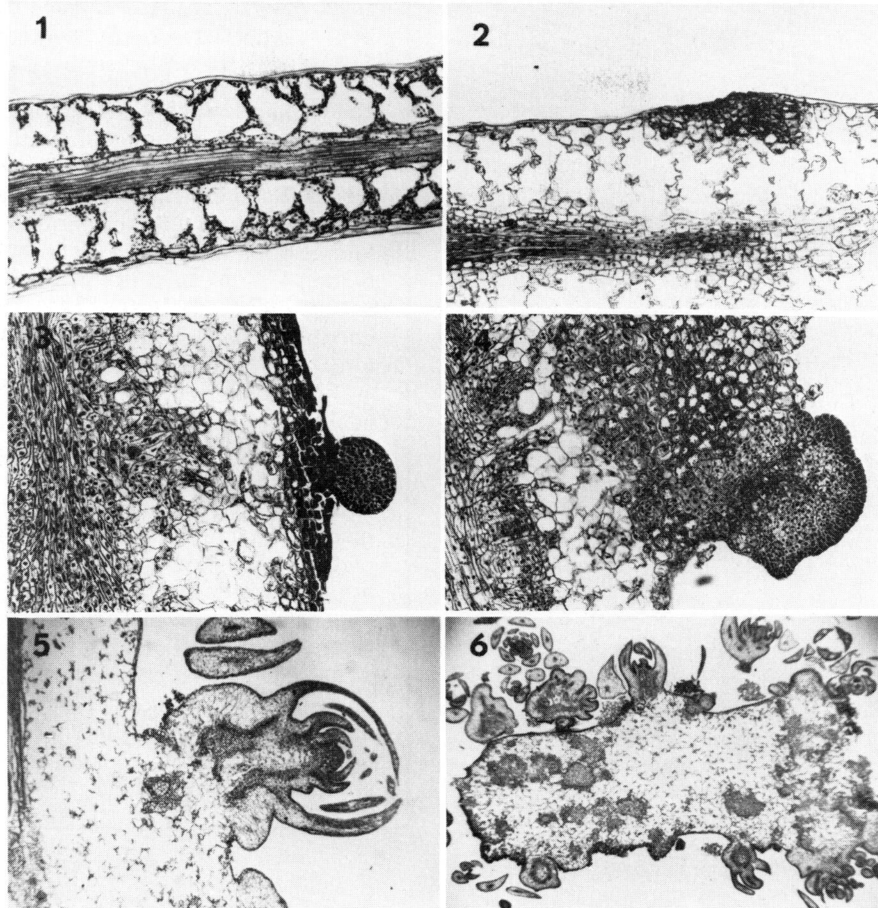

FIGURE 17–17 Adventitious shoot initiation on Douglas-Fir *(Pseudotsuga menziesii)* cotyledons. *Top left:* Cross section of cotyledon before culture. *Top right:* Initiation of a meristemoid on surface of cotyledon. *Center left:* Development of meristemoid into a tiny globular mass of tissue. *Center right:* Differentiation into a shoot primordium. *Lower left:* A completely developed shoot tip. *Lower right:* Cotyledon showing masses of adventitious shoot tips. (Courtesy Dr. Tsai-Ying Cheng.)

CALLUS, CELL, AND PROTOPLAST CULTURE SYSTEMS

Callus provides an important tissue culture system that can be subcultured and maintained more or less indefinitely. Several important pathways of development may follow.

1. New roots, shoots, or complete plantlets may develop from individual cells or clumps of cells **(organogenesis)**.
2. Cells disassociate when grown in liquid culture medium on a rotary shaker and develop into a **cell suspension**. One interesting type of suspension culture is the **nodule culture** observed in several woody plants (*115*). Nodules are round aggregates of cells with a distinct morphology that can become organogenic when moved to a stationary culture.
3. **Somatic embryogenesis** may take place among individual or clumps of cells in the callus or cell suspension.
4. Cells may be treated to produce a **protoplast culture** by removing the cell wall around the cytoplasm and thus enhancing the ability to introduce specific materials into the cell or to

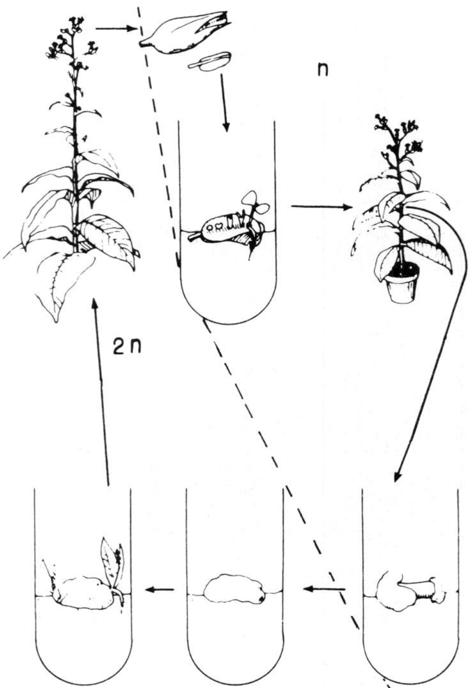

FIGURE 17–18 Anther culture is a procedure for obtaining haploid plants from normally diploid plants. A flower bud from *Nicotiana tabacum* is excised just as the petals are emerging from the bud *(upper left).* The immature anther is removed aseptically and planted on agar with a standard nutrient medium without hormones. Pollen should be at the uninucleate, microspore stage of development. Somatic embryos develop from callus derived from the haploid microspores. Haploid (1n) embryos germinate to form plantlets.

SHOOT-TIP GRAFTING IN VITRO

Citrus embryos are excised from rootstock seeds, surface sterilized, and planted in standard inorganic salt medium with 1 percent agar (*133*). Embryos germinate in the dark in two weeks. Seedlings are then removed and decapitated to a 1 to 1.5 cm length; cotyledons and lateral buds are excised with a mounted razor blade. A 0.14- to 0.18-mm tip with three leaf primordia is used as the scion. This gives reasonable success and the shoot tip eliminates viruses.

An inverted T-bud cut is made in the seedling rootstock, cutting 1 mm down the stem, followed by a horizontal cut on the bottom. The excised shoot tip is placed inside the flap next to the cambium. Grafted plants are placed in a liquid medium. A filter paper bridge with a center hole supports the stem. Cultures are kept in the light for three to five weeks to heal. When two expanded leaves appear on the scion, the grafted plant is transplanted.

A similar procedure has been used for apple (*82*) and plum (*134*). Rootstocks are either seedling plants or rooted stems. Shoot-tip scions taken from cultured plants reduce contamination problems.

join two cells together in a so-called **cybrid.** Protoplasts are significant because many manipulations with the cells are possible (*161, 176, 184*) once freed of the enclosing cell walls. For example, viruses can be more easily incorporated into protoplasts. In plant breeding, the fusion of protoplasts of two different genotypes, such as two species, which combine two nuclei and two cytoplasms, has been accomplished in a process called **somatic** (or **parasexual**) **hybridization** (*20*). Protoplast culture is particularly important in genetic engineering, since protoplasts can absorb DNA, proteins, and other large macromolecules. New genetic material can be incorporated directly into cells of an organism. This procedure is called **transformation**. Isolated protoplasts are also capable of taking up nuclei and chloroplasts (**organelle transfer**).

Callus and (particularly) cell cultures are potentially useful as methods of commercial propagation because of the high rates of multiplication and the possibility of industrialization. In practice, these methods have not been used directly, however, because significant amounts of genetic and/or epigenetic aberrations tend to develop during cell multiplication (*121*).

This class of culture system has been used extensively in the production of new and novel genotypes. Genetic variation induced in plants produced from populations of cells in culture is known as **somaclonal variation** (see Chapter 9). Variation may either be natural (*44, 159*) or may be induced by mutagenic treatments or through specific genetic engineering techniques.

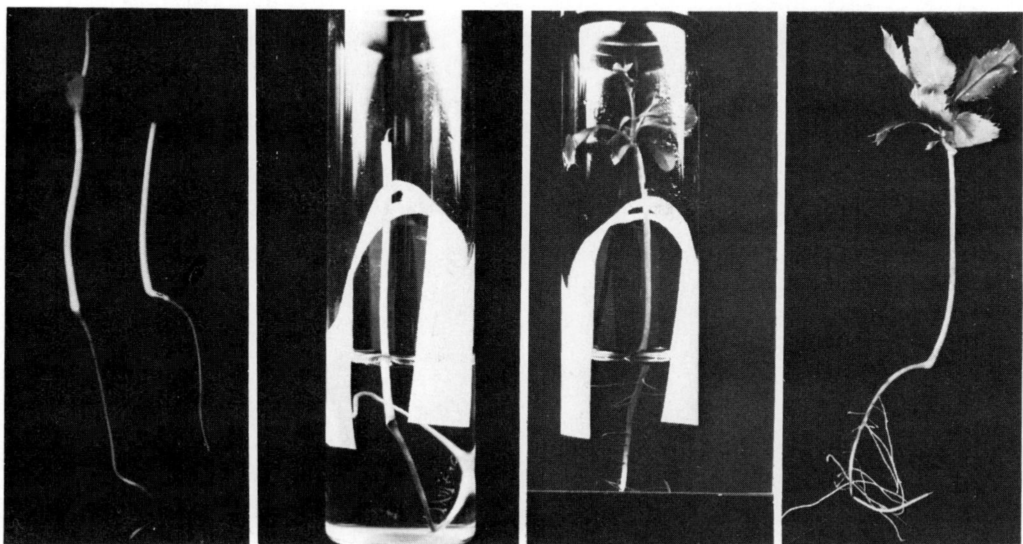

FIGURE 17–19 Micrografting apple meristems. *Left:* Seedling hypocotyl sections before and after decapitation. *Left center:* Grafted plant one week after grafting. *Right center:* Grafted plant six weeks after grafting with scion growing. *Right:* Eight weeks after grafting, ready for transplanting. (Courtesy S. C. Huang and D. F. Millikan (*82*).)

Cell suspensions are also used for the industrial production of secondary compounds, such as pharmaceuticals (*9*).

Callus Culture

Callus is produced on explants *in vitro* as a response to wounding and growth substances, either endogenous or supplied in the medium. Explants from almost any plant structure or part—seeds, stems, roots, leaves, storage organs, or fruits—can be excised, disinfested, and induced to form callus (*176, 184*). Continued subculture at three- to four-week intervals of small cell clusters taken from these callus masses can maintain the callus culture for long periods (Figure 17–20).

The most common culture medium is Murashige-Skoog (MS) (*132*), which established optimum rates for inorganic compounds, plus the Linsmaier and Skoog (*110*), which optimized organic supplements. Both media are rich in macroelements, particularly nitrogen, including both nitrate (NO_3) and ammonium ions (NH_4), sucrose, and certain vitamins. Initiation of cell division and subsequent callus production requires a supply of cytokinin and auxin in the medium at the proper proportion (*163*). Auxin at a moderate to high concentration is the primary growth substance used to produce callus. The principal auxins include indoleacetic acid (IAA), naphthaleneacetic acid (NAA), and 2,4-dichlorophenoxyacetic acid (2,4-D), in increasing order of effectiveness. Cytokinin, (such as kinetin, or benzyladenine (BA)) is supplied in a lesser amount if not adequate within the explant.

Although callus tissue cultures may appear outwardly to be uniform masses of cells, in reality their structure is relatively complex with considerable morphological, physiological, and genetic variation within the callus. Growth follows a typical logarithmic pattern. There is (a) a slow initial cell division *induction* period requiring auxin, (b) a rapid cell *division* phase involving active synthesis of DNA, RNA, and protein, followed by (c) a gradual *cessation* of cell division along with (d) *differentiation* into larger parenchyma and vascular-type cells. Cell division does not take place throughout the culture mass but is located primarily in a meristematic layer on the outer periphery of cells. The inner parts of the callus remains as an undivided mass of older tissue and, in time, may differ physiologically and genetically from cells of the outer layer. Division in the exterior layer decreases and the appearance of the callus may become "knobby" as cell division becomes restricted to specific islands of cells. Thus, variations in cell age and type may occur within the tissue culture mass. The inner cells are older and the exterior cells are younger as a meristematic

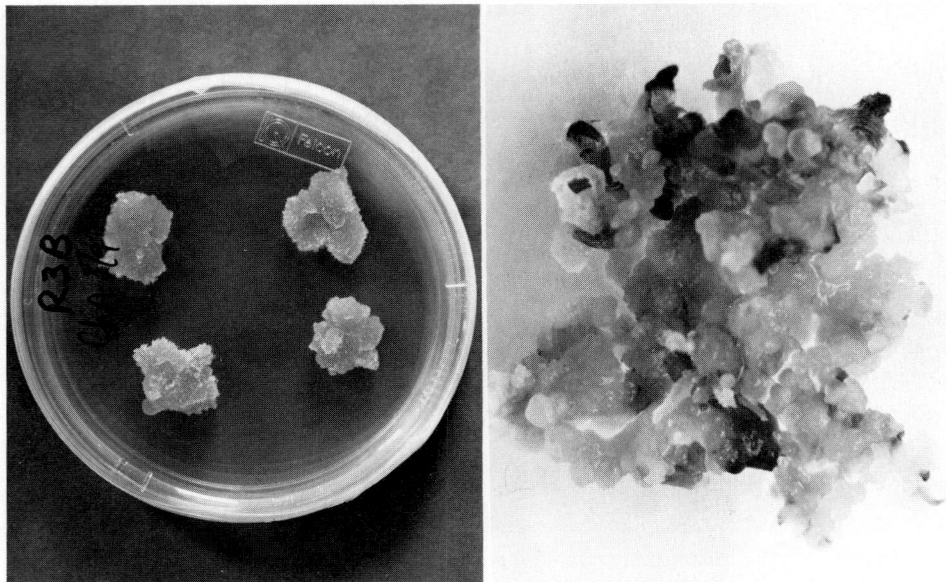

FIGURE 17–20 Callus cultures. *Left:* Undifferentiated callus of tomato *(Lycopersicon esculentum)* growing *in vitro. Right:* Callus of corn *(Zea mays)* showing dark spots that are green meristemoid-producing areas that will initiate shoots. (Courtesy Carole Meredith.)

region persists around the periphery of the callus mass.

Organogenesis begins with **dedifferentiation** of parenchyma cells to produce centers of meristematic activity **(meristemoids)** (*43, 183, 187*). In early studies by Skoog, tobacco callus produced shoots (*163*) if a relatively high cytokinin/auxin ratio was supplied. If the ratio was reversed, roots tended to form. An intermediate ratio produced both. Adenine was synergistic with cytokinin and increased inorganic phosphate (PO_4) was useful. Although the same basic pattern tends to follow with most other plants, an exact formula for optimizing conditions for regeneration is needed for each species or cultivar.

Cell Suspensions

A suspension culture is started by placing a piece of friable callus or homogenized tissue in liquid medium so that the cells disassociate from each other (Figure 17–21). In **batch cultures,** cells are grown in a flask placed on a shaking device that allows air and liquid to mix. Rotating devices that result in continuous bathing of the tissue are available. Another device, called a **chemostat** or **turbidostat,** continuously cycles the media through the cell culture essentially in the same manner as in the culture of microorganisms. In a third method, cells on a **filter paper layer** are placed on a shallow liquid medium in a petri dish with no agitation.

Growth of cells follows a typical pattern based on changes in rates of cell division. Cells first divide slowly *(lag phase),* then more rapidly *(exponential),* increasing to a steady state *(linear),* followed by a declining rate *(deceleration)* until a *stationary* state is reached. When cells are transferred to a new liquid medium, the process will be repeated. Under proper environmental conditions with media control, the process can go on indefinitely.

Devices known as **bioreactors** are used to grow cell suspensions on a large commercial scale (*179*) (Figure 17–22). These were originally developed to grow microorganisms or other living cells for fermentation or to produce various secondary products for industrial use. These devices include provision for the introduction of fresh medium and removal of spent medium, along with proper environmental controls.

Culture media are usually similar to those for callus tissue culture and include a complete range of ingredients: inorganic salts, sucrose, vitamins, and a proper balance of growth substances.

FIGURE 17–21 Suspension culture of soybean *(Glycine)* containing proembryogenic masses.

Protoplast Culture

Protoplasts are the living parts of plant cells, containing the nucleus, cytoplasm, vacuole, and various cellular structures surrounded by a semipermeable membrane (plasmalemma), but with the cell wall removed. The plant cell, in contrast to the animal cell, is surrounded by a firm nonliving cell wall composed of cellulose and hemicellulose and held together by pectin materials. The cell wall is digested by microbial enzymes to yield the protoplast. Protoplasts can be obtained from cells in suspension or can be derived directly from mesophyll leaf cells.

The major advance that permitted protoplast cultures to be made (*26, 176*) was the discovery that the walls of plant cells could be removed by enzymes that digest the pectin and allow the protoplast surrounded by the cellular membrane to survive (Figure 17–23). Commercial enzyme preparations are available for this purpose. Maintaining an adequate osmotic pressure to prevent disruption of membranes is necessary; mannitol (0.45 to 0.8 *M*) has been used for this purpose. Protoplasts are cultured in media similar to those for cells except for the presence of the osmoticum. During subsequent culture, regeneration of the cell walls takes place rapidly within several days. When new cell walls are produced, cells resume cell division and can be used to start tissue cultures or cell suspensions.

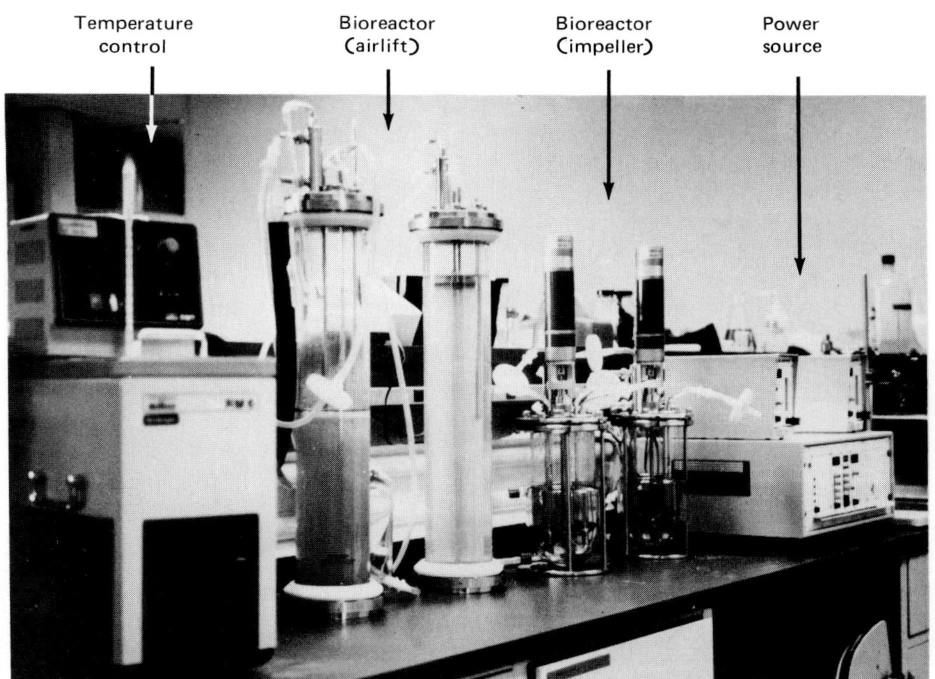

FIGURE 17–22 Laboratory version of two kinds of bioreactors. These are used to propagate plants from somatic embryos. (Courtesy Steven Strickland, Plant Genetics, Inc., Davis, Calif.)

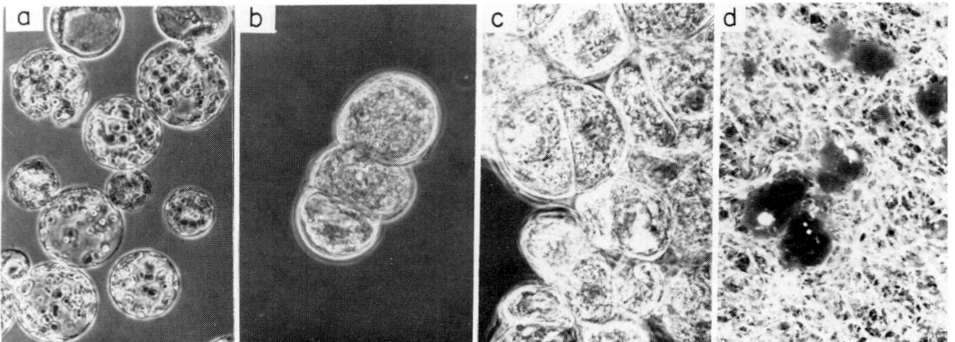

FIGURE 17–23 Protoplasts of Douglas-Fir *(Pseudotsuga menziesii)* cotyledons. (a) Freshly isolated protoplasts with cell walls removed. (b) Four-cell stage after protoplast has resynthesized a cell wall and divided twice. (c) Cells forming a colony. (d) Callus formation. (Courtesy Dr. Tsai-Ying Cheng.)

SOMATIC EMBRYOGENESIS AND SYNTHETIC SEED PRODUCTION

Somatic embryogenesis is the development of embryos from cells and tissues within *in vitro* systems (*160*). Early in the study of cell suspension systems, F. C. Steward et al. (*170*) discovered that carrot cells, when treated with coconut milk, stopped multiplying and differentiated into miniature embryolike structures that were called **embryoids.** At about the same time, Reinert (*155*) independently discovered the same phenomenon in carrot grown on agar using high auxin concentrations as the inducing agent. Since then, specific tissues in various species have been found to either have a capacity (competence) for somatic embryogenesis in culture systems or can be induced to develop competency in culture by specific treatments to the medium. Somatic embryos develop through stages similar to zygotic embryos as described in Chapter 5 (Figure 17–24). However, the final size of the cotyledons are usually reduced and there is no development of endosperm or seed coat (*59*). The genes involved in the competence to form somatic embryos and the regulation of genes common to both somatic and zygotic embryo development are an active area of basic research (*201*). Somatic embryos can arise from two different pathways: (a) adventitious embryogenesis, and (b) induced embryogenesis.

Adventitious somatic embryogenesis. Somatic embryos develop directly from cells or callus which are associated with the explant. These cells have been embryonically predetermined prior to their excision as explants and can be referred to as **embryogenic.** This type can originate from three fundamentally different kinds of explants:

Type 1. Includes nucellus or integuments of young ovules of some polyembryonic species including citrus, mangosteen, and

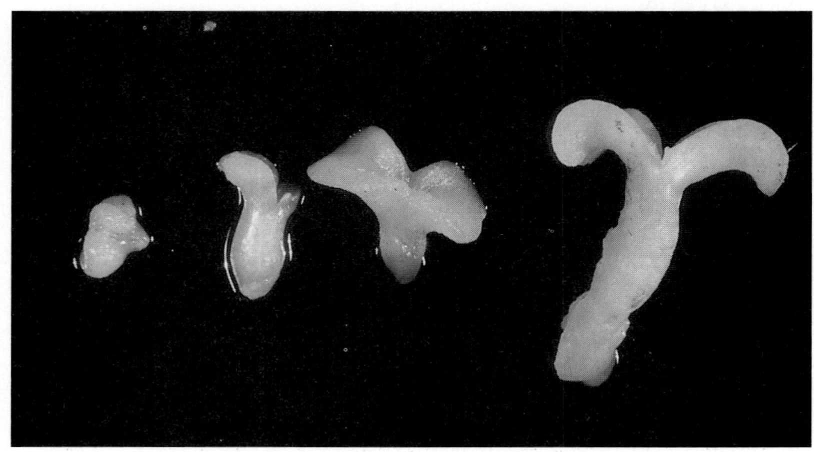

FIGURE 17–24 Somatic embryos at various stages of development. Precotyledonary stage (left) to mature cotyledonary stage at the right.

mango, as well as some monoembryonic species (loquat, rubber, apple, pear, cacao, and grape). This type reproduces the genotype of the mother plant as in apomixis (see pp. 137).

Type 2. Includes the **embryonal-suspensor mass (ESM)** at the very earliest stage that precedes embryo development (*41, 63*) (Figure 17–25). This tissue is prominent in conifers where its embryonic potential leads to polyembryony *in vivo* but it also occurs in angiosperms. Its appearance is white and mucilaginous, stains red with acetocarmine (0.10 percent w/v), fluoresces weakly under UV light as compared to green for embryonic cells and yellow for moribund cells (*41*).

Type 3. Includes the developing zygotic embryo at various stages of development or seedling tissue as the explant source (Figure 17–26). This is a common pattern of somatic embryogenesis in a wide variety of plants.

Induced somatic embryogenesis. This type of embryogenesis results from callus and cell suspen-

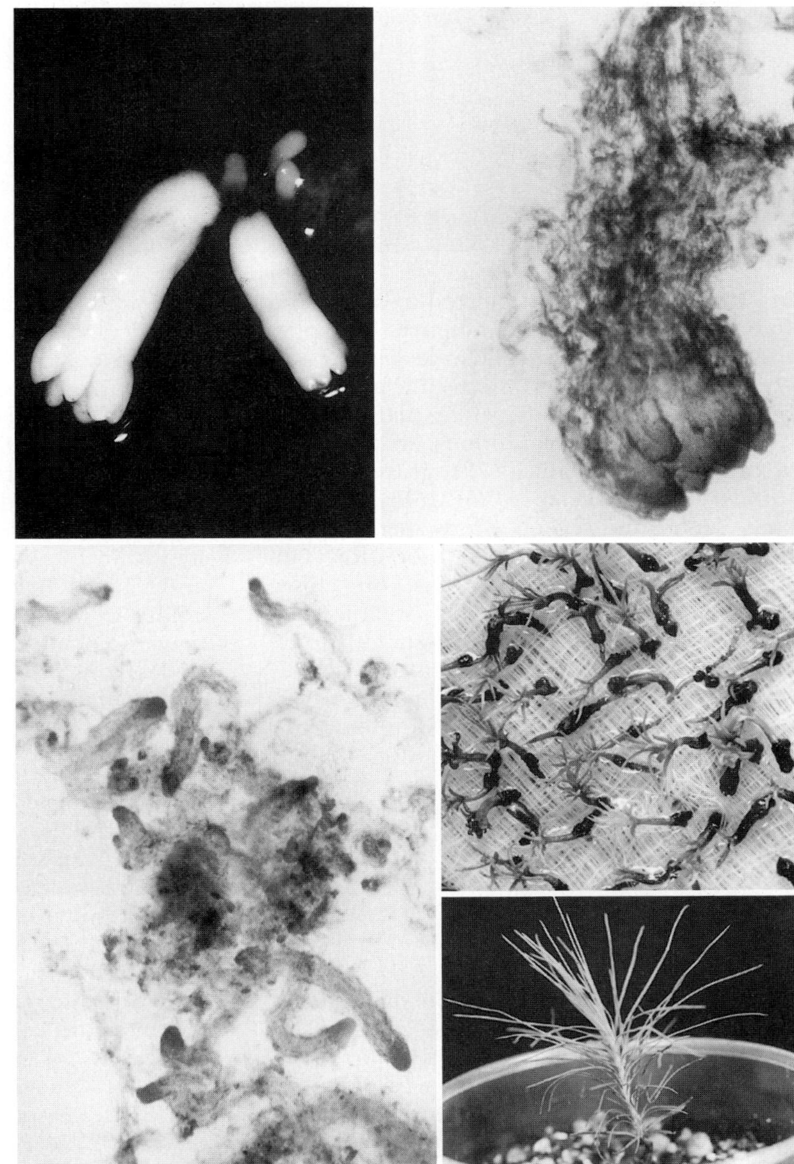

FIGURE 17–25 Somatic polyembryogenesis in conifer cell suspension cultures. *Upper left:* Transplanted embryonal-suspensor mass (ESM) of sugar pine with rescued zygotic embryos and secondary embryos developing from the ESM on agar plant. *Upper right:* Cell suspension cultures of ESM of Douglas-Fir showing cleavage polyembryony. Proembryonal cells in the cultures' ESM divide longitudinally to produce a polyembryonic mass. *Lower left:* Cell suspension cultures of ESM of loblolly pine showing multiple embryos with suspensors developing mainly by budding polyembryogenesis. *Middle right:* Recovered plantlets from Norway spruce *(Picea)* cell suspension cultures. Plantlets have cotyledons, new emerging shoots, and a root. *Lower right:* In soil, loblolly pine plantlet derived by somatic polyembryogenesis. (Courtesy D. J. Durzan.)

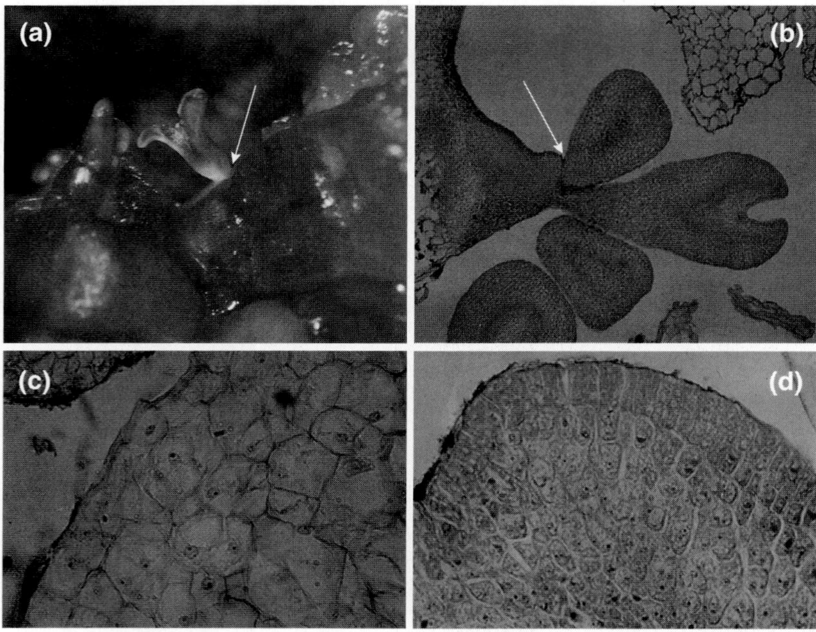

FIGURE 17–26 Somatic embryo forming directly from cotyledon tissue in eastern redbud *(Cercis canadensis)* (*53*). (a and b) Somatic embryo showing attachment to original explant at the radicle tip. (c and d) Cells in subepidermal tissue of the cotyledon dividing to form a meristematic layer from which somatic embryos will arise.

sions but only if the tissue is subjected to conditioning to induce embryogenic competence. The phenomenon seems to be biologically widespread (*103, 104, 140, 197*) but may require systematic analysis of the embryonic potential of different explant sources and the required culture conditions, as has been done for carrot (*111*), alfalfa (*179*), grasses (*191*), coffee (*167*), palm species (*185, 186*), soybean (*23*), celery (*188*), and several woody species (*86*).

Although embryogenic cell suspension cultures outwardly appear to be callus, on closer inspection these cell masses are well organized as proembryogenic masses (PEMs) (Figure 17–27). PEMs continue to develop in suspension cultures until they are moved to a stationary medium (agar-based development medium) to develop into mature somatic embryos. PEMs are often passed through sizing screens to get uniformity and synchrony of development.

Applications of Somatic Embryogenesis to Propagation

Mass Propagation

Embryos produced as a result of induced somatic embryogenesis as well as Type 1 of adventitious embryogenesis originate from somatic tissue and have the genotype of plant from which they were obtained. This creates the potential for mass clonal propagation of normally seed-propagated cultivar crops analogous to the production of apomictic seedlings. Potential crops include field crops (rice, alfalfa, orchard grass, soybean), vegetables (carrot, celery, lettuce), plantation crops (oil and date palm, coffee), and forest trees (conifers). To make this potential a reality, technical problems of engineering the entire process of embryogenesis to seedling production and planting need to be addressed. Problems with inherent genetic variability within the system need to be better understood and careful field testing and monitoring for variability is required. The concerns of monoculture may also need to be addressed (see pp. 240). Nevertheless, this procedure has considerable promise for having major applications to various species.

Genetic improvement programs of plant cultivars. Somatic embryogenesis could be useful for isolating somaclonal genetic variation within populations of cells. This is possible because of the single-cell origin of somatic embryos. Somaclonal variability is sometimes inherently present within the source plant, results from the callusing system, is induced by mutagenic agents, or can be produced by genetic engineering (see pp. 244).

Somatic embryos develop from a few cells (often single cells) (*197*). This makes them attractive targets for genetic transformation. Figure 17–28 shows PEM tissue transformed with a marker gene (GUS). Blue spots indicate integration of the new gene into the PEM cells. Transformation occurred

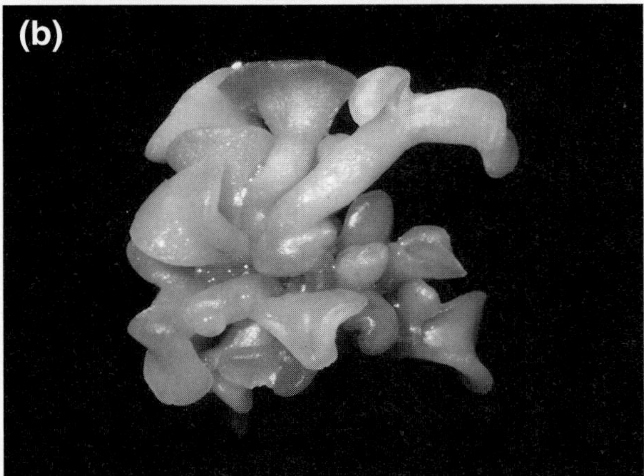

FIGURE 17–27 Somatic embryo formation in soybean *(Glycine)* cultures. (a) Proembryogenic masses (PEMs) formed from suspension cultures. (b) Formation of embryos after transferring PEMs to agar medium.

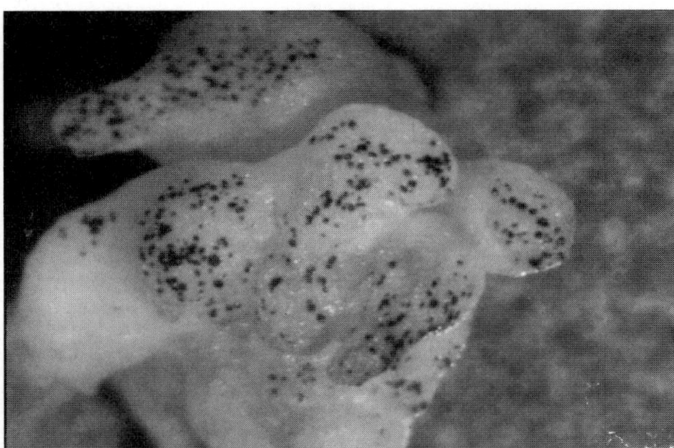

FIGURE 17–28 GUS expression (dark spots) after particle bombardment of PEMs using gene gun (biolistics).

by using the gene gun (see Chapter 2). Regeneration of somatic embryos from these cells would yield a genetically transformed plant.

In tree improvement programs, seven to ten years or more might be required to evaluate individual plants derived from culture. To solve this problem, long-term storage to preserve the original embryogenic cell masses by cryopreservation (*65*) has been suggested.

Protocols for Somatic Embryogenesis

Protocols for the induction of embryogenically determined cells or the exploitation of preembryogenically determined cells need to be established for each genotype. Nevertheless, certain generalizations can be made. Consequently, the system has the following stages:

*1. **Selection and culture of appropriate explant material.*** The selection of explant source material is the most critical decision and may require a systematic analysis of the embryogenic potential of different explant sources within the plant. The first step is the production of callus, cell-suspension, or protoplast material by methods that have already been described.

*2. **Induction of embryogenic potential (conditioning) in the cell explants.*** Induction is necessary for nonembryogenically determined cells and explants. Induction is achieved through the transfer of cells to a basal medium with a high concentration of auxin. The most effective auxins are (a) 2,4-dichlorophenoxyacetic acid (2,4-D) or (b) coconut milk plus a low concentration of naphthaleneacetic acid (NAA). After one to two weeks some small proembryos may appear. The larger clumps and proembryos may be separated by different-sized screens for transferring to the differentiation medium. Smaller cells can be subcultured for continued production of somatic embryos.

*3. **Differentiation and maturation of somatic embryos.*** After induction of embryogenic potential on an auxin-containing medium, proembryo masses are shifted to an auxin-free basal medium, high in ammonium nitrogen. Somatic embryos arise from single cells in clumps or small masses, develop polarity, and follow a pattern mimicking normal zygotic embryogenesis (see Figure 17–24) (*1, 2, 194*). Development may be variable in rate and

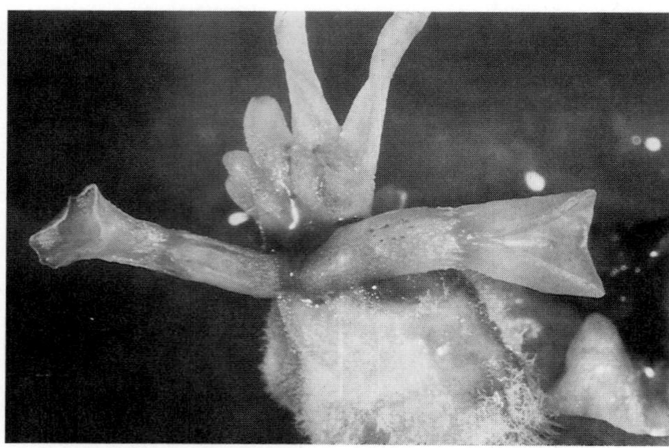

FIGURE 17–29 Malformed somatic embryo formation in cultures of eastern redbud *(Cercis canadensis)*. Typical malformations include fasciated and fused cotyledons and poor apical meristem formation.

abnormal in appearance (Figure 17–29), with secondary embryos forming on primary embryos. Synchrony and normalized development may be obtained by separating PEMs into different sizes by screening or density gradients. Adding abscisic acid (ABA) to the medium has improved uniformity and promoted normal development (*59*). Dehydration may also induce embryo maturity similar to that produced by *in vivo* embryo development (see pp. 140) (*45, 46, 58, 95*).

*4. **Plantlet formation.*** Matured somatic embryos that have reached a "normal" size can be plated onto an agar medium devoid of any auxin but containing a low level of cytokinin.

*5. **Transplanting.*** After leaves and roots have formed, the plantlet can be transplanted to a medium in a container and handled as with any seedling plant or plantlet. Mass propagation of somatic plants by embryogenesis to produce "synthetic seeds" requires the development of procedures: first, generate the required "true-to-type" embryos; second, control the development of the proembryos through "normal" patterns of embryo development and maturation; and third, provide a delivery system to make it possible to plant the seeds under greenhouse or field conditions to produce "true-to-type" plants. Considerable progress has been made in the development and utilization of large reactor vessels (*179*) to generate embryos similar to those used in fermenters or mass propagation of microbes and cells for industrial production of pharmaceuticals and other products. Fluid drilling (*59*) has some potential as a planting procedure (Chapter 7). Development of synthetic seed coverings to encapsulate the embryos in an "artificial" seed coat is another possible procedure (*139, 153, 154*).

CONTROL OF THE TISSUE CULTURE ENVIRONMENT *(88, 100, 152, 165)*

Temperature. The temperature range used most often for tissue culture is between 20 to 27° C (68 to 81° F). There is some evidence to suggest that high temperatures should be avoided and can reduce cytokinin-induced shoot formation (*48*). Most tissue culture labs maintain a constant temperature; however, for some species (like lily), bulblet formation is enhanced by using alternating day/night temperature cycles (*172*).

Commerical growers also use temperature to suspend growth when demand for plantlets is down. Cultures are placed in refrigerated storage to reduce growth and eliminate the need and cost of subculturing. These cultures are removed from refrigeration to resume multiplication of shoots as the demand for plants increases.

Light. Light effects can be separated into the impact of light irradiance (photosynthetic active radiation), duration (photoperiod), and quality on growth and development (see Chapter 3). Typical irradiance levels for micropropagation are between 40 and 80 $\mu mol \cdot m^{-2} \cdot sec^{-1}$ at culture height, but irradiance inside the vessel may be much lower. Also, the type of closure used can reduce light transmitted into the culture vessel (*49*). Although species differ in their light requirements, this range is a very low irradiance compared to outdoor and greenhouse irradiances (600 to 1200 $\mu mol\ m^{-2} \cdot sec^{1}$). This is because sucrose is provided in the medium of **heterotrophic** cultures rather than relying on photosynthesis for energy. Shoots in heterotrophic culture show low photosynthetic rates because of low CO_2 in the culture vessel and feedback inhibition of photosynthesis by the sugar in the medium. Higher light irradiance in heterotrophic cultures can result in loss of chlorophyll and leaf necrosis.

Photoperiod and light quality (wavelength). Very few studies have systematically studied the impact of photoperiod on culture development, but longer daylengths (12 to 16 hours) are most often used. Light quality is a function of the lamps used to illuminate the cultures and the type of vessels being

An alternative to heterotrophic growth is **autotrophic** or **photoautotrophic** growth (*88, 102*). Various investigators have argued that many if not most of the problems in commercial micropropagation are related to the use of closed heterotrophic systems for producing plants. Toyoki Kozai and colleagues have investigated conditions that would provide a photoautotropic alternative to conventional micropropagation (*102*). These conditions rely on photosynthesis as the major energy source rather than sugar in the medium. Model systems using potato shoot-tip cultures have been very productive. These cultures are characterized by having little or no sucrose, increased CO_2 levels, and increased light irradiance (100 to 200 μmol $\cdot$ m^{-2} $\cdot$ sec^{-1}). This provides enhanced photosynthesis for culture, while reducing costs and contamination. Photoautotrophic culture offers a high potential for automation for some species.

used. In general, tissue culture labs use cool white fluorescent (white light) or growlight (added red light) lamps in the culture room. Light quality is also affected by the transmission quality of the vessel and the type of closure for the vessel (*35*). Glass vessels do not transmit light wavelengths shorter than 290 nm, while polycarbonate vessels do not transmit light shorter than 390 nm. Although this is a measured difference, the important wavelengths for photosynthesis and photomorphogenesis are between 400 and 800 nm.

Light quality can also alter the growth response of shoots *in vitro.* For example, geranium *(Pelargonium)* cultures exposed to incandescent (red) light showed increased shoot elongation compared to fluorescent (white) light, while blue light reduced shoot elongation (*5*). In contrast, photosynthetic capacity was highest when birch *(Betula)* shoots were exposed to blue light compared to either white or red light (*157*). Blue light also stimulated chlorophyll production and increased leaf size. It was suggested that these qualities might be important during acclimatization of plantlets and could be promoted by a blue light treatment prior to removing plants from culture.

Light quality can also indirectly impact shoot development by causing changes to the media components altering culture development (*72*).

In general, light inhibits root growth and there is some indication that excluding light from the rooting zone is beneficial to root formation.

Photoperiodic response of flowering can be produced *in vitro.* For example, carnation shoots can be induced to flower by treatment with 16 hours of light but remain vegetative if kept at 12 hours (*171*). It is likely that other plants can respond similarly.

Gases that impact shoot development in culture include oxygen, carbon dioxide, and ethylene (*152*). Generally, commercial growers do not attempt to alter the levels of gases in culture. However, all closures and caps used for tissue culture are permeable to gases to some degree. There is usually some enhancement in growth by venting closures or providing filtered air exchange. This is due to increased CO_2 available for photosynthesis (even in heterotrophic cultures) and in some cases, reduced ethylene levels. Of course, elevated CO_2 concentrations are a key component of proposed photoautotrophic systems.

SPECIAL PROBLEMS ENCOUNTERED BY IN VITRO CULTURE

Hyperhydricity. **Hyperhydricity** (also termed **vitrification**) is characterized by a translucent, water-soaked, succulent appearance that can result in cultures that deteriorate and fail to proliferate (*28, 29, 50*). Physiologically, expression involves excess water uptake ("waterlogging") and inhibition of lignin and cellulose synthesis. This condition appears to be a consequence of the water relationships between the matric potential in the medium and developing shoots, as well as a low nitrate to ammonium ratio. Hyperhydricity is more prevalent if plants are grown in liquid media or with low agar concentrations, high humidity, and high ammonium concentrations (as in the MS medium). Some of the factors that have provided some relief from this problem include increasing the agar concentration, changing brands of agar, modification of inorganic ingredients in the medium, changing the cytokinin concentration, and the addition of antivitrification agents (*140, 204*).

Internal Pathogens. Some explants are found to have internal pathogens in the medium that are not apparent until after the culture has been grown for some time.

Excessive exudation. For discussion see pp. 557.

Shoot-tip necrosis and internal browning. Actively growing shoot tips sometimes develop "tip die-back" and subsequently die (Figure 17–30). This condition is usually caused by calcium deficiency in the medium and correction requires refining the basic medium (*114*).

Tissue proliferation (*109*). Tissue proliferation (TP) is the formation of gall-like growths on the stem of micropropagated plants. This may occur while plants are still in culture or may not be seen until after plants have been moved to the greenhouse or nursery. TP is most common in species in the heath family (Ericaceae) like *Rhododendron* and mountain laurel *(Kalmia)*. The cause of TP is not known but apparently is not pathogen-related. Similar gall-like structures have been seen naturally on *Rhododendron* and it has been suggested that the tissue culture environment may promote the expression of TP in species having this tendency naturally. Some cultivars have a greater tendency to show TP and care should be taken in micropropagating these plants.

Habituation. This term has been given to a phenomenon in which callus and cell systems lose their requirement for specific growth regulators. This apparently epigenetic trait tends to develop after long exposure to high concentrations of a particular growth regulator (*119, 120, 171, 175*). For example, shoot cultures may become habituated to cytokinin and continue to proliferate even after the culture has been transferred to a medium without any growth regulator.

Tissue that has been infected (transformed) with various strains of bacteria *(Agrobacterium tumefaciens* or *A. rhizogenes)* can also show growth on tissue culture medium without any growth regulators. *Agrobacterium* integrates a segment of bacterial DNA (T-DNA) into a chromosome of the plant's cell (*97*). Wild type *Agrobacterium* integrates segments of T-DNA that code for auxin and cytokinin, so infected tissue makes its own growth hormones needed for callus, shoot, or root development. This is the basis for plant transformation techniques using *Agrobacterium* (see Chapter 2) and hairy root cultures to produce secondary compounds in culture (*70*).

VARIATION IN MICROPROPAGATED PLANTS

Standards for micropropagated vegetatively propagated plant cultivars are the same as standard propagated plants, i.e., true-to-cultivar, true-to-type, and free from serious pathogens (see Chapter 9). The problem of deviation from these standards may be even more acute with micropropagation, however, because the very high multiplication rates magnify any problem, and potential economic losses can be greatly increased before detection and correction (Figure 17–31).

FIGURE 17–30 Shoot-tip necrosis in cultures of eastern redbud *(Cercis canadensis)*. (a) Comparison of normal (right) and a culture with shoot-tip necrosis (left). (b) Degrees of severity in shoot-tip necrosis.

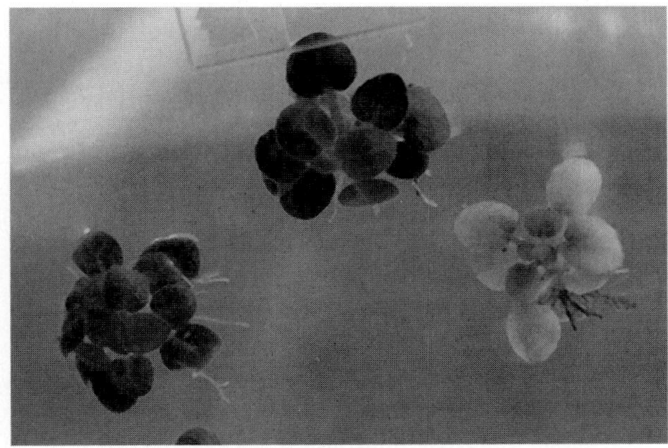

FIGURE 17–31 A variant produced during the culture of African violet. It is not uncommon to find albino individuals that lack chlorophyll production forming among normal green plants.

Much field testing has been carried out with micropropagated products and considerable commercial experience has now been obtained with various crops (*71, 171, 182, 202*). Early experience with commercial products resulting from micropropagation showed specific instances of unexpected aberrations, and off-type performance in some cases resulted in some public resistance to these products. **On the whole, most plants from micropropagation systems have been true-to-type and uniform in phenotypic appearance.**

Variability can be categorized as in Chapter 9, although it is not always possible to assign a specific variant to a given category. Furthermore, variation may be species- or genotype-specific. The basic principle is that trueness-to-type must be tested through the same phenotypic and genotypic selection procedures as described in Chapter 9.

Disease Problems

Even though micropropagation is considered to be a pathogen-free system, complete absence should not be taken for granted. The original source material should be indexed for specific viruses even if the original selection is from shoot tip cultures or micrografts. Some internal bacteria can be very persistent and hard to eliminate.

For example, in the initial enthusiasm for orchid micropropagation, the importance of initial meristem treatment was not appreciated and shoot tip culture was assumed to control viruses. As a result, many of the commercial sources of orchid cultivars were soon found to be infected, causing considerable economic loss.

Transient Phenotypic Variation

One of the first observations with micropropagated plants was that in some cases, growth patterns tended to be different in the initial post-culture period from those plants propagated by conventional methods. Usually, these were temporary and disappeared as the plants continued to grow and/or were repropagated. A particular variant may or may not be desirable, depending upon the standards for trueness-to-type of the particular cultivar and its pattern of growth and development.

Vigor. In many species, one common characteristic of micropropagated plants is enhanced vigor. For example, strawberry plants produced as a result of meristem tip propagation invariably show enhanced vigor. The reason is not completely clear. One explanation has been that the unknown viruses or other pathogens have been eliminated. Another explanation is that the plants have been rejuvenated to a more juvenile state, as described in a later section.

Change in developmental stage. Along with enhanced vigor, some species may show a specific developmental phase (*19, 181*). Strawberry *(Fragaria)* provides an example of variation in growth patterns. Micropropagated plants used directly for fruit production tended to produce somewhat smaller fruits and lower yields. The same plants tended to develop into an enhanced runnering stage and were thus useful for nursery plant production (see Chapter 19). Consequently, meristem-tip culture to control disease and shoot-tip multiplication is a useful combination to obtain "mother plants" for conventional nursery production.

Branching. A common characteristic of many herbaceous ornamentals propagated by micropropagation (*171*) is enhanced branching as compared to conventionally propagated plants. This characteristic is considered highly useful for some species, e.g., *Hosta, Begonia,* aster, chrysanthemum, rose (*171*), *Syngonium,* ferns, and other foliage plants (*26*). The same characteristics are considered undesirable for *Gypsophila, Kalanchoe, Gerbera,* and *Saintpaulia* (*171*). This characteristic has also occurred in strawberry, blackberry, and apple (*202*).

The effect has been explained as a carry-over effect of cytokinin in the multiplication medium. Another explanation could be an epigenetic effect related to phase change. Note that in some species (see Chapter 9), propagules taken from the more mature parts of a plant tend to be more branched with enhanced lateral orientations.

Epigenetic (Rejuvenation) Effects

Variations due to epigenetic effects primarily involve phase changes involved with the juvenile-mature cycle (*68*). Note the characteristics of vegetatively propagated plants in the clonal cycle described in Chapter 9 and the effect on reproductive capacity (flowering), trueness-to-type morphology, and regeneration (rooting potential). Micropropagated plants and the system of *in vitro* culture are not only sensitive to the initial status of the explant but conditions during culture can bring about changes of rejuvenation.

Rejuvenation. Many, if not all, woody plants that are consecutively recultured in shoot-tip culture show some degree of rejuvenation in rooting potential that can also be expressed in leaf morphology and/or performance. This phenomenon has been observed in blackberry, gooseberry, grape, and apple (*129, 202*). Enhanced rooting potential may persist after micropropagation such that the plants produced may be useful later as stock plants in conventional cutting propagation (*178*).

Effect on flowering. Flowering problems have been reported as occurring in some herbaceous crops, such as chrysanthemum, *Dianthus, Kalanchoe,* and *Gypsophila* (*171*). This is manifested in unevenness and reluctance or delay in flowering. Scheduling of flowering times becomes a problem and excessive variation can limit the use of micropropagated material of some crops. Cultivar differences exist in these responses and selection of cultivars that resist this problem may be the long-term solution. Future developments may include selection of cultivars specifically adapted to micropropagation procedures. The explanation for this phenomenon is variation in juvenility status among individual micropropagated plants.

The effect of micropropagation on the age of flowering varies with genotype, method of handling, and source of explant. For example, fruit and nut cultivars propagated from shoot-tip culture have not demonstrated a significant increase in the age of bearing as compared to conventionally grafted trees (*202*), as might be expected if the plants are juvenile. In fact, examples have occurred where particularly precocious cultivars of micropropagated walnut have flowered and fruited at an unusually early age, even in the nursery row. This pattern would be particularly favorable for the production of fruit and nut cultivars and for forest species grown for seed orchards, but it is a disadvantage for trees grown for timber.

True-to-type juvenile phenotypes. The production of a true-to-type juvenile phenotype as compared to an adult is important in the propagation of forest species to be used for lumber, biomass production, and other cases where the vegetative structure is the product. In general, production of the mature phenotype persists from vegetative propagation, whether by conventional procedures or by those micropropagation procedures that have been successful. Selecting a superior ("elite") genotype of conifers for forest uses must be done at a mature age, but this creates problems not only in propagation but in the production of the true-to-type phenotype (*12, 40*). It is possible that this problem can be solved by explant selection from juvenile material, reversion of mature tissues, or development of suitable micropropagation techniques that will capture the genetic gains of clonal propagation, retain a juvenile phenotype, and provide an efficient economical mass-propagation system (*12, 16*).

Genetic and Chimeral Effects

Genetic changes prior to, or during micropropagation are the most serious types of aberration because they are permanent and may be difficult to detect during the operation. As with conventional propagation, a problem may not appear until after the plants have been grown in a permanent location. It is important therefore to establish whether the problem is a preexisting one in the source material or due to conditions that occur during the propagation procedure.

Causes of genetic variation. Many important horticultural crops are chimeras as described in Chapter 9. Several cases were described in relation to conventional propagation concerning root and leaf cuttings which manifested reversions to underlying cell layers with different genotypes. Similarly, micropropagation with any method that includes adventitious shoot initiation will tend to revert and result in **chimeral breakdown** (Figure 17–32).

Systems that proliferate callus on the explant or on the microplant during any of the stages of development are apt to result in mutant sectors or individual growing points. The reason is that callus proliferation tends to result in polyploid or aneuploid cells that can produce aberrant mutant plants.

FIGURE 17–32 Micropropagation of chimeral plants requires judicious observation to maintain plants with the proper phenotype. (a) Range of variants that can be recovered in a chimeral hosta cultivar micropropagation system. (b) Selection of plants with appropriate phenotype. (c) Plants must be observed for trueness-to-type prior to sale.

Shoots being cultured often develop callus at the nodes and the base from which adventitious shoots can arise. Usually, these are low in frequency but the rate can be increased by extended subculturing or increased levels of auxin and cytokinin. This problem was recognized in Boston fern *(Nephrolepsis)* early in the development of micropropagation research (*131*). Often off-types plants are characteristic of specific genotypes. Some strawberry cultivars have been reported to produce chimeral sectors (0.1 to 7 percent), white streaks (1 to 8 percent), and dwarfs (1 percent).

Control of genetic variation. Like conventional propagation, micropropagation systems that multiply by maintaining an intact shoot tip remain stable. For this reason, axillary shoot proliferation culture is recommended for commercial micropropagation. Here it may be important to limit the number of subcultures from any single explant, as is recommended for Boston fern. Lowering the growth regulator levels during subculturing to reduce the amount of callusing may also be indicated. It may be that the best conditions for inducing rapid growth and high propagation ratios may not be the best conditions for production of true-to-type plants (*42*).

Systems that utilize adventitious shoot production or the multiplication of callus and cells may result in considerably more variation in the end product than the axillary shoot system. Where adventitious shoots are part of the system, the potential for variability of the specific genotype needs to be carefully evaluated through specific progeny tests of the end product (Figure 17–32). This precaution would apply to production from bulb scales, cotyledons, leaves, and flower scapes. Chimeras would be difficult to duplicate by these methods.

In any micropropagation system, visual inspections should be practiced throughout the production cycle and any off-type or aberrant plant rogued prior to distribution. Maintaining progeny source records should be a regular aspect of production so that the origin of a specific problem can be traced to its source. Also, long-term evaluation is needed before introducing any plant into widespread distribution.

REFERENCES

1. Ammirato, P.V. 1985. Patterns of development in culture. In *Tissue culture in forestry and agriculture,* R.R. Henke, K.W. Hughes, M.J. Constantin, and A. Hollaender, eds. New York: Plenum Press, pp. 9–29.
2. Ammirato, P.V. 1987. Organizational events during somatic embryogenesis. In *Plant tissue and cell culture,* C.E. Green, D.A. Somers, W.P. Hackett, and D.D. Biesboer, eds. New York: Alan R. Liss, pp. 57–81.
3. Anderson, W.C. 1975. Propagation of rhododendrons by tissue culture. 1. Development of a culture medium for multiplication of shoots. *Comb. Proc. Inter. Plant Prop. Soc.* 25:129–35.
4. Anderson, W.C. 1980. Mass propagation by tissue cultures: Principles and techniques. In *Proc. conf. on nursery production of fruit plants: Applications and feasibility,* R.H. Zimmerman, ed. U.S. Dept. of Agr. Sci. and Education Administration ARR-NE-11, pp. 1–14.
5. Appelgren, M. 1991. Effects of light on stem elongation of *Pelargonium* in vitro. *Scientia Hort.* 45:345–51.
6. Bajaj, Y.P.S., ed. 1986. *Biotechnology in agriculture and forestry,* Vol. I, *Trees.* Berlin: Springer-Verlag.
7. Barlass, M., and K.G.M. Skene. 1980. Studies on the fragmented shoot apex of grapevine. I. The regenerative capacity of leaf primordial fragments in vitro. *Jour. Exp. Bot.* 31:483–88.
8. ———. 1980. Studies on the fragmented shoot apex of grapevine. II. Factors affecting growth and differentiation *in vitro. Jour. Exp. Bot.* 31: 489–95.
9. Barz, W., E. Reinhard, and M.H. Zenk, eds. 1977. *Plant tissue culture and its biotechnological applications.* Berlin: Springer-Verlag.
10. Beasley, C.A. 1984. Culture of cotton ovules. In *Cell culture and somatic cell genetics of plants,* Vol. I. *Laboratory procedures and their applications,* I.K. Vasil, ed. New York: Academic Press, pp. 232–40.
11. Bilkey, P.C., B.H. McCown, and A.C. Hildebrandt. 1978. Micropropagation of African violet from petiole cross-sections. *HortScience* 13:37–8.
12. Bonga, J.M. 1986. Vegetative propagation in relation to juvenility, maturity, and rejuvenation. In *Cell and tissue culture in forestry* (2nd ed.), J.M. Bonga and D.J. Durzan, eds. Dordrecht: Martinus Nijhoff Publishers, pp. 387–411.
13. Boulay, M. 1987. In vitro propagation of tree species. In *Plant tissue and cell culture,* C.E. Green, D.A. Somers, W.P. Hackett, and D.D. Biesboer, eds. New York: Alan R. Liss, pp. 367–82.
14. Bourgin, J.P., and J.P. Nitsch. 1967. Obtention de *Nicotiana* haploides à partir d'elamines cultivées *in vitro. Ann. Phys. Veg.* 9:377–82.
15. Boxus, P., and P. Druart. 1986. Virus-free trees through tissue culture. In *Biotechnology in agriculture and forestry,* Vol. I, *Trees,* Y.P.S. Bajaj, ed. Berlin: Springer-Verlag, pp. 24–30.
16. Boxus, P.H., M. Quoirin, and J.M. Laine. 1977. Large scale propagation of strawberry plants from tissue culture. In *Applied and fundamental aspects of plant cell, tissue and organ culture,* J. Reinert and Y.P.S. Bajaj, eds. Berlin: Springer-Verlag, pp. 131–43.
17. Brainerd, K.E., L.H. Fuchigami, S. Kwiatkowski, and C.S. Clark. 1981. Leaf anatomy and water stress of aseptically cultured 'Pixy' plum grown under different environments. *HortScience* 16:173–75.
18. Burk, L.G., G.R. Gwynn, and J.R. Chaplin. 1972. Diploidized haploids from aseptically cultured anthers of *Nicotiana tabacum. Jour. Hered.* 63:355–60.
19. Cameron, J.S., and J.F. Hancock. 1986. Enhanced vigor in vegetative progeny of micropropagated strawberry plants. *HortScience* 21:1225–26.
20. Carlson, P.S., H.H. Smith, and R.D. Dearing. 1972. Parasexual interspecific plant hybridization. *Proc. Nat. Acad. Sci. USA* 69:2292–94.
21. Cassels, A.C. 1991. Problems in tissue culture: culture contamination. In. P.C. Debergh and R.H. Zimmerman (eds.). *Micropropagation: technology and application.* Dordrecht: Kluwer Academic Pub. pp. 31–44.
22. Cheah, Kheng-Tuan, and Tsai-Ying Cheng. 1978. Histological analysis of adventitious bud formation in cultured Douglas fir cotyledon. *Amer. Jour. Bot.* 65:845–49.
23. Christianson, M.L. 1985. An embryogenic culture of soybean: towards a general theory of somatic embryogenesis. In *Tissue culture in forestry and agriculture,* R.R. Henke, K.W. Hughes, M.J. Constantin, and A. Hollaender, eds. New York: Plenum Press, pp. 83–103.
24. Christie, C.B. 1978. Rapid propagation of aspens and silver poplars using tissue culture techniques. *Comb. Proc. Inter. Plant Prop. Soc.* 28:255–60.
25. Chu, I.Y.E. 1986. The application of tissue culture to plant improvement and propagation in the ornamental horticulture industry. In *Tissue culture as a plant production system for horticultural crops,* R.H. Zimmerman, R.J. Griesbach, F.A. Hammerschlag, and R.H. Lawson, eds. Dordrecht: Martinus Nijhoff Publishers, pp. 15–33.
26. Cocking, E. C. 1986. The tissue culture revolution. In *Plant tissue culture and its agricultural applications,* L.A. Withers, and P.G. Alderson, eds. London: Butterworth, pp. 3–20.
27. Collins, G.B., and J.W. Grosser. 1984. Culture of embryos. In *Cell culture and somatic cell genetics of plants,* Vol. 1, I.K. Vasil, ed. New York: Academic Press, pp. 241–57.

28. Debergh, P., J. Aitken-Christie, D. Cohen, B. Grout, S. von Arnold, R. Zimmerman and M. Ziv. 1992. Reconsideration of the term 'vitrification' as used in micropropagation. *Plant Cell Tiss. Organ Cult.* 30:135–40.
29. ———. 1987. Recent trends in the application of tissue culture to ornamentals. In *Plant tissue and cell culture,* C.E. Green, D.A. Somers, W.P. Hackett, and D.D. Biesboer, eds. New York: Alan R. Liss, pp. 383–93.
30. ———, and P.E. Read. 1991. Micropropagation. In P.C. Debergh and R.H. Zimmerman. (eds.). *Micropropagation: technology and application.* Dordrecht: Kluwer Academic Pub. pp. 1–14.
31. de Fossard, R.A. 1976. *Tissue culture for plant propagators.* Armidale, N.S.W., Australia: Univ. of New England Printery.
32. ———. 1986. Principles of plant tissue culture. In *Tissue culture as a plant production system for horticultural crops,* R.H. Zimmerman, R.J. Griesbach, F.A. Hammerschlag, and R.H. Lawson, eds. Dordrecht: Martinus Nijhoff Publishers, pp. 1–13.
33. Dirr, M.A., and C.W. Heuser, Jr. 1987. *The reference manual of woody plant propagation: From seed to tissue culture.* Athens, Ga.: Varsity Press.
34. Dodds, J.H., D. Silva-Rodriguez and P. Tovar. 1992. Micropropagation of potato (*Solanum tuberosum* L.). In Y.P.S. Bajaj (ed.). *Biotechnology in agriculture and forestry.* Vol 19. *High tech and micropropagation III.* Springer-Verlag, Berlin.
35. Dooly, J.H. 1991. Influence of lighting spectra on plant tissue culture. *ASAE Paper No. 917530.* Amer. Soc. Agr. Eng., MI.
36. Driver, J.A., 1986. Method for acclimatizing and propagating plant tissue culture shoots. Patent No. *4,612,725.* Washington, D.C.: U.S. Patent Office.
37. ———, and G.R.L. Suttle. 1986. Nursery handling of propagules. In *Cell and tissue culture in forestry.* (2nd ed.), J.M. Bonga and D.J. Durzan, eds. Dordrecht: Martinus Nijhoff Publishers, pp. 320–31.
38. Dunstan, D.I., and K.E. Turner. 1984. The acclimatization of micropropagated plants. In *Cell culture and somatic cell genetics of plants,* Vol. I, *Laboratory procedures and their applications,* I.K. Vasil, ed. New York: Academic Press, pp. 123–29.
39. Dunwell, J.M. 1986. Pollen, ovule and embryo culture as tools in plant breeding. In *Plant tissue culture and its agricultural applications,* L.A. Withers, and P.G. Alderson. London: Butterworth, pp. 375–404.
40. Durzan, D.J. 1986. Special problems: adult vs. juvenile explants. In *Handbook of plant cell culture,* Vol. 4, D.A. Evans, W.R. Sharp, and P.V. Ammirato, eds. New York: Macmillan, pp. 471–503.
41. ———, and P.K. Gupta. 1988. Somatic embryogenesis and polyembryogenesis in conifers. In *Biotechnology in agriculture,* Mizrahl, A., ed. New York: Alan R. Liss, pp. 53–81.
42. Evans, D.A., and J.E. Bravo. 1986. Phenotypic and genotypic stability of tissue cultured plants. In *Tissue culture as a plant production system for horticultural crops,* R.H. Zimmerman, R.J. Griesbach, F.A. Hammerschlag, and R.H. Lawson, eds. Dordrecht: Martinus Nijhoff Publishers, pp. 73–94.
43. ———, W.R. Sharp, and C.E. Flick. 1981. Plant regeneration from cell cultures. *Hort. Rev.* 3: 214– 314.
44. ———, W.R. Sharp, and H.P. Medina-Filho. 1984. Somaclonal and gametoclonal variation. *Amer. Jour. Bot.* 71:759–74.
45. Finkelstein, R.R., and M.L. Crouch. 1984. Precociously germinating rapeseed embryos retain characteristics of embryogeny. *Planta* 162:125–31.
46. ———, and M.L. Crouch. 1987. Hormonal and osmotic effects on developmental potential of maturing rapeseed. *HortScience* 22:797–800.
47. Fiorino, P., and F. Loreti. 1987. Propagation of fruit trees by tissue culture in Italy. *HortScience* 22:353–58.
48. Fonnesbech, M. 1974. Temperature effects on shoot and root development from *Begonia* × *Cheimantha* petiole segments grown in vitro. *Physiol. Plant.* 32:282–86.
49. Fujiwara, K., T. Kozai, Y. Nakajo and I. Watanabe. 1989. Effects of closures and vessels on light intensities in plant tissue culture vessels. *Jour. Agric. Meteorol.* 45:143–49.
50. Gaspar, T., C. Kevers, P. Debergh, L. Maene, M. Paques, and P. Boxus. 1986. Vitrification: Morphological, physiological, and ecological aspects. In *Cell and tissue culture in forestry,* Vol. 1, *General principles and biotechnology,* J.M. Bonga and D.J. Durzan, eds. Dordrecht: Martinus Nijhoff Publishers, pp. 152–66.
51. Gautheret, R.J. 1985. History of plant tissue and cell culture: A personal account. In *Cell culture and somatic cell genetics of plants,* Vol. 2, I.K. Vasil, ed. New York: Academic Press, pp. 1–59.
52. Geneve, R.L., S.T. Kester and S. El-Shall. 1990. In vitro shoot initiation in Kentucky coffeetree (*Gymnocladus dioicus* L.). *HortScience* 25:578.
53. ———, and S.T. Kester. 1990. The initiation of somatic embryos and adventitious roots from developing zygotic embryo explants of *Cercis canadensis* L. cultured in vitro. *Plant Cell Tiss. Organ Cult.* 22:71–6.
54. Gengenbach, B.G. 1984. In vitro pollination, fertilization, and development of maize kernels. In *Cell culture and somatic cell genetics of plants,* Vol. 1, *Laboratory procedures and their applications,* I.K. Vasil, ed. New York: Academic Press, pp. 276–82.
55. George, E.F. 1993. *Plant propagation by tissue culture.* Part I. The Technology. Edington, Wilts. Eng.: Exergetics Ltd.
56. ———, and P.D. Sherrington. 1984. *Plant propagation by tissue culture: Handbook and directory of commercial laboratories.* Eversley, England: Exegetics Ltd.

57. Goldy, R.G., and U. Amborn. 1987. In vitro culturability of ovules from 10 seedless grape clones. *HortScience* 22(5):952.

58. Gray, D.J. 1987. Quiescence in monocotyledonous and dicotyledonous somatic embryos induced by dehydration. *HortScience* 22:810–14.

59. ———, and A. Purohit. 1991. Somatic embryogenesis and development of synthetic seed technology. *Critical Rev. Plant Sci.* 10:33–61.

60. Green, C.E., D.A. Somers, W.P. Hackett, and D.D. Biesboer, eds. 1987. *Plant tissue and cell culture.* Proc. 6th international congress plant tissue and cell culture. New York: Alan R. Liss.

61. Grout, B.W.W. and M.J. Aston, 1977. Transplanting of cauliflower plants regenerated from meristem culture. I. Water loss and water transfer related to changes in leaf wax and to xylem regeneration. *Hort. Res.* 17:1–7.

62. Guha, S., and S.C. Maheshwari. 1964. In vitro production of embryos from anthers of *Datura. Nature* (London) 204:497.

63. Gupta, P.K., and D.J. Durzan. 1986. Somatic polyembryogenesis from callus of mature sugar pine embryos. *Biotechnology* 4:643–45.

64. ———, and D.J. Durzan. 1987. Micropropagation and phase specificity in mature, elite Douglas fir. *Jour. Amer. Soc. Hort. Sci.* 112:969–71.

65. ———, D.J. Durzan, and B.J. Finkle. 1987. Somatic polyembryogenesis in embryogenic cell masses of *Picea abies* (Norway spruce) and *Pinus taeda* (loblolly pine) after freezing in liquid nitrogen. *Can. Jour. For. Res.* 17:1130–34.

66. Haberlandt, G. 1902. Kulturversuche mit isolierten Pflanzenzellen. *Sitzungsber. Akad. der Wiss. Wien, Math.-Naturwiss.* Kl. 111:69–92.

67. Hackett, W.P. 1966. Applications of tissue culture to plant propagation. *Comb. Proc. Inter. Plant Prop. Soc.* 16:88–92.

68. ———. 1985. Juvenility, maturation, and rejuvenation in woody plants. *Hort. Rev.* 7:109–55.

69. ———, and J.M. Anderson. 1967. Aseptic multiplication and maintenance of differentiated carnation shoot tissue derived from shoot apices. *Proc. Amer. Soc. Hort. Sci.* 90:365–69.

70. Hamill, J.D. and S.F. Chandler. Use of transformed roots for root development and metabolism studies and progress in charecterizing root-specific gene expression. In T.D. Davis and B.E. Haissig (eds.). *Biology of adventitious root formation.* Plenum Press, NY. pp. 163–180.

71. Hammerschlag, F.A. 1986. Temperate fruits and nuts. In *Tissue culture as a plant production system for horticultural crops,* R.H. Zimmerman, R.J. Griesbach, F.A. Hammerschlag, and R.H. Lawson, eds. Dordrecht: Martinus Nijhoff Publishers, pp. 221–36.

72. Hangarter, R.P. and T.C. Stasinopoulos. 1991. Repression of plant tissue culture growth by light is caused by photochemical change in the culture medium. *Plant Science* 79:253–57.

73. Henny, R.J. 1980. *In vitro* germination of *Maranta leuconeura* embryos. *HortScience* 15:198–200.

74. Henny, R.J., J.F. Knauss, and A. Donnan, Jr. 1981. Foliage plant tissue culture. In *Foliage plant production,* J. Joiner, ed. Englewood Cliffs, N.J.: Prentice-Hall.

75. Hesse, C., and D.E. Kester. 1955. Germination of embryos of *Prunus* related to degree of embryo development and method of handling. *Proc. Amer. Soc. Hort. Sci.* 65:251–64.

76. Heuser, C.W., and J. Horker. 1976. Tissue culture propagation of day lilies. *Comb. Proc. Inter. Plant Prop. Soc.* 26:269–72.

77. Hicks, G.S. 1980. Patterns of organ development in plant tissue culture and the problem of organ determination. *Bot. Rev.* 46:1–23.

78. Holdgate, D.P., and J.S. Aynsley. 1977. The development and establishment of a commercial tissue culture laboratory. *Acta Hort.* 78:31–36.

79. Holley, W.D., and R. Baker. 1963. *Carnation production.* Dubuque, Iowa: William C. Brown.

80. Hollings, M. 1965. Disease control through virus-free stock. *Ann. Rev. Phytopath.* 3:367–96.

81. Hu, C., and P. Wang. 1986. Embryo culture: Technique and applications. In *Handbook of plant cell culture,* Vol. 4, D.A. Evans, W.R. Sharp, and P.V. Ammirato, eds. New York: Macmillan, pp. 43–96.

82. Huang, S., and D.F. Millikan. 1980. *In vitro* micrografting of apple shoot-tips. *HortScience* 15:741–43.

83. Hussey, G. 1976. *In vitro* release of axillary shoots from apical dominance in monocotyledonous plantlets. *Ann. Bot.* 40:1323–25.

84. ———. 1977. *In vitro* propagation of some members of the Liliaceae, Iridaceae and Amaryllidaceae. *Acta Hort.* 78:303–9.

85. ———, and A. Falavigna. 1980. Origin and production of in vitro adventitious shoots in the onion, *Allium cepa* L. *Jour. Exp. Bot.* 31:1675–86.

86. Jain, S.M., P.K. Gupta and R.J. Newton. 1994. *Somatic embryogenesis in woody plants.* Dordrecht: Kluwer Academic Pub.

87. James, D.J. and I.J. Thurbon. 1981. Shoot and root initiation in vitro in the apple rootstock M9 and the promotive effects of phloroglucinol. *Jour. Hort. Sci.* 56:15–20.

88. Jeong, B.R., K. Fujiwara and T. Kozai. 1995. Environmental control and photoautotrophic micropropagation. *Hort. Rev.* 17:125–72.

89. Jonard, R. 1986. Micrografting and its applications to tree improvement. In *Biotechnology in agriculture and forestry,* Vol. 1, *Trees,* Y.P.S. Bajaj, ed. Berlin: Springer-Verlag, pp. 31–48.

90. Jones, J.B. 1986. Determining markets and market potential. In *Tissue culture as a plant production system*

for horticultural crops, R.H. Zimmerman, R.J. Griesbach, F.A. Hammerschlag, and R.H. Lawson, eds. Dordrecht: Martinus Nijhoff Publishers, pp. 175–82.

91. Jones, R.O., R.L. Geneve and S.T. Kester. 1994. Micropropagation of gas plant (*Dictamnus albus* L.). *Jour. Environ. Hort.* 12:216–18.

92. Kartha, K.K. 1984. Elimination of viruses. In *Cell culture and somatic cell genetics of plants,* Vol. 1, I.K. Vasil, ed. New York: Academic Press, pp. 577–85.

93. ———. 1986. Production and indexing of disease-free plants. In *Plant tissue culture and its agricultural applications,* L.A. Withers, and P.G. Alderson, eds. London: Butterworth, pp. 219–38.

94. Keller, W.A. 1984. Anther culture of *Brassica.* In *Cell culture and somatic cell genetics of plants,* Vol. 1. I.K. Vasil, ed. New York: Academic Press, pp. 302–10.

95. Kermode, A.R., J.D. Bewley, J. Dasgupta, and S. Misra. 1986. The transition from seed development to germination: A key role for desiccation? *HortScience* 21:1113–18.

96. Kester, D.E., and C.O. Hesse. 1955. Embryo culture of peach varieties in relation to season of ripening. *Proc. Amer. Soc. Hort. Sci.* 65:265–73.

97. Klee, H., R. Horsch and S. Rogers. 1987. Agrobacterium-mediated plant transformation and its further applications to plant biology. *Ann. Rev. Plant Physiol.* 38:467–81.

98. Knauss, J.F. 1976. A tissue culture method of producing *Dieffenbachia picta* cv. Perfection free of fungi and bacteria. *Proc. Fla. State Hort. Soc.* 89:293–96.

99. Knauss, J.F., and J.W. Miller. 1978. A contaminant, *Erwinia carotovera,* affecting commercial plant tissue culture. *In Vitro* 14:754–56.

100. Kohlenback, H.W. 1984. Culture of isolated mesophyll cells. In *Cell culture and somatic cell genetics of plants,* Vol. 1, *Laboratory procedures and their applications,* I.K. Vasil, ed. New York: Academic Press, pp. 204–12.

101. Kosuge, T., C.P. Meredith, and A. Hollaender, eds. 1983. *Genetic engineering of plants: An agricultural perspective.* New York: Plenum Press.

102. Kozai, T., K. Fujiwara, M. Hayashi, and J. Aitken-Christie. 1992. In K. Kurata and T. Kozai (eds.). *Transplant production systems.* Dordrecht: Kluwer Academic Pub. pp. 247–282.

103. Krikorian, A.D. 1982. Cloning higher plants from aseptically cultured tissues and cells. *Biol. Rev.* 57:151–218.

104. ———, R.P. Kann, S.A. O'Connor, and M.S. Fitter. 1986. Totipotent suspensions as a means of multiplication. In *Tissue culture as a plant production system for horticultural crops,* R.H. Zimmerman, R.J. Griesbach, F.A. Hammerschlag, and R.H. Lawson, eds. Dordrecht: Martinus Nijhoff Publishers, pp. 61–72.

105. Kyte, L. 1987. *Plants from test tubes: An introduction to micropropagation* (rev. ed.). Portland, Oreg.: Timber Press.

106. Lammerts, W.E. 1942. Embryo culture, an effective technique for shortening the breeding cycle of deciduous trees and increasing germination of hybrid seed. *Amer. Jour. Bot.* 29:166–171.

107. Langhans, R.W., R.K. Horst, and E.D. Earle. 1977. Disease-free plants via tissue culture propagation. *HortScience* 12:149–50.

108. Lee, C.W., R.M. Skirvin, A.I. Soltero, and J. Janick. 1977. Tissue culture of *Salpiglossis sinuata* L. from leaf discs. *HortScience* 12:547–49.

109. Linderman, R.G. 1993. Tissue proliferation. *Amer. Nurseryman* 178:57–67.

110. Linsmaier, E.M., and F. Skoog. 1965. Organic growth factor requirements of tobacco tissue cultures. *Physiol. Plant* 18:101–27.

111. Lutz, J.D., J.R. Wong, and J. Rowe. 1985. Somatic embryogenesis for mass cloning of crop plants. In *Tissue culture in forests and agriculture,* R.R. Henke, K.W. Hughes, M.J. Constantin, and A. Hollaender, eds. New York: Plenum Press, pp. 105–16.

112. Mapes, M.O. 1973. Tissue culture of bromeliads. *Comb. Proc. Inter. Plant Prop. Soc.* 23:47–55.

113. McClelland, M.T., M.A.L. Smith and Z.B. Carothers. 1990. The effects of in vitro and ex vitro root initiation on subsequent microcutting root quality in three woody plants. *Plant Cell Tissue Organ Cult.* 23:115–23.

114. McCown, B.H. 1986. Woody ornamentals, shade trees, and conifers. In *Tissue culture as a plant production system for horticultural crops,* R.H. Zimmerman, R.J. Griesbach, F.A. Hammerschlag, and R.H. Lawson, eds. Dordrecht: Martinus Nijhoff Publishers, pp. 333–42.

115. ———, E.L. Zeldin, H.A. Pinkalla and R.R. Dedolph. 1987. Nodule culture: a developmental pathway with high potential for regeneration, automated micropropagation, and plant metabolite production from woody plants. In J.W. Hanover and D.E. Keathley (eds.). *Genetic manipulation of woody plants.* New York: Plenum Press. pp. 149–166.

116. McCown, D.D., and B.H. McCown. 1986. North American hardwoods. In *Cell and tissue culture in forests,* Vol. 3, *Case histories: Gymnosperms, angiosperms and Palms* (2nd ed.), J.M. Bonga and D.J. Durzan, eds. Dordrecht: Martinus Nijhoff, pp. 247–60.

117. McGarrity, G.L. 1975. Control of microbiological contamination. *Tissue Culture Assn. Man.* 1:181–85.

118. McGrew, J.R. 1980. Meristem culture for production of virus-free strawberries. In *Proc. conf. on nursery production of fruit plants through tissue culture: Applications and feasibility,* R.H. Zimmerman, ed. U.S. Dept. of Agr. Sci. and Education Administration ARR-NE-11, pp. 80–5.

119. Meins, F., Jr., and A.N. Binns. 1978. Epigenetic clonal variation in the requirements of plant cells for cytokinins. In *The clonal basis of development,* S. Subtelny and I.M. Sussex, eds. New York: Academic Press, pp. 185–201.

120. ———. 1979. Cell determination in plant development. *BioScience* 29:221–25.

121. Meredith, C., and P.S. Carlson. 1978. Genetic variation in cultured plant cells. In *Propagation of higher plants through tissue culture,* K.W. Hughes, R. Henke, and M. Constantin, eds. Springfield, Va.: U.S. Dept. of Energy, Tech. Inform. Center, pp. 166–76.

122. Meyer, M.M., Jr. 1976. Culture of *Paeonia* embryos by *in vitro* techniques. *Amer. Peony Soc. Bul.* 217:32–35.

123. ———. 1980. *In vitro* propagation of *Hosta sieboldiana. HortScience* 15:737–38.

124. ———, L.H. Fuchigami, and A.N. Roberts. 1975. Propagation of tall bearded irises by tissue culture. *HortScience* 10:479–80.

125. ———, and G.M. Milbrath. 1977. In vitro propagation of horseradish with leaf pieces. *HortScience* 12:544–45.

126. Morel, G.M. 1960. Producing virus-free cymbidiums. *Amer. Orch. Soc. Bul.* 29:495–97.

127. ———. 1964. Tissue culture: A new means of clonal propagation of orchids. *Amer. Orch. Soc. Bul.* 33:473–78.

128. Mudge, K.W., C.A. Borgman, J.C. Neal, and H.A. Weller. 1986. Present limitations and future prospects for commercial micropropagation of small fruits. *Comb. Proc. Inter. Plant Prop. Soc.* 36:538–43.

129. Mullins, M.G., G.Y. Nair, and P. Sampet. 1979. Rejuvenation *in vitro:* Induction of juvenile characters in an adult clone of *Vitis vinifera* L. *Ann. Bot.* 44:623–27.

130. Murashige, T. 1966. Principles of in vitro culture. *Comb. Proc. Inter. Plant Prop. Soc.* 16:80–8.

131. ———. 1974. Plant propagation through tissue cultures. *Ann. Rev. Plant Phys.* 25:135–66.

132. ———, and F. Skoog. 1962. A revised medium for rapid growth and bioassays with tobacco tissue cultures. *Phys. Plant.* 15:473–97.

133. Navarro, L., and J. Juarez. 1977. Tissue culture techniques used in Spain to recover virus-free citrus plants. *Acta Hort.* 78:425–35.

134. Negueroles, J., and O.P. Jones. 1979. Production *in vitro* of rootstock/scion combinations of *Prunus* cultivars. *Jour. Hort. Sci.* 54:279–81.

135. Ng, T.J. 1986. Use of tissue culture for micropropagation of vegetable crops. In *Tissue culture as a plant production system for horticultural crops,* R.H. Zimmerman, R.J. Griesbach, F.A. Hammerschlag, and R.H. Lawson, eds. Dordrecht: Martinus Nijhoff Publishers, pp. 259–70.

136. Nitsch, J.P., and C. Nitsch. 1969. Haploid plants from pollen grains. *Science* 163:85–7.

137. Oglesby, R.P., and J.L. Griffis, Jr. 1986. Commercial in vitro propagation and plantation crops. In *Tissue culture as a plant production system for horticultural crops,* R.H. Zimmerman, R.J. Griesbach, F.A. Hammerschlag, and R.H. Lawson, eds. Dordrecht: Martinus Nijhoff Publishers, pp. 253–57.

138. Oglesby, R.P., and R.E. Strode. 1979. Commercial tissue culturing at Oglesby Nursery. *Comb. Proc. Inter. Plant Prop. Soc.* 29:341–44.

139. Onishi, N., Y. Sakamoto and T. Hirosawa. 1994. Synthetic seeds as an application of mass production of somatic embryos. *Plant Cell Tissue Organ Cult.* 39:137–45.

140. Pierik, R.L.M. 1987. *In vitro culture of higher plants.* Dordrecht: Martinus Nijhoff Publishers.

141. Pierik, R.L.M., J.L.M. Jansen, A. Maasdam, and C.M. Binnendijk. 1975. Optimalization of *Gerbera* plantlet production from excised capitulum explants. *Sci. Hort.* 3:351–57.

142. ———, and H.H.M. Steegmans. 1976. Vegetative propagation of Freesia through the isolation of shoots *in vitro. Neth. Jour. Agr. Sci.* 24:274–77.

143. Preece, J.E. and E.G. Sutter. 1991. Acclimatization of micropropagated plants to the greenhouse and field. In P.C. Debergh and R.H. Zimmerman. (eds.). *Micropropagation: technology and application.* Dordrecht: Kluwer Academic Pub., pp. 71–94.

144. Quak, F. 1977. Meristem culture and virus-free plants. In *Applied and fundamental aspects of plant cell, tissue and organ culture,* J. Reinert and Y.P.S. Bajaj, eds. Berlin: Springer-Verlag, pp. 598–615.

145. Raju, B.C., and J.C. Trolinger. 1986. Pathogen indexing in large-scale propagation of florist crops. In *Tissue culture as a plant production system for horticultural crops,* R.H. Zimmerman, R.J. Griesbach, F.A. Hammerschlag, and R.H. Lawson, eds. Dordrecht: Martinus Nijhoff Publishers, p. 138.

146. Ramming, D.W. 1985. In ovulo embryo culture of early-maturing *Prunus. HortScience* 20:419–20.

147. Randolph, L.F., and L.G. Cox. 1943. Factors influencing the germination of *Iris* seed and the relation of inhibiting substances to embryo dormancy. *Proc. Amer. Soc. Hort. Sci.* 43:284–300.

148. Rangan, T.S. 1984. Culture of ovaries. In *Cell culture and somatic cell genetics of plants,* Vol. 1, I.K. Vasil, ed. New York: Academic Press, pp. 221–26.

149. Rangan, T.S. 1984. Culture of ovules. In *Cell culture and somatic cell genetics of plants,* Vol. 1, I.K. Vasil, ed. New York: Academic Press, pp. 227–31.

150. Rangaswamy, N.S. 1977. Application of *in vitro* pollination and *in vitro* fertilization. In *Applied and fundamental aspects of plant cell, tissue and organ culture,* J. Reinert and Y.P.S. Bajaj, eds. Berlin: Springer-Verlag, pp. 412–25.

151. Ranjit, M., and D.E. Kester. 1988. Micropropagation of cherry rootstocks: II. Invigoration and enhanced rooting of '46-1 Mazzard' by co-culture with 'Colt.' *Jour. Amer. Soc. Hort. Sci.* 113:150–54.

152. Read, P.E. 1992. Environmental and hormonal effects in micropropagation. In K. Kurata and T. Kozai (eds.). *Transplant production systems.* Dordrecht: Kluwer Academic Pub. pp. 231–46.

153. Redenbaugh, K., D. Slade, P. Viss, and J.A. Fujii. 1987. Encapsulation of somatic embryos in synthetic seed coats. *HortScience* 22:803–9.

154. ———, P. Viss, D. Slade, and J.A. Fujii. 1987. Scale-up: artificial seeds. In *Plant tissue and cell culture,* C.E. Green, D.A. Somers, W.P. Hackett, and D.D. Biesboer, eds. New York: Alan R. Liss, pp. 473–93.

155. Reinert, J. 1959. Über die Kontrolle die Morphogenese und die Induktion von Adventiveembryonen an gewebekulturen aus karotten. *Planta* 53:318–33.

156. Roest, S., and G.S. Bokelmann. 1973. Vegetative propagation of *Chrysanthemum cinerariaefolium in vitro. Scientia Hort.* 1:120–22.

157. Saebo, A., T. Krekling and M. Appelgren. 1995. Light quality affects photosynthesis and leaf anatomy of birch plantlets in vitro. *Plant Cell Tissue Organ Cult.* 41:177–85.

158. Sagawa, Y., T. Shoji, and T. Shoji. 1966. Clonal propagation of cymbidium through shoot meristem culture. *Amer. Orchid Soc. Bul.* 35:118–22.

159. Scowcroft, W.R., R.I.S. Brettell, S.A. Ryan, P.A. Davies, and M.S. Pallotta. 1987. Somaclonal variation and genomic flux. In *Plant tissue and cell culture,* C.E. Green, D.A. Somers, W.P. Hackett, and D.D. Biesboer, eds. New York: Alan R. Liss, pp. 275–88.

160. Sharp, W.R., M.R. Sondahl, L.S. Caldas, and S.B. Maraffa. 1980. The physiology of in vitro asexual embryogenesis. *Hort. Rev.* 2:268–309.

161. Shepard, J.F., D. Bidney, and E. Shakin. 1980. Potato protoplasts in crop improvement. *Science* 208:17–24.

162. Sinha, S., R.P. Roy, and K.K. Jha. 1979. Callus formation and shoot bud differentiation in anther culture of *Solanum surattense. Can. Jour. Bot.* 57: 2524–27.

163. Skoog, F., and C.O. Miller. 1957. Chemical regulation of growth and organ formation in plant tissues cultured *in vitro. Symp. Soc. for Exp. Biol.* 11:118–31.

164. Slack, S.A. 1980. Pathogen-free plants: Meristem-tip culture. *Plant Disease* 64(1):15–7.

165. Smith, M.A.L. and M.T. McClelland. 1991. Gauging the influence of in vitro conditions on in vivo quality and performance of woody plants. In *Vitro Cell. Dev. Biol.* 27P:52–6.

166. Smith, M.A.L. and A.A. Obeidy. 1991. Micropropagation of a mature, male Kentucky coffee tree. *HortScience* 26:14–26.

167. Sondahl, M.R., and L.C. Monaco. 1981. In vitro methods applied to coffee. In *Plant tissue culture: Methods and applications in agriculture,* T.A. Thorpe, ed. New York: Academic Press, pp. 325–47.

168. Spiegel-Roy, P., N. Sahar, J. Baron, and U. Lavi. 1985. In vitro culture and plant formation from grape cultivars with abortive ovules and seeds. *Jour. Amer. Soc. Hort. Sci.* 110:109–12.

169. Sriskandarajah, C., and M.G. Mullins. 1981. Micropropagation of Granny Smith apple: Factors affecting root formation *in vitro. Jour. Hort. Sci.* 56:71–6.

170. Steward, F.C., M.O. Mapes, and K. Mears. 1958. Growth and organized development of cultured carrots. II. Organization in cultures from freely suspended cells. *Amer. Jour. Bot.* 454:705–8.

171. Stimart, D.P. 1986. Commercial micropropagation of florist flower crops. In *Tissue culture as a plant production system for horticultural crops,* R.H. Zimmerman, R.J. Griesbach, F.A. Hammerschlag, and R.H. Lawson, eds. Dordrecht: Martinus Nijhoff Publishers, pp. 301–15.

172. ———, and P.D. Ascher. 1981. Developmental response of *Lilium longiflorum* bulblets to constant or alternating temperatures in vitro. *Jour. Amer. Soc. Hort. Sci.* 106:450–54.

173. Stoltz, L.P. 1971. Agar restriction of the growth of excised mature iris embryos. *Jour. Amer. Soc. Hort. Sci.* 96:611–84.

174. Stone, O.M. 1978. The production and propagation of disease-free plants. In *Propagation of higher plants through tissue culture,* K.M. Hughes, R. Henke, and M. Constantin, eds. Springfield, Va.: U.S. Dept of Energy, Tech. Inform. Center, pp. 25–34.

175. Stoutemyer, V.T., and O.K. Britt. 1969. Growth and habituation in tissue cultures of English ivy, *Hedera helix. Amer. Jour. Bot.* 56:222–26.

176. Street, H.E., ed. 1977. *Plant tissue and cell culture.* Berkeley, Calif.: Univ. of Calif. Press.

177. Strode, R.A., and G. Abner. 1986. Large scale tissue culture production for horticultural crops. In *Tissue culture as a plant production system for horticultural crops,* R.H. Zimmerman, R.J. Griesbach, F.A. Hammerschlag, and R.H. Lawson, eds. Dordrecht: Martinus Nijhoff Publishers, pp. 367–71.

178. Struve, D.K. and R.D. Lineberger. 1988. Restoration of high adventitious root regeneration potential in mature *Betula papyrifera* Marsh. Softwood stem cuttings. *Can. Jour. For. Res.* 18:265–69.

179. Stuart, D.A., S.G. Strickland, and K.A. Walker. 1987. Bioreactor production of alfalfa somatic embryos. *HortScience* 22:800–3.

180. Sunderland, N. 1984. Anther culture of *Nicotiana tabacum.* In *Cell culture and somatic cell genetics of plants,* Vol. 1, I.K. Vasil, ed. New York: Academic Press, pp. 283–92.

181. Swartz, H.J. 1991. Post culture behavior: genetic and epigenetic effects and related problems. In P.C. Debergh and R.H. Zimmerman. (eds.). *Micropropagation: technology and application.* Dordrecht: Kluwer Academic Pub., pp. 95–122.

182. ———, and J.T. Lindstrom. 1986. Small fruit and grape tissue culture from 1980 to 1985: Commercialization of the technique. In *Tissue culture as a plant production system for horticultural crops,* R.H. Zimmerman, R.J. Griesbach, F.A. Hammerschlag, and R.H. Lawson, eds. Dordrecht: Martinus Nijhoff Publishers, pp. 201–20.

183. Thorpe, T.A. 1979. Regulation of organogenesis *in vitro.* In *Propagation of higher plants through tissue culture,* K.W. Hughes, R. Henke, and M. Constantin, eds. Springfield, Va.: Natl. Tech. Inf. Ser. U.S. Dept. of Commerce, pp. 87–101.

184. ———, ed. 1981. *Plant tissue culture: Methods and applications in agriculture.* New York: Academic Press.

185. Tisserat, B. 1981. Date palm tissue culture. *USDA Agr. Res. Ser., Adv. in Agr. Tech., Western Ser. 17,* pp. 1–50.

186. ———, E.B. Esan, and T. Murashige. 1979. Somatic embryogenesis in angiosperms. *Hort. Rev.* 1:1–78.

187. Torrey, J.G. 1977. Cytodifferentiation in cultured cells and tissues. *HortScience* 12:14–5.

188. Toth, K.F. and M.C. Lacy. 1992. Micropropagation of celery (*Apium graveolens* var. Dulce). In Y.P.S. Bajaj (ed.). *Biotechnology in agriculture and forestry. Vol 19. High tech and micropropagation III.* Springer-Verlag, Berlin.

189. Tran Thanh Van, K., and T.H. Trinh. 1986. Fundamental and applied aspects of differentiation in vitro and in vivo. In *Handbook of plant cell culture,* Vol. 4, *Techniques and applications,* D.A. Evans, W.R. Sharp, and P.V. Ammirato, eds. New York: Macmillan, pp. 316–35.

190. van Overbeek, J. 1942. Hormonal control of embryo and seedling. *Cold Spring Harbor Symp. Quantitative Biol.* 10:126–34.

191. Vasil, I.K. 1985. Somatic embryogenesis and its consequences in the Gramineae. In *Tissue culture in forestry and agriculture,* R.R. Henke, K.W. Hughes, M.J. Constantin, and A. Hollaender, eds. New York: Plenum Press, pp. 31–47.

192. Wenzel, G., and B. Foroughi-Wehr. 1984. Anther culture of *Solanum tuberosum.* In *Cell culture and somatic cell genetics of plants,* Vol. 1, I.K. Vasil, ed. New York: Academic Press, pp. 293–301.

193. ———, and B. Foroughi-Wehr. 1984. Anther culture of cereals and grasses. In *Cell culture and somatic cell genetics of plants,* Vol. 1, I.K. Vasil, ed. New York: Academic Press, pp. 31–127.

194. Wetherell, D.F. 1978. *In vitro* embryoid formation of cells derived from somatic plant tissues. In *Propagation of higher plants through tissue culture,* K.M. Hughes, R. Henke, and H. Constantin, eds. Springfield, Va.: U.S. Dept. of Commerce, Natl. Tech. Inform. Ser., pp. 102–24.

195. Wetzstein, H.Y. and H.E. Sommer. 1982. Leaf anatomy of tissue-cultured *Liquidambar styraciflua* (Hamamelidaceae) during acclimatization. *Amer. Jour. Bot.* 69:1579–86.

196. White, P.R. 1963. *The cultivation of animal and plant cells* (2nd ed.). New York: Ronald Press.

197. Williams, E.G., and G. Maheswaran. 1986. Somatic embryogenesis: factors influencing coordinated behaviour of cells as an embryonic group. *Ann. Bot.* 57:443–62.

198. Yeung, E.C., T.A. Thorpe, and C.J. Jensen. 1981. In vitro fertilization and embryo culture. In *Plant tissue culture: Methods and applications in agriculture,* T.A. Thorpe, ed. New York: Academic Press, pp. 253–71.

199. Yusnita, S., R.L. Geneve and S.T. Kester. 1990. Micropropagation of a white flowering form of eastern redbud. *Jour. Environ. Hort.* 8:177–79.

200. Zilis, M., D. Zwagerman, D. Lamberts, and L. Kurtz. 1979. Commercial propagation of herbaceous perennials by tissue culture. *Comb. Proc. Inter. Plant Prop. Soc.* 29:404–13.

201. Zimmerman, J.L. 1993. Somatic embryogenesis: a model for early development in higher plants. *Plant Cell* 5:1441–23.

202. Zimmerman, R.H. 1986. Propagation of fruit, nut and vegetable crops—overview. In *Tissue culture as a plant production system for horticultural crops,* R.H. Zimmerman, R.J. Griesbach, F.A. Hammerschlag, and R.H. Lawson, eds. Dordrecht: Martinus Nijhoff Publishers, pp. 183–200.

203. Ziv, M. 1986. *In vitro* hardening and acclimatization of tissue culture plants. In *Plant tissue culture and its agricultural applications,* L.A. Withers, and P.G. Alderson, eds. London: Butterworth, pp. 187–96.

204. ———. 1991. Vitrification: morphological and physiological disorderers of in vitro plants. In P.C. Debergh and R.H. Zimmerman. (eds.). *Micropropagation: technology and application.* Dordrecht: Kluwer Academic Pub. pp. 45–70.

205. ———, A.H. Halevy, and R. Shilo. 1970. Organs and plantlets regeneration of *Gladiolus* through tissue culture. *Ann. Bot.* 34:671–76.

SUPPLEMENTARY READING

AHUJA, M.R. ed. 1993. *Micropropagation of woody plants.* Dordrecht: Kluwer Academic Pub.

BAJAJ, Y.P.S., ed. 1986–96. *Biotechnology in agriculture and forestry,* Vol. 1–27, Berlin: Springer-Verlag.

BONGA, J.M., and D.J. DURZAN. 1986. *Cell and tissue culture in forestry,* 4 volumes (2nd ed.). Dordrecht: Martinus Nijhoff Publishers.

COCKING, E.C. 1986. The tissue culture revolution. In *Plant tissue culture and its agricultural applications,* L.A. Withers and P.G. Alderson, eds. London: Butterworth, pp. 3–20.

DEBERGH, P.C. and R.H. ZIMMERMAN. 1991. *Micropropagation. Technology and Application.* Dordrecht: Kluwer Academic Pub.

DONNELLY, D.J., and W.E. VIDAVER. 1988. *Glossary of plant tissue culture.* Portland, Oreg.: Dioscorides Press.

EVANS, D.A., W.R. SHARP, and P.V. AMMIRATO, eds. 1986. *Handbook of plant cell culture,* Vol. 4. New York: Macmillan.

GAUTHERET, R.J. 1985. History of plant tissue and cell culture: A personal account. In *Cell culture and somatic cell genetics of plants,* Vol. 2, I.K. Vasil, ed. New York: Academic Press, pp. 1–59.

GEORGE, E.F. 1993. *Plant propagation by tissue culture. Part I. The technology.* Edington, Wilts: Exergetics Ltd.

HUETTERMAN, C.A. and J.E. PREECE. 1993. *Thidiazuron: a potent cytokinin for woody plants tissue culture.* Plant Cell Tissue Organ Culture 33:105–19.

JAIN, S.M., P.K. GUPTA and R.J. NEWTON. 1994. *Somatic embryogenesis in woody plants.* Dordrecht: Kluwer Academic Pub.

K. KURATA and T. KOZAI (eds.). 1992. *Transplant production systems.* Dordrecht: Kluwer Academic Pub.

KRIKORIAN, A.D. 1982. Cloning higher plants from aseptically cultured tissues and cells. *Biol. Rev.* 57:151–218.

THORPE, T.A., ed. 1978. *Frontiers of plant tissue culture.* Calgary: Univ. of Calgary Press.

TISSUE CULTURE ASSOCIATION. *In Vitro* (Series).

VASIL, I.K., ed. 1984. *Cell culture and somatic cell genetics of plants,* Vol. 1, *Laboratory procedures and their applications.* New York: Academic Press.

VASIL, I.K., ed. 1985. *Cell culture and somatic cell genetics of plants,* Vol. 2, *Cell growth, nutrition, cytodifferentiation, and cryopreservation.* New York: Academic Press.

VASIL, I.K. and T.A. Thorpe, ed. 1994. *Plant cell and tissue culture.* Dordrecht: Kluwer Academic Pub.

WITHERS, L.A., and P.G. ALDERSON, eds. 1986. *Plant tissue culture and its agricultural applications.* London: Butterworth.

18

Techniques of In Vitro Culture for Micropropagation and Biotechnology

INTRODUCTION

Tissue culture is an aseptic laboratory procedure that requires unique facilities and special skills. Plant parts are cultured under controlled environments in small containers where all the nutrients required for growth are provided. Specific growth regulators are supplied to control growth and development.

There are a number of reasons to produce plants in tissue culture. These include: (a) micropropagation; (b) somatic embryogenesis; (c) natural products synthesis for such chemicals as natural flavors, colors, pharmaceuticals (medicines), and even plastics; (d) embryo rescue (see pp. 553); (e) anther (microspore) culture to produce haploid plants for crop breeding (see pp. 566); (f) micrografting (see pp. 566); and (g) crop improvement through biotechnology. This includes gene transfer (see Chapter 2), somaclonal variation (see pp. 246), and protoplast fusion (see pp. 568). Of these techniques, the two most directly related to plant propagation are micropropagation and somatic embryogenesis. This chapter will describe the general kinds of facilities required to reproduce plants by *in vitro* culture. Secondly, specific procedures used for micropropagation and somatic embryogenesis will be discussed with some applications to specific crops as case histories. Additional information on micropropagation is included in Chapters 19, 20, and 21. These chapters contain propagation methods for specific plants.

CHARACTERISTICS OF MICROPROPAGATION

Micropropagation has become an important part of commercial propagation for many plants (*15, 27, 29, 85*). The advantages of micropropagation as a propagation system have been reviewed by many authors (*12, 13, 31, 41, 43, 45, 53, 58, 70, 78*) and can be summarized as:

1. Mass propagation of specific clones.
2. Production of pathogen-free plants.
3. Clonal propagation of parental stock for hybrid seed production.

4. Year-round nursery production.
5. Germplasm preservation.

1. Mass propagation of specific clones. The objective of commercial propagation is to reproduce (either sexually or asexually) copies of an original parent plant. The controlled aspects of micropropagation permit the rapid propagation of individuals from a single plant. Multiplication rates can be very high, since plants in culture can theoretically be multiplied at an exponential rate by consecutive subculturing (e.g., one month apart). Although such theoretical rates are not usually maintained in practice, the actual rates that can be attained under proper management are impressive. A multiplication rate of fourfold in cultures subcultured every four weeks will theoretically produce over one million plants in ten months. Many commercial laboratory nurseries in operation have the capacity to produce millions of micropropagated plants a year (*21, 57*). Commercial micropropagation is particularly useful for the following:

a. Plants whose natural rate of increase is relatively slow—orchids, bulbs, Boston fern, gerbera, many foliage plants, herbaceous perennials, palm species. A good example can be seen in the herbaceous perennial *Hosta*. Traditionally, this plant is propagated by division. Division yields fewer than ten plants from the original mother plant per year. Micropropagated *Hosta* can yield a similar multiplication rate every month!

b. New cultivars where commercial demands require getting to market in as short a time as possible. This has been important for a number of crops. The introduction of a new apple rootstock illustrates this concept well. Apple rootstocks are propagated by stooling (mound layering—see pp. 509). However, relying on stooling to propagate the thousands of rootstocks necessary to supply the market would take many years. Micropropagation allows the fruit breeder to quickly multiply a new rootstock in a short time. These micropropagated plants, in turn, are used by the commercial propagator to create new stool beds to propagate rootstocks for distribution to growers.

c. Cultivars whose high value makes micropropagation a viable alternative to conventional methods. Micropropagation is relatively expensive compared to other methods of propagation. However, in some cases, propagation costs are minimal compared to the value of the finished crop. In these cases, micropropagation may be the method of choice for reasons other than cost. Examples include specialty woody plants such as valuable rootstocks (*18, 38, 49*), rhododendrons, and specialty perennials such as daylily.

d. Propagation of difficult-to-root plants. Many woody plants fail to root from cuttings (see Chapter 10). Traditionally, grafting has been the only method to clonally propagate these plants. Micropropagation offers a viable alternative for propagation of these plants on their own roots. An excellent example are the many cultivars of mountain laurel *(Kalmia latifolia)* that have only become available to the nursery trade because of micropropagation.

e. Conservation of endangered species. Micropropagation can provide a rapid method for the multiplication of endangered or threatened species. In these cases, only a few plants may be available as stock plants or the collection of plants and seeds from wild plants needs to be minimized. Micropropagation can also be an alternative to collecting plants from natural habitats to supply commercial markets. An example is the over-collection of spring snowflake *(Leucojum)* from its natural populations. Harvesting bulbs for commercial sale has seriously reduced populations and has been a driving force for the propagation of these plants from micropropagation. Several organizations like the Conservation and Reproduction of Endangered Wildlife (CREW) at the Cincinnati Zoo and The Botanical Garden and the Center for Plant Conservation at the Missouri Botanical Garden are involved in developing and supporting efforts to conserve native plants, including the use of micropropagation.

2. Production of pathogen-free plants. Production of propagation material free of fungal, bacterial, and systemic viruses and viruslike pathogens has become an essential aspect of propagation (see Chapter 9). Micropropagation provides a method to rid a clone of external pathogens and can also provide a system in which plants are kept free of reinfection until delivery. Propagators should not automatically assume systemic pathogens are absent unless culture indexes for bacteria, fungi, or viruses are made. Specific aspects of pathogen control are as follows:

a. Maintain germplasm and source material in a pathogen-free condition.

b. Include micropropagation as the initial step in production for Foundation Clone Systems (*57, 61, 76*).

c. Allow movement of germplasm materials across quarantine barriers (*57*).

d. Facilitate the distribution of commercial material in international trade. This has allowed the importation of specific plants between countries (*58, 80*). For example, specialty producers of ferns in Israel and Australia supply most of the fern liners to U.S. growers.

3. Clonal propagation of parental stocks for hybrid seed production. This procedure is used for marigolds, asparagus, tomato, cucurbits, and broccoli (*5, 17, 54*).

4. Provide year-round nursery production. Most nursery operations are seasonal. Micropropagation has the potential for continuous year-round operation with production scheduled according to market demands. High-volume production requires high-volume distribution and the facilities to stockpile items. Combining production with cold storage facilities makes it possible to hold material for peak marketing periods (*57, 73*).

5. Germplasm preservation. The major method of germplasm preservation is seeds (see Chapter 6). However, seed storage requires a relatively large facility with specialized storage conditions. Also, seed storage is not an option for recalcitrant seeds that do not tolerate dry storage conditions. For these reasons, preservation of vegetative tissue as explants is an attractive alternative (*36*). Cryopreservation, often using antifreeze materials as an aid, has been used to ultrafreeze vegetative tissue in a fashion similar to that of seeds (see Chapter 6). Frozen material can be thawed and subsequently used to micropropagate the original plants. Proper freezing and storage conditions as well as evaluation of stored material for genetic purity are active areas of current research.

DISADVANTAGES OF MICROPROPAGATION

Micropropagation on a commercial scale has particular characteristics that may create problems that could limit use (*52, 58*). These include:

1. Expensive and sophisticated facilities, trained personnel, and specialized techniques.
2. High cost of production results from expensive facilities and high labor input. For example, shoot-tip propagation requires much hand labor to transfer individual propagules.
3. A high-volume, more or less continuous distribution system or adequate storage facilities to stockpile products is required.
4. Contamination or insect infestation can cause high losses in a short time.
5. Variability and production of off-type individuals can be a risk in the products emerging from micropropagation. Careful roguing, prior field testing of new products, and continuing research and development are essential to decrease this risk (see Chapter 9).
6. Economics and marketing are key to the success of commercial operations.

GENERAL LABORATORY FACILITIES AND PROCEDURES

Facilities and Equipment

Facilities for aseptic procedures may be placed into three categories as determined by their scope, size, sophistication, and cost. These categories are: (a) research laboratories where precise work requires highly sophisticated equipment (*77*), (b) large commercial propagation facilities where several millions of plants can be mass produced annually (*8, 9, 10, 11, 27, 34, 35, 57, 81, 83*), and (c) limited facilities for small research laboratories, individual nurseries, or hobbyists where a relatively small volume of material is handled (*32, 46, 72*). No matter what size, the facility should include three basic components: (a) preparation area, (b) transfer area, and (c) growing area. In addition there may be a need for service areas, office, and cold storage. This facility should be separated from the regular nursery and greenhouse production area and restricted entry should be maintained to avoid introducing contaminants into the culture rooms. Floors, benches, and table tops should be kept scrupulously clean. In some commercial labs, technicians use clean lab coats, foot coverings, and hairnets.

There is much variation in the kinds of facilities and opportunities for cost cutting, but the basic principles of *in vitro* culture must be followed (*14, 27, 32, 41, 72*). In any case, careful consideration must be given to the costs and labor required (*5, 8, 16, 46, 57, 72*).

Preparation Area

The preparation area is essentially a kitchen with three basic functions: (a) cleaning glassware, (b) preparation and sterilization of media, and (c) storage of glassware and supplies (Figure 18–1).

An efficient method of washing is required, either by hand or by machine. Normal washing is followed by rinsing in distilled or deionized water. A sink, running water, and electrical or gas outlets for heating are necessary; air or vacuum outlets are often useful. Table surfaces should be made of a material that can be cleaned easily.

The following equipment items are used in the preparation area:

1. Refrigerator to store chemicals, stock solutions, and small batches of media.

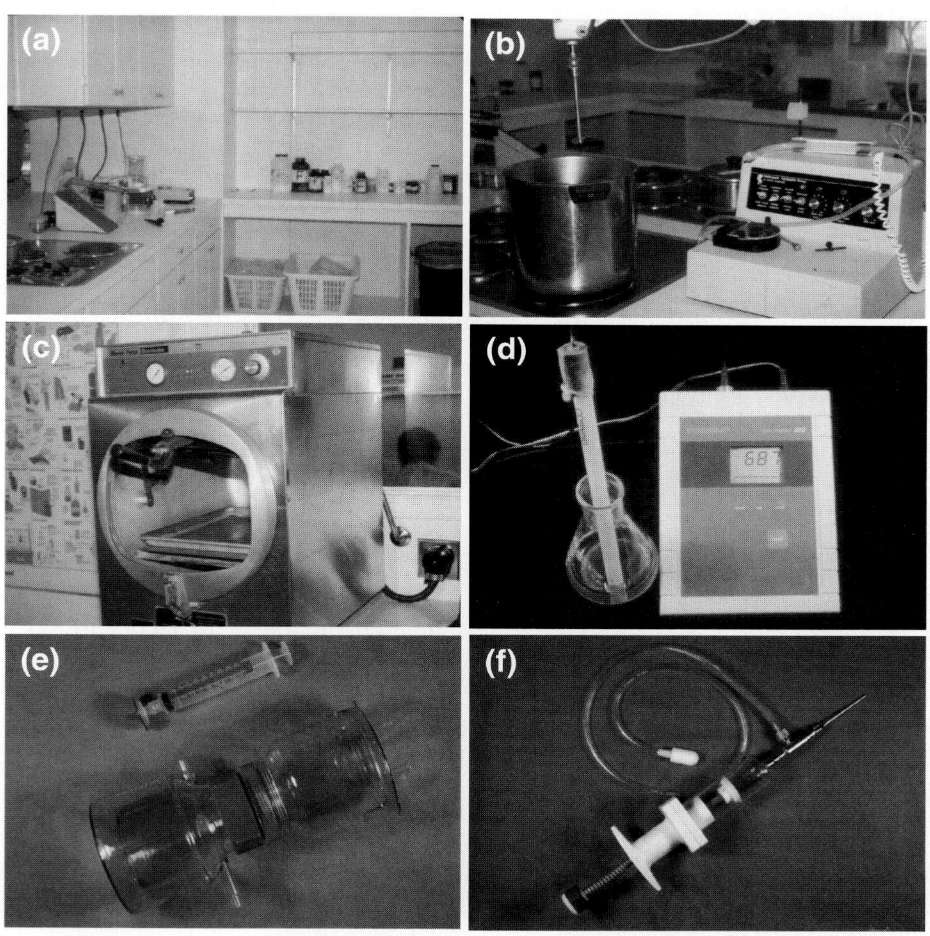

FIGURE 18–1 Common features found in the tissue culture preparation area. (a) Kitchen facilities including places to heat medium and wash dishes. (b) A peristaltic pump can significantly reduce the time to dispense medium. A mixer incorporates agar into the medium uniformly. (c) An autoclave or stove-top pressure canner are essential to sterilize medium and glassware. (d) Every lab must have a pH meter. (e) Filter apparatus should be available to sterilize chemicals that can not be autoclaved. These may be disposable, as the presterilized 25 ml syringe and Acrodisc syringe tip filter, or permanent autoclavable units with replaceable filters to sterilize larger quantities of fluids. The permanent unit must use a vacuum to move fluid through the disposable paper filter. (f) For smaller operations that can not afford a peristaltic pump, small plunger dispensers can be effective labor savers.

2. Scales or analytical balance, preferably top loading.
3. Autoclave capable of reaching 120° C (250° F). A household pressure cooker can be used to sterilize small batches of media.
4. A pH meter. Indicator papers can be used as a substitute in less precise work.
5. Gas or heating plate.
6. Stirrer and mixing device.
7. Filters for sterilizing nonautoclavable ingredients (particularly in research laboratories). Filters are used to sterilize chemicals that are modified by autoclaving. These consist of special funnel-like units fitted with a bacteria-proof membrane through which a solution is pulled by vacuum or passed under positive pressure. Various commercial types are available.
8. Equipment to purify water, a glass still, an ion exchanger, reverse osmosis system, or a combination. In smaller operations, distilled water may be purchased and stored in plastic containers.
9. A vacuum pump or an ultrasonic cleaner. These specific items can be used to decontaminate explants.
10. A media dispenser, valuable as a labor-saving device.
11. Storage for flasks, bottles, and other supplies.

Transfer Area

The transfer area is the place where explants are inserted into culture, and where transfers of subcultures to fresh media take place. The key requirement is that the environment of the transfer area **must** be sterile and free from any contaminating organisms. Transfer is most conveniently done in an open-sided **laminar airflow hood** (Figure 18–2) where filtered air is passed from the rear of the hood outward on a positive pressure gradient. Air passes through a prefilter to remove dust. Then it is drawn through a high-efficiency particulate air (HEPA) filter that removes microbial spores. Turn on the hood at least 30 minutes before use. With such equipment the transfer may be carried out in a room with other activities. Less expensive alternatives, such as an enclosed walk-in room or enclosed transfer boxes, can be used if they are carefully sterilized before use.

Ultraviolet (UV) germicidal lamps can be used to sterilize the interior of the transfer chamber. These are turned on about two hours prior to using the chamber but must be turned off during operation. The light should be directed inward, not toward the worker's face, because UV may adversely affect the eyes.

Growing Area

Cultures should be grown in a separate, lighted facility where both daylength and light irradiance can be controlled and where specific tem-

FIGURE 18–2 Laminar flow transfer hoods are used for all sterile operations. Left: Operator is sterilizing end of a test tube by flaming. Right: Transferring cultures in a commercial orchid operation.

perature regimes can be provided (Figure 18–3). If different kinds of plants are to be propagated, it is useful to have several rooms, each programmed to meet the temperature and light needs of specific kinds of plants. Light requirements vary from about 3 to 30 W/M^2 PAR. Cool-white or Gro-lux fluorescent lights are generally used. A 16/8 (light/dark) photoperiod is most common. Temperatures of 21 to 30° C (70 to 86° F) are generally adequate, although some kinds of plants may need lower temperatures. The relative humidity at the temperature given is about 30 to 50 percent, but depends on the type of closure. Dehydration of the medium and increase in salt concentration might occur during long culture periods if the relative humidity is too low. If too high, contamination may develop.

Various kinds of rolling drum culture devices or shakers are sometimes used. Their advantage is providing aeration in liquid culture systems.

Containers for Growing Cultures

Test tubes, Erlenmeyer flasks, petri dishes, and shell vials of various sizes are routinely used in research laboratories (Figure 18–4). These may be Pyrex glass but are often less expensive soda glass and borosilicate, all of which can be sterilized and flamed during transfer operations. Various kinds of commercial plastic containers are also available that are less expensive, less likely to break, and often disposable.

FIGURE 18–3 Cultures can be grown in (a) test tubes, (b) flasks, and (c) tubs.

FIGURE 18–4 Containers used for tissue culture can include test tubes, petri dishes, plastic or glass jars. Specially designed closures are available to cover all types of containers. Caps allow air to enter and leave the tissue culture vessel, but prevent contamination. Different color caps can be useful to keep track of different types of media or species.

Petri dishes are glass or plastic, reusable or disposable. Less expensive larger glass containers commonly used in the past for home canning or commercial canning are widely used for the later stages of development in large-scale propagation. Most of these supplies are available from commercial supply houses or from food industry supply houses.

Various kinds of closures are necessary for the containers. Nonabsorbent cotton plugs have been used in research laboratories but are inconvenient in large commercial operations. Metal or plastic covers or polyurethane or polypropylene plugs are more convenient. A second cover of aluminum foil or various plastic films is desirable for holding moisture and reducing infection but allowing exchange of CO_2, O_2, and ethylene *(58)*. Covering wide-mouthed jars with petri dish halves that fit works well. Parafilm (paraffined paper in a sterile roll) is used to seal petri dishes and other containers. Parafilm or plastic films can also help reduce the spread of culture mites or thrips. Culture mite or thrip infestations can be devastating in a commercial or research lab. Mites move from one culture vessel to another spreading contamination. Mite "trails" are evident as spots of fungal contamination on the surface of the agar. Thrips can enter the lab through unscreened windows or doorways. They can also be present on the original explant. A magnifying glass may be required to see the actual mites or thrips. To eliminate pests, destroy infested cultures in the autoclave and disinfect the culture room area.

Media Preparation

Ingredients

Ingredients of the culture medium vary with the kind of plant and the propagation stage at which one is working. In general, certain standard mixtures are used for most plants, but exact formulations may need to be established by testing. Media can be made from the pure chemicals or purchased as commercial premixed culture media (Figure 18–5). Empirical trials may be necessary to test available combinations of ingredients when dealing with a new kind of plant. These ingredients can be grouped into categories of (a) inorganic salts, (b) organic compounds, (c) complex natural ingredients, and (d) inert supports.

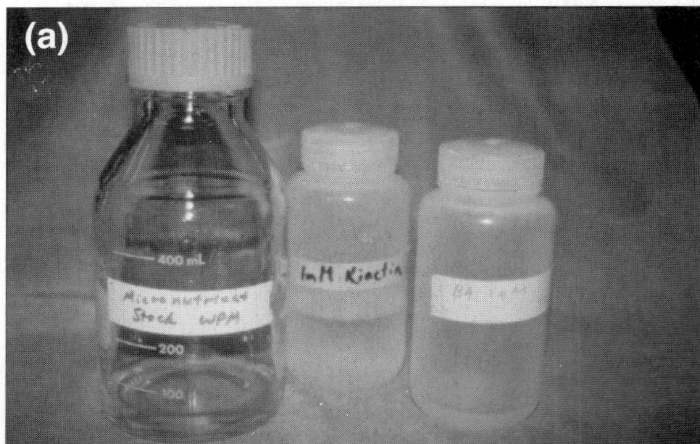

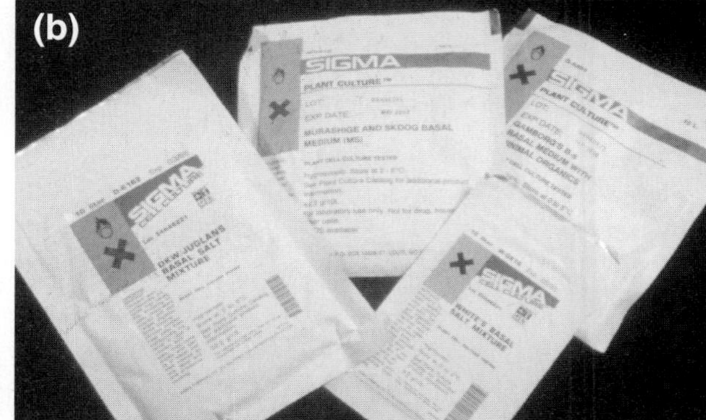

FIGURE 18–5 Media are usually prepared from (a) stock solutions or (b) premeasured commercial mixes.

Inorganic Salts (Groups A to E, Table 18–1)

Inorganic salts provide the macroelements (nitrogen, phosphorus, potassium, calcium, and magnesium) and microelements (boron, cobalt, copper, manganese, iodine, iron, and zinc). These salts can be made up in stock solutions by group and stored in a refrigerator. A stock solution is a concentrated solution (usually 100 times the concentration found in the final medium) from which portions are removed to make the final medium. Stock solutions save time because dry salts need only be weighed once. The reasons for preparing different stock solutions is that certain kinds of chemicals when mixed together will precipitate and not remain in solution. Stock solutions should be stored in the refrigerator (not frozen) and are good for several months. Cloudy stock solution should be discarded.

There are many different formulations reported in the literature with detailed descriptions (*27, 41, 58*). In this chapter, we describe four representative media. The Murashige-Skoog (MS) and Linsmaier-Skoog media (*42, 56*) have been used extensively for a range of culture types and species, particularly herbaceous plants. For woody plants, a dilution of three to ten times in inorganic salts or a shift to other inorganic media is desirable. The Woody Plant Medium (WPM) (*44*) was developed for woody plants and the Anderson (AND) medium was developed for rhododendrons (*3, 41*). The Gamborg B5 medium has been used widely for cell and tissue cultures (*23*).

Organic Compounds (Group F and G, Table 18–1)

Carbohydrates. Sucrose at 2 to 4 percent is used for most cultures, but concentrations as high as 12 percent might be used in some cases, as for young embryos. Glucose has sometimes been used for monocots and selected dicots (like strawberry). Fructose, maltose, and starch have been used occasionally. These materials are added at the time of making the cultures.

TABLE 18–1

Stock solutions (grams/liter) used in preparing various media for micropropagation[a]

Group	Compound	Murashige and Skoog (MS)	Woody Plant Medium (WPM)	Anderson (AND)	Gamborg B5
A	NH_4NO_3	165.00	40.00	40.00	—
	KNO_3	190.00	—	48.00	250.00
	$Ca(NO_34H_2O)$	—	55.6	—	—
B	K_2SO_4	—	99.00	—	—
	$MgSO_4 \bullet 7H_2O$	37.00	37.00	37.00	25.00
	$MnSO_4 \bullet H_2O$	1.69	2.23	1.69	1.00
	$ZnSO_4 \bullet 7H_2O$	0.86	0.86	0.86	0.2
	$CuSO_4 \bullet 5H_2O$	0.0025	0.0025	—	0.0025
	NH_4SO_4	—	—	—	13.4
C	$CaCl_2 \bullet 2H_2O$	44.00	9.6	44.00	15.00
	KI	0.083	—	0.083	0.075
	$CoCl_2 \bullet 6H_2O$	0.0025	—	0.083	0.0025
D	KH_2PO_4	17.00	17.00	—	—
	H_3BO_3	0.62	0.62	0.62	0.30
	$Na_2MoO_4 \bullet 2H_2O$	0.025	0.025	0.025	0.025
	$NaH_2PO_4 \bullet H_2O$	[b]	—	38.00	15.00
E	$FeSO_4 \bullet 7H_2O$	2.784	2.78	5.57	2.78
	$Na_2 \bullet EDTA$	3.724	3.73	7.45	3.725
F	Thiamin•HCl	0.10	0.10	0.04	1.00
	Nicotinic acid	0.05	0.05	—	0.10
	Pyridoxine•HCl	0.05	0.05	—	0.10
	Glycine	0.20	0.20	—	—
G	Myo-inositol	10.00	10.00	10.00[c]	10.00

[a]These are 100 × the final solution. Use 10 ml of each stock solution for preparing 1 liter of the culture medium.

[b]Commonly added as a supplement directly to the culture medium at 85 to 255 mg/liter.

[c]Adenine sulfate added, 80 mg/liter.

Vitamins. Thiamin (0.1 to 0.5 mg/l) is almost always considered essential, and nicotinic acid (0.5 mg/l) and pyridoxine (0.5 mg/l) are usually added. Inositol at 100 mg/l is beneficial in many cultures and is usually added routinely. Other vitamins that are sometimes beneficial include pantothenic acid (0.1 mg/l) and biotin (0.1 mg/l). All of these substances are soluble in water and should be prepared as stock solutions, ready for dilution at 100 times the final concentration. Stock solutions should be stored in a refrigerator.

Hormones and growth regulators. The two most important classes that are used to control organ and tissue development are the auxins and cytokinins. Gibberellins have sometimes been used to promote shoot elongation.

Auxins. The natural auxin, indoleacetic acid (IAA), is usually used at concentrations of 1 to 50 mg/l. Synthetic auxins are usually more stable and include naphthaleneacetic acid (NAA) (0.1 to 10 mg/l), 3-indolebutyric acid (IBA)(0.1 to 10 mg/liter), and 2, 4-dichlorophenoxyacetic acid (2,4-D) (0.05 to 0.5 mg/l).

Cytokinins. The compounds include N^6-benzyladenine (BA), kinetin, N^6-isopentenyl-adenine (2iP), and zeatin. They are used at a range of 0.01 to 10 mg/l. Thidiazuron (*20, 37*) and N-2-chloro-4-pyridyl N-phenylurea (CPPU) (*20, 37, 51*) have cytokinin activity and are often used in combination with traditional cytokinins (like BA), but at ⅒ the concentration. Adenine sulfate is often added to 40 to 120 mg/l.

Gibberellins. Gibberellic acid (GA_3) is sometimes used but must be filter-sterilized.

All of these materials should be prepared in advance and maintained as stock solutions in a refrigerator at near freezing temperature. Stock solutions of auxins and other organic materials deteriorate with time.

Miscellaneous. Citric acid (150 mg/l) or ascorbic acid (10 mg/l) can be used as an antibrowning agent (antioxidant). Malic acid (100 mg/l) is added in some embryo culture media. These should be filter-sterilized, since autoclaving causes decomposition. These organic acids may be incorporated in the medium or used in prewashing steps. Fine-grade activated charcoal, preferably prewashed, is used in some cases (0.1 to 1 percent) to adsorb and counteract inhibiting substances released by some tissues. The use of activated charcoal can reduce the effectiveness of growth regulators.

Complex natural ingredients. Various materials of unknown composition have been used to establish cultures when known substances fail to do so. Protein hydrolysates from casein or other proteins (30 to 3,000 mg/l) are sometimes helpful, primarily for providing organic nitrogen and amino acids. Coconut milk (endosperm of either green or ripe coconuts, 10 to 20 percent by volume) has been used but must be filter-sterilized for some uses. Malt (500 mg/l), yeast extract (50 to 5,000 mg/l), and such substances as tomato juice (30 percent v/v) and orange juice (30 percent v/v) also have been used. These substances are dissolved in water and added during preparation of the medium. Most of these materials are available from commercial sources.

Preparation of Stock Solutions

Inorganic substances and many of the organic materials can be maintained under refrigeration in stock solutions of 100 times the required concentration (Figure 18–5). Ten ml of the stock solution can be added to each liter of medium to provide the required concentration. From the list in Table 18–1, the following materials can be combined without forming precipitates: (a) nitrates, (b) sulfates, (c) halides (i.e., chlorides and iodides), (d) phosphates, borates, and molybdates. Iron can be purchased as a chelated iron solution, or it can be prepared separately by combining the substances in group E, or alternatively supplied as inorganic iron—$FeCl_36H_2O$ (1 mg/l), $FeSO_4$(2.5 mg/l), or Fe tartrate (10 mg/l). Iron solutions should be stored in a dark bottle.

Some organic compounds are relatively insoluble in water. A small quantity of organic solvent (not more than 0.5 percent of the final medium) are effective in dissolving most organic substances. Cytokinins are weak bases and can be dissolved in a dilute acid, then diluted to a final volume with water. Auxins are weak acids and can be dissolved in a dilute base or in alcohol. Use 0.3 ml of 1 *N* HCl for cytokinin and NaOH (for auxin) for each 10 mg of the compound. Stock solutions and organic substances should be refrigerated.

Supports

Agar. Agar is obtained from certain species of red algae and is commercially available in a powdered form. Its value in culture systems is (a) ability to melt when heated, (b) change to a semisolid gel at room temperatures, and (c) relative biological inertness. It may contain certain impurities, notably salts, in its natural state. Most agars are prewashed or purified (USP grade). Quality may vary among commercial producers. Agar performance may be affected by concentration and by pH. Usually a pH of 5.0 to 6.0 is obtained, but the pH of the medium may need to be adjusted through the addition of acids or bases. As the pH is lowered (4.5 or less), the agar may fail to solidify.

The lowest concentration of agar that will support the explants is best, which is about 0.5 to 0.6 percent. This allows good contact between explant and medium and adequate uptake of nutrients. At lower concentrations the explant may sink into the agar and aeration is impaired. As the concentration increases (up to 1.0 to 1.2 grams per liter), growth may be depressed because of higher osmotic pressure.

Agar substitutes. Gelrite is a polysaccharide derived from a *Pseudomonas* bacteria (*27, 30, 41, 58*). It makes a very clear gel upon heating in combination with cations, e.g., magnesium, calcium, potassium, and sodium for gelation. Its required concentration is less than agar but is often used in combination with agar, such as in a ratio of 3:1 Gelrite to agar, with the most effective combination varying with the brand of agar (*41*).

Liquid. Nutrient solutions without agar have an advantage for some plants. Some type of support is required to keep the explants and cultures from sinking, or a shaker or rotating drum must be used to provide aeration. Filter paper bridges may be inserted into the liquid to support the culture with the paper acting as a wick. Plastic fabric supports made of 100 percent polyester fleece are available commercially. Liquid media are useful for some plant species whose explants exude toxic substances from the cut surfaces. The liquid medium can be exchanged without reculturing.

Preparation of the Medium

High water quality (i.e., low salts and/or low amounts of organic or chemical impurities) is important for growing cultures, and tests of the water supply used for growing cultures should be determined. Single distilled water may be adequate in commercial micropropagation, but it may be desirable to double distill the water or to pass it through deionizers and equipment to further purify it.

To prepare media, use a large Pyrex flask or beaker, and add a portion (one-half to two-thirds) of the final volume of water. Sugar may be added at this time or not until the end. Stir. Add by pipettes or a graduated cylinder proper amounts of each ingredient from stock bottles. Agar is added next, and then the solution is made up to the prescribed volume by adding water. Adjust to the desired pH (usually about 5.7) with drop-by-drop additions of 1 *N* HCl or 1 *N* NaOH using a pH meter (or pH indicator strips) to measure.

The solution is heated to melt the agar. This may be done in an autoclave at 121° C (218° F) for three to seven minutes or on a hot plate, stirring continuously to prevent the agar from settling and burning. The hot solution is then dispensed into containers and sterilized in an autoclave for 20 minutes at 121° C. Occasional agitation of the solution while dispensing will ensure even distribution of medium components into each container. The greater the volume of media to be sterilized, the longer the sterilization time (7).

Substances that are unstable during autoclaving must be filter-sterilized. Commercial devices equipped with a cellulose acetate or cellulose nitrate membrane filter that is porous enough to allow liquid to pass but nonporous enough to prevent the movement of organisms such as bacteria are available from commercial sources (Figure 18–1). These require pressure from a syringe plunger or a gentle vacuum to pass solutions through the filter.

The appropriate concentration of the material is added to the medium after autoclaving when the medium has cooled to 35 to 40° C (95 to 104° F) before dispensing into the final containers. The medium should be used within two to four weeks and refrigerated if kept longer.

Aseptic Procedures

Disinfestation

Disinfestation is the process of removing contaminants from the surface of the organ rather than from within the organ. Primary disinfestants include alcohol (ethyl, methyl, or isopropyl, usually about 80 percent) and bleach (calcium or sodium hypochlorite, usually with 5.25 to 6.0 percent active ingredient). Bleach is usually diluted to 10 to 20 percent before use. A few drops of a surfactant or detergent should be added to the bleach to improve surface coverage. The effectiveness of the bleach treatment is a time-dosage response; the disinfestation effectiveness increases with an increase in both time and concentration but the damage to living tissues also increases. Thus a compromise must be developed through testing for the kind of explant to be disinfested.

Additional disinfectants include hydrogen peroxide, silver nitrate, benzalkonium chloride, and mercuric chloride. Care should be taken, especially with mercuric chloride, to handle these materials safely and to properly dispose of them. Request the MSDS sheet for these chemicals from your dealer for additional information about care in using these compounds.

A typical procedure for shoot-tip propagation is to cut the shoot in short pieces several centimeters long, and wash in tap water with a detergent. For woody pieces, a quick dip in alcohol may be helpful. Wrapping the shoots with small squares of sterile gauze will hold them intact during the treatment. In a hood, the plant parts are placed into the disinfesting solution (with a surfactant or detergent added), using a solution of 1 to 10 percent of the prepared material, for 5 to 15 minutes. The solution is then poured off and the material rinsed two or three times with sterile water. Modifications of this basic procedure that may improve effectiveness include prewashing, mechanical agitation, vacuuming, and multiple treatments.

Microbial contaminants, consisting of yeasts and various species of fungi and bacteria, usually appear on the agar surface within a few days to a week as white or opaque slime or as variously colored colonies, sometimes with black spores and mycelium (*22*). When these contaminants can be identified visually, the culture vessels containing them should be discarded quickly by autoclaving. Some bacterial contaminants, such as *Bacillus subtilis, Erwinia* (*33*), or *Pseudomonas* (*4*), sometimes remain inside the plant material without being detected for several months after the initial culture or incubation. Procedures such as culture indexing to detect or to identify these contaminants may be included (see also Chapter 9). Antibiotics are sometimes used in the medium to improve control of bacterial growth (*23, 58*).

Microorganisms occur primarily as spores resting on surfaces of tables, hands, arms, clothing, and various objects and settle out of the air or are blown in the dust on air currents. Aseptic transfer procedures are done in some kind of transfer hood or box to minimize the movement of dust and contaminants. These should be equipped with a HEPA filter with a positive air pressure blowing outward from the rear of the hood to prevent movement of airborne spores into the chamber. The sides and surfaces of the chamber should be washed with a disinfestant such as sodium or calcium hypochlorite or 95 percent alcohol. Hands and arms should be washed thoroughly, and gloves and a clean lab coat should be worn.

A **culture indexing test** for contamination by specific pathogens in containers is as follows: Cut off a piece of plant tissue, dice it, and place the pieces into 6 ml of nutrient broth that is optimum for bacterial and fungal growth. Positive results are shown in cloudiness of the medium and should appear within four to five days. If made at the initial explant stage, however, this procedure does not always identify all potential contaminants (*33*). Additional tests may be necessary because of latent contamination.

Supplies needed for a transfer operation are as follows:

1. Burner for sterilizing dissecting tools and containers. This can be an automatic gas burner or an inexpensive glass alcohol lamp with a wick. A glass bead sterilizer (Figure 18–6) and certain types of burner-incinerators are available commercially and can be used to avoid the dangers of alcohol or gas fumes with their possible attendant fires.
2. Forceps to hold and manipulate tissue. Three useful sizes (*72*) include 300 mm, 200 mm straight-tipped, and 115 mm curve-tipped.
3. Dissecting needles with wood handles, or a metal needle holder with replaceable tips.
4. Dissecting scalpels. Metal scalpel holders with replaceable knife blades are most convenient.

FIGURE 18–6 A glass bead sterilizer is an alternative to burners for sterilization of tools.

A piece of broken razor blade glued to a holder can be used for fine work, or various kinds of microtools can be made.

5. Sterile petri dishes can be used for holding explants and sectioning them. Small squares of sterile moist paper toweling inside each petri dish are convenient to assist cleaning later (*41*). Sterilized paper towels placed directly on the surface of the lamina flow hood are commonly used as a working surface. A moist surface can keep small explants from drying out.
6. A dissecting microscope and a top-loading balance. These may be important for specific operations. These should be wiped with a disinfectant. The dissecting microscope should have a plastic or glass tunnel over the viewing area to protect against contamination.

Scalpels, needles, and other dissecting tools are moved into position, usually on holders. Sterile petri dishes are placed into the chamber to use as working trays. Sterile water containers are included, as are containers of medium, as needed. These containers are generally kept to the front of the working area so that reaching beyond a particular "sterile" area to the rear of the hood can be avoided. Outside surfaces of containers should be wiped or sprayed with a disinfectant.

In preparing explants for transfer, the implements must be sterilized. Although this can be done by dipping them into alcohol and flaming them before use, many propagators avoid this because of the fire hazard. Furthermore, simply dipping and flaming the scalpels may not kill the most resistant pathogens or spores. Longer exposure to the flame, or use of a chemical sterilant, may be required (*64, 71*). A Bacti-incinerator is a commercial device that heats forceps and tools to a very high temperature (about 1,600° F) when inserted into the core (*41*). However, the manufacturer warns that sharp instruments should not be inserted into the core. An alternative to burners and the Bacti-incinerator is the glass bead sterilizer (Figure 18–6). Instruments placed in the heated beads at 250° C are sterilized in 10 to 15 seconds. Tools should be dry and free of agar. Tools should not remain in glass beads for more than one minute. Glass beads need to be replaced periodically. Regardless of how tools are sterilized, needles, scalpels, and forceps should be allowed to cool 30 to 60 seconds before using to prevent damage to tender tissue. When working in an open area, the propagator usually flames the mouth of the container before opening it and after inserting the material to avoid entrance of contaminants.

MICROPROPAGATION PROCEDURES

As discussed in Chapter 17, there are basically four stages to the micropropagation process (Figures 18–7, 18–8). These are:

- Stage I. Establishment and stabilization
- Stage II. Shoot multiplication
- Stage III. Root formation
- Stage IV. Acclimatization

Common Terms Related to Micropropagation

Explant. Piece of plant used to initiate a culture.

Subculture. The act of subdividing the developing explant into smaller pieces and moving them to a new medium.

Transfer. Moving the entire culture to a new medium. There is no subdivision as in subculturing.

Microshoots. Shoots developed during tissue culture.

Microcuttings. Single microshoots moved to a medium to induce rooting. Microcuttings can be rooted either *in vitro* or *ex vitro.*

Plantlets. A plant derived from micropropagation that has both shoots and roots.

In vitro. Growth and development in the tissue culture environment.

Ex vitro. Growth and development outside the tissue culture environment.

STAGE I—ESTABLISHMENT AND STABILIZATION

The function of this stage is to disinfest the explant, establish it in culture, and then stabilize the explants for multiple shoot development.

Preparation for Establishing Cultures

Handling stock plants. Two principal considerations are involved in handling stock plants. One is reducing the potential for contamination by fungi, viruses, and other pathogens. These

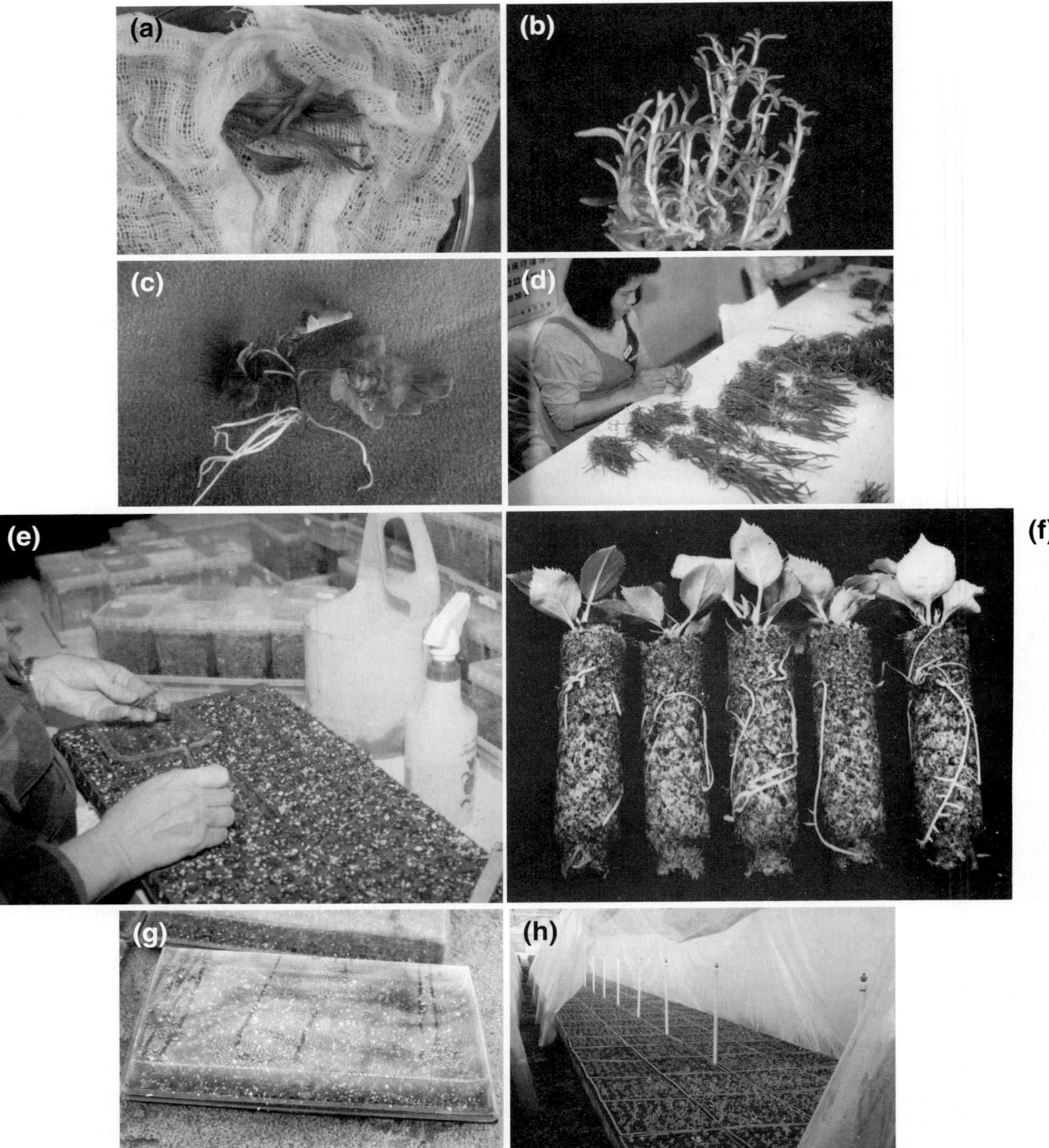

FIGURE 18–7 The basic process for micropropagation includes: (a) *Stage I.* Tissue is disinfested with a solution of bleach (usually 10 to 20 percent); (b) *Stage II.* Multiplication of shoots takes place on a medium with a high cytokinin to auxin ratio; Stage III. Root formation can take place in an *in vitro* or *ex vitro* environment; (c) Roots formed in vitro; (d) Orchids rooted *in vitro* are ready to transplant. The technician is grading plants by size; (e) Microcuttings can be directly rooted (*ex vitro*) in a greenhouse medium. Note the spray bottle used to frequently mist these rhododendron microcuttings during sticking; (f) Peat plugs commercially treated with a binder are popular for rooting microcuttings (apple cuttings photo courtesy of R. H. Zimmerman); (g) Plastic lids fit standard sized flats and keep a high humidity for the rooting environment; *Stage IV.* (h) Large commercial growers purchase rooted or unrooted cuttings from tissue culture specialists and acclimatize them under plastic tents, often with mist.

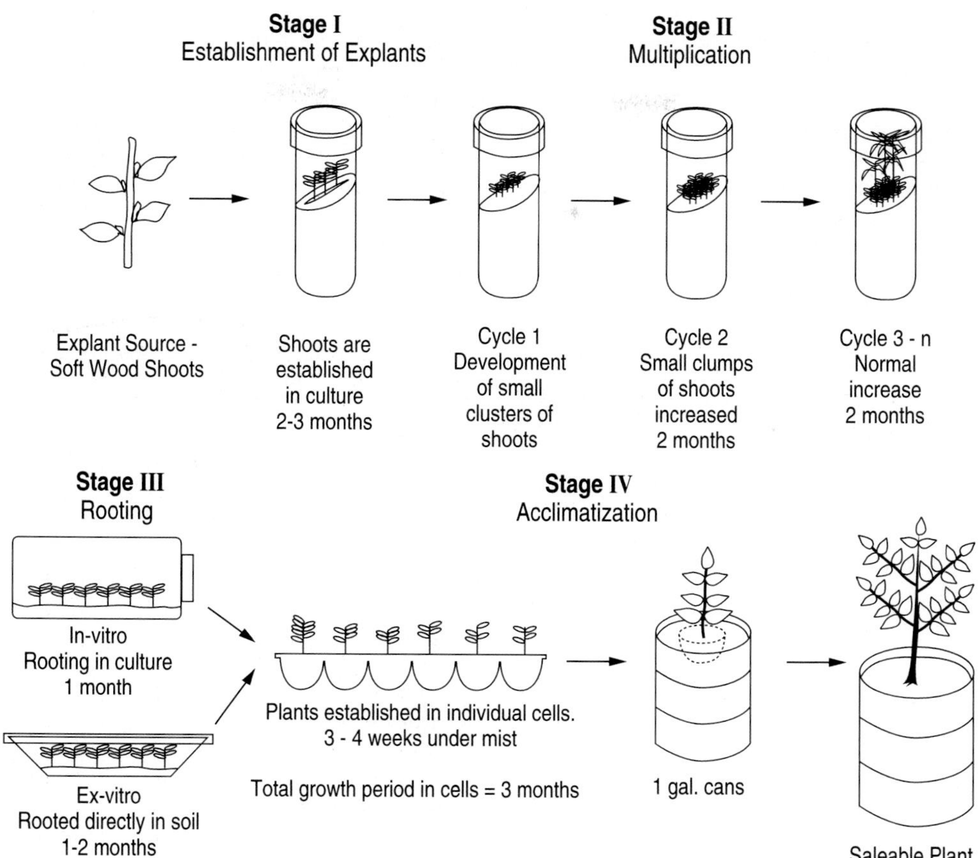

FIGURE 18–8 Schematic representation of micropropagation system applied to rhododendron (courtesy of W. Anderson).

pathogens are not automatically eliminated by *in vitro* technique unless this goal is built into the system. Preselected "virus-free" or pathogen-indexed stock plants should be used. Otherwise special techniques, such as meristem-tip culture, heat treatment, and various culture-indexing tests, may be incorporated into the procedure.

Choice of explant. The kinds of explant and where and how they are collected varies with the purpose of the culture, the species, and often the cultivar (see Chapter 17). Shoot tips should be collected from plants in relatively active growth. The size of the explant may vary from as small as a 1 to 5 mm long meristem tip for meristem culture to a piece of shoot several centimeters long. A single node bearing a lateral bud also can be used. For woody plants, a shoot tip from a dormant (but not resting) bud may be utilized but is often difficult to disinfest. Removal of the bud scales, plus cutting away the leaf scar, may give sterile tissue.

For woody plants, it may be helpful to collect branches in early spring before growth has begun and to bring the cut branches into a growth chamber or greenhouse to force growth. A florist's cut-flower preservative in the forcing water can improve growth. Collect explants from the expanding new growth. Explants collected in this way are easier to disinfest than those collected directly from new growth outdoors.

Pieces of leaves with veins present, bulb scales, flower scapes, and cotyledons also may be used for obtaining explants, as described in Chapter 17 for adventitious shoot initiation.

Pre-treatment. Shoots are disinfested by procedures described in the previous paragraphs. Sometimes a preliminary treatment step is included. In a petri dish, the explants are placed on the surface of an agar medium of basic salts and sugar, but without growth regulators. The purpose is to detect contamination of individual materials and to control exudation of phenolics and other substances from the cut surface that inhibits development. Addition of activated charcoal, ascorbic acid, or citric acid to the medium can be used to overcome the problem. Using liquid media in the initial stage or frequent transfers on agar medium for several days may leach out the toxic materials. Likewise, using antioxidants, such as ascorbic acid and citric acid, in the preliminary washing may be useful.

Initial Shoot Development

Depending on the type of explant, shoots will initiate from (a) the stimulation of axillary shoots; (b) the initiation of adventitious shoots on excised shoots, leaves, bulb scales, flower scapes, cotyledons, and other organs, as described in Chapter 17; or (c) the initiation from callus on the cut surfaces. The medium selected varies with the species, cultivar, and kind of explant to be used. A basal medium (BM) includes the ingredients listed in Table 18–1, plus sucrose and sometimes other supplements. Control of development is most often achieved by manipulating the levels of auxin and cytokinin. If axillary shoots are desired, a moderate level of cytokinin (BA, kinetin, or 2iP) is used (0.5 to 1 mg/l). The auxin level is kept very low (0.01 to 0.1 mg/l) or omitted altogether.

To develop adventitious shoots on explants (excised shoots, leaves, bulb scales, cotyledons, flower scapes, etc.), a somewhat higher cytokinin level is needed. For callus formation, increased auxin levels are needed but must complement an appropriate level of cytokinin. Usually four to six weeks are required to complete Stage I and produce explants ready to be transplanted to Stage II.

Some woody plant species can take up to a year to complete Stage I (*47*). This is termed **stabilization.** A culture is stabilized when explants produce a consistent number of "normal" shoots after subculturing. "Abnormal" shoots usually are arrested in development and fail to elongate.

STAGE II—SHOOT MULTIPLICATION

In the multiplication stage each explant has expanded into a cluster of microshoots arising from the base of the explant (Figure 18–9). This structure is divided into separate microshoots, which are transplanted into a new culture medium. This is called **subculturing**. During the multiplication stage, cultures are subcultured every four to eight weeks.

The kind of medium used depends on the species, cultivar, and type of culture. Usually, the basal medium is the same as in Stage I, but often the cytokinin and mineral supplement level is increased. Adjustments may need to be determined by experimentation (*4, 14, 55*).

Different species vary in the optimum size of the microshoot and method of cutting apart. There is generally a minimum mass of tissue that is necessary to produce uniform rapid multiplication in the next transfer. New cultures are initiated in two ways from these developing cultures. The elongated shoots are cut from the original culture and are subcultured as nodal explants. These may have the leaves removed and are typically two to four nodes in length. These shoots can be inserted into the medium in a vertical position or laid horizontally on the surface. Avoid pushing shoots too deeply into the agar and submerging the nodes. Elongated shoots laid horizontally on the agar surface perform as a type of layering and stimulate lateral shoots to develop. After the elongating shoots have been removed and subcultured, the original explant can be further subdivided and recultured. Division may be made by vertically cutting the tissue mass into sections, keeping some of the base with each piece. Sometimes one dominant shoot develops and inhibits elongation of others. If this happens, the shoot may be cut off and the base recultured. For clump-forming plants (like hosta or day lily), all shoots form from a common base. These crowns are separated in a similar way to standard division (see Chapter 15) and each division is subcultured.

The number of microshoots produced ranges from 5 to 25 or more depending upon the species and conditions of culture. Multiplication may be repeated several times to increase the supply of material to a predetermined level for subsequent rooting and transplanting. Sometimes microshoots deterio-

FIGURE 18–9 Multiple shoot formation in gas plant *(Dictamnus)* cultures grows on MS medium containing 5 μM BA.

rate with time, lose leaves, fail to grow, develop tip-burn, go dormant, or lose the potential to regenerate (*4*) (see pp. 558 for a discussion of stabilization.)

During multiplication, off-type propagules sometimes appear, depending on the kind of plant and method of regeneration. In some plants, such as Boston fern, restricting the multiplication phase to only three propagations before reinitiating the cultures is recommended to avoid development of off-type microshoots. Other procedures that can reduce the potential for variability include reducing growth regulator concentration and avoiding basal callus that may give rise to adventitious shoots in subculturing (*25*).

STAGE III—ROOT FORMATION

Shoots developed during the multiplication stage do not usually have roots. There are some exceptions (like African violet) that spontaneously root on the multiplication medium. For most other species, single shoots (microcuttings) must be moved to a medium or suitable environment to induce roots. Therefore, the purpose of Stage III is to prepare the microplants for transplanting from the artificial heterotrophic environment of the test tube to an autotrophic free-living existence in the greenhouse and on to their ultimate location (Figure 18–10). This preparation may not only involve rooting, but it also may involve conditioning of the microplant to increase its potential for acclimation and survival during transplanting. For example, the agar and/or sucrose concentration might be increased. Light intensity is sometimes increased during this period.

Microcuttings can be rooted either *in vitro* or *ex vitro* (Figures 18–7, 18–8). For *in vitro* rooting, individual microcuttings are subcultured into new containers in a sterile medium with reduced or omitted cytokinin, an increased auxin concentration, and often a reduced inorganic salt concentration. For some plants, rooting is best if the microcutting is kept in the auxin medium for only one or two days and then transferred to an auxin-free medium. Or the microcutting may simply be dipped into a rooting (auxin) solution and inserted directly into an auxin-free medium. Some plants root best if placed in the dark during the auxin treatment period.

Some plants respond to an "elongation" phase between Stages II and III by placing the microshoot into an agar medium for two to four weeks without cytokinins (or at very low levels) and, in some cases, adding (or increasing) gibberellic acid (*68*). This reduces the influence of cytokinin. The microshoots are then rooted in a cytokinin- and GA-free medium.

Selection for uniformity and roguing of obviously abnormal, aberrant, or diseased microshoots should be made prior to Stage III (*4*). Plants of species having an inherent dormancy or rest period may need to be chilled to stimulate new growth and elongation (*4*).

Microcuttings can also be rooted directly under *ex vitro* conditions. Microcuttings can be treated with auxin as a quick-dip (see Chapter 11)

FIGURE 18–10 Apple rootstock propagation by shoot-tip proliferation. (a) Proliferated shoots in Stage II. (b) Rooted plant ready for transplanting. (c) Apple plant after transplanting (Courtesy I. Snir and A. Erez).

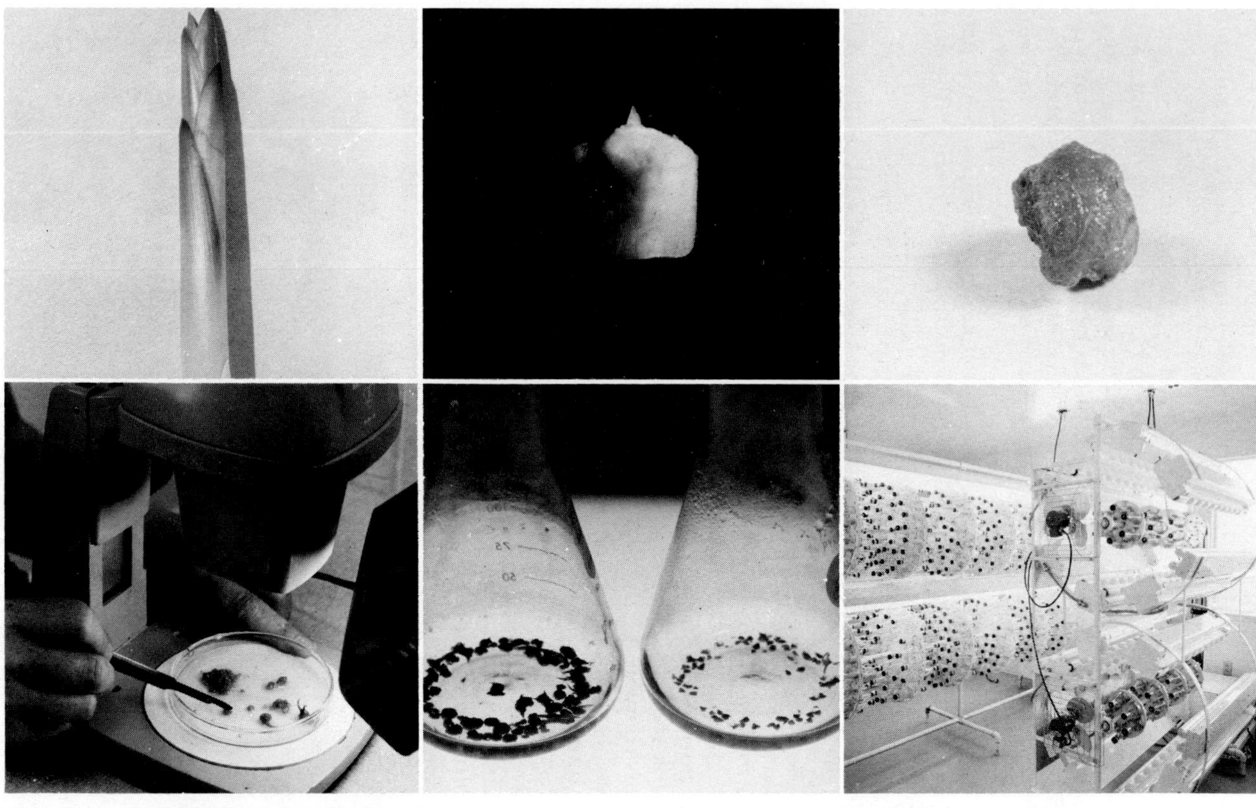

FIGURE 18–11 Steps in the micropropagation of Cymbidium orchids (66). Top row (left to right): Growing shoot-source of meristem; excised shoot apical meristem; protocormlike bodies after enlarging in nutrient medium and before dividing. Second row (left to right): Dividing each enlarged protocormlike structure into several pieces; before and after dividing; rotating wheels with divided mericlones in nutrient solutions for enlargement and prevention of root-shoot polarity.

FIGURE 18–11 (cont.) Third row (left to right): Protocormlike material placed on nutrient agar for polarized growth and development of roots and shoots; greenhouse filled with flasks for growing on and into plants; well-developed plants large enough for potting. Bottom row (left to right): Well-developed orchid plants ready for potting; potted Cymbidium orchid originating from a divided protocormlike body; orchid house with plants coming into flowering. (Courtesy Richard Smith, Rod Mclellan Co., South San Francisco.)

or "pulsed" with auxin by *in vitro* exposure to a medium with relatively high levels of auxin for five to ten days. After auxin treatment, microcuttings are "stuck" into plugs (*48, 84*) containing a greenhouse rooting medium (e.g., two parts peat to one part vermiculite) and placed in a high humidity environment to form roots (Figure 18–7). This procedure is the same as that used during standard cutting propagation (see Chapter 11) except care must be taken to prevent the fragile leaves on the microcutting from drying out during the sticking operation. Frequent misting prevents this type of damage (Figure 18–7).

STAGE IV—ACCLIMATIZATION TO GREENHOUSE CONDITIONS

Once plantlets are well rooted, they must be acclimatized to the normal greenhouse environment (*60, 82*). *In vitro* rooted plants are removed from the culture vessel and the agar is washed away completely to remove a potential source of contamination. Plantlets are transplanted into a standard pasteurized rooting or soil mix in small pots or cells in a more or less conventional manner. Initially, microplants should be protected from desiccation in a shaded, high-humidity tent or under mist or fog. Several days may be required for new functional roots to form (*67, 8*).

Plantlets should be gradually exposed to a lower relative humidity and a higher light intensity. Any dormancy or resting condition that develops may need to be overcome as part of the establishment process.

SPECIFIC PROTOCOLS FOR ASEPTIC CULTURE

Micropropagation

Micropropagation involves the vegetative propagation of plants through the (1) establishment, (2) multiplication, (3) rooting, and (4) transplanting of clones *in vitro* starting with small explants and ending with a rooted plant established in a container or in the ground. The following protocols are step-by-step procedures by which specific representative species can be propagated. For more details about additional plant species, see individual genera in Chapters 19, 20, and 21.

Carnation.* *(75) Carnation is typically started by meristem-tip culture to ensure that the material is pathogen-free. Once established, "clean" material can be readily propagated through axillary shoot proliferation.

Stage I. Establishment. Use shoots of about 12 nodes, removing leaves that are over 3 mm long. Wash with detergent. Place in 10 percent solution of bleach (0.525 percent sodium hypochlorite) for five minutes. Rinse two times for 3 minutes each in sterile water. Remove stem tip with one to two leaves 1 mm or larger and place on surface of modified MS agar medium (see Table 18–1) plus 0.2 mg/l kinetin and 0.2 mg/l NAA. Place in growth chamber at 22° C in continuous light.

Stage II. Multiplication. After approximately one month or when the culture has grown sufficiently, remove from the container, and cut into segments approximately 1 cm long. Transfer each to separate containers on agar with Mod MS + 0.5 mg/l kinetin + 0.1 mg/l NAA. Continue to segment and transfer to fresh medium.

Stage III. Rooting. Transfer microcuttings to a medium without kinetin or NAA. Roots form easily *in vitro. Ex vitro* rooting is an alternative. Microcuttings from Stage II are treated with IBA and placed in a rooting medium as a typical softwood cutting would be.

Stage IV. Acclimatization. Remove rooted microshoots from container. Wash away solid media if present. Transplant into growing medium and hold under mist or under high-humidity tent. New functional roots need to form to replace the older roots. Plug containers of the type used for seeds are useful. Once established, microplants are gradually exposed to lower humidity and higher light intensity.

***Hosta*.** An example of an easily micropropagated herbaceous perennial (*50, 81*).

Stage I. Initiation. Select top florets as an explant after the bottom florets have opened. Surface disinfest the entire flower scape in 10 percent bleach for 15 minutes followed by several rinses with sterile water. Remove florets and place on modified MS medium with BA at 0.5 to 2.5 mg/l and NAA at 0.5 to 2.5 mg/l. Callus and new shoots develop from the flower tissue in 8 to 12 weeks.

Stage II. Multiplication. Transfer individual shoots to the same medium with 0.1 mg/l BA and 0.5 mg/l NAA to induce further shoot formation. Clumps of shoots can be subcultured every four to six weeks.

Stage III. Root Formation. Shoots root readily *in vitro* on a medium without BA or NAA.

Stage IV. Acclimatization. Wash agar from plantlets and transfer to greenhouse medium under high humidity. Gradually reduce humidity to acclimatize plants. Evaluate variegated cultivars for trueness-to-type.

Boston Fern (Nephrolepsis) (*59, 66*). Micropropagation is very effective because natural increase by rhizomes is slow.

Stage I. Initiation. Use 5- to 10-cm long actively growing runners. Surface disinfest in 20 percent bleach (1.05 percent sodium hypochlorite) for 20 minutes; rinse three times in sterile water. Use 1/2 or 3/4 strength salts MS medium with 1.0 mg/l kinetin and 0.1 mg/l NAA. Cut runners into 2.5 cm. sections. Growth should appear within six weeks.

Stage II. Multiplication. Use the same medium. Establish at six to eight weeks by subdividing clumps into 1 cm^2 sections. Subculture every six to eight weeks but do not do more than three times to avoid excessive mutation from proliferating cultures.

Stage III. Rooting. Microcuttings form roots *in vitro* on ½ strength MS medium without growth regulators, or microcuttings can be rooted *ex vitro* directly into a rooting medium under high humidity or mist.

Stage IV. Acclimatization. Gradually reduce humidity after plantlets are well rooted.

Cherry (Prunus) (*69*). Represents a woody plant that is sometimes difficult to propagate.

Stage I. Initiation. Collect 7- to 10-cm young shoots in spring. Wash vigorously in 10 percent bleach for 10 minutes. Place in running water for two minutes. Disinfest in 10 to 20 percent bleach for 20 minutes; rinse three times with sterile water. Cut above and below node and place upright in agar. Medium is ½ strength MS with 0.5 to 1.0 mg/l BA.

Stage II. Multiplication. Place one or two node microshoots on ½ strength MS salts with BA (0.5 to 2.5 mg/l) and IBA (0.01 mg/l). Subculture microshoots every six to eight weeks.

Stage III. Root formation. Root single microcuttings on a ½ strength MS salts medium with 0.2 mg/l IBA and no cytokinin.

Stage IV. Acclimatization. When roots are well formed and new growth is evident from the plantlet, wash agar off roots and move to greenhouse potting mix under high humidity. Gradually reduce humidity to acclimatize to greenhouse conditions.

Rhododendron (*2, 74*). Woody ornamental that is extensively grown commercially by micropropagation.

Stage I. Initiation. Wash 5-cm shoot tips in 10 to 20 percent bleach; remove leaves and terminal bud; dip in 70 percent alcohol for 10 seconds, rinse; soak in 10 percent bleach for 20 minutes; rinse three times for one minute each. Trim 1 cm from base leaving a 2- to 4-cm explant that is laid horizontally on an agar slant. The medium can vary depending on the species but a typical protocol includes: Anderson medium plus 2iP (5.0 mg/l), IAA (1.0 mg/l), and adenine (80 mg/l). Transfer to new medium every two weeks until growth starts, then every five weeks, but do not subculture until shoots are well developed.

Stage II. Multiplication. Subculture individual shoots onto Anderson medium with 2iP.

Stage III. Root formation. Microcuttings can be rooted in vitro on ½ MS with 2 to 7 mg/l IBA plus activated charcoal or rooted *ex vitro* using 1:1 peat-perlite after treatment with IBA.

Stage IV. Acclimatization. Reduce humidity after plantlets are well rooted.

SOMATIC EMBRYOGENESIS AND THE PRODUCTION OF SYNTHETIC SEEDS

Somatic embryogenesis

Somatic embryogenesis is the production of embryos to produce seeds or seedlings directly from vegetative tissue as a form of cloning (*28*). There are several potential advantages of somatic embryogenesis as a method for propagation. These include:

1. Clonal propagation for species traditionally propagated from seeds such as vegetables and forest species.
2. Extremely high regeneration rates.
3. The development of a synthetic seed industry (see pp. 608).
4. The potential for automation that could dramatically reduce costs relative to seeds.
5. A good system to regenerate transgenic plants because somatic embryos develop from only a few cells (often a single cell).

Several factors have limited the commercialization of plants propagated from somatic embryos. Most important is the relative cost of the procedure compared to traditional seed propagation. Improvements in automating somatic embryo systems are reducing costs, particularly in important forestry species (*1*). A second problem is the poor storability of somatic embryos and synthetic seed. Encapsulated somatic embryos have a storage life of only about 30 days (*28*). This limits commercial transport of synthetic seeds, limits production flexibility, and adds to overall costs.

This procedure is an alternate method to micropropagation and has the potential of mass propagation under industrialized conditions at low cost per unit. In order for extensive commercial use to occur, detailed knowledge of the biological steps in embryogenesis must be known for specific species. Considerable potential for somatic variation exists and must be understood and evaluated for specific crop plants.

Nevertheless, the general steps in the protocol for propagating plants through somatic embryogenesis are known (Figure 18–11) and include the following:

1. selection of explant material.
2. conditioning cells to express embryogenic potential.
3. differentiation and maturation of somatic embryos.
4. conversion (germination) of the somatic embryo into a seedling.

1. Selection and culture of appropriate explant material. Explant tissue for somatic embryogenesis includes reproductive structures (ovules or nucellus) and seedling tissue. There are two patterns of differentiation that lead to the production of somatic embryos. **Adventitious somatic embryos** form directly on the explant tissue without an intervening callus stage. The second pattern of embryogenesis **(induced somatic embryos)** requires a step in the procedure to produce callus (often in the form of suspension culture cells) that must be induced to develop into embryos as described in Chapter 17. The pattern of differentiation determines the procedures selected by the propagator.

2. Conditioning the cells to express their embryonic potential. Cells are transferred to a basal medium with a high concentration of auxin. The most effective auxins have been 3,4-dichlorophenoxyacetic acid (2,4-D), although other synthetic auxins (NAA, Picloram) and cytokinins (BA) can induce somatic embryogenesis.

3. Differentiation and maturation of somatic embryos (Figure 18–12). In many cases, 2,4-D is essential to initiate somatic embryogenesis but most species must be removed from the 2,4-D medium to permit further development of the somatic embryos. Proembryo masses (PEMs) are shifted to an auxin-free medium, usually high in ammonium nitrogen. In some cases, the tissue is sized by passing through screens (sieves) of various sizes to give the highest number of PEMs and synchronous development. In cases where somatic embryos form directly without callus formation, the entire original explant is moved to an auxin-free medium to permit further embryo development. Somatic embryos arise from clumps or small masses, develop polarity, and follow a pattern mimicking normal embryo development within the ovule (*28*). Development may be variable in rate and abnormal in appearance with secondary embryos developing from primary embryos. ABA may be added to the development medium to promote more normal somatic embryo development (*6,24*).

4. Conversion (germination) of the somatic embryo to a seedling. Once a mature somatic embryo has reached an appropriate size, it is plated onto an agar medium without auxin. Shoots and roots may then develop. Some somatic embryos show poor ability to convert to seedlings (*28*). Treatment with cytokinin can promote germination in these embryos. However, the most common treatment is to partially desiccate somatic embryos prior to germination. A slow rate of drying enhances survival of embryos during desiccation.

5. Transplanting. Once an adequate shoot and root has developed, the seedling can be transplanted to a greenhouse medium as with any seedling plant or treated with a "synthetic seed coat" and handled as a synthetic seed.

Synthetic Seeds

A **synthetic seed** contains an embryo produced by somatic embryogenesis enclosed within an artificial medium that supplies nutrients and is encased in an artificial seed covering. Technology to carry out each of these steps is available for some crop seeds although not available as a commercial operation (*28, 39, 63*). Many of these operations are protected by plant patents.

The proposed uses for synthetic seeds include:

1. Clonal propagation to replace traditional seed propagation.

2. A replacement to hand-pollinated hybrid plants.
3. Carriers for beneficial microorganisms, pesticides, and growth regulators (see Chapter 5).

The process of developing synthetic seeds starts with the development of somatic embryos as previously described. Synchronous development is important to get a large number of somatic embryos at the same stage of development for further treatment. Somatic embryos are then either directly encapsulated or partially dehydrated before encapsulation. Materials that have been used for encapsulation include sodium alginate (a soluble hydrogel), carrageenan gum, Gelrite, and polyoxyethylene (polyox wafers) (*28, 39, 63*). Synthetic seeds can be then sown for germination like traditional seeds. One major difference between synthetic and most natural crop seeds is the short storage life of synthetic seeds.

METHODS FOR MICROPROPAGATING REPRESENTATIVE HORTICULTURAL CROPS

Many horticultural crops are being micropropagated on a regular basis. Tables 18–2 to 18–5 provide information about key steps in the propagation of representative species being grown in commercial propagation. Except where noted, methods are compiled from published information in Dirr and Heuser (*15*), George (*27*), Kyte (*41*), and Zimmerman (*85*) and are not comprehensive. Further details and additional crop information needs to come from surveys of manuals and texts listed at the end of this chapter or Chapter 17. Abbreviations and explanations used in the tables are listed below Table 18–2:

General methods described in this chapter include:

I. Establishment
II. Multiplication
III. Root formation
IV. Acclimatization

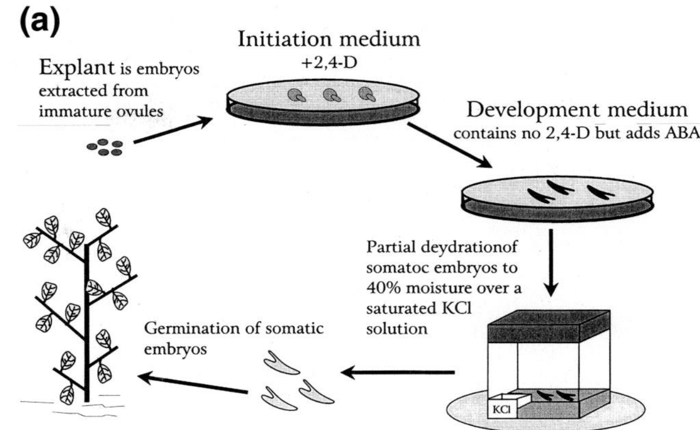

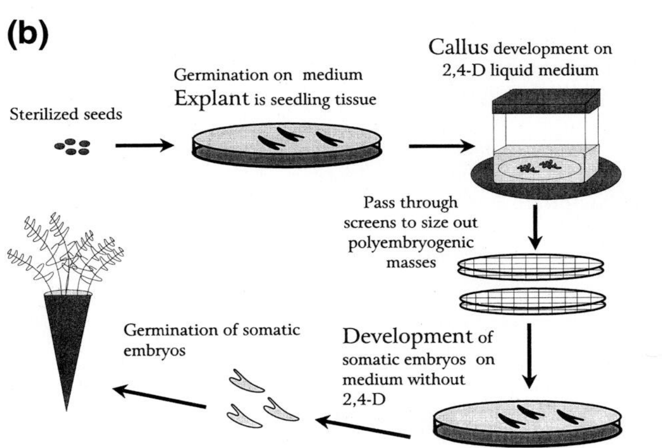

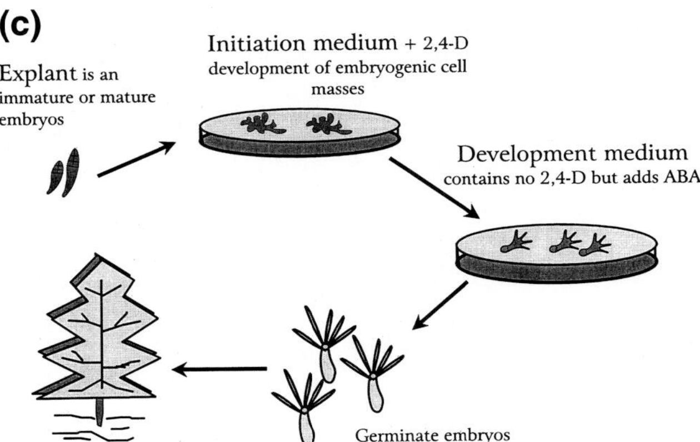

FIGURE 18–12 Three common protocols for the development of somatic embryos. (a) Direct somatic embryogenesis from developing zygotic tissue in redbud *(Cercis)* (*26*). (b) Induced somatic embryogenesis in carrot cell suspensions; (c) Induced somatic embryogenesis from embryogenic cell masses in conifers.

TABLE 18–2

Micropropagation methods for horticultural crops—foliage plants

Species	*Method*	*Explant*	*Sterilization*
1. *Anthurium*	AS, ASP	Single node with bud; excise 2 mm GP; leaf	20 min1:10 BL + DT; GP 1:10 BL 30–45 min; rinse 5 min
2. *Dieffenbachia* 'Perfection'	ASP	Active growing shoot tips	Standard
3. *Ficus benjamina*	ASP	3–5 mm ST	Standard
4. *Nephrolepsis* (Boston fern)	ASP, AS	5–10 mm tip of actively growing runner	1/10 BL; alcohol; rinse 3×
5. *Peperomia*	AS	Leaf sections	Standard
6. *Saintpaulia* (African violet)	AS	1 cm^2 leaf; 2-mm cross section of petiole	Wash leaf or petiole; agitate 10 min; 1:10 BL; rinse 3×; cut
7. *Spathiphyllum*	ASP	Terminal or lateral ST from crown	Standard
8. *Syngonium*	ASP	1–2 cm young plant stems; 1–2 mm MT	15 min 1:10 BL; rinse 3×

AC	activated charcoal
AD	adenine
AND	Anderson's rhododendron medium
AS	adventitious shoots
ASP	axillary shoot proliferation
BA	benzyladenine
BL	"bleach" such as sodium hypochlorite
CH	casein hydrolysate
CM	coconut milk (fresh coconuts)
DT	detergent such as Tween 20
GA	gibberellic acid
GP	growing point
IAA	indoleacetic acid
IBA	indolebutyric acid
KIN	kinetin
LS	lateral shoot
mod	modified
MS	standard Murashige-Skoog; includes standard supplements
MT	meristem tip—0.1 to 0.5 mm; primarily for virus control
NAA	naphthaleneacetic acid
ST	shoot tip; usually 1 cm or more long
TS	terminal shoot
2iP	iso-pentenyl adenine
WPM	woody plant medium
rinse:	sterile distilled water
dip:	momentary immersion
wash:	longer period in tap water; nonsterile, usually with a detergent

Light requirements: converted to watts per sq meter PAR

Propagation	*Rooting; acclimation*	*Environmental conditions*
I. MS + vitamins 15% coconut milk, 20% sucrose; rotating liquid 6 weeks II. Two-node section, same – CM, + BA (0.2)	III. None IV. Standard	Continuous light at 16 W/m^2; 25 to 28° C
I. MS; IAA (2.0); 2iP (16.0) separate base from tip in 30 days; repeat II. Use base; 30–60 days; liquid MS + KIN (2.0)	III. None IV. Root in sterilized soil	27° C; 16 hrs; 3 W/m^2
I. MS; IAA (0.3); 2iP (30.0) II. same; subculture for 4–8 wks	III. MS—hormones or supplement; AC IV. Standard	I, II. 33 W/m^2 III, IV. increase 21 to 229 W/m^2
I. 1/2 or 3/4 MS; KIN (1.0) NAA (0.1) II. same as 1	III. MS—hormones IV. Standard	3–6 W/m^2; 6/8; 25° C
I. MS; KIN (4.0); NAA (2.0) II. Same	III. Same IV. Standard	25° C; continuous light
I. MS; BA (0.1); NAA (0.1) for petiole; mod MS + IAA (2.0) + BA (0.08) for leaf II. same	III. Same IV. Standard	25° C; 10 W/m^2 for I, II; 1.5–3 W/m^2 for III
I. 2/3 MS, 3% sucrose II. Same	III. Same but add IBA (15.0) IV. Standard	15 W/m^2
I. Liquid, MS + 2iP (3.0) + IAA (1.0) II. same	III. MS salts only IV. Standard	25° C; 10 W/m^2 for I, II; 25 W/m^2 for III

Remarks

1. Procedure is particularly useful to avoid serious contamination problems with species. Also can use leaf explants to get adventitious shoots for later ASP culture.
2. Be sure to use source plants free of fungi and bacteria as disease control is a major benefit. Most commercial plants are grown from micropropagation.
3. Procedures also work with *F. decora and F. pandurata.*
4. Limit to no more than three subcultures to reduce variation; potential to produce one million plants per year. Most commercial plants are grown from micropropagation (see Figure 18–7).
5. Propagates readily from leaf disks and follows classical hormone responses. Cultures in spring or summer yield more than in winter.
6. One of easiest to grow. Propagates from leaves or petioles as in *in vivo* propagation.
7. Roots develop in four weeks. Most commercial plants are grown from micropropagation. May need to restart from new explants after one year to maintain quality.
8. Propagates readily from tissue culture. Most commercial plants are grown from micropropagation.

TABLE 18–3

Micropropagation methods for horticultural crops—florists' crops

Species	*Method*	*Explant*	*Sterilization*
1. *Cattleya* orchid	ASP	2 mm lateral buds or MT from 2- to 5-cm shoots	Wash + DT; alcohol dip 10 sec; 1:10 BL 15 min; rinse dry; dissect 2-mm ST in sterile antioxidant
2. *Chrysanthemum*	ASP	MT or ST	See procedure of *Dianthus*
3. *Cymbidium* orchid (see Figure 18–12)	ASP	Meristem: apical dome, 2 leaves, cube of tissue 0.5 mm	Remove outer leaves; dip in alcohol 2 sec; rinse in 1:10 BL 15 min; rinse; remove inner leaves
4. *Dianthus* (carnation)	ASP	1 mm ST + 1–2 leaves, or use larger ST	Remove shoot of ca. 12 nodes; remove leaves over 3 mm; wash with DT; add 1:10 BL 5 min; rinse 2 × for 3 min each
5. *Gerbera*	ASP	3–10 mm ST from crown; emerging bud with 1-mm base	Wash running water; add water + DT 3× 10 min; rinse 3× 2 min each; 5 min 1:20 DL; excise; dip 1 min 1:20 BL and plant
6. *Kalanchoe*	ASP, AS	Leaf blade 1.5 cm; stem 2 cm; ST 2 mm	Wash 3–5 cm cuttings; rinse; add 1:20 BL 15 min; rinse 3× 1 min each; trim
7. *Hemerocallis* (daylily)	AS	2 mm sections; young (10-cm) inflorescence scapes; petals and bracts from 1-mm flower buds	Discard bracts; wash; rinse; 1:10 BL 20 min; rinse 3×; slice into 2-mm sections with sterile antioxidant; place 2-mm segment upside down
8. *Lilium* (lily)	AS (scale) AS (leaf) AS (fl bud)	4–5-mm cubes from scale base; trim inner to include basal plate	Wash; rinse running water 15 min; rinse in sterile water; 1:10 DT 10–20 min; rinse 3× in sterile water

Propagation	Rooting; acclimation	Environmental conditions
Use *Cymbidium* method + coconut milk (100 mg/l)	Same as *Cymbidium*	Same as *Cymbidium*
I. Mod MS + IAA (0.5) + KIN (10) II. Same	III. omit KIN and adenine; IAA (10.0); 2 weeks IV. Standard	I, II. 27° C; 3 W/m^2 III. 30 W/m^2
I. Liquid or agar (Knudson's 2% sucrose) II. Same; keep dividing	III. Shift to agar for shoot and root development	Continuous light; 6 W/m^2; 22° C
I. Mod MS + KIN (0.2) + NAA (0.2); II. Mod MS + KIN (0.5) + NAA (0.1)	III. Omit KIN and NAA IV. Standard or root *in vivo*	22° C; 6 W/m^2 continuous
I. Mod MS + KIN (1.0) + IAA (1.0) II. Same	III. Omit KIN and IAA: IBA (1.0) IV. Standard	3–9 W/m^2
I. Mod MS, i.e., KH_2 PO4, CH, malt extract; AD, NAA (10.0), KIN (0.1) II. Same, but less	III. NAA (0.5) for plantlet to form IV. Standard	Dark 4–8 weeks for callus; continuous light for plants
I. Mod MS, KIN (1.0); IAA (1.0); II. Same	III. Omit BA and IAA: IBA (1.0) IV. Standard	3 W/m^2; 16/8 hours; 22° C
I. Mod MS (1/2) NAA (0.03); use 2iP for Easter lily; buds, MS + BA and IAA at 0.1 or 1.0 each	III. Same IV. Transfer to vermiculite	Continuous dark better for bulbs; give *in vitro* chilling

Remarks
1. Liquid medium better than agar in initial culture; same as *Cymbidium* but more complex medium. Growing in liquid on a shake continues proliferation and overcomes polarity. Cutting into sections has same effect. Placing on stationary medium or stopping cutting allows shoot and root formation.
2. Used primarily as a disease-controlling procedure in nuclear stock program. If culture prolonged, cytokinin can stunt growth; avoid by removing for several cultures or use gibberellic acid. High light intensity improves branching. May show uneven flowering in post-culture period.
3. Initially forms protocormlike structures; keeping in liquid on rotating wheel offsets polarity and maintains proliferation (*71*); transplant to agar for plantlet formation, or grow on agar and keep cutting into sections (Figure 18–7). Seeds are also propogated in aseptic culture (Fig. 18–11).
4. Used primarily as a disease-controlling procedure in nuclear stock program; makes excellent material for teaching demonstration. Same comments as *Chrysanthemum* on quality of product.
5. Contamination is a persistent problem; extensively used to propagate selected forms and specific colors. Higher cytokinin may be required initially with lower concentration later. Requires renewal of stock cultures after one year.
6. Meristems hard to clean; uses principle of callus production + differentiation of plantlets, then ASP. May have uneven flowering in post-culture period particularly in some cultivars.
7. Utilizes flower buds from which adventitious shoots develop followed by ASP. Highly successful plant in culture. Relatively stable in long-term continuous culture.
8. Used mostly for Asiatic hybrids and oriental species. Easter lily for disease protection; similar methods can be used on a wide range of bulbous species. Also can use slices of young elongated stems and axillary buds for explants.

TABLE 18–4

Micropropagation methods for horticultural crops—fruit and nut plants

Species	Method	Explant	Sterilization
1. *Actinidia* (kiwifruit)	MT;ASP ASP	Store shoots 1° C; use 0.2–0.5 mm MT (dormant) Three-cm shoot tip (spring)	Wash 20 min; rinse; dip in alcohol; rinse; 1:10 BL; 15 min rinse 3×; dip in alcohol; rinse; dissect 0.6% BL 20 min; rinse 3×; remove 1 cm from base
2. *Fragaria* (strawberry)	MT ASP	0.5 mm tip from apical and lateral buds of young runner	3-cm runner before leafing; wash DT 5 min; 1:10 BL 10 min; wash 2× 10 sec each; excise under dissecting microscope
3. *Malus* (apple)	ASP	I. One- to two-node active shoot II. Use preemergent buds	I. Same as cherry II. Use swollen, green axillary bud: cut like "chip" bud; wash vigorously DT 10 min; stir 1:2 or 1:10 BL 10–20 min; rinse 3×; remove scales; ST
4. *Prunus* (cherry)	ASP	One-to two-node active shoot	Collect 7- to 10-cm first growth in spring; wash vigorously BL 10 min; running water 2 min; 1:5 to 1:10 BL 20 min; rinse 3×; cut above and below node; plant upright
5. *Prunus* (peach)	ASP	1.5-cm active shoot	15% BL 15 min; rinse 4×; trim to 0.5 cm
6. *Rubus* (blackberry)	ASP	5 cm tip; retain young leaves	Agitate-wash in DT 5 min 4×; 1:20 BL 5 min; rinse 3× 5 min each
7. *Vaccinium* (blueberry)	ASP	One- to four-cm shoot; use node with 5 mm each side of bud	2–5 cm active shoot; trim; wash; rinse 5 min; place in fresh Ca hypochlorite 10 min; rinse 2× 10 sec; dip antioxidant; rinse 2 min
8. *Vitis* (grape)	ASP	Three-note shoot tip; 2-mm MT for virus	5-cm shoot tip; remove leaves; stir in 70% alcohol 1 min or BL 20 min; rinse 4×

Propagation	Rooting; acclimation	Environmental conditions
I. 3/4 MS, BA (2.0) IBA (0.02) II. Rotating liquid works very well	III. Omit IV. Direct rooting after IBA (2.0) dip	6–9 W/m^{2}16/8 hr; 23° C
I. MS + BA (1.0) II. Same	III. Replace BA with IBA (1) + charcoal IV. Standard or use *in vivo* rooting	25° C; 3–9 W/m^2
I. 1/2 MS, full Fe II. full MS but vary BA/IBA (0.1 to 1.0)	III. 1/2 MS; BA (0.2) or omit IV. Standard	
I. 1/2 MS, BA (0.5 to 1.0) II. Same, but BA (0.5 to 2.5) + IBA (0.01)	III. Omit BA: IBA (0.2), or omit IV. Standard	3–9 W./m^2 16/8 hr; 25–27° C
I. Mod MS + BA (1.0) + GA (0.5); liquid or agar II. Same	III. omit BA: IBA (1.0) IV. Standard; or *in vivo* rooting	25° C; 6–12 W/m^2
I. MS minor element but reduce major; 2 × Fe; IAA (4.0), 2iP (15) II. Same	III. Omit hormones; IBA (5.0), or omit IV. *In vivo* rooting with IBA dip	6–12 W/m^2 16/8 hr
I. Mod MS, 2 × Fe; 2iP (15) + IAA (4) II. Same	III. Omit 2iP IV. *In vivo* rooting	25° C; 6–12 W/m^2 hr
I. 3/4 MS, BA (0.1) II. BA (0.2), $NaH_2 PO_4$ (170) AD (80.0)	III. Reduce additives; IAA (0.1) or omit IV. *In vivo* rooting slow to establish	9 W/m^2 16/8 hr; 23° C; 30° C for rooting

Remarks

1. Appears to lend itself to rapid multiplaction. Increasingly important fruit crop around the world.
2. Extensively used for nuclear stock production system for disease control programs. Microplants tend to be vigorous, runner extensively,and make excellent nursery bed plants. Micropropagated plants number in the hundreds of thousands in the world.
3. Own-rooted cultivars, clonal rootstocks and ornamental crabs are being produced with varying degrees of success. Stabilization to an easy-to-root juvenile-like form appears to develop after repeated subculturing. However, field tests show early productivity and promising performance (see FIgure 18–10).
4. Rootstock clones and own-rooted cultivars can be produced with varying degrees of ease. Also, various rootstock plants have promise of production.
5. Peach cultivars can be micropropagated with varying degrees of ease. Preliminary field tests show comparable performance to conventional trees. Best success is with specific peach × almond hybrid rootstock clones of which millions have now been propagated.
6. Subculture microshoots with short piece of original explant; highly successful procedure, particularly useful in virus control programs. Field performance is favorable. Can multiply at very high rates at reasonable price.
7. Microshoots from mature clones undergo reversion to juvenile like form prior to rooting. Variation in requirements among species and cultivars. Micropropagation comparable in time and cost to conventional propagation but growing to plant takes more time *(68)*. Micropropagated plants in field show higher yield because of greater branching *(59)*.
8. Inconsistent results requiring frequent transfers. Useful to multiply plants rapidly after meristem culturing. Difficult to compete with conventional propagation methods.

TABLE 18–5

Micropropagation methods for horticultural crops—woody perennials

Species	Method	Explant	Sterilization
1. *Eucalyptus*	ASP	Seedlings; use nodes, basal shoots, shoot tips	Remove leaves; retain 1 cm petiole; wash running water 1 hr; wash in DT 5 min; agitate in fresh, filtered Ca hypochlorite 20 min; rinse 3×; trim to 1 cm; incubate on agar 24 hr
2. *Kalmia*	ASP	2–3 cm shoots	Remove leaves > 1 cm; dip 70% alcohol; 1:10 BL 15 min; 1 min rinse 3×; remove any damaged tissue
3. *Nandina*	ASP	Lateral and terminal buds	5-cm shoots, remove leaves; wash DT 10 min; rinse; agitate 1:10 BL 20 min; excise under dissecting microscope
4. *Pinus radiata* (Monterey pine)	AS	Cotyledons of germinating embryos	Seeds in 1:1 BL 15–20 min; rinse 24 hr running water; 6% H_2O 5–10 min; rinse 2×; refrigerate 2 days; resterilize; rinse; excise embryo and place on agar medium
5. *Populus* (poplar)	ASP	Chilled buds; active shoots	Rinse in alcohol; sterilize in BL; rinse 3×
6. *Rhododendron* (rhododendron)	ASP	Two- to four-cm actively growing shoots	Wash 5-cm tip in BL; remove leaves and terminal bud; dip 70% alcohol 10 sec; rinse; 1:10 BL 20 min; rinse 3× 1 min each; trim off 1-cm base and lay on agar slant
7. *Rosa* (rose)	ASP	One- to two-cm active shoot	Remove lateral leaves; wash; 1:10 BL 20 min; rinse 3× 2 min each
8. *Sequoia* (redwood)	ASP	One- to three-cm ST	Wash; rinse; alcohol rinse 1 min, 0.1% fungicide 20 min; rinse 1:20 BL 20 min; rinse 1% BL; alcohol dip 10 sec; rinse 3×

Propagation	*Rooting; acclimation*	*Environmental conditions*
I. 1/2 MS; 2% sucrose; BA (0.2); IBA (0.16) II. Same	III. IBA (2.0) hard to acclimate; deep shade, gradually increase light	Start in dark in 24 hr; shift to 3–6 W/m^2
I. Agitated liquid WPM 1 week; change daily; stationary liquid; 2iP (1.0) II. After 2 months change to agar, 4-week interval	III. Optional IV. Root directly in 100% peat; no 2iP; high humidity	28–30° C Continuous light 3–6 W/m^2
I. Mod Gamborg B5, MS Fe, NAA (0.1); BA (1.0); AC II. Same; omit BA	III. 1/3 MS; AC; NAA (3.0)	19 W/m^2; 26° C
I. Low salts, 2% sucrose; BA (5.0); NAA (2.0); 3 weeks II. Same but BA omitted; transfer each 4–6 weeks	III. Single shoots, water (0.8%), agar, NAA (0.5), IBA (1.0) 5 days; rooted in trays in rooting medium	25° C/20° C with 16/8 hr light; high RH' mist in rooting
I. Mod MS, BA (0.2–0.5)	III. Same, IBA (0.2–0.5); rooting in 7–14 days	25° C; 3–9 W/m^2,
II. Same; BA (0.5)	III. omit 2iP; NAA (5.0); AC	3–9 W/m^2; 25° C
I, II. AND: 2iP (5.0), IAA (1.0); adenine (*80*); shift every 2 weeks until growth starts, then every 5 weeks; gradually remove explant	IV. Can root treated shots in 1:1 peat-perlite	
I. Mod WPM; long explants vertical in liquid; short explants on agar II. Agar; BA (1.0)	III. Omit IV. Root directly; aided by being in dark.	3–9 W/m^2; 25° C
I. MS, KIN (2.0); IAA (0.5) II. Same	III. Omit KIN; IAA (2.0), IBA (3.0); ACIV. Root 4-cm cuttings after treatment with hormone and fungicide	25° C

Remarks

1. Difficult subject; exudate and explant selection problems; hard to establish after rooting.
2. Highly successful methods have been developed. Commercial.
3. Easy culture should produce large numbers of plants; solves virus problems with this procedure.
4. Commercial production occurs in New Zealand; similar technique being used with various other pine species; Reference: Aiken-Christie and T. A. Thorpe. 1984. In *Cell culture and somatic cell genetics of plants*, I. K. Vasil, ed. Orlando: Academic Press.
5. Utilized for biomass production. Poplars are relatively easy to propagate vegetatively. Relatively easy to "stabilize" in culture.
6. Widely propagated in commercial production. One of most successful. Unique because of specific use of 2iP as a cytokinin. Proper medium is important, particularly lower inorganic content; AC important for rooting. Essentially involves rapid subculturing that may produce reversion to more juvenile form (see Figure 18–8).
7. Commercial production by tissue culture is a viable alternative to grafting. Requires some time to achieve "stabilization." May show seasonal patterns in proliferation. May require stock renewal at frequent intervals or will slow down in multiplication rate.
8. The first mass produced conifer. Relatively easy to micropropagate.

TABLE 18–6

Approximate start-up costs for a tissue culture facility. Costs are per item and presented in 1996 dollars. Item costs are provided in a range with specific costs dependent on the size of the operation. Larger scale operations would require more than one of many of these items.

Item	Approximate cost ($)	Approximate cost for small size operation*
Basic Equipment Required by most Labs		
Autoclave	3,000 to > 20,000	8,000
Pressure cooker	300 to 450	
Laminar flow hood	2,000 to > 4,000	3@2,500 = 7,500
Glassware	1,000 to > 5,000	1,500
Test tubes and containers	5,000 to > 10,000	6,000+
Refrigerator/freezer	500 to > 2,000	750
Water purification system	2,000 to > 10,000	4,500
Dishwasher	400 to > 6,000	1,000
Dispensor for media	500 to > 1,000	1,000
Chemicals	500 to > 1,000	750+
Lights and racks for cultures	500 to > 1,000	700
Sterilizer	250 to 400	3@300 = 900
pH meter	200 to > 1,000	450
Balance	150 to > 2,000	Top loader 850 Analytical 2,000
Safety equipment	300 to 500	300
Carts	200 to 500	4@350 = 1400
Electronic stirrer	150 to 500	150
Stir plate/hot plate	150 to 300	2@200 = 600
Optional Equipment		
Incubators	2,000 to > 15,000	
Rotary shakers	1,500 to > 3,000	
Microscopes	800 to > 5,000	
Microwave	150 to 400	300

*small scale operation at less than 500,000 units per year

TABLE 18–7

Yearly production (1996) of plants in the United States by micropropagation

Crop Type	***Numbers in thousands***	***Percentage of total***
Foliage plants	63,695	52.7
Trees and shrubs	15,294	12.7
Vegetables	12,862	10.6
Greenhouse crops	11,297	9.3
Herbaceous perennials	9,448	7.8
Fruits	3,721	3.1
Miscellaneous	4,545	3.8

Source: R. Zimmerman, USDA, Beltsville, MD

REFERENCES

1. Aitken-Christie, J. 1991. Automation. In *Micropropagation technology and application.* P.C. Debergh and R.H. Zimmerman (eds.). Dordrecht: Martinus Nijhoff Publishers. pp. 363–88.
2. Anderson, W.C. 1975. Propagation of rhododendrons by tissue culture. 1. Development of a culture medium for multiplication of shoots. *Comb. Proc. Inter. Plant Prop. Soc.* 25:129–35.
3. ______. 1978. Rooting of tissue cultured rhododenderons. *Comb. Proc. Inter. Plant Prop. Soc.* 28:135–39.
4. ______. 1980. Mass propagation by tissue culture: Principles and practice. In *Proc. of the conference on nursery production of fruit plants through tissue culture. Applications and feasibility,* R.H. Zimmerman, ed. U. S. Dept. of Agr. Sci. and Education Administration ARR-NE-11, pp. 1–10.
5. ______, and J.B. Carstens, 1977. Tissue culture propagation of broccoli, *Brassica oleracea* (italica group), for use in F_1 hybrid seed production. *Jour. Amer. Soc. Hort. Sci.* 102:69–73.
6. Attree, S.M., D. Moore, V.K. Sawhney and L.C. Fowke. 1991. Enhanced maturation and desiccation tolerance in white spruce [*Picea glauca* (Moench) Voss] somatic embryos: effects of a non-plasmolysing water stress and abscisic acid. *Ann. Bot.* 68:519–25.
7. Burger, D. 1988. Guidelines for autoclaving liquid media used in plant tissue culture. *HortScience* 23: 1066.
8. Boxus, P., and P. Druart. 1980. Micropropagation, an industrial propagation method of quality plants true to type and at a reasonable price. In *Plant cell cultures: Results and perspectives,* F. Sala et al., eds. Amsterdam: Elsevier/North-Holland Biomedical Press.
9. Bridgen, M.P., and J.W. Bartok, Jr. 1987. Designing a plant micropropagation laboratory. *Comb. Proc. Inter. Plant Prop. Soc.* 37:462–67.
10. Broome, O.C. 1986. Laboratory design. In *Tissue culture as a plant production system for horticultural crops.* R.H. Zimmerman, R.J. Griesbach, F.A. Hammerschlag, and R.H. Lawson, eds. Dordrecht: Martinus Nijhoff Publishers, pp. 351–65.
11. Brown, D.C.W., and T. Thorpe. 1984. Organization of a plant tissue culture laboratory. In *Cell culture and somatic cell genetics of plants.* Vol. 1, I. K. Vasil, ed. New York: Academic Press, pp. 1–12.
12. Chu, I.Y.E. 1986. The application of tissue culture to plant improvement and propagation in the ornamental horticulture industry. In *Tissue culture as a plant production system for horticultural crops.* R.H. Zimmerman, R.J. Griesbach, F.A. Hammerschlag, and R.H. Lawson, eds. Dordrecht: Martinus Nijhoff Publishers, pp. 15–33.
13. Debergh, P.C. 1987. Recent trends in the application of tissue culture to ornamentals. In *Plant tissue and cell culture,* C.E. Green, D.A. Somers, W.P. Hackett, and D.D. Biesboer, eds. New York: Alan R. Liss, pp. 383–93.
14. de Fossard, R.A. 1976. *Tissue culture for plant propagators.* Armidale, Australia: Univ. of New England Printery.
15. Dirr, M.A., and C.W. Heuser, Jr. 1987. *The reference manual of woody plant propagation: From seed to tissue culture.* Athens, Ga.: Varsity Press.
16. Donnan, A., Jr. 1986. Determining and minimizing production costs. In *Tissue culture as a plant production system for horticultural crops,* R.H. Zimmerman, R.J. Griesbach, F.A. Hammerschlag, and R.W. Lawson, eds. Dordrecht: Martinus Nijhoff Publishers, pp. 167–73.
17. Dore, C. 1987. Application of tissue culture to vegetable crop improvement. In *Plant tissue and cell culture,* C.E. Green, D.A. Somers, W. P. Hackett, and D.D. Biesboer, eds. New York: Alan R. Liss, pp. 419–32.
18. Driver, J.A., and G.R.L. Suttle. 1986. Nursery handling of propagules. In *Cell and tissue culture forestry* (2nd ed.), J.M. Bonga and D.J. Durzan, eds. Dordrecht: Martinus Nijhoff Publishers, pp. 320–31.
19. Durzan, D.J. and P.E. Durzan. 1991. Future technologies: model-reference control systems for the scale-up of embryogenesis and polyembryogenesis in cell suspension cultures. In *Micropropagation technology and application.* P.C. Debergh and R.H. Zimmerman (eds.). Kluwer Dordrecht. pp. 389–423.
20. Fellman, C.D., P.E. Reed, and M.A. Hosier. 1987. Effects of thidiazuron and CPPU on meristem formation and shoot proliferation. *HortScience* 22: 1197–1200.
21. Fiorino, P., and F. Loreti. 1987. Propagation of fruit trees by tissue culture in Italy. *HortScience* 22:353–58.
22. Fogh, J., ed. 1973. *Contamination in tissue culture.* New York: Academic Press.
23. Gamborg, O.L., and L.R. Wetter, eds. 1975. *Plant tissue culture methods.* Saskatoon, Canada: Prairie Regional Laboratory.
24. Fuji, J.A.A., D. Slade, R. Olsen, S.E. Ruzin and K. Redenbaugh. 1990. Alfalfa somatic embryo maturation and conversion to plants. *Plant Sci.* 72:93–100.
25. Geneve, R.L. 1989. Variations within a clone in tissue culture production. *Comb. Proc. Intern. Plant Prop. Soc.* 39:458–62.
26. ______, and S.T. Kester. 1990. The initiation of somatic embryos and adventitous roots from developing zygotic embryo explants of *Cercis canadensis* L. *Plant Cell Tissue Organ Cult.* 22:71–6.
27. George, E.F., and P.D. Sherrington. 1984. *Plant Propagation of tissue culture: Handbook and directory of commercial laboratories.* Eversley, England: Exegetics Ltd.

28. Gray, D.J. and A. Purohit. 1991. Somatic embryogenesis and development of synthetic seed technology. *Critical Rev. Plant Sci.* 10:31–61.
29. Griesbach, R.J. 1986. Orchid tissue culture. In *Tissue culture as a plant production system for horticultural crops.* R.H. Zimmerman, R.J. Griesbach, F.A. Hammerschlag, and R.H. Lawson, eds. Dordrecht: Martinus Nijhoff Publishers, pp. 343–51.
30. Halquist, J.L., M.A. Hosier, and P.E. Read. 1983. A comparison of several gelling agents and concentrations on callus growth and organogenesis in vitro. *In Vitro* 19:248.
31. Hammerschlag, F.A. 1986. Temperate fruits and nuts. In *Tissue culture as a plant production system for horticultural crops.* R.H. Zimmerman, R.J. Griesbach, F.H. Hammerschlag, and R.H. Lawson, eds. Dordrecht: Martinus Nijhoff Publishers, pp. 221–36.
32. Hartmann, H.T., and J. Whisler. 1977. Micropropagation exercises in teaching plant propagation. *Comb. Proc Inter. Plant. Prop. Soc.* 27:407–13.
33. Henny, R.J., J.F. Knauss, and A. Donnan, Jr. 1981. Foliage plant tissue culture. In *Foliage plant production,* J. Joiner, ed. Englewood Cliffs, N.J.: Prentice-Hall.
34. Holdgate, D.P., and J.S. Aynsley. 1977. The development and establishment of a commercial tissue culture laboratory. *Acta Hort.* 78:31–36.
35. Jones, J.B. 1986. Determining markets and market potential. In *Tissue culture as a plant production system for horticultural crops.* R.H. Zimmerman, R.J. Griesbach, F.A. Hammerschlag, and R.H. Lawson, eds. Dordrecht: Martinus Nijhoff Publishers, pp. 175–82.
36. Kartha, K.K. 1985. Cryopreservation of plant cells and organs. CRC Press, Boca Raton FL.
37. Kerns, H.R., and M.M. Meyer, Jr. 1986. Tissue culture propagation of *Acer* × *freemanii* using thidiazuron to stimulate shoot tip proliferation. *HortScience* 21: 1209–10.
38. Kester, D.E., and C. Grasselly. 1987. Almond rootstocks. In *Rootstocks for fruit crops,* R.C. Rom and R.F. Carlson, eds. New York: John Wiley, pp. 265–93.
39. Kitto, S.L. and J. Janick. 1985. Production of synthetic seeds by encapsulating asexual embryos of carrot. *Jour. Amer. Soc. Hort. Sci.* 110:277–80.
40. Knudson, L. 1951. Nutrient solution for orchids. *Bot. Gaz.* 112:528–32.
41. Kyte, L. 1987. *Plants from test tubes; an introduction to micropropagation.* (rev. ed.) Portland, Oreg.: Timber Press.
42. Linsmaier, E.M., and F. Skoog. 1965. Organic growth factor requirements of tobacco tissue cultures. *Physiol. Plant.* 18:101–27.
43. Litz, R.E. 1987. Application of tissue culture to tropical fruits. In *Plant tissue and cell culture,* C.E. Green, D.A. Somers, W.P. Hackett, and D.D. Biesboer, eds. New York: Alan R. Liss, pp. 407–18.
44. Lloyd, G., and B. McCown. 1980. Commercially feasible micropropagation of mountain laurel, *Kalmia latifolia,* by use of shoot-tip culture. *Comb. Proc. Inter. Plant Prop. Soc.* 30:421–27.
45. Macdonald, B. 1986. *Practical woody plant propagation for nursery growers.* Portland, Oreg.: Timber Press.
46. Matsuyama, J. 1980. Overview of tissue culture at K.M. Nursery. *Comb. Proc. Inter. Plant Prop. Soc.* 30:40–2.
47. McCown, B.H. 1986. Woody ornamentals, shade trees, and conifers. In *Tissue culture as a plant production system for horticultural crops,* R.H. Zimmerman, R.J. Griesbach, F.A. Hammerschlag, and R.H. Lawson, eds. Dordrecht: Martinus Nijhoff Publishers, pp. 333–42.
48. McCown, D.D. 1986. Plug systems for micropropagules. In *Tissue culture as a plant production system for horticultural crops,* R.H. Zimmerman, R.J. Griesbach, F.A. Hammerschlag, and R.H. Lawson, eds. Dordrecht: Martinus Nijhoff Publishers, pp. 53–60.
49. McGranahan, G.H., and P.B. Catlin. 1987. *Juglans* rootstocks. In *Rootstocks for fruit crops,* R.C. Rom and R.F. Carlson, eds. New York: John Wiley, pp. 411–50.
50. Meyer, M.M. Jr. 1980. In vitro propagation of Hosta sieboldiana. *HortScience* 15:737–38.
51. Mok, M.C., D.W.S. Mok, J.E. Turner, and C.V. Mufer. 1987. Biological and biochemical effects of cytokinin-active phenylurea derivatives in tissue culture systems. *HortScience* 22:1194–97.
52. Mudge, K.W., C.A. Borgman, J.C. Neal, and H.A. Weller. 1986. Present limitations and future prospects for commercial micropropagation of small fruits. *Comb. Proc. Inter. Plant Prop. Soc.* 36:538–43.
53. Mullins, M.G. 1987. Propagation and genetic improvement of temperate fruits, the role of tissue culture. In *Plant tissue and cell culture,* C.E. Green, D.A. Somers, W.P. Hackett, and D.D. Biesboer, eds. New York: Alan R. Liss, pp. 395–406.
54. Murashige, T. 1974. Plant propagation through tissue cultures. *Ann. Rev. Plant Phys.* 25:135–66.
55. ______. 1977. Plant cell and organ cultures as horticultural practices. *Acta Hort.* 78:17–30.
56. ______, and F. Skoog. 1962. A revised medium for rapid growth and bioassays with tobacco tissue cultures. *Physiol. Plant.* 15:473–97.
57. Oglesby, R.P., and J.L. Griffis, Jr. 1986. Commercial in vitro propagation and plantation crops. In *Tissue culture as a plant production system for horticultural crops,* R.H. Zimmerman, R.J. Griesbach, F.A. Hammerschlag, and R.H. Lawson, eds. Dordrecht: Martinus Nijhoff Publishers, pp. 253–57.
58. Pierik, R.L.M. 1987. *In vitro culture of higher plants.* Dordrecht: Martinus Nijhoff Publishers.
59. Pedhya, M.A. and A.R. Mehta. 1982. Propagation of fern *(Nephrolepis)* through tissue culture. *Plant Cell Reports* 1:261–63.

60. Preece, J.E. and E.G. Sutter. 1991. Acclimatization of micropropagated plants to the greenhouse and field. In *Micropropagation technology and application.* P.C. Debergh and R.H. Zimmerman (eds.). Dordrecht: Martinus Nijhoff Publishers. pp. 71–93.

61. Raju, B.C., and J.C. Trolinger, 1986. Pathogen indexing in large-scale propagation of florist crops. In *Tissue culture as a plant production system for horticultural crops,* R.H. Zimmerman, R.J. Griesbach, F.A. Hammerschlag, and R.H. Lawson, eds. Dordrecht: Martinus Nijhoff Publishers, p. 138.

62. Read, P.E., C.A. Hartley, J.R. Sandahl, and D.K. Wildung. 1987. Field performance of in vitro propagated blueberries. *Comb. Proc. Inter. Plant Prop. Soc.* 37:450–52.

63. Redenbaugh, K., D. Slade and J.A. Fujii. 1987. Encapsulation of somatic embryos in synthetic seed oats. *HortScience* 22:803–09.

64. Singha, S., G.K. Bissonnette, and M.L. Double. 1987. Methods for sterilizing instruments contaminated with *Bacillus* spp. from plant tissue cultures. *HortScience* 22:659.

65. Smith, R.J. 1972. Orchid propagation by in vitro culture techniques. *Comb. Proc. Inter. Plant Prop. Soc.* 22:174–77.

66. Smith, R.H. 1992. *Plant tissue culture. Techniques and Experiments.* Academic Press, NY.

67. Smith, W.A. 1981. The aftermath of the test tube. *Comb. Proc. Inter. Plant Prop. Soc.* 31:47–49.

68. Snir, I., and A. Erez. 1980. In vitro propagation of Malling Merton apple rootstocks. *HortScience* 15:597–98.

69. ______. 1982. In vitro propagation of sweet cherry cultivars. *HortScience* 17:192–93.

70. Stimart, D.P. 1986. Commercial micropropagation of florist flower crops. In *Tissue culture as a plant production system for horticultural crops,* R.H. Zimmerman, R.J. Griesbach, F.A. Hammerschlag, and R.H. Lawson, eds. Dordrecht: Martinus Nijhoff Publishers, pp. 301–15.

71. Stokes, M.J. 1980. Current aspects of commercial micropropagation. *Comb. Proc. Inter. Plant Prop. Soc.* 30:249–54.

72. Stoltz, L.P. 1979. Getting started in tissue culture: Equipment and costs. *Comb. Proc. Inter. Plant Prop. Soc.* 29:375–81.

73. Strode, R.A., and G. Abner. 1986. Large scale tissue culture production for horticultural crops. In *Tissue culture as a plant production system for horticultural crops,* R.H. Zimmerman, R.J. Griesbach, F.A. Hammerschlag, and R. H. Lawson, eds. Dordrecht: Martinus Nijhoff Publishers, pp. 367–71.

74. ______, P.A. Travers and R.P. Oglesly. 1979. Commercial micropropagation of rhododendrons. *Comb. Proc. Intern. Plant Prop. Soc.* 29:439–43.

75. Sutter, E. and R.W. Langhans. 1979. Epicuticular wax and its effect on formation in carnation plantlets regenerated from tip culture. *Journ. Amer. Soc. Hort. Sci.* 104:493–96.

76. Swartz, H.J., and J.T. Lindstrom. 1986. Small fruit and grape tissue culture from 1980 to 1985: Commercialization of the technique. In *Tissue culture as a plant production system for horticultural crops,* R.H. Zimmerman, R.J. Griesbach, F.A. Hammerschlag, and R.H. Lawson, eds. Dordrecht: Martinus Nijhoff Publishers, pp. 201–20.

77. Thorpe, T.A., ed. 1981. *Plant tissue culture: Methods and applications in agriculture.* New York: Academic Press.

78. Wetherell, D.F. 1982. *Introduction to in vitro propagation.* Avery Pub. Group, Wayne NJ.

79. Wimber, D.E. 1963. Clonal multiplication of Cymbidium through tissue culture. *Amer. Orch. Soc. Bul.* 32: 105–7.

80. Wochok, Z.S. 1986. Status of crop improvement through tissue culture. *Comb. Prop. Inter. Plant Prop. Soc.* 36:72–77.

81. Zilis, M., D. Zwagerman, D. Lamberts, and L. Kurtz. 1979. Commercial propagation of herbaceous perennials by tissue culture. *Comb. Proc. Inter. Plant Prop. Soc.* 29:404–13.

82. Zimmerman, R.H. 1988. Micropropagation of woody plants: Post tissue culture aspects. *Acta Hort.* 227:489–499.

83. ______, 1979. The laboratory of micropropagation at Cesena, Italy. *Comb. Proc. Inter. Plant Prop. Soc.* 29:398–400.

84. ______, and 1. Fordham. 1985. Simplified method for rooting apple cultivars in vitro. *Jour. Amer. Soc. Hort. Sci.* 110:34–8.

85. ______, R.J. Griesbach, F.A. Hammerschlag, and R.H. Lawson, eds. 1986. *Tissue culture as a plant production system for horticultural crops.* Dordrecht: Martinus Nijhoff Publishers.

SUPPLEMENTARY READING

AITKEN-CHRISTIE, J., T. KOZAI, and M.A.L. SMITH, ed. 1995. *Automation and environmental control in plant tissue culture.* Dordrecht: Kluwer Academic Publ.

BAJAJ, Y.P.S. ed., *Biotechnology in agriculture and forestry series.* Springer-Verlag, NY.

BONGA, J.M., and D.J. DURZAN. 1986. *Cell and tissue culture in forestry,* 4 volumes (2nd ed.). Dordrecht: Martinus Nijhoff Publishers.

BRIDGEN, M.P. and J.W. BARTOK, JR. 1988, Designing a plant micropropagation laboratory. *Comb. Proc. Inter. Plant Prop. Soc.* 37:462–67.

DEBERGH, P.C. and R.H. ZIMMERMAN. 1991. *Micropropagation. Technology and Application.* Dordrecht: Kluwer Academic Publishers.

DEFOSSARD, R.A. 1976. *Tissue culture for plant propagators.* Armidale, Australia: Univ. of New England Printery.

GEORGE, E.F. 1993. *Plant propagation by tissue culture. Part I. The technology.* Edington, Wilts. Eng.:Exergetics Ltd.

HUTCHINSON, J.F., and R.H. ZIMMERMAN. 1989. Tissue culture of temperate fruit and nut trees. *Hort. Rev.* 9:273–349.

INGRAM, D.S., and J.P. HELGESON. 1981. *Tissue culture methods for plant pathologists.* Oxford: Blackwell Scientific Publ.

INTERNATIONAL PLANT PROPAGATORS SOCIETY. Proceedings of annual meetings.

INTERNATIONAL PLANT TISSUE CULTURE SOCIETY. Newsletters.

JAIN, S.M., P.K. GUPTA AND R.J. NEWTON eds. 1994. *Somatic embryogenesis in woody plants.* Dordrecht: Martinus Nijhoff Publishers.

KOZAI, T., R.H. ZIMMERMAN, Y. KITAYA and K. FUJIWARA, ed. 1995. *Environmental effects and their control in plant tissue culture.* Acta Hort. 393. 230 pp.

KYTE, L. 1987. *Plants from test tubes: An introduction to micropropagation* (Rev. ed.). Portland, Oreg.: Timber Press.

NISHIO, T. and C. DORE. 1995. *Genetic improvement of horticultural crops by biotechnology.* Acta Hort. 392. 274 pp.

PREECE, J.E. and M.E. COMPTON. 1991. Problems with explant exudation. In: Y.P.S. Bajaj (ed.). *Biotechnology in Agriculture and Forestry.* Vol 17. High-Tech and Micropropagation I. Springer-Verlag, Berlin. pp 168–189.

TORRES, K.C. 1989. *Tissue culture techniques for horticultural crops.* New York: Van Nostrand Reinhold.

VASIL, I.K. and T.A. THORPE, ed. 1994. *Plant cell and tissue culture.* Dordrecht: Kluwer Academic Publishers. 593 pp.

19

Propagation Methods and Rootstocks for Important Fruit and Nut Species

Few fruit or nut crop cultivars reproduce true-to-type when propagated by seed. Propagation by some vegetative method is necessary. Most tree fruit and nut species are propagated by budding or grafting on rootstocks—seedlings, rooted cuttings, or layered plants. Rootstocks are very important for two main reasons. First, most fruit and nut tree cultivars are difficult to root by cuttings, except for a few species such as the grape, fig, olive, quince, currant, gooseberry, pomegranate, blackberry, and raspberry. Some, such as the filbert and various tropical species, can be propagated by layering. Secondly, rootstocks are essential in many of the species for resisting critical soil pathogens and for adapting plants to specific areas and management conditions.

Clonal rootstocks are very important and rooting is by cuttings, layering, and micropropagation. In general, micropropagation is used mostly to produce virus-tested sources as part of a program of Registration and Certification or to produce stock plants with enhanced rooting potential.

Fruit and nut cultivars are examples of monocultures with the attendant problems of vulnerability to introduced pathogens. For this reason, commercial propagation of fruit and nut crops has become very sophisticated and specialized, requiring inputs of research and development to overcome developing problems (*68*).

Nevertheless, large nurseries specializing in fruit and nut propagation provide many hundreds of thousands of fruit and nut trees each year by various vegetative methods (*20, 47, 67, 123, 167*).

Achras zapota. Sapodilla, Nispero. Tropical tree originating in Mexico and South America but grown throughout the tropics (*15*). Many cultivars exist in the world. 'Prolific' and 'Brown Sugar' are grown in Florida. Seedling rootstocks are preferred. Fruits are depulped, washed, dried. Soaking seed overnight is desirable before sowing in seed beds or pots. Germination is in about four weeks, and the seedlings are transferred to small pots when 15 cm high. Inarching is the most important graft method, but side grafting and budding can be done.

Actinidia deliciosa. Kiwifruit (*23, 62*). Chinese gooseberry. Large-fruited dioecious subtropical vine requiring male and female cultivars to ensure fruiting. *A. arguta* is a related species producing small fruits in a cluster. Kiwi vines are propagated commercially mostly by budding or grafting cultivars to seedling rootstocks, but they may also be propagated by leafy, semi-hardwood cuttings under mist, hardwood cuttings, and root cuttings. Grafted plants are believed to be more vigorous and to come into bearing sooner than those started as rooted cuttings.

Seed. Seedling plants have a long juvenile period and their sex cannot be determined until fruiting at seven years or more. Seed should be taken from soft, well-ripened fruit, dried and stored at 5° C (41° F). After at least two weeks at this temperature, subject seed to fluctuating temperatures—10° C (50° F) night and 20° C (68° F) day—for two or three weeks before planting.

Cuttings. Leafy semi-hardwood cuttings taken from apical and central parts of current season's growth in late spring and midsummer may be rooted under mist in coarse vermiculite with 6,000 ppm IBA. Hardwood cuttings, taken in midwinter and planted in a greenhouse, require higher IBA concentrations.

Grafting. Seedlings are grafted successfully by the whip graft in late winter using dormant scionwood. T-budding in late summer is also successful.

Micropropagation. Kiwifruit can be propagated by using short meristem tips or longer apical shoots as the initial explants.

Almond. *See Prunus dulcis.*

Anacardium occidentale. Cashew (*6, 156*). These tender tropical evergreen trees are usually grown as seedlings. Clonal selections have been made, but vegetative propagation and transplanting are difficult.

Seed. Two to three are planted directly in place in the orchard and later thinned to one tree per location. Dormancy does not occur, but seeds may lack embryos. Test by placing seeds in water and discard those that float. Seeds germinate in 15 to 20 days. Growing in biodegradable containers will improve transplanting success.

Grafting. Rootstocks should be one to two months old but not tall and spindly. In softwood grafting, the scions are 8 to 10 cm (3 to 4 in.) long, pencil thick, with bulging tips (*15*). Scions are green to brown in color with terminal leaves dark green. Leaf blades are clipped off seven to ten days before grafting, and grafting is done when the petioles abscise. One to two leaves are retained below the grafting which is done on the soft portion. A longitudinal cut about 3 to 4 cm (1 to 1.5 in.) long is made. The scion wedge is inserted and the graft tied with polyethylene grafting tape. Grafts are kept in shade for a week before transferring to the open.

Miscellaneous Asexual Methods. Cashew can also be propagated by rooting leafy cuttings, stooling (*164*), approach-grafting (*180*), and air layering (*161, 164*). Other methods, such as T-budding and patch budding also have been used successfully in limited trials.

Ananas comosus. Pineapple. Tropical terrestrial monocot of the Bromiliadaceae family which originated in South America. A large number of cultivars are grown in Hawaii, South America, Africa, the Philippines, Australia, and Southeast Asia, much of them in large industrial plantings (*37*). The plant consists of a thick stem with the flowering rosette on the top.

Offshoots. Three main types of vegetative propagules include offshoots (offsets) from the main stem either from below or above ground *(suckers),* lateral offshoots from the fruiting stem (peduncle) just below the fruit *(slips),* and the vegetative shoot emerging above the fruit *(crowns).* All three types can be used for propagation by removing from the plant and planting either by machine or by hand (see Figure 15–16). Drying (curing) for several weeks to allow callusing on the cut surface will reduce decay after planting. Suckers are somewhat limited in number, producing a mature fruit in 15 to 17 months, and are rarely used as planting material. If used, they are removed about one month after peak fruit harvest. Slips are preferred and are most abundant. They are cut from the mother plant two to three months after peak harvest. A slip-produced plant will produce fruit in about 20 months. They can be stored for a relatively long time and retain vigor for replanting. Slips of about the same size and age will produce uniform flowering. Crowns are removed at about the same time as fruit harvest and require about 24 months to produce fruit.

Micropropagation. Explants of axillary buds from crowns of mature fruits can be propagated in an MS medium with 24 percent cocoanut water followed by subculturing in reduced cocoanut water supplemented by BA (*155, 238*).

Annona cherimola. Cherimoya, Chirimoya, Custard Apple. Sugar apple *(A. squamosa).* Atemoyas *(A. squamosa × A. cherimola).* Cherimoya is evergreen, sugar apple is deciduous, and atemoyas are semi-deciduous and somewhat tolerant of frost (*15, 73*).

These tropical plants grow as small trees in Spain, Mexico, Ecuador, Peru, Chile, Southern Africa—with limited production in California, India, and Sri Lanka. In many areas, seedlings are grown and some selections in Mexico and South America are reported to come nearly "true." Seedling plants are vigorous, have long juvenile periods, and inferior fruits.

Seed. Viability is retained for many years if kept dry. Germination takes place in a few weeks after planting. Seeds have been successfully propagated in containers transplanted from flats to small pots when 7.5 to 10 cm (3 to 4 in.) high, then to small pots at 20 cm (8 in.), and to larger pots or to open ground.

Grafting. Cultivars can be cleft grafted or T-budded. Topworking is successful with bark or cleft grafting carried out at the end of a dormant period. Seedling rootstocks are cherimoya, or the related sugar apple *(Annona squamosa),* which gives a dwarf plant, or custard apple (*Annona* hybrids). The latter two species are susceptible to cold, and root-rot is a problem on poorly drained soils.

Layering (15). Stooling (mound layering) has given promising results. Three-year-old seedlings were headed to ground level, shoots were girdled, and IBA applied.

Apple. *See Malus × domestica.*

Apricot. *See Prunus armeniaca.*

Artocarpus heterophyllus. Jackfruit. A medium-sized tropical tree with fruit of unique flavor. It has many related genera and species including breadfruit *(Artocarpus altilis).* Grown mostly by seed, but cultivars are available. The seed has limited viability and must be germinated within one month. Presoaking the seed improves germination. Can be propagated by cuttings when stock plants are etiolated, and cuttings treated with IBA and rooted under mist. Grafting, including inarching, epicotyl grafting, and various budding methods are successful. Can be propagated by air layering.

Atemoyas. *See Annona.*

Averrhoa carambola. Carambola (*15*). A medium-sized evergreen tree for warm tropical and subtropical areas. It produces a five-cornered fruit, used for fresh fruit, juice, or decoration. It is mostly propagated by seed, but vegetatively propagated cultivars are available. In Florida, grafting is done with a side veneer graft, otherwise the wedge, whip-and-tongue, and approach grafts are successful. Standard air layering is highly successful.

Avocado. *See Persea.*

Banana. *See Musa.*

Berry (Youngberry, Boysenberry, Loganberry, Dewberry). *See Rubus.*

Blackberry. *See Rubus.*

Blueberry. *See Vaccinium.*

Butternut. *See Juglans cinerea.*

Cacao. *See Theobroma.*

Carambola. *See Averrhoa.*

Carica papaya. Papaya. Pawpaw (*87, 208*). These are large tropical fast-growing, short-lived herbaceous perennials. They come into fruiting within about five months and live four to five years.

Seed. Propagation is by seed from selected parents or controlled crosses. Plants are dioecious, require cross-pollination, and must be screened for sex form after coming into bearing. Seeds do equally well if taken from stored or fresh fruit. Viability can be retained up to six years at 5° C (41° F) in sealed, moisture-proof bags (*11*). Seeds are sown in flats of soil or in seed beds in the open—and germinate in two to three weeks. Seedlings are first transplanted at about 10 cm (4 in.) tall and then generally retransplanted once or twice before they are put in their permanent location. Another technique is to plant four to eight seeds per container and thin to two to four of the strongest when about 10 cm (4 in.) tall. Soil pasteurization is recommended, since young papaya seedlings are very susceptible to damping-off organisms.

The usual practice in Florida is to plant seeds in midwinter and set the young plants in the field by early spring. Growth occurs during spring and summer, first fruits are mature by fall, and the plants bear all winter and the following season.

Miscellaneous Asexual Methods. Cuttings are used for specific cultivars with parthenocarpic fruit (*15*). Cuttings of branches with a "heel" are placed under mist with bottom heat. Budding is another method (*200*).

Micropropagation. Papaya can be micropropagated (*136*).

Carob. *See Ceratonia.*

Carya illinoensis. Pecan (*84, 141*). Pecans are native to the southwest U.S. and northern Mexico where many natural seedling groves exist. Commercial groves use selected cultivars grafted to pecan seedling rootstocks.

Seed. Pecan seeds of certain cultivars may show **vivipary** (p. 142), which ruins the pecan crop. They start to germinate in the hulls before harvesting and lose their viability in warm, dry storage. To maintain viability, seeds should be stored at 0° C (32° F) at 5 percent moisture immediately after harvest and until planted. Pecan seeds should be stratified for 12 to 16 weeks at about 1 to 5° C (34 to 41° F) to ensure good, rapid germination. Growth restrictions of the shell have been reported to be reduced by germination at high temperatures—30 to 35° C (86 to 95° F) (*51, 142, 215*).

Midwinter planting, with seedling emergence in the spring, is a successful procedure. Deep, well-drained, sandy soil should be used for growing pecan nursery trees. Young seedlings are tender and should be shaded against sunburn. In the summer, toward the end of the second growing season, the seedlings are large enough to bud.

Grafting (198). Cultivars are propagated by budding or grafting to two-year-old pecan seedling rootstocks. Patch budding in the nursery is the usual method. After the budded top grows for one or two seasons, the nursery tree is transplanted to a permanent location. Young pecan trees have a long taproot and must be handled carefully in digging and replanting.

Seedlings may be crown grafted with the whip graft in late winter or early spring. Pecan seedling trees, growing in place in the orchard, can be top-worked by inlay bark grafting limbs 4 to 9 cm (1 ½ to 3 ½ in.) in diameter (see Figure 13–17).

Rootstocks (84). Pecan seedlings from the 'Apache,' 'Riverside,' 'Elliott,' and 'Curtis' cultivars are reported to produce excellent grafted plants. Pecan roots are susceptible to verticillium wilt. Experimental trials have indicated that *C. aquatics* might be a possible rootstock for wet soils. Although pecan cultivars will grow on hickory species rootstocks, the nuts generally do not attain normal size.

Miscellaneous Asexual Methods. It is extremely difficult to root pecans from cuttings. Some rooting occurs with hardwood cuttings taken in the fall and treated with 10,000 ppm IBA (*195*). Mound (stool) layering for clonal rootstock production is done in Peru (see Figure 15–9) (*150*). Plants have been propagated by root cuttings and by air layering when treated with IBA.

Micropropagation. Seedling shoots of pecans have been micropropagated and rooted *ex vitro* (*86*).

Carya ovata. Shagbark hickory (*140*). Shagbark hickory is native to North America and grown as a nut tree and landscape ornamental.

Seed. Most trees are grown as seedlings because of difficulties of grafting and transplanting and slow growth of the trees. Some variation in seed requirements exists. Hickory seed will germinate when planted in the spring but should be kept in moist, cool storage until planting. However, seed can be stored dry at low temperatures. Seed should be stratified for up to seven months (*235*). Fall planting is successful if the soil in temperate climates is well mulched to prevent excessive freezing and thawing.

Grafting. Patch budding is done by some nurseries in the commercial propagation of hickories, usually in late summer. Seedling rootstocks of *C. ovata* or pecan *(C. illinoensis)* are grown for two years or more before they are large enough to bud. The inlay bark graft is another option. Transplanting is a problem, so root pruning to force out lateral root growth is sometimes done.

Cashew. *See Anacardium.*

Castanea spp. Chestnut (*118, 147, 218*). The chestnut of commerce in edible nuts is *C. sativa*—the European (Spanish) sweet chestnut. It is produced in quantity from very old groves and, more recently, new plantations throughout Europe import large quantities to the U.S. *C. dentata* was a large native American tree of eastern North America that was very important for timber as well as nuts until the early 1900s when the fungal blight, *Endothia parasitica,* was accidentally introduced from the Orient. As a result, essentially all mature trees of the American chestnut have been killed to near the ground in natural woodlands of the eastern U.S. The European chestnut is also susceptible but has not been as severely affected. Blight-resistant chestnut species exist in China *(C. mollissima)* and Japan *(C. crenata)* and provide an important source of resistance. Breeding programs both for cultivars and rootstocks are in progress and new improved blight-resistant materials are becoming available. In addition, a virus attacking the blight fungus has been discovered that has the promise of providing biocontrol of the disease (*34*).

Seed. Seeds are used for rootstocks, breeding, and occasionally for crops. Seeds are large and fleshy and should be prevented from drying following harvest and until planting. The nuts (seeds) are gathered as soon as they drop and are either planted in the fall or kept in moist storage one or two months at 0 to 2° C (32 to 40° F) for spring planting. Seeds are satisfactorily stored in tight tin cans, with one or two very small holes for ventilation, at 0° C (32° F) or slightly higher; this storage temperature also aids in overcoming embryo dormancy. Weevils in the nuts, which will destroy the embryo, can be killed by hot-water treatment at 49° C (120° F) for 30 minutes. After one year's growth, the seedlings should be large enough to transplant to their permanent location or be grafted to the desired cultivar.

Grafting and budding (104). Clonal selections are usually propagated onto rootstocks using the splice, chip, whip, or inverted T-buds. Regular T-buds tend to "drown" from the excessive sap flow.

Cuttings. Chinese chestnut can be propagated by rooting leafy cuttings under mist if treated with IBA (*110*) but the method has not been done on a large scale.

Nut grafting (117). In this procedure, a nut that has started to germinate is cut off just above the root. A dormant scion with a wedge cut on the bottom is inserted into the inverted nut which is then placed in a closed heated frame containing peat moss and vermiculite. After about four weeks of warm temperature the grafts are hardened-off and removed.

Stooling. This typical method of propagation is used in Europe for obtaining plants on their own roots but it is expensive.

Rootstocks. Seedling rootstocks are usually used, using seed from the cultivar being propagated. Because of hybridization, variability often results, which has been associated with graft incompatibility.

Micropropagation. A chestnut hybrid *(C. sativa* × *C. castanea)* has been micropropagated using explants from mature trees (*216*).

Ceratonia siliqua. Carob. This subtropical evergreen tree is usually propagated by seeds, which germinate without difficulty when freshly harvested. However, if seeds dry out and the seed coat becomes hard, they should be softened by hot-water or sulfuric acid treatment. Transplanting of bare-root seedlings gives poor results, so seeds are best planted in their permanent location or started in containers for later transplanting. Selected cultivars are chip budded in late spring, which is faster than grafting (*52*). Cuttings can be rooted if taken in mid-spring and treated with 7,500 ppm IBA. Air layering in late summer is successful.

Micropropagation using explants from both seedling and mature trees has been done successfully (*190*).

Cherry (Sour). *See Prunus cerasus.*

Cherry (Sweet). *See Prunus avium.*

Chinese gooseberry. *See Actinidia.*

Cherimoya. *See Annona.*

Citron. *See Citrus.*

Citrus spp. Citrus (*174, 184*). This includes cultivars of *C. aurantifolia* (lime), *C. limon* (lemon), *C. maxima* (pumulo), *C. medica* (citron), *C. reticulata* × *C. sinensis* (tangor), *C.* × *paradisi* (grapefruit), *C. reticulata* (mandarin orange), *C. sinensis* (sweet orange), *C. paradisi* × *C. regiculata* (tangelo), and related citrus species used for rootstock. In general, propagation is the same for all species of citrus. Members of this genus are readily intergrafted and grafted to other closely related genera such as *Fortunella* (kumquat) and *Poncirus* (trifoliate orange).

Seeds. Polyembryony occurs in seeds of most citrus species used as rootstocks due to nucellar embryony (see p. 134). The sexual seedlings present within the embryo tend to be weak, variable, and are usually rogued out. The apomictic seedlings, which arise from the nucellus, are usually uniform and have the same genotype as the seed tree. However, as a commercial plant, nucellar seedlings are vigorous, thorny, upright-growing, slow to come into bearing, and completely undesirable as an orchard tree. These characteristics (vigor, thorniness, and delayed bearing) become less pronounced in consecutive vegetative generations of propagation, particularly if propagules are taken from the upper part of older nucellar seedling trees. Nucellar seedlings are largely free of viruses and other systemic pathogens. Foundation clones of nucellar cultivars have been developed for all the commercial citrus cultivars and are the basis of the "new-line" citrus sources (*24*) (see Chapter 9). The vigor and uniformity of nucellar seedlings makes them ideal as rootstocks, and most rootstock species or cultivars are nucellar.

Citrus seeds generally have no dormancy but are injured if allowed to dry. Consequently, seeds should be planted immediately after being extracted from the ripe fruit. Seeds may be stored moist, in polyethylene bags at a low temperature of (4° C; 40° F); before storage they should be soaked for ten minutes in water at 49° C (120° F) to aid in eliminating seed-borne diseases. Treatment of the seed with a fungicide is beneficial.

Grafting. The most important method used in commercial production of citrus is to bud or graft onto nucellar seedling rootstocks. T-buds, inverted T-buds, or modified cleft grafts can be used. In microbudding, a very small sliver of stem bearing the bud is inserted under the bark of the rootstock.

An alternate method used in commercial production of dwarf citrus is a cutting-graft in which the scion and the rootstock are grafted together, then placed in a mist bed so that the rootstock produces roots and the graft heals at the same time. The rooted grafted unit is potted into a container for further production as a salable product (*50*).

Source Selection. Source selection of cultivars and rootstocks is essential in modern citrus nursery production. About 12 known viruses, a viroid causing "exocortis," and a mycoplasmalike organism causing "stubborn" disease can infect citrus. The most serious worldwide problem is *tristeza*, causing "quick decline" and/or "slow decline" in specific cultivar-rootstock combinations. Good indexing procedures have been worked out for most of them (*21, 124, 184*). In addition, citrus is subject to the production of low-producing, inferior "budsport limbs." Budwood should be taken only from known high-producing, disease-free trees, preferably *foundation clones*. It is desirable to select the budsticks from a single tree, avoiding any off-type "sporting" branches.

Rootstock Selection. Next to scion source selection, the choice of rootstock is critical in commercial citrus production. Rootstocks vary greatly in their relative susceptibility to diseases (*Phytopthora* sp.), viruses (particularly *tristeza*), interactions with scion, quality of fruit, size and vigor, and tolerance to soil problems, such as chlorosis, salt, boron excess, etc.

Nursery Production Systems.

Field Production of Nursery Plants. Avoid using soils infested with citrus nematodes *(Tylenchulus semipenetrans)* or burrowing nematodes *(Radopholus similis)*, or soil-borne diseases, although citrus is resistant to verticillium wilt. For nurseries, it is preferable to use virgin soil or at least a soil that has not been formerly planted with citrus. For small operations, raised seed beds enclosed by 12-in. boards can be used. A soil mixture of ¾ sandy loam and ¼ peat moss is satisfactory. Treating the soil with a fumigant such as DD (dichloropropane—dichloropropene) will minimize the chances of nematode infestations. To reduce the hazard of fungus infection, methyl bromide has been used to fumigate the seed bed and nursery site. Following treatment, planting should be delayed for six to ten weeks to allow the fumigant to dissipate. Some soils in California, however, have remained toxic to citrus for a year following such treatment. The seed bed should be in a lathhouse, or some other provision should be made for screening the young seedlings from the full sun. The fruit of certain species, such as the trifoliate orange or its hybrids, matures in the fall. If the seeds are to be planted at that time, they should be held in moist storage at -1 to 4° C (30 to 40° F) four weeks before planting. Trifoliate orange seedlings, if grown during times of the year having short days, will respond strongly with increased growth when supplementary light is provided to lengthen the day. The same is true for seedlings of certain citrange and sweet orange cultivars (*219*).

The best time to plant the seeds is in the spring after the soil has warmed [above 15° C (60° F)]. Seeds should be planted in rows 5 to 7.5 cm (2 to 3 in.) apart, and 2.5 cm (1 in.) apart in the row. They are pressed lightly in the soil and covered with a 2 cm (¾ in.) layer of clean, sharp river sand. The sand prevents crusting and aids in the control of "damping-off" fungi. The soil should be kept moist at all times until the seedlings emerge. Either extreme, allowing the soil to become dry and baked or overly wet, should be avoided. PVC pipe with circulating hot water is placed below the seed bed to maintain a temperature of 27 to 29° C (80 to 85° F)—which hastens germination. By this method, seeds may be planted in the winter months, and the seedlings will be large enough to line-out in the nursery in the spring. Many can be budded by fall or the following spring. This technique often shortens the propagation time by 6 to 12 months.

After the seedlings are 20 to 30 cm (8 to 12 in.) tall, they are ready to be transplanted from the seed bed to the nursery row, preferably in the spring after danger of frost has passed. The seedlings are dug with a spading fork after the soil has been wet thoroughly to a depth of 45 cm (18 in.). They can then be loosened and removed with little danger of root injury. All stunted or off-type seedlings or those with crooked, misshapen roots should be discarded.

The nursery site should be in a frost-free, weed-free location on a medium-textured, well-drained soil at least 61 cm (24 in.) deep and with irrigation water available. Old citrus soils should be avoided unless heavily fumigated with DD or methyl bromide before planting. The seedlings should be planted at the same depth as in the seed bed and spaced 25 to 30 cm (10 to 12 in.) apart in 90 to 120 cm (3 to 4 ft) rows.

Citrus seedlings are usually budded in the fall in Florida and California, starting in mid-September, early enough so that warm weather will ensure a good bud union, yet late enough so that bud growth does not start and the wound callus does not grow over the bud itself.

The best type of budwood is that next to the last flush of growth, or the last flush after the growth hardens. A round bud-stick gives more good buds than an angular one. The best buds are those in the axils of large leaves. The bud-sticks are usually cut at the time of budding, the leaves removed, and the bud-sticks protected against drying. Bud-sticks may be stored for several weeks if kept moist and held under refrigeration at 4 to 13° C (40 to 55° F).

The T-bud method is satisfactory for citrus. The bud piece is cut to include a sliver of wood beneath the bud. Fall buds are unwrapped in six to eight weeks after budding, spring buds in about three weeks. In California and Texas the buds are usually inserted at a height of 30 to 45 cm (12 to 18 in.), but in Florida the buds are inserted very low on the stock—2.5 to 5 cm (1 to 2 in.) above the soil. Such low budding is often necessitated by profuse branching in rough lemon and sour orange seedlings, which is caused by partial defoliation due to scab and anthracnose spot.

Buds inserted in the fall are forced into growth in the spring by "lopping" or "crippling" the top of the seedlings 5 to 7.5 cm (2 to 3 in.) above the bud. This is done just before spring growth starts, and consists of partly severing the top, allowing it to fall over on the ground. The top thus continues to nourish the seedling roots, but the bud is forced into growth. Lopping of spring buds is done when the bud wraps are removed—about three weeks after budding. If possible, the "lops" should be left until late summer, at which time they are cut off just above the bud union. Although lopping is satisfactory, it may make irrigation and cultivation difficult. An alternative practice is to first cut the seedling completely off 30.5 to 35.5 cm (12 to 14 in.) above the bud, then later cut it back immediately above the bud. However, this practice does not force the bud as well as lopping or cutting the seedling just above the bud.

Young citrus nursery trees may be dug "balled and burlapped" or bare-root. Bare-rooted trees should have the tops pruned back severely before digging. Transplanting of such trees is best done in early spring, but balled trees can be moved any time during spring before hot weather starts.

Container Production of Nursery Plants (*1, 56*). Citrus seeds are mechanically extracted and washed. They are then germinated in special elongated square plastic pots in a greenhouse. After 3 or 4 months the seedling rootstock plants are large enough for inverted T-budding; finished budded trees can be developed in about 12 months when grown in plastic containers (*27*). Sometimes cleft grafting is used on the young seedlings after they are growing in their final containers (*149*). Citrus nursery trees apparently grow well in containers and develop into good orchard trees upon field planting, providing they are not held in the container so long that root binding occurs (*157*).

Miscellaneous Asexual Methods. Citrus plants can be propagated by a variety of vegetative methods. Many citrus species can be propagated by rooting leafy cuttings, or by leaf-bud cuttings, although nursery trees are not commonly propagated in this manner (*48*). The persian lime *(C. aurantifolia)* is propagated to some extent by air layering, as is the pummelo *(C. grande)* in southeast Asian countries.

Micropropagation. Citrus can be micropropagated but it is not a commercial method. Primary use is production of virus-free clones through micrografting procedures (*162*).

Rootstocks for Citrus (26, 48, 221, 233). The principal characteristics of citrus are listed in the following chart. *Sweet orange (C. sinensis)* is a good rootstock for all citrus cultivars, producing large, vigorous trees but, due to its high susceptibility to gummosis (*Phytophthora* spp.), is not grown much today.

Sour Orange (*C. aurantium*). This is historically the most important rootstock species, used all over the world due to its vigor, hardiness, deep root system, resistance to gummosis diseases, and ability to produce high-quality, smooth, thin-skinned, and juicy fruit. Its high susceptibility to the *tristeza* virus (*13*) which produces quick decline has eliminated its use in California and reduced its use in Florida. This rootstock is still used in Texas, Mexico, Cuba, Venezuela, Honduras, Sicily, and Israel. Tolerant strains to *tristeza* have been tested in Australia (*203*). The scion top itself may be tolerant to the virus, but a hyperreaction at the graft union kills the phloem tissues and the roots starve.

Trait	*Sweet Orange*	*Sour Orange*	*Rough Lemon*	*Citrange*	*Trifoliata*	*Cleopatra Mandarin*
Nucellar	70–90%		90–100%	nearly 100%	60%	80%
Incompatibility	OK	OK	some	OK	OK	'Eureka'
Vigor	large, vigorous	large, vigorous	very vigorous	semi-dwarf	dwarfing	
Fruit quality	very good	very good	tends to be low	good	very good	high
Tristeza	tolerant	very susceptible				resistant
Phytopthora	very susceptible	resistant	susceptible		resistant	susceptible
Nematodes					resistant	
Soil	light, well drained		lighter soils, well drained		medium	salt-tolerant
Hardiness	medium	hardy	tender	hardy	hardy	

Rough Lemon (C. jambhiri). This has been the most important stock in Florida because its high vigor is well adapted to warm, humid areas with deep sandy soils. Its primary drawback is the low quality of fruit that is produced, evidently a function of the high vigor. It is also very cold-sensitive. Because of its susceptibility to blight and *Phytopthora,* its use is declining there. It remains an important rootstock in Arizona.

Trifoliate Orange (Poncirus trifoliata). This is a deciduous species that is important because of its cold tolerance and tendency to dwarf the scion. It is the principal citrus rootstock in Japan and it is used in China as a stock for Satsumas and kumquats. In northern Florida and along the Gulf coast to Texas, it has long been used as a stock for Satsuma oranges and kumquats, for which it is excellent. Trifoliate orange is commonly used as a stock for ornamental citrus and in home orchards for dwarfed trees. Trees on this stock yield heavily and produce high-quality fruits.

Trifoliate orange fruits produce large numbers of plump seeds that germinate easily. The upright-growing, thorny seedlings are easy to bud and handle in the nursery, but their slow growth often necessitates an extra year in the nursery before salable trees are produced.

'Cleopatra' Mandarin (C. reshni). This is a good rootstock, particularly in Florida, as a stock for oranges, grapefruit, and other mandarin types, and has come into use in California and Texas as a replacement for sour orange rootstock. Its resistance to gummosis, comparative salt tolerance, and resistance to *tristeza* seemingly justify its greater use. Its chief disadvantages are the slow growth of the seedlings, slowness in coming into bearing, and susceptibility to *Phytophthora parasitica* root rot.

Citranges (Trifoliate Orange × Sweet Orange Hybrids). These include several named cultivars—'Savage,' 'Morton,' 'Troyer,' and 'Carrizo,' whose seedlings are used as rootstocks in California and Florida. 'Savage' is especially suitable as a dwarfing rootstock for grapefruit and the mandarins. 'Morton' is not as dwarfing but produces heavy crops of excellent-quality fruit. Seed production is so low, however, that getting quantities of nursery trees started is difficult. Sweet orange trees on 'Troyer' are vigorous, cold-hardy, resistant to gummosis, and produce high-quality fruit. This rootstock is widely used commercially in California particularly for replanting on old citrus soils. 'Troyer' is relatively fruitful and produces 15 to 20 plump seeds per fruit, facilitating the propagation of nursery trees.

Nucellar seedlings of citrange cultivars develop strong, single trunks, easily handled in the nursery. As with trifoliate orange, only exocortis-free buds should be used on citrange rootstocks; otherwise, dwarfing—and eventual low production—will result. 'Troyer' and 'Carrizo' are the two commercially most important citrus rootstocks in California and Arizona. 'Carrizo' is an important rootstock in Florida.

Rangpur Lime (C. aurantifolia × C. reticulate). This is the most widely used citrus rootstock in Brazil, producing vigorous, fruitful trees, which are resistant to *tristeza.* It is highly susceptible to *exocortis.* In Texas, it has been more salt-tolerant than other citrus rootstocks. Some strains of Rangpur lime are susceptible to *Phytophthora.*

Alemow (Citrus macrophylla). This is used in California as a rootstock for lemons in high-boron areas, owing to its boron-tolerance. It is susceptible to *tristeza* when sweet orange scion cultivars are used and to a rootstock necrosis first detected about 1960.

Coconut. *See Cocos.*

Cocos nucifera. Coconut (*32, 176*). This tropical monocot is probably the most important nut crop in the world. There are two forms—tall and dwarf—with named seed-propagated cultivars of each. These are based on their origin in specific "seed gardens" collected from isolated groves. Some named cultivars maintain their characteristics quite dependably but some of these are also susceptible to the highly lethal *golden disease.* Seeds should be collected from trees that produce large crops of high-quality nuts.

The nuts are usually germinated in seed beds. The nuts, still in the husk, are set at least 30 cm (12 in.) apart in the bed and laid on their sides with the stem end containing the "eyes" slightly raised. The sprout emerges through the eye on the side that has the longest part of the triangular hull. As soon as this occurs (about a month after planting), the sprout sends roots downward through the hull and into the soil. In 6 to 18 months the seedlings are large enough to transplant to their permanent location.

Coffea arabica. Coffee (*226*).

Seed. The most common method of propagation is by seed, preferably obtained from selected superior trees. Coffee seeds lose viability quickly and are subject to drying through the seed coverings. Seeds held at a moisture content of 40 to 50 percent and at 4 to 10° C (40 to 50° F) will keep for several months. There is no seed dormancy. Seeds are usually planted in seed beds under shade (*204*). Sometimes the seedlings are started in soil in containers formed from leaves, or in polyethylene bags, to facilitate transplanting. Germination takes place in four to six weeks. When the first pair of true leaves develop, the seedlings are transplanted to the nursery and set 30 cm (12 in.) apart. After 12 to 18 months in the nursery the plant has formed six to eight pairs of laterals and is ready to set out in the plantation.

Miscellaneous Asexual Methods. Coffee can be propagated vegetatively by almost all methods. Leafy cuttings have potential for commercial use, but the resulting plants show strong **topophysis** (see p. 256). Cutting material should be taken only from upright shoot terminals in order to produce the desired upright-growing tree. Leafy cuttings of partially hardened wood treated with auxin can be rooted fairly easily with mist and partial shade. (*193*).

Micropropagation. Coffee plants have been propagated by the formation of adventitious embryos in aseptic callus cultures (*101*).

Coffee. *See Coffea.*

Cola nitida. Kola. This tropical tree produces a caffein-containing nut used to make refreshing beverages.

Seed. Plants are grown from seed planted directly in the field or in nurseries for later field planting. Seeds for planting should be harvested only when completely mature. Seedling plants have shown pronounced juvenility; seven to nine years being required for the seedlings to flower.

Miscellaneous Asexual Methods. Vegetative propagation using mature-growth propagules can produce early production but propagation is difficult. Rooting of terminal leafy cuttings in polyethylene-covered frames without auxin have been successful (*213*). Air layering with IBA treatment is successful—commercial nut production is induced 12 to 18 months after planting the rooted layers. Patch budding is also a successful method of vegetative propagation (*8*).

Corylus spp. Filbert, Hazelnut (*127, 148*). Cultivars of hazelnut *(C. avellana)* are widely grown in Europe, but are called filberts in the Willamette valley of Oregon where they are mostly grown in the U.S. Native species in North America (e.g., *C. americana*) also occur.

Seed. Freshly harvested seeds of *C. avellana* are not dormant but develop dormancy with drying in a complex interplay of inhibitors and growth promoters (see page 210). Seed dormancy can be overcome with stratification for three months at 0 to 4° C (32 to 40° F). Seeds are mostly planted in the nursery in the fall.

Layering. The usual method of commercial filbert propagation is by simple layering in the spring using suckers arising from the base of vigorous young trees four to eight years old. Special stool mother plants can be maintained in a bush form just for layering purposes (see Figure 15–3). After one season's growth, a well-rooted plant 0.6 to 1.8 in. (2 to 6 ft) tall may be obtained, ready to set out in the orchard. Old orchard trees are not suitable to use for layering. Sometimes suckers arising from the roots are dug and grown in the nursery row for a year or, if well rooted, planted directly in place in the orchard.

Grafting. Grafting in midwinter using a whip graft has been successful with the aid of a "hot grafting" procedure, as shown in Figure 13–37 (*106, 202*). Commercial cultivars are grafted on seedling *C. avellana* rootstock.

Cuttings. Propagation is difficult but some reports indicate that with careful timing in mid-June to mid-July while in active growth rooting can occur (*52*). Cuttings are treated with a 5,000 to 10,000 ppm IBA quick-dip or talc, and rooted under mist in a well-drained medium. Too much water results in decline, which is why fog is preferred to mist.

Micropropagation. Corylus can be micropropagated but is not a commercial procedure (*5*).

Crab apple (Flowering). *See Malus in Chapter 20.*

Cranberry. *See Vaccinium.*

Currants. *See Ribes.*

Custard apple. *See Annona.*

Cydonia oblonga. Quince. The quince is a small bushy tree that is grown for the jelly-making qualities of the fruit, due to high pectin content, and as a dwarfing rootstock for pears; see the rootstock section for *Pyrus communis.*

Cuttings. Quince roots readily by hardwood cuttings using "heel" cuttings attached to one-year-old wood. Cuttings from two- to three-year-old wood also root easily. The burrs or knots found on this older wood are masses of adventitious root initials. Cuttings usually make sufficient growth in one season to transplant to their permanent location.

Grafting. Quince cultivars can be T-budded on rooted cuttings (such as 'Angers') or sometimes on quince seedlings.

Layering. Quince can be propagated readily by mound layering (see p. 509) which has been a traditional method of propagation.

Micropropagation. Methods for large-scale clonal production of 'Provence' quince has been described for use as a dwarfing pear rootstock (*4*).

Date palm. *See Phoenix.*

Diospyros kaki. Japanese persimmon (*177, 186, 196*). Subtropical tree species with many cultivars grown widely in the Orient and to a limited extent in the U.S. for its large fruit (*186*). Cultivars are propagated by grafting on seedling rootstocks. American persimmon *(D. virginiana)* grows throughout the midwest and southern part of the U.S. Some named cultivars are grown in home yards.

Seed. Seeds of *D. lotus, D. kaki,* and *D. virginiana* are stratified for 120 days at about 10° C (50° F). Dry seeds should be soaked in warm water for two days before stratification. Excessive drying of the seed is harmful, especially for *D. kaki.* Seeds are planted either in flats or in the nursery row. Young persimmon seedlings require shading.

Cuttings (15). Softwood cuttings with bottom heat, under mist, and treated with IBA will root, whereas hardwood cuttings do not root successfully.

Grafting. Crown grafting by the whip-and-tongue method or cleft can be done in early spring when both scion-wood and rootstock are still dormant. Bench grafting may also be used. Budding is less successful than grafting (*177*). *D. virginiana* is primarily produced by budding and grafting onto *D. virginiana* seedlings.

Rootstocks for Japanese Persimmon (105)

Diospyros lotus. This rootstock widely used in California is very vigorous and drought-resistant. It produces a fibrous type of root system, which transplants easily. This rootstock is susceptible to crown gall and verticillium, will not tolerate poorly drained soils, and is highly resistant to oak root fungus. 'Hachiya' does not produce well on *D. lotus* stock because of excessive shedding of fruit in all stages (*188*). 'Fuyu' scions usually do not form a good union with *D. lotus,* although 'Fuyu' topworked on a compatible *D. kaki* interstock on *D. lotus* roots makes a satisfactory tree.

D. kaki. This rootstock is most favored in Japan and is probably best for general use. It develops a good union with all cultivars, is resistant to crown gall and oak root fungus but susceptible to verticillium. Seedlings have a long taproot with few lateral roots, making transplanting somewhat difficult.

D. virginiana. Seedlings of this species are utilized in the southern U.S. due to their wide soil adaptation, but they have not proven satisfactory in some localities. 'Hachiya' growing on this rootstock in California is dwarfed, has sparse bloom, and yields poorly (*188*), but most other Oriental cultivars make a good union with this rootstock. However, diseases of an unknown nature carried in *D. kaki* scions will cause death of the *D. virginiana* roots. Otherwise, this rootstock seems to be tolerant of both drought and excess soil moisture. Its fibrous type of root system makes transplanting easy but increases suckering.

Micropropagation. Oriental persimmon cultivars can be micropropagated (*40, 205*).

Eriobotrya japonica. Loquat. This subtropical evergreen pome fruit is grown for its fruit which is used fresh, or in jams, jellies, etc. The species is also used as a landscape plant. Many of the trees in use are unselected seedlings. Improved cultivars exist but are difficult to propagate.

Seed. No treatment is apparently required, but seeds tend to germinate spontaneously. Should be sown soon after extraction as heat and light tends to reduce quality of germination.

Grafting. Scions are T-budded or side grafted to loquat seedlings. Established seedling trees may be topworked, using the cleft graft. Quince can also be used as a rootstock, producing a dwarfed tree. Inarching is also used. Topworking is easily done with bark grafting.

Layering. Air layering is successful and can be improved by auxin application (*192*). Three-month-old shoots are ringed and layered (*15*).

Euphoria longan. Longan (*15*). A tropical tree from southwest Asia and China. Seedlings have a long juvenile period and are variable. There are many vegetatively propagated cultivars in southern China.

Seed. Easily grown from seed.

Grafting. Cleft, whip, side, and inarch grafting are successful. The scions are semi-hard terminal shoots—about 0.5 cm (¼ in.) in diameter, with leaves still attached. Grafts are made into a node subtended by a leaf and enclosed with a polyethylene bag, which is shaded. Seedling rootstocks of the cultivar to be grafted are used.

Cuttings. Leafy cuttings with thick hard terminal wood, about 25 cm (10 in.) long, are taken in winter and rooted with bottom heat and mist.

Layering. Marcottage (air layering) is the preferred method. This is carried out in spring with well-matured, recently flushed growth. See the procedure described in Chapter 15. The layer is removed after two to four months when roots have turned from white to creamy brown. Establish in plastic bags or pots and transplant to the field after 6 to 12 months.

Feijoa sellowiana. Feijoa. Pineapple Guava (*114*). This small subtropical evergreen tree is grown for its small edible fruit useful for jelly and juice.

Seed. Seeds germinate without difficulty. They are started in flats of soil and later transplanted to the nursery row.

Cuttings. Leafy softwood cuttings treated with auxin and started under closed frames can be rooted. Rooting under mist is usually quite difficult. Otherwise layering can be successful.

Grafting. Scions are grafted to Feijoa seedlings using a whip graft or a cleft graft (*60*).

Ficus carica. Fig. The cultivated fig is an ancient fruit crop from the Mediterranean area which grows into small trees (*38, 125*). Specific cultivars have been propagated by hardwood cuttings for centuries.

Seed. Seeds are used only for breeding new cultivars. The fruit is a fleshy receptacle bearing many small achenes inside. Some of these are sterile and others fertile. The heavier fertile seeds can be separated from the lighter ones by floating them in water. Seeds can be germinated easily in flats of soil mix.

Cuttings. Hardwood cuttings of two- or three-year-old wood or basal parts of vigorous one-year shoots with a minimum of pith are suitable for cuttings. For best results cuttings should be prepared in early spring, well before bud-break, with the bases allowed to callus for about ten days at about 24° C (75° F) in slightly moist bark or peat media before planting. IBA will enhance rooting. The cuttings are grown for one or two seasons in the nursery, then transplanted to their permanent location. Figs can also be grown in containers in the nursery. A method in European countries is to plant long cuttings, 0.9 to 1.2 meters (3 to 4 ft), with their full length in the ground where the tree is to be located permanently. Sometimes two cuttings are set in one location to increase the possibility that at least one will survive. Pregirdling of fig shoots 30 days before rooting is beneficial (*15*).

Grafting. The fig can be budded, using either T-buds inserted in vigorous one-year-old shoots on heavily pruned trees, or patch budded on older shoots.

Layering. Figs are easily air layered. One-year-old branches, if layered in early spring, are usually well-rooted by midsummer.

Micropropagation. Figs can be micropropagated (*175*).

Fig. *See Ficus.*

Fragaria × ananassa. Strawberry. (*72, 225*). The strawberry is an herbaceous perennial whose basic vegetative structure is a thickened vertical stem, known as a "crown." It produces a rosette of leaves and inflorescences in the spring followed by runners (see p. 513) in the summer. In the normal annual cycle, the plant is induced to initiate flowering by warm days and shortened photoperiod in the fall. This is followed by chilling of the crown during winter to produce vegetative growth and flowering for about six weeks in spring. In the long days of summer, runners are formed (induced) and grow out from the main crown to start new plantlets at every other node. In some strawberry cultivars, runner formation is inhibited and the plant expands by increasing its crown ("everbearing types"). In recent years, "day-neutral" genotypes have been produced by breeding which makes flowering possible at any time of the year. Natural vegetative reproduction is by runner development or by crown division.

Modern commercial strawberry production has manipulated the strawberry plant by breeding, nursery and plant management, and environmental controls into an integrated system. Production has increased annual production more than tenfold, and expanded the season from a few weeks in the spring to literally year-round production, depending on cultivar, management, and climate.

Strawberry has many problems that must be controlled during propagation and production. These include fungal soil and foliage diseases, nematodes, viruses spread by aphids and nematodes, and mycoplasma diseases spread by aphids. Control involves cultivar resistance, pathogen-free propagation and planting material, chemical control, and sanitation.

Seeds. Seed propagation is used only in breeding programs for the development of new cultivars. The strawberry fruit is an expanded receptacle upon which are borne the true "fruit" or achenes. These are extracted usually with a blender. The achenes are quite hard and scarifying with sulfuric acid for 15 to 20 minutes prior to planting improves germination (*189*).

Nursery Production. Pathogen-free sources are developed and maintained through a combination of meristem culture, thermotherapy, and certification schemes (*39*). Nursery ground is located in a favorable location, and isolated from other crops, including wild strawberries. Land is fumigated the previous fall (*197*). Nursery plants are planted in the spring and subsequently maintained under irrigation for rapid growth and rooting of runners. Plants are dug in the fall after complete dormancy is produced, packed into plastic lined containers, and stored at −1 to −2° C (28 to 30° F) until planting time (*79*).

Field nurseries are planted at relatively high elevations where there is little rainfall but lots of sunlight to maximize starch reserves. Plants develop runners. Runner plants are dug in the fall and winter when mature using special machines. Immediately thereafter, the leaves and petioles are removed and the plants are packed with roots to the center in vented cardboard or wood boxes lined with polyethylene (0.75 to 1.5 mil) to keep plants from drying out.

Fruit Production. Plants are removed from storage, planted in the appropriate location, and timed to a particular planting date depending upon cultivar, nursery history, storage timing, and location. Boxes of 1,000 or 2,000 plant units can be stored and kept in good condition up to one year at −2 and −1° C (28 to 30° F). During storage, the plants not only remain dormant but physiological changes occur that overcome endodormancy and enhance vigor when the plants are later removed and planted into the production area. The time schedule for digging, storage, and planting is geared to the individual cultivar, the production schedule at the production fields, and the environment at the growing site.

Crown Division. Everbearing cultivars that produce few runners are propagated by crown division. Certain cultivars, such as the 'Rockhill,' may produce 10 to 15 strong crowns per plant by the end of the growing season. In the spring, such plants are dug and carefully cut apart; each crown may then be used as a new plant.

Micropropagation. Micropropagation has become an integral part of commercial strawberry production. Certain commercial laboratories, particularly in Europe, produce millions of plants through shoot-tip culture for direct sale to customers (*16, 46, 145*). In most U.S. production systems, meristem culture is used to produce virus-free "mother" plants which are then used as the foundation clones to be multiplied for producing commercial material as part of a Registration and Certification system (*72*).

Garcinia mangostana. Mangosteen. A slow-growing evergreen tree of the tropics (*15*). It produces a soft, sweet, and delicious fruit. Most plants are nucellar seedlings which may take 8 to 15 years to produce fruit. Seeds have low viability but can germinate in four to six days. Grafting and budding are successful, but the rootstock must be compatible. Layering is improved by IBA treatment.

Gooseberries. *See Ribes.*

Grape. *See Vitis.*

Grapefruit. *See Citrus.*

Guava (Common or Lemon; Cattley or Strawberry). *See Psidium.*

Guava (Pineapple). *See Feijoa.*

Hickory (Shagbark). *See Carya.*

Jackfruit. *See Artocarpus.*

Juglans cinerea. Butternut. This tree is native to the eastern U.S. where it is grown mostly as a timber tree. Several butternut cultivars have been selected for nut production.

Seed. The very hard-shelled seeds are enclosed within a husk that must be removed after collection. Nuts may be stored at 3 to 4° C (37 to 39° F) at 80 to 90 percent humidity. Seeds have dormant embryos and should be either planted in the fall out-of-doors or stratified for several months and planted in the spring.

Grafting. Cultivars are propagated by chip budding or bench grafting (using the whip graft) on Persian *(J. regia)* or black *(J. nigra)* walnut seedling rootstocks. Butternut is also bark grafted onto *J. nigra* seedlings. Excessive sap "bleeding" is a problem in grafting, and necessary steps must be taken to overcome it.

Juglans nigra. Black Walnut (*71*). This species is native to eastern north America where it is an important timber tree. It also produces excellent nuts, although it is very hard-shelled.

Seed. The very hard-shelled nuts apparently require a period of warm stratification to soften the shells followed by cold stratification to overcome embryo dormancy. Early fall planting could provide both conditions. Seedlings are started in nurseries and only the strongest, most vigorous trees are set out in the plantation. Careful planting is necessary to obtain the essential rapid, early growth.

Cuttings. Cuttings are very difficult to root.

Grafting. Black walnut cultivars selected for their nut qualities are propagated by patch or ring budding or side grafting on *J. nigra* rootstocks (*29, 194*). This method is suitable for seedling stocks up to about 2.5 cm (1 in.) in diameter.

Juglans regia. Persian or English walnut (*66*). It is the most important commercial nut producer and is grown in California, Oregon, and in the Mediterranean area. They originated in the area of present-day Iran. "Carpathian walnuts" are a cold-hardy strain of *J. regia* which originated in cold winter areas of the Carpathian Mountains of Poland and the Kiev and Poltava regions of the Ukraine. Carpathian cultivars have been developed that are winter-hardy in the eastern United States (*77*).

Seed. Nuts of most of the *Juglans* species used as seedling rootstocks for *J. regia* should be either fall-planted, or stratified for about two to four months at 2 to 4° C (36 to 40° F) before they are planted in the spring. It is better to plant the seeds before they start sprouting in the stratification boxes, but with care, sprouted seeds can be planted successfully. At the end of one season, the seedlings should be large enough to bud. Seedlings of Juglans species are very sensitive to waterlogged soil conditions under which they will not survive (*28*).

Grafting. *J. regia* is propagated by patch budding, T-budding, or whip grafting on one-year-old seedling rootstocks. Topworking-established seedling trees in the orchard that are one to four years old can be done by bark grafting in late spring. Patch budding in the spring or summer works well when topworking smaller size material.

Rootstocks for Persian (English) Walnut (146).

Northern California Black Walnut (J. hindsii). This rootstock is the most commonly used in California. The seedlings are vigorous and make a strong graft union. They are somewhat resistant to crown rot, oak root fungus *(Armillaria mellea),* and root-knot nematode (*Meloidogyne* spp.)—but are susceptible to crown and root rot (*Phytophthora* spp.), and the root-lesion nematode *(Pratylenchus vulnus).* Persian walnut trees grafted on this rootstock are subject to a serious problem known as "blackline." The latter has appeared in California, Oregon, England, Italy, and France. It is characterized by a breakdown of the tissues in the cambial region at the graft union, leading to a lethal girdling of the trees. The cause of this problem is a pollen-transmissible virus, cherry leaf roll virus (*152, 153*). The Persian walnut top above the graft union eventually dies, leaving the *J. hindsii* stock alive.

Persian Walnut (J. regia). Worldwide, this is probably the most common rootstock for Persian walnut cultivars due to availability of seeds. Seedlings of this species as a rootstock produce good trees with an excellent graft union. The roots are susceptible to crown gall, oak root fungus, and salt accumulation in the soil, and are not as resistant to root-knot nematodes as *J. hindsii.* Nurserymen object to the slow initial growth of the seedlings. Trees on this rootstock should be satisfactory in soils free of oak root fungus and salt accumulation. Its use is recommended in Oregon, where blackline is very serious. Seedlings of the 'Manregian' clone (of the Manchurian race of *J. regia*), imported by the USDA as P. I. No. 18256, are used as rootstocks. They are vigorous, cold-hardy, and—in Oregon—no more susceptible to oak root fungus than *J. hindsii* seedlings (*168*).

Paradox Walnut (J. hindsii × J. regia). First-generation (F_1) hybrid seedlings are obtained from seed taken from *J. hindsii* trees, whose pistillate (female) flowers have been wind-pollinated with pollen from nearby *J. regia* trees. When seeds from such a *J. hindsii* tree are planted, some of the seedlings may be the hybrid progeny. The amount varies widely, from none to almost 100 percent, depending upon the individual mother tree. The hybrids are easily distinguished by their large leaves in comparison with the smaller-leaved, self-pollinated *J. hindsii* seedlings (see Fig. 4–10). *Seedlings from Paradox trees* themselves *should not* be used for producing rootstocks, because of their great variability in all characteristics. Although first-generation (F_1) seedlings are variable in some characteristics, most of them exhibit hybrid vigor and make excellent vigorous rootstocks for the Persian walnut. They are more resistant than either parent to crown rot, and are tolerant of saline and heavy, wet soils. Paradox seedlings are susceptible to crown gall. Persian walnut on Paradox roots are just as susceptible to the blackline virus as *J. hindsii* seedlings. Trees on Paradox rootstocks grow and yield as well as, or better than, those on *J. hindsii* roots, and may produce large-size nuts with better kernel color. In heavy or low-fertility soils, trees on Paradox rootstock grow faster than those on *J. hindsii.*

Since Paradox seeds are difficult to secure in quantity, vegetative propagation methods would be very desirable in order to establish superior Paradox clones from which large numbers of rootstock plants could be obtained. Propagation by leafy cuttings under mist, hardwood cuttings, and trench layering can be done but with difficulty and are more of a problem to transplant.

Eastern Black Walnut (*J. nigra*). This has been recommended in some cases as a rootstock for Persian walnut cultivars in Europe and the Soviet Union—but not in California. As compared to *J. regia* roots, it is reported to have greater tolerance to nematodes, oak root fungus, *Phytophthora,* and crown gall, but in California rootstock trials, trees on this stock showed poor yields.

Micropropagation. J. regia cultivars can be micropropagated (*146*).

Jujube. *See Zizyphus.*

Kiwifruit. *See Actinidia.*

Kola. *See Cola.*

Lemon. *See Citrus.*

Lime. *See Citrus.*

Litchi (Hairy). *See Nephelium.*

Litchi chinensis. Litchi. Tropical tree fruit with delicious juicy fruit. Many vegetatively propagated cultivars exist.

Seed. The seeds are used in breeding, but have a long juvenile period of 10 to 15 years (*15, 35*). Litchi seeds germinate in two to three weeks if planted immediately upon removal from the fruit. They lose their viability within a month if not planted.

Cuttings. Generally difficult to root from cuttings. Tip cuttings from a flush of growth in the spring have been rooted in fairly high percentages under mist in the full sun. Auxin is beneficial. Semi-hardwood cuttings from an active flush of new growth root more readily than those from dormant hardwood cuttings.

***Grafting* (*15*).** Some success has been reported for splice grafting using subterminal portions of shoots for the scion. It is best to retain leaves on the rootstock. Approach grafting is also successful, as is chip and T-budding.

Layering (Marcottage). Air layering is the most important method of propagation, since cuttings root with difficulty. Layering can be done at any time of the year, but best results are obtained in spring or summer (*138*). Large limbs with mature vegetative growth air layer easier than small ones with recently flushed wood. Auxin applications are beneficial. Stooling (mound layering) has also been reported to produce good results (*15*).

Loganberry. *See Rubus.*

Longan. *See Euphoria.*

Loquat. *See Eriobotrya.*

Macadamia spp. Macadamia nut. *M. integrifolia* and *M. tetraphylla* are the two principal species (*83*). Trees of this subtropical evergreen nut tree may be grown as seedlings, but clonal cultivars are preferred. Macadamia is highly resistant to *Phytophthora cinnamomi* and will tolerate heavy clay soil.

Seed. Fresh seeds should be planted in the fall as soon as they mature, either directly in the nursery or in sand boxes in a lathhouse. Seedlings are transplanted to the nursery or into poly containers after growing 10 to 15 cm (4 to 6 in.) tall. Seeds should not be cracked because they are readily attacked by fungi. Only seeds that sink when placed in water should be used (*83*). Seeds retain viability for about 12 months at 4° C (40° F), but at room temperature, viability starts decreasing after about four months. Scarifying or soaking the seed in hot water hastens germination.

Cuttings. Macadamia can be propagated by rooting leafy, semi-hardwood cuttings of mature, current season's growth. Tip cuttings 8 to 10 cm (3 to 4 in.) work best. Treatment with IBA at 8,000 to 10,000 ppm is beneficial. Cuttings should be placed in a closed propagating frame or under mist for rooting. Bottom heat at 24° C (75° F) is beneficial. Cultivar differences exist in ease of rooting. *M. tetraphylla* cuttings root more readily than those of *M. integrifolia.* Rooted cuttings are not widely used in commercial plantings because of their shallow root system.

Grafting. One of the side graft methods is used. Leaves should be retained for a time on the rootstock. Rapid healing of the union is promoted if prior to grafting the rootstock is checked in growth by water or nitrogen deficiency to permit carbohydrate accumulation. Ringing the branches that are to be the source of the scions several weeks before they are taken increases their carbohydrate content and promotes healing of the union. Budding has generally been unsuccessful (*53*). Seedling rootstock of *M. tetraphylla* is preferred for grafting *M. integrifolia* cultivars (*83, 201*). Macadamia can also be propagated by the cutting-graft method (see p. 477).

Malus spp. Crab apple (Flowering). See Chapter 20.

Malus × domestica. Apple. This deciduous fruit crop is one of the oldest and most widely grown fruit crops in the world. Thousands of cultivars exist, many of which are chimeras and bud-sports. Viruses and viruslike organisms are serious problems for apples (see Chapter 9) such that essentially all commercial nurseries utilize virus-tested foundation clones and produce trees in a program of Registration and Certification (*20, 47, 67*).

Seed. Seeds are used for breeding and are the rootstock for many older orchards in the U.S. At present, seedling rootstocks are used in limited parts of the world. The principal source of commercial seeds has been the pomace from processed apples. For spring sowing, seeds require stratification for 60 to 90 days at 2 to 7° C (35 to 45° F) to germinate. Some nurseries fall-plant seeds so that they receive the natural winter chilling. Soil should be raked over in the spring to avoid seedlings that become crooked in breaking through the crust. To obtain a branched root system, the seedlings may be undercut while small to prevent the development of a taproot. However, a straight root may be preferred for bench grafting. Seedlings that do not grow to a satisfactory size in one year should be eliminated.

Layering. The principal method of propagating the widely used clonal rootstocks is by mound layering which is described in Chapter 15. Large nursery areas in the Pacific Northwest, the eastern U.S., and in northern Europe and England are devoted to stool beds.

Cuttings. Propagation of cultivars by hardwood cuttings is not successful, but certain clonal rootstocks can be propagated in this manner. For this, special methods described in Chapter 11 are used. Softwood cuttings can be rooted under mist, but this method is not used commercially.

Grafting. Root grafting either by the whip graft or by machine has been the traditional method of propagation (see Chapter 13). Seedling or layered liners are obtained from rootstock nurseries during the dormant season and stored at low temperatures until it is time to graft. Grafting is done during the dormant season and the plants are lined-out in the nursery in the spring as the outside temperatures become warm.

T-budding is a traditional method of propagation, either in the fall or in the spring. Dormant liners are obtained from rootstock nurseries and planted in late winter. In a mild winter area, such as California, rootstocks are budded (T- or chip) in April or early May with nursery trees produced by the end of the same season (*20*). In other areas, the rootstock may grow for the remainder of that season and are fall-budded to grow an additional year. In recent years, fall chip budding has become widely adopted in different parts of the world.

Micropropagation. Methods to micropropagate apple scions and rootstock have been developed (*116, 128, 199, 239*). Nevertheless, the commercial application is quite limited. Part of the reason is that micropropagated plants are more expensive than conventionally propagated ones. Another reason is that plants on their own roots are more vigorous and larger in size than those desired in commercial orchards. The possibility of rooting spur-type and dwarf genotypes is in the future (*121*). Clonal rootstock propagation is more feasible, and some production occurs, particularly in Europe. Enhanced rootability of source material produced by micropropagated plants has a possibility of application to nursery production of apple rootstocks.

Rootstocks (64). All apple rootstocks, either seedling or clonal, are in the genus *Malus,* although the apple will grow for a time and even come into bearing while grafted on pear *(Pyrus communis)* rootstock. At one time, apple seedling rootstocks were widely used in the western and southeastern United States but the trend has been almost entirely towards using clonal-size-controlling rootstocks in the midwestern, northeastern, south-central U.S. and in Europe.

Seedling Rootstocks

Apples on seedling roots produce large-sized trees. Seedlings of 'Delicious,' 'Golden Delicious,' 'McIntosh,' 'Winesap,' 'Yellow Newtown,' and 'Rome Beauty' (but particularly 'Delicious') have been the most successful, being uniform with no incompatibility problems. In purchasing seed it is sometimes difficult to determine the seed source. In the colder portions of the United States—the Dakotas and Minnesota—the hardy Siberian crab apple *(Malus baccata)* and seedlings of such cultivars as 'Antonovka,' containing some *M. baccata* parentage, are favored. In Poland, 'Antonovka' seedlings are the chief apple rootstock. Some nurseries in British Columbia, Canada, have used 'McIntosh' seedlings because of their winter-hardiness, upright growth in the nursery, and early fall shedding of leaves. They tend to be somewhat susceptible to *hairy root,* a form of *crown gall,* and to *powdery mildew,* and the resulting trees tend to be variable in size and performance.

Apples with the triploid number of chromosomes, such as 'Gravenstein,' 'Baldwin,' 'Stayman,' 'Winesap,' 'Arkansas,' 'Rhode Island Greening,' 'Bramley's Seedling,' 'Jonagold,' 'Mutsu,' and 'Tompkins King,' produce seeds that are of low viability and are not recommended as a seed source. Seeds of 'Wealthy,' 'Jonathan,' or 'Hibernal' have given unsatisfactory results.

Apple roots are resistant to root-knot and root-lesion nematodes, moderately susceptible to oak root fungus, and highly resistant to verticillium wilt.

Clonal Rootstocks

Numerous clonal, asexually propagated apple rootstocks have been developed. Most are in the species *Malus* × *domestica.* In various apple-producing countries, rootstock breeding and testing programs are underway to develop new dwarfing and semi-dwarfing rootstocks to replace them.

Malling and Malling-Merton Series. Beginning in 1912, the East Malling Research Station in England began a pioneer program of selecting and classifying a series of vegetatively propagated size-controlling apple rootstocks. Size of the grafted cultivar ranged from very dwarfed to very invigorated (*236*), although size control was also a function of the scion cultivar. Fruit size was large, especially on young dwarfed trees, often larger than on standard-sized trees. The Malling rootstocks are compatible with most apple cultivars and trees grafted to these rootstocks have been planted in varying amounts all over the world. In 1928, a second phase of this program was instigated jointly by (*179*) the John Innes Horticultural Institution and the East Malling Research Station. A new series of apple rootstocks were produced to add resistance to woolly aphids, and to provide a further range in tree vigor. These were referred to as Malling-Merton (or MM).

Other improved cultivar characteristics associated with these rootstocks include high yield, precocity of flowering (with some rootstocks), well-anchored trees (with some rootstocks), freedom from suckering, and good propagation qualities. All of the Malling and Malling-Merton stocks are readily propagated, mostly by stool-bed layering. A number of rootstock clones, notably 'Malling 26,' 'MM 106,' and 'MM 111,' propagate well by hardwood cuttings (*96*). Virus-tested material of both of these groups of rootstocks has been developed by a further joint effort by East Malling and Long Ashton Research Stations in England. "Clean" material is distributed under the designations EMLA 7, EMLA 9, EMLA 26, EMLA 27, EMLA 106, and EMLA 111. In many cases, these "clean" rootstocks produce trees with 10 to 15 percent or more increased vigor than the older virus-infected stocks of the same clone.

A summary of the characteristics of the most useful of these two groups of clonal rootstocks follows, with the rootstocks grouped according to their effect on the vigor of the scion cultivar (*64, 167*). However, the particular scion cultivar used has a definite influence on the size of the composite tree. The most popular of these rootstocks in North America are 'Malling 7, 9, 26' and 'Malling-Merton 106 and III' (See Figure 12–3).

Dwarfing Clonal Rootstocks

'Malling 27.' This is the most dwarfing of all the Malling stocks, producing trees 1 ½ to 2 m (4 to 6 ft) tall, one-half to two-thirds the size of those on 'Malling 9,' making it useful for high-density plantings. Virus-tested propagating material was released by the East Malling Research Station in 1970. It can also be used as an interstock to give a dwarfing effect.

'Malling 9' ('Jaune de Metz'). This rootstock originated as a chance seedling in France in 1879 and has been used widely in Europe for many years as an apple rootstock. It is a dwarfed tree itself and a valuable dwarfing rootstock much in demand for producing small trees for the home garden or for

commercial high-density plantings. Recommended tree spacing is 1.8 × 3.6 m (6 × 12 ft). Such trees are seldom over 3 m (9 ft) tall when mature, usually starting to bear in the first year or two after planting. It is the most important rootstock in Europe and has been planted extensively in the U.S. except for California. There are a number of clonal selections of 'Malling 9,' all dwarfing but somewhat different otherwise.

'Malling 9' has numerous thick, fleshy, brittle roots and requires a fertile soil; the trees require staking or trellising for support. It is resistant to collar rot *(Phytophthora cactorum)* but susceptible to mildew, crown gall, fire blight, and woolly apple aphid. Roots are sensitive to low winter temperatures. It is propagated by stooling.

'Malling 9' can be used as an *interstock* in double-working trees, but the dwarfing of the scion cultivar is less than when it is used as the rootstock (*212*).

'Malling 26.' This rootstock was introduced in 1959 at the East Malling Research Station, originating from a cross between 'Malling 16' and 'Malling 9.' Better anchored than 'M 9', it produces a tree somewhat larger and sturdier than 'Malling 9' but less so than 'Malling 7' or 'MM 106.' It still requires staking. Suggested tree spacing is 3 × 4.2 m (10 × 14 ft). It can be propagated by softwood cuttings under mist or by hardwood cuttings (*96*), but produces poorly in stool beds. Although quite winter-hardy, it does not tolerate heavy or poorly drained soils, and is especially susceptible to fire blight but less to collar rot.

Semi-dwarfing Clonal Rootstocks

'Malling 7a.' Malling 7 was originally selected at East Malling from a group of French traditional rootstocks known as Doucin. This stock produces trees somewhat larger than those on 'Malling 26' roots. The 'Malling 7a' designation indicates a clonal selection free of certain viruses present in the original 'Malling 7,' the viruses having been removed by heat therapy. A virus-indexed EMLA clone was introduced in 1974. 'Malling 7a' has a stronger, deeper, root system than 'Malling 9' and produces an early-bearing, semi-dwarf tree. It is tolerant of excessive soil moisture, but susceptible to crown gall. It is very well anchored, requiring staking for the first few years. Suggested tree spacing is 4.2 × 4.8 m (14 × 16 ft). This rootstock has the undesirable characteristic of suckering badly, and the trees are not very winter-hardy. 'Malling 7a' is easily propagated by stooling or by leafy cuttings under mist.

'Malling-Merton 106.' This clone originated as a cross between 'Northern Spy' and 'Malling 1,' producing trees two-thirds to three-fourths the size of trees on seedling rootstocks. Although once popular, its planting has decreased because of crown rot problems. On good soils, with some scion cultivars, it can produce a large tree. The roots are well anchored and do not sucker. Suggested planting distances are 4.2 × 5.4 m (14 × 18 ft). In some areas 'MM 106' is susceptible to mildew and highly susceptible to collar rot, which may be its chief weakness. It has not been affected by fire blight. It grows well in the nursery but drops its leaves later in the fall than most other understocks and is not resistant to early fall freezes. Hardwood and softwood cuttings root easily and stool beds are quite productive.

Vigorous Clonal Rootstocks

'Malling-Merton 111.' This stock originated as a cross between 'Northern Spy' and 'Merton 793' ('Northern Spy' × 'Malling 2'). A virus-indexed EMLA clone was introduced in 1969. Grafted trees are about 75 to 80 percent the size of seedlings in most orchards where they are grown. It produces more precocious bearing than those trees on seedlings but less than those on 'MM 106,' 'M 7,' and 'M 26.' It does well on a wide range of soil types. It is susceptible to mildew but not to collar rot or woolly aphid. Stool beds are highly productive with heavy root systems developing. Hardwood cuttings root well with proper treatment (*96*) and softwood cuttings under mist root well. 'MM 111' is more winter-hardy than 'Mailing 7' or 'MM 106.' Suggested planting distances are 4.8 × 6 m (16 × 20 ft). It shows excessive vigor in some situations.

Clonal Rootstocks from Miscellaneous Sources

'Alnarp 2.' This rootstock developed at the Alnarp Fruit Tree Station in Sweden is widely used there for its winter-hardy properties. The roots are well anchored and give about 20 percent size reduction as compared to trees on seedling roots. It is susceptible to woolly apple aphid and to fire blight.

'Robusta No. 5' [M. robusta (M. baccata × M. prunifolia)]. This vigorous, very hardy clonal apple rootstock, propagated by stooling or stem cuttings, originated in 1928 at the Central Experimental Farm, Ottawa, Canada. It is apparently resistant to fire blight and crown rot and seems to be compatible with most apple cultivars. It is extensively used as an apple rootstock in eastern Ontario and Quebec, as well as in the New England states. It is the best rootstock for use where extreme winter-hardiness is required. However, its low chilling requirement produces the unfavorable habit of starting growth too early in the spring following three or four warm days in late winter. It is not a dwarfing rootstock.

'Mark.' This rootstock was developed at Michigan State University as an open-pollinated

seedling of 'Malling 9.' Trees on this stock are similar in size to those on 'Malling 9' or slightly larger and about the same in productivity. This rootstock does not sucker, is well anchored but requires staking. It propagates easily in stool beds. Trees on this stock have an open, spreading growth habit. Some trees have developed proliferation just below the soil line but the cause is not known.

Polish Series. These rootstocks were the result of a breeding program started in Poland in 1954. Six clones (P-1, P-2, P-14, P-16, P-18, and P-22) are available (*64, 167*) with a range of tree size potential and other characteristics. As a group, they were selected for hardiness.

Budagovsky Series. This material originated from a breeding program in Russia to produce very hardy apple rootstocks (*64, 167*). 'Budagovski 9' is very dwarfing and is recommended as an interstock in cold regions. 'Budagovski 118' (75 percent size), 'Budagovski 490' (65 percent), and 'Budagovski 491' (20 percent) are other promising rootstocks of the series.

Mandarin. *See Citrus.*

Mangifera indica. Mango (*30*). Most plantings of this tropical evergreen fruit tree are seedlings which may be sexual (monoembryonic) or nucellar (polyembryonic) in origin—both conditions may occur simultaneously in the seed. Growth of several shoots from one seed does not necessarily indicate nucellar embryos, since certain cultivars develop shoots from below ground, arising in the axils of the cotyledons of one embryo, which may or may not be of zygotic origin (*7*). Polyembryonic cultivars commonly occur in mango. Monoembryonic cultivars should not be propagated by seed, as they do not come true.

Seed. Mango seeds are used for rootstocks although seedlings of polyembryonic cultivars are apomictic. Seedlings have long juvenile periods and are overly vigorous. Seeds should be planted within a week of maturation. Storage is possible within the fruits or in polyethylene bags at about 21° C (70° F) for at least two months. Low-temperature (below 10° C; 50° F) storage and excessive drying should be avoided. By removing the tough endocarp that surrounds the seed and planting in a sterilized medium, good germination should occur within two to three weeks. Seedlings should be transplanted to pots or into nursery rows soon after they start to grow.

Cuttings. Cuttings are difficult but some success is reported from India using etiolation, IBA treatment, mist, and bottom heat (*144*).

Grafting. Mangos are commonly propagated in Florida by veneer grafting or by chip budding. A week after budding, the top of the rootstock is removed two to three nodes above the bud—with final removal back to the bud when the bud shoot is 7.5 to 10 cm (3 to 4 in.) long. The best budwood is prepared from hardened terminal growth 6 to 10 mm (¼ to ⅜ in.) in diameter. The leaves are removed, with the exception of two or three terminal ones. The buds swell in two to three weeks, and are then ready to use. If the buds are to be used on stocks older than three weeks, ringing the base of the shoots from which the buds are to be taken about ten days before they are used increases their carbohydrate supply and seems to promote graft union formation.

Budding is best done when the rootstock seedlings are two to three weeks old—in the succulent red stage. Four to six weeks after budding, the inserted bud should start growth (*138*). T-budding also has been successful. Patch, shield, and Forkert budding is done in India (*144*).

Approach grafting, termed *inarching* in India, has been used since ancient times in propagating the mango. Veneer grafting also is successful (*159*), as is saddle grafting (*209*). Veneer grafting is recommended in India (*144*) either for nursery propagation or topworking. The scion should be a terminal nonflowering shoot of 3 to 4 months' maturity. The scions are defoliated seven to ten days before they are removed for grafting, keeping a part of the petioles attached which helps the buds to swell. This is done from March to September in northern India. In India, assorted rootstock seedlings are mostly used. Monoembryonic rootstocks can be multiplied by air layering or by cuttings and subsequently clonally propagated by stooling (*158*).

A method called epicotyl/stone grafting has been described in India (*144*). Seeds are germinated in a sand bed covered with leaf mold. Eight to 15-day-old germinated seedlings are lifted, decapitated about 5 cm above the stone, and wedge-shaped scions are inserted into a vertical split. Buds are wrapped with poly tape and grafted plantlets are planted in poly bags or outdoors in a trench. They are then shifted to the field.

Layering. Air layering is successful, especially when etiolated shoots are treated with IBA at 10,000 ppm; such treatments have given high rooting and survival (*160*). Pot-layering and stooling are also suggested in India (*144*).

Mango. *See Mangifera.*

Mangosteen. *See Garcinia.*

Morus spp. Mulberry. Mulberry cultivars grown for their fruit are mostly *M. alba* or *M. nigra,* but some of *M. rubra* occur. Seeds are easily grown but produce weedy trees.

Cutting propagation is readily started by hardwood cuttings 20 to 30 cm (8 to 12 in.) long of wood and planted in early spring. Leafy softwood or semi-hardwood cuttings treated with 8,000 ppm root easily under mist (*52*).

Mulberry can be commercially micropropagated (*112*).

Mulberry. *See Morus.*

Musa spp. Banana. The banana "tree" is a large herbaceous perennial. The "stem" consists of compressed, curved leaf stalk bases arranged spirally in strips. The bases of the leaf stalks are attached to the true stem which is a rhizome (see Chapter 15). New "suckers" grow from buds on the corm and soon develop their own roots and a base as large as the parent plant, which dies and deteriorates shortly after the fruit bunch is harvested. A banana plant may live for a considerable time but it is really a succession of new plants, each arising as a sucker from a rhizomatous bud; any given sucker fruits only once, then dies.

Asexual Methods. Since the edible types rarely produce seeds, commercial propagation of the banana is asexual, consisting of division of the rhizome and replanting of the pieces or the suckers. A large rhizome is cut into pieces, which are termed "heads," weighing 3 to 4.5 kg (7 to 10 lb), depending on the cultivar. Each head should contain at least two buds capable of growing into suckers. Each sucker produces two branches in the first crop. Fairly large "sword" suckers 90 to 180 cm (3 to 6 ft) high with well-developed roots also are used, but the leaves must be shortened considerably to reduce water loss after the sucker is cut from the parent plant. These suckers are removed with a sharp cutting tool inserted vertically about halfway between the parent stalk and stem of the sucker. These sword suckers produce only one bunch of fruit in the first crop, but they are often preferred, because of the large size of the bunch.

Micropropagation (111). To propagate planting material from sources free of *Fusarium* wilt, micropropagation can be used starting with explants obtained from a decapitated shoot apex from "clean" banana suckers. In one year about 1 million pathogen-free plantlets can be produced for commercial plantings. Similarly, mosaic-free plants were obtained after meristem tip culture and thermotherapy (*15*).

Nectarine. *See Prunus persica.*

Nephelium lappaceum. Hairy Litchi. A tropical evergreen fruit tree similar to litchi and longan (*15*). Seedlings are variable and have a long juvenile period. Cultivars are grafted to seedling rootstocks. Seeds are sown immediately upon ripening because drying causes loss of viability. Seeds are removed from the fruit, washed, and sown horizontally. Two leaved seedlings are produced within two weeks. The greatest success is with patch budding or Forkert budding onto one- to two-year-old rootstocks of the same cultivar. Inarching is successful, as is air layering.

Olea europaea. Olive. (*63, 98*). Many cultivars of this ancient crop exist. They are mostly propagated by budding or grafting onto seedling or clonal rootstocks, hardwood or semi-hardwood cuttings, or "suckers" from old trees.

Seed. Seeds are used primarily for rootstocks. Germination is somewhat difficult due to the hard stony endocarp ("pit") that surrounds the seed. Seeds of small-fruited cultivars germinate more easily than those of large-fruited ones. Since germination is sometimes prolonged over one or two years, a common practice is to plant more seeds than will be needed as seedlings to offset the low germination percentage. Clipping the end of the endocarp will hasten germination. Softening the endocarp with sulfuric acid or sodium hydroxide, then warm-stratifying at 15° C (59° F) will improve germination (*43*) as will stratifying the seeds (still in the endocarp) at 10° C (50° F), removing the endocarp and planting at 20° C (68° F) (*217*). Removing the seed from the pit, removing the seed coats and germinating at 20° C in petri dishes on moist filter paper will give high germination if the fruits are collected shortly after pit hardening.

Cuttings. Hardwood cuttings may be made from two- to three-year-old wood about 2.5 cm (1 in.) in diameter and 20 to 30 cm (8 to 12 in.) long. All leaves are removed. IBA aids rooting. An older technique is to soak the basal ends of the cuttings in 15 ppm IBA for 24 hours, followed by storage in moist sawdust at 15 to 21° C (60 to 70° F) for a month preceding spring planting in the nursery (*91*).

Semi-hardwood cuttings are made from vigorous, one-year-old wood about 6 mm (¼ in.) in diameter. Cuttings are taken in early summer or midsummer, and trimmed to 10 to 15 cm (4 to 6 in.) long with two to six leaves retained on the upper portion of the cutting. Cuttings are treated with a 4,000 ppm IBA quick-dip and rooted under mist. Cultivars vary greatly in case of rooting (*97*).

Grafting. Seedlings tend to grow slowly and may take a year or two to become large enough to be grafted or budded. Successful methods include T-budding, patch budding, whip grafting, and side-tongue grafting. A widely used method in Italy is bark grafting small seedlings in the nursery row in the spring. The stocks are cut off several centimeters above ground, and one small scion is inserted in each seedling, followed by tying and waxing. After grafting or budding, one or two more years are required to produce a tree large enough for transplanting to the orchard.

Rootstocks. *Olea europaea* seedlings are commonly used although they vary considerably in tree vigor and size (*92*). Using rooted cuttings of a strong-growing cultivar such as 'Mission' may be more desirable because of uniformity and vigor. Because of susceptibility to *Verticillium* wilt, use of resistant clonal rootstocks such as 'Oblonga' or 'Swan Hill' (*99*) may be better. Other *Olea* species do not make satisfactory rootstocks for the edible olive.

Olive. *See Olea.*

Orange (Mandarin, Sour, Sweet). *See Citrus.*

Papaya. *See Carica.*

Passiflora edulis. Passion fruit (*54*). The genus has about 60 species that produce edible fruit grown in tropical and subtropical areas. In Hawaii, Brazil, Sri Lanka, Kenya, and New Zealand, production is based on seedlings of the yellow or purple passion fruit. In Australia and South Africa, hybrid cultivars are grafted on specific rootstocks.

Seed. Must be extracted from freshly harvested fruit, and removed from fruit and juice after fermentation for about three days. Seeds are washed, dried, and kept in sealed containers in a refrigerator where they will remain viable for at least three months. Germination takes place in two to three weeks after planting in pots whereupon they are transplanted in the two- to three-leaf stage.

Grafting. Cleft-grafting is done when seedlings are 30 to 40 cm high. Rootstock is cut off about 15 to 20 cm above ground and split. The scion is a small piece of vine wood about 8 to 10 cm long with larger leaves removed. The wedge at the scion base is inserted into the rootstock cleft and the union bound with plastic budding tape. Callusing takes four to six weeks, at which time plants are transplanted to the field. Scions of purple-fruited hybrid cultivars are grafted to seedling rootstocks of golden passion fruit (*P. edulis* forma *flavicarpa*), which are resistant to *Fusarium* and to nematodes (*207*).

Passion fruit. *See Passiflora.*

Pawpaw. *See Carica.*

Peach. *See Prunus persica.*

Pear. *See Pyrus.*

Pecan. *See Carya.*

Persea americana. Avocado (*17, 173*). This tender tropical tree is susceptible to *Phytopthora* and must be propagated onto resistant clonal or seedling rootstocks.

Seed. Seeds are used for the production of seedling rootstocks. To avoid sun-blotch viroid, the seeds (and budwood) must be taken from source trees that have been registered by state certifying agencies as free of the disease. Seeds are generally planted shortly after removal from fruit taken from the tree (not picked from the ground), care being taken not to allow them to dry out. Seeds can be stored six to eight months if packed in dry peat moss and held at 5° C (41° F) with 90 percent relative humidity.

To eliminate infection from avocado root rot *(Phytophthora cinnamomi),* seeds are dipped in hot water at 49 to 52° C (120 to 125° F) for 30 minutes before planting (*237*). Germination of the seeds is hastened by removing the brown seed coats and cutting a thin slice from the apical and basal end of each seed before planting. The seed coats can be removed by wetting the seeds and allowing them to dry in the sun. Seeds are then sown in nursery rows or for container production.

Cuttings. Semi-hardwood stem cuttings from mature trees of the Mexican race have been rooted under intermittent mist using bottom heat at 25° C (77° F), provided seven or more leaves are retained on a 20 cm (8 in.) cutting. If the leaves drop, rooting ceases (*183*). Cuttings taken from mature trees of the Guatemalan or West Indian race are difficult to root. Some rooting success has been achieved by *etiolating* the basal portion of a leafy shoot while it is still attached to the stock plant—prior to making the cutting. By covering the shoot base with Velcro or black electrical tape, the basal portion grows in complete darkness, whereas the terminal portion develops in the light until three to five leaves have formed. Those shoots with etiolated bases can then be detached and rooted in a propagating case (*69, 70*). See Chapters 10 and 11 for discussions on banding and etiolation systems for stock plants. Auxin does not enhance rooting. Avocados started as rooted cuttings eventually make satisfactory trees, but in general, grow poorly in their initial stages.

Grafting. Nursery trees are commercially propagated in California by T-budding, tip grafting (splice or whip graft) (*206*), or wedge or cleft grafting on avocado seedlings or clonal rootstocks (*18*). T-budding is used occasionally in Florida and some of the Caribbean countries—but the usual nursery practice is to graft mature tip scions, either as side-veneer or cleft grafts on young succulent rootstock seedlings.

Budwood must be selected properly. The best buds are usually near the terminal ends of completed growth cycles with fully matured, leathery leaves. To prevent drying, the leaves should be removed when the budwood is taken.

Nursery Production Systems

Field Production of Nursery Plants. Seeds of the Mexican race, which ripen their fruit in the fall, may be planted in beds in late fall or early winter. They should be placed with the large, basal end down, just deep enough to cover the tips. If grown in a warm area, the sprouted seeds are ready to line-out in the nursery row the following spring. By summer or fall, the seedlings are usually large enough to permit T-budding; if not, they can be budded the following spring.

Four to six weeks after budding, the seedling rootstock is cut off 20 to 25 cm (8 to 10 in.) above the bud, or bent over a few inches above the bud. The remaining portion of the seedling above the bud is not cut off until the bud shoots have completed a cycle of growth. The new shoots are usually staked and tied. In digging from the nursery, the trees are "balled and burlapped" for removal to their permanent location following the first or second growth cycle or just as the first flush starts.

Container Production of Nursery Plants. Seeds are planted in polyethylene bags and the resulting seedlings are cleft or wedge grafted about 9 cm above the seed two to four weeks after germination. Scions are taken from freshly cut terminal shoots showing strong plump buds. The plastic containers, with the grafts, are placed on raised benches in plastic-covered houses, which facilitates sanitation procedures to avoid *P. cinnamomi*. After four to six weeks, the grafts are moved to a 50 percent shade house to acclimatize, then transplanted into large containers for moving to an outdoor area (*172*).

In Florida, young, succulent West Indian seedlings (*138*), grown in gallon containers, are grafted when they are 15 to 25 cm (6 to 10 in.) high and 6 to 10 mm (¼ to ⅜ in.) in diameter. The scions are shoot terminals, 5 to 7.5 cm (2 to 3 in.) long, with a plump terminal bud, taken just as it resumes growth.

Rootstocks for Avocado

Mexican Race (P. americana var. drymifolia). Seedlings of this race are preferred in California for their cold-hardiness and their partial resistance to *Phytophthora cinnamomi*, lime-induced chlorosis, and *Dothiorella* and *Verticillium*. They are, however, susceptible to injury from high salinity. In Florida, where seedlings of large diameter are preferred for grafting, the Mexican types are little used, owing to their thin shoots.

Guatemalan Race (P. americana). These are occasionally used in California when there is a scarcity of Mexican seeds. Guatemalan seedlings are often more vigorous initially than Mexican seedlings but are more susceptible to diseases and to injury from cold.

West Indian Race (P. americana). These seedlings are susceptible to frost injury in California but are widely used in Florida. The large seed produces a pencil-size shoot suitable for side grafting in two to four weeks after germination.

Persimmon (American, Japanese). *See Diospyros.*

Phoenix dactylifera. Date palm. (*163*). This monocotyledonous plant is propagated by seeds, offshoots, and more recently by micropropagation (somatic embryogenesis). Date palms are dioecious. In a commercial planting, most of the trees are female, but a few male trees are necessary for cross-pollination.

Seed. Date seeds germinate readily. However, seedling populations have a long juvenile period and produce about half of the trees' males. Seedling female trees produce fruits of variable and generally inferior types.

Offshoots. Superior clonal cultivars exist, some since ancient times. The natural method of vegetative propagation is from offshoots which arise from axillary buds near the juvenile base of the tree. Roots develop on the base after three to five years and are ready to be removed. To promote rooting, soil should be added to the base of the offshoot.

Large, well-rooted offshoots, weighing 18 to 45 kg (40 to 100 lb), are more likely to grow than smaller ones. Offshoots higher on the trunk can be induced to root by placing moist rooting medium against the base of the offshoot in a box or polyethylene tube. Unrooted offshoots arising even higher on the stem can be cut off and rooted in the nursery, but percentages may be reduced. A single date palm may yield 10 to 25 offshoots during its first 10 to 15 years.

Considerable skill is required to cut off the date offshoot properly. Soil is dug away from the rooted offshoot, but a ball of moist earth as thick as possible should remain attached to the roots. The connection with the parent tree should be exposed on each side by removing loose fiber and old leaf bases. A special chisel, with a blade flat on one side and beveled on the other, is used to sever the offshoot. The first cut is made to the side of the base of the offshoot close to the main trunk. The beveled side of the chisel is toward the parent tree, which gives a smooth cut on the offshoot. A single cut may be sufficient, but usually one or more cuts from each side are necessary to remove the offshoot. The offshoot should never be pried loose; it should be cut off cleanly. After removal, it should be handled carefully, replanted as soon as possible, and roots prevented from drying out (see Figure 15–15).

Micropropagation. Commercial micropropagation has become an important alternative because of its greater efficiency and speed (*166, 210*). Two general methods can be used. One is production of somatic embryos from callus generated from shoot tips. This technique produces a larger number of plants but has greater danger of variability. The second method is rooting of axillary shoots produced from the shoot apex of offshoots. The general procedure is described in Chapters 17 and 18.

Pineapple. *See Ananas.*

Pistacia vera. *Pistachio* (*82, 120*).

Seed. Seeds are used in commercial production for rootstocks. Obtain fruits when the hulls turn blue-green. Pinkish seed is blank. The hulls contain a germination inhibitor and should be removed. *P. terebinthus* seed should be held in moist sand at 5 to 100° C (41 to 50° F) for six weeks before sowing, then germinated at 20° C (68° F). With early planting of seed, the more vigorous seedlings may reach budding size by fall.

Grafting. Buds of *P. vera* cultivars are quite large, requiring a fairly large seedling to accommodate them. Budding is possible over a considerable period of time, but if done before mid-spring, when sap flow may be excessive, the percentage of takes is low. A marked improvement in bud union occurs as budding is extended through the summer and fall. Because of their long taproot, best results are usually obtained by planting seedling rootstocks in their permanent orchard location, then T-budding them 0.6 m (2 ft) or more above ground after the seedlings are well established. Seedlings can be grown in long tubular, biodegradable pots ready for planting in the orchard.

Rootstocks. *P. atlantica* and *P. terebinthus* seedlings were used in earlier commercial orchards in California but have been highly susceptible to verticillium wilt. *P. integerrima* shows considerable resistance to this soil-borne disease and is being used in areas having verticillium wilt problems.

The F_1 hybrid seedlings between *P. atlantica* and *P. integerrima* have superior cold resistance as compared to *P. integerrima* and also show verticillium resistance and considerable vigor. These have been produced by hand pollination under the name of "UCB-1".

P. terebinthus seedling rootstocks may have some resistance to *Armillaria,* are cold tolerant but grow slowly.

Micropropagation. Cultivars of *P. vera* have been micropropagated (*9, 169*).

Pistachio. *See Pistacia.*

Plum (European). *See Prunus domestica.*

Plum (Japanese). *See Prunus salicina.*

Pomegranate. *See Punica.*

Prunus spp. Stone-fruit and nut-producing trees include: sweet cherry *(P. avium),* sour cherry *(P. cerasus),* apricot *(P. armeniaca),* almond *(P. dulcis),* peach and nectarine *(P. persica),* European plum *(P. domestica),* and Japanese plum *(P. salicina).* In addition, several hundred related species and species hybrids exist worldwide.

Seed. Seed propagation is important in breeding and in rootstock production. *Prunus* seeds have a hard protective endocarp (shell, pit, stone) that surrounds the seed (see p. 128). The embryo is nonendospermic (Figure 5-1) and exhibits embryo dormancy that requires one to three months' cold stratification to germinate (see p. 201). Seeds of different species and sometimes individual cultivars have specific time requirements. In addition, the stone may be inhibiting either because of substances or because of hardness and/or lack of water uptake (cherry). Presoaking, stone removal, leaching in running water, or warm stratification preceding cold stratification may be helpful.

Cuttings. Propagating *Prunus* cultivars or rootstocks varies greatly by species and individual cultivar. Producing own rooted peach trees of specific cultivars by hardwood cuttings has been used to produce high density orchards (*41*). See cutting propagation methods for ornamental *Prunus* species in Chapter 20. For the most part, propagation by cuttings is more successful in producing clonal rootstocks which have been selected specifically for ease of rooting.

Direct rooting of hardwood cuttings in the nursery is a preferred method and variations of this method are described in Chapter 11. Clones of many species can be rooted by softwood cuttings under mist, but the method is not usually commercially practiced. Hard-to-root rootstock clones have been rooted commercially by trench layering ('F-12' cherry) or mound layering ('Colt').

Grafting. The most important method of propagation of stone fruits is to bud or graft cultivars to rootstocks in the nursery and transplant to the orchard. Seedling rootstocks are prepared either by stratifying the seeds during the winter and planting early in the spring, or fall planting and allowing natural chilling in the ground to bring about the moist-chilling process. Seedling liners which have been produced during the previous year may be purchased and lined-out in the nursery row for later budding.

T-budding is the usual practice and may be done in the fall, early spring or late spring, or summer depending on the length of the growing season. Chip budding may also be utilized in areas with a relatively short growing season. Trees are produced within one, two, or three years depending upon the nursery site and conditions, the species, and the desires of the propagator.

Micropropagation (55, 65, 113, 181). Much research and development has been devoted to micropropagation of *Prunus* cultivars and rootstocks. Successful methods have been achieved with plum, cherry, peach, and peach × almond hybrids. European micropropagation labs have produced millions of commercial plants of cultivars such as 'GF677' peach × almond. Production has not been economical in the U.S., and tissue-culture-based nursery systems have not replaced conventional propagation except for early introductions and disease control.

Embryo rescue and ovule culture are effective tools in breeding programs which make possible the propagation of early-ripening cultivars or interspecific hybrids whose embryos abort when the fruits are ripe. Meristem culture is used for removing viruses from infected plants in initiating foundation clones.

Prunus avium. Sweet Cherry.

Prunus cerasus. Sour cherry. Cultivars of both species are propagated by T-budding or chip budding on either seedling or clonal rootstocks.

Nursery Production. Because of sensitivity to virus and MLO problems, only virus-tested sources should be used for scions, clonal rootstocks, and seed orchards (see Chapter 9).

Rootstock seedlings can be grown in a closely planted seed bed or nursery row for one year, dug, and then lined-out in the nursery row about 10 cm (4 in.) apart and grown a second season before budding. Under favorable growing conditions, stratified seeds may be planted directly in the nursery in early spring, and the seedlings will be large enough for budding by late summer or early fall.

Rootstocks for Cherry (171). Historically, the two most common stocks for sweet cherry and sour cherry cultivars are Mazzard *(Prunus avium)* and Mahaleb *(P. mahaleb)* seedlings. Recently a number of clonal rootstocks became available. Both rootstocks are susceptible to verticillium wilt. *P. avium* is moderately resistant to oak root fungus, but *P. mahaleb* is susceptible.

Mazzard (P. avium). There is considerable variation among the sources of Mazzard seeds, some undoubtedly better than others. Mazzard seedlings used in the United States are available from sources in Oregon and Washington. In germinating Mazzard seeds it is beneficial to soak them in water that is changed daily for about eight days prior to stratification. A warm, moist stratification period at 21° C (70° F) for four to six weeks prior to cold stratification should improve germination. Finally, a cold (4° C; 40° F) stratification period of 150 days is used. When a percentage of the seeds shows cracking of the endocarp with root tips emerging, they should be removed and planted (*130*).

Sweet cherry cultivars make an excellent graft union with Mazzard roots, giving trees that are vigorous and long-lived, but may be so large that harvest is uneconomical. Mazzard roots are not particularly suitable for heavy, poorly aerated, wet soils, but will tolerate such conditions better than Mahaleb. Mazzard seedlings have a shallow, horizontal root system.

Mazzard roots are immune to one species of root-knot nematode *(Meloidogyne incognita)* and resistant to *M. javanica,*but susceptible to the root-lesion nematode *(Praylenchus vulnus).*

Mahaleb (P. mahaleb). This species is heterogeneous with considerable variability in many characteristics. The seeds should be soaked in water for 24 hours, then stratified for about 100 days at 4° C (40° F). Leafy Mahaleb cuttings are easily rooted under mist if treated with IBA, thus permitting the establishment of clonal Mahaleb stocks. Mahaleb is the principal rootstock in the United States for 'Montmorency,' the leading sour cherry cultivar. Mahaleb rootstocks may produce a somewhat dwarfed tree, particularly if budded high, e.g., 38 to 50 cm (15 to 20 in.) on the rootstock. In good soils the trees may become as large as those on Mazzard roots. Mahaleb roots are resistant to the buckskin virus such that it has been used to produce a high-worked multitrunked tree with sweet cherry scions. If any branch becomes infected, it can be regrafted below the union without losing the whole tree.

Mahaleb-rooted trees do not grow satisfactorily in heavy, wet soils with high water tables, being susceptible to *Phytophthora* root rot. Under dry, unirrigated, drought conditions, Mahaleb is more likely to survive than Mazzard, presumably because of the deep, vertical rooting habit of Mahaleb. Trees on Mahaleb roots are more cold-hardy than those on Mazzard. Sweet cherries often grow faster for the first few years on Mahaleb than on Mazzard roots, and some cultivars start heavy bearing rather early, which may result in some dwarfing of the tree. There is evidence, especially in England, that trees on Mahaleb roots are relatively short-lived, but in the United States, good productive trees known to be more than 50 years old are grown on this rootstock. Mahaleb roots are more resistant to root-lesion nematode *(Praylenchus vulnus)* than Mazzard. They are also resistant to one species of the root-knot nematode *(Meloidogyne incognita)* but susceptible to *M. javanica.* Mahaleb roots are more resistant to bacterial canker than Mazzard roots.

Clonal rootstocks

'Malling F12/1.' This clone of *P. avium* was selected because of its high resistance to bacterial canker. It was developed at the East Malling Research Station, England. Propagation is primarily by trench layering (see Chapter 15) but it has been produced by micropropagation. It is widely used in the United Kingdom, Australia, New Zealand, and South Africa. In Oregon, this stock has worked well for the root, trunk, and primary scaffold system onto which the scion cultivar is topworked.

'Colt' (Prunus avium × P. pseudocerasus). This was released as a dwarfing clonal cherry rootstock in 1958 at the East Malling Research Station. A virus-tested strain, EMLA Colt, was released in 1977. This rootstock can be used for both sweet and sour cherries and is easily propagated by hardwood cuttings and has also been commercially produced by mound layering. Tree size is about 60 to 70 percent of trees on F12-1 and 80 percent of trees on Mazzard seedlings. It is reported to produce early heavy cropping, and resistance to bacterial canker, "stem pitting" virus, and replant problems. It is susceptible to crown gall.

M × M Clones (P. avium × p. mahaleb). A series of clones were selected from large seedling populations in Oregon. 'M × M 14' is widely used in Europe.

G.M. Series. Three clonal rootstocks named G.M. 9. *(P. incisa × P. serrula),* G.M. 61/1 *(Prunus dawyckansis),* and G.M. 79 *(P. canescens)* were introduced by Fruit and Vegetable Research Station, Gembloux, Belgium (*167, 171*). These produce tree sizes ranging from about 30 percent to 60 percent of trees on F12-1. They are propagated by softwood cuttings or micropropagation.

Gisela Series. Five clonal rootstocks known by number (148-1, 148-2, 148-8, 148-9, and 195-1) were developed at the Justus Liebeg University, Giessen, Germany (*167, 171*). These are progenies of crosses among various *Prunus* species including *P. canescens, P. fruticosa,* and *P. avium.* Tree size ranges from 40 to 60 percent of trees on Mazzard. Trees on these rootstocks are reported to be precocious, heavy bearing, well anchored, and hardy.

St. Lucie (P. mahaleb). Clonal selection made at INRA, Grande Ferrade, France (*61*). Propagates from softwood and semi-hardwood cuttings. Grown in Europe. Adapts to a wide range of soils, including calcareous, but requires good drainage. Trees are compact and intermediate in vigor.

Prunus armeniaca. Apricot. Cultivars of apricot are propagated commercially by T-budding or chip budding on various seedling *Prunus* rootstocks. Fall budding is the usual practice, but spring and June budding may be used. Bench grafting also has been successful (*49*).

***Rootstocks for Apricot* (45).** Rootstocks include apricot seedlings, peach seedlings, and myrobalan plum *(P. cerasifera)* seedlings and recent clonal hybrids. Seeds of all these species require low-temperature 5° C (41° F) stratification before planting.

Apricot (P. armeniaca). Seedlings of local apricot cultivars are best on good, well-drained soils. Seeds of 'Royal' or 'Blenheim' produce excellent rootstock seedlings in California. In the eastern

United States, seedlings of 'Manchurian,' 'Goldcot,' and 'Curtis' are recommended. In Europe local selections are used.

Apricot root is almost immune to the root-knot nematode (*Meloidogyne* spp.). In addition, it is somewhat resistant to the root-lesion nematode. It is susceptible to crown rot (*Phytophthora* spp.) and intolerant of poor soil-drainage conditions. Apricot roots are not as susceptible to crown gall *(Agrobacterium tumefaciens),* as are peach and plum roots. Apricot seedling roots are susceptible to oak root fungus and highly susceptible to verticillium wilt.

Peach (P. persica). In California, 'Nemaguard,' 'Nemared,' and 'Lovell' peach seedlings are satisfactory as rootstocks for apricot cultivars. See under **Peach** for characteristics. In the eastern United States and Canada, some apricot cultivars show incompatibility on peach seedlings (*25, 129*).

Myrobalan Plum (P. cerasifera). Although successful high-yielding apricot orchards grow on this rootstock, some instances have been observed where the trees have broken off at the graft union in heavy winds, and die-back conditions occur. Nurseries often have trouble starting apricots on myrobalan roots. Some of the trees fail to grow rapidly and upright or else have weak or rough unions. After these weaker ones are culled out, the remaining trees seem to grow satisfactorily. This rootstock is useful for apricot when the trees are to be planted in heavy soils or under excessive soil moisture conditions.

'Marianna 2624.' This clonal rootstock has become an alternative to myrobalan plum seedlings on which apricots do well.

'Citation'. This clonal rootstock is being used in new plantings in California. See page 652.

Prunus dulcis. Almond (*61, 122*). The almond is an ancient crop in the Mediterranean, southwest Asia, and North Africa where it was traditionally grown as seedlings. Commercial almond orchards in the Mediterranean region have been traditionally propagated by T-budding onto almond seedling rootstocks by fall budding, but more recently by budding to 'GF 677' (almond × peach). Older California almond orchards were budded to almond seedlings, but as the industry shifted to irrigated systems, rootstock selection shifted to peach seedlings, to a limited amount to 'Marianna 2624' plum, and more recently to peach × almond hybrids. Almond stem cuttings have given but slight success.

Rootstocks for Almond

Almond (P. dulcis) Seeds of bitter types or certain commercial cultivars are used. In California, 'Texas' ('Mission'), seeds are commonly used. In the Mediterranean, seeds of 'Garrigues,' 'Atocha,' and 'Desmayo Royo' are typically used. Almond seeds require limited stratification of three to four weeks before planting. Almond roots are adapted to deep, well-drained soil. In poorly drained soils, almond roots are often unsatisfactory, owing to their susceptibility to waterlogging, infection by crown rot (*Phytophthora* spp.), and susceptibility to crown gall *(Bacterium tumefasciens).* Their deep-rooting tendency is an advantage in orchards grown on unirrigated soils or where drought conditions occur. Almond seedlings are tolerant of high-lime soils and, of the rootstocks available for almonds, are the least affected by excess boron salts. Almond seedlings are susceptible to root-knot and root-lesion nematodes as well as to oak root fungus *(Armillarea mellea).* Nematode-resistant almond selections 'Alnem 1,' 'Alnem 88,' and 'Alnem 201' have been selected in Israel.

Peach (P. persica). Peach seedlings are used as rootstocks for almond where irrigation is practiced, such as in California. 'Lovell' and 'Halford' are typically used in nonnematode soils but 'Nemaguard' and sometimes 'Nemared' are used in nematode-infested areas or where nematode concern exists. For characteristics of peach rootstocks see under **Peach.**

'Marianna 2624' Plum. This clonal rootstock is used in California where dense soils occur or where oak root fungus is present (*122*). Almond cultivars vary in their relative size and in their degree of incompatibility on this rootstock (see p. 419). In general, trees are about one-third smaller than those on the other available rootstocks. Not all almond cultivars are compatible with this rootstock (see p. 419). (See under **Plum** for characteristics.)

Peach-almond Hybrids (*61, 122*). These first-generation hybrids of peach and almond can be produced as seedling populations by natural crossing between adjoining trees representative of the two species. Combinations of 'Titan' almond and 'Nemaguard' have been used. Seedling hybrid plants are identified in the nursery row by their high vigor, intermediate appearance between the parents and red foliage color in certain combinations using a redleaf red parent (see p. 118).

Clonal hybrids have been developed. 'GF 677,' discovered as a chance seedling in France, is resistant to alkaline soils and is widely used in the calcareous soils of the Mediterranean region for both almonds and peaches. Commercial propagation by micropropagation is widely practiced in that area.

'Hansen 536' and 'Hansen 2168' are hybrid rootstock clones released by the University of California in 1983. Rooting is possible by hardwood cuttings treated with IBA plus a fungicide, planted directly in the nursery row in the fall. Peach-almond hybrids can be rooted by semi-hardwood cuttings under mist if given IBA treatments (*151*). These rootstocks are noted for their vigor and excellent compatibility with scion cultivars. 'Hansen 536' and 'Hansen 2138' have some susceptibility to crown rot and show somewhat reduced survival in commercial orchards.

Prunus persica. Peach and Nectarine (*33, 131*).

Peach and nectarine cultivars are universally budded to peach seedling rootstocks. Most other Prunus species (almond, plum, apricot) tend to produce incompatibility.

Cuttings. Some peach cultivars can be propagated by leafy, succulent softwood cuttings taken in spring or summer, treated with auxin, and rooted in a mist-propagating bed (*41, 95*). Also, in areas with mild winters, some peach cultivars can be started from hardwood cuttings if they are treated with a 4,000 ppm IBA quick-dip, then set out in the nursery in the fall (*85*). Propagation of peaches by cuttings is used commercially with high-density tree production systems, requiring a large number of inexpensive nursery trees (*58*).

Grafting. Nursery trees of peaches are propagated by T-budding or chip budding on seedling rootstocks. In California and other areas with a long growing season, trees are produced in one year by June budding. Seeds are planted in early fall and allowed to stratify over winter, germinating in early spring (March) to produce vigorous large plants by May. Fall budding may be used but a two year production cycle is required.

Rootstocks for Peach (20, 47, 131)

Peach (P. persica). Peach seedlings are the most satisfactory rootstock for peach and should be used unless certain special conditions warrant other stocks. Seeds of 'Halford,' 'Lovell,' 'Elberta,' or 'Rutgers Red Leaf' are usually used, since they germinate well and produce vigorous seedlings. Seedlings of these cultivars are not resistant to root-knot nematodes, however, and where this is a problem, resistant peach stocks should be considered. Seeds from peach cultivars whose fruits mature early in the season should not be used because their germination percentage is usually low. It is best to obtain seeds from the current season's crops, since viability decreases with storage. Seeds are sometimes obtained from wild types such as, in the United States, the Tennessee Naturals or Indian peaches.

Peach roots are quite susceptible to the root-knot nematode (*Meloidogyne* spp.), especially in sandy soils, and are susceptible to the root-lesion nematode *Pratylenchus vulnus.* 'Nemaguard,' a *Prunus persica* × *P. davidiana* hybrid rootstock introduced by the USDA in 1959, produces seedlings that are uniform and resistant to both *M. incognita* and *M. javanica,* but not to the ring nematode, *Criconemella xenoplax.* Nemaguard roots give strong, well-anchored, high-yielding trees, although certain peach and plum cultivars have not done well on them and they have shown susceptibility to bacterial canker and to crown rot. 'Nemaguard' is not hardy in the colder peach-growing areas. 'Nemared' is a red-leaved selection from 'Nemaguard.'

'Siberian C' is a winter-hardy peach rootstock—withstanding 11° C (12° F) soil temperatures—introduced in 1967 by the Canadian Department of Agriculture Research Station at Harrow, Ontario. It does well on light, sandy soils and will give about 15 percent dwarfing to the scion cultivar. It is, however, susceptible to root-knot and root-lesion nematodes and has shown high tree mortality in tests in the southeastern United States.

Nursery trees on peach roots often make unsatisfactory growth when planted on soils previously planted to peach trees. Peach roots are susceptible to oak root fungus, crown rot, crown gall, and verticillium wilt.

Plum. Some plum rootstocks are used for peach cultivars to produce somewhat smaller trees and to grow on heavy dense soil. These include: *Prunus insititia* ('St. Julien d'Orleans,' 'St. Julien Hybrid No. 1,' 'St. Julien GF 655.2'); *P. cerasifera* (myrobalan plums); *P. domestica* ('GF 43'); and *P. domestica* × *P. spinosa* ('Damas GF 1869'). In the United Kingdom, rootstocks such as these have been compatible with all peach and nectarine cultivars worked on them, and produce medium-sized to large trees. Plum rootstocks are especially adapted to wet, waterlogged soils.

Apricot (P. armeniaca). Apricot seedlings are occasionally used as a rootstock for the peach. The graft union is not always successful, but numerous trees and commercial orchards of this combination have produced fairly well for many years. Seedlings of the 'Blenheim' apricot seem to make better rootstocks for peaches than those of 'Tilton.' The apricot root is highly resistant to root-knot but not to root-lesion nematodes.

Almond (P. amygdalus). Almond seedlings have been used with limited success as a rootstock for peaches. There are trees of this combination growing well, but in general this is not a satisfactory combination. The trees are often dwarfed and tend to be short-lived.

Peach/Almond Hybrids. These are clonal rootstocks, such as 'GF556' and 'GF667,' that are important in France and other Mediterranean countries. They are excellent rootstocks for peach cultivars and can be propagated by rooting semi-hardwood cuttings under mist (*171*). Micropropagation is used commercially for 'GF667.'

Western Sand Cherry (P. besseyi). When this stock was used in limited experiments as a dwarfing rootstock for several peach cultivars, it was found that although the bud unions were excellent, about 40 percent of the nursery trees failed to survive. The remainder grew well, however, and developed into typical dwarf trees with healthy, dark-green foliage. The trees bore normal-sized fruit in the second or third year after transplanting to the orchard.

Nanking Cherry (P. tomentosa). This may be suitable for some peach cultivars as a dwarfing rootstock.

Prunus domestica. European plums and prunes. **P. salicina.** Japanese plums. Plum cultivars are propagated by T-budding or chip budding in the fall on seedling rootstocks or, with certain rootstocks, on rooted cuttings or layers. Budding can also be done in the spring. Plums tend to be easier to root than other species, being propagated by hardwood cuttings (*94*) and some by leafy, softwood cuttings under mist (*95*). Bench grafting in winter and planting the grafts in the nursery in spring gives good results using the whip graft (*49*). Japanese plum (*P. salicina*) nursery trees have been produced by micropropagation (*55, 185*).

Rootstocks for Plum (175)

Myrobalan Plum (P. cerasifera). This widely used plum rootstock is particularly desirable for the European plums, *P. domestica* (which includes the commercially important prune cultivars). It is also satisfactory for Japanese plums. Some cultivars of plum—'President,' 'Kelsey,' 'Stanley,' and 'Robe de Sergeant'—are not entirely compatible with this stock, however. Cultivars that are Japanese-American hybrids (*P. salicina* × *P. americana*) are best worked on American plum seedlings (*P. americana*).

Myrobalan roots are adapted to a wide range of soil and climatic conditions. They will endure fairly heavy soils and excess moisture, and are resistant to crown rot but susceptible to root-knot nematodes and oak root fungus. They grow well on light sandy soils.

Myrobalan seeds require stratification for about three months at 2 to 4° C (36 to 40° F). They may then be planted thickly in a seed bed for one season, then transplanted to the nursery and grown for a second season before budding; or the seeds may be planted directly in the nursery row and grown there for one season, the seedlings being budded in late summer or fall.

Certain vigorous myrobalan selections are propagated by hardwood cuttings. One of these, 'Myro 29C,' is immune to root-knot nematodes. In England, a selection, 'Myrobalan B,' was developed at the East Malling Research Station and is propagated by hardwood cuttings. It is particularly valuable in producing vigorous trees, although there are cultivars not completely compatible with it. Most plum cultivars in England belong to the *P. domestica* species.

Marianna Plum (P. cerasifera × P. munsoniana). This clonal rootstock originated in Texas as an open-pollinated seedling of the myrobalan plum, apparently a hybrid with *P. munsoniana.* It is propagated by hardwood cuttings (*94*). Some plums have grown well on it; others have not. An exceptionally vigorous seedling selection of the parent 'Marianna' plum, made by the California Agricultural Experiment Station in 1926, is widely used under the identifying name 'Marianna 2624.' It is adaptable to heavy, wet soils and is immune to root-knot nematodes, resistant to crown rot, crown gall, oak root fungus, and verticillium wilt, but is susceptible to bacterial canker. It suckers badly from roots and is sensitive to "brownline" virus. Since it is propagated as rooted cuttings, it has a shallow root system for the first few years, but as the trees grow older they develop a deeper root system. Hardwood cuttings root well if fall-planted (in mild winter climates) after IBA treatments.

Peach (P. persica). Some plum and prune orchards in California are on peach seedling rootstocks. This stock has proved satisfactory for light, well-drained soils or where bacterial canker has been a problem. However, peach rootstocks should be avoided if the trees are to be planted on a site formerly occupied by a peach orchard. In some areas, plum on peach roots tend to overbear and develop a die-back condition. Peach is not satisfactory as a rootstock for some plum cultivars, including 'Sugar' prune and 'Robe de Sergeant.'

Apricot (P. armeniaca). Apricot seedlings can be used as a plum stock in nematode-infested sandy soils for those cultivars that are compatible with apricot. Japanese plums tend to do better than European plums on apricot roots.

Almond (P. dulcis). Some plum cultivars can be grown successfully on almond seedlings. The 'French' prune does very well on this rootstock. The trees grow faster and bear larger fruit than when myrobalan roots are used. Plum cultivars on almond roots tend to overbear, sometimes to the detriment of the tree. Except where plantings are to be made on well-drained, sandy soils, high in lime or boron, this stock probably should not be used for plums.

'Brompton' and 'Common' plum (P. domestica). These two clonal plum rootstocks are used chiefly in England. 'Brompton' seems to be compatible with all plum cultivars and tends to produce medium to large trees. 'Common' plum, which produces small to medium trees, has shown incompatibility with some cultivars. Propagation is by hardwood cuttings.

'St. Julien,' 'Common Mussel,' 'Damas,' and 'Damson' (P. insititia). The first three stocks are used mostly in England. 'Damas C' produces medium to large trees, whereas 'Common Mussel' generally produces small to medium trees. The latter stock seems to be compatible with all plums. 'St. Julien A' produces small to medium trees for all compatible scion cultivars. Results have been variable with these stocks, since they vary widely in type. At the East Malling Research Station, 'St. Julien,' 'Mussel,' and 'Damas' clones have been selected and listed as A, B, C, D, and so forth. 'St. Julien' has been used to some extent as a plum stock for the *P. domestica* cultivars in the United States. 'Pixy' has been released by the East Malling Research Station as a dwarfing plum rootstock.

In England, the following clonal rootstocks are recommended for plums: for *vigorous* trees, 'Myrobalan B'; for *semi-vigorous* trees, 'Brompton'; for *intermediate* trees, 'Marianna,' 'Pershore'; for *semi-dwarf* trees, 'Common' plum, 'St. Julien A'; for *dwarf* trees, 'St. Julien K.'

Florida Sand Plum (P. angustifolia). This species may be useful as a dwarfing rootstock for compatible plum cultivars. In California, after 16 years, 'Giant,' 'Burbank,' and 'Beauty' on this stock were healthy and very productive, with a dwarf type of growth.

Japanese Plum (P. salicina). Seedlings of this species are used as plum rootstocks in Japan but apparently not elsewhere. European plums *(P. domestica),* when topworked on Japanese plum rootstocks, result in very short-lived trees; the reverse combination, though, produces compatible unions.

Western Sand Cherry (P. besseyi). This rootstock has produced satisfactory dwarfed plum trees of the Japanese and European types, but poor bud unions and shoot growth developed when it was used as a rootstock for cultivars of *P. insititia.*

Hybrid Prunus Rootstocks

A number of clonal rootstocks originating as interspecific hybrids of a number of species have been developed and introduced. Among these are the following:

'Citation' (Interspecific of Peach and Plum) (20, 47). Compatible with apricot and plum. Induces early bearing in the orchard and early defoliation, and dormancy, in the nursery. Tolerant of waterlogging and resistant to root-knot nematodes. Susceptible to crown gall and oak root fungus *(Armillaria).* Not suitable for peach and almond.

'Viking' (Interspecific among Peach, Almond, Plum, and Apricot) (47). Extremely vigorous and precocious. May have resistance to root-knot nematodes. Intolerant of wet soil conditions.

'Myran' (Prunus cerasifera × P. salicina × 'Yunnan' Peach) (61, 131). Compatible with almond and peach cultivars. Tolerates wet soil conditions and has some tolerance to *Armillaria.* Propagated by hardwood cuttings.

'Ishtara' (P. myrobalan × P. myrobalan × P. persica). Introduced by INRA, France. Used for prunes in Europe but compatible for apricot, almond, plum, and some cultivars of peach (*61*).

Pummulo. *See Citrus.*

Psidium spp. Guava. Cattley or Strawberry guava *(P. cattleianum),* and Common or Lemon Guava *(P. guajava)* (*15, 138, 144*). The Cattley guava has no named cultivars, and most nursery plants are propagated by seed. This species comes nearly true from seed—large-fruited, superior trees are used as the seed source. It is difficult to propagate by vegetative methods. The common guava *(P. guajava)* is widely grown in the world as a tropical and subtropical fruit tree. It is indigenous to tropical America. The common guava has several cultivars, but they are not grown extensively because of difficulties in asexual propagation.

Seed. Seeds are short-lived and must be planted immediately. They are grown in poly bags or in a nursery, and transplanted to the field when six to eight weeks old. Most trees of these two species are propagated by seeds, which germinate easily and in high percentages. Seedlings are somewhat susceptible to damping-off organisms, and should be started in sterilized soil or otherwise treated with fungicides. When about 4 cm (1½ in.) high, the seedlings should be transplanted into individual containers. In six months the plants should be about 30 cm (12 in.) high and can be transplanted to their permanent location.

Cuttings. Cuttings are difficult to root. Best results are obtained by treating softwood and semi-hardwood cuttings with IBA and rooting under mist (*170*).

Grafting. For large-scale propagation of common guava cultivars, grafting or budding is necessary. Chip budding has been successful, done any time during the summer using greenwood buds from selected cultivars inserted into seedling rootstocks about 5 mm in thickness. Plastic wrapping tape is used to cover the buds. The rootstock is cut off above the inserted bud after about three weeks (*115*). Inarching is an important method of propagation, but labor is expensive. Side-veneer grafting is also possible using scions from terminal growth flushes with well-developed axillary buds. Several species of *Psidium* are used as seedling rootstock: *P. cujanillis, P. molle, P. cattleianum,* and *P. guineense.* They tend to sucker from the base.

Layering. For the common guava, air layering is one of most important commercial propagation methods, as described in Chapter 15. With both species, simple and mound layering are effective methods of starting new plants; the layers may be tightly wrapped with wire just below the point where roots are wanted, or can be ringed, and treated with 500 ppm IBA in lanolin prior to earthing up. Two series of shoots can be produced in one year (*143*).

Punica granatum. Pomegranate. Small tree or shrub of subtropical and tropical Mediterranean and southwest Asian areas. Both deciduous and semi-deciduous types exist and are adapted to semi-arid climates. Seedlings are grown but cultivars can be readily propagated by hardwood cuttings. Cuttings are taken from fully mature one-year-old wood or suckers from the base of the tree. Cuttings are trimmed to 20 to 25 cm (8 to 10 in.) long, and treated with IBA. Softwood cuttings root successfully under mist. Suckers are produced. These may be dug during the dormant season, leaving a piece of root attached, and then planted.

Pummelo. *See Citrus.*

Pyrus spp. Pear. Over 20 species of pears exist around the world. They are divided into four groups (*137*). The most important edible types are the European pear (*P. communis* and a number of related species in Europe) and the Asiatic pears (mostly *P. pyrifolia* but also complex hybrids). Both are ancient fruit crops in Europe, North America, China, and Japan respectively. Many named cultivars are hundreds of years old. Many of the related species are involved in rootstock selection.

The history of the pear illustrates the problems of monocultures and the changing patterns of disease and insect vectoring. European pear cultivars are highly susceptible to the bacteria fire blight *(Bacillus amyl)*. "Pear decline" is a graft union failure caused by a mycoplasmalike organism (*103*) which was spread by the pear psylla *(Psylla pyricola)* first in Europe (*191*) and then in western North America (*12, 14*). Certain pear species are highly susceptible and when used as a rootstock they cause decline or death of the tree. Other rootstock species produce a serious defect on the fruit known as "black-end" or "hard-end." Serious viruses and viruslike diseases affect the pear, and production must be carried through a virus control program (*187*).

Propagation Methods

Seed. Seeds are used for rootstocks and in breeding programs. Pear seeds must be stratified for 60 to 100 days at about 4° C (40° F). They are then planted thickly about 13 mm (½ in.) deep in a seed bed, where they are allowed to grow during one season. The following spring they are dug, the roots and top are cut back, and then they are transplanted to the nursery row, where they are grown a second season, ready for budding in the fall.

Grafting. Essentially all pear cultivars are propagated by grafting to rootstocks, using fall budding, (T- or chip budding). Pear trees can also be started by whole-root grafting, using the whip-and-tongue method.

Cuttings. Some pear cultivars, such as 'Old Home' and 'Bartlett,' can be propagated by hardwood cuttings or by leafy cuttings under mist if treated with IBA (*93, 94, 227*). Own-rooted 'Bartlett' trees have shown excellent production with large, well-shaped fruit—they are resistant to pear decline (*74, 93*) and with age become partially dwarfed—a desirable attribute for high-density plantings. However, rooting percentages are not high and cutting propagation is not widely practiced.

Micropropagation. Pear cultivars and rootstocks can be micropropagated (*31, 238*). Commercial production has occurred primarily in Italy and France where own-rooted 'Bartlett' ('Williams') have been grown for commercial orchards. Micropropagation has been used to enhance the rooting potential of hard-to-root clones in the establishment of stock blocks.

Rootstocks for Pear (76, 90, 137, 228)

Pear Seedlings. Seed selection is very important if rootstocks of a specific species are desired.

Various *Pyrus* species hybridize freely, some bloom at the same time, and cross-pollination is necessary for seeds to develop. Use isolated groups of trees as the seed source of a known species and avoid collecting seeds from a collection of several different species, since hybrids are likely to be produced.

French Pear (P. communis). French pear seedlings are generally grown from seeds of 'Winter Nelis' or 'Bartlett.' This rootstock is moderately vigorous and winter-hardy, produces moderately productive, uniform trees with a strong, well-anchored root system, and is resistant to pear decline. It forms an excellent graft union with all pear cultivars and will tolerate relatively wet (but not waterlogged) and heavy soils. French pear roots are resistant to verticillium wilt, oak root fungus *(Armillaria mellea),* root-knot and root-lesion nematodes, and crown gall.

French pear rootstocks are susceptible to pear root aphid, *Eriosoma pyricola,* and fire blight, *Erwinia amylovora.* Millions of pear trees on French pear roots have died from fire blight, owing to the high susceptibility of this stock. Seedlings of the 'Kieffer' pear—a hybrid between *P. communis* and *P. pyrifolia*—have been satisfactorily used for many years in Australia as a pear rootstock.

Blight-Resistant French Pear Rootstocks and Interstocks. Pear selections resistant to blight include 'Old Home' (*75, 93*) and 'Farmingdale' (*182*), and have been used as an intermediate stock grafted on seedling rootstocks. The desired cultivar is topworked after the trunk and primary scaffold branches develop. A blight attack in the top of the tree will only go to the resistant interstock, which can then be regrafted after blight has been cut out. Thirteen clonal selections from an 'Old Home' × 'Farmingdale' progeny have been made in Oregon to be propagated by hardwood cuttings. These provide blight resistance and give a range of dwarfing of potential value in pear production (*167*).

P. calleryana. This stock is blight-resistant and produces vigorous trees with a strong graft union; fruit quality is good, with no black-end. 'Bartlett' orchards with this rootstock have produced well in California, and it is popular in the southern part of the United States for 'Keiffer' and other hybrid pears. In areas with cold winters it lacks winter-hardiness. Where good pear psylla control has been practiced, trees on *P. calleryana* roots are resistant to pear decline. However, trees with *P. calleryana* roots show less resistance to oak root fungus than those with *P. communis* roots. Seedlings of a selection of this species, known as 'D-6,' are widely used as a pear rootstock in Australia.

P. ussuriensis. This Oriental stock has been used in the past to some extent; many pear cultivars on this root develop black-end, although not to the extent found with *P. pyrifolia* roots. It also produces small trees that are susceptible to pear decline, and should not be used in areas where this may occur.

P. betulaefolia. This species has vigorous seedlings, resistance to leaf spot and pear root aphid, tolerance to alkali soils, an adaptability to a wide range of climatic conditions, good resistance to pear decline, and produces large high-yielding trees.

Cydonia Oblonga. Quince (229). This species has been used for centuries as a dwarfing stock for pear and is widely used as a pear rootstock in Western Europe. Some cultivars fail to make a strong union directly on the quince, hence double-working with an intermediate stock such as 'Hardy' or 'Old Home' is necessary. Cultivars that require such a compatible interstock when worked on quince roots include 'Bartlett,' 'Bosc,' 'Winter Nelis,' 'Seckel,' 'Easter,' 'Clairgeau,' 'Conference,' 'Guyot,' 'Clapp's Favorite,' 'Farmingdale,' and 'El Dorado.' A selection of 'Bartlett' ('Swiss Bartlett'), compatible with 'Quince A' rootstock, originated in Switzerland, presumably as a bud mutation from an incompatible form (*178*). The following pears also appear to be compatible when worked directly on quince: 'Anjou,' 'Old Home,' 'Hardy,' 'Packham's Triumph,' 'Gorham,' 'Comice,' 'Flemish Beauty,' 'Duchess,' and 'Maxine.'

Quince roots are resistant to pear root aphids and nematodes, but are susceptible to oak root fungus, fire blight, and calcareous soil. They are not winter-hardy in areas where extremely low temperatures occur. In some areas, trees on quince roots have developed pear decline, but in others they have not, possibly because different quince rootstocks were used. The black-end trouble has not developed with pears on quince roots. There are a number of quince cultivars, most of which are easily propagated by hardwood cuttings or layering. 'Angers' quince is commonly used as a pear rootstock because its cuttings root readily, it grows vigorously in the nursery, and it does well in the orchard. 'Provence' quince roots produce larger pear trees than 'Angers' and are more winter-hardy.

The East Malling Research Station has selected several clones of quince suitable as pear rootstocks and designated them as 'Quince A,' 'B,' and 'C.' 'Quince A' ('Angers') has proved to be the most satisfactory stock. 'Quince B' (Common quince) is somewhat dwarfing, whereas 'Quince C' produces very dwarfing but highly productive trees. It is important to use only virus-tested quince stock.

Provence quince (LePage Series C and BA-29C) originated in France. These are winter-hardy and, when used as rootstocks for pears, give trees one-half to two-thirds the size of standard pear trees. The BA-29C series is a virus-free selection of LePage Series C.

Pyrus pyrifolia *(P. serotina).* Asiatic pear. Cultivars of Asiatic pear are propagated on seedling rootstocks of this species in Japan. This rootstock was once widely used in the United States as a pear rootstock from about 1900 to 1925, but is no longer recommended due to its high susceptibility to pear decline and to the physiological defect black-end (or "hard-end") which may occur in 'Bartlett,' 'Anjou,' 'Winter Nelis,' and other cultivars when they are propagated on it (*100*).

Quince. *See Cydonia.*

Raspberry. *See Rubus.*

Ribes spp. Currants and gooseberries. Specific species, hybrids, and cultivars are found in the northern hemisphere, grow as bushes and produce small berries used in making jams, jellies, and pies. Cultivars of red and white currants are in *Ribes sativum;* black currants are in *R. nigrum.* Flowering current is *R. odoratum.* American gooseberries include *R. grossularia* and its hybrids, while European gooseberries are in *R. uva-crispa.* All of the species are alternate hosts of white pine blister rust and their planting is restricted by law. Propagation is similar for cultivars of all of these species.

Cuttings. Currants are propagated by collecting hardwood cuttings 20 to 25 cm (8 to 10 in) long in late fall, precallusing them at low temperatures over winter, and planting in the spring. Cultivars of gooseberry can also be rooted but with more difficulty.

Layering. Mound layering is used for American gooseberry. Shoots usually root well after one season. They are then cut off and transferred to the nursery row for a second season's growth before they are set out in their permanent location. The slower-rooting layers of European cultivars may have to remain attached to the parent plant for two seasons before they develop enough roots to be detached.

Micropropagation. Gooseberries can be easily micropropagated and the rooted plantlets stored under refrigeration for as long as 130 days with 100 percent survival (*224*).

Rubus spp. Raspberry, Blackberry, Youngberry, Boysenberry, Loganberry, Dewberry. This group includes a complex of species growing worldwide from which a large number of commercially important cultivars have been produced. Commercial propagation requires vegetative methods. Seeds are used in breeding programs. Raspberry and blackberry produces clusters of drupelets surrounding a hard-coated achene.

Asexual Methods

Upright Type. Blackberries (*Rubus;* subgenus *Eubatus*). Reproduce by suckers removed in the spring with roots attached and replanted to a new location or to a nursery row for an additional year (*119*). Also propagate by root cuttings, conventional stem cuttings, one-node cuttings, and leaf-bud cuttings under mist with auxin (*22*). They are commercially micropropagated (see below).

Trailing Types. Youngberry, Boysenberry, Loganberry, or Dewberry (*Rubus* spp.). These do not produce many suckers but reproduce naturally by tip layering (see Chapter 15). They can be propagated by root cuttings, but some thornless forms are periclinal chimeras and revert to the nonmutated (thorny) type. Thornless cultivars which do not revert to thorny types have been developed through biotechnology (see p. 218). Boysenberry types can also be propagated by conventional stem cuttings, one-node stem cuttings (*240*), or leaf-bud cuttings rooted under mist with auxin hormones. They are commercially micropropagated (see below).

Black Raspberry (R. occidentalis). Usual method is by tip layering (see p. 513) but can be rooted from leaf-bud cuttings in about three weeks under mist.

Purple Raspberry (*R. occidentalis* × *R. idaeus*) (*119*). Tip layering is the usual method of propagation but roots less easily than the black raspberry.

Red Raspberry (R. strigosus × R. idaeus). Propagated usually by removing one-year-old suckers in early spring with piece of old root attached (*119*). Young, green suckers of new wood may also be dug in the spring shortly after they appear above ground. Leave piece of old root attached. Survival requires removal during cool weather and irrigation following transplanting. Sucker production is stimulated by inserting a spade deeply at intervals in the vicinity of old plants to cut off roots. Or mulch with straw or sawdust.

Viruses and crown gall can be problems with red raspberries, so suckers should be taken only from clean plants obtained from nurseries specializing in certified plants. Root cuttings (*102, 211*) should be thick; use 15 cm (6 in.) root lengths; for thin roots use 5 cm (2 in.) long pieces. Cuttings are planted very shallow—13 mm (½ in.) (*102, 109*). Strong nursery plants may be produced in one year.

Cuttings. Leafy softwood cuttings can be made in early spring from young sucker shoots just emerging from the soil. These should have a 2.5 to 5 cm (1 to 2 in.) etiolated section from below the surface. Place in a propagating frame (*230*). Shoot cuttings taken directly from the canes are difficult to root, although success has been reported with pre-etiolated shoots, treated with IBA, and rooted under ventilated high-humidity fog (*107*).

Micropropagation. Blackberry and raspberry (*19, 223*) cultivars are commercially micropropagated by shoot-tip cultures on a large enough scale to allow mass propagation. In practice, micropropagation is used as part of a system to produce virus-free stock plants in virus control programs (see Chapter 9).

Sapodilla (Nispero). *See Achras.*

Strawberry. *See Fragaria × ananassa.*

Tangelo. *See Citrus.*

Tangerine. *See Citrus.*

Theobroma cacao. Cacao. The dried, partly fermented fatty seeds of this native South American tree are used in the production of cocoa, chocolate, and cocoa butter. Large commercial cacao plantings have been established in West Africa and South America (*214*). Production is mostly from seedling trees, which are highly variable.

Seed. Seeds are obtained from selected high-yielding clones as well as from hybrid seeds resulting from cross-pollination between parents which are propagated vegetatively. Freshly harvested mature seeds are planted immediately, since cacao seeds quickly deteriorate after harvesting, normally losing all capacity to germinate within a week after removal from the pod. Longer seed life can be produced by keeping seeds from drying, and storing at 24 to 29° C (75 to 85° F) (*10*).

Cacao is often direct (seeded) field planted, three to four seeds to a planting site, retaining all seedlings surviving as branches of one tree. Seedlings may be started in a nursery bed and later transplanted to their permanent site. Seedlings may also be started in poly bags, baskets, bamboo or paper cylinders, or clay pots, or containers from which they are removed later and planted. The germination temperature should be about 26° C (80° F).

Miscellaneous Asexual Methods. Patch budding, and sometimes T-budding, is done on seedling rootstock (*59*). Air layering is quite successful.

Vaccinium spp. Blueberry and Cranberry. Lowbush Blueberry *(V. angustifolium),* Rabbiteye Blueberry *(V. ashei),* Highbush blueberry *(V. corymbosum L* and *V. austral),* Cranberry *(Vaccinium macrocarpon).* The blueberry and cranberry are native to North America and are grown, respectively, as a small bush or trailing vine.

Seed. Seed propagation is used for breeding. There is no pregermination treatment for blueberry, but with cranberry, optimum seed germination occurs after a three-month cold stratification (*57*). Seeds are removed from ripe berries and spread over a well-drained, acid-type soil mix containing one-third peat moss. Seeds are covered with a layer of finely ground sphagnum moss and kept moist until they germinate, usually in three to four weeks. Seedlings 2 cm (¾ in.) tall are transferred to peat pots.

Miscellaneous Asexual Methods

Lowbush Blueberry (V. angustifolium) (*81*). Cultivars of this species are probably best propagated by leafy softwood cuttings under intermittent mist with bottom heat, using sand and peat moss (1:1) as a rooting medium. Cuttings taken in late spring and early summer from actively growing shoots root well, some clones giving almost 100 percent rooting. Rooted cuttings are transferred to peat pots for further growth and overwintering. Cuttings made from rhizomes also can be rooted. These are best taken in early spring or late summer and fall, avoiding the midsummer rest period of the rhizome buds.

Rabbiteye Blueberry (V. ashei). Cultivars of this species can be propagated by hardwood cuttings as well as by leafy softwood cuttings taken in midsummer, treated with a 10,000 ppm IBA talc, and rooted under mist. Using supplementary light to give a 16-hour daylength may improve root production (*42*). Micropropagation is also possible (*36*).

Highbush blueberry (V. corymbosum L. *and V. austral)* (*57*). Dormant hardwood stem cuttings are used. The blueberry can also be started by leaf-bud cuttings. Cuttings should be spaced about 5 cm (2 in.) apart in the rooting bed and set with the top bud just showing. When leaves appear, the frame should be raised slightly to allow for ventilation. Either mist or frequent watering to maintain a high humidity is required. Roots start to form in about two months. Both softwood and hardwood cuttings treated with a 8,000 ppm IBA talc root well (*52*). To take advantage of different soil types, *V. corymbosum* can be grafted on *V. ashei* and *V. arboreum* rootstock. Grafting has been by cleft, whip, side graft, and T-budding (*52*).

Cranberry (V. macrocarpon) (*44, 80*). This vine type of evergreen plant produces trailing runners upon which are numerous short upright branches. Propagation is by cuttings made from either runners or upright branches. Cutting material is obtained by mowing the vines in early spring before new growth has started. The cuttings are then set directly in place in their permanent location without previous rooting at distances of 6 to 18 in. apart each way. Two to four cuttings are set in sand in each "hill." The cuttings are 13 to 25 cm (5 to 10 in.) long and set deep enough so that only an inch is above ground. A more rapid method of starting a cranberry bog is to scatter the cuttings over the ground and work them into the soil with a special disk-type planter. This is justifiable when there is an abundance of cutting material and a scarcity of labor for setting the cuttings by hand. Water is applied to the bog immediately after planting. The cuttings root during the first year and make some top growth, but the plants do not start bearing until three or four years later.

Vitis spp. Grape (*220, 231, 232*). Grapevines are propagated by seeds, cuttings, layering, budding, or grafting (*2*). Grape propagation methods have been modernized by the use of virus-indexed "clean" planting stock, mist propagation techniques for leafy cuttings, and rapid machine-grafting procedures. Traditionally, commercial propagation has been by dormant hardwood cuttings. For types difficult to root, such as the Muscadine *(Vitis rotundifolia)*, layering or leafy cuttings under mist is possible (*78*). Budding or grafting on rootstocks is used to increase vine life, plant vigor, and yield and where noxious soil organisms, such as phylloxera *(Dacylosphaera vitifoliae)* or root-knot nematodes (*Meloidogyne* spp.), are present. Cultivars of susceptible species such as *V. vinifera* must be grafted or budded onto a resistant rootstock. Root-knot nematodes can be eradicated from grapevine rootings by dipping them in hot water (51.5 to 54.5° C; 125 to 130° F) for five to three minutes, respectively (*132*).

Methods of Propagation

Seeds. Seeds are used in breeding programs to produce new cultivars. Grape seeds are not difficult to germinate. Best results with vinifera grape seeds are obtained after a moist stratification period at 0.5 to 4° C (33 to 40° F) for about three months before planting (*88*).

Hardwood Cuttings. Grape cultivars and clonal rootstocks have traditionally been propagated by dormant hardwood cuttings, which root readily. Cutting material should be collected during the winter from healthy, vigorous, mature vines. Well-developed current season's canes should be used; which are medium in size and have moderately short internodes. Cuttings 8 to 13 mm (⅓ to ½ in.) in diameter and 36 to 46 cm (14 to 18 in.) long are generally used, and planted in the spring deep enough to cover all but one bud. One season's growth in the nursery should produce plants large enough to transplant to the vineyard. Auxins have not been needed to root hardwood grape cuttings.

Leafy Cuttings. Leafy greenwood grape cuttings root profusely under mist in about ten days if given relatively high (27 to 30° C; 80 to 85° F) bottom heat and treated with IBA. Scarce planting stock (such as virus-indexed material) can be increased very rapidly using one-budded stem cuttings, then consecutively taking additional cuttings from the shoot arising from the bud on the rooted cutting, and so on.

Layers. Grape cultivars difficult to start by cuttings can be propagated by simple, trench, or mound layering (see Chapter 15).

Micropropagation. Grapes have been produced by several *in vitro* techniques, including embryoid formation and fragmented shoot-tip cultures (*126*).

Grafting. Bench grafting (*89, 222*) is widely used; scions are grafted on either rooted or unrooted disbudded rootstock cuttings by the whip graft or, better, by machine grafting (see Figures 13–31, 13–32, 13–33). The grafts are made in late winter or early spring from completely dormant scion and rootstock material. The stocks are cut to 31 to 36 cm (12 to 14 in.) with the lower cut just below a node and the top cut 2.5 cm (1 in.) or more above a node. All buds are removed from the rootstock to prevent subsequent suckering. Scionwood should have the same diameter as the stock.

After grafting, using a one-bud scion, the union is stapled together or wrapped with budding rubber. The grafts should be held for three to four weeks in well-aerated, moist wood shavings or peat moss at about 26.5° C (80° F) for callusing.

The callused grafts are removed from the callusing boxes or plastic bags and any roots and the scion shoot are carefully trimmed back to an 18 mm (½ in.) stub. The scion is dipped quickly into a temperature-controlled container of melted (low melting point) paraffin or rose wax to a depth of 2.5 cm (1 in.) below the graft union and then quickly into cool water. The paraffined bench grafts are then planted into 5.0 × 5.0 × 25 cm (2 × 2 × 10 in.) milk cartons, containers, or planting tubes, that contain a mixture of perlite and pumice, or perlite and peat moss. The containers or tubes are placed upright in flats. The flats are set on pallets and moved into a heated greenhouse for six to eight weeks. The growing bench grafts are then transferred to a 50 percent shade screen house for about two weeks for hardening-off prior to planting in the nursery or vineyard.

The bench grafts are planted deeply so that the tops of the cartons or planting tubes are at least 2 to 3 in. below the soil level to ensure that water will get inside the cartons or tubes. At no time, however, should bench grafts be planted in the vineyard with the graft union at or below ground level.

Greenwood Grafting. Greenwood grafting is a simple and rapid procedure for propagating vinifera grapes on resistant rootstocks (*88*). A one-budded greenwood scion is splice-grafted during the active growing season on new growth arising either from a one-year-old rooted cutting or from a cutting during midseason of the second year's growth.

Field Budding (135). An older method of establishing grape cultivars on resistant rootstocks is to field bud on rapidly growing, well-rooted cuttings that were planted in their permanent vineyard location the previous winter or spring. T-budding can be done in late spring using dormant budwood held under refrigeration. Shortly after budding, the trunks should be cut with diagonal slashes at the base to allow the "bleeding" to take place there rather than where the bud was inserted (*3*). An alternate method is to chip bud in late summer or early fall as soon as fresh mature buds from wood with light brown bark can be obtained and before the stock goes dormant. In areas where mature buds cannot be obtained early in the fall, growers may store under refrigeration bud-sticks collected in the winter and bud them in late spring or early summer.

The bud is inserted in the stock 5 to 10 cm (2 to 4 in.) above the soil level, preferably on the side adjacent to the supporting stake, tied in place with budding rubber, or poly budding tape, but not waxed. The bud is then covered with 13 to 25 cm (5 to 10 in.) of well-pulverized, *moist* soil to prevent drying. In areas of hot summers, or in soils of low moisture, variable results are likely to be obtained, and bench or nursery grafted vines should be used. If the buds are tied with white 13 mm (½ in.) plastic tape, it is unnecessary to mound them over with soil.

Top-Grafting Grapevines. Cultivars of mature grapevines can be changed by cutting off the tops of the vines in early spring 30 to 53 cm (12 to 21 in.) below the lower wire. Stocks are side whip-grafted, using a two-bud scion of the desired cultivar. The scions are wrapped with 1-in. white plastic tape and the cut surfaces are covered with grafting wax.

The vines may also be T-budded with the inverted T and wrapped with white plastic tape after the bark "slips" in late spring. No grafting compound is needed. Both methods give highly successful takes.

Rootstocks for American Hybrid Grapes (108)

'Ramsey' ('Salt Creek') and 'Dog Ridge' are widely used as rootstocks for grapes in the southern US. They have also been used successfully to resist nematodes on low fertility sites in California.

The northeastern grape-growing regions are largely own-rooted, but in areas where *V. vinifera* cultivars are grown, or where more vigor is desired scions are primarily grafted to 'Couderc 3309,' 'Teleki 5C,' and '101–14 Mgt.' Other stocks that are occasionally used include 'Couderc 1616,' and 'Kober 5BB.' Some inland areas still derive scion vigor from American hybrid rootstocks, such as 'Cynthiana' or 'Lenoir.'

Rootstocks for Vinifera Grapes (133, 134, 233, 234)

The rootstock AXR#1 (*V. vinifera* × *V. rupestris*), although widely used in California through the early 1980s, does not possess sufficient resistance to phylloxera and its use is declining within the state. The replanting effort is proceeding with a wide range of new rootstocks, although current evaluation data indicating which rootstocks are best suited for which areas is not complete. The following list describes the most widely used rootstocks at this time. Most of these rootstocks were developed in Europe about 100 years ago and were designed to resist phylloxera and grow well under European conditions.

'Rupestris St. George.' This pure *V. rupestris* rootstock was one of the original rootstocks developed to address the phylloxera crisis in France. It is only moderately resistant to phylloxera, but has not collapsed to this pest. It is very susceptible to nematodes, but roots and grafts well. 'St. George' is best suited to shallow soils, or soils with moderate fertility, and can produce overly vigorous scions on fertile soils or on sites with excessive soil moisture.

'Teleki 5C.' This widely used *V. berlandieri* × *V. riparia* rootstock (*108*) is low to moderate in vigor and well adapted to fertile soils. It resists phylloxera and many nematode species, but is sensitive to water stress and performs best with adequate irrigation. Its cuttings root relatively easily and its low to moderate vigor produces good fruit quality.

'3309 Couderc.' This *V. riparia* × *V. rupestris* hybrid is widely used in California, as is the similar '101–14 Mgt.' These two rootstocks are suited to fertile soils and need adequate irrigation. Both are phylloxera resistant, but are susceptible to nematodes. They root and graft well and produce vines of moderate vigor (scions on '101–14' are relatively more vigorous) with good fruit quality.

'110 Richter.' This *V. berlandieri* × *V. rupestris* rootstock is widely used for drought tolerance on sites with shallow soils and limited rainfall. The similar rootstock '1103 Paulsen' is also used on these sites, and is slightly more vigorous. Both rootstocks are phylloxera resistant, but susceptible to nematodes. They are tolerant of limestone soils and root and graft well.

'Ramsey.' This pure *V. champinii* rootstock is used on soils with low fertility and nematode problems. It resists most root knot nematode strains, but can produce an overly vigorous vine on more fertile soils leading to problems with fruit set and fruit quality. Its phylloxera resistance is only moderate, but phylloxera pressure is usually low on the sandy soils 'Ramsey' is best suited for. 'Dog Ridge' is a *V. champinii* rootstock with similar characteristics, but slightly less vigor. Both rootstocks root with difficulty, but graft well once rooted.

'Freedom.' This widely used rootstock was developed in the 1960s at the USDA-Fresno station as a cross between a seedling of open pollinated 'Couderc 1613' × a seedling of open pollinated 'Dog Ridge' (both of these rootstock parents are female vines and the pollen source is unknown). 'Couderc 1613' is ¼ *V. vinifera* and is moderately susceptible to phylloxera; this fact and the open pollinated nature of this cross cast suspicion on the phylloxera resistance of 'Freedom' and the similar rootstock 'Harmony.' However, 'Freedom' has excellent nematode resistance and is used on many sites throughout California. 'Freedom' imparts relatively high vigor to the scion and is best suited for sandy or low fertility soils. It can be difficult to root and graft well.

'039–16.' This *V. vinifera* × *Muscadinia rotundifolia* rootstock was released by UC/Davis in 1989 to resist fanleaf degeneration (*233*). This virus disease is vectored by the dagger nematode, *Xiphinema index.* Although this rootstock allows the nematode to vector the virus it moderates the virus' negative effect on fruit set. Because it is half *V. vinifera,* 039–16's phylloxera resistance is questionable. It is only recommended for sites where fanleaf degeneration is severe, and no alternative exists. '039–16' can impart excessive vigor to the scion and is very sensitive to water stress. Its *M. rotundifolia* percentage makes this rootstock difficult to root, but once rooted it will graft moderately well.

Walnut (Black). *See Juglans nigra.*

Walnut (Paradox). *See J. hindsii* × *J. regia.*

Walnut (Persian or English). *See Juglans regia.*

Zizyphus jujuba. Jujube, Chinese date. Deciduous temperate-zone tree adapted to hot, arid regions. Other evergreen tropical species with cultivars also exist in India. The leading jujube cultivars in the U.S. are 'Lang' and 'Li.' Seeds should be stratified at about 4° C (40° F) for several months before planting. Can be rooted by hardwood stem cuttings and root cuttings. T-budding is done on jujube seedling rootstock.

REFERENCES

1. Adams, W.G. 1985. Improved techniques used in producing budded citrus nursery trees for commercial fruit production. *Comb. Proc. Intl. Plant Prop. Soc.* 34:496–99.
2. Alley, C.J. 1980. Propagation of grapevines. *Calif. Agr.* 34(7):29–30.
3. Alley, C.J., and A.T. Koyama. 1978. Vine bleeding delays growth of T-budded grapevines. *Calif. Agr.* 32(8):6.
4. Al Maarri, K., Y. Arnaud, and E. Miginiac, 1986. *In vitro* micropropagation of quince *(Cydonia oblonga Mill.). Scientia Hort.* 28:315–21.
5. Anderson, W.C. 1984. Micropropagation of filberts, *Corylus avellana. Comb. Proc. Intl. Plant Prop. Soc.* 33:132–37.
6. Argles, G.K. 1976. *Anacardium occidentale* (cashew). In *The propagation of tropical fruit trees,* R.J. Garner and S.A. Chaudri, eds. Hort. Rev. 4. East Malling, England: FAO and Commonwealth Agricultural Bureaux.
7. Arndt, C.H. 1935. Mango polyembryony and other multiple shoots. *Amer. J. Bot.* 22:26.
8. Ashiru, G.A., and T. Quarcoo. 1971. Vegetative propagation of kola *(Cola nitida). Trop. Agr.* 48(1):85–92.

9. Barghchi, M., and P.G. Alderson. 1985. In vitro propagation of *Pistacia vera* L. and the commercial cultivars, Ohadi and Kalleghochi. *J. Hort. Sci.* 60(3): 423–30.
10. Barton, L.V. 1965. Viability of seeds of *Theobroma cacao* L. *Contr. Boyce Thomp. Inst.* 23(4):109–22.
11. Bass, L.N. 1975. Seed storage of *Carica papaya* L. *HortScience* 10(3):232–33.
12. Batjer, L.P., and H. Schneider. 1960. Relation of pear decline to rootstocks and sievetube necrosis. *Proc. Amer. Soc. Hort. Sci.* 76:85–97.
13. Bitters, W.P., and E.R. Parker. 1953. Quick decline of citrus as influenced by top-root relationships. *Calif. Agr. Exp. Sta. Bul. 733.*
14. Blodgett, E.C., H. Schneider, and M.D. Aichele. 1962. Behavior of pear decline disease on different stock-scion combinations. *Phytopathology* 52:679–84.
15. Bose, T.K. and S.K. Mitra. 1990. *Fruits: Tropical and Subtropical.* Calcutta, India: Naya Prokash.
16. Boxus, P. 1974. The production of strawberry plants by in vitro micropropagation. *J. Hort. Sci.* 49:209–10.
17. Brokaw, W.H. 1977. Subtropical fruit tree production; avocado as a case study. *Comb. Proc. Intl. Plant Prop. Soc.* 27:113–21.
18. Brokaw, W.H. 1988. Avocado clonal rootstock propagation. *Comb. Proc. Intl. Plant Prop. Soc.* 87:97–102.
19. Broome, O.C., and R.H. Zimmerman. 1978. In vitro propagation of blackberry. *HortScience* 13(2):151–53.
20. Burchell Nursery. 1994. *Orchard varieties and information guide.* Modesto, CA; Burchell Nursery, Inc. 40 pp.
21. Calavan, E.C., E.F. Frolich, J.B. Carpenter, C.N. Roistacher, and D.W. Christiansen. 1964. Rapid indexing of exocortis of citrus. *Phytopathology* 54:1359–62.
22. Caldwell, J.D. 1984. Blackberry propagation. *HortScience* 19(2):193–95.
23. Caldwell, J.D., D.C. Coston, and K.H. Brock. 1988. Rooting of semi-hardwood 'Hayward' kiwifruit cuttings. *HortScience* 23(4):714–17.
24. Cameron, J.W., and R.K. Soost. 1953. Nucellar lines of citrus. *Calif. Agr.* 7(1):8, 15, 16.
25. Carlson, R.F. 1965. Growth and incompatibility factors associated with apricot scion/rootstock in Michigan. *Mich. Quart. Bul* 48(1):23–29.
26. Castle, W.S. 1987. Citrus rootstocks. In *Rootstocks for fruit crops,* R.C. Rom and R.F. Carlson, eds. New York: John Wiley.
27. Castle, W.S., W.G. Adams, and R.L. Dilley. 1979. An indoor container system for producing citrus nursery trees in one year from seed. *Proc. Fla. State Hort. Sci.* 92:3–7.
28. Catlin, P.B., and E.A. Olsson, 1986. Response of eastern black walnut and northern California black walnut to waterlogging. *HortScience* 21(6):1379–80.
29. Chase, S.B. 1947. Budding and grafting eastern black walnut. *Proc. Amer. Soc. Hort. Sci.* 49:175–80.
30. Chaudri, S.A. 1976. *Mangifera indica—Mango.* In *The propagation of tropical fruit trees,* R.J. Garner and S.A. Chaudri, eds. Hort. Rev. 4. East Malling, England: FAO and Commonwealth Agricultural Bureaux.
31. Chevreau, E., B. Thibault, and Y. Arnaud. 1992. Micropropagation of Pear (*Pyrus communis* L.) In Bajaj, Y.P.S. ed. *Biotechnology in Agriculture and Forestry* 18:244–261. Berlin: Springer-Verlag.
32. Child, R. 1964. *Coconuts.* London: Longmans, Green.
33. Childers, N., ed. 1988. *The peach: World cultivars to marketing.* Gainesville, Fla.: Horticultural Publications.
34. Choi, G.H. and C.L. Nuss. 1992. A viral gene confers hypovirulence-associated traits to the chestnut blight fungus. *EMBO J.* 11:473–477.
35. Cobin, M. 1954. The lychee in Florida. *Fla. Agr. Exp. Sta. But. 546.*
36. Cohen, D., and D. Elliott. 1979. Micropropagation methods for blueberries and tamarillos. *Comb. Proc. Intl. Plant Prop. Soc.* 29:177–79.
37. Collins, J.L. 1960. *The pineapple—history, cultivation, utilization.* London: Leonard-Hill.
38. Condit, I.J. 1969. *Ficus, the exotic species.* Berkeley: Univ. Calif. Div. Agr. Sci.
39. Converse, R.A. 1972. The propagation of virus-tested strawberry stocks. *Comb. Proc. Intl. Plant Prop. Soc.* 22:73–76.
40. Cooper, P.A. and D. Cohen. 1984. Micropropagation of Japanese persimmon *(Diospyros kaki). Comb. Proc. Intl. Plant Prop. Soc.* 34:118–124.
41. Couvillon, G.A., and A. Erez. 1980. Rooting, survival, and development of several peach cultivars propagated from semi-hardwood cuttings. *HortScience* 15(1):41–43.
42. Couvillon, G.A., and F.A. Pokorny. 1968. Photoperiod, indolebutyric acid, and type of cutting wood as factors in rooting of rabbiteye blueberry (*Vaccinium ashei Reade*), cv. Woodward. *HortScience* 3(2): 74–75.
43. Crisosto, C., and E. Sutter. 1985. Improving 'Manzanillo' seed germination. *HortScience* 20(1)100–102.
44. Cross, C.E., I.E. Demoranville, K.H. Deubert, R.M. Devlin, J.S. Norton, W.E. Tomlinson, and B.M. Zuckerman. 1969. *Modern cultural practice in cranberry growing. Mass. Agr. Exp. Sta. Ext. Ser. Bul. 39.*
45. Crossa-Raynaud, P., and J.M. Anderson. 1987. Apricot rootstocks. In *Rootstocks for fruit crops,* R.C. Rom and R.F. Carlson, eds. New York: John Wiley.
46. Damiano, C. 1980. Strawberry micropropagation. In *Proc. Conf. Nurs. Prod. Fruit Plants through Tissue Cult.,* R.H. Zimmerman, ed. USDA, SEA, ARR-NE-11.
47. Dave Wilson Nursery. 1994. *Catalogue and Reference Guide.* Hickman, CA: Dave Wilson Nursery. 28 pp.

48. Davies, F.S. 1986. The navel orange. In *Horticultural reviews,* Vol. 8, J. Janick, ed. Westport, Conn.: AVI Publ. Co., pp. 129–80.

49. Davis, B., II. 1976. Bench grafting plum and apricot as compared to T-budding. *Comb. Proc. Intl. Plant Prop. Soc.* 26:253–55.

50. Dillon, D. 1967. Simultaneous grafting and rooting of citrus under mist. *Comb. Proc. Intl. Plant Prop. Soc.* 17:114–18.

51. Dimalla, G.G., and J. Van Staden. 1978. Pecan nut germination: A review for the nursery industry. *Scientia Hort.* 8:1–9.

52. Dirr, M.A., and C.W. Heuser, Jr. 1987. *The reference manual of woody plant propagation.* Athens, Ga.: Univ. Press.

53. Doggrell, L.M. 1976. Commercial propagation of macadamias. *Comb. Proc. Intl. Plant Prop. Soc.* 26:391–93.

54. Doncaster, T. 1981. The whys and hows of passion fruit growing in Queensland. *Comb. Proc. Intl. Plant Prop. Soc.* 30:616–17.

55. Druart, P. 1992. In vitro culture and micropropagation of plum (Prunus sp.) In Bajaj, Y.P.S., *Biotechnology in Agriculture and Forestry* 18:279–303. Berlin: Springer-Verlag.

56. Durling, D.L. 1990. International citrus nursery production as it relates to *Phytopthora,* virus and growing media. *Comb. Proc. Intl. Plant Prop. Soc.* 16:51–54.

57. Eck, P. 1988. *Blueberry science.* New Brunswick, N.J.: Rutgers Univ. Press.

58. Erez, A., and Z. Yablowitz. 1981. Rooting of peach hardwood cuttings for the meadow orchard. *Scientia Hort.* 15:137–44.

59. Evans, H. 1951. Investigations on the propagation of cacao. *Trop. Agr.* 28:147–203.

60. Fankhauser, I. 1985. Propagating feijoa by bench grafting. *Comb. Proc. Intl. Plant Prop. Soc.* 34:401–3.

61. Felipe, A. 1989. *Patrones para frutales de pepita y huesco.* Barcelona, Spain: Ediciones Tecnicas Europeas.

62. Ferguson, A.R. 1984. Kiwifruit: a botanical review. In *Horticultural reviews,* Vol. 6, J. Janick, ed. Westport, Conn., AVI Publ. Co., pp. 1–53.

63. Ferguson, L., et. al. eds. *Olive Production Manual.* Berkeley, CA: Div. of Agr. and Nat. Res., University of California, Publ. 3353.

64. Ferree, D.C., and R.F. Carlson. 1987. Apple rootstocks. In *Rootstocks for fruit crops,* R.C. Rom and R.F. Carlson, eds. New York: John Wiley.

65. Fiorino, P., and F. Loreti. 1987. Propagation of fruit trees by tissue culture in Italy. *HortScience* 22(3): 353–58.

66. Forde, H.I., 1979. Persian walnut in the western United States. In *Nut tree culture in North America.* R.A. Jaynes, ed. Hamden. Conn. Northern Nut Growers Assn.

67. Fowler Nursery. 1992. *Rootstock and planting guide.* Newcastle, CA: Fowler Nurseries, inc. 27 pp.

68. Fridlund, P.R. 1980. Maintenance and distribution of virus-free fruits. In *Proc. conf. on nursery production of fruit plants through tissue culture.* R.H. Zimmerman ed. USDA Sci. and Education Administration, ARR-NE-II, pp 11–22.

69. Frolich, E.F. 1966. Rooting citrus and avocado cuttings. *Comb. Proc. Intl. Plant Prop. Soc.* 16:51–54.

70. Frolich, E.F., and R.G. Platt. 1972. Use of the etiolation technique in rooting avocado cuttings. *Calif. Avoc. Soc. Yearbook, 1971–1972,* pp. 97–109.

71. Funk, D.T. 1979. Black walnuts for nuts and timber. In *Nut tree culture in North America,* R.A. Jaynes, ed. Hamden, Conn.: North American Nut Growers Assn.

72. Galleta, G.J. and D.G. Himelrick. 1990. *Small Fruit Crop Management.* Englewood Cliffs, N.J.: Prentice Hall.

73. George, A.P., and R.J. Nissen. 1987. Propagation of *Annona* species: A review. *Scientia Hort.* 33:75–85.

74. Griggs, W.H., J.A. Beutel, W.O. Reil, and B.T. Iwakiri. 1980. Combination of pear rootstocks recommended for new Bartlett plantings. *Calif. Agr.* 34(10):20–24.

75. Griggs, W.H., and H.T. Hartmann. 1960. Old Home pears show resistance to decline when on own roots. *Calif. Agr.* 14:8, 10.

76. Griggs, W.H., D.D. Jensen, and B.T. Iwakiri. 1968. Development of young pear trees with different rootstocks in relation to psylla infestation, pear decline, and leaf curl. *Hilgardia* 39:153–204.

77. Grimo, E. 1979. Carpathian (Persian) walnuts. In *Nut tree culture in North America,* R.A. Jaynes, ed. Hamden, Conn.: North American Nut Growers Assn.

78. Goode, D.Z., Jr., and R.P. Lane. 1983. Rooting leafy muscadine grape cuttings. *HortScience* 18(6):944–46.

79. Guttridge, C.G., D.T. Mason, and E.G. Ing. 1965. Cold storage of strawberry runner plants at different temperatures. *Exp. Hort.* 12:38–41.

80. Hall, I.V. 1969. Growing cranberries. *Can. Dept. Agr. Publ. 1282* (rev.).

81. Hall, I.V., L.E. Aalders, L.P. Jackson, G.W. Wood, and C.L. Lockhart. 1975. Lowbush blueberry production. *Can. Dept. Agr. Publ. 1477.*

82. Hall, W.J. 1975. Propagation of walnuts, almonds, and pistachios in California. *Comb. Proc. Intl. Plant Prop. Soc.* 25:53–57.

83. Hamilton, R.Z., and W. Yee. 1974. Macadamia: Hawaii's dessert nut. *Univ. Hawaii Cir. 485.*

84. Hanna, J.D. 1987. Pecan rootstocks. In *Rootstocks for fruit crops,* R.C. Rom and R.F. Carlson, eds. New York: John Wiley.

85. Hansen, C.J., and H.T. Hartmann. 1968. The use of indolebutyric acid and captan in the propagation of clonal peach and peach-almond hybrid rootstock. *Proc. Amer. Soc. Hort. Sci.* 92:135–40.

86. Hansen, K.C., and J.E. Lazarte. 1984. In vitro propagation of pecan seedlings. *HortScience* 19(2): 237–39.

87. Harkness, R.W. 1967. Papaya growing in Florida. *Fla. Agr. Exp. Sta. Cir. S-180.*

88. Harmon, F.N. 1954. A modified procedure for greenwood grafting of *Vinifera* grapes. *Proc. Amer. Soc. Hort. Sci.* 64:255–58.

89. Harmon, F.N., and J.H. Weinberger. 1959. Effects of storage and stratification on germination of vinifera grape seeds. *Proc. Amer. Soc. Hort. Sci.* 73:147–50.

90. Hartman, Henry. 1961. Historical facts pertaining to root and trunkstocks for pear trees. *Oreg. Agr. Exp. Sta. Misc. Paper 109.*

91. Hartmann, H.T. 1952. Further studies on the propagation of the olive by cuttings. *Proc. Amer. Soc. Hort. Sci.* 59:155–60.

92. ———. 1958. Rootstock effects in the olive. *Proc. Amer. Soc. Hort. Sci.* 72:242–51.

93. Hartmann, H.T., W.H. Griggs, and C.J. Hansen. 1963. Propagation of own-rooted Old Home and Bartlett pears to produce trees resistant to pear decline. *Proc. Amer. Soc. Hort. Sci.* 82:92–102.

94. Hartmann, H.T., and C.J. Hansen. 1958. Rooting pear, plum rootstocks. *Calif. Agr.* 12(10):4, 14–15.

95. ———. 1955. Rooting of softwood cuttings of several fruit species under mist. *Proc. Amer. Soc. Hort. Sci.* 66:157–67.

96. Hartmann, H.T., C.J. Hansen, and F. Loreti. 1965. Propagation of apple rootstocks by hardwood cuttings. *Calif. Agr.* 19(6):4–5.

97. Hartmann, H.T., and F. Loreti. 1965. Seasonal variation in rooting leafy olive cuttings under mist. *Proc. Amer. Soc. Hort. Sci.* 87:194–98.

98. Hartmann. H.T. and K. Opitz. *Olive production in California.* Univ. Calif. Div. Agr. Sfi. Leaflet 2464.

99. Hartmann, H.T., W.C. Schnathorst, and J. Whisler. 1971. Oblonga, a clonal olive rootstock resistant to verticillium wilt. *Calif. Agr.* 25(6):12–15.

100. Heppner, M.J. 1927. Pear black-end and its relation to different rootstocks. *Proc. Amer. Soc. Hort. Sci.* 24:139–42.

101. Herman, E.B., and G.J. Hass. 1975. Clonal propagation of *Coffea arabica* L. from callus culture. *HortScience* 10(6):588–89.

102. Heydecker, W., and Margaret Marston. 1968. Quantitative studies on the regeneration of raspberries from root cuttings. *Hort. Res.* 8(2):142–46.

103. Hibino, H., G.H. Kaloostian, and H. Schneider. 1971. Mycoplasmalike bodies in the pear psylla vector of pear decline. *Virology* 43:34–40.

104. Hilton, H. 1986. Budding European (Spanish) chestnut (*Castanea sativa* Mill.) *Proc. Inter. Plant Prop. Soc.* 35:72–76.

105. Hodgson, R.W. 1940. Rootstocks for the Oriental persimmon. *Proc. Amer. Soc. Hort. Sci.* 37: 338–39.

106. Holden, V.L. 1984. Propagation of filbert trees by layering and grafting. *Comb. Proc. Intl. Plant Prop. Soc.* 33:51–52.

107. Howard, B.H., E. Tal, and S.K. Miles. 1987. Red raspberry propagation from leafy summer cuttings. *J. Hort. Sci.* 62(4):485–92.

108. Howell, G.S. 1987. *Vitis* rootstocks. In *Rootstocks for fruit crops,* R.C. Rom and R.F. Carlson, eds. New York: John Wiley.

109. Hudson, J.P. 1954. Propagation of plants by root cuttings I. Regeneration of raspberry root cuttings. *J. Hort. Sci.* 29:27–43.

110. Hunter, A.G., and J.D. Norton. 1985. Rooting stem cuttings of Chinese chestnut. *Scientia Hort.* 26:43–45.

111. Hwang, S.C., C.L. Chen, J.C. Lin, and H.L. Lin. 1984. Cultivation of banana using plantlets from meristem culture. *HortScience* 19(2):231–33.

112. Ivanicka, J. 1987. In vitro micropropagation of mulberry, *Morus nigra* L. *Scientia Hort.* 32:33–39.

113. Ivanicka, J. 1992. Micropropagation of cherry. In Bajaj, Y.P.A. ed. *Biotechnology in agriculture and forestry* 18:304–322. Berlin: Springer-Verlag.

114. Ivey, I.D. 1979. Feijoas: Selection and propagation. *Comb. Proc. Intl. Plant Prop. Soc.* 29:161–68.

115. Jaffee, A. 1970. Chip grafting guava cultivars. *The Plant Propagator* 16(2):6.

116. James, D.J., and I.J. Thurbon. 1981. Shoot and root initiation *in vitro* in the apple rootstock M.9 and the promotive effects of phloroglucinol. *J. Hort. Sci.* 56(1):15–20.

117. Jaynes, R.A. 1961. Buried-inarch technique for rooting chestnut cuttings. *North. Nut Grow. Assoc. 52nd Ann. Rpt.,* pp. 37–39.

118. ———. 1979. Chestnuts. In *Nut tree culture in North America,* R.A. Jaynes, ed. Hamden, Conn.: Northern Nut Growers Assn.

119. Jennings, D.L. 1988. *Raspberries, blackberries, their breeding, diseases, and growth.* London: Academic Press.

120. Joley, L.E. 1979. Pistachios. In *Nut tree culture in North America,* R.A. Jaynes, ed. Hamden, Conn.: Northern Nut Growers Assn.

121. Jones, O.P. 1993. Propagation of apple in vitro. In Ahuja, A., ed. *Micropropagation of woody plants.* Dordrecht: Berlin.

122. Kester, D.E., and C. Grasselly. 1987. Almond rootstocks. In *Rootstocks for fruit crops,* R.C. Rom and R.F. Carlson, eds. New York: John Wiley.

123. Kidd, E.L., Jr. 1987. Asexual propagation of fruit and nut trees at Stark Brothers Nurseries. *Comb. Proc. Intl. Plant Prop. Soc.* 36:427–30.

124. Klotz, L.J., E.C. Calavan, and L.G. Weathers. 1972. Virus and virus-like diseases of citrus. *Calif. Agr. Exp. Sta. Circ. 559.*

125. Krezdorn, A.H., and G.W. Adriance. 1961. *Fig growing in the South.* U.S. Dept. Agr. Handbook 196. Washington, D.C.: U.S. Govt. Printing Office.

126. Krul, W.R., and J. Meyerson. 1980. In vitro propagation of grape. In *Proc. Conf. Nurs. Prod. Fruit Plants through Tissue Culture,* R.H. Zimmerman, ed. USDA, SEA, ARR-NE-11.

127. Lagerstedt, H.B. 1979. Filberts. In *Nut tree culture in North America,* R.A. Jaynes, ed. Hamden, Conn.: Northern Nut Growers Assn.

128. Lane, W.D., 1992. Micropropagation of Apple (*Malus domestica* Burkh.). In Bajaj, Y.P.S. Ed., *Biotechnology in Agriculture and Forestry* 18:231–244. Berlin: Springer-Verlag.

129. Lapins, K. 1959. Some symptoms of stock-scion incompatibility of apricot varieties on peach seedling rootstock. *Can. J. Plant Sci.* 39:194–203.

130. Lawyer, E.M. 1978. Seed germination of stone fruits. *Comb. Proc. Intl. Plant Prop. Soc.* 28:106–9.

131. Layne, R.E.C. 1987. Peach rootstocks. In *Rootstocks for fruit crops,* R.C. Rom and R.F. Carlson, eds. New York: John Wiley.

132. Lear, B., and L. Lider. 1959. Eradication of root-knot nematodes from grapevine rootings by hot water. *Plant Dis. Rpt.* 43:314–17.

133. Lider, L. 1958. Phylloxera-resistant grape rootstocks for the coastal valleys of California. *Hilgardia* 27:287–318.

134. ———. 1960. Vineyard trials in California with nematode-resistant grape rootstocks. *Hilgardia* 30:123–52.

135. ———. 1963. Field budding and the care of the budded grapevine. *Calif. Agr. Ext. Ser. Leaflet 153.*

136. Litz, R.E. and V.S. Jaiswal. 1991. Micropropagation of tropical and subtropical fruits. In deBergh, P.C. and R.H. Zimmerman, ed. *Micropropagation: Technology and application.* Dordrecht, Neth.; Kluwer Publ. 247–264.

137. Lombard, P.B., and M.N. Westwood. 1987. Pear rootstocks. In *Rootstocks for fruit crops,* R.C. Rom and R.F. Carlson, eds. New York: John Wiley.

138. Lynch, S.J., and R.O. Nelson. 1956. Current methods of vegetative propagation of avocado, mango, lychee, and guava in Florida. *Cieba* (Escuela Agricola Panamericana, Tegucigalpa, Honduras) 4:315–77.

139. Lyrene, P.M. 1980. Micropropagation of rabbiteye blueberries. *HortScience* 15(1):80–81.

140. MacDaniels. L.H. 1979. Hickories. In *Nut tree culture in North America,* R.A. Jaynes, ed. Hamden, Conn.: Northern Nut Growers Assn.

141. Madden, G.E. 1979. Pecans. In *Nut tree culture in North America,* R.A. Jaynes, ed. Hamden, Conn.: Northern Nut Growers Assn.

142. Madden, G.E., and H.W. Tisdale. 1975. Effects of chilling and stratification on nut germination of northern and southern pecan cultivars. *HortScience* 10(3):259–60.

143. Majumdar, P.K., and S.K. Mukherjee. 1968. Guava: A new vegetative propagation method. *Indian Hort.,* Jan.-Mar.

144. Majumdar, P.K., and D.K. Sharma. 1990. Mango. In: Bose, T.K. and S.K. Mitra, ed. *Fruits: tropical and subtropical.* pp. 1–62. Calcutta, India: Naya Prokash.

145. Martinelli, A. 1992. Micropropagation of strawberry (*Fragaria* spp.). In Bajaj, Y.P.S., ed. *Biotechnology in Agriculture and Forestry.* Berlin: Springer-Verlag. pp. 354–370.

146. McGranahan, G.H., and P.B. Catlin. 1987. *Juglans* rootstocks. In *Rootstocks for fruit crops,* R.C. Rom and R.F. Carlson, eds. New York: John Wiley.

147. McKay, J.W. and R.A. Jaynes. 1969. Chestnuts. In R.A. Jaynes, Ed., *Handbook of North American nut trees,* Knoxville, TN: Northern Nut Growers Assoc.

148. McMillan-Browse, P.D.A. 1970. Vegetative propagation of *Corylus. Comb. Proc. Intl. Plant Prop. Soc.* 20:356–58.

149. Maxwell, N.P., and C.G. Lyons. 1979. A technique for propagating container-grown citrus on sour orange rootstock. *HortScience* 14(1):56–57.

150. Medina, J.P. 1981. Studies of clonal propagation of pecans at Ica, Peru. *Plant Propagator* 27:10–11.

151. Mehlenbacher, S.A. 1986. Rooting of interspecific peach hybrids by semi-hardwood cuttings. *HortScience* 21(6):1374–76.

152. Mircetich, J.S. M., J. Refsguard, and M.E. Matheron. 1980. Blackline of English walnut trees traced to graft-transmitted virus. *Calif. Agr.* 34(11, 12):8–10.

153. Mircetich, J.S.M., and A. Rowhani. 1984. The relationship between cherry leafroll virus and blackline disease of English walnut trees. *Phytopathology* 74:426–28.

154. Mitra, S.K. and T.K. Bose. 1990. Guava. In: Bose, T.K. and S.K. Mitra ed. *Fruits: tropical and subtropical.* pp. 280–303. Calcutta, Ind.: Naya Prokash.

155. Moore, G.A., M.G. Dewald and M.H. Evans. Micropropagation of Pineapple (*Ananas comosus* L.). In Bajaj, Y.P.S., ed. *Biotechnology in Agriculture and Forestry* 18:460–470. Berlin: Springer-Verlag.

156. Morton, J.F., and F.D. Venning. 1972. Avoid failures and losses in the cultivation of cashew. *Econ. Bot.* 26(3):245–54.

157. Moss, G.I., and R. Dalgleish. 1981. Rapid propagation of citrus in containers. *Comb. Proc. Intl. Plant Prop. Soc.* 31:262–74.

158. Mukherjee, S.K., and P.K. Majumdar. 1963. Standardization of rootstock of mango. I. Studies on the propagation of clonal rootstocks by stooling and layering. *Indian J. Hort.* 20:204–9.

159. ———. 1964. Effect of different factors on the success of veneer grafting in mango. *Indian J. Hort.* 21:46–50.

160. Mukherjee, S.K., and N.N. Bid. 1965. Propagation of mango. II. Effect of etiolation and growth regulator treatment on the success of air layering. *Indian J. Agr. Sci.* 35:309–14.

161. Nagabhushanam, S., and M.A. Menon. 1980. Propagation of cashew (*Anacordium occidentale* L.) by etiolation, girdling, and stooling. *Plant Propagator* 26(1):11–13.

162. Navarro, L. 1992. Citrus shoot-tip grafting. In Bajaj, Y.P.S. *Biotechnology in Agriculture and Forestry* 18:327–326. Berlin: Springer-Verlag.

163. Nixon, R.W. 1969. Growing dates in the United States. *USDA Agr. Inf. Bul. 207.*

164. Northwood, P.J. 1964. Vegetative propagation of cashew *(Anacardium occidentale)* by the air layering method. *E. Afr. Agr. For. Jour.* 30:35–37.

165. Okie, W.R. 1987. Plum rootstocks. In *Rootstocks for fruit crops,* R.C. Rom and R.F. Carlson, eds. New York: John Wiley.

166. Omar, M.S., M.K. Hameed and M.S. Al-Rawi. 1992. Micropropagation of date palm (*Phoenix dactypifera* L.). In Bajaj, Y.P.S. ed. in *Biotechnology in agriculture and forestry* vol. 18:470–492.

167. Oregon Rootstock and Tree Co. Inc. 1991. *Catalog and reference manual for rootstock and nursery varieties.* 54 pp. Treco: Woodburn, OR.

168. Painter, J.H. 1967. Producing walnuts in Oregon. *Ore. Agr. Ext. Bul. 795.*

169. Parfitt, D.E., and A.A. Almehdi. 1994. Use of high CO_2 atmosphere and medium modifications for the successful micropropagation of pistachio. *Sci. Hort.* 56:321–29.

170. Pennock, W. and G. Maldonado. 1963. The propagation of guavas from stem cuttings. *Jour. Agr. Univ. Puerto Rico* 47:280–290.

171. Perry, R.L. 1987. Cherry rootstocks. In *Rootstocks for fruit crops,* R.C. Rom and R.F. Carlson, eds. New York: John Wiley.

172. Platt, R.G. 1976. Current techniques of avocado propagation. In *Proc. 1st Inter. Trop. Short Course: The avocado,* J.W. Sauls, R.L. Phillips, and L.K. Jackson, eds. Gainesville, Fla.: Fruit Crops Dept., Univ. of Fla.

173. Platt, R.G., and E.F. Frolich. 1965. Propagation of avocados. *Calif. Agr. Exp. Sta. Cir. 531.*

174. Platt, R.G., and K. Opitz. 1973. The propagation of citrus. In *The citrus industry,* Vol. 3, W. Reuther, ed. Berkeley: Univ. Calif. Div. Agr. Sci.

175. Pontikis, C.A., and P. Melas. 1986. Micropropagation of *Ficus carica* L. *HortScience* 21(1):153.

176. Popenoe, J. 1970. Coconut varieties and propagation. In *Coconuts: Production, processing, products,* J.G. Woodroof, ed. Westport, Conn.: AVI Publ. Co.

177. Porter, G. 1988. Non-astringent persimmon propagation in southeast Queensland. *Comb. Proc. Intl. Plant Prop. Soc.* 37:148–50.

178. Posnette, A., and R. Cropley. 1962. Further studies on a selection of Williams Bon Chretien pear compatible with Quince A rootstocks. *J. Hort. Sci.* 37:291–94.

179. Preston, A.P. 1966. Apple rootstock studies: Fifteen years' results with Malling-Merton clones. *J. Hort. Sci.* 41:349–60.

180. Rao, V.N.M., and I.K.S. Rao. 1957. Studies on the vegetative propagation of cashew (*Anacardium occidentale* L.). Approach grafting with and without the aid of plastic film wrappers. *Indian J. Agr. Sci.* 27:267–75.

181. Reeves, D.W. and G.A. Couvillon. 1992. Micropropagation of peach (*Prunus persica* L.) Batsch. In Bajaj, Y.P.S., ed. *Biotechnology in Agriculture and Forestry* 18:262–273. Berlin: Springer-Verlag.

182. Reimer, R.C. 1925. Blight resistance in pears and characteristics of pear species and stocks. *Oreg. Agr. Exp. Sta. Bul. 214.*

183. Reuveni, O., and M. Raviv. 1981. Importance of leaf retention to rooting of avocado cuttings. *J. Amer. Soc. Hort. Sci.* 106(3):127–30.

184. Reuther, W., ed. *The citrus industry,* Vol. 1, 1967; Vol. 2, 1968; Vol. 3, 1973; Vol. 4, 1978. Berkeley: Univ. Calif. Div. Agr. Sci.

185. Rosati, P., G. Marino, and C. Swierczewski. 1980. In vitro propagation of Japanese plum (*Prunus salicina* Lindl. cv. Calita.). *J. Amer. Soc. Hort. Sci.* 105(1):126–29.

186. Ryugo, K., C.A. Schroeder, A. Sugiura, and K. Yonemori. 1988. Persimmons for California. *Calif. Agr.* 42(4):7–9.

187. Schneider, H. 1970. Graft transmission and host range of the pear decline causal agent. *Phytopathology* 60:204–7.

188. Schroeder, C.A. 1947. Rootstock influence on fruit set in the Hachiya persimmon. *Proc. Amer. Soc. Hort. Sci.* 209.

189. Scott, D.H., and D.P. Ink. 1948. Germination of strawberry seed as affected by scarification treatment with sulfuric acid. *Proc. Amer. Soc. Hort. Sci.* 51:299–300.

190. Sebastian, K.T., and J.A. McComb. 1986. A micropropagation system for carob. *Scientia Hort.* 28: 127–31.

191. Shalla, T., and L. Chiarappa. 1961. Pear decline in Italy. *Bul. Calif. Dept. Agr.* 50:213–17.

192. Singh, S., and D.V. Chugh. 1961. Marcotting with some plant regulators in loquat *(Eriobotrya japonica Lind.). Indian J. Hort.* 18:123–29.

193. Singh-Dhaliwal, T., and A. Torres-Sepulveda. 1961. Recent experiments on rooting coffee stem cuttings in Puerto Rico. *Univ. Puerto Rico Agr. Exp. Sta. Tech. Paper 33.*

194. Sitton, B.G. 1931. Vegetative propagation of the black walnut. *Mich. Agr. Exp. Sta. Tech. Bul. 119.*

195. Smith, I.E., B.N. Wolstenholme, and P. Allan. 1974. Rooting and establishment of pecan (*Carya illinoensis* [Wang] K. Koch.) stem cuttings. *Agroplantae* 6:21–28.

196. Smith, J.C., D.J. Jordon, and F.H. Wood. 1986. Persimmon propagation: research highlights from Ruakura. *Comb. Proc. Intl. Plant Prop. Soc.* 35:323–28.

197. Smith, S.H., R.E. Hilton, and N.W. Frazier. 1970. Meristem culture for elimination of strawberry viruses. *Calif. Agr.* 24(8):8–10.

198. Smith, M.W. 1987. Grafting and cutting propagation of pecans. *Comb. Proc. Intl. Plant Prop. Soc.* 36: 414–18.

199. Snir, I. and A. Erez. 1980. *In vitro* propagation of Malling-Merton apple rootstocks. *HortScience* 15(5): 597–98.

200. Sookmark, S., and E.A. Tai. 1975. Vegetative propagation of papaya by budding. *Acta Hort.* 49:85–90.

201. Storey, W.B. 1976. *Macadamia tetraphylla*—the preferred rootstock. *Calif. Macadamia Soc. Yearbook* 22:101–7.

202. Strametz, J.R. 1984. Hot callus grafting of filbert trees. *Comb. Proc. Intl. Plant Prop. Soc.* 33:52–53.

203. Stubbs, L.L. 1963. Tristeza-tolerant strains of sour orange. *F.A.O. Plant Prot. Bul* 11:8–10.

204. Suarez de Castro, F. 1961. Semilleros o germinadores de cafe (Coffee seed beds or germinators). *Agri. Trop.* (Bogota) 17:317–24.

205. Sugiura, A., R. Tao, H. Murayama, and T. Tomana. 1986. In vitro propagation of Japanese persimmon. *HortScience* 21(5):1205–7.

206. Teague, C.P. 1966. Avocado tip-grafting. *Comb. Proc. Intl. Plant Prop. Soc.* 16:50–51.

207. Teulon, J. 1971. Propagation of passion fruit *(Passiflora edulis)* on a fusarium-resistant rootstock. *Plant Propagator* 17(3):4–5.

208. Theakston, F.E. 1976. Carica papaya. In *The propagation of tropical fruit trees,* R.J. Garner and S.A. Chaudri, eds. Hort. Rev. 4. East Malling, England: FAO and Commonwealth Agricultural Bureaux.

209. Thomas, C.A. 1981. Mango propagation by saddle grafting. *J. Hort. Sci.* 56(2):173–75.

210. Tisserat, B. 1984. Propagation of date palms by shoot tip cultures. *HortScience* 19(2):230–31.

211. Torre, I.C., and B.H. Barritt. 1979. Red raspberry establishment from root cuttings. *J. Amer. Soc. Hort. Sci.* 104(1):28–31.

212. Tukey, H.B., and K.D. Brase. 1943. The dwarfing effect of an intermediate stempiece of Malling IX. *Proc. Amer. Soc. Hort. Sci.* 42:357–64.

213. van Eignatten, C.L.M. 1969. Propagation of kola *(Cola nitida). Communm. NEDERF* (Amsterdam), Sept.

214. Urquhart, D.H. 1961. *Cocoa* (2nd ed.). London: Longmans, Green.

215. van Staden, J., B.N. Wolstenholme, and G.G. Dimala. 1976. Effect of temperature on pecan seed germination. *HortScience* 11(3):261–62.

216. Vietez, A.M., A. Ballester, M.L. Vieitez, and E. Vieitez. 1983. In vitro plantlet regeneration of mature chestnut. *J. Hort. Sci.* 58(9):45763.

217. Voyiatzis, D.C., and I.C. Porlingis. 1987. Temperature requirements for the germination of olive seeds (*Olea europaea* L.) *J. Hort. Sci.* 62(3): 405–11.

218. Wallace, R.D. and L.G. Spinella, ed. 1952. *World chestnut industry conference.* Aluchua, FL; Chestnut Marketing Association.

219. Warner, R.M., Z. Worku, and J.A. Silva. 1979. Effect of photoperiod on growth responses of citrus rootstocks. *J. Amer. Soc. Hort. Sci.* 104(2):232–35.

220. Weaver, R.J. 1976. *Grape growing.* New York: John Wiley.

221. Webber, H.J. 1948. Rootstocks: their character and reactions. In *The citrus industry,* Vol. II, H.J. Webber and L.D. Batchelor, eds. Berkeley: Univ. of Calif. Press.

222. Weinberger, J.H., and N.H. Loomis. 1972. A rapid method for propagating grapevines on rootstocks. *USDA Agr. Res. Ser. ARS-W-2.*

223. Welander, M. 1985. In vitro culture of raspberry *(Rubus ideaus)* for mass propagation. *J. Hort. Sci.* 60(4):493–99.

224. Welander, M. 1985. Micropropagation of gooseberry, *Ribes grossularia. Scientia Hort.* 26:267–72.

225. Welch, N.C., R. Bringhurst, A.S. Greathead, V. Voth, W.S. Seyman, N.F. McCalley, and H.W. Otto. 1982. *Strawberry production in California.* Leaflet 2959. Berkeley: Univ. Calif. Div. Agr. Sci.

226. Wellman, F.L. 1961. *Coffee: Botany, cultivation, utilization.* London: Leonard-Hill.

227. Westwood, M.N., and L.A. Brooks. 1963. Propagation of hardwood pear cuttings. *Comb. Proc. Intl. Plant Prop. Soc.* 13:261–68.

228. Westwood, M.N., H.R. Cameron, P.B. Lombard, and C.B. Cordy. 1971. Effects of trunk and rootstock on decline, growth, and performance of pear. *J. Amer. Soc. Hort. Sci.* 96(2):147–50.

229. Westwood, M.N., and A.N. Roberts. 1965. Quince root used for dwarf pears. *Oreg. Orn. and Nurs. Dig.* 9(1):1–2.

230. Williams, M.W., and R.A. Norton. 1959. Propagation of red raspberries from softwood cuttings. *Proc. Amer. Soc. Hort. Sci.* 74:401–6.

231. Winkler, A.J., J.A. Cook, L.A. Lider, and W.M. Kliewer. 1974. Propagation. Chapter 9 in *General viticulture* (2nd ed.). Berkeley: Univ. of Calif. Press.

232. Wolpert, J.P., M.A. Walker, E. Weber, and D. Roberts. ed. 1995. *Proc. Intern. Symp. on Clonal Selection.* Davis, CA.: Amer. Soc. Enol. and Vit.

233. Wolpert, J.A., M.A. Walker, and E. Weber. eds. 1992. *Proceedings of rootstock seminar. A worldwide perspective.* Davis, CA: ASEV Press.

234. Wolpert, J.A., M.A. Walker, E. Weber, L. Bettiga, R. Smith and P. Verdegal. 1994. Use of phylloxera-resistant rootstocks in California: Past, present and future. *Grape Grower,* Part 1 January: 16-19, Part 2 February: 4–7, Part 3. March: 10–17.

235. Young, J.A. and C.G. Young, 1986. *Seeds of Wildland Plants.* Portland, OR: Timber Press.

236. Zeiger, D., and H.B. Tukey. 1960. A historical review of the Malling apple rootstocks in America. *Mich. State Univ. Circ. Bul. 226.*

237. Zentmyer, G.A., A.O. Paulus, and R.M. Burns. 1962. Avocado root rot. *Calif. Agr. Exp. Sta. Circ. 511.*

238. Zimmerman, R.H. 1991. Micropropagation of temperate zone fruit and nut crops. In DeBergh, P.C. and R.H. Zimmerman, ed. *Micropropagation: technology and application.* Dordrecht, Neth.: Kluwer Publ.

239. Zimmerman, R.H., and D.C. Broome. 1980. Apple cultivar micropropagation. *USDA, Agr. Research Results ARR-NE-11,* pp. 54–63.

240. Zimmerman, R.H., G.J. Galleta, and O.C. Broome. 1980. Propagation of thornless blackberries by one-node cuttings. *J. Amer. Soc. Hort. Sci.* 105(3): 405–7.

SUPPLEMENTARY READING

Carlson, R.F. 1971. Fruit trees—dwarfing and propagation. *Mich. State Univ. Hort. Rpt. 1* (rev.).

Day, L.H. 1953. Rootstocks for stone fruits. *Calif. Agr. Exp. Sta. Bul. 736.*

Garner, R.J., and S.A. Chaudri.1976. *The propagation of tropical fruit trees.* Hort. Rev. 5. East Malling, England: FAO and Commonwealth Agricultural Bureaux.

International Dwarf Fruit Tree Association. *Compact tree fruit and Proceedings of Annual Meetings.*

Jaynes, R.A., ed. 1979. *Nut tree culture in North America.* Hamden, Conn.: Northern Nut Growers Assn.

Morton, Julia F. 1987. *Fruits of warm climates.* Winterville, N.C.: Creative Resource Systems.

Purseglove, J.W. 1968. *Tropical crops—dicotyledons.* New York: John Wiley & Sons.

Purseglove, J.W. 1992. *Tropical crops—monocotyledons.* New York: John Wiley & Sons.

Roach, F.A. 1969. *Fruit tree raising: Rootstocks and propagation.* Bul. 135 (5th ed.) London: Ministry of Agriculture, Fisheries, and Food.

Rom, R.C., and R.F. Carlson,eds. 1987. *Rootstocks for fruit crops.* New York: John Wiley.

Ryugo, K. 1988. *Fruit culture: Its science and art.* Somerset, N.J.: John Wiley.

Tukey, H.B. 1964. *Dwarfed fruit trees.* New York: Macmillan.

World Wide Web. The electronic yellow pages of the future. Information is available from plant propagation to nurseries to plant species.

Young, J.A. and C.G. Young. 1992. *Seeds of Woody Plants in North America.* Portland, Oreg.: Dioscorides Press.

20

Propagation of Ornamental Trees, Shrubs, and Woody Vines

In any commercial propagation system it is import to conduct small trials before propagating on a large scale. The propagation techniques and references listed are to serve as a guide. Propagators must develop their own procedures and chemical treatments that work best for their particular propagation systems.

Abelia × grandiflora. Abelia, Glossy Abelia. Commercially propagated with semi-hardwood cuttings. Can be rooted easily under mist in spring, summer or fall. Rooting is enhanced by applying talc or quick-dips of 1,000 to 2,000 ppm IBA, or IBA-NAA combinations totaling 1,000–2,000 ppm have produced superior results (*115*). Hardwood cuttings also may be rooted in fall or late winter, but less successfully than with semi-hardwood cuttings. Abelia is commercially micropropagated.

Abies spp. Fir.

Seed. Seed propagation is not difficult, but fresh seed should be used, since most species lose their viability after one year in ordinary storage. Embryo dormancy is generally present; fall planting or stratification at about 4° C (40° F) for one to three months is required for good germination. Bulk presoaking *A. procera* (noble fir) seed in water should be avoided because of imbibition damage; it is best to allow seeds to slowly uptake water on moist filter paper and then stratify at 4° C (40° F) for a minimum of three weeks (*209*). Alternatively, seeds can be placed for five to ten days in moist perlite for imbibition, and then cold stratified. Abies seedlings are very susceptible to damping-off. They should be given partial shade during the first season, since they are injured by excessive heat and sunlight.

Cuttings. Fir cuttings are considered difficult-to-root, but *Abies fraseri* hardwood cuttings—selected from young trees, and basally wounded and treated with IBA—can be rooted in high percentages (*187*). First order laterals root in higher percentages than primary axes (*40*). Cuttings taken from lateral branches root readily, but tend to become plagiotropic. A quick-dip of 5,000 ppm IBA and maintaining a bottom heat of 18 to 24° C (65 to 75° F) is best for rooting hardwood cuttings (*40*).

This species, along with white fir *(A. concolor)*, red fir *(A. magnifica)*, and the California 'Silver Tip,' are important Christmas tree species.

Grafting. The side-veneer graft is commonly used. Japanese Momi fir *(A. firma)* is one of the few firs that will tolerate the heavy clay, wet soil conditions (low soil oxygen), and heat of the southeastern U.S. Consequently, researchers at North Carolina State University recommend grafting desirable fir cultivars on *A. firma* rootstock, rather than the less tolerant *A. fraseri* or *A. balsamea* (*100*).

> In the U.S., it is not legal for producers/growers to use technical-grade auxins. Only end-use formulations are permitted, such as Dip N' Grow, Woods Rooting Compound, and other commercial liquid and talc auxin formulations. Information is presented in Chapters 19, 20, and 21 for rooting with technical-grade auxins, since one can legally make recommended auxin concentrations by diluting end-use formulations. See Chapter 11 for examples of making recommended auxin concentrations with end-use formulation products for quick-dip applications.

Abutilon spp. Abutilon, Flowering Maple, Chinese Bell Flower. Seeds germinate without difficulty. Can also be rooted by leafy cuttings under mist from summer through fall.

Acacia spp. Acacia.

Seed. Generally propagated by seed. The impervious seed coats must be softened before planting by soaking in concentrated sulfuric acid for 20 minutes to 2 hours or by pouring boiling water over the seeds and allowing them to soak for 12 hours in the gradually cooling water. *A. cyanophylla* (beach acacia, golden willow), *A. farnesiana* (sweet acacia, West Indian blackthorn), and *A. koa* (koa acacia) seeds are scarified by soaking in warm water overnight. All but the youngest plants are difficult to transplant because of a pronounced taproot.

Cuttings. Leafy cuttings of partially matured wood can be rooted under mist if treated with 8,000 ppm IBA talc. *A. redolens* and *A. subprosa* root during the fall with 3,000 ppm IBA liquid. Cuttings with heels form vigorous new wood (*85*). *A. koa* is difficult to root from stem cuttings, and better results have occurred with micropropagation.

Micropropagation. Micropropagation of *A. melanoxylon* has been reported (*207, 259*).

> Extreme care should be used when handling concentrated acids for scarification. Protective clothing, including safety glasses or face shields, should be worn. Use of the acid ideally should be under a laboratory hood, or at least in a well-ventilated area. The acid should be properly disposed of after use. It is important that seeds be *thoroughly washed* after acid treatment before storage or handling.

Acer spp. Maple. (*359*). Various methods of propagation are used—seeds, grafting, budding, cutting, layering, and micropropagation (*53, 215*).

Seeds. Most maples produce seed in the fall and have one or more types of dormancy. Collecting seed before it is completely mature and not subjecting it to drying will minimize germination delay by avoiding a hard seed coat and the possibility of secondary dormancy. Excising embryos from fresh seed and incubating in cytokinin helps circumvent testa-imposed dormancy and the need to stratify—and results in germination within one week (*384*).

Two species—*A. rubrum* and *A. saccharinum*—produce seed in the spring. Such spring-ripening seeds should be gathered promptly when mature and sown immediately without drying, since the seed does not store well. Typically, seed will germinate within a seven to ten day period.

For other species, stratification, usually for 90 days at 4° C (40° F), followed by spring planting, gives good germination. Fall planting out-of-doors may be done if the seeds are first soaked for a week, changing the water daily; seed can also be soaked under running water. Seeds of the Japanese maple, *A. palmatum,* germinate satisfactorily if they are placed in warm water (about 43° C; 110° F) and allowed to soak for two days, followed by stratification for 60 to 120 days. Soaking seeds of *A. rubrum* and *A. negundo* in cold running water for five days and two weeks, respectively, before planting may increase germination. Seeds of *A. negundo, A. barbatum,* and *A. floridanum* are treated with a cold-warm-cold stratification. *Acer* seeds should not be allowed to dry out.

***Cuttings* (*115, 377*).** Leafy Japanese maple (*A. palmatum*) cuttings, as well as those of other Asiatic maples, will root if they are made from tips of vigorous pencil-sized shoots in late spring and placed under mist. IBA at 8,000 to 20,000 ppm in talc or quick-dips has given good results in rooting leafy cuttings of various *Acer* species under mist and over bottom heat in the greenhouse. In general, it is best to wait until the teminal bud has formed, but before the last set of leaves has fully expanded. Semi-hardwood cuttings of *A. floridanum* are propagated successfully with 8,000 ppm IBA talc. Softwood cuttings of *A. griseum* from stock blocks are rooted successfully with 8,000 ppm IBA talc (*191*) or 5,000 ppm IBA quick-dip. Etiolation of stock plants can improve rooting of softwood basal stem cuttings from five-year-old plants of paper bark maple (*A. griseum*) (*249*). *A. macrophyllum* roots well from softwood cuttings (June and July in Washington) treated with a talc formulation of 8,000 ppm IBA (*189*).

A. negundo semi-hardwood cuttings are rooted with 8,000 ppm IBA talc. Cuttings of selected *A. rubrum* and *A.* × *freemanii* cultivars can be rooted from single-node cuttings treated with 3,000 to 8,000 ppm IBA talc applications (*385*). Single-node cuttings of *A. glabrum* subspecies *Douglasii* (Douglas maple) root with 8,000 ppm IBA talc; rooted cuttings should be overwintered before transplanting (*353*). It is often difficult to overwinter rooted maple cuttings. To overcome this problem, new growth should be induced on the cuttings, shortly after rooting, by using supplementary lighting and supplying fertilizers. Rooting earlier in the season also allows more time to encourage new growth.

In Alabama, cuttings of dwarf Japanese maple (*A. palmatum*) are taken in early spring, as soon as the new growth has hardened, and quick-dipped in 6,250 ppm IBA/2,500 ppm NAA (*201*). Liners are overwintered in a cool house, where temperatures do not drop below −3° C (25° F) and planted in the following spring. Hardwood cuttings of *Acer palmatum* taken in midwinter have been rooted successfully in the greenhouse after wounding and treating with IBA (*72*). *A. truncatum* softwood cuttings taken from trunk sprouts of a ten-year-old tree had 79 percent rooting with a 5,000-ppm IBA ten-second quick-dip (*274*). Softwood cuttings taken in August (Kansas) from three-year-old seedling stock root well with a 5,000 ppm IBA quick-dip. Growth after rooting via long-day manipulation is important for winter survival of transplanted rooted cuttings and cuttings left in the propagation bed before lifting in spring.

Sugar maple (*A. saccharum*) cuttings are best taken in late spring after cessation of shoot elongation, then rooted under mist. They are more difficult to root—IBA treatments give variable results (*118*).

Softwood cuttings (June to August in Kansas) of *A. truncatum* root best with an IBA quick-dip of 1,000 ppm. Cuttings taken in August root readily, but are difficult to overwinter unless kept under long days to encourage some growth before winter (*277*).

Grafting. *Acer palmatum* seedlings are used as the rootstock for Japanese maple cultivars. In Oklahoma, seedling rootstock of *A. palmatum* is selected from the more vigorous green-leaved cultivars, rather from red-leaved cultivars (*148*). Seedling rootstocks of *A. saccharum* are used for sugar maples, *A. rubrum* for red maple cultivars, *A. pseudoplatanus* for striped maple (*A. pennsylvanicum*), and *A. platanoides* for such Norway maple clones as 'Crimson King,' 'Schwedler,' and the pyramidal forms. There is some evidence of delayed incompatibility in using *A. saccharinum* as a rootstock for red maple cultivars.

In Oregon, Japanese maples are direct seeded in liner pots in February, and by August the seedling rootstock is of graftable size. In other systems, seedling rootstock plants are grown for one year in a seed bed. Then in the fall or early spring, they are dug and transplanted into small pots and grown in propagating frames through the second summer. In late winter, the rootstock plants are brought into the greenhouse preparatory to grafting. As soon as roots show signs of growth, the rootstock is ready for grafting. Dormant scions are taken from outdoor plants. The side-veneer graft is ordinarily used with *A. palmatum* 'Dissectum,' *A. pensylvanicum, A. pseudoplatanus,* and *A. saccharinum* (*164*).

In Southern California, summer-grafted *A. palmatum* cultivars are side grafted and wrapped with budding rubbers or poly tape for three weeks with 90 percent grafting success. It is important to completely remove the leaves of the scion, prior to grafting in the summer. In Canada, an apical splice graft is preferred to the traditional side-veneer graft for bench grafting cultivars of *A. palmatum* with containerized seedling rootstock (*200*).

See the published list of *Acer* rootstocks that can be used for interspecific grafting of rare and common *Acer* species (*360*). In general, scions of "milky" sap groups will not graft on "non-milky" rootstock, and conversely, "non-milky" scions will not take on "milky" rootstock, i.e., use *A. platanoides* rootstock for scions of *A. cappadocicum, A. catalpifolium, A. lobelii* and other cultivars. These scions will not take on *A. pseudoplatanus* rootstock, which is "non-milky."

Maples are also T-budded and chip budded on one-year seedlings in the nursery row, the buds being inserted from mid- to late summer. In T-budding on active rootstock, the wood (xylem) is removed from the bud shield, which then consists only of the actual bud and attached bark. The seedling is cut back to the bud the following spring just as growth is starting. *A. platanoides* 'Crimson King' is more successfully chip budded than T-budded.

Micropropagation. Maple species are commercially micropropagated.

Actinidia spp. Fast-growing vigorous vines with fragrant flower and edible fruit. *A. hemsleyana* is a vigorous, narrow-leaved climbing vine that grows to 10 m (35 ft), and has attractive red soft bristles on the new growth. Gooseberry *(A. arguta)* can be propagated by seed, which is cold stratified for three months. It is best to use cuttings selected from female plants for flower and fruit characteristics. Can be propagated from single-node cuttings with talc applications of 8,000 ppm IBA (*244*). *Actinidia* is commercially micropropagated (*387*).

Aesculus spp. Buckeye, Horse chestnut (*115*).

Seed. This tree may be propagated by seeds, but prompt sowing or stratification after gathering in the fall is necessary. If the seeds lose their waxy appearance and become wrinkled, their viability will be reduced. For best germination, seeds of *A. arguta* (Texas buckeye), *A* × *carnea* (red horse chestnut), *A. hippocastanum* (common horse chestnut), *A. indica, A. octandra, A. sylvatica,* and *A. turbinata* should be stratified for four months at about 4° C (40° F) immediately after collecting. *A. californica* and *A. parvifolia* seed need no pretreatment. Though no dormancy has been reported, for *A. pavia,* a one-month cold stratification is recommended (*115*). In southern California, seeds of *A. carnea* and *A. hippocastanum* are leached in a bag with running water for ten days; seeds germinate within this leaching period. Seeds of buckeye are poisonous if eaten.

Cuttings. Cuttings of *A. parviflora* are taken from new growth on stock plants from June to July (Chicago). Leafy single node cuttings with two leaves and 2.5+cm (1+in.) of stem tissue are treated with 1,250 to 2,000 ppm IBA + 625 to 1,000 ppm NAA—the higher IBA/NAA level is used if the cutting wood is more lignified. The cuttings are inserted with buds set at below the top surface of the propagation medium. Cuttings will not root under very moist medium conditions (*217*). Terminal stem cuttings 15 cm (6 in.) taken in May rooted better than those taken in July (North Carolina); a 2,500 ppm IBA quick-dip in ethanol gave best results (*36*). *A. parviflora* can also be propagated by underground stem pieces treated with 0 to 5,000 ppm IBA quick-dip and propagated by root cuttings.

Grafting. T-budding, cleft grafting, or bench grafting, using the whip graft, can be used for selected cultivars grafted on *A. hippocastanum* as the rootstock. Although normally propagated by seed, grafting is done to: (1) perpetuate superior sterile forms (*A. hippocastanum* 'Baumannii'), (2) more quickly produce a large-caliper tree, (3) avoid suckering problems, and (4) produce a plant with better flowers and/or foliage.

Agarita. *See Berberis trifoliata (Mahonia trifoliata).*

Ailanthus altissima. Ailanthus, Tree of Heaven. Seed propagation is easy, and self-sowing usually occurs when both male and female trees are grown close together. Embryo dormancy is present in freshly harvested seed. Stratification at about 4° C (40° F) for two months aids germination. Seed propagation produces both types of trees, but planting male trees should be avoided, since the staminate flowers produce an obnoxious odor. The more desirable female trees can be propagated by root cuttings planted in the spring. *A. malabarica* (*122*) and *A. altissima* (*389*) have been micropropagated.

Alberta magna. An attractive, flowering evergreen species, native to South Africa. It is used as a small landscape tree and has potential as flowering pot plant. Seed propagation is poor and the seedlings have a long juvenile period. Cuttings taken from mature trees root poorly. Mature-phase scionwood has been successfully wedge grafted onto micropropagated juvenile plants, which also speeds up flowering (*26*).

Albizia julibrissin. Albizzia, Silk tree, Mimosa. This species is propagated by seed. The impermeable seed coat is scarified by a 30-minute soak in sulfuric acid. Seed selection is important for obtaining plants with good flower color. Stem cuttings do not root, but root cuttings about 7.5 cm (3 in.) long and 12 mm (0.5 in.) or more in diameter taken and planted in early spring are successful (*143*). Juvenile shoots arising from root pieces can be removed and rooted with 1,000 to 3,000 IBA quick-dip or talc (*108*). Seedling material has been successfully micropropagated (*316*).

Alder. *See Alnus.*

Allamanda cathartica. Yellow allamanda, Golden Trumpet. Easily propagated using semi-hardwood cuttings in summer and fall (Texas) with 1,000 to 2,000 ppm IBA liquid.

Almond (Dwarf, Flowering). *See Prunus.*

Alnus spp. Alder.

Seed. *A. incana* (gray alder), *A. serrulata* (tag alder), *A. rhombifolia* (sierra alder), *A. rugosa* (speckled alder), *A. cordata* (Italian alder), and *A. glutinosa* (European alder). Seeds of this small deciduous tree should be thoroughly cleaned and can be planted immediately after fall harvest, but three to five months of stratification improves germination. Seed will also germinate if given a 24-hour soak in 0.2 percent potassium nitrate. Light appears to be beneficial in germinating *Alnus* seed.

Cuttings. Softwood cuttings of *A. glutinosa* root with 3,000 ppm IBA talc. *A. incana* stem cuttings are rooted with 8,000 ppm IBA talc, and can also be grafted on potted seedling understock. Cultivars of *A. glutinosa* can also be side-veneer grafted (*197*).

Micropropagation. Alder is commercially micropropagated.

Amelanchier spp. Serviceberry. (*109, 115*). These trees and shrubs are generally seed-propagated.

Seed. Seeds show embryo dormancy which can be overcome by stratification at 2° C (36° F) for three to six months. Seeds should not be allowed to dry out. Seeds of *A. alnifolia* (Saskatoon) and *A. laevis* (Allegheny serviceberry) require 15 to 30 minutes of sulfuric acid scarification prior to three months of cold stratification.

Cuttings. Many species are readily propagated by leafy softwood cuttings taken before the terminal bud has formed and when the new growth is several inches long—and then rooted under mist. Talc and quick-dips of IBA from 1,000 to 10,000 ppm enhance rooting. At 40° N latitude, softwood to semi-hardwood cuttings of *A. laevis* taken in mid-May to mid-June root best with a K-IBA of 2,500 ppm or greater (*339*).

Micropropagation. The newer cultivars 'Prince Charles' and 'Princess Diana,' which are selections of *A. laevis* and *A.* × *grandiflora,* respectively, are commercially micropropagated.

Amorpha spp. Indigo bush (*108*). A large deciduous shrub in the legume family. They have an impermeable seed coat which requires acid scarification. *A. canescens* (leadplant amphora) seed is acid-scarified for 15 minutes followed by two to eight weeks of stratification. Seed of *A. fruticosa* (indigo bush amorpha) can be fall planted or given ten minutes of acid scarification; easily roots from untreated softwood cuttings. *A. canescens* is rooted with cuttings treated with 8,000 ppm IBA talc.

Anacho Orchid tree. *See Bauhinia.*

Angelica tree. *See Aralia.*

Anise. *See Illicium.*

Apricot (Japanese flowering). *See Prunus.*

Aralia spp. Walking stick, Angelica tree. Suckering shrubs and trees. Seeds should be sown in early fall. Scarify with sulfuric acid for 30 to 40 minutes, and cold stratify for three months to overcome the double dormancy requirement. *A. spinosa* has a three-month cold stratification requirement. Asexual propagation is with 10 to 13 cm (4 to 5 in.) root cuttings, which are taken in fall and stored in cool storage until spring planting (*108*).

Araucaria spp. These large evergreen trees are seed propagated.

Seed. Seeds of *A. bidwillii* (bunya-bunya), *A. columnaris* (Cook or New Caledonia pine), *A. araucana* (monkey puzzle tree), and *A. cunninghamii* (colonial pine) must be cleaned and then sown within a month after collection. There are no dormancy problems and germination time is two to four weeks. Plant seed of *A. heterophylla* (Norfolk Island pine) directly in pots, placing the seed vertical and burying to half the seed height. Keeping seeds moist is critical for the first 10 to 14 days after planting and can be accomplished by misting once each hour (Texas).

Cuttings. Can also be propagated by semi-hardwood cuttings. Cuttings of side branches will root but produce horizontally growing plants. Plants produced from terminal cuttings grow upright.

Micropropagation. *A. cunninghamii* and other species can be micropropated (*61*).

Arborvitae. *See Thuja.*

Arbutus menziesii. Pacific Madrone. This Pacific coast evergreen tree is usually propagated by seeds, which are stratified for 2 to 3 months at 2 to 4° C (35 to 40° F) (*245*). Seedlings are started in flats, then transferred to pots. They are difficult to transplant and should be set in their permanent location when not over 18 in. tall. Propagation can be done by cuttings, layering, and also grafting. *Arbutus* is commercially micropropagated.

Arctostaphylos spp. Manzanita (*A.* spp.), bearberry *(A. uva-ursi)*.

Seed. There is a double dormancy requirement which makes seed propagation difficult. Seed scarification can be done by treating with sulfuric acid from three to six hours (*133*). Seed can also be scarified by inserting in boiling water—then immediately turning off the heat and soaking the seed in the gradually cooling water for 24 hours.

Cuttings. Cutting propagation is much more practical. Terminal cuttings are taken from November to February (California), submerged in 5 to 10 percent Clorox, given a 1,000 ppm IBA plus 500 ppm NAA quick-dip or 8,000 ppm IBA talc, and rooted under intermittent mist with bottom heat [21° C (70° F)].

Micropropagation. *Arctostaphylos* is commercially micropropagated.

Ardisia spp. Coralberry, spiceberry *(A. crenata)*, and Japanese *(A. japonica)*. These medium-sized evergreen shrubs and ground covers *(A. japonica)* can be seed-propagated without stratification. *A. crenata* can be propagated by softwood cuttings. *A. japonica* is easily rooted with a 1,000 ppm IBA quick-dip, and this stoloniferous plant can also be divided (*108*).

Aronia arbutifolia. Red Chokeberry *(Aronia arbutifolia)* and black *(A. melanocarpa)*. Propagated by seed, which should be collected in autumn, then stratified for about three months at 5° C (41° F) before planting. Can also be started by leafy cuttings treated with a 4,000 ppm IBA quick-dip under mist, or by suckers.

Asarum europaeum. European wild ginger. A glossy-leaved, spreading ground cover. Can be propagated by division in spring or fall (*334*).

Ash. *See Fraxinus.*

Asimina triloba. Common paw-paw and dwarf pawpaw *(A. parviflora)*. This small tree and deciduous shrub are propagated primarily by seed which is stratified at 4° C (40° F) for two to three months. A southern ecotype of *A. triloba* easily germinates without stratification (*115*). Successful cutting propagation has not been reported.

Aspen. *See Populus.*

Aspen (Quaking). *See Populus.*

Avocado. *See* Chapter 19.

Australian Pine. *See Casuarina.*

Azalea. *See Rhododendron.*

Baccharis spp. False willow, narrowleaf baccharis *(B. angustifolia)*, and eastern baccharis *(B. halimifolia)*. These small evergreen shrubs have excellent salt tolerance. Baccharis is seed-propagated with no seed pretreatment required. Germination occurs within one to two weeks. In Arizona, cuttings of male plants of *B. sarothodes* (Desert broom) are trimmed to 18 cm (7 in.), dipped in 8,000 ppm IBA talc, stuck in a peat:perlite medium, and rooted under mist (*319*). *B. halimifolia* roots easily from softwood cuttings treated with 2,500 ppm K-IBA.

Bald Cypress. *See Taxodium.*

Bamboo. *See Bambusa.*

Bambusa spp. Bamboo (*Arundinaria, Dendrocalamus, Gigantochloa, Phyllostachys, Pseudoasa, Sasa, Schizostachym, Sinarundinaria, Thamnocalamus, Thyrsostachys* spp.) (*299, 324*). Bamboo has about 1,000 species in some 50 genera. Two classes of bamboo are commercially propagated in containers (Louisiana) to eliminate much of the hand labor and extensive field stock space needed in the old traditional method of hand digging and separating established clumps.

Division. Clump-forming bamboo (*pachymorphs*—with constricted rhizomes) are propagated by dividing the young, peripheral rhizomes in late spring or early summer; the culms (above-ground stem with branches and leaves) are cut back and the rhizome is containerized in a 5 bark to 1 sand medium with slow-release fertilizer, and watered as needed. The following spring, rhizomes from the containerized stock plant are further divided and transplanted with attached culms and roots to new containers for further propagation or finishing-off as a marketable plant.

In Oregon, bamboo from five-gallon containers are sawed vertically into quarters, then replanted into five-gallon containers with 100 percent survival and faster reestablishment.

The running-type bamboos (*leptomorphs*—with vigorous rhizomes extending beyond parent plants) are more cold-tolerant but less desirable in the garden. They are propagated by the division technique described above for rhizomes less than 1 cm (½ in.), or by using 30-cm (1-ft) rhizomes 2 cm (¾ in.) thick which are containerized and grown for two years as stock plants. Rhizomes 15 cm (6 in.) long are then containerized and grown until they are a finished crop. Prevention of drying during transplanting is important. Best results are obtained from rhizomes taken and planted in late winter or early spring before the buds begin to elongate.

Micropropagation. Some 15 genera and 54 species of bamboo have been micropropagated, which offers potential for those bamboo species which are currently difficult to propagate or transplant (*291*). Somatic embryogenesis and successful transplanting of plantlets has been reported (*307*).

Banksia spp. (27). Banksia are native Australian shrubs and small trees in the *Proteaceae* family. Two species with promise as a plantation-grown flower crop with export potential are *B. coccinea* (scarlet banksia) and *B. menziesii* (raspberry frost banksia).

Seed. These species are propagated by seed, although they are difficult to remove from the seed-containing structures. One method is to soak seed several days in water and then dry quickly. There are no pregermination treatments; however, 15° C is the optimum germination (20 days) temperature for *B. coccinea*.

Cuttings. This species can also be propagated from semi-hardwood cuttings using a quick-dip of 8,000 to 12,000 ppm IBA in 50 percent ethanol. Cuttings should be wounded and inserted in a porous mix under intermittent mist with 25° C (77° F) bottom heat. It is best to root cuttings directly in liner pots and avoid root disturbance. Graft incompatibility limits successful grafting of *Banksia* spp. to rootstock—resistant to *phytohphora* and tolerant of high soil phosphorus and poorly drained sites.

Barberry. *See Berberis.*

Basswood. *See Tilia.*

Bauhinia congesta. Anacho Orchid Tree (*195*). This multistemmed deciduous shrub has emerald leaves, which are cleft to the petiole, and produces abundant white flowers from March through summer (Texas). Seeds germinate easily without pregermination treatment. *B. purpurea* has been micropropagated (*223*).

Bayberry. *See Myrica.*

Bay Laurel. *See Laurus nobilis.*

Beaucarnea recurvata. Ponytail palm. *See* palms.

Beech. *See Fagus.*

Berberis spp. Barberry.

Seed. Propagated without difficulty by fall-sowing or by spring-sowing seeds that have been stratified for two to six weeks at 4° C (40° F). Some species require up to two to three months of stratification or seed can be sown in fall. It is important to remove all pulp from seeds. Seedlings are susceptible to damping-off and some to black stem rust (Oregon).

Cuttings. Berberis are generally propagated by cuttings. Semi-hardwood cuttings taken from spring to fall can be rooted under mist (*373*). Some propagators have success with hardwood cuttings (leaves still attached) taken from October to November. IBA from 2,000 to 8,000 ppm aids rooting in some species. *B. thunbergii* 'Atropurpurea nana' is propagated with 5 to 6 in. (14 cm) semi-hardwood cuttings in May and June (Texas), which are collected when the new growth is firm at the cutting base (*19*). The stem of the cutting should have a greenish-yellow color; avoid using brown wood. Cuttings are quick-dipped into 1,870 ppm IBA with 50 percent alcohol and direct stuck into 5 cm (2¼ in.) liner pots. Timing, application of mist, and the hardening-off process are critical. In Georgia, cuttings of *B. thunbergii* cultivars and other *Berberis* species are taken from May 15 through July; the 8 to 10 cm (3 to 4 in.) cuttings are wounded 6 to 12 mm (¼ to ½ in.) from the base and quick-dipped with 1,500 to 3,500 ppm IBA. Greenhouse grafting of some selected types is also practiced, and layering is done occasionally. Division of crowns is useful for small quantities of plants.

Micropropagation. *B. thunbergii* 'Atropurpurea' is commercially micropropagated (*355*).

Berberis trifoliata. *(Mahonia trifoliata),* Agarita (*195*). A dense evergreen rounded shrub with holly-like leaves that is an early spring bloomer in Texas with yellow fragrant flowers in March. This drought-resistant plant also has edible red berries and grows best in full sun. Seeds require a two- to three-month cold stratification.

Betula spp. Birch.

Seed. Seeds should be either fall-planted or spring-planted following stratification at about 4° C (40° F) for one to three months. *Betula nigra* seeds mature in the spring; however, they do not store well. If planted promptly, the seeds will germinate at once without pretreatment. Fresh birch seeds can also be soaked for 24 hours with 0.2 percent potassium, and cold stratification avoided.

Cuttings. Birch is considered difficult to propagate by cuttings, but leafy, semi-hardwood cuttings root under mist if taken in midsummer. Softwood cuttings of *B. alleghaniensis, B. lenta, B. nigra, B. papyrifera,* and *B. pendula* are rooted with 8,000 ppm IBA talc. Softwood cuttings (July in Ohio) of *B. nigra* root best with a 1,000 ppm IBA dip (*189*). Softwood one-node cuttings of 18-year-old *B. papyrifera* collected in mid-July (Vermont) treated with 4,000 to 6,000 ppm IBA in 50 percent ethanol had 40 to 60 percent rooting; problems occurred with survival of post-rooted cuttings (*280*).

Grafting. Some of the selected, weeping forms are grafted on *Betula pubescens* or *B. pendula* seedlings (*102*). T-budding and chip budding have been used in field-grown rootstocks of *B. pendula* (*96*). In the selection of birches that will tolerate the wet soils and high temperatures of the southeast U.S., whitespire Japanese birch *(B. platyphylla)*—which has heat and drought resistance, but is intolerant of poorly drained soils—has proven ideal when grafted to rootstock of river birch *(B. nigra)* or European birch rootstock *(B. pendula)* (*295, 296*).

Micropropagation. Birch is commercially micropropagated (*53, 252*).

Birch. *See Betula.*

Bittersweet. *See Celastrus.*

Black Gum. *See Nyssa.*

Black Haw. *See Bumelia.*

Black Tupelo. *See Nyssa.*

Boston Ivy. *See Parthenocissus.*

Bottlebrush. *See Callistemon.*

Bougainvillea spp. Bougainvillea. This showy tropical, woody, evergreen vine grows outdoors only in mild climates, and is propagated by leafy cuttings taken at any time of the year. Rooting is aided by supplemental bottom heat. Difficult-to-root cultivars should be treated with IBA. In Australia, semi-hardwood (wood changing from green to brown) 4- to 6-node cuttings are cut to 10 to 13 cm (4 to 5 in.) length (*190*). IBA at 4,000 to 16,000 ppm is applied, depending on the cultivar and the time of year. Bottom heat is used to maintain a propagation bed surface temperature of 25 to 27° C (77 to 81° F). Intermittent mist is used, but bougainvilleas do not tolerate a wet environment. Can be micropropagated by shoot-tip culture (*80*).

Boxwood. *See Buxus.*

Boxwood (Oregon). *See Paxistima.*

Breath of Heaven. *See Diosma (Ericoides).*

Broom. *See Cytisus.*

Buckeye. *See Aesculus.*

Buckeye (Mexican). *See Ungnadia.*

Buckthorn. *See Rhamnus.*

Buddleia spp. Butterfly Bush. Seeds require no pretreatment. Seeds started in the greenhouse in early spring will provide flowering plants by fall, although reproduction is not genetically true by seed. Generally propagated by softwood or semi-hardwood cuttings taken in the summer, treated with 8,000 ppm IBA, and rooted under mist. *Buddleia* is commercially micropropagated.

Buffaloberry. *See Shepherdia.*

Bumelia lanuginosa. Bumelia, Black Haw, or False Buckthorn. This deciduous small tree is propagated by seed. Seed is fall-collected and stratified at 4° C (40° F) for two months. Seed is sown in the spring in outdoor beds.

Bunya-Bunya. *See Araucaria.*

Butia capitata. Pindo palm. *See* palms.

Butterfly Bush. *See Buddleia.*

Buxus spp. Boxwood (*Buxus sempervirens* L. and *B. microphylla*). Seeds are rarely used, because of the very slow growth of the seedlings. Cuttings are commonly used—either softwood taken in spring or summer, semi-hardwood taken in late summer and fall, or hardwood cuttings taken in the winter (Texas). Quick-dips of 2,500 to 5,000 ppm IBA enhance rooting (*366*). An Oregon nursery uses a light application of liquid fertilizer (about 50 ppm N) after callus formation to help speed root development. Young liner plants should always be grown in containers or transplanted with balls of soil around their roots.

Callistemon spp. BottleBrush. Although seeds germinate without difficulty, seedlings should be avoided because many of them prove worthless as ornamentals. The preferred method of propagation is by semi-hardwood cuttings taken from selected cultivars. These root under mist quite easily.

Calluna vulgaris. Heather. *See Erica* (Heath).

Calocedrus decurrens. Incense Cedar. Propagated by seed. Germination is promoted by a stratification period of about eight weeks at 0 to 4° C (32 to 40° F). Can be rooted from cuttings treated with quick-dips of 2,500 ppm NAA, or 2,500 ppm NAA plus 2,500 ppm IBA. *C. decurrens* can be grafted on Thuja.

Calycanthus spp. ***(108).*** Sweetshrub, Allspice. Stoloniferous shrubs. *C. floridus* is common to the eastern and southern United States. *C. floridus* 'Athens' ('Katherine') is very fragrant with yellow flowers, while *C. occidentalis* has reddish flowers. Seed should be collected when receptacles change from green to brown and immediately planted. Seed collected later should be cold stratified for three months. Good rooting success is reported with summer semi-hardwood cuttings using 2,000 to 3,000 ppm IBA, or a 10,000 ppm K-IBA quick-dip.

Camellia spp. Camellias can be propagated by seed, cuttings, grafting, layering, and micropropagation (*242, 314*). They do not come true from seed. Seedlings are used in breeding new cultivars, as rootstocks for grafting, or in growing hedges where foliage is the only consideration. To perpetuate cultivars—cuttings, grafts and micropropagation are used.

Seed. In the fall when the capsules begin to turn reddish brown and split, seeds should be gathered before the seed coats harden and the seeds become scattered. The seeds should not be allowed to dry out and should be planted before the seed coats harden. If the seeds must be stored for long periods, they will keep satisfactorily mixed with ground charcoal and stored in an airtight container placed in a cool location. After the hard seed coats develop, scarification is accomplished by pouring boiling water over the seeds and allowing them to remain in the cooling water for 24 hours. It takes four to seven years to bring camellias into flowering from seeds.

Cuttings. Most *C. japonica, C. oliefera, C. sasanqua, C. reticulata,* and *C. sinensis* cultivars and hybrids are produced commercially from cuttings. Cuttings are best taken from midsummer to mid-fall from the flush of growth after the wood has matured somewhat and changed from green to light brown in color. Tip cuttings are used, 8 to 15 cm (3 to 6 in.) long, with two or three terminal leaves. Rooting is much improved if the cuttings are treated with IBA quick-dips of 3,000 to 5,000 ppm, 6,000 to 8,000 ppm IBA plus 2,500 ppm NAA, or 3,000 to 8,000 ppm IBA talc (*21, 30, 115*). *C. reticulata* hybrids root best with 8,000 to 20,000 ppm IBA talc (*308*). Wounding the base of the cuttings before they are treated is also likely to improve rooting. Cuttings root best either in a polyethylene closed frame or under mist.

Camellias also may be started as leaf-bud cuttings which are handled as stem cuttings. In this case, excessive auxin concentrations should be avoided because this may inhibit development of the single bud.

Grafting. Camellias are frequently grafted—since some cultivars (e.g., 'Pink Pagoda') are poor rooters and grow poorly on their own roots—for multiplying new cultivars faster and for changing cultivars of older established plants. Vigorous seedlings or rooted cuttings of either *C. japonica* or *C. sasanqua* can be used as rootstocks for grafting. Any of the side-graft methods is suitable. *C. japonica, C. sinensis,* and *C. reticulate* can be container-grafted by using a whip graft.

Micropropagation. Camellias are commercially micropropagated.

Camphor Tree. *See Cinnamomum.*

Campsis spp. Trumpet Creeper. This vine is usually propagated by cuttings, but seeds can also be used. With the latter method, stratification for two months at 4 to 10° C (40 to 50° F) hastens but does not increase germination. Both softwood and hardwood cuttings root readily and can be quick-dipped with 1,000 ppm IBA. *C. radicans* can be started by root cuttings. One Oregon nursery takes root pieces [5 cm length × 0.6 cm width (2 × ¼ in.)], which are direct-stuck vertically with 0.3 cm (⅛ in.) of the root showing. Watch the polarity of the root cutting.

Cape Jasmine. *See Gardenia.*

Caragana pygmaea. Pygmy (Dwarf) Pea shrub. Shrub with bright yellow, pealike pendulous flowers. Propagated by softwood tip cuttings, which are treated with 1,000 ppm IBA + 500 ppm NAA quick-dip. Rooted cuttings are transplanted the following spring.

Carob. *See Ceratonia.*

Carpinus spp. Hornbeam.

Seed. For seed propagation, collect seeds while the wings are still soft and pliable. Check to make sure that seeds have embryos, which may be absent after stressful growing conditions. Do not allow seeds to dry out. Sow outdoors in autumn or stratify (three to four months), overwinter, and sow in spring. If the seed dries, a hard seed coat germination block develops, and this double dormancy requires scarification before stratification; two months of warm stratification followed by two months of cold stratification has been used with *C. caroliniana* (*52*).

Cuttings. Summer stem cuttings of *C. betulus* 'Fastiagata' root with a quick-dip of 20,000 ppm IBA, *C. Carolina* 'Pyramidalis' with 16,000 ppm IBA, and *C. japonica* with a talc application of 3,000 ppm IBA (*115*). Wounding cuttings has been recommended. Softwood cuttings may root better than semi-hardwood cuttings. Stock plant etiolation, shading, and stem banding improved rooting of *C. betulus* softwood cuttings (*250*). Cultivars may be side-veneer grafted or budded on seedlings of the same species.

Carya spp. Hickory, Water Hickory *(Carya aquatica),* bitternut hickory *(C. cordiformis),* pignut hickory *(C. glabra),* shellbark *(C. laciniosa),* nutmeg hickory *(C. myristicae-formis),* shagbark hickory *(C. ovata),* and Mockernut hickory *(C. tomentosa).* Some of these tree species have edible fruits. Seeds (nuts) are fall-collected and will erratically germinate without pretreatment. Most species have a three- to four-month cold stratification requirement for uniform germination. Few reports exist on successful rooting or grafting of hickory. Cutting propagation is difficult. However, Medina (*258*) reported high rooting success with pecans using mounding and stooling techniques. Pecan *(C. illinoiensis)* is successfully grafted by patch budding, the inlay bark graft, and four-flap (banana graft). Some of these asexual propagation techniques could be applied to other *Carya* species (see Chapter 19).

Castanea spp. Chestnut. American chestnut *(Castanea dentata),* Chinese chestnut *(C. mollissima),* and sweet chestnut *(C. sativa).*

Seed. These deciduous medium-sized trees are predominantly seed propagated. Seeds are harvested in early fall and stratified for two to three months. It is important that seeds not dry out.

Cuttings. Mature cuttings are very difficult to root. Rooting success has occurred with juvenile cuttings of *C. mollissima* (8,000 ppm IBA talc or 7,000 ppm IBA in 95 percent alcohol) (*115*). Limited success has occurred with hardwood cuttings quick-dipped with 12,000 ppm IBA and with air layering. Leafy softwood summer cuttings have rooted with 1,250 ppm IBA five-second quick-dip in a high-humidity ventilated fog system; pretreatment with a blanching tape also improved rooting (*286*). Grafting success has occurred with cleft graftage, inlay bark graftage, and chip budding.

Micropropagation. Juvenile (*305*) and mature (*361*) explants of *Castanea* have been successfully micropropagated (*327*).

Casuarina spp. Ironwood *(C. cunninghamiana),* Australian pine *(C. equisetifolia),* and longleaf Casuarina *(C. glauca).* These large evergreen trees from Australia are easily propagated from seeds. Seeds require no pretreatment, and after harvesting and cleaning, seeds are stored and spring sown in outdoor beds.

Catalpa spp. Catalpa.

Seed. Catalpa seeds germinate readily without any pretreatment. They are stored dry, overwintered at room temperature, and planted in late spring. Catalpa species can also be propagated in summer by softwood cuttings rooted under mist. Hardwood cuttings of *C. bignonioides* and *C. speciosa* root with 8,000 ppm IBA talc (*115*). The terminal bud of the cuttings should be removed.

Grafting. *C. bignonioides* 'Nana' is often budded or grafted high on stems of *C. speciosa,* giving the "umbrella tree" effect. A strong shoot is forced from a one-year-old seedling rootstock, which is then budded with several buds in the fall at a height of 1.8 m (6 ft).

Ceanothus spp. Ceanothus can be propagated by seed, cuttings, layering, and sometimes grafting.

Seed. Seeds must be gathered shortly before the capsules open or they will be lost. Those of *C. arboreus, C. cuneatus, C. jepsoni, C. megacarpus, C. oliganthus, C. rigidis,* and *C. thyrsiflorus* have only seed coat dormancy. Germination is aided by placing seeds in hot water [82 to 87° C (180 to 190° F)] and allowing them to cool for 12 to 24 hours. To obtain germination in other *Ceanothus* species, which have both seed-coat and embryo dormancy, the seed should be immersed in hot water (as described above), then stratified at 2 to 4° C (35 to 40° F) for two to three months.

Cuttings. Ceanothus hybrid cultivars 'Concha,' 'Frosty Blue,' 'Joyce Coulter,' and 'Victoria' are all cutting-grown. Semi-hardwood cuttings can be rooted under mist at any time from spring to fall. Rooting is enhanced with 1,000 to 3,000 ppm (and up to 5,000 ppm) IBA/NAA quick-dips (*333*). Terminal softwood cuttings taken from vigorously growing plants in containers root when treated with 1,000 ppm IBA.

Cedar. *See Cedrus.*

Cedar (Incense). *See Calocedrus.*

Cedrus spp. Cedar.

Seed. Seeds germinate if not permitted to dry out. No dormancy conditions occur, but soaking the seeds in water several hours before planting may be helpful. Also, a one-month cold stratification period will improve the germination rate.

Cuttings. *C. ailantica* and *C. libani* are difficult to root. *C. deodara* can be rooted by wounding cuttings and using bottom heat; cuttings are collected in late fall to early winter and quick-dipped with 5,000 ppm IBA (*269*).

Grafting. Side-veneer grafting of selected forms on one- or two-year-old potted seedling stocks may be done in the spring (or winter in southern California). Scions should be taken from vigorous terminal growth of current season's wood rather than from lateral shoots. In winter grafting, both lateral and terminal scions from the previous season's growth can be used, producing identical results. *C. atlantica* selections and other species are grafted on *C. deodara* seedling rootstock. Rooted cuttings of *C. deodara,* which are a year or older, also make acceptable rootstock material.

Celastrus spp. Bittersweet. These dioecious twining vines have male flowers on one plant and female on another, and the two types must be near each other to produce attractive berries.

Seed. To propagate, seeds must be removed from the berries and then fall-planted or stratified for about three months at 4° C (40° F) before planting.

Cuttings. Clones of known sex can be propagated by softwood cuttings taken in midsummer, or by hardwood cuttings taken in winter. Semi-hardwood summer cuttings root well when treated with 8,000 ppm IBA talc (*115*). IBA-treated softwood cuttings can also be used.

Celtis spp. Hackberry. Seeds are ordinarily used, sown either in the fall or stratified for two or three months at about 4° C (40° F) and planted in the spring (*255*). Clones of two species, *C. occidentalis* and *C. laevigata* (sugarberry), can be started by cuttings, but the rooting percentage is low. Grafting and chip budding also have been used with *C. occidentalis* and *C. laevigata* as rootstock for other species.

Cephalotaxus harringtonia. Japanese plum yew. This ornamental species also has anticarcinogenic attributes. Can be propagated by seed, which requires a three-month cold stratification period. Rooting by cuttings is difficult. An IBA talc or quick-dip of 5,000 to 10,000 ppm is reported to enhance rooting (*115*). Rooting of cuttings varies with cultivars. Can be micropropagated. Micropropagated shoots can be rooted *ex vitro* under mist with or without auxin (*202*). See *Taxus spp.*

Ceratonia siliqua. Carob, locust bean. This medium-sized evergreen tree is propagated primarily by seed. Seed is fall harvested, cleaned, scarified, and sown in greenhouse beds. The taproot is easily injured, so it is best to sow seeds in air-pruning flats. This species can be layered and grafted.

Cercis spp. Redbud.

Seed. Seed propagation is successful, but seed treatments are necessary because of dormancy resulting from an impervious seed coat plus a dormant embryo. Probably the most satisfactory treatment is a 30- to 60-minute soaking period in concentrated sulfuric acid, or an 82° C (180° F) hot-water soak; scarification treatments must be followed by stratification for three months at 2 to 4° C (35 to 40° F). A one-hour sulfuric acid scarification, followed by 35 days of cold stratification is optimum for Mexican redbud *(C. canadensis* var. *mexicana)* (*350*). Gibberellic acid and ethephon can overcome dormancy in nonchilled dormant seed (*156*). Outdoor fall-sowing of untreated seeds also may give good germination. Seed provenance is important, since ecotypic variation affects survival, dormancy, and plant growth potential (e.g., a Florida versus a Canadian seed source).

Cuttings. Propagating *Cersis* by cuttings can be difficult. Leafy softwood cuttings of some *Cercis* species root under mist if taken in spring or early summer. *C. canadensis* var. *mexicana* roots readily when 10 to 15 cm (4 to 6 in.) long, leafy terminal cuttings are taken four weeks after bud-break and treated with a 21,000 ppm K-IBA 5-second dip (*349*).

Grafting. T-budding in midsummer on *C. canadensis* seedlings is used commercially for *Cercis* cultivars such as *C. canadensis* var. *texensis 'Oklahoma'* (*371, 372*). Stick budding is used in California.

Micropropagation. *Cercis* is commercially micropropagated. Both *C. canadensis* and *C. canadensis* var. *mexicana* have been micropropagated and somatic embryos have been produced from micropropagated zygotic embryo explants of *C. canadensis* (*157, 159, 388*).

Chaenomeles spp. Quince (Flowering). Flowering quince is easily started by seeds, which should be fall-planted or stratified for two or three months at 4° C (40° F) before sowing. Clean the seeds from the fruit. Softwood cuttings treated with 1,000 to 5,000 ppm IBA root readily. Quince is commercially micropropagated.

Chamaecyparis spp. Chamaecyparis, False cypress. *C. thysoides* (Atlantic white cedar).

Seed. In the fall, cones are collected, dried, and the seeds knocked free. Seeds are then stratified at about 4° C (40° F) for two to three months. Some species such as *C. nootkatensis, C. thyoides,* and *C. praecox* have a double dormancy and a warm-cold stratification is required.

Cuttings. Cuttings of most species are not difficult to root, particularly if juvenile forms are used. They may be taken in fall and rooted in a cold frame or in winter and rooted under mist or a polytent using bottom heat. Quick-dips of 3,000 to 8,000 ppm IBA are used (*226*). *C. thyoides* roots well with hardwood or softwood cuttings 24 cm (9 in.) long; IBA up to 15,000 ppm was unnecessary for rooting, but increases root number (*188*). At high auxin levels, more primary roots are produced, while at lower to moderate auxin levels, more secondary roots are produced, which is more desirable.

Micropropagation. *C. nootkatensis* (Alaska yellow cedar) is micropropagated (*239*).

Chamelaucium spp. Waxflower. Small to medium-sized shrubs with waxy flowers. In Australia, terminal stem cuttings 8 to 10 cm (3 to 4 in.) long with growing points intact are dipped in a 2 percent sodium hypochlorite solution for disease prevention. Leaves are stripped from the basal 2 cm of the stem and cuttings quick-dipped in 2,000 ppm IBA in ethanol solution (*166*).

Cherry (Flowering). *See Prunus.*

Chestnut. *See Castanea.*

Chilopsis linearis. Desert willow. This small tree is a member of the *Bignoniaceae* family, and has funnel-form flowers in open terminal panicles that range in color from white to lavender to burgundy. Can be propagated by seed or cuttings (*351*). Seeds will germinate immediately after harvest or can be stored under refrigeration. Germination takes one to three weeks. Hardwood cuttings root easily. Softwood and semi-hardwood cuttings treated with less than 5,000 ppm IBA can be rooted either under mist or in a high-humidity chamber.

Chionanthus spp. Fringe Tree *(Chionanthus virginicus* and *C. retusus).*

Seed. Seed propagation can be used, but is very slow. Manually nicking the seed coat helps. There is a double dormancy requirement, with a three-month warm stratification period (watch for the radicle to emerge), followed by a three-month cold stratification period. Seeds can then be sown (*115*).

Cuttings. Cutting propagation of the Chinese fringe tree *(C. retusus)* generally has been considered almost impossible, but by taking softwood cuttings in late spring, rooting under mist, treating with 8,000 to 10,000 ppm IBA talc, and using a mixture of sand and vermiculite as the rooting medium, excellent rooting percentages can be obtained. Rooting improves when stock plants are kept juvenile or when serial propagation is used.

Cuttings of *C. virginicus* are very difficult to root, but some success has been reported in Alabama by taking cuttings in early spring from hardened-off new growth and quick-dipping with 1,250 ppm IBA (*201*). Wounding appears to be helpful, and soft-tip cuttings have been reported to root in limited percentages (*18*). Serially propagating those cuttings which root can increase rooting success.

Micropropagation. *Chionanthus* is commercially micropropagated.

Chokeberry, Red. *See Aronia.*

Cinquefoil (Potentilla). *See Potentilla.*

Cinnamomum camphora. Camphor Tree. Seed propagation is the most common propagation method, unless cultivars have been selected. Can also be propagated by semi-hardwood cuttings taken in spring and rooted under mist.

Citrus spp. *See Chapter 19.* There are a number of citrus cultivars for landscapes and home gardens, all of which are asexually grown by cuttings or grafted or budded onto seedling rootstock. *Citrus sinensis* 'Valencia' and 'Washington' are grown by cuttings (California).

Grafting. In Texas, citrus varieties are grafted on *Poncirus trifoliata* 'Rubidoux' rootstock. Some grafted cultivars include: 'Improved Myer's Limon' *(C. limon),* which is grown by cuttings in California, 'Satsuma' *(C. reticulata), C. sinensis* 'Tomango' and 'Louisiana Sweet,' *C.* × *paradisi* 'Bloomsweet,' 'Rio Red' and 'Star Ruby' (cutting-grown), and *Fortunella miewa* 'Kumquat.'

T-budding in Texas is done in the spring and fall. Dormant budwood is collected in February and refrigerated. Budding begins as soon as the bark begins slipping and ends in mid-April when the weather gets hot. In the fall, budwood is taken from stems that are round—compared to the normal triangular shape of citrus stems. The fall budding is from October until mid-November when the bark stops slipping. Budding rubbers and clear poly tape are used for tying.

Clematis spp. Clematis can be propagated by seed, cuttings, grafting (*313*), division of roots, or layering (*135*).

Seed. Seeds of some clematis species have embryo dormancy, so stratification for one to three months at about 4° C (40° F) is needed. Some species require a warm-cold stratification.

Cuttings. Clematis is probably best propagated by cuttings taken from young plants; these root under mist in about five weeks. Young wood with short internodes taken in the spring gives satisfactory results, but partially matured wood taken in late spring to late summer is more commonly used. Leaf-bud cuttings taken in midsummer will also root readily under mist. Semi-hardwood and hardwood cuttings can be treated with 3,000 to 8,000 ppm IBA talc, and softwood cuttings with 3,000 to 5,000 ppm IBA (*189*). Internodal cuttings (leaf-bud cuttings with one leaf intact) and the stem node inserted above the rooting medium—root best with 1,000 ppm IBA quick-dip or 4,000 ppm talc (*293*).

Micropropagation. Clematis is commercially micropropagated.

Clethra spp. Clethra are low-growing shrubs or small trees with attractive foliage, bark, and fragrant flowers. Can be propagated by seed which is sown in flats under mist. *C. alnifolia* is rooted from softwood cuttings (June to September in Delaware, USA) and treated with 1,000 ppm K-IBA (*59, 113*). Other species are rooted with 1,000 to 5,000 ppm IBA.

Coccoloba spp. *(279)*. Some of these species are useful landscape plants with high salt tolerance. *C. uvifera* (sea grape) and *C. diversifolia* (pigeon plum) are primarily seed-propagated. Seeds are collected from August to November (Florida), the seed coats peeled off, and no other pregermination is needed. It is important that seeds not dry out before planting. Some nursery producers propagate *C. uvifera* by cuttings and air layering.

Coralberry. *See* Ardisia.

Cornus spp. Dogwood.

Seed. Seeds have various dormancy conditions; those of the popular flowering dogwood *(C. florida)* require either fall planting or a stratification period of about four months at 4° C (40° F). Best germination is obtained if the seeds are gathered as soon as the fruit starts to color, and sown or stratified immediately. If allowed to dry out it is best to remove seeds from the fruit and soak in water. Seed germination of *Cornus canadensis* is enhanced with a four-month cold stratification (*246*). Low precipitation during seed formation reduces seed germination of red-osier dogwood *(C. stolonifera),* which must also be stratified (*1*). Other species require, in addition, a treatment to soften the seed coat. Two months in moist sand at diurnally fluctuating temperatures (21 to 30° C; 70 to 85° F), followed by four to six months at 0 to 4° C (32 to 40° F), is effective. With some species, the warm stratification period may be replaced by mechanical scarification or soaking in sulfuric acid.

Cuttings. Some dogwoods can be started easily by cuttings. *C. kousa* cuttings collected in mid-June through July (Massachusetts) root well when wounded on one side and treated with either 8,000 ppm IBA talc or quick-dips of 2,500 ppm IBA and NAA (*146*). Summer softwood cuttings of *C. alba* root readily when treated with a 1,000 to 3,000 IBA quick-dip or talc. Cuttings of *C. florida* are best taken in late spring or early summer from new growth after flowering, then rooted under mist (*165*). In Alabama, two-node cuttings of *C. florida* 'Stokes Pink' and 'Weaver's White' are taken from slightly stiff new-growth early in the spring and quick-dipped with 10,000 ppm IBA (*201*). *C. florida* 'Spring Grove' roots well from softwood cuttings (June, July in Ohio) treated with 8,000 ppm IBA in talc (*189*). In Florida, semi-hardwood cuttings taken in May and late August are quick-dipped in 10,000 ppm K-IBA for three or ten seconds, respectively (*93*). *C. florida* 'Rubra' can be rooted successfully if cuttings are taken in early summer after the second growth flush, treated with 3,000 ppm IBA, and propagated under mist (*320*).

To ensure survival through the following winter in cold climates, the potted cuttings should be kept in heated cold frames or polyhouses to hold the temperature between 0 and 7° C (32 and 45° F). Rooted cuttings that had shoot growth in the fall, but were not fertilized, had the best overwinter survival (*165*).

Grafting. Selected types such as the red flowering dogwood, *C. florida* 'Rubra,' and the weeping forms (difficult to start by cuttings) are often propagated by T-budding in late summer or by whip grafting in the greenhouse in winter on *C. florida* seedling rootstock. *C. kousa* and *C. florida* can be reciprocally grafted (*146*).

Micropropagation. Dogwood is commercially micropropagated (*129*).

Corylopsis spp. Winterhazel. Difficult to propagate by seeds, which have a warm and cold stratification requirement. Softwood to semi-hardwood cuttings can be treated with a 1,000 ppm IBA quick-dip and rooted under mist. IBA talc at 3,000 ppm has also been used. Buttercup winterhazel *(C. pauciflora)* is commercially micropropagated.

Corylus spp. Filbert, Hazelnut. European filbert *(C. avellana)* is grown for fruit production and ornamental characteristics (*C. avellana* 'Contorta').

Seed. Seeds should be stored under refrigerated conditions immediately after cleaning (*115*). Seeds have a two- to six-month cold stratification requirement.

Cuttings. In Oregon, cuttings in an active stage of growth are taken in mid-June to mid-July and treated with 5,000 to 10,000 ppm IBA quick-dip or talc. Cuttings are very sensitive to overwatering, and should be kept in an active growth stage after rooting. Etiolation of stock plants and use of softwood cuttings enhanced rooting of *C. maxima* (*248*). Stooling and collection of hardwood cuttings in mid-February (England), and treating cuttings with 10,000 ppm IBA at 21° C (70° F) can enhance rooting.

Grafting. Commercial fruit and ornamental cultivars are grafted using a whip graft on *C. avellana* seedling rootstock. Chip budding is successful and budded plants are more vigorous than grafted plants (*115*). See Chapter 13 for the discussion on the hot grafting technique.

Micropropagation. Corylus are commercially micropropagated (*5*).

Cotinus coggygria. Smoke Tree. Smoke trees should not be propagated by seeds, since many of the seedlings are male plants, lacking the showy flowering panicles. Only vegetative methods should be used, with propagating wood taken from plants known to produce large quantities of the desirable fruiting clusters. Also, the purple foliage color will only come true from vegetative reproduction.

Cuttings. Tip cuttings taken from spring growth (May in Oregon) should be very soft, 5 to 8 cm (2 to 3 in.) long. Treated with a 1,000 to 3,000 ppm IBA quick-dip, cuttings should root in about five weeks. After the mist is discontinued, the rooted cuttings are left in place, undisturbed, and transplanted the following spring.

Cotoneaster spp. Cotoneaster.

Seed. Seeds of most Cotoneaster species should be soaked for about 90 minutes in concentrated sulfuric acid, rinsed, and then stratified for three to four months at about 4° C (40° F).

Cuttings. Leafy cuttings of many species taken in spring or summer will root under mist without much difficulty. Evergreen/semi-evergreen types root better than deciduous, and are treated with a 1,000 to 3,000 ppm IBA quick-dip. Semi-hardwood cuttings of *C. buxifolius* root best with a 4,000 to 6,000 ppm IBA quick-dip in 100 percent perlite medium.

Grafting. Cotoneaster can be budded high onto pear trees to produce a "tree" cotoneaster. A blight-resistant pear rootstock, such as 'Old Home,' should be used. *C. bullatus* and *C. actifollus* rootstocks are more commonly used.

Cottonwood. *See Populus.*

Crab Apple (Flowering). *See Malus.*

Crape Myrtle. *See Lagerstroemia.*

Crataegus spp. Hawthorn.

Seed. Hawthorns tend to reproduce true by seed. Pronounced seed dormancy is present because of a combination of an impermeable seed coat and embryo conditions. Scarify seed with sulfuric acid, and then cold stratify for five months at about 4° C (40° F). Planting the seed in early summer will provide these conditions naturally, resulting in germination the following spring. Seeds of some *Crataegus* species do not have impermeable seed coats, so they need only be cold stratified. Since hawthorn develops a long taproot, transplanting is successful only with very young plants. Air pruning of seedling roots may be beneficial.

Grafting. Selected clones may be T-budded or root-grafted on seedlings of *C. crus-galli,* or *C. coccinea* for the American (entire-leaf) types and on seedlings of *C. laevigata* or *C. monogyna* for the European (cut-leaf) types.

Cryptomeria japonica. Japanese Cryptomeria, Sugi. Can be propagated either by seeds or by cuttings.

Seed. Seeds should not dry out. Seeds should be cold stratified for three months, or given a three-month warm stratification followed by a three-month cold stratification.

Cuttings. Chinese and Japanese foresters have been commercially propagating Sugi by cuttings for more than 500 years, and detailed methods for taking and rooting cuttings were published in the 1600s (*303*). Cuttings, 5 to 15 cm (2 to 6 in.) long, should be taken from greenwood at a stage of maturity at which they break with a snap when bent. Root with bottom heat; keep the cuttings shaded and cool. After roots start to form, in about two weeks, give more light; transplant to pots when roots are about 13 mm (½ in.) long. Stem cuttings of *C. japonica* 'Yoshino' can be rooted year-round. Hardwood cuttings taken in January (North Carolina) can be rooted when tips of first-order laterals or proximal halves of first-order laterals (Fig. 10–29)are treated with 3,000 to 9,000 IBA quick-dip applications (*210, 211*).

× **Cupressocyparis leylandii.** Leyland Cypress. This bigeneric hybrid of *Cypressus macrocarpa* and *Chamaecyparis nootkatensis* is propagated by cuttings, and rooted under mist with bottom heat (*381*). Cuttings can be taken any time from late winter to autumn and should be treated with IBA at 3,000 ppm, or 8,000 ppm in December-January (Georgia) (*289*). In California, 6,000 ppm IBA was optimal (*45*). Leyland cypress is micropropagated (*342*).

Cupressus spp. Cypress. Seeds have embryo dormancy, so stratification for about four weeks at 2 to 4° C (35 to 40° F) is used. Cuttings can be rooted if taken during winter months. Treatments with 2,000 to 8,000 ppm IBA enhance rooting. In California, quick-dips of 6,000 to 8,000 ppm IBA or 6,000 IBA + 6,000 NAA aid rooting (*45*). Side-veneer grafting of selected forms on seedling *Cupressus* rootstocks in the spring is often practiced.

Currant (Red Flowering). *See Ribes.*

Cycads. *See* Cycas.

Cycas spp. Cycads. Species of *Cycas, Zamia,* and *Encephalartos.*

Seed. Many are seed-propagated. The seed of most cycads germinate without difficulty, but sometimes slowly over a span of several months. In general, soaking the seed in water for a few days and subsequently in 1,000 ppm gibberellic acid for 24 to 48 hours enhances germination (*104*). Species of *Cycas, Zamia,* and *Encephalartos* have seed germination problems that include an impermeable seed coat (the sclerotesta), an immature embryo, and physiological dormancy. Scarification with sulfuric acid, followed by a gibberellic acid treatment resolves most of these dormancy problems. Scarifying the seeds by cracking them is not recommended. Seeds of *Zamia floridana* germinate well when treated with scarified acid for one hour, soaked in gibberellic acid for 48 hours, and then placed under intermittent mist for six weeks; *Z. furfuracea* seeds germinate best with a 15-minute acid treatment, 24-hour gibberellic acid soak, and then being placed under mist for six weeks (*104*).

Seeds of *Cycas* float in water whether viable or nonviable. The embryo is immature at the time of seed collection. For improved germination, Sago palm *(Cycas revoluta)* seed should be stored at room temperature and scarified for one hour with concentrated sulfuric acid (*151*).

Cycad seedlings should be root-pruned by severing the roots at the root-shoot juncture, and dipping the cut end of the leafy portion in IBA. They are then potted in liner pots, and placed under mist for two weeks. This encourages branched root systems and more rapid growth and development of the seedlings.

Vegetative propagation. Slow growth rates limit the potential for vegetative propagation. The American *Zamia* can be regenerated from its underground tuberous root and stem tissue. Cycads do not have lateral buds to produce side shoots. However, when taxa such as *Cycas* are injured, side shoots known as "pups" regenerate from callus. The pups are removed and their severed base is quick-dipped in 2,000 to 10,000 ppm IBA; rooting can take up to one year (*104*).

Micropropagation. The South African cycad, *Stangeria eriopus,* has been micropropagated (*273*). However, as of 1996, there is no commercial micropropagation of cycads.

Cypress. *See Cupressus.*

Cytisus spp. Broom.

Seed. Seeds of many of the species germinate satisfactorily if gathered as soon as mature and treated with sulfuric acid for 30 minutes to soften the hard seed coats before planting. Since the various *Cytisus* species crossbreed readily, stock plants for seed sources should be isolated.

Cuttings. Hardwood cuttings taken in late February or early March, and treated with 2,500 ppm IBA liquid, root in good percentages. Cuttings can be rooted rather easily under mist in midsummer if treated with 3,000 to 8,000 ppm IBA and given bottom heat.

Daphne spp. Daphne can be propagated by seed, stem and leaf-bud cuttings, layering, and grafting (*70, 77*).

Seed. Seeds should be sown at once after harvest or, if dried, then scarify the seed and give a moist-chilling period before sowing. Berries are very poisonous if eaten.

Cuttings. Daphne is probably best propagated by leafy cuttings in perlite and peat moss (4:1) mix under mist; a well-drained 100 percent medium perlite is also recommended for cuttings (*119*). In England, cuttings are taken in mid- to late May, towards the end of flowering. Cuttings are also taken in summer from partially matured current season's growth. Rooting can be enhanced with a 1,000 ppm IBA + 500 ppm NAA, or a 2,500 ppm IBA quick-dip (*119*). The daphnes do not transplant easily and should be moved only when young.

Micropropagation. *D. odora* is commercially micropropagated (*301*).

Dawn Redwood. *See Metasequoia.*

Davidia involucrata. Dove tree. Seeds have a double dormancy requirement of five to six months' warm stratification followed by three months of cold stratification (*147*). Polyethylene bags containing 1 sand:1 peat is a suggested system for the pretreatment of *Davidia* seeds. Rooting is variable. Leaf-bud cuttings treated with 3,000 ppm IBA talc rooted 85 percent. Wounded cuttings with four leaves that were treated with 8,000 ppm IBA talc and propagated under mist had 50 percent rooting (*115*). Disturbance and overwintering of rooted cuttings can be a problem.

Delonix regia. Royal Poinciana. This spectacular tropical flowering tree is propagated by seed. Germination is rapid when seeds are treated to soften seed coats, as by pouring boiling water over seeds or by soaking in concentrated sulfuric acid for one hour.

Deutzia spp. Deutzia is easily propagated either by hardwood cuttings lined-out in the nursery row in spring, or by softwood cuttings taken in the spring or summer and treated with IBA talc or quick-dip at 1,000 to 3,000 ppm.

Diosma ericoides. Diosma, Breath of Heaven. Propagated by leafy cuttings taken in summer and rooted under mist.

Diospyros texana. Texas persimmon. Japanese persimmon *(D. kaki)*. The former is a small multi-trunked tree of 15 to 30 ft. It has intricate branching and smooth gray bark. Propagation of both species is by seed and no pregermination treatment is necessary (*195*). *D. kaki* is also propagated by shield or chip budding and side-veneer or side-tongue grafted on *D. virginiana* rootstock. *D. kaki* is commercially micropropagated (*91*).

Disanthus cercidifolius *(108)*. A multistemmed shrub of potential in the Hammamelidaceae family with red fall color and purple flowers. Reported to have a double dormancy requirement requiring that seed be warm stratified for five months and cold stratified for three months. Cuttings collected in July (Georgia) root best when treated with 10,000 ppm IBA alcohol quick-dip and propagated under mist with peat-perlite. Successfully overwintered in a polyhouse if potted up immediately after rooting, fertilized, and allowed to harden-off naturally in the fall. *Disanthus* is commercially micropropagated.

Dogwood. *See Cornus.*

Douglas-Fir. *See Pseudotsuga.*

Dove tree. *See Davidia.*

Elaeagnus spp. Elaeagnus, Russian olive, Silverberry, Silverthorn. Seeds planted in the spring germinate readily following a stratification period of three months at 4° C (40° F). Removal of the pit (endocarp) for silverberry seeds *(E. commutata)* resulted in about 90 percent germination of unstratified seeds, since a germination inhibitor is apparently present in the pit; if not fall-planted, seeds should be cold stratified for three months. Seeds of the Russian olive, *E. angustifolia,* should be treated with sulfuric acid for 30 to 60 minutes before fall planting or stratification; this deciduous species is commercially seed-propagated, and only limited rooting success (28 percent) has been obtained with 3,000 ppm IBA talc (*115*). Leafy cuttings of the evergreen species root readily. Hardwood and semi-hardwood cuttings of *E. pungens* root well when treated with 8,000 to 20,000 ppm IBA talc (*30*).

Elderberry. *See Sambucus.*

Elm. *See Ulmus.*

English Ivy. *See Hedera.*

Enkianthus spp. Enkianthus are attractive shrubs with pronounced fall color. Seed propagation requires no pretreatment, and germination occurs within two to three weeks after sowing (*115, 270*). Cuttings root quite easily. Leafy cuttings taken in mid-June (Massachusetts) are treated with talc or quick-dips of IBA at 5,000 to 8,000 ppm, and rooted under mist. Rooted cuttings should be allowed to harden-off and overwinter before they are disturbed. An extended photoperiod is helpful in ensuring overwintering success. Enkianthus are commercially micropropagated.

Epigaea repens. Mayflower, trailing arbutus. This creeping, evergreen shrub has fragrant white to pink flowers and is propagated by seed collected in late spring (Massachusetts). Immature seed is white, and dark when mature. The cleaned seed is stored at 4° C (40° F) until sowing; there is no stratification requirement (*304*).

Erica spp. Heath. The propagation of the closely related genera, Heather *(Calluna vulgaris)*, is quite similar (*117*).

Seed. Seeds may be germinated in flats in the greenhouse in winter or in a shaded outdoor cold frame in spring.

Cuttings. Leafy, partially matured cuttings taken at almost any time of year, but especially in early summer, root readily under mist in a glasshouse or polyethylene-covered cold frame. While most species root readily, IBA (1,000 ppm quick-dip or 4,000 ppm talc) speeds up rooting.

Micropropagation. *Erica* and *Calluna* are commercially micropropagated.

Eriobotrya japonica. Loquat. These large evergreen shrubs or small trees can be propagated by seed, requiring no pretreatment. Considered difficult-to-root. Side grafting and budding is another form of propagation. *E. japonica* has been successfully micropropagated.

Eriostemon spp. Eriostemon. An Australian evergreen shrub with long-lasting white or pink flowers and fragrant foliage. Can be propagated by seed and stem cuttings. Stem cuttings can be slow to root or root poorly, and ease of rooting is clonally variable. *E. myoporoides* and *E.* 'Stardust' can be micropropagated (*12*).

Escallonia spp. Escallonia is easily started by leafy cuttings taken after a flush of growth. Cuttings root well under mist and respond markedly to treatment with 1,000 ppm IBA. It is best to direct root in liner pots, since transplanting is difficult.

Eucalyptus spp. Eucalyptus.

Seeds. Eucalyptus is largely propagated by seeds planted in the spring (*92*). Mature capsules are obtained just before they are ready to open. No dormancy conditions occur in most species, so seeds are able to germinate immediately following ripening. Seeds of some species, however—for example, *E. dives, E. niphophila,* and *E. pauciflora*— require stratification for about two months at 4° C (40° F) for best germination. Eucalyptus seedlings are very susceptible to damping-off. Seeds are usually planted in flats of pasteurized soil placed in a shady location or flats covered with a white plastic (*37*). From flats, they are transplanted into small pots, from which they are later lined-out in a nursery row. The roots of young trees will not tolerate drying, so the young plants should be handled as container-grown stock. Seeds may be sown directly into containers in which the seedlings are grown until planting in their permanent location (*169*).

Cuttings. Eucalyptus is difficult to start from cuttings, but good rooting can be obtained from some species. For example, leafy cuttings of *E. camaldulensis* root when taken in early spring from shoots arising from the base of young trees, wounded, treated with a 4,000 ppm IBA + 4,000 ppm NAA, and propagated under mist with bottom heat at 21° C (70° F) (*137*). *E. grandis* stock plants are hedged and new developing shoots used as cuttings. Four-node cuttings are prepared and treated with IBA. Leaf retention is very important, so avoid using older shoots with leaves that will abscise before rooting occurs (*71*). Clones of *E. cladocalyx, E. tereticornis, E. grandis, E. camaldulensis* and *E. trabutii* will root when three to four node cuttings are treated with a quick-dip of 4,000 to 8,000 K-IBA (*312*). Rooting is best when cuttings are taken from rapidly growing stock plants (April to October in California).

Grafting. *E. ficifolia* has been grafted successfully by a side-wedge method, using young vigorous *Eucalyptus* seedling rootstocks growing in containers and placed under very high humidity following grafting. Use of scions taken from shoots that had been girdled at least a month previously increased success (*189*).

Micropropagation. Eucalyptus are routinely micropropagated, even with mature explant tissue (*28, 60, 103, 208, 236*).

Euonymus spp. Euonymus.

Seed. Stratification for three to four months at 0 to 10° C (32 to 50° F) is required for satisfactory seed germination (*153*). Remove seeds from fruit and prevent drying.

Cuttings. Euonymus is easily started by cuttings. Deciduous species are started from hardwood cuttings in late winter/early spring, while evergreen types are propagated with leafy semi-hardwood cuttings. *E. alatus* roots best from stem cuttings collected before growth is fully hardened in the spring. IBA can be applied as a 1,000 to 3,000 ppm quick-dip or from 3,000 to 8,000 ppm talc (*115*). The production cycle of *E. fortunei* 'Sarcoxie' is significantly shortened when summer-rooted cuttings are exposed to a minimum of 500 chilling hours and then winter-forced in a heated greenhouse (*82*).

Micropropagation. *E. alataus* is commercially micropropagated (*330*).

Euphorbia spp. Poinsettia *(E. pulcherrima)*. This species is generally propagated by leafy cuttings under mist (*125, 231*). A quick-dip of 500 to 1,000 ppm IBA + NAA enhances rooting. Stock plants of self-branching cultivars should be used as a source of cuttings. It is best to root cuttings in small container liners or root cubes, so that the roots of rooted cuttings are not disturbed during transplanting to larger containers. Cuttings can be rooted in the greenhouse from spring to fall. Specialists or their licensed propagators will sell rooted liners to greenhouse producers for either stock plants or for shifting up to a larger pot for finishing. Scarlet plum *(E. fulgens),* which is a medium-sized shrub from Mexico used for cut-flower production in greenhouses, is generally propagated by cuttings. Micropropagation of this species is faster and requires less greenhouse space than conventional cuttage (*390*).

Fagus spp. Beech.

Seed. Seeds germinate readily in the spring from fall planting or after being stratified for three months at about 4° C (40° F). Seeds should not be allowed to dry out. *F. sylvatica* seeds are sometimes hollow and nonviable.

Cuttings. Cuttings taken from seedling plants will root at different percentages due to clonal influences. Overwintering survival of rooted cuttings (Denmark) is a problem (*230*). Etiolation of stock plants and covering shoot bases with black adhesive tape enhances rooting.

Grafting. Selected clones are grafted by either the cleft, whip, or side-veneer method on seedling rootstock of European beech (*F. sylvatica*). In Oregon, *F. sylvatica* cultivars are field-grafted on seedling rootstock using a "stick bud" method; the scionwood generally contains two buds and is inserted into a T-cut in the rootstock (*131*).

False Cypress. *See Chamaecyparis.*

× Fatshedera *(Fatsia × Hedera). See Hedera helix.*

Feijoa sellowiana. *See Feijoa* in Chapter 19.

Ficus spp. Fig, rubber plant. A wide range of ornamental trees, vines, and ornamental potted plants. For the edible, common fig, see *Ficus carica* in Chapter 19. Many species form aerial roots, and are easily rooted by cuttings and layering.

Cuttings. Single-node propagation of *Ficus* spp. produces many more plants per stock plant than the common method of air layering (*288*). *F. benjamina* (weeping fig) is easily started by leafy semi-hardwood cuttings taken in spring or early summer and rooted under mist. *F. elastica* (rubber plant) is propagated by cuttings taken from 5 to 27 cm (6 to 12 in.) shoots; single buds or "eyes" can be removed and rooted. These cuttings are made in spring, inserted in sand or a similar medium, and held in a warm greenhouse. *F. pumila* (creeping fig) is propagated with 10 to 13 cm (4 to 5 in.) long cuttings. The juvenile form, which is a climbing vine with aerial rootlets, roots easily all year round; 1,000 to 1,500 ppm IBA quick-dips enhance rooting (*99*). The mature form lacks aerial roots, but can be rooted successfully with 2,000 to 3,000 ppm IBA. *F. lyrata* (fiddleleaf fig) can be propagated by cuttings, or by air layering (*206*).

Air layering. Shoots of trees growing outdoors in the tropics are air layered and the rooted air layers shipped to wholesale nurseries for finishing off the crop. Indoor plants that become too "leggy" also can be air layered

Micropropagation. *Ficus* are micropropagated commercially (*287*).

Fig. *See Ficus.*

Fig (Common). *See Ficus carica* in Chapter 19.

Filbert. *See Corylus.*

Fir. *See Abies.*

Firethorn. *See Pyracantha.*

Forsythia spp. Forsythia is easily propagated by hardwood cuttings set in the nursery row in early spring or by leafy softwood cuttings taken during late spring and rooted under intermittent mist.

Fothergillia gardenii. Dwarf Fothergillia. Generally difficult to propagate by seed—based on lack of availability, and a six-month warm stratification, followed by a three-month cold stratification requirement. Easy to root by cuttings, which are treated with 4,000 ppm IBA (*115*). Cultivars of Fothergillia are commercially micropropagated.

Franklinia alatamaha. Franklinia can be propagated by seed which is sown immediately (if fresh). Otherwise, seed needs a one-month stratification period if stored. Roots easily by cuttings treated with 1,000 ppm IBA. Root early in the season to allow growth of rooted liners prior to overwintering. Franklinia is commercially micropropagated.

Fraxinus spp. Ash.

Seed. Seeds of most species germinate if stratified for two to four months at about 4° C (40° F). Seed of *F. anomala* and *F. pennsylvanica* germinate with ease after a three-month cold stratification. Some seeds have a one- to three-month warm stratification followed by a cold stratification requirement of five to six months at 4° C (40° F), i.e., *F. excelsior, F. nigra,* and *F. quadrangulata.*

Cuttings. Ashes are difficult to propagate from stem cuttings. *F. greggii* (littleleaf ash) roots readily when 10 to 15 cm (4 to 6 in.) long—leafy terminal cuttings are taken 16 weeks after bud-break and treated with a 17,000 ppm K-IBA quick-dip (*349*). Rooting is enhanced with early, softwood cutting material. *F. pennsylvanica* 'Summit' has been rooted in low percentages when taken from softwood cuttings in April and May in Pennsylvania (*18*). *F. anomala* cuttings taken from containerized seedling stock plants had high rooting when treated with auxins from 5,000 to 20,000 ppm.

Grafting. *F. excelsior, F. ornus* and *F. Pennsylvanica* seedlings are used as rootstocks for grafting or budding ash cultivars.

Micropropagation. Ashes are commercially micropropagated (*8, 23, 173*).

Fringe Tree. *See Chionanthus.*

Fuchsia spp. Fuchsia is easily rooted by leafy cuttings maintained under humid conditions (*364*). Roots develop in two to three weeks.

Gardenia jasminoides. Gardenia, Cape jasmine. Leafy terminal cuttings are treated with 3,000 ppm IBA and rooted under mist from spring to fall. Gardenias are difficult to transplant and should be moved only when small. Gardenia is commercially micropropagated (*127*).

Garrya elliptica ***(300).*** Garrya 'James Roof' is an ornamental tree producing catkins 300 mm long. It is generally considered difficult to propagate asexually. Tip nodal cuttings 10 cm long on side shoots taken with a heel with well-developed terminal buds are collected from late summer to December (England). Wounding is optional and cuttings are treated with 8,000 ppm IBA talc and propagated under mist or plastic film. Direct rooting in small liner pots is desirable to avoid root disturbance problems. Rooted cuttings should complete their first spring flush of growth before transplanting. *Garrya* is commercially micropropagated.

Genista spp. Genista. Landscape plants that range from creeping ground covers to massing shrubs. These species are not difficult to propagate.

Seed. Seed propagation is similar to that of other legumes with hard seed coats. Scarify with a 30-minute treatment of concentrated sulfuric acid, or a boiling-water treatment. Seeds are placed in boiling water which is immediately allowed to cool to room temperature (*142*). Some species have no scarification requirement. Seeds of *G. tinctoria* and some other species also require a three-month cold stratification of 5° C (41° F). When in doubt, immerse seeds in water overnight, then separate those that have imbibed water and swelled from those that did not. Cold stratify the swollen ones for three months or until radicle growth first becomes visible. Soak those seeds that did not swell in concentrated sulfuric acid for 30 minutes, then wash them thoroughly, and cold stratify in the same way (*142*).

Cuttings. Softwood cuttings taken in early summer (Indiana) root easily without auxin, while semi-hardwood cuttings taken in late summer and hardwood cuttings taken in fall root best with 8,000 ppm IBA in talc (*142*). Bottom heat should not be allowed to drop below 18° C (65° F). Seedlings or rooted cuttings should be grown in containers, since field-dug plants do not transplant well.

Ginger (European, wild). *See Asarum.*

Ginkgo biloba. Ginkgo. Seed propagation should not be used except for rootstock production. The "fruits" are collected in mid-fall, and the pulp removed. A warm and cold stratification of one to two months, each, is required. Seedlings produce either male or female trees, but the sex cannot be determined until the trees flower—after about 20 years. The plumlike "fruits" on the female trees have a very disagreeable odor, so only male trees are used for ornamental planting. Cutting-produced male liners do not grow as quickly as grafted plants. Softwood cuttings taken in early summer are treated with 8,000 ppm IBA talc or quick-dip and rooted under mist. Commercial propagation is by T-budding or chip budding, using buds from male trees inserted into ginkgo seedlings. Male cultivars can also be cleft or whip grafted on container seedling rootstock during winter.

Gleditsia triacanthos. Common Honey locust. Readily propagated either by seeds planted in the spring or by cuttings. In seed propagation, soaking the seed in sulfuric acid for one to two hours, followed by stratification at 2° C (36° F) for three months, gives good germination. Very difficult to propagate by cuttings—can be rooted by root cuttings. The thornless honey locust, *G. triacanthos,* var. *inermis,* and the thornless and fruitless 'Moraine' locust are usually propagated by T-budding on seedlings of the thorny type.

Golden Chain. *See Laburnum.*

Goldenrain Tree. *See Koelreuteria.*

Gold Tree. *See Tabebuia argentea.*

Golden Trumpet. *See Allamanda.*

Gordonia spp. (115). Loblolly bay, Black laurel. These medium-sized trees can be propagated by seed, which germinate readily without pregermination requirements. Semi-hardwood cuttings are easily rooted with a 2,500 to 3,000 ppm IBA quick-dip or 3,000 ppm IBA talc. These species transplant easily after rooting.

Grevillea spp. These Australian native shrubs or trees are propagated by seed, cuttings, or graftage. Cuttings of the low-growing species root readily, but larger-growing species, such as *G. robusta,* silk oak, are best propagated by seed. *G. johnsonii* is difficult to root, but fall cuttings (Australia) treated with 4,000 ppm IBA-ethanol five-second quick-dip have 70 percent rooting (*132*). With *G. asplenifolia,* single-node cuttings with reduced leaf areas work best, whereas retaining all leaves on larger two- to three-node cuttings is optimal for *G. juniperina.* A number of Grevillea species can also be grafted using a whip graft (*47*). *G. robusta* is the primary rootstock, since it has resistance to *Phytophthora cinnamomi.* Specialized weeping forms of Grevillea are produced by using approach grafting of two independent, containerized Grevillea plants (*95*). *G. scapigera* (*58*) and G. robusta (*294*) can be micropropagated.

Gymnocladus dioicus. Kentucky Coffeetree. An attractive landscape tree, which is dioecious. Male trees are more desirable than female trees, which have long seed pods that abscise and detract from the ornamental value of the plant. Can be propagated by seed, but undesirable female plants are produced. Cutting propagation is very difficult. Seedlings (*158*) and mature male trees have been successfully micropropagated (*332*).

Hackberry. *See Celtis.*

Halesia spp. Silverbell. A small, valuable landscape tree, native to the southeastern United States. It has a striped bark, bell-shaped flowers, and interesting fruit (two- to four-winged drupe). Seed propagation of *Halesia* requires complex stratification regimes and success is often limited (*160*). *H. carolina* is warm stratified for two to four months, followed by two to three months of cold stratification. Spring and summer cuttings root well when treated with 1,000 to 10,000 ppm IBA quick-dips (*108*). Silverbell is commercially micropropagated (*49*).

Hamamelis spp. Witch Hazel. Propagated by seed, budding, or grafting.

Seed. Seeds are gathered and planted outdoors in early fall. *H. japonica* seeds should be soaked for a week with water changed daily. Alternatively, seeds warm stratified for four to six months, followed by a three-month cold moist stratification, will germinate well.

Cuttings. Cutting propagation is difficult but possible. Leafy cuttings of *H. mollis, H. virginiana, H. japonica,* and *H. vernalis* are treated with 8,000 ppm IBA talc or a 10,000 IBA quick-dip and rooted under mist. Cuttings of *H.* × *intermedia* 'Arnold Promise' should be taken as early as possible in spring—cuttings taken in late spring and summer will root, but not survive the winter. Collect three- to four-node cuttings with basal portion firm, quick-dip for five seconds in 10,000 ppm K-IBA, or for softer tissue, 5,000 ppm IBA (*116*). To improve overwintering survival, it is best to induce a growth flush after rooting and prior to winter. The rooted cuttings need to be left undisturbed until after their first dormancy (*130, 227*).

Grafting. Cultivars of *H. mollis, H.* × *intermedia,* and *H. japonica* are propagated by budding or grafting (side-veneer graft) on *H. virginiana* seedlings; however, rootstocks tend to sucker and compete with scions (*292*). In Oregon, chip budding works well.

Micropropagation. Witch hazel is commercially micropropagated (*53*).

Hawthorn. *See Crataegus.*

Heath. *See Erica.*

Heather. *Calluna. See Erica* (Heath).

Hebe spp. Hebe, Veronica. Propagated by seed, by leafy cuttings in summer under mist, or by layering. *Hebe* is commercially micropropagated.

Hedera helix. English Ivy. English ivy is readily propagated by rooting cuttings of the juvenile (nonfruiting, lobed-leaf) form. It is also sometimes grafted onto *Fatshedera (Fatsia japonica* × *Hedera helix)* as a rootstock. English ivy is commercially micropropagated.

Hemlock. *See Tsuga.*

Heptacodium miconioides. Seven Son Flower. A small deciduous flowering tree in the Caprifoliaceae family, which is native to China. It has potential as a new nursery crop because of its exfoliating bark, vigorous growth, fragrant white, late-summer flowers, and beautiful rose to crimson samaras (a dry, simple fruit with winglike appendages) that clothe the tree in the fall. Best rooting occurred with basal and middle softwood cuttings and basal semi-hardwood cuttings (*232*). A quick-dip of 5,000 ppm K-IBA enhanced rooting. Shoots from lateral cuttings appear to exhibit plagiotrophic growth, but vertical shoots will rapidly grow from the base of the plant. Plagiotrophic growth can be reversed by growing rooted cuttings for a season or two and then cutting back to one bud.

Heteromeles arbutifolia. Toyon, Christmas Berry. Usually propagated by seed, which is stratified for three months, or sown in the fall to obtain outdoor winter chilling of the seed. Softwood tip cuttings taken in mid-spring, treated with 8,000 ppm IBA talc, and placed under mist will root (*168*). It can also be started by layering.

Hibiscus rosa-sinensis. Hibiscus (Chinese).

Cuttings. Softwood and hardwood cuttings are not difficult to root; however, there are cultivar differences (*213*). Softwood and semi-hardwood cuttings are rooted under mist in liner pots with peat:perlite, or Oasis Horticubes to prevent root damage during transplanting. In Australia, soft-tip cuttings 11 cm (4 in.) long are taken from December to April and treated with 5,000 ppm IBA, and rooted under mist (*128*). Cultivars that root rapidly receive maximum benefit from propagation medium temperatures of 26 to 30° C (79 to 86° F) (*66*).

Grafting. Vigorous cultivars which are resistant to soil pests and can be started easily by cuttings—such as 'Single Scarlet,' 'Dainty,' 'Euterpe,' or 'Apple Blossom'—are also used as rootstocks. Some clones develop into much better plants when grafted on these rootstocks than when on their own root system from cutting propagation. Whip grafting in the spring or cleft grafting or side grafting in late spring or early summer is successful. Scions of current season's growth, about pencil size, are grafted on rooted cuttings of about the same size (*322*).

In New Zealand, simultaneous bench grafting and rooting is done with a grafting tool that makes matching "V" cuts in the scion and unrooted rootstock. After wrapping the graft union with floral tape, the grafted rootstock is quick-dipped with 2,000 ppm IBA and rooted under mist (*24*).

Micropropagation. Hibiscus is commercially micropropagated.

Hibiscus syriacus. Hibiscus, Shrub-althea, Rose of Sharon. Easily propagated, either by hardwood cuttings in the nursery row in spring, or by softwood cuttings in midsummer under mist. Lateral shoots make good cutting material. Softwood cuttings respond well to treatment with 1,000 ppm IBA quick-dip. Deciduous, hardwood cuttings have rooted well when treated with 8,000 ppm IBA talc.

Hickory. *See Carya.*

Holly. *See Ilex.*

Honey locust (Common). *See Gleditsia.*

Honeysuckle. *See Lonicera.*

Hornbeam. *See Carpinus.*

Horse chestnut. *See Aesculus.*

Hovenia dulcis. Japanese Raisin Tree. Used for ornamental, medicinal, and fruit production. Seeds should be collected in the fall (Delaware). Fruits are dried, cracked, and added to water to separate the seeds. Discard seeds that float in water. The remaining seed are scarified with sulfuric acid for 45 minutes, stratified by storage at 5° C (41° F) for 90 days, and planted in seedling flats (*150*).

Huckleberry (Evergreen). *See* Vaccinium.

Hydrangea spp. Hydrangea are generally easy to root. Softwood and semi-hardwood cuttings are treated with 1,000 ppm IBA, and hardwood cuttings and more difficult-to-root species with 3,000 to 5,000 ppm IBA as a talc or quick-dip (*115*). In Oregon, soft-tip cuttings of *H. pedularis* are taken late May to early June, treated with 1,000 ppm IBA, and rooted under mist. In Alabama, *H. quercifolia* roots well from spring to early summer. Half of the leaf surface on one-node cuttings are trimmed off and the cuttings are dipped in 2,500 ppm IBA (*201*). Rooted liners do not overwinter well in a warm greenhouse, so they must be kept cool during the winter. Florists' hydrangea have been rooted with 5,000 to 10,000 ppm IBA talc under mist (*14*). Hydrangea is commercially micropropagated (*13, 340*).

Hypericum spp. St. Johnswort. Easily started by softwood cuttings taken in late summer from the tips of current growth and rooted under high humidity or mist. *Hypericum* is commercially micropropagated.

Ilex spp. Holly. Can be propagated by seeds, cuttings, grafting, budding, layering, and division (*141, 325*). Most hollies are dioecious. The female plants produce the very desirable decorative berries if male plants are nearby for pollination. For most hollies, any male will pollinate the female flower as long as the two are in bloom at the same time and planted fairly close together. In seed propagation, sex cannot be determined until the seedlings start blooming—at 4 to 12 years.

Seeds. The only reason to grow seedlings is for rootstock for graftage, or for breeding and selection work. *Cutting propagation is the preferred method for producing superior clones.* Germination of holly seed is very erratic; species such as *I. crenata, I. cassine, I. glabra, I. vomitoria, I. amelanchier,* and *I. myrtifolia* germinate promptly and should be planted as soon as they are gathered. Seeds of other species, *I. aquifolium* (English holly), *I. cornuta* (Chinese holly), *I. verticilliata, I. decidua,* and most *I. opaca* (American holly), should be collected and cleaned as soon as the fruit is ripe in the fall and then stored at about 4° C (40° F) until spring in a mixture of moist sand and peat moss; they do not germinate until a year or more after planting even though stratified, due to rudimentary embryos at the time of harvest. For deciduous hollies *(I. verticilliata, I. decidua),* stratify seeds for three to four months, and if no germination occurs, place seeds under warm stratification at 20 to 30° C (68 to 86° F) for after-ripening and embryo development for two to three months—followed by a two- to three-month cold stratification period (*110*).

Cuttings. This is the method most used commercially, permitting large-scale production of superior clones. Semi-hardwood tip cuttings from well-matured current season's growth produce the best plants. Cuttings taken from flat, horizontal branches of *I. crenata* tend to produce prostrate plants (plagiotropic) and those from upright growth produce upright (orthotropic) plants.

Timing is important—best rooting is usually obtained from mid- to late summer, but cuttings may be successfully taken into the following spring. Wounding the base of the cuttings helps induce root formation. The wounding induced by stripping off the lower leaves may be sufficient.

The use of auxins, particularly IBA at relatively high concentrations (8,000 to 20,000 ppm), is essential in obtaining rooting of some cultivars, such as *I. opaca* 'Savannah' (10,000 ppm IBA liquid), whereas 2,500 ppm IBA is sufficient for medium-difficult species such as *I. cornuta* (*30*). Semi-hardwood cuttings of *I. crenata* root best with 8,000 ppm talc, whereas hardwood cuttings should be treated with a 10,000 ppm IBA dip (*189*). Deciduous hollies such as *I. serrata, I. decidua* and *I. verticillata* are best rooted in early July (Indiana) with 10 to 13 cm (4 to 5 in.) cuttings having four to five nodes; the bottom leaf is stripped off and the cutting quick-dipped in 5,000 ppm IBA. Hollies can be sensitive to solvents used as IBA carriers. They usually exhibit chlorosis, followed by defoliation, e.g., *I. vomitoria* is sensitive to alcohol quick-dips. It is best to switch to water-based K-IBA products.

I. vomitoria 'Nana' are direct-stuck into 6-cm (2 ¼-in.) liner pots using 3,000 ppm K-IBA spring-summer (Alabama) and 5,000 ppm K-IBA fall-winter (*123*). Semi-hardwood cuttings taken in early fall

(Alabama) are strip-wounded at the base by tearing off the lowest branch and given a three-second dip with 2,000 ppm IBA (*170*). Using NAA on *I. vomitoria* will burn the stem and cause defoliation. Bottom heat at 21 to 24° C (70 to 75° F) is beneficial. A table of liquid auxin solutions for *Ilex taxa* has been published (*32*).

Micropropagation. Ilex are commercially micropropagated.

Illicium spp. Anise. Evergreen, medium-sized shrubs which are fragrant. Seed propagation requires no pretreatment; however, most species are propagated by semi-hardwood cuttings, with good to poor rooting, depending on the species. Summer cuttings treated with a 3,000 ppm IBA quick-dip are rooted in peat-perlite media with mist (*108*).

Incense Cedar. *See Calocedrus.*

Indigo bush. *See Amorpha.*

Ironwood. *See Casuarina.*

Jacaranda mimosifolia. Jacaranda. Easily propagated by seed taken from capsules after blooming (*374*). Vegetative propagation is normally not used; however, a white flowering form is grafted (southern California).

Japanese raisin tree. *See Hovenia.*

Jasminum spp. Jasmine. Propagated without difficulty by leafy semi-hardwood cuttings taken in late summer and rooted under mist. *J. sambac* 'Maid of Orleans,' 'Sambac Jasmine,' and 'Grand Duke' are propagated by single- or double-node cuttings that root within four to five weeks. Cuttings of *J. nudiflorum* root best when treated with a 3,000 ppm IBA quick-dip (*101*). Layers and suckers also can be used.

Jasmine (Asiatic). *See Trachelospermum.*

Jasmine ('Maid of Orleans,' 'Sambac Jasmine,' and 'Grand Duke'). *See Jasminum.*

Jasmine (Star, Confederate). *See Trachelospermum.*

Juniperus spp. Juniper. Junipers are divided into ground cover types, bushes, and upright pyramidal types. They are generally propagated by cuttings, but in some cases difficult-to-root species such as the upright types are grafted on seedlings or select cutting-grown species.

Seeds. Seedlings of the red cedar, *Juniperus virginiana,* or of *J. chinensis,* are ordinarily used as rootstocks for grafting ornamental clones. Seeds should be gathered in the fall as soon as the berry-like cones become ripe. For best germination, seeds should be removed from the fruits, then treated with sulfuric acid for 30 minutes before being stratified for about four months at 4° C (40° F). Rather than the acid treatment, two to three months of warm 21 to 30° C (70 to 85° F) stratification, or summer planting, could be used. As an alternative for cold stratification, the seed may be sown in the fall. Germination is delayed at temperatures above 15° C (60° F). Viability of the seed varies considerably from year to year and among different lots, but it never germinates much more than 50 percent. Treated seed is usually planted in the spring, either in outdoor beds or in flats in the greenhouse. Two or three years are required to produce plants large enough to graft.

Cuttings. The spreading, prostrate types of junipers are more easily rooted than upright kinds. Cuttings are made 5 to 15 cm (2 to 6 in.) long from new lateral-growth tips stripped off older branches. Sometimes a small piece of old wood—a heel—is thus left attached to the base of the cutting. In other cases, good results are obtained when the cuttings are just clipped without the heel from the older wood. Terminal growth of current season's wood also roots well.

Cuttings to be rooted in the greenhouse can be taken at any time during the winter (*186*) or rooted outdoors on heated beds (southern California). In more temperate areas, exposing the stock plants to several hard freezes seems to give better rooting. Optimum time for taking cuttings is when stock plants have ceased growth (e.g., late fall-winter propagation period is more successful than summer). For propagating in an outdoor cold frame, cuttings are taken in late summer or early fall. There may be advantages to using bottom heat.

Lightly wounding the base of the cuttings is sometimes helpful, as is the use of IBA. Recommendations have included 2,500 IBA quick-dip (Alabama) for medium-difficult juniper species to 3,000 to 8,000 ppm IBA liquid or talc. For upright junipers, one large California nursery uses combinations of 3,000 to 6,000 ppm K-IBA (*88*).

Maintaining a minimum bottom heat of 16 to 18° C (60 to 65° F) is critical during the first six weeks of propagation—to allow the basal wound of cuttings to callus. In southern California, heat is withheld for six weeks to allow *callusing.* Bottom heat is then raised to 21 to 24° C (70 to 75° F) to encourage *rooting* (*88*). In the southern U.S. (Gulf Coast), many growers use no heat at all during the rooting process, and take advantage of the ambient temperatures. Hardwood cuttings can be rooted in outdoor field beds. Optimal rooting of dwarf Japanese juniper (*J. procumbens* 'Nana') occurred with an IBA quick-dip at 8,000 ppm (or 16,000 ppm talc); rooted cuttings were potted up after 15 weeks, and unrooted, but callused cuttings were retreated with IBA and restuck in a peat:perlite medium to allow for further rooting (*46*).

In North Carolina, optimum rooting of eastern red cedar (*J. virginiana*) occurred with hardwood cuttings collected in January, then wounded, treated with 5,000 ppm IBA, and rooted under mist in a propagation house (*184*). Unlike many coniferous species, eastern red cedar has no problem with plagiotropism (undesirable prostrate growth); cuttings from lateral branches retain an upright growth habit after rooting.

***Grafting* (15, 285).** Vigorous seedling rootstocks with straight trunks, about pencil size, are dug in the fall from the seedling bed and potted in small pots set in peat moss in a cool, dry greenhouse. Seedlings potted earlier—in the spring—may also be used for grafting in the wintertime. About two to three weeks before grafting, the rootstock is brought into the greenhouse to induce root activity before grafting takes place. Seedlings can give variable graft results, and later, variable growth of the grafted plant. In Oregon, *J. virginiana* 'Skyrocket,' which is clonally propagated from cuttings and has a very columnar form resembling a skyrocket, is widely used as a rootstock or standard for patio tree stock grafts.

The scions should be selected from current season's growth taken from vigorous, healthy plants and preferably of the same diameter as the rootstock to be grafted. Scion material can be stored under high humidity at –1 to 4° C (30 to 40° F) for several weeks until used.

Side-veneer or other side-graft methods are ordinarily used. The unions are best tied with budding rubber strips. The grafted plants are set in a greenhouse bench filled with peat moss deep enough to cover the union. The temperature around the graft union should be kept as constant as possible at 24° C (75° F) with a relative humidity of 85 percent or more around the tops of the plants. A lightly shaded greenhouse should be used to avoid injuring the grafts. Adequate healing will take place in two to eight weeks, after which the temperature and humidity can be lowered. The rootstock plant is then cut off above the graft union to allow the scion to develop.

Kalmia spp. Laurel. Can be propagated readily by seed germinated at about 20° C (68° F). Germination of *K. latifolia* seed is enhanced by cold stratification for eight weeks, or by a 12-hour soak in 200 ppm gibberellic acid (*203, 204*). Seeds of *K. latifolia* should not be covered during sowing, since they require light for germination (continuous light is optimal) and the seeds are extremely small [1.4 million seeds/28 g (1 oz)] (*247*). Rooting of Kalmia is highly variable among species and cultivars. Some respond to higher IBA concentration (8,000 + ppm), whereas others do not (*144*). Can also be propagated by cleft or side grafting, or by layering.

K. latifolia cultivars are commercially micropropagated, which alleviates rooting problems and is probably the most efficient propagation method (*53, 240*).

Kentucky Coffeetree. *See Gymnocladus.*

Koelreuteria spp. Goldenrain Tree. This tree is usually propagated by seed. Supposedly the seeds have double dormancy, germinating best if the seed coats are scarified for about 60 minutes in sulfuric acid or by mechanical scarification—followed by stratification for about 90 days at 2 to 4° C (35 to 45° F) to overcome embryo dormancy. In southern California, stratification is avoided by pouring 83° C (180° F) water on seed and soaking overnight. This would suggest that *Koelreuteria* does not have a double dormancy requirement, or the ecotypes used in California do not. Softwood cuttings taken in the spring are treated with 1,000 to 3,000 ppm IBA and rooted under mist.

Laburnum spp. Laburnum, Golden Chain. Propagated by seed which are scarified for two hours in sulfuric acid. Seeds are poisonous if eaten. Some cultivars are propagated by grafting or budding on laburnum seedling rootstock.

Lagerstroemia indica. Crape Myrtle. Seeds will germinate readily without pregermination treatment. However, this species is commercially propagated by cuttings because it is easily rooted from softwood or hardwood cuttings. Most softwood cuttings are much easier to root prior to flower initiation. An IBA quick-dip of 1,000 to 1,250 ppm will aid root formation. Most cultivars are easy to transplant; however, some dwarf cultivars must be transplanted with a ball of soil. Hardwood cuttings of field-grown plants are gathered after the first hard frost, sawed into 8-in. (20-cm) cuttings, graded, bunched, stored over winter, and then planted in open fields in spring (March in Alabama) without auxin treatment. Eighty percent rooting is achieved (*63*). Crape myrtle is commercially micropropagated (*391*).

Larch. *See Larix.*

Larix spp. Larch. Most of these deciduous conifers are propagated easily by fall-planted seeds (*183*). Cones should be collected before they dry and open on the tree. Several species have empty or improperly developed seeds. Seeds of some species have some embryo dormancy, so for spring planting, stratification for one month at about 4° C (40° F) is advisable. Cutting propagation is best done by rooting leafy, semi-hardwood cuttings in late summer under mist. IBA at a 8,000 ppm quick-dip or 2,000 ppm talc promotes rooting—but only at a low percentage. Cutting material should be taken from young trees only.

Adult larch trees have been successfully micropropagated by shoot-tip culture and plantlets acclimatized (220).

Laurel (Cherry, English). *See Prunus.*

Laurus nobilis. Bay Laurel. It is normally propagated by seed. Seed germination is best with mechanical scarification, followed by a 30-day cold stratification. Rooting of stem cuttings should be carried out in the summer (Israel), which is the period of most active growth. Semi-hardwood cuttings from hedged trees root better than softwood cuttings (*298*).

Leptospermum spp. Tea Tree. Some species of this Australian native must be propagated by seed. Cultivars of other species, such as *L. scaparium,* can be readily propagated by cuttings.

Leucothoe spp. Switch Ivy, Drooping Leucothoe *(L. fontaneisana),* Coast Leucothoe *(L. axillaris).* Evergreen shrubs with white, pitcher-shaped flowers. Because of the small seed size and light requirement to germinate, seeds should not be covered with media during propagation. Seeds can be germinated at 25° C with continuous light (*42*). Shrubs can be asexually propagated by stem cuttings from June to December (southern U.S.). Treating with 1,000 to 3,000 ppm IBA will hasten rooting, but is optional (*115*). Terminal cuttings 10 to 13 cm (4 to 5 in.) long are taken, and leaves stripped from the basal end 2.5 cm (1 in.). Cuttings root in 10 to 12 weeks and can be transplanted after rooting. *Leucothoe* is commercially micropropagated.

Leucophyllum spp. *L. frutescens, L. candidum* (Texas Sage, Purple Sage). Easily propagated from seed, but cuttings are preferred (*383*). Softwood, succulent cuttings from 8 to 10 cm (3 to 4 in.) are treated with 8,000 ppm IBA talc, and propagated under mist in a well-drained propagation medium (e.g., 100 percent perlite). Cuttings will root in three to four weeks under high light and a propagation medium temperature of 20 to 22° C (70 to 72° F). Diseases after rooting cause major problems with survivability. It is difficult to transplant, so it is best if direct-stuck (rooted) in small liner pots.

Leyland Cypress. *See × Cupressocyparis.*

Leucospermum spp. Leucospermum, Protea. *L. conocarpodendron* × *L. cuneiforme* 'Hawaii Gold,' which is a cut-flower protea. Leucospermum is micropropagated (*225*). See *Grevillea* for other proteaceous plants that can be micropropagated.

Ligustrum spp. Privet. Seed propagation is easily done. The cleaned seed should be stratified for two to three months at 0 to 10° C (32 to 50° F) before planting. Hardwood cuttings of most species planted in the spring root easily, as do softwood cuttings in summer under mist. Japanese privet *(L. japonicum)* is somewhat difficult to start from cuttings, but good results were obtained with actively growing terminal shoots rather than more mature wood. Treat with a 2,500 ppm IBA quick-dip (*30*).

Lilac. *See Syringa.*

Linden. *See Tilia.*

Lindera obtusiloba. Japanese spicebush. Propagated by cuttings quick-dipped with 4,000 ppm NAA or 8,000 ppm IBA (*115*). *Lindera* is commercially micropropagated.

Liquidambar styraciflua. Liquidambar, American Sweet Gum. Propagation is usually by seeds, which are collected in the fall. Seeds are stratified for one to three months at about 4° C (40° F) to overcome seed dormancy. Sweet gum can also be propagated by stem cuttings treated with 4,000 to 8,000 ppm IBA—overwintering may be a problem. Girdling of ten-year-old *L. formosana* prior to taking cuttings, and IBA treatment enhanced rooting (*177*). Selected clones are grafted or T-budded from spring through fall onto *L. styraciflua* seedlings. Sweet gum is commercially micropropagated (*50, 51, 238, 346*).

Liriodendron tulipifera. Tulip Tree, Yellow Poplar. Seed propagation is somewhat difficult. Seeds of this species are often devoid of embryos, so cutting tests of each seed lot should be made. Seeds should be stratified for about two months before planting and should not be allowed to dry out. Fall planting, with outdoor stratification through the winter, also has given good germination.

Although considered difficult to root, leafy stem cuttings taken in summer that are not fully hardened have been rooted in fairly good percentages. Can be propagated by root cuttings. Propagation by both budding and grafting is successful. Cultivars can also be chip budded.

Loblolly Bay. *See Gordonia.*

Locust (Black). *See Robinia.*

Lonicera spp. Honeysuckle. Seeds show considerable variation in their dormancy conditions, some species having both seed-coat and embryo dormancy, while others have only embryo dormancy, or no dormancy inhibition. This variability also occurs among different lots of seeds of the same species. In *L. tatarica,* some lots have no seed dormancy. In general, however, for prompt germination, stratification for two to three months at about 4° C (40° F) is recommended. Seeds of *L. hirsute* and *L. oblongifolia* should have two months of warm stratification [21 to 30° C (70 to 85° F)], followed by two to three months of cold stratification at about 4° C (40° F).

Most honeysuckle species are propagated easily by hardwood cuttings, softwood cuttings (spring), or semi-hardwood (summer) cuttings treated with a 1,000 ppm IBA talc and propagated under mist (*189*). Layering of vine types, such as 'Hall's' honeysuckle, is very easy, since roots form wherever the canes become buried under moist soil.

Loquat. *See Eriobotrya.*

Loropetalum chinense. Chinese Witch Hazel. An evergreen shrub that averages 3 m (10 ft). 'Rubrum' has ruby-red, purplish-red to greenish-purple leaves and salmon-pink flowers. Propagated by hardwood cuttings treated with a 3,000 ppm K-IBA quick-dip (*112*). It is commercially micropropagated.

Maclura pomifera. Osage-orange. A tree for difficult sites. Can be seed-propagated with a 30-day stratification treatment (*275*). A two-day water soak overcomes dormancy and permits germination without stratification (*115*). Male and thornless cultivars 'Altamont,' 'Park,' and 'Wichita' can be asexually produced by softwood cuttings or hardwood cuttings (January, Kansas) treated with 5,000 to 10,000 ppm IBA. This species can also be budded. Osage-orange can also be micropropagated (*218*).

Madrone, Pacific. *See Arbutus.*

Magnolia spp. Magnolia. Seeds, cuttings, grafting, layering, and micropropagation are utilized in propagating magnolias (*54,79*).

Seeds. Magnolia seeds are gathered in the fall as soon as possible after the fruit is ripe, when the red seeds are visible all over the fruit. The red pulp on *M. grandiflora* seed acts as an inhibitor to germination, so cleaning is very important. After cleaning, the seeds should either be sown immediately in the fall or—prior to spring planting—stratified for two to three months at about 4° C (40° F). Allowing the seeds to dry out during storage seems to be harmful. *M. grandiflora* seeds, and some other species, lose their viability if stored through the winter at room temperature. If prolonged storage is necessary, the seeds should be held in sealed containers at 0 to 4° C (32 to 40° F). Magnolia seedlings grow rapidly, and generally are large enough to graft by the end of the first season. Transplanting should be kept at a minimum, since this retards the plants.

Cuttings. There is no one best way to root a magnolia cultivar. Some species, such as *M. soulangeana* and *M. stellata,* are commercially propagated by leafy softwood cuttings. Cuttings may be taken from late spring to late summer after terminal growth has stopped and the wood has become partly matured.

With *M. stellata,* 6 to 10 cm (2 to 4 in.) softwood leaf-bud cuttings are wounded on the opposite side of the bud and the leaf blade is reduced 50 to 60 percent. The cutting is treated with a 8,000 ppm IBA talc or quick-dip (*229*).

Leafy cuttings of *M. grandiflora* (with two to four leaves per cutting) are taken from late spring to late summer. Terminal buds should be hardened, and not making new growth. Wounding encourages rotting, but enhances rooting in some cultivars.

Auxins are needed to root *M. grandiflora.* Generally, a range of *3,000 to 8,000 ppm* IAA and IBA combinations, or combinations of K-IBA and K-NAA give best results; cultivars are treated with as little as 1,000 ppm to a high of 15,000 ppm auxin combinations (*182*). *M. grandiflora* root best with mist under warm conditions, i.e., maintain a bottom heat of 24° C (75° F)—minimum 16° C (60° F) in winter (*182*).

With *M. grandiflora,* there is considerable cultivar variation in rooting. Brown-black cultivars are harder to root, with the hairy felt on the leaves that makes moisture control more difficult. Other reports indicate that semi-hardwood cuttings root well when quick-dipped in 5,000 to 10,000 ppm NAA in 50 percent alcohol (*108*). To obtain survival of the rooted cuttings through the following winter, they should be rooted early enough in the season so that some resumption of growth will occur before fall. In Scotland, the leaf blade of leaf-bud or nodal-tip cuttings is partially trimmed to reduce transpiration, save propagation space, and allow sufficient air circulation; cuttings are treated with 3,000 ppm IBA in talc or a 1,000 ppm quick-dip, and propagated under mist with a basal rooting temperature of 18° C (64° F) (*106*).

In South Carolina, *Magnolia grandiflora* 'Little Gem' are stuck in August and rooted by February of the next year. Cuttings are maintained at an initial 35° C (95° F) air temperature[minimum 16° C (60° F) bottom heat in late fall/winter], single wounded, and quick-dipped in 5,000 ppm IBA + 2,500 ppm NAA (*182*).

Grafting. *Magnolia kobus* is probably the best rootstock for the Oriental magnolias, whereas *M. acuminata* can be used as a rootstock for either Oriental or American species. *M. sprengeri* 'Diva' makes excellent, comparably vigorous rootstocks for the large Asian species and their hybrids. *Magnolia grandiflora* seedlings are used for *M. grandiflora* cultivars.

Winter or bench graftage using side-veneer grafts are satisfactory, with the union and scion taped or waxed after grafting. Some propagators pot the seedlings in the fall, then bring them into the greenhouse and do the grafting in midwinter. The newly grafted plants may be set on open benches in the greenhouse or placed in closed propagating frames, where they stay for seven to ten days while the union is healing.

In Alabama and Mississippi, cleft grafting *M. grandiflora* with containerized seedling rootstock has been reported (*31*). Chip budding is possible throughout the growing season (*354*) and has been used with more difficult-to-root cultivars budded onto container-grown rootstock (*106, 229*) and with field-grown rootstock (*193*). Seedling rootstock tends to sucker and must be maintained (*253*). Clonal rootstock (from rooted cuttings) are also being utilized to better accommodate caliper differences that frequently occur between scions and seedling rootstocks (*193*). Clonal rootstocks are being selected for dwarfing characteristics so that Magnolias can be sold in bloom and better accommodate smaller-sized, urban gardens.

Micropropagation. Magnolias are commercially micropropagated.

Mahogany. *See Swietenia.*

Mahonia spp. Mahonia. Seed propagation is generally easy for most cultivars, while cuttings can be difficult (*89*). *M. bealei, M. lomarifolia,* and *M. japonica* are easy to grow from seed and do not require special treatments. *M. aquifolia* and *M. repens* seed must be separated from the fruit pulp, leached of inhibitors (for one week with a slow continuous flow of water), and later stratified for either three months (Georgia) or a total of five months (California). Cuttings of *M. aquifolium* 'Compacta' are collected in fall or winter, quick-dipped with 3,000 ppm IBA, and rooted under mist with 25° C (77° F) bottom heat; 8,000 ppm IBA talc has also been used with cuttings of *M. bealei, M. nervosa, M. pinnata, M. repens,* and *M. wagneri* (*115*).

Malus spp. Crab Apple (Flowering). Crab apples can be produced from seed, rooted as cuttings, grafted or budded, and some cultivars are commercially produced from micropropagation (*370*).

Seed. Four species of crab apples—*M. toringoides, M. hupehensis, M. sikkimensis,* and *M. florentina*—will reproduce true from seed.

Cuttings. Selected forms of all other crab apple species, such as *M. sargentii, M. floribunda,* and *M.* 'Dolgo,' should be *clonally propagated by asexual methods,* which results in a stronger, more uniform crop. *M.* 'Hopa,' 'Almey,' and *M.* × *eleyi* are commercially propagated with July semi-hardwood cuttings (Florida) which are wounded at the base, quick-dipped in 5,000 ppm K-IBA, and rooted under mist within four to six weeks (*93*). Hardwood cuttings of crab apples are difficult to root. Establishing selected crab apple cultivars and species on their own root system with softwood cuttings is cheaper and avoids rootstock suckering, crooks in the trunk, and graft incompatibility associated with traditional grafting and budding systems (*56*). Softwood cuttings taken in late spring root when treated with 3,000 to 8,000 ppm IBA in talc (*370*) or 2,500 to 10,000 ppm IBA liquid under mist (*108*).

Grafting. Nursery trees are commonly propagated either by root grafting, using the whip graft, or by T-budding and chip budding (see Figure 14–8) seedlings in the nursery row. Budding is done either in spring or fall (*328*). Fall budding (late summer in Oregon) is considered by most nursery people to be the faster and most desirable method of propagating crab apples. Various seedling rootstocks have been used, such as *M.* × *domestica* (common apple), *M. baccata, M. ioensis,* and *M. coronaria.* Hardiness and suckering can be a problem with some seedling rootstock.

To eliminate suckering, clonal apple rootstock—'EMLA 111' and semi-dwarfing 'M7a'—are used. They also give better anchorage than own-rooted crabs and give a moderate improvement in root-hardiness (Oregon). While T-budding generally works well on domestic apple rootstock, chip budding is preferable with 'EMLA 111' (*370*). Crab apples are commercially micropropagated (*329*).

Mandevilla Spp. Mandevilla 'Alice DuPont.' Take one-node cuttings and trim off half of the leaf surface and dip cuttings in 2,500 ppm IBA. In Alabama, cuttings are rooted in July, rooted liners planted in October, and plants finished off and sold in 3.7-liter (1-gal) containers in April (*201*).

Manzanita. *See Arctostaphylos.*

Maple. *See Acer.*

Melaleuca spp. The seeds of these native Australian species are almost dustlike but germinate easily and can be handled like eucalyptus seed.

Mesquite. *See Prosopis.*

Metasequoia glyptostroboides. Metasequoia, Dawn Redwood. Seeds germinate without difficulty, and both softwood and hardwood cuttings will root (*84*). The leafless hardwood cuttings may be lined-out in the nursery row in early spring. Leafy cuttings root easily under mist if taken in summer and treated with 3,000 to 8,000 IBA talc or quick-dip.

Mimosa. *See Albizzia.*

Mock Orange. *See Philadelphus.*

Monkey Puzzle Tree. *See Araucaria.*

Morus alba. Fruitless or White Mulberry. Some mulberry trees produce only male flowers and hence do not bear fruits. Seeds need to be separated from fruits; some seeds have dormant embryos and impermeable seed coats. Seeds collected in early summer, cleaned, and sown have high percent germination. Stored seeds, cold stratified for one to three months, can be sown in the spring. Roots easily from semi-hardwood (summer cuttings) and hardwood cuttings treated with 8,000 ppm talc or 10,000 to 15,000 ppm IBA quick-dip. *Morus alba* 'Pendula' is grafted on a standard of *M. alba* 'Tatarica' to produce a small weeping tree (*115*). The fruitless *Morus alba* 'Chaparral' is also grafted. Mulberry is commercially micropropagated.

Mountain Ash. *See Sorbus.*

Mountain Laurel. *See Kalmia.*

Mulberry (Fruitless). *See Morus.*

Myrica spp. Bayberry, Wax myrtle. Propagated by seed or by cuttings. With northern wax myrtle *(M. pennsylvanicum)* and southern wax myrtle *(M. cerifera),* the wax must be removed from the seed for germination to occur. The wax coating may be removed by rubbing the seed over a screen. One propagator uses a hot water soak with several drops of mild detergent to help remove the wax. A three-month cold stratification is recommended. Kinetin and gibberellic acid treatments are reported to enhance germination of northern wax myrtle *(M. pennsylvanicum)* (*172*).

M. cerifera is fairly easy to root under mist using a 2,000 ppm IBA quick-dip (Texas). *M. pennsylvanicum* can be easily rooted from softwood cuttings (Pennsylvania) taken early before the terminal bud has formed—a 5,000 ppm IBA quick-dip is beneficial.

Myrtle. *See Myrtus.*

Myrtus spp. Myrtle. Crop production of *M. communis* takes longer by seed than by cuttings. Seedling plants produce a special swollen root (lignotuber) that permits myrtle to survive drought conditions, disease, and insect damage. Plants produced by cuttings do not produce the lignotuber, and are not as drought-resistant as seed-propagated plants. The species is most commonly propagated from 8 to 15 cm (3 to 6 in.) shoot-tip cuttings taken in August

(Pennsylvania). A quick-dip of 5,000 ppm IBA enhances rooting. Micropropagation has been obtained from mature field-grown plants (*271*).

Nandina domestica. Nandina, Heavenly Bamboo.

Seed. Can be propagated by seed. The embryos in the mature fruits are rudimentary, but will develop in cold storage. Seeds can be collected in late fall, held in slightly moist storage at 4° C (40° F), then planted in late summer. Germination occurs in about 60 days. Germination tends to take place in autumn regardless of planting date. No cool moist stratification period is necessary.

Cuttings. Dwarf nandina ('Compacta nana,' 'Purpurea Nana,' 'Gulf Stream,' 'Harbour Dwarf,' and 'Moon Bay') are easy to root (*20*). Shoot tips (without brown wood) are trimmed to 4 cm (1.5 in.) lengths, stripped of bottom leaves, quick-dipped in 1,250 ppm IBA and 500 ppm NAA, and stuck directly into 6-cm (2.2-in.) liner pots. Cuttings can be rooted any time of the year (Texas), except during spring flush. Winter rooting requires bottom heat. 'Harbour Dwarf' and 'Compacta' can be propagated by separation of suckers at the base. This often results with suckering in the liner pot with new shoots, more rapidly producing a salable plant. Suckering and more rapid plant development in the liner pot can also be promoted by rooting a cutting with one node under soil; when the liner is transplanted, it is buried deep enough to cover the next node up the stem, which results in suckering from one or both of the nodes (*201*).

Micropropagation. Nandinas are commercially micropropagated (*53*).

Nerium oleander. Oleander. Seedlings reproduce fairly true-to-type, although a small percentage of plants with different flower colors will appear. The seeds should be collected in late fall after a frost has caused the seed pods to open. Rubbing the seeds through a coarse mesh wire screen removes most of the fuzzy coating. The seeds are then planted immediately in the greenhouse in flats without further treatment. Germination occurs in about two weeks. Leafy cuttings root easily under mist if taken from rather mature wood during the summer and treated with a 3,000 ppm IBA quick-dip. Plant parts are very poisonous. Can be micropropagated (*272*).

Norfolk Island Pine. *See* Araucaria.

Nyssa sylvatica. Black Gum, Black Tupelo, Pepperidge Tree. One of the best trees for dependable fall color, even in mild climates. Propagated by seed. A high seed germination occurs with a three-month cold stratification. Softwood cuttings root when treated with 8,000 ppm IBA talc and 21° C (70° F) bottom heat.

Oak. *See Quercus.*

Olea europaea. Olive. A fruitless cultivar, 'Swan Hill,' is available for use as a patio or street tree; it is usually grafted on *O. oblonga.* Cuttings are difficult to root. They should be placed under mist and treated with IBA at 2,000 to 3,000 ppm, or they can be grafted on easily rooted cultivars used as a rootstock (*179*). *O. europaea* 'Wilsonii' and 'Majestic Beauty' are also fruitless and root more easily.

Oleander. *See Nerium.*

Olive. *See Olea.*

Orchid (Anacho) Tree. *See Bauhinia.*

Osage-Orange. *See Maclura.*

Osmanthus spp. Sweet olive, Fragrant tea olive (*O. fragrans*). Seeds are slow and difficult to propagate. Semi-hardwood cuttings or those with firm wood are used and a 3,000 to 8,000 IBA talc or quick-dip enhances rooting. Softwood cuttings of *Osmanthus* × *fortunei* treated with 2,500 IBA root well (*38*). In Alabama, *O. fragrans* cuttings root best when taken from semi-hardwood, new-growth stock in early August and quick-dipped in 15,000 ppm IBA—the bigger the caliper of the cutting, the better it roots (*201*). Semi-hardwood and hardwood cuttings of *O. heterophyllus* 'Ilicifolius' and hardwood cuttings of 'Rotundifolius' root in high percentages without auxin (*39*).

Oxydendrum arboreum. Sourwood. Sourwood is propagated commercially by seeds. Fall-harvested seeds need no pretreatment and flats containing seeds are generally placed under continuous light. Stratification of two to three months speeds germination and decreases the light required for germination (*22*). Propagation by cuttings is extremely difficult. However, soft-tip cuttings treated with a 2,000 to 3,000 ppm IBA quick-dip can be rooted (Pennsylvania). Sourwood is commercially micropropagated (*16*).

Pachysandra terminalis. Japanese spurge. This evergreen ground cover for shady areas spreads naturally by rhizomes. Can be propagated easily by division or by leafy cuttings under mist without auxin. *Pachysandra* is commercially micropropagated.

Paeonia suffruticosa. Tree Peony. Seed propagation is complicated by epicotyl dormancy. Seeds are sown in pots and exposed to a warm-moist stratification of 18 to 21° C (65 to 75° F) for root/hypocotyl growth, than cold stratified to break epicotyl dormancy. Further epicotyl growth occurs under warm conditions. Selected cultivars are propagated by grafting in late summer on herbaceous peony *(P. lactiflora)* roots as the rootstock. The grafts are covered and callused in a sand-peat medium in a greenhouse until fall, when they are potted. In Oregon, peonies are cleft grafted in early August with a fresh, single-edge razor blade for each graft. The newly grafted units are packed in plastic-lined crates containing *slightly* moist peat and maintained at 27° C (80° F) for two to four weeks until the graft union has formed (*265*). This species has been micropropagated (*48*).

Palms (57, 74). There are numerous species and genera of ornamental palms.

Seed. They are propagated with fully mature seed, as indicated by color changes from green to red, yellow, black, etc., depending on species. It is best to use only fresh seed, which should be planted as soon as possible after harvesting and not allowed to dry out. Generally, palm seeds remain viable for only a short period. Seed should be collected from trees and not taken from the ground. It is recommended that the fleshy seed coat be removed. The coat often contains an inhibitor. Soak the cleaned seed for three days prior to planting, changing the water daily (*297*). Any seeds that float in water should be discarded. Presoaking seed for one to two days in water containing a fungicide and insecticide can also speed germination time and the percentage of germination. Palm seeds are susceptible to surface molds and should be protected by dusting with a fungicide.

For storing palm seed, fresh, cleaned seed should be conditioned for two days to 85 to 90 percent RH, then dusted with a seed protectant fungicide (e.g., thiram), tightly sealed in heavy polyethylene bags, and stored at 23° C (73° F) (*120*).

Seed germination of some species can be accelerated by scarification, followed by soaking in gibberellic acid at 1,000 ppm for 72 hours—and seed flats placed over bottom heat at 27° C (81° F) (*267*). *Rhapidophyllum hystrix* seed can be scarified under sterile conditions by removing the seed coat and cap that cover the cavity containing the embryo (*68*). High germination was obtained from scarified seeds of *R. hystrix*, which were stored for 12 months at 5° C (41° F) and 100 percent relative humidity; high viability of the stored seeds was obtained when the seeds' moisture content was 14 to 36 percent (*67*). Successful germination also occurs with unscarafied seeds of *R. hystrix*—but the trick is to fully hydrate the seeds and expose them to alternating temperatures. Seeds are soaked for seven days with aerated and/or running tap water at 30° C (86° F), then seeds are germinated in moist peat moss with alternating 40° C (104° F) for 6 hours and 25° C (77° F) for 18 hours daily; the fully hydrated embryos can readily penetrate the thick-walled seed coats, but need the high temperature stimulus to promote embryo growth (*67, 68*).

Seeds of most species germinate in one to three months, especially if bottom heat, 28° C (80° F), is maintained, but some may take as long as one to two years. Some palm species have limited optimum temperatures for seed germination, and temperatures above and below the optimum level contribute to irregular and low percentage germination. Four native Florida palms had optimum germination temperatures of 35° C (95° F) (*64*). As a general rule, germination temperatures should be maintained between 29 to 35° C (85 to 95° F). With *Butia capitata*, seeds require 90 to 150 days' after-ripening at 5 to 25° C (41 to 77° F) before sowing (*65*). After-ripening, followed by a 40° C (104° F) germination temperature, increased seed germination and reduced germination time.

Division. Multiclumping palms, such as some date palm species, can be asexually propagated by removing and rooting offshoots or basal suckers.

Micropropation. The date palm *(Phoenix dactylifera)* (*352*), and ponytail palm *(Beaucarnea recurvata)* (*315*) can be micropropagated.

Parrotia persica. Persian Parrotia. Lack of available seed is a problem, as is a long warm/cold stratification requirement. Easily propagated from semi-hardwood cuttings treated with 1,000 to 4,000 ppm IBA. It is commercially micropropagated.

Parthenocissus spp. Virginia Creeper *(P. quinquefolia)*, Boston Ivy *(P. tricuspidata)*. These two ornamental vines can be propagated by seeds planted in the fall or stratified for two months at about 4° C (40° F) before planting in the spring. Leafy softwood cuttings taken in late summer root easily under mist when treated with a 3,000 ppm IBA talc. So do hardwood cuttings planted outdoors in early spring.

Grafting of named cultivars on *P. quinquefolia* seedlings or rooted cuttings is done by some nurseries using the whip or cleft graft.

Passiflora × alatocaerulea. Passion Vine. This tender, subtropical vine is propagated by leafy cuttings under glass or mist.

Pawpaw. *See Asimina.*

Paxistima myrsinites. Oregon Boxwood. An evergreen shrub native to the Pacific Northwest. Readily propagated from semi-hardwood cuttings from midsummer until bud-break in spring (Vancouver, B.C.). Cuttings are treated with 8,000 ppm IBA talc, placed under mist, and maintained with a minimum basal temperature of 21° C (70° F). Can also be rooted in fall and winter under contact polyethylene film (*353*).

Pear (Callery cvs. Bradford, Captial, Whitehouse, etc.). *See Pyrus.*

Pear (Evergreen). *See Pyrus.*

Peony (Tree). *See Paeonia.*

Penstemon fruticosus. Purple Haze Penstemon. A low, compact, evergreen to semi-evergreen ground cover (subshrub) with tubular, mauve-purple flowers. Propagated with softwood and semi-hardwood cuttings (June through September in Vancouver); 3,000 ppm IBA talc is optional (*189*). Requires a well-drained container soil. Penstemon is commercially micropropagated.

Persian Parrotia. *See Parrotia.*

Persimmon (Texas). *See Diospyros.*

Philadelphus spp. Mock Orange. The many cultivars of mock orange are best propagated by cuttings, which root easily. Hardwood cuttings (late winter) treated with 2,500 to 8,000 ppm IBA or softwood cuttings treated with a 1,000 ppm IBA quick-dip can be rooted under mist. Removing rooted suckers arising from the base of old plants is an easy means of obtaining a few new plants.

Phoenix dactylifera. Date palm. *See* Palms.

Photina arbutifolia. *See Heteromeles.*

Photinia spp. Photinia. Large evergreen shrubs with small pome fruits. *P.* × *fraseri* (Fraser or red-tip photinia) can be propagated by seed exposed to a two-month cold stratification. Semi-hardwood cuttings are best rooted with a 10,000 ppm IBA quick-dip (*30*) or 8,000 ppm talc. Wounding cuttings, trimming leaves of cuttings, and 3,000 to 8,000 ppm IBA talc or quick-dips are optional for *P.* × *fraseri* 'Red Robin,' which is a hybrid of *P. glabra* and *P. serrulata* (*83, 107*). Photinia is commercially micropropagated.

Picea spp. Spruce.

Seed. They are ordinarily propagated without difficulty by seed. Most species have embryo dormancy, requiring a one- to three-month cold stratification at 4° C (40° F) for good germination. They can also be fall-planted and naturally stratified. Seeds of *P. abies, P. engelmannii, P. glauca* var. *albertiana,* and *P. omorika* germinate well without stratification. Colorado blue spruce (*P. pungens* 'Glauca') grown from seed produces trees with a slight bluish cast. Only a small percentage of the seedlings have the very desirable bright blue color. Several exceptionally fine blue seedling specimens have been selected as clones and are perpetuated by grafting. Three of the best-known grafted blue spruces are Koster blue spruce (*P. pungens* 'Koster'), the compact 'Moerheim' blue spruce (*P. pungens* 'Moerheimii'), and *P. pungens 'Hoopsii'*—which is considered the bluest form.

Cuttings. Selected clones of spruce are difficult to propagate by cuttings, but there are instances in which good percentages of cuttings have rooted (*163, 199*), especially from selected young source trees. In general, pyramidal forms are difficult-to-root, and are usually grafted or seed propagated. Dwarf and more prostrate, spreading forms are easier to root from cuttings.

Taking cuttings from vigorous containerized trees gives good results. Cuttings taken in spring, midsummer, and mid-autumn have been rooted. For cuttings, it is best to use only shoot terminals, which should be gathered in early morning when the wood is turgid. Wounding, mist, and high light irradiance during rooting are helpful. *P. glauca* and *P. pungens* have been successfully rooted with 3,000 to 10,000 ppm IBA treatments (*115*). In making cuttings of upright-growing types, terminal shoots should be selected rather than lateral branches, since the latter, if rooted, tend to produce prostrate, sprawling plants rather than the desired upright form.

Grafting. The Koster blue spruce is propagated commercially by grafting scions on Norway spruce *(P. abies)* seedlings. In Canada, bench grafting of *P. abies* on seedling rootstock is done with a side-veneer or apical-wedge graft. Most grafting is done in winter, but there are advantages to late summer graftage. In Oregon, blue spruce (*P. Pungens* 'Hoopsi') is grafted on containerized, dormant *P. abies* rootstock and overwintered in unheated structures; soaking the scion bases in 200 ppm IBA for three minutes increased grafting success (*25*).

Micropropagation. Clones of *P. glauca* and *P. pungens* can be vegetatively micropropagated by somatic embryogenesis (*75*). Meristem micrografting has been done with *P. abies in vitro* (*261*).

Pieris spp. Pieris. *P. floribunda* and *P. japonica* reproduce readily by seed with no treatment necessary (*145*). During sowing, seeds should not be covered with media, since they require light to maximize germination and are relatively small [210,000 seeds per 28g (1 oz)] (*338*). Some Pieris species can be started easily by cuttings with a 1,000 ppm IBA quick-dip, but *P. floribunda* cuttings are difficult to root. Rooting under mist or polyethylene-covered frames is enhanced by wounding, IBA talc, or quick-dips at 5,000 to 8,500 ppm, and bottom heat. Rockwool propagating sheets are reported to reduce transplant shock (*11*). Pieris is commercially micropropagated (*337*).

Pileostegia viburnoides. A self-clinging evergreen climber in the Hydrangeaceae family. Produces panicles of creamy-yellow flowers. Can be rooted from cuttings using talc applications of 8,000 ppm IBA.

Pine. *See Pinus.*

Pinus spp. Pine.

Seed. Pines are ordinarily propagated by seed (*347*). Considerable variability exists among the species in regard to seed dormancy conditions. Seeds of many species have no dormancy and will germinate immediately upon collection, whereas others have embryo dormancy. With the latter, stratification at 0 to 4° C (32 to 40° F) for one to three months will increase germination. Moist perlite as a stratification medium will enhance overall germination. *P. cembra* may have immature embryos, so a warm stratification of 21 to 27° C (70 to 80° F) for two to three months, followed by a cool stratification for three months at 2° C (36° F) will aid germination.

Species whose seeds have no dormancy conditions and can be planted without treatment include *Pinus aristata, P. banksiana, P. canariensis, P. caribaea, P. clausa, P. contorta, P. coulteri, P. edulis, P. halepensis, P. jeffreyi, P. latifolia, P. monticola, P. mugo, P. nigra, P. palustris, P. pinaster, P. ponderosa, P. pungens, P. radiata, P. resinosa, P. roxburghii, P. sylvestris, P. thunbergii, P. virginiana,* and *P. wallichiana.* However, if seeds of the above species have been stored for any length of time, it is advisable to give them a cold stratification period before planting. Pine seeds can be stored for considerable periods of time without losing viability if held in sealed containers between–15 to 0° C (5 and 32° F). Seeds should not be allowed to dry out.

Cuttings. Pine cuttings are difficult to root, although those of mugo pine *(Pinus mugo)* root easily if taken in early summer (*216*), and selected clones of *P. radiata* are commercially rooted. Success is more likely if cutting material is taken in winter from low-growing lateral shoots on young trees. Treatment with IBA is beneficial (*237*).

Considerable study has been given to the rooting of cuttings of Monterey pine *(P. radiata)* because of its importance as a timber crop in New Zealand and Australia, and as a Christmas tree species. There are clonal variations in the commercial rooting of this species. Best rooting is from cuttings taken in early winter. Wounding, plus a 4,000 ppm IBA quick-dip enhances rooting. A more symmetrical root system could be induced by clipping the ends of the original roots and allowing the root system to develop from the secondary roots.

Rooting of *P. strobus* was improved when stock plants were etiolated and shoot bases covered with black adhesive tape. Accelerated growth techniques of supplementary lighting, elevated CO_2, temperature manipulation, and optimum fertility enhance rooting in *P. echinata, P. thunbergiana,* and *P. elliottii,* (*176*).

Rooting needle fascicles. Pines can be propagated asexually by rooting needle fascicles (needle leaves held together by the scale leaves, containing a base and a diminutive shoot apex) (*87*). Rooting is best when the fascicles are taken from trees younger than four years old. IBA treatments are helpful. By selecting certain seedlings whose cuttings root easily and by using critical timing in taking cuttings, it is possible to select clones in which cutting propagation is commercially feasible. This was shown to be true for the mugo pine (*162*). Cuttings of the Scotch pine have been rooted by a unique method of forcing out interfascicular shoots from young stock plants. These shoots, when made into cuttings, root in high percentages.

Grafting. Side-veneer grafting is used for propagating selected clones. Well-established two-year-old seedlings of the same or closely related species should be used as rootstocks. Scions should be of new growth, taken from firm, partly matured wood. Winter grafting in greenhouses works well in Oregon. *P. cembra* is side-veneer grafted on *P. strobus,* and selected forms of *P. densiflora* are grafted on seedling *P. densiflora* or *P. sylvestris.*

Micropropagation. Selected pine species can be micropropagated (*53, 326, 382, 386*).

Pistache (Chinese). *See Pistacia.*

Pistacia chinensis. Chinese Pistache. Seeds should be collected from relatively large fruits, blue-green in color, in mid-fall. Pulp must be removed. Soak fruits in water, then rub over a screen. Seeds that float in water have aborted embryos and should be discarded. Stratification at 4 to 10° C (40 to 50° F) for ten weeks gives good germination. Seedlings exhibit a wide range of variability. Propagation by cuttings is very difficult. T-budding selected clones on seedling *P. chinensis* rootstocks in late summer is used to produce uniform, superior trees (*205*). Limited success has occurred with the micropropagation of *P. atlantica* (*283*).

Pittosporum spp. Pittosporum. Can be started by seeds or cuttings. The seeds are not difficult to germinate; dipping a cloth bag containing the seeds for several seconds in boiling water may hasten germination. Cultivars are propagated with leafy semi-hardwood cuttings. An IBA quick-dip or talc application of 1,000 to 3,000 ppm is beneficial to rooting of *P. tobira*—and 6,000 to 8,000 ppm IBA for *P. tenuifolium.*

Platycladus orientalis. *See Thuja.*

Plane Tree. *See Platanus.*

Platanus spp. Plane Tree, Sycamore. Seeds are ordinarily used in propagation, but they should not be allowed to dry out. The best procedure is to allow the seeds to overwinter in the seed balls right on the tree. They may then be collected in late winter or early spring and planted immediately. Germination usually occurs promptly. If the seeds are collected in the fall, then stratification at about 4° C (40° F) should be used. The hybrid London plane tree, *P × acerifolia,* can be propagated by hardwood cuttings, taken and planted in the nursery in autumn or by leafy softwood cuttings taken in midsummer, and rooted under mist. Auxin treatments do not always enhance rooting (*266*). *Platanus × acerifolia* (London Plane tree) cultivars, such as as 'Bloodgood,' 'Columbia,' and 'Liberty,' can be budded on seedling rootstock; the same cultivars can also be rooted from cuttings treated with 8,000 ppm IBA (*115*).

Plums (Flowering, Mexican). *See Prunus.*

Plumbago spp. Plumbago. Seeds sown in late winter usually germinate easily. Leafy cuttings taken from partially matured wood can be rooted without difficulty under mist. Root cuttings also can be used, and old plants can be divided. *P. rosea* has been micropropagated (*224*).

Plumeria spp. Plumeria. Leafy cuttings 15 to 20 cm (6 to 8 in.) long of this tender tropical, Hawaiian shrub will root readily under mist—if treated with 2,500 ppm IBA.

Podocarpus spp. Podocarpus. These evergreen trees and shrubs have foliage resembling the related yews (*Taxus*) and make good container plants. Generally they are seed-propagated. They can be propagated by stem cuttings taken in late summer and early fall. Rooting of *P. marophyllus* is slow, even with a 3,000 to 8,000 ppm IBA talc application.

Poinsettia. *See Euphorbia.*

Poplar. *See Populus.*

Populus spp. Poplar, Cottonwood, Aspen (*282*).

Seed. These trees can be propagated by seeds; they should be collected as soon as the capsules begin to open in spring, and planted at once, because they lose viability rapidly, and should not be allowed to dry out. However, if held in sealed containers near 0° C (32° F), seeds of some species can be stored for as long as three years. There are no dormancy conditions and seeds germinate within a few days after planting. The seedlings are highly susceptible to damping-off fungi and will not tolerate excessive heat or drying. Poplars are difficult to propagate in quantity by seed.

Cuttings. Hardwood cuttings of *Populus* root easily (except the aspens) when propagated in the spring. Treatment with IBA is likely to improve rooting. Leafy softwood cuttings (of some species at least) taken in midsummer also root well. *P. tremuloides,* quaking aspen, can be propagated by removing root pieces, root suckers, and layering—inducing adventitious shoots to form from these propagules in vermiculite. Then rooting these adventitious shoots as stem cuttings under mist with IBA treatments.

Micropropagation. Many *Populus* species are micropropagated (*376*).

Potentilla spp. Potentilla, Cinquefoil. Propagation is usually by cuttings, but seed and division can be used. Cuttings are taken from early summer through fall (*149*). Rooting is best under light mist with bottom heat; mist should be reduced as soon as rooting occurs. Softwood cuttings of *P. fruticosa* root readily with 1,000 ppm IBA. Potentilla is commercially micropropagated.

Privet. *See Ligustrum.*

Prosopis spp. Mesquite. A landscape shrub or small tree for arid and semi-arid regions. Seed germination is hindered by external seed dormancy due to a water-impermeable seed coat. Mechanical scarification and hot-water treatment enhance seed germination of *P. alba* and *P. flexuosa* (*73*). Auxin has been used to enhance the rooting of *P. alba* cuttings (*219*). A thornless variety of Chilean mesquite (*P. chilensis*) is propagated by cuttings. *Prosopis* has been micropropagated (*7*).

Protea spp. Protea. This South African native, popular for its cut flowers, is propagated by seed, cuttings, and grafting. Cuttings are difficult to root but some species respond to auxin treatments (*281*).

Prunus spp. Japanaese Flowering Apricot (*P. mume*), Flowering Cherries (cultivars of *P. campanulata, P. sargentii, P. serrulata, P. sieboldi, P. subhirtella* and *P. yedoenis*), Plums (*P. cerasifera, P. mexicana*), Almonds (*P. dulcis, P. glandulosa, P. tenella*), Cherry Laurel, English Laurel (*P. caroliniana, P. laurocerasus*). Prunus are classified as evergreen and deciduous shrubs, flowering and fruiting trees.

Seeds. Seeds of *P. mume* require a one- to three-month cold-moist stratification for maximum germination. If cross-pollination with other species can be avoided, *P. sargentii, P. campanulata,* and *P. yedoensis* will reproduce true from seed.

Cuttings. As a general rule, ornamental Prunus shrubs are propagated by cuttings (e.g., *P.* × *cistena,* cultivars of *P. laurocerasus, P. lusitanica*), while ornamental Prunus trees are grafted (e.g., cultivars of *P. cerasifera, P. serrulata*). Leafy cuttings of some of the flowering cherry species can be rooted under mist in high percentages if treated with IBA, but subsequent survival and overwintering are sometimes difficult. Rooting is enhanced in *P. tenella* semi-hardwood cuttings collected in July (Ireland), quick-dipped in 1,250 ppm K-IBA + 1,250 K-NAA; IBA alone was ineffective (*214*). *P. serrulata* 'Kwanzan' is successfully own-rooted from softwood cuttings by wounding and using a 10,000 ppm IBA and 5,000 ppm NAA quick-dip (*241*). *Dwarf flowering almond* (*P. glandulosa*) is easily rooted from softwood summer cuttings with 1,000 ppm IBA. In Georgia, cuttings of *P. mume* are collected early in the season, when growth is just starting to firm, and treated with 3,000 ppm K-IBA (*111*); with later, firmer wood cuttings, no auxin or a low dosage of 1,000 ppm is used. Mexican plums (*P. mexicana*) are propagated by softwood cuttings (May and June in Texas), treated with 1,000 to 1,500 ppm IBA, or they are produced as seedlings. *P. caroliniana* are rooted from softwood or semi-hardwood cuttings treated with 3,000 to 8,000 ppm IBA. *P. laurocerasus* are rooted from semi-hardwood cuttings treated with 1,000 to 3,000 ppm IBA.

Grafting. Cultivars of oriental cherry (*P. serrulata*) are grafted onto seedling or clonal rootstocks of the Mazzard cherry (*P. avium*); T-budding, either in the fall or in the spring is also done. *Prunus dropmoreana* is a suitable rootstock for *P. serrulata* 'Kwanzan.' In England, *P. mume* is bench grafted in February and chip budded in August onto *P. cerasifera* rootstock (*111*). Seedling 'Nemaguard' peach rootstock is used in California.

Micropropagation. Prunus species are commercially micropropagated (*53, 335*).

Pseudotsuga menziesii. Douglas-Fir. This important lumber and Christmas tree species is normally seed-propagated, but clonal regeneration with cuttings is becoming important with Christmas tree selection and the production of elite timber trees (302). Limited seed from elite, control-pollinated plants (full-sib families) are collected and sown. Each seedling is used as a stock plant (ortet), and is sheared and cuttings (ramets) are taken and rooted. The rooted cuttings are combined so that many clones are mixed together, thus avoiding monoculture production. **Bulking** refers to the mixing of many elite clones together in the nursery and plantation—e.g. a few clonal copies are made of a large number of genotypes versus **cloning** where a large number of copies are produced from one or a few genotypes (302, 303).

Seeds. Seeds exhibit varying degrees of embryo dormancy (*34*). For prompt germination, it is best to sow the seeds in the fall or stratify them in moist perlite for two months at about 4° C (40° F).

Cuttings. Douglas-Fir cuttings are rather difficult to root, but by taking them in late winter, treating them with IBA, and rooting them in a sand/peat-moss mixture, it is possible to obtain fairly good rooting. Cuttings from young trees root much more easily than do those from old trees, and cuttings from certain source trees are easier to root than those from others. Cuttings are taken in December and January from sheared or hedged stock blocks, and treated with 5,000 ppm NAA + 5,000 ppm IBA. One timber company uses a three-year propagation regime with greenhouse production of stock plants (ortets) from elite seed, then roots cuttings of the ortets under fog, and finishes the rooted cuttings in an outdoor bare-root nursery (*302*). Rooting of cuttings for Douglas-Fir Christmas trees is strongly related to clonal variation and seasonal fluctuation. NAA is more effective than IBA in stimulating rooting (*290*).

Micropropagation. Micropropagation was successful with cotyledon explants (*78*).

Pygmy Pea shrub. *See* Caragana.

Pyracantha spp. Pyracantha, Firethorn. Seed require a three-month cold stratification. However, propagation is almost always by cuttings. Partially matured, leafy, current-season's growth is taken from late spring to late fall and rooted either in the greenhouse or under mist for good results. Treatments with a 2,500 ppm IBA quick-dip are beneficial (*30*). Semi-hardwood cuttings collected in fall rooted well when wounded and treated with 8,000 ppm IBA talc or quick-dip (*115*).

Pyrus calleryana. Bradford Pear (*P. calleryana* 'Bradford'). Can be rooted with 8,000 ppm IBA talc, quick-dips of 10,000 ppm IBA + 5,000 ppm NAA, or a 20,000 ppm K-IBA quick-dip. Semi-hardwood cuttings are taken June to August (after the terminal bud has formed), and rooted under mist (*2, 115*). There are some reports of poor performance of 'Bradford,' which is being replaced by better-performing *P. calleryana* cultivars, such as 'Capital,' 'Cleveland Select,' 'Redspire,' and 'Whitehouse.' These cultivars are grafted on seedling *P. calleryana* rootstock by T-budding. Bench grafting can also be done with a whip-and-tongue graft. This species is commercially micropropagated (*53*).

Pyrus kawakamii. Evergreen Pear. Propagated by cuttings or, more commonly, by grafting on *P. calleryana* seedling rootstock. Bench grafting is done in midwinter using the cleft graft. The grafts can be planted in containers or in the nursery row.

Quercus spp. Oak.

Seed. Seed propagation is generally practiced. Wide variations exist in the germination requirements of oak seed, particularly between the black or red oak (acorns maturing the second year) and white oak (acorns maturing the first year) groups. Seeds of the white oak group have little or no dormancy and, with few exceptions, are ready to germinate as soon as they mature in the fall. Seeds of the following species will germinate without a low-temperature stratification period: *Quercus agrifolia, Q. alba, Q. arizonica, Q. bicolor, Q. chrysolepis, Q. douglassii, Q. garryana, Q. lobata, Q. macrocarpa, Q. montana, Q. petraea, Q. prinus, Q. robur, Q. stellata, Q. suber, Q. turbinella,* and *Q. virginiana.* Germination of *Q. nigra* is maximized when the seed coat is removed (*3*). Seeds of most species of the black oak group have embryo dormancy, requiring either stratification [0 to 2° C (32 to 35° F)] for one to three months, or fall planting to allow natural stratification.

Acorns are often attacked by weevils. Soaking in water held at 49° C (120° F) for 30 minutes will rid the acorns of this damaging pest. However, research with *Q. virginiana* indicates no commercial advantage with heat treatment for weevil control. In fact, there is a loss of seed viability.

Seeds are usually floated in water, and those seeds that float are discarded. With seeds of stratified *Q. rubra,* the presence of a split pericap following a ten-day aerated water-soak treatment is a non-destructive method for identifying high-quality seed before sowing (*90*). Acorns of many species tend to lose their viability rapidly when stored dry at room temperature. Seeds of some northern species can be stored for several years without losing viability by holding them at 1 to 3° C (34 to 37° F) in polyethylene bags. The seeds should have a 60 to 70 percent moisture level at the start of storage (*136*). Consider treating seeds with an insecticide or heat treatment to prevent weevil damage during long-term storage.

Oaks are strongly taprooted and seedling roots will quickly encircle containers. To obtain lateral root branching—which makes the seedlings more adaptable to transplanting—the acorns can be planted in bottomless flats or small liner containers. The bottoms of the containers are covered by a screen mesh or placed on an open wire bench, which allows for air-pruning of the taproot. The tip of the taproot, upon contacting this mesh, is air-pruned and killed, forcing development of many lateral roots. Copper-based paints are being used to coat the inside of liner pots and larger production containers to prevent root girdling (*344*). A system for speeding up the production of red oak whip production in containers has been described (*343*).

Cuttings. Attempts to propagate oaks by cuttings or layering is frequently difficult, and species-specific. Some success has been obtained in rooting leafy softwood cuttings of *Quercus robur* 'Fastigata' under outdoor mist in midsummer after treatment with IBA at 20,000 ppm (*139*). Rooting of oaks depends on the species, with no consistent rooting patterns between red and white oak groups (*121*).

Commercial propagation of *Q. virginiana* and *Q. laurifolia* has been done by taking July semi-hardwood cuttings (Florida) 6 cm (2 to 3 in.), wounding, quick-dipping in 12,000 ppm K-IBA salt, and rooting under mist; rooting takes seven to nine weeks (*93*). *Q. shumardii* cuttings taken in July (Florida) are quick-dipped in 10,000 ppm K-IBA salt and also root in seven to nine weeks. Some success has occurred with serial propagation of rooted *Q. Virginiana* which are maintained under accelerated growth techniques in a greenhouse and used as stock plants for future propagules (*263*). *Q. Virginiana* can be rooted from rhizomic shoots produced at the crown of the tree, which are treated with a five-second quick-dip of IBA in 50 percent ethanol (*368*).

Girdling of stock-plant shoots prior to collecting cuttings and treating cuttings with a rooting powder of auxins, sucrose, and fungicide increased rooting (76 percent) and survival of 19- to 57-year-old water oak (*178*).

Grafting. Bench grafting of potted seedling rootstocks in the greenhouse in late winter or early spring is moderately successful. Side or whip grafting is ordinarily used, with dormant one-year-old wood for scions. Seedlings in place in the nursery row are occasionally crown grafted in the spring, after the rootstock plants start to leaf out. Scions are taken from wood gathered when dormant and stored under cool, moist conditions until used. Various grafting methods are satisfactory-whip, cleft, or bark. A modified bark graft has been successfully used for topworking established rootstock in the field (*348*). Budding generally has been unsatisfactory. In grafting, only seedlings of the black oak group should be used for scion cultivars of the same group and, in the same manner, only seedlings of the white oak group should be used as rootstocks for other members of the white oaks. The use of seedlings of the same species is preferable. Although some distantly related species of oak will unite satisfactorily, incompatibility symptoms usually appear later.

Quercus virginiana is the recommended rootstock for the southeastern U.S. Researchers at North Carolina State University report that it tolerates the heavier wet clay soils, and temperature of the South.

Root grafting with a white oak *Q. macrocarpa* scion on a *Q. robur* root piece (rootstock) has been reported (*235*). Root grafting with a side-veneer graft on *Q. robur* has been reported to produce plants with more uniform growth and no suckering; this technique was not successful with *Q. rubra* or *Q. palustris* (*233*).

Micropropagation. Single-node-stem sections of *Q. shumardii* were successfully micropropagated (*29*). Juvenile and mature explants of *Q. robur* have been micropropagated and rooted successfully *in vitro* (*362, 363*). Embryoid germination and plant regeneration has been done with *Q. bicolor* micropropagated from male catkins (*161*).

Quince, Flowering. *See Chaenomeles.*

Redbud. *See Cercis.*

Redwood (Coast redwood, Giant Sequoia, Sierra Redwood). *See Sequoia.*

Redwood (Dawn). *See Metasequoia.*

Raphiolepis indica. Raphiolepis, India Hawthorn. Seed germination is easy after the pulp has been removed. The cultivars of *R. indica* are commercially propagated by cuttings. *R. indica* 'Jack Evans' is best rooted when quick-dipped in 50 percent ethanol solutions of 5,000:2,000 ppm or 9,000:1,000 ppm IBA:NAA (*228*). Auxin response will vary with cultivar. Semi-hardwood cuttings taken from lower portions of stems rooted slightly better than those from upper portions (California). The patio tree form, *R. indica* 'Monrey,' is top grafted on clonal rootstock propagated from cuttings.

Rhamnus spp. Buckthorn. Can be propagated by planting seed out-of-doors in the fall. Macerate fruits and clean seeds. Some species may germinate better if scarified for 20 minutes with sulfuric acid before sowing.

Rhapidophyllum hystrix. Needle palm. *See* Palms.

Rhododendron spp. Rhododendron, Azalea (*358, 378*). Azaleas are no longer a separate genus, and are incorporated within the genus *Rhododendron*. *Rhododendron* are normally propagated by cuttings or tissue culture, but seeds and grafting are an option. All parts of rhododendron and azalea plants are poisonous if ingested.

Seeds. Seedlings may be used as rootstocks for grafting or for propagation of ornamental species. *Rhododendron ponticum* is one of the principal rootstocks for grafting. The seed should be collected just when the capsules are beginning to dehisce, and may be stored dry and planted in late winter or early spring in the greenhouse. Seed to be kept for long periods should be put in sealed bottles and held at about 4° C (40° F). A good germination medium is a layer of shredded sphagnum moss or vermiculite over a mixture of sand and peat. The very small seeds are sifted on the surface of the medium and watered with a fine spray. The flats should be covered with glass or propagated under mist (*124, 379*). Careful attention must be given to provide adequate moisture and ventilation as well as even heat: 15 to 21° C (60 to 70° F). The plants grow slowly, taking three months to reach transplanting size. After two or three true leaves form, they are moved to another flat and spaced 2.5 to 5 cm (1 to 2 in.) apart, where they remain through the winter in a cool greenhouse or in cold frames. In the spring, the plants are set out in the field in an acid soil, and by fall they are ready to be dug and potted preparatory to grafting in the winter. Seeds of *R. carolinianum* should not be covered during propagation because of their extremely small size [825,000 seeds per 28 g (1 oz)] and because light is required for germination (*43*). Seeds of *R. catawbiense* and *R. maximum* should be subjected to light to maximize germination (*41*).

Cuttings (*155, 167, 358*). Rooting cuttings is the chief method of rhododendron propagation. These are best taken, midsummer into fall, from stock plants grown in full sun. However, stem cuttings—or leaf-bud cuttings—of some hybrids taken in midwinter will root well. Any flower buds should be removed from the cuttings. Treatments with IBA at relatively high concentrations are required—an IBA talc or quick-dip of 10,000 to 20,000 ppm works well. Wounding the base of the cutting on both sides is a strong stimulus to rooting in rhododendrons (Figure 11–27). A rooting medium of two-thirds sphagnum peat moss and one perlite or 1 peat:1 perlite (v/v) is suitable. Bottom heat at 24° C (75° F) should be used. Rhododendron cuttings are best rooted under mist in the greenhouse or in a closed polytent, and should be lifted soon after roots are well formed (about three months) or the roots will deteriorate.

In California, Rhododendrons are direct rooted into liners (small rose pots) (*317*). After rooting and transplanting (into peat moss, with added fertilizers), the cuttings should be held at 4° C (40° F) for about 20 days, after which the night temperature can be raised to a minimum of 18° C (65° F). Supplementary light at this stage to extend the daylength will give good growth response. Plants started from cuttings usually develop rapidly and are free of the disadvantage of suckering from rootstock, which occurs with grafted plants.

In Alabama, spring cuttings of Azalea cultivars are cut to 7 to 10 cm (3 to 4 in.) and tender tops removed. Basal ends are quick-dipped in 3,000 ppm K-IBA or 5,000 ppm K-IBA for more difficult-to-root cultivars (*170*).

Grafting. A side-veneer graft is most successful. The best scionwood is taken from straight, vigorous current season's growth. After grafting, the plants are kept in closed frames under high humidity and a temperature near 21° C (70° F) until the union has healed. The plants should then be moved to cooler conditions—10 to 15° C (50 to 60° F)—and the top of the rootstock removed above the graft union. After the plant has hardened, it is transplanted to the nursery row and grown for two years, after which it is ready to dig as a salable nursery plant. A modified chip-bud method using leaf buds in early summer (Rhode Island) was successful for grafting difficult-to-root rhododendron cultivars (*256*).

Micropropagation. Many rhododendron, evergreen, and deciduous azaleas are commercially micropropagated (*4, 126, 134, 198, 212*). There have been some micropropagation problems with reproducing Rhododendrons that are true-to-type. See Chapters 17 and 18 for discussion.

Rhus spp. Sumac.

Seed. Commonly propagated by seeds, which are collected in the fall. For prompt germination the seeds should be scarified in concentrated sulfuric acid for one to six hours, depending upon the species—then either fall-planted out-of-doors or stratified for two months at about 4° C (40° F) before planting. Seeds of *R. virens* (evergreen sumac) need to be acid scarified with concentrated sulfuric acid for 50 minutes, and then cold stratified for 73 days (*195, 350*). To ensure fruiting, only plants bearing both male and female flowers should be propagated asexually.

Cuttings. Leafy softwood cuttings, of some species, such as *R. aromatics,* taken in midsummer, root well if treated with 10,000 ppm IBA. For those species that sucker freely, such as *R. typhina* and *R. copallina,* root cuttings are planted in the nursery row in early spring. In Minnesota, root cuttings of *R. typhina* are trimmed to 10 cm (4 in)., with a 6 to 19 mm (¼ to ¾ in.) diameter and the ends dipped in talc containing 3,000 ppm IBA; the root cuttings are allowed to callus at 10 to 16° C (50 to 60° F)—then are stored and field-planted (*94*). It is important to maintain proper polarity of the root cuttings.

Ribes sanguineum Red Flowering Currant. An attractive flowering deciduous shrub native to the Pacific Northwest. Easily propagated by softwood cuttings taken during summer to early September (Vancouver, B.C.), or by hardwood cuttings made with a "heel" of two-year-old wood (*353*). Both cutting types do best with basal heat [21° C (70° F)] and 8,000 ppm IBA talc. Currant is commercially micropropagated.

Robinia pseudoacacia. Black Locust. Readily propagated by seeds, which are soaked in concentrated sulfuric acid for one hour, followed by thorough rinsing in water, before planting. A hot-water scarification followed by a 24-hour soak prior to sowing can also be used. This species can be propagated by root cuttings and by grafting using a whip, side-veneer, cleft, or wedge ("V") graft (*278, 369*). Black locust is micropropagated (*76*).

Rosa spp. Rose. All rose cultivars selected are propagated by asexual methods (*181, 222*). T-budding on vigorous rootstocks is most common, although the use of softwood or hardwood cuttings, chip budding, simultaneous budding and rooting (stenting), layering, the use of suckers, and micropropagation are also practiced.

Seed. Seed propagation is used in breeding new cultivars, in producing plants in large numbers for conservation projects or mass landscaping, and in growing seedling rootstock of certain species, such as *R. canina,* and *R. multiflora.* Some commercial rose-bush producers prefer propagating rootstock by seed to avoid virus transmission through conventional asexual techniques (*69*).

As soon as the rose fruits ("hips") are ripe but before the flesh starts to soften, they should be collected and the seeds extracted. It is best to stratify them immediately at 2 to 4° C (35 to 40° F). Six weeks is sufficient for *Rosa multiflora,* but others—*R. rugosa* and *R. hugonis*—require four to six months, and *R. blanda* ten months. *Rosa canina* germinates best if the seeds are held at room temperature for two months in moist vermiculite and then transferred to 0° C (32° F) for an additional two months.

Hybrid rose seeds usually respond best to a stratification temperature of 1 to 4° C (34 to 40° F) for 60 to 90 days, although some seeds may germinate with no cold stratification treatment. Germination is probably prevented in rose seeds by inhibitors occurring in the seed coverings, as well as by the mechanical restriction imposed by the massive pericarp (fruit wall). The seeds may be planted either in the spring or in the fall in seed beds or in the nursery row. In areas of severe winters, seedlings are likely to be winter-killed if they are smaller than 10 cm (4 in.) by the onset of cold weather.

Hardwood Cuttings. Hardwood cuttings are widely used commercially in the propagation of rose rootstocks—and to some extent for propagating the strong-growing polyanthas, pillars, climbers, and hybrid perpetuals on their own root systems. The hybrid teas and other similar ever-blooming roses also can be started by cuttings, but more winter-hardy and nematode-resistant plants are produced if they are budded on selected vigorous rootstocks.

In mild climates (Texas, California, Spain), the cuttings are taken and field-planted in the nursery in the fall (*171*). In areas with severe winters, cuttings may be made in late fall or early winter, tied in bundles, and stored in damp peat moss or sand at about 4° C (40° F) until spring, when they are planted in the nursery row. The rootstocks are ready to bud by the following spring, summer, or fall. The cuttings are made into 15 to 20 cm (6 to 8 in.) lengths from previous season's canes of 6 to 9 mm (¼ to ⅛ in.) diameter. Commercially, large bundles of canes are run through band saws to cut them to the correct length. Disbudding ("de-eying") is done in rootstock propagation; all axillary buds except the top two or three are removed to prevent subsequent sucker growth in the nursery row. The use of auxins and other rooting pretreatments are of little benefit in the commercial propagation of *Rosa multiflora* hardwood cuttings (*97*).

Softwood Cuttings. Softwood cuttings are made from current season's growth, from early spring to late summer, depending upon the time the wood becomes partially mature. Rooting is fairly rapid, occurring in 10 to 14 days. At the end of the season the cuttings may be transplanted to their permanent location, potted, and overwintered in a cold frame, or transferred to the nursery row for another season's growth. Some desired cultivars are budded to rooted rootstock. Cultivars of most miniature roses are easily propagated by softwood or semi-hardwood cuttings under mist.

Budding. T-budding is the method ordinarily used. The buds are inserted into 5 to 10 mm (³⁄₁₆ to ⅜ in.) diameter rootstock plants. In mild climates, budding can be done during a long period, from late winter until fall, but mostly in the spring. Early buds will make some growth during the summer and produce a salable plant by fall. Some propagators partially break over the top or "cripple" the rootstock about two weeks after budding to force the bud out. After the bud has reached a length of 10 to 20 cm (4 to 8 in.), the top of the rootstock is entirely removed. In areas with shorter growing seasons, budding is done during the summer. Buds inserted late in the summer either make little growth or remain dormant until the following spring. In this case, the rootstock is cut off just above the bud in late winter or early spring, forcing the inserted bud into growth. Shoots from buds started in the fall are cut back to 13 mm (½ in.) in the spring. The shoot then grows through the following entire summer, producing a well-developed plant by fall. After the shoot has grown about 15 cm (6 in.), it is generally cut back to 5 to 7.5 cm (2 or 3 in.) to force out side branches.

In California, Texas, and Spain, commercial landscape rose bush producers collect budwood in late fall (prior to digging two-year-old rose bushes), store the budwood at 29° F (−2° C), and then T-bud in the following spring (*98*). The budwood is collected in late fall after the flowers are shed and the thorns become dark. The leaves are removed by hand, or "sweated-off" under high humidity, and the thorns are left intact. Sticks 15 to 25 cm (6 to 10 in.) long are put up in bundles of 30 or 40 each. The bundles are wrapped as tightly as possible in moist paper and then placed in polyethylene bags.

From spring to early summer, the budwood is taken out of storage, the thorns are removed, and shield buds from the bud-wood are T-budded by inserting them into the medial, de-eyed portion of the original cuttings (rather than into new growth arising from the rootstock).

During recent years, there has been an interest in simultaneously grafting (budding) and rooting landscape roses through bench chip budding (see Figure 14–7) and direct rooting in the field (*98*)—and using the stenting system in grafting and rooting greenhouse cut-flower roses (see Figure 13–45) (*356*).

Rootstocks and Interstocks for Rose Cultivars (55). Most rose rootstock clones have been in use for many years, propagated by cuttings. Many of the clones are virus-infected, thus infecting the cultivar top after budding. However, these clonal rootstocks are available with the viruses eliminated by heat treatments. Holding potted rose plants at a dry heat of 37 to 38° C (98 to 100° F) for four to five weeks will rid infected rootstocks of viruses. Again, seed propagation of rootstocks helps avoid most virus problems.

Rosa multiflora. This is a useful rootstock, especially in its thornless forms, for landscape roses. Several cultivars have been developed, some giving better bud unions and bud development than others. Cuttings of this species root readily, develop a vigorous root system, and do not sucker excessively. It is adaptable to a wide range of soil and climatic conditions. Seedlings are used in the eastern part of the United States, Great Britain, and Australia, and cuttings are used on the Pacific coast and Texas. The bark often becomes so thick late in the season that budding is impossible.

Rosa canina. Dog rose. Although this species has not done well under American conditions, the rootstock is commonly used in Europe. It is usually propagated by seed, since the cuttings do not root easily; however, the seeds are difficult to germinate. The prominent thorns make it difficult to handle. It also tends to sucker. Young plants on this rootstock grow slowly, but they are long-lived. *Rosa canina* is adaptable to drought and alkaline soil conditions, and is used as greenhouse rootstock for cut-flower rose production in the Netherlands.

Rosa chinensis. 'Gloire de Rosomanes,' 'Ragged Robin.' This old French rootstock is popular in California for outdoor roses, resisting heat and dry conditions well. It is also resistant to nematodes and does not sucker if the lower buds on the cuttings are removed. This rootstock grows steadily through the summer, permitting budding at any time, under irrigated conditions. The fibrous root system is easy to transplant but requires good soil drainage. In some areas, however, it is difficult to propagate and is injured by leaf spot.

Rosa 'Dr. Huey.' This is the principal rootstock in Arizona and the southern San Joaquin Valley, California, rose districts, replacing 'Ragged Robin' to a large extent. It has also performed well in Australia (*306*). It is useful for late season budding because of its thin bark. It is very vigorous and well adapted to irrigated conditions, and its cuttings root readily. It is very good as a rootstock for weak-growing cultivars. Defects include injury from subzero temperatures and susceptibility to black spot, mildew, and verticillium.

Rosa × noisettiana 'Manetti.' This is an old rootstock, very popular for greenhouse forcing roses. It is also of value for dwarf roses and for planting in sandy soils. It is easily propagated by cuttings, produces a plant of moderate vigor, and is resistant to some strains of verticillium.

Rosa odorata (Odorata 22449). Tea rose. This rootstock is excellent for greenhouse forcing roses. Cuttings root easily under suitable conditions, and produce a large symmetrical root system. It is adapted to both excessively dry and wet soil conditions. Since it is not cold-hardy, it should be used only in areas with mild winters. Some propagation stock of this clone is badly diseased and does not root well. The plants are not adaptable to cold-storage handling. It is more susceptible than *R. manetti* to verticillium.

Rosa IXL (Tausendschon/Veilchenblau). This interstock is used primarily as a trunk for tree roses. It is very vigorous and has no thorns. The canes tend to sunburn and are somewhat susceptible to low-temperature injury.

Rosa Multiflore de la Grifferaie. This interstock is useful as a trunk for tree roses, producing desirable straight canes. It is vigorous, extremely hardy, and resistant to borers, but very susceptible to mite injury.

Rosa rugosa. This form, which is used as a rootstock, bears single, purplish-red flowers. For bush roses it is propagated by cuttings, and for tree roses by seed. The root system is shallow and fibrous and tends to sucker badly, but the plants are very long-lived. It is also used as the upright stem in producing standard (tree) roses.

Propagating Tree (Standard) Roses. A satisfactory method of producing this popular form of rose is to use *Rosa multiflora* as the rootstock, which is budded in the first summer to IXL or, preferably, the *Grifferaie* interstock (Figure 12–5). These are trained to an upright form and kept free of suckers. In the second summer, at a height of about 0.9 m (3 ft), three or four buds of the desired flowering cultivar are inserted into the interstock trunk. During the winter, the cane above the inserted buds is removed. The buds develop the following summer, as do buds from the rootstock, which must be removed. In the fall, the plants may be dug and moved to their permanent location. Tree roses are sometimes dug as balled and burlapped plants because an extensive root system is formed during the two years in which the top is being developed.

Propagation of Miniature Roses (262). Soft or semi-hardwood cuttings are taken the year round and rooted under mist, after dipping in IBA talc. Miniature rose cultivars are especially bred for their ease of rooting.

Micropropagation. Selected rose cultivars are commercially micropropagated (*180, 365*).

Rose of Sharon. *See Hibiscus.*

Rosemary. *See Rosmarinus.*

Rosmarinus officinalis. Rosemary. This ground cover with aromatic leaves is easily propagated by seed and leafy cuttings under mist.

Rosewood, Arizona. *See Vauquelinia.*

Royal Poinciana. *See Delonix.*

Rubber Plant. *See Ficus.*

Russian Olive. *See Elaeagnus.*

Sage (Texas, Purple). *See Leucophyllum.*

Salix spp. Willow. Seeds must be collected as soon as the capsules mature (when they have turned from green to yellow) and planted immediately, since they retain their viability for only a few days at room temperature. Even under the most favorable conditions, maximum storage is four to six weeks. Willows are difficult to propagate in quantity by seed. Willows root so readily by either stem or root cuttings that there is little need to use other methods. Hardwood cuttings planted in early spring root promptly. Summer cuttings from hardened new growth also performs well. *Salix* can be micropropagated (*268*).

Sambucus spp. Elderberry. Seed propagation is difficult because of complex dormancy conditions involving both the seed coat and embryo. Probably the best treatment is a warm [21 to 30° C (70 to 85° F)], moist stratification period for two months, followed by a cold (4° C; 40° F) stratification period for three to five months. These conditions could be obtained naturally by planting the seed in late summer, after which germination should occur the following spring. Since softwood cuttings root easily if taken in spring or summer, this method is generally practiced.

Sapindus drummondii. Soapberry (Western). Western soapberry is a deciduous landscape tree with a bright yellow fall color and a low water requirement. It performs well on highly calcareous soils. For maximum germination, fall or winter collection of seeds followed by a 60-minute acid scarification and three-month cold stratification is recommended (*264*). May and June cuttings can be rooted when treated with 16,000 ppm IBA (*115*).

Sarcococca spp. Sarcococca, Fragrant Sweetbox (*S. ruscifolia*). Attractive, broad-leaf evergreen ground cover that tolerates shade conditions. Sarcococca species can be propagated from firm cuttings. A

3,000 ppm IBA quick-dip enhanced rooting of *S. hookerana* (*114*).

Sea Grape. *See Coccoloba.*

Schizophragma hydrangeoides. A vine in the Hydrangeaceae family that climbs to 12 m (40 ft) on the bark of trees and bears a mass of creamy-white flowers with reddish new shoots. Can be propagated from seed stratified at 1 to 3° C (34 to 38° F) for 10 to 12 weeks. The seed are very small and can be difficult to handle. Can be propagated asexually by nodal tip cuttings or single nodal cuttings with two opposite buds using talc applications of 5,000 ppm IBA (*244*). The most easily rooted cuttings are those obtained from nonflowering shoots near the base of the plant, which often have preformed root initials. *Schizophragma* is commercially micropropagated.

Senicio cineraria. *See Senicio* (Chapter 21).

Sequoiadendron giganteum. *See Sequoia.*

Sequoia spp. Coast Redwood *(Sequoia sempervirens),* Giant Sequoia or Sierra Redwood *(Sequoiadendron giganteum).* Both genera are ordinarily propagated by seed; however, some cultivars are propagated by cuttings.

Seed. Seeds of *Sequoia sempervirens* are mature at the end of the first season, but those of *Sequoiadendron giganteum* require two seasons for maturity of the embryos. Cones are collected in the fall and allowed to dry for two to four weeks, after which the seeds can be separated. Seeds may be kept for several years in sealed containers under 4° C (40° F) storage without losing viability. Stratification for ten weeks before planting at about 4° C (40° F) promotes germination of *Sequoiadendron giganteum* seed. Seeds of *Sequoia sempervirens* will germinate without a stratification treatment.

Fall planting also may be done, sowing the seed about 3 mm (⅛ in.) deep in a well-prepared seed bed. The young seedlings should be given partial shade for the first 60 days.

Cuttings. Introduction of vegetatively propagated ornamental cultivars of *Sequoia sempervirens* has resulted in vast improvement over highly variable seed-propagated specimens. Hardwood cuttings (February in southern California) were trimmed to 12 cm (5 in.) in length so that the outer tissue on the main stem of the cutting was brown at the base and green above. Cuttings were propagated in 9 perlite:1 peat media, placed in cutting flats on outdoor heated beds [21° C (70° F)], and rooted with mist under full sun for five months. The cultivar 'Majestic Beauty' rooted best with 3,000 ppm IBA and 3,000 ppm NAA liquid; 'Santa Cruz' rooted best with 16,000 ppm IBA talc, and 'Soquel' rooted best with 6,000 ppm IBA and 6,000 ppm NAA quick-dip (*44*). *Sequoiadendron giganteum,* grown for Christmas trees, roots easily if cuttings are taken from young trees and treated with IBA under mist (*138*).

Micropropagation. Both genera are micropropagated (*33, 196*).

Serviceberry. *See Amelanchier.*

Seven Son Flower (Tree). *See Heptacodium.*

Shepherdia spp. Buffaloberry. Hardy deciduous and evergreen shrubs. Seeds of *S. canadensis* and *S. rotundifolia* have been reported to exhibit seed-coat and internal dormancy. Seeds should be scarified in sulfuric acid for 30 minutes, followed by a minimum 30-day cold stratification. *S. canadensis* can be rooted with hardwood cuttings treated with 3,000 ppm IBA. *S. rotundifolia* can be micropropagated with WPM (*221*).

Shrub-althea. *See* Hibiscus.

Silk Tree. *See* Albizzia.

Silverbell. *See Halesia.*

Smoke Tree. *See Cotinus.*

Snowberry. *See Symphoricarpos.*

Soapberry (Western). *See Sapindus.*

Sophora spp. Sophora.

For uniform seed germination, seed of Texas Mountain Laurel *(S. secundiflora)* must be scarified, regardless of the seed age (young, soft seeds with yellow seed coats or mature, firm seeds with red seed coats) (*367*). Scarifying with sulfuric acid from 35 to 120 minutes, or nicking the seed coat with a saw, works well (*309, 341*). *S. japonica* (Japanese pagoda tree) is generally seed-propagated. Seed must be cleaned by soaking fruit for 48 hours; generally, no scarification or stratification treatments are needed (*115*). Cuttings are not a successful method for propagating *S. secundiflora* and *S. japonica.* However, cultivars of *S. microphylla* are rooted successfully by taking semi-hardwood cuttings in winter (New Zealand), which are stripped of lower leaves, wounded, dipped in 8,000 ppm IBA talc, and inserted into 1 bark:1 pumice medium under mist with 20° C (68° F) bottom heat (*62*). *S. microphylla* can be grafted by T-budding or with a side graft. *S. secundiflora* can be micropropagated, which offers potential for clonal selection (*152*).

Sorbus spp. Sorbus, Mountain Ash.

Seed. Seeds should be collected as soon as the fruits mature; fleshy parts are removed to eliminate inhibitors. Some species have double dormancy and require a three- to five-month warm stratification, followed by a three-month cold stratification of 5° C (41° F) for good seed germination.

Cuttings. Cuttings are generally difficult to root; however, 6 to 10 cm (2 to 4 in.) long softwood cuttings of *S. aucuparia* and *S. hybrida*—which had been severed from forced stock plants at the junction between the growth of the current year and that of the previous year—root well (*175*). A talc formulation of 5,000 to 20,000 ppm IBA improved rooting of the softwood cuttings.

Grafting. Either fall chip budding or bench grafting (whip grafting) is successful. Selected cultivars are best worked on seedlings of their own species, although *S. aria, S. aucuparia,* and *S. cuspidata* (European mountain ash) seedlings seem satisfactory as a rootstock for other species.

Micropropagation. *S. aucuparia* (*9, 76*) and *S. domestica* are micropropagated (*10*).

Spiceberry. *See Ardisia.*

Spicebush (Japanese). *See Lindera.*

Spiraea spp. Spirea. Usually propagated by cuttings, although some species, such as *S. thunbergi,* are more easily started by seeds, which should not be allowed to dry out. Leafy softwood cuttings taken in midsummer are generally successful. Treatments with auxin at 1,000 ppm aids rooting. Some species, such as *S.* × *vanhouttei,* can be started readily by hardwood cuttings, planted in early spring. Spirea is commercially micropropagated.

Spruce. *See Picea.*

Spurge (Japanese). *See Pachysandra.*

St. Johnswort. *See Hypericum.*

Stangeria eriopus. *See Cycas.*

Stewartia spp. Stewartia. Double dormancy is overcome by a three- to five-month warm stratification followed by a three-month cold stratification (*108*).

Semi-hardwood cuttings collected in early summer root well when treated with 5,000 to 8,000 ppm IBA, or 2,500 ppm IBA + 2,500 ppm NAA. Problems occur with winter survival. Rooted cuttings should be hardened-off and not transplanted during the fall. It is best not to fertilize until the cuttings leaf out the following spring. In Oregon, 15–20 cm (6–8 in.) cuttings taken in mid-July are treated with 2,000 ppm IBA + 1,000 ppm NAA and direct-stuck in small liner pots under mist. Rooted cuttings are not transplanted until after one full year of growth (*131*).

Sumac. *See Rhus.*

Sweetshrub. *See Calycanthus.*

Swietenia mahagoni. Mahogany. A salt-tolerant, semi-deciduous landscape tree for southern Florida. Seeds are collected when mature in late winter and germinate easily without soaking or pregermination treatments (*279*). While seed propagation is the preferred propagation method, propagation by cuttings is important for clonal propagation and as a research tool. The inclusion of older growth in stem cuttings is necessary for successful rooting. Good rooting was obtained from a twenty-year-old tree (south Florida). Cuttings are trimmed to 10 to 20 cm (4 to 8 in.) in length with four to five fully expanded leaves, succulent terminal growth removed, treated with 1,000 ppm IBA + 300 ppm NAA, and rooted under mist (*194*).

Sycamore. *See Platanus.*

Symphoricarpos spp. Snowberry. Seed propagation is difficult because of a hard, impermeable endocarp and a partially developed embryo at harvest. Give seeds a three- to four-month warm, moist stratification followed by cold stratification at 5° C (41° F) for six months before planting. Cuttings root easily when treated with 1,000 to 3,000 ppm IBA.

Syringa spp. Lilac. *Syringa vulgaris* cvs. and other Syringa species are either micropropagated or propagated from cuttings.

Seeds. Seedlings are used mostly as rootstock for grafting or in hybridization. Lilac cultivars will not reproduce true from seed. Seeds require fall planting out-of-doors or a stratification period of 40 to 60 days at about 4° C (40° F) for good germination.

Cuttings. Cuttings can be rooted, if attention is given to proper timing. Ordinarily, good rooting of lilacs can be obtained only if terminal leafy cuttings are taken within a narrow period shortly after growth commences in the spring. When the new, green shoots have reached a length of 10 to 15 cm (4 to 6 in.) they should be cut off and trimmed into cuttings. Since they are very succulent at this state it is difficult to prevent wilting. Rooting can be obtained in a polyethylene-covered bed in the greenhouse with bottom heat, or in indoor and outdoor mist beds.

Chinese lilac *(S. × chinensis)* roots well in late spring (before the terminal bud has set), when treated with a 4,000 ppm IBA quick-dip. Korean lilac *(S. patula)* cuttings rooted well when taken from plants forced in a greenhouse and treated with 8,000 ppm talc (*115*). Softwood cuttings of *S. henryi*, *S. josikaea*, and *S. villosa* root easily with a 3,000 ppm IBA talc. With *S. vulgaris*, optimal rooting occurs with softwood cuttings treated with 1,500 to 3,000 ppm IBA—depending on the cultivar.

Cuttings of hybrid *S. vulgaris* cultivars showed improved rooting with prior etiolation of stock plants, and the period over which lilac cuttings could be propagated successfully was lengthened considerably. IBA talc of 3,000 to 8,000 ppm enhanced rooting.

Grafting. Today, grafting is less popular due to the expense, graft incompatibility problems, suckering of the rootstock, reduced plant life, and poor growth renewal. Also, tissue culture production of Syringa is much more broadly utilized. Grafting or budding on rootstock of California privet *(Ligustrum ovalifolium)* or Amur privet *(L. amurense)* cuttings, or on lilac or green ash *(Fraxinus pennsylvanica)* seedlings has been used in lilac propagation (*86, 174, 375*).

Micropropagation. Lilacs are commercially micropropagated (*185, 254*).

Tabebuia argentea. Gold Tree, Silver Trumpet Tree. Has a cork-like light-colored bark with spectacular golden yellow flowers. It is generally seed-propagated without any pregermination treatment (*279*). It is important that the propagation medium be drenched with fungicides to control damping-off. This species can also be produced by cuttings and air layering.

Tamarisk. *See Tamarix.*

Tamarix spp. Tamarisk. These are easily rooted by hardwood cuttings, which are usually made about 30 cm (12 in.) long and planted deeply. Softwood cuttings taken in early summer also will root readily under mist.

Taxodium distichum. Bald Cypress. Bald cypress is propagated by seeds, which should be fall-planted or stratified at 5° C (41° F) for 90 days. Soaking the seeds for 24 to 48 hours enhances germination.

Taxus spp. Yew. These ornamental trees and shrubs are also valued for the anticancer drug, taxol, which is found in various parts of the plant.

Seeds. Seedling propagation is little used, because of variation in the progeny, complicated seed dormancy conditions, and the slow growth of seedlings. This method is confined in commercial practice almost entirely to the Japanese yew *(T. cuspidata)*, which comes fairly true from seed if isolated plants can be located as sources of seed. Seed imported from Japan is believed to produce uniform offspring. Plant growth habit and basal branching is better with seedling-grown plants than cutting-grown.

For good germination, seeds should be given a warm 20° C (68° F) stratification period in moist peat moss or other medium for three months, followed by four months at a lower temperature [5° C (41° F)]. Seedling growth is very slow. Two years in the seed bed, followed by two years in a lining-out bed, then three or four years in the nursery row are required to produce a salable-size plant of Japanese yew.

Cuttings. Most clonal selections of yews are propagated by cuttings, which root without much difficulty (*311, 323*). *Taxus* cuttings can be rooted outdoors in cold frames or in the greenhouse under mist, the latter giving much faster results. For the cold frame, fairly large cuttings, 20 to 25 cm (8 to 10 in.) long, are made in early fall from new growth with a section of old wood at the base. Many species and cultivars respond best when stripped of needles at the base of the cutting (wounding) and treated with 8,000 ppm IBA talc or an IBA quick-dip of 5,000 to 10,000 ppm. Cuttings may be kept in closed frames through the winter. Rooting takes place slowly during the following spring and summer.

In Ohio (*345*) and Rhode Island (*257*), spring-rooted Taxus were equal to or superior to cuttings taken in late fall, rooted on bottom heat, and spring-planted. With bottom heat, there is no need to strip needles from the basal end of the cuttings, and cuttings propagated on bottom heat are ready to plant sooner than those without heat. Cuttings are initially quick-dipped in 2,000 to 2,500 ppm IBA and 1,000 to 1,250 ppm NAA combinations.

For greenhouse propagation, cuttings should be taken in early winter, after several frosts have occurred, and rooted under mist in 90 percent sand:10 percent peat with bottom heat at about 21° C (70° F) and an air temperature of 10 to 13° C (50 to 55° F). Rooting in the greenhouse takes only about two months, but cuttings should not be dug too soon. Allow time for secondary roots to develop from the first-formed primary roots. There may be advantages of a two-month cold period after cuttings have rooted before planting out liners. There is evidence that cuttings from male plants (at least in *T. cuspidata expansa*) root more readily than cuttings from female plants (those that produce fruits).

Side or side-veneer grafting is practiced for those few cultivars that are especially difficult to start by cuttings, with easily rooted cuttings used as rootstock. Taxus can be micropropagated (*81*).

Telopea speciosissima. Waratah. Seeds of this Australian native shrub with beautiful chrysanthemum-like flowers germinate easily, but the seedlings are difficult to transplant and grow outside their native environment. Waratah requires soil of extremely low phosphorus content.

Thuja spp. Arborvitae. American *(Thuja occidental)*, Oriental [*Platycladus orientalis (Thuja orientalis)*], Korean *(T. koraiensis)*, and Western Arborvitae *(T. plicata)*.

Seeds. Germination is relatively easy, but stratification of seeds for 60 days at about 4° C (40° F) may be helpful. *T. plicata* seed generally does not require stratification. *T. occidentalis* are fall-planted and *Platycladus orientalis* are spring-planted.

Cuttings. Hardwood cuttings of *T. occidentalis* can be rooted in midwinter under mist in the greenhouse. Best rooting is often found with cuttings taken from older plants no longer making rapid growth. The cuttings should be about 20 cm (6 in.) long and may be taken either from succulent, vigorously growing terminals or from more mature side growth several years old. Wounding and treating with 3,000 to 8,000 ppm IBA quick-dip or talc is beneficial. No shading should be used. Cuttings may also be made in midsummer and rooted out-of-doors in a shaded, closed frame.

Cuttings of *P. orientalis* are often more difficult to root than those of *T. occidentalis*. Small, soft cuttings several centimeters long, taken in late spring, can be rooted in mist beds if treated with 8,000 ppm IBA talc. Cuttings 10 to 15 cm (4 to 6 in.) can be taken in summer or winter, treated with a 5,000 ppm IBA quick-dip, and rooted under mist (Texas). *T. plicata* roots easily from fall cuttings treated with 3,000 ppm IBA talc (*108*). In southern California, conifers are rooted outdoors in full sun; cuttings are taken in November and December before spring rains, since these species have cultural problems with too much water.

Grafting. The side graft is used in propagating selected clones of *T. orientalis*, with two-year-old potted *P. orientalis* seedlings as the rootstock. Grafting is done in late winter in the greenhouse. After making the grafts, the potted plants are set in open benches filled with moist peat moss just covering the union. The grafted plants should be ready to set out in the field for further growth by mid-spring. Selected cultivars of *T. occidentalis* can also be side-veneer grafted. *P. orientalis* is generally own-rooted, rather than grafted.

Tilia spp. Linden, Basswood.

Seed. Seeds are used primarily for seedling rootstock for budding and grafting. This is a difficult plant to get good germination and seedling production. The seeds have a dormant embryo plus an impermeable seed coat which, in some species, is surrounded by a hard, tough pericarp. Such seeds are slow and difficult to germinate (*140, 357*). Removing the pericarp—either mechanically or by soaking the seeds in concentrated nitric acid for one-half to two hours, rinsing thoroughly and drying, then soaking the seeds for about 15 minutes in concentrated sulfuric acid to etch the seed coat, followed by stratification for four months at 2° C (35° F), may give fairly good germination; otherwise, warm [15 to 27° C (60 to 80° F)] stratification for four to five months, followed by an equal period of cold stratification at 2 to 4° C (35 to 40° F) can be used. Collecting the seed from the tree just as the seed coats turn completely brown (but before the seeds drop and the seed coats become hard and dry), followed by immediate planting has given good germination.

Cuttings. *T. americana* and *T. cordata* softwood and semi-hardwood cuttings, wounded, and treated with 20,000 to 30,000 IBA quick-dips will root successfully (Kansas). Cuttings should be very soft and taken before the terminal bud hardens (Pennsylvania). Leaf abscission under mist is a problem. Suckers arising around the base of trees cut back to the ground have been successfully mound layered, and softwood cuttings taken from stump sprouts have been rooted.

Budding. T-budding or chip budding (England) in late summer on seedling rootstocks of the same species gives good results (*140*).

Micropropagation. *T. cordata* is commercially micropropagated (*76*).

Toyon. *See Heteromeles.*

Trachelospermum asiaticum. Asiatic jasmine. An important ground cover that can be rooted any time of the year (Texas). Cuttings are quick-dipped in 3,500 ppm IBA and rooted under mist or under heavy shade in outdoor beds. They root easily from the nodes and can be direct-stuck in any suitable rooting media. In Alabama, four to five cuttings are stuck in the center of a small pot, which results in less production time to produce a full liner plant (*201*). *T. asiaticum* can be micropropagated (*6*).

Trachelospermum jasminoides. Star or Confederate Jasmine. Leafy cuttings of partially matured wood root easily, especially when placed under mist and treated with 1,000 to 3,000 ppm IBA quick-dip or talc.

Tree of Heaven. *See Ailanthus.*

Tree Tea. *See Leptospermum.*

Trumpet Creeper. *See Campsis.*

Tsuga spp. Hemlock.

Seed. Hemlocks are propagated by seed without difficulty. Seed dormancy is variable; some lots exhibit embryo dormancy, whereas others do not. To ensure good germination it is advisable to stratify the seeds for two to four months at about 4° C (40° F). Fall planting outdoors generally gives satisfactory germination in the spring. The seedlings should be given partial shade during the first season.

Cuttings. The preferred method of propagating hemlock cultivars is by cuttings. *T. canadensis* cultivars can be rooted with either hardwood cuttings (bottom heat plus 5,000 ppm each of IBA and NAA—or a 20,000 ppm IBA quick-dip) or softwood cuttings treated with 8,000 ppm IBA talc—plus wounding the cuttings. Softwood cuttings have a lower rooting percentage, but once rooted have a greater growth rate than hardwood cuttings (*105*). Moisture stress will cause needles to drop, so do not store the cuttings prior to rooting. Rooting varies widely among cultivars.

Tulip Tree. *See Liriodendron.*

Ulmus spp. Elm.

Seeds. Seed propagation is commonly used. Elm seed loses viability rapidly if stored at room temperature, but it can be kept for several years in sealed containers at 0 to 4° C (32 to 40° F). Seed that ripens in the spring should be sown immediately, and germination usually takes place promptly. For those species that ripen their seed in the fall, either fall planting or stratification for two months at about 4° C (40° F) should be used. To obtain tree uniformity, selected clones are propagated by budding on seedling rootstocks of the same species.

Cuttings. Softwood cuttings of several elm species can be rooted under mist when taken in early summer. Semi-hardwood cuttings have been rooted without mist or auxin treatments (*318*). Softwood cuttings taken from new growth arising from cut-off stumps root readily if treated with IBA and placed under mist (*321*). Ulmus 'Frontier' *(U. carpinifolia × U. parvifolia), U. parvifolia* 'Pathfinder,' and *U. wilsoniana* 'Prospector'—all which are highly adaptable to tough urban sites and have resistance to Dutch elm disease—are propagated with softwood cuttings, treated with an 8,000 ppm IBA quick-dip (*189*).

U. parvifolia can be rooted by softwood cuttings—treated with 5,000 ppm IBA or 1,250 ppm, each, of IBA + NAA. However, hardwood cuttings treated with 10,000 ppm IBA are preferred, since there is greater growth of the rooted cutting in the same season, and winter protection is not required, as with softwood cuttings (Kansas) (*276*). *U. hollandica* can be propagated by hardwood cuttings taken in late winter, treated with IBA at 1,500 ppm, then placed in a bin over bottom heat for six weeks before planting (*380*).

Micropropagation. Ulmus is commercially micropropagated.

Ungnadia speciosa. Mexican buckeye. An outstanding native of central Texas and northern Mexico which grows as a tree or a large multistemmed shrub. Propagation is by seed, which is collected in late summer and needs no pregermination treatment (*195*). Seeds turn from green to dark brown and become mature in July (Texas). Seedlings require full sun (*336*).

Vaccinium spp. Blueberry, Cranberry, Huckleberry. Evergreen huckleberry *(V. ovatum)* is similar to Japanese holly, but with reddish-colored new growth and pale pink flowers in profusion.

Seed. Most Vaccinium species have no seed pregermination requirements. Exceptions are with lingonberry or mountain cranberry *(V. vitis-idaea)*—its seed is separated from the fruit and cleaned, then cold stratified for 90 days at 4° C (40° F). A higher percentage of seed will germinate with stratification than without (*304*).

Cuttings. *V. ovatum* is propagated by cuttings from fully matured shoots taken in fall and winter; cuttings made from previous year's growth taken the third week in April root readily (Vancouver, B.C.). Basal heat [21° C (70° F)] and 3,000 to 4,000 ppm IBA talc enhance rooting (*353*). Rabbiteye Blueberry *(V. ashei)* and Highbush Blueberry *(V. corymbosum)* are propagated by terminal softwood cuttings treated with 8,000 ppm IBA talc or 4,000 ppm NAA. American cranberry *(V. macrocarpon)* is easily rooted with 1,000 ppm IBA talc (*115*). Many blueberry species root best when very soft cuttings are taken, and not treated with auxins.

Micropropagation. Blueberry cultivars are commercially micropropagated.

Vauquelinia spp. Arizona rosewood *(V. california)*. This is a drought-tolerant, rosaceous evergreen shrub, which is one of the most popular landscape plants used in semi-arid regions of the U.S. southwest. Seed propagation is highly variable (*331*). Cuttings taken from six- to ten-year-old stock plants root best when propagated from May to June (Arizona) under mist and bottom heat [32° C (90° F)]. An IBA quick-dip at 8,000 ppm is of some benefit, but clonal variability (0 to 95 percent rooting) and season timing are the most important factors.

Veronica. *See Hebe.*

Viburnum spp. *Viburnum.* This large group of desirable shrubs can be propagated by a number of methods, including seeds, cuttings, grafting, and layering (*35*). At least one species *(V. dentatum)* is readily started by root cuttings.

Seeds. The viburnums have rather complicated seed dormancy conditions. Seeds of some species, such as *V. sieboldi,* will germinate after a single ordinary low-temperature [4° C (40° F)] stratification period, but for most species, a period of two to nine months at high temperatures [20 to 30° C (68 to 86° F)], followed by a two- to four-month period at low temperatures (4° C; 40° F) is required. The initial warm temperatures cause root formation, and the subsequent low temperature causes shoot development. Cold stratification alone will not result in germination. Such rather exacting treatments may best be given by planting the seeds in summer or early fall (at least 60 days before the onset of winter), thus providing the initial high-temperature requirement; the subsequent winter period fulfills the low-temperature requirement. After this, the seeds should germinate readily in the spring. Often, collecting the seeds early, before a hard seed coat has developed, will hasten germination. Viburnum seed can be kept up to ten years if stored dry in sealed containers and held just above freezing. *V. lantana, V. opulus,* and *V. rhytidophyllum* are commonly propagated by seed.

Cuttings. Although some viburnum species (*V. opulus, V. dentatum,* and *V. trilobum*) can be propagated by hardwood cuttings, softwood cuttings rooted under mist are successful for most species. Soft, succulent cuttings taken in late spring root faster than those made from more mature tissue in midsummer, but the latter are more likely to grow on into sturdy plants that will survive through the following winter. Treatments with 8,000 ppm IBA talc and a 2,500 ppm IBA quick-dip have been recommended (*30*). In Illinois, cuttings of *V. carlesii* taken in June root best when treated with 10,000 ppm K-IBA. Cuttings are direct-rooted in small liner pots in a heated Quonset house, left in place, and allowed to go dormant and maintained throughout the winter at a minimum temperature of 2° C (28° F) (*284*).

One of the chief problems with viburnum cuttings is to keep them growing after rooting. Cuttings made from succulent, rapidly growing material often die in a few weeks after being potted. This problem may be overcome by not transplanting the cuttings too soon, and allowing a secondary root system to form, which will better stand the transplanting shock. It may help to fertilize the rooted cuttings with a nutrient solution about ten days before the cuttings are to be removed. It is best not to fertilize rooted cuttings of *V. carlesii* until new growth has started. Placing the rooted cuttings under supplementary lighting to increase daylength also is helpful. Cuttings of some species root more easily than others. *Viburnum carlesii,* for example, is difficult to root, while *V. burkwoodii, V. rhytidophylloides, V. lantana, V. sargentii,* and *V. plicatum* root readily.

***Grafting* (*17*).** Selected types of viburnum are often propagated by grafting on rooted cuttings, layers, or seedlings of *V. dentatum* or *V. lantana.* Often, grafted viburnums will develop into vigorous plants more quickly than those started as cuttings. *V. opulus* 'Roseum' (Snowball) is dwarfed when grafted onto *V. opulus* 'Nanum' cuttings. It is important that all buds be removed from the rootstock so that subsequent suckering from the rootstock does not occur. The rootstocks are potted in the fall and brought into the greenhouse, where they are grafted in midwinter by the side-graft method, using dormant scion-wood. After grafting, the potted plants are placed in a closed, glass-covered or poly frame with the unions buried in damp peat moss.

Grafting can be done in late summer also, using potted rootstock plants and scion material that has stopped growing and become hardened. *V. carlesii* 'Compactum' is grafted on *V. dentatum* rootstock during mid-September (Rhode Island) using a conventional side-vereer graft (*192*). The grafted plants are buried in slightly damp peat moss in closed frames in the greenhouse until the unions heal, after which they are moved to outdoor, poly-covered cold frames for hardening-off for the winter.

Micropropagation. Viburnums are commercially micropropagated.

Virginia Creeper. *See Parthenocissus.*

Walking Stick. *See Aralia.*

Waratah. *See Telopea.*

Waxflower. *See Chamelaucium.*

Wax Myrtle. *See Myrica.*

Weigela spp. Weigela. This shrub is easily propagated either by hardwood cuttings planted in early spring or by softwood tip cuttings under mist taken any time from late spring into fall. Rooting is promoted by treating softwood cuttings with a 1,000 to 3,000 ppm IBA quick-dip, and hardwood cutttings with a 8,000 ppm IBA talc (England). Weigela is commercially micropropagated.

Willow, False Willow. *See Baccharis.*

Willow. *See Salix.*

Winterhazel. *See Corylopsis.*

Wisteria spp. Wisteria. It may take eight to twelve years for seed-propagated wisteria to flower, which is why it is vegetatively propagated. Seeds and pods are also poisonous. Wisterias may be started by softwood cuttings under mist taken in midsummer. IBA often aids rooting. Some species can be started by hardwood cuttings set in the greenhouse in the spring. *W. sinensis* can be propagated with 6 to 8 cm (2.5 to 3 in.) long root cuttings without auxins (*251*). It is important to maintain polarity of root cuttings and to plant vertically with tops of cuttings level with the surface of a porous rooting medium. Simple layering of the long canes is quite successful. Choice cultivars are often grafted on rooted cuttings of less desirable types. Suckers arising from roots of such grafted plants should be removed promptly. Wisteria is commercially micropropagated.

Witch Hazel. *See Hamamelis.*

Xylosma congestum. This is propagated by rooting leafy cuttings, taken in late summer or early fall, using the first and second subterminal cuttings on the shoot. Cuttings are treated with a 5,000 ppm IBA quick-dip and propagated under mist with bottom heat (21° C; 70° F). Rooted cuttings should be hardened in a cool, humid greenhouse.

Yellow Poplar. *See Liriodendron.*

Yew. *See Taxus.*

Zamia. *See Cycas.*

REFERENCES

1. Acharya, S.N., C.B. Chu, R. Hermesh and G.B. Schaalje. 1992. Factors affecting red-osier dogwood seed germination. *Can. J. Bot.* 70:1012–1016.
2. Ackerman, W.L., and G.A. Seaton. 1969. Propagating the Bradford pear from cuttings. *Amer. Nurs.* 130(7):8.
3. Adams, J.C. and K.W. Farrish. 1992. Seed coat removal can enhance the germination rate of water oak seed. *Amer. Nurser.* 176(12):141.
4. Anderson, W.C. 1978. Rooting of tissue-cultured rhododendrons. *Comb. Proc. Intl. Plant Prop. Soc.* 28:135–39.
5. ———. 1983. Micropropation of filberts, *Corylus avellana. Comb. Proc. Intl. Plant Prop. Soc.* 33:132–137.
6. Apter, R.C., F.T. Davies, Jr., and E.L. McWilliams. 1993. In vitro and ex vitro adventitious root formation of Asian jasmine *(Trachelospermum asiaticum).* II Physiological comparisons. *J. Amer. Soc. Hort. Sci.* 118(6):906–909.
7. Arce, P. and O. Balboa. 1991. Seasonality in rooting of *Prosopis chilensis* cuttings and in vitro micropropagation. *Forest Ecol. Manag.* 40:163–73.
8. Arrillaga, I., V. Lerma, and J. Segura. 1992. Micropropagation of juvenile and mature ash. *J. Amer. Soc. Hort. Sci.* 117(2):346–350.
9. Arrillaga, I., T. Marzo, and J. Segura. 1991. Micropropagation of juvenile and adult *Sorbus domestica* L. *Plant Cell, Tissue, Org. Cult.* 27:341–348.

10. ———. 1992. Embryo culture of *Fraxinus ornus* and *Sorbus domestica* removes seed dormancy. *HortScience* 27(4):371.
11. Artlett, E.G. 1985. Propagation of *Pieris japonica*. *Comb. Proc. Intl. Plant Prop. Soc.* 35:128–29.
12. Ault, J.R. 1994. In vitro propagation of *Eriostemon myoporoides* and *Eriostemon* 'stardust.' *HortScience.* 29(6):686–688.
13. Bailey, D.A., G.R. Seckinger, and P.A. Hammer. 1986. In vitro propagation of florists' hydrangea. *HortScience* 21:525–26.
14. Bailey, D.A., and T.C. Weiler. 1984. Rapid propagation and establishment of florists' hydrangea. *HortScience* 19:850–52.
15. Bakker, D. 1992. Grafting of Junipers. *Comb. Proc. Intl. Plant Prop. Soc.* 439–441.
16. Banko, T.J. and M.A. Stefani. 1989. In vitro propagation of *Oxydendrum arboreum* from mature trees. *HortScience* 24(4):683–685.
17. Barnes, H.W. 1992. Grafting viburnums: new ideas and techniques. *Comb. Proc. Intl. Plant Prop. Soc.* 42:436–438.
18. ———. 1988. Rooting responses and possibilities of Fraxinus, Osmanthus, Chionanthus. *Plant Propagator.* 34(2):5–6.
19. Barr, B. 1985. Propagation of *Berberis thunbergii 'Atropurpurea Nana.' Comb. Proc. Inter Plant Prop. Soc.* 35:711–12.
20. ———. 1987. Propagation of dwarf *nandina* cultivars. *Comb. Proc. Intl. Plant Prop. Soc.* 37:507–8.
21. ———. 1994. Propagation of Camellias by cuttings. *Comb. Proc. Intl. Plant Prop. Soc.* 44:454–456.
22. Barton, S.S., and V.P. Bonaminio. 1986. Influence of stratification and light on germination of sourwood [*Oxydendrum arboreum* (L.) DC]. *J. Environ. Hort.* 4:8–11.
23. Bates, S., J.E. Preece, N.E. Navarrete, J.W. Van Sambeek and G.R. Gaffney. 1992. Thidiazuron stimulates shoot organogenesis and somatic embryogenesis in white ash (*Fraxinus americana* L.). *Plant Cell, Tissue Org. Cult.* 31:21–29.
24. Bayly, R.B. 1989. Propagating Hibiscus by cuttings and grafting. *Comb. Proc. Intl. Plant Prop. Soc.* 39: 278–280.
25. Beeson, R.C., Jr. and W.M. Proebsting. 1991. Propagation tips for blue spruce. *Amer. Nurs.* 172:86–90.
26. Ben-Jaacov, J., A. Ackerman and E. Tal. 1991. Vegetative propagation of *Alberta magna* by tissue culture and grafting. *HortScience* 26(1):74.
27. Bennell, M., and G. Barth. 1986. Propagation of *Banksia coccinca* by cuttings and seed. *Comb. Proc. Intl. Plant Prop. Soc.* 36:148–52.
28. Bennett, I.J., J.A. McComb, C.M. Tonkin, and D.A.J. McDavid. 1994. Alternating cytokinins in multiplication in media stimulated in vitro shoot growth and rooting of *Eucalyptus globulues* Labill. *Annals Bot.* 74:53–58.
29. Bennett, L.K., and F.T. Davies, Jr. 1986. In vitro propagation of *Quercus shumardii* seedlings. *HortScience* 21:1045–47.
30. Berry, J.B. 1984. Rooting hormone formulations: A chance for advancement. *Comb. Proc. Intl. Plant Prop. Soc.* 34:486–91.
31. ———. 1991. Cleft grafting of *Magnolia grandiflora*. *Comb. Proc. Intl. Plant Prop. Soc* 41:345–346.
32. ———. 1994. Propagation and production of *Ilex* species in the southeastern United States. *Comb. Proc. Intl. Plant Prop. Soc.* 44:425–429.
33. Berthon, J.Y., S.B. Tahar, G. Gasper and N. Boyer. 1990. Rooting phases of shoots of *Sequoiadendron giganteum* in vitro and their requirements. *Plant Physiol. Biochem:* 28(5):631–638.
34. Bhella, H.S. 1975. Some factors affecting propagation of Douglas fir. *Comb. Proc. Intl. Plant Prop. Soc.* 25:420–24.
35. ———. 1980. Vegetative propagation of viburnum, cvs. Alleghany, Mohican, and Onondaga. *Plant Propagator* 26(3):5–9.
36. Bir, R.E. and H.W. Barnes. 1994. Stem cutting propagation of bottlebrush buckeye. *Comb. Proc. Intl. Plant Prop. Soc.* 44: 499–502.
37. Blakemore, S.D. 1993. Propagation of acacias and eucalypts. *Comb. Proc. Intl. Plant Prop. Soc.* 43: 317–319.
38. Blazich, F.A., and J.R. Acedo. 1987. Propagation of *Osmanthus × fortunei* by softwood cuttings. *J. Environ. Hort.* 5:70–71.
39. ———. 1989. Propagation of *Osmanthus heterophyllus* 'llicifolius' and 'Rotundifolius' by stem cuttings. *J. Environ. Hort.* 7(4):133–135.
40. Blazich, F.A., and L.E. Hinesley. 1995. Fraser fir. *Amer. Nurser.* 181(5):54–67.
41. Blazich, F.A., S.L. Warren, J.R. Acedo and W.M. Reece. 1993. Seed germination of *Rhododendron catawbiense* and *Rhododendron maximum:* influence of light and temperature. *J. Environ. Hort.* 9(1):5–8.
42. Blazich, F.A., S.L. Warren, J.R. Acedo and R.O. Whitehead. 1991. Seed germination of *Leucothoe fontanesiana* as influenced by light and temperature. *J. Environ. Hort.* 9(2):72–75.
43. Blazich, F.A., S.L. Warren, M.C. Starrett, and J.R. Acedo. 1993. Seed germination of *Rhododendron carolinianum:* influence of light and temperature. *J. Environ. Hort.* 11(2):55–58.
44. Blythe, G. 1984. Cutting propagation of *Sequoia sempervirens. Comb. Proc. Intl. Plant Prop. Soc.* 34:204–11.
45. ———. 1989. Cutting propagation of *Cupressus* and *Cupressocyparis. Comb. Proc. Intl. Plant Prop. Soc.* 39:154–160.
46. ———. 1994. Cutting propagation of *Juniperus procumbens* 'Nana' *Comb. Proc. Intl. Plant Prop. Soc.* 44:409–413.
47. Boorman, D. 1991. How to commercially graft grevilleas for profit or preservation. *Comb. Proc. Intl. Plant Prop. Soc.* 41:57–58.

48. Bouza, L., M. Jacques, E. Miginiac. 1994. In vitro propagation of *Paeonia suffruticosa* Andr. cv. 'Mme de Vatry': developmental effects of exogenous hormones during the multiplication phase. *Scientia Hort.* 57:241–245.

49. Brand, M.H., and R.D. Lineberger. 1986. In vitro propagation of *Halesia Carolina* L. and the influence of explanation timing on initial shoot proliferation. *Plant Cell Tissue Org. Cult.* 7:103–13.

50. ———. 1988. In vitro adventitious shoot formation on mature-phase leaves and petioles of *Liquidambar styraciflua* L. *Plant Sci.* 57:173–179.

51. ———. 1991. The effect of leaf source and development stage on shoot organogenic potential of sweetgum (*Liquidambar styraciflua* L.) leaf explants. *Plant Cell Tissue Org. Cult.* 24:1–7.

52. Bretzloff, L.V., and N.E. Pellet. 1979. Effect of stratification and gibberellic acid on the germination of *Carpinus caroliniana* Walt. *HortScience* 14: 621–22.

53. Briggs, B.A., and S.M. McCulloch. 1983. Progress in micropropagation of woody plants in the United States and Canada. *Comb. Proc. Intl. Plant Prop. Soc.* 33:239–48.

54. Browse, P.M. 1986. Dormancy control in *Magnolia* seed germination. *Comb. Proc. Intl. Plant Prop. Soc.* 36:116–20.

55. Buck, G.J. 1951. Varieties of rose understocks. *Amer. Rose Ann.* 36:101–116.

56. Bunge, B. 1988. Propagation and production of crab apples on their own roots. *Comb. Proc. Intl. Plant Prop. Soc.* 38:542–544.

57. Bunker, E.J. 1975. Germinating palm seed. *Comb. Proc. Intl. Plant Prop. Soc.* 25:377–78.

58. Bunn, E. and K.W. Dixon. 1992. In vitro propagation of the rare and endangered *Grevillea scapigera* (Proteaceae). *HortScience* 27(3):261–262.

59. Bunting, B.A. 1985. Propagation of *Clethra alnifolia*. *Comb. Proc. Intl. Plant Prop. Soc.* 35:712–13.

60. Burger, D.W. 1987. In vitro micropropagation of *Eucalyptus sideroxylon*. *HortScience* 22:496–97.

61. Burrows, G.E., D.D. Doley, R.J. Haines and D.G. Nikles. 1988. In vitro propagation of *Araucaria cunninghamii* and other species of Araucariaceae via axillary meristems. *Austral. J. Bot.* 36:665–676.

62. Butcher, S.M., and S.M.N. Wood. 1984. Vegetative propagation and development of *Sophora microphylla* Ait. *Comb. Proc. Intl. Plant Prop. Soc.* 34:407–16.

63. Byers, D. 1983. Selection and propagation of crape myrtle. *Comb. Proc. Intl. Plant Prop. Soc.* 33:542–45.

64. Carpenter, W.J. 1988. Temperature affects seed germination of four Florida palm species. *HortScience* 23:336–37.

65. ———. 1988. Seed after-ripening and temperature influence *Butia capitata* germination. *HortScience* 23:702–703.

66. ———. 1989. Medium temperature influences the rooting response of *Hibiscus rosa-inensis* L. *J. Environ. Hort.* 7(4):143–146.

67. Carpenter, W.J., E.R. Ostmark, and J.A. Cornell. 1993. Embryo cap removal and high temperature exposure stimulate rapid germination of needle palm seeds. *HortScience* 28(9):904–907.

68. Carpenter, W.J., E.R. Ostmark, and K.C. Ruppert. 1995. Promoting the rapid germination of needle palm seeds. *N. Amer. Plant Propagator* 7(1):20–23.

69. Carter, A.R. 1969. Rose rootstocks—performance and propagation from seed. *Comb. Proc. Intl. Plant Prop. Soc.* 19:172–80.

70. ———. 1979. Daphne propagation. *Comb. Proc. Intl. Plant Prop. Soc.* 29:248–51.

71. Carter, A.R. and M.U. Slee. 1992. The effects of shoot age on root formation of cuttings of *Eucalyptus grandis* W. Hill ex Maiden. *Comb. Proc. Intl. Plant Prop. Soc.* 42:43–47.

72. Carville, L. 1975. Propagation of *Acer palmatum* cultivars from hardwood cuttings. *Comb. Proc. Intl. Plant Prop. Soc.* 25:39–42.

73. Catalan, L.A. and R.E. Macchiavelli. 1991. Improving germination in *Prosopis flexuosa* D.C. and *P. alba* Griseb. with hot water treatments and scarification. *Seed Sci. Technol.* 19:253–262.

74. Caulfield, H.W. 1976. Pointers for successful germination of palm seed. *Comb. Proc. Intl. Plant Prop. Soc.* 26:402–5.

75. Cervelli, R. 1994. Propagation of ornamental varieties of spruce (*Picea* spp.) through somatic embroyogenesis. *Comb. Proc. Intl. Plant Prop. Soc.* 44:300–3.

76. Chalupa, V. 1987. Effect of benzylaminopurine and thidiazuron on in vitro shoot proliferation of *Tilia cordata* Mill., *Sorbus aucuparia* L. and *Robinia pseudoacacia* L. *Biolog. Plantar. (Praha)* 29:425–29.

77. Chandler, G.P. 1969. Rooting daphnes from cuttings. *Comb. Proc. Intl. Plant Prop. Soc.* 19:205–6.

78. Chang, T.Y., and T.H. Vogue. 1977. Regeneration of Douglas-Fir plantlets through tissue culture. *Science* 198:306–7.

79. Chase, H.H. 1964. Propagation of Oriental magnolias by layering. *Comb. Proc. Intl. Plant Prop. Soc.* 14:67–69.

80. Chaturvedi, A., K. Sharma, and P.N. Prasad. 1978. Shoot apex culture of *Bougainvillea glabra 'Magnifica.' HortScience* 13(1):36.

81. Chee, P.B. 1994. In vitro culture of zygotic embryos of Taxus species. *HortScience* 29(6): 695–697.

82. Chong, C.B. 1988. Dormancy requirement and greenhouse forcing of three Euonymus cultivars. *Comb. Proc. Intl. Plant Prop. Soc.* 38:580–583.

83. Christie, C.B. 1986. Factors affecting root formation on *Photinia* 'Red Robin' cuttings. *Comb. Proc. Intl. Plant Prop. Soc.* 36:490–94.

84. Chu, K., and W.S. Cooper. 1950. An ecological reconnaissance in the native home of *Metasequoia glyptostroboides*. *Ecology* 31:260–78.

85. Coate, B. 1983. Vegetative propagation of *Acacia iteaphylla*. *Comb. Proc. Intl. Plant Prop. Soc.* 33:118–20.

86. Coggeshall, R.C. 1977. Propagating French hybrid lilacs by softwood cuttings. *Comb. Proc. Intl. Plant Prop. Soc.* 27:442–44.

87. Cohen, M.A. 1975. Vegetative propagation of *Pinus strobus* by needle fascicles. *Comb. Proc. Intl. Plant Prop. Soc.* 25:413–19.

88. Connor, D.A. 1985. Propagation of upright junipers. *Comb. Proc. Intl. Plant Prop. Soc.* 35:719–21.

89. ———. 1985. Propagation of *Mahonia* species and cultivars. *Comb. Proc. Intl. Plant Prop. Soc.* 35:279–81.

90. Cooper, C., D.K. Struve and M.A. Bennett. 1991. Pericarp splitting after aerated water soak can be used as an indicator of red oak seed quality. *Can. J. For. Res.* 21:1694–1697.

91. Cooper, P.A. and D. Cohen. 1984. Micropropagation of Japanese persimmon *(Diospyros kaki)*. *Comb. Proc. Intl. Plant Prop. Soc.* 34:118–124.

92. Costin, J.J. 1977. Production of Eucalyptus. *Comb. Proc. Intl. Plant Prop. Soc.* 27:44–48.

93. Covan, D.A. 1986. Softwood cutting propagation of oaks, magnolias, crab apples and dogwoods. *Comb. Proc. Intl. Plant Prop. Soc.* 36:419–21.

94. Cross, R.E. 1988. Persistence pays. *Amer. Nurs.* 168(12):63–67.

95. Crossen, T. 1990. Approach grafting Grevilleas. *Comb. Proc. Intl. Plant Prop. Soc.* 40:69–72.

96. Currie, R. 1990. Betula propagation. *Comb. Proc. Intl. Plant Prop. Soc.* 40:315–317.

97. Davies, F.T., Jr. 1985. Adventitious root formation in *Rosa multiflora* 'Brooks 56' hardwood cuttings. *J. Environ. Hort.* 3:55–56.

98. Davies, F.T., Jr., Y. Fann, J.E. Lazarte, and D.R. Paterson. 1980. Bench chip budding of field roses. *HortScience* 15:817–18.

99. Davies, F.T., Jr., and J.N. Joiner. 1980. Growth regulator effects on adventitious root formation in leaf bud cuttings of juvenile and mature *Ficus pumila*. *J. Amer. Soc. Hort. Sci.* 105:91–95.

100. Davis, R.G. 1995. A consortium of conifers. Amer. Nurser. 182(10):20–27.

101. Davis, T. 1995. *Jasminum nudiflorum*. *Nurser. Manag. Prod.* 11(2):26.

102. Deering, T. 1979. Bench grafting of *Betula* species. *Plant Propagator* 25(3):8–9.

103. de Fossard, R.A., and H. de Fossard. 1988. Micropropagation of some members of the Myrtaceae. *Acta Hort.* 227:346–51.

104. Dehgan, B. 1996. Permian permanance: Cycads. *Amer. Nurs.* 183(2):66–81.

105. Del Tredici, P. 1985. Propagation of *Tsuga canadensis* cultivars: Hardwood versus softwood cuttings. *Comb. Proc. Intl. Plant Prop. Soc.* 35:565–69.

106. Dick, L. 1990. Magnolia production. *Comb. Proc. Intl. Plant Prop. Soc.* 40:318–320.

107. Dirr, M.A. 1983. Comparative effects of selected rooting compounds on the rooting of *Phoitinia* × *Fraseri*. *Comb. Proc. Intl. Plant Prop. Soc.* 33:536–40.

108. ———. 1985. Ten woody plants that deserve a longer look. *Comb. Proc. Intl. Plant Prop. Soc.* 35:728–34.

109. ———. 1987. Native amelanchiers: a sampler of northeastern species. *Amer. Nurs.* 166:64–83.

110. ———. 1988. To know them is to love them. *Amer. Nurs.* 163:23–41.

111. ———. 1991. The Japanese apricot: flowers in the winter and style in the landscape. *Nurs. Manag.* 7(4):40–42.

112. ———. 1993. *Loropetalum chinense* var. Rubrum: as a specimen, in a grouping in masses or in a container, the shub has exciting potential for the landscape. *Nurs. Manag.* 9(1):24–25.

113. ———. 1994. Confessions of a Clethra-phile. *Nurs. Manag.* 10(11):14–21.

114. ———. 1994. Lustrous leaves, fragrant flowers and shiny red fruit make *Sarcococca* a sweet addition to the landscape. *Nurs. Manag.* 10(5):24–25.

115. Dirr, M.A., and C.W. Heuser, Jr. 1987. *The reference manual of woody plant propagation.* Athens, Ga.: Univ. Press.

116. Dirr, M.A. and A.E. Richards. 1989. Cutting propagation of *Hamamelis* × *intermedia* 'Arnold Promise.' *N. Amer. Plant Prop.* 1(2):9–10.

117. Dodge, M.H. 1986. Heath and heather propagation. *Comb. Proc. Intl. Plant Prop. Soc.* 36:563–67.

118. Donnelly, J.R., and H.W. Yawney. 1972. Some factors associated with vegetatively propagating sugar maple by stem cuttings. *Comb. Proc. Intl. Plant Prop. Soc.* 22:413–31.

119. Donovan, D.M. 1990. *Daphne cneorum:* its propagation by softwood cuttings. *N. Amer. Reg. Plant Prop.* 2(2):13.

120. Donselman, H. 1990. Ornamental palm propagation and container production. *Comb. Proc. Intl. Plant Prop. Soc.* 40:236–241.

121. Drew, J.J. and M.A. Dirr. 1989. Propagation of *Quercus* L. species by cuttings. *J. Environ. Hort.* 7(3):115–117.

122. D'Silva, I. and L. D'Souza. 1992. Micropropagation of *Ailanthus malabarica* D.C. using juvenile and mature tree tissues. *Silvae Genet.* 41(6):333–339.

123. Duck, P.H. 1985. Propagation of *Ilex vomitoria* 'Nana.' *Comb. Proc. Intl. Plant Prop. Soc.* 35:710–11.

124. Duncan, P.J., and T.E. Bilderback. 1982. Effect of irrigation systems, gibberellic acid and photoperiod on seed germination of *Kalmia latifolia* L. and *Rhododendron maximum* L. *HortScience* 17:916–17.

125. Ecke, P., and O.A. Matkin. 1976. *The poinsettia manual.* Encinitas, Calif.: Paul Ecke Poinsettias.

126. Economou, A.S., and P.E. Read. 1984. In vitro shoot proliferation of Minnesota deciduous azaleas. *HortScience* 19:60–61.

127. Economou, A.S., and K.M.J. Spanoudaki. 1985. In vitro propagation of *Gardenia. HortScience* 20:213.

128. Eden, W.S. 1985. Hibiscus propagation in cool climates. *Comb. Proc. Intl. Plant Prop. Soc.* 35:132–34.

129. Edson, J.L., D.L. Wenny, and A. Leege-Brusven. 1994. Micropropagation of Pacific dogwood. *HortScience* 29(11):1355–56.

130. Effner, W.S. 1992. Successful cutting propagation of *Hamamelis* × *intermedia* 'Arnold Promise' and *Hamamelis mollis* 'Brevipetala.' *Comb. Proc. Intl. Plant Prop. Soc.* 42:497–498.

131. Ekstrom, D.E. and J.A. Ekstrom. 1988. Propagation of cultivars of *Stewartia, Acer palmatum,* and *Fagus sylvatica* for open ground production. *Comb. Proc. Intl. Plant Prop. Soc.* 38:180–184.

132. Ellyard, R.K., and P.J. Ollerenshaw. 1984. Effect of indolebutyric acid, medium composition and cutting type on rooting of *Grevillea johnsonii* cuttings at two basal temperatures. *Comb. Proc. Intl. Plant Prop. Soc.* 34:101–108.

133. Emery, D.E. 1985. *Arctostaphylos* propagation. *Comb. Proc. Intl. Plant Prop. Soc.* 35:281–84.

134. Ettinger, T.L., and J.E. Preece. 1985. Aseptic micropropagation of *Rhododendron* P. J. M. hybrids. *J. Hort. Sci.* 60:269–74.

135. Evison, R.J. 1977. Propagation of clematis. *Comb. Proc. Intl. Plant Prop. Soc.* 27:436–40.

136. Farmer, R.E., Jr. 1975. Long term storage of northern red and scarlet oak seed. *Plant Propagator* 20(4) and 21(1):11–14.

137. Fazio, S. 1964. Propagating *Eucalyptus* from cuttings. *Comb. Proc. Intl. Plant Prop. Soc.* 14:288–90.

138. Fins, L. 1980. Propagation of giant sequoia by rooting cuttings. *Comb. Proc. Intl. Plant Prop. Soc.* 30:127–32.

139. Flemer, W., III. 1962. The vegetative propagation of oaks. *Comb. Proc. Intl. Plant Prop. Soc.* 12:168–71.

140. ———. 1980. Linden propagation review. *Comb. Proc. Intl. Plant Prop. Soc.* 30:333–36.

141. Fleming, R.A. 1978. Propagation of holly in southern Ontario. *Comb. Proc. Intl. Plant Prop. Soc.* 28:553–57.

142. Flint, H. 1992. Genista—golden treasures for the landscape. *Amer. Nurser.* 76(11):57–61.

143. Fordham, A.J. 1972. Vegetative propagation of *Albizia. Amer. Nurs.* 128:(4):7, 63.

144. ———. 1977. Propagation of *Kalmia latifolia* by cuttings. *Comb. Proc. Intl. Plant Prop. Soc.* 27:479–83.

145. ———. 1977. *Pieris floribunda* and its propagation. *Comb. Proc. Intl. Plant Prop. Soc.* 27:495–97.

146. ———. 1984. *Cornus kousa* and its propagation. *Comb. Proc. Intl. Plant Prop. Soc.* 34:598–602.

147. ———. 1987. *Davidia involucrata* var. *vilmoriniana*—dove tree and its propagation by seeds. *Comb. Proc. Intl. Plant Prop. Soc.* 37:343–45.

148. Foster, S. 1992. Propagation of Japanese maples by softwood cutting and grafting. *Comb. Proc. Intl. Plant Prop. Soc.* 42:373–374.

149. Freeland, K.S. 1977. Propagation of potentilla. *Comb. Proc. Intl. Plant Prop. Soc.* 27:441–42.

150. Frett, J.J. 1989. Germination requirements of *Hovenia dulcis* seeds. *HortScience* 24(1):152.

151. ———. 1987. Seed germination of *Cycas revoluta. J. Environ. Hort.* 5(3):105–106.

152. Froberg, C.A. 1985. Tissue culture propagation of *Sophora secundiflora. Comb. Proc. Intl. Plant Prop. Soc.* 35:750–54.

153. Fuller, C.W. 1979. Container production of *Euonymus alata* 'Compacta.' *Comb. Proc. Intl. Plant Prop. Soc.* 29:360–62.

154. George, C. 1993. Rhododendron propagation—methods and techniques carried out in the pacific northwest of the USA. *Comb. Proc. Intl. Plant Prop. Soc.* 43:178–182.

155. Galle, F. 1988. *Azaleas.* Portland, Oreg.: Timber Press.

156. Geneve, R.L. 1991. Seed dormancy in eastern redbud *(Cercis canadensis). J. Amer. Soc. Hort. Sci.* 116(1):85–88.

157. Geneve, R.L. and S.T. Kester. 1990. The initiation of somatic embryos and adventitious roots from developing zygotic embryo explants of *Cercis canadensis* L. cultured in vitro. *Plant Cell Tissue Org. Cult.* 22:71–76.

158. Geneve, R.L., S.T. Kester, and S. El-Shall. 1990. In vitro shoot initiation in Kentucky coffeetree. *HortScience.* 25(5):578.

159. Geneve, R.L. and S.T. Kester and S. Yusnita. 1992. Micropropagation of Eastern Redbud (*Cercis canadensis* L.). *Comb. Proc. Intl. Plant Prop. Soc.* 42:417–420.

160. Giersbach, J., and L.V. Barton. 1932. Germination of seeds of the silverbell, *Halesia carolina. Contrib. Boyce Thomp. Inst.* 4:27–37.

161. Gingas, V.M. 1991. Asexual embryogenesis and plant regeneration from male catkins of *Quercus. HortScience* 26(9):1217–1218.

162. Girouard, R.M. 1971. Vegetative propagation of pines by means of needle fascicles—a literature review. *Inform. Rpt. Dept. Environ., Can. For. Ser., Quebec.*

163. ———. 1973. Rooting, survival, shoot formation, and elongation of Norway spruce stem cuttings as affected by cutting types and auxin treatments. *Plant Propagator* 19(2):16–17.

164. Goddard, A.N. 1970. Grafting Japanese maples. *Plant Propagator* 16(4):6.

165. Goodman, M.A., and D.P. Stimart. 1987. Factors regulating overwinter survival of newly propagated

tip cuttings of *Acer palmatum* Thumb 'Bloodgood' and *Cornus florida* L. var. *rubra. HortScience* 22:1296–98.

166. Gordon, I. 1992. Cutting propagation of *Chamelaucium* cultivars. *Comb. Proc. Intl. Plant Prop. Soc.* 42:196–199.

167. Goreau, T. 1980. Rhododendron propagation. *Comb. Proc. Intl. Plant Prop. Soc.* 30:532–37.

168. Greever, P.T. 1979. Propagation of *Heteromeles arbutifolia* by softwood cuttings or by seed. *Plant Propagator* 25(2):10–11.

169. Grossbechler, F. 1981. Mass production of eucalyptus seedlings by direct sowing method. *Comb. Proc. Intl. Plant Prop. Soc.* 31:276–79.

170. Gwaltney, T. 1992. Dwarf yaupon, weeping yaupon and azalea propagation at Flowerwood Nursery, Inc. *Comb. Proc. Intl. Plant Prop. Soc.* 42:369–372.

171. Hambrick, C.E., F.T. Davies, Jr. and H.B. Pemberton. 1991. Seasonal changes in carbohydrate/nitrogen levels during field rooting of *Rosa multiflora* 'Brooks 56' hardwood cuttings. *Scientia Hort.* 46:137–146.

172. Hamilton, D.F., and P.L. Carpenter. 1977. Seed germination of *Myrica pennsylvanicum* L. *HortScience* 12(6):565–66.

173. Hammatt, N. 1994. Shoot initiation in the leaflet axils of compound leaves from micropropagated shoots of juvenile and mature common ash (*Fraxinus excelsior* L.). 1994. *J. Exp. Bot.* 45(275):871–875.

174. Hand, N.P. 1978. Propagation of lilacs. *Comb. Proc. Intl. Plant Prop. Soc.* 28:348–50.

175. Hansen, O.B. 1990. Propagating *Sorbus aucuparia* L. and *Sorbus hybrida* L. by softwood cuttings. *Scientia Hort.* 42:169–175.

176. Hare, R.C. 1974. Chemical and environmental treatments promote rooting of pine cuttings. *Can. J. For. Res.* 4:101–6.

177. ———. 1976. Rooting of American and Formosan sweetgum cuttings taken from girdled and nongirdled cuttings. *Tree Plant. Notes* 27:6–7.

178. ———. 1977. Rooting of cuttings from mature water oak. *S. Jour. Appl. For.* 1:24–25.

179. Hartmann, H.T. 1967. 'Swan Hill': A new fruitless ornamental olive. *Calif. Agr.* 21(1):4–5.

180. Hasegawa, P.M. 1979. Factors affecting shoot and root initiation from cultured rose shoot tips. *J. Amer. Soc. Hort. Sci.* 105(2):216–20.

181. Hasek, R.F. 1980. Roses. In *Introduction to floriculture,* R.A. Larson, ed. New York: Academic Press.

182. Head, R.H. 1995. Propagation of *Magnolia grandiflora* cultivars. *Comb. Proc. Intl. Plant Prop. Soc.* 45:In press.

183. Heit, C.E. 1972. Propagation from seed: Growing larches. *Amer. Nurs.* 135(8):14–15, 99–110.

184. Henry, P.H., F.A. Blazich, and L.E. Hinesley. 1992. Vegetative propagation of eastern redcedar by stem cuttings. *HortScience* 27(12):1272–1274.

185. Hildebrandt, V., and P.M. Harney. 1983. In vitro propagation of *Syringa vulgaris* 'Vesper.' *HortScience* 18:432–34.

186. Hill, J.B. 1962. The propagation of *Juniperus chinensis* in greenhouse and mist bed. *Comb. Proc. Intl. Plant Prop. Soc.* 12:173–78.

187. Hinesley, L.E., and F.A. Blazich. 1980. Vegetative propagation of *Abies fraseri* by stem cuttings. *HortScience* 15(1):96–97.

188. Hinesley, L.E., F.A. Blazich, and L.K. Snelling. 1994. Propagation of Atlantic white cedar by stem cuttings. *HortScience* 29(3):217–219.

189. Higginbotham, J.S. 1992. What's new in '92—new introductions from non-commercial sources. *Amer. Nurs.* 175(4):37–62.

190. Higginbotham, R. 1992. Bougainvillea propagation. *Comb. Proc. Intl. Plant Prop. Soc.* 42:37–38.

191. Hoogendoorn, D.P. 1984. Propagation of *Acer griseum* from cuttings. *Comb. Proc. Intl. Plant Prop. Soc.* 34:570–73.

192. ———. 1988. Grafting *Viburnum carlesii* 'Compactum.' *Comb. Proc. Intl. Plant Prop. Soc.* 38: 563–566.

193. Hooper, V. 1990. Selecting and using Magnolia clonal understock. *Comb. Proc. Intl. Plant Prop. Soc.* 40:343–346.

194. Howard, F.W., S.D. Verkade, and J.D. DeFilippis. 1988. Propagation of West Indies, Honduran and hybrid mahoganies by cuttings, compared with seed propagation. *Proc. Fla. State Hort. Soc.* 101:296–298.

195. Hubbard, A.C. 1986. Native ornamentals for the U.S. southwest. *Comb. Proc. Intl. Plant Prop. Soc.* 36:347–50.

196. Hunter, S.A. and N. O'Donnell. 1988. Outline of a system for in vitro propagation of *Sequoia sempervirens. Comb. Proc. Intl. Plant Prop. Soc.* 38:268–272.

197. Huss-Danell, K., L. Eliasson, and I. Ohberg. 1980. Conditions for rooting of leafy cuttings of *Alnus incana. Physiol. Plant.* 49:113–16.

198. Iapichino, G., T.H.H. Chen and L.H. Fuchigami. 1991. Adventitious shoot production from a *Vireya* hybid of *Rhododendron. HortScience* 26(5):594–596.

199. Iseli, J., and D. Howse. 1981. New cultivars of *Picea pungens glauca*—their attributes and propagation. *Plant Propagator* 27(1):5–8.

200. Intven, W.J. and T.J. Intven. 1989. Apical graftage of *Acer palmatum* and other deciduous plants. *Comb. Proc. Intl. Plant Prop. Soc.* 39:409–412.

201. Jacobs, R.M., J. Berry and P. Duck. 1990. New propagation techniques. *Comb. Proc. Intl. Plant Prop. Soc.* 40:394–396.

202. Janick, J., A. Whipkey, S.L. Kitto, and J. Frett. 1994. Micropropagation of *Cephalotaxus harringtonia. HortScience* 29(2):120–122.

203. Jaynes, R.A. 1988. *Kalmia, the laurel book II.* Portland, Oreg.: Timber Press.

204. ———. 1976. Mountain laurel selections and how to propagate them. *Comb. Proc. Intl. Plant Prop. Soc.* 26:233–36.

205. Joley, L. 1960. Experiences with propagation of the genus *Pistacia. Comb. Proc. Intl. Plant Prop. Soc.* 10:287–92.

206. Jona, R., and I. Gribaudo. 1987. Adventitious bud formation from leaf explants of *Ficus lyrata. HortScience* 22:651–53.

207. Jones, C. and D. Smith. 1988. Effect of 6-benzylaminopurine and 1-naphthylacetic acid on in vitro axillary bud development of mature *Acacia Melanoxylon. Comb. Proc. Intl. Plant Prop. Soc.* 38:389–393.

208. Jones, N.B. and J. Van Staden. 1994. Micropropagation and establishment of *Eucalyptus grandis* hybrids. *S. Afr. J. Bot.* 60(2):122–126.

209. Jones, S.K., Y.K. Samuel and P.G. Gosling. 1991. The effect of soaking and prechilling on the germination of noble fir seeds. *Seed Sci. Technol.* 19:287–293.

210. Jull, L.G., S.L. Warren, F.A. Blazich and J.C. Warren. 1993. Effects of growth stage, branch order and IBA treatment on rooting stem cuttings of 'Yoshino' Cryptomeria. *Comb. Proc. Intl. Plant Prop. Soc.* 43:404–407.

211. Jull, L.G., S.L. Warren, and F.A. Blazich. 1994. Rooting of 'Yoshino' Cryptomeria stem cuttings as influenced by growth stage, branch order and IBA treatment. *HortScience* 29(12):1532–1535.

212. Kavanagh, J.M., S.A. Hunter, and P.J. Crossan. 1986. Micropropagation of catawba hybrid *Rhododendron* 'Nova Zembla,' 'Cynthia' and 'Pink Pearl.' *Comb. Proc. Intl. Plant Prop. Soc.* 36:264–72.

213. Kelety, M.M. 1984. Container-grown hibiscus: Propagation and production. *Comb. Proc. Intl. Plant Prop. Soc.* 34:480–86.

214. Kelly, J.C. 1986. Propagation of *Prunus tenella* 'Firehill' from cuttings. *Comb. Proc. Intl. Plant Prop. Soc.* 36:279–81.

215. Kerns, H.R., and M.M. Meyer. 1986. Tissue culture propagation of *Acer × freemanii* using thidiazuron to stimulate shoot tip proliferation. *HortScience* 21:1209–10.

216. Kiang, Y.T., O.M. Rogers, and R.B. Pike. 1974. Rooting mugho pine cuttings. *HortScience* 9(4):350.

217. King, P.C. 1993. Fog and air circulation techniques to propagate *Aesculus parviflora* and trifoliate maples. *Comb. Proc. Intl. Plant Prop. Soc.* 43:470–473.

218. King, S.M. 1988. Tissue culture of osage-orange. *HortScience* 23:613–15.

219. Klass, S., J. Wright and P. Felker. 1987. Influence of auxins, thiamine and fungal drenches on the rooting of *Prosopis alba* clone B2V50 cuttings. *J. Hort. Sci.* 62:97–100.

220. Kretzschmar, U. and D. Ewald. 1994. Vegetative propagation of 140-year-old *Larix decidua* trees by different in-vitro techniques. *J. Plant Physiol.* 144: 627–630.

221. Krishnan, S. and H. Hughes. 1991. Asexual propagation of *Shepherdia canadensis* and *S. rotundifolia. J. Environ. Hort.* 9(4):218–220.

222. Krussmann, G. 1981. *The complete book of roses.* Portland, Oreg.: Timber Press.

223. Kumar, A. 1992. Micropropagation of a mature leguminous tree—*Bauhinia purpurea. Plant Cell Tissue Org. Cult.* 31:257–259.

224. Kumar, K.S. and K.V. Bhavanandan. 1988. Micropropagation of *Plumbago rosea* Linn. *Plant Cell Tissue Org. Cult.* 15:275–278.

225. Kunisaki, J.T. 1989. In vitro propagation of *Leucospermum* hybrid, 'Hawaii Gold.' *HortScience* 24(4): 686–687.

226. Lamb, J.G.D. 1970. Trials on propagation of *Chamaecyparis* at Kinsealy. *Comb. Proc. Intl. Plant Prop. Soc.* 20:334–38.

227. ———. 1976. The propagation of understocks for *Hamamelis. Comb. Proc. Intl. Plant Prop. Soc.* 26: 127–30.

228. Lane, B.C. 1987. The effect of IBA and/or NAA and cutting wood selection on the rooting of *Rhaphiolepis indica* 'Jack Evans.' *Comb. Proc. Intl. Plant Prop. Soc.* 37:77–82.

229. Lane, C.G. 1993. Magnolia propagation. *Comb. Proc. Intl. Plant Prop. Soc.* 43:163–166.

230. Larsen, O.N. 1985. Clonal propagation of *Fagus sylvatica* L. by cuttings. *Comb. Proc. Intl. Plant Prop. Soc.* 35:438–42.

231. Larson, R.A., J.W. Love, D.L. Strider, R.K. Jones, J.R. Baker, and K.F. Horn. 1978. *Commercial poinsettia production.* Raleigh, N. C.: N. C. State Univ. Dept. Agr. Inf.

232. Lee, C.C. and T.E. Bilderback. 1990. Propagation of *Heptacodium miconioides,* seven son flower, by softwood and semi-hardwood cuttings. *Comb. Proc. Intl. Plant Prop. Soc.* 40:397–401.

233. Leiss, J. 1984. Root grafting of oaks. *Comb. Proc. Intl. Plant Prop. Soc.* 34:526–27.

234. ———. 1985. Seed treatments to enhance germination. *Comb. Proc. Intl. Plant Prop. Soc.* 35:495–99.

235. ———. 1988. Piece root grafting of oaks: an update. *Comb. Proc. Intl. Plant Prop. Soc.* 38:531–532.

236. Le Roux, J.J. and J. Van Staden. 1991. Micropropagation of *Eucalyptus* species. *HortScience* 26(2): 199–200.

237. Libby, W.J., and M.T. Conkle. 1966. Effects of auxin treatment, tree age, tree vigor, and cold storage on rooting young Monterey pine. *For. Sci.* 12:484–502.

238. Lin, X., B.A. Bergmann, and A.M. Stomp. 1995. Effect of medium physical support, shoot length and genotype on in vitro rooting and plantlet morphology of sweetgum. *J. Environ. Hort.* 13(3): 117–121.

239. Ling, C.H. and L.K.C. Clay. 1988. Commercial conifer micropropagation. *Comb. Proc. Intl. Plant Prop. Soc.* 38:209–215.
240. Lloyd, G., and B. McCown. 1980. Commercially-feasible micropropagation of mountain laurel, *Kalmia latifolia,* by use of shoot-tip culture. *Comb. Proc. Intl. Plant Prop. Soc.* 30:421–27.
241. Lohnes, J.P., B.C. Van Duyne, and C.H. Case. 1988. Propagating 'Kwanzan' cherries. *Amer. Nurs.* 167: 69–85.
242. Macdonald, A.B. 1974. Camellia propagation. *Comb. Proc. Intl. Plant Prop. Soc.* 24:152–54.
243. ———. 1989. Bench grafting Colorado blue spruce—criteria for budding success. *Comb. Proc. Intl. Plant Prop. Soc.* 39:131–134.
244. ———. 1990. Ornamental Asian climbing vines for the Pacific Northwest. *Comb. Proc. Intl. Plant Prop. Soc.* 40:464–466.
245. Maleike, R. and R.L. Hummel. 1991. Germination of Madrona seed. *Comb. Proc. Intl. Plant Prop. Soc.* 41:283–285.
246. ———. 1991. Germination of *Cornus canadensis* seed. *Comb. Proc. Intl. Plant Prop. Soc.* 41:286–289.
247. Malek, A.A., F.A. Blazich, S.L. Warren and J.E. Shelton. 1989. Influence of light and temperature on seed germination of mountain laurel. *J. Environ. Hort.* 7(4):161–162.
248. Maynard, B.K., and N.L. Bassuk. 1987. Stock plant etiolation and blanching of woody plants prior to cutting propagation. *J. Amer. Soc. Hort. Sci.* 112: 273–76.
249. ———. 1990. Rooting softwood cuttings of *Acer griseum:* promotion by stockplant etiolation, inhibition by catechol. *HortScience* 25(2):200–202.
250. ———. 1992. Stock plant etiolation, shading, and banding effects on cutting propagation of *Carpinus betulus. J. Amer. Soc. Hort. Sci.* 117(5):740–744.
251. McConnell, J.F. 1993. Wisteria propagation by root cuttings. *Comb. Proc. Intl. Plant Prop. Soc.* 43:304.
252. McCown, B., and R. Amos. 1979. Initial trials with micropropagation of birch selections. *Comb. Proc. Intl. Plant Prop. Soc.* 29:387–93.
253. McCracken, P. 1995. The controversy continues: comparisons of propagation techniques of *Magnolia grandiflora. Comb. Proc. Intl. Plant Prop. Soc.* 45:In press.
254. McCulloch, S.M. 1989. Tissue culture propagation of French hybrid lilacs. *Comb. Proc. Intl. Plant Prop. Soc.* 39:105–108.
255. McDaniel, J.D. 1964. A look at some hackberries. *Comb. Proc. Intl. Plant Prop. Soc.* 14:143–46.
256. McGuire, J.J., W. Johnson, and C. Dawson. 1987. Leaf-bud or side-graft nurse root grafts for difficult to root *Rhododendron cultivars. Comb. Proc. Intl. Plant Prop. Soc.* 37:447–49.
257. ———. 1991. Cold storage of rooted *Taxus* cuttings and subsequent summer regrowth. *J. Environ. Hort.* 9(1): 36–37.
258. Medina, J.P. 1981. Studies of clonal propagation of pecans at Ica, Peru. *Plant Propagator* 27:10–11.
259. Meyer, H.J. and J. van Staden. 1987. Regeneration of *Acacia melanoxylon* plantlets in vitro. *S. Afr. Tydskv. Planik.* 53(3):206–209.
260. Monroe, J.E. 1992. Propagation of *Corylus avellana* "Contorta' from root suckers. *Comb. Proc. Intl. Plant Prop. Soc.* 42:499.
261. Monteuuis, O. 1994. Effect of technique and darkness on the success of meristem micrografting of *Picea abies. Silvae Genet.* 43:2–3.
262. Moore, R.S. 1981. Miniature rose production. *Comb. Proc. Intl. Plant Prop. Soc.* 30:54–60.
263. Morgan, D.L. 1985. Propagation of *Quercus virginiana* cuttings. *Comb. Proc. Intl. Plant Prop. Soc.* 35:716–19.
264. Munson, R.H. 1984. Germination of western soapberry as affected by scarification and stratification. *HortScience* 19:712–13.
265. Murphy, W.R. 1995. A unique approach to mass propagation. *Comb. Proc. Intl. Plant Prop. Soc.* 45: In press.
266. Myers, J.R., and S.M. Still. 1979. Propagating London plane tree from cuttings. *Plant Propagator* 25(3):8–9.
267. Nagao, M.A., K. Kanegawa, and W.S. Sakai. 1980. Accelerating palm seed germination with gibberellic acid, scarification, and bottom heat. *HortScience* 15(2):200–201.
268. Neuner, H. and R. Beiderbeck. 1993. In vitro propagation of *Salix caprea* L. by single node explants. *Silvae Genet.* 42(6):308–310.
269. Nicholson, R. 1984. Propagation notes on *Cedrus deodara* 'Shalimar' and *Calocedrus decurrens. Plant Propagator* 30:5–6.
270. ———. 1987. Enkianthus—a worthy genus waits to emerge from companion-plant status. *Amer. Nurs.* 166:91–97.
271. Nobre, J. 1994. In vitro shoot proliferation of *Myrtus communis* L. from field-grown plants. *Scientia Hort.* 58:253–258.
272. Oliphant, J.L. 1992. Micropropagation of *Nerium oleander. Comb. Proc. Intl. Plant Prop. Soc.* 42:288–289.
273. Osborne, R., and J. Van Staden. 1987. In vitro regeneration of *Stangeria eriopus. HortScience* 22: 13–26.
274. Pair, J.C. 1986. Propagation of *Acer truncatum,* a new introduction to the southern great plains. *Comb. Proc. Intl. Plant Prop. Soc.* 36:403–13.
275. ———. 1986. New plant introductions with great stress tolerance to conditions in the southern great plains. *Comb. Proc. Intl. Plant Prop. Soc.* 36:351–55.

276. ———. 1992. Evaluation and propagation of lacebark elm selections by hardwood and softwood cuttings. *Comb. Proc. Intl. Plant Prop. Soc.* 42:431–435.
277. ———. 1995. *Acer tuncatum. Nurs. Manag. Prod.* 11(2):27.
278. Parr, G. 1983. Summer grafting of golden robinia. *Comb. Proc. Intl. Plant Prop. Soc.* 33:164–67.
279. Patel, S.I. 1983. Propagation of some rare tropical plants. *Comb. Proc. Intl. Plant Prop. Soc.* 33:573–80.
280. Pelleh, N.E., and K. Alpert. 1985. Rooting softwood cuttings of mature *Betula papyrifera.Comb. Proc. Intl. Plant Prop. Soc.* 35:519–25.
281. Perry, D.K. 1987. Successfully growing Proteaceae. *Comb. Proc. Intl. Plant Prop. Soc.* 37:112–115.
282. Phipps, H.M., D.A. Belton, and D.A. Netzer. 1977. Propagating cuttings of some Populus clones for tree plantations. *Plant Propagator* 23(4):8–11.
283. Picchioni, G.A. and F.T. Davies, Jr. 1990. Micropropagation of *Pistacia atlantica* shoots from axillary buds. *N. Amer. Plant Propagator.* 2(1):14–15.
284. Pickerill, J.D. 1992. Overwintering rooted cuttings of *Viburnum carlesii. Comb. Proc. Intl. Plant Prop. Soc.* 42:468–469.
285. Pinney, J.J. 1970. A simplified process for grafting junipers. *Amer. Nurs.* 131(10):7, 82–84.
286. Ponchia, G., and B.H. Howard. 1988. Chestnut and hazel propagation by leafy summer cuttings. *Acta Hort.* 227:236–41.
287. Pontikis, C.A., and P. Melas. 1986. Micropropagation of *Ficus carica* L. *HortScience* 21:153.
288. Poole, R.T., and C.A. Conover. 1984. Propagation of ornamental *Ficus* by cuttings. *HortScience* 19:120–21.
289. Powell, J.C. 1985. Production of × *Cupressocyparis Leylandii. Comb. Proc. Intl. Plant Prop. Soc.* 35:722–23.
290. Proebsting, W.E. 1984. Rooting of Douglas-fir stem cuttings: Relative activity of IBA and NAA. *HortScience* 19:845–56.
291. Prutpongse, P. and P. Gavinlertvatana. 1992. In vitro micropropagation of 54 species from 15 genera of bamboo. *HortScience* 27:453–454.
292. Purcell, G.V. 1973. The budding of *Hamamelis. Comb. Proc. Intl. Plant Prop. Soc.* 23:129–32.
293. Putland, L. 1994. The best way to propagate *Clematis. Nursery Manag.* 10(8):55–58.
294. Rajesekaran, P. 1994. Production of clonal plantlets of *Grevillea robusta* in in vitro culture via axillary bud activation. *Plant Cell Tissue Org. Cult.* 39:277–279.
295. Ranney, T.G. and R.E. Bir. 1994. Comparative flood tolerance of birch rootstocks. *J. Amer. Soc. Hort. Sci.* 119:43–48.
296. Ranney, T.G. and E.P. Whitman. 1995. Growth and survival of 'Whitespire' Japanese birch grafted on rootstocks on five species of birch. *HortScience* 30(3):521–522.
297. Rauch, F.D. 1994. Palm seed germination. *Comb. Proc. Intl. Plant Prop. Soc.* 44:304–307.
298. Raviv, M. and E. Putievsky. 1989. The effect of intree position and type of cutting on rooting of bay laurel (*Laurus nobilis* L.) stem cuttings. *N. Amer. Plant Propagator.* 1(3):14–15.
299. Richard, M.A. 1984. Propagation of bamboo by vegetative means. *Comb. Proc. Intl. Plant Prop. Soc.* 34:440–43.
300. Ridgway, D. 1984. Propagation and production of *Garrya elliptica. Comb. Proc. Intl. Plant Prop. Soc.* 34:261–65.
301. Ripphausen, F. 1989. Propagation and nutrition of daphne cuttings and tissue culture plantlets. *Comb. Proc. Intl. Plant Prop. Soc.* 39:305–308.
302. Ritchie, G.A. 1993. Production of Douglas-fir, *Pseudotsuga menziesii* (Mirb.) Franco, rooted cuttings for reforestation by Weyerhaeuser company. *Comb. Proc. Intl. Plant Prop. Soc.* 43:284–288.
303. ———. 1994. Commercial application of adventitious rooting to forestry. In *Biology of adventitious root formation,* T.D. Davis and B.E. Haissig, eds. New York: Plenum Press.
304. Rogers, C.S. 1994. Perennial Preservation. *Amer. Nurs.* 179(6):64–67.
305. Rodríquez, R. 1982. In vitro propagation of *Castanea sativa* Mill. through meristem-tip culture. *HortScience* 17:888–89.
306. Ross, D.M. 1977. Rose rootstock, 'Dr. Huey,' in South Australia. *Comb. Proc. Intl. Plant Prop. Soc.* 27:562–63.
307. Rout, G.R. and P. Das. 1994. Somatic embryogenesis and in vitro flowering of 3 species of bamboo. *Plant Cell Reports* 13:683–686.
308. Rumbal, J. 1992. The development of cutting propagation of *Camellia reticulata* hybrids. *Comb. Proc. Intl. Plant Prop. Soc.* 42:295–296.
309. Ruter, J.M. and D.L. Ingram. 1991. Germination and morphology of *Sophora secundiflora* seeds following scarification. *HortScience* 26(3):256–257.
310. Ryan, G.F. 1966. Grafting *Eucalyptus ficifolia. Plant Propagator* 12(2)4–6.
311. Sabo, J.E. 1976. Propagation of *Taxus* in Northern Ohio. *Comb. Proc. Intl. Plant Prop. Soc.* 26:174–76.
312. Sachs, R.M., C. Lee, J. Ripperda and R. Woodward. 1988. Selection and clonal propagation of *Eucalyptus. Calif. Agr.* 42:27–31.
313. Salter, C.E. 1970. *Clematis armandii* grafting. *Comb. Proc. Intl. Plant Prop. Soc.* 20:330–32.
314. Samartin, A. 1984. In vitro propagation of *Camellia japonica seedlings. HortScience* 19:225–26.
315. Samyn, G.L.J. 1993. In vitro propagation of ponytail palm: producing multiple-shoot plants. *HortScience* 28(3):225.

316. Sankhla, D., T.D. Davis, and N. Sankhla. 1994. Thidiazuron-induced in vitro shoot formation from roots of intact seedlings of *Albizzia julibrissin. J. Plant Growth Reg.* 14:267–272.
317. Santana, C. 1993. Direct stick Rhododendron production. *Comb. Proc. Intl. Plant Prop. Soc.* 43:293.
318. Saul, G.H., and L. Zsuffa. 1978. Vegetative propagation of elms by green cuttings. *Comb. Proc. Intl. Plant Prop. Soc.* 28:490–94.
319. Sallee, K. 1989. Desert broom—drought-tolerant shrub of Southwest deserves respect in the landscape. Nurs. Manag, 5(3):106–109.
320. Savella, L. 1981. Propagating pink dogwoods from rooted cuttings. *Comb. Proc. Intl. Plant Prop. Soc.* 30:405–7.
321. Schreiber, L.R., and M. Kawase. 1975. Rooting of cuttings from tops and stumps of American elm. *HortScience* 10(6):615.
322. Scott, A. 1976. A successful technique for grafting hibiscus. *Comb. Proc. Intl. Plant Prop. Soc.* 26:389–91.
323. Shugert, R. 1985. Taxus production in the U.S.A. *Comb. Proc. Intl. Plant Prop. Soc.* 35:149–53.
324. Simon, R.A. 1986. A survey of hardy bamboos: Their care, culture and propagation. *Comb. Proc. Intl. Plant Prop. Soc.* 36:528–31.
325. Simpson, R.C. 1981. Propagating deciduous holly. *Comb. Proc. Intl. Plant Prop. Soc.* 30:338–42.
326. Sen, S., M.E. Magallanes-Cedeno, R.H. Kamps, C.R. McKinley and R.J. Newton. 1994. In vitro micropropagation of Afghan pine. *Can. J. For. Res.* 24:1248–1252.
327. Serres, R., P. Read, W. Hackett and P. Nissen. 1990. Rooting of American chestnut microcuttings. *J. Environ. Hort.* 8(2):86–88.
328. Simpson, T.R. 1988. Propagating and growing crabapples by budding. *Comb. Proc. Intl. Plant Prop. Soc.* 38:537–541.
329. Singha, S. 1982. In vitro propagation of crab apple cultivars. *HortScience* 17:191–92.
330. Smith, C.C. and J.A. Jernstedt. 1989. In vitro development of adventitious shoots in *Euonymus alatus* (Celastraceae). *Scientia Hort.* 41:161–169.
331. Smith, E.Y., and C.W. Lee. 1983. Propagation of Arizona rosewood by stem cuttings. *HortScience* 18:764–65.
332. Smith, M.A.L. and A.A. Obdeidy. 1991. Micropropagation of a mature, male Kentucky coffeetree. *HortScience.* 26(11):1426.
333. Smith, M.N. 1985. Propagating *Ceanothus. Comb. Proc. Intl. Plant Prop. Soc.* 35:301–5.
334. Smith, R.C. 1990. *Asarum europaeum. Amer. Nurs.* 172(8):166.
335. Snir, I. 1982. In vitro propagation of sweet cherry cultivars. *HortScience* 17:192–93.
336. Stanford, G. 1982. *Ungnadia speciosa* (Mexican buckeye). *Plant Propagator* 28(2):5–6.
337. Starrett, M.C., F.A. Blazich, J.R. Acedo, and S.L. Warren. 1993. Micropropagation of *Pieris floribunda. J. Environ. Hort.* 11(4):191–195.
338. Starrett, M.C., F.A. Blazich, and S.L. Warren. 1993. Seed germination of *Pieris floribunda:* influence of light and temperature. *J. Environ. Hort.* 10(2): 121–124.
339. Still, S.M. and S. Zanon. 1991. Effects of K-IBA rates and timing on rooting percentage and root quality of *Amelanchier laevis. J. Environ. Hort* 9(2):86–88.
340. Stoltz, L.P. 1984. In vitro propagation and growth of *Hydrangea. HortScience* 19:717–19.
341. Strong, M.E., and F.T. Davies, Jr. 1982. Influence of selected vesicular-arbuscular mycorrhizal fungi on seedling growth and phosphorus uptake of *Sophora secundiflora. HortScience* 17:620–21.
342. Sturrock, J.W. and J.D. Ferguson. 1989. Macro and micro propagation of Leyland cypress. *Comb. Proc. Intl. Plant Prop. Soc.* 39:285–290.
343. Struve, D.K., M.A. Arnold, and D.H. Chinery. 1987. Red oak whip production in containers. *Comb. Proc. Intl. Plant Prop. Soc.* 37:415–20.
344. Struve, D.K., M.A. Arnold, R. Beeson, Jr., J.M. Ruter, S. Svenson, and W.T. Witte. 1994. The copper connection: the benefits of growing woody ornamentals in copper-treated containers. *Amer. Nurs.* 179(4): 52–61.
345. Studebaker, D.W., D.M. Maronek and M. Oberly. 1988. Propagation methods affect Taxus cuttings and liner quality. *Comb. Proc. Intl. Plant Prop. Soc.* 38:550–554.
346. Sutter, E., and P. Barker. 1985. In vitro propagation of mature *Liquidambar styraciflua. Plant Cell Tissue Org. Cult.* 5:13–21.
347. Ticknor, R.L. 1969. Review of the rooting of pines. *Comb. Proc. Intl. Plant Prop. Soc.* 19:132–37.
348. Tietje, W.D., J.H. Foott and E.L. Labor. 1990. Grafting California oaks. *Calif. Agr.* 44(2):30–31.
349. Tipton, J.L. 1990. Vegetative propagation of Mexican redbud, larchleaf goldenweed, littleleaf ash, and evergreen sumac. *HortScience* 25(2):196–198.
350. ———. 1992. Requirements for seed germination of Mexican redbud, evergreen sumac, and mealy sage. *HortScience.* 27(4):313–316.
351. ———. 1995. Easy-to-grow *Chilopsis* lends an ethereal look to desert vistas. *Nurs. Manag. Prod.* 11(5):14–15.
352. Tisserat, B. 1979. Propagation of the date palm (L.) in vitro. *J. Exp. Bot.* 30:1275–1283.
353. Tubesing, C.E. 1985. Cutting propagation of *Paxistima myrsinites, Vaccinium ovatum, Ribes sanguineum* and *Acer glabrum subsp. Douglasii. Comb. Proc. Intl. Plant Prop. Soc.* 35:293–96.

354. ———. 1987. Chip budding of magnolias. *Comb. Proc. Intl. Plant Prop. Soc.* 37:377–79.

355. Uno. S., and J.E. Preece. 1987. Micro- and cutting propagation of 'Crimson Pygmy' barberry. *HortScience.* 22:488–91.

356. Van de Pol, P.A., and A. Breukelaar. 1982. Stenting of roses: A method for quick propagation by simultaneously cutting and grafting. *Scientia Hort.* 17:187–96.

357. Vanstone, D.E. 1978. Basswood *(Tilia americana)* seed germination. *Comb. Proc. Intl. Plant Prop. Soc.* 28:566–69.

358. Van Veen, T. 1971. The propagation and production of rhododendrons. *Amer. Nurs.* 133(8):15–16, 52–58.

359. Vertrees, J.D. 1987. *Japanese maples.* Portland, Oreg.: Timber Press.

360. ———. 1991. Understock for rare *Acer* species. *Comb. Proc. Intl. Plant Prop. Soc.* 41:272–275.

361. Vieitez, A.M., A. Ballester, M.L. Vieitez, and E. Vieitez. 1983. In vitro plantlet regeneration of mature chestnut. *J. Hort. Sci.* 58:457–63.

362. Vieitez, A.M., M.C. San-Jose, and E. Vieitez, 1985. In vitro plantlet regeneration from juvenile and mature *Quercus robur* L. *J. Hort. Sci.* 60:99–106.

363. Vieitez, A.M., M. Concepción Sanchez, J.B. Amo-Marco and A. Ballester. 1994. Forced flushing of branch segments as a method for obtaining reactive explants of mature *Quercus robur* trees for micropropagation. *Plant Cell Tissue Org Cult.* 37:287–295.

364. Wallis, J.S. 1976. The propagation and training of standard fuchsias. *Comb. Proc. Intl. Plant Prop. Soc.* 26:346–48.

365. Wang, H.C. and N.A. Reichert. 1992. In vitro propagation of modern roses. *Comb. Proc. Intl. Plant Prop. Soc.* 42:398–403.

366. Wang, Y.T. 1989. Effect of water salinity, IBA concentration and season of rooting on Japanese boxwood cuttings. *Acta. Hort.* 246:191–198.

367. ———. 1991. Enhanced germination of *Sophora secundiflora* seeds. *Subtrop. Plant Sci.* 44:37–40.

368. Wang, Y.T. and R.E. Rouse. 1989. Rooting live oak rhizomic shoots. *HortScience* 24(6):1043.

369. Ward, P.W. 1993. The production of *Robinia pseudoacacia* 'Frisia.' *Comb. Proc. Intl. Plant Prop. Soc.* 43:357–358.

370. Warren, K. 1989. A crab apple system. *Amer. Nurser.* 170(5):31–35.

371. Warren, P. 1973. Propagation of *Cercis* cultivars by summer budding. *Plant Propagator* 19(3): 16–17.

372. ———. 1995. *Cercis canadensis* ssp. *texensis* 'Oklahoma.' 1995. *Nurs. Manag. & Product.* 11(4):13.

373. Wasley, R. 1979. The propagation of *Berberis* by cuttings. *Comb. Proc. Intl. Plant Prop. Soc.* 29:215–16.

374. Watkins, J.V. 1972. Jacaranda. *Horticulture* 50(5): 22–23.

375. Wedge, D. 1977. Propagation of hybrid lilacs. *Comb. Proc. Intl. Plant Prop. Soc.* 27:432–36.

376. Welander, M., E. Jansson and H. Lindqvist. 1989. In vitro propagation of *Populus* × *wilsocarpa*—a hybrid of ornamental value. *Plant Cell Tissue Org. Cult.* 18:209–219.

377. Wells, J.S. 1980. How to propagate Japanese maples. *Amer. Nurs.* 151(9):14, 117–20.

378. ———. 1981. A history of rhododendron cutting propagation. *Amer. Nurs.* 154(9):14–15, 114–26.

379. ———. 1985. *Plant propagation practices.* Chicago: American Nurseryman Publishing Co.

380. Whalley, D.N. 1975. Propagation of Commelin elm by hardwood cuttings in heated bins. *Plant Propagator* 21(3):4–6.

381. ———. 1979. Leyland cypress—rooting and early growth of selected clones. *Comb. Proc. Intl. Plant Prop. Soc.* 29:190–202.

382. Whitten, M. 1994. A better Virginia pine—send in the clones. *Nurs. Manag.* 10(7):39–42.

383. ———. 1994. Plan ahead: here's a southwest shrub to add to your 1996 catalogs. *Nurs. Mang.* 10(10):20–23.

384. Wiegrefe, S.J. 1994. Overcoming dormancy in maple seeds without stratification. *N. Amer. Plant Prop.* 6(3):8–11.

385. Wilkins, L.C., W.R. Graves, and A.M. Townsend. 1995. Development of plants from single-node cuttings differs among cultivars of red maple and Freeman maple. *HortScience.* 30(2):360–362.

386. Wisniewski, L.A., L.J. Frampton, and S.E. McKeand. 1986. Early shoot and root quality effects on nursery and field development of tissue-cultured loblolly pine. *HortScience* 21:1185–86.

387. Woodward, S. 1993. Micropropagation, rooting and survival of *Actinidia kolomikta. N. Amer. Plant Prop.* 5(3):9–12.

388. Yusnita, S., R.L. Geneve and S.T. Kester. 1990. Micropropagation of white flowering eastern redbud (*Cercis canadensis* var. *alba* L.). *J. Environ. Hort.* 8(4):177–179.

389. Zenkteler, M. and B. Stefaniak. 1991. The de novo formation of buds and plantlets from various explants of *Ailanthus altissima* Mill. culture in vitro. *Biolog. Plant.* 33(4):332–336.

390. Zhang, B. and L.P. Stoltz. 1989. Shoot proliferation of *Euphorbia fulgens* in vitro affected by medium components. *HortScience* 24(3):503–504.

391. Zhang, Z.M., and F.T. Davies, Jr. 1986. In vitro culture of crape myrtle. *HortScience* 21:1044–45.

SUPPLEMENTARY READING

American Nurseryman Magazine. Chicago: American Nurseryman Publishing Co.

Burger, D.W. and C.I. Lee 1994. Development of a computer database for vegetative propagation of trees and shrubs. *J. Environ. Hort.* 12(2):87–89.

DIRR, M.A. 1990. *Manual of woody landscape plants,* 4th ed. Champaign, Ill.: Stipes Publ. Co.

DIRR, M.A., and C.W. HEUSER, JR. 1987. *The reference manual of woody plant propagation.* Athens, Ga.: Univ. Press.

DUNMIRE, J.R., ed. 1988. *Western garden book.* Menlo Park, Calif.: Lane.

FORDHAM, A.J., and L.S. SPRAKER. 1977. Propagation manual of selected gymnosperms. *Arnoldia* 37:188.

INTERNATIONAL PLANT PROPAGATORS' SOCIETY. *Combined Proceedings of annual meetings.*

KRUSSMAN, G. 1984. *Manual of cultivated broadleaved trees and shrubs.* Portland, Oreg.: Timber Press.

LAMB, J.G.D., J.C. KELLY, and P. BOWBRICK. 1985. *Nursery stock manual.* London: Grower Books.

MACDONALD, B. 1986. *Practical woody plant propagation for nursery growers,* Vol 1. Portland, Oreg.: Timber Press.

MCCLINTOCK, E., and A.T. LEISER. 1979. An annotated checklist of woody ornamental plants of California, Oregon and Washington. *Univ. Calif. Div. Agr. Sci. Priced Publ. 4091.*

MCMILLAN-BROWSE, P.D.A. 1979. *Hardy woody plants from seed.* London: Grower Books.

New Zealand Journal of Forestry Sciences. 1974. Special issue on vegetative propagation of conifers and hardwoods, 4 (*2*).

RECHT, C. and M.F. WETTERWALD. 1992. *Bamboos.* Portland, Oreg.: Timber Press.

SCHOPMEYER, C.S., ed. 1974. *Seeds of woody plants in the United States.* U.S. Dept. Agr. For. Ser. Handbook 450. Washington, D.C.: U.S. Govt. Printing Office.

STEWART, L. 1994. *A guide to palms and cycads of the world.* New York, N.Y.: Harper Collins, Inc.

WRIGLEY, J.W., and M. FAGG. 1979. *Australian native plants: Propagation, cultivation, and use in landscaping.* Sydney, Australia: William Collins Publishers.

WORLD WIDE WEB. The electronic yellow pages of the future. Information is available from plant propagation to nurseries to plant species.

YOUNG, J.A. and C.G. YOUNG. 1992. *Seeds of Woody Plants in North America.* Portland, Oreg.: Dioscorides Press.

ZIMMERMAN, R.H., R.J. GRIESBACH, F.A. HAMMERSCHLAG, and R.H. LAWSON. 1986. *Tissue culture as a plant production system for horticultural crops.* Dordrecht: Martinus Nijhoff Publishers.

21

Propagation of Selected Annuals and Herbaceous Perennials Used as Ornamentals

Herbaceous plants are classified as *annuals, biennials,* or *perennials,* although the differences among these types may not be obvious. They may also be classified as *hardy, semi-hardy,* or *tender.* In general, the propagation procedures for such plants depend upon their categories and the locality where they are to be grown. In the following list of plants, seed germination data are given for some species, including suggested approximate temperatures that should give the most rapid and complete germination, along with the expected germination time. A single figure indicates a constant temperature. The propagation methods indicated will serve as a guide, but some variation from these methods may be necessary with individual cultivars (*18,21*).

Achillea spp. Yarrow. Hardy perennial. Seeds germinate in one to two weeks at 21° C (70° F), but seeds of *A. filipendulina,* which is used as a field-grown cut flower crop (*12*), sometimes produce inferior plants. Other species can be seed-propagated. Fast propagation is common from summer softwood cuttings. Propagation by division is easy and necessary for good garden performance (*190*).

Achimenes spp. Tender perennial (*67*). Seeds germinated in a warm greenhouse can be used for propagating this species. Plants grow from small scaly rhizomes, which can be divided for propagation. Softwood cuttings in spring or leaf cuttings in summer can be rooted. Partially dried leaf scales can be planted and rooted.

Aconitum spp. Monkshood (*190*). Hardy perennial. Seeds often show dormancy and must be moist-chilled below 5° C (41° F) for six weeks before planting. Considered difficult to propagate by seed. Plants have tuberous roots that can be divided, but once established they should not be transplanted. All parts of the plants are poisonous.

Adiantum. *See Fern.*

Aegopodium spp. Goutweed. A hardy perennial used as a groundcover. Propagated by division.

African Violet. *See Saintpaulia ionantha.*

Agapanthus spp. Lily-of-the-Nile. Tender perennial. Grown for blue lilylike flowers. The thick rhizomes can be divided to produce new plants.

Agave spp. Many species of succulents, including the century plant. Perennial. Seeds should be sown in sandy soil when mature. Reproduces vegetatively by offsets from base of plant, or from the flower stalk of some species; these are removed along with roots and repotted in spring. Some species produce bulbils that can be used for propagation. Agave are commercially mass-produced by tissue culture.

Ageratum houstonianum. Ageratum (*146*). Half-hardy annual. Blue and white flowering bedding plants. Taller forms are grown as cut flowers. Seeds germinate in one to two weeks at 26 to 29° C (80 to 85° F). Ageratum may also be propagated by cuttings.

Aglaonema spp. An important foliage plant that is easily propagated by canes (long stems), shoot cuttings, division, or seeds. Canes should be treated carefully as delicate leaf cuttings. Rooting is enhanced with IBA and bottom heat [24 to 28° C (75 to 80° F)] (*156*).

Agrostemma githago. Corncockle. Hardy annual. Seeds germinate in two to three weeks at 20° C (68° F).

Ajuga spp. Bugle flower. Hardy perennial. Blue or pink flowers on a spreading groundcover. Naturally layers itself by stolons. Can be propagated by seeds, cuttings, divisions, or tissue culture. Variegated foliage types are from cuttings (*190*).

Allium spp. Ornamental onion; also onion, chives, and garlic. Propagated by seed. Plants grow from bulbs, which produce offsets. Clumps can also be divided. Many species produce bulbils.

Aloe spp. Succulents of the lily family. Propagated by seed in well-drained sandy soil. Germination takes place in three to four weeks at 20 to 24° C (68 to 75° F). Plants produce offshoots that can be detached and rooted. Plants with long stems can be made into cuttings, which should be exposed to air for a few hours to allow cut surfaces to suberize. These species are commercially micropropagated. *Aloe barbadensis* can be micropropagated from shoot explants (*136,144*).

Alstroemeria spp. Parrot lily (*67*). Half-hardy perennial. Grown commercially as a cut-flower and pot plant. Propagated by division of the fleshy rhizome. Rhizomes are also multiplied in tissue culture to produce disease-free plants (*30*). *Alstroemeria* as pot plants are propagated from seed. Fresh seeds are dormant, but moistened one-year-old seeds germinate after four weeks of 18 to 25° C (65 to 75° F) followed by four weeks of 7° C (45° F) conditions (*111*).

Alcea rosea. Hollyhock. Half-hardy biennial. Seeds germinate in two to three weeks at 20° C (68° F). Where winters are not too severe, sow seeds in summer, transplant in fall for bloom the following year, or sow seeds in warm greenhouse in winter and transplant outdoors.

Sweet alyssum. *See Lobularia.*

Alyssum saxatile. Goldentuft (*146*). Hardy perennial. Seeds germinate in three to four weeks at 15 to 21° C (60 to 70° F). Sow in summer for bloom the following year. Germination may be stimulated by light or exposure of moist seeds at 15° C (50° F) for five days (*14*). Propagate by division or by softwood cuttings in spring. Double forms must be propagated by cuttings or division.

Alyxia olivaeformis. Maile. An important foliage plant which is native to Hawaii. It is propagated almost exclusively by seeds which need to be depulped (*196*). Rooting of single-node maile stem cuttings is improved by removing one-half leaf surface area, placing greenhouse-grown cuttings in water prior to treatment, and propagating in a shady cloth-covered greenhouse. Cuttings were irrigated once daily with spray stakes. A five-second quick-dip in 3,000 to 8,000 ppm IBA was most effective (*197*).

Amaranthus caudatus. Love-lies-bleeding (*146*). Half-hardy annual. Seeds germinate in one to two weeks at 21 to 24° C (70 to 75° F). Light may increase germination (*14*). Sow in warm greenhouse for later transplanting or sow out-of-doors when frost danger is past.

***A. gangeticus tricolor,* Joseph's coat.** Same as for *A. caudatus*. Sensitive to excess water.

Amaryllis belladonna. Belladonna lily (*67*). Perennial. Grows from bulbs outdoors in mild areas or in pots in cold climates. Propagate by bulb cuttings, separation of bulbs, or tissue culture (*51, 64*).

Amsonia tabernaemontana. Willow amsonia. Hardy perennial. Pale blue flowers above willowlike foliage. Propagation is from seed or summer softwood cuttings. Seed should be stratified at 2 to 5° C (34 to 40° F) for four to six weeks.

Anchusa capensis. Bugloss. Hardy annual or biennial. Seeds germinate in two to three weeks at 20 to 30° C (68 to 86° F). Sow seeds in summer for bloom next year or plant in greenhouse in winter for later transplanting to garden.

A. azurea. Perennial. Selected clones best propagated by root cuttings or clump division (*69*).

Anemone spp. Poppy anemone ***(A. coronaria)***. Tender perennials. Seeds germinate in five to six weeks at 20° C (68° F) and may be sensitive to higher temperatures (*14*). Plants develop clusters of small, clawlike tuberous roots.

A. blanda. Greek windflower (*67*). Hardy perennial produced from a tuber. Propagated from seed or by division of the tuber into sections.

A. japonica. Japanese anemone. Hardy perennial. Since seeds do not come true, cultivars are propagated by division or by root cuttings. Roots are dug in fall and cut into 5-cm (2-in.) pieces, which are laid in flats or in a cold frame, then covered with 2.5 cm (1 in.) of soil. Plants can be potted after shoots appear.

A. pulsatilla. Pasque flower. Hardy perennial. Seeds germinate in five to six weeks at 20° C (68° F), but may be sensitive to high temperature. Plants can be divided or propagated from root cuttings taken in spring. There is an export market for *Anemone* tubers (*82*).

Anigozanthus spp. Kangaroo paw (*123*). This native perennial Australian genus is used for cut-flower production and for containerized plants. Seed supplies are often scarce and germination rates of available seed are usually low and variable for many species. Hot-water and chemical pretreatment can be used to improve germination. Some hybrids are sterile and do not set seed at all. Clumps of rhizomes can be divided, but the rate of multiplication is low and unreliable. The most effective means of commercial propagation is through micropropagation (*67*).

Anthemis spp. Golden Marguerite, camomile. Hardy perennial. Seeds germinate in one to three weeks at 20° C (68° F). Plants can be divided or propagated by stem cuttings.

Anthurium andraeanum. Anthurium. Remove offshoots with attached roots from the parent plant or root two- or three-leaved terminal cuttings under mist. Anthurium can be propagated by *in vitro* methods using a vegetative bud explant (*118*). Seed propagation is a lengthy process requiring 1½ to 3 years for flowering, and cultivars do not come true from seed (*90*).

Antirrhinum majus. Snapdragon. Tender perennial, treated as an annual. Seeds germinate in one to two weeks with 27° C (80° F) days and 24° C (75° F) nights (*45*). Chilling seeds at 5° C (40° F) can improve germination. Light seeds for the first three days, then provide dark to allow radicle growth. Move to light when seedlings emerge. Start indoors for later outdoor planting (in fall in mild climates or in spring in severe winter areas). Softwood cuttings root readily. This species is tissue cultured (*147, 153*).

Aquilegia spp. Columbine. Hardy perennials. Seeds germinate in three to four weeks at 21 to 24° C (70 to 75° F). A short period of stratification for three to four weeks, moist-chilling at 5° C (41° F), can improve germination (*77*), but may not be necessary for all species. Many of the columbines come true from seed.

Arabis spp. Rockcress. Hardy perennials. Seeds germinate in three to four weeks at 18 to 21° C (65 to 70° F) and may respond to light (*14*). Softwood cuttings taken from new growth immediately after bloom root readily. Plants can be divided in spring or fall.

Arctotis stoechadifolia. African daisy. Half-hardy annual. Seeds germinate in two to three weeks at 20° C (68° F). Sow indoors for later transplanting.

Armeria spp. Thrift. Hardy evergreen perennials. Seeds germinate in three to four weeks at 15 to 21° C (60 to 70° F). Can also be propagated by clump division in spring or fall.

Artemisia spp. Hardy perennial. *A. ludoviciana* can be used as a foliage plant and is propagated by division or stem cuttings. *A. schmidtiana* (wormwood) is a hardy perennial used as a specimen plant and is propagated by stem cuttings, rather than by division.

Arum italicum 'Pictum.' Painted arum. Hardy perennial produced from a tuber. Evergreen foliage and naked red seed heads are attractive for the perennial garden. Propagated by division of the tuber. Seed requires stratification.

Aruncus dioicus (A. sylvester). Goat's beard. A hardy perennial used as a specimen or border plant. Seeds have a cold stratification requirement of 5° C (40° F) for four weeks. Usually propagated by division.

Asclepias tuberosa. Butterfly weed. Hardy perennial. Seeds germinate in three to four weeks at 21 to 24° C (70 to 75° F). Fresh seed may need chilling. Vegetative propagation is from 3-cm-long root cuttings (*72*). Plants should not be disturbed once established.

A. curassavica. Bloodflower. Tropical perennial. Propagated by seed or by rooting softwood cuttings. Long taproot makes division difficult.

Asparagus asparagoides. *A. plumosus,* fern asparagus, *A. sprengeri,* Sprenger asparagus. Tender perennials. Propagated by seeds, which germinate in three to four weeks at 20 to 30° C (68 to 86° F). Sow seeds soon after they ripen, since seeds are short-lived (*21*). Cuttings can be made of young side shoots taken from old plants in spring; clumps can be divided.

Asplenium nidus. Bird's nest fern. *See Fern.*

Aster spp. Hardy perennials. Seeds germinate in two to three weeks at 18 to 21° C (65 to 70° F). Cultivars are propagated by lifting clumps in fall and dividing into rooted sections, discarding the older parts or stem cuttings. *A. frikartii* 'Monch' is micropropagated commercially.

Astilbe spp. Astilbe. Hardy perennials. Propagated by division in early spring when 2.5 cm (1 in.) tall and then again following flowering (*23*). Seed germination is slow and produces a mixed progeny. Germination takes three to four weeks at 16 to 21° C (60 to 70° F) (*190*).

Athrium spp. Painted fern. *See Fern.*

Astrantia spp. Masterwort. Perennial with unusual and attractive flowers. Generally propagated by division. Seeds require a cold stratification.

Aubrieta deltoidea. Aubrieta. Hardy perennials, sometimes treated as annuals. Seeds germinate in two to three weeks at 13° C (55° F). Clumps are difficult to divide; cuttings may be taken immediately after blooming.

Aucuba japonica 'Variegata.' Gold Dust Plant. Propagated by leafy stem cuttings under mist, which root easily, or by root cuttings. Grows well in shade.

Baptisia spp. False Indigo. Hardy perennial. Seed germination at 20° C (68° F) is slow and uneven. *Baptisia* is a hard-seeded legume and requires scarification. Gather seeds when ripe, scarify, and sow outdoors to overwinter. Clumps are difficult to divide because of a long taproot.

Begonia spp. Begonia (*173, 174*). Tropical perennials. Seeds, which are very fine and need light, germinate in two to four weeks at 22° C (72° F). Best seed germination is at 28° C (82° F) for five to seven days followed by 25° C (78° F) until seedlings emerge (*146*). Sow on moist, light medium with little or no covering. Begonia species, tuberous begonias, and wax begonias are propagated by seed, but other types are propagated vegetatively.

Tuberous Begonias. In addition to seed propagation, these can be grown from tuberous stems, which are divided into sections, each bearing at least one growing point. Leaf, leaf-bud, and short-stem cuttings (preferably with a piece of tuberous stem attached) will root readily. Tissue culture propagation is also possible (*195*).

Fibrous-Rooted Begonias. Wax begonias, Christmas begonias, and others are propagated by leaf cuttings or softwood cuttings taken from young shoots in spring or summer. The cytokinin PBA was more effective in bud and shoot development from leaf cuttings than BA or kinetin (*63*).

Rhizomatous Types. Various species and cultivars, including *Rex begonia* plants, are divided or their rhizomes are cut into sections. Propagation is usually by leaf cuttings, but stem cuttings also will root. Treatment of leaf cuttings with a cytokinin increases the number of plantlets produced per leaf (*215*). *B. evansiana* produces small tubercles, which are detached and planted. Begonia can also be micropropagated using leaf petioles (*139*), petiole explants (*159, 175, 176, 215*), and somatic embryos (*150*).

Belamcanda chinensis. Blackberry lily. Hardy perennial. Summer-blooming orange blossoms are held above irislike foliage. Commercially propagated by seed at 24° C (75° F). Division is also possible while plants are dormant.

Bellis perennis. English daisy. Hardy perennial often treated as annual or biennial. Seeds germinate in one to two weeks at 21 to 24° C (70 to 75° F) and may respond to light (*14*). Clumps should be divided every year to prevent crowding.

Bergenia cordifolia. Hardy perennial. Evergreen cabbagelike foliage and attractive spring flowers. Can be propagated by seed or division. Seeds require chilling stratification of 5° C (41° F) for six to eight weeks. Germination is in two to three weeks at 21 to 24° C (70 to 75° F) in the light. *Bergenia* is commercially micropropagated.

Bleeding heart. *See Dicentra.*

Boltonia spp. Boltonia. Hardy perennial. Autumn-blooming perennial resembling asters. Seeds germinate in two to three weeks at 20° C (68° F). Commercially propagated by division while plants are dormant.

Bouvardia ternifolia (*97*). Scarlet bouvardia. An outstanding perennial with scarlet tubular flowers which bloom from midsummer to frost in Texas and New Mexico. It is propagated by semi-hardwood cuttings throughout the growing season.

Bromeliads. About 2,000 species of tropical herbs or subshrubs in 45 genera. The pineapple *(Ananas)* is the best-known. Propagation is mainly by seeds or by asexual division of lateral shoots, but micropropagation has been used successfully with some species (*93*). Conditions vary for the successful micropropagation of *Guzmania, Tillandsia,* and *Vriesea* species (*135*).

Brodieia spp. (Also *Triteleia*). Perennial plants grown as cut flowers and produced from a corm. Brodiaea is propagated from seed that requires stratification for eight weeks at 3° C (37° F) (*84*), or from cormels. Plants can also be propagated through liquid tissue cultures to develop corms (*100*).

Browallia spp. Amethyst flower (*146*). Tender, blue-flowered perennial often treated as an annual. Seeds germinate in two to three weeks at 24° C (75° F). Softwood cuttings can be taken in fall or spring. Can be used as flowering pot plant indoors in winter.

Brunnera macrophylla. Siberian bugloss. Hardy perennial. Blue forget-me-not flowers appear in the spring followed by large green leaves. Can be propagated by seed, division, or root cuttings.

Cactus (*49,151*). Large group of many genera, species, and some cultivars. Tender to semi-hardy perennials. Seed propagation can be used for most species, but seeds often germinate slowly. Sow fungicide-treated seed in well-drained sterile mixture and water sparingly, but do not allow medium to dry out. Pieces of stem can be broken off and rooted as cuttings (*204*), or small offsets, which root readily, can be removed. Allow offsets to dry for a few days to heal (suberize) cut surfaces before rooting. Cuttings require two to three weeks to heal. High humidity during rooting is unnecessary, but bottom heat is beneficial. Grafting is used to provide a decay-resistant stock for certain kinds and to produce unusual growth forms. For example, the pendulous *Zygocactus truncatus* is sometimes grafted on tall erect stems of *Pereskia aculeata.* Intergeneric grafts are usually successful. A type of cleft graft is used. The stem of the stock is cut off, and a wedge-shaped piece is removed. The scion is prepared by removing a thin slice from each side of the base; this is fitted into the opening in the stock. The scion is held in place with a pin or a thorn. A grafting adhesive can be used to adhere scions to stocks of transversally cut (tip-grafted) cactus (*218*). The completed graft is held in a warm greenhouse until healed (*46, 73, 218*) The development of cacti shoots by micropropagation can be extremely rapid in comparison with greenhouse-germinated seedlings (*16*) and where poor branching limits propagation by traditional vegetative propagation methods (*54, 98, 124*).

Caladium hybrids (*67*). This tropical perennial, grown for its strikingly colorful foliage, produces tubers. Propagation is by removing the tubers from the parent plant at the end of the four- to five-month dormancy period just before planting. Commercially, tubers are cut into 2-cm pieces (chips), each containing at least two buds ("eyes"). Caladiums do best out-of-doors when planted after the minimum night temperature is above 18° C (65° F) or as pot plants maintained with night temperatures of 18 to 21° C (65 to 70° F) and day temperatures of 24 to 29.5° C (75 to 85° F). Dried caladium seed has a short storage life. Seeds require light and temperatures between 25 to 30° C (68 to 86° F) (*37*).

Calamagrostis acutiflora. Feather reed grass. Hardy perennial. Upright grass with attractive flowering plumes. Propagation is by division in late spring.

Calceolaria spp. Pocketbook plant. Tender perennials are often grown as annuals. Seeds germinate in two to three weeks at 21° C (70° F). Propagation is also possible by softwood cuttings.

Calendula officinalis. Pot marigold. Hardy annual; gives winter bloom in mild climates from seed sown in late summer. Seeds germinate in one to two weeks at 21° C (70° F). Thin plants 30 cm (12 in.) apart.

Calla. *See Zantedeschia* spp.

Callistephus chinensis. China aster. Half-hardy annual. Seeds germinate in two to three weeks at 20° C (68° F). Plant only wilt-resistant types.

Caltha palustris. Marsh marigold. A hardy perennial used around water gardens or ponds. Propagation is by division.

Campanula carpatica. Tussock, Bellflower. Hardy perennial. Seeds germinate in two to three weeks at 20 to 30° C (68 to 86° F) and may respond to light (*14*).

C. lactiflora. Bellflower. Hardy perennial. Seeds germinate best after two to four weeks of stratification at 4° C (40° F).

C. medium. Canterbury bells. Hardy biennial. Seeds, which germinate in two to three weeks at 21° C (70° F), are sown in late spring or early summer for bloom the following year.

C. persicifolia. Peach bells. Hardy perennial. Seeds germinate in two to three weeks at 13 to 32° C (55 to 90° F) and may respond to light (*14*). Small offsets can be detached and rooted.

C. pyramidalis. Chimney bellflower. Hardy perennial, often treated as a biennial. Seeds germinate in two weeks at 20 to 30° C (68 to 86° F).

Many of the named cultivars of *Campanula* species cannot be produced by seed, so division or cuttings are used. Cuttings are produced from the rhizomatous growth of stock plants and rooting occurs from the etiolated base. Cuttings are placed in peat-perlite media, and given basal heat under glass or in a tunnel (England) (*178*).

Canna spp. Canna (*67*). Tender perennial. Cultivars do not come true from seed. Seeds, which have hard coats and must be scarified before planting, are germinated in a warm greenhouse. Cultivars are propagated by dividing the rhizome, keeping as much stem tissue as possible for each growing point. In mild climates this is done after the shoots die down in the fall or before growth starts in the spring. In cold climates, the plants are dug in fall, stored over winter, divided in spring, then started in sand or sandy soil for transplanting outdoors when frost danger is over.

Carnation. *See Dianthus caryophyllus.*

Catananche caerulea. Cupid's dart. Hardy perennial. Seeds germinate in two to four weeks at 20 to 30° C (68 to 86° F). Plants may be divided in the fall.

Catharanthus roseus. Vinca (*45*). Tender annual. Vinca is a major bedding plant grown from seed. Optimum temperature for germination is from 24 to 27° C (75 to 78° F) in the dark. Do not keep seeds too moist. Vinca is also the commercial source of alkaloids used for cancer research.

Celosia argentea. Cockscomb. Tender annual. Both plumed and cockscomb (fasciated) cultivars are available as bedding plants and cut flowers. Seeds germinate in one to two weeks at 26 to 30° C (80 to 86° F) and may respond to light (*14*).

Centaurea spp. Tender and hardy perennials. Seeds germinate in two to four weeks at 24 to 26° C (75 to 80° F). Cuttings can be rooted. Seeds of *C. cyanus* (cornflower or bachelor button) and *C. moschata* (sweet sultan) germinate in three to four weeks at 15 to 21° C (60 to 70° F). *C. hypoleuca* (knapweed), *C. macrocephala* (Globe centaurea), and *C. montana* are hardy perennials propagated by division or seed.

Centranthus ruber. Red valerian. Hardy perennial. Rose-colored flowers are produced throughout the summer. Propagated from seed. Seeds emerge in two to three weeks at 15 to 18° C (60 to 65° F). Stem cuttings are also possible.

Cephalotus follicularis. Australian pitcher plant. Perennial carnivorous plant. Usually propagated from IBA-treated leaf or stem cuttings. Also easily propagated by tissue culture (*1*).

Cerastium tomentosum. Snow-in-summer. Hardy perennial. Seeds germinate in two to four weeks at 20° C (68° F). This is easily propagated by division in the fall or by softwood cuttings in summer.

Cheiranthus cheiri (synonym is *Erysimum asperum*). Wallflower. Semi-hardy perennial often treated as a biennial. Seeds germinate in two to three weeks at 13° C (54° F) and may respond to light (*14*). Choice plants may be increased by cuttings taken in early summer.

Chelone spp. (*190*). Turtlehead. A hardy perennial used in wet areas. Propagation is by division or by cuttings. Seeds may require cold stratification for germination.

Chionodoxa spp. Glory-of-the-snow. Hardy perennial bulb. Primarily grown from seed. Ripe seeds are stored through the summer at 17° C (63° F) and sown outdoors in September to stratify over winter (*67*). Bulb cuttage (see *Hyacinthus*) and offsets are successful for vegetative propagation.

Chlorophytum comosum. Spider plant. Propagated mainly by planting miniature plants developing at ends of stolons. Stolon formation is under day-length control; short days (12 hours or less daily) promote stolon production (*83*). It can also be propagated by division.

Chrysanthemum carinatum, C. coronarium, and C. segetum Hybrids (Synonym is Dendranthemum). Many cultivars. Hardy annuals and perennials. Seeds germinate in two to four weeks at 20° C (68° F).

C. coccincum. Painted daisy. Propagate by seed as described above or by division.

C. parthenium. Feverfew. Hardy perennial usually grown as an annual. Start from seeds, as described above. Plants easily self-seed and can be divided.

C. superbum (maximum). Shasta daisy. Hardy perennial but often treated as a biennial, since it is short-lived. Propagated by seeds at 18 to 21° C (65 to 70° F). Division is from side shoots that have roots. Commercially micropropagated.

C. morifolium. Garden and greenhouse chrysanthemum; and *C. frutescens,* Marguerite. Hardy and semi-hardy perennials. After flowering, lateral shoots develop from the base of the flowering stems, particularly if the tops are cut back. When the new side shoots are 9 to 10 cm (3½ to 4 in.) long and firm but not woody, they are cut off and rooted as softwood cuttings under mist and with a treatment of indolebutyric acid rooting hormone. The best source of new cuttings is a mother block (or increase block) grown in an isolated area away from the producing area. Such plants are grown in programs designed to keep them pathogen- and virus-free and true-to-type (*21*). It is becoming more common for stock plants to be periodically replaced with new disease-indexed stock plants that have been regenerated through micropropagation (*34*). Softwood cuttings which are disease-indexed are then propagated by conventional means. Unrooted cuttings can be held for as long as 30 days at 0.5° C (33° F). For outdoor planting, cuttings may be taken in the same way. In areas with mild winters, cuttings can be taken in late winter for rooting and later transplanting to the garden. In cold-winter areas, the plants should be dug in the fall and brought into the greenhouse or cold frame; cuttings should be made in winter. The plants may also be left in place and divided in spring or fall. If cuttings are taken from the ends of stems high above ground, they are not likely to be infected with soil-borne insects and diseases.

C. nipponicum. Nippon daisy. Propagated by spring tip cuttings. Division of clumps is difficult.

Chrysanthemum is readily micropropagated by shoot-tip and petal-segment explants (*20, 110, 168*).

Clarkia spp. *See Godetia.*

Cleome spinosa. Spiderflower. Tender annual. Seeds germinate in one to two weeks at 26° C days and 21° C nights (80 to 70° F). Seeds may respond to light (*14*). Can also be propagated by division.

Codiaeum variegatum. Croton (*162*). Tropical perennial. Propagated by leafy terminal cuttings in spring or summer. Tall "leggy" plants can be propagated by air layering. Stem cutting root number (root initiation) was unaffected by (28° C; 83° F) bottom heat or increasing light intensity; however, root length was increased (*210*). During shipping of cuttings, exposure to light results in shorter roots. Unrooted cuttings can be shipped for five to ten days at 15 to 30° C (60 to 86° F) (*209*).

Colchicum autumnale. Autumn crocus, saffron. Hardy perennial that grows from corms (see *Crocus*). Seeds are sown as soon as they are ripe in the fall but may require chilling over winter to germinate. Several years are required for plants to reach flowering size.

Coleus blumei. (*21*) Tender perennials. Bedding plants for the shade. Grown for their colorful foliage. Seeds germinate in two to three weeks at 21 to 24° C (70 to 75° F) but seedlings can be variable. Selected individuals are propagated by softwood cuttings, which root easily.

Convallaria majalis. Lily-of-the-valley. Hardy perennial that grows as a rhizome, whose end develops a large underground bud, commonly called a "pip." In fall, the plants are dug, and the pip, with attached roots, is removed and used as the planting stock. Digging should take place in early autumn, with replanting completed by late autumn. Single pips may be stored in plastic bags in the refrigerator, then planted in late winter for spring bloom. Micropropagation has also been successful (*205*).

Cordyline spp. Ti *(C. terminalis)* is easily propagated by cuttings and by micropropagation (*65, 117*). Other *Cordyline* species can be seed-propagated. A vegetative propagation technique for *C. australis, C. kaspar,* and *C. pumillo* by division of the underground stems of stock plants has been described (*155*).

Coreopsis spp. Hardy annuals and perennials. Seeds, which germinate in two to three weeks at 20° C (68° F), may respond to light (*14*). Seed germination can be improved by growth regulator treatment or seed priming (*44, 169*). Perennial clumps can be divided in spring or fall. *C. verticillata* can be propagated by cuttings and is hardy to zone 3 (*57*).

Cortaderia selloana. Pampas grass. Best feathery plumes are found on female plants. Propagated by clump division. Micropropagation is possible from immature flower parts (*165*).

Cosmos bipinnatus and C. sulphureus. Half-hardy annuals. Seeds germinate in one to two weeks at 21° C (70° F) and may respond to light (*14*).

Crassula argentea. Jade plant. Can be propagated at any time by leaf or stem cuttings.

Crocosmia spp. Crocosmia. Hardy perennial used for cut flowers or as a border plant. Offsets that form at the base of corms are propagated.

Crocus spp. (*67*). Hardy perennials that grow from corms. Seeds germinate as soon as ripe in summer; several years are required for plants to flower. When leaves die in fall, plants are dug and corms and cormels are separated and planted. Tissue culture propagation is from corm fragments, isolated buds, or flower parts (*75, 92*).

Cucurbita pepo var. ovifera. Ornamental gourds. Tender annuals. Seeds germinate in one to three weeks at 20 to 30° C (68 to 86° F).

Cyclamen spp. (*67, 125, 130, 133*). Tender perennials. Plants grow from a large tuberous underground stem. Cyclamen is propagated best by seeds that germinate in three to four weeks in the dark at temperatures about 20° C (68° F), no higher than 22° C. Seeds are planted from midsummer to midwinter. Germination is best in a medium of peat moss to which pulverized limestone and mineral nutrients have been added (*216*). Seedlings require one to several years to flower and should be shaded in spring and summer to prevent leaf scald. The tubers can be divided for the production of a few plants identical to the parent. Short shoots of cyclamen with two to three leaves are easily rooted in three weeks when given a ten-second dip of 3,000 to 5,000 ppm K-IBA under intermittent mist, 21° C (70° F) basal heat, and 1 perlite:1 vermiculite rooting medium (*125*). Tissue culture is used to multiply F_1 hybrids (*171*).

Cymbalaria muralis. Kenilworth ivy. Semi-hardy perennial. Seeds germinate in one to four weeks at 12° C (54° F). Self-seeds readily. Softwood cuttings or clump division may be used.

Cynoglossom amabile. Chinese forget-me-not. Hardy biennial grown as an annual. Seeds germinate in two to three weeks at 20° C (68° F) and may respond to light (*14*).

Cypripedium. Lady slipper. *See Orchids.*

Dahlia. Tender perennials consisting of hundreds of cultivars. Commercially propagated by seed or stem cuttings. Seeds germinate in one to two weeks at 26 to 29° C (80 to 85° F) when planted indoors for later transplanting outdoors. Large plants can be dug in the fall before frost and stored over winter at 2 to 10° C (30 to 50° F), and covered with a material such as soil or vermiculite to prevent shriveling. In spring, when new sprouts begin to appear, divide the clumps so that each root section has at least one sprout. Plant outdoors when danger of frost is over. Dahlias can also be propagated by softwood or leaf-bud cuttings. Tissue culture is used to recover virus-free plants (*214*).

Day lily. *See Hemerocallis* spp.

Delphinium spp. Hardy perennials, usually propagated by seeds, which germinate in three to four weeks at 18 to 24° C (65 to 75° F). 'Giant Pacific' germinates better with alternating 26° C (75° F) day and 21° C (70° F) night temperatures. There may be some benefit to chilling dry seed at 3° C (35° F) for one week prior to sowing seeds (*39, 146*). Seeds are short-lived and should be used fresh, or stored in containers at low temperature and reduced moisture. Seeds are usually sown outdoors in spring or summer to produce plants that flower the following year. Delphiniums can be propagated by softwood cuttings (*68*). Clumps can be divided in spring or fall, but such plants tend to be short-lived.

Dianthus caryophyllus (*25*). Carnation. Tender to semi-hardy perennial grown as an annual that has many cultivars used in the florist's trade. Seeds germinate readily but are used primarily for breeding. Carnations are readily propagated by softwood cuttings (*91*). With mist, and with growth regulator treatment, rooting can be done any time of the year. The best source of cuttings is a mother (or stock) block isolated from the producing area. This block originates from cuttings taken from stock plants maintained under a complex program designed to keep them pathogen- and virus-free and true-to-type. As with chrysanthemum, carnation stock blocks are periodically replenished with meristem-tip culture for disease-free plants. Conventional cutting propagation then proceeds with these clean stock blocks (see Chapter 11). Such rooted cuttings are produced to a large extent by specialist growers, but commercial growing benches may be another source, if careful disease control and selection is practiced in the blocks. Lateral shoots ("breaks") that arise after flowering are removed and used as cuttings. Cuttings root in two to four weeks and may be planted directly to a greenhouse bench or transplanted to peat pots or to a nursery bed. Carnations can also be micropropagated on a large scale using shoot-tip explants (*71*). Tissue culture is also possible from petal explants (*143*).

D. chinensis, D. plumarius, and related species. Garden pinks. Hardy perennials, although some kinds are grown as annuals or biennials. Seeds germinate easily in two to three weeks at 15 to 21° C (60 to 70° F). Softwood cuttings are taken in early summer and rooted to produce next year's plants. Layering and division also can be used.

D. gratinanopolitanus. These are propagated by cuttings or division.

D. barbatus. Sweet William. Perennial but grown as a biennial. Start by seed planted outdoors in spring. In mild-winter areas, transplant to permanent location in fall. In cold-winter areas, overwinter in a cold frame and transplant in spring.

Dicentra spp. Bleeding heart. Hardy perennials. Seeds are sown in late summer or fall for overwintering at low temperatures; alternatively, seeds should be stratified for six weeks below 5° C (41° F) before planting. Seeds will germinate in three to four weeks at 10 to 13° C (50 to 55° F). Divide clumps in spring or fall. Stem cuttings can be rooted if taken in spring after flowering. Root cuttings about 7.5 cm (3 in.) long can be taken from large roots after flowering. *D. exima* is commercially propagated from seeds or division. *D. spectabilis* is usually propagated by division of the woody rhizome or from stem cuttings. It can also be micropropagated.

Dictamnus albus. Gas plant. Hardy perennial. Seed germination is difficult and inconsistent. Stratify seeds at 1 to 5° C (34 to 41° F) for three to four months. Cultivars can be propagated from root cuttings. Gas plant is easily micropropagated from shoot explants (*106*). Plants should not be disturbed after establishment; thus division is not done. Some people are allergic to plants of this species.

Dieffenbachia spp. Dumbcane. Tropical perennial. Cut stem into 5-cm (2-in.) segments, with one or two nodes per section, and place horizontally, half exposed in sand. New shoots and roots will develop from nodes. If plant gets tall and "leggy," the top may be cut off and rooted as a cutting, or the plant may be air layered. Leaves and stem are poisonous and may cause rashes on skin.

Digitalis spp. Foxglove. Hardy biennial or perennial plants. Seeds germinate in two to three weeks at 15 to 18° C (60 to 65° F) and may respond to light (*14*). Sow seeds outdoors in spring, transplant to a nursery row at 9-in. spacing, then transplant to a permanent location in fall. Perennial species increased by clump division.

Dimorphotheca spp. Cape marigold. Half-hardy annual with daisylike flowers. Seeds germinate in two to three weeks at 21° C (70° F).

Dionaea muscipula. Venus fly trap. Carnivorous plants which have unique appearance, unusual mode of life, and are in demand by plant collectors. Can be propagated by tissue culture, from leaves, adventitious buds, and peduncle explants (*24, 140*).

Dodecatheon maedia. Shooting star. Hardy perennial. Unique flowers in white or purple resemble tiny darts. Easily divided when plants are dormant. Seed is sown in autumn to stratify over winter for spring emergence.

Doronicum spp. Leopard bane. Hardy perennial. Yellow daisylike flowers in spring. Seeds germinate in two to three weeks at 20° C (68° F). Divide plants in spring or fall.

Dracaena spp. Variable group of tropical perennial foliage plants which are available in bush, cane, tree, and stump forms. Seeds germinate in three to four weeks at 30° C (86° F). Some species are propagated from leaf-bud cuttings which are treated with IBA and rooted under intermittent mist. *D. fragrans,* which is an important cane form used for interiorscapes, is propagated from cane stem cuttings which are cut into 30- to 183-cm (1- to 6-ft) sections, waxed on the distal (top) end; basal ends are treated with IBA and placed in a porous medium without intermittent mist under shade (indoors or field-propagated). Branching of canes during field propagation is done by cutting one-third to one-half way through the cane, which results in the development of lateral buds anywhere from directly below to 15 cm (6 in.) below the cut (*53*). Micropropagation is from stem explants (*52, 65, 206*).

Drosera spp. Sundew. Easily grown carnivorous plants which produce leaves with conspicuous glandular hairs that trap insects. Drosera can be propagated by seed, or root cuttings (*48*), or easily from tissue culture (*5*). Non-tuber-forming plants can also be propagated from IBA-treated leaf cuttings.

Dusty Miller. *See Senecio* spp.

Dryoptris spp. *See Fern.*

Dyssodia tenuiloba. Dahlberg daisy. Tender annual. Yellow daisylike flowers on compact edging plants. Propagated from seed. Seeds germinate in two to three weeks at 18 to 21° C (65 to 70° F). Germination can be erratic due to dormant seeds.

D. pentacheta. Perennial. A low-growing Texas wildflower suitable for xeriscapes. Propagated from softwood cuttings treated with IBA under mist or by tissue culture (*219*).

Echeveria. *See Succulents.*

Echinacea purpurea. Purple coneflower. Hardy perennial. Purple or white flowers typical of the sunflower family are attractive as garden plants or as cut flowers. Seeds germinate in two to three weeks at 21 to 25° C (70 to 78° F). Germination can be erratic and improved by seed priming or chilling stratification at 15° C (59° F) (*169, 212, 213*). Garden plants can be divided.

Echinops exaltatus. Globe thistle (*146*). Hardy perennials. Unique metallic blue flowers on thistle-like plants. Seeds germinate in one to four weeks at 15 to 18° C (60 to 65° F). Plants may be divided in spring. Root cuttings, 5- to 7.5-cm (2- to 3-in.) long, may be made in the fall and planted in sandy soil in a cold frame.

Epimedium spp. Barrenworts. Hardy perennials. Popular ground cover for shady areas. Propagated by division in the spring.

Epiphyllum spp. Leaf-flowering cactus. Tender perennial. Seeds do not germinate well when fresh but will after 6 to 12 months' storage if planted in a warm greenhouse. Propagated readily by leaf cuttings (botanically, modified stems called *phyllocades*) or by grafting to Optunia. Can be micropropagated (*124*). *See Cactus.*

Epipremnum aureum. Golden pothos. Among the most important commercially produced foliage plants. Producers cut the long vines into single-node, leaf-bud cuttings for propagation. Leaf-bud cutting propagation is enhanced with light intensity of about 2,000 ft-c and basal heat of 28° C (83° F) (*208*). Stock plants should be maintained at four to five nodes (14 to 15 leaves), and a 3-cm or longer internode section below the node and a fraction of the old aerial root should be retained on the cuttings for most rapid axillary shoot development (*211*). Maintained in the juvenile phase by cutting propagation. Mature phase has a much larger leaf and flowers.

Eranthis hyemalis. Winter aconite (*67*). Hardy perennial produced from tubers. Early yellow flowering plants popular in the rock garden. Commercially propagated from seed (see *Chionodoxa*).

Eremurus bungei. Foxtail lily (*67*). Tender perennial produced from a tuberous root. Tall flowering spikes are commercially grown as a cut flower. Propagated from seed (see *Chionodoxa*) or division.

Erigeron spp. Fleabanes. Hardy perennial. Usually produces blue daisylike flowers with yellow centers. Can be propagated by seed, division, and stem cuttings. Usually commercially propagated by seed germinated at 21 to 24° C (70 to 75° F).

Eryngium spp. Sea-holly, eryngo. A diverse species of perennials, some of which are used as specimen and border plants. Propagation by division is possible, but a long taproot makes transplanting difficult. Root cutting propagation is the commercial method for species not coming true from seed (*190*). Seeds of *E. bourgatii* have a warm-cold stratification requirement of four weeks at 21° C (70° F), followed by six weeks of 3° C (38° F) and then a warm temperature of 18 to 23° C (65 to 75° F).

Eschscholzia californica. California poppy. Hardy annual. Sow seeds outdoors in fall in mild climates or in early spring in colder areas. Tends to self-sow. Seedlings are difficult to transplant because of a long taproot.

Eucomis spp. Pineapple lily. Summer-blooming perennial bulb. Propagation is by offsets or from seeds. Micropropagation by twin scaling offers commercial potential (*15*).

Euphorbia spp. Euphorbia, spurge. Perennials used as border and sometimes specimen plants. Plants are also grown for latex production. Propagation is normally done by division. The thick seeds of *E. epithymoides (polychroma)* take 15 to 20 days to germinate at 18 to 21° C (65 to 70° F) (*190*). *In vitro* techniques have also been developed from stem explants (*157*).

Eustoma grandiflorum. Lisianthus (*146*). Annual. Grown as a pot plant or cut flower. Seeds germinate in two to three weeks at 24° C (75° F).

Exacum affine. Exacum. A popular greenhouse-grown pot plant with fragrant flowers, multiple blooms, and good postharvest quality. Can be propagated from seed or cuttings. Seeds germinate in two to three weeks at 21° C (70° F). Exacum can also be micropropagated (*200*).

× Fatshedera lizei. Tree ivy. Cross between *Hedera helix* and *Fatsia japonica.* Propagated by stem cuttings or by air layering.

Festuca spp. Blue fescue. Hardy perennial grass. A clump-forming grass with blue foliage. A warm season grass, fescue should be divided in the spring or fall. *F. ovina glauca* comes relatively true from seed.

Fern (*152, 166, 194*). Many genera and species. Spores are collected from the spore cases on lower sides of fronds. Examine these sporangia with a magnifying glass to be sure they are ripe but not empty. Place fronds with the spores in a manila envelope and dry for a week at 21° C (70° F). Screen them to separate spores from the chaff. Transfer to a vacuum-tight bottle and store in a dry, cool place. Sow spores evenly on top of sterilized moist soil mixture (e.g., two-thirds peat moss, one-third perlite in flats), paying particular attention to sanitation. Leave 1 in. space on top and cover with a pane of glass. Use 18 to 24° C (65 to 75° F) air temperature; bottom heat may be helpful. Keep moist, preferably using distilled water to avoid salt injury.

Spores germinate and produce mosslike growth ⅛ in. thick, composed of many prothallia. Fertilization of the archegonium on the underside of the prothallus occurs in three to six months. In a first transplanting, a small piece of prothallus is removed with tweezers and transplanted to wider spacing in a new flat of soil mixture. The prothallia expand to about ½ in. in diameter and produce tiny sporophyte plants with primary leaves and roots. The fern plant will grow from a second transplanting. Procedures have been developed for propagating ferns from spores *in vitro* using nutrient agar solutions (*113, 122, 199*).

Several vegetative propagation methods are possible. Ferns grow from thick rhizomes, which can be divided. Certain species (e.g., *Cystopteris bulbifera*) produce small "bulblets," about the size of a pea, on the underside of the leaf. These drop when mature, are planted, and produce a fern plant by the second year. Other species produce small vegetative buds on the upper surface or edge of the leaves; these detach and form new plants. *Polystichum setiferum* also produces bulbils in the axils of fronds which will root and produces young plants when the fronds are pegged down on propagation media (*194*).

Cultivars of the sterile (nonpropagatable by spores) Boston fern group *(Nephrolepsis)* are now largely micropropagated starting with rhizome tips (*3, 33*). This technique is also applicable to other fern genera, such as *Adiantum* (maidenhair fern), *Alsophila* (Australian tree fern), *Pteris* (brake fern), *Microlepia, Playcerium* (staghorn fern), and *Woodwardia* (chain fern).

Freesia spp. Tender perennials produced from a corm. Commercially produced as a cut flower. Seeds planted in fall germinate in four to six weeks and will bloom the next spring. A germination temperature of about 18.5° C (65° F) is best *(81)*. Plants are commercially propagated from cormels that are planted in spring and dug in fall. Micropropagation is also used to produce corms as well as disease-free plants (*154*).

Fritillaria spp. Checker lily (*67*). Perennials produced from nontunicate bulbs. Interesting group of spring flowering bulbs of which *F. imperialis* (crown imperial) and *F. melaegris* are the best-known. Propagated from offsets, bulb scaling, and bulb cuttage (chipping). Micropropagation is also successful (*119*).

Fuchsia × hybrida. *Fuchsia magellanica* hybrids. Fuchsia are tender perennials treated as annuals that are utilized as hanging baskets, containers, or trained to tree form on standards. The most effective way to propagate cultivars is by cuttings taken in spring or late summer. For optimum cutting production it is best to maintain stock plants on ten-hour photoperiods (*207*).

Gaillardia spp. Blanketflower (*146*). Annual and hardy perennials. Seeds germinate in two to three weeks at 21 to 24° C (70 to 75° F) and may respond to light (*14*). Perennial kinds are planted in spring to bloom the following year. These may also be started from root cuttings or may be divided in spring or fall but are not long-lived.

Galanthus spp. Snowdrop. Hardly perennial produced from an annual tunicate bulb. Bulbs are planted in the fall for bloom the following spring. Offsets are removed when bulbs are dug. *G. nivalis* and *G. elwesii* are mostly propagated by seed sown as soon as they are ripe in the spring. Bulb cuttage (chipping) and twin scale micropropagation can also be used (*51*).

Gasteria. *See Succulents.*

Gazania spp. (*146*). Tender perennial often grown as an annual. Propagated by seeds sown in spring or fall, or by softwood cuttings taken in late summer that are rooted in a cold frame, then transplanted in spring. Seeds germinate in one to two weeks at 21° C (70° F). Divide clumps after three or four years.

Gentiana spp. Gentian. Many species, mostly hardy perennials, although some are annuals and biennials. Plant fresh seed in the fall to overwinter outdoors. Seeds germinate in one to four weeks at 20° C (68° F). However, seeds need to be precooled for three weeks at 2° C (36° F) or treated with 300 ppm GA_3 (*26*). Seedlings are very delicate and should not be transplanted until roots are established during the first month. Cutting propagation is used for some white cultivars which have poor seed germination. Micropropagated liners are now becoming available.

Geranium. *See Pelargonium × hortorum.*

Geranium spp. Cranesbill. Hardy perennials. True geraniums are popular herbs and perennial garden plants. Geraniums can be propagated by seed, division, stem or root cuttings. Seeds germinate in two to four weeks at 21° C (70° F), but may have a hard seed coat that requires scarification. Commercial propagation is most often by division when plants are dormant or from root cuttings taken in the winter. Root cuttings are sensitive to rot if overwatered. One approach is to allow buds to form on root pieces by holding them in "sweat" boxes (polyethylene tents) at near 100 percent humidity prior to planting in a potting medium.

Gerbera jamosonii. Transvaal daisy. Tender perennial. Seeds germinate in two to three weeks at 20° C (68° F); it is important to use fresh seed. Or remove basal shoots from the rhizome and use as cuttings. Micropropagation from shoot tips can be used for rapid, large-scale multiplication (*121, 142*).

Geum spp. Avens. Hardy perennials. Common perennial easily propagated by seeds. Seeds germinate in three to four weeks at 18 to 21° C (65 to 70° F). Propagate also by clump division in spring or fall.

Gladiolus (*67*). Tender perennial grown from a corm. Popular cut flower. Seed propagation is used for developing new cultivars. Seeds are planted in spring either indoors for later transplanting or outdoors when danger of frost is over (see Chapter 16). Commercial propagation is from cormels or division of the corm leaving at least one bud (eye) per piece (*132*). *In vitro* techniques using buds or liquid-shake culture have improved multiplication rates (*19, 127, 220*).

Gloriosa spp. Gloriosa lily (*67*). Tender perennial. Vines are produced from tuberous stems. Unique flowers with recurved petals are grown as cut flowers or container plants. Propagation is from daughter tubers that form at the shoot base of the original tuber. Micropropagation is possible from tuber explants (*78*).

Gloxinia. *See Sinningia speciosa.*

Godetia spp. *(also Clarkia)* Hardy annuals. Attractive plants grown as cut flowers or pot plants (*4*). Sow seeds in early spring; these germinate in one to two weeks at 21° C (70° F).

Gomphrena globosa. Globe amaranth (*67*). Buttonlike white or purple flowers make this a popular cut flower and bedding plant. Propagate from seeds that emerge in one to two weeks at 22° C (72° F).

Gypsophila spp. *(G. elegans).* Baby's breath. Annual. Grown as a cut flower. Seed germinates in two to three weeks at 21 to 26° C (70 to 80° F).

G. paniculata. Hardy perennial. Started by seed as above. Plants can be divided in spring and fall. Double-flowered cultivars are grafted on seedling. *G. paniculata* (single-flowering) roots. Grafting can be done in summer and fall, using outdoor-grown plants for rootstocks and placing them in a cold frame for healing of the graft; grafting is also done in winter and early spring, using greenhouse-grown stock plants. *G. paniculata* can be micropropagated using shoot-tip explants (*120*).

Haworthia. *See Succulents.*

Helenium autumnale. Sneezeweed. Hardy perennial. Sunflowerlike blooms in unique colors produced in late summer in the perennial garden. Seeds germinate in one to two weeks at 21° C (70° F). Cultivars are increased by division. Separate rooted shoots in spring, line-out in nursery, then transplant in fall and winter.

Helianthemum nummularium. Sunrose. Hardy perennial. Drought-tolerant spring blooming groundcover. Seeds germinate in two to three weeks at 21 to 24° C (70 to 75° F). Cultivars are propagated by softwood cuttings taken from young shoots in spring. Transplant to pots and place in permanent location the following winter or spring. Division of clumps is also possible, but plants tend to be short-lived.

Helianthus annuus. Sunflower. Hardy annual. Popular as a cut flower. Seeds germinate in two to three weeks at 20 to 30° C (68 to 86° F). *H. decapetalus* and other hardy perennial species are increased by division.

Heliconia spp. (*60*). These tropical ornamental herbaceous perennials are prized for their showy inflorescences. Commercially produced as a cut flower. They are easily propagated by division of the rhizomes. Micropropagated from rhizome buds (*145*).

Helichrysum bracteatum. Strawflower (*146*). Annual. Popular cut flower for drying. Propagated by seeds that emerge in one to two weeks at 21 to 24° C (70 to 75° F).

Heliopsis spp. Heliopsis. Hardy perennial. Seeds germinate in one to two weeks at 18 to 21° C (65 to 70° F). Divide clumps in fall.

Heliotropium spp. Heliotrope. Tender perennial usually grown as an annual. Seed germinates in three to four weeks at 20 to 30° C (68 to 86° F) and may respond to light (*14*). Start indoors for later spring planting. Take softwood cuttings of side shoots in fall or spring and root at low temperatures [10° C (50° F)] in slightly moist conditions.

Helleborus spp. Hellebore, Christmas and lenten rose. Perennials used as border plants. One of the earliest plants to bloom in the perennial garden. Propagation by seed is very slow. *H. lividus* has a double dormancy requirement of warm stratification (21° C; 70° F) for eight to ten weeks followed by cool stratification of 3° C (37° F) for eight to ten weeks; seeds still may take up to two years to germinate. Division is the most common method by carefully separating the crown (*190*). Roots and leaves are poisonous.

Hemerocallis spp. (*62*). Daylily. Hardy perennial. Seeds require about six weeks of moist-chilling for good germination. Seed propagation is used only to develop new cultivars and requires six weeks of stratification; germination takes three to seven weeks at 16 to 21° C (60 to 70° F) (*190*). Divide clumps in fall or spring, separating into rooted sections, each with about three offshoots. Clones can also be micropropagated using flower petals and sepals as explants (*6, 138*).

Heuchera spp. American alumroot, coralbells. Perennials used as border plants for their foliage and flowers. Propagated by clump division or by leaf cuttings. Seeds germinate in three weeks at 18° C (65° F). Commercially micropropagated from stem explants (*183*).

Hibiscus spp. Mallow. Hardy perennials. Large saucer-shaped flowers in vibrant colors. Cultivars are propagated by division while dormant. Seeds germinate in one to two weeks at 21 to 26° C (70 to 75° F).

Hippeastrum spp. Amaryllis. Tender bulbous perennial. Garden plant and popular indoor flowering bulb. Propagated by offsets or micropropagation. Bulb offsets will flower the second year. Bulb cuttings can be made in late summer. Dry membranous seeds are borne in dehiscing capsules. Seeds germinate under warm conditions 20 to 30° C (68 to 86° F). Seedlings take two to four years to produce flowers. Micropropagated by twin scaling method (*51, 96*).

Hosta spp. Plantain-lily. Herbaceous perennials which are used for massed plantings or as specimen plants for their foliage and flowers. Propagated by clump division in spring. One producer removes the apical dominance of the crown (terminal) bud and slices (divides) the remaining clumps into quarters which are then placed outdoors in trays (England) which are winter protected and then planted when they begin to shoot. It takes three years to produce a flowering plant from seed. Micropropagation is being used with new cultivars to speed up propagation (*137*).

Hunnemannia fumariifolia. Goldencup. Tender perennial often grown as an annual. Seeds germinate in two to three weeks at 20° C (68° F). For bloom first year, sow seeds early indoors then transplant outdoors when danger of freezing is over.

Hyacinthus spp. Hyacinth (*67*). Hardy, spring-flowering perennial; bulbs are planted in the fall. Removal of offset bulbs gives small increase. For commercial propagation, new bulbs are obtained by scoring or scooping mature bulbs (see Chapter 16). Micropropagation, using segments of the bulb, leaf, inflorescence, or stem as an explant, is successful (*17, 99*). Seeds may be planted outdoors in fall, but up to six years are required to produce blooms.

Hypoestes phyllostachya. Polka dot plant (*146*). Tender annual. Grown as a bedding plant or indoor plant because of unique spotted foliage. Propagated from seeds that emerge in one to two weeks at 21 to 24° C (70 to 75° F).

Iberis spp. Candytuft. Hardy annual and perennial species. Seeds germinate in one to two weeks at 15 to 18° C (60 to 65° F) but may need light. Cultivars are propagated by softwood cuttings in summer or plants are divided in fall.

Impatiens spp. Snapweed, touch-me-not, balsam (*42, 45*). Perennials and half-hardy annuals. An important bedding plant. Seeds germinate in two to three weeks at 24 to 25° C (75 to 78° F). Give seeds light for the first three days then move to darkness until seedlings emerge. New Guinea impatiens can be started by cuttings, and by micropropagation (*187, 188*).

Incarvillea spp. Incarvillea. Hardy perennial. Seeds germinate in one to two weeks at 20° C (68° F). Divide in fall or, preferably, in spring.

Ipomoea spp. Morning glory. Tender perennial grown as an annual. Seeds germinate in one to three weeks at 20 to 30° C (68 to 86° F). Notch seed coats or soak seeds overnight in warm water before planting.

Iresine spp. Bloodleaf. Tender perennial. Softwood cuttings root easily. Keep stock plants over winter in greenhouse and take cuttings in late winter or spring.

Iris spp. Perennials. There are several different groups of hardy or semi-hardy iris, which grow either from rhizomes or from bulbs. Rhizomes are divided after bloom. Discard the older portion and use only the vigorous side shoots. Leaves are trimmed to about 15 cm (6 in.).

Bulbous species follow a typical spring-flowering, fall-planting sequence. The old bulb completely disintegrates, leaving a cluster of various-size new bulbs. These are separated and graded, the largest size being used to produce flowers, the smaller for further growth.

Seeds, which are used to propagate species and to develop new cultivars, should be planted as soon as ripe after being given a moist-chilling period; germination is often irregular and slow. Removal of embryo from the seed and growing it in artificial culture has given prompt germination in some cases. Iris can be micropropagated, which greatly hastens production of new cultivars over the customary division of rhizomes (*99, 103, 202*).

Ixia spp. Corn lily. Tender, summer- or fall-flowering perennials grown from corms. In cold climates, these are dug in fall and stored over winter. Small cormels are removed and planted in the ground or in flats to reach flowering size, as is done with gladiolus.

Ixora spp. (*160*). Ixora. Several species are used in Hawaii as landscape and flowering pot plants. Many species are easy to propagate by cuttings. Difficult-to-root *L. acuminata* three-node cuttings had optimal rooting when given a five-second dip of IBA-NAA, both at 2,500 ppm.

Kalanchoe spp. Tropical perennials. *See also Succulents.*

Kangaroo Paw. *See Anigozanthus spp.*

Kniphofia Hybrids *(K. tritoma)* (*190*). Torch lily or poker plant. Perennials used as specimen plants, borders, and cut flowers. Normally propagated by division. Seed propagation takes two to three years to produce a flowering plant.

Lamiastrum galeobdolan. Yellow archangel. Hardy perennial. Ground cover for shade gardens. Grown for the attractive variegated foliage. Only cultivars are grown and these are propagated by softwood cuttings. Division is also possible.

Lamium maculatum. Spotted deadnettle. Hardy perennial. Ground cover for shade. Propagation is same as *Lamiastrum.*

Lantana sellowiana, L. camara. Lantana. Tender perennials. Seeds germinate in six to seven weeks at 20° C (68° F). Softwood cuttings root easily.

Lathyrus latifolius. Perennial pea vine. Hardy perennial. Seeds germinate in two to three weeks at 20 to 30° C (68 to 86° F). Clumps may be divided.

L. odoratus. Sweet pea. Hardy annual. Seed germinates in two weeks at 20° C (68° F). Notching seed or soaking in warm water may hasten germination. Plant outdoors in fall where winters are mild, in spring where winters are severe.

Lavandula spp. Lavender. Half-hardy perennial native to Mediterranean. Seeds, which may be planted in winter, germinate in two to three weeks at 18 to 24° C (65 to 75° F). Take cuttings from side shoots in later summer or fall; plant in soil and cover or start in cold frame. Divide clumps in the fall. Micropropagation is from hypocotyl explants (*36*).

Leontopodium alpinum. Edelweiss (*146*). Hardy perennial. Mounded plants with silvery leaves and unique flowers. Seeds germinate in three to four weeks at 20 to 22° C (68 to 72° F). Division is also possible.

Lespedeza spp. Bushclover. Considered a herbaceous perennial in more northern latitudes or a semi-woody shrub in the southern United States. Excellent landscape shrub for massing and screening with its blue-grass foliage and purple flowers. Seeds can be direct-sown after harvest or scarified with a 15-minute acid treatment if stored. Roots easily from softwood cuttings with 1,000 ppm IBA in 50 percent ethanol quick-dip.

Leucojum spp. (*67*). Spring snowflake. Hardy perennial bulb. Bulbs have been collected for sale from native populations to the point of endangerment. Plants produced from seed take four to five years to flower. Propagation is normally done by separating bulbs, which is done by digging bulbs after foliage has turned brown. Can be micropropagated (*51, 182*).

Liatris spp. Gayfeather. Herbaceous perennials that are being utilized for cut-flower crops (*12*). Seeds require stratification at 4° C (34° F) for six weeks and will germinate in three to four weeks at 21° C (70° F), but flower development can take two years. Asexual propagation is by woody corms or rhizomes which are divided in the spring (*190*). *L. spicata* has been micropropagated (*191*).

Lilium spp. (*67, 134*). Lily. Hardy perennials. These are spring- and summer-flowering plants grown from scaly bulbs; most have a vertical axis, but in some species growth is horizontal with a rhizomatous structure. Lilies include many species, hybrids, and named cultivars. Seed propagation is used for species and for new cultivars. Seeds of different lily species have different germination requirements (*167*).

Immediate seed germinators include most commercially important species and hybrids (*L. amabile, L. concolor, L. longiflorum, L. regale, L. tigrinum,* Aurelian hybrids, Mid-Century hybrids, and others). Germination is epigeous; shoots generally emerge three to six weeks after planting at moderately high temperatures. Treat seeds with a fungicide to control *Botrytis.* Sow ¾ in. deep in flats during winter or outdoors in a seed bed in early spring. Dig the small bulblets in fall, sort for size, and replant with similar sizes together. Plants normally grow two years in a seed bed and two years in a nursery row before producing good-sized flowering bulbs.

Another group consists of the *slow seed germinators* of the epigeal type (*L. candidum, L. henryi,* Aurelian hybrids, and others) in which seed germination is slow and erratic; the procedures used are essentially the same as described above. The most difficult group to propagate are the *slow seed germinators* of the hypogeous type (*L. auratum,* L. *bolanderi,* L. *canadense,* L. *martagon,* L. *parvum,* L. *speciosum,* and others). Seeds of this group require three months under warm conditions for the root to grow and produce a small bulblet, then a cold period of about six weeks, followed by another warm period in which the leaves and stem begin to grow. This sequence can be provided by planting the seeds outdoors in summer as soon as they are ripe, or planting seeds in flats and then storing under appropriate conditions to provide the required temperature sequence.

Vegetative methods of propagation include natural increase of the bulbs, such as bulblet production on underground stems (either naturally or artificially), aerial stem bulblets (bulbils), or scaling. Outer and middle scales are used for scale propagation to increase the number of forcible commercial bulbs (*131*). These procedures are described in Chapter 16.

Lilies can be micropropagated from bulb scales (*193*) and pedicels (*129*). *L. longiflorum* can be propagated by leaf cuttings.

Limonium spp. Statice. *L. sinuatum* is a perennial herb native to the eastern Mediterranean which is grown commercially around the world as a cut flower for both fresh- and dry-flower arrangements. Plants are propagated by seeds that germinate in one to two weeks at 21° C (70° F). Statice has recently been micropropagated (*35, 87*).

Linaria spp. Toadflax. Hardy annual and perennials. Seeds germinate in two to three weeks at 12° C (54° F). Perennial species take two years to produce bloom from seed. Clumps can be divided in spring or fall.

Linum spp. Flax. Hardy annual and perennial species. Seeds germinate in three to four weeks at 18 to 24° C (65 to 75° F). Divide clumps of perennial species in fall or spring.

Lisianthus. *See Eustoma.*

Liriope spp. Lily turf. Hardy perennial. Vigorous ground cover. Can be propagated by seed, but commercially propagated by division in spring or autumn.

Lobelia erinus. Lobelia. Tender perennial grown as an annual. Seeds germinate in two to three weeks at 24 to 26° C (75 to 80° F), but seedling growth is slow. May respond to light (*14*). Start indoors 10 to 12 weeks before transplanting outdoors after last frost. Mature plants, if potted in the fall and kept in greenhouse over winter, can be used to provide new growth for cuttings to be taken in late winter.

Lobelia spp. Hardy perennials. Seeds germinate in three to four weeks at 20 to 30° C (68 to 86° F). Species self-seeds. Divide clumps in fall or spring.

Lobularia maritima. Sweet alyssum (*146*). Perennial grown as a hardy annual. Seed germinates in one to two weeks at 26 to 28° C (75 to 82° F) and blooms appear in six weeks. Often seeded with more than one seed per plug for better pack development.

Lunaria annua. Honesty. Biennial, sometimes grown as an annual. Seeds germinate in two to three weeks at 20° C (68° F).

L. rediviva. Hardy perennial. Propagated by seed as described above. Also increased by division.

Lupinus hartwegii, L. nanus, and others. Hardy annuals. Seeds germinate in two to three weeks at 20° C (68° F).

L. polyphyllus, L. 'Russell Hybrid.' Perennial. Seeds germinate in three to four weeks at 20° C (68° F). Seed coats may be hard and should be scarified.

L. arboreus. Tree lupine. Hardy perennial. Start from seeds indoors and transplant to permanent location.

L. texensis. Seeds of L. *texensis* (Texas bluebonnet) require scarification. L. *texensis* has been micropropagated from cotyledonary node explants (*201*). Sow seeds in spring or summer, or propagate by cuttings taken in early spring with small piece of root or crown attached.

Lychnis spp. Campion. Mostly hardy perennials but some are grown as annuals or biennials. Seeds germinate in three to four weeks at 20° C (68° F). Clumps can be divided in spring or fall.

Lycoris spp. Spider lily, surprise lily (*67*). Tender and semi-hardy perennials from a tunicate bulb. Propagation is by bulb offsets, which are removed when the dormant bulbs are dug. These are replanted to grow larger. Bulb cuttings can also be used for increase. Micropropagation by twin scaling has also been developed (*95*).

Lysimachia spp. Loosestrife. Hardy perennial. Vigorous plants with striking white or yellow flowers.

Can be propagated from seed, division, or softwood cuttings. Commercially propagated by softwood cuttings in the summer.

Lythrum salicaria. Purple loosestrife. Hardy perennial. Long-blooming plants produce purple flowering spikes. Can be a pest in states with wet habitats. Growing lythrum is restricted by some states. Propagation is easy from softwood cuttings. Division is possible but difficult because the roots are woody.

Malva alcea. Hollyhock mallow. Drought-tolerant perennial native to Italy; utilized in border plantings. Propagated by division. Seed is another method, since this species self-seeds.

Maranta leuconeura. Synonym *Calathea.* Prayer plant. Perennial herbaceous plants that are used in terrariums or as small potted plants. They are normally propagated asexually by stem cuttings because seeds rarely germinate. Cuttings are rooted in humidity tents for four to six weeks. *M. leuconeura* 'Kerchoviana' can be micropropagated (*70*).

Marigold. *See Tagetes spp.*

Matteuccia. Ostrich fern. *See Fern.*

Matthiola incana. Common stock (*146*). *M. longipetala bicornis.* Evening scented stock. Perennial that is grown as biennial or annual. Seed germinates in two weeks at 18 to 21° C (65 to 70° F) and may respond to light (*14*). Seeds are sown in summer or fall for winter bloom, in late winter indoors for spring bloom, or outdoors in spring for summer bloom.

Melampodium paludosum. Annual. Rounded plants produce yellow daisylike flowers throughout the growing season. Seeds germinate in one to two weeks at 18° C (65° F). Self-sowing can be a problem in the garden. Will root from softwood cuttings.

Mesembryanthemum spp. *See Succulents.*

Mimulus spp. Monkey flower. Includes many species of tender to hardy plants. Mostly perennials, but sometimes grown as annuals. Seeds germinate in one to two weeks at 15 to 21° C (60 to 70° F). Softwood cuttings taken from young shoots can be rooted.

Miscanthus sinensis. Maiden grass. There are many cultivars of this popular perennial grass. They make excellent specimens in the landscape with their showy feathery inflorescences. Most cultivars are sterile and are propagated by division. *Miscanthus* is a warm-season grass and should be divided in late spring. *Miscanthus* can be micropropagated (*79*).

Molluccella laevis. Bells of Ireland. Half-hardy annual. Seed germinates in three to five weeks at 10° C (50° F). Difficult to transplant; sow seeds in place.

Monarda didyma. Bee balm (*190*). A perennial garden plant native to eastern North America. Can be propagated by seed, which germinates in two to three weeks at 16 to 21° C (60 to 70° F). Can also be propagated by softwood cuttings or by clump division.

Monstera deliciosa. Often misnamed cut-leaf philodendron. Easily propagated by rooting sections of the main stem, by stem cuttings, or by air layering.

Muscari spp. Grape hyacinth (*67*). Hardy perennials from tunicate bulbs. Plants bloom in spring; bulbs become dormant in fall when they are lifted and divided. Increase by removing bulb offsets. Seed propagation can also be used. Bulb scooping and scoring produce bulb offsets. Micropropagation is easy from leaves, scales, or flower parts (*148*).

Myosotis sylvatica. Forget-me-not. Hardy biennial. Seeds germinate in two to three weeks at 20° C (68° F). Sow in summer and transplant to permanent location the following spring. *M. scorpioides* is a perennial started from seed; division in spring is also used.

Narcissus. Daffodil (*67*). Hardy perennials; spring-flowering tunicate bulbs. Vegetative propagation procedures are described in Chapter 16. Vegetative techniques include twin scaling, chipping (almost the same method as twin scaling, except that the bulb is cut across the root plate into 16 pieces with up to two bulbils developing per section), and micropropagation (*51, 85, 94*).

Nasturtium. *See Tropaeolum majus.*

Nelumbo lutea. American lotus. An aquatic plant for water gardens in the urban landscape. Usually propagated vegetatively through rhizome division. Rhizome cultures were established *in vitro* from excised embryos (*108*).

Nemesia strumosa. Half-hardy annual. Seeds germinate in two to three weeks at 13 to 18° C (55 to 65° F). Temperatures above 18° C (55° F) inhibit germination. Sow outdoors in spring in cold climates, in fall in mild-winter areas.

Nepenthes spp. A large group of carnivorous plants producing pitchers to trap insects. Traps are formed at the tips of tendrils that extend from the leaves. Propagation is from leaf or stem cuttings (*48*). Tissue culture is also successful (*163*).

Nepeta mussinii. Nepeta, catmint. Hardy perennial. Seeds germinate in two to three weeks at 20° C (68° F). Softwood cuttings of nonflowering side shoots taken in early summer root readily. Cuttings may be inserted directly into soil if protected. Plants may be divided in spring, using newest parts and discarding older portion of clumps.

Nerine spp. (*67*). Perennial tunicate bulb. There has been increasing interest in nerine as a cut flower. Propagation is from offsets or bulb cuttage, but mostly by twin scaling. Tissue culture is possible from scales, lateral buds, and the young flower stalk (*61*).

Nicotiana spp. Flowering tobacco. Half-hardy annuals. Seeds germinate in one to two weeks at 24° C (75° F) and may respond to light (*14*). Sow indoors four to six weeks before last frost then transplant out-of-doors.

Nierembergia spp. Cupflower (*146*). Tender perennial, sometimes grown as an annual. Seeds germinate in two to three weeks at 21° C (70° F). Softwood cuttings removed from new growth in spring root readily. Clumps can be divided.

Nymphaea spp. Water lily. Consists of numerous species and many named cultivars. Plants grow as rhizomes. Clumps are divided in spring. Seeds are used to grow species and to develop new cultivars. Tropical water lily hybrids and species grow from tubers. Seeds do not reproduce hybrids. Both kinds of seeds are planted 2.5 cm (1 in.) deep in sandy soil, then immersed in water 3 to 4 in. deep. Hardy species should be started at 16° C (60° F), tropical species at 21 to 27° C (70 to 80° F). Vegetative propagation is either from small tubers that can be removed from old tubers in fall or from small epiphyllous plantlets growing from the leaf (*177*). Micropropagation is also possible by using epiphyllous plantlets as explants (*104*). N. 'Gladstone' production can be extended by photoperiod control (*109*).

Oenothera spp. Evening-primrose. Hardy perennials, but some kinds are biennial. Seeds germinate in one to three weeks at 21 to 26° C (70 to 80° F). Plants can also be increased by division of clumps in the fall.

Ophiopogon japonicus. Mondo grass. Hardy perennial. Evergreen ground cover. Propagation is the same as *Liriope.*

Opuntia spp. Prickly pear cactus. Hardy perennial. The only cactus hardy to the northern U.S. Propagation is from seed, division, or cuttings. *See Cactus.*

Orchids (*141, 172, 189*). Many genera, hybrids, and cultivars are cultivated, and many more are found in nature. Some, such as *Aerides, Arachnis, Phalaenopsis, Renanthera,* and *Vanda,* exhibit a *monopodial* habit of growth. This means they are erect and grow continuously from the shoot apex and can be propagated by tip cuttings. Adventitious roots are produced along the stem and inflorescences are produced laterally from leaf axils. Most others, including *Brassovola, Calanthe, Cattleya, Cymbidium, Laelia, Miltonia, Odontoglossom, Oncidium,* and *Phalus* have a *sympodial* habit of growth, are procumbent, and do not grow continuously from the apex. Their main axis is a rhizome in which new growth arises from offshoots, or "breaks." Pseudobulbs are usually present on plants of this type. Many orchids are *epiphytes* (i.e., air plants), typically growing on branches of trees. Others *(Cymbidium, Cypripedium, Paphiopedilum)* are terrestrial and grow in the ground.

Epiphytic Orchids

Seed propagation is mainly used for hybridization. Many important cultivars are seedling hybrids, either from species or between genera, resulting from controlled crosses of carefully selected parents. Many such important crosses are between tetraploid and diploid parents to produce triploids. These offspring are sterile and are not in turn usable as parents. Seedling variation occurs, since orchids are heterozygous. Five to seven years is required for a seedling plant to bloom. Orchid flowers are hand-pollinated. A seed capsule requires 6 to 12 months to mature. A single capsule will contain many thousands of tiny seeds with relatively underdeveloped embryos. *In vitro* culture is universally used for seed propagation. (The procedure is described in Chapter 18.) Knudson's C medium is usually used. Arditti (*8, 9*) has summarized the many experiences of testing various nutritional and other factors for orchid seed germination. Orchid seed can be stored for many years if held in sealed containers over calcium chloride at about 2° C (36° F).

Vegetative methods for orchids are generally slow, difficult for many genera, and usually too low-yielding for extensive commercial use. Sympodial species are increased by division of the rhizome while it is dormant or just as new growth begins. Four or five pseudobulbs are included in each section. "Back-bulbs" and "greenbulbs" can be used for some genera.

Orchids with long canelike stems, such as *Dendrobium* and *Epidendrum,* sometimes produce offshoots ("keiki") that produce roots. Offshoots can also be produced if the stem is cut off and laid horizontally in moist sphagnum or some other medium. Flower stems of *Phaius* and *Phalaenopsis* can be cut off after blooming and handled in the same way. A drastic method of inducing offshoots is to cut out or mutilate the growing point of *Phalaenopsis,* remove the small leaves, and treat the injured portion with a fungicide. Offshoots may then be produced. Monopodial species can be propagated by long (30 to 37 cm) tip cuttings with a few roots already present. Air layering also is possible. Vegetative propagation by proliferation of shoot-tip (meristem) cultures *in vitro* has revolutionized orchid propagation, particularly for *Cymbidium, Cattleya,* and some other genera (*10, 141*). The shoot growing point is dissected from the plant and grown on a special, sterile medium; a proliferated mass of tissue and small protocorms develops, which can be divided periodically. Many thousands of separate protocorms can be developed in this way within a matter of months, each of which will eventually differentiate shoots and roots to produce an orchid plant (*10*). The procedure is described in Chapter 18. *Vanilla planifolia,* which is an orchid essential for its oil that grows as a vine, is normally propagated commercially by cuttings; nodal stem explants of this species were successfully micropropagated with BA and microshoots rooted *ex vitro* (*114*).

Terrestrial Orchids

These orchids can be difficult to propagate because they require a symbiotic relationship with an appropriate mycorrhizal fungus. *In vitro* seed germination using techniques similar to those used with epiphytic orchids has been successful both with and without fungal assistance (*7*). However, these procedures have not been extensively used commercially. Sowing seeds in pasteurized potting mixes containing mycorrhizal fungi has been successful and offers commercial potential (*158*).

Osmunda spp. *See Fern.*

Pachysandra spp. *See* Chapter 20.

Paeonia spp. Paeonia hybrids *(P. hybryda),* fernleaf peony *(P. tenuifolia),* tree peony [*P. suffructicosa (P. arborea)*]. Hardy perennials native to China used for specimen plants in borders and as cut flowers. Seed propagation is difficult, taking five to seven years to produce a flowering plant from seed (*190*). Germination may take one to two years to meet epicotyl dormancy requirement. Seeds are sown in fall for cold stratification requirement during the winter. Roots develop during the first summer, and shoots develop the second spring. Plants developed from seed are generally not true-to-type. Another method is to collect seeds before they become black and completely ripe. Do not allow them to dry out; sow in pots, which should be buried in the ground for six to seven weeks. Roots will develop; dig up and plant in a protected location or under mulch over winter. Best propagation method for herbaceous peonies is to divide clumps in fall; each tuberous root should have at least one bud or "eye," preferably three to five.

P. suffruticosa is wedge grafted in late summer on *P. lactiflora* understock. Peonies can be micropropagated (*29, 32*).

Pansy. *See Viola cornuta.*

Papaver nudicaule. Iceland poppy. Hardy perennial grown as a biennial. Seeds germinate in one to two weeks at 18 to 24° C (65 to 75° F). Sow in permanent location in summer for bloom next year.

P. orientale. Oriental poppy. Hardy perennial. Very fine seeds, which may respond to light, should be covered very lightly (*14*). Seeds germinate in same time as *P. nudicaule.* Cultivars are propagated by root cuttings. Dig when leaves die down in fall, cut into 7.5- to 10-cm (3- to 4-in.) sections, and lay horizontally in a flat covered by 2.5 cm (1 in.) of sandy soil. Root cuttings are transplanted in spring. Or dig plants in spring, prepare root cuttings, and plant directly in a permanent location.

P. rhoeas. Corn poppy, Shirley poppy. Hardy annual. Seed germinates in one to two weeks at 13° C (55° F). Sow in late summer for early spring bloom or in early spring for summer bloom. Do not transplant.

Pelargonium × hortorum. Geranium. Started by cuttings and by seed. Traditionally propagated by cuttings, which root easily with bottom heat, but *Pythium* and *Botrytis* infection can be serious problems. Pathogen-free stock, identified by culture indexing, should be used and can be supplied by specialists (*198*). There may be practical value in applying ABA in the shipment and storage of geranium cuttings (*13*).

In the mid-1970s, large-scale seed propagation of geraniums began with the introduction of certain cultivars that would grow from seed to flower in 14 to 16 weeks. Seeds germinate best at about 21 to 24° C (70 to 75° F) in an artificial medium (*11*). Chlormequat (cycocel) is utilized in greenhouse geranium production to produce compact, early-flowering, well-branched plants; there may be advantages in applying this growth retardant in seed-propagated garden geraniums (*170*). *In vitro* propagation has been developed but is not used extensively for commercial propagation (*47*).

Pennisetum spp. Fountain grass. Perennial grasses with feathery inflorescences. They are propagated by division in late spring to early summer (*57*). Seeds germinate easily and volunteers in the garden can be a nuisance.

Penstemon spp. Beardtongue. Semi-hardy to hardy perennials, sometimes handled as annuals. Seeds germinate in two to three weeks at 18 to 21° C (65 to 70° F) but growth is slow and uneven; seeds may respond to light (*14*). Some species benefit from eight weeks of stratification at 15° C (59° F) (*2*). Plants started indoors in early spring and transplanted outdoors later may bloom the first year. Plants are usually short-lived. Softwood cuttings taken from nonflowering side shoots of old plants root readily. Make cuttings in fall to obtain plants for next season. Clumps may be divided. Penstemon can also be micropropagated from lateral buds used as explants (*128*).

Peperomia spp. Tender perennial. Softwood stem, leaf-bud, or leaf cuttings root readily. Plants can be divided. Peperomia can also be micropropagated from excised leaf explants (*89*) and petioles (*159*).

Periwinkle. *See Vinca minor.*

Pervoskia atriplicifolia. Russian sage. Hardy perennial. Large shrublike perennial with silvery foliage and blue flowering spikes. Propagated from softwood cuttings.

Petunia Hybrids. Petunia (*45*). Tender perennial often grown as an annual. Seeds germinate in one to two weeks at (24 to 25° C; 75 to 78° F) for good germination. Give light for the first three days then move to darkness until seedlings emerge. It is best to start plants indoors for later outdoor planting. Softwood cuttings taken in late summer or fall from side shoots root easily. Petunia is easily micropropagated from leaf segments (*192*).

Philodendron spp. Tropical vines. Seeds germinate readily at about 25° C (77° F) if sown as soon as they are ripe and before they become dry. Best propagation methods are use of leaf-bud or stem cuttings, rooting sections of main stem, or by air layering.

Phlox divaricata. Sweet William. Hardy perennial. Expose seeds to cold during winter before planting. Softwood cuttings taken in spring root easily. Divide clumps in spring or fall.

Phlox drummondii. Annual phlox (*45*). Hardy annual. Seed germinates in two to three weeks at 15 to 18° C (60 to 65° F). Initial seed germination occurs in either light or darkness, but light inhibits radicle growth, so it is common to germinate phlox in the dark (*43*). Start indoors for later outdoor planting, or outdoors after frost.

Phlox paniculata. Garden phlox. Hardy perennial. Plants do not come true from seed. Sow seeds as soon as ripe in fall to germinate the next spring. They will germinate in three to four weeks at 20° C (68° F). Grow plants one season and transplant in fall. Softwood cuttings taken from young shoots in spring or summer root easily, but they are subject to damping-off if kept too wet. It is best to propagate from root cuttings. Dig clumps in fall; remove all large roots to within 5 cm (2 in.) of crown (which is replanted). Cut roots into 2 in. lengths and place in flats of sandy soil; cover 13 mm (½ in.) deep. Transplant next spring. Divide clumps in fall or spring. Phlox can be micropropagated using shoot explants.

Phlox subulata. Moss pink. Hardy perennial. Evergreen ground cover with profusion of blooms in early spring. Propagation is by division or softwood stem cuttings.

Physalis alkekengi. Chinese lantern. Hardy perennial. Grown for showy fruits shaped like orange paper lanterns. Seeds emerge in three to four weeks at 15° C (60° F). Can be divided in the autumn.

Physostegia virginiana. False dragonhead or lionsheart. A herbaceous perennial that is used as a field-grown cut flower crop (*12*). Seeds will germinate in two to three weeks at 18 to 21° C (65 to 70° F). This species can also be propagated by division.

Pinguicula spp. Butterwort. Perennial carnivorous plants. Glandular hairs are produced on the leaves of these rossette plants to trap insects. An attractive flowering carnivorous plant. Propagation is from seed or tissue culture (*1*).

Platycodon grandiflorus. Balloon flower. Hardy perennial. Long-lasting blue flowers that resemble balloons when in bud. Seed germinates in one to two weeks at 15 to 21° C (60 to 70° F). Can be divided in spring.

Poinsettia *(Euphorbia pulcherrima).* See Chapter 20.

Polemonium spp. Jacob's ladder. Hardy perennial. In spring, blue or white flowers appear on plants with a leaflet pattern that resembles a ladder. Seeds germinate in three to four weeks at 21° C (70° F). Divide clumps or root stem cuttings.

Polianthes tuberosa. Tuberose. Tender bulbous perennial. Propagated by removing offsets at planting time. The small bulbs take more than one year to flower. Divide clumps every four years.

Poliomintha longiflora (*97*). Mexican oregano. Has small evergreen leaves which smell like the spice oregano and is a striking landscape plant which produces light lavender flowers throughout the summer. It is easily propagated by semi-hardwood cuttings.

Polygonatum spp. Solomon's seal. Hardy perennials. Used in wildflower and perennial gardens. Propagation is by seed or division. Cultivars are propagated by dividing rhizomes in the spring.

Polygonum spp. *(190).* Fleeceflower. Versatile perennials and annuals used as ground covers, and in rock gardens, planters, and hanging baskets. Generally propagated by division or seed. *P. capitatum* 'Magic Carpet,' which is an excellent annual ground cover, is propagated by seed at 21 to 27° C (70 to 80° F) and germinates in three weeks.

Portulaca grandiflora. Moss rose. Half-hardy annual. Seeds germinate in two to three weeks at 26 to 29° C (80 to 85° F); they respond to light (*14*). Reseeds itself. Can be propagated by cuttings.

Potentilla spp. Cinquefoil. Hardy perennial. Creeping perennials with attractive blooms. Can be propagated by seed, division, or stem cuttings. *P. nepalensis* is a common garden perennial and is grown from seed germinated at 21° C (70° F).

Pothos. *See Epipremnum.*

Primula obconica. Primrose (*146*). Tender perennial grown as an annual. Seeds germinate well in a cool greenhouse or after three to four weeks at 20 to 21° C (68 to 70° F). The very tiny seeds respond to light and should not be covered.

Primula malacoides. Fairy primrose. Seeds will germinate in two to three weeks at 15 to 18° C (59 to 65° F).

Primula sinensis. Chinese primrose. Seeds will germinate in three to four weeks at 20° C (68° F). In these species, double-flowering cultivars do not produce seeds but are propagated by cuttings taken in spring, or by division.

P. × polyantha* and *P. vulgaris. Hardy perennials grown outdoors. Seeds germinate in three to six weeks at 20° C (68° F), but some species may require lower temperatures (*56, 74*). It is best to collect and sow seeds as soon as they are ripe in the fall. Cuttings taken in spring root easily. Clumps can be divided just after flowering.

Pulmonaria spp. Lungwort. Hardy perennial. Attractive spotted or silvery foliage on plants adapted to the shade garden. Common companion plant to hostas. Can be divided, or propagated by tissue culture. Tissue culture propagation and the introduction of new cultivars has increased the popularity of lungworts.

Puschkinia scilloides. Striped squil. Hardy perennial bulb. Propagation is the same as *Scilla.*

Ranunculus asiaticus. Turban or Persian ranunculus. Tender perennial. Seeds germinate in one to four weeks at 20° C (68° F). Grows from tuberous roots which can be divided.

Ranunculus spp. Buttercup. Hardy perennial. Seeds germinate in one to four weeks at 20° C (68° F). Plant seeds as soon as they are ripe. Divide plants in spring or fall.

Reseda odorata. Mignonette. Hardy annual. Seeds should not be covered; they germinate in two to three weeks at 12° C (54° F) and respond to light (*14*).

Rhipsalidopsis spp. Easter cactus. Propagation same as *Epiphyllum.*

Ricinus communis. Castor bean. Soak seeds in water for 24 hours or nick with file before planting. Plant each seed in an individual pot, and after frost danger is past, transplant to outdoor location. Seeds are poisonous.

Rodgersia pinnata. Featherleaf, Rodgersflower. Perennial for moist border with attractive foliage and ornamental flower. Propagation is by division.

Rudbeckia spp. Black-eyed Susan, coneflower. Hardy annual, biennial, and perennial species. Seeds germinate in two to three weeks at 21 to 24° C (70 to 75° F). Perennial kinds are propagated by clump division. Will reseed naturally. Osmotic priming can improve germination (*76*).

Saintpaulia ionantha. African violet (*101, 116, 217*). Tropical perennial. Propagation is by seed, division, or cuttings. The very fine seeds, which germinate in three to four weeks at 30° C (86° F), should not be covered. Seedlings are subject to damping-off. Vegetative methods are necessary to maintain cultivars. Plants can be divided. Leaf cuttings (blade and petiole) are easily rooted, either in a rooting medium or in water. Rapid, large-scale propagation can be accomplished by *in vitro* culture techniques using leaf petiole sections (*27, 59, 107, 185*).

Salpiglossis sinuata. Painted tongue. Semi-hardy annual. Seeds are difficult to germinate but some will start in one to two weeks at 21 to 22° C (70 to 72° F). Start indoors in peat pots. Can be micropropagated (*126*).

Salvia spp. Sage (*45*). Annual, biennial, and perennial species. Salvia are used as borders or as cut flowers (*12*). Seeds germinate in two to three weeks at 24 to 25° C (75 to 78° F) and may respond to light (*14*). Softwood cuttings of young shoots 3 to 4 in. long root readily.

S. officinalis basal cuttings root better than apical cuttings. Flowering reduces rooting and removal of flowers enhances propagation. Rooting is enhanced with basal dips of 1,000 ppm K-IBA salt and 100 ppm dithiothreitol (DTT) (*161*). Rooting ability was highest in spring (Israel).

S. greggii is easily propagated by semi-hardwood cuttings (*97*). Plants can be divided, but such divisions are slow to recover.

S. splendens. Scarlet sage. Tender perennial grown as an annual. Germinate seeds at 24 to 25° C (75 to 78° F), then grow at 13° C (55° F) night temperature. Seeds soaked for six days at 6° C (43° F) can promote germination (*38*). Softwood cuttings taken in fall root readily.

Sansevieria trifasciata and S. 'hahnii' (Dwarf Form). Bowstrip hemp. Snakeplant. Tropical perennial. Plants grow from a rhizome, which can be readily divided. Leaves may be cut into sections, several inches long, and inserted into a rooting medium; a new shoot and roots will develop from base of leaf cutting. The variegated form, *S. t.* 'Laurentii,' is a chimera, which can be maintained only by division. *S. trifasciata* has been micropropagated (*28*).

Santolina chamaecyparissus *(incana)*. Lavender cotton. A perennial native to the Mediterranean that is used as carpet bedding or a low hedge. It needs to be trimmed to maintain compact growth. It is normally propagated by cuttings.

Sanvitalia procumbens. Creeping zinnia. Hardy annual. Seeds germinate in one to two weeks at 20° C (68° F) and may respond to light (*14*). Sow in place in spring, or in fall in mild climates.

Saponaria officinalis. Bouncing Bet. Hardy perennial. Seeds germinate in two to three weeks at 20° C (68° F). The plant spreads rapidly by an underground creeping stem which can be divided.

Saponaria vaccaria. Soapwort. Hardy annual. Seeds germinate in two to three weeks at 20° C (68° F).

Sarracenia spp. Pitcher plant. Perennial carnivorous plants native to bog ecosystems. Leaves are modified to form spectacular pitchers that entrap insects. Also grown for cut foliage for the florist industry. Propagation is by division of the rhizome (*48*).

Saxifraga spp. Many interesting unusual species and hybrids. Mostly hardy perennials. Seeds germinate easily; they are sown preferably when ripe. Some hybrids and cultivars are maintained only by vegetative methods and are propagated once flowering is finished in June or July (England) (*178*). Cuttings are small and slow growing—it takes one year to grow a liner. Cuttings are rooted in a cold frame or seed tray. Most plants grow as small rosettes and are easily propagated by making small cuttings involving single rosettes. Small plants from stolons root readily. Plants can be divided in spring or fall. *S. stolonifera,* strawberry geranium, is a tender perennial that reproduces by runners.

Scabiosa spp. Pincushion flower. Annual and hardy perennial species. Seeds germinate in two to three weeks at 15 to 18° C (60 to 65° F). Perennial kinds can be divided.

Scarlet Sage. *See Salvia splendens.*

Schefflera arboricola (*86*). Schefflera is an important foliage plant that can be propagated easily by seeds, cuttings, or air layering. Basal cuttings develop more roots and longer shoots and require less time to break lateral buds than do apical cuttings. As single-node cutting length increased to 20 cm (8 in.) so did rooting, bud-break, and shoot growth. Seeds germinate in three to four weeks at 22 to 24° C (72 to 75° F).

Schizanthus spp. Butterfly flower. Tender annual. Seeds germinate in one to two weeks at 15 to 21° C (60 to 70° F) in the dark. Seeds are sensitive to high temperatures (*14*). Sow seeds in fall for early spring blooms indoors, or sow in early spring to be transplanted outdoors for summer blooms.

Schlumbergera truncata. Christmas cactus. Propagation same as *Epiphyllum.*

Scilla spp. Squill (*67*). Includes several kinds of bulbous hardy spring-flowering perennials. Dig plants when leaves die down in summer and remove the bulblets. *S. autumnalis* is planted in spring and blooms in fall.

Sedum spp. (*178*). Sedum is composed of a wide range of species including herbaceous perennials, evergreens, and monocarps. Many of the *Sedum* species can be raised by seed, but generally this method is limited to herbaceous perennials only; seeds germinate at 15 to 18° C (65° F). The mat-forming species are propagated by division, since the creeping shoots root into the ground as they travel and mats are easily pulled apart. Direct sticking of cuttings into containers is done, since many species root so readily.

Sempervivum. *See Succulents.*

Senecio spp. Dusty miller *(S. cineraria)* (*146*). These tender perennials are grown as annuals. Seeds germinate in two to three weeks at 24 to 26° C (75 to 80° F). Stem-tip cuttings root rapidly if treated with IBA and placed under mist over bottom heat. Florists' cineraria *(S. cruentus)* is a cool-season crop with true blue or lavender flowers. Seeds can be obtained from self-pollinated plants, but size and quality of flowers deteriorate after one to two generations. Plants produced vegetatively by cuttings have reduced vigor and flower size. This species is micropropagated (*55,80*).

Sidalcea malviflora. Sidalcea (*190*). A native western U.S. perennial used as a border plant. Cultivars are commonly propagated by division. Can be seed-propagated, but cultivars do not come true from seed.

Sinningia speciosa. Gloxinia. Tropical perennial. Commonly grown from seeds, which are very fine and require light. Sow in uncovered, well-drained peat moss medium; they germinate in two to three weeks at 20° C (68° F). Vegetative methods are required to reproduce cultivars. Plant grows from a tuberous root on which a rosette of leaves is produced. The root can be divided as described for tuberous begonia. Softwood cuttings or leaf cuttings taken in spring from young shoots starting from the tubers root easily. Gloxinia can also be micropropagated using leaf explants (*105*).

Snapdragon. *See Antirrhinum majus.*

Solidago spp. Goldenrod. Hardy perennial. Grown as a cut flower and for the perennial garden. Can be grown from seed but is usually propagated from cuttings. × *Solidaster* is an intergeneric hybrid between *Solidago* and *Aster* that is also grown as a cut flower from stem cuttings. Plants can also be divided.

Spider Plant. *See Chlorophytum comosum.*

Stachys spp. Lamb's ear *(S. byzantina),* big betony *(S. grandiflora).* Hardy perennials used as border plants or ground covers. Propagated by clump division or by seed. Seed germinates in three to four weeks at 21° C (70° F).

Stock. *See Matthiola.*

Stokesia laevis. Stokes aster. Hardy perennial. Seeds germinate in four to six weeks at 21 to 24° C (70 to 75° F) and the plants bloom the first year. Commonly propagated from root cuttings or division.

Strelitzia reginae. Bird-of-paradise. Seed propagation of this tropical perennial is undesirable due to juvenility and genetic variation. Seeds are sown under warm conditions, and freshly harvested seed should be used to avoid seed coat impermeability. Bottom heat of 37° C (98° F) aids germination, which occurs in six to ten weeks *(50)*. This species grows from a rhizome, which can be divided in the spring; however, division is limited by a low rate of multiplication with 0.5 to 1.5 divisions per branch per year. A technique has been developed to overcome the strong apical dominance which inhibits branching of axillary buds into propagules. A triangular incision with a knife is made at the base of a separated branch 8 to 12 mm above the basal plate to reach and remove the apex from adult plants (*203*). After two to six months, 2 to 30 lateral shoots develop from each fan (separated branch). During the next six months, newly formed laterals root and can be divided into individual plants (see Chapter 16).

Streptocarpus × hybridus. Cape primrose is a herbaceous perennial in the Gesneriacae. It is generally propagated by seed or leaf cuttings. Many of the new commercial cultivars bloom throughout the year with large blooms on compact plants, but are sterile hybrids and must be vegetatively propagated. Cape primrose has been successfully micropropagated using explants of leaf discs, shoot apices, stem, petiole, pedicel segments, and corolla flower parts (*149*).

Succulents (*22, 73, 151, 186*). This loosely defined horticultural group includes many genera, such as *Agave, Aloe, Crassula, Echevaria, Euphorbia, Gasteria, Haworthia, Hoya, Kalanchoe, Mesembryanthemum, Portulacaria, Sedum, Sempervirens,* and *Yucca.* These are plants with fleshy stems and leaves that store water, or plants that are highly drought-resistant. Most are half-hardy or tender perennials. Seed propagation is possible, although young plants are often slow to develop and to produce flowers. It is best to germinate seeds indoors at high day temperatures (29 to 35° C; 85 to 95° F). Seedlings are susceptible to damping-off.

Cuttings of most species root readily—either stem, leaf-bud, or leaf—in a 1:1 peat-perlite medium. They should be exposed to the open air or inserted into dry sand for a few days to allow callus to develop over the cut end. Some protection from drying is needed during rooting. Some species can be reproduced by removing offsets. Grafting is possible as described for cacti. Many of the succulents can be micropropagated (*164, 179, 181, 192*).

Sweet Alyssum. *See Lobularia maritima.*

Sweet Pea. *See Lathyrus.*

Symphytum spp. Comfrey. Hardy perennial. Grown as an herb or ornamental garden plant. Cultivars are propagated mainly from root cuttings. Micropropagation is from stem explants (*88*).

Tagetes spp. Marigold (*146*). Tender annuals. Seeds germinate readily in one week at 24 to 26° C (75 to 80° F) and sometimes respond to light (*14*). Can be sown in place in spring after frost in mild climates. *T. erecta* (African marigold) has been micropropagated from leaf segments (*115*).

Thalictrum spp. Meadow rue. Hardy perennial. Seeds germinate in four to six weeks at 20° C (68° F). Sometimes hard seeds are present. Plants can be divided in spring or fall.

Thermopsis caroliniana. Hardy perennial. Seeds have hard seed coats and require scarification. Seeds germinate in two to three weeks at 21° C (70° F). Plants can be divided, but it is best to leave them undisturbed.

Thunbergia spp. Clockvine. Tender perennials grown as annuals. Seeds germinate in two to three weeks at 21 to 24° C (70 to 75° F), but seedlings grow slowly. Softwood cuttings taken from new shoots root readily.

Thymus spp. Thyme. Hardy perennials. Seeds germinate in one to two weeks at 12 to 32° C (54 to 90° F). Germination may be promoted by light. Can also be increased by division or by softwood cuttings taken in summer.

Ti. *See Cordyline terminalis.*

Tiarella cordifolia. Foamflower. Hardy perennial. Grown as a ground cover in the shade. Can be propagated from seed or division. *Tiarella* is commercially micropropagated with shoot explants (*112*).

Tigridia pavonia. Tiger flower. Tender bulbous perennials. Plant bulbs in spring and dig in fall when leaves die. Increase by removal of small bulblets. Easily started by seed.

Tithonia rotundifolia. Mexican sunflower. Tender perennial grown as an annual. Seeds germinate in two to three weeks at 21° C (70° F).

Tolmiea menziesii. Piggyback plant. New plantlets form on upper surface of leaves. Such leaves are removed and the petiole is stuck in rooting medium to the depth of new plantlet, which then resumes growth.

Torenia fournieri. Wishbone flower (*180*). Torenia is a tender perennial grown as an annual. Seeds germinate at 21 to 23° C (70 to 75° F) media temperature in 10 to 14 days.

Tradescantia spp. Spiderwort. Hardy perennial. Long blooming blue, pink, or white flowers above grasslike foliage. Propagated from spring or autumn divisions.

Trachymene coerulea. Laceflower. Tender perennial grown as an annual. Seeds germinate in two to three weeks at 20° C (68° F).

Trollius spp. Globeflower. Hardy perennial. Plant seeds in fall. This will produce plants that flower by next spring (*31*). Also increases by clump division.

Tropaeolum majus. Nasturtium. Tender perennial grown as an annual. Plant seeds in place; they germinate in one to two weeks at 20° C (68° F), but are difficult to transplant. Double-flowering kinds must be propagated vegetatively, usually by softwood cuttings.

Tricyrtis spp. Toad lily. Hardy perennial. Unique flowers produced in the early autumn. Plants prefer shade. Propagation is from seed, division, or stem cuttings. Seeds require stratification. *Tricyrtis* has become widely available partly due to its ease of micropropagation.

Tulipa spp. and Hybrid Cultivars. Tulip (*67*). Hardy perennials from tunicate bulbs. Plant bulbs in fall for spring blooms. Seeds are used to reproduce species and for breeding new cultivars. They germinate readily after stratification. Vegetative methods include removal of offset bulbs in the fall. Different bulb sizes are planted separately, since the time required to produce flowers varies with size. For details of procedure, see Chapter 16. Tissue culture of tulip is possible but results are variable and success rates are low. Additional work is needed to develop a commercial micropropagation system.

Valeriana officinalis. Valerian. Hardy perennial. Seeds germinate in three to four weeks at 12 to 32° C (54 to 90° F). New plants can be obtained by clump division.

Venidium fastuosum. Venidium. Half-hardy annual. Seeds germinate in four to six weeks at 20 to 30° C (68 to 86° F); or sow outdoors at 10 to 13° C (50 to 55° F).

Verbascum spp. Mullein. Hardy perennials and biennials. Seeds are slow to germinate; best temperature is 30° C (86° F). Propagate named cultivars by root cuttings taken in early spring.

Verbena spp. Verbena [*Verbena* × *hybrids* (*V.* × *hortensis*)]. Tender perennial grown as an annual. Seeds germinate in three to four weeks at 24° C (75° F) days and 15° C (60° F) nights, but germination can be erratic; sometimes promoted by light (*45*). Seeds are sensitive to overwatering (*41*). May be propagated by softwood cuttings taken in summer.

V. canadensis. Clump verbena. Hardy perennial that blooms first year from seed. Seeds germinate in two to four weeks at 12 to 32° C (54 to 90° F). Plants can be propagated by division or by softwood cuttings.

Veronica spp. Speedwell. Hardy perennials. Seeds germinate in two weeks at 12 to 26° C (54 to 80° F). Plants are increased by division in spring or fall or by softwood cuttings taken in the spring or summer.

Vinca. *See Catharanthus.*

Vinca major. Tender perennial. Propagate by division or by softwood cuttings taken in summer.

V. minor. Periwinkle. Hardy perennial. Seeds germinate in two to three weeks at 20° C (68° F). Easily propagated by softwood cuttings or by division. *V. minor* is also micropropagated (*184*).

Viola spp. Many hardy perennial kinds. These are grown by seeds as above, but germination may be slow and seeds are best exposed to cold before planting. Many species produce seeds in inconspicuous, enclosed (cleistogamous) flowers near the ground, whereas the conspicuous, showy flowers produce few or no seeds. These plants can also be reproduced by cuttings or by division (*56*).

Viola cornuta. Horned violet, tufted pansy. Hardy perennial. Seeds germinate in two to three weeks at 12 to 32° C (54 to 90° F). Seeds of some cultivars need light. Vegetative propagation is by cuttings taken from new shoots obtained by heavy cutting back in the fall. Clumps may also be divided.

V. odorata. Sweet violet. Tender to semi-hardy perennials. Grows by rhizomelike stems, which can be separated from others on the crown and treated as a cutting with some roots present.

V. tricolor. Johnny-jump-up. Hardy or semi-hardy, short-lived perennial. Usually propagated by seeds as described for *V. cornuta* but may also be increased by division.

Viola × wittrockiana. Pansy. Short-lived perennial grown as an annual. Popular as an autumn and early spring bedding plant. Propagation is by seed. Seeds germinate at 18 to 21° C (65 to 70° F). Pansy seed experiences thermoinhibition at temperatures above 30° C (86° F) and fails to germinate. These temperatures are common in summer greenhouses when pansy seed is normally sown. Seed priming alleviates this problem (*40*). Pansy seed is commonly sold as primed seed from the seed company.

Yucca spp. Yucca. Tender to semi-hardy perennials. Seeds germinate at 20° C (68° F) but rather slowly and require four to five years to flower. Plants are monocots; some are essentially stemless and grow as a rosette, while others have either long or short stems. Offsets growing from the base of the plant can be removed and handled as cuttings; sometimes entire branches or the top of the plant can be detached a few inches below the place where leaves are borne and replanted in sandy soil. Sections of old stems can be laid on sand or other medium in a warm greenhouse, and new side shoots that develop can be detached and rooted. *Y. elephantipes* is rooted by long canes (*156*).

Zantedeschia spp. Calla. Several species, which have similar propagation requirements. Tropical perennials. Plants grow by thickened rhizomes that produce offsets or rooted side shoots; these are removed and planted. Calla are also micropropagated (*102*).

Zebrina pendula. Wandering Jew. Tender perennial. Easily propagated at any season by stem cuttings.

Zinnia elegans and Other Species. Zinnia. Half-hardy hot-weather annuals. Seeds germinate outdoors in one week at 26 to 29° C (80 to 85° F). Sometimes seeds respond to light (*14*).

REFERENCES

1. Adams, R.M., S.S. Koenigsberg and R.W. Langhans. 1979. In vitro propagation of *Cephalotus follicularis* (Australian pitcher plant). *HortScience* 14:512–3.
2. Allen, P.S. and S.E. Meyer. 1990. Temperature requirements of three *Penstemon* species. *HortScience* 25:191–93.
3. Amaki, A. and H. Higuchi. 1992. Micropagation of Boston ferns (*Nephrolepsis* spp.). *Biotechnology in Agriculture and Forestry. Vol. 20. High-Tech and micropropagation IV.* Y.P.S. Bajaj (ed.). Springer-Verlag, Berlin. pp. 484–94.
4. Anderson, R.A., R.L. Geneve and L. Utami. 1991. Gotta cut godetia. *Greenhouse Grower* 9(1):46–8.
5. Anthony, J.L. 1992. In vitro propagation of *Drosera* spp. *HortScience* 27:850.
6. Apps, D.A., and C.W. Heuser. 1975. Vegetative propagation of Hemerocallis—including tissue culture. *Comb. Proc. Inter. Plant Prop. Soc.* 25:362–67.
7. Arditti, J. 1967. Factors affecting the germination of orchid seeds. *Bot. Rev.* 33:1–97.
8. ———. 1977. *Orchid biology.* Ithaca, N.Y.: Cornell Univ. Press.
9. ———. 1982. Orchid seed germination and seedling culture—a manual. In J. Arditti (ed.). *Orchid biology: reviews and perspectives* II. Cornell Univ. Press, London. pp. 242–370.
10. ———. and R. Ernst. 1993. *Micropropagation of orchids.* John Wiley, NY.
11. Armitage, A.M. 1986. *Seed propagated geraniums.* Portland Oreg.: Timber Press.
12. ———. 1987. The influence of spacing on field grown perennial crops. *HortScience* 22:904–7.
13. Arteca, R.N., D.S. Tsai, and C. Schiangnhauler. 1985. Abscisic acid effects on photosynthesis and transpiration in geranium cuttings. *HortScience* 20:370–72.
14. Assn. Off. Seed Anal. 1993. Rules for seed testing. *Jour. Seed Tech.* 16:1–113.
15. Ault, J.R. 1995. In vitro propagation of *Eucomis autumnalis, E. comosa* and *E. zambesiaca* by twin scaling. *HortScience* 30:1441–42.
16. ———, and W.J. Blackmon. 1987. In vitro propagation of *Ferocactus acanthodes* (Cactaceac). *HortScience* 22:126–27.
17. Bach, A. 1992. Micropagation of Hyacinths. *(Hyacinthus orientalis* L.*). Biotechnology in Agriculture and Forestry. Vol. 20. High-Tech and micropropagation* IV. Y.P.S. Bajaj (ed.). Springer-Verlag, Berlin. pp. 144–59.
18. Bailey, L.H., E.Z. Bailey, and Staff of Bailey Hortorium. 1976. *Hortus third.* New York: Macmillan.
19. Bajaj, Y.P.S., M.M.S. Sidhu and A.P.S. Gill. 1992. Micropagation of *Gladiolus. Biotechnology in Agriculture and Forestry. Vol. 20. High-Tech and micropropagation* IV. Y.P.S. Bajaj (ed.). Springer-Verlag, Berlin. pp. 135–43.

20. ———. 1992. Micropagation of *Chrysanthemum. Biotechnology in Agriculture and Forestry. Vol. 20. High-Tech and micropropagation* IV. Y.P.S. Bajaj (ed.). Springer-Verlag, Berlin. pp. 69–80.

21. Ball, V., ed. 1991. *The Ball red book: Greenhouse growing* (15th ed.). Reston, Va.: Reston Publ. Co.

22. Bayer, M.B. 1982. *The new Haworthia handbook*. Pretoria: National Botanic Gardens of South Africa.

23. Beattie, D.J. 1993. Propagation of *Astilbe. Comb. Proc. Intern. Plant Prop. Soc.* 43:509–10.

24. Beebee, J.D. 1980. Morphogenetic response of seedlings and adventitious buds of the carnivorous plant *Dionaea muscipula* to aseptic culture. *Bot. Gaz.* 141:396–400.

25. Besemer, S. 1980. Carnations. In *Introduction to floriculture,* R.A. Larson, ed. New York: Academic Press.

26. Bicknell, R.A. 1984. Seed propagation of *Gentiana scabra. Comb. Proc. Inter. Plant Prop. Soc.* 34:396–401.

27. Bilkey, P.C., B.H. McCown, and A.C. Hildebrandt. 1978. Micropropagation of African violet from petiole cross-sections. *HortScience* 13:37–8.

28. Blazich, F.A., and R.T. Novitzky. 1984. In vitro propagation of *Sansevieria trifasciata. HortScience* 19:122–23.

29. Bouza, L., M. Jacques and E. Miginiac. 1994. In vitro propagation of *Paeonia suffruticosa* Andr. cv. 'Mme. De Vatry': developmental effects of exogenous hormones during the multiplication phase. *Scientia Hort.* 57:241–51.

30. Bridgen, M.P., J.J. King, C. Pedersen, M.A. Smith and P.J. Winski. 1992. Micropropagation of *Alstroemeria* hybrids. *Comb. Proc. Intern. Plant Prop. Soc.* 42: 427– 430.

31. Brumback, W.E. 1985. Propagation of wildflowers. *Comb. Proc. Inter. Plant Prop. Soc.* 35:542–48.

32. Buchheim, J.A.T. and M.M. Meyer, Jr. 1992. Micropropagation of peony (*Paeonia* spp.). *Biotechnology in Agriculture and Forestry. Vol. 20. High-Tech and micropropagation* IV. Y.P.S. Bajaj (ed.). Springer-Verlag, Berlin. pp. 269–85.

33. Burr, R.W. 1975. Mass production of Boston fern through tissue culture. *Comb. Proc. Inter. Plant Prop. Soc.* 25:122–24.

34. Bush, S.R., E.D. Earle, and R.W. Langhans. 1976. Plantlets from petal segments, petal epidermis, and shoot tips of the periclinal chimera, *Chrysanthemum morifolium* 'Indianapolis,' *Amer. Jour. Bot.* 63:729–37.

35. Butcher, S.M., R.A. Bicknell, J.F. Seelye, and N.K. Borst. 1986. Propagation of *Limonium peregrinum. Comb. Proc. Inter. Plant Prop. Soc.* 36:448–50.

36. Calvo, M.C. and J. Segura. 1989. In vitro propagation of lavender. *HortScience* 24:375–76.

37. Carpenter, W.J. 1990. Light and temperature govern germination and storage of *Caladium* seed. *HortScience* 25:71–4.

38. Carpenter, W.J. 1989. *Salvia splendens* seed pregermination and priming for rapid and uniform plant emergence. *Jour. Amer. Soc. Hort. Sci.* 114:247–50.

39. Carpenter, W.J. and J.F. Boucher. 1992. Temperature requirements and germination of *Delphinium* × *cultorum* seed. *HortScience* 27:989–92.

40. Carpenter, W.J. and J.F. Boucher. 1991. Priming improves high-temperature germination of pansy seed. *HortScience* 26:541–44.

41. Carpenter, W.J. and S. Maekawm. 1991. Substrate moisture level governs the germination of verbena seed. *HortScience* 26:1468–72.

42. Carpenter, W.J., E.R. Ostmark and J.A. Cornell. 1994. Light governs the germination of *Impatiens wallerana* Hook. F. Seed. *HortScience* 29:854–57.

43. Carpenter, W.J., E.R. Ostmark and J.A. Cornell. 1993. The role of light during *phlox drummondii* Hook. Seed germination. *HortScience* 28:786–88.

44. Carpenter, W.J. and E.R. Ostmark. 1992. Growth regulator and storage temperature govern germination of *Coreopsis* seed. *HortScience* 27:1190–93.

45. Carpenter, W.J. and S.W. Williams. 1993. Keys to successful seeding. *Grower Talks* 57:34–44.

46. Carter, F.M. 1973. Grafting cacti. *Horticulture* 51:34–5.

47. Cassells, A.C. 1992. Micropagation of commercial *Pelargonium* species and hybrids. *Biotechnology in Agriculture and Forestry. Vol. 20. High-Tech and micropropagation* IV. Y.P.S. Bajaj (ed.). Springer-Verlag, Berlin. pp. 286–306.

48. Cheers, G. 1992. *Letts guide to carnivorous plants of the world.* London. Charles Lett and Co.

49. Chidamian, C. 1984. *Book of cacti and other succulents.* Portland, Oreg.: Timber Press.

50. Chopping, N. 1986. Replacing the bird: Pollination in the genus *Strelitzia. Comb. Proc. Inter. Plant Prop. Soc.* 36:204–7.

51. Christie, C.B. 1985. Propagation of amaryllids: A brief review. *Comb. Proc. Inter. Plant Prop. Soc.* 35:351–57.

52. Chua, B.U., J.T. Kunisaki and Y. Sagawa. 1981. In vitro propagation of *Dracaena marginata* 'Tricolor'. *HortScience* 16:494.

53. Cialone, J. 1984. Developments in *Dracena* production. *Comb. Proc. Inter. Plant Prop. Soc.* 34:491–94.

54. Clayton, P.Q., J.F. Hubstenberger and G.C. Phillips. 1990. Micropropagation of members of the Cactaceae subtribe Cactinae. *Jour. Amer. Soc. Hort. Sci.* 115:337–43.

55. Cockrel, A.D., G.L. McDaniel, and E.T. Graham. 1986. In vitro propagation of florist's cineraria. *HortScience* 21:139–40.

56. Colborn, L.N. 1986. Primroses and violets: What's new. *Comb. Proc. Inter. Plant Prop. Soc.* 36:245–49.

57. Connor, D.M. 1985. Plants for the discriminating propagator. *Comb. Proc. Inter. Plant Prop. Soc.* 35:274–77.
58. Conover, C.A., and R.T. Poole. 1978. Production of *Ficus elastica* 'Decora' standards. *HortScience* 13:707–8.
59. Cooke, R.C. 1977. Tissue culture propagation of African violets. *HortScience* 12:549.
60. Criley, R.A. 1988. Propagation of tropical cut flowers: *Strelitzia, Alpinia,* and *Heliconia. Acta Hort.* 226:509–17.
61. Custers, J.B.M. and J.H.W. Bergervoet. 1992. Differences between *Nerine* hybrids in micropropagation potential. *Scientia Hort.* 52:247–56.
62. Darrow, G.M., and F.G. Meyer, eds. 1968. *Day lily handbook.* American Horticultural Society, Vol. 47, No. 2.
63. Davies, F.T., Jr., and B.C. Moser. 1980. Stimulation of bud and shoot development of Rieger begonia leaf cuttings with cytokinins. *Jour. Amer. Soc. Hort. Sci.* 105:27–30.
64. De Bruyn, M.H., D.I. Ferreira, M.M. Slabbert and J. Pretorius. 1992. In vitro propagation of *Amaryllis belladonna. Plant Cell Tissue Organ Cult.* 31:179–84.
65. Debergh, P. 1976. An in vitro technique for the vegetative multiplication of chimaeral plants of *Dracaena* and *Cordyline. Acta Hort.* 64:17–9.
66. Debergh, P., and J. DeWall. 1977. Mass propagation of *Ficus lyrata. Acta Hort.* 78:361–64.
67. DeHertogh, A. and M. Le Nard. 1993. *The physiology of flowering bulbs.* Elsevier, Amsterdam.
68. Dodge, M.H. 1978. Propagation of named delphinium cultivars. *Comb. Proc. Inter. Plant Prop. Soc.* 28:496–98.
69. Dodge, M.H. 1985. Propagation of herbaceous perennials by root cuttings. *Comb. Proc. Inter. Plant Prop. Soc.* 35:548–55.
70. Dunston, S., and E. Sutter. 1984. In vitro propagation of prayer plants. *HortScience* 19:511–12.
71. Earle, E.D., and R.W. Langhans. 1975. Carnation propagation from shoot tips cultured in liquid medium. *HortScience* 10:608–10.
72. Ecker, R. and A. Barzilay. 1993. Propagation of *Asclepias tuberosa* from short root segments. *Scientia Hort.* 56:171–74.
73. Edinger, P., ed. 1970. *Succulents and cactus.* Menlo Park, Calif.: Lane.
74. Erickson, D.M. 1985. Propagating and growing primroses in the Pacific Northwest. *Comb. Proc. Inter. Plant Prop. Soc.* 35:219–21.
75. Fakhrai, F. and P.K. Evans. 1989. Morphogenetic potential of cultured explants of *Crocus chrysanthus* Herbert cv. E.P. Bowles. *Jour. Exp. Bot.* 40:809–12.
76. Fay, A.M., M.A. Bennett and S.M. Still. 1994. Osmotic seed priming of *Rudbeckia fulgida* improves germination and expands germination range. *HortScience* 29:868–70.
77. Finnerty, T.L., J.M. Zajicek and M.A. Hussey. 1992. Use of seed priming to bypass stratification requirements of three *Aquilegia* species. *HortScience* 27:310–13.
78. Finnie, J.F. and J. Van Staden. 1989. In vitro propagation of *Sandersonia* and *Gloriosa. Plant Cell Tissue Organ Cult.* 19:151–58.
79. Gawel, N.J., C.D. Robacker and W.L. Corley. 1990. In vitro propagation of *Miscanthus sinensis. HortScience* 25:1291–93.
80. Gertsson, U.E. 1992. Micropagation of cineraria (*Senecio* × *hybridus* Hyl.). *Biotechnology in Agriculture and Forestry. Vol. 20. High-Tech and micropropagation* IV. Y.P.S. Bajaj (ed.). Springer-Verlag, Berlin. pp. 396–406.
81. Gilbertson-Ferriss, T.L., and H.F. Wilkins. 1977. Factors influencing seed germination of *Freesia refracta* Klatt cv. Royal Mix. *HortScience* 12:572–73.
82. Gill, L.M. 1984. Anemone tuber production in southwest England. *Comb. Proc. Inter. Plant Prop. Soc.* 34:290–94.
83. Hammer, P.A. 1976. Stolon formation in *Chlorophytum. HortScience* 11:570–72.
84. Han, S.S. 1993. Chilling, ethephon and photoperiod affect cormel production of *Brodiaea. HortScience* 28:1095–97.
85. Hanks, G.R. and A.R. Rees. 1979. Twin-scale propagation of *Narcissus.* A review. *Scientia Hort.* 10:1–14.
86. Hansen, J. 1986. Influence of cutting position and stem length on rooting of leaf-bud cuttings of *Schefflera arboricola. Scientia Hort.* 28:177–86.
87. Harazy, A., B. Leshem, A. Cohen, and H.D. Rabinowitch. 1985. In vitro propagation of statice as an aid to breeding. *HortScience* 20:361–62.
88. Harris, P.J.C., C.G. Grove and A.J. Harvard. 1989. In vitro propagation of *Symphytum* species. *Scientia Hort.* 40:275–81.
89. Henny, R.J. 1978. In vitro propagation of *Peperomia* 'Red Ripple' from leaf discs. *HortScience* 13:150–51.
90. Higaki, T., and D.P. Watson. 1973. Anthurium culture in Hawaii. *Univ. Hawaii Coop. Ext. Ser. Circ. 420.*
91. Holley, W.D., and R. Baker. 1963. *Carnation* production. Dubuque, Iowa: William C. Brown Co., Publishers.
92. Homes, J., M. Legeos and M. Jaziri. 1987. In vitro multiplication of *C. sativus* L. *Acta Hort.* 212:675–76.
93. Hosoki, T., and T. Asahira. 1980. In vitro propagation of bromeliads in liquid culture. *HortScience* 15:603–4.
94. Houghton, W.J. 1984. New narcissus and their propagation. *Comb. Proc. Inter. Plant Prop. Soc.* 34:294–96.
95. Huang, F.H., G.L. Klingaman, and H.H. Chen. 1992. Micropropagation of surprise lily *(Lycoris*

squamigera). Biotechnology in Agriculture and Forestry. Vol. 20. High-Tech and micropropagation IV. Y.P.S. Bajaj (ed.). Springer-Verlag, Berlin. pp. 198–212.

96. Huang, C.W., H. Okubo and S. Uemoto. 1990. Comparison of bulblet formation from twin scales and single scales on *Hippeastrum hybridum* cultured in vitro. *Scientia Hort.* 46:151–60.

97. Hubbard, A.C. 1986. Native ornamentals for the U.S. southwest. 1986. *Comb. Proc. Inter. Plant Prop. Soc.* 36:347–50.

98. Hubstenberger, J.F., P.W. Clayton and G.C. Phillips. 1992. Micropropagation of cacti *(Cactaceae). Biotechnology in Agriculture and Forestry. Vol. 20. High-Tech and micropropagation* IV. Y.P.S. Bajaj (ed.). Springer-Verlag, Berlin. pp. 49–68.

99. Hussey, G. 1975. Totipotency in tissue explants of some members of the Liliaceae, Iridaceae, and Amaryllidaceae. *Jour. Exp. Bot.* 26:253–62.

100. Ilan, A., M. Ziv and A.H. Halevy. 1995. Propagation of corm development of *Brodiaea* in liquid cultures. *Scientia Hort.* 63:101–12.

101. Jackson, H.C. 1975. Propagation and culture of African violets. *Comb. Proc. Inter. Plant Prop. Soc.* 25:269–71.

102. Jamieson, A.C. 1988. New Zealand callas. *Grower Talks* 51:56–60.

103. Jeham, H., D. Courtois, C. Ehret, K. Lerch and V. Petiard. 1994. Plant regeneration in *Iris pallida* Lam. and *Iris germanica* L. via somatic embryogenesis from leaves, apices and young flowers. *Plant Cell Rpts.* 13:671–75.

104. Jenks, M., M. Kane, F. Morousky, D. McConnell and T. Sheehan. 1990. In vitro establishment and epiphyllous plantlet regeneration of *Nymphaea* 'Daubeniana.' *HortScience* 25:1664.

105. Johnson, B.B. 1978. In vitro propagation of gloxinia from leaf explants. *HortScience* 13:149–50.

106. Jones, R.O., R.L. Geneve and S.T. Kester. 1994. Micropropagation of gas plant (*Dictamnus albus* L.). *Jour. Environ. Hort.* 12:216–18.

107. Jungnickel, F. and S. Zaid. 1992. Micropropagation of African violets (*Saintpaulia* spp. and cvs.). *Biotechnology in Agriculture and Forestry. Vol. 20. High-Tech and micropropagation* IV. Y.P.S. Bajaj (ed.). Springer-Verlag, Berlin. pp. 357–395.

108. Kane, M.E., T.J. Sheehan and F.H. Ferwerda. 1988. In vitro growth of American lotus embryos. *HortScience* 23:611–13.

109. Kelly, J.W., and J.J. Frett. 1986. Photoperiodic control of growth in water lilies. *HortScience* 21: 151.

110. Khehra, K.C. Lowe, M.R. Davey and J.B. Power. 1995. An improved micropropagation system for chrysanthemum based on Pluronic F-68-supplemented media. *Plant Cell Tissue Organ Cult.* 41:87–90.

111. King, J.J. and M.P. Bridgen. 1990. Environmental and genotypic regulation of *Alstroemeria* seed germination. *HortScience* 25:1607–09.

112. Kitto, S.L. and A. Hoopes. 1992. Micropropagation and field establishment of *Tiarella cordifolia. Jour. Environ. Hort.* 10:171–74.

113. Knauss, J.F. 1976. A partial tissue culture method for pathogen-free propagation of selected fern from spores. *Proc. Fla. State Hort. Soc.* 89:363–65.

114. Kononowicz, H., and J. Janick. 1984. In vitro propagation of *Vanilla planifolia. HortScience* 19:58–9.

115. Kothari, S.L., and N. Chandra. 1984. In vitro propagation of African marigold. *HortScience* 19:703–5.

116. Kramer, J. 1971. *How to grow African violets* (4th ed.). Menlo Park, Calif.: Lane.

117. Kunisaki, J.T. 1975. In vitro propagation of *Cordyline terminalis* (L.) Kurth. *HortScience* 10:601–2.

118. ———. 1980. In vitro propagation of *Anthurium andreanum* Lind. *HortScience* 15:508–9.

119. Kukulczanka, K., K. Kromer, B. Cvzanstka. 1989. Micropropagation of *Fritillaria melengris* L. through tissue culture. *Acta Hort.* 251:147–53.

120. Kusey, W.E., Jr., P.A. Hammer, and T.C. Weiler. 1980. In vitro propagation of *Gypsophila paniculata* L. 'Bristol Fairy'. *HortScience* 15(5):600–601.

121. Laliberté, S., L. Chretien, and J. Vieth. 1985. In vitro plantlet production from young capitulum explants of *Gerbera jamesonii. HortScience* 20: 137–39.

122. Lane, B.C. 1981. A procedure for propagating ferns from spore using a nutrient agar solution. *Comb. Proc. Inter. Plant Prop. Soc.* 30:94–7.

123. Lawson, G.M., and P.B. Goodwin. 1985. Commercial production of kangaroo paws. *Comb. Proc. Inter. Plant Prop. Soc.* 35:57–65.

124. Lazarte, J.E., M.S. Gaiser, and O.R. Brown. 1982. In vitro propagation of *Epiphyllum chrysocardium. HortScience* 17:84.

125. Lee, C.I., and H.C. Kohl. 1985. Note on vegetative reproduction of *Cyclamen indicum.* The Plant Propagator 31:4.

126. Lee, C.W., R.M. Skirvin, A.I. Soltero, and J. Janick. 1977. Tissue culture of *Salpiglossis sinuata* L. from leaf discs. *HortScience* 12:547–49.

127. Lilien-kipnis, H. and M. Kochba. 1987. Mass propagation of new *Gladiolus* hybrids. *Acta Hort.* 212: 631–38.

128. Lindgren, D.T. and B. McCown. 1992. Multiplication of four penstemon species in vitro. *HortScience* 27:182.

129. Liu, L., and D.W. Burger. 1986. In vitro propagation of Easter lily from pedicels. *HortScience* 21:1437–38.

130. Lyons, R.E., and R.E. Widmer. 1980. Origin and historical aspects of *Cyclamen persicum* Mill. *HortScience* 15(2):132–35.

131. Matsuo, E., and J.M. van Tuyl. 1986. Early scale propagation results in forcible bulbs of Easter lily. *HortScience* 21:1006–7.

132. McKay, M.E., D.E. Bythe and J.A. Tommerup. 1981. The effects of corm size and division of the mother corm in gladioli. *Aust. Jour. Expt. Agric. Animal Husbandry* 21:343–48.

133. McMillan, R.N. 1980. Cyclamen production problems. *Comb. Proc. Inter. Plant Prop. Soc.* 29:173–76.

134. McRae, E.A. 1978. Commercial propagation of lilies. *Comb. Proc. Inter. Plant Prop. Soc.* 28:166–69.

135. Mekers, O. 1977. In vitro propagation of some Tillandsiodeae (Bromeliaceae). *Acta Hort.* 78:311–20.

136. Meyer, H.J. and J. Van Stadden. 1991. Rapid in vitro propagation of *Aloe barbandensis* Mill. *Plant Cell Tissue Organ Cult.* 26:167–71.

137. Meyer, M.M., Jr. 1980. In vitro propagation of *Hosta sieboldianum. HortScience* 15:737–38.

138. ———. 1976. Propagation of daylilies by tissue culture. *HortScience* 11:485–87.

139. Mikkelsen, E.P., and K.C. Sink, Jr. 1978. In vitro propagation of Rieger Elatior begonias. *HortScience* 13(3):242–44.

140. Minocha, S.C. 1985. In vitro propagation of *Dionaea muscipula. HortScience* 20:216–17.

141. Morel, G.M. 1966. Meristem culture: Clonal propagation of orchids. *Orchid Digest* 30:45–49.

142. Murashige, T., M. Serpa, and J.B. Jones. 1974. Clonal multiplication of gerbera through tissue culture. *HortScience* 9:175–80.

143. Nakano, M., Y. Hoshino and M. Mii. 1994. Adventitious shoot regeneration from cultured petal explants of carnation. *Plant Cell Tissue Organ Cult.* 36:15–20.

144. Natali, L., I.C. Sanchez and A. Cavallini. 1990. In vitro culture of *Aloe barbadensis* Mill.: Micropropagation from vegetative meristems. *Plant Cell Tissue Organ Cult.* 20:71–4.

145. Nathan, M.J., C.J. Goh and P.P. Kumar. 1992. In vitro propagation of *Heliconia psittacorum* by bud culture. *HortScience* 27:450–52.

146. Nau, J. 1993. *Ball culture guide: the encyclopedia of seed germination.* Ball Pub. Batavia, IL.

147. Newbury, H.J., E.A.B. Aitken, N.J. Atkinson and B.V. Ford-Lloyd. 1992. Micropropagation of snapdragon (*Antirrhinum majus* L.). *Biotechnology in Agriculture and Forestry. Vol. 20. High-Tech and micropropagation* IV. Y.P.S. Bajaj (ed.). Springer-Verlag, Berlin. pp. 19–33.

148. Peck, D.E. and B.G. Cumming. 1986. Beneficial effects of activated charcoal on bulblet production in cultures of *Muscari armeniacum. Plant Cell Tissue Organ Cult.* 6:9–14.

149. ———. 1984. In vitro vegetative propagation of cape primrose using the corolla of the flower. *HortScience* 19:399–400.

150. ———. 1984. In vitro propagation of *Begonia × tuberhydra* from leaf sections. *HortScience* 19:395–97.

151. Perl, P. 1978. *Cacti and succulents.* Alexandria, Va.: Time-Life Books.

152. ———. 1977. Ferns. Alexandria, Va.: Time-Life Books.

153. Pfister, J.M., and J.M. Widholm. 1984. Plant regeneration from snapdragon tissue culture. *HortScience* 19:852–54.

154. Pierik, R.L.M. and H.H.M. Steegmans. 1976. Vegetative propagation of *Freesia* through isolation of shoots in vitro. *Netherlands Jour. Agric. Science* 24:274–77.

155. Platt, G.C. 1985. Propagation of the cordylines by vegetative means. *Comb. Proc. Inter. Plant Prop. Soc.* 35:364–66.

156. Poole, R.T., and C.A. Conover. 1987. Vegetative propagation of foliage plants. *Comb. Proc. Inter. Plant Prop. Soc.* 37:503–7.

157. Preece, J.E. and K.P. Ripley. 1992. In vitro culture and micropropagation of *Euphorbia* sp. In *Biotechnology in Agriculture and Forestry. Vol. 20. High-Tech and micropropagation* IV. Y.P.S. Bajaj (ed.). Springer-Verlag, Berlin. pp. 91–112.

158. Quay, L., J.A. McComb and K.W. Dixon. 1995. Methods for ex vitro germination of Australian terrestrial orchids. *HortScience* 27:182.

159. Ramachandra, S. And H. Khatamian. 1989. Micropropagation of *Peperomia* and *Begonia* using petiole segments. *HortScience* 24:153.

160. Rauch, F.D., and R.M. Yamakawa. 1980. Effects of auxin on rooting of *Ixora acuminata. HortScience* 15:97.

161. Raviv, M., E. Putieusky, and D. Sanderovich. 1984. Rooting stem cuttings of sage (*Salvia officinalis* L.). *The Plant Propagator* 30:8–9.

162. Raward, I.D. 1975. Propagation of *Codiaeum* (croton) by tip cuttings. *Comb. Proc. Inter. Plant Prop. Soc.* 25:386.

163. Redwood, G.N. and J.C. Bowling. 1990. Micropropagation of *Nepenthes* species. *Bot. Gardens Microprop. News* 1:19–20.

164. Richwine, A.M., J.L. Tipton and G. Thompson. 1995. Establishment of *Aloe, Gasteria,* and *Haworthia* shoot cultures from inflorescence explants. *HortScience* 30:1443–44.

165. Robacker, C.D. and W.L. Corley. Plant regeneration of pampas grass from immature inflorescences cultured in vitro. *HortScience* 27:841–43.

166. Roberts, D.J. 1965. Modern propagation of ferns. *Comb. Proc. Inter. Plant Prop. Soc.* 15:317–21.

167. Rockwell, F.F., G.C. Grayson, and J. de Graff. 1961. *The complete book of lilies.* Garden City, N.Y.: Doubleday.

168. Roest, S., and G.S. Bokelmann, 1975. Vegetative propagation of *Chrysanthemum morifolium* Ram in vitro. *Scientia Hort.* 3:317–30.

169. Samfield, D.M., J.M. Zajicek and B.G. Cobb. 1988. The effects of osmoconditioning on herbaceous perennial seed germination at different temperatures. *HortScience* 23:750.

170. Schwartz, M.A., R.N. Payne, and G. Sites. 1985. Residual effects of chlormequat on garden performance in sun and shade of seed and cutting propagated cultivars of geraniums. *HortScience* 20:368–70.

171. Schwenkel, H.G. and J. Grumwaldt. 1988. In vitro propagation of *Cyclamen persicum* Mill. *Acta Hort.* 226:659–62.

172. Sheehan, T.J. 1983. Recent advances in botany, propagation and physiology of orchids. In *Horticultural reviews,* Vol. 5, J.A. Janick ed., Westport, Conn., AVI Publ. Co., pp. 279–315.

173. Sheerin, P. 1974. Propagation of various types of begonia. *Comb. Proc. Inter. Plant Prop. Soc.* 24:292–93.

174. Shoemaker, C.A. and W.H. Carlson. 1992. Temperature and light affect seed germination of *Begonia semperflorens-cultorum. HortScience* 27:181.

175. Simmonds, J. 1992. Micropropagation of *Begonia* spp. *Biotechnology in Agriculture and Forestry. Vol. 20. High-Tech and micropropagation* IV. Y.P.S. Bajaj (ed.). Springer-Verlag, Berlin. pp. 34–48.

176. ———, and T. Werry. 1987. Liquidshake culture for improved micropropagation of *Begonia* × *hiemalis. HortScience* 22:122–24.

177. Slocum, P.D. 1985. Propagating water lilies and aquatics. *Brooklyn Botanic Garden Record* 41:25–28.

178. Small, D.J. 1986. Propagation of choice alpines. *Comb. Proc. Inter. Plant Prop. Soc.* 36:241–44.

179. Smith, R.H., and A.E. Nightingale. 1979. In vitro propagation of *Kalanchoe. HortScience* 14(1):20.

180. Solem, M. 1988. Torenia. *Grower Talks* 52:18.

181. Standifer, L.C., E.N. O'Rourke, and R. Porche-Sorbet. 1984. Propagation of *Haworthia* from floral scapes. *The Plant Propagator* 30:4–6.

182. Stanilova, M.I., V.P. Ilcheva and N.A. Zagorska. 1994. Morphogenetic potential and in vitro micropropagation of endangered plant species *Leucojum aestivum* L. and *Lilum rhodopaeum* Delip. *Plant Cell Rpts.* 13:451–53.

183. Sapfer, R.E. and C.W. Heuser. 1986. Rapid multiplication of *Heuchera sanguinea* Engelm. 'Rosamundi' propagated in vitro. *HortScience* 21:1043–44.

184. ———. 1985. In vitro propagation of periwinkle. *HortScience* 20:141–42.

185. Start, N.D., and B.G. Cumming 1976. In vitro propagation of *Saintpaulia ionantha* Wendl. *HortScience* 11:204–6.

186. Stefanis, J.P., and R.W. Langhans. 1980. Factors affecting the production and propagation of xerophytic succulent species. *HortScience* 15:504–5.

187. Stephens, L.C., J.L. Weigle, S.L. Krell and K. Han. Micropropagation of *Impatiens. Biotechnology in Agriculture and Forestry. Vol. 20. High-Tech and micropropagation* IV. Y.P.S. Bajaj (ed.). Springer-Verlag, Berlin. pp. 160–172.

188. ———, S.L. Krell, and J.L. Weigle. 1985. In vitro propagation of Java, New Guinea and Java × New Guinea Impatiens. *HortScience* 20:362–63.

189. Stewart, J., and E. Hennessy. 1981. *Orchids of Africa.* Boston: Houghton Mifflin.

190. Still, S.M. 1994. *Herbaceous ornamental plants* (3rd ed.). Champaign, Ill.: Stipes Publ. Co.

191. Stimart, D.P. and J.F. Harbage. 1989. Shoot proliferation and rooting in vitro propagation of *Liatris spicata. HortScience* 24:835–36.

192. ———. 1986. Commercial micropropagation of florist flower crops. In *Tissue culture as a plant production system for horticultural crops,* R.H. Zimmerman et al., eds. Dordrecht: Martinus Nijhoff Publishers.

193. ———, and P.D. Ascher. 1978. Tissue culture of bulb scale sections for asexual propagation of *Lilium longiflorum* Thumb. *Jour. Amer. Soc. Hort. Sci.* 103:182–84.

194. Stokes, P. 1984. Hardy ferns. *Comb. Proc. Inter. Plant Prop. Soc.* 34:332–33.

195. Takayama, S. 1990. Begonia. In *Handbook of plant cell culture.* P.V. Ammirato, D.A. Evans, W.R. Sharp and Y.P.S. Bajaj (eds.). McGraw-Hill. pp. 253–283.

196. Tanabe, M.J. 1980. Effect of depulping and growth regulators on seed germination of *Alyxia olivaeformis. HortScience* 15:199–200.

197. ———. 1982. Single node stem propagation of *Alyxia olivaeformis. HortScience* 17:50.

198. Thorn-Horst, A., R.K. Horst, S.H. Smith, and W.A. Oglevee. 1977. A virus-indexing tissue culture system for geraniums. *Flor. Rev.* 160(4148):28–29, 72–74.

199. Tjosvold, S. 1978. Uniform spore dispersal on warm nutrient agar solution. *Univ. Calif. Nursery and Flower Rpt.,* summer issue.

200. Torres, K.C., and N.J. Natarella. 1984. In vitro propagation of *Exacum. HortScience* 19:224–25.

201. Upadhyaya, A., T.D. Davis, D. Sankhla, and N. Sankhla. 1992. Micropropagation of *Lupinus texensis* from cotyledonary node explants. *HortScience* 27:1222–23.

202. Van der Linde, P.C.G. and J.A. Schipper. Micropagation of iris with special reference to *Iris* × *hollandica* Tub. *Biotechnology in Agriculture and Forestry. Vol. 20. High-Tech and micropropagation* IV. Y.P.S. Bajaj (ed.). Springer-Verlag, Berlin. pp. 173–197.

203. Van de Pol, P.A., and T.F. van Hell. 1988. Vegetative propagation of *Strelitzia reginae. Acta Hort.* 226:581–86.

204. Van Dyk, M., and R. Currah. 1983. Vegetative propagation of prairie forbs native to southern Alberta, Canada. *The Plant Propagator* 28:12–14.

205. Verron, P., M. Le Nard and J. Cohat. 1995. In vitro organogenic competence of different organs and tissues of lily of the valley 'Grandiflora of Nantes'. *Plant Cell Tissue Organ Cult.* 40:237–42.

206. Vinterhalter, D.V. 1989. In vitro propagation of green-foliaged *Dracaena fragrans* Ker. *Plant Cell Tissue Organ Cult.* 17:13–19.

207. Von Hentig, W.U., M. Fisher, and K. Köhler. 1984. Influence of daylength on the production and quality of cuttings from *Fuchsia* mother plants. *Comb. Proc. Inter. Plant Prop. Soc.* 34:141–49.

208. Wang, Y.T. 1987. Effect of warm-medium, light intensity, BA and parent leaf on propagation of golden pothos. *HortScience* 22:597–99.

209. ———. 1987. Effect of temperature, duration and light during simulated shipping on quality and rooting of croton cuttings. *HortScience* 22:1301–2.

210. ———. 1987. Influence of light and heated medium on rooting and shoot growth of two foliage plant species. *HortScience* 23:346–47.

211. ———, and C.A. Boogher. 1988. Effect of nodal position, cutting length, and root retention on the propagation of golden pothos. *HortScience* 23:347–49.

212. ———. 1994. Source and seed quality influence germination in purple coneflower *(Echinacea purpurea). HortScience* 29:1443–44.

213. Wartidininghsih, N. And R.L. Geneve. 1994. Osmotic priming or chilling stratification improve seed germination of purple coneflower *(Echinacea purpurea). HortScience* 29:1445–48.

214. Watelet-Gonad, M.C. and J.M. Favre. 1981. Miniaturization et rajeunissement chez *Dahlia* variabilis (variete Television) cultive in vitro. *Ann. Sciences Naturelles Bot.,* Paris pp. 51–67.

215. Welander, T. 1978. In vitro organogenesis in explants from different cultivars of *Begonia* × *hiemalis. Physiol. Plant.* 41:142–45.

216. Widmer, R.E. 1980. Cyclamens. In *Introduction to floriculture,* R.A. Larson, ed. New York: Academic Press.

217. Wilson, H.V. 1980. *Saintpaulia* species. *Amer. Hort.* 58:35–39.

218. Zieslin, N., and A. Keren. 1980. Effects of rootstock on cactus grafted with an adhesive. *HortScience* 21:153–54.

219. Zimmerman, T.W., F.T. Davies, Jr. and J.M. Jajicek. 1991. In vitro and macropropagation of the wildflower *Dyssodia pentacheta* (D.C.) Robins. *HortScience* 26:1555–57.

220. Ziv, M. and and J. Lilien-Kipnis. 1990. *Gladiolus.* In *Handbook of plant cell culture.* P.V. Armmirato, D.A. Evans, W.R. Sharp and Y.P.S. Bajaj. McGraw-Hill, NY. pp. 461–78.

SUPPLEMENTARY READING

Aden, P. 1988. *The hosta book.* Portland, Oreg.: Timber Press.

Armitage, A.M. 1986. *Seed propagated geraniums.* Portland, Oreg.: Timber Press.

Y.P.S. Bajaj (ed.). *Biotechnology in Agriculture and Forestry. Vol. 20. High-Tech and micropropagation* IV. Springer-Verlag, Berlin.

Ball, V. 1991. *Ball red book: Greenhouse growing* (15th ed.) Reston, Va.: Reston Publ. Co.

Benson, L. 1983. *The cacti of the U.S. and Canada.* Palo Alto, Calif.: Stanford Univ. Press.

Dehertogh, A. and M. Le Nard. 1993. *The physiology of flowering bulbs.* Elsevier, Amsterdam.

Chidamian, C. 1984. *Book of cacti and other succulents.* Portland, Oreg.: Timber Press.

George, A.S. 1985. *The banksia book.* Portland, Oreg.: Timber Press.

Giles, F.A., R. McIntosh, and D.C. Saupe. 1980. *Herbaceous perennials.* Reston, Va.: Reston.

Graf, A.B. 1986. *Exotica 4: Pictorial encyclopedia of exotic plants from tropical and near-tropical regions* (12th ed.). East Rutherford, N.J.: Roehrs Company.

Growers Books. 1980. *New cut flower crops. Grower guide 18.* London: Grower Books. International Plant Propagators' Society. Proceedings of annual meetings.

Joiner, J.N., ed. 1981. *Foliage plant production.* Englewood Cliffs, N.J.: Prentice-Hall.

Kohlein, F. 1988. *Iris.* Portland, Oreg.: Timber Press.

Larson, R.A., ed. 1980. *Introduction to floriculture.* New York: Academic Press.

Laurie, A., D.C. Kiplinger, and K.S. Nelson. 1979. *Commercial flower forcing* (8th ed.). New York: McGraw-Hill.

Levy, M. 1988. Perennial productions grower's notebook for summer. *Grower Talks* 52:72–78.

Mastalerz, J.W., ed. 1976. *Bedding plants* (2nd ed.). University Park, Pa.: Pennsylvania State Univ.

Nau, J. 1993. *Ball culture guide: the encyclopedia of seed germination.* Ball Pub. Batavia, IL.

Neel, P.L. 1979. Macropropagation of tropical plants as practiced in Florida. *Comb. Proc. Inter. Plant Prop. Soc.* 29:468–80.

Padilla, V. 1986. *Bromeliads.* New York: Crown Publ. Co.

POOLE, R.T., and C.A. CONOVER. 1988. Vegetative propagation of foliage plants. *Comb. Proc. Inter. Plant Prop. Soc.* 37:503–7.

PROFESSIONAL PLANT GROWERS ASSOCIATION. Annual proceedings.

REILLY, A. 1978. *Park's success with seeds.* Greenwood, S.C.: Geo. W. Park Seed Co., Inc.

RICE, G. 1986. *A handbook of annuals and bedding plants.* Portland, Oreg: Timber Press.

STILL, S.M. 1994. *Herbaceous ornamental plants* (3rd ed.). Champaign, Ill.: Stipes Publ. Co.

STIMART, D.P. 1986. Commercial micropropagation of florist flower crops. In *Tissue culture as a plant production system for horticultural crops,* R.H. Zimmerman, et al., eds. Dordrecht: Martinus Nijhoff Publishers.

WELLS, J.S. 1988. *Modern miniature daffodils.* Portland, Oreg.: Timber Press.

YOUNG, J.A.C. 1986. *Collecting, processing and germinating seeds of wildland plants.* Portland, Oreg.: Timber Press.

ZILIS, M, D. ZWAGERMAN, D. LAMBERTS, and L. KURTZ. 1979. Commercial propagation of perennials by tissue culture. *Comb. Proc. Inter. Plant Prop. Soc.* 29:404–13.

Indexes

SUBJECT INDEX

Abscisic acid (ABA), 20, 25, 142, 205, **208,** 291
Accelerated growth techniques (AGT), **83,** 317
Acclimation, 235
Acclimatization, 562, 608
Acrylic, 58
Additive variation, 116
Adenine, 17
Adventitious bud formation, 12, 556, **563,** 564, **565, 566**
Adventitious buds, 284, **285**
  origin, 283–84
Adventitious embryony, 19, 31, 134
Adventitious roots, 12, 19, **277–79, 283–85,** 289
  de novo, **278,** 281, **285,** 289
  preformed, 277, **278, 279**
  ex vitro, 560, **561, 602, 603,** 605
  in vitro, 560, **561, 602, 603,** 605
Adventitious shoots, 12, 19, 284, **285, 290,** 556, 563–66
Aeroponic systems, 44
Aerial stem bulblets, 529
After-ripening, 177, 195, 199
Agar, 598
  substitutes, 599
Agrobacterium, 17, 93, 296, 363
Algal growth, and cutting propagation, 91, 373
Alternation of generations, 10
Amylase, 180
American Association of Nurserymen, 6
American Seed Trade Association, Inc., 6
American Society for Horticultural Science, 6
AMO 1618 (Carvadan), 25
Ancymidol, 25
Anderson (AND) medium, 597
Annuals, 26
Anther culture, 566, 568
Antigibberellin, 25
Antioxidants, 557
Antisense technology, 20
Aplospory, 31
Apomixis, 108, 116, 134–39, 137
  faculative, 108
  life cycle, apomictic, 31
  obligate, 108, 137
  seedlings, apomictic, 108, 629
  significance of, 31, 138
  vegetative, apomixis, 139
Aseptic procedures, 599
Asexual reproduction, 10, **12**
Association of Official Seed Analysts, 6, 154
Association of Official Seed Certifying Agencies, 6, 120, 121, 154
Aster yellows, 259
Autotrophic (autotrophism), 577
Auxins, 20, 22, 288, 289, 309, 310, 355–62
  activity, 20, 22, 141
  and physiology of in cuttings, 289
  list of commercial compounds, 356
  methods of application, 24, 358, 359, **360, 361,** 561
  storage of solutions, 24, **357**
  structural formulas, **23**
Auxin synergists, 291
Axillary shoot proliferation, **563**

Bacillus thuringiensis (BT), 88
Back bulbs, 545
Bailey, Liberty Hyde, 4, 166
Banding, 303, 348, **350,** 503
BA (BAP), 24, 25, 560, 598
Bare-root nursery stock, handling, 383, **384, 385**
Bark, 73, 351
Basal plate, 520, **522**
Bedding Plants, Inc. (B.P.I.), 7
Benches, movable, **48**
Bench grafting (bench working), 470, 471
Benzyladenine (BA or BAP), 24, 25, 289, 560, 598
Best management practices (BMP), 81, 93, **94**
Biochemical markers, 16, 260
Biennials, 26
Biocontrol (*see* integrated pest management), 85, 88, 163, 363
Biofungicides, 85, 88
Biological control in IPM, 85, 88, 363
Bioreactors, 570, **571**
Biotechnology, 15, 244, 549
Biotic factors, 40, **41,** 83
Black-end of pears, 653
Black-line of walnuts, 637
Blanching, 303, 348–50, 503
Botanical classification, 32, 33
Botanical Nomenclature, International Code of, 32
Botanical varieties, 32, 117
Botrytis, 84, 88
Bottom (basal) heat, for cuttings, **366**
Breeder's seed, 112, 120
Breeding, 15, 16
Breeding system, 106, 114
Bryophyta, 32

Boldface numbers indicate figures.

Bud formation, 556, **563,** 564, **565, 566**
Budding, 392, 393
  bud-forcing methods, 412, **413**
  bark slipping 406, **408,** 409, 411
  chip, **408,** 487, **488, 489, 490**
  double-working by budding, **499**
  fall budding, 482, 483, **484, 487**
  flute, **498**
  I, 496, **498**
  June budding, 485, **486, 487**
  microbudding, 568
  patch, 494, **495, 496, 497,** 498
  ring, **498**
  rootstocks for, 482
  spring budding, 484, 485, 487
  T, or shield, **408,** 491, **492, 493,** 494
  top-budding (topworking), **499**
  types of budding, 487–98
Bud-mutations, 34, 244
Buds, effects on rooting, 306
  proper type for budding, **483**
Bud-sport, **244**
Bulb chipping (fractional scale-stem cuttage), 532
Bulbils, **528,** 529, **530**
Bulblet formation, **528,** 529, **530**
Bulbs,
  defined, 520
  growth cycles, 520
  micropropagation of, **533**
  nontunicate (scaly), 522, **524**
  offset, 527
  structure of, **523**
  tunicate (laminate), **522, 523**
  types, 522
Bulb scales, 529, **530, 531**
Burr knots, 277, 279

Calcined clay, 165
Calcium hypochlorite, 553, 557, 599, 560
Callus, **12,** 552
  culture, **563,** 566, 567, 569
  in grafting, 394, **409**
  root formation, 278, 281, **283,** 335
Callus bridge, in grafting, **401,** 402, **403**
Cambium, vascular, 281, 394, **401**
Canadian Seed Trade Association, 7
Captan, 163
Carbon dioxide, 44
  enrichment, 82, 577
Carbohydrates,
  in micropropagation, 587
  in rooting, 304, 305, 317
Carpal tunnel syndrome, in grafting, 469
Cell competency-to-root, 294, **295,** 296, 297
Cell recognition, in grafting, 422
Cell structure, 10, 11
Cell suspensions, 570, **571**
Certified seed, 112, 120
Chemical control of disease, 85, **91,** 163, 258, **362, 363**
Chemical fumigation, 85, **86**
Chemical revolution, 258
Cherry leaf roll virus, 637
Chimeral breakdown, 250, 580
Chimeras, 244
  anatomical origin of, 246, **248, 249**
  detection of, 250
  graft, 245
  in micropropagation, 250, 580
  reversions, **249**
  stability, 248, 580, **581**
Chlorine, 92, **93,** 362
Chlormequat (CCC), 25
Chloropicrin, 87
Chromosomes, homologous, 10
  structure, 17, **18**
Classification of plants, 33
Cline, 32, 117
Clonal life cycle, **30,** 251
  repositories, 269
  seed sources, 118
  selection, 267
  variation, genetic, **244**
    nongenetic, **30, 31**
Clone, 12, 34, 239
  cultivars, origin, **242**
  history, 239
  reasons for using, 240
Cloning, 12, 239, **241**
Clorox, 362, 553, 557, 560
Closed-case propagation, 58, **59, 60,** 62, 81, 314, **364,** 365
Coconut milk, 25
Codon, 17
Cofactors, in rooting, 291, 292, **293, 294,** 295
Cold frames, 59, **60**
Cold storage, of rooted and unrooted cuttings, 380, 381, **382, 383,** 384, 385
Color breaks, 258
Competency (*see* rooting), 19, **22**
Compound layering, 505
Computerized controllers, for propagation, environmental systems, **43,** 55, 370
Computer modeling in propagation, 83, **371**
Cone of juvenility, 251, 256
Cones, processing, 153
Contact polyethylene propagation system, 315
Containers, 62–68
  clay pots, 63
  fiber pots, 65
  in-ground fabric containers (grow bags), 97, **98**
  plastic containers (pots), **64, 67, 69**
  pot-in-a-pot container systems, 97, **98**
  metal containers, 65
  paper pots, 65
  propagation flat (trays), 62, 228, **353**
  wood containers, 68, **69**
Container growing (production), 94
  Ohio Production System (OPS), 97
Contractile roots, **525**
Contracts, 36
Copper hydroxide, containers treated with, 64, **66,** 98
Copyright, 36
Cormels, **535**
Corms, **534, 535**
Cornell peat-lite mixes, 74
Crinkle, cherry, 257
Crop Improvement Association, 112, 120
Cross-pollination, 106, 114, **115**
Crown, 286, **541**
  division, 515, **516–18**
Cruciferin, 140
Cryopreservation, of seeds, 169, **170**
Culm, 453
Cultivar, 14, 34
  asexually reproduced, 34, **242**
  sexually reproduced, 34, 111
Culture, shoot-tip, **265**
Culture indexing, 262, 557, 579
Cuttings (*see* rooting)
  accelerated growth techniques, **83,** 317
  advantages of, 329
  algal growth problems, 91, 373,
  auxin effects on, 288, 289, 310, 311
  bottom (basal) heat, **52,** 335, 336, **366,** 378
  broad-leaved evergreens, 308
  care of during rooting, 378, 379
  costing variables, 375
  direct sticking (rooting), 40, **76,** 352, **353**
  disease/pathogen control and treatment (*see* IPM), 352, **362, 363**
  environmental conditions for rooting, 312–17, 363, 378, 379
  environmental modeling, 83, **371**
  fog systems, 370–73
  fungicides and rooting, 362, 363
  handling field-propagated hardwood cuttings, **335**
  handling cutting material, 308–11
  hardening off and post-propagation care, 379–83
  hardwood cuttings, 331–33, **334–36**
  herbaceous cuttings, **339, 340**
  herbicide use and effects, 379
  humidity effects, 314
  photosynthesis effects, 317, **318**
  temperature effects, 315, 316
  leaching of nutrients, 311
  leaf cuttings, 282, 285, 340, **341, 342**
  leaf and bud effects, 287, **288,** 306
  leaf-bud (single-eye, single node cutting), 341, **342**
  light influence, 316, 317
  microcuttings, 560, **561, 602, 603,** 605
  mist propagation controllers, **57,** 366, **367, 369**
  narrow-leaved evergreens, 308
  nutrition of, 45, 310, 311, 377, 378
  piece-work systems 375
  polyethylene sheeting for propagation, 364
  preparing for rooting, 308–11, 329–40
  preventing mist operation problems, 373
  record keeping, 374, **375, 376**
  root cuttings, 342, **343,** 344
  rooting medium, 68–75, 349–52
  seasonal timing, **307**
  semi-hardwood cuttings, **337**
  softwood cuttings, **304,** 337, **338, 339**
  sources of cutting material, 344, 345
  split base, hardwood cuttings, **312**
  stem cuttings, 330, 331
  sticking (inserting) cuttings, **374**
  storage temperature and duration, 308, 380–85, **382, 383**

Cuttings, (*cont.*)
stripping, 353, **354**
temperature effects, 315, **316**
timing and scheduling, 377, **378**
treatment with growth regulators, 288–92, 309, 310, 355–62
types of, 330–44
water quality problems, 373
water relations during rooting, 312–15, **313**
wounding, 311, **312,** 313, 353, **354,** 355
Cutting material, 305, **306**
apical, medial, and basal portions of shoot, 306
flowering and vegetative wood, 306
heel vs. nonheel cuttings, 331
lateral and terminal growth, 305
rejuvenation techniques, 579, 580
source of, 305, 306
type of wood selected, 305, **306,** 308
Cyclocel (CCC), 25
Cyclophysis, 28
Cylindrocladium, 84
Cyroflex, 58
Cytosine, 5, 17
Cytokinin, 20, 25, 142, 206, 289, 560, 598
and organ formation, 25, **285,** 289, **290,** 558
and shoot development, 25, **285,** 289, **290, 560**

Damping-off pathogenic fungi, 84, 193, 352
Dedifferentiation, in rooting, 277, 278
Deionization, 594
Deoxyribose nucleic acid (DNA), 17, 139, 190
Desiccation, seed, 140, 141
Determinism, 20, **22**
Dichogamy, 108, 114
Dihybrid cross, 15
Dioecy, 108, **109,** 114
Diphenylurea, 25, 598
Diploid, 10, 128
Diplospory, 31
Direct seeding in nursery row, 223
Direct sticking, of cuttings, 40, **76,** 352, **353**
Disease and propagation, 258
Disease-free, concept of, 261
Disinfectants, **91,** 92, 553, 557, 599, 560
Distal, 286, 414
DNA, 17, **18**
fingerprinting, **160**
Domestication (plants), 2
Dominance, 14
Dormancy, seed, 141, 194–210
chemical, 198
concepts of, 194
double, 195, 204
endogenous, 195, 198–204
epicotyl, 204
exogenous, 195–197
intermediate, 195, 201–203
morphological, 198, 199
nondeep (shallow), 195, 199
photodormancy, 199, **200,** 201
physical, 195, **196, 197**
physiological, 195, 199
primary, 194
secondary, 195, 204
thermodormancy, 205
Double cross, hybrid seed production, 111
Double dormancy, 195, 204
Double fertilization, 128
Double working, *see* budding, grafting, **399, 499**
Drip irrigation, 94, **95**
Droppers, 523
Dynaglas, 58

Ecodormancy, 194
Ecotype, 32
Edaphic factors, 40
management of in propagation, **41,** 68–75
Electrolytic leakage, 161
Electrophoresis, 160
ELISA test, 84, 263, **265**
Elite plants, **42,** 116, 117
Embryo,
abortion, 129
culture, 555
development in conifers, 133, **134, 136**
development in dicots, 130
development in monocots, 132
dormancy, 199
globular (shaped), **132, 133**
heart-shaped, **132, 133**
maturation, 140
rudimentary, 199
undeveloped, 199
proembryo, **132, 135**
scutellar embryo, 132, **134**
suspensor, **132, 133**
torpedo-shaped, **132, 133**
Embryogenesis, 132–36
somatic, **574, 575, 580,** 609, 610
Embryoids, **572**
Embryonal-suspensor mass (ESM), **573**
Enclosed mist systems, (*see* propagation systems), **59**
Enclosure systems (closed-case propagation), 58, **59, 60, 364,** 365
Endodormancy, 195
Endoplasmic reticulum, in grafting, **407**
Endosperm, 128, 130, **134,** 185, 186
Endotha parasitica, 628
Environmental conditions, for rooting cuttings, 312–17
Enzyme-linked immunosorbent assay, 263, **265**
Epicormic shoots, **251,** 256
Epicotyl, 184
Epicotyl dormancy, 195
Epigenetic process, 19, 29, 252, 254, 580
Epigeous germination, 184
Ergonomics, in budding, 493
Ethephon, 26
Ethylene (C_2H_4), **20,** 26, 143, 206, 291
Etiolation, 303, **304,** 348, **349, 350,** 503
Excised embryo seed test, 155, **158**
Excised leaf virus testing, **264**
Exolite, 58
Exon, 17
Exudation, 557, 577
Ex vitro rooting, **560, 603,** 605

F_1 hybrid, 111, 112, **113**
F_2 strains, 111
Facultative apomicts, 108, 138
Family selection, seeds, 118
Fastidious bacteria, 259
Federal Insecticide, Fungicide and Rodenticide Act (FIFRA), 87
Federal Seed Act, 120, 153
Female gametophyte, 130, 133
Fern,
reproduction, **14**
spore cultures, 735
Fertigation, 45
Fertilization, 45, 74, 75
double fertilization, ovule, 128, 129
Fertilizers, during and after propagation,74–78, 95
controlled-release (slow-release), 74, 76–78
granular, 74
liquid, 74, 75, 77
pre-plant (preincorporated), 74
post-plant, 74, 77, 78
topdressing (broadcasting), 77
Fiberglass, 58
Fiber pots, 65
Field production (nursery), 97, 98, 383–87
Finger printing, 160, 260
"Fixing"
apomixis, 31
in clones, 240
phase change, 254
self pollination, 15, 106, **107**
Flats (trays), 62, 228, **353**
Float beds, **234**
Floor ebb and flood system (flood floor), 46, **49**
Flower structure, **129**
Fluid drilling, **166,** 222
Fogging controllers, 371
Fogging systems, 315, 370–73
Foundation,
block, 112
clone, 267
seed, 120
system, **270**
Fragmented shoot apex culture, 565
Frameworking, 473
Fruit and seed development, 126–44
growth, in lettuce, **131**
morphology, 126, 127
Fumigation with chemicals, 85
Fungicidal soil drenches, 87
Fusarium, 84, 87, 88

Gametes, 10, 12
Gene
expression, 19, 20, 257
pattern genes, 257
structure, 17
Genetic engineering, 15, **575,** 578
Genetic drift, 110
Genetic shift, 110
Genetic variability, 105
concepts of, 14
control of, 110, 115
Genotype, 9, 14
Genotype x environment interaction, 242, 250
Genotypic selection, 116, 261
Germination,
of barley seeds, **186**
of cocklebur seeds, **206**
curve, **188**
environmental factors, 188
seed stratification, 218
seed treatments to facilitate, 162
percentage, 154, 187
rate, 187
speed, 187
stages of, 179
testing, 153
Germplasm resources information network (GRIN), 171
Gibberellins, 20, 25, 142, 205, 290, 598
Girdling, as related to rooting, **347**
Gladiolus, corm development of, **535**
Glasshouses (*see* greenhouses), 45
Gliocladium virens, 88
Golden disease, 633
Grafting,
aftercare of grafted plants,

in bench grafting, 470, 471, **472**
in field/nursery grafting, 472
apical graftage, 439
banana (four-flap) graft, 444, **446, 447**
cleft (split) graft, 441, **443**
saddle graft, 444, **446**
splice (whip) graft, 439, **442**
wedge (saw-kerf) graft, 443, **444, 445**
wedge graft (with equal sized scion and rootstock), 464, **465**
whip-and-tongue graft, 439, **440, 441**
approach graft, **457,** 458
inlay approach, **458**
spliced approach, **457, 458**
tongued approach, **458**
bark graft (rind graft), 452, **453**
inlay bark (veneer crown graft), **454**
bark slipping, 406, **408,** 409, 411
bench grafting (bench working), 470, 471
bud-forcing methods, 412, **413**
crippling, **413**
crown grafting, 472
compatibility, 409
between clones within a species, 415
between families, 416
between genera, 415
between species within a genus, 415
within a clone, 415
cutting-grafts, 477
stenting, **478**
detached scion graftage, 439–57
double-working, 397, **399,** 474, **476, 499**
environmental conditions during and following grafting, **410,** 411, **452**
failure to graft, 416, 417
frameworking, 473
growth regulators, 412
herbaceous grafting, 476, **477**
hot pipe callusing, 470, **471**
incompatibility, 409, 412, 416–25
causes and mechanisms of, 412, 418, 420, **421,** 422, **423**
cellular recognition, 422, 423
nontranslocatable (localized), 418, **421**
pathogen-induced incompatibility, 419, 420
predicting incompatible combinations, 423, 424
symptoms of, **416–20**
translocatable, 419
interstock (intermediate stock), 397
limits of grafting, 414–16
machines, 464, **465, 466**
automation/robotics, 465, **467, 468**
micrografting, 263, 478, 566, 568, **569**
natural grafting, 400
origins of, 393
polarity, 413, **414**
production processes of grafting, 461–72
reasons for grafting and budding, 395–400
record keeping, 470
repair graftage, 458–61
bracing, 461, **463**
bridge graft, **414,** 459, **461, 462**
inarching, 458, **459, 460**
root grafting, 455–57
nurse root grafting, **456**
piece-root grafting, 455, **456**
whole-root grafting, **455**
rootstock (stock, understock), **393,** 394, **396**
clonal, 394, 395
growth activity of, 411
handling for bench grafting, 469
seedling, 394
serial graftage for rooting, **302**
scion, **393**
scion-rootstock (shoot-root) relationships, 425–30
mechanisms of, 427–30
scionwood selection, handling, storage, 466–68
side graftage
side-stub graft, **448**
side-tongue graft, 449, **450**
side-veneer graft, 449, **451, 452**
temperature effects, **410**
techniques and ergonomics, 469, **493**
tools and accessories for grafting, 462–64
tying, wrapping, and sealing materials, 463–64
budding rubbers, 463
knives, 462, **463, 494**
raffia, 463
nursery adhesive tape, 463
Parafilm tape, 463
polyvinyl chloride (PVC) budding strips, 463
waxed string (twine), **463**
waxes, 464
topworking, 397, **473–76, 499**
top-budding, **499**
top-grafting, **473, 474**
types of grafting, 438–62
Graft-chimera, 244, **245**
Graft union,
adhesion of cells, 403, 423
callus bridge, **401,** 402, **403, 409**
cellular recognition, 422
dictyosomes, **404**
endoplasmic reticulum, **407**
formation of, 401–9, **401, 403, 404, 406, 407**
formation in T- and Chip-budding, 406, 407, **408,** 409
plasmodesmata, 403, **407**
symplastic and apoplastic connections, 403, **407**
wound periderm, 402
wounding response, 402
wound-repair xylem and phloem, 401–05
Green bulbs, 545
Greenhouses, 45
covering materials, 55, 58
environmental controllers and sensors, **43,** 54, 55, **56, 57**
gutter-connected, **46**
heating and cooling systems, 47, **50,** 51, **52–54**
plastic covered, **47**
Growth phases, 28, 254
Growth regulators, 22
to induce rooting, 288–92
structural formulas, **23**
Growth retardants, 291
Growth rooms
seed germination, **229,** 231
tissue culture, **595**
Guanine, 17
Gymnosperm, sexual cycle in, 136

Habituation, 578
Haploid, 10
Hardening-off,
post-propagation care of rooted cuttings, 379–83
care of liner plants, 235
Hard seed coats, 195, **196, 197**
Hardwood cuttings, 331–36
deciduous species, 331, **334–36**
heel cuttings, 331, **334**
machine planting of, **335**
mallet, 331, **334**
narrow-leaved evergreen species, **336**
sawing, **335**
straight cuttings, 331, **334**
types of, 331, **334**
Heat treatments, **86,** 89, **90,** 163, 263
Hedging, **251,** 256
Heeling-in, of bare-rooted nursery stock, 383, **385**
Heirloom varieties, 105, 111
Herbaceous cuttings, **339, 340**
Herbaceous perennials, 36
Herbicide use in propagation, 379
Heterotrophic, 576, 577
Heterozygous plant, 14, 106
Hilum, 141, 178
Histodifferentiation, 130
Histogens, 246
History
clones, 239
grafting, 393
propagation, 1–5
seed cultivars, 105
Homologous chromosomes, **13**
Homozygous plant, 14, 106
Hormones (phytohormones), 20, **23**
Hotbeds (hot frames), 58
Humidity control in rooting, 312, **313, 314,** 315
Hybrid cultivars, 111
Hybridization analysis, 263
Hybrid lines, 34, 111, 112
Hybrids, woody plants, 118
Hybrid seed production, 112, **114**
in corn, **113**
in paradox walnut, **117,** 118
seed sources, 118
Hybrid vigor, 107
Hydrotime germination model, 188
Hyperhydricity, 577
Hypocotyl, 184
Hypogeous germination, 184

Imbibition, 78, **179–81**
Inbred lines, 34, 111
Inbreeding, **107**
Inbreeding depression, 107
Incompatibility in grafting, 409, 412, 416–25
Increase or registered block, 112
Indexing, culture, **262,** 557, 579
for virus transmission, 263, **264**
Indolebutyric acid (IBA), 24, 289, 309, 310, 561, 598
Indole-3-acetic acid (IAA), 22, 24, 288, 289, 309, 310, 598
Infrared heaters, 50
Inheritance, **15**
Inherited disorder, 257
Inositol, 26
Integrated pest management (IPM), 84–93
biocontrol, 85, 163

Integrated pest management (IPM), (*cont.*)
biological control, 85, 88, 163, 363
chemical control, 85–88, **86, 362, 363**
cultural control, 85, **86,** 88, 89, 362
Intermittent mist, 81, 315, **353,** 365–70
International Association for Plant Tissue Culture, 7
International Code of Botanical Nomenclature, 32
International Code of Nomenclature for Cultivated Plants, 32
International Convention for Plant Variety and Protection, 35
International Dwarf Fruit Tree Association, 7
International Horticultural Society, 7
International Plant Propagators' Society, 6
International Seed Testing Association, 7, 154
Interstock, 397
Invigoration, 254
In vitro culture, 549
Irradiance, 576
Irrigation, 94
capillary mat subirrigation, **95, 233**
cyclic (interval, pulse irrigation), 94
drip irrigation, **95**
ebb and flood (flood floor) irrigation, 46, **49**
subirrigation, 233, **234**
Isolation, in seed production, 110
Isopentenyladenine (2iP), 25, 289, 598
Isozyme analysis, 160, 260, **261**
Ivy, phase changes in, **28, 30**

Juvenile phase, 28, 251
rooting, 580
variation due to, **253,** 254, **255**

Keikies, 513
Kinetin, 25, 289, 598

Laminar flow transfer hood, **594**
Landraces, 105, 110
Lathhouses, **61**
Layering, 502
advantages of, 502
air, 502, **506, 507**
avoiding plant stress, 503
compound, or serpentine, 505
containerized mound (stool), 511
drop layering, 512
management, 503
mound (stool), 508
mounding, 508
natural plant modifications for, 513
physiology, 503
rejuvenation, 503
simple layering, 503
stool beds, 509
stooling, **508**
tip, **513**
trench, 511, **512**
Leaching
of nutrients, from cuttings, 311
of seeds, 198
Legal protection, 35, 120
Leptomorph, **541,** 672
Lexan, 58
Life cycles, 26
Light, **55,** 316, 317
banding, 303, 348, **350,** 503
blanching, 303, 348–50, 503
duration (photoperiod), 41, **56,** 82, 316
etiolation, 303, **304, 349, 350,** 448, 503
exposure, effect on seed germination, 192, 199
irradiance, 41, 42, **53,** 82, 316, 576
manipulation for rooting, 82, 316, 317
shading, **193,** 303, **304,** 348–50, 503
spectral quality (wave length), 41, 44, 82, 317, 576
tissue culture, 576
Line, 34, 111
Liners (liner plants), 40, **63, 76,** 94
Linsmaier and Skoog, 597
Local seed, 115
Lucite, 58

Macrosclereid cells, **196**
Mass clonal system, 268
Matric potential, 164, 180
Matriconditioning, 164, 165, 190
Matrix seed priming, 164, 165, 190
Maturation, drying in seeds, 140
ontogenetic, 251
Mechanical dormancy, 198
Media, for liner and container growing, 74, 75
Media, for propagation, 68–75, 71, 349, **351,** 352
bark, 73, 351
calcined clay, 72
compost, 73
mineral (inorganic) media components, 68, 349, 597
organic components, 68, 349, 597
peat (moss), 70
peat-lite mixes, 74, 75
perlite, 72
pumice, 73
rockwool (mineral wool), 65, 73
salinity, 78
sand, 70
sphagnum moss peat, 72
synthetic plastic aggregates, 73
vermiculite, 72
water management of, 352
Media, tissue culture, 597
stock solutions, **596,** 597
Megapascal (MPa), 165, 178
Meiosis, 12, **13**
Mendel, Gregor, 4
Mericlinal chimera, 245, **246**
Meristems, 277, 282
Meristem-tip culture, **265,** 563, **564**
Mesomorphs, 542
Messenger RNA, 20, 182, 190
Methyl bromide, 85, **86,** 87
Microclimatic conditions (in propagation), 40
Micropropagation, 16, 556, 601
apple, 605
Cymbidium orchids, 606
ferns, 607
general methods, 556, 601
genetic and chimeral effects, 580, 581
juvenility effects, 580
methods for specific crops, 612
shoot multiplication, 558–60
stages, 556, 601
tissue proliferation, 578
transient phenotypic variation, 579
variation in, 578
Microspore culture, 566, **568**
Mineral nutrition, 45
rooting of cuttings, 45, 310, 311, 377, 378
Mist boom, **233**
Mist controls, **57,** 366, **367, 369**
Mist nozzles, 365, **368**
Mist (intermittent) systems, 81, 315, **353,** 365–70
Mitosis, 10, 11
Modeling, environment during propagation), 83, **371**
Moist-chilling, 202, 219
Monoculture, 240
Monoecy, 108, 114
Monohybrid cross, **15**
Morphological dormancy, 198, 199
Mound layering, **508–10**
mRNAs, **20, 21,** 182, 190
Multiline, 111
Murashige and Skoog, 553
culture medium, 597
Mutations, 12, **243,** 580
Mycoplasma-like organisms (MLOs), 259, 419, 420
Mycorrhizal fungi, 88, 258
Myo-inositol, 26, 597, 598

Naphthaleneacetic acid (NAA), 289, 309, 310, 598,
National Seed Storage Laboratory, 171
Nodule culture, 567
Nomenclature, plant, 31
Nomenclature for Cultivated Plants, International Code of, 33
Nonchimeral variation within clones, 250
Noninfectious bud failure, 257, 268
Nonrecurrent apomixis, 31
Nucellar budding, 31
Nucellar embryony, 31, 137
Nuclear seed production, 112
Nuclear seedlings (cultivars), 629
Nucleotide, 17
Nucellus, 130, 137
Nuclear stock, pedigree system, 112, 258
Nurse-root grafting, 456
Nursery row selection, 116, **117**
Nursery production (field),
balled and burlapped (B&B), 385, **386**
bare-root, 383, **384, 385,** 387
digging machines, **384, 386**
in-ground containers (grow bags), **98,** 385
in-ground plastic containers, **98,** 385
production in containers, 94, 97, **98**
Nutrition, of cuttings, 310, 311, 377, 378

Obligate apomictics, 34, 108, 138
Organization for Economic Cooperation and Development (OECD), 120
Offset bulbs, 527
Offsets (offshoot), 502, 514, **515**
Off-type, 110
Ontogenetic aging, 28
Ontogeny, 19
Organogenesis, 550
Ornamental virus, 258
Ortet, 242, 252, 300
Orthotropism, 256, **257**
Osmoconditioning, 164, **165,** 190
Osmotic potential, 180
Osmotic priming, 164, **165,** 190
Outbreeding, 106

Outdoor planting, for moist chilling, 219
Ovary, 129
Ovary culture, 555
Ovule, 129
Ovule culture, 555

Pachymorph, 540, **542,** 672
Paclobutrazol, 25, 563
Paradormancy, 194
Parafilm tape, 596
Parthenocarpy, 143
Pasteurization of propagation media, 89
Patents, 35
Pathogen-free, concept of, 106, 259, 261
Pathogens (infectious agents),
  elimination of, 263–66
  in micropropagation, 557, 600
  integrated pest management, 84–93
  seed protection against, 163
PCR, 260
Pear decline, 259
Peat, 70
Peat-lite mixes, Cornell, 74
Pedigree system,
  for seed selection, **112**
  for source selection, 267
Pelleted seed, **164**
Perennials, 26
Pericarp, 126, **127**
Periclinal chimera, 245
Perisperm, **127**
Perlite, 72
Phase change, 19, **27, 28**
Phase variation, 251, 252, **253**
Phenolics, 557, 577
Phenology, 19
Phenotype, 9
Phenotypic correlation, 116
Phenotypic selection, 116, 250, 261
Phenotypic variation, **251**
Photodormancy, 199, **200,** 201
Photoelectric cells, for mist propagation, **369**
Photoperiod, 27, 316, 317
Photosynthesis of cuttings, 317, **318**
Photosynthetic photon flux (PPF), 42
Phylloxera, 258
Physan, 90, 362
Physical dormancy, 195, **196, 197**
Physiological dormancy, 195, 199
Physiological dwarfs, 203, **204**
Physiologically deep dormancy, 201
Physiological maturity, seeds, 140
Phytophthora, 84, 87, 88, 94, 193, 258
Phytoplasma, 259
Pip, **541**
Pierce's disease, 259
Plagiotropism, 256, **257**
Plant
  breeding, 1, **16**
  exchange, 3
  exploration, 3
  nomenclature, 31
Planting density, 22, 226
Plant growth regulators, 22, **23,** 288–92, 598
  methods of application, 24
  solvents (carriers), 24
Plant hormones (phytohormones), 20–26, 288–92, 598
Plant improvement, 1
Plant propagation,
  defined, 1
  history, 2–5
  industry, 5
  organizations, 6
Plant taxonomy, 32
Plant Variety Protection Act, 35, 121
Plasmid, 17, 296
Plasmodesmata, in grafting, 403, **407**
Plexiglass, 58
Plug propagation, of seeds, 228, **229,** 230
  forest tree seedlings, 230, **231**
  requirements for germination, 230
Polarity,
  in grafting, 413, **414**
  in root regeneration, **286**
Pollen culture, 566, **568**
Pollen incompatibility, **108**
Pollen sterility, 108
Pollination, 108, **114,** 128, 149
Polyamines, 26, 291
Polycarbonate, 58
Polyembryogenesis, 134
  somatic, **573**
Polyembryony, 31, **137**
Polyethylene (polythene, poly), 55, **59, 63**
  bags, 64, 67
Polygal, 58
Polymorphic seed, 207
Polystyrene flakes, 73
Potassium nitrate, 206
Pots (*see* containers), 62–68
Precocious germination, 142
Preformed
  primary meristems, 283
  roots, 279
Pregerminated seed (pregermination), 165, **166**
Primary dormancy, 194
  in seed development, 141, 179
Proembryo, **132, 135**
Professional Plant Growers Association, Inc., 7
Progeny test, seedling, 110, **115,** 116
  clonal selection, 261, 269
Propagation,
  beds, indoor, outdoor, **353, 366, 367**
  blocks (peat, fiber, expanded foam, Rockwool), 65, **67**
  media, 68–75
  propagation flats (trays), 62, 228, **353**
  source, 266–68
Propagation systems, for cuttings,
  enclosed (closed case propagation), 58, **59, 60,** 62, 81, 314, **364,** 365
  enclosed mist, 58, **59, 63, 314,** 315, **364,** 365
  fogging systems, 370–73, **372**
  indoor polytent, **364**
  intermittent mist, 81, 315, **353,** 365–70
  nonmisted enclosed systems, **60**
  open mist, **314, 315, 353, 366**
Protein, 17
  storage 139, **140,** 184
  synthesis **18, 21**
Protoplast culture, 571, **572**
Provenances, 34, 115, 117
Proximal, 286, 414
Pseudobulbs, 543, **545**
Pure line, 111
Pure stand, 118
Putrescine, 26
Pythium, 84, 87, 88, 193

Quarantines, 269
Quick decline, in citrus, 630

Radicle (seed), 130–34
Ramet, **242,** 252, 300
RAPD, 160, 260
Recalcitrant seeds, 143
Recessive, 14
Recombinant DNA technology, 16, 17, 244
Recurrent apomixis, 31, 137
Registered seed, 120
Registration and certification, 120
  clones, 268
  seeds, 120
Rejuvenation, 256
  in layering, 503
  in micropropagation, 580
  of stock plants, 299
  techniques, 299
Reproduction, sexual, 12
Residual mRNA, 182
Rest period, 287
Reversion, 254, 256
RFLP, 160, 260
Rhizocaline, 287, 292, 295
Rhizoctonia, 84, 87, 88, 193
Rhizomes, 540, **541, 542**
Rhizosphere, 88
Ribose nucleic acid (RNA), 17, **18,** 20
Ribosomal RNA, 20
RNA detection, 263
Rockwool, 65
Roguing, 110
Rolled towel seed test, 155
Root cuttings, 285, 342, **343,** 344
  chimeral plants, 285
  preparation of, 342–44
  species propagated by, 344
Root development, in containerized plants, 64, **66,** 96, **97**
Rootgerms, 279
Root inducing (RI) plasmid, 296, 297, 578
Root pruning,
  air pruning roots, **64, 66,** 97
  chemical pruning, 64, **66**
Rooting,
  ancillary rooting compounds, 291
  bioassays, 292
  biochemical changes in development, 294
  boron influence, **294**
  bud and leaf effects, 287, **288,** 306
  carbohydrate effects, 304, 305, 307
  chronological age of cuttings, 299
  classification of plants by rooting success, 280, 282, 292
  cofactors (auxin synergists) in rooting, 291, 292, **293, 294,** 295
  competency of cells to root, 294, **295,** 296, 297
  compounds, list of, 356
  de novo adventitious roots, **278,** 279, 281
  developmental stages of, 278, 279, 282, **294**
  direct root formation, 280, **282**
  ex vitro rooting, 560, **561, 603,** 605
  herbaceous plants, 279, **280**
  humidity effects, 312–15
  indirect root formation, 280, **282**
  inhibitors of, 292, 295
  in vitro rooting, 560, **561, 603,** 605
  light effects on, 316, 317
  manganese influence, **311**
  media, 68–75
  molecular/bioltechnological advances, 296, 297, **298**
  photosynthesis of cuttings, 317, **318**
  physiological (biological) age of cuttings, 299, 301
  rhizocaline, 287, 292, 295
  rooting inhibitors, 291, 295

Rooting, (*cont.*)
  temperature effects, 315, **316**
  vascularization and rooting, 297
  woody perennial plants, 280
Roots,
  adventitious, **277, 278,** 279, **284, 288**
  de novo, **278,** 279, 281
  developmental stages of, 278, 279, 282, 294
  origin of, 279–82
  fleshy, **538, 539**
  latent root initials, 277, **279**
  preformed, 277, **278, 279**
  primordia, **278,** 279, **280, 282, 283**
  sclerenchyma ring, **280**
  tuberous, **538, 539**
  wound-induced, 277, 278, 281
Root spiraling, **97**
Rootstocks (*see* grafting), 393–96
Rudimentary embryos, 127, 198, 199
Runners, 502, 513, **514**

Salinity,
  media and soil, 78
  water, 78
Sanitation in propagation, 89, 90
Scaling, of bulbs, 529, **530,** 531
Scarification, **197, 218,** 219
Scion, 393
Scion-stock relationships, 425–30
Sclerenchyma tissue and rooting, 281
Scooping, 531
Scoring, 531
Scouting system in IPM, 84
Screen balance, for mist propagation, 368, **369**
Scutellar embryo, 132, **134**
Sectorial chimera, 245, **246**
Seed,
  accelerated aging, 161
  banks, 207
  certification, 120
  chilling injury, 182,
  classification, 126, **127**
  coat dormancy, 196
  cold test, 161
  collecting, 149
  collection zones, 118, **119**
  conditioning, 148, 151
  development, stages, **130**
  dispersal, 143
  dormancy, 194–210
    kinds of, 195
    embryo, 198
  film-coated seed, **164**
  food reserves, 139–40
  germination,
    aeration effects, 192
    conserved mRNA, 182
    curve, 188
    disease control during, 193
    hormones, 205, 206
    hydrotime germination model, 188
    imbibition, 178, **179–81**
    light effects, 192
    percentage (germination), 154, 187
    phases of, 179
    rate (germination), 187
    tests, 153
      excised-embryo test, 155, **158**
      direct germination test, 154
      purity, 159
      standard germination test, 154, **156**
      tetrazolium, 156, **158**
      vigor testing, 161, **162**
      x-ray analysis, **158**
    types of, **184**
    temperature effects, 190
    termal time germination model, 191
    thermohydrotime germination model, 191
    water effects, 188
  germinator, **229**
  harvesting and processing, 149–53
  health, 161
  longevity, 167–72
  matric potential, 180
  matriconditioning, 164, 165
  maturation drying, 154
  moisture determination, 150
  morphological types, 126, **127**
  orchards, 119, 149
  origin, 115, 117
  osmotic potential, 180
  pelleted seed, **164**
  physiological dwarfing, 203, **204**
  polymorphic, 207
  pregerminated seed (pregermination), 165, 166
  priming, 164, 165
  production, 112
  production areas, 118, 149
  propagation, 216
    field seeding, 217
    indoor sowing for transplants, 227
    nursery row production, 223
  purity, 159, 160
  sampling, **154**
  seeders, **221, 232**
    belt seeders, 222
    cylinder seeders, **232, 233**
    drum seeders, **232, 233**
    mechanical seeders, **232, 233**
    needle seeders, **232, 233**
    plate seeders, 222
    precession seeders, 221
    random seeders, 221
    template seeders, **232**
    vacuum seeders, 222
    wheel seeders, 222
  selection of, 105, 106
  sources of, **147**
  storage of, 170–72
  testing, 153
    electrolytic leakage, 161
    purity, 159, 160
    standard germination test, 154
    tetrazolium testing, 156, 159
    vigor, 161, **162**
Seedlessness, 143
Seedling,
  cycle, 26, 251
  defined, 26
  life cycle, 26, 251
  progeny test, 110
Seed Storage Laboratory, U.S. National, 171
Self pollination, 106, **107,** 114
Self-sterility, **109**
Semihardwood cuttings, **337**
Separation, 520
Serial rooting of cuttings, 299
Serology, 263
Sexual cycle, **13**
Sexual reproduction, 12, 128
Shading, **193,** 303, **304,** 348–50, 503
Shirofugen cherry indexing, 263, **264**
Shoot apex culture, 263, **265,** 563
Shoot culture, adventitious, 564
Shoots, adventitious, 564
Shoot tip, in micropropagation, 563
Shoot-tip necrosis, 578
Simple layering, 503, **504, 505**
Single cross, hybrid seed production, 111
Single-eye (single-node) cuttings, 341, **342**
Society for In Vitro Biology, 7
Sodium hypochlorite, **91,** 553, 557, 560, 599
Softwood cuttings, 337, **338, 339**
  etiolation of, **304**
Soil, 68, 70
Somaclonal variation, 246, **247,** 580
Somaclones, **247**
Somatic embryogenesis, 572–76, 609–11
Somatic (parasexual) hybridization, 571
Somatic polyembryogenesis, **573**
Somatic variation, 246, 580
Source-identified seed, 121
Source selection, 259
Species, 14, 32
Spermidine, 26
Spermine, 26
Sphaeroblasts, **251,** 256, 301
Sphagnum moss peat , 70, 72
Spore culture, ferns, 735
"Sports", 243, **244**
Stabilization, 558
  in clonal propagation, 243
Standard germination test, 154, 155, **156**
Stem bulblets, 528
Stenting, of roses, 478
Stock blocks, **267**
Stock plants, 298–306, 344, **345, 346,** 347–50, 556
  carbohydrates in, 304
  carbon dioxide enrichment, 304
  girdling, 304, **347**
  hedging, **299,** 301, 346
  light effects, 303, 348, **349, 350**
  mineral nutrition of, 305
  rejuvenation, **299,** 301, **302**
  shearing, **299, 510**
  sources of cutting material, 301
  temperature effects on, 303
  effects of pruning, 300, 345–47
  water effects on, 302
Stolons, 502, 514
Stooling, 509, **510**
Stool layering, **509**
Stoolshoots, 256, 564
Stratification of seeds, 202, 219
Stripping in cutting preparation, **354**
Strophiole, 197
Stubborn disease, 259
Stump sprouts, **251,** 256
Suberization, 278
Subspecies, 32
Suckers, 285, 502, 517
Sun tunnels (heated and unheated), 58, **59**
Suspension culture, 570, **571**
Suspensor, **132, 133**
Synthetic cultivar, 111
Synthetic lines, 34
Synthetic somatic seeds, 109, 576, **610**
Systemic pathogens, 258

T-budding, 491–94
T-DNA, 578
Temperature, 44
  effects on seed germination, 190
  effects on rooting, 315, **316**
Testa, 128
Tetrazolium seed test, 156, **158**

Thermal time germination model, 191
Thielaviopsis, 84, 87
Thymine, 17
Thermal curtain, **53**
Thermodormancy, 205
Thermohydrotime germination model, 191
Thermotherapy chamber, 263, **266**
of seeds, 163
Thidiazuron (TDZ), 25, 289
Thin layer epidermal strips, 565
Thiourea, 25
Three-way cross, hybrid seed production, 111
Threshold water potential, 188
Timers, for mist propagation, 366, 367
Tip layering, 502, **512**
Tissue culture,
equipment and facilities, 592
genetic variation in, 580
principles of, 550
exudation, 557, 558
tissue proliferation, 578
Topcross, hybrid seed production, 111
Topophysis, 256, 633
Topworking, 473–76
Totipotency, 17, **19,** 553
Tracheary nests, **283**
Trademark, 35
Transcription, 21
Transfer RNA, 20
Transformation, 16
Transformed plants, 16, **575**
Translation, 20, **21**
Transplanting seedlings, 235
Transposable elements, 246
Tree and shrub seeds, germination requirements, 157
Tree seed certification, 121
Triploid, 128
Trueness-to-name, 106, 259
Trueness-to-type, 106, 259, 260
True-to-type, 110, 254, **255**
Tubercles, 537
Tuberous roots, **538, 539**
definition and structure, 538
growth pattern, 539
propagation of, 540
Tuberous stems, **538, 539**
definition and structure, 538
propagation of, 540
Tubers,
definition and structure, 536
growth patterns, 536
potato tubers, **536,** 537,
potato vegetative "seed" propagule, **537**
Turgor (pressure potential), 180, 312
Twin scaling, **533**

U.S. Clonal Repository System, 269
USDA IR-2 Repository, 269
Ultrafreezing of seeds, 169, **170**
Understock, 393–94
Utility patent, 36

Vapor pressure deficit, 314
Variation, in clones
genetic, 242–250
nongenetic, 250–58
in micropropagated plants, 578
Variegation, **244,** 248
Variety, botanical, 32
Vegetative progeny test, 261, 269
Vegetative propagation, 10
Vermiculite, 72
Vernalization, 26
Viability, seed, 154
Viability test, excised embryo, 155, **158**
Vigor testing, seeds, **161**
Viroids, 259
Virus, 258
diseases, studying, 258
free concept, 261
indexing, 262, 557, 579
Vitrification, in tissue culture, 577
Vivipary, 142

Wardian cases, 3, 4
Water potential, 44
seeds, 180
threshold water potential, 188
turgor potential, 44, 180
Water recycling, 79, **80,** 81
Water quality, 78, **79, 80**
electrical conductivity (EC), 78, **80**
pH, 78, **80**
salinity, 78
Water relations, 44, 180
humidity control, 44, 81, 312, **313, 314,** 315,
rooting, 312–15
Watersprouts, **251,** 256
Weed seeds, 153, 159
Winter protection, 95, **96**
Woody perennials, 26
Woody plant medium (WPM), 597
Worker protection standard (WPS), 85
Wounding,
cuttings, 311, **312,** 313, 353, **354,** 355
Wounding response, 278
Wound-induced,
secondary meristems, 283
wound-induced roots, 277, 278, 281
Wound-repair xylem and phloem, in grafting, 401–05

X-disease, 259
X-ray analysis of seeds, 158

Yellows
strawberry, 257

Zeatin, 25, 598
Zeatin riboside, 25
Zygote, 17, 130

PLANT INDEX, SCIENTIFIC NAMES

Abelia x grandiflora, 667
Abies spp., 29, 35
Abutilon spp., 258, 668
Acacia spp., 253, **668**
Acer spp., 29, 668
Achillea spp., 725
Achimenes spp., 725
Achras zapota, 625
Aconitum spp., 725
Actinidia spp. 625, 670
Adiantum spp., 735
Aegopodium spp., 725
Aesculus spp., 670
Agapanthus spp., 726
Agave spp., 726
Ageratum houstonianum, 112, 726
Aglaonema spp., 726
Agrostemma githago, 726
Ailanthus altissima, 670
Ajuga spp., 513, 514, 726
Alberta magna, 670
Albizia julibrissin, 670
Alcea rosea, 726
Allamanda cathartica, 670
Allium spp., 31, 726
Alnus spp., 671
Aloe spp., 726
Alstroemeria spp. 726
Alyssum saxatile, 726
Alyxia olivaeformis, 726
Amaranthus spp., 726
Amaryllis belladonna, 726
Amelanchier spp., 671
Amorpha spp., 671
Amsonia tabernaemontana, 726
Anacardium occidentale, 626
Ananas comosus, 626
Anchusa spp., 727
Anemone blanda, 727
Anigozanthus spp., 727
Annona cherimola, 626
Anthemis spp., 727
Anthurium andraeanum, 727
Antirrhinum majus, 727
Aquilegia spp., 727
Arabidopsis spp., 29
Arabis spp., 727
Aralia spp., 671
Araucaria spp., 257, 671
Arbutus menziesii, 671
Arctotis stoechadifolia, 727
Arctostaphylos spp., 672
Ardisia spp., 672
Armeria spp., 727
Aronia spp., 672
Artemisia spp., 727
Artocarpus spp. 627
Arum italicum, 727
Aruncus spp., 728
Arundinaria spp., 672
Asclepias spp., 728
Asimina triloba, 672
Asparagus spp., 728
Asplenium spp. 735
Aster spp., 728
Astilbe spp., 728
Astrantia spp., 728
Athrium spp. 735
Aubrieta deltoidea, 728
Aucuba japonica 'Variegata', 728
Averrhoa carambola 627

Baccharis spp., 672
Bambusa spp., 672
Banksia spp., 673
Baptisia spp., 728
Bauhinia congesta, 673
Beaucarnea recurvata, 696
Begonia spp., 728
Belamcanda chinensis, 728
Bellis perennis, 112, 728
Berberis spp., 512, 673
Bergenia cordifolia, 729
Betula spp., 29, 673
Boltonia spp., 729
Bougainvillea spp., 674
Bouvardia ternifolia, 729
Brodieia spp. (*also Triteleia*), 729
Browallia spp., 729
Brunnera macrophylla, 729
Buddleia spp., 674
Bumelia lanuginosa, 674
Butia capitata, 696
Buxus spp., 674

Caladium bicolor, 729
Calamagrostis acutiflora, 729
Calceolaria spp., 729
Calendula officinalis, 730
Callistemon spp., 674
Callistephus chinensis, 730

Boldface numbers indicate figures.

Calluna vulgaris, 683
Calocedrus decurrens, 675
Caltha palustris, 730
Calycanthus spp., 675
Camellia spp. 675
Campanula spp., 730
Campsis spp., 675
Canna spp., 730
Caragana pygmaea, 675
Carica papaya, 627
Carpinus spp., 676
Carya spp., 676
  C. illinoensis, 628
  C. ovata, 628
Castanea spp., 628, 676
Casuarina spp., 676
Catalpa spp., 676
Catananche caerulea, 730
Catharanthus roseus, 730
Cattleya, 742
Ceanothus spp., 676
Cedrus spp., 677
Celastrus spp., 677
Celosia argentea, 730
Celtis spp., 677
Centaurea spp., 730
Cephalotaxus harringtonia, 677
Centranthus ruber, 730
Cephalotus follicularis, 730
Cerastium tomentosum, 730
Ceratonia siliqua, 629
Cercis spp., 33, 677
Chaenomeles spp., 678
Chamaecyparis spp., 35, 254, 678
Chamelaucium spp., 678
Cheiranthus cheiri, 730
Chelone spp., 730
Chilopsis linearis, 678
Chionodoxa spp., 731
Chionanthus spp., 678
Chlorophytum comosum, 513, **514,** 731
Chrysanthemum spp., **262,** 263, 266, 269, **339, 383,** 731
Cinnamomum camphora, 678
Citrus spp. 29, 206, 245, 629–32, 678
Clarkia spp., 736
Clematis spp., 506, 679
Cleome spinosa, 731
Clethra spp., 679
Coccoloba spp., 679
Cocos nucifera, 632
Codiaeum variegatum, 731
Coffea arabica, 633
Cola nitida, 633
Colchicum autumnale, 731
Coleus blumei, 731
Convallaria majalis, 731
Cordyline spp., 732
Coreopsis spp., 732
Cornus spp., 514, 679
Cortaderia selloana, 732
Corylopsis spp., 633, 680
Corylus spp., 633, 680
  C. avellana, 680
Cosmos spp., 732
Cotinus coggygria, 680
Cotoneaster spp., 680
Crassula argentea, 732
Crataegus spp., 245, 680
Crocosmia spp., 732
Crocus vernus, 732
Cryptomeria japonica, 240, 254, 681
Cucurbita spp., 732
Cunninghamia lanceolata, 240
x Cupressocyparis leylandii, 33, 681
Cupressus spp., 681
Cycas spp., 681
Cyclamen spp., 732
Cydonan spp., 502
Cydonia oblonga, **334,** 634, 654
Cymbalaria muratis, 732
Cymbidium, 742
Cynodon spp., 502, 514
Cynoglossom amabile, 732
Cypripedium spp. 742
Cytisus spp., 681

Dahlia spp., 732
Daphne spp., 682
Davidia involucrata, 682
Delonix regia, 682
Delphinium spp., 732
Dendrobium spp., 513, 742
Dendrocalamus spp., 672
Deutzia spp., 682
Dianthus spp., 733
Dicentra spp., 733
Dictamnus albus, 733
Dieffenbachia spp., **340,** 502, **504,** 733
Digitalis spp., 733
Dimorphotheca spp., 733
Dionaea muscipula, 733
Dioscorea spp., 239
Diosma ericoides, 682
Diospyros spp., 634, 682
Disanthus cercidifolius, 682
Dodecatheon maedia, 733
Doronicum spp., 733
Dracaena spp., **340,** 734
  D. marginata, **506**
Drosera spp., 734
Dryoptris spp. 735
Duchesnea indica, 513
Dyssodia spp. 734

Echinacea purpurea, 734
Echinops exaltatus, 734
Elaeagnus spp., 682
Encephalartos spp., 681
Enkianthus spp., 683
Epigaea repens, 683
Epimedium spp., 734
Epiphyllum spp., 734
Epipremnum aureum, 734
Eranthis hyemalis, 734
Eremurus bungei, 734
Erica spp., 683
Erigeron spp., 734
Eriobotrya japonica, 634, 683
Eriostemon spp., 683
Eryngium spp., 734
Escallonia spp., **355,** 683
Eschscholzia californica, 735
Eucalyptus spp., 683
Eucomis spp., 735
Euonymus spp., **337,** 684
Euphorbia spp., 684, 735
Euphoria longan, 635
Eustoma grandiflorum, 735
Exacum affine, 735

Fagus spp., 29, 684
x Fatshedera (Fatsia x Hedera), 687, 735
Feijoa sellowiana, 635
Festuca spp. 735
Ficus spp., 254, 506, 684
  F. benghatensis, 277
  F. benjamina, 684
  F. carica, 635
  F. elastica, 502, **507,** 684
  F. lyrata, 684
  F. pumila, **278, 283,** 684
Forsythia spp., 685
Fortunella spp., 629
Fothergillia gardenii, 685
Fragaria x ananassa, 502, 635
Franklinia alatamaha, 685
Fraxinus spp., 29, 685
Freesia spp., 735
Fritillaria spp., 735
Fuchsia spp., 685
Fuchsia x hybrida, 736

Gaillardia spp., 736
Galanthus spp., 736
Garcinia mangostina
Gardenia jasminoides, 685
Garrya elliptica, 683
Gazania spp., 736
Genista spp., 685
Gentiana spp., 736
Geranium spp., 736
Gerbera jamesonii, 736
Geum spp., 736
Ginkgo biloba, 686
Gladiolus spp., 736
Gleditsia triacanthos, 686
Gloriosa spp., 736
Gloxinia spp., 736
Godetia spp., 736
Gomphrena globosa, 737
Gordonia spp., 686
Grevillea spp., 686
Gymnocladus dioicus, 686
Gypsophila spp., 737

Halesia spp., 686
Hamamelis spp, 686
Haworthia spp., 748
Hebe spp., **337,** 687
Hedera helix, 28, 29, 30, 254, 687
Helenium autumnale, 737
Helianthemum nummularium, 737
Helianthus annuus, 737
Heliconia spp., 737
Helichrysum bracteatum, 737
Heliopsis scabra, 737
Heliotropium spp., 737
Helleborus spp., 737
Hemerocallis spp., 737
Heptacodium miconioides, 687
Heteromeles arbutifolia, 687
Heuchera spp., 737
Hibiscus rosa-sinensis, 687, 737
  H. syriacus, 687
Hippeastrum spp., 738
Hosta spp., 515, 738
Hovenia dulcis, 688
Hunnemannia fumariifolia, 738
Hyacinthus spp., 738
Hydrangea spp., 688
Hypericum spp., 688
Hypoestes phyllostachya, 738

Iberis spp., 738
Ilex spp., 688
Illicium spp., 689
Impatiens spp., 738
Incarvillea spp., 738
Ipomoea spp., 239, 738
Iresine spp., 738
Iris spp., 738
Ixia spp., 738
Ixora spp., 738

Jacaranda mimosifolia, 689
Jasminum spp., 689
Juglans spp. 438
  J. cinerea, 637
  J. hindsii x J. regia, 116, **117,** 118
  J. nigra, 637
  J regia, 116, 637
Juniperus spp., 689

Kalanchoe spp., 342, 748
Kalmia spp., 690
Kniphofia hybrids,739
Koelreuteria spp., 690

Laburnum spp, 690
Lagerstroemia indica, 690
Lamiastrum galeobdolan, 739
Lamium maculatum, 739
Lantana spp., 739
Larix spp., 29, 118, 691
Laurus nobilis, 691
Lathyrus spp., 739
Lavandula spp., 739
Leontopodium alpinum, 739
Leptospermum spp. 691
Lespedeza spp., 739
Leucojum spp., 739
Leucothoe spp., 691
Leucophyllum spp., 691
Leucospermum spp., 691
Liatris spp., 739
Ligustrum spp., 691
Lilium spp., 739
Limonium spp., 740
Lindera obtusiloba, 691
Linaria spp., 740
Linum spp., 740

Liquidambar spp., 116, 691
Liriodendron tulipifera, 691
Liriope spp. **345,** 740
Litchi chinensis, 638
Lobelia spp., 740
Lobularia maritima, 740
Lonicera spp., 692
Loropetalum chinense, 692
Lunaria spp., 740
Lupinus spp., 740
Lychnis spp., 740
Lycoris spp., 740
Lycopersicon spp., 245
Lysimachia spp., 740
Lythrum salicaria, 741

Macadamia spp., 638
Maclura pomifera, 692
Magnolia spp., 692
Mahonia spp., 506, 693
M. trifoliata, 673
Malus spp., 29, 639, 693
M. hupehensis, 255, 693
Malva alcea, 741
Mandevilla spp., 694
Mangifera indica, 642
Maranta leuconeura, 741
Matteuccia spp. 735
Matthiola spp., 735
Mentha spp., 514
Mespilus germanica, 245
Melaleuca spp., 694
Melampodium paludosum, 741
Mesembranthemum spp., 748
Metasequoia glyptostroboides, 694
Mimulus spp., 741
Miscanthus spp., 741
Molluccela laevis, 741
Monarda didyma, 741
Monstera deliciosa, 506, 741
Morus spp., 643
M. alba, 694
Musa spp., 239, 643
Muscari spp., 741
Myosotis sylvatica, 741
Myrica spp., 694
Myrtus spp., 338, 694

Nandina domestica, 695
Narcissus, 741
Nelumbo lutea, 741
Nemesia strumosa, 742
Nepenthes spp., 742
Nepeta mussinii, 742
Nephelium larraceum, 643
Nephrolepsis spp., 22, 735
Nerine spp., 742
Nerium oleander, **338,** 695
Nicotiana spp., 22, 742
Nierembergia spp., 742
Nymphaea spp., 742
Nyssa sylvatica, 695

Oenothera spp., 742
Olea europaea, 643, 695
Ophiopogon japonicus, 742
Opuntia spp., 742
Osmunda spp. 735
Osmanthus spp., 695
Oxydendrum arboreum, 695

Pachysandra terminalis, 695
Paeonia spp., 696, 743
Parrotia persica, 696
Papaver spp., 743
Parthenocissus spp., 696
Paspalum notatum, 108
Passiflora edulis, 644
P. x alatocaerulea, 697
Paxistima myrsinites, 697
Pelargonium x hortorum, 743
Pennisetum spp., 108, 744
Penstemon fructicosus, 697
Peperomia spp., 744
Persea americana, 644
Pervoskia atriplicifolia, 744
Petunia spp., 744
Philadelphus spp., 697
Philodendron spp., 744
Phlox spp., 744
Phoenix dactylifera, 645, 696
Photinia spp., 697
Phyllostachys spp., 672
Physalis alkekengi, 744
Physostegia virginiana, 744
Picea spp., 29, 35, 116, 117, 697
Pieris spp., 698
Pileostegia viburnoides, 698
Pinguicula spp. 745
Pinus spp., 29, 117, 118, 254, **284,** 698
Pistacia atlantica, 699
P. chinensis, 110, 699
P. vera, 115, 646
Pittosporum spp., 699
Platanus spp., 699
Platycerium spp., 735
Platycladus orientalis, 710
Plumbago spp., 699
Plumeria spp., 699
Poa spp. 31, 108
Podocarpus spp., 699
Polemonium spp., 745
Polianthes tuberosa, 745
Poliomintha longiflora, 745
Polygonatum spp. 745
Polygonum spp., 745
Polystichum spp., 735
Poncirus trifoliata, 629, 632
Populus spp., 699
Portulaca grandiflora, 745
Potentilla spp., 700, 745
Primula spp., 745
Prosopis spp. 700
Protea spp., 700
Prunus spp. 11, 29, 646, 700
P. angustifolia, 652
P. armeniaca, 648, 651
P. avium, 257, 647
P. besseyi, 651, 652
P. cerasifera, 648, 649, 651
P. cerasifera x munsoniana, 651
P. cerasus, 647
P. domestica, 33, 651, 652
P. dulcis, 244, 257, 649, 652
P. insititia, 650, 652
P. mahaleb, 648
P. persica, 650
P. salicina, 651, 652
P. tomentosa, 651
Pseudoasa spp., 672
Pseudotsuga menziesii, 29, **42,** 700
Psidium cattleianum, 652
P. guajava, 652
Pteris spp., 735
Pulmonaria spp., 745
Punica granatum, 653
Puschkinia scilloides, 745
Pyracantha spp., **338,** 701
Pyrus, 29, 653
P. betulaefolia, 654
P. calleryana, 654
P. calleryana 'Bradford', 654
P. communis, 654
P. kawakamii, 701
P. pyrifolia, 653, 655
P. ussuriensis, 654

Quercus spp., 29, 701

Ranunculus spp., **745**
Raphiolepis indica, 702
Reseda odorata,
Rhamnus spp., 702
Rhapidophyllum hystrix, 696
Rhipsalidopsis spp., 745
Rhododendron spp., **342, 354,** 702
Rhus spp., 703
Ribes spp., 655
R. sanguineum, 704
R. sativum, **286**
Ricinus communis, **280,** 746
Robinia pseudoacacia, 704
Rodgersia pinnata, 746
Rosa spp., 29, **399,** 704
R. multiflora, **336,** 704
Rosmarinus officinalis, 706
Rubus spp. 31, 50, 245, **513,** 655
R. occidentalis, 655
R. occidentalis x R. idaeus, 655
Rudbeckia spp., 746

Saccharum spp., 239
Saintpaulia ionantha, **341,** 746
Salix spp., 706
Salpiglossis sinuata, 746
Salvia spp., 746
Sambucus spp., 706
Sansevieria spp., **341,** 746
Santolina chamaecyparissus, 746
Sanvitalia procumbens, 746
Sapindus drimmondii, 706
Saponaria officinalis, 746
Sarcococca spp., 706
Sarracenia spp., 746
Sasa spp., 672
Saxifraga spp., 513, 746
Scabiosa spp., 747
Schefflera arboricola, 747
Schizanthus spp., 747
Schizophragma hydrangeoides, 707
Schlumbergera truncata, 747
Scilla spp., 747
Sedum spp., 747
Sempervirens spp., 748
Senecio spp., 747
Sequoia spp., 29, 707
Sequoiadendron giganteum, 29, 707
Shepherdia spp. 707
Sidalcea malviflora, 747
Sinarundinaria spp, 672
Sinningia speciosa, 747
Solanum spp. 245
S. tuberosum, 234
Solidago spp., 747
x Solidaster, 747
Sophora spp., 707
Sorbus spp., 708
Spiraea spp., 708
Stachys spp., 524, 747
Stangeria spp., 681
Stewartia spp., 708
Stokesia laevis, 747
Strelitzia reginae, **544,** 747
Streptocarpus x hybridus, 748
Swietenia mahagoni, 708
Symphoricarpos spp., 708
Symphytum spp., 748
Syringa spp., 34, **304,** 708

Tabebuia argentea, 709
Tagetes spp., 748
Tamarix spp., 709
Taxodium distichum, 709
Taxus spp., 709
Telopea speciosissima, 710
Thalictrum spp., 748
Thamnocalamus spp., 672
Theobroma cacao, 656
Thermopsis caroliniana, 748
Thuja spp., 29, **354,** 710
Thunbergia spp., 748
Thymus spp., 748
Thyrsostachys spp., 672
Tiarella cordifolia, 748
Tigridia pavonia, 748
Tilia spp., 710
Tithonia rotundifolia, 748
Tolmiea menziesii, **341,** 748
Torenia fournieri, 749
Trachelospermum asiaticum, 711
T. jasminoides, 711
Trachymene coerulea, 749
Tradescantia spp., 749
Tricyrtis spp., 749
Trollius spp., 749
Tropaeolum majus, 749
Tsuga spp., 29, 711
Tulipa spp., 749

Ulmus spp., 711
Ungnadia speciosa, 711

Vaccinium spp. 656, 711
V. angustifolium, 656
V. ashei, 656

Vaccinium spp. (*cont.*)
V. australe, 656
V. corymbosum, 656
V. macrocarpon, 656, 657
V. ovatum, 711
Valeriana officinalis, 749
Vauquelinia spp., 712
Venidium fastuosum, 749
Verbascum spp., 749
Verbena spp., 749
Veronica spp., 687
Viburnum spp., 33, 712
Vinca spp., 749
Viola spp., 749
Vitis spp., 26, **286,** 657, 658
V. rotundifolia, 502
Weigela spp., 713
Wisteria spp., 506, 713
Woodwardia spp., 735

Xylosma congestum, 713

Yucca spp., 750

Zamia spp., 681
Zantedeschia spp., 750
Zebrina pendula 750
Zingiber officinale, 542
Zinnia spp., 750
Zizyphus jujuba, 659
Zygocactus spp., 729

PLANT INDEX, COMMON NAMES

Abelia, 667
Abutilon, 668
Acacia, **253**
Achimenes, 725
African daisy, 727
African violet, 249, **341**
Agapanthus, 726
Agarita, 673
Agave, 726
Ageratum, 726
Aglaonema, 726
Ailanthus, 670
Albizzia, 670
Alder, 108, 671
Alemow, 632
Allamanda, 670
Allspice, 675
Almond, **244,** 257, 649, 650, 652
Dwarf flowering, 700
Almond x peach, 649, 651
Aloe, 726
Amaryllis, 263, 726
American alumroot, 737
Amethyst flower, 729
Anacho orchid tree, 673
Angelica tree, 671
Anise, 689
Annual phlox, 744
Anthurium, 727
Apple, 29, 239, 243, 254, 263, **396, 398, 413, 414, 489,** 502, **504, 509,** 511, 639
Apricot, 239, 648, 650, 651
Japanese flowering, 700
Araucaria, 671
Arborvitae, 29, 710
Arbutus, trailing, 683
Armeria, 727
Artemisia, 727
Ash, 29, 685
Mountain ash, 708
Asparagus, **109,** 516
Aspen, 699
Atemoya, 626
Asters, 728
Astilbe, 728
Aubrieta, 728
Australian pine, 676
Australian pitcher plant, 730
Autumn crocus, 731
Avens, 736
Avocado, 644
Azalea, 702
Baccharis, 672
Bachelor button, 730
Bald cypress, 709
Balloon flower, 745
Balsam, 738
Bamboo, 672
Banana, 243, 263, 515, 643
Banksia, 673
Banyan tree, **277**
Barberry, 673
Barrenworts, 734
Basswood, 710
Bayberry, 694
Bay laurel, 691
Bearberry, 672
Beardstongue, 744
Bee balm, 741
Beech, 29, 684
Begonia, 728
Belladonna lily, 726
Bells of Ireland, 741
Bermuda grass, 502, 514
Big betony, 747
Birch, 29, 673
Bird-of-paradise, **544, 747**
Bistort, 745
Bittersweet, 677
Blackberry, 31, 245, 248, 502, 513, 655
Blackberry lily, 728
Black-eyed Susan, 746
Black gum (Black tupelo), 695
Black haw, 674
Blanketflower, 736
Bleedingheart, 733
Bloodflower, 728
Bloodleaf, 738
Blue fescue, 735
Blueberry, 656
Boltonia, 729
Boston fern, 513, 735
Boston ivy, 696
Bottlebrush, 675
Bougainvillea, 674
Bouncing Bet, 746
Boxwood, 512, 674
Oregon boxwood, 697
Boysenberry, **513,** 658
Breadfruit, 239
Breath of Heaven, 682
Bromeliads, 729
Broom, 681
Buckeye, 670
Mexican buckeye, 711
Buckthorn, 702
False buckthorn, 674
Buffaloberry, 707
Bugle flower, 513, 726
Bugloss, 727
Bumelia, 674
Bunya-bunya, 671
Bushclover, 739
Buttercup, 745
Butterfly bush, 674
Butterfly flower, 747
Butterfly weed, 728
Butternut, 637
Butterwort, 745

Cacao, 656
Cactus, **400, 406,** 629
Caladium, 729
Calceolaria, 729
California poppy, 735
Calla, 750
Camellia, 245, 675
Camphor tree, 678
Campion, 740
Candytuft, 738
Canna, 730
Canterbury bells, 730
Cape jasmine, 685
Cape marigold, 733
Cape primrose, 745
Carnation, 263, **265,** 266, 269, 733
Carob, 629
Cashew, 626
Cassava, 263
Castor bean, **280,** 746
Catalpa, 676
Catmint, 742
Ceanothus, 676
Cedar, 677
Chamaecyparis, 678
Checker lily, 735
Cherimoya, 626
Cherry, 239, 257, 511, 700
Flowering cherry, 700
Nanking cherry, 651
Sweet cherry, 646
Sour cherry, 647
Western sand cherry, 651
Chestnut, 676
Chimney bellflower, 730
China aster, 730
Chinese date, 659
Chinese forget-me-not, 732
Chinese Lantern, 744
Chokeberry, 672
Christmas berry, 687
Christmas cactus, 747
Christmas rose, 737
Chrysanthemum, **262,** 263, 266, 269, **339, 383,** 731
Cinquefoil, 700, 745
Citrange, 632
Citrus, 29, **31,** 254, 263, 266, 629, 678
Clarkia, 736
Clematis, 679
Clethra, 679
Clockvine, 748
Cockscomb, 730
Coconut, 632
Coffee, **257,** 633
Coleus, 731
Colonial pine, 671
Colt cherry, 648
Columbine, 727
Comfrey, 748
Common stock, 741
Coneflower, 734
Coralbells, 737
Coralberry, 672
Coreopsis, 732
Corncockle, 726
Cornflower, 730
Cosmos, 732
Cotoneaster, 680
Cottonwood, 108, 699
Crabapple, flowering, 693
Cranberry, 656
Mountain cranberry, 711
Cranesbill, 736
Crape myrtle, 690
Creeping zinnia, 746
Crocosmia, 732
Croton, 502, 506, 731
Cryptomeria, Japanese, 681
Cup flower, 742
Cupidsdart, 730
Currant, **286,** 509, 655
Red flowering currant, 704
Custard apple, 626
Cycads, 681
Cyclamen, 732
Cypress, 681
Leyland cypress, 681
Lawson cypress, 254

Daffodil, 741
Dahlia, 732
Daphne, 682
Daylily, 737

Delphiniums, 732
Deutzia, 682
Dewberry, 513, 655
Dieffenbachia, 733
Diosma, 682
Disanthus, 682
Dogwood, 214, 679
Douglas fir, 29, **42, 115,** 117, 700
Dove tree, 682
Dracaena, 734
Drooping Leucothoe, 691
Durian, 240
Dusty miller, 747
Dutch crocus, 732

Edelweiss, 739
Elaeagnus, 682
Elderberry, 706
Elm, 711
English daisy, 728
Enkianthus, 683
Eriostemon, 683
Escallonia, 683
Eucalyptus, **241, 253, 257,** 683
Euonymous, **337,** 684
Euphorbia, spurge, 735
Evening scented stock, 741
Exacum, 735

Falsecypress, 678
False dragonhead, 744
False indigo, 728
False willow, 672, 726
Featherleaf, 672, 746
Feather reed grass, 729
Feijoa, 635
Fern, **14,** 247, 735
Fern, asparagus, 728
Feverfew, 731
Fig, 239, 684
  Creeping fig, 239, 684
  Fiddle leaf fig, 684
  Strangling fig, 254
  Weeping fig, 684
Filbert, 502, 505, 633, 680
Fir, 29, 667
  Chinese fir, 240
Flax, 740
Fleabane, 734
Fleeceflower, 745
Florists' cineraria, 747
Foam flower, 748
Forget-me-not, 741
Forsythia, 685
Fountain grass, 744
Fothergillia (dwarf), 685
Foxglove, 733
Foxtail lily, 734
Franklinia, 685
Freesia, 263, 735
Fringe tree, 678
Fuchsia, 685, 736

Gardenia, 685
Garden phlox, 744
Garden pinks, 733
Garrya, 685
Gasplant, 733
Gayfeather, 739
Gazania, 736
Genista, 685
Gentian, 736
Geranium, **114,** 263, 269, 736
  Strawberry geranium, 513
Ginger,
  European wild ginger, 672
  Tropical ginger, 542
Ginkgo, 686
Gladiolus, 263, 736
Globe amaranth, 737
Globe centaurea, 730
Globeflower, 749
Globe thistle, 734
Gloriosa lily, 736
Glory-of-the-snow, 731
Gloxinia, 747
Goat's beard, 728
Godetia, 736
Gold dust plant, 728
Golden chain, 690
Goldencup, 726
Golden Marguerite, camomile, 727
Golden pothos, 734
Golden rod, 747
Goldenrain tree, 690
Goldentuft, 726
Gold tree, 709
Gooseberry, 263, 509, 655, 670
  Chinese gooseberry, 625
Goutweed, 725
Grape, 29, 239, 242, 258, 263, 286, 657
  Muscadine grape, 502, 506
Grapefruit, 244, 629
Grape hyacinth, 741
Greek windflower, 727
Grevillea, 686
Guava, 652
  Pineapple guava, 635
Gypsohilia, 737

Hackberry, 677
Hawthorn, 245, 680
Hazelnut, 633, 680
Heath, 683
Heather, 683
Hebe, 687
Heliconia, 737
Heliopsis, 737
Heliotrope, 737
Hellebore, 737
Hemlock, 29, 711
Hibiscus, **404,** 687
  Chinese hibiscus, 687, 737
Hickory, 628, 676
Holly, 688
Hollyhock, 726
Hollyhock mallow, 741
Honesty, 740
Honeylocust, common, 686
Honeysuckle, 692
Hornbeam, 676
Horned violet, 749
Horsechestnut, 670
Horseradish,
Huckleberry, 711
  Evergreen huckleberry, 711
Hyacinth, 738
Hydrangea, 688

India hawthorn, 702
Incarvillea, 738
Incense cedar, 675
Indigo bush, 671
Iris, 263, 738
Ironwood, 676
Ivy, 29, **30**
  Boston ivy, 696
  English ivy, **347,** 687
  Kenilworth ivy, 732
  Switch ivy, 691
Ixia, 738
Ixora, 738

Jacaranda, 689
Jackfruit, 627
Jacob's ladder, 745
Jade plant, 732
Japanese anemone, 727
Japanese hydrangea vine, 707
Japanese pagoda tree, 707
Japanese raison tree, 688
Japanese spice bush, 691
Japanese spurge, 695
Jasmine, 689
  Asiatic jasmine, 711
  Star (confederate) jasmine, 711
Joseph's coat, 726
Jujube, 659
Juniper, 254, **336,** 689

Kangaroo paw, 727
Kentucky coffee tree, 686
Kiwifruit, 625
Knapweed, 730
Kola, 633

Laburnum, 245, 690
Laceflower, 749
Lamb's ear, 747
Lantana, 739
Larch, 29, **256,** 691
Laurel, 690
  Bay laurel, 691
  Black laurel (Loblolly bay), 686
  Cherry laurel, 700
  English laurel, 700
  Mountain laurel, Texas, 707
Lavender, 739
Lavender cotton, 746
Lemon, 244, **288,** 629
Lenton rose, 737
Leopard bane, 733
Leucospermum, 691
Ligonberry, 711
Lilac, **304,** 708
Lily, 263, 739
Lily turf, 740
Lily-of-the-Nile, 726
Lily-of-the-valley, 731
Lime, 506, 629
  Persian lime, 506
  Rangpur lime, 632
Linden, 710
Lionsheart, 744
Liquidambar, 691
Lisianthus, 734
Litchi, 240, 502, 506, 638
  Hairy litchi, 643
Lobelia, 740
Loblolly bay (black laurel), 686
Locust, black, 704
Locust bean, 629, 677
Loganberry, 655
Longan, 240, 506
Loosestrife, 740
Loquat, 634, 683
Lotus, American, 741
Love-lies-bleeding, 726
Lungwort, 745
Lupinus, 740

Macadamia, 638
Madrone, Pacific, 671
Magnolia, 254, 692
Mahaleb, 648
Mahogany, 708
Mahonia, 693
Maiden grass, 741
Maile, 726
Mallow, 737
Mandarin, 629, 632
Mandevilla, 694
Mango, 240, 502, 642
Manzanita, 672
Maple, 29, **490,** 668
Marigold, 748
Marsh marigold, 730
Masterwort, 728
Mayflower, 683
Meadow rue, 748
Medlar, 245
Melaleuca, 694
Mesquite, 700
Metasequoia, 694
Mexican oregano, 745
Mexican sunflower, **748**
Mignonette, 745
Mimosa, 670
Mint, 514
Mock orange, 697
Mondo grass, 742
Monkey flower, 741
Monkey puzzle tree, 671
Monkshood, 725
Morning glory, 738
Moutan, 743
Mulberry, 643
  fruitless mulberry, 694
Mullein, 749
Myrtle, 694

Nandina, 695
  Dwarf nandina, 695
Narcissus, 263, 741
Nasturtium, 749
Nectarine, 263
Nerine, 263
Nightshade, 245
Nispero, 625
Norfolk Island pine, 671

Oak, 29, 108, 701
Oleander, **338,** 695
Olive, 239, 643
  Fragrant tree olive, 695
  Fruitless olive, 695
  Russian olive, 682
  Sweet olive, 695
Orange, 245, **246**
  Sweet orange, 629, 631
  Trifoliate orange, 632
Orchid, 254, 263, 742
Ornamental gourds, 732
Ornamental onion, 726
Osage-orange, 692
Osmanthus, 695

Painted arum, 728
Painted daisy, 731
Painted tongue, 746
Palms, 696
  Date palms, **515,** 645
Pampas grass, 732
Pansy, 749
Papaya, 627
Parrot lily, 726
Parrotia (Persian), 696
Pasque flower, 727
Passion fruit, 644
Passion vine, 697
Pawpaw, 627, 672
Peach, **15,** 239, **249,** 649–650
Peach bells, 730
Pear, 29, 239, 242, 254, **414,** 653
  Asian pear, 655
  'Bradford' pear, 701
  Evergreen pear, 701
  'Old Home', 288, **395**
Pecan, **510,** 628
Penstemon, purple haze, 697
Peony, 743
  Tree peony, 696, 743
Peperomia, 744
Perennial pea vine, 739
Persimmon, 634
  Texas persimmon, 682
Petunia, 744
Philodendron, 744
Photinia, 697
Pieris, 698
Pigeon plum, 679
Piggyback plant, **341,** 748
Pincushion flower, 747
Pine, 29, 117, **284,** 698
  Monterey pine, 240, 241
Pineapple, 263, 515, **516,** 626
Pineapple lily, 735
Pistache, Chinese, 699
Pistachio, **115,** 646
Pitcher plant, 746
Pittosporum, **345,** 699
Plane tree, 699
Plantain-lily, 738
Plum, 239, 632, 652
  'Brompton' plum, **278**
  'Citation' plum, 649, 652
  'Marianna 2624' plum, 649, 651
  Mexican flowering plum, 700
  'Myrobalan' plum, 649–651
  'Viking' plum, 652
Plumbago, 699
Plumeria, 699
Pocketbook plant, 729
Podocarpus, 699
Poinsettia, **318, 345,** 684
Polianthes tuberose, 745
Polka dot plant, 738
Polygonatum, 745
Pomegranate, 653
Poplar, 239, 241, 699
  Yellow poplar, 692
Poppy, 743
Potato, 239, 258, 266, 514
Potentilla, 700, 745
Pot marigold, 730
Prayer plant, 741
Prickly pear cactus, 712
Privet, 691
Protea, 691, 700
Pumulo, 629
Purple loosestrife, 741
Pygmy (dwarf) pea shrub, 675
Pyracantha, 701

Quince, **334, 395,** 511, 634, 654
  Flowering quince, 678

Rambutan, 240
Ranunculus, **745**
Raphiolepis, 702
Raspberry, 263
  Black raspberry, 502
Red valerian, 730
Redbud, 33, 677
Redwood, 29, 707
  Coast redwood, 707
  Dawn redwood, 694
  Sierra redwood (Giant sequoia), 707
Rhododendron, **342, 354,** 702
Rockcress, 727
Rodgersflower, 746
Rose, 29, **336, 399, 413, 489, 490, 493,** 704
Rosemary, 706
Rose moss,
Rose of Sharon, 687
Rosewood, Arizona, 712
Rough lemon, 632
Royal poinciana, 682
Rubber plant, 502, 684

Sage, 746
  Purple sage, 691
  Texas sage, 691
  Russian sage, 744
Saffron, 731
Sapodilla, 625
Sarcococca, 706
Saxifraga, 513, 746
Scarlet bouvardia, 729
Schefflera, 747
Sea grape, 679
Sea-holly, eryngo, 734
Sedum, **404,** 747
Sequoia, Giant, 707
Serviceberry, 671
Seven son flower, 687
Shasta daisy, 731
Shooting star, 733
Shrub-althea, 687
Siberian bugloss, 729
Sidalcea, 747
Silk oak, 686
Silk tree, 670
Silverbell, 686
Silverberry, 682
Silverthorn, 682
Silver trumpet tree, 709
Smoke tree, 680
Snakeplant, 746
Snapdragon, 246, 727
Sneezeweed, 737
Snowberry, 708
Snowdrop, 737
Snow-in-winter, 730
Soapberry, Western, 706
Solomon's seal, 745
Sophora, 707
Sorbus, 708
Sourwood, 695
Speedwell, 749
Spiceberry, 672
Spiderflower, 731
Spiderwort, 744
Spider lily, 731
Spider plant, 513, **514**
Spirea, 708
Spotted deadnettle, 739
Sprenger asparagus, 728
Spring snowflake, 739
Spruce, 29, **403,** 697
  Sitka spruce, **255**
Squill, 747
Statice, 740
Stewartia, 708
St. Johnswort, 688
Stokes aster, 747
Strawberry, 253, 263, 269, 502, 513, 635
Strawflower, 737
Striped squill, 745
Succulents, 748
Sugar cane, 239, 263
Sugi, 240, 681
Sumac, 703
Sundew, 734
Sunflower, 737
Sunrose, 737
Sweet alyssum, **740**
Sweet gum, 691
Sweet pea, 739
Sweet potato, 239
Sweetbox, Fragrant, 706
Sweetshrub, 675
Sweet William, 733
Switch ivy

Tamarisk, 709
Tangelo, 629
Tangor, 629
Tangerine, 629
Taro, 263
Tea tree, 691
Thermopsis, 749
Thrift, 727
Thyme, 748
Ti, 732
Tiger flower, 748
Toadflax, 740
Toad lily, 749
Tobacco, flowering, 22, 742
Tomato, 245
Torch lily, 739
Touch-me-not, 738
Toyon (Christmas berry), 687
Transvaal daisy, 736
Tree of heaven, 670
Tree ivy, 735
Trumpet creeper, 675
Tufted pansy, 749
Tulip, 258
Tulip tree, 692
Turtlehead, 731
Tussock, bellflower, 730

Valerian, 749
Venidium, 749
Venus fly trap, 733
Verbena, 749
Veronica, 687, 749
Viburnum, 712
Vinca, 730, 749
Virginia creeper, 696

Walking stick, 671
Wallflower, 730
Walnut, 108, 118, 254, **438, 497,** 511
  Black walnut, 637
  Carpathian walnut, 637
  English walnut, 637
  Paradox walnut, 116, **117,** 637
  Persian walnut, 637
Wandering Jew, 750
Waratah, 710
Water lily, 742
Waxflower, 678
Wax myrtle, 694
Weigela, 713
Willow, 239, 241, 706
  Desert willow, 678
Willow amsonia, 726
Winter aconite, 734
Winterhazel, 680
Wishbone flower, 749
Wisteria, 686
Witch hazel, 686
  Chinese witch hazel, 692

Xylosma, 713

Yam, 239
Yarrow, 725
Yellow archangel, 739
Yew, 709
  Japanese plum yew, 677
Youngberry, 655
Yucca, 750

Zinnia, 750

Grafting and Budding Propagation

Whip and tongue graft.

Side graft for Japanese maple.

Field grafting ornamental cherry trees.

Chip bud ready to wrap.

Healed chip bud about to grow out in the Spring.

Bulb Propagation

Seed germination in lily seed following stratification to relieve dormancy.

Daffodil is propagated from naturally produced offsets.

Small bulblets form at the base of scales in lily.